HANDBOOK
OF
NATURE STUDY

ANNA BOTSFORD COMSTOCK

SEPTEMBER 1, 1854 – AUGUST 24, 1930

HANDBOOK
OF
NATURE STUDY

BY

NNA BOTSFORD COMSTOCK, B.S., L.H.D.

LATE PROFESSOR OF NATURE STUDY IN
CORNELL UNIVERSITY

With a Foreword by
VERNE N. ROCKCASTLE

Comstock Publishing Associates

A DIVISION OF

CORNELL UNIVERSITY PRESS

ITHACA AND LONDON

Revised edition published 1939 by Comstock Publishing Company, Inc. Twenty-second printing of the revised edition published 1974 by Cornell University Press. Published in cloth and paperback editions, with a new foreword, 1986. Bibliographic materials deleted in this reissue.

LIBRARY OF CONGRESS CATALOGING-IN-PUBLICATION DATA

Comstock, Anna Botsford, 1854–1930.
Handbook of nature study.

Reprint. Originally published: Ithaca, N.Y.: Comstock Pub. Co., 1939. With new foreword.
"Bibliographical materials deleted in this reissue"—T.p. verso.
Includes index.
1. Nature study. I. Title
QH51.C735 1986 508 85-29144
ISBN 978-0-8014-9384-3 (pbk.: alk. paper)

PRINTED IN THE UNITED STATES OF AMERICA

Cornell University Press strives to use environmentally responsible suppliers and materials to the fullest extent possible in the publishing of its books. Such materials include vegetable-based, low-VOC inks and acid-free papers that are recycled, totally chlorine-free, or partly composed of nonwood fibers. For further information, visit our website at www.cornellpress.cornell.edu.

Paperback printing 20 19 18 17 16 15 14

TO

LIBERTY HYDE BAILEY

UNDER WHOSE WISE, STAUNCH, AND INSPIRING LEADERSHIP
THE NATURE-STUDY WORK AT CORNELL UNIVERSITY
HAS BEEN ACCOMPLISHED

AND TO MY CO-WORKER

JOHN WALTON SPENCER

WHOSE COURAGE, RESOURCEFULNESS, AND UNTIRING ZEAL
WERE POTENT FACTORS IN THE SUCCESS OF THE CAUSE

THIS BOOK IS DEDICATED

FOREWORD TO THE 1986 REISSUE

matchless handbook for decades, this ic has been the natural history bible for itless teachers and others who seek in- ation about their environment. Written nally for those elementary school ners who knew little of common plants animals, and even less about the earth ath their feet and the skies overhead, book is for the most part as valid and ful today as it was when first written in . After all, dandelions, toads, robins, constellations have changed little since ! And modern society's concern with the ity of life and the impact of people on water, and wildlife makes this book even e relevant. It is because of the current ving concern with our natural environ- it that this volume, including the revi- s of 1939, has been brought back into t.

ature-study, as used in this handbook, ompasses all living things except hu- is, as well as all nonliving things such rocks and minerals, the heavens, and ther. Of the living things described, most common in the northeastern states, and ny, such as the dandelion, milkweed, and llein, and the house mouse, muskrat, red fox, are so widespread that people ng outside the United States will recog- e them easily.

)ther, less common but still easily recog- ed organisms such as bloodroot, teasel, ristmas fern, and dogwood, as well as nmon mosses, ferns, and mushrooms, nplete an amazing coverage of plants. d descriptions of insects of many kinds, es, salamanders, frogs and toads, rep- s, birds, and mammals give the animal gdom broad representation. Missing from book are only the plants and animals igenous to the sea; this is a book about d organisms—the ones most of us see ry day.

The most common species of plants and animals in the Northeast are described in detail. Any relatives of these which live in other states will differ mostly in appearance; their life histories will be largely similar to the ones described in the handbook. It is true that though many people know the most familiar organisms by name, the details of their lives are not at all well known. Anna Botsford Comstock very appropriately took the view that we should know first and best the things closest to us. Only then, when we have an intimate knowledge of our neigh- bors, should we journey farther afield to learn about more distant things.

Teachers and children will find the ma- terial in this book invaluable in that regard. Details of the most common, but in some ways the most interesting, things are brought out, first by careful, nontechnical descrip- tions of the things themselves and later by thoughtful questions and study units. Be- cause the most common things are treated in greatest detail, materials for study are easy to find. Whether the reader lives in the inner city or in the rural outback, the handbook is a treasure trove of information.

A teacher does not need to know much about nature to use this handbook. The in- formation is there for the novice and the ex- pert alike. All that is needed is an inquiring mind, senses to observe, and a willingness to think about nature on a personal level. To enter this book in search of information about any common organism, stone, or ob- ject in the sky is to open the door to a fresh and lively acquaintance with one's environ- ment. People may change in the ways they live and dress, but Comstock's handbook is one of those timeless treasures that every family and every classroom should have at the ready. It is a source of enrichment and enjoyment that has not been duplicated be- tween any other covers.

As you study this book, there are some limitations that you should bear in mind.

Some sections, such as the one about weather, simply cannot reflect the sweeping scientific advances that have been made since this descriptive material was written. Weather forecasting today depends almost entirely on computerized information. The computer was not even on the drawing boards when this book was written. However, the clouds of today differ little from the clouds of yesteryear, and frost forms for the same reasons today that it did earlier in the century.

Like the clouds, objects in the sky have changed little with time, even though the ways in which we study them today differ dramatically from those of decades ago. When reading about the skies, then, read with some charity toward the author, who, had she written in the computer age, might have passed over completely the delightful insights about the more basic relations in the natural world overhead.

The suggested readings and the bibliography have been deleted because many of the books are out of print or unavailable to most readers. The questions in the lessons following the various accounts, however, have been retained. They provide observational starting points for the study of organisms in the lessons.

A few of the organisms that were comr in Mrs. Comstock's day have now all disappeared from our land. This is tru the American elm, a gorgeous specime which is pictured in the book. A few o resistant elm trees can still be found, but text as written in 1911 does not match reality of our nearly elmless countryside

At the same time, some species consid in Mrs. Comstock's time to be on the v of extinction seem now to be on the v of a comeback. Such is true of the Ameri chestnut, nearly eradicated by the chest blight. One nonfatal strain of the blight seems to render its host tree immune to fatal strain—hope for the future.

These exceptions aside, the reissue Comstock's *Handbook of Nature Stud* most welcome. You need only open its ers and turn to almost any page to be los an adventure with nature—the nature is closest to you and an adventure that i interesting as a safari to wildest Africa.

VERNE N. ROCKCAS

ITHACA, NEW YORK
January 1986

PUBLISHER'S FOREWORD

'he publication of the twenty-fourth ion of the *Handbook of Nature-Study* ned an appropriate time to make cer- revisions which had become press- y necessary, to replace and improve illustrations, and to incorporate sug- ions which had been received from ıy interested friends. Accordingly, the re text has been carefully scrutinized, has been corrected or elaborated in light of the most recent knowledge. ere the earlier treatment seemed in- quate new material has been added, Part IV in particular has been much anded. New subjects, such as soil con- ation, have been introduced. We k it is safe to say that the *Handbook* been well modernized.

ut by far the greater part of Mrs. nstock's work proved to be as accurate timely in 1939 as in 1911, a striking ute to the scientific genius of the hor. In such cases the language of the ier text has been preserved, for no rovement could be made on the rming style that has won friends in the ; of thousands. And a careful attempt been made throughout to preserve the hod of treatment adopted by Mrs. nstock. Perhaps some justification of policy is needed. Some readers of the ıdbook have suggested that the new ion be oriented away from the nature- ly approach, and be made instead to e as an introduction to the natural nces. For the convenience of readers wish preparation for the academic lies, some scientific classifications and ninology have been introduced. But nature-study approach has been pre- ed. The kernel of that method of tment is the study of the organism in environment, its relation to the world ut it, and the features which enable it unction in its surroundings. This study takes the individual organism, rather than an abstract phylum or genus, as the point of departure. Mrs. Comstock believed that the student found in such a study a fresh, spontaneous interest which was lacking in formal textbook science, and the phenomenal success of her work seems to prove that she was right. Moreover, nature-study as Mrs. Comstock conceived it was an aesthetic experience as well as a discipline. It was an opening of the eyes to the individuality, the ingenuity, the personality of each of the unnoticed life-forms about us. It meant a broadening of intellectual outlook, an expansion of sympathy, a fuller life. Much of this Mrs. Comstock succeeded in conveying into her work; and perhaps it is this informing spirit that is the chief virtue of the book.

But it should not be thought that nature-study is not a science. The promising science of ecology is merely formalized nature-study; indeed it might be said that nature-study is natural science from an ecological rather than an anatomical point of view. The truth is that nature-study is a science, and is more than a science; it is not merely a study of life, but an experience of life. One realizes, as he reads these pages, that with Mrs. Comstock it even contributed to a philosophy of life.

Only the generous efforts of many specialists made possible the thoroughgoing revision of the book. Dr. Marjorie Ruth Ross assumed in large part the responsibility for editorial supervision and co-ordination, and performed most of the labor of revision and replacement of illustrations. Professor A. H. Wright and Mrs. Wright made valuable suggestions and criticisms of the book in general, provided hitherto unpublished photographs for the sections on reptiles and amphibians, and read proof on those sections.

Professor Glenn W. Herrick, Professor J. G. Needham, and Dr. Grace H. Griswold made suggestions for the revision of the material on insects, and supplied illustrations for that section. Professor E. F. Phillips contributed criticism for the lesson on bees. Professor A. A. Allen kindly made suggestions and provided illustrations for the material on birds. Professor B. P. Young gave assistance in the treatment of aquatic life; Dr. W. J. Koster made suggestions for improving the section on fish; and Dr. Emmeline Moore selected photographs of fish, and on behalf of the New York State Department of Conservation gave permission to use them.

Thanks are due to Professor W. J. Hamilton, Jr., for criticism of the section on mammals and for supplying several photographs; to Professor E. S. Harrison for aid in revising the lesson on cattle and supplying illustrations. Mrs. C. N. Stark made helpful suggestions for the revision of the lesson on bacteria. Miss Ethel Belk suggested many revisions in the part on plants. Professor W. C. Muenscher made useful criticisms of the section on weeds, and supplied illustrations. Professor C. H. Guise revised the portion dealing with the chestnut tree and Professor Ralph W. Curtis gave valuable assistance in the revision of the whole section on trees, and furnished pictures. Professor Joseph Oskamp suggested several improvements in the text on the apple tree. Mr. Will Marcus Ingram, Jr. prepared the capti for the illustrations of shells.

Professor H. Ries made extensive visions and additions in the lessons re ing to geology. Professor H. O. Buckr revised the lesson on soil. Professor A Gustafson revised the lesson on brook, and added material on soil con vation. Professor S. L. Boothroyd not revised the old text on the sky, but he provided new material and supplied m and photographs to illustrate it. Dr. H Geren made valuable suggestions for revision of the text on weather. Eva L. Gordon made numerous sug tions for revision of parts of the text provided some of the illustrations.

Dr. F. D. Wormuth acted as lite editor of the manuscript. Dr. John Raines composed many of the capt for the new illustrations, and, with Raines, read proof of the entire book

Many teachers throughout the cou offered constructive criticisms; an atte has been made to put them into ef To all of these persons the publishers to express most cordial and sincere tha

THE PUBLIS

ITHACA, NEW YORK
January 1, 1939

Anna Comstock's *Handbook of Nature Study*, written in 1911, was revised in the spirit of its author by a group of persons in 1939. Since then more than 100,000 copies have been sold, and many orders come to the Press each month. This vigorous, continuing life is a tribute to the durable charm and correctness of the original text, written more than fifty years ago.

Little has been outmoded in Anna's love affair with nature. The screech owl's song, the bat's wings, yellow jackets lapping juice, all are fascinating and wonderful still. The poets quoted still sp tellingly of nature. Where knowledge advanced, Mrs. Comstock's stance scientific care have saved her from bl dering. The illustrations remain pe nent: rabbits playing in the moonli the toad from egg to adult, an ice sto spider webs, flowers, trees, rocks, stars.

"In the early years," she suggests, q ing Liberty Hyde Bailey, "we are no teach nature as science, we are not teach it primarily for method or for d we are to teach it for loving."

January 1970

PREFACE

e Cornell University Nature-Study ganda was essentially an agricultural ment in its inception and its aims; inaugurated as a direct aid to better ods of agriculture in New York During the years of agricultural de- on 1891–1893, the Charities of New City found it necessary to help many e who had come from the rural dis- — a condition hitherto unknown. philanthropists managing the Associ- for Improving the Condition of the asked, "What is the matter with and of New York State that it can- upport its own population?" A con- ce was called to consider the situa- to which many people from different of the State were invited; among was the author of this book, who realized that in attending that meet- he whole trend of her activities would thereby changed. Mr. George T. ell, who had been a most efficient Di- or of Farmers' Institutes of New York e, was invited to the conference as an ert to explain conditions and give ad- as to remedies. The situation seemed erious that a Committee for the Pro- ion of Agriculture in New York State appointed. Of this committee the norable Abram S. Hewitt was Chair- , Mr. R. Fulton Cutting, Treasurer, Wm. H. Tolman, Secretary. The er members were Walter L. Suydam, n. E. Dodge, Jacob H. Schiff, George Powell, G. Howard Davidson, Howard wnsend, Professor I. P. Roberts, C. Namee, Mrs. J. R. Lowell, and Mrs. B. Comstock. Mr. George T. Powell s made Director of the Department of ricultural Education.

At the first meeting of this committee r. Powell made a strong plea for inter- ing the children of the country in ming as a remedial measure, and main- tained that the first step toward agricul- ture was nature-study. It had been Mr. Powell's custom to give simple agricul- tural and nature-study instruction to the school children of every town where he was conducting a farmers' institute, and his opinion was, therefore, based upon experience. The committee desired to see for itself the value of this idea, and experi- mental work was suggested, using the schools of Westchester County as a labo- ratory. Mr. R. Fulton Cutting generously furnished the funds for this experiment, and work was done that year in the West- chester schools which satisfied the com- mittee of the soundness of the project.

The committee naturally concluded that such a fundamental movement must be a public rather than a private enterprise; and Mr. Frederick Nixon, then Chairman of the Ways and Means Committee of the Assembly, was invited to meet with the committee at Mr. Hewitt's home. Mr. Nixon had been from the beginning of his public career deeply interested in im- proving the farming conditions of the State. In 1894, it was through his influ- ence and the support given him by the Chautauqua Horticultural Society under the leadership of Mr. John W. Spencer, that an appropriation had been given to Cornell University for promoting the horticultural interests of the western counties of the State. In addition to other work done through this appropriation, horticultural schools were conducted un- der the direction of Professor L. H. Bailey with the aid of other Cornell instructors and especially of Mr. E. G. Lodeman; these schools had proved to be most use- ful and were well attended. Therefore, Mr. Nixon was open-minded toward an educational movement. He listened to the plan of the committee and after due con- sideration declared that if this new measure would surely help the farmers of the State, the money would be forthcoming. The committee unanimously decided that if an appropriation were made for this purpose it should be given to the Cornell College of Agriculture; and that year eight thousand dollars were added to the Cornell University Fund, for Extension Teaching and inaugurating this work. The work was begun under Professor I. P. Roberts; after one year Professor Roberts placed it under the supervision of Professor L. H. Bailey, who for the fifteen years since has been the inspiring leader of the movement, as well as the official head.

In 1896, Mr. John W. Spencer, a fruit grower in Chautauqua County, became identified with the enterprise; he had lived in rural communities and he knew their needs. He it was who first saw clearly that the first step in the great work was to help the teacher through simply written leaflets; and later he originated the great plan of organizing the children in the schools of the State into Junior Naturalists Clubs, which developed a remarkable phase of the movement. The members of these clubs paid their dues by writing letters about their nature observations to Mr. Spencer, who speedily became their beloved "Uncle John"; a button and charter were given for continued and earnest work. Some years, 30,000 children were thus brought into direct communication with Cornell University through Mr. Spencer. A monthly leaflet for Junior Naturalists followed; and it was to help in this enterprise that Miss Alice G. McCloskey, the able Editor of the present *Rural School Leaflet*, was brought into the work. Later, Mr. Spencer organized the children's garden movement by forming the children of the State into junior gardeners; at one time he had 25,000 school pupils working in gardens and reporting to him.

In 1899, Mrs. Mary Rogers Miller, who had proven a most efficient teacher when representing Cornell nature-study in the State Teachers' Institutes, planned and started the Home Nature-Study Course Leaflets for the purpose of helping the teachers by correspondence, a work fell to the author in 1903 when Miller was called to other fields.

For the many years during which York State has intrusted this imp work to Cornell University, the tea of nature-study has gone steadily on University, in teachers' institutes, in summer schools, through various pu tions and in correspondence co Many have assisted in this work, nc Dr. W. C. Thro, Dr. A. A. Allen Miss Ada Georgia. The New York cation Department with Charles R. ner as Commissioner of Education Dr. Isaac Stout as the Director of T ers' Institutes co-operated heartily the movement from the first. Later the co-operation of Dr. Andrew D as Commissioner of Education, ma the Cornell leaflets have been w with the special purpose of aidin carrying out the New York State Syll in Nature-Study and Agriculture.

The leaflets upon which this volu based were published in the Home ture-Study Course during the years 1 1911, in limited editions and were out of print. It is to make these les available to the general public that volume has been compiled. While subject matter of the lessons herein g is essentially the same as in the lea the lessons have all been rewritten fo sake of consistency, and many new sons have been added to bridge gaps make a coherent whole.

Because the lessons were written ing a period of so many years, each le has been prepared as if it were the one, and without reference to other there is any uniformity of plan in the sons, it is due to the inherent qualitie the subjects, and not to a type plan in mind of the writer; for, in her opin each subject should be treated indiv ally in nature-study; and in her long perience as a nature-study teacher she never been able to give a lesson twice a on a certain topic or secure exactly same results twice in succession. It sho also be stated that it is not because

: undervalues physics nature-study : has been left out of these lessons, ecause her own work has been always biological lines.

: reason why nature-study has not complished its mission, as thought- or much of the required work in our : schools, is that the teachers are as ole untrained in the subject. The en are eager for it, unless it is spoiled e teaching; and whenever we find a er with an understanding of out-of- life and a love for it, there we find e-study in the school is an inspira- and a joy to pupils and teacher. It is ise of the author's sympathy with ntrained teacher and her full com- ension of her difficulties and help- ess that this book has been written. e difficulties are chiefly three-fold: teacher does not know what there is e in studying a planet or animal; she vs little of the literature that might her; and because she knows so little e subject, she has no interest in giving son about it. As a matter of fact, the ature concerning our common ani- and plants is so scattered that a her would need a large library and al- t unlimited time to prepare lessons in extended nature-study course.

he writer's special work for fifteen s in Extension teaching has been the ping of the untrained teacher through onal instruction and through leaflets. ny methods were tried and finally e was evolved the method followed in volume: All the facts available and tinent concerning each topic have been embled in the "Teacher's story" to ke her acquainted with the subject; this ollowed by an outline for observation the part of the pupils while studying object. It would seem that with the cher's story before the eyes of the cher, and the subject of the lesson be- e the eyes of the pupils with a number questions leading them to see the es- tial characteristics of the object, there ould result a wider knowledge of nature in is given in this or any other book.

That the lessons are given in a very informal manner, and that the style of writing is often colloquial, results from the fact that the leaflets upon which the book is based were written for a correspondence course in which the communications were naturally informal and chatty. That the book is meant for those untrained in science accounts for the rather loose terminology employed; as, for instance, the use of the word *seed* in the popular sense whether it be a drupe, an akene, or other form of fruit; or the use of the word *pod* for almost any seed envelope, and many like instances. Also, it is very likely, that in teaching quite incidentally the rudiments of the principles of evolution, the results may often seem to be confused with an idea of purpose, which is quite unscientific. But let the critic labor for fifteen years to interest the untrained adult mind in nature's ways, before he casts any stones! And it should be always borne in mind that if the author has not dipped deep in the wells of science, she has used only a child's cup.

For many years requests have been frequent from parents who have wished to give their children nature interests during vacations in the country. They have been borne in mind in planning this volume; the lessons are especially fitted for field work, even though schoolroom methods are so often suggested.

The author feels apologetic that the book is so large. However, it does not contain more than any intelligent country child of twelve should know of his environment; things that he should know naturally and without effort, although it might take him half his life-time to learn so much if he should not begin before the age of twenty. That there are inconsistencies, inaccuracies, and even blunders in the volume is quite inevitable. The only excuse to be offered is that, if through its use, the children of our land learn early to read nature's truths with their own eyes, it will matter little to them what is written in books.

The author wishes to make grateful acknowledgment to the following people: To Professor Wilford M. Wilson for his chapter on the weather; to Miss Mary E. Hill for the lessons on mould, bacteria, the minerals, and reading the weather maps; to Miss Catherine Straith for the lessons on the earthworm and the soil; to Miss Ada Georgia for much valuable assistance in preparing the original leaflets on which these lessons are based; to Dean L. H. Bailey and to Dr. David S. Jordan for permission to quote their writings; to Mr. John W. Spencer for the use of his story on the movements of the sun; to Dr. Grove Karl Gilbert, Dr. A. C. Gill, Dr. Benjamin Duggar, Professor S. H. Gage and Dr. J. G. Needham for reading and criticizing parts of the manuscript; to Miss Eliza Tonks for reading the proof; to the Director of the College of Agriculture for the use of the engravings made for the original leaflets; to Miss Martha Van Rensselaer for the use of many pictures from *Boys and Girls*; to Professor Cyrus Crosby, and to Messrs. J. T. Lloyd, Allen and R. Matheson for the their personal photographs; to the Geological Survey and the U. S. Service for the use of photograp Louis A. Fuertes for drawings of bi Houghton Mifflin & Company for t of the poems of Lowell, Harte an com, and various extracts from Burr and Thoreau; to Small, Maynard & pany and to John Lane & Compa the use of poems of John T. Bab Doubleday, Page & Company for th of pictures of birds and flowers; and American Book Company for the electrotypes of dragon-flies and a omy. Especially thanks are extend Miss Anna C. Stryke for numerous ings, including most of the initials.

Anna Botsford Coms

Ithaca, New York
July, 1911

CONTENTS

PART I
THE TEACHING OF NATURE-STUDY

PART II
ANIMALS

PART III
PLANTS

PART IV

EARTH AND SKY

FULL-PAGE PLATES

PART I
THE TEACHING OF NATURE-STUDY

THE TEACHING OF NATURE-STUDY

Ralph W. Curtis

What Nature-Study Is

ature-study is, despite all discussions perversions, a study of nature; it con- of simple, truthful observations that , like beads on a string, finally be aded upon the understanding and held together as a logical and har- ious whole. Therefore, the object of nature-study teacher should be to cul- te in the children powers of accurate ervation and to build up within them lerstanding.

What Nature-Study Should Do for the Child

irst, but not most important, nature- dy gives the child practical and help- knowledge. It makes him familiar with nature's ways and forces, so that he is not so helpless in the presence of natural misfortune and disasters.

Nature-study cultivates the child's imagination, since there are so many wonderful and true stories that he may read with his own eyes, which affect his imagination as much as does fairy lore; at the same time nature-study cultivates in him a perception and a regard for what is true, and the power to express it. All things seem possible in nature; yet this seeming is always guarded by the eager quest of what is true. Perhaps half the falsehood in the world is due to lack of power to detect the truth and to express it. Nature-study aids both in discernment and in expression of things as they are.

Nature-study cultivates in the child a

love of the beautiful; it brings to him early a perception of color, form, and music. He sees whatever there is in his environment, whether it be the thunder-head piled up in the western sky, or the golden flash of the oriole in the elm; whether it be the purple of the shadows on the snow, or the azure glint on the wing of the little butterfly. Also, what there is of sound, he

Louis Agassiz Fuertes Council, Boy Scouts of America

A nature hike

hears; he reads the music score of the bird orchestra, separating each part and knowing which bird sings it. And the patter of the rain, the gurgle of the brook, the sighing of the wind in the pine, he notes and loves and becomes enriched thereby.

But, more than all, nature-study gives the child a sense of companionship with life out-of-doors and an abiding love of nature. Let this latter be the teacher's criterion for judging his or her work. If nature-study as taught does not make the child love nature and the out-of-doors, then it should cease. Let us not inflict permanent injury on the child by turning him away from nature instead of toward it. However, if the love of nature is in the teacher's heart, there is no danger; such a teacher, no matter by what m takes the child gently by the han walks with him in paths that lead seeing and comprehending of wh may find beneath his feet or abo head. And these paths, whether the among the lowliest plants, or wheth the stars, finally converge and brin wanderer to that serene peace and ful faith that is the sure inheritance those who realize fully that they are ing units of this wonderful universe.

Nature-Study as a Help to Hea

Perhaps the most valuable practica son the child gets from nature-study personal knowledge that nature's law not to be evaded. Wherever he look discovers that attempts at such eva result in suffering and death. A knowl thus naturally attained of the imm bility of nature's " must " and " shall r is in itself a moral education. The rea tion that the fool as well as the trans sor fares ill in breaking natural laws m for wisdom in morals as well as in hygi

Out-of-door life takes the child a and keeps him in the open air, which only helps him physically and occu his mind with sane subjects, but k him out of mischief. It is not only du childhood that this is true, for love nature counts much for sanity in later This is an age of nerve tension, and relaxation which comes from the comf ing companionship found in woods fields is, without doubt, the best rem for this condition. Too many men w seek the out-of-doors for rest at the pres time, can only find it with a gun in ha To rest and heal their nerves they m go out and try to kill some unfortun creature — the old, old story of sacrifi blood. Far better will it be when, throu properly training the child, the man sh be enabled to enjoy nature through see how creatures live rather than watch them die. It is the sacred privilege nature-study to do this for future gene tions and for him thus trained, shall t words of Longfellow's poem to Agas apply:

he wandered away and away, with
Nature the dear old nurse,
o sang to him night and day, the
rhymes of the universe.
when the way seemed long, and his
heart began to fail,
sang a more wonderful song, or told
a more wonderful tale.

AT NATURE-STUDY SHOULD DO FOR THE TEACHER

uring many years, I have been watch- teachers in our public schools in their scientious and ceaseless work; and so as I can foretell, the fate that awaits n finally is either nerve exhaustion or ve atrophy. The teacher must become er a neurasthenic or a "clam."

have had conversations with hundreds eachers in the public schools of New k State concerning the introduction ature-study into the curriculum, and t of them declared, "Oh, we have not e for it. Every moment is full now!" ir nerves were at such a tension that one more thing to do they must fall t. The question in my own mind dur- these conversations was always, how can she stand it! I asked some of n, "Did you ever try a vigorous walk he open air in the open country every rday or every Sunday of your teach- year?" "Oh no!" they exclaimed in pair of making me understand. "On day we must go to church or see our ds and on Saturday we must do our pping or our sewing. We must go to dressmaker's lest we go unclad, we t mend, and darn stockings; we need rday to catch up."

es, catch up with more cares, more ries, more fatigue, but not with more wth, more strength, more vigor, and e courage for work. In my belief, there two and only two occupations for Sat- y afternoon or forenoon for a teacher. e is to be out-of-doors and the other o lie in bed, and the first is best. t in this, God's beautiful world, there everything waiting to heal lacerated ves, to strengthen tired muscles, to ase and content the soul that is torn to shreds with duty and care. To the teacher who turns to nature's healing, nature-study in the schoolroom is not a trouble; it is a sweet, fresh breath of air blown across the heat of radiators and the noisome odor of overcrowded small humanity. She who opens her eyes and her heart nature-ward even once a week finds nature-study in the schoolroom a delight and an abiding joy. What does such a one find in her schoolroom instead of the terrors of discipline, the eternal watching and eternal nagging to keep the pupils quiet and at work? She finds, first of all, companionship with her children; and second, she finds that without planning or going on a far voyage, she has found health and strength.

WHEN AND WHY THE TEACHER SHOULD SAY "I DO NOT KNOW"

No science professor in any university, if he be a man of high attainment, hesitates to say to his pupils, "I do not know," if they ask for information beyond his knowledge. The greater his scientific reputation and erudition, the more readily, simply, and without apology he says this. He, better than others, comprehends how vast is the region that lies beyond man's present knowledge. It is only the teacher in the elementary schools who has never received enough scientific training to reveal to her how little she does know, who feels that she must appear to know everything or her pupils will lose confidence in her. But how useless is this pretense, in nature-study! The pupils, whose younger eyes are much keener for details than hers, will soon discover her limitations and then their distrust of her will be real.

In nature-study any teacher can with honor say, "I do not know"; for perhaps the question asked is as yet unanswered by the great scientists. But she should not let lack of knowledge be a wet blanket thrown over her pupils' interest. She should say frankly, "I do not know; let us see if we cannot together find out this mysterious thing. Maybe no one knows it as yet, and I wonder if you will discover it before I do." She thus conveys the right impression, that only a little about the intricate life of plants and animals is yet known; and at the same time she makes her pupils feel the thrill and zest of investigation. Nor will she lose their respect by doing this, if she does it in the right spirit. For three years, I had for comrades in my walks afield two little children and they kept me busy saying, "I do not know." But they never lost confidence in me or in my knowledge; they

Leonard K. Beyer

Long-spurred violet

simply gained respect for the vastness of the unknown.

The chief charm of nature-study would be taken away if it did not lead us through the border-land of knowledge into the realm of the undiscovered. Moreover, the teacher, in confessing her ignorance and at the same time her interest in a subject, establishes between herself and her pupils a sense of companionship which relieves the strain of discipline, and gives her a new and intimate relation with her pupils which will surely prove a potent element in her success. The best teacher is always one who is the good comrade of her pupils.

Nature-Study, the Elixir of Youth

The old teacher is too likely to become didactic, dogmatic, and "bossy" if she does not constantly strive with herself. Why? She has to be thus five days in the week and, therefore, she is likely to be so seven. She knows arithmetic, grammar, and geography to their uttermost, she is never allowed to forget that knows them, and finally her interests come limited to what she knows.

After all, what is the chief sign growing old? Is it not the feeling we know all there is to be known? not years which make people old; ruts, and a limitation of interests. W we no longer care about anything ex our own interests, we are then old matters not whether our years be tw or eighty. It is rejuvenation for teacher, thus growing old, to stand norant as a child in the presence of of the simplest of nature's miracle the formation of a crystal, the evolu of the butterfly from the caterpillar, exquisite adjustment of the silken in the spider's orb web. I know how "make magic" for the teacher wh growing old. Let her go out with youngest pupil and reverently watch him the miracle of the blossoming vi and say: "Dear Nature, I know na of the wondrous life of these, your sm est creatures. Teach me!" and she suddenly find herself young.

Nature-Study as a Help in School Discipline

Much of the naughtiness in scho a result of the child's lack of interes his work, augmented by the physical action that results from an attempt t quietly. The best teachers try to ob both of these causes of misbehav rather than to punish the naughtiness results from them. Nature-study is an in both respects, since it keeps the c interested and also gives him somet to do.

In the nearest approach to an school that I have ever seen, for chil of second grade, the pupils were allo as a reward of merit, to visit the aq or the terrarium for periods of five utes, which time was given to the bli observation of the fascinating priso The teacher also allowed the readin stories about the plants and animals der observation to be regarded as ward of merit. As I entered the sc

eight or ten of the children were
e windows watching eagerly what
appening to the creatures confined
in the various cages. There was a
aquarium for the frogs and sala-
ers, an aquarium for fish, many
aquaria for insects, and each had
r two absorbedly interested specta-
who were quiet, well-behaved, and
getting their nature-study lessons
ideal manner. The teacher told me
the problem of discipline was solved
is method, and that she was rarely
ed to rebuke or punish. In many
schools, watching the living crea-
in the aquaria or terraria has been
as a reward for other work well done.

HE RELATION OF NATURE-STUDY TO SCIENCE

ature-study is not elementary science
taught, because its point of attack
t the same; error in this respect has
ed many a teacher to abandon nature-
y and many a pupil to hate it. In
entary science the work begins with
simplest animals and plants and pro-
ses logically through to the highest
s; at least this is the method pursued
ost universities and schools. The ob-
of the study is to give the pupils an
ook over all the forms of life and their
tion one to another. In nature-study
work begins with any plant or crea-
which chances to interest the pupil.
egins with the robin when it comes
k to us in March, promising spring;
it begins with the maple leaf which
ters to the ground in all the beauty of
autumnal tints. A course in biological
nce leads to the comprehension of
kinds of life upon our globe. Nature-
dy is for the comprehension of the
ividual life of the bird, insect, or plant
t is nearest at hand.

Nature-study is perfectly good science
hin its limits, but it is not meant to
more profound or comprehensive than
capabilities of the child's mind. More
n all, nature-study is not science be-
led as if it were to be looked at through
reversed opera glass in order to bring
it down small enough for the child to play with. Nature-study, as far as it goes, is just as large as is science for "grown-ups." It may deal with the same subject matter and should be characterized by the same accuracy. It simply does not go so far.

To illustrate: If we are teaching the science of ornithology, we take first the Archæopteryx, then the swimming and scratching birds, and finally reach the song birds, studying each as a part of the whole. Nature-study begins with the robin because the child sees it and is interested in it, and notes the things about the habits and appearance of the robin that may be perceived by intimate observa-

Hugh Spencer

An aquarium

tion. In fact, he discovers for himself all that the most advanced book of ornithology would give concerning the ordinary habits of this one bird; the next bird studied may be the turkey in the barnyard, or the duck on the pond, or the screech owl in the spruces, if any of these happen to impinge upon his notice and interest. However, such nature-study makes for the best of scientific ornithology, because by studying the individual birds thus thoroughly, the pupil finally studies a sufficient number of forms so that his knowledge, thus assembled, gives him a better comprehension of birds as a whole than could be obtained by the routine study of them. Nature-study does not start out with the classification given in books, but in the end it builds up in the child's mind a classification which is based on fundamental knowledge; it is a classification like that evolved by the first naturalists, because it is built on careful personal observations of both form and life.

Nature-Study Not for Drill

If nature-study is made a drill, its pedagogic value is lost. When it is properly taught, the child is unconscious of mental effort or that he is suffering the act of teaching. As soon as nature-study becomes a task, it should be dropped; but how could it ever be a task to see that the sky is blue, or the dandelion golden, or to listen to the oriole in the elm!

Stanley Mulaik

A young entomologist

The Child Not Interested in Nature-Study

What to do with the pupil not interested in nature-study subjects is a problem that confronts many earnest teachers. Usually the reason for this lack of interest is the limited range of subjects used for nature-study lessons. Often the teacher insists upon flowers as the lesson subject, when toads or snakes would prove the key to the door of the child's interest. But whatever the cause may be, there is only one right way out of this difficulty: The child not interested should be kept at his regular school work and not admitted as a member of the nature-study class, where his influence is always demoralizing. He had much better be learni spelling lesson than learning to ha ture through being obliged to stud jects in which he is not intereste general, it is safe to assume that th pil's lack of interest in nature-stu owing to a fault in the teacher's me She may be trying to fill the child's with facts when she should be le him to observe these for himself, v is a most entertaining occupation fo child. It should always be borne in that mere curiosity is always imperti and that it is never more so than exercised in the realm of nature. A ine interest should be the basis o study of the lives of plants and animals. Curiosity may elicit facts, only real interest may mold these into wisdom.

When to Give the Lesson

There are two theories concerning time when a nature-study lesson sh be given. Some teachers believe th should be a part of the regular rou others have found it of greatest val reserved for that period of the sc day when the pupils are weary and less, and the teacher's nerves straine the snapping point. The lesson on a insect, or flower at such a moment aff immediate relief to everyone; it is a n tal excursion, from which all return freshed and ready to finish the dutie the day.

While I am convinced that the us the nature-study lesson for mental freshment makes it of greatest value, I realize fully that if it is relegated such periods, it may not be given at It might be better to give it a reg period late in the day, for there is stren and sureness in regularity. The teac is much more likely to prepare herself the lesson, if she knows that it is requ at a certain time.

The Length of the Lesson

The nature-study lesson should short and sharp and may vary from minutes to a half hour in length. Th

d be no dawdling; if it is an observa-
.esson, only a few points should be
and the meaning for the observa-
made clear. If an outline be sug-
d for field observation, it should be
in an inspiring manner which shall
each pupil anxious to see and read
ruth for himself. The nature story
properly read is never finished; it
ways at an interesting point, "con-
d in our next."

e teacher may judge as to her own
ess in nature-study by the length
me she is glad to spend in reading
nature's book what is therein writ-
As she progresses, she finds those
s spent in studying nature speed
r, until a day thus spent seems but
our. The author can think of nothing
would so gladly do as to spend days
months with the birds, bees, and flow-
with no obligation to tell what she
ld see. There is more than mere ination in hours thus spent. Lowell
ribes them well when he says:

se old days when the balancing of a
yellow butterfly o'er a thistle bloom
s spiritual food and lodging for the
whole afternoon.

The Nature-Study Lesson Always New

nature-study lesson should not be
ated unless the pupils demand it. It
uld be done so well the first time that
e is no need of repetition, because it
thus become a part of the child's con-
usness. The repetition of the same les-
in different grades was, to begin with,
opeless incubus upon nature-study.
e disgusted boy declared, "Darn ger-
ation! I had it in the primary and last
r and now I am having it again. I
w *all about germination.*" The boy's
itude was a just one; but if there had
n revealed to him the meaning of
mination, instead of the mere process,
would have realized that until he had
nted and observed every plant in the
rld he would not know all about ger-
nation, because each seedling has its
own interesting story. The only excuse for repeating a nature-study lesson is in recalling it for comparison and contrast with other lessons. The study of the violet will naturally bring about a review of the pansy; the dandelion, of the sunflower; the horse, of the donkey; the butterfly, of the moth.

Nature-Study and Object Lessons

The object lesson method was introduced to drill the child to see a thing accurately, not only as a whole but in detail, and to describe accurately what he saw. A book or a vase or some other object was held up before the class for a

Leonard K. Beyer

A mountain brook

moment and then removed; afterwards the pupils described it as perfectly as possible. This is an excellent exercise and the children usually enjoy it as if it were a game. But if the teacher has in mind the same thought when she is giving the nature-study lesson, she has little comprehension of the meaning of the latter and the pupils will have less. In nature-study, it is not desirable that the child see all the details, but rather those details that have something to do with the life of the creature studied; if he sees that the grasshopper has the hind legs much longer than the others, he will inevitably note that there are two other pairs of legs and he will in the meantime have come into an illuminating comprehension of the reason the insect is called "grasshopper." The child should see definitely and accurately all that is necessary for the recognition of a plant or animal; but in nature-study, the observation of form is for the purpose of better understanding life. In fact, it is form linked with life, the relation of "being" to "doing."

Nature-Study in the Schoolroom

Many subjects for nature-study lessons may be brought into the schoolroom. Whenever it is possible, the pupils should themselves bring the material, as the collecting of it is an important part of the

A. I. Root Co.

An observation beehive

lesson. There should be in the schoolroom conveniences for caring for the little prisoners brought in from the field. A terrarium and breeding cages of different kinds should be provided for the insects, toads, and little mammals. Here they may live in comfort, when given their natural food, while the children observe their interesting ways. The ants' nest and the observation hive yield fascinating views of the marvelous lives of the insect socialists, while the cheerful prisoner in the bird cage may be made a constant illustration of the adaptations and habits of all birds. The aquaria for fishes, tadpoles, and insects afford the opportunity for continuous study of these water creatures and are a never-failing source of interest to the pupils, while the window garden may be made not only an ornament and an æsthetic delight, but a basis for intere study of plant growth and developı

A schoolroom thus equipped is a of delight as well as enlightenme the children. Once, a boy whose luxu home was filled with all that money buy and educated tastes select, said little nature-study laboratory which in the unfinished attic of a school b ing, but which was teeming with lif think this is the most beautiful roo the world."

Nature-Study and Museum Specimens

The matter of museum specime another question for the nature-s teacher to solve, and has a direct be on an attitude toward taking life. T are many who believe the stuffed bi the case of pinned insects have no p in nature-study; and certainly t should not be the chief material. let us use our common sense; the sees a bird in the woods or field and not know its name; he seeks the bir the museum and thus is able to pla and read about it and is stimulate make other observations concerning Wherever the museum is a help to study of life in the field, it is well good. Some teachers may give a live son from a stuffed specimen, and o teachers may stuff their pupils with f about a live specimen; of the two, former is preferable.

There is no question that making a lection of insects is an efficient way developing the child's powers of c observation, as well as of giving him m ual dexterity in handling fragile thi Also it is a false sentiment which att utes to an insect the same agony at ing impaled on a pin that we might su at being thrust through by a stake. insect nervous system is far more c veniently arranged for such an ordeal t ours; and, too, the cyanide bottle bri immediate and painless death to the sects placed within it; moreover, the sects usually collected have short li anyway. So far as the child is concern

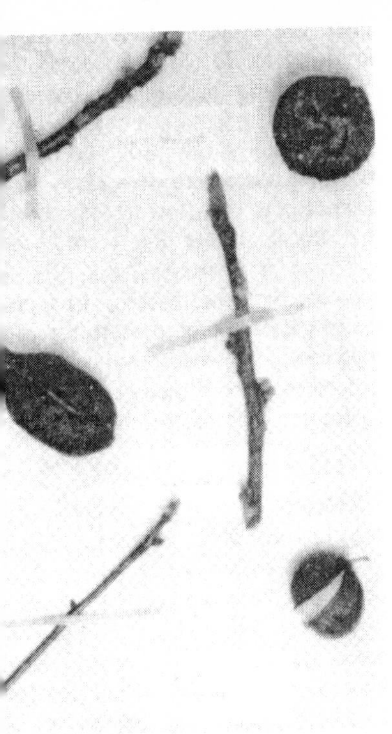

unted twigs and nuts. These may be put bottom of a shallow box with a sheet of hane pasted over the top

thinking of his collection of moths tterflies and not at all of taking life; is not teaching him to wantonly oy living creatures. However, an in-iminate encouragement of the mak-of insect collections cannot be ad-. There are some children who will t by it and some who will not, and iestionably the best kind of study of ts is watching their interesting ways e they live.

o kill a creature in order to prepare r a nature-study lesson is not only ng but absurd, for nature-study has to with life rather than death, and the 1 of any creature is interesting only n its adaptations for life are studied. again, a nature-study teacher may be opportunist; if without any volition her part or the pupils', a freshly killed imen comes to hand, she should ke the most of it. The writer remem-s most illuminating lessons from a par-ge that broke a window and its neck simultaneously during its flight one winter night, a yellow hammer that killed itself against an electric wire, and a musk-rat that turned its toes to the skies for no understandable reason. In each of these cases the creature's special physical adaptations for living its own peculiar life were studied, and the effect was not the study of a dead thing, but of a successful and wonderful life.

The Lens, Microscope, and Field Glass as Helps in Nature-Study

In elementary grades, nature-study deals with objects which the children can see with the naked eye. However, a lens is a help in almost all of this work because it is such a joy to the child to gaze at the wonders it reveals. There is no lesson given in this book which requires more than a simple lens for seeing the most minute parts discussed. An excellent lens may be bought for a dollar, and a fairly good one for fifty cents or even twenty-five cents. The lens should be chained to a table or desk where it may be used by the pupils at recess. This gives each an opportunity for using it and obviates the danger of losing it. If the pupils themselves own lenses, they should be fastened by a string or chain to the pocket.

A microscope has no legitimate part in nature-study. But if there is one available, it reveals so many wonders in the commonest objects that it can ofttimes be

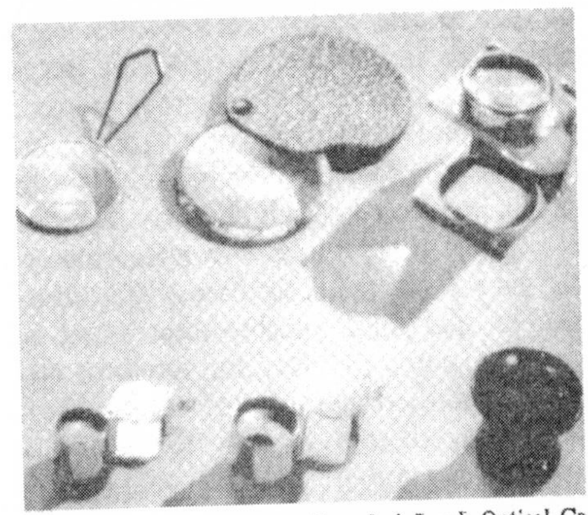

Bausch & Lomb Optical Co.

Hand lenses

Bausch & Lomb Optical Co.

A field glass

made a source of added interest. For instance, thus to see the scales on the butterfly's wing affords the child pleasure as well as edification. Field or opera glasses, while indispensable for bird study, are by no means necessary in nature-study. However, the pupils will show greater interest in noting the birds' colors if they are allowed to make the observations with the help of a glass.

Uses of Pictures, Charts, and Blackboard Drawings

Pictures alone should never be used as the subjects for nature-study lessons, but they may be of great use in illustrating and illuminating a lesson. Books well illustrated are more readily comprehended by the child and are often very helpful to him, especially after his interest in the subject is thoroughly aroused. If charts are used to illustrate the lesson, the child is likely to be misled by the size of the drawing, which is also the case in blackboard pictures. However, this error may be avoided by fixing the attention of the pupil on the object first. If the pupils are studying the ladybird and have it in their hands, the teacher may use a diagram representing the beetle as a foot long and it will still convey the idea accurately; but if she begins with the picture, she probably can never convince the children that the pict
anything to do with the insect.

In making blackboard drawing
trative of the lesson, it is best, if p
to have one of the pupils do the d
in the presence of the class; or,
teacher does the drawing, she shoul
the object in her hand while do
and look at it often so that the c
may see that she is trying to repre
accurately. Taking everything int
sideration, however, nature-study
and blackboard drawings are of lit
to the nature-study teacher.

The Uses of Scientific Nam

Disquieting problems relative to
tific nomenclature always confro
teacher of nature-study. My own p
has been to use the popular names
cies, except in cases where confusion
ensue, and to use the scientific nam
anatomical parts. However, this ma
of little importance if the teacher b
mind that the purpose of nature-
is to know the subject under obser
and to learn the name incidentally.

Common tree frog or tree toad, Hyla v
color versicolor. *Another species,* Hyla
cifer, *is also often called the tree frog and*
toad. Common names, then, will not di
guish these amphibians one from anot
the scientific names must be applied

he teacher says, "I have a pink he-. Can anyone find me a blue one?" ildren, who naturally like grownup , will soon be calling these flowers icas. But if the teacher says, "These s are called hepaticas. Now please one remember the name. Write it ur books as I write it on the black-, and in half an hour I shall ask you what it is," the pupils naturally look the exercise as a word lesson and its gnificance is lost. This sort of nature-is dust and ashes and there has been much of it. The child should never quired to learn the name of any-; in the nature-study work; but the should be used so often and so rally in his presence that he will it without being conscious of the ess.

HE STORY AS A SUPPLEMENT TO THE NATURE-STUDY LESSON

any of the subjects for nature lessons be studied only in part, since but one e may be available at the time. Often, cially if there is little probability that pupils will find opportunity to com-e the study, it is best to round out r knowledge by reading or telling the y to supplement the facts which they e discovered for themselves. This y should not be told as a finality or complete picture but as a guide and piration for further study. Always e at the end of the story an interroga-mark that will remain aggressive and stent in the child's mind. To illus-e: Once a club of junior naturalists ught me rose leaves injured by the leaf-ter bee and asked me why the leaves re cut out so regularly. I told them the ry of the use made by the mother bee these oval and circular bits of leaves d made the account as vital as I was le; but at the end I said, "I do not ow which species of bee cut these ves. She is living here among us and ilding her nest with your rose leaves, hich she is cutting every day almost der your very eyes. Is she then so much more clever than you that you cannot see her or find her nest?" For two years following this lesson I received letters from members of this club. Two carpenter bees and their nests were discovered by them and studied before the mysterious leaf-cutter was finally ferreted

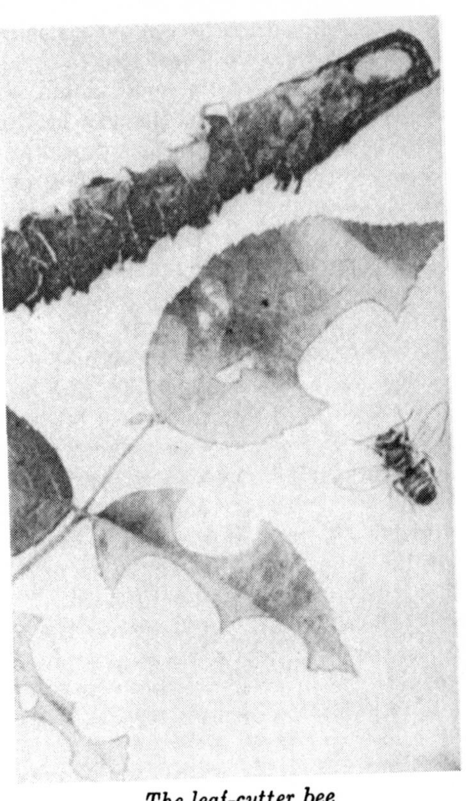

The leaf-cutter bee

out. My story had left something interesting for the young naturalists to discover. The children should be impressed with the fact that the nature story is never finished. There is not a weed or an insect or a tree so common that the child, by observing carefully, may not see things never yet recorded in scientific books; therefore the supplementary story should be made an inspiration for keener interest and further investigation on the part of the pupil. The supplementary story simply thrusts aside some of the obscuring underbrush, thus revealing more plainly the path to further knowledge.

The Nature-Study Attitude toward Life and Death

Perhaps no greater danger besets the pathway of the nature-study teacher than the question involved in her pupils' attitude toward life and death. To inculcate in the child a reverence for life and yet to keep him from becoming mawkish and morbid is truly a problem. It is almost inevitable that the child should become sympathetic with the life of the animal or plant studied, since a true understanding of the life of any creature creates an interest which stimulates a desire to protect this particular creature and make its life less hard. Many times, within my own experience, have I known boys, who began by robbing birds' nests for egg collections, to end by becoming most zealous protectors of the birds. The humane qualities within these boys budded and blossomed in the growing knowledge of the lives of the birds. At Cornell University, it is a well-known fact that those students who turn aside so as not to crush the ant, caterpillar, or cricket on the pavement are almost invariably those that are studying entomology; and in America it is the botanists themselves who are leading the crusade for flower protection.

Thus, the nature-study teacher, if she does her work well, is a sure aid in inculcating a respect for the rights of all living beings to their own lives; and she needs only to lend her influence gently in this direction to change carelessness to thoughtfulness and cruelty to kindness. But with this impetus toward a reverence for life, the teacher soon finds herself in a dilemma from which there is no logical way out, so long as she lives in a world where lamb chop, beefsteak, and roast chicken are articles of ordinary diet; a world in fact, where every meal is based upon the death of some creature. For if she places much emphasis upon the sacredness of life, the children soon begin to question whether it be right to slay the lamb or the chicken for their own food. It would seem that there is nothing for the consistent nature-study teacher to do but become a vegetarian, and even there might arise refinements in this tion of taking life; she might have t sider the cruelty to asparagus in c it off in plump infancy, or the eth devouring in the turnip the food la by the mother plant to perfect her In fact, a most rigorous diet woul forced upon the teacher who shoul fuse to sustain her own existence a cost of life; and if she should attem teach the righteousness of such a she would undoubtedly forfeit her tion; and yet what is she to do! She soon find herself in the position of a tain lady who placed sheets of stick paper around her kitchen to rid her h of flies, and then in mental anguish pi off the buzzing, struggling victims sought to clean their too adhesive w and legs.

In fact, drawing the line between w to kill and what to let live requires use of common sense rather than l First of all, the nature-study teacher, w exemplifying and encouraging the mane attitude toward the lower creat and repressing cruelty which want causes suffering, should never mag the terrors of death. Death is as nat as life and is the inevitable end of phy life on our globe. Therefore, every s and every sentiment expressed wh makes the child feel that death is terr is wholly wrong. The one right way teach about death is not to emphasiz one way or another, but to deal wit as a circumstance common to all; it sho be no more emphasized than the fact t creatures eat or fall asleep.

Another thing for the nature-st teacher to do is to direct the interest the child so that it shall center upon hungry creature rather than upon the which is made into the meal. It is v to emphasize that one of the conditi imposed upon every living being in woods and fields is that if it is cle enough to get a meal it is entitled to when it is hungry. The child natura takes this view of it. I remember w that as a child I never thought parti about the mouse which my cat ating; in fact, the process of trans- ıg mouse into cat seemed altogether ɪr, but when the cat played with the e, that was quite another thing, and ever permitted. Although no one ap- ıtes more deeply than I the debt h we owe to Thompson Seton and ɾs of his kind, who have placed be- the public the animal story from the al point of view and thus set us all ıinking, yet it is certainly wrong to ess this view too strongly upon the g and sensitive child. In fact, this ess should not begin until the judg- t and the understanding are well de- ɔed, for we all know that although ıg the other fellow's standpoint is a ce of strength and breadth of mind, living the other fellow's life is, at , an enfeebling process and a futile e of energy.

OULD THE NATURE-STUDY TEACHER TEACH HOW TO DESTROY LIFE?

is probably within the proper scope he nature-study teacher to place em- sis upon the domain of man, who, be- the most powerful of all animals, as- s his will as to which ones shall live in midst. From a standpoint of abstract ice, the stray cat has just as much ıt to kill and eat the robin which lds in the vine of my porch as the ɪin has to pull and eat the earth- ɾms from my lawn; but the place is ne, and I choose to kill the cat and pre- ve the robin.

When emphasizing the domain of n, we may have to deal with the kill- ; of creatures which are injurious to : interests. Nature-study may be tribu- y to this, in a measure and indirectly, t the study of this question is surely ɪt nature-study. For example, the child ıdies the cabbage butterfly in all its ıges, the exquisitely sculptured yellow g, the velvety green caterpillar, the ırysalis with its protecting colors, the hite-winged butterfly, and becomes in- rested in the life of the insect. Not ıder any consideration, when the attention of the child is focused on the insect, should we suggest a remedy for it when it becomes a pest. Let the life story of the butterfly stand as a fascinating page of nature's book. But later, when the child enters on his career as a gardener, when he sets out his row of cabbage plants and waters and cultivates them, and does his best to bring them to maturity, along comes the butterfly, now an arch enemy, and begins to rear her progeny on the product of his toil. Now the child's interest is focused on the cabbage, and the question is not one of killing insects so much as of saving plants. In fact, there is nothing in spraying the plants with Paris green which suggests cruelty to innocent caterpillars, nor is the process likely to harden the child's sensibilities.

To gain knowledge of the life story of insects or other creatures is nature-study. To destroy them as pests is a part of agriculture or horticulture. The one may be of fundamental assistance to the other, but the two are quite separate and should never be confused.

THE FIELD NOTEBOOK

A field notebook may be made a joy to the pupil and a help to the teacher. Any kind of blank book will do for this, except that it should not be too large to be carried in the pocket, and it should always have the pencil attached. To make the notebook a success the following rules should be observed:

(a) The book should be considered the personal property of the child and should never be criticized by the teacher except as a matter of encouragement; for the spirit in which the notes are made is more important than the information they cover.

(b) The making of drawings to illustrate what is observed should be encouraged. A graphic drawing is far better than a long description of a natural object.

(c) The notebook should not be regarded as a part of the work in English. The spelling, language, and writing of the notes should all be exempt from criticism.

(d) As occasion offers, outlines for ob-

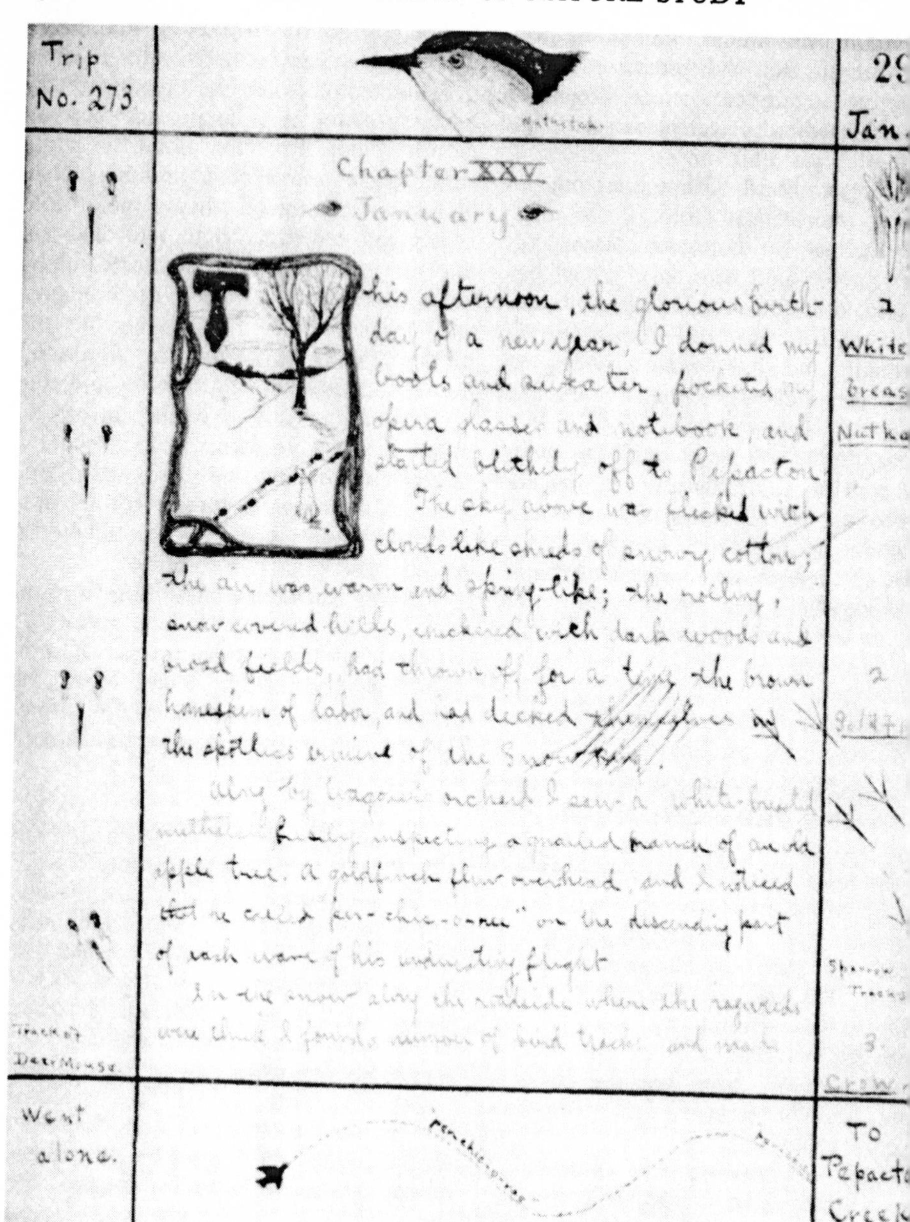

Trip No. 273

Nuthatch.

Chapter XXV

January

This afternoon, the glorious birth-day of a new year, I donned my boots and sweater, pocketed my opera glasses and notebook, and started blithely off to Pepacton.

The sky above was flecked with clouds like shreds of snowy cotton; the air was warm and spring-like; the rolling, snow covered hills, checkered with dark woods and broad fields, had thrown off for a time the brown homespun of labor, and had decked themselves in the spotless ermine of the Snow King.

Along by Wagoner's orchard I saw a white-breasted nuthatch busily inspecting a gnarled branch of an old apple tree. A goldfinch flew overhead, and I noticed that he called "per-chic-o-ree" on the descending part of each wave of his undulating flight.

In the snow along the roadside where the ragweeds were thick I found numbers of bird tracks and ma

Track of Deer Mouse

Went alone.

Jan.

White breas Nuthat

Sparrow Tracks

Crow

To Pepact Creek

A page from the field notebook of a boy of fourteen who read Thoreau and admired the boo of Ernest Thompson Seton

serving certain plants or animals may be placed in the notebook previous to the field excursion so as to give definite points for the work.

(e) No child should be compelled to have a notebook.

The field notebook is a veritable gold mine for the nature-study teacher to work in securing voluntary and happy obser tions from the pupils concerning th out-of-door interests. It is a friendly g which admits the teacher to a knowled of what the child sees and cares f Through it she may discover where t child's attention impinges upon t realm of nature and thus may kno

A brook in winter

: to find the starting point for cul-
ng larger intelligence and wider in-
.

ave examined many field notebooks by pupils in the intermediate grades have been surprised at their pleni- of accurate observation and graphic ration. These books ranged from : account books furnished by the ly grocer up to a quarto, the pages of h were adorned with many marginal rations made in passionate admira- of Thompson Seton's books and l with carefully transcribed text that ved the direct influence of Thoreau. se books, of whatever quality, are pre- s beyond price to their owners. And not? For they represent what cannot ought or sold, personal experience in happy world of out-of-doors.

The Field Excursion

Many teachers look upon the field ex- sion as a precarious voyage, steered be- en the Scylla of hilarious seeing too ch and the Charybdis of seeing noth- at all because of the zest which comes m freedom in the fields and wood. is danger can be obviated if the teacher ns the work definitely before starting, l demands certain results.

It is a mistake to think that a half day necessary for a field lesson, since a very efficient field trip may be made during the ten or fifteen minutes at recess, if it is well planned. Certain questions and lines of investigation should be given the pupils before starting and given in such a manner as to make them thoroughly interested in discovering the facts. A certain teacher in New York State has studied all the common plants and trees in the vicinity of her school by means of these recess excursions and the pupils have been enthusiastic about the work.

The half-hour excursion should be preceded by a talk concerning the purposes of the outing and the pupils must know that certain observations are to be made or they will not be permitted to go again. This should not be emphasized as a punishment; but they should be made to understand that a field excursion is only, naturally enough, for those who wish to see and understand outdoor life. For all field work, the teacher should make use of the field notebook which should be a part of the pupils' equipment.

Pets as Nature-Study Subjects

Little attention has been given to making the child understand what would be the lives of his pets if they were in their native environment, or to relating their habits and lives as wild animals. Almost any pet, if properly observed, affords an admirable opportunity for understanding the reasons why its structure and peculiar habits may have made it successful among other creatures and in other lands.

Moreover, the actions and the daily

W. J. Hamilton, Jr.

Young woodchucks

life of the pet make interesting subject matter for a notebook. The lessons on the dog, rabbit, and horse as given in this volume may suggest methods for such study, and with apologies that it is not better and more interesting, I have placed with the story of the squirrel a few pages from one of my own notebooks regarding my experiences with "Furry." I include this record as a suggestion for the children that they should keep notebooks of their pets. It will lead them to closer observation and to a better and more natural expression of their experiences.

The Correlation of Nature-Study with Language Work

Nature-study should be so much a part of the child's thought and interest that it will naturally form a thought core for other subjects quite unconsciously on his part. In fact, there is one safe rule for correlation in this case — it is legitimate and excellent training as long as the pupil does not discover that he is correlating. But there is something in human nature which revolts against doing one thing to accomplish quite another. A boy once said to me, "I'd rather never go on a field excursion than to have to write it up for English," a sentiment I sympathized with keenly; ulterior motive is sickening to the honest spirit. But if that same boy had been a member of a field class and had enjoyed all the new experiences and had witnessed the interesting things discovered on this excursion, and if later his teacher had asked him to write for her an account of some part of it, because *she wished to know what he had discovered*, the chances are that he would have written his story joyfully and with a certain pride that would have counted much for achievement in word expression.

When Mr. John Spencer, known to so many children in New York State as "Uncle John," was conducting the Junior Naturalist Clubs, the teachers allowed letters to him to count for language exercises; and the eagerness with which these letters were written should given the teachers the key to the method of teaching English. Mr. Sp requested the teachers not to corre letters, because he wished the ch to be thinking about the subject rather than the form of expression so anxious were many of the pup make their letters perfect that the nestly requested their teachers to them write correctly, which was an condition for teaching them En Writing letters to Uncle John was a joy to the pupils that it was use privilege and a reward of merit in schools. One rural teacher reduce percentage of tardiness to a minimu giving the first period in the morni the work in English which consist letters to Uncle John.

Why do pupils dislike writing En exercises? Simply because they are interested in the subject they are to write about, and they know tha teacher is not interested in the info tion contained in the essay. But they are interested in the subject write about it to a person who is ested, the conditions are entirely char If the teacher, overwhelmed as she work and perplexities, could only ke mind that the purpose of a languag after all, merely to convey ideas, son her perplexities would fade away. A veyance naturally should be fitted fo load it is to carry, and if the pupi quires the load first he is very likel construct a conveyance that will be quate. How often the conveyance is m perfect through much effort and poli through agony of spirit and the load tirely forgotten!

Nature-study lessons give much e lent subject matter for stories and es but these essays should never be critic or defaced with the blue pencil. T should be read with interest by teacher; the mistakes made in them transformed as to be unrecognizable, be used for drill exercises in grammat construction. After all, grammar and sp ing are only gained by practice and th royal road leading to their acquire-

Correlation of Nature-Study and Drawing

e correlation of nature-study and ng is so natural and inevitable that eds never be revealed to the pupil. n the child is interested in studying object, he enjoys illustrating his ob- tions with drawings; the happy ab-

. mounted fern. A pressed dry fern placed a layer of cotton batting backed by card- rd is covered with a sheet of cellophane l is slipped into an envelope from which a el has been cut

ption of children thus engaged is a ight to witness. At its best, drawing is erfectly natural method of self-expres- n. The savage and the young child, th untutored, seek to express them- ves and their experiences by this means. is only when the object to be drawn foreign to the interest of the child that awing is a task.

Nature-study offers the best means for idging the gap that lies between the kindergarten child who makes drawings because he loves to and is impelled to from within, and the pupil in the grades who is obliged to draw what the teacher places before him. From making crude and often meaningless pencil strokes, which is the entertainment of the young child, to the outlining of a leaf or some other simple and interesting natural object, is a normal step full of interest for the child because it is still self-expression.

Miss Mary E. Hill, formerly of the Goodyear School of Syracuse, gave each year an exhibition of the drawings made by the children in the nature-study classes; and these were universally so excellent that most people regarded them as an exhibition from the art department; and yet many of these pupils never had had lessons in drawing. They had learned to draw because they liked to make pictures of the living objects which they had studied. One year there were in this exhibit many pictures of toads in various stages, and although their anatomy was sometimes awry in the pictures, yet there was a certain vivid expression of life in their representation; one felt that the toads could jump. Miss Hill allowed the pupils to choose their own medium, pencil, crayon, or water color, and said that they seemed to feel which was best. For instance, when drawing the outline of trees in winter they chose pencil, but when representing the trillium or iris they preferred the water color, while for bittersweet and crocuses they chose the colored crayons.

It is through this method of drawing that which interests him that the child retains and keeps as his own what should be an inalienable right, a graphic method of expressing his own impressions. Too much have we emphasized drawing as an art; it may be an art, if the one who draws is an artist; but if he is not an artist, he still has a right to draw if it pleases him to do so. We might as well declare that a child should not speak unless he put his words into poetry, as to declare that he should not draw because his drawings are not artistic.

The Correlation of Nature-Study with Geography

Life depends upon its environment. Geographical conditions and limitations have shaped the mold into which plastic life has been poured and by which its form has been modified. It may be easy for the untrained mind to see how the deserts and oceans affect life. Cattle may not roam in the former because there is

U. S. Geological Survey — Photo by W. G. Pierce

A meandering stream

nothing there for them to eat, nor may they occupy the latter because they are not fitted for breathing air in the water. And yet the camel can endure thirst and live on the scant food of the desert; and the whale is a mammal fitted to live in the sea. The question is, how are we to impress the child with the "have to" which lies behind all these geographical facts? If animals live in the desert they *have to* subsist on scant and peculiar food which grows there; they *have to* get along with little water; they *have to* endure heat and sand storms; they *have to* have eyes that will not become blinded by the vivid reflection of the sunlight on the sand; they *have to* be of sand color so that they may escape the eyes of their enemies or creep upon their prey unperceived.

All these "have to's" are not mere chance, but they have existed so long that the animal, by constantly coming in contact with them, has attained its present form and habits.

There are just as many "have to's" in the stream or the pond back of the schoolhouse, on the dry hillside behind it, or in the woods beyond the creek as there are in desert or ocean; and when the child gets an inkling of this fact, he has a great step into the realm of geog When he realizes why water lilie grow only in still water that is nc deep and which has a silt bottom why the cattails grow in swamps there is not too much water, and wh mullein grows in the dry pasture, why the hepatica thrives in the damp woods, and why the daisies in the meadows, he will understand this partnership of nature and geog illustrates the laws which govern Many phases of physical geograph long to the realm of nature-study: brook, its course, its work of erosion sedimentation; the rocks of many k the soil, the climate, the weather, a legitimate subjects for nature-study sons.

The Correlation of Nature-St with History

There are many points where na study impinges upon history in a that may prove the basis for an insp lesson. Many of our weeds, cultiv plants, and domestic animals have introduced from Europe and are a pa our colonial history; while many of most commonly seen creatures have pl their part in the history of ancient ti For instance, the bees which gave to the only means available to him for sw ening his food until the 17th century, closely allied to the home life of anc peoples. The buffalo which ranged western plains had much to do with life of the red man. The study of the g hopper brings to the child's atten stories of the locusts' invasion mentio in the Bible, and the stars which witnes our creation and of which Job sang the ancients wrote, shine over our he every night.

But the trees, through the lengthy s of their lives, cover more history indivi ally than do other organisms. In glanc across the wood-covered hills of N York one often sees there, far above other trees, the gaunt crowns of old w pines. Such trees belonged to the fo

val and may have attained the age
ɔ centuries; they stand there look-
ɪt over the world, relics of another
vhen America belonged to the red
and the bear and the panther played
ɪght beneath them. The cedars live

The Arnold Arboretum

he Endicott pear tree. This tree was ɪted by Governor John Endicott in his len in Salem, Massachusetts, in 1630. rge Washington, Abraham Lincoln, and ɪiel Webster enjoyed the fruit of this riarchal tree. Sprouts, shown above, from old tree still bear

ger than do the pines, and the great rlet oak may have attained the age of ɪr centuries before it yields to fate.

Perhaps in no other way can the atten- n of the pupil be turned so naturally past events as through the thought ɪt the life of such a tree has spanned much of human history. The life his- ry of one of these ancient trees should made the center of local history; let e pupils find when the town was first ttled by the whites and where they came om, and how large the tree was then; hat Indian tribes roamed the woods be- re that and what animals were common the forest when this tree was a sapling. hus may be brought out the chief events the history of the county and town- ɪip, when they were established and for whom or what they were named; and a comparison of the present industries may be made with those of a hundred years ago.

The Correlation of Nature-Study with Arithmetic

The arithmetical problems presented by nature-study are many; some of them are simple and some of them are complicated, and all of them are illuminating. Seed distribution especially lends itself to computation; a milkweed pod contains 140 seeds; there are five such pods on one plant; each milkweed plant requires at least one square foot of ground to grow on; how much ground would be required to grow all of the seeds from this one plant? Or, count the seeds in one dandelion head, multiply by the number of flower heads on the plant and estimate how many plants can grow on a square foot, then ask a boy how long it would take for one dandelion plant to cover his

W. C. Muenscher

A red cedar and its seedlings

father's farm with its progeny; or count the blossoms on one branch of an apple tree, later count the ripened fruit; what percentage of blossoms matured into fruit? Measuring trees, their height and thickness and computing the lumber they will make combines arithmetic and geometry, and so on ad *infinitum*.

As a matter of fact, the teacher will find in almost every nature lesson an arithmetic lesson; and when arithmetic is used in this work, it should be vital and inherent and not "tacked on"; the pupils should be really interested in the answers to their problems; and as with all correlation, the success of it depends upon the genius of the teacher.

Gardening and Nature-Study

Erroneously, some people maintain that gardening is nature-study; this is not so necessarily nor ordinarily. Gardening may be a basis for nature-study, but it is rarely made so to any great extent. Even the work in children's gardens is so conducted that the pupils know little or nothing of the flowers or vegetables which they grow except their names, their uses to man, and how to cultivate them. They are taught how to prepare the soil, but the reason for this from the plant's standpoint is never revealed; and if the child becomes acquainted with the plants in his garden, he makes the discovery by himself. All this is nothing against gardening! It is a wholesome and valuable experience for a child to learn how to make a garden even if he remains ignorant of the interesting facts concerning the plants which he there cultivates. But if the teachers are so inclined, they may find in the garden and its products the most interesting material for the best of nature lessons. Every plant the child grows is an individual with its own peculiarities as well as those of its species in manner of growth. Its roots, stems, and leaves are of certain form and structure; and often the special uses to the plant of its own kind of leaves, stems, and roots are obvious. Each plant has its own form of flower and even its own tricks for securing pollination; and its own manner of developing and scattering its seeds. Every weed of the garden has developed some special method of winning and holding its place among the cultivated plants; and in no other way can the child so fully and naturally come into a comprehension of that term "the survival of the fittest" as by studying the ways of the fit as e
plified in the triumphant weeds c
garden.

Every earthworm working belov
soil is doing something for the ga
Every bee that visits the flowers th
on an errand for the garden as well
herself. Every insect feeding on le
root is doing something to the ga
Every bird that nests near by or that
visits it, is doing something which a
the life and the growth of the gar
What all of these uninvited guests
doing is one field of garden nature-st
Aside from all this study of indivi
life in the garden, which even the yo
est child may take part in, there are
more advanced lessons on the soil. W
kind of soil is it? From what sort of
was it formed? What renders it me
and fit for the growing of plants? M
over, what do the plants get from it? H
do they get it? What do they do v
what they get?

This leads to the subject of plant ph
ology, the elements of which may
taught simply by experiments carried
by the children themselves, experime
which should demonstrate the sap c
rents in the plant; the use of water

carry food and to make the plant rig
the use of sunshine in making the pla
food in the leaf laboratories; the nouris
ment provided for the seed and its gern
nation, and many other similar lessons.

A child who makes a garden, and th
becomes intimate with the plants he cu
tivates, and comes to understand the i
terrelation of the various forms of li

he finds in his garden, has pro- far in the fundamental knowledge ure's ways as well as in a practical edge of agriculture.

ATURE-STUDY AND AGRICULTURE

kily, thumb-rule agriculture is be- ushed to the wall in these enlight- days. Thumb rules would work better if nature did not vary her mances in such a confusing way. nment experiment stations were es- ıed because thumb rules for farm- vere unreliable and disappointing; ll the work of all the experiment ns has been simply advanced nature- and its application to the practice riculture. Both nature-study and ag- ure are based upon the study of life the physical conditions which en- ge or limit life; this is known to the as the study of the natural sciences; f we see clearly the relation of nature- to science, we may understand r the relation of nature-study to ag- ure, which is based upon the sciences. ture-study is science brought home. a knowledge of botany, zoology, and gy as illustrated in the dooryard, the field or the woods back of the house. e people have an idea that to know sciences one must go to college; do not understand that nature has ished the material and laboratories very farm in the land. Thus, by be- ing with the child in nature-study we him to the laboratory of the wood arden, the roadside or the field, and materials are the wild flowers or the

Marion E. Wesp

A wheat shock

Dept. of Agronomy, N. Y. State College of Agriculture

A meadow at harvest time

weeds, or the insects that visit the golden-rod or the bird that sings in the maple tree, or the woodchuck whistling in the pasture. The child begins to study living things anywhere or everywhere, and his progress is always along the various tracks laid down by the laws of life, along which his work as an agriculturist must always progress if it is to be successful.

The child through nature-study learns the way a plant grows, whether it be an oak, a turnip, or a pigweed; he learns how the roots of each are adapted to its needs; how the leaves place themselves to get the sunshine and why they need it; and how the flowers get their pollen carried by the bee or the wind; and how the seeds are finally scattered and planted. Or he learns about the life of the bird, whether it be a chicken, an owl, or a bobolink; he knows how each bird gets its food and what its food is, where it lives, where it nests, and its relation to other living things. He studies the bumblebee and discovers its great mission of pollen-carrying for many flowers, and in the end would no sooner strike it dead than he would voluntarily destroy his clover patch. This is the kind of learning we call nature-study and not science or agriculture. But the country child can never learn anything in nature-study that has not something to do with science, and that has not its own practical lesson for him, when he shall become a farmer.

Some have argued, "Why not make nature-study solely along the lines of agriculture? Why should not the child begin nature-study with the cabbage rather than with the wild flowers?" This argument carried out logically provides recreation for a boy in hoeing corn rather than in playing ball. Many parents in the past have argued thus and have, in consequence, driven thousands of splendid boys from the country to the city with a loathing in their souls for the drudgery which seemed all there was to farm life. The reason the wild flowers may be selected for beginning the nature-study of plants is that every child loves these woodland posies, and his happiest hours are spent in gathering them. Never yet have we known of a case where a child, having gained his knowledge of the way a plant lives through studying the plants he loves, has failed to be interested and delighted to find that the wonderful things he discovered about his wild flower may be true of the vegetable in the garden, or the purslane which fights with it for ground to stand upon.

Some have said, "We, as farmers, care only to know what concerns our pocketbooks; we wish only to study those things which we must, as farmers, cultivate or destroy. We do not care for the butterfly, but we wish to know the plum weevil; we do not care for the trillium, but we are interested in the onion; we do not care for the meadowlark, but we cherish the gosling." This is an absurd argument since it is a mental impossibility for any human being to discriminate between two things when he knows or sees only one. In order to understand the important economic relations to the world of one plant or animal, it is absolutely necessary to have a wide knowledge of other plants and animals. One might as well say, "I will see the approaching cyclone, but never look at the sky; I will look at the clover, but not see the dandelion; I will look for the sheriff when he comes over the hill, but will not see any other team on the road."

Nature-study is an effort to make the individual use his senses instead of losing them; to train him to keep his eyes open to all things so that his powers crimination shall be based on wi The ideal farmer is not the man w hazard and chance succeeds; he man who loves his farm and all th rounds it because he is awake t beauty as well as to the wonders are there; he is the man who under as far as may be the great forces of which are at work around him, and fore he is able to make them wo him. For what is agriculture save a sion of natural forces for the bene man! The farmer who knows these only when restricted to his paltry and has no idea of their larger applic is no more efficient as a farmer than a who knew only how to start and st engine would be as an engineer.

In order to appreciate truly his the farmer must needs begin as a with nature-study; in order to be su ful and make the farm pay, he must continue in nature-study; and to mak declining years happy, content, fu wide sympathies and profitable tho he must needs conclude with na study; for nature-study is the alphab agriculture and no word in that grea cation may be spelled without it.

Nature-Study Clubs

The organizing by the pupils of a for studying out-of-door life is a great and inspiration to the work in nature-s in the classroom. The essays and the before the club prove efficient aid in lish composition; and the varied inte of the members of the club furnish and vital material for study. A button badge may be designed for the club of course, it must have a constitution bylaws. The proceedings of the club m ings should be conducted accordin parliamentary rules; but the field ex sions should be entirely informal.

The meetings of the Junior Natura Clubs, as organized in the schools of N York State by Mr. John W. Spen were most impressive. The school ses would be brought to a close, the tea stepping down and taking a seat with

. The president of the club, some
ıl boy or slender slip of a girl,
take the chair and conduct the
ng with a dignity and efficiency
y of a statesman. The order was per-
the discussion much to the point.
fess to a feeling of awe when I at-
d these meetings, conducted so seri-
and so formally, by such youngsters.
ubtedly, the parliamentary training
xperience in speaking impromptu are
g the chief benefits of such a club.
ese clubs may be organized for spe-
tudy. In one bird club of which I
there have been contests. Sides
chosen and the number of birds seen
from May 1 to 31 inclusive was the test of supremacy. Notes on the birds were taken in the field with such care that, when at the end of the month each member handed in his notes, they could be used as evidence of accurate identification. An umpire decided the doubtful points with the help of bird manuals. The contest was always close and exciting.

The programs of the nature club should be varied so as to be continually interesting. Poems and stories concerning the objects studied help make the program attractive. Observing nature, however, should be the central theme of all meetings.

HOW TO USE THIS BOOK

rst and indispensably, the teacher
ld have at hand the subject of the
n. She should make herself familiar
the points covered by the questions
read the story before giving the les-
If she does not have the time to go
the observations suggested before
g the lesson, she should take up the
tions with the pupils as a joint inves-
ion, and be boon companion in dis-
ring the story.

he story should not be read to the
ils. It is given as an assistance to the
her, and is not meant for direct in-
nation to the pupils. If the teacher
ws a fact in nature's realm, she is then
position to lead her pupils to dis-
er this fact for themselves.

lake the lesson an investigation and
e the pupils feel that they are in-
igators. To tell the story to begin
h inevitably spoils this attitude and
nches interest.

he "leading thought" embodies
e of the points which should be in
teacher's mind while giving the les-
; it should not be read or declared to
pupils.

The outlines for observations herein
en by no means cover all of the ob-
vations possible; they are meant to suggest to the teacher observations of her own, rather than to be followed slavishly.

The suggestions for observations have been given in the form of questions, merely for the sake of saving space. The direct questioning method, if not employed with discretion, becomes tiresome

Marion E. Wesp

to both pupil and teacher. If the questions do not inspire the child to investigate, they are useless. To grind out answers to questions about any natural object is not nature-study, it is simply "grind," a form of mental activity which is of much greater use when applied to spelling or the multiplication table than to the study of nature. The best teacher will cover the points suggested for observations with few direct questions. To those who find the questions inadequate I will say that, although I have used these outlines once, I am sure I should never be able to use them again without making changes.

Marion E. Wesp

A hickory tree

The topics chosen for these lesson not be the most practical or the interesting or the most enlight that are to be found; they are s those subjects which I have used i classes, because we happened to find at hand the mornings the lessons given.

While an earnest attempt has made to make the information in book accurate, it is to be expected a be hoped that many discrepancies be found by those who follow the les No two animals or plants are just and no two people see things exactl same way. *The chief aim of this vo is to encourage investigation rather to give information.* Therefore, if takes are found, the object of the will have been accomplished, and author will feel deeply gratified. If teacher finds that the observations by her and her pupils do not agree the statements in the book, I earn enjoin upon her to trust to her own rather than to any book.

No teacher is expected to teach al lessons in this book. A wide rang subjects is given, so that congenial ch may be made.

PART II
ANIMALS

ANIMAL GROUPS

For some inexplicable reason, the word *animal*, in common parlance, is restricted to the mammals. As a matter of fact, the bird, the fish, the insect, and the snake have as much right to be called animals as the squirrel or the deer. And while I believe that much freedom in the matter of scientific nomenclature is permissible in nature-study, I also believe that it is well for the child to have a clearly defined idea of the classes into which the animal kingdom is divided; I would have him gain this knowledge by noting how one animal differs from another rather than by st ing the classification of animals in b He sees that the fish differs in many from the bird and that the toad di from the snake; and it will be easy him to grasp the fact that the mam differ from all other animals in that t young are nourished by milk from breasts of the mother; when he app ates this, he will understand that s diverse forms as the whale, the cow, bat, and man are members of one g class of animals.

BIRDS

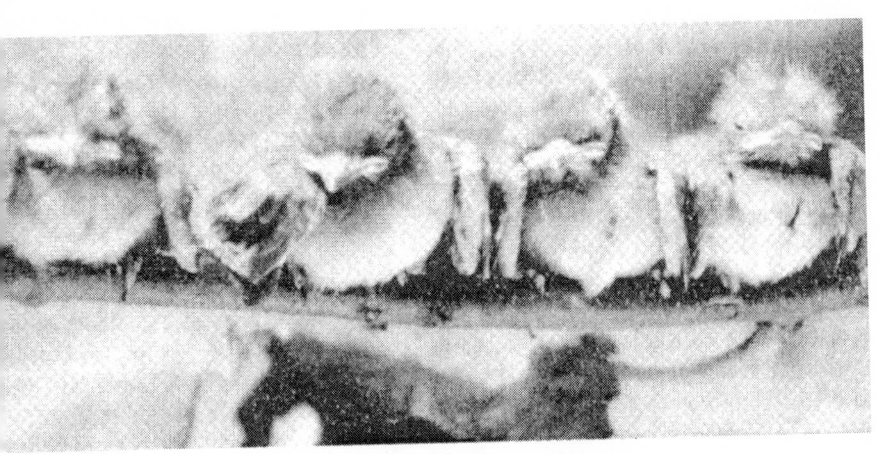

Young phoebes that have just left the nest

e reason for studying any bird is to
tain what it does; in order to accom-
this, it is necessary to know what
bird is, learning what it is being
y a step that leads to a knowledge
at it does. But, to hear some of our
devotees talk, one would think that
able to identify a bird is all of bird
. On the contrary, the identification
rds is simply the alphabet to the real
, the alphabet by means of which
nay spell out the life habits of the
To know these habits is the ambition
e true ornithologist, and should like-
be the ambition of the beginner,
though the beginner be a young
.

everal of the most common birds have
selected as subjects for lessons in
book; other common birds, like the
be and the wrens, have been purposely
tted; after the children have studied
birds, as indicated in the lessons, they
enjoy working out lessons for themes with other birds. Naturally, the sence of these lessons does not follow
ntific classification; in the first lessons,
attempt has been made to lead the
child gradually into a knowledge of bird
life. Beginning with the chicken there follow naturally the lessons with pigeons and
the canary; then there follow the careful
and detailed study of the robins and constant comparison of them with the bluebirds. This is enough for the first year
in the primary grades. The next year the
work begins with the birds that remain
in the North during the winter, the

Leonard K. Beyer

A family of cedar waxwings

chickadee, nuthatch, and downy woodpecker. After these have been studied carefully, the teacher may be an opportunist when spring comes and select any of the lessons when the bird subjects are at hand. The classification suggested for the woodpeckers and the swallows is for more advanced pupils, as are the lessons on the geese and turkeys. It is to be hoped that these lessons will lead the child directly to the use of the bird books, of which there are many excellent ones.

Beginning Bird Study in the Primary Grades

The hen is especially adapted as an object lesson for the young beginner of bird study. First of all, she is a bird, notwithstanding the adverse opinions of two of my small pupils who stoutly maintained that "a robin is a bird, but a hen is a hen." Moreover, the hen is a bird always available for nature-study; she looks askance at us from the crates of the world's marts; she comes to meet us in the country barnyard, stepping toward us sedately; looking at us earnestly with one eye, then

Leonard K. Beyer

A redstart at her nest

turning her head so as to check up her observations with the other; meantime she asks us a little question in a wheedling, soft tone, which we understand perfectly to mean, "Have you perchance brought me something to eat?" Not only is the hen an interesting bird in herself, but she is a bird with problems; a studying her carefully we may be duced into the very heart and cen bird life.

This lesson may be presented i ways: First, if the pupils live in the try, where they have poultry at hom whole series of lessons may best be a plished through talks by the teache lowed on the part of the children b servations to be made at home. Th sults of these observations should be in school in oral or written lessons ond, if the pupils are not familiar fowls, a hen and a chick, if possible, s be kept in a cage in the schoolroom few days, and a duck or gosling shou brought in one day for observation. crates in which fowls are sent to m make very good cages. One of the tea of the Elmira, N. Y. schools introd into the basement of the schoolho hen, which there hatched her broo chicks, much to the children's deligh edification. After the pupils have bee thoroughly interested in the hen an familiar with her ways, after they hav her and watched her, and have for sense of ownership, the following le may be given in an informal manner, they were naturally suggested to teacher's mind through watching the

FEATHERS AS CLOTHING

ıe bird's clothing affords a natural ıning for bird study because the wear- ›f feathers is a most striking character- distinguishing birds from other crea-

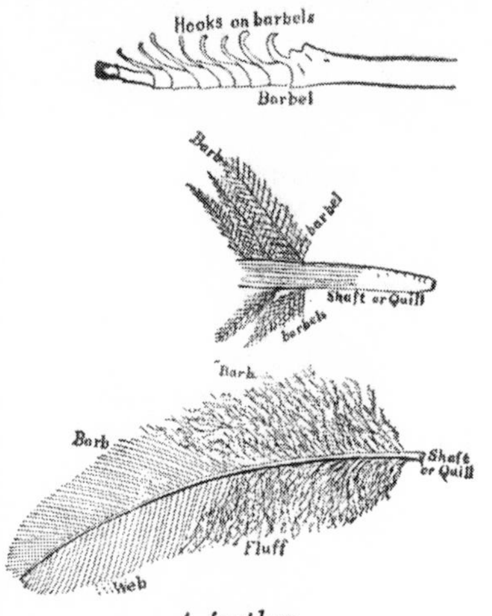

A feather

:s; also, feathers and flying are the first ıgs the young child notices about birds. The purpose of all these lessons on hen are: (a) To induce the child to ke continued and sympathetic observa- ıs on the habits of the domestic birds.) To cause him involuntarily to com- e the domestic with the wild birds.) To induce him to think for himself w the shape of the body, wings, head, ık, feet, legs, and feathers are adapted in :h species to protect the bird and assist ın getting its living.

The overlapping of the feathers on a hen's back and breast is a pretty illustration of nature's method of shingling, so that the rain, finding no place to enter, drips off, leaving the bird's underclothing quite dry. It is interesting to note how a hen behaves in the rain; she droops her tail and holds herself so that the water finds upon her no resting place, but simply a steep surface down which to flow to the ground.

Each feather consists of three parts, the shaft or quill, which is the central stiff

Feathers help birds to endure the cold

stem of the feather, giving it strength. From this quill come off the barbs which, toward the outer end, join together in a smooth web, making the thin, fanlike portion of the feather; at the base is the fluff, which is soft and downy and near to the body of the fowl. The teacher should put on the blackboard this figure so that incidentally the pupils may learn the parts of a feather and their structure. If a microscope is available, show both the web and the fluff of a feather under a three-fourths objective.

The feathers on the back of a hen are longer and narrower in proportion than those on the breast and are especially fitted to protect the back from rain; the breast feathers are shorter and have more of the fluff, thus protecting the breast from the cold as well as the rain. It is plain to any child that the soft fluff is comparable to our underclothing while the smooth, overlapping web forms a rain- and wind-proof outer coat. Down is a feather with no quill; young chicks are covered with down. A pin-feather is simply a young feather rolled up in a sheath, which bursts later and is shed, leaving the feather free to assume its form. Take a large pin-feather and cut the sheath open and show the pupils the young feather lying within.

When a hen oils her feathers it is a process well worth observing. The oil gland is on her back just at the base of the tail feathers; she squeezes the gland with her beak to get the oil and then rubs the beak over the surface of her

Young pelicans are born naked, but are soon covered with white down

feathers and passes them through it; she spends more time oiling the feathers on her back and breast than those on the other parts, so that they will surely shed water. Country people say that when the hen oils her feathers it is a sure si rain. The hen sheds her feathers o year and is a most untidy looking meanwhile, a fact that she seems to ize, for she is as shy and cross as a y lady caught in company with her h curlers; but she seems very pleased

J. E.

Feathers of a rooster, showing their rela size, shape, and position

1, neck hackle; 2, breast; 3, wing shoulder cover wing flight covert; 5, wing primary; 6, wing seconc 7, wing covert; 8, back; 9, tail covert; 10, main 11, fluff; 12, thigh; 13, saddle hackle; 14, the sick feather of beauty; 15, lesser sickle

herself when she finally gains her feathers.

LESSON 1

Feathers as Clothing

Leading Thought — Feathers gr from the skin of a bird and protect t bird from rain, snow, wind, and co Some of the feathers act as cloaks intoshes and others as undercloth-

ETHOD — The hen should be at close for this lesson where the children observe how and where the different of feathers grow. The pupils should study separately the form of a feather the back, from the breast, from the r side of the body, and a pin-feather.

BSERVATIONS FOR PUPILS — 1. How he feathers arranged on the back of hen? Are they like shingles on the

How does a hen look when standing e rain?

How are the feathers arranged on breast?

Compare a feather from the back one from the breast and note the rence.

Are both ends of these feathers alike? ot, what is the difference?

Is the fluffy part of the feather on the outside or next to the bird's skin? What is its use?

7. Why is the smooth part of the feather (the web) on the outside?

8. Some feathers are all fluff and are called "down." At what age was the fowl all covered with down?

9. What is a pin-feather? Why do you think it is so called?

10. How do hens keep their feathers oily and glossy so they will shed water?

11. Where does the hen get the oil? Describe how she oils her feathers; which ones does she oil most? Does she oil her feathers before a rain?

"How beautiful your feathers be!"
The Redbird sang to the Tulip-tree
New garbed in autumn gold.
"Alas!" the bending branches sighed,
"They cannot like your leaves abide
To keep us from the cold!"

— JOHN B. TABB.

FEATHERS AS ORNAMENT

he ornamental plumage of birds is of the principal illustrations of a great ciple of evolution. The theory is that male birds win their mates because heir beauty, those that are not beauti- being doomed to live single and leave progeny to inherit their dullness. On other hand, the successful wooer ds down his beauty to his sons. How- r, another quite different principle acts n the coloring of the plumage of the ther birds; for if they should develop ht colors themselves, they would at- t the eyes of the enemy to their pre- us hidden nests; only by being incon- cuous are they able to protect their s and nestlings from discovery and th. The mother partridge, for instance, o nearly the color of the dead leaves on ground about her that we may almost p upon her before we discover her; if were the color of the male oriole or ager she would very soon be the center attraction to every prowler. Thus it has come about that among the birds the male has developed gorgeous colors which attract the female, while the female has kept modest, unnoticeable plumage.

Olin Sewall Pettingill, Jr.

Not a candidate for a beauty contest. A young belted kingfisher clothed in pin feathers

The curved feathers of the rooster's tail are weak and mobile and could not possibly be of any use as a rudder; but they give grace and beauty to the fowl and cover the useful rudder feathers underneath by a feather fountain of iridescence. The neck plumage of the cock

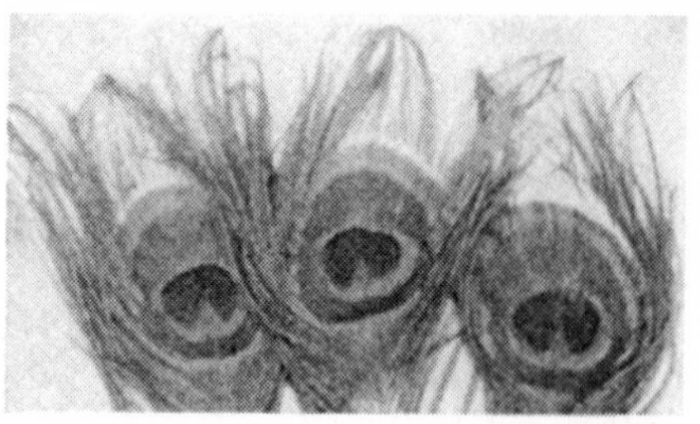

Peacock feathers. Is beauty useful?

is also often luxurious and beautiful in color and quite different from that of the hen. Among the Rouen ducks the brilliant blue-green iridescent head of the drake and his wing bars are beautiful, and make his wife seem Quaker-like in contrast.

As an object lesson to instill the idea that the male bird is proud of his beautiful feathers, I know of none better than that presented by the turkey gobbler, for he is a living expression of self-conscious vanity. He spreads his tail to the fullest extent and shifts it this way and that to show the exquisite play of colors over the feathers in the sunlight, meanwhile throwing out his chest to call particular attention to his blue and red wattles; and to keep from bursting with pride he bubbles over in vainglorious "gobbles."

The hen with her chicks and the turkey hen with her brood, if they follow their own natures, must wander in the fields for food. If they were bright in color, the hawks would soon detect them and their chances of escape would be small; this is an instance of the advantage to the young of adopting the colors of the mother rather than of the father; a fact equally true of the song birds in cases where the males are brilliant in color at maturity. The male Baltimore oriole does not assist his mate in brooding, but he sits somewhere on the home tree and cheers her by his glorious song and by glimpses of his gleaming orange coat. Some have accused him of being lazy; on the contrary, he is a wise householder, for, instead of att ing the attention of crow or squirrel t nest, he distracts their attention fro by both color and song.

A peacock's feather should really lesson by itself, it is so much a thin beauty. The brilliant color of the pu eye-spot, and the graceful flowing b that form the setting to the central are all a training in æsthetics as we in nature-study. After the children studied such a feather let them see peacock, either in reality or in picture, give them stories about this bird of — a bird so inconspicuous, except fo great spread of tail, that a child se him for the first time cried, "Oh, oh, this old hen all in bloom!"

The whole question of sexual selec may be made as plain as need be for little folks, by simply telling them the mother bird chooses for her mate one which is most brightly and beautif dressed; make much of the comb and tles of the rooster and gobbler as additi to the brilliancy of their appearance.

LESSON 2

Feathers as Ornament

Leading Thought — The color feathers and often their shape make so birds more beautiful; while in others, color of the feathers serves to prot them from the observation of their e mies.

Method — While parts of this less relating to fowls may be given in prim grades, it is equally fitted for pupils w have a wider knowledge of birds. Be with a comparison of the plumage of hen and the rooster. Then, if possib study the turkey gobbler and a peacock life or in pictures. Also the plumage o Rouen duck and drake, and if possib the Baltimore oriole, the goldfinch, scarlet tanager, and the cardinal.

Observations — 1. Note the differen in shape and color of the tail feathers hen and rooster.

Do the graceful curved tail feathers rooster help him in flying? Are they nough to act as a rudder?
If not of use in flying what are they Which do you think the more beauti-e hen or the rooster?
In what respects is the rooster a more iful fowl?
What other parts of the rooster's age are more beautiful than that of en?
If a turkey gobbler sees you looking n he begins to strut. Do you think es this to show off his tail feathers? how he turns his spread tail this way that so the sunshine will bring out eautiful changeable colors. Do you he does this so you can see and ad-him?
Describe the difference in plumage een the hen turkey and the gobbler. the hen turkey strut?
Note the beautiful blue-green irides-head and wing patches on the wings e Rouen ducks. Is the drake more tiful than the duck?
What advantage is it for these fowls ve the father bird more beautiful and t in color than the mother bird?
. In the case of the Baltimore oriole, e mother bird as bright in color as the er bird?
. Study a peacock's feather. What

Peacocks

color is the eye-spot? What color around that? What color around that? What color and shape are the outside barbs of the feather? Do you blame a peacock for being proud when he can spread a tail of a hundred eyes? Does the peahen have such beautiful tail feathers as the peacock?

The bird of Juno glories in his plumes;
Pride makes the fowl to preene his feathers so.
His spotted train fetched from old Argus' head,
With golden rays like to the brightest sun,
Inserteth self-love in the silly bird;
Till midst its hot and glorious fumes
He spies his feet and then lets fall his plumes.

— "THE PEACOCK,"
ROBERT GREENE (1560)

HOW BIRDS FLY

o convince the children that a bird's gs correspond to our arms, they should a fowl with its feathers off, prepared market or oven, and they will infer fact at once.
he bird flies by lifting itself through sing down upon the air with its wings. re are several experiments which are ded to make the child understand this. difficult for children to conceive that air is really anything, because they can-see it; so the first experiment should to show that the air is something we push against or that pushes against us. Strike the air with a fan and we feel there is something which the fan pushes; we feel the wind when it is blowing and it is very difficult for us to walk against a hard wind. If we hold an open umbrella in the hand while we jump from a step, we feel buoyed up because the air presses up against the umbrella. The air presses up against the wings of the birds just as it does against the open umbrella. The bird flies by pressing down upon the air with its wings just as a boy jumps high by pressing down with his hands on his vaulting pole.

Study wing and note: (a) That the wings open and close at the will of the bird. (b) That the feathers open and shut on each other like a fan. (c) When the wing is open the wing quills overlap, so

Olin Sewall Pettingill, Jr.

Common tern. While we are having winter this bird spends the summer in South America. It will return to spend our summer with us

that the air cannot pass through them. (d) When the wing is open it is curved so that it is more efficient, for the same reason that an umbrella presses harder against the atmosphere when it is open than when it is broken by the wind and turned wrong side out.

A wing feather has the barbs on the front edge lying almost parallel to the quill, while those on the hind edge come off at a wide angle. The reason for this is easy to see, for this feather has to cut the air as the bird flies; and if the barbs on the front side were like those of the other side, they would be torn apart by the wind. The barbs on the hind side of the feather form a strong, close web so as to press down on the air and not let it through. The wing quill is curved; the convex side is up and the concave side below during flight. The concave side, like the umbrella, catches more air than the upper side; the down stroke of the wings is forward and down; while on the up stroke, as the wing is lifted, it bends at the joint like a fan turned sidewise, and offers less surface to resist the air. Thus, the up stroke does not push the bird down.

Observations should be made on the use of the bird's tail in flight. The hen spreads her tail like a fan when sh[e] to the top of the fence; the robin likewise when in flight. The fact th[at] tail is used as a rudder to guide the in flight, as well as to give more su[rface] for pressing down upon the air, is ha[rd for] the younger pupils to understand, perhaps can be best taught by w[atch]ing the erratic unbalanced flight of y[oung] birds whose tail feathers are no[t] grown.

The tail feather differs from the [wing] feather in that the quill is not curved the barbs on each side are of about [equal] length and lie at about the same ang[le on] each side of the quill. See Fig. p. 30.

How Birds Fly

Leading Thought — A bird flies [by] pressing down upon the air with its wi[ngs] which are made especially for this [pur]pose. The bird's tail acts as a rudder [dur]ing flight.

Method — The hen, it is hoped, [will] by this time be tame enough so that [the] teacher may spread open her wings [for] the children to see. In addition, ha[ve a] detached wing of a fowl such as is use[d in] farmhouses instead of a whisk-broom.

Observations — 1. Do you thin[k a] bird's wings correspond to our arms[? If] so why?

2. Why do birds flap their wings w[hen] they start to fly?

3. Can you press against the air w[ith] a fan?

4. Why do you jump so high wit[h a] vaulting pole? Do you think the bird [uses] the air as you use the pole?

How are the feathers arranged on the so that the bird can use it to press ıe air?

If you carry an umbrella on a windy ing, which catches more wind, the r or the top side? Why is this? Does urved surface of the wing act in the way?

Take a wing feather. Are the barbs ıg on one side of the quill as on the ? Do they lie at the same angle from uill on both sides? If not why?

Which side of the quill lies on the side and which on the inner side of ving?

Is the quill of the feather curved?

. Which side is uppermost in the , the convex or the concave side? a quill in one hand and press the gainst the other hand. Which way it bend more easily, toward the convex or the concave side? What has this to do with the flight of the bird?

11. If the bird flies by pressing the wings against the air on the down stroke, why does it not push itself downward with its wings on the up stroke?

12. What is the shape and arrangement of the feathers which prevent pushing the bird back to earth when it lifts its wings?

13. Why do you have a rudder to a boat?

14. Do you think a bird could sail through the air without something to steer with? What is the bird's rudder?

15. Have you ever seen a young bird whose tail is not yet grown, try to fly? If so, how did it act?

16. Does the hen when she flies keep the tail closed or open like a fan?

17. Compare a tail feather with a wing feather and describe the difference.

MIGRATION OF BIRDS

ıe travelogues of birds are as fascinat- as our favorite stories of fairies, ad- ure, and fiction. If we could accom- certain birds, such as the Arctic s, on their spring and autumn trips, logs of the trips would be far more ex- g than some recorded by famous avia- The Arctic tern seems to hold the rd for long-distance flight. Its nest is e within the bounds of the Arctic cir- nd its winter home is in the region of Antarctic circle. The round-trip mile- for this bird during a year is about oo miles. Wells W. Cooke, a pioneer lent of bird migration, has called atten- to the interesting fact that the Arctic "has more hours of daylight than any r animal on the globe. At the north- nesting-site the midnight sun has ady appeared before the birds' arrival, it never sets during their entire at the breeding grounds. During two nths of their sojourn in the Antarctic birds do not see a sunset, and for the of the time the sun dips only a little below the horizon and broad daylight is continuous. The birds, therefore, have twenty-four hours of daylight for at least eight months in the year, and during the other four months have considerably more daylight than darkness." It is true that few of our birds take such long trips as does the Arctic tern; but most birds do travel for some distance each spring and fall.

Each season brings to our attention certain changes in the bird population. During late summer, we see great flocks of swallows; they are on telephone or telegraph wires, wire fences, clothes lines, or aerial wires. They twitter and flutter and seem all excited. For a few days, as they prepare for their southern journey, they are seen in such groups, and then are seen no more until the following spring. Some birds do not gather in flocks before leaving for the winter; they just disappear and we scarcely know when they go. We may hear their call notes far over our heads as they wing their way to their winter homes. Some birds migrate only during the day, others go only during the night, and others may travel by either day or night.

Those birds that do not migrate are called permanent residents. In the eastern United States chickadees, jays, downy

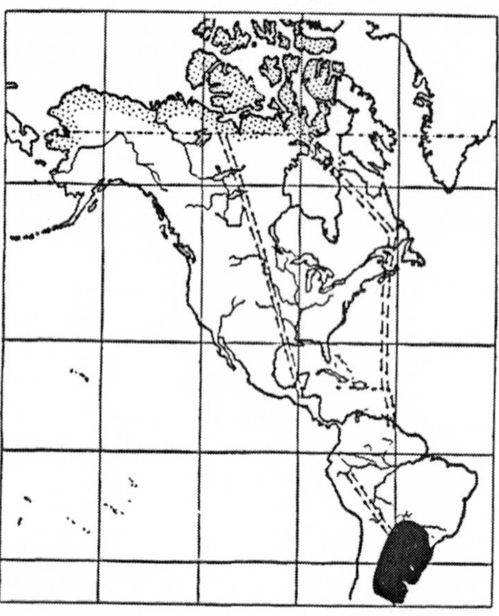

After Cooke

The migration routes of the golden plover. The dotted area is the summer home and nesting place; the black area is the winter home. Migration routes are indicated by arrows. On the southern route the plover makes a flight of 2,400 miles from Labrador to South America

woodpeckers, nuthatches, grouse, and pheasants are typical examples of the permanent resident group. These birds must be able to secure food under even the most adverse conditions. Much of their food is insect life found in or about trees; some fruits and buds of trees, shrubs, and vines are also included in their diet.

Birds that travel are called migratory birds. If the spring migrants remain with us for the summer, we call them our summer residents. Fall migrants that remain with us for the winter are called winter residents. The migrants that do not remain with us but pass on to spend the summer or winter in some other area are called our transients or visitors. Of course, we must remember that the birds which visit us only for a short time are summer residents and winter residents in other parts of the country. Our summer dents are the winter residents of other area.

In spring we await with interest arrival of the first migrants. These are, in general, those which have the winter only a comparatively short tance away. In the eastern United St we expect robins, red-winged blackb song sparrows, and bluebirds among earliest migrants. In many species males arrive first; they may come as m as two weeks ahead of the females. immature birds are usually the last t rive. The time of arrival of the first grants is determined somewhat by wea conditions; their dates cannot be dicted with as much accuracy as can t of birds which, having spent the wint a greater distance from us, arrive when the weather is more favorable some places, for example at Ithaca, York, bird records have been kept season for more than thirty years. the information from these records, possible to indicate almost to a day w certain birds, such as barn swallows, oles, or hummingbirds, may be expe to arrive. Usually the very first birds kind to arrive are those individuals w will within a few days continue t northward journey. The later arrivals usually those that remain to become s mer residents. In some species all i viduals are migrants; for southern York the white-throated sparrow is re sentative of such a group. It winters ther south and nests farther north t southern New York.

Why do birds migrate? This quest has often been asked; but in answer it we must say that while we know m about where birds go and how fast t travel, we still know actually very li about the reasons for their regular seaso journeys.

As the airplane pilot has man-made struments to aid him in reaching a cert airport, so the birds have a well-develo sense of direction which guides them their destination. Each kind of b seems, in general, to take the route d by its ancestors; but this route be varied if for any reason food ld become scarce along the way. Such s are so exactly followed year after that they are known as lanes of mi-on. Persons desiring to study a cer-species of bird can have excellent op-unities to do so by being at some vantage point along this lane. Some-s undue advantage has been taken of in birds, especially hawks. Persons ing to kill these birds have collected trategic points along the lanes and tonly killed many of them. As a result uch activities sanctuaries have been blished at certain places along the s to give added protection to birds.

he routes north and south followed a given species of bird may lead entirely different parts of the country; e are called double migration routes. y may vary so much that one route lead chiefly over land while the other lead over the ocean. The golden ver is an example of such a case. See migration map.

Iuch valuable information as well as sure can be gained from keeping a ndar of migration and other activities birds. It is especially interesting dur-the spring months when first arrivals recorded if daily lists are made of all cies observed. In summer, nesting ac-ties and special studies of an individual cies provide something of interest for h day. More pleasure can be derived m the hobby if several people take it and compare their findings. Interests photography, sketching, or nature-story ting are natural companions of such d study.

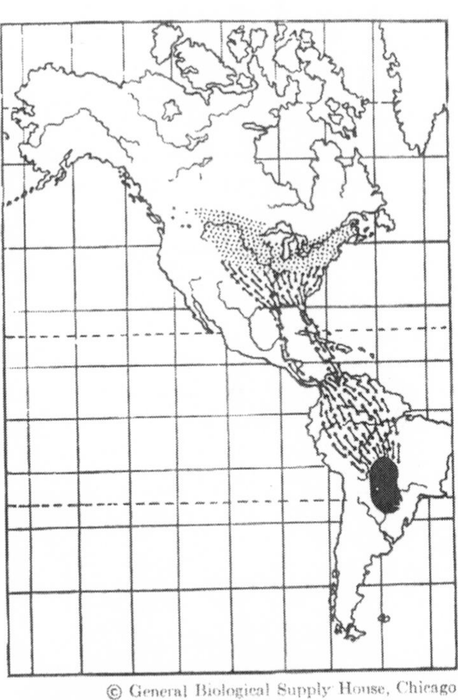

The travels of the bobolink. The migration routes of the bobolink are shorter than those of the plover and follow land more closely

EYES AND EARS OF BIRDS

The hen's eyes are placed at the side of the head so that she cannot see the same object with both eyes at the same time, and thus she has the habit of looking at us first with one eye and then the other to be sure she sees correctly. The position of the hen's eyes gives her a command of her entire environment. All birds have much keener eyes than we have; and they can adjust their eyes for either near or far vision much more effectively than we can; some hawks, flying high in the air, can see mice on the ground.

A wide range of colors is found in the eyes of birds: white, red, blue, yellow, brown, gray, pink, purple, and green are found in the iris of different species. The

Olin Sewall Pettingill, Jr.

A duck hawk. Notice the strong hooked beak, the keen eye, and the prominent nostril

hen's eye consists of a black pupil at the center, which must always be black in any eye, since it is a hole through which enters the image of the object. The iris of the hen's eye is yellow; there is apparently no upper lid, but the lower lid comes up during the process of sleeping. When the bird is drowsy the little film lid comes out from the corner of the eye and spr over it like a veil; just at the corner of own eye, next the nose, is the remain this film lid, although we cannot mo as the hen does.

The hearing of birds is very acute though in most cases the ear is sin a hole in the side of the head, and is n or less covered with feathers. The h ear is like this in many varieties of ch ens; but in others and in the roosters t are ornamental ear lobes.

LESSON 4

Eyes and Ears of Birds

Leading Thought — The eyes and of birds are peculiar and very efficien

Method — The hen or chicken and rooster should be observed for this less notes may be made in the poultry yar in the schoolroom when the birds brought there for study.

Observations — 1. Why does the turn her head first this side and then t as she looks at you? Can she see an obj with both eyes at once? Can she see w

2. How many colors are there in a he eye? Describe the pupil and the iris.

3. Does the hen wink as we do? she any eyelids?

4. Can you see the film lid? Does come from above or below or the inne outer corner? When do you see this f lid?

5. Where are the hen's ears? How they look? How can you tell where rooster's ears are?

6. Do you think the hen can see a hear well?

THE FORM AND USE OF BEAKS

ce the bird uses its arms and hands
ing, it has been obliged to develop
organs to take their place, and of
work the beak does its full share. It
l to emphasize this point by letting
hildren at recess play the game of
; to eat an apple or to put up their
, and pencils with their arms tied
d them; such an experiment will
how naturally the teeth and feet
to the aid when the hands are use-

e hen feeds upon seeds and insects
h she finds on or in the ground; her
is horny and sharp and acts not only
pair of nippers, but also as a pick as
trikes it into the soil to get the seed

A. A. Allen

A red-eyed vireo repairing her nest

nsect. She has already made the place
by scratching away the grass or sur-
of the soil with her strong, stubby
. The hen does not have any teeth,
does she need any, for her sharp beak
bles her to seize her food; and she
; not need to chew it, since her gizzard
s this for her after the food is swal-
ed.

he duck's bill is broad, flat, and much
er than the hen's beak. The duck feeds
n water insects and plants; it obtains
e by thrusting its head down into the
er, seizing the food, and holding it fast while the water is strained out through the sieve at the edges of the beak; for this use, a wide, flat beak is necessary. It would be quite as impossible for a duck to pick up hard seeds with its broad, soft bill as it would for the hen to get the duck's food

Leonard K. Beyer

These holes were made by a pileated woodpecker in search of insects

out of the water with her narrow, horny bill.

Both the duck and hen use their bills for cleaning and oiling their feathers and for fighting also; the hen strikes a sharp blow with her beak, making a wound like a dagger, while the duck seizes the enemy and simply pinches hard. Both fowls also use their beaks for turning over the eggs when incubating, and also as an aid to the feet when they make nests for themselves.

The nostrils are very noticeable and are situated in the beak near the base. However, we do not believe that birds have a keen sense of smell, since their nostrils are not surrounded by a damp, sensitive, soft surface as are the nostrils of the deer and dog. This arrangement aids these animals to detect odor in a marvelous manner.

LESSON 5

The Beak of a Bird

Leading Thought — Each kind of bird has a beak especially adapted for getting its food. The beak and feet of a bird are its chief weapons and implements.

Method — Study first the beak of the hen or chick and then that of the duckling or gosling.

Observations — 1. What kind of food does the hen eat and where and how does she find it in the field or garden? How is her beak adapted to get this food? If her beak were soft like that of a duck could she peck so hard for seeds and worms? Has the hen any teeth? Does she need any?

2. Compare the bill of the hen with that of the duck. What are the differences in shape? Which is the harder?

3. Note the saw teeth along the edge of the duck's bill. Are these for chewing? Do they act as a strainer? Why does the duck need to strain its food?

4. Could a duck pick up a hen's food from the earth or the hen strain out a duck's food from the water? For what other things than getting food do these fowls use their bills?

5. Can you see the nostrils in the bill of a hen? Do they show plainer in the duck? Do you think the hen can smell as keenly as the duck?

It is said that nature-study tea should be accurate, a statement that good teacher will admit without d but accuracy is often interpreted to completeness, and then the state cannot pass unchallenged. To study dandelion," "the robin," with emp on the particle "the," working ou complete structure, may be good la tory work in botany or zoology fo vanced pupils, but it is not an eleme educational process. It contributes ing more to accuracy than does the na order of leaving untouched all phases of the subject that are out o child's reach; while it may take ou life and spirit of the work, and the spi quality may be the very part that is worth the while. Other work may pr the formal "drill"; this should suppl quality and vivacity. Teachers often s me that their children have done exce work with these complete methods, they show me the essays and draw but this is no proof that the work is mendable. Children can be made t many things that they ought not to d that lie beyond them. We all need t to school to children. — "The Out to Nature," L. H. Bailey

Weather and wind and waning moo
Plain and hilltop under the sky,
Ev'ning, morning and blazing noon,
Brother of all the world am I.
The pine-tree, linden and the maize,
The insect, squirrel and the kine,
All — natively they live their days —
As they live theirs, so I live mine,
I know not where, I know not wha
Believing none and doubting non
Whate'er befalls it counteth not, —
Nature and Time and I are one.

— L. H. Ba

THE FEET OF BIRDS

Obviously, the hen is a digger of the soil; her claws are long, strong, and slightly hooked, and her feet and legs are covered with horny scales. These scales protect her feet from injury when they are use scratching the hard earth to lay bare seeds and insects hiding there. The is a very good runner indeed. She

wings a little to help, much as an
:tic runner uses his arms, and so can
r ground with amazing rapidity, her
ıg toes giving her a firm foothold. The
: she makes is very characteristic; it
ists of three toe-marks projecting for-
l and one backward. A bird's toes are
bered thus: the hind toe is number
the inner toe number two, the mid-
oe three, and the outer toe four.

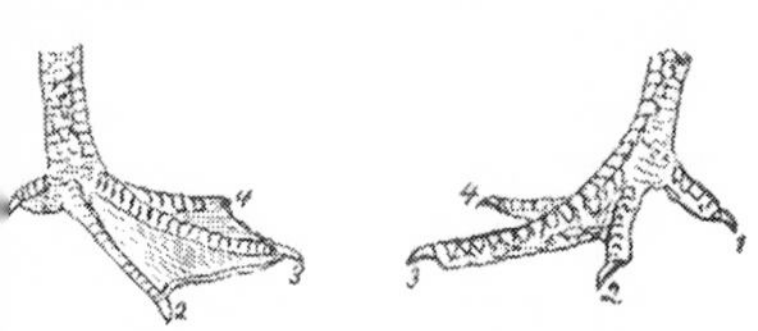

Duck's foot and hen's foot with toes numbered

duck has the same number of toes as
hen, but there is a membrane, called
web, which joins the second, third,
fourth toes, making a fan-shaped foot;
first or hind toe has a little web of
own. A webbed foot is first of all a
dle for propelling its owner through
water; it is also a very useful foot on
shores of ponds and streams, since its
ıdth and flatness prevent it from sink-
into the soft mud.

'he duck's legs are shorter than those
the hen and are placed farther back
wider apart. They are essentially
nming organs and are not fitted for
.tching or for running. They are
:ed at the sides of the bird's body so
t they may act as paddles, and are
her back so that they may act like the
:el of a propeller in pushing the bird
ıg. We often laugh at a duck on land,
:e its short legs are so far apart and so
back that its walk is necessarily an awk-
d waddle; but we must always remem-
that the duck is naturally a water bird,
l on the water its movements are grace-
Think how a hen would appear if
attempted to swim! The duck's body
o poorly balanced on its short legs that
:annot run rapidly; and if chased even
ıort distance it will fall dead from the
ırt, as many a country child has dis-
'ered to his sorrow when he tried to
ve the ducks home from the creek or
pond to coop. The long hind claw of the
hen enables her to clasp a roost firmly
during the night; a duck's foot could not
do this and the duck sleeps squatting on

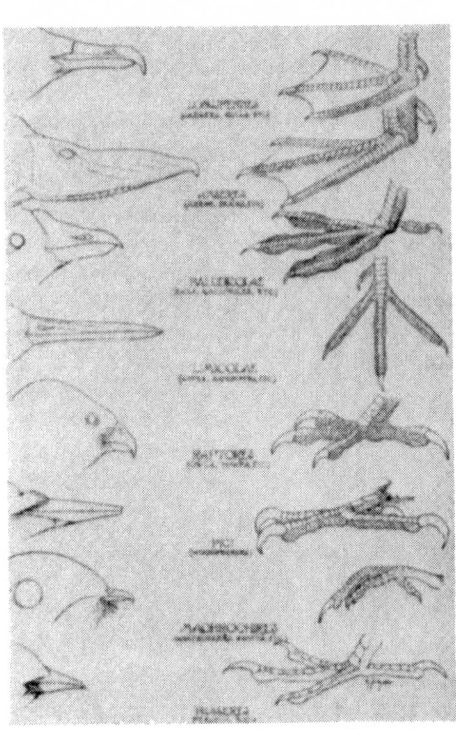

© General Biological Supply House, Chicago

Types of bills and feet

the ground. However, the Muscovy ducks, which are not good swimmers, have been known to perch.

LESSON 6

The Feet of Birds

Leading Thought — The feet of birds are shaped so as to assist the bird in getting its food as well as for locomotion.

Method — The pupils should have opportunity to observe the chicken or hen and a duck as they move about; they should also observe the duck swimming.

OBSERVATIONS — 1. Are the toes of the hen long and strong? Have they long, sharp claws at their tips?

2. How are the legs and feet of the hen covered and protected?

3. How are the hen's feet and legs fitted for scratching the earth, and why does she wish to scratch the earth?

4. Can a hen run rapidly? What sort of track does she make?

5. You number your fingers with the thumb as number one and the little finger as five. How do you think the hen's toes are numbered?

6. Has the duck as many toes a hen? What is the chief difference bet the feet of the duck and of the hen?

7. Which of the duck's toes are nected by a web? Does the web exte the tips of the toes? How does the help the duck?

8. Are the duck's legs as long a hen's? Are they placed farther forwa farther back than those of the hen? they farther apart?

9. Can a duck run as well as a hen? the hen swim at all?

10. Where does the hen sleep and does she hold on to her perch? Coul duck hold on to a perch? Does the need to perch while sleeping?

SONGS OF BIRDS

Anyone who attempts to recognize birds by sight alone misses much of the pleasure that comes to those who have

taken the time and pains to learn bird songs and use them as a means of bird recognition. It is true that not all people have a talent for music; but anyone interested in birds can learn to identify the songs and most of the call notes of common birds.

The observer will notice that in cases only the male bird sings, but a exceptions are recorded, notably th male rose-breasted grosbeak and carc grosbeak, which sing under some co tions. Birds do most of their singing in early morning and during the spring early summer months. The male b have not only a favorite time of day a particular season of the year du which they do most of their singing, they even have a certain perch or narro defined territory from which they sin

Each person will need to decide hov can best remember bird songs. Most ple will doubtless use such method were used by earlier bird students. L literary descriptions were given for e song. Alexander Wilson, for instance, scribes the call of the male blue jay "repeated creakings of an ungrea wheelbarrow." Often the call of a part lar bird is put into words; in many c these words have come to be accepte the common name of the bird, such bobwhite and whip-poor-will. The im nation of students may suggest cert words to represent the song or call n of a bird. These are often more easily membered than the song itself.

Some ornithologists have develo

plicated systems of recording bird s as musical scores. Wilson Flagg and . Mathews are well-known names in field. Such a method has its limita- ; because many variations of bird s cannot be indicated by the charac- used in writing music. The song of a written as music is not usually recog- ble when played on a musical instru- t. Other ornithologists have devel- l more graphic methods of recording songs. One leader in this field, A. A. nders, has proposed and used a system loying lines, dots, dashes, and sylla- . This system is very interesting and is eful one to a person who has a good for music. One of the latest methods ecording bird songs has been devel- d by the Department of Ornithology, nell University, Ithaca, New York. By this method bird songs are photographed on moving picture film and later may be recorded on phonograph records; these records can be played over and over again to give the student practice in identifying bird songs. Sound pictures have also been produced; the pictures of the various birds are shown on the screen as their songs are being heard by the audience.

ATTRACTING BIRDS

f suitable and sufficient food, water, lter, and nesting sites are provided, and rotection is given from such enemies cats and thoughtless men, it is possi- to attract many kinds of birds to ne grounds or gardens. The most logi- time to begin to attract birds is during winter months; but the best time is enever one is really interested and is ling to provide the things most needed the birds. Certain types of food, such suet or sunflower seeds, are sought by ds at any season. During the summer nths water for drinking and bathing y be more desired than food, but in e winter almost any seeds, fruits, or ty foods are welcome.

In the spring nesting boxes properly nstructed and placed will do much to ract some kinds of birds, especially ose that normally nest in holes in trees. n abundance of choice nesting materials ll entice orioles, robins, or chipping arrows to nest near by. Straws, sticks, athers, cotton, strings, or even hairs om old mattresses may be put out as in- cements to prospective bird tenants.

An invitation to our garden friends to partake of suet and peanuts in addition to their regular fare

The spring is also a good time to plant fruit-bearing trees, shrubs, and vines; these

A bird bath in the author's garden

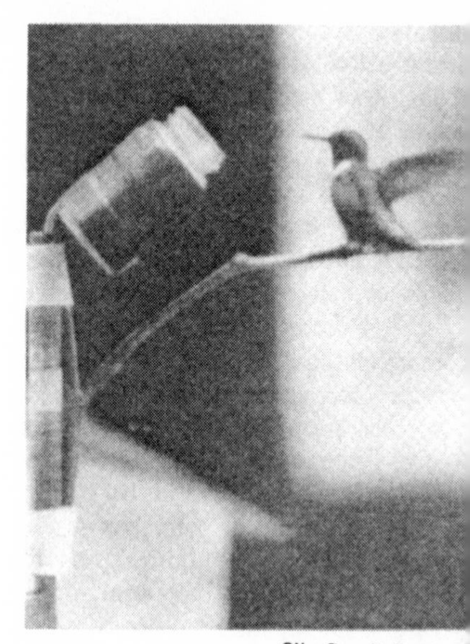

Olin Sewall Pettingi

Ruby-throated hummingbird attracted vial containing sweetened water

natural food counters become more attractive each year as they grow larger and produce more fruit and better nesting places for birds.

Autumn is the ideal time to establish feeding centers to which the birds may be attracted during the winter months. Food, such as suet or seeds, should be put at a great many places throughout the area in which one wishes to attract birds. The birds will gradually work their way one of these feedings points to ano soon it will be possible to concentrat feeding at one point, and the birds continue to come to that point as as food is provided there.

VALUE OF BIRDS

l you ever try to calculate in dollars leasure that you receive from seeing aring the first spring migrants? The , bluebird, and meadowlark bring to thousands of people every year. d, it would be difficult to find any- except perhaps in large cities, who not notice the arrival of at least spring birds — the robins on the the honk of the wild geese overhead, e song sparrows as they sing from the of a shrub. Birds are interesting to people because of their mere pres- their songs, their colors, or their s. Persons engaged in nature-study d outdoors and thus have opened to many other nature fields.

ne needs to observe a bird for only rt time to discover for himself what been known by scientists for many , that birds are of great economic rtance. Watch a chickadee or nut- as it makes its feeding rounds on nter day. Note how carefully each branch is covered by the chickadee what a thorough examination of the s and trunks is made by the nuthatch. ntless insect eggs as well as insects onsumed. On a sunny day in spring, rve the warblers as they feed about the y opened leaves and blossoms of the . See them as they hunt tirelessly for quota of the tiny insects so small they are generally overlooked by r birds. It must be remembered too some birds do, at times, take a toll ultivated crops; this is especially true ne seed-eating and insectivorous birds. they deserve some pay for the work do for man, and so in reality he should begrudge them a little fruit or grain. ome of the birds of prey are active all time; the hawks work in the daytime the owls come on duty for the night t. Countless destructive small mam- s and insects are eaten by them; thus tend to regulate the numbers of nerous small pests of field and wood, thereby preventing serious outbreaks of such animals. There has been much discussion of the real economic status of hawks and owls; many food studies have been made and the general conclusion is that most species are more useful than harmful. It is true that some species do take a toll of game birds, song birds, and poultry; but they include also in their diet other animal forms, many of which are considered harmful. One individual bird may be especially destructive and thus give a bad name to an entire species.

Leonard K. Beyer

A red-eyed vireo on her nest. Vireos live largely on insects gleaned from the under surfaces of leaves and from crevices in bark

There are even garbage gatherers among the birds; vultures, gulls, and crows serve in this capacity. The vultures are commonly found in the warmer parts of the country and serve a most useful purpose by their habit of devouring the unburied bodies of dead animals. The gulls are the scavengers of waterways and shore lines. The crow is omnivorous — that is, it eats both plant and animal food; but it seems to like carrion as well as fresh meat.

The farmer and the gardener owe quite a debt of thanks to the birds that eat weed seeds. Of course there are still bountiful crops of weeds each year; but there would

Verne Morton

A goldfinch nest in winter

be even more weeds if it were not for the army of such seed-eating birds as sparrows, bobwhites, and doves.

The game birds, such as grouse, pheasant, and bobwhite are important today, chiefly from the standpoint of the recreation they afford sportsmen and other lovers of the outdoors. The food habits of game birds do not present much of an economic problem; the birds are not numerous enough at the present time to be an important source of meat for man as they were in pioneer days.

Thus, a brief consideration of a few types of birds will show even a casual observer that birds have economic importance and that each species seems to have a definite work to perform.

LESSON 7
The Study of Birds' Nests in Winter

There are very good reasons for not studying birds' nests in summer, since the birds misinterpret familiarity on the part of eager children and are likely, in sequence, to abandon both nest a cality. But after the birds have go sunnier climes and the empty nes the only mementos we have of them, we may study these habitations car and learn how to appreciate pr the small architects which made I think that every one of us who fully examines the way that a nest is must have a feeling of respect f clever little builder.

I know of certain schools wher children make large collections of winter nests, properly labeling each thus gain a new interest in the bir of their locality. A nest when coll should be labeled in the following ner:

The name of the bird which buil nest.

Where the nest was found.

If in a tree, what kind?

How high from the ground?

After a collection of nests has made, let the pupils study them ac ing to the following outline:

1. Where was the nest found?

(a) If on the ground, describe t cality.

(b) If on a plant, tree, or shru the species, if possible.

(c) If on a tree, tell where it w a branch — in a fork, or hanging b end of the twigs.

Leonard K

A homemade wren house and its occu

(d) How high from the ground, and was the locality?

(e) If on or in a building, how situ-

Did the nest have any arrangement otect it from rain?

Give the size of the nest, the di- er of the inside and the outside; also lepth of the inside.

What is the form of the nest? Are ides flaring or straight? Is the nest ed like a cup, basket, or pocket?

What materials compose the out- of the nest and how are they ar- ed?

Of what materials is the lining made, how are they arranged? If hair or feathers are used, on what creature did they grow?

7. How are the materials of the nest held together, that is, are they woven, plastered, or held in place by environment?

8. Had the nest anything peculiar about it either in situation, construction, or material that would tend to render it invisible to the casual glance?

U. S. Department of Agriculture

Chicks, a few days old

CHICKEN WAYS

Dame Nature certainly pays close at- tion to details. An instance of this is little tooth on the tip of the upper ndible of the young chick, which aids n breaking out of its egg-shell prison; ce a tooth in this particular place of no use later, it disappears. The chil- n are delighted with the beauty of a ffy little chick with its bright, question- eyes and its life of activity as soon as it is freed from the shell. What a contrast to the blind, bare, scrawny young robin, which seems to be all mouth! The difference between the two is fundamental since it gives a means for distinguishing ground birds from perching birds. The young partridge, quail, turkey, and chick are clothed and active and ready to go with the mother in search of food as soon as they are hatched; while the young of

An anxious stepmother. The ducklings pay her little heed

the perching birds are naked and blind, being kept warm by the brooding mother, and fed and nourished by food brought by their parents, until they are large enough to leave the nest. The down which covers the young chick differs from the feathers which come later; the down has no quill but consists of several flossy threads coming from the same root; later on, this down is pushed out and off by the true feathers which grow from the same sockets. The pupils should see that the down is so soft that the little, fluffy wings of the chick are useless until the real wing feathers appear.

We chew food until it is soft and fine, then swallow it, but the chick swallows it whole; after being softened by juices from the stomach the food passes into a little mill, in which is gravel that the chicken has swallowed. This gravel helps to grind up the food. This mill is called the gizzard and the pupils should be taught to look carefully at this organ the next time they have chicken for dinner. A chicken has no muscles in the throat, like ours, to enable it to swallow water as we do. Thus, it has first to fill its beak with water, then hold it up so the water will flow down the throat. As long as the little chick has its mother's wings to sleep under, it not need to put its head under its wing; but when it grows up and sp the night upon a roost, it usually its head under its wing while sleepi

The conversation of the barnyard covers many elemental emotions a easily comprehended. It is well fo children to understand from the first the notes of birds mean something nite. The hen clucks when she is ing her chicks afield so that they know where she is in the tall grass; chicks follow "cheeping" or "peep as the children say, so that she will k where they are; but if a chick feels lost its "peep" becomes loud and consolate; on the other hand, there i sound in the world so full of cosy tentment as the low notes of the c when it cuddles under the mother's w When a hen finds a bit of food she u rapid notes which call the chicks hurry, and when she sees a hawk she a warning "q-r-r" which makes e chick run for cover and keep quiet. W hens are taking their sun and dust b together, they seem to gossip and we almost hear them saying, "Didn't think Madam Dorking made a great over her egg today?" Or, "That grown young rooster has got a crow match his legs, hasn't he?" Con these low tones with the song of the as she issues forth in the first warm

Poultry Dept., N. Y. State College of Agric

White leghorns are prolific layers

ring and gives to the world one of the
joyous songs of all nature. There is
a different quality in the triumphant
le of a hen telling to the world that
has laid an egg and the cackle which
es from being startled. When a hen
tting or is not allowed to sit, she is
ous and irritable, and voices her
tal state by scolding. When she is
y afraid, she squalls; and when seized
an enemy, she utters long, horrible
wks. The rooster crows to assure his
that all is well; he also crows to show
er roosters what he thinks of himself
of them. The rooster also has other
s; he will question you as you ap-
ch him and his flock, and he will
a warning note when he sees a hawk;
n he finds some dainty tidbit, he calls
flock of hens to him and they usually
ve just in time to see him swallow the
sel.

Vhen roosters fight, they confront each
er with their heads lowered and then
to seize each other by the back of the
k with their beaks, or strike each other
a the wing spurs, or tear with the leg
rs. Weasels, skunks, rats, hawks, and
ws are the most common enemies of
fowls, and often a rooster will attack
of these invaders and fight valiantly;
hen also will fight if her brood is dis-
bed.

LESSON 8

Chicken Ways

Leading Thought — Chickens have
eresting habits of life and extensive
nversational powers.

Method — For this lesson it is neces-
y that the pupils observe the inhabit-
ts of the poultry yard and answer these
estions a few at a time.

Observations — 1. Did the chick get
t of the egg by its own efforts? Of what
is the little tooth which is on the tip
of the upper part of a young chick's beak? Does this remain?

2. What is the difference between the down of the chick and the feathers of the hen? The little chick has wings; why can it not fly?

3. Why is the chick just hatched so pretty and downy, while the young robin is so bare and ugly? Why is the young chick able to see while the young robin is blind?

4. How does the young chick get its food?

5. Does the chick chew its food before swallowing? If not, why?

6. How does the chick drink? Why does it drink this way?

7. Where does the chick sleep at night? Where will it sleep when it is grown up?

8. Where does the hen usually put her head when she is sleeping?

9. How does the hen call her chicks when she is with them in the field?

10. How does she call them to food?

11. How does she tell them there is a hawk in sight?

12. What notes does the chick make when it is following its mother? When it gets lost? When it cuddles under her wing?

13. What does the hen say when she has laid an egg? When she is frightened?

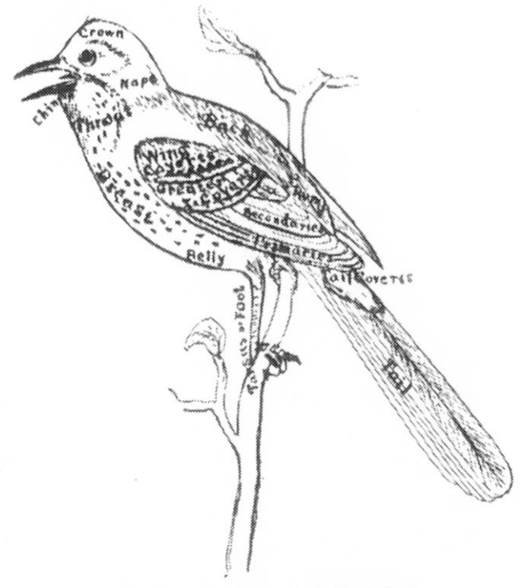

Parts of the bird labeled

This figure may be placed on the blackboard where pupils may consult it when studying colors and markings of birds.

When she is disturbed while sitting on eggs? When she is grasped by an enemy? How do hens talk together? Describe a hen's song.

14. When does the rooster crow? What other sounds does he make?

15. With what weapons does rooster fight his rivals and his enen

16. What are the natural enemie the barnyard fowls and how do the cape them?

J. H. Com

Pigeon houses of the upper Nile

PIGEONS

There is mention of domesticated pigeons by writers three thousand years ago; and Pliny relates that the Romans were fervent pigeon fanciers at the beginning of the Christian era. All of our domestic varieties of pigeons have been developed from the Rock pigeon, a wild species common in Europe and Asia. The carrier pigeon was probably the first to be specially developed because of its usefulness; its love and devotion to its mate and young and its homesickness when separated from them were used by man for his own interests. When a knight of old started off on a Crusade or to other wars, he took with him several pigeons from the home cote; and after riding many days he wrote a letter and tied it to the neck or under the wing of one of his birds, which he then set free, an flew home with its message; later he wo set free another in like manner. The dr back to this correspondence was tha went only in one direction; no bird fr home brought message of cheer to wandering knight. Nowadays mail rou telegraph wires, and wireless currents mesh our globe, and the pigeon as a rier is out-of-date; but fanciers still perf the homer breed and train pigeons very difficult flight competitions, so of them over distances of hundreds miles. Recently a homer made one th sand miles in two days, five hours, a fifty minutes.

The natural food of pigeons is gra we feed them cracked corn, wheat, p Kafir corn, millet, and occasionally he

Verne Morton

Homing pigeons

; it is best to feed mixed rations as birds tire of a monotonous diet. Pi-s should be fed twice a day; the pi-s and their near relatives, the doves, the only birds which can drink like orse, that is, with the head lowered. walk of a pigeon is accompanied by a liar nodding as if the head were in e way attached to the feet, and this ement sends waves of iridescent rs over the bird's plumage. The flight he pigeon is direct without soaring, wings move rapidly and steadily, the s circling and sailing as they start or ht. The crow flaps hard and then for a distance when it is inspecting ground, while the hawk soars on mo-less wings. It requires closer attention nderstand the language of the pigeon n that of the hen, nor has it so wide nge of expression as the latter; how-r, some emotions which the children understand are voiced in the cooing. he nest is built of grass and twigs; the ther pigeon lays two eggs for a sitting; in some breeds a pair will raise from en to twelve broods per year. The eggs ch in from sixteen to eighteen days, both parents share the labors of in-ating. In the case of the homer the ner bird sits from about 10 A.M. to .M. and the mother the remainder of day and night. The devotion of pigeons to their mates and to their young is great, and has been sung by the poets and praised by the philosophers during many ages; some breeds mate for life. The young pigeons or squabs are fed in a peculiar manner; in the crops of both parents is secreted a cheesy substance, known as pigeon milk. The parent seizes the beak of the squab in its own and pumps food from its own crop into the stomach of the young. This nutritious food is given to the squab for about five days and then replaced by grain which has been softened in the parents' stomachs, until the squabs are old enough to feed themselves. Rats, mice, weasels, and hawks are the chief enemies of the pigeons; since pigeons cannot fight, their only safety lies in flight.

As the original Rock pigeon built in caves, our domesticated varieties naturally build in the houses we provide for them. A pigeon house should not be built for more than fifty pairs; it should be well ventilated and kept clean; it should face the south or east and be near a shallow, running stream if possible. The nest boxes should be about twelve inches square and nine inches in height with a door at one side, so that the nest may remain hidden. In front of each door there should be a little shelf to act as a balcony on which the resting parent bird may sit and coo to relieve the monotony of the sitter's task. Some breeders make a double compart-

J. Demary

Pouter pigeons

ment instead of providing a balcony, while in Egypt branches are inserted in the wall just below the doors of the very ornamental pigeon houses. The houses should be kept clean and whitewashed with lime to which carbolic acid is added in the proportion of one teaspoonful of acid to two gallons of the wash; the leaf stems of tobacco may be given to the pigeons as material for building their nests, so as to help keep in check the

Hugh Spencer

Domestic pigeon

bird lice. There should be near the pigeon house plenty of fresh water for drinking and bathing; also a box of table salt, and another of cracked oyster shell and one of charcoal as fine as ground coffee. Salt is very essential to the health of pigeons. The house should be high enough from the ground to keep the inmates safe from rats and weasels.

LESSON 9

Pigeons

Leading Thought — The pigeon
fer from other birds in appearance
also in their actions. Their nesting h
are very interesting and there are
things that may be done to make
pigeons comfortable. They were, i
cient days, used as letter carriers.

Method — If there are pigeons
in the neighborhood, it is best to en
age the pupils to observe these birds
of-doors. Begin the work with an int
ing story and with a few questions w
will arouse the pupils' interest in
birds.

Observations — 1. For an out-of
exercise during recess let the pupil
serve the pigeon and tell the colors o
beak, eyes, top of the head, back, b
wings, tail, feet, and claws. This exe
is excellent training to fit the pupi
note quickly the colors of wild bird.

2. On what do pigeons feed? Are
fond of salt?

3. Describe how a pigeon drinks.
does it differ in this respect from c
birds?

4. Describe the peculiar movemen
the pigeon when walking.

5. Describe the pigeon's flight.
rapid, high in the air, do the wings
constantly, etc.? What is the chief di
ence between the flight of pigeons
that of crows or hawks?

6. Listen to the cooing of a pigeon
see if you can understand the diffe
notes.

7. Describe the pigeon's nest. H
many eggs are laid at a time?

8. Describe how the parents share
labors in hatching the eggs. How lon
it after the eggs are laid before the yo
hatch?

9. How do the parents feed their yo
and on what material?

10. What are some enemies of pige
and how do they escape from them? H
can we protect the pigeons?

11. Describe how a pigeon ho
should be built.

What must you do for pigeons to them healthy and comfortable?

How many breeds of pigeons do know? Describe them.

r my own part I readily concur with n supposing that housedoves are de- from the small blue rock-pigeon, mba livia, for many reasons. . . . what is worth a hundred arguments e instance you give in Sir Roger yn's housedoves in Cærnarvonshire; h, though tempted by plenty of food gentle treatment, can never be pre- d on to inhabit their cote for any ; but as soon as they begin to breed, ke themselves to the fastnesses of shead, and deposit their young in y amidst the inaccessible caverns and ipices of that stupendous promon- . "You may drive nature out with a hfork, but she will always return": turam expellas furca . . . tamen us- que recurret." irgil, as a familiar occurrence, by way imile, describes a dove haunting the rn of a rock in such engaging numbers, that I cannot refrain from quoting the passage.

Qualis spelunca subito commota Columba,
Cui domus, et dulces latebroso in pumice nidi,
Fertur in arva volans, plausumque exterrita pennis
Dat tecto ingentem, mox aere lapsa quieto,
Radit iter liquidum, celeres neque commovet alas.

(Virg. Aen. v. 213–217)

As when a dove her rocky hold forsakes,
Roused, in a fright her sounding wings she shakes;
The cavern rings with clattering: — out she flies,
And leaves her callow care, and cleaves the skies;
At first she flutters: — but at length she springs
To smoother flight, and shoots upon her wings.

(Dryden's Translation)

— White of Selbourne

THE CANARY AND THE GOLDFINCH

ı childhood the language of birds and nals is learned unconsciously. What d, who cares for a canary, does not erstand its notes which mean loneli- ;, hunger, eagerness, joy, scolding, ht, love, and song!

'he pair of canaries found in most es are not natural mates. The union is : de convenance, forced upon them by ple who know little of bird affinities. : could hardly expect that such a mat- would be always happy. The singer, the male is called, is usually arbitrary l tyrannical and does not hesitate to chastising beak upon his spouse. The ression of affection of the two is usu- very practical, consisting of feeding h other with many beguiling notes l much fluttering of wings. The singer y have several songs; whether he has many or few depends chiefly upon his education; he usually shows exultation when singing by throwing the head back like a prima donna, to let the music well

Leonard K. Beyer

A goldfinch on her nest in a hawthorn

forth. He is usually brighter yellow in color with more brilliantly black markings than his mate; she usually has much gray in her plumage. But there are about fifty varieties of canaries and each has distinct color and markings.

Canaries should be given a more varied diet than most people think. The seeds we buy or that we gather from the plantain or wild grasses, they eat eagerly. They like fresh, green leaves of lettuce and chickweed and other tender herbage; they enjoy bread and milk occasionally. There should always be a piece of cuttlefish bone or sand and gravel where they can get it, as they need grit for digestion. Above all, they should have fresh water. Hard-boiled egg is given them while nesting. The canary seed which we buy for them is the product of a grass in the Canary Islands. Hemp and rape seed are also sold for canary food.

The canary's beak is wide and sharp and fitted for shelling seeds; it is not a beak fitted for capturing insects. The canary, when drinking, does not have to lift the beak so high in the air in order to swallow the water as do some birds. The nostrils are in the beak and are easily seen; the ear is hidden by the feathers. The canary is a fascinating little creature when it shows interest in an object; it has such a knowing look, and its perfectly round, black eyes are so intelligent and cunning. If the canary winks, the act is so rapid as to be seen with difficulty, but when it is drowsy, the little inner lid appears at the inner corner of its eye and the outer lids close so that we may be sure that they are there; the lower lid covers more of the eye than the upper.

The legs and toes are covered with scale armor; the toes have long, curved claws that are neither strong nor sharp but are especially fitted for holding to the perch; the long hind toe with its stronger claw makes complete the grasp on the twig. When the canary is hopping about on the bottom of the cage we can see that its toes are more fitted for holding to the perch than for walking or hopping on the ground.

When the canary bathes, it duc head and makes a great splashing wi wings and likes to get thoroughly Afterward, it sits all bedraggled "humped up" for a time and then ally preens its feathers as they dry. V going to sleep, it at first fluffs ou feathers and squats on the perch, c back its head, and looks very dro Later it tucks its head under its win the night and looks like a little ba feathers on the perch.

Canaries make a great fuss when b ing their nest. A pasteboard box is us given them with cotton and string lining; usually one pulls out what other puts in; and they both industri tear the paper from the bottom of cage to add to their building mat Finally, a makeshift of a nest is c pleted and the eggs are laid. If the si is a good husband, he helps incubate eggs and feeds his mate and sings to frequently; but often he is quite th verse and abuses her abominably. nest of the caged bird is very diffe in appearance from the neat nests of g plant down, and moss which the wild cestors of these birds made in some retreat in the shrubs or evergreens of Canary Islands. The canary eggs are blue, marked with reddish-brown. incubation period is 13 to 14 days. young are as scrawny and ugly as n little birds and are fed upon food tially digested in the parents' stoma Their first plumage usually resem that of the mother.

In their wild state in the Canary Isla and the Azores, the canaries are c green above with golden yellow brea When the heat of spring begins, t move up the mountains to cooler le and come down again in the winter. T may rear three or four broods on th way up the mountains, stopping at s cessive heights as the season advanc until finally they reach the high peaks

The Goldfinch or Thistle Bird

The goldfinches are small birds their songs are so sweet and reedy t seem to fill the world with music effectually than many larger birds. are fond of the seeds of wild grass, especially of thistle seed; and they ng the pastures and fence corners e the thistles hold sway. In summer, male has bright yellow plumage with le black cap "pulled down over his " like that of a grenadier. He has also ck tail and wings with white-tipped rts and primaries. The tail feathers white on their inner webs also, which not show when the tail is closed. head and back of the female are n and the under parts yellowish e, with wings and tail resembling e of the male except that they are not vidly black. In winter the male dons ss more like that of his mate; he loses black cap but keeps his black wings tail.

he song of the goldfinch is exquisite he sings during the entire period of golden dress; he sings while flying as as when at rest. The flight is in itself tiful, being wavelike up and down, graceful curves. Mr. Chapman says on the descending half of the curve male sings "Per-chick or-ree." The

Audubon Educational Leaflet No. 17

A pair of goldfinches

A. A. Allen

The nest and eggs of a goldfinch in an elm tree

goldfinch's call notes and alarm notes are very much like those of the canary.

Since the goldfinches live so largely upon seeds of grasses, they stay with us in small numbers during the winter. During this period both parents and young are dressed in olive green, and their sweet call notes are a surprise to us of a cold, snowy morning, for they are associated in our memory with summer. The male dons his winter suit in October.

The goldfinch nest is a mass of fluffiness. These birds make feather beds for their young, or perhaps we should say beds of down, since it is the thistledown which is used for this mattress. The outside of the nest consists of fine shreds of bark or fine grass closely woven; but the inner portion is a mat of thistledown — a cushion an inch and a half thick for a nest which has an opening of scarcely three inches; sometimes the outside is ornamented with lichens. The nest is usually placed in some bush or tree, often in an evergreen, and ordinarily not more than five or six feet from the ground; but sometimes it is placed thirty feet high. The eggs are from four to six in number and bluish white in color. The female builds the nest, her mate cheering her with song meanwhile; he feeds her while she is incubating and helps feed the young. A strange thing about the nesting habits of the goldfinches is that the nest is not built until August. It has been surmised that this nesting season is delayed until there is an abundance of thistledown for building material.

LESSON 10

The Canary and the Goldfinch

Leading Thought—The canary is a close relative of the common wild goldfinch. If we compare the habits of the two we can understand how a canary might live if it were free.

Method—Bring a canary to the schoolroom and ask for observations. Ask the pupils to compare the canary with the goldfinches which are common in the summer. The canary offers opportunity for very close observation, which will prove excellent training for the pupils for beginning bird study.

Observations—1. If there are two canaries in the cage, are they always pleasant to each other? Which one is the "boss"? How do they show displeasure or bad temper? How do they show affection for each other?

2. Which one is the singer? Does the other one ever attempt to sing? What other notes do the canaries make besides singing? How do they greet you when you bring their food? What do they say when they are lonesome and hungry?

3. Does the singer have more than one song? How does he act while singing? Why does he throw back his head like an opera singer when singing?

4. Are the canaries all the same color? What is the difference in color between the singer and the mother bird? Describe the colors of each in your notebook as follows: top and sides of head, back, tail, wings, throat, breast, and under parts.

5. What does the canary eat? What sort of seeds do we buy for it? What seeds do we gather for it in our garden? D goldfinches live on the same seeds? does the canary do to the seeds b eating them? What tools does he u take off the shells?

6. Notice the shape of the car beak. Is it long and strong like a rob Is it wide and sharp so that it can seeds? If you should put an insect ir cage would the canary eat it?

7. Why do we give the canary c bone? Note how it takes off pieces o bone. Could it do this if its beak were sharp?

8. Note the actions of the birds v they drink. Why do they do this?

9. Can you see the nostrils? Where they situated? Why can you not see ear?

10. When the canary is intereste looking at a thing how does it act? I closely at its eyes. Does it wink? does it close its eyes? When it is dr can you see the little inner lid come f the corner of the eye nearest the b Is this the only lid?

11. How are the legs and feet cove Describe the toes. Compare the lengt the claw with the length of the toe. W is the shape of the claw? Do you t that claws and feet of this shape are b fitted for holding to a branch than walking? Note the arrangement of toes when the bird is on its perch. Is hind toe longer and stronger? If so, w Do the canaries hop or walk about bottom of the cage?

12. What is the attitude of the car when it goes to sleep at night? How c it act when it takes a bath? How doe get the water over its head? Over its ba What does it do after the bath? If forget to put in the bath dish how d the bird get its bath?

Nesting Habits to be Observed in the Spring

13. When the canaries are ready build a nest, what material do we furn them for it? Does the father bird h the mother to build the nest? Do t strip off the paper on the bottom of

or nest material? Describe the nest it is finished.

Describe the eggs carefully. Does ather bird assist in sitting on the Does he feed the mother bird when sitting?

How long after the eggs are laid the young ones hatch? Do both ts feed the young? Do they swallow od first and partially digest it before it to the young?

How do the very young birds look? t is their appearance when they the nest? Does the color of their age resemble that of the father or nother?

Where did the canaries originally from? Find the place on the map.

The Goldfinch

ADING THOUGHT — Goldfinches are at their best in late summer or ember, when they appear in flocks ever the thistle seeds are found in dance. Goldfinches so resemble the ries in form, color, song, and habits they are called wild canaries.

ETHOD — The questions for this les- may be given to the pupils before the of school in June. The results may be rted to the teacher in class when the ol begins in the autumn.

BSERVATIONS — 1. Where do you find goldfinches feeding? How can you nguish the father from the mother s and from the young ones in color?

2. Describe the colors of the male goldfinch and also of the female as follows: crown, back of head, back, tail, wings, throat, breast, and lower parts. Describe in particular the black cap of the male.

3. Do you know the song of the goldfinch? Is it like the song of the canary? What other notes has the goldfinch?

4. Describe the peculiar flight of the goldfinches. Do they fly high in the air? Do you usually see them singly or in flocks?

5. Where do the goldfinches stay during the winter? What change takes place in the coat of the male during the winter? What do they eat during the winter?

6. At what time of year do the goldfinches build their nests? Describe the nest. Where is it placed? How far above the ground? How far from a stream or other water? Of what is the outside made? The lining? What is the general appearance of the nest? What is the color of the eggs?

Sometimes goldfinches one by one will drop
From low-hung branches; little space they stop,
But sip, and twitter, and their feathers sleek,
Then off at once, as in a wanton freak;
Or perhaps, to show their black and golden wings;
Pausing upon their yellow flutterings.

— John Keats

THE ROBIN

Most of us think we know the robin , but very few of us know definitely habits of this, our commonest bird. object of this lesson is to form in the ils a habit of careful observation, and nable them to read for themselves the resting story of this little life which ved every year before their eyes. More- r, a robin notebook, if well kept, is a sure for any child; and the close obser- ion necessary for this lesson trains the pupils to note in a comprehending way the habits of other birds. It is the very best preparation for bird study of the right sort.

A few robins occasionally find a swamp where they can obtain food to nourish them during the northern winter, but for the most part they go in flocks to our southern states, where they settle in swamps and cedar forests and live chiefly upon fruits and berries. The robins do not

Leonard K. Beyer

A robin and its hungry young

nest or sing while in Southland. When the robins first come to us in the spring they feed on wild berries, being especially fond of those of the Virginia creeper. As soon as the frost is out of the ground they begin feeding on earthworms, cutworms, white grubs, and other insects. The male robins come first, but do not sing much until their mates arrive.

The robin is ten inches long and the English sparrow is only six and one-third inches long; the pupils should get the sizes of these two birds fixed in their minds for comparison in measuring other birds. The father robin is much more decided in color than his mate; his beak is yellow, there is a yellow ring about the eye and a white spot above it. The head is black and the back slaty-brown; the breast is brilliant reddish brown or bay and the throat is white, streaked with black. The mother bird has paler back and breast and has no black upon the head. The wings of both are a little darker than the back; the tail is black with the two outer feathers tipped with white. These white spots do not show except when the bird is flying and are "call colors" — that is, they enable the birds to see each other and thus keep together when flying in flocks during the night. The white patch made by th der tail-coverts serves a similar pur The feet and legs are strong and da color.

The robin has many sweet songs ar may be heard in the earliest dawn and in the evenings; if he wishes to chee mate he may burst into song at any He feels especially songful before summer showers, when he seems to "I have a theory, a theory, it's g to rain." And he might well say he also has a theory, based on ex ence, that a soaking shower will many of the worms and larvæ in the up to the surface where he can get t Besides these songs the robins have a variety of notes which the female sh although she is not a singer. The ag ing, angry cries they utter when the a cat or squirrel must express their fee fully; they give a very different war note when they see crow or hawk. note is hard to describe; it is a long, very loud squeak.

A robin can run or hop as pleases best, and it is interesting to see one, w hunting earthworms, run a little dista then stop to bend the head and li and look; when he finally seizes the e worm he braces himself on his strong and tugs manfully until he sometime

Herbert E.

Four blue eggs in a nest on a rail fenc

falls over backward as the worm lets hold. The robins, especially at nest-me, eat many insects as well as earth-s.

e beginning of a robin's nest is very sting; much strong grass, fine straw, s, and rootlets are brought and placed secure support. When enough of this rial is collected and arranged, the bird to the nearest mud puddle or stream in and fills its beak with soft mud; en goes back and "peppers" it into est material; after the latter is soaked, bird gets into it and molds it to the by turning around and around. In case which the author watched the ier bird did this part of the building, ugh the father worked industriously inging the other materials. After the is molded but not yet hardened, it is l with fine grass or rootlets. If the n is very dry and there is no soft at hand, the robins can build without aid of this plaster. Four eggs, which n exquisite greenish blue in color, are lly laid.

oth parents share the monotonous ness of incubating, and in the instance er the eyes of the author the mother was on the nest at night; the period ncubating is from eleven to fourteen s. The most noticeable thing about

A. A. Allen

Young robins. Their spotted breasts show their relationship to the thrushes

a very young robin is its wide, yellow-margined mouth, which it opens like a satchel every time the nest is jarred. This wide mouth cannot but suggest to anyone who sees it that it is meant to be stuffed, and the two parents work very hard to fill it. Both parents feed the young and often the father feeds the mother bird while she is brooding. Professor Treadwell experimented with young robins and found that each would take 68 earthworms daily; these worms if laid end to end would measure about 14 feet. Think of 14 feet of earthworm being wound into the little being in the nest; no wonder that it grows so fast! I am convinced that each pair of robins about our house has its own special territory for hunting worms, and that any trespasser is quickly driven off. The young birds' eyes are opened when they are from six to eight days old, and by that time the feather tracts, that is, the places where the feathers are to grow, are covered by the spinelike pin-feathers; these feathers push the down out and it often clings to their tips. In eleven days the birds are pretty well feathered; their wing feathers are fairly developed, but alas, they have no tail feathers! When a young robin flies from the nest he is a very uncertain and tippy youngster, not having any tail to steer him while flying, or to balance him when alighting.

It is an anxious time for the old robins when the young ones leave the nest, and

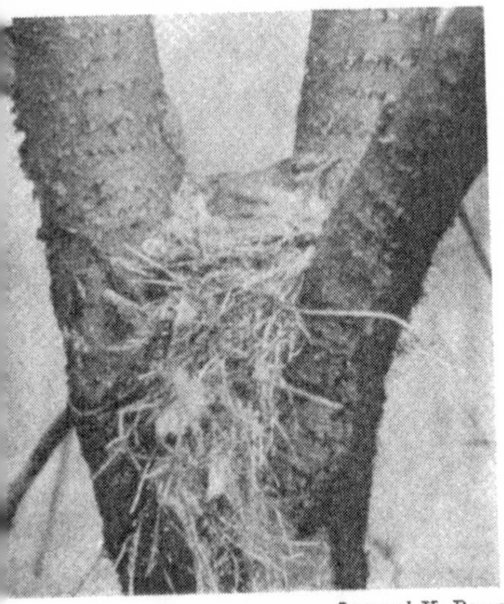

Leonard K. Beyer

A robin on its nest

they flutter about and scold at anyone who comes in sight, so afraid are they that injury will come to their inexperienced young ones: for some time the parents care for the fledglings, solicitously feeding them and giving them warnings of danger. The young robin shows in its plumage its relation to the thrush family, for it is yellowish and very spotted and speckled, especially on the breast. The parents may raise several broods, but they rarely use the

Leonard K. Beyer

This robin became so entangled in material it had gathered for its nest that it was unable to fly

same nest for two consecutive broods, both because it may be infested with parasites and because it is more or less soiled, although the mother robin works hard to keep it clean; she carries away all waste matter in her beak and drops it at some distance from the nest. Robins do not sing much after the breeding season is over until after they have molted. They are fond of cherries and other pulp fruits and often do much damage to such crops. The wise orchardist will plant a few Russian mulberry trees at a reasonable distance from his cherry trees, and thus, by giving the robins a fruit which they like better, and which ripens a little earlier, he may save his cherries. It has been proved conclusively that the robins are far more beneficial than damaging to the farmer; they destroy many noxious insects, two- of their food throughout the year co ing of insects; during April and May do a great work in destroying cutworn

The robins stay in the North later most migrating birds, often not le us entirely before November. Occas stragglers may remain all winter, in protected areas. Their chief enemi northern climates are cats, crows, squirrels. Cats should be taught t birds alone (see lesson on cat) or sh be killed. The crows have driven robins into villages where they can their nests under the protection of If crows venture near a house to attac robins, firing a gun at them once or t will give them a hint which they are slow to take. The robins of an e neighborhood will attack a nest-rob crow, but usually too late to save the lings. The robins can defend thems fairly well against the red squirrel u he steals the contents of the nest w the owners are away. There can be doubt that the same pair of robins re to the same nesting place year after On the Cornell University campu robin lacking the white tip on one of his tail was noted to have returne the same particular feeding ground several years; and we are very certain the same female bird built in the vine our piazza for seven consecutive year took two years to win her confidence, after that she seemed to feel as if she w a part of the family and regarded us as friends. We were sure that during fifth year she brought a new young band to the old nesting site; proba her faithful old husband had met w some mischance during the winter.

LESSON 11
The Robin

:ading Thought — To understand all :an about the life and ways of the n.

[ethod — For first and second grades work may be done by means of an a blackboard, or what is far better, ts of ordinary, buff, manila wrapping :r fastened together at the upper end, aat they may be hung and turned over a calendar. On the outside page make cture of a robin in colored chalk or ons, coloring according to the chil-'s answers to questions of series "b." vote each page to one series of ques-s, as given below. Do not show these stions to the pupils until the time is : for the observations. Those pupils ng accurate answers to these questions uld have their names on a roll of honor the last page of the chart.

'or third or higher grades the pupils y have individual notebooks in which h one may write his own answers to : questions of the successive series, ich should be written on the black-ard at the proper time for the observa-ns. This notebook should have a page out 6 x 8 inches and may be made of any nk paper. The cover or first page should ow the picture of the robin colored by e pupil, and may contain other illus-tive drawings, and any poems or other erature pertinent to the subject.

Observations by Pupils — Series a o be given in March in the northern ates).

1. At what date did you see the first bin this year?
2. Where did the robin spend the win-r? Did it build a nest or sing when in its inter quarters?
3. What does it find to eat when it rst comes in the spring? How does this iffer from its ordinary food?
4. Does the robin begin to sing as soon as it comes north?

Series b (to be given the first week of April).

1. How large is the robin compared with the English sparrow?
2. What is the color of the beak? The eye? Around and above the eye?
3. The color of the top of the head? The back? The throat? The breast?
4. Do all the robins have equally bright colors on head, back, and breast?
5. What is the color of the wing feathers?
6. What is the color of the tail feathers? Where is the white on them? Can the white spots be seen except during flight of the bird? Of what use to the robin are these spots?
7. Is there white on the underside of the robin as it flies over you? Where?
8. What is the color of the feet and legs?

Series c (to be given the second week of April).

1. At what time of day does the robin sing? Is it likely to sing before a rain? How many different songs does a robin sing?
2. What note does a robin give when it sees a cat?
3. What sounds do the robins make when they see a crow or a hawk?
4. Does a robin run or walk or hop?
5. Do you think it finds the hidden earthworm by listening? If so, describe the act.
6. Describe how a robin acts as it pulls a big earthworm out of the ground.
7. Do robins eat other food than earthworms?

Series d (to be given in the middle of April or a little later).

1. At what date did your pair of robins begin to build their nest?
2. Where was the nest placed and with what material was it begun?
3. Can you tell the difference in colors between the father and mother birds? Do both parents help in making the nest?

4. How and with what material is the plastering done? How is the nest molded into shape? Do both birds do this part of the work?

5. Where is the mud obtained and how carried to the nest?

6. How is the nest lined?

Series e (to be given a week after series *d*).

1. What is the number and color of the eggs in the nest?

2. Do both parents do the sitting? Which sits on the nest during the night?

3. Give the date when the first nestling hatches.

4. How does the young robin look? The color and size of its beak? Why is its beak so large? Can it see? Is it covered with down? Compare it to a young chick and describe the difference between the two.

5. What does the young robin do if it feels any jar against the nest? Why does it do this?

6. Do the young robins make any noise?

7. What do the parents feed their young? Do both parents feed them? Are the young fed in turns?

8. Do you believe each pair of robins has a certain territory for hunting worms which is not trespassed upon by other robins?

Series f (to be given three days after series *e*).

1. How long after hatching befor young robin's eyes are open? Can yo where the feathers are going to How do the young feathers look?

2. How long after hatching befor young birds are covered with feathe

3. Do their wing or tail feathers first?

4. How is the nest kept clean?

5. Give the date when the young r leave the nest. How do the old robin at this important crisis?

6. Describe the young robin's f Why is it so unsteady?

7. How do the young robins diff colors of breast from the parents?

8. Do the parents stay with the y for a time? What care do they give th

9. If the parents raise a second br do they use the same nest?

Series g (to be given for summer r ing and observations).

1. Do the robins sing all sum Why?

2. Do the robins take your berries cherries? How can you prevent them f doing this?

3. How does the robin help us?

4. How long does it stay with us in fall?

5. What are the chief enemies of robin and how does it fight or esc them? How can we help protect it?

6. Do you think the same robins c back to us each year?

THE BLUEBIRD

Stern as were our Pilgrim Fathers, they could not fail to welcome certain birds with plumage the color of June skies, whose sweet voices brought hope and cheer to their homesick hearts at the close of that first, long, hard winter of 1621. The red breasts of these birds brought to memory the robins of old England, and so they were called "blue robins"; and this name expresses well the relationship implied, because the bluebirds and robins of America are both members of the thrush family, a family noted for exqui song.

The bluebirds are usually ahead of robins in the northward journey and of arrive in New York amid the blizzards early March, their soft, rich "curly" no bringing, even to the doubting mind, g convictions of coming spring. There i family resemblance between voices bluebird and robin, a certain rich qua of tone; but the robin's song is far m assertive and complex than is the so ng" song of the bluebird, which en vocalized as "tru-al-ly, tru-al-ly." love songs cease with the hard work ling the nestlings, but may be heard is a prelude to the second brood in The red breast of the bluebird is its color resemblance to the robin, alh the young bluebirds and robins are spotted, showing the thrush colors. obin is so much larger than the bluehat commonly the relationship is oticed. This is easily explained bethere is nothing to suggest a robin exquisite cerulean blue of the bluehead, back, tail, and wings. This is most brilliant when the bird is e wing, in the sunshine. However, is a certain mirror-like quality in blue feathers; and among leaf shador even among bare branches they neasure reflect the surroundings and render the bird less noticeable. female is paler, being grayish blue e and with only a tinge of red-brown

This bluebird is nesting in a cavity drilled by a woodpecker the previous year

Leonard K. Beyer

hollow fence post is a common home of bluebird. The young are fed chiefly on cts

on the breast; both birds are white beneath.

The bluebirds haunt open woods, fields of second growth, and especially old orchards. They flit about in companies of three or four until they mate for nesting. While feeding, the bluebird usually sits on a low branch keeping a keen eye on the ground below, now and then dropping suddenly on an unsuspecting insect and then returning to its perch; it does not remain on the ground hunting food as does the robin. The nest is usually built in a hole in a tree or post and is made of soft grass. A hollow apple tree is a favorite nesting site.

In building birdhouses we should bear in mind that a cavity about ten inches deep and six inches in height and width will give a pair of bluebirds room for building a nest. The opening should not be more than two or two and one-half inches in diameter and there should be no threshold; this latter is a very particular point. If there is a threshold or place to alight upon, the sparrows are likely to dispute with the bluebirds and drive them away, but the sparrow does not care for a place which has no threshold. The box for the bluebird may be made out of old boards or may be a section of an old tree trunk; it should be fastened from six to fifteen feet above the ground, and should be in nowise noticeable in color from its surroundings. To protect the nest from cats, barbed wire should be wound around the tree or post below the box. If the box for the nest is placed upon a post, the barbed wire will also protect it from the squirrels. The eggs are bluish white; the young birds in their first feathers are spotted on the back and have whitish breasts mottled with brown. The food of the nestlings is almost entirely insects. In fact, this bird during its entire life is a great friend to man. The food of the adult is more than three-fourths insects and the remainder is wild berries and fruits, the winter food being largely mistletoe berries. It makes a specialty of beetles, caterpillars, and grasshoppers, and seems never to touch any of our cultivated fruits. We should do everything in our power to encourage and protect these birds from their enemies, which are chiefly cats, squirrels, and English sparrows.

The migration takes place in flocks during autumn, but it is done in a most leisurely manner with frequent stops where food is plenty. The bluebirds we see in September are probably not the ones we have had with us during the summer, but are those which have come from farther north.

They winter largely in the Gulf states; the writer has often heard them singing in midwinter in southern Mississippi. The bluebirds seem to be the only ones that sing while at their winter resorts. They live the year round in the Bermudas, contrasting their heavenly blue plumage with the vivid red of the cardinals. The bluebird should not be confused with the indigo bunting; the latter is darker blue and has a blue breast.

Winged lute that we call a bluel
You blend in a silver strain
The sound of the laughing waters,
The patter of spring's sweet rai
The voice of the winds, the sunsh
And fragrance of blossoming th
Ah! You are an April poem,
That God has dowered with wi
— "The Bluebird," Re

LESSON 12

The Bluebird

Leading Thought — The bluebi related to the robins and thrushes a as beneficial as it is beautiful. We sl study its habits and learn how to nesting boxes for it, and protect it ways.

Method — The observations of lesson must be made in the field an the pupils individually. Give to eac outline of questions to answer thr seeing. There should follow reading sons on the bluebird's value to us an winter migrations, and the lesson sh end in discussions of the best way to l boxes for its use in nesting season, its tection from cats and other enemies.

Observations — 1. Which comes n earlier in spring, the robin or the l bird?

2. How do the two resemble each o and differ from each other?

3. Describe the bluebirds' song. they sing all summer?

4. Describe the colors of the blue as follows: the head, back, breast, ur parts, wings, tail. How does the male b bird differ from his mate in colors?

5. Where were the bluebirds you s What were they doing? If feeding, l did they act?

6. Can you see the color of the b

s plainly when it is in a tree as when
ying? If not, why?
Where do the bluebirds build their
Of what material are the nests
? Do both parents work at the nest
ng?
What is the color of the eggs? How
e young birds look, when old enough
ve the nest, as compared with their
ts?
What do the bluebirds eat? How do
benefit us?
What can we do to induce the blue-
to live near our houses? How can we
ct them?
. Where do the bluebirds spend the
er?

12. Make a colored picture of a bluebird. How can we tell the bluebird from the indigo bunting?

13. What are the bluebirds' chief enemies?

Hark! 'tis the bluebird's venturous strain
High on the old fringed elm at the gate —
Sweet-voiced, valiant on the swaying bough,
Alert, elate,
Dodging the fitful spits of snow,
New England's poet-laureate
Telling us Spring has come again!

— Thomas Bailey Aldrich

THE WHITE-BREASTED NUTHATCH

The busy nuthatch climbs his tree
Around the great bole spirally,
Peeping into wrinkles gray,
Under ruffled lichens gay,
Lazily piping one sharp note
From his silver mailèd throat.

— Maurice Thompson

lithe and mellow is the ringing "ank,
" note of the nuthatch, and why need
allude to its nasal timbre! While it
ot a strictly musical note, it has a most
icing quality and translates into sound
picture of bare-branched trees and the
ing of enchantment which permeates
forest in winter; it is one of the most
oodsy" notes in the bird repertoire.
d while the singer of this note is not
bewitching as his constant chum the
ckadee, yet he has many interesting
ys quite his own. Nor is this "ank,
k" his only note. I have often heard
air talking to each other in sweet confi-
ntial syllables, "wit, wit, wit," very difent from the loud note meant for the
rld at large. The nuthatches and chicka-
es hunt together all winter; it is no mere
siness partnership but a matter of con-
nial tastes. The chickadees hunt over
e twigs and smaller branches, while the
thatches usually prefer the tree trunks and the bases of the branches; both birds like the looks of the world upside down, and while the chickadee hangs head down from a twig, the nuthatch is quite likely to alight head down on a tree bole, holding itself safely in this position by thrusting its toes out at right angles to the body, thus getting a firm hold upon the bark. Sometimes its foot will be twisted completely around, the front toes pointed up the tree. The foot is well adapted for clinging to the bark as the front toes are strong and the hind toe is very long and is armed with a strong claw. Thus equipped, this bird runs about on the tree so rapidly that it has earned the name of "tree mouse." It often ascends a tree trunk spirally but is not so hidebound in this habit as is the brown creeper. It runs up or down freely, head first, and never flops down backwards like a woodpecker.

In color the nuthatch is bluish gray above with white throat and breast and reddish underparts. The sides of the head are white; the black cap extends back upon the neck but is not " pulled down " to the eyes as with the chickadees. The wing feathers are dark brown edged with pale gray. The upper middle tail feathers are bluish like the back; the others are dark brown and tipped with white in such a manner that the tail when spread shows a broad white border on both sides. The most striking contrast between the chickadee and nuthatch in markings is that the latter lacks the black bib. However, its entire shape is very different from that of the chickadee and its beak is long and slender, being as long as its head or longer, while the beak of the chickadee is a short, sharp little pick. The bill of the nuthatch is fitted to reach in crevices of the bark and pull out hiding insects, or to hammer open the shell of nut or acorn and get both the meat of the nut and the grub feeding upon it. It will wedge an acorn into a seam in the bark and throw back its head, woodpecker fa and drive home its chisel beak. But i not always use common sense in habit. I have often seen one cut off a of suet, fly off and thrust it into crevice, and hammer it as hard as were encased in a walnut shell. Th

S. A. Gr

A family of white-breasted nuthatches

ways seems bad manners, like carrying fruit from *table d'hôte;* but the nutha is polite enough in using a napkin, after eating the suet, it invariably wipes bill on a branch most assiduously, f one side then the other, until it is p fectly clean.

The nuthatches are a great benefit our trees in winter, for then is when th hunt for hiding pests on the trun Their food consists of beetles, caterpilla pupæ of various insects, also seeds of r weed, sunflowers, acorns, etc. While t nuthatch finds much of its food on tre yet Mr. Torrey has seen it awkwardly tu ver fallen leaves hunting for insects, Mr. Baskett says it sometimes catches s on the wing and gets quite out of n from this unusual exercise.

s only during the winter that we comy see the nuthatches, for during the ng season they usually retire to the woods, where they may occupy a y in a tree used by a woodpecker last or may make a hole for themselves their sharp beaks. The nest is lined leaves, feathers, and hair; from five ine creamy, speckled eggs are the ure of this cave.

L. H. Bailey

A characteristic pose

LESSON 13

THE NUTHATCH

EADING THOUGHT — The nuthatch is n a companion of the chickadees and odpeckers. It has no black bib, like the ckadee, and it alights on a tree trunk d downward, which distinguishes it n woodpeckers.

METHOD — This bird, like the chicka and downy, gladly shares the suet banquet we prepare for them and may be observed at leisure while "at table." The contrast between the habits of the nuthatch and those of its companions makes it a most valuable aid in stimulating close and keen observation on the part of the pupils.

Leonard K. Beyer

The nuthatch runs head first down tree unks in search of insects. Here he is eating et which has been fastened to the tree

OBSERVATIONS — 1. Where have you seen the nuthatches? Were they with other birds? What other birds?

2. Does a nuthatch usually alight on the ends of the branches of a tree or on the trunk and larger limbs? Does it usually alight head down or up? When it runs down the tree, does it go head first or does it back down? When it ascends the tree, does it follow a spiral path? Does it use its tail for a brace when climbing, as does the downy?

3. How does the arrangement of the nuthatch's toes assist it in climbing? Are the three front toes of each foot directed downward when the bird alights head downward? How does it manage its feet when in this position?

4. What is the general color of the nuthatch above and below? The color of the top and sides of head? Color of back? Wings? Tail? Throat? Breast?

5. Does the black cap come down to

the eyes on the nuthatch as on the chickadee? Has the nuthatch a black bib?

6. What is the shape of the beak of the nuthatch? For what is it adapted? How does it differ from the beak of the chickadee?

7. What is the food of the nuthatch? Where is it found? Does it open nuts for the grubs or the nut meat? Observe the way it strikes its beak into the suet; why does it strike so hard?

8. How would you spell this note? Have you heard it give more one note?

9. How does the nuthatch benefi trees? At what season does it benefit most? Why?

10. Where do the nuthatches their nests? Why do we see the hatches oftener in winter than in mer?

Acadian chickadees

Olin Sewall Pettingi

THE CHICKADEE

He is the hero of the woods; there are courage and good nature enough in compact little body, which you may hide in your fist, to supply a whole grov of May songsters. He has the Spartan virtue of an eagle, the cheerfulness of a thr the nimbleness of Cock Sparrow, the endurance of the sea-birds condensed into tiny frame, and there have been added a pertness and ingenuity all his own. His c osity is immense, and his audacity equal to it; I have even had one alight upon barrel of the gun over my shoulders as I sat quietly under his tree.

— ERNEST INGERSC

However careless we may be of our bird friends when we are in the midst of the luxurious life of summer, even the most careless among us give pleased attention to the birds that bravely endure with us the rigors of winter. And when this winged companion of winter proves to be the most fascinating little ball of feathers ever created, constantly overflowing with cheerful song, our pleased attention changes to active delight. Thus it is, that in all the lands of snowy winters the chickadee is a loved comrade of the country wayfarer; that happy song "chick-a-dee-dee-dee" finds its way to the dullest consciousness and the most callous he

The chickadees appear in small flo in the winter and often in company w the nuthatches. The chickadees work the twigs and ends of branches, while t nuthatches usually mine the bark of t trunk and larger branches, the forn hunting insect eggs and the latter, inse tucked away in winter quarters. When t chickadee is prospecting for eggs, it fi looks the twig over from above and th hangs head down and inspects it from b low; it is a thorough worker and doesn't tend to overlook anything whatever; a however busily it is hunting, it always fin

›r singing; whether on the wing or d upon a twig or hanging from it acrobat, head down, it sends forth ›ppy "chickadeedee" to assure us his world is all right and good n for anybody. Besides this song, it in February to sing a most seductive ›ee," giving a rising inflection to the ›llable and a long, falling inflection last, which makes it a very different from the short, jerky notes of the :her called phœbe, which cuts the ›llable short and gives it a rising in- ›n. More than this, the chickadee ›me chatty conversational notes, and nd then performs a bewitching little , which is a fit expression of its own ous personality.

e general effect of the colors of the adee is grayish brown above and h white below. The top of the head ck, the sides white, and it has a se- ve little black bib under its chin. back is grayish, the wings and tail are gray, the feathers having white mar- The breast is grayish white changing ıff or brownish at the sides and below. often called the "Black-capped Tit- se," and it may always be distin-

S. A. Grimes

ıck-capped chickadees. The friendly chick-adee is easily tamed

Leonard K. Beyer

A "banded" chickadee

guished by black cap and black bib. It is smaller than the English sparrow; its beak is a sharp little pick just fitted for taking insect eggs off twigs and from under bark. Insects are obliged to pass the winter in some stage of their existence, and many of them wisely remain in the egg until there is something worth doing in the way of eating. These eggs are glued fast to the food trees by the mother insect and thus provide abundant food for the chickadees. It has been estimated that one chickadee will destroy several hundred insect eggs in one day, and it has been proved that orchards frequented by these birds are much more free from insect pests than other orchards in the same locality. They can be enticed into orchards by putting up beef fat or bones and thus we can secure their valuable service. In summer these birds attack caterpillars and other insects.

When it comes to nest building, if the chickadees cannot find a house to rent they proceed to dig out a proper hole from some decaying tree, which they line with moss, feathers, fur, or some other soft material. The nest is often not higher than six to ten feet from the ground. One which I studied was in a decaying fence post. The eggs are white, sparsely speckled and spotted with lilac or rufous. The young birds are often eight in number. How these fubsy birdlings manage to pack themselves in such a small hole is a wonder; it probably gives them good discipline in bearing hardships cheerfully.

LESSON 14

The Chickadee

Leading Thought — The chickadee is as useful as it is delightful; it remains in the North during winter, working hard to clear our trees of insect eggs and singing cheerily all day. It is so friendly that we can induce it to come even to the window sill by putting out suet to show our friendly interest.

Method — Put beef fat on the trees near the schoolhouse in December and replenish it about every two or three weeks. The chickadees will come to the feast and may be observed all winter. Give the questions a few at a time and let the children read in the bird books a record of the benefits derived from this bird.

Observations — 1. Where hav seen the chickadees? What were t ing? Were there several together?

2. What is the common song chickadee? What other notes has it you heard it yodel? Have you h sing "fee-bee, fee-bee"? How do song differ from that of the phœbe? it sing on the wing or when at rest?

3. What is the color of the chic top and sides of head, back, wing. throat, breast, under parts?

4. Compare the size of the chic with that of the English sparrow.

5. What is the shape of the chick bill and for what is it adapted? W the food in winter? Where does th find it? How does it act when feedin hunting for food?

6. Does the chickadee usually alig the ends of the branches or on the portions near the trunk of the tree?

7. How can you distinguish the c dees from their companions, the hatches?

8. Does the chickadee ever seen couraged by the snow and cold wea Do you know another name for chickadee?

9. Where does it build its nest what material? Have you ever wat one of these nests? If so, tell about i

10. How does the chickadee benef orchards and shade trees? How ca induce it to feel at home with us and for us?

THE DOWNY WOODPECKER

Friend Downy is the name this attractive little neighbor has earned, because it is so friendly to those of us who love trees. Watch it as it hunts each crack and crevice of the bark of your favorite apple or shade tree, seeking assiduously for cocoons and insects hiding there, and you will soon, of your own accord, call it friend; you will soon love its black and white uniform, which consists of a black coat speckled and barred with white, and whitish gray vest and trousers. The front of the head is black and there is a b streak extending backward from the with a white streak above and also be it. The male has a vivid red patch on back of the head, but his wife shows such giddiness; plain black and white good enough for her. In both sexes throat and breast are white, the mi tail feathers black, while the side tail fe ers are white, barred with black at t tips.

The downy has a way of alighting

on a tree trunk or at the base of a branch and climbing upward in a ashion; it never runs about over the or does it turn around and go down irst, like the nuthatch; if it wishes down a short distance it accoms this by a few awkward, backward but when it really wishes to descend, s off and down. The downy, like woodpeckers, has a special arrangeof its physical machinery which enit to climb trees in its own manner. grasp the bark on the side of a tree firmly because its fourth toe is d backward and works as a comon with the thumb. Thus it is able utch the bark as with a pair of niptwo claws in front and two claws be; and as another aid, the tail is ar-d to prop the bird, like a bracket. tail is rounded in shape and the mideathers have rather strong quills; but secret of the adhesion of the tail to bark lies in the great profusion of s which, at the edge of the feathers, bristling tips, and when applied to side of the tree act like a wire brush all the wires pushing downward. s explains why the woodpecker can go backward without lifting the tail. ut even more wonderful than this is mechanism by which the downy and y woodpeckers get their food, which sists largely of wood-borers or larvæ king under the bark. When the woodker wishes to get a grub in the wood, eizes the bark firmly with its feet, uses tail as a brace, throws its head and uppart of the body as far back as posle, and then drives a powerful blow h its strong beak. The beak is adapted just this purpose, as it is wedge-shaped the end, and is used like a mason's drill netimes, and sometimes like a pick. hen the bird uses its beak as a pick, it ikes hard, deliberate blows and the ips fly; but when it is drilling, it strikes pidly and not so hard and quickly drills small, deep hole leading directly to the rrow of the grub. When finally the grub reached, it would seem well-nigh imposble to pull it out through a hole which is

L. W. Brownell

Friend Downy

too small and deep to admit of the beak being used as pincers. This is another story and a very interesting one; the downy and hairy can both extend their tongues far beyond the point of the beak, and the tip of the tongue is hard and horny and covered with short backward-slanting hooks acting like a spear or harpoon; and thus when the tongue is thrust into the grub it pulls it out easily. The bones of the tongue have a spring arrangement; when not in

Friend Downy's foot

use, the tongue lies soft in the mouth, like a wrinkled earthworm, but when in use, the bones spring out, stretching it to its full length, and it is then slim and small. The process is like fastening a pencil to the tip of a glove finger; when drawn back the finger is wrinkled together, but when thrust out, it straightens. This spring arrangement of the bones of the woodpecker's tongue is a marvelous mechanism and should be studied through pictures.

Since the food of the downy and the

hairy is where they can get it all winter, there is no need for them to go south; thus they stay with us and work for us the entire year. We should try to make them feel at home with us in our orchards and shade trees by putting up pieces of beef fat, to convince them of their welcome. No amount of free food will pauperize these birds, for as soon as they have eaten of the fat, they commence to hunt for grubs on the tree and thus earn their feast. They never injure live wood.

James Whitcomb Riley describes the drumming of the woodpecker as "weed-

A. A. Allen

Part of the tree has been cut away to show Downy's nest

ing out the lonesomeness" and that is exactly what the drumming of the woodpecker means. The male selects some dried limb of hard wood and there beats out his well-known signal which advertises far and near, "Wanted, a wife." And after he wins her, he still drums on for a time to cheer her while she is busy with her family cares. The woodpecker has no voice for singing, like the robin or thrush; and luckily, he does not insist on singing, like the peacock, whether he can or not. He chooses rather to devote his voice to terse and business-like conversation; and when he is musically inclined, he turns drummer. He is rather particular about his instrument, and having found one that is sufficiently resonant he returns to after day. While it is ordinarily the that drums, I once observed a f drumming. I told her that she was a minx and ought to be ashamed of self; but within twenty minutes sh drummed up two red-capped suitors chased each other about with great mosity, so her performance was evid not considered improper in woodp society. I have watched a rival pair of downies fight for hours at a time, but duel was of the French brand — n fuss and no bloodshed. They adva upon each other with much haughty ing and many scornful bobs of the h but when they were sufficiently ne stab each other they beat a mutual circumspect retreat. Although we hea male downies drumming every spri doubt if they are calling for new wiv believe they are, instead, calling the at tion of their lawful spouses to the fact it is time for nest building to begin. I h come to this conclusion because downies and hairies which I have watc for years have always come in pairs to take of suet during the entire winter; while only one at a time sits at meat the lord and master is somewhat bossy, they seem to get along as well as most r ried pairs.

The downy's nest is a hole, usually partly decayed tree; an old apple tree favorite site and a fresh excavation is m each year. There are from four to six wh eggs, which are laid on a nice bed of ch almost as fine as sawdust. The door to nest is a circle about an inch and a qua across.

The hairy woodpecker is fully one-th larger than the downy, measuring n inches from tip of beak to tip of tail, wh the downy measures only about six inch The tail feathers at the side are white the entire length, while they are barred the tips in the downy. There is a bla "parting" through the middle of the r patch on the back of the hairy's head. T two species are so much alike that it difficult for the beginner to tell the apart. Their habits are very similar, exce he hairy lives in the woods and is not mmonly seen in orchards or on shade The food of the hairy is much like of the downy; it is, therefore, a ficial bird and should be protected.

LESSON 15

THE DOWNY WOODPECKER

EADING THOUGHT — The downy dpecker remains with us all winter, ling upon insects that are wintering in ices and beneath the bark of our trees. s fitted especially by shape of beak, gue, feet, and tail to get such food and "friend in need" to our forest, shade, orchard trees.

METHOD — If a piece of beef fat be ened upon the trunk or branch of a which can be seen from the school-m windows, there will be no lack of in-est in this friendly little bird; for the wny will sooner or later find this feast ead for it and will come every day to take. Give out the questions, a few at a e, and discuss the answers with the pils.

OBSERVATIONS — 1. What is the gen-l color of the downy above and below? e color of the top of the head? Sides of e head? The throat and breast? The lor and markings of the wings? Color d markings of the middle and side tail athers?

2. Do all downy woodpeckers have the d patch at the back of the head?

3. What is the note of the downy? oes it make any other sound? Have you er seen one drumming? At what time of e year? On what did it drum? What did use for a drumstick? What do you sup-ose was the purpose of this music?

4. How does the downy climb a tree trunk? How does it descend? How do its actions differ from those of the nut-hatch?

5. How does the arrangement of the woodpecker's toes help it in climbing a tree trunk? How does this arrangement of toes differ from that of other birds?

6. How does the downy use its tail to assist it in climbing? What is the shape of the tail and how is it adapted to assist?

7. What does the downy eat and where does it find its food? Describe how it gets at its food. What is the shape of its bill and how is it fitted for getting the food? Tell how the downy's tongue is used to spear the grub.

8. Why do you think the downy does not go south in winter?

9. Of what use is this bird to us? How should we protect it and entice it into our orchards?

10. Write an account of how the downy builds its nest and rears its young.

A few seasons ago a downy woodpecker, probably the individual one who is now my winter neighbor, began to drum early in March in a partly decayed apple-tree that stands in the edge of a narrow strip of woodland near me. When the morning was still and mild I would often hear him through my window before I was up, or by half-past six o'clock, and he would keep it up pretty briskly till nine or ten o'clock, in this respect resembling the grouse, which do most of their drumming in the fore-noon. His drum was the stub of a dry limb about the size of one's wrist. The heart was decayed and gone, but the outer shell was loud and resonant. The bird would keep his position there for an hour at a time. Between his drummings he would preen his plumage and listen as if for the response of the female, or for the drum of some rival. How swift his head would go when he was delivering his blows upon the limb! His beak wore the surface percep-tibly. When he wished to change the key, which was quite often, he would shift his position an inch or two to a knot which gave out a higher, shriller note. When I climbed up to examine his drum he was much disturbed. I did not know he was in the vicinity, but it seems he saw me from a near tree, and came in haste to the neighboring branches, and with spread plumage and a sharp note demanded plainly enough what my business was with his drum. I was invading his privacy, desecrating his shrine, and the bird was much put out. After some weeks the female appeared; he had literally drummed up a mate; his urgent and oft-repeated advertisement was answered. Still the drumming did not cease, but was quite as fervent as before. If a mate could be won by drumming she could be kept and entertained by more drumming; courtship should not end with marriage. If th felt musical before, of course he felt more so now. Besides that, the g deities needed propitiating in beha the nest and young as well as in beh the mate. After a time a second fe came, when there was war between two. I did not see them come to b but I saw one female pursuing the about the place, and giving her no res several days. She was evidently tryi run her out of the neighborhood. and then she, too, would drum brief if sending a triumphant message to mate. — "WINTER NEIGHBORS," J BURROUGHS

THE SAPSUCKER

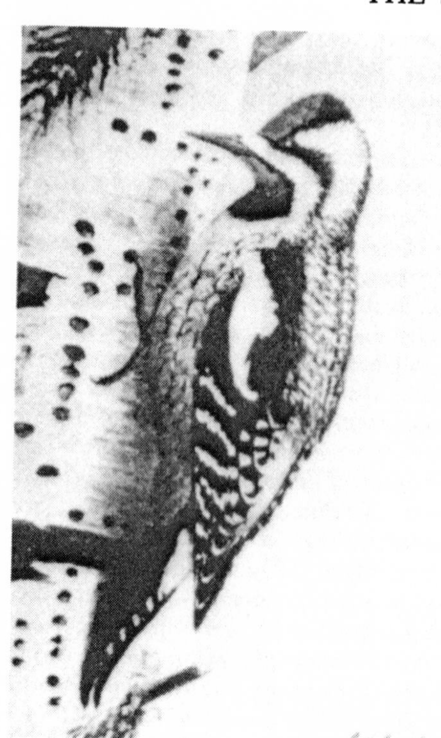

L. A. Fuertes

The yellow-bellied sapsucker

The sapsucker is a woodpecker that has strayed from the paths of virtue; he has fallen into temptation by the wayside, and instead of drilling a hole for the sake of the grub at the end of it, he drills for drink. He is a tippler, and sap is his erage; and he is also fond of the soft, i bark. He often drills his holes in reg rows and thus girdles a limb or a t and for this is pronounced a rascal by who have themselves ruthlessly cut f our land millions of trees that should be standing. It is amusing to see a sucker take his tipple, unless his sal happens to be one of our prized yo trees. He uses his bill as a pick and ma the chips fly as he taps the tree; then goes away and taps another tree. Aft time he comes back and holding his b close to the hole for a long time seems be sucking up the sap; he then thr back his head and "swigs" it down w every sign of delirious enjoyment. T avidity with which these birds come to bleeding wells which they have made, in it all the fierceness of a toper crazy drink; they are particularly fond of sap of the mountain ash, apple, thorn ple, canoe birch, cut-leaf birch, red map red oak, white ash, and young pines. Ho ever, the sapsucker does not live solely sap; he also feeds upon insects whene he can find them. When feeding th young, the sapsuckers are true flycatch snatching insects while on the wing. T male has the crown and throat crimsc edged with black with a black line exten ck of the eye, bordered with white and below. There is a large, black r patch on the breast which is bor- at the sides and below with lemon . The female is similar to the male as a red forehead, but she has a bib instead of a red one beneath the The distinguishing marks of the sap- should be learned by the pupils. ed is on the front of the head instead the crown, as is the case with the y and hairy; when the bird is flying road, white stripes extending from oulders backward, form a long, oval , which is very characteristic.

e sapsuckers spend the winter in the ern states where they drill wells in hite oak and other trees. From Vir- to northern New York and New and, where they breed, they are seen during migration, which occurs in ; then the birds appear two and three her and are very bold in attacking e trees, especially the white birch. y nest only in the northern United s and northward. The nest is usually le in a tree about forty feet from the nd, and is likely to be in a dead birch.

LESSON 16

The Sapsucker

EADING THOUGHT — The sapsucker a red cap, a red bib, and a yellow ast; it is our only woodpecker that does ry to trees. We should learn to distin- sh it from the downy and hairy, as the er are among the best bird friends of trees.

METHOD — Let the observations begin h the study of the trees (common al- st everywhere) which have been at- ked by the sapsucker, and thus lead an interest in the culprit.

OBSERVATIONS — 1. Have you seen the rk of the sapsucker? Are the holes illed in rows completely around the tree? If there are two rows or more, are the holes set evenly one below another?

2. Do the holes sink into the wood, or are they simply through the bark? Why does it injure or kill a tree to be girdled with these holes? Have you ever seen the sapsuckers making these holes? If so, how did they act?

3. How many kinds of trees can you find punctured by these holes? Are they likely to be young trees?

4. How can you distinguish the sap- sucker from the other woodpeckers? How have the hairy and downy which are such good friends of the trees been made to suf- fer for the sapsucker's sins?

5. What is the color of the sapsucker: forehead, sides of head, back, wings, throat, upper and lower breast? What is the difference in color between the male and female?

6. In what part of the country do the sapsuckers build their nests? Where do they make their nests and how?

In the following winter the same bird (a sapsucker) tapped a maple-tree in front of my window in fifty-six places; and, when the day was sunny and the sap oozed out he spent most of his time there. He knew the good sap-days, and was on hand promptly for his tipple; cold and cloudy days he did not appear. He knew which side of the tree to tap, too, and avoided the sunless northern exposure. When one series of well-holes failed to supply him, he would sink another, drilling through the bark with great ease and quickness. Then, when the day was warm, and the sap ran freely, he would have a regular sugar-maple debauch, sitting there by his wells hour after hour, and as fast as they became filled sipping out the sap. This he did in a gentle, caressing manner that was very suggestive. He made a row of wells near the foot of the tree, and other rows higher up, and he would hop up and down the trunk as they became filled. — "WINTER NEIGHBORS," JOHN BURROUGHS

THE REDHEADED WOODPECKER

The redhead is well named, for his helmet and visor show a vivid glowing crimson that stirs the sensibilities of the color lover. It is readily distinguished from the other woodpeckers because its entire head and bib are red. For the rest, it is a beautiful dark metallic blue with the lower back, a band across the wing, and the under parts white; its outer tail feathers are tipped with white. The female is colored like the

L. A. Fuertes

The redheaded woodpecker

male, but the young have the head and breast gray, streaked with black and white, and the wings barred with black. It may make its nest by excavating a hole in a tree or a stump or even in a telegraph pole; the eggs are glossy white. This woodpecker is quite different in habits from the hairy and downy, as it likes to flit along from stump to fence post and catch insects on the wing, like a flycatcher. The only time that it pecks wood is when it is making a hole for its nest.

As a drummer, the redhead is most adept and his roll is a long one. H
adaptable fellow, and if there is n
nant dead limb at hand, he has
known to drum on tin roofs and lig
rods; and once we also observed hi
cuting a most brilliant solo on th
of a barbed fence. He is especially f
beechnuts and acorns, and being a t
fellow as well as musical, in time of
he stores up food against time of nee
places his nuts in crevices and forks
branches or in holes in trees or any
hiding place. He can shell a bee
quite as cleverly as can the deer m
and he is own cousin to the carp
woodpecker of the Pacific Coast, v
is also redheaded and which drills
in the oak trees wherein he drives a
like pegs for later use.

LESSON 17

The Redheaded Woodpecker

Leading Thought — The redhe
woodpecker has very different habits f
the downy and is not so useful to u
lives upon nuts and fruit and such in
as it can catch upon the wing.

Method — If there is a redhead in
vicinity of your school the children wi
sure to see it. Write the following q
tions upon the blackboard and off
prize to the first one who will make a
on where the redhead stores his wi
food.

Observations — 1. Can you tell
redhead from the other woodpeck
What colors especially mark his pl
age?

2. Where does the redhead nest?
scribe eggs and nest.

What have you observed the red- eating? Have you noticed it storing nd acorns for the winter? Have you d it flying off with cherries or other

What is the note of the redhead? you ever seen one drumming? t did he use for a drum? Did he come often to this place to make his music?

other trait our woodpeckers have endears them to me, and that has been pointedly noticed by our orni- gists, is their habit of drumming in pring. They are songless birds, and yet e musicians; they make the dry limbs uent of the coming change. Did you k that loud, sonorous hammering h proceeded from the orchard or the near woods on that still March or l morning was only some bird getting reakfast? It is downy, but he is not rap- at the door of a grub; he is rapping at door of spring, and the dry limb thrills eath the ardor of his blows. Or, later in season, in the dense forest or by some ote mountain lake, does that meas- d rhythmic beat that breaks upon the ice, first three strokes following each er rapidly, succeeded by two louder s with longer intervals between them, that has an effect upon the alert ear f the solitude itself had at last found a e — does that suggest anything less n a deliberate musical performance? In t, our woodpeckers are just as charac- istically drummers as is the ruffed grouse, and they have their particular limbs and stubs to which they resort for that purpose. Their need of expression is apparently just as great as that of the song-birds, and it is not surprising that they should have found out that there is music in a dry, seasoned limb which can be evoked beneath their beaks.

The woodpeckers do not each have a particular dry limb to which they resort at all times to drum, like the one I have described. The woods are full of suitable branches, and they drum more or less here and there as they are in quest of food; yet I am convinced each one has its favorite spot, like the grouse, to which it resorts, especially in the morning. The sugar-maker in the maple woods may notice that this sound proceeds from the same tree or trees about his camp with great regularity. A woodpecker in my vicinity has drummed for two seasons on a telegraph-pole, and he makes the wires and glass insulators ring. Another drums on a thin board on the end of a long grape-arbor, and on still mornings can be heard a long distance.

A friend of mine in a Southern city tells me of a redheaded woodpecker that drums upon a lightning-rod on his neighbor's house. Nearly every clear, still morning at certain seasons, he says, this musical rapping may be heard. "He alternates his tapping with his stridulous call, and the effect on a cool, autumn-like morning is very pleasing." — "Birds, Bees and Sharp Eyes," John Burroughs

THE FLICKER OR YELLOW-HAMMER

The first time I ever saw a flicker I said, What a wonderful meadowlark and at is it doing on that ant hill?" But an- ier glance revealed to me a red spot on e back of the bird's neck, and as soon I was sure that it was not a bloody gash, knew that it marked no meadowlark. ie top of the flicker's head and its back e slaty-gray, which is much enlivened by bright red band across the nape of the neck. The tail is black above and yellow tipped with black below; the wings are black, but have a beautiful luminous yellow beneath, which is very noticeable during flight. There is a locket adorning the breast; it is a thin, black crescent, much narrower than that of the meadowlark. Below the locket, the breast is yellowish white thickly marked with circular, black spots. The throat and sides of the head

Olin Sewall Pettingill, Jr.

A brood of seven young flickers

are pinkish brown, and the male has a black mustache extending backward from the beak with a very fashionable droop. Naturally enough the female, although she resembles her spouse, lacks his mustache. The beak is long, strong, somewhat curved and dark colored. This bird is distinctly larger than the robin. The white patch on the rump shows little or not at all when the bird is at rest. This white mark is known as a "color call" — for it has been said that it serves as a rear signal by means of which the flock of migrating birds are able to keep together in the night. The yellow-hammer's flight is wave-like and jerky — quite different from that of the meadowlark; it does not stay so constantly in the meadows, but often frequents woods and orchards.

The flicker has many names, such as golden-winged woodpecker, yellow-hammer, highhole, yarup, wake-up, clape, and many others. It earned the name of highhole because of its habit of excavating its nest high up in trees, usually between ten and twenty-five feet from the ground. It especially loves an old apple tree as a site for a nest, and most of our large old orchards can boast of a pair of these handsome birds during the nesting season of May and June. The flicker is not above renting any house he finds vacant, vated by some other birds last yea earned his name of yarup or wake-up his spring song, which is a rollicking "wick-a, wick-a, wick-a-wick" — a commonly heard the last of March or April. The chief insect food of the is ants, although it also eats beetles, and wild fruit; it does little or no da to planted crops. Its tongue has be modified, like that of the anteater; long and is covered with a sticky stance; and when it is thrust into a hill, all of the little citizens, disturb their communal labors, at once br attack the intruder and become glue to it; they are thus withdrawn and t ferred to the capacious stomach of bird. It has been known to eat three t sand ants at a single meal.

Those who have observed the fl during the courting season declare to be the most silly and vain of all wooers. Mr. Baskett says: "Whe wishes to charm his sweetheart he mo a small twig near her, and lifts his w spreads his tail, and begins to nod and left as he exhibits his mustache t charmer. He sets his jet locket first on side of the twig and then on the o He may even go so far as to turn his half around to show her the pretty

A. A.

The male flicker has a black mustach

is back hair. In doing all this he pers the most ludicrous antics and has illiest expression of face and voice as losing his heart, as some one phrases e had lost his head also."

ne nest hole is quite deep and the e eggs are from four to ten in num- The feeding of the young flickers is a ess painful to watch. The parent takes food into its own stomach and par- digests it, then thrusts its own bill n the throat of the young one and ps the soft food into it "kerchug, hug," until it seems as if the ng one must be shaken to its foun- ons. The young flickers as soon as leave the nest climb around freely he home tree in a delightful, playful ner.

Stanley Mythaler

The homes of flickers

Olin Sewall Pettingill, Jr.

The female flicker

LESSON 18

The Flicker

Leading Thought — The flicker is a true woodpecker but has changed its habits and spends much of its time in meadows hunting for ants and other insects; it makes its nest in trunks of trees, like its relatives. It can be distinguished from the meadowlark by the white patch above the tail which shows during flight.

Method — This is one of the most important of the birds of the meadow. The work may be done in September, when there are plenty of young flickers which have not learned to be wary. The observations may be made in the field, a few questions being given at a time.

Observations — 1. Where do you find the flicker in the summer and early autumn? How can you tell it from the meadowlark in color and in flight?

2. What is it doing in the meadows? How does it manage to trap ants?

3. What is the size of the flicker as compared to the robin? What is its general color as compared to the meadowlark?

4. Describe the colors of the flicker as follows: top and sides of the head, back of the neck, lower back, tail, wings, throat, and breast. Describe the color and shape of the beak. Is there a difference in markings between the males and females?

5. Does the patch of white above the tail show, except when the bird is flying? Of what use is this to the bird?

6. What is the flicker's note? At what time of spring do you hear it first?

7. Where does the flicker build its nest and how? What is the color of the eggs? How many are there?

8. How does it feed its young? How do the young flickers act?

9. How many names do you know for the flicker?

The high-hole appears to drum more promiscuously than does the downy. He utters his long, loud spring call, whick-whick-whick, and then begins to rap his beak upon his perch before th note has reached your ear. I have seen drum sitting upon the ridge of the The log-cock, or pileated woodpecke largest and wildest of our Northern cies, I have never heard drum. His b should wake the echoes.

When the woodpecker is searchin food, or laying siege to some hidden g the sound of his hammering is dea muffled, and is heard but a few yards. only upon dry, seasoned timber, free its bark, that he beats his reveille to sp and woos his mate. — "Birds, Bees Sharp Eyes," John Burroughs

THE MEADOWLARK

L. A. Fuertes

The meadowlark

The first intimation we have in early spring that the meadowlark is again with us comes to us through his soft, sweet, sad note which Van Dyke describes so graphically when he says it "leaks slowly upward from the ground." One wonders how a bird can express happiness in these melancholy, sweet, slurred notes, and yet undoubtedly it is a song expressing joy, the joy of returning home, the happi of love and of nest building.

The meadowlark, as is indicated by name, is a bird of the meadow. It is of confused with another bird of the mea which has very different habits, the flic The two are approximately of the sa size and color and each has a black c cent or locket on the breast and e shows the "white feather" during flig The latter is the chief distinguishing c acteristic; the outer tail feathers of meadowlark are white, while the tail fea ers of the flicker are not white at all, bu has a single patch of white on the run The flight of the two is quite differe The lark lifts itself by several sharp mo ments and then soars smoothly over t course, while the flicker makes a contin ous up-and-down, wavelike flight. T songs of the two would surely never confused, for the meadowlark is amo our sweetest singers, to which class t flicker with his "flick-a-flick" hardly b longs.

The colors of the meadowlark are m harmonious shades of brown and yello well set off by the black locket on breast. Its wings are light brown, ea feather being streaked with black ar brown; the line above the eye is yello bordered with black above and below;

ne extends from the beak backward ıe crown. The wings are light brown ave a mere suggestion of white bars; ns of the outer feathers on each side tail are white, but this white does ıow except during flight. The sides throat are greenish, the middle part oreast are lemon-yellow, with the black crescent just below the throat. oeak is long, strong, and black, and eadowlark is decidedly a low-browed the forehead being only slightly r than the upper part of the beak. It ittle larger than the robin, which it in plumpness.

e meadowlark has a particular liking neadows which border streams. It when on the ground, on the bush nce and while on the wing; and it during the entire period of its north-stay, from April to November, ex-while it is moulting in late summer. Mathews, who is an eminent author-n bird songs, says that the meadow-of New York have a different song those of Vermont or Nantucket, al-gh the music has always the same ral characteristics. The western spe-has a longer and more complex song ours of the East. It is one of the few fornia birds that is a genuine joy to eastern visitor; during February and ch its heavenly music is as pervasive ıe California sunshine.

The meadowlark's arched nest

R. W. Hegner

A father prairie horned lark at his nest. These birds nest in early March, and often snow falls on the nest and brooding bird

The nest is built in a depression in the ground near a tuft of grass; it is constructed of coarse grass and sticks and is lined with finer grass; there is usually a dome of grass blades woven above the nest; and often a long, covered vestibule leading to the nest is made in a similar fashion. This is evidently for protection from the keen eyes of hawks and crows. The eggs are laid about the last of May and are usually from five to seven in number; they are white, speckled with brown and purple. The young meadowlarks are usually large enough to be out of the way before haying time in July.

The food of the meadowlark during the entire year consists almost exclusively of insects which destroy the grass of our meadows. It eats great quantities of grasshoppers, cutworms, chinch bugs, army worms, wireworms, and weevils, and also destroys some weed seeds. Each pupil should make a diagram in his notebook showing the proportions of the meadowlark's different kinds of food. This may be copied from *Audubon Leaflet* 3. Everyone should use his influence to the uttermost to protect this valuable bird. It has been estimated that the meadowlarks save to every township where hay is produced, twenty-five dollars each year on this crop alone.

LESSON 19

The Meadowlark

Leading Thought — The meadowlark is of great value in delivering the grass of our meadows from insect destroyers. It has a song which we all know; it can be identified by color as a large, light brown bird with white feathers on each side of the tail, and in flight by its quick up-and-down movements finishing with long, low, smooth sailing.

Method — September and October are good months for observations on the flight, song, and appearance of the meadowlark, and also for learning how to distinguish it from the flicker. The notes must be made by the pupils in the field, and after they know the bird and its song let them, if they have opportunity, study the bird books and bulletins, and prepare written accounts of the way the meadowlark builds its nest and of its economic value.

Observations — 1. Where have you seen the meadowlark? Did you ever see it in the woods? Describe its flight. How can you identify it by color when it is flying? How do its white patches and its flight differ from those of the flicker?

2. Try to imitate the meadowlark's notes by song or whistle. Does it sing while on the ground, or on a bush or fence, or during flight?

3. Note the day when you hear its last song in the fall and also its first song in the spring. Does it sing during August and September? Why? Where does it spend the winter? On what does it feed while in the South?

4. Is the meadowlark larger or s than the robin? Describe from you observations, as far as possible, the of the meadowlark as follows: top of line above the eye, back, wings, throat, breast, locket, color and sha beak. Make a sketch of your own copy from Louis Fuertes' excellent p of the meadowlark in the *Audubon let*, and color it accurately.

5. When is the nest built; where placed; of what material is it built? H it protected from sight from above? this protection? How many eggs are in the nest? What are their colors markings?

6. What is the food of the mea

Sweet, sweet, sweet! O happy that I a
(Listen to the meadow-larks, acros
fields that sing!)
Sweet, sweet, sweet! O subtle breat
balm,
O winds that blow, O buds that g
O rapture of the spring!

Sweet, sweet, sweet! O happy world
is!
Dear heart, I hear across the fields
mateling pipe and call
Sweet, sweet, sweet! O world so fu
bliss,
For life is love, the world is love,
love is over all!

— Ina Coolb

S. A. Grimes

English sparrows at a feeding station

THE ENGLISH SPARROW

So dainty in plumage and hue,
A study in grey and in brown,
How little, how little we knew
The pest he would prove to the town!
From dawn until daylight grows dim,
Perpetual chatter and scold.
No winter migration for him,
Not even afraid of the cold!
Scarce a song-bird he fails to molest,
Belligerent, meddlesome thing!
Wherever he goes as a guest
He is sure to remain as a King.

— Mary Isabella Forsyth

The English sparrow, like the poor and housefly, is always with us; and since is here to stay, let us make him useful ve can devise any means of doing so. ere is no bird that gives the pupils a re difficult exercise in describing colors l markings than does he; and his wife almost equally difficult. I have known ly skilled ornithologists to be misled some variation in color of the hen sparv, and it is safe to assert that the maity of people " do not know her from am." The male has the top of the head y with a patch of reddish brown on her side; the middle of the throat and upper breast is black; the sides of the throat white; the lower breast and under parts grayish white; the back is brown streaked with black; the tail is brown, rather short, and not notched at the tip; the wings are brown with two white bars and a jaunty dash of reddish brown. The female has the head grayish brown, the breast, throat, and under parts grayish white; the back is brown streaked with black and dirty yellow, and she is, on the whole, a " washed out " looking lady bird. The differences in color and size between the English sparrow and the chippy are quite noticeable, as the chippy is an inch shorter and far more slender in appearance, and is especially marked by the reddish brown crown.

When feeding, the English sparrows are aggressive, and their lack of table manners make them the "goops" among all birds; in the winter they settle in noisy flocks on the street to pick up the grain undigested by the horses, or in barnyards where the grain has been scattered by the farm animals. They only eat weed seeds when other food fails them in the winter, for they are civilized birds even if they do not act so, and they much prefer the cultivated grains. It is only during the nesting season that they destroy insects to any extent; over one-half the food of nestlings is insects, such as weevils, grasshoppers, cutworms, etc.; but this good work is largely offset by the fact that these same nestlings will soon give their grown-up energies to attacking grain fields, taking the seed after sowing, later the new grain in the milk, and later still the ripened grain in the sheaf. Wheat, oats, rye, barley, corn, sorghum, and rice are thus attacked. Once I saw on the upper Nile a native boat loaded with millet which was attacked by thousands of sparrows; when driven off by the sailors they would perch on the rigging like flies, and as soon as the men turned their backs they would drop like bullets to the deck and gobble the grain before they were again driven off. English sparrows also destroy for us the buds and blossoms of fruit trees and often attack the ripening fruit.

The introduction of the English sparrow into America is one of the greatest arguments possible in favor of nature-study; for ignorance of nature-study methods in this single instance costs the United States millions of dollars every year. The English sparrow is the European house sparrow, and people had a theory that it was an insect eater, but never took the pains to ascertain if this theory were a fact. About 1850, some people with more zeal than wisdom introduced these birds into New York, and for twenty years afterwards there were other importations of the sparrows. In twenty years more, people discovered that they had taken pains to establish in our country one c worst nuisances in all Europe. In add to all the direct damage which the En sparrows do, they are so quarrelsome they have driven away many of our n beneficial birds from our premises, now vociferously acclaim their presen places which were once the haunts of with sweet songs. After they drive of other birds they quarrel among t selves, and there is no rest for tired ea their vicinity. There are various n made by these birds which we can u stand if we are willing to take the p the harassing chirping is their song; squall when frightened and peep p tively when lonesome, and make a agreeable racket when fighting.

But to "give the devil his due" must admit that the house sparrow clever as it is obnoxious, and its succe doubtless partly due to its superior cl ness and keenness. It is quick to ta hint, if sufficiently pointed; firing a s gun twice into a flock of these birds driven them from our premises; and ing down their nests assiduously f month seems to convey to them the that they are not welcome. Another stance of their cleverness I witnessed day: I was watching a robin, worn nervous with her second brood, ferve hunting earthworms in the lawn to fill gaping mouths in the nest in the Virg creeper shading the piazza. She fin pulled up a large, pink worm, and a sparrow flew at her viciously; the ro dropped the worm to protect herself, the sparrow snatched it and carried it triumphantly to the grape arbor wh she had a nest of her own full of gap mouths. She soon came back, and a safe distance watched the robin pull another worm, and by the same tac again gained the squirming prize. Th times was this repeated in an hour, then the robin, discouraged, flew up i a Norway spruce and in a monologue sullen cluckings tried to reason out w had happened.

The English sparrow's nest is quite

ing with the bird's other qualities; it ually built in a hole or box or in some ected corner beneath the eaves; it is often built in vines on buildings and sionally in trees. It is a good example fuss and feathers"; coarse straw, or other kind of material, and feathers of or of other birds, mixed together out fashion or form, constitute the . In these sprawling nests the whitish, vn or gray-flecked eggs are laid and oung reared; several broods are reared ne pair in a season. The nesting begins ost as soon as the snow is off the nd and lasts until late fall.

uring the winter, the sparrows gather ocks in villages and cities, but in the ng they scatter out through the coun- where they can find more grain. The place where this bird is welcome is sibly in the heart of a great city, where other bird could pick up a livelihood. a true cosmopolite and is the first bird reet the traveler in Europe or northern ica. These sparrows will not build in es suspended by a wire; and they do like a box where there is no resting ce in front of the door leading to the t.

A. A. Allen

The sprawling nest of the English sparrow

LESSON 20

The English Sparrow

Leading Thought — The English spar- v was introduced into America by peo- who knew nothing of its habits. It has ally overrun our whole country, and to reat extent has driven out from towns d villages our useful American song ds; it should be discouraged and not owed to nest around our houses and grounds. As a sparrow it has interesting habits which we should observe.

Method — Let the pupils make their observations in the street or wherever they find the birds. The greatest value of this lesson is to teach the pupils to observe the coloring and markings of a bird accurately and describe them clearly. This is the best of training for later work with the wild birds.

Observations — 1. How many kinds of birds do you find in a flock of English sparrows?

2. The ones with the black cravat are naturally the men of the family, while their sisters, wives, and mothers are less ornamented. Describe in your notebook or from memory the colors of the cock sparrow as follows: top of head, sides of the head, the back, the tail, the wings, wing bars, throat and upper breast, lower breast and under parts.

3. Describe the hen sparrow in the same manner and note the difference in markings between the two. Are the young birds, when they first fly, like the father or the mother?

4. Compare the English sparrow with the chippy and describe the differences in size and color.

5. Is the tail when the bird is not flying square across the end or notched?

6. What is the shape of the beak? For what sort of food is it adapted?

7. What is the food of the English sparrows and where do they find it? Describe the actions of a flock feeding in the yard or street. Are the English sparrows kindly or quarrelsome in disposition?

8. Why do the English sparrows stay in the North during the coldest of winters? Do they winter out in the country or in villages?

9. Describe by observation how they try to drive away robins or other native birds.

10. Describe the nest of this sparrow. Of what material is it made? How is it supported? How sheltered? Is it a well-built nest?

11. Describe the eggs. How many broods are raised a year? What kin food do the parents generally give nestlings?

12. If you have ever seen these spar do anything interesting, describe the cumstance.

13. In what ways are these birds a sance to us?

14. How much of English sparrow do you understand?

15. How can we build bird-boxe that the English sparrows will not tr take possession of them?

Do not tire the child with questi lead him to question you, instead. Be s in any case, that he is more intereste the subject than in the questions ab the subject.

THE CHIPPING SPARROW

Leonard K. Beyer

A chipping sparrow on its nest

This midget lives in our midst, and yet among all bird kind there is not another which so ignores us as does the chippy. It builds its nest about our houses, it hunts for food all over our premises, it sings like a tuneful grasshopper in our ears, it brings up its young to disregard us, and every hour of the day it "t tsips" us to scorn. And, although it well earned the name of "doorstep s row," since it frugally gathers the cru about our kitchen doors, yet it rarely comes tame or can be induced to from the hand, unless it is trained sc do as a nestling.

Its cinnamon-brown cap and tiny bl forehead, the gray streak over the eye the black through it, the gray cheeks the pale gray, unspotted breast disting it from the other sparrows, although brown back streaked with darker co and brown wings and blackish tail, hav very sparrowish look; the two whitish w bars are not striking; it has a bill fitted shelling seeds, a characteristic of all sparrows. Despite its seed-eating bill, chippy's food is about one-third inse and everyone should know that this li bird does good to our gardens and tr It takes in large numbers cabbage ca pillars, pea lice, the beet leaf-miners, l hoppers, grasshoppers, and cutworms, a does its share in annihilating the ca pillars of the terrible gypsy and brown moths. In fact, it works for our ben even in its vegetable food, as this cons

y of the seeds of weeds and unde-
e grasses. It will often fly up from
rch after flies or moths, like a fly-
er; and the next time we note it, it
be hopping around hunting for the
bs we have scattered for it on the
floor. The song of the chippy is
interesting to it than to us; it is a
nuous performance of high, shrill,
notes, all alike so far as I can detect;
it utters many of these in rapid suc-
on it is singing, but when it gives
singly they are call notes or mere
ersation.
ne peculiarity of the nest has given
sparrow the common name of hair-
for the lining is almost always of
coarse hair, usually treasure trove
the tails of horses or cattle, switched
gainst boards, burs, or other obstacles.
he many nests I have examined, black
ehair was the usual lining; but two
s in our yard show the chippy to be
ourceful bird; evidently the hair mar-
was exhausted and the soft, dead
lles of the white pine were used in-
l and made a most satisfactory lining.
nest is tiny and shallow; the outside
f fine grass or rootlets carefully but
closely woven together; it is placed

A. A. Allen

"The breadline." Young chipping sparrows being fed by one of their parents

in vine or tree, usually not more than ten or fifteen feet from the ground; a vine on a house is a favorite nesting site. Once a bold pair built directly above the entrance to our front door and mingled cheerfully with other visitors. Usually, however, the nest is so hidden that it is not discovered until after the leaves have fallen. The eggs are light blue tinged with green, with fine, purplish brown specks or markings scrawled about the larger end.

The chippy comes to us in early spring and usually raises two broods of from three to five "piggish" youngsters, which even after they are fully grown follow pertinaciously their tired and "frazzled out" parents and beg to be fed; the chippy parents evidently have no idea of discipline but indulge their teasing progeny until our patience, at least, is exhausted. The young differ from the parents in having streaked breasts and lacking the reddish crown. In the fall the chippy parents lose their red-brown caps and have streaked ones instead; and then they fare forth in flocks for a seed-harvest in the fields. Thereafter our chippy is a stranger to us; we do not know it in its new garb, and it dodges into the bushes as we pass, as if it had not tested our harmlessness on our own door-stone.

A. A. Allen

*A cowbird laid the large egg in this chip-
g sparrow's nest. The cowbird depends
on other birds to brood its eggs and care
its young*

LESSON 21

The Chipping Sparrow

Leading Thought — The chipping sparrow is a cheerful and useful little neighbor. It builds a nest, lined with horsehair, in the shrubbery and vines about our homes and works hard in ridding our gardens of insect pests and seeds of weeds.

Method — Begin this lesson with a nest of the chippy, which is so unmistakable that it may be collected and identified in the winter. Make the study of this nest so interesting that the pupils will wait anxiously to watch for the birds which made it. As soon as the chippies appear, the questions should be asked, a few at a time, giving the children several weeks for the study.

The Nest

Observations — 1. Where was this nest found? How high from the ground?

2. Was it under shelter? How was it supported?

3. Of what material is the outside of the nest? How is it fastened together? How do you suppose the bird wove this material together?

4. Of what material is the lining? Why is the bird that built this nest called the "hair-bird"? From what animal do you think the lining of the nest came? How do you suppose the bird got it?

5. Do you think the nest was well hidden when the leaves were about it? Measure the nest across and also its depth; do you think the bird that made it is as large as the English sparrow?

The Bird

6. How can you tell the chippy from the English sparrow?

7. Describe the colors of the chippy as follows: beak, forehead, crown, marks above and through the eyes, cheeks, throat, breast, wings, and tail. Note if the wings have whitish bars and how many.

8. Describe the shape of the be compared with that of the robin. is this shaped bill adapted for?

9. What is the food of the ch Why has it been called the doc sparrow?

10. Note whether the chippy ca flies or moths on the wing like the ph

11. Why should we protect the c and try to induce it to live nea gardens?

12. Does it run or hop when se food on the ground?

13. How early in the season doe chippy appear and where does it s the winter?

14. Can you describe the chi song? How do you think it won the of chipping sparrow?

15. If you have the luck to find a of chippies nesting, keep a diary of observations in your notebook cov the following points: Do both pa build the nest? How is the frame laid? How is the finishing done? Wh the number and color of the eggs? both parents feed the young? How young chippies act when they first l the nest? How large are the young before the parents stop feeding th What are the differences in color markings between parents and young

THE FIELD-SPARROW

A bubble of music floats, the slope of
hillside over;
A little wandering sparrow's notes;
the bloom of yarrow and clover,
And the smell of sweet-fern and the
berry leaf, on his ripple of song
stealing,
For he is a cheerful thief, the wealt
the fields revealing.

One syllable, clear and soft as a raindr
silvery patter,
Or a tinkling fairy-bell; heard aloft, in
midst of the merry chatter
Of robin and linnet and wren and jay,
syllable, oft repeated;
He has but a word to say, and of that
will not be cheated.

singer I have not seen; but the song
' arise and follow
brown hills over, the pastures green,
nd into the sunlit hollow.
a joy that his life unto mine has
'ent, I can feel my glad eyes glisten,
gh he hides in his happy tent, while
I stand outside, and listen.

This way would I also sing, my dear little
hillside neighbor!
A tender carol of peace to bring to the
sunburnt fields of labor
Is better than making a loud ado; trill on,
amid clover and yarrow!
There's a heart-beat echoing you, and
blessing you, blithe little sparrow!

— LUCY LARCOM

THE SONG SPARROW

He does not wear a Joseph's coat of many colors, smart and gay
His suit is Quaker brown and gray, with darker patches at his throat.
And yet of all the well-dressed throng, not one can sing so brave a song.
It makes the pride of looks appear a vain and foolish thing to hear
In " Sweet, sweet, sweet, very merry cheer."

A lofty place he does not love, he sits by choice and well at ease
In hedges and in little trees, that stretch their slender arms above
The meadow brook; and then he sings till all the field with pleasure rings;
And so he tells in every ear, that lowly homes to heaven are near
In " Sweet, sweet, sweet, very merry cheer."

— HENRY VAN DYKE

:hildren may commit to memory the m from which the above stanzas were :n; seldom in literature have detailed ırate observation and poetry been so pily combined as in these verses. The on might begin in March when we all listening eagerly for bird voices, and children should be asked to look out a little, brown bird which sings, weet, sweet, sweet, very merry cheer," as Thoreau interprets it, " Maids! ids! Maids! Hang on the teakettle, kettle-ettle-ettle." In early childhood arned to distinguish this sparrow by its 'eakettle" song. Besides this song, it others quite as sweet; and when rmed it utters a sharp " T'chink, ıink."

The song sparrow prefers the neighbor-od of brooks and ponds which are bor-red with bushes, and also the hedges nted by nature along rail or other field ces, and it has a special liking for the rubbery about gardens. Its movements d flight are very characteristic; it usually sits on the tip-top of a shrub or low tree when it sings; when disturbed, however, it never rises in the air but drops into a low

Leonard K. Beyer

The song sparrow usually builds its nest on the ground

flight and plunges into a thicket with a defiant twitch of the tail which says plainly, " Find me if you can."

A. A. Allen

The eggs are bluish white with many brown markings

The color and markings of this bird are typical of the sparrows. The head is a warm brown with a gray streak along the center of the crown and one above each eye, with a dark line through the eye. The back is brown with darker streaks. The throat is white with a dark spot on either side; the breast is white spotted with brown with a large, dark blotch at its very center; this breast blotch distinguishes this bird from all other sparrows. The tail and wings are brown and without buff or white bars or other markings. The tail is long, rounded, and very expressive of emotions, and makes the bird look more slender than the English sparrow.

The nest is usually placed on the ground or in low bushes not more than five feet from the ground; it varies much in both size and material; it is sometimes constructed of coarse weeds and grasses; and sometimes only fine grass is used. Sometimes it is lined with hair, and again, with fine grass; sometimes it is deep, but occasionally is shallow. The eggs have a whitish ground-color tinged with blue or green, but are so blotched and marked with brown that they are safe from observation of enemies. The nesting season begins in May, and there are usually three and sometimes four broods; but so far as I have observed, a nest is never used for two consecutive broods. The song s rows stay with us in New York State late in the fall, and a few stay in shelt places all winter. The quality in this which endears him to us all is the s of song which stays with him; his s trill may be heard almost any mont the year, and he has a charming hab singing in his dreams.

The song sparrow is not only the dea of little neighbors, but it also works lu for our good and for its own food at same time. It destroys cutworms, p lice, caterpillars, canker-worms, gro beetles, grasshoppers, and flies; in wi it destroys thousands of weed seeds, w otherwise would surely plant themse to our undoing. Every boy and girl sho take great pains to drive away stray and to teach the family puss not to me with birds; for cats are the worst of the song sparrow's enemies, destroy thousands of its nestlings every year.

LESSON 22

The Song Sparrow

Leading Thought — The beaut song of this sparrow is usually heard ear in the spring than the notes of blueb or robin. The dark blotch in the center its speckled breast distinguishes this sp row from all others; it is very benefi and should be protected from cats.

"Sweet, sweet, sweet, very merry cheer"

ƖETHOD — All the observations of the ; sparrow must be made in the field, they are easily made because the bird ds near houses, in gardens, and in the bbery. Poetry and other literature ıt the song sparrow should be given ıe pupils to read or to memorize.

BSERVATIONS — 1. Have you noticed ttle brown bird singing a very sweet ; in the early spring? Did the song ıd as if set to the words "Little Maid! le Maid! Little Maid! Put on the tea-le, teakettle-ettle-ettle"?

Where was this bird when you heard singing? How high was he perched ve the ground? What other notes did hear him utter?

. Describe the colors and markings of song sparrow on head, back, throat, ıst, wings, and tail. Is this bird as large he English sparrow? What makes it : more slim?

. How can you distinguish the song rrow from the other sparrows? When urbed does it fly up or down? How s it gesture with its tail as it disappears he bushes?

. Where and of what material does song sparrow build its nest?

. What colors and markings are on eggs? Do you think these colors and rkings are useful in concealing the eggs en the mother bird leaves the nest?

. How late in the season do you see song sparrows and hear their songs?

. How can we protect these charming le birds and induce them to build near houses?

9. What is the food of the song sparrows and how do they benefit our fields and gardens? Name some of the injurious insects that they eat.

THE SING-AWAY BIRD

Have you ever heard of the Sing-away bird,
That sings where the Runaway River
Runs down with its rills from the bald-headed hills
That stand in the sunshine and shiver?
"Oh, sing! sing-away! sing-away!"
How the pines and the birches are stirred
By the trill of the Sing-away bird!

And the bald-headed hills, with their rocks and their rills,
To the tune of his rapture are ringing;
And their faces grow young, all the gray mists among,
While the forests break forth into singing.
"Oh, sing! sing-away! sing-away!"
And the river runs singing along;
And the flying winds catch up the song.

'Twas a white-throated sparrow, that sped a light arrow
Of song from his musical quiver,
And it pierced with its spell every valley and dell
On the banks of the Runaway River.
"Oh, sing! sing-away! sing-away!"
The song of the wild singer had
The sound of a soul that is glad.

— Lucy Larcom

THE MOCKINGBIRD

Among all the vocalists in the bird rld, the mockingbird is seldom rivaled he variety and richness of his repertoire. e mockingbirds go as far north as south-New England, but they are found at ir best in the Southern states and California. On the Gulf Coast the ckers begin singing in February; in rmer climates they sing almost the r through. During the nesting season, the father mocker is so busy with his cares and duties during the day that he does not have time to sing, and so he devotes the nights to serenading; he may sing almost all night long if there is moonlight, and even on dark nights he gives now and then a happy, sleepy song. Not all mockingbirds are mockers; some sing their own song, which is rich and beautiful; while others learn, in addition, not only the

L. A. Fuertes

The mockingbird

songs of other birds, but their call notes as well. One authority noted a mocker which imitated the songs of twenty species of birds during a ten-minute performance. When singing, the mocker shows his relationship to the brown thrasher by lifting the head and depressing and jerking the tail. A good mocker will learn a tune, or parts of it, if it is whistled often enough in his hearing; he will also imitate other sounds and will often improve on a song he has learned from another bird by introducing frills of his own; when learning a song, he sits silent and listens intently, but will not try to sing it until it is learned.

Although the mockingbirds live in wild places, they prefer the haunts of men, taking up their home sites in gardens and cultivated grounds. Their flight is rarely higher than the tree tops and is decidedly jerky in character with much twitching of the long tail. For nesting sites, they choose thickets or the lower branches of trees, being especially fond of orange trees; the nest is usually from four to twenty feet from the ground. The foundation of the nest is made of sticks, grasses, and weed stalks interlaced and crisscrossed; on these is built the nest of softer materials, such as rootlets, horsehair, cotton, or in fact anything suitable which is at hand. T nest is often in plain sight, since mocker trusts to his strength as a fig to protect it. He will attack cats with g ferocity and vanquish them; he will o kill snakes; good-sized black snakes h been known to end thus; he will also d away birds much larger than himself making his attack, the mocker ho above his enemy and strikes it at the b of the head or neck.

The female lays from four to six p greenish or bluish eggs blotched w brown which hatch in about two we then comes a period of hard work for parents, as both are indefatigable in cat ing insects to feed the young. The mock by the way, is an amusing sight as chases a beetle on the ground, lifting wings in a pugnacious fashion. The mo ers often raise three broods a season; young birds have spotted breasts, show their relationship to the thrasher.

As a wooer, the mocker is a bird much ceremony and dances into his lad graces. Mrs. F. W. Rowe, in describ this, says that the birds stand facing e other with heads and tails erect and wi drooping; "then the dance would beg and this consisted of the two hopp sideways in the same direction and rather a straight line a few inches a

A. A. A

A mockingbird on her nest in a thicket

always keeping directly opposite other and about the same distance They would *chassez* this way four e feet, then go back over the same n the same manner." Mrs. Rowe bserved that the male mockers have ng preserves of their own, not allowny other males of their species in precincts. The boundary was sus-l by tactics of both offense and se; but certain other species of were allowed to trespass without of.

urice Thompson describes in a de-ul manner the "mounting" and pping" songs of the mocker which during the wooing season. The r flits up from branch to branch of e, singing as he goes, and finally on opmost bough gives his song of tri-n to the world; then, reversing the ess, he falls backward from spray to , as if drunk with the ecstasy of his song, which is an exquisitely soft gling series of notes, liquid and sweet, seem to express utter rapture."

ne mockingbirds have the same colors oth sexes; the head is black, the back hy-gray; the tail and wings are so brown that they look black; the tail ry long and has the outer tail feathers ely white and the two next inner ones white for more than half their length; wings have a strikingly broad, white which is very noticeable when the is flying. The under parts and breast grayish white; the beak and legs are kish. The food of the mockingbirds bout half insects and half fruit. They largely on the berries of the red cedar, tle, and holly, and we must confess are n too much devoted to the fruits in our ards and gardens; but let us put down heir credit that they do their best to rminate the cotton boll caterpillars moths, and also many other insects rious to crops.

'he mocker is full of tricks and is dis-tly a bird of humor. He will frighten er birds by screaming like a hawk and n seem to chuckle over the joke.

idney Lanier describes him well:

Leonard K. Beyer

The brown thrasher, a close relative of the mockingbird, is also an accomplished musician

Whate'er birds did or dreamed, this bird could say.
Then down he shot, bounced airily along
The sward, twitched in a grasshopper, made song
Midflight, perched, prinked, and to his art again.

LESSON 23

The Mockingbird

Leading Thought — The mockingbird is the only one of our common birds that sings regularly at night. It imitates the songs of other birds and has also a beautiful song of its own. When feeding their nestlings, the mockers do us great service by destroying insect pests.

Method — Studies of this bird are best made individually by the pupils through watching the mockers which haunt the houses and shrubbery. If there are mockingbirds near the schoolhouse, the work can be done in the most ideal way by keeping records in the school of all the observations made by the pupils, thus bringing out an interesting mockingbird story.

Observations — 1. During what months of the year and for how many months does the mockingbird sing in this locality?

2. Does he sing only on moonlight nights? Does he sing all night?

3. Can you distinguish the true mockingbird song from the songs which he has learned from other birds? Describe the actions of a mocker when he is singing.

4. How many songs of other birds have you heard a mocker give and what are the names of these birds?

5. Have you ever taught a mocker a tune by whistling it in his presence? If so, tell how long it was before he learned it and how he acted while learning.

6. Describe the flight of the mockingbirds. Do they fly high in the air like crows?

7. Do these birds like best to live in wild places or about houses and gardens?

8. Where do they choose sites for their nests? Do they make an effort to hide the nest? If not, why?

9. Of what material is the nest made? How is it lined? How far from the ground is it placed?

10. What are the colors of the eggs? How many are usually laid? How long before they hatch?

11. Give instances of the parents' devotion to the young birds.

12. Have you seen two mockingbirds dancing before each other just before the nesting season?

13. In the spring have you heard a mocker sing while mounting from the lower to the upper branches of a tree and then after pouring forth his best song fall backward with a sweet, gurgling song as if intoxicated with his music?

14. How many broods does a pair of mockers raise during one season? How does the color of the breast of the young differ from that of the parent?

15. How does the father bird protect the nestlings from other birds, cat snakes?

16. Does the mocker select c places for his own hunting grounc drive off other mockers which tresp

17. Describe the colors of the mo bird as follows: beak, head, back wings, throat, breast, under parts an

18. What is the natural food c mockingbirds and how do they bene farmer? How does the mocker act attacking a ground beetle?

19. Have you seen mockin frighten other birds by imitating t of a hawk? Have you seen them play tricks?

20. Tell a story which includes own observations on the ways of mo birds which you have known.

Soft and low the song began: I sc
caught it as it ran
Through the melancholy trill of the
tive whip-poor-will,
Through the ringdove's gentle wail,
tering jay and whistling quail,
Sparrow's twitter, catbird's cry, red
whistle, robin's sigh;
Blackbird, bluebird, swallow, lark,
his native note might mark.

Oft he tried the lesson o'er, each
louder than before;
Burst at length the finished song, lou
clear it poured along;
All the choir in silence heard, hushe
fore this wondrous bird.
All transported and amazed, sca
breathing, long I gazed.
Now it reached the loudest swell; l
lower, now it fell, —
Lower, lower, lower still, scarce it sou
o'er the rill.

— Joseph Rodman D

THE CATBIRD

The Catbird sings a crooked song, in minors that are flat,
And, when he can't control his voice he mews just like a cat,
Then nods his head and whisks his tail and lets it go at that.
— OLIVER DAVIE

; a performer, the catbird distinctly ngs to the vaudeville, even going so s to appear in slate-colored tights. His ialties range from the most exquisite ; to the most strident of scolding s; his nasal "n-y-a-a-h, n-y-a-a-h" is so very much like the cat's mew after but when addressed to the intruder eans "get out"; and not in the whole ut of bird notes is there another which quickly inspires the listener with this re. I once trespassed upon the terri- of a well-grown catbird family and squalling that ensued was ear-splitting; retreated, the triumphant youngsters owed me for a few rods with every of triumph in their actions and voices; y obviously enjoyed my apparent ht. The catbirds have rather a pleasant uck, cluck" when talking to each er, hidden in the bushes, and they also e a variety of other notes. The true g of the catbird, usually given in the ly morning, is very beautiful. Mr. thews thinks it is a medley gathered m other birds, but it seems to me very lividual. However, true to his vaude- e training, this bird is likely to intro- ce into the middle or at the end of his quisite song some phrase that suggests cat call. He is, without doubt, a e mocker and will often imitate the oin's song, and also if opportunity offers rns to converse fluently in chicken guage. One spring morning I heard tside my window the mellow song of e cardinal, which is a rare visitor in ew York, but there was no mistaking the tor-re-do, tor-re-do." I sprang from my d and rushed to the window, only to e a catbird singing the cardinal song, d thus telling me that he had come om the sunny South and the happy companionship of these brilliant birds. Often when the catbird is singing, he sits on the topmost spray of some shrub lifting his head and depressing his tail, like a brown thrasher; and again, he sings completely hidden in the thicket.

In appearance the catbird is tailor-made, belonging to the same social class as the cedar-bird and the barn swallow.

Robert Matheson

A catbird on its nest

However, it affects quiet colors, and its well-fitting costume is all slate-gray except the top of the head and the tail which are black; the feathers beneath the base of the tail are brownish. The catbird is not so large as the robin, and is of very different shape; it is far more slender and has a long, emotional tail. The way the catbird twitches and tilts its tail, as it hops along the ground or alights in a bush, is very characteristic. It is a particularly alert and nervous bird, always on the watch for intruders, and the first to give warning to all other birds of their approach. It is a good fighter in defending its nest, and there are several observed instances where it has fought to defend the nest of other species of birds; and it has gone even

The catbird lays three to five eggs of a rich greenish blue in a well constructed nest in a dense thicket

further in its philanthropy, by feeding their orphaned nestlings.

The catbird chooses a nesting site in a low tree or shrub or brier, where the nest is built usually about four feet from the ground. The nest looks untidy, but is strongly made of sticks, coarse grass, weeds, bark strips, and occasionally paper; it is lined with soft roots and is almost always well hidden in dense foliage. The eggs are from three to five in number and are dark greenish blue. Both parents work hard feeding the young and for this purpose destroy many insects which we can well spare. Sixty-two per cent of the food of the young has been found in one instance to be cutworms, showing what a splendid work the parents do in our gardens. In fact, during a large part of the summer, while these birds are rearing their two broods, they benefit us greatly by destroying the insect pests; and although later they may attack our fruits and berries, it almost seems as if they had earned the right to their share. If we only had the wisdom to plant along the fences some elderberries or Russian mulberries, the catbirds as well as the robins would feed upon them instead of the cultivated fruits.

The catbirds afford a striking exa for impressing upon children that species of birds haunts certain kin places. The catbirds are not often f in deep woods or in open fields, but ally near low thickets along streams in shrubbery along fences, in tangl vines, and especially do they like to about our gardens, if we protect t They are very fond of bathing, ar fresh water is given them for this pur we may have opportunity to witness most thorough bath a bird can tak catbird takes a long time to bathe preen its feathers and indulges in luxurious sun baths and thus deserv earns the epithet of "well-groomed is one of the most intelligent of all birds and soon learns "what is what," repays in the most surprising way the ble of careful observation.

LESSON 24

The Catbird

Leading Thought — The catbird a beautiful song as well as the h "miou," and can imitate other birds though not so well as the mockingb It builds in low thickets and shrubl and during the nesting season is of g benefit to our gardens.

Method — First, let the pupils st and report upon the songs, scoldings, other notes of this our northern mocki bird; then let them describe its appeara and habits.

Observations — 1. Do you think squall of the catbird sounds like the m of a cat? When does the bird use this n and what for? What other notes have heard it utter?

2. Describe as well as you can the c bird's true song. Are there any harsh no in it? Where does he sit while singi Describe the actions of the catbird wh he is singing.

3. Have you ever heard the catbird i

the songs of other birds or other
:s?

Describe the catbird as follows: its
and shape compared to the robin; the
r and shape of head, beak, wings, tail,
st, and under parts.

Describe its peculiar actions and its
acteristic movements.

Where do catbirds build their nests?
v high from the ground? What ma-
l is used? Is the nest compact and
fully finished? Is it hidden?

. What is the color of the eggs? Do
h parents care for the young?

. What is the food of the catbird?
ay is it an advantage to us to have cat-
ls build in our gardens?

. Do you ever find catbirds in the deep
ods or out in the open meadows?
nere do you find them?

10. Put out a pan of water where the catbirds can use it and then watch them make their toilets and describe the process. Describe how the catbirds take sun baths.

He sits on a branch of yon blossoming bush,
This madcap cousin of robin and thrush,
And sings without ceasing the whole morning long;
Now wild, now tender, the wayward song
That flows from his soft, gray, fluttering throat;
But often he stops in his sweetest note,
And, shaking a flower from the blossoming bough,
Drawls out, " Mi-eu, *mi-ow!* "

— "THE CATBIRD," EDITH M. THOMAS

Olin Sewall Pettingill, Jr.

A family of seven young belted kingfishers that were posed for the camera

THE BELTED KINGFISHER

This patrol of our streams and lake
nores, in his cadet uniform, is indeed a
nilitary figure as well as a militant per-
onality. As he sits upon his chosen branch
verhanging some stream or lake shore,
is crest abristle, his keen eye fixed on the water below, his whole bearing alert, one must acknowledge that this fellow puts " ginger " into his environment, and that the spirit which animates him is very far from the " *dolce far niente* " which permeates the ordinary fisherman. However,

Olin Sewall Pettingill, Jr.

A moment between diggings. This male belted kingfisher hesitates on the doorstep of the nesting burrow which he is digging. To him, rather than to his mate, falls the task of home-building

he does not fish for fun but for business; his keen eye catches the gleam of a moving fin and he darts from his perch, holds himself for a moment on steady wings above the surface of the water, to be sure of his quarry, and then there is a dash and a splash and he returns to his perch with the wriggling fish in his strong beak. Usually he at once proceeds to beat its life out against a branch and then to swallow it sensibly, head first, so that the fins will not prick his throat nor the scales rasp it. He swallows the entire fish, trusting to his internal organs to select the nourishing part; and later he gulps up a ball of the indigestible scales and bones.

The kingfisher is very different in form from an ordinary bird; he is larger than a robin, and his head and fore parts are much larger in proportion; this is the more noticeable because of the long feathers

Kingfisher's foot. This shows the weak toes; the third and fourth are joined together, which undoubtedly assists the bird in pushing out soil when excavating

of the head which he lifts into a crest, and because of the shortness of the tail. The beak is very long and strong, enabling the kingfisher to seize the fish and ho
fast, but the legs are short and weak.
third and fourth toes are grown toge
for a part of their length; this is of
to the bird in pushing earth from the
row, when excavating. The kingfishe
no need for running and hopping,
the robin, and therefore does not
the robin's strong legs and feet. His c
are beautiful and harmonious; the u
parts are grayish blue, the throat and c
white, as is also the breast, which h
bluish gray band across the upper
this giving the name of the Belted K
fisher to the bird. The feathers of
wings are tipped with white and the
feathers narrowly barred with white.
under side of the body is white in
males, while in the females it is somev
chestnut in color. There is a striking w
spot just in front of the eye.

The kingfisher parents build their
in a burrow which they tunnel hori
tally in a bank; sometimes there is a v
bule of several feet before the nes
reached, and at other times it is b
very close to the opening. Both parents
industrious in catching fish for their n
lings, but the burden of this duty
heaviest upon the male. Many fish bo
are found in the nest, and they seem
clean and white that they have been
garded as nest lining. Wonderful tales
told of the way the English kingfishers

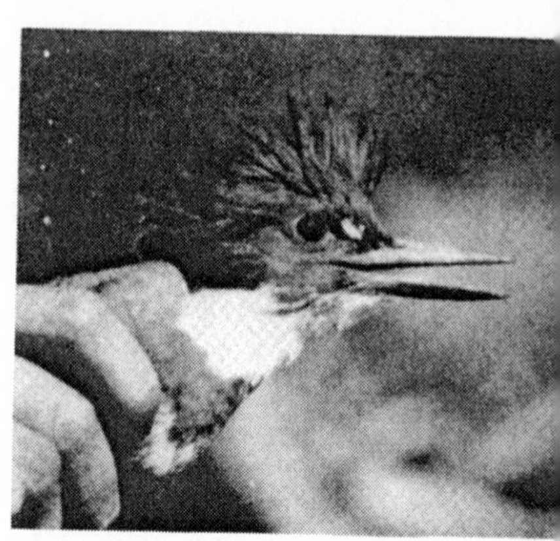

Olin Sewall Pettingill,

A large sharply pointed bill and a good e behind it is all the equipment this feathe fisherman needs to catch his food

ɔnes to support the earth above their and tributes have been paid to their ectural skill. But it is generally con- that the lining of fish bones in the of our kingfisher is incidental, since ood of the young is largely fish, al- h frogs, insects, and other creatures ten eaten with relish. It is interesting te the process by which the young sher gets its skill in fishing. I have seen one dive horizontally for a yard o beneath the water and come up nant and sputtering because the fish scaped. It was fully two weeks more e this one learned to drop like a t on its quarry.

e note of the kingfisher is a loud rat- ot especially pleasant close at hand, ot unmusical at a little distance. It is ious coincidence that it sounds very like the clicking of the fisherman's it is a sound that conjures visions of e-dappled streams and the dancing, waters of tree-fringed lakes and ls.

ere seems to be a division of fishing nd among the kingfishers, one bird y trespassing upon its neighbor's pre- s. Unless it be the parent pair work- near each other for the nestlings, or nestlings still under their care, we sel- see two kingfishers in the same im- iate locality.

LESSON 25

The Kingfisher

EADING THOUGHT — The kingfisher is d by form of body and beak to be a erman.

IETHOD — If the school be near a am or pond the following observations be made by the pupils; otherwise let boys who go fishing make a study of bird and report to the school.

OBSERVATIONS — 1. Where have you the kingfisher? Have you often seen it a certain branch which is its favorite ch? Is this perch near the water? What is the advantage of this position to the bird?

2. What does the kingfisher feed upon? How does it obtain its food? Describe the actions of one of these birds while fishing.

3. With what weapons does the king- fisher secure the fish? How long is its beak compared with the rest of its body? How does it kill the fish? Does it swallow the fish head or tail first? Why? Does it tear off the scales or fins before swallowing it? How does it get rid of these and the bones of the fish?

4. Which is the larger, the kingfisher or the robin? Describe the difference in shape of the bodies of these two birds; also in the size and shape of feet and beaks, and explain why they are so differ- ent in form. What is there peculiar about the kingfisher's feet? Do you know which two toes are grown together?

5. What are the colors of the kingfisher in general? The colors of head, sides of head, collar, back, tail, wings, throat, breast, and under parts? Is there a white spot near the eye? If so, where? Do you know the difference in colors between the parent birds?

6. Where is the nest built? How is it lined?

7. What is the note of the kingfisher? Does it give it while perching or while on the wing? Do you ever find more than one kingfisher on the same fishing grounds?

THE KINGFISHER (OF ENGLAND)

For the handsome Kingfisher, go not to the tree,
No bird of the field or the forest is he;
In the dry river rock he did never abide,
And not on the brown heath all barren and wide.

He lives where the fresh, sparkling waters are flowing,
Where the tall heavy Typha and Loose- strife are growing;
By the bright little streams that all joyfully run
Awhile in the shadow, and then in the sun.

He lives in a hole that is quite to his mind,
With the green mossy Hazel roots firmly entwined;
Where the dark Alder-bough waves gracefully o'er,
And the Sword-flag and Arrow-head grow at his door.

There busily, busily, all the day long,
He seeks for small fishes the shallows among;
For he builds his nest of the pearly fish-bone,
Deep, deep, in the bank, far retired, and alone.

Then the brown Water-Rat from h
row looks out,
To see what his neighbor Kingfi
about;
And the green Dragon-fly, flitting
away,
Just pauses one moment to bid him
day.

O happy Kingfisher! What care sho
know,
By the clear, pleasant streams, as he
to and fro,
Now lost in the shadow, now bright
sheen
Of the hot summer sun, glancing s
and green!

— Mary Ho

THE SCREECH OWL

Disquiet yourselves not: 'Tis nothing but a little, downy owl. — Shelley

Of all the sounds to be heard at night in the woods, the screech owl's song is surely the most fascinating; its fascination does not depend on music but upon the chills which it sends up and dow
spine of the listener, thus attacking a
different set of nerves than do other
songs. The weird wail, tremulous and
drawn out, although so blood-curdli
from the standpoint of the owlet the
beautiful music in the world; by mea
it he calls to his mate, cheering her
the assurance of his presence in the w
evidently she is not a nervous crea
The screech owls are likely to sing at
during any part of the year; nor shoul
infer that when they are singing the
not hunting, for perchance their n
frightens their victims into fatal act
Although the note is so unmistakable
there is great variation in the songs c
dividuals; the great variety of quave
the song offers ample opportunity fo
expression of individuality. More
these owls often give themselves ov
tremulous whispering and they emph
excitement by snapping their beaks i
alarming manner.

Country Life in America

Screech owls

Any bird that is flying about and sin
in the night time must be able to
where it is going, and the owls have
cial adaptations for this. The eyes

large and the yellow iris opens and s about the pupil in a way quite simi- ɔ the arrangement in the cat's eye, ɔt that the pupil in the owl's eye is d when contracted instead of elon- ; in the night this pupil is expanded it covers most of the eye. The owl not need to see behind and at the , since it does not belong to the birds h are the victims of other birds and als of prey. The owl is a bird that s instead of being hunted, and it s only to focus its eyes on the creature chasing. Thus, its eyes are in the front e head like our own; but it can see nd, in case of need, for the head turns n the neck as if it were fitted on a ball- ing joint. I have often amused my- by walking around a captive screech which would follow me with its eyes urning the head until it almost made circle; then the head would twist back such lightning rapidity that I could lly detect the movement. It seemed ost as if the head were on a pivot and ld be moved around and around in- nitely. Although the owl, like the cat, eyes fitted for night hunting, it can see fairly well during the daytime.

beak with the upper mandible end- in a sharp hook signifies that its owner s upon other animals and needs to d and tear flesh. The owl's beak thus ned is somewhat buried in the feathers he face, which gives it a striking resem- nce to a Roman nose. This, with the at, staring, round eyes, bestows upon owl an appearance of great wisdom. t it is not the beak which the owl uses a weapon of attack; its strong feet and rp, curved claws are its weapons for king the enemy and also for grappling h its prey. The outer toe can be moved k at will, so that in grasping its prey its perch, two toes may be directed for- rd and two backward, thus giving a onger hold.

The ear is very different in form from e ear of other birds; instead of being a re hole opening into the internal ear, it nsists of a fold of skin forming a chan- l which extends from above the eye

S. A. Grimes

A barn or monkey-faced owl

around to the side of the throat. Thus equipped, while hunting in the dark the owl is able to hear any least rustle of mouse or bird and to know in which direction to descend upon it. There has been no relation established between the ear tufts of the screech owl and its ears, so far as I know, but the way the bird lifts the tufts when it is alert always suggests that this movement in some way opens up the ear.

In color there are two phases among the screech owls, one reddish brown, the other gray. The back is streaked with black, the breast is marked with many shaft-lines of black. The whole effect of the owl's plumage makes it resemble a branch of a tree or a part of the bark, and thus it is protected from prying eyes during the daytime when it is sleeping. Its plumage is very fluffy and its wing feathers, instead of being stiff to the very edge, have soft fringes which cushion the stroke upon the air. The owl's flight is, therefore, noiseless; and the bird is thus able to swoop down upon its prey without giving warning of its approach.

The screech owls are partial to old apple orchards for nesting sites. They will often use the abandoned nest of a woodpecker; the eggs are almost as round as marbles and as white as chalk; it is well that they are laid within a dark hole, for otherwise their color would attract the

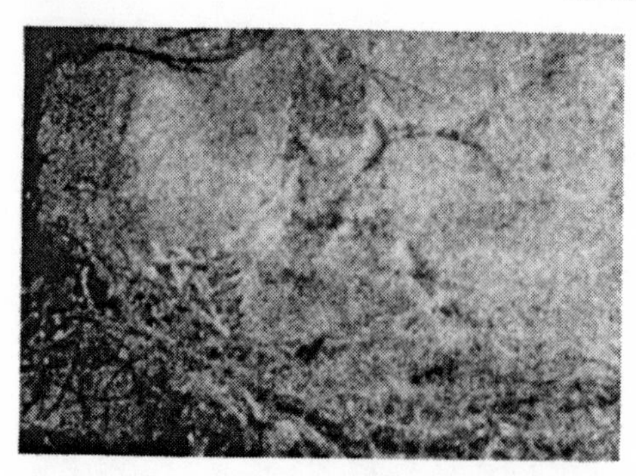

S. A. Grimes

The great horned owl

eyes of enemies. There are usually four eggs; the fubsy little owlets climb out of their home cave by the end of May and are the funniest little creatures imaginable. They make interesting but decidedly snappy pets; they can be fed on insects and raw beef. It is most interesting to see one wake up late in the afternoon after its daytime sleep. All day it has sat motionless upon its perch with its toes completely covered with its fluffy feather skirt. Suddenly its eyes open, the round pupils enlarging or contracting with great rapidity as if adjusting themselves to the amount of light. When the owl winks it is like a moon in eclipse, so large are the eyes, and so entirely are they obscured by the lids, which seem like circular curtains. When it yawns, its wide bill absurdly resembles a human mouth, and the yawn is very human in its expression. It then stretches its wings; it is astonishing how far this wing can be extended below the feet. It then begins its toilet. It dresses its feathers with its short beak, nibbling industriously in the fluff; it scratches its under parts and breast with its bill, then cleans the bill with its foot, meanwhile moving the head up and down as if in an attempt to see its surroundings better.

The owls are loyal lovers and are said to remain mated through life, the twain being very devoted to their nests and nestlings. Sometimes the two wise-looking little parents sit together on the eggs, a most happy way to pass the wearisome incubation period.

The screech owls winter in the north and are distinctly foresighted in paring for winter. They have often observed catching mice, during the fall, and placing them in some hollow for cold storage, whence they may be t in time of need. Their food consis some extent of insects, especially n flying moths and beetles, and also c pillars and grasshoppers. However, larger part of their food is mice; s times small birds are caught, and the lish sparrow is a frequent victim. Chic are rarely taken, except when small, s this owlet is not as long as a robin. It lows its quarry as whole as possible, t ing to its inner organs to do the sifting selecting. Later it throws up pellets of indigestible bones, hair, etc. By the s of these pellets, found under owl ro the scientists have been able to detern the natural food of the bird, and the unite in assuring us that the screech does the farmer much more good t harm, since it feeds so largely upon c tures which destroy his crops.

LESSON 26

The Screech Owl

Leading Thought — This owl is es cially adapted to get its prey at night feeds largely on field mice, grasshopp caterpillars, and other injurious insects a is therefore the friend of the farmer.

Method — This lesson should be when the children first hear the cry of t owl; and an owl in captivity is a fascin ing object for the children to obser However, it is so important that the c dren learn the habits of this owl that t teacher is advised to hinge the lesson observation whatever made by the pu- and illustrate it with pictures and es.

BSERVATIONS — 1. Have you ever d the screech owl? At what time of day or night? Why was this? Why the owl screech? How did you feel n listening to the owl's song?

Describe the owl's eyes. Are they oted to see by night? What changes place in them to enable the owl to by day also? In what way are the s eyes similar to the cat's? Why is it essary for an owl to see at night? Are owl's eyes placed so that they can at the sides like other birds? How s it see an object at the sides or be- d it?

. Note the owl's beak. For what pur- e is a hooked beak? How does the owl its beak? Why do we think that the looks wise?

. Describe the feet and claws of the ech owl. What are such sharp hooked ws meant for? Does an owl on a perch ays have three toes directed forward one backward?

. Describe the colors of the screech . Are all these owls of the same color? w do these colors protect the bird from enemies?

. How is the owl's plumage adapted to nt flight? Why is silent flight advan- eous to this bird?

7. How does the owl's ear differ from e ears of other birds? Of what special ad- ntage is this? As the owl hunts during e night, what does it do in the daytime? ow and by what means does it hide it- f?

8. Where does the screech owl make its st? Do you know anything about the votion of the parent owls to each other d to their young? How many eggs are laid? What is their color? At what time of year do the little owls appear?

9. Where does the screech owl spend the winter? What do the screech owls feed upon? Do they chew their food? How do they get rid of the indigestible portion of their food? How does this habit help sci- entists to know the food of the owls?

10. How does the screech owl work in- jury to the farmers? How does it benefit them? Does not the benefit outweigh the injury?

11. How many other kinds of owls do you know? What do you know of their habits?

TWO WISE OWLS

We are two dusky owls, and we live in a tree;
Look at her, — look at me!
Look at her, — she's my mate, and the mother of three
Pretty owlets, and we
Have a warm cosy nest, just as snug as can be.

We are both very wise; for our heads, as you see,
(Look at her — look at me!)
Are as large as the heads of four birds ought to be;
And our horns, you'll agree,
Make us look wiser still, sitting here on the tree.

And we care not how gloomy the night-time may be;
We can see, — we can see;
Through the forest to roam, it suits her, it suits me;
And we're free, — we are free
To bring back what we find, to our nest in the tree.

— ANONYMOUS

S. A. C

The fish hawk or osprey. This hawk builds its large nest from twenty to fifty feet above ground. It subsists almost entirely on fish

THE HAWKS

Above the tumult of the cañon lifted, the gray hawk breathless hung,
Or on the hill a winged shadow drifted where furze and thornbush clung.

— Bret Harte

It is the teacher's duty and privilege to try to revolutionize some popular misconceptions about birds, and two birds, in great need in this respect, are the so-called hen hawks. They are most unjustly treated, largely because most farmers consider that a " hawk is a hawk," and should always be shot to save the poultry, although there is as much difference in the habits of hawks as there is in those of men. The so-called hen hawks are the red-shouldered and the red-tailed species, the latter being somewhat the larger and rarer of the two. Both are very large birds. The red-shouldered has cinnamon brown epaulets; the tail is blackish, crossed by five or six narrow white bars, and the wing feathers are also barred. The red-tailed species has dark brown wings; the feath are not barred, and it is distinguished its tail which is brilliant cinnamon co with a black bar across it near the end is silvery white beneath. When the ha is soaring, its tail shows reddish as it whe in the air. Both birds are brown above a whitish below, streaked with brown.

The flight of these hawks is similar a is very beautiful; it consists of soaring outstretched wings in wide circles high the air, and is the ideal of graceful aer motion. In rising, the bird faces the wi and drops a little in the circle as its ba turns to the leeward, and thus it clim an invisible winding stair until it is a m speck in the sky. When the bird wishes drop, it lifts and holds its wings above and comes down like a lump of lead, o catch itself whenever it chooses to again to climb the invisible spiral. ll this is done without fatigue, for birds have been observed to soar for hours together without coming th. When thus soaring the two spe- ay be distinguished from each other eir cries; the red-tailed gives a high ering scream, which Chapman likens sound of escaping steam; while the ouldered calls in a high not unmusi- ote "kee-you, kee-you" or "tee-ur, ."

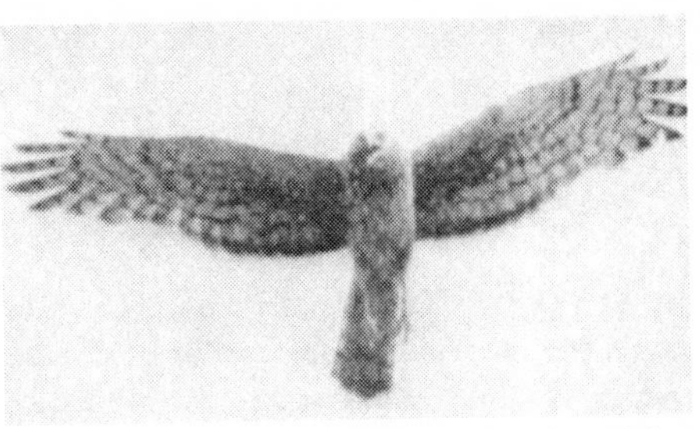

S. A. Grimes

The marsh hawk. This is a bird of the open fields. It flies low in search of rodents, reptiles, frogs, and insects. It may be identified by a white spot on the rump

e popular fallacy for the teacher to ct about these birds is that they are ies of the farmers. Not until a hawk ctually been seen to catch chick- hould it be shot, for very few of them ilty of this sin. Sixty-six per cent of ood of the red-tailed species consists jurious animals, i.e., mice and go- , etc., and only seven per cent con- of poultry; the victims are probably disabled fowls, and fall an easy prey; bird much prefers mice and reptiles to try. The more common red-shouldered hawk feeds generally on mice, snakes, frogs, fish, and is very fond of grasshoppers. Ninety per cent of its food consists of creatures which injure our crops or pastures and scarcely one and one-half per cent is made up of poultry and game. These facts have been ascertained by the experts in the Department of Agriculture at Washington who have examined the stomachs of hundreds of these hawks taken from different localities. Furthermore, Dr. Fisher states that a pair of the red-shouldered hawks bred for successive years within a few hundred yards of a poultry farm containing 800 young chickens and 400 ducks, and the owner never saw them attempt to catch a fowl.

A. A. Allen

Red-tailed hawk

However, there are certain species of hawks which are to be feared; these are the Cooper's hawk and the sharp-shinned hawk, the first being very destructive to poultry and the latter killing many wild birds. These are both somewhat smaller than the species we are studying. They are both dark gray above and have very long tails, and when flying they flap their wings for a time and then glide a distance. They do not soar on motionless outspread pinions by the hour.

When hawks are seen soaring, they are likely to be hunting for mice in the meadows below them. Their eyes are remarkably keen; they can see a moving creature from a great height, and can suddenly drop upon it like a thunderbolt out of a clear sky. Their wonderful eyes are far-sighted when they are circling in the sky,

BIRDS OF PREY AND SCAVENGERS

1. Sparrow Hawks. *In summer these birds will be found from northern Canada south to the Gulf states except in peninsular Florida and the arid regions of the Southwest; in winter from the northern United States to Panama. About eleven inches in length, this pretty little hawk has readily adapted itself to civilization and in densely populated areas makes its nest about buildings and even in birdhouses. The sparrow hawk should be protected everywhere, for it is useful to man; it feeds chiefly on mice and insects. (Photo by Dorothy M. Compton)*

2. Snowy Owl. *One of the largest and most handsome of owls, the snowy owl, is at home in the northern part of the Northern Hemisphere; it breeds as far north as land is found and as far south as northern Quebec, Manitoba, and British Columbia. In winter it migrates southward in search of food if mice and lemmings become scarce in the North. In North America the winter range may extend as far south as the Gulf states, in Europe as far south as France and Switzerland, and in Asia to northern India and Japan. This owl is seldom seen in trees, preferring the open country, probably because the rodents which are its principal food are found there. (Photo by Olin Sewall Pettingill, Jr.)*

3. A Young Screech Owl. *The range of these birds extends from southern Canada to the southern United States. They breed over most of this area. The screech owl is not quite so long as a robin. It often nests in a small cavity in a tree or even in a birdhouse. It is not unusual for the owl to use the same nesting place year after year. It feeds largely on mice, other small mammals, insects, and small birds. This owl is unique in that it has two color phases; both male and female may be either gray or reddish brown. (Photo by Dorothy M. Compton)*

4. Herring Gull. *These birds are engers found along the coasts and waters of the Northern Hemisphere. nest in colonies, usually on islands but near the water. The nest of seaweed, g or moss is generally built on the ground. of herring gulls are often seen near pier wharves where they perform a valuable by feeding on garbage and refuse. It is ger this bird that follows coastwise boats u for refuse to be thrown overboard. (Ph Olin Sewall Pettingill, Jr.)*

5. An Adult Screech Owl. *Perchec tree, the screech owl is difficult to dete he is easily mistaken for branches and (Photo by A. A. Allen)*

6. A Black Vulture at the Entr to Its Nest. *This is a scavenger of the Though it rarely breeds north of Mar it is occasionally seen in some of the c states. The value of these birds in rem health-menacing garbage and carrion great that they are protected by law and sentiment. They are quite numerous in South and are often seen in towns and The black vulture does not build a nest eggs are laid in cavities in trees or rock hollow stumps, or on the ground beneath br (Photo by S. A. Grimes)*

7. Audubon's Caracara. *This bird's range is from Lower California, Arizona, T and southern Florida southward to Ecu it has been reported as an accidental visit far north as Ontario. The nest is a bulky s ture of sticks, branches, roots, grass, and le usually placed in trees or on bushes or le Caracaras are often seen in the compan vultures, feeding on carrion, and they capture and eat snakes, frogs, and lizards. caracara's flight is direct and rapid, not like that of the vulture, which sails and soar spirals. (Photo by S. A. Grimes)*

1
2
3
4
5
6
7

Leonard K. Beyer

Nest and eggs of the marsh hawk

but as they drop, the focus of the eyes changes automatically with great rapidity, so that by the time they reach the earth they are nearsighted, a feat quite impossible for our eyes unless aided by glasses or telescope.

These so-called hen hawks will often sit motionless, for hours at a time, on some dead branch or dead tree; they are probably watching for something eatable to stir within the range of their keen vision. When seizing its prey, a hawk uses its strong feet and sharp, curved talons. All hawks have sharp and polished claws, even as the warrior has a keen, bright sword; the legs are covered by a growth of feathers extending down from above, looking like feather trousers. The beak is hooked and very sharp and is used for tearing apart the flesh of the quarry. When a hawk fights some larger animal or man, it throws itself over upon its back and strikes its assailant with its strong claws as well as with its beak; but the talons are its chief weapons.

Both species build a large, shallow nest of coarse sticks and grass, lined with moss, feathers, etc.; it is a rude, rough structure, and is placed in tall trees from fifty to seventy-five feet from the ground. Only two to four eggs are laid; these are whitish, spotted with brown. These hawks are said to remain mated for life and are devoted to each other and to their young. H and eagles are very similar in form habits, and if the eagle is a noble bir is the hawk.

LESSON 27

The Hawks

Leading Thought — Uninformed ple consider all hawks dangerous n bors because they are supposed to exclusively on poultry. This idea is and we should study carefully the h of hawks before we shoot them. The nary large reddish "hen hawks," w circle high above meadows, are doing g good to the farmer by feeding upon mice and other creatures which steal grain and girdle his trees.

Method — Begin by observations the flight of one of these hawks and plement this with such observations as pupils are able to make, or facts wh they can discover by talking with hun or others, and by reading.

Observations — 1. How can you t

Leonard K.

Young marsh hawks

., when flying, from a crow or other bird? Describe how it soars. Does ove off in any direction? If so, does it e off in circles? How often does it e strokes with its wings? Does it rise n it is facing the wind and fall as it s its back to the wind?

Have you seen a hawk flap its wings y times and then soar for a time? If what hawk do you think it was? How it differ in habits from the "hen ks"?

Have you noticed a hawk when soar-drop suddenly to earth? If so, why did this?

How does a hawk hunt? How, when so high in the air that it looks like a ing speck in the sky, can it see a mouse meadow? If it is so farsighted as , how can it be nearsighted enough to h the mouse when it is close to it? uld you not have to use field glasses elescope to do this?

. When a hawk alights what sort of e does it choose? How does it act?

. Do hawks seize their prey with their vs or their beaks? What sort of feet l claws has the hawk? Describe the k. What do you think a beak of this pe is meant for?

. Why do people shoot hawks? Why t a mistake for people to wish to shoot hawks?

. What is the food of the red-shoul-ed hawk as shown by the bulletin of U. S. Department of Agriculture or by Audubon leaflets?

9. Where does the hawk place its nest? Of what does it build its nest?

10. Compare the food and the nesting habits of the red-shouldered and red-tailed hawks?

11. How devoted are the hawks to their mates and to their young? Does a hawk, having lost its mate, live alone ever after?

12. Describe the colors of the hen hawks and describe how you can tell the two species apart by the colors and markings of the tail.

13. What is the cry of the hawk? How can you tell the two species apart by this cry? Does the hawk give its cry only when on the wing?

14. Why should an eagle be considered so noble a bird and the hawk be so scorned? What difference is there between them in habits?

Yet, ere the noon, as brass the heaven turns,
The cruel sun smites with unerring aim,
The sight and touch of all things blinds and burns,
And bare, hot hills seem shimmering into flame!

On outspread wings a hawk, far poised on high,
Quick swooping screams, and then is heard no more:
The strident shrilling of a locust nigh
Breaks forth, and dies in silence as before.

—"SUMMER DROUGHT," J. P. IRVINE

THE SWALLOWS AND THE CHIMNEY SWIFT

These friendly little birds spend their ne darting through the air on swift ngs, seeking and destroying insects ich are foes to us and to our various ps. However, it is safe to assume that ey are not thinking of us as they skim ove our meadows and ponds, hawking r tiny foes; for like most of us, they are nply intent upon getting a living. 'ould that we might perform this nec-sary duty as gracefully as they!

In general, the swallows have a long, slender, graceful body, with a long tail which is forked or notched, except in the case of the eave swallow. The beak is short but wide where it joins the head; this enables the bird to open its mouth wide and gives it more scope in the matter of catching insects; the swift flight of the swallows enables them to catch insects on the wing. Their legs are short, the feet are weak and fitted for perching; it would be quite im-

L. A. Fuertes

Swallows and swifts

possible for a swallow to walk or hop like a robin or blackbird.

THE EAVE OR CLIFF SWALLOWS — These swallows build under the eaves of barns or in similar locations. In early times they built against the sides of cliffs; but when man came and built barns, they chose them for their dwelling sites. The nest is made of mud pellets and is somewhat globular in shape, with an entrance at one side. When the nest is on the side of a cliff or in an unprotected portion of a barn, a covered passage is built around the door, which gives the nest the shape of a gourd or retort; but when protected beneath the eaves the birds seem to think

A. A. Allen

Nests of cliff swallows

this vestibule is unnecessary. The nest is warmly lined with feathers and materials, and often there are many built so closely together that they to The eave swallow comes north about 1, and soon after that may be seen a streams or other damp places gathe mud for the nests. It seems necessary the bird to find clay mud in order to der the nest strong enough to support eggs and nestlings. The eggs are wh blotched with reddish brown. The par cling to the edge of the nest when fee

Leonard K.

Barn swallow and nest

the young. Both the barn and eave sw lows are blue above, but the eave swal has the forehead cream white and rump of pale brick-red, and its tail square across the end as seen in flight. T barn swallow has a chestnut forehead a its outer tail feathers are long, makin distinct fork during flight, and it is not upon the rump.

THE BARN SWALLOWS — These bi choose a barn where there is a hole in gable or where the doors are kept open the time. They build upon beams or ters, making a cup-shaped nest of layers pellets of mud, with grass between; it well lined with feathers. The nest is u ally the shape of half of a shallow c which has been cut in two lengthwise, le being plastered against the side of fter. Sometimes the nests are more s supported upon a beam or rafter; gs are white and dotted with reddish n. The barn swallows, aside from constant twittering, have also a song. Both parents work at building the nest and feeding the young; there ely to be several pairs nesting in the building. The parents continue to the young long after they have left est; often a whole family may be seen g on a telegraph wire or wire fence, parents still feeding the well-grown gsters. This species comes north in

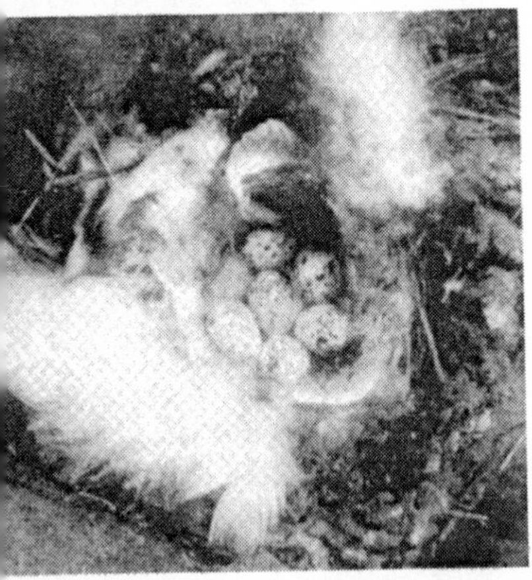

is barn swallow's nest is well feathered

latter part of April and leaves early in ember. It winters as far south as il.

he barn swallow has a distinctly tailore appearance; its red-brown vest and scent blue coat, with deeply forked at tails" give it an elegance of style ch no other bird, not even the chic ar waxwing, can emulate.

HE BANK SWALLOW — When we see a ly bank apparently shot full of holes as mall cannon balls, we may know that have found a tenement of bank swals. These birds always choose the perdicular banks of creeks or of railroad or of sand pits for their nesting sites; y require a soil sufficiently soft to be neled by their weak feet, and yet not oose as to cave in upon the nest. The nel may extend from one to four feet

Leonard K. Beyer

The band of color across the breast is the distinguishing mark of the bank swallow

horizontally in the bank with just enough diameter to admit the body of the rather small bird. The nest is situated at the extreme end of the tunnel and is lined with soft feathers and grasses.

The bank swallows arrive late in April and leave early in September. They may be distinguished from the other species by their grayish color above; the throat and breast are white with a broad, brownish band across the breast; the tail is slightly forked. The rough-winged swallow, which is similar in habits to the bank swallow, may be distinguished from it by its gray breast which has no dark band.

THE TREE SWALLOW — This graceful little bird builds naturally in holes in trees, but readily accepts a box if it is provided. It begins to build soon after it comes north in late April, and it is well for us to encourage the tree swallows to live near

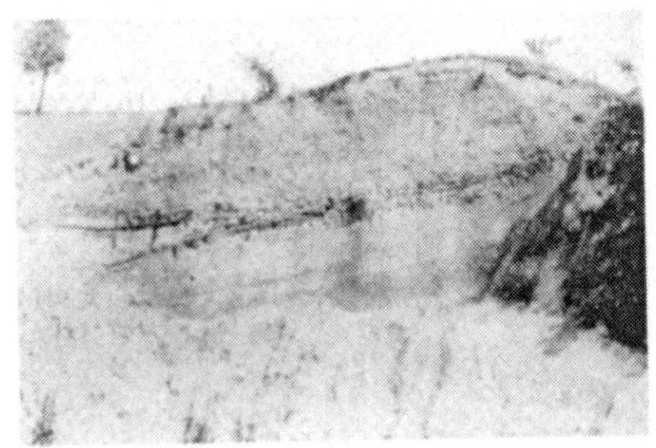

Leonard K. Beyer

Nesting site of a colony of bank swallows

George Fiske, Jr.

A tree swallow

our houses by building houses for them and driving away the English sparrows. The tree swallows live upon many insects which annoy us and injure our gardens and damage our orchards; they are, therefore, much more desirable neighbors than the English sparrows. The tree swallows congregate in great numbers for the southern migration very early in the season,

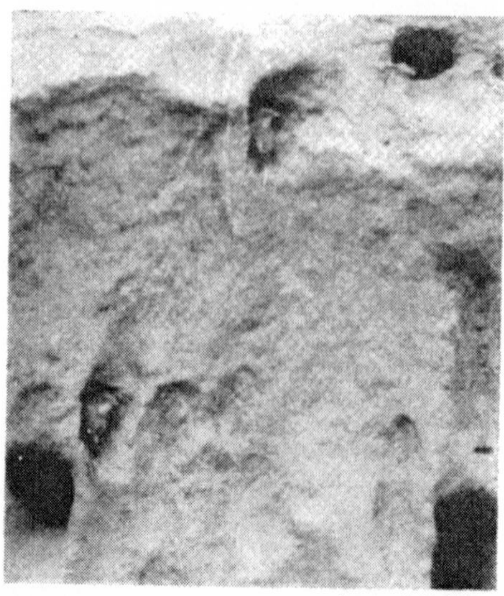

Leonard K. Beyer

Two bank swallows at the entrances to their burrows

often in early August. They are lik congregate in marshes, as are als other swallows. In color the tree sw has a green metallic back and head a pure white breast with no band a it; these peculiarities distinguish it all other species.

THE PURPLE MARTIN — The mar a larger bird than any other swallov ing eight inches in length, while the swallow does not measure quite s The male is shining, steel-blue above below; the female is brownish above a gray throat, brownish breast and is beneath. The martins originally nest hollow trees but for centuries have cared for by man. The Indians were to put out empty gourds for them to in; and as soon as America was settle Europeans, martin boxes were buil

A. A.

Nest of chimney swifts

tensively. But when the English spar came, they took possession of the bo and the martins have to a large extent appeared; this is a pity since they are b ficial birds, feeding upon insects w are injurious to our farms and gard They are also delightful birds to h around, and we may possibly induce th to come back to us by building ho for them and driving away the sparr

THE CHIMNEY SWIFT

When the old-fashioned firepl went out of use and were walled up, l ing the great old chimneys useless, th sociable birds took possession of th

they built their nests and reared their
;, and twittered and scrambled about,
ened all sleepers in the neighbor-
at earliest dawn, and in many ways
themselves a distinct part of family
With the disappearance of these old
neys and the growing use of the
er chimney, the swifts have been
or less driven from their close asso-
n with people; and now their nests
ften found in hay barns or other
ded buildings, although they still
r in chimneys when opportunity
.

e chimney swifts originally built
in hollow trees and caves; but with
coming of civilization they took pos-
n of the chimneys disused during the
ner, and here is where we know them
The nests are shaped like little wall
ets; they are made of small sticks of
y uniform size which are glued to-
er and glued fast to the chimney wall
neans of the saliva secreted in the
th of the bird. After the nesting sea-
the swifts often gather in great flocks
live together in some large chimney;
rd nightfall they may be seen cir-
; about in great numbers and drop-
into the mouth of the chimney, one
ne, as if they were being poured into
nnel. In the morning they leave in
se manner, each swift flying about
idening circles as it leaves the chim-
The swifts are never seen to alight
where except in hollow trees or chim-
or similar places; their tiny feet have
p claws for clinging to the slightest
hness of the upright wall; the tail
as a prop, each tail feather ending in
ne which is pressed against the chim-
side when the bird alights, thus
bling it to cling more firmly. In this
ion the swifts roost, practically hung
against a wall.

he swift has a short beak and wide
th which it opens broadly to engulf
cts as it darts through the air. Chim-
swifts have been known to travel at the
of 110 miles an hour.

his bird should never be confused
h the swallows, for when flying, its tail seems simply a sharp point, making the whole body cigar-shaped. This characteristic alone distinguishes it from the long-tailed swallows. In color it is sooty brown, with a gray throat and breast; the wings are long and narrow and apparently curved. The manner of flight and appearance in the air make it resemble the bat more than it does the swallow.

LESSON 28

The Swallows and Swifts

Leading Thought — The swallows are very graceful birds and are exceedingly swift fliers. They feed upon insects which they catch upon the wing. There are five native swallows which are common — the eave, or cliff, the barn, the bank, the tree swallow, and the purple martin. The chimney swift, although often called so, is not a swallow; it is more nearly related to the hummingbird than to the swallows.

Method — The questions should be given as an outline for observation, and may be written on the blackboard or placed in the field notebook. The pupils should answer them individually and from field observation. We study the swifts and swallows together to teach the pupils to distinguish them apart.

Observations — 1. What is the general shape of the swallow? What is the color of the forehead, throat, upper breast, neck, rump, and tail?

2. Is the tail noticeably forked, especially during flight?

Leonard K. Beyer

Nest of bank swallows. The bank has been cut away so that the nest and eggs could be photographed

3. Describe the flight of the swallow. What are the purposes of its long, swift flight? How are the swallow's wings fitted for carrying the bird swiftly?

4. Describe the form of the beak of the swallow. How does it get its food? What is its food?

5. In what particular locations do you see the swallows darting about? At what time of day do they seem most active?

6. Describe the swallow's legs and feet and explain why they look so different from those of the robin and blackbird.

The Eave or Cliff Swallow

7. Where do the eave swallows build their nests? Of what material is the outside? The lining? Describe the shape of the nest and how it is supported.

8. How early in the spring do the eave swallows begin to make their nests? Where and by what means do they get the material for nest building? Are there a number of nests usually grouped together?

9. Describe the eave swallow's egg. Where do the parents sit when feeding the young? What is the note of the eave swallow?

10. What are the differences between the barn and the eave swallow in and shape of tail?

The Barn Swallow

11. Where does the barn swallow its nest? What is the shape of the Of what material is it made?

12. What is the color of the eggs scribe the feeding of the young and sounds made by them and their par Do both parents work together to the nest and feed the young?

13. Is there usually more than one in the same locality? When the y swallows are large enough to leave nest, describe how the parents cont to care for them.

14. Have you ever heard the barn lows sing? Describe their conversati notes.

15. When do the barn swallows grate and where do they go during winter? How can you distinguish the swallow from the eave swallow?

The Bank Swallow

16. Where do the bank swallows b What sort of soil do they choose?

17. How does a bank which is tena by these birds look?

18. How far do the bank swal tunnel into the earth? What is the ameter of one of these tunnels? Do extend straight or do they rise or def

A. A.

Nest and eggs of tree swallows

. With what tools is the tunnel exca-l? Where is the nest situated in the el and how is it lined?

. How can you distinguish this spe- from the barn and eave and tree lows? At what time do the bank swal- leave us for migration south?

THE TREE SWALLOW

. Where does the tree swallow make nest? How does its nest differ from of the barn, eave, or bank swallow? en does it begin to build?

. How can we encourage the tree low to build near our houses? Why e tree swallow a much more desirable to have in birdhouses than the Eng- sparrow?

3. Describe the peculiar migrating its of the tree swallow. How can you this species from the barn, the eave, the bank swallows?

THE PURPLE MARTIN

4. Compare the purple martin with swallows and describe how it differs in and color.

25. Where did the martins build their nests before America was civilized? Where do they like to nest now? How do the purple martins benefit us and how can we induce them to come to us?

THE CHIMNEY SWIFT

26. Where do the chimney swifts build their nests? Of what materials is the nest made? What is its shape and how is it supported? Where does the chimney swift get the glue which it uses for nest building?

27. Describe how the chimney swifts enter their nesting place at night. Where and how do they perch? Describe the shape of the swift's tail and its use to the bird when roosting.

28. On what does the chimney swift feed and how does it procure this food? Describe how its beak is especially fitted for this.

29. How can you distinguish the chimney swift from the swallows? In what respect does the chimney swift resemble the swallows? In what respects does it differ from them?

THE HUMMINGBIRD

Formerly it was believed that this dain-st of birds found the nectar of flowers ple support for its active life; but the er methods of discovering what birds by examining the contents of their machs, show that the hummingbird is insect eater of most ravenous appetite. ot only does it catch insects in mid , but undoubtedly takes them while ey are feasting on the nectar of the bular flowers which the hummingbird ves to visit. Incidentally, the humming-rd carries some pollen for these flowers d may be counted as a friend in every spect, since usually the insects in the ctaries of those flowers with long tubu- corollas are stealing nectar without ving in return any compensation to the ower by carrying its pollen. Such insects ay be the smaller beetles, ants, and flies. The adaptations of the hummingbird's beak and long, double-tubed tongue, are especially for securing this mingled diet of insects and nectar. It is interesting to note that the young hummingbirds have the beak much shorter than the mature birds. The hummingbird's beak is exactly fitted to probe those flowers where the bird finds its food. The tongue has the outer edges curved over, making a tube on each side. These tubes are provided with minute brushes at the tips and thus are fitted both for sucking nectar and for sweeping up the insects.

The natural home of the hummingbird seems to have been in the American tropics. The male of our one species east of the Rocky Mountains has a ruby throat. This bird comes to us after a very long journey each year. One species on the Pa-

A. A. Allen

Ruby-throated hummingbird turning her eggs

cific Coast is known to travel three thousand miles to the north for the summer and back again in winter.

Hummingbirds are not supposed to sing, but to use their voices for squeaking when angry or frightened. However, I once had the privilege of listening to a true song by a hummingbird on the Pacific Coast. The midget was perched upon a twig and lifted up his voice with every appearance of ecstasy in pouring forth his lay. To my uncultured ear this song was a fine, shrill, erratic succession of squeaks, "as fine as a cambric needle," said my companion.

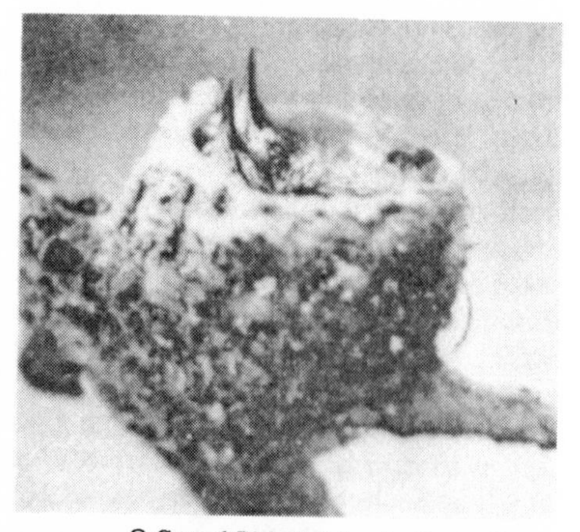

© General Biological Supply House, Chicago

Two young hummingbirds. They remain in nest for about three weeks

The nest of the hummingbird most exquisite structure; it is about t fourths of an inch in diameter on th side and about half an inch deep. in shape, a symmetrical cup; the ou is covered with lichens, so that it ex resembles the branch on which it r the inside is lined with the down of p seeds and plant fibers. The lichens often fastened to the outside with

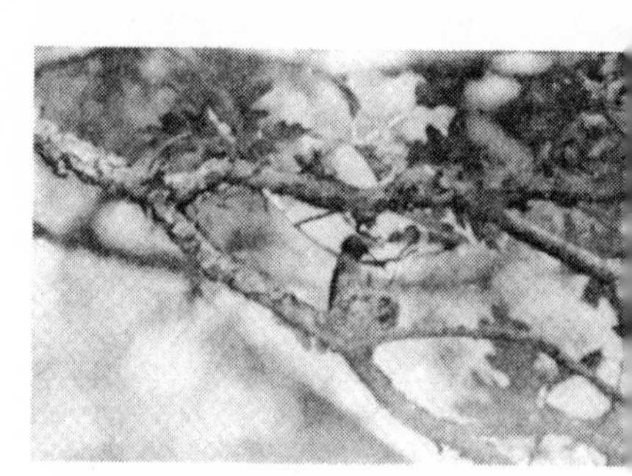

© General Biological Supply House, C

Not much larger than a walnut, the h mingbird's nest looks like a knot on a br

silk web of spiders or caterpillars. The is usually saddled on a branch of a from ten to fifty feet above the grou The eggs are two in number and wh they look like tiny beans. The young black and look, at first glance, more insects than like birds.

LESSON 29

The Hummingbird

Leading Thought — The hummi bird in flight moves its wings so rapi that we cannot see them. It can hold it poised above flowers while it thrusts long beak into them for nectar and sects.

Method — Give the questions to

and let them make the observations
they have the opportunity.
SERVATIONS — 1. Where did you
he hummingbird? What flowers was
ting? At what time of day? Can you
whether it is a hummingbird or a
moth which is visiting the flowers?
hat time of day do the hawkmoths
r?
Did you ever see the hummingbird
to rest? Describe its actions while
g.

3. What are the colors of the back, throat, breast, and under parts? How do you distinguish the mother hummingbird from her mate?

4. How does the hummingbird act when extracting the nectar? How does it balance itself in front of a flower? Have you ever seen hummingbirds catch insects in the air? If so, describe how they did it.

5. Describe the hummingbird's nest. How large is it in diameter? What is the covering outside? With what is it lined?

THE RED-WINGED BLACKBIRD

e blackbirds are among our earliest
rs in the spring; they come in flocks
beset our leafless trees like punctua-
marks, meanwhile squeaking like mu-
wheelbarrows. What they are, where
come from, where they are going and
they are going to do, are the ques-
that naturally arise at the sight of
sable flocks. It is not easy to distin-
grackles, cowbirds, and rusty black-
at a glance, but the redwing pro-
ns his identity from afar. The bright
epaulets, margined behind with pale
w, make up a uniform which catches
admiring eye. The bird's glossy black
nage brings into greater contrast his
ht decorations. No one who has seen
actions can doubt that he is fully
re of his beauty: he comes sailing
n at the end of his strong, swift flight,
balances himself on some bending
; then, dropping his long tail as if
vere the crank of his music box, and
ding both wings lifted to show his scar-
decorations, he sings his "quong-quer-
e." Little wonder that such a hand-
e, military-looking fellow should be
now and then to win more than
share of feminine admiration. But
n though he become an entirely suc-
sful bigamist or even trigamist, he has
ved himself to be a good protector
each and all of his wives and nestlings;
wever, he often has but one mate.
"The redwing flutes his O-ka-lee" is Emerson's graphic description of the sweet song of the redwing; he also has many other notes. He clucks to his mates and clucks more sharply when suspicious, and has one alarm note that is truly alarm-

Male and female red-winged blackbirds

ing. The male redwings come from the South in March; they appear in flocks, often three weeks before their mates arrive. The female looks as though she belonged to quite a different species. Al-

Nest and eggs of the red-winged blackbird

though her head and back are black, the black is decidedly rusty; it is quite impossible to describe her, she is so inconspicuously speckled with brown, black, whitish buff, and orange. Most of us never recognize her unless we see her with her spouse. She probably does most of the nest building, and her suit of salt, pepper, and mustard renders her invisible to the keen eyes of birds of prey. Only when she is flying does she show her blackbird characteristics — her tail being long and of obvious use as a steering organ; and she walks with long, stiff strides. The redwings are ever to be found in and about swamps and marshes. The nest is usually built in May; it is made of grasses and stalks of weeds and is lined with finer grass or reeds. It is bulky and is placed in low bushes or among the reeds. The eggs are pale blue, streaked and spotted with purple or black. The young resemble the mother in color, the males being obliged to wait a year for their epaulets. As to the food of the redwings here in the North, Mr. Forbush has said:

"Although the red-wings almost i ably breed in the swamp or marsh, have a partiality for open fields plowed lands; however, most of the birds that nest in the smaller swam jacent to farm lands get a large sh their food from the farmer's fields. forage about the fields and meadows they first come north in the spring. they follow the plow, picking up g worms and caterpillars; and should be an outbreak of canker-worms in th chard, the blackbirds will fly at leas a mile to get canker-worms for their y Wilson estimated that the red-win the United States would in four m destroy sixteen thousand two hu

A. A.

The mother arrives with food for her yo

million larvæ. They eat the caterpilla the gypsy moth, the forest tent-caterp and other hairy larvæ. They are am the most destructive birds to weevils, beetles, and wire-worms. Grasshopp ants, bugs, and flies form a portion of red-wing's food. They eat comparati little grain in Massachusetts although get some from newly sown fields in spr as well as from the autumn harvest; they feed very largely on the seeds weeds and wild rice in the fall. In h they join with the bobolink in tating the rice fields, and in the t they are often so numerous as to oy the grain in the fields; but here [in North and East] the good they do far eighs the injury, and for this reason are protected by law."

LESSON 30

THE RED-WINGED BLACKBIRD

ADING THOUGHT — The red-winged kbird lives in the marshes where it ds its nest. However, it comes over our plowed lands and pastures and s the farmer by destroying many in- which injure the meadows, crops, trees.

METHOD — The observations should be le by the pupils individually in the l. These birds may be looked for in ks early in the spring, but the study ald be made in May or June when they be found in numbers in almost any mp. The questions may be given to the ils a few at a time or written in their l notebooks and the answers discussed en discovered.

OBSERVATIONS — 1. How can you dis- guish the red-winged blackbird from other blackbirds? Where is the red his wings? Is there any other color be- es black on the wings? Where? What he color of the rest of the plumage of bird?

. What is there peculiar in the flight the redwing? Is its tail long or short? w does it use its tail in flight? What is its position when the bird alights on a reed?

3. What is the song of the redwing? Describe the way he holds his wings and tail when singing, balanced on a reed or some other swamp grass. Does he show off his epaulets when singing? What note does he give when he is surprised or suspicious? When frightened?

4. When does the redwing first appear in the spring? Does he come alone or in flocks? Does his mate come with him? Where do the redwings winter? In what localities do the red-winged blackbirds live? Why do they live there? What is the color of the mother redwing? Would you know by her looks that she was a blackbird? What advantage is it to the pair that the female is so dull in color?

5. At what time do these birds nest? Where is the nest built? Of what material? How is it concealed? What is the color of the eggs?

6. Do the young birds resemble in color their father or their mother? Why is this an advantage?

7. Is the redwing ever seen in fields adjoining the marshes? What is he doing there? Does he walk or hop when looking for food? What is the food of the redwings? Do they ever damage grain? Do they not protect grain more than they damage it?

8. What great good do the redwings do for forest trees? For orchards?

9. At what time in the summer do the redwings disappear from the swamps? Where do they gather in flocks? Where is their special feeding ground on the way south for the winter?

L. A.

The Baltimore oriole

THE BALTIMORE ORIOLE

I know his name, I know his note,
That so with rapture takes my soul;
Like flame the gold beneath his throat,
His glossy cope is black as coal.
O Oriole, it is the song
You sang me from the cottonwood,
Too young to feel that I was young,
Too glad to guess if life were good.
— William Dean Howells

Dangling from the slender, drooping branches of the elm in winter, these pocket nests look like some strange persistent fruit; and, indeed, they are the fruit of much labor on the part of the oriole weavers, those skilled artisans of the bird world. Sometimes the oriole "For the summer voyage his hammock swings" in a sapling, placing it near the main stem and near the top; otherwise it is almost invariably hung at the end of branches and is rarely less than twenty feet from the ground. The nest is pocket-shaped, and usually about seven inches long, and four and a half inches wide at the largest part, which is the bottom. The top is attached to forked twigs at th so that the mouth or door will be k open to allow the bird to pass in and when within, the weight of the causes the opening to contract somew and protects the inmate from prying e Often the pocket hangs free so that breezes may rock it, but in one case found a nest with the bottom stayed a twig by guy lines. The bottom is m more closely woven than the upper for a very good reason, since the o meshes admit air to the sitting bird. nest is lined with hair or other soft terial, and although this is added last, inside of the nest is woven first. The ike to build the framework of twine, t is marvelous how they will loop this nd a twig almost as evenly knotted as ocheted; in and out of this net the ner bird with her long, sharp beak es bits of wood fiber, strong, fine , and scraps of weeds. The favorite g is horsehair, which simply cushions bottom of the pocket. Dr. Detwiler a pet oriole which built her nest of air, which she pulled from his head; possible that orioles get their supply orsehair in a similar way? If we put ht-colored twine or narrow ribbons in venient places, the orioles will weave n into the nest, but the strings should be long lest the birds become entan- . If the nest is strong the birds may it a second year.

hat Lord Baltimore found in new erica a bird wearing his colors must cheered him greatly; and it is well is that this brilliant bird brings to our ds kindly thoughts of that tolerant, -minded English nobleman. The ori- s head, neck, throat, and part of the k are black; the wings are black but the hers are margined with white; the tail lack except that the ends of the outer hers are yellow; all the rest of the bird olden orange, a luminous color which kes him seem a splash of brilliant sun- ne. The female, although marked much same, has the back so dull and mot- l that it looks olive-brown; the rump, ast, and under parts are yellow but by means showy. The advantage of these et colors to the mother bird is obvious, ce it is she that makes the nest and in it without attracting attention to location. In fact, when she is sitting, brilliant mate places himself far ough away to distract the attention of ddlers, yet near enough for her to see flash of his breast in the sunshine and hear his rich and cheering song. He a good spouse and brings her the ma- ials for the nest which she weaves in, nging head downward from a twig and ng her long sharp beak for a shuttle. d his glorious song is for her alone. me hold that no two orioles have the same song, and I know of two individuals at least whose songs were sung by no other birds: one gave a phrase from the Wald- vogel's song in *Siegfried*; the other whis- tled over and over, "Sweet birdie, hello, hello." The orioles can chatter and scold as well as sing.

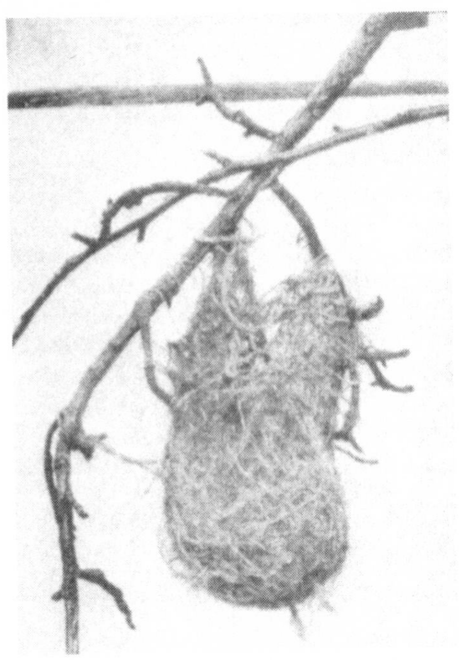

C. R. Crosby

An oriole's nest, anchored to the windward

The oriole is a brave defender of his nest and a most devoted father, working hard to feed his ever-hungry nestlings; we can hear these hollow mites peeping for more food, "Tee dee dee, tee dee dee," shrill and constant, if we stop for a mo- ment under the nest in June. The young birds dress in the safe colors of the mother, the males not donning their bright plum- age until the second year. A brilliant col- ored fledgling would not live long in a world where sharp eyes are in constant quest for little birds to fill empty stom- achs.

The food of the oriole places it among our most beneficial birds, since it is al- ways ready to cope with the hairy cater- pillars avoided by most birds; it has learned to abstract the caterpillar from his spines and is thus able to swallow him minus his

Young orioles just out of the nest

"whiskers." The orioles are waging a great war against the terrible brown-tail and gypsy moths; they also eat click beetles and many other noxious insects. Once when we were breeding big caterpillars in the Cornell University Insectary, an oriole came in through the open windows of

Leonard K. Beyer

An orchard oriole

the greenhouse, and thinking he had found a bonanza proceeded to work it, carrying off our precious crawlers before we discovered what was happening.

The orioles winter in Central America and give us scarcely four months of their company. They do not usually appea fore May and leave in early Septemb

LESSON 31

The Oriole

Leading Thought — The oriole is most skillful of all our bird archite It is also one of our prized song b and is very beneficial to the farmer and fruit grower because of the insect p which it destroys.

Method — Begin during winter early spring with a study of the nest, wh may be obtained from the elms of roadsides. During the first week in M give the questions concerning the b and their habits. Let the pupils keep questions in their notebooks and ans them when they have opportunity. T

ations should be summed up once
k.

SERVATIONS — 1. Where did you
he nest? On what species of tree?
it near the trunk of the tree or the
the branch?

What is the shape of the nest? How
is it? How wide? Is the opening as
as the bottom of the nest? How is
ng to the twigs so that the opening
ns open and does not pull together
the weight of the bird at the bottom?
e bottom of the nest stayed to a
or does it hang loose?

With what material and how is the
fastened to the branches? Of what
rial is the outside made? How is it
n together? Is it more loosely woven
top than at the bottom? How many
of material can you find in the out-
of the nest?

With what is the nest lined? How
p is it lined? With what tool was the
woven? If you put out bright-colored
of ribbon and string do you think
orioles will use them? Why should
not put out long strings?

At what date did you first see the
more oriole? Why is it called the
more oriole? How many other names
t? Describe in the following way the
rs of the male oriole: top of head,
, wings, tail, throat, breast, under
s. What are the colors of his mate?
would it endanger the nest and nest-
if the mother bird were as bright
red as the father bird?

Which weaves the nest, the father
he mother bird? Does the former as-
in any way in nest building?

Where does the father bird stay and
what does he do while the mother bird
is sitting on the eggs?

8. What is the oriole's song? Has he
more than one song? What other notes
has he? After the young birds hatch, does
the father bird help take care of them?

9. By the middle of June the young
birds are usually hatched; if you know
where an oriole nest is hung, listen and
describe the call of the nestlings for food.

10. Which parent do the young birds
resemble in their colors? Why is this a
benefit?

11. What is the oriole's food? How is
the oriole of benefit to us in ways in which
other birds are not?

12. Do the orioles use the same nest
two years in succession? How long does
the oriole stay in the North? Where does
it spend its winters?

Hush! 'tis he!
My oriole, my glance of summer fire,
Is come at last, and, ever on the watch,
Twitches the packthread I had lightly
wound
About the bough to help his house-
keeping, —
Twitches and scouts by turns, blessing his
luck,
Yet fearing me who laid it in his way,
Nor, more than wiser we in our affairs,
Divines the Providence that hides and
helps.
Heave, ho! Heave, ho! he whistles as the
twine
Slackens its hold; once more, now! and a
flash
Lightens across the sunlight to the elm
Where his mate dangles at her cup of felt.

— "UNDER THE WILLOWS," LOWELL

THE CROW

Thoreau says: " What a perfectly New England sound is this voice of the crow! If you stand still anywhere in the outskirts of the town and listen, this is perhaps the sound which you will be most sure to hear, rising above all sounds of human industry and leading your thoughts to some far-away bay in the woods. The bird sees the white man come and the Indian withdraw, but it withdraws not. Its untamed voice is still heard above the tinkling of the forge. It sees a race pass away, but it passes not away. It remains to remind us of aboriginal nature."

The crow is probably the most intelligent of all our native birds. It is quick to learn and clever in action, as many a farmer will testify who has tried to keep it out of corn fields with various devices, the harmless character of which the crow soon understood perfectly. Of all our birds, this one has the longest list of virtues and of sins, as judged from our standpoint; but we should listen to both sides of the case before we pass judgment. I find with crows, as with people, that I like some more than I do others. I do no at all the cunning old crow which s the suet I put on the trees in winte the chickadees and nuthatches; and I hired a boy with a shotgun to protec eggs and nestlings of the robins and birds in my neighborhood from the ages of one or two cruel old crows have developed the nest-hunting h On the other hand, I became a sir admirer of a crow flock which worke a field close to my country home, a have been the chosen friend of se tame crows who were even more i esting than they were mischievous.

The crow is larger than any othe our common black birds; the nort raven is still larger, but is very rarely s Although the crow's feathers are b yet in the sunlight a beautiful purpl descence plays over the plumage, e cially about the neck and back; it h compact but not ungraceful body, long, powerful wings; its tail is med sized and is not notched at the end feet are long and strong; the track sh toes directed forward and one long directed backward. The crow does ail through the air as does the hawk, progresses with an almost constant ing of the wings. Its beak is very g and is used for tearing the flesh ; prey and for defense, and in fact lmost anything that a beak could be for; its eye is all black and is very and intelligent. When hunting for in the field, it usually walks, but times hops. The raven and the fish s are the nearest relatives of the rican crow, and next to them the jays. should hardly think that the blue jay he crow were related to look at them, when we come to study their habits, h is to be found in common.

ne crow's nest is usually very large; it ade of sticks, of grape vines and bark, horsehair, moss, and grasses. It is ed in trees or in tall bushes rarely less twenty feet from the ground. The are pale bluish green or nearly white brownish markings. The young crows h in April or May. Both parents are ted in the care of the young, and ain with them during most of the mer. I have often seen a mother crow ing her young ones which were following her with obstreperous caws, although they were as large as she.

While the note of the crow is harsh when close at hand, it has a musical quality in the distance. Mr. Mathews says: "The crow when he sings is nothing short of a clown; he ruffles his feathers, stretches his neck, like a cat with a fish bone in her throat, and with a most tremendous effort delivers a series of hen-like squawks." But aside from his caw, the crow has some very seductive soft notes. I have held long conversations with two pet crows, talking with them in a high, soft tone, and finding that they answered readily in a like tone in a most responsive way. I have also heard these same tones among the wild crows when they were talking together; one note is a guttural tremolo, most grotesque.

Young crows are a noisy lot

Herbert E. Gray

A crow's nest and eggs

Crows gather in flocks for the winter; these flocks number from fifty to several hundred individuals, all having a common roosting place, usually in pine or hemlock forests or among other evergreens. They go out from these roosts during the day to get food, often making a journey of many miles. During the nesting season they scatter in pairs, and they do not gather again in flocks until the young are fully grown.

When crows are feeding in the fields there is usually, if not always, a sentinel posted on some high point so that he can give warning of danger. This sentinel is

Verne Morton

The story of a take-off. With the third wing beat the crow is away

an experienced bird and is keen to detect a dangerous from a harmless intruder. I once made many experiments with these sentinels; I finally became known to those of a particular flock and I was allowed to approach within a few yards of where the birds were feeding, a privilege not accorded to any other person in the neighborhood.

The crow is a general feeder and will eat almost any food; generally, however, it finds its food upon the ground. The food given to nestlings is very largely insects, and many pests are thus destroyed. The crows do harm to the farmer by pulling the sprouting corn and by destroying the eggs and young of poultry. They also do much harm by destroying the eggs and nestlings of other birds which are beneficial to the farmer; they also do some harm by distributing the seeds of poison ivy and other noxious plants. All these must be set down in the account against the crow, but on the credit side must be placed the fact that it does a tremendous amount of good work for the farmer by eating injurious insects, especially the grubs and cutworms which work in the ground, destroying the roots of grasses and grains. It also kills many mice other rodents which are destructiv crops.

One of the best methods of preve crows from taking sprouting corn treat the seed corn with some st smelling substance, such as tar.

If any of the pupils in your school had any experience with tame crows will relate interesting examples of love of the crow for glittering ob I once knew a tame crow which sto of the thimbles in the house and b them in the garden; he would w for a thimble to be laid aside when sewing was dropped, and would sei almost immediately. This same crow sisted in taking the clothespins off line and burying them, so that he finally imprisoned on wash-days. He fond of playing marbles with a little of the family. The boy would sho marble into a hole and then Billy, crow, would take a marble in his and drop it into the hole. The bird see to understand the game and was hi indignant if the boy played out of and made shots twice in succession.

LESSON 32

The Crow

Leading Thought — The crow has keenest intelligence of all our comn birds. It does good work for us and a does damage. We should study its w before we pronounce judgment, for some localities it may be a true friend a in others an enemy.

Method — This work should begin winter with an effort on the part of

to discover the food of the crows
: snow is on the ground. This is a
time to study their habits and their
s. The nests are also often seen in
er, although usually built in ever-
ıs. The nesting season is in early
l, and the questions about the nests
ld be given then. Let the other ques-
; be given when convenient. The
t, the notes, the sentinels, the food,
benefit and damage may all be taken
:parate topics.

he following topics may be given to
elate with work in English: "What
et crow of my acquaintance did";
idences of crow intelligence"; "A
a crow might make in self-defense to
farmer who wished to shoot him";
ıe best methods of preventing crows
ı stealing planted corn."

BSERVATIONS — 1. How large is the
v compared with other black birds?

Describe its colors when seen in the
ight.

Describe the general shape of the
v.

. Are its wings long and slender or
rt and stout?

. Is the tail long or short? Is it notched
traight across the end?

. Describe the crow's feet. Are they
e and strong or slender? How many
; does the track show in the snow or
d? How many are directed forward and
, many backward?

7. Describe a crow's flight compared with that of the hawk.

8. Describe its beak and what it is used for.

9. What is the color of the crow's eye?

10. When hunting for food does the crow hop or walk?

11. Which are the crow's nearest relatives?

12. Where and of what material do the crows build their nests?

13. Describe the eggs. At what time of the year do the young crows hatch? Do both parents take care of and feed the young? How long do the parents care for the young after they leave the nest?

14. What are the notes of the crow? If you have heard one give any note besides "caw," describe it.

15. Where and how do crows live in winter? Where do they live in summer?

16. Do they post sentinels if they are feeding in the fields? If so, describe the action of the sentinel on the approach of people.

17. Upon what do the crows feed? What is fed to the nestlings?

18. How do the crows work injury to the farmer? How do they benefit the farmer? Do you think they do more benefit than harm to the farmer and fruitgrower?

19. Have you known of instances of the crow's fondness for shining or glittering articles, like pieces of crockery or tin?

THE CARDINAL GROSBEAK

There never lived a Lord Cardinal who
ssessed robes of state more brilliant in
or than the plumage of this bird. By
: way, I wonder how many of us ever
nk when we see the peculiar red called
dinal, that it gained its name from the
ss of this high functionary of the
ırch? The cardinal grosbeak is the best
me for the redbird because that de-
ibes it exactly, both as to its color and
chief characteristic, since its beak is
ck and large; the beak is also red, which
a rare color in beaks, and in order to make its redness more emphatic it is set in a frame of black feathers. The use of such a large beak is unmistakable, for it is strong enough to crush the hardest of seed shells or to crack the hardest and driest of grains.

What cheer! What cheer!
That is the grosbeak's way,
With his sooty face and his coat of red

sings Maurice Thompson. Besides the name given above, this bird has been

After Audubon Leaflet 18

The cardinal grosbeak

called in different localities the redbird, Virginia redbird, crested redbird, winter redbird, Virginia nightingale, the red corn-cracker; but it remained for James Lane Allen to give it another name in his masterpiece, *The Kentucky Cardinal*.

The cardinal is a trifle smaller than the robin and is by no means slim and graceful, like the catbird or the scarlet tanager, but is quite stout and is a veritable chunk of brilliant color and bird dignity. The only bird that rivals him in redness is the scarlet tanager, which has black wings; the summer tanager is also a red bird, but is not so vermilion and is more slender and lacks the crest. The cardinal surely finds his crest useful in expressing his emotions; when all is serene, it lies back flat on the head, but with any excitement, whether of joy or surprise or anger, it lifts until it is as peaked as an old-fashioned nightcap. The cardinal's mate is of quiet color; her back is greenish gray and her breast buffy, while her crest, wings, and tail reflect in faint ways the brilliancy of his costume.

The redbird's song is a stirring succession of syllables uttered in a rich, ringing tone, and may be translated in a variety of ways. I have heard him sing a thousand times " tor-re'-do, tor-re'-do, tor-re'-do," but Dr. Dawson has heard him sing " che'-pew, che'-pew, we'-woo, we'-w[illegible] " bird-ie, bird-ie, bird-ie; tschew, tsc[illegible] tschew "; and " chit-e-kew, chit-e-kew[illegible] weet, he-weet." His mate breaks the [illegible]tom of other birds of her sex and sin[illegible] sweet song, somewhat softer than [illegible] Both birds utter a sharp note " tsip, ts[illegible]

The nest is built in bushes, vines, or [illegible] trees, often in holly, laurel, or other [illegible] evergreens, and is rarely more than si[illegible] eight feet above the ground. It is mad[illegible] twigs, weed stems, tendrils, the bark of [illegible] grapevine, and coarse grass; it is lined [illegible] fine grass and rootlets; it is rather loo[illegible] constructed but firm and is well hid[illegible] for it causes these birds great anguish [illegible] have their nest discovered. Three or f[illegible] eggs are laid, which are bluish white [illegible] grayish, dully marked with brown. [illegible] father cardinal is an exemplary husb[illegible] and father; he cares for and feeds his m[illegible] tenderly and sings to her gloriously w[illegible] she is sitting; and he works hard catch[illegible] insects for the nestlings. He is also a b[illegible] defender of his nest and will attack [illegible] intruder, however large, with undaun[illegible] courage. The fledglings have the dull c[illegible] of the mother and have dark-colored b[illegible] Until the young birds are able to take c[illegible] of themselves, their dull color somew[illegible] protects them from the keen eyes of th[illegible] enemies. If the male fledglings were [illegible] color of their father, probably not [illegible] would escape a tragic death. While [illegible] mother bird is hatching the secc[illegible]

Leonard K. Be[illegible]

The cardinal builds its nest in thick bushes [illegible] vines

the father keeps the first brood him and cares for them; often ıole family remains together during inter, making a small flock. How- he flocking habit is not characteris- these birds, and we only see them in lerable numbers when the exigencies :king food in the winter naturally them together.

: cardinals are fond of the shrubbery hickets of river bottoms near grain or where there is plenty of wild and they only visit our premises driven to us by winter hunger. Their consists of the seeds of rank weeds, wheat, rye, oats, beetles, grasshop- flies, and to some extent, wild and :n berries; but they never occur in ient numbers to be a menace to our . The cardinals may often be seen in :ornfields after the harvest, and will an overlooked ear of corn and crack kernels with their beaks in a most erous manner. During the winter we coax them to our grounds by scatter- :orn in some place not frequented by thus, we may induce them to nest us, since the cardinal is not naturally grant but likes to stay in one locality mer and winter. It has been known to ie as far north as Boston and southern v York, but it is found in greatest ıbers in our Southern states.

LESSON 33

The Cardinal Grosbeak

LEADING THOUGHT — The cardinal is : most brilliantly colored of all our ds, and one of our most cheerful sing- . We should seek to preserve it as a utiful ornament to our groves and unds.

METHOD — This work must be done personal observation in the field. The ld notes should be discussed in school.

OBSERVATIONS — 1. Do you know the rdinal? Why is it so called?

2. How many names do you know for is bird?

Leonard K. Beyer

The cardinal sings a beautiful song

3. Is the cardinal as large as the robin? Is it graceful in shape?

4. Is there any color except red upon it? If so, where?

5. What other vividly red birds have we and how can we distinguish them from the cardinal?

6. Describe the cardinal's crest and how it looks when lifted. Why do you think it lifts it?

7. Describe its beak as to color, shape, and size. What work is such a heavy beak made for?

8. Is the cardinal's mate the same color as he? Describe the color of her head, back, wings, tail, breast.

9. Can you imitate the cardinal's song? What words do you think he seems to sing? Does his mate sing also? Is it usual for mother birds to sing? What other notes besides songs do you hear him utter?

10. Where does the cardinal usually build its nest? How high from the ground? Of what materials? Is it compact or bulky? How many eggs are there and what are their colors?

11. How does the father bird act while his mate is brooding? How does he help take care of the young in the nest?

12. How do the fledglings differ in color from their father? From their mother? Of what use to the young birds is their sober color?

13. What happens to the fledglings of the first brood while the mother is hatching the eggs of the second brood?

14. In what localities do you most often see the cardinals? Do you ever see them in flocks?

15. What is the food of the cardinals? What do they feed their nestlings?

16. How can you induce the cardinals to build near your home?

17. What do you know about th protecting birds? Why should suc be observed?

Along the dust-white river road,
The saucy redbird chirps and tri
His liquid notes resound and ris
Until they meet the cloudless ski
And echo o'er the distant hills.

—N.

GEESE

To be called a goose should be considered most complimentary, for of all the birds the goose is probably the most intelligent. An observant lady who keeps geese on her farm assures me that no animal, not even dog or horse, has the intelligence of the goose. She says that these birds learn a lesson after a few repetitions, and surely

Canada geese in a field of grain

her geese were patterns of obedience. While I was watching them one morning, they started for the brook via the cornfield; she called to them sharply, "No, no, you mustn't go that way!" They stopped and conferred; she spoke again and they waited, looking at her as if to make up their minds to this exercise of self-sacrifice; but when she spoke the third time they left the cornfield and took the other path to the brook. She could bring her geese into their house at any time c by calling to them, "Home, home soon as they heard these words, would start and not stop until the las was housed.

In ancient Greece maidens made p geese; and often there was such a dev between the bird and the girl that the latter died her statue with that o goose was carved on her burial tablet. loyalty of a pet goose came under th servation of Miss Ada Georgia. A gander was the special pet of a small in Elmira, New York, who took sole of him. The bird obeyed commands a dog but would never let his little m out of his sight if he could avoid it; sionally he would appear in the sc yard, where the pupils would tease by pretending to attack his master at risk of being so severely whipped the bird's wings that it was a tes bravery among the boys so to challe him. His fidelity to his master was treme; once when the boy was ill in the bird wandered about the yard hon disconsolately and refused to eat; he driven to the side of the house where master could look from the window he immediately cheered up, took his fc and refused to leave his post beneath window while the illness lasted.

The goose is a stately bird whether land or water; its long legs give it good p portions when walking, and the neck, ing so much longer than that of the du gives an appearance of grace and digni The duck on the other hand is beauti when on the water or on the wing; ort legs, placed far back and far out at ides, make it a most ungraceful r. The beak of the goose is harder in re and is not flat like the duck's; no er the bird was a favorite with the an- Greeks, for the high ridge from the to the forehead resembles the fa- Grecian nose. The plumage of geese ry beautiful and abundant and for reason they are profitable domestic . They are picked late in summer the feathers are nearly ready to be ed; at this time the geese flap their s often and set showers of loose feath- lying. A stocking or a bag is slipped the bird's head and she is turned st side up with her head firmly be- n the knees or under the arm of the er. The tips of the feathers are seized the fingers and come out easily; only breast, the under parts, and the feath- beneath the wings are plucked. Geese not seem to suffer while being plucked ept through the temporary inconven- e and ignominy of having their heads ust into a bag; their dignity is hurt re than their bodies.

The wings of geese are very large and utiful; although our domestic geese e lost their powers of flight to a great ent, yet they often stretch their wings l take little flying hops, teetering along if they can scarcely keep on earth; this ist surely be reminiscent of the old in- nct for traveling in the skies. The tail the goose is a half circle and is spread en flying; although it is short, it seems be sufficiently long to act as a rudder. e legs of the goose are much longer an those of the duck; they are not set so back toward the rear of the body, and erefore the goose is the much better nner of the two. The track made by the ose's foot is a triangle with two scallops n one side made by the webs between the ree front toes; the hind toe is placed igh up; the foot and the unfeathered por- on of the leg, protected by scales, are sed as oars when the bird is swimming. When she swims forward rapidly, her feet xtend out behind her and act on the principle of a propeller; but when swimming around in the pond she uses them at almost right angles to the body. Although they are such excellent oars they are also efficient on land; when running, her body may waddle somewhat, but her head and neck are held aloft in stately dignity.

The Toulouse are our common gray geese; the Embdens are pure white with orange bill and bright blue eyes. The African geese have a black head with a large black knob on the base of the black bill; the neck is long, snakelike, light gray, with a dark stripe down the back; the wings and tail are dark gray; there is a dewlap at the throat. The brown Chinese geese have also a black beak and a black knob at the base of the bill. The neck is light brown with a dull yellowish stripe down the neck. The back is dark brown; breast, wings, and tail are grayish brown. The white Chinese are shaped like the brown Chinese, but the knob and bill are orange and the eyes light blue.

The Habits of Geese

Geese are monogamous and are loyal to their mates. Old-fashioned people declare that they choose their mates on Saint Valentine's Day, but this is a pretty myth; when once mated, the pair live together year after year until one dies; an interesting instance of this is one of the traditions in my own family. A fine pair of geese belonging to my pioneer grandfather had been mated for several years and had reared handsome families; but one spring a conceited young gander fell in love with the old goose, and as he was young and lusty, he whipped her legitimate lord and master and triumphantly carried her away, although she was manifestly disgusted with this change in her domestic fortunes. The old gander sulked and refused to be comforted by the blandishments of any young goose whatever. Later the old pair disappeared from the farmyard and the upstart gander was left wifeless. It was inferred that the old couple had run away with each other into the encompassing wilderness and much sympathy was felt for them because of this sacrifice of their lives for loyalty. However, this was misplaced sentiment, for later in the summer the happy pair was discovered in a distant "slashing" with a fine family of goslings, and all were brought home in triumph. The old gander, while not able to cope with his rival, was still able to trounce any of the animal marauders which approached his home and family.

The goose lines her nest with down and the soft feathers which she plucks from her breast. The gander is very devoted to his goose while she is sitting; he talks to her in gentle tones and is fierce in her defense. The eggs are about twice as large as those of the hen and have the ends more

A. A. Allen

A pair of Canada geese. While one broods the eggs the other stands guard

rounded. The period of incubation is four weeks. The goslings are beautiful little creatures, covered with soft down, and have large, bright eyes. The parents give them most careful attention from the first. One family which I studied consisted of the parents and eighteen goslings. The mother was a splendid African bird; she walked with dignified step, her graceful neck assuming serpentine curves; and she always carried her beak "lifted," which gave her an appearance of majestic haughtiness. The father was just a plebeian white gander, probably of Embden descent, but he was a most efficient protector. The family always formed a procession in going to the creek, the majestic mother at the head, the goslings following her and the gander bringing up the rear to be sure there were no stragglers; if ling strayed away or fell behind, the went after it, pushing it back int family circle. When entering the co night he pushed the little ones in g with his bill; when the goslings took first swim, both parents gently pu them into the water, "rooted them as the farmer said. Any attempt to liberties with the brood was met bristling anger and defiance on the pa the gander; the mistress of the farm me that he had whipped her black blue when she tried to interfere with goslings.

The gander and goose always show picion and resentment by opening mouth wide and making a hissing n showing the whole round tongue mocking defiance. When the gande tacks, he thrusts his head forward, with or below the level of his back, se his victim firmly with his hard, too bill so that it cannot get away, and t with his strong wings beats the life ou it. I remember vividly a whipping whi gander gave me when I was a child, h ing me fast by the blouse while he laid the blows.

Geese feed much more largely u land vegetation than do ducks; a g growth of clover and grass makes excell pasture for them; in the water, they f upon water plants but do not eat ins and animals to any extent.

Undoubtedly goose language is va and expresses many things. Geese talk each other and call from afar; they sh in warning and in general make such turmoil that people do not enjoy it. T goslings, even when almost grown, k up a constant "pee wee, pee wee," wh is nerve-racking. There is a good opp tunity for some interesting investigati in studying out just what the differe notes of the geese mean.

The goose is very particular about h toilet; she cleans her breast and back a beneath her wings with her bill; and s cleans her bill with her foot; she al cleans the top of her head with her fo and the under side of her wing with t of that side. When oiling her feath-he starts the oil gland flowing with beak, then rubs her head over the l until it is well oiled; she then uses head as a "dauber" to apply the oil to the feathers of her back and breast. When thus polishing her feathers, she twists the head over and over and back and forth to add to its efficiency.

The Jack Miner Migratory Bird Foundation, Inc.

e corner of Jack Miner's Bird Sanctuary, Kingsville, Ontario, Canada, where Canada geese find food, shelter, and protection

WILD GEESE

There is a sound, that, to the weather-e farmer, means cold and snow, even ugh it is heard through the hazy atmos-ere of an Indian summer day; and that he honking of wild geese as they pass their southward journey. And there is t a more interesting sight anywhere in e autumn landscape than the wedge-aped flock of these long-necked birds th their leader at the front apex. "The ld goose trails his harrow," sings the et; but only the aged can remember e old-fashioned harrow which makes this nile graphic. The honking which reveals us the passing flock, before our eyes can scern the birds against the sky, is the call of the wise old gander who is the leader, to those following him, and their return salute. He knows the way on this long thousand-mile journey, and knows it by instinct and in part by the topography of the country. If ever fog or storm hides the earth from his view, he is likely to be-come confused, to the dismay of his flock, which follows him to the earth with many lonely and distressful cries.

The northern migration takes place in April and May, and the southern from October to December. The journey is made with stops for rest and refreshment at certain selected places, usually some se-cluded pond or lake. The food of wild geese consists of water plants, seeds and corn, and some of the smaller animals living in water. Although the geese come to rest on the water, they go to the shore to feed. In California, the wild geese are dreaded visitors of the cornfields, and men with guns are employed regularly to keep them off.

The nests are made of sticks lined with down, usually along the shores of streams, sometimes on tree stumps and sometimes in deserted nests of the osprey. There are

A. R. Dugmore

Wild geese flying in even ranks

only four or five eggs laid and both parents are devoted to the young, the gander bravely defending his nest and family from the attacks of any enemies.

Although there are several species of wild geese on the Atlantic Coast, the one called by this name is usually the Canada goose. This bird is a superb creature, brown above and gray beneath, with head, neck, tail, bill, and feet of black. These black trimmings are highly ornamental and, as if to emphasize them, there is a white crescent-shaped "bib" extending from just back of the eyes underneath the head. This white patch is very striking, and gives one the impression of a bandage for sore throat. It is regarded as a call-color, and is supposed to help keep the flock together; the side tail-coverts are also white and may serve as another guide to follow.

Often some wounded or wearied of the migrating flock spends the w in farmyards with domestic geese. morning a neighbor of mine found during the night a wild gander, inj in some way, had joined his flock. stranger was treated with much cou by its new companions as well as by farmer's family and soon seemed fectly at home. The next spring he m with one of the domestic geese. In the summer, my neighbor, mindful of geese habits, clipped the wings of the der so that he would be unable to join passing flock of his wild relatives. As migrating season approached, the ga became very uneasy; not only was he easy and unhappy always but he insi that his wife share his misery of un He spent days in earnest remonstr with her and, lifting himself by cropped wings to the top of the barn fence, he insisted that she keep him c pany on this, for webbed feet, uneasy ing place. Finally, after many day tribulation, the two valiantly started sc on foot. News was received of their p ress for some distance and then they lost to us. During the winter our neigh visited a friend living eighteen miles to southward and found in his barnyard errant pair. They had become tired of grating by tramping and had joined farmer's flock; but we were never abl determine the length of time required this journey.

LESSON 34

Geese

Leading Thought — Geese are most intelligent of the domesticated bi and they have many interesting habits.

Method — This lesson should not

unless there are geese where the
may observe them. The questions
d be given a few at a time and an-
d individually by the pupils after the
vations are made.

SERVATIONS — 1. What is the chief
ence between the appearance of a
and a duck? How does the beak of
oose differ from that of the duck in
and in texture? Describe the nostrils
heir situation.

What is the difference in shape be-
n the neck of the goose and that of
uck?

What can you say about the plum-
of geese? How are geese "picked"?
hat time of year? From what parts
e body are the feathers plucked?

Are the wings of the goose large com-
d with the body? How do geese exer-
their wings? Describe the tail of the
e and how it is used.

How do the legs and feet of the
e differ from those of the duck? De-
e the goose's foot. How many toes are
bed? Where is the other toe? What is
shape of the track made by the goose's
? Which portions of the legs are used
oars? When the goose is swimming
vard where are her feet? When turning
nd how does she use them? Does the
se waddle when walking or running as
uck does? Why? Does a goose toe in
n walking? Why?

. Describe the shape and color of the
owing breeds of domestic geese: The
ulouse, the Embden, the African, and
Chinese.

Habits of Geese

1. What is the chief food of geese? What do they find in the water to eat? How does their food differ from that of ducks?

2. How do geese differ from hens in the matter of mating and nesting? At what time of year do geese mate? Does a pair usually remain mated for life?

3. Describe the nest and compare the eggs with those of hens. Describe the young goslings in general appearance. With what are they covered? What care do the parents give to their goslings? Describe how the parents take their family afield. How do they induce their goslings to go into the water for the first time? How do they protect them from enemies?

4. How does the gander or goose fight? What are the chief weapons? How is the head held when the attack is made?

5. How does the goose clean her feathers, wings, and feet? How does she oil her feathers? Where does she get the oil and with what does she apply it?

6. How much of goose language do you understand? What is the note of alarm? How are defiance and distrust expressed? How does a goose look when hissing? What is the constant note which the gosling makes?

7. Give such instances as you may know illustrating the intelligence of geese, their loyalty and bravery.

8. "The Canada Goose, its appearance, nesting habits, and migrations," would be an interesting topic for discussion.

GAME BIRDS

1. Ring-necked Pheasants. *These birds, native to China, have been introduced into many other parts of the world. They were first brought to the United States in* 1881 *and since then have become common in many of the states. The cock is handsome and brightly colored, the hen an inconspicuous brown. These pheasants are found in fields and in hedgerows or brush-covered areas rather than in forested sections. They feed chiefly on the ground, eating weed seeds, insects, ungarnered grain, and wild or waste fruit. In winter, whenever the ground is covered with crusted snow or ice, it is hard for them to get food and many of them starve unless man feeds them. Another difficulty of theirs in winter is that their long tail feathers get loaded with snow and ice, which keeps them from going about after food and even from seeking shelter.* (*Photo by courtesy of* Country Life in America)

2. Wild Turkey. *This game bird was once common from New England southward and west to the Rocky Mountains. It has been exterminated in the North, but it is still found locally in the South and West. Because the wild turkey thrives upon a variety of foods and because it can adapt itself to varied conditions of climate, it is again being introduced in many sections of the country.* (*Photo by L. W. Brownell*)

3. Nest of the Ruffed Grouse. *The ruffed grouse, a much prized game bird, is native to the eastern and central United States. It is a very hardy bird, being able to withstand extreme cold, and to live on the buds and twigs of trees when insects, berries, and seeds are not available. In winter ruffed grouse take shelter at night in a "pocket" of snow or beneath brush; in summer they usually roost in trees. In appearance this bird is not unlike the dusky grouse* (*No.* 5). (*Photo by Marjorie Ruth Ross*)

4. Eastern Bobwhite *or* Quail. *in the eastern United States, except sular Florida, and as far west as Co except New Mexico and southern Texas white or quail are permanent residents. like open fields with brushy fence-corn low bushes near at hand for protection storm and enemies. The pretty song is translated* bob-white *or* buck-wheat. *Th is made upon the ground under a bun grass or some bush, and in it are laid eighteen white eggs. The family or cove remain together until spring, and at will squat close together in a circle with together and heads out ready to scatter directions at the slightest indication of ger. In winter when quail are in this f tion, they may be covered with snow; if a crust of sleet or ice which they are u to break should form, the entire covey smother or starve.* (*Photo by L. W. Brown*

5. Dusky Grouse. *A relative of the r grouse, this species is found in the R Mountain regions of the United States Canada.* (*Photo by L. W. Brownell*)

6. A Woodcock on Its Nest. *Excep the Far West the woodcock is found spread over the United States. It winter the South. It lives largely on earthworms grubs for which it probes moist soft with a long, sensitive bill. The courtship s flights of the male are unique: with a ca his mate he rises into the air; by a serie loops he flies higher and higher until fro height of about two hundred feet he d suddenly to a place on the ground very where he started. The young quickly l to fly, but until they do they are freque carried from place to place by their mo who holds them between her legs with her* (*Photo by Olin Sewall Pettingill, Jr.*)

1
2
3
4
5
6

The beginning of the strut. These gobblers are strutting before the camera, hidden by br in an endeavor to attract the hen turkey whose mating call the camera man is imitating

THE TURKEY

That the turkey and not the eagle should have been chosen for our national bird, was the conviction of Benjamin Franklin. It is a native of our country, it is beautiful as to plumage, and like the American Indian, it has never yielded entirely to the influences of civilization. Through the hundreds of years of domestication it still retains many of its wild habits. In fact, it has many qualities in common with the red man. Take for instance its sun dance, which anyone who is willing to get up early enough in the morning and who has a flock of turkeys at hand can witness. Miss Ada Georgia made a pilgrimage to witness this dance and describes it thus: "While the dawn was still faint and gray, the long row of birds on the ridge-pole stood up, stretched legs and wings and flew down into the orchard beside the barnyard and began a curious, high-stepping, 'flip-flop' dance on the frosty grass. It consisted of little, awkward, up-and-down jumps, ied by forward springs of about a f with lifted wings. Both hens and m danced, the latter alternately strutting hopping and all 'singing,' the hens cal a 'Quit, quit,' the males accompany with a high-keyed rattle, sounding lil hard wood stick drawn rapidly alon picket fence. As the sun came up and sky brightened, the exhibition ended s denly when 'The Captain,' a great th pound gobbler and leader of the flo made a rush at one of his younger bre ren who had dared to be spreading a too near to his majesty."

The bronze breed resembles m closely our native wild turkey and is th fore chosen for this lesson. The colors markings of the plumage form the bro turkey's chief beauty. Reaching from skin of the neck halfway to the middle the back is a collar of glittering bro with greenish and purple iridescence, e er tipped with a narrow jet band. remainder of the back is black except each feather is edged with bronze. breast is like the collar and at its er is a tassel of black bristles called beard which hangs limply downward n the birds are feeding; but when the bler stiffens his muscles to strut, this d is thrust proudly forth. Occasionally hen turkeys have a beard. The long ls, or primaries, of the wings are barred ss with bands of black and white; secondaries are very dark, luminous wn, with narrower bars of white. Each her of the fan-shaped tail is banded black and brown and ends with a k bar tipped with white; the tail-cov- are lighter brown but also have the k margin edged with white. The colors the hen are like those of the gobbler ept that the bronze brilliance of breast, k, and wings is dimmed by the faint of white which tips each feather.

The heads of all are covered with a ty wrinkled skin, bluish white on the wn, grayish blue about the eyes, and other parts are red. Beneath the throat hanging fold called the wattle, and ve the beak a fleshy pointed knob led the caruncle, which on the gobbler prolonged so that it hangs over and be- the beak. When the bird is angry se carunculated parts swell and grow re vivid in color, seeming to be gorged th blood. The color of the skin about head is more extensive and brilliant in gobblers than in the hens. The beak is ghtly curved, short, stout, and sharp- inted, yellowish at the tip and dark at e base.

The eyes are bright, dark hazel with a in red line of iris. Just back of the eye is e opening of the ear, seemingly a mere le, yet leading to a very efficient ear, on which every smallest sound im- nges.

The legs of the young turkeys are nearly ack, fading to a brownish gray when ma- re. The legs and feet are large and stout, e middle toe of the three front ones be- g nearly twice the length of the one on ther side; the hind toe is the shortest of the four. On the inner side of the gobbler's legs, about one-third the bare space above the foot, is a wicked-looking spur which is a most effective weapon. The wings are large and powerful; the turkey flies well for such a large bird and usually roosts high, choosing trees or the ridge-pole of the barn for this purpose.

In many ways the turkeys are not more than half domesticated. They insistently prefer to spend their nights out of doors instead of under a roof. They are also great wanderers and thrive best when allowed to forage in the fields and woods for a part of their food.

The gobbler is the most vainglorious bird known to us; when he struts to show his flock of admiring hens how beautiful he is, he lowers his wings and spreads the stiff primary quills until their tips scrape the ground, lifting meanwhile into a semicircular fan his beautiful tail feathers; he protrudes his chest, and raises the iridescent plumage of his neck like a ruff to make a background against which he throws back his red, white, and blue decorated head. He moves forward with slow and mincing steps and calls attention to his grandeur by a series of most aggressive "gobbles." But we must say for the gobbler that although he is vain he is also a brave fighter. When beginning a fight he advances with wings lowered and sidewise as if guarding his body with the spread wing. The neck and the sharp beak are outstretched and he makes the attack so suddenly that it is impossible to see whether he strikes with both wing and beak or only with the latter, as with fury he pounces upon his adversary apparently striving to rip his neck open with his spurs.

Turkey hens usually begin to lay in April in this latitude (southern New York) and much earlier in more southern states. At nesting time each turkey hen strays off alone, seeking the most secluded spot she can find to lay the large, oval, brown-speckled eggs. Silent and sly, she slips away to the place daily, by the most roundabout ways, and never moving in the direction of the nest when she thinks herself observed. Sometimes the sight of any person near her nest will cause her to desert it. The writer has spent many hours when a child, sneaking in fence corners and behind stumps and tree trunks, stalking turkeys' nests. Incubation takes four weeks. The female is a most persistent sitter and care should be taken to see that she gets a good supply of food and water at this time. Good sound corn or wheat is the best food for her at this period. When sitting she is very cross and will fight most courageously when molested on her nest.

Turkey nestlings are rather large, with long, bare legs and scrawny, thin necks; they are very delicate during the first six weeks of their lives. Their call is a plaintive "peep, weep," and when a little turkey feels lost its cry is expressive of great fear and misery. But if the mother is freely ranging she does not seem to be much affected by the needs of her brood; she will fight savagely for them if they are near her, but if they stray, and they usually do, she does not seem to miss or hunt for them, but strides serenely on her way, keeping up a constant crooning "kr-rit, kr-rit," to encourage them to follow. As a consequence, the chicks are lost, or get draggled and chilled by struggling through wet grass and leaves that are no obstacle to the mother's strong legs, and thus many die. If the mother is confined in a coop it should be so large and roomy that she can move about without trampling on the chicks, and it should have a dry floor, since dampness is fatal to the little ones.

For the first week the chicks should be fed five times a day, and for the next five weeks they should have three meals a day. They should be given only just about enough to fill each little crop and none should be left over to be trodden under their awkward little feet. Their quarters should be kept clean and free from vermin.

LESSON 35

Turkeys

Leading Thought — The turkey is a native of America. It was introduced into Spain from Mexico about 1518, and s then has been domesticated. How there are still in some parts of the c try flocks of wild turkeys. It is a beau bird and has interesting habits.

Method — If the pupils could vis flock of turkeys, the lesson would be g to a better advantage. If this is imposs ask the questions a few at a time and those pupils who have opportunities observing the turkeys give their ans before the class.

Observations — 1. Of what breed the turkeys you are studying: Bro Black, Buff, White Holland, or Narra sett?

2. What is the general shape and of the turkey? Describe its plumage, ing every color which you can see in Does the plumage of the hen turkey fer from that of the gobbler?

3. What is the covering of the hea the turkey, what is its color and how does it extend down the neck of the bi Is it always the same color; if not, w causes the change? Is the head cover alike in shape and size on the male a the female? What is the part called t hangs from the front of the throat bel the beak? From above the beak?

4. What is the color of the beak? I short or long, straight or curved? Wh are the nostrils situated?

5. What is the color of the turke eyes? Do you think it is a keen-sight bird?

6. Where are the ears? Do they sh as plainly as a chicken's ears do? Are t keys quick of hearing?

7. Do turkeys scratch like hens? A they good runners? Describe the feet a legs as to shape, size, and color. Has t male a spur on his legs, and if so, where it situated? For what is it used?

8. Can turkeys fly well? Are the win small or comparatively large and stro for the weight of the body? Do turke prefer high or low places for perchi when they sleep? Is it well to house an confine them in small buildings and par as is done with other fowls?

9. Tell, as nearly as you can discover close observation, how the gobbler se

part of his plumage when he is "showing off" or strutting. What do you think is the bird's purpose in thus exhibiting his fine feathers? Does the "king of the flock" permit any such action by other gobblers in his company?

. Are turkeys timid and cowardly or independent and brave, ready to meet and fight anything which they think is threatening to their comfort and safety?

. When turkeys fight, what parts of their bodies seem to be used as weapons? Does the male "gobble" during a fight, only as a challenge or in triumph when victorious? Do the hen turkeys ever fight, or only the males?

. How early in the spring does the turkey hen begin to lay? Does she nest about the poultry yard and the barns or is she likely to seek some secret and distant place where she may hide her eggs? Describe the turkey's egg, as well as you can, as to color, shape, and size. Can one tell it by the taste from an ordinary hen's egg? About how many eggs does the turkey hen lay in her nest before she begins to "get broody" and want to sit?

13. How many days of incubation are required to hatch the turkey chick? Is it as downy and pretty as other little chicks? How often should the young chicks be fed, and what food do you think is best for them? Are turkey chicks as hardy as other chicks?

14. Is the turkey hen generally a good mother? Is she cross or gentle when sitting and when brooding her young? Is it possible to keep the mother turkey as closely confined with her brood as it is with the mother hen? What supplies should be given to her in the way of food, grits, dust-baths, etc.?

BIRDS OF MARSH AND SHORE

1. Shoveller, Spoonbill, or Broadbill. *The range of the shoveller extends from Alaska in summer to Colombia, South America, in winter. With its uniquely long, broad bill, this shallow-water "dabbler" gathers up water and ooze; by means of the comblike teeth with which the bill is equipped it strains out the insects and vegetable matter which are its favorite food.* (Photo by L. W. Brownell)

2. The Mallard. *The range of the mallard in North America extends in summer south of the Arctic circle, east to Hudson Bay, and south to Lower California and Texas. In winter it is found from the Aleutian Islands south to Panama. Being a "dabbler" the mallard generally feeds in shallow water, but it is very adaptable as to food and environment. From the economic standpoint it is the most important duck in the world, since it is the ancestor of most domestic ducks, is widely distributed, and produces meat of good quality.* (Photo by L. W. Brownell)

3. Lesser Scaup Ducks. *This is one of the most common ducks in the open waters of rivers, larger lakes and bays, and along seacoasts. Its food, consisting chiefly of insects, crustaceans, water snails, tadpoles, and aquatic plants, it secures by diving. In the Gulf states, the lesser scaup is often called the "raft duck" because of the great numbers that collect into flocks and move about on the water. These rafts are sometimes a mile long.* (Photo by S. A. Grimes)

4. Pied-billed Grebe on Its Nest. *The summer range of this grebe is from southern Canada to the southern United States; its winter range extends to Mexico and Cuba. It moves south when ice forms on northern streams, and returns when it breaks up in spring. Its food consists chiefly of aquatic animals and some water plants. To escape danger it dives rather than flies. This grebe, like others, often carries its young on its back, thus hiding them from observers; the mother can even dive with the young and when she comes again to the surface keep them still concealed.* (Photo by Olin Sewall Pettingill, Jr.)

5. Spotted Sandpiper Approaching Its Nest. *The sandpiper* (also called tip-up or tip-tail), *said to be the most widely and commonly distributed shore bird in North America, is found in regions about both fresh and salt water. Although it can swim and dive readily, its food consists chiefly of grasshoppers, cutworms, grubs, and pests of cultivated lands. The nest, a hollow in the ground, may be along shores or even in cultivated fields far from water; it is built by the united efforts of the pair.* (Photo by L. W. Brownell)

6. Chicks of Wilson's Plover. *These newly hatched chicks were picked up on a sandy beach and "posed" in a shell.* (Photo by Olin Pettingill, Jr.)

7. Wilson's Plover at Its Nest. (Se No. 6.) *Wilson's plover is found in the c regions of southern North America and tral America. It feeds on the tiny sea cre that the falling tide leaves strewn along flats and sandy beaches. The nest, u placed above high water on a sandy bea a hollowed out place in the sand. The and eggs blend so with the sand as to be a unnoticeable. In the one pictured here, one egg beneath the female, one in front o and newly hatched chick behind her.* (by S. A. Grimes)

8. King Rail on Its Nest. *The range o bird is in the central and southern portio the eastern half of the United States. Its consists largely of insects of cultivated which it secures from the edges of sw areas in uplands. Rails are found chief grassy marshes. The legs are strong an wings are weak, and hence when pursued will run or hide, but will fly only as a last r* (Photo by S. A. Grimes)

9. The Common Tern at Its Nest. *T live in both the Eastern and Western H spheres.*

Terns nest in colonies, usually on the sand of an island beach. They can be d guished from gulls by their more pointed narrower wings, and by their habit of divi swimming to catch their food, which consis small fish, aquatic worms, and insects. (Phot S. A. Grimes)

10. American Egret, Great White E or White Heron. *The summer range of egret is chiefly from the southern United St south to Patagonia. In late summer it migr northward to Maine. Its winter range is C rado, Texas, and South Carolina southw The egrets and other herons are comm found about the shores of lakes, rivers, or b They usually nest in flocks. Once in dange extinction, they are now under protection are increasing in numbers.* (Photo by S Grimes)

11. An American Bittern on the Def sive. *This inhabitant of the marshes range summer across the North American conti from central Canada to the southern Un States. In winter it is found from the south United States to Panama. When approac bitterns fall into a rigid pose which they h until the intruder retires or frightens them flight. The cry of this bird is most arresting unusual. It is compared to the sound of dri a stake or the sound of a pump in action. Fr snakes, small fish, mice, and insects comp its food.* (Photo by S. A. Grimes)

1
2
3
4
5
6
7
8
9
10
11

FISHES

It remains yet unresolved whether the happiness of a man in this world doth sist more in contemplation or action. Concerning which two opinions I shall fore to add a third by declaring my own, and rest myself contented in telling you both of these meet together, and do most properly belong to the most honest, in ious, quiet and harmless art of angling. And first I tell you what some have obser and I have found to be a real truth, that the very sitting by the riverside is not only quietest and the fittest place for contemplation, but will invite an angler to it.

— Isaak Wal

Dear, human, old Isaak Walton discovered that nature-study, fishing, and philosophy were akin and as inevitably related as the three angles of a triangle. And yet it is surprising how little the fish have been used as subjects for nature lessons. Every brook and pond is a treasure to the teacher who will find what there is in it and who knows what may be got out of it.

Almost any of the fishes found in a brook or pond may be kept in an aquarium for a few days of observation in the schoolroom. A large water pail or a bucket does very well if there is no glass aquarium. The water in an aquarium should be changed whenever it becomes foul. The practice should be established, once for all, of putting these finny prisoners back into the identical body of water from which they were taken. Much damage has been done by liberating fish in bodies of water where they do not belong. Many fish have cannibalistic traits: black bass, for instance, if they are either the comers or the original inhabitants, be likely to attack and destroy o fish. Besides, even if the new home vides suitable living conditions for newcomers, they may upset the bala existing among the various forms of p and animal life already there.

THE GOLDFISH

Once upon a time, if stories are true, there lived a king called Midas, whose touch turned everything to gold. Whenever I see goldfish, I wonder if, perhaps, King Midas were not a Chinese and if he perchance did not handle some of the little fish in Orient streams. But common man has learned a magic as wonderful as that of King Midas, although it does not act so immediately, for it is through agency in selecting and breeding t we have gained these exquisite fish our aquaria. In the streams of China goldfish, which were the ancestors of th effulgent creatures, wore safe green col like the shiners in our brooks; and if a goldfish escape from our fountains a run wild, their progeny return to th e olive-green color. There are many ch dull-colored goldfish in the lakes rivers of our country. It is almost in- eivable that one of the brilliant-col- fishes, if it chanced to escape into our ls, should escape the fate of being n by some larger fish attracted by such ering bait.

he goldfish, as we see it in the aquar- is brilliant orange above and pale on-yellow below; there are many speci- s that are adorned with black patches. as if this fish were bound to imitate precious metals, there are individuals ch are silver instead of gold; they are ized silver above and polished silver w. The goldfish are closely related he carp and can live in waters that stale. If water plants and scavengers, as water snails, are kept in the arium, the water does not become foul. water, then, need not be changed; but ess the aquarium is covered, it will be essary to add water to replace that ch evaporates. Goldfish should not be fed too lavishly. An inch square of one of the sheets of prepared fish food we have found a fair daily ration for five medium sized fish; these fish are more likely to

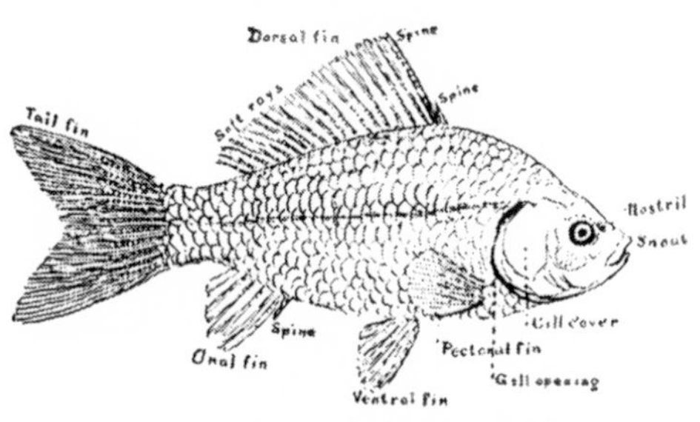

Goldfish with parts named

die from overfeeding than from starving. Goldfish are naturally long-lived; Miss Ada Georgia kept them until seven years old in a school aquarium; and there is on rec- ord one goldfish that lived nine years.

Helen F. Hill

Fish in a hatchery pond

Too often the wonderful common things are never noticed because of their commonness; and there is no better in- stance of this than the form and move- ments of a fish. It is an animal in many ways similar to animals that live on land; but its form and structure are such that it is perfectly adapted to live in water all its life; there are none of the true fishes which live portions of their lives on land as do the frogs. The first peculiarity of the fish is its shape. Looked at from above, the broader part of the body is near the front end, which is rounded or pointed so as to cut the water readily. The long, narrow, hind portion of the body with the tail acts as a propeller in the sense that it pushes the body forward; this movement is not at all similar to the action of an airplane propeller or a ship's screw. Seen from the side, the body is a smooth, graceful oval and this form is especially adapted to move through the water swiftly, as can be demonstrated to the pupil by cutting a model of the fish from wood and trying to move it through the water sidewise.

Normally, the fish has seven fins, one along the back called the dorsal, one at the end of the tail called the tail or caudal fin, one beneath the rear end of the body called the anal, a pair on the lower side

N. Y. State Conservation Dept.

Large-mouthed black bass
Aplites salmoides

of the body called the ventrals, and a pair just back of the gill openings called the pectorals. All these fins play their own parts in the movements of the fish. The dorsal fin is usually higher in front than behind and can be lifted or shut down like a fan. This fin when it is lifted gives the fish greater height and it can be twisted to one side or the other and thus be made a factor in steering. The anal fin on the lower side acts in a similar manner. The tail fin is the propeller and sends the body forward by pressing backward on the water, first on one side and then on the other, being used like a scull. The tail fin varies in shape very much in different species. In the goldfish it is fanlike, with a deeply notched hind edge, but in some it is rounded or square. The paired fins correspond anatomically to our arms and legs, the pectorals representing the arms, the ventrals the legs.

Fishes' eyes have no eyelid but the eyeball is movable, and this often gives the impression that the fish winks. Fishes are necessarily nearsighted since the lens of the eye has to be spherical in order to see in the water. The sense of smell is located in a little sac to which the nostril leads; the nostrils are small and often partitioned and may be seen on either side of the snout. The nostrils of a fish have no connection whatever with breathing.

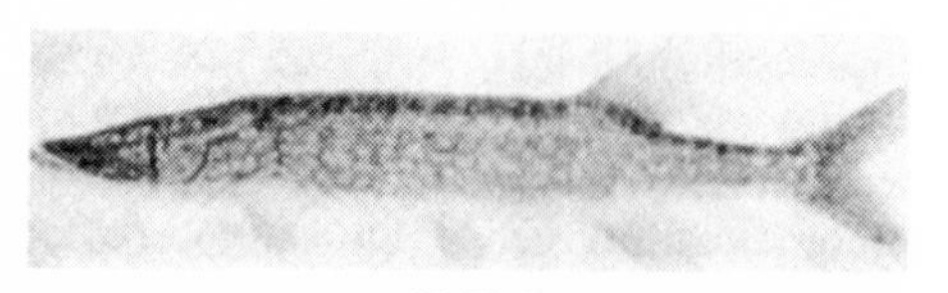

N. Y. State Conservation Dept.

A chain pickerel
Esox niger

The tongue of the fish is very bon
gristly and immovable. Very little s
of taste is developed in it. The sh
number, and position of the teeth var
cording to the food habits of the fish.
commonest teeth are fine, sharp, and s
and are arranged in pads, as seen in
bullhead. Some fish have blunt teeth
able for crushing shells. Some herbivo
fishes have sharp teeth with serrated ec
while those living upon crabs and sn
have incisor-like teeth. In some specie
find several types of teeth; in others, s
as goldfish or minnows in general,
teeth may be entirely absent. The te
are borne not only on the jaws but
in the roof of the mouth, on the tong
and in the throat.

The ear of the fish has neither out
form nor opening and is very imper
in comparison with that of man. Exte
ing along the sides of the body from h
to tail is a line of modified scales cont
ing small tubes connecting with ner

N. Y. State Conservation D

A yellow perch
Perca flavescens

this is called the lateral line and it is
lieved that it is in some way connec
with the fish's senses, perhaps with
sense of hearing.

The covering of fishes varies: most fi
such as the yellow perch and black ba
are sheathed in an armor of scales; othe
such as the bullhead, have only a smoo
skin. All fish are covered with a sli
substance which somewhat reduces fr
tion as they swim through the water.

In order to understand how the f
breathes we must examine its gills.
front, just above the entrance to the gull
are several bony ridges which bear t
rows of pinkish fringes; these are the g
arches and the fringes are the gills. T

re filled with tiny bloodvessels, and water passes over them, the impu- of the blood pass out through the skin of the gills and the life-giving n passes in. Since most fish cannot use of air unless it is dissolved in , it is very important that the water e aquarium provide a sufficient sur- area to enable the fish to secure air. gill arches also bear a series of bony sses called gill-rakers. Their function prevent the escape of food through ills while it is being swallowed, and vary in size according to the food s of the fish. We note that the fish in quarium constantly opens and closes mouth; this action draws the water the throat and forces it out over the and through the gill openings; this, , is the act of breathing.

LESSON 36

A Study of the Fish

eading Thought — A fish lives in the er where it must breathe, move, and its food. The water world is quite erent from the air world and the fish e developed forms, senses, and habits ch fit them for life in the water.

Method — The goldfish is used as a ject for this lesson because it is so veniently kept where the children may it. However, a shiner or other minnow ıld do as well.

Before the pupils begin the study, place diagram shown on p. 145 on the black- rd, with all the parts labeled; thus pupils will be able to learn the parts the fish by consulting it, and not be npelled to commit them to memory itrarily. It would be well to associate goldfish with a geography lesson on ina.

Observations — 1. Where do fish live?

2. What is the shape of a fish when seen from above? Where is the widest part? What is its shape seen from the side? Think if you can in how many ways the shape of the fish is adapted for moving swiftly through the water.

3. How many fins has the fish? Make a sketch of the goldfish with all its fins and name them from the diagram on the blackboard.

4. How many fins are there in all? Four of these fins are in pairs; where are they situated? What are they called? Which pair corresponds to our arms? Which to our legs?

5. Describe the pectoral fins. How are they used? Are they kept constantly moving? Do they move together or alternately? How are they used when the fish swims backward?

6. How are the ventral fins used? How do they assist the fish when swimming?

7. Observe a dorsal fin and an anal fin. How are these used when the fish is swimming?

8. With what fin does the fish push itself through the water? Make a sketch of the tail. Note if it is square, rounded, or notched at the end.

9. Watch the goldfish swim and describe the action of all the fins while it is in motion. In what position are the fins when the fish is at rest?

10. What is the nature of the covering of the fish? Are the scales large or small? In what direction do they seem to overlap? Of what use to the fish is this scaly covering?

11. Can you see a line which extends from the upper part of the gill opening, along the side to the tail? This is called the lateral line. Do you think it is of any use to the fish?

12. Note carefully the eyes of the fish. Describe the pupil and the iris. Are the eyes placed so that the fish can see in all directions? Can they be moved so as to see better in any direction? Does the fish wink? Has it any eyelids? Do you know why fish are nearsighted?

13. Can you see the nostrils? Is there a little wartlike projection connected

with the nostril? Do you think fishes breathe through their nostrils?

14. Describe the mouth of the fish. Does it open upward, downward, or directly in front? What sort of teeth have fish? How does the fish catch its prey? Does the lower or upper jaw move in the process of eating?

15. Is the mouth kept always in motion? Do you think the fish is swallowing water all the time? Do you know why it does this? Can you see a wide opening along the sides of the head behind the gill cover? Does the gill cover move with the movement of the mouth? How does a fish breathe?

16. What are the colors of the goldfish above and below? What would happen to our beautiful goldfish if they were put in a brook with other fish? Why could they not hide? Do you know what happens to the colors of the goldfish when they run wild in our streams and ponds?

17. Can you find in books or cyclopedias where the goldfish came from? Are they gold and silver in color in the streams where they are native? Do you think that they had originally the long, slender, swallow-tails which we see sometimes in goldfish? How have the beautiful colors and graceful forms of the gold and silver fishes been developed?

I have my world, and so have you,
A tiny universe for two,
A bubble by the artist blown,
Scarcely more fragile than our own,
Where you have all a whale could wi
Happy as Eden's primal fish.
Manna is dropt you thrice a day
From some kind heaven not far aw
And still you snatch its softening cru
Nor, more than we, think when
comes.
No toil seems yours but to explore
Your cloistered realm from shore to s
Sometimes you trace its limits roun
Sometimes its limpid depths you sour
Or hover motionless midway,
Like gold-red clouds at set of day;
Erelong you whirl with sudden whi
Off to your globe's most distant rim,
Where, greatened by the watery le
Methinks no dragon of the fens
Flashed huger scales against the sky,
Roused by Sir Bevis or Sir Guy;
And the one eye that meets my view,
Lidless and strangely largening, too,
Like that of conscience in the dark,
Seems to make me its single mark.
What a benignant lot is yours
That have an own All-out-of-doors,
No words to spell, no sums to do,
No Nepos and no parlyvoo!
How happy you, without a thought
Of such cross things as Must
Ought —
I too the happiest of boys
To see and share your golden joys!

— "THE ORACLE OF THE GOLDFISH

LOW

THE BULLHEAD

The bull-head does usually dwell and hide himself in holes or amongst stone clear water; and in very hot days will lie a long time very still and sun himself and be easy to be seen on any flat stone or gravel; at which time he will suffer an angle put a hook baited with a small worm very near into his mouth; and he never ref to bite, nor indeed, to be caught with the worst of anglers. — ISAAK WALTON

When one looks a bullhead in the face one is glad that it is not a real bull, for its barbels give it an appearance quite fit for the making of a nightmare; and yet from the standpoint of the bullhead, how truly beautiful those fleshy feelers For without them how could it feel way about searching for food in the m Two of these barbels stand straight the two largest ones stand out on e

State of New York Conservation Department

Common bullhead
Ameiurus nebulosus

of the mouth, and two pairs of short
; adorn the lower lip, the smallest pair
ıe middle.

s the fish moves about, it is easy to
:hat the large barbels at the side of the
ıth are of the greatest use; it keeps
n in a constantly advancing move-
ıt, feeling of everything it meets. The
er ones stand straight up, keeping
ch for whatever news there may be
n above; the two lower ones spread
rt and follow rather than precede the
, seeming to test what lies below. The
er and lower pairs seem to test things
hey are, while the large side pair deal
h what is going to be. The broad
ıth seems to be formed for taking in
things eatable, for the bullhead lives
almost anything alive or dead that it
:overs as it noses about in the mud.
vertheless, it has its notions about its
d, for I have repeatedly seen one draw
terial into its mouth through its breath-
motion and then spew it out with a
ıemence one would hardly expect from
h a phlegmatic fish.

Although it has feelers which are very
cient, it also has perfectly good eyes
ich it uses to excellent purpose; note
w promptly it moves to the other side
the aquarium when we are trying to
dy it. The eyes are not large; the pupils
black and oval and are rimmed with
ıarrow band of shiny pale yellow. The
:s are prominent so that when moved
ckward and forward they gain a view
of the enemy in the rear or at the front while the head is motionless. It seems strange to see such a pair of pale yellow, almost white eyes in such a dark body.

The general shape of the front part of the body is flat, in fact, it is shaped decidedly like a tadpole; this shape is especially fitted for groping about muddy

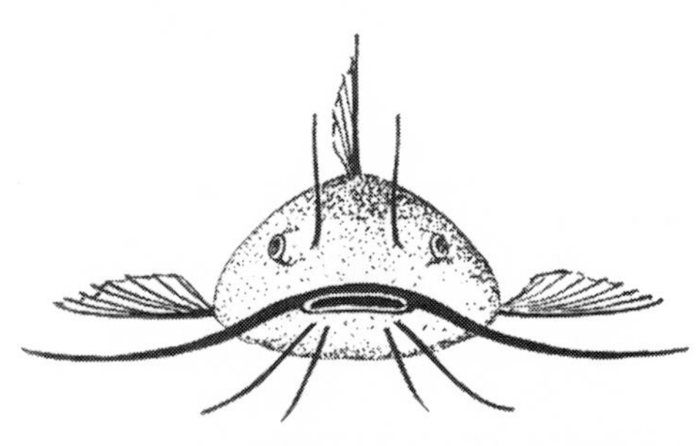

bottoms. The flat effect of the body is emphasized by the gill covers opening below rather than at the sides, every pulsation widening the broad neck. The pectoral fins also open out on the same plane as the body, although they can be turned at an angle if necessary; they are thick and fleshy and the sharp tips of their spines offer punishment to whosoever touches them. The dorsal fin is far forward and not large; it is usually raised at a threatening angle.

Near the tail there is a little fleshy dorsal fin which stands in line with the body, and one wonders what is its special use. The ventral fins are small. The anal fin is far back and rather strong, and this with the long, strong tail gives the fish good motor power; it can swim very rapidly if occasion requires.

The bullhead is mud-colored and has no scales. The skin is very thick and leathery so that it is always removed before the fish is cooked. The bullhead burrows deep into the mud in the fall and remains there all winter; when the spring freshets come, it emerges and is hungry for fresh meat.

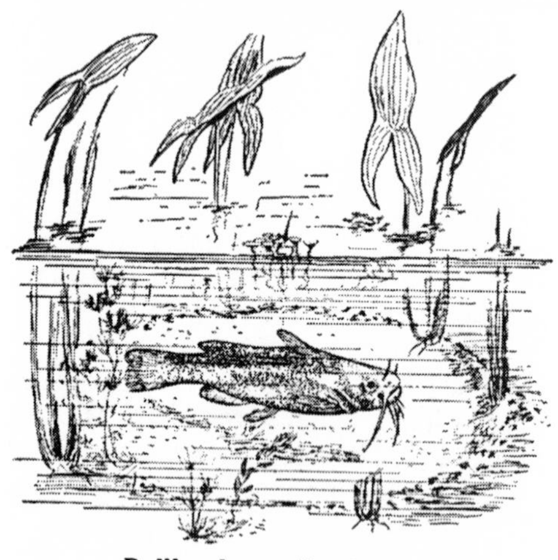

Bullhead guarding his nest

The family life of the bullheads and other catfishes seems to be quite ideal. Dr. Theodore Gill tells us that bullheads make their nests by removing stones and gravel from a more or less irregularly circular area in shallow water, and on sandy or gravelly ground. The nest is somewhat excavated, both parents removing the pebbles by sucking them into the mouth and carrying them off for some distance. After the eggs are laid, the male watches over and guards the nest and seems to have great family responsibilities. He is the more active of the two in stirring and mixing the young fry after they are hatched. Smith and Harron describe the process thus: "With their chins on the bottom, the old fish brush the corners where the fry were banked, and with the barbels all directed forward, and flexed where they touch the bottom, thoroughly agitate the mass of fry, bringing the deepest individuals to the surface. This act is usually repeated several times in quick succession.

"The nests are usually made ben logs or other protecting objects an shallow water. The paternal care is tinued for many days after the birth o young. At first these may be crowde gether in a dense mass, but as time p they disperse more and more and sp around the father. Frequently, espec when the old one is feeding, some — or more — of the young are taken int mouth, but they are instinctively rated from the food and spit out. At the young swarm venture farther their birthplace, or perhaps they are away by their parents."

LESSON 37

The Bullhead, or Horned Pou

Leading Thought — The bullh lives in mud bottoms of streams ponds and is particularly adapted for in such locations.

Method — A small bullhead may placed in a small aquarium jar. At firs the water be clear and add a little p weed so as to observe the natural tende of the fish to hide. Later add mud gravel to the aquarium and note the havior of the fish.

Observations — 1. What at the glance distinguishes the bullhead f other fish? Describe these strange "w kers" growing about the mouth; many are there and where are they s ated? Which are the longest pair? the fish move them in any direction will?

2. Where do we find bullheads? what do they feed? Would their eyes h them to find their food in the mud? H do they find it?

3. Explain, if you can, why the b head has barbels, or feelers, while trout and bass have none.

4. What is the shape of the mout

What is the general shape of the y? What is its color? Has it any scales?

Why should the bullhead be so flat zontally while the sunfish is so flat he opposite direction?

. Describe the bullhead's eyes. Are large? What is their color? Where they placed?

. Describe the dorsal fin, giving its parative size and position. Do you see ther dorsal fin? Where is this peculiar and how does it differ from all of the ers?

. Describe the tail fin. Does it seem g and strong? Is the bullhead a good mmer?

o. Is the anal fin large or small as comed with that of the goldfish?

1. How do the pectoral fins move as npared with those of the sunfish? Why he position of the pectoral and dorsal of benefit to this fish?

.2. How does the bullhead inflict unds when it is handled? Tell how se spines may protect it from its natural mies.

3. When is the best season for fishing bullheads? Does the place where they found affect the flavor of their flesh? hy?

14. What is the spawning season? Do u know about the nests the bullheads ild and the care they give their young?

15. Write an essay on the nest-making bits of the bullheads and the care given e young by the parents.

And what fish will the natural boy natally take? In America, there is but one h which enters fully into the spirit of e occasion. It is a fish of many species cording to the part of the country, and as many sizes as there are sizes of boys. his fish is the horned pout, and all the st of the species of Ameiurus. Horned out is its Boston name. Bullhead is good ough for New York; and for the rest of e country, big and little, all the fishes of this tribe are called catfish. A catfish is a jolly blundering sort of a fish, a regular Falstaff of the ponds. It has a fat jowl, and a fat belly, which it is always trying to fill. Smooth and sleek, its skin is almost human in its delicacy. It wears a long mustache, with scattering whiskers of other sort. Meanwhile it always goes armed with a sword, three swords, and these it has always on hand, always ready for a struggle on land as well as in the water. The small boy often gets badly stuck on these poisoned daggers, but, as the fish knows how to set them by a muscular twist, the small boy learns how, by a like untwist, he may unset and leave them harmless.

The catfish lives in sluggish waters. It loves the millpond best of all, and it has no foolish dread of hooks when it goes forth to bite. Its mouth is wide. It swallows the hook, and very soon it is in the air, its white throat gasping in the untried element. Soon it joins its fellows on the forked stick, and even then, uncomfortable as it may find its new relations, it never loses sight of the humor of the occasion. Its large head and expansive forehead betoken a large mind. It is the only fish whose brain contains a Sylvian fissure, a piling up of tissue consequent on the abundance of gray matter. So it understands and makes no complaint. After it has dried in the sun for an hour, pour a little water over its gills, and it will wag its tail, and squeak with gratitude. And the best of all is, there are horned pouts enough to go around.

The female horned pout lays thousands of eggs, and when these hatch, she goes about near the shore with her school of little fishes, like a hen with myriad chicks. She should be respected and let alone, for on her success in rearing this breed of "bullying little rangers" depends the sport of the small boy of the future.

— "FISH STORIES," CHARLES FREDERICK HOLDER AND DAVID STARR JORDAN

State of New York Conservation Depar

The common sucker
Catostomus commersonnii

THE COMMON SUCKER

He who loves to peer down into the depths of still waters, often sees upon the sandy, muddy, or rocky bottom several long, wedge-shaped sticks lying at various angles one to another. But if he thrust down a real stick, behold, these inert, water-logged sticks move off deftly! And then he knows that they are suckers. He may drop a hook baited with a worm in front of the nose of one, and if he waits long enough before he pulls up he may catch this fish, not by its gills but by the pit of its stomach; for it not only swallows the hook completely but tries to digest it along with the worm. Its food is made up of soft-bodied insects and other small water creatures; it is also a mud eater and manages to make a digestive selection from the organic material of silt. For this latter reason it is not a desirable food fish, although its flesh varies in flavor with the locality where it is found. The suckers taken when the waters are cold, are tasty but somewhat more bony than most fishes, while those taken from warm waters are very inferior in flavor and often unpalatable.

Seen from above, the sucker is wedge-shaped, being widest at the eyes; seen from the side it has a flat lower surface and an ungracefully rounded contour above, which tapers only slightly toward the tail. The profile of the face gives the imp
sion of a Roman nose. The young sp
mens have an irregular scale-mosaic
tern of olive-green blotches on a p
ground color, while the old ones are q
brown above and on the sides. The s
ers differ from most other fishes in ha
the markings of the back extend do
the sides almost to the belly. This
help in concealing the fish, since its s
show from above quite as distinctly as
back because of its peculiar form. T
scales are rather large and are noticea
larger behind than in the region of
head. Like other fish it is white below

The dorsal fin is placed about mid
the length of the fish as measured fr
nose to tail. The tail is long and str
and deeply notched; the anal fin exte
back to where the tail begins. The ven
fins are small and are directly opposite
hind half of the dorsal fin. The pecto
are not large but are strong and are pla
low down. The sucker has not a lav
equipment of fins, but its tail is str
and it can swim swiftly; it is also very
citable; in its efforts to escape, it will ju
from the aquarium more successfully th
any other fish. When resting on the b
tom, it is supported by its extended p
toral and ventral fins, which are strong
though not large.

e eyes are fairly large but the iris is
ainy; they are placed so that the fish
asily see above it as well as at the
the eyes move so as to look up or
and are very well adapted to serve
that lives upon the bottom. The
ls are divided, the partition project-
ntil it seems a tubercle on the face.
mouth opens below and looks like
uckered opening of a bag. The lips
hick but are very sensitive; it is by
cting these lips, in a way that re-
s one of a very short elephant's
, that it is enabled to reach and find
ood in the mud or gravel; so al-
gh the sucker's mouth is not a beauti-
ature, it is doubly useful. The sucker
he habit of remaining motionless for
periods of time. It breathes very
ly and appears sluggish; it never seizes
od with any spirit but simply slowly
lfs it; and for this reason it is consid-
poor game. It is only in the spring
n they may be speared through the ice
there is any fun in catching suckers;
at this season of the year that they
e upstream to shallow riffles to
n. Even so lowly a creature as the
er seems to respond to influences of
springtime, for at that period the
has a faint rosy stripe along his sides.
the winter these fish retire to the
hs of the rivers or ponds.

here are many species of suckers and
vary in size from six inches to three
in length. They inhabit all sorts of
ers, but they do not like a strong cur-
and are, therefore, found in still
s. The common sucker (*Catostomus*
mersonii), which is the subject of this
on, sometimes attains the length of
nty-two inches and the weight of five
nds. The ones under observation were
ut eight inches long, and proved to be
acrobats of the aquarium, since they
e likely at any moment to jump out;
eral times I found one on the floor.

LESSON 38

The Common Sucker

Leading Thought — The sucker is especially adapted by shape for lying on the bottom of ponds under still water where its food is abundant.

Method — If still-water pools along rivers or lakesides are accessible, it is far more interesting to study a sucker in its native haunts, as an introduction to the study of its form and colors when it is in the aquarium.

Observations — 1. Where do you find suckers? How do you catch them? Do they take the hook quickly? What is the natural food of the sucker?

2. What is the shape of this fish's body when seen from above? From the side? What is the color above? On the sides? Below? Does the sucker differ from most other fishes in the coloring along its sides? What is the reason for this? What do suckers look like on the bottom of the pond? Are they easily seen?

3. Describe or sketch a sucker, showing the position, size, and shape of the fins and tail. Are its scales large or small? How does it use its fins when at rest? When moving? Is it a strong swimmer? Is it a high jumper?

4. Describe the eyes; how are they especially adapted in position and in movement to the needs of a fish that lives on the bottom of streams and ponds?

5. Note the nostrils. Are they used for breathing?

6. Where is the mouth of the sucker situated? What is its form? How is it adapted to get food from the bottom of the stream and from crevices in the rocks?

7. Tell all you know about the habits of the suckers. When do you see them first in the spring? Where do they spend the winter? Where do they go to spawn? How large is the largest one you have ever seen? Why is their flesh sometimes considered poor in quality as food? Is there a difference in the flavor of their flesh depending upon the temperature of the water in which they live?

State of New York Conservation Depa

Common shiner or redfin
Notropis cornutus

THE SHINER

This is a noteworthy and characteristic lineament, or cipher, or hieroglyphi type of spring. You look into some clear, sandy bottomed brook where it spreads a deeper bay, yet flowing cold from ice and snow not far off, and see indistinctly p over the sand on invisible fins, the outlines of the shiner, scarcely to be distinguis from the sands behind it as if it were transparent. — THOREAU

There are many species of shiners and it is by no means easy to recognize them or to distinguish them from chub, dace, and other minnows, since all these belong to one family; they all have the same arrangement of fins and live in the same water; and the plan of this lesson can with few changes be applied to any of them.

Never were seen more exquisite colors than shimmer along the sides of the common shiner (*Notropis cornutus*). It is pale olive-green above, just a sunny brook-color; this is bordered at the sides by a line of iridescent blue-purple, while the shining silver scales on the sides below flash and glimmer with the changing hues of the rainbow. Most of the other minnows are darker than the shiners.

The body of the shiner is ideal for slipping through the water. Seen from above it is a narrow wedge, rounded in front and tapering to a point behind; from the side, it is long, oval, lance-shaped. The scales are large and beautiful, and the lateral line looks like a series of dots embroidered at the center of the diamond-shaped scales.

The dorsal fin is placed just back o center of the body and is not very l it is composed of soft rays, the first being stiff and unbranched. The ta long, large, graceful and deeply notc The anal fin is almost as large as the sal. The ventral pair is placed on the l side, opposite the dorsal fin; the pect are set at the lower margin of the b just behind the gill openings. The sh and its relatives use the pectoral fin aid in swimming, and keep them stantly in motion when moving thr the water. The ventrals are moved now and then and evidently help in k ing the balance. When the fish m rapidly forward, the dorsal fin is raise that its front edge stands at right angl the body and the ventral and anal fins expanded to their fullest extent. But w the fish is lounging, the dorsal, anal, ventral fins are more or less closed, though the tip of the dorsal fin sw with every movement of the fish.

The eyes are large, the pupils b very large and black; the iris is pale ye

hining; the whole eye is capable of
movement forward and back. The
il is divided by a little projecting par-
which looks like a tubercle. The
th is at the front of the head; to see
apabilities of this mouth, watch the
r yawn, if the water of the aquarium
mes stale. Poor fellow! He yawns just
do in the effort to get more oxygen.
e shiners are essentially brook fish
ugh they may be found in larger
es of water. They lead a precarious
ence, for the larger fish eat them in all
stages. They hold their own only by
g countless numbers of eggs. They
chiefly on water insects, algæ, and
eggs, including their own. They are
ty and graceful little creatures and
be seen swimming up the current in
middle of the brook. They often oc-
in schools or flocks, especially when
ng.

LESSON 39

The Shiner

eading Thought — The shiners are ong the most common of the little fish our small streams. They are beautiful orm and play an important part in the of our streams.

Method — Place in the aquarium shin- and as many as possible of the other cies of small fish found in our creeks d brooks. The aquarium should stand ere the pupils may see it often. The fol- ving questions may be asked, giving the ildren time for the work of observation.

Observations — 1. What is the shape the shiner's body when seen from ove? When seen from the side? Do you ink that its shape fits it for moving rap- y through the water?

2. What is the coloring above? On the les? Below?

3. Are the scales large and distinct, or ry small? Can you see the lateral line? here are the tiny holes which make this e placed in the scales?

4. Describe or sketch the fish, showing position, relative size, and shape of all the fins and the tail.

5. Describe the use and movements of each of the fins when the fish is swimming.

6. Describe the eyes. Do they move?

7. Describe the nostrils. Do you think each one is double?

8. Does the mouth open upward, downward, or forward? Have you ever seen the shiner yawn? Why does it yawn? Why do you yawn?

9. Where do you find the shiners living? Do they haunt the middle of the stream or the edges? Do you ever see them in flocks or schools?

MINNOWS

How silent comes the water round that bend;
Not the minutest whisper does it send
To the o'er-hanging sallows; blades of grass
Slowly across the chequer'd shadows pass,
Why, you might read two sonnets, ere they reach
To where the hurrying freshnesses aye preach
A natural sermon o'er their pebbly beds;
Where swarms of minnows show their little heads,
Staying their wavy bodies 'gainst the streams,
To taste the luxury of sunny beams
Tempered with coolness. How they ever wrestle
With their own sweet delight, and ever nestle
Their silver bellies on the pebbly sand!
If you but scantily hold out the hand,
That very instant not one will remain;
But turn your eye, and there they are again.
The ripples seem right glad to reach those cresses,
And cool themselves among the em'rald tresses;
The while they cool themselves, they freshness give,
And moisture, that the bowery green may live.

— John Keats

State of New York Conservation Depa

The brook trout
Salvelinus fontinalis

THE BROOK TROUT

Up and down the brook I ran, where beneath the banks so steep,
Lie the spotted trout asleep. — WHITTIER

But they were probably not asleep, as Mr. Whittier might have observed if he had cast a fly near one of them. There is in the very haunts of the trout a suggestion of where it gets its vigor and wariness: the cold, clear streams where the water is pure; brooks that wind in and out over rocky and pebbly beds, here shaded by trees and there dashing through the open — it makes us feel vigorous even to think of such streams. Under the overhanging bank or in the shade of some fallen log or shelving rock, the brook trout hides where he may see all that goes on in the world above and around him without being himself seen. Woe to the unfortunate insect that falls upon the surface of the water in his vicinity or even flies low over it, for the trout will easily jump far out of the water to seize its prey It is this habit of taking the insect upon and above the water's surface which has made trout fly-fishing the sport that it is. Man's ingenuity is fairly matched against the trout's cunning in this contest. I know of one old trout that has kept fishermen in the region around on the *qui vive* for years; and up to date he is still alive, making a dash now and then at a tempting bait, showing himself enough to tant his would-be captors with his sple size, but always retiring at the sight of line.

The brook trout varies much in c depending upon the soil and the rock the streams in which it lives. Its bac marbled with dark olive or black, r ing it just the color of shaded water. marbled coloration also marks the dc and the tail fins. The sides, which much in color, are marked with beau vermilion spots, each placed in the ce of a larger, bluish spot. In some insta the lower surface is reddish, in ot whitish. All the fins on the lower of the body have the front edges cre or yellowish white, with a darker st behind.

The trout's head is quite large somewhat blunt. The large eye is a li in front of the middle of the head. dorsal fin is at about the middle of body, and when raised is squarish in c line. Behind the dorsal fin and near tail is the little, fleshy adipose fin, so ca because its tissue is more or less adip in nature. The tail is fan-shaped, sligl notched at the end and is large and str

anal fin is rather large, being shaped
n like the dorsal fin, only slightly
ler. The ventral fins are directly be-
the dorsal fin and a little behind its
lle. The pectorals are low down, being
w and just behind the gill arches.
size the brook trout may reach four-
inches, but the majority of those
ht are seldom longer than seven or
t inches. It does not flourish in water
ch is warmer than 70° Fahrenheit,
prefers a temperature of about 50°
renheit. It must have the pure water
nountain streams and cannot endure
water of rivers which is polluted by
s or the refuse of cities. Where it has
ess to streams that flow into the ocean,
orms the salt-water habit, going out
sea and remaining there during the
ter. Such specimens become very
e.

'he trout can lay eggs when about six
nes in length. The eggs are laid from
tember until late November in most
ts of the United States. One small
ther trout lays from 400 to 600 eggs,
the large-sized ones lay more. The pe-
l of hatching depends upon the tem-
ature of the water. In depositing their
s the trout seek water with a gravelly
tom, often where some spring enters
o a stream. The nest is shaped by the
of the fish, the larger stones being car-
d away in the mouth. To make the pre-

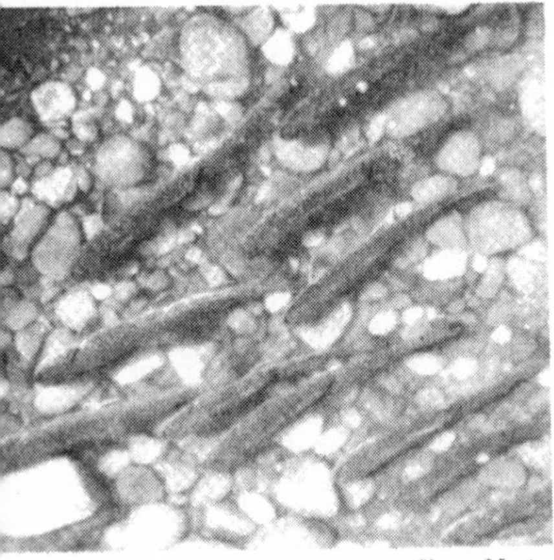

Verne Morton

'hen resting on a stream bed trout face into the current

Where the trout live

cious eggs secure they are covered with gravel.

Strict laws have been enacted by almost all of our states to protect the brook trout and preserve it in our streams. While it is true that brook trout spawn when five to six inches in length, the legal size in most states is six to seven inches; this gives them a chance to spawn at least about once before being caught. It is the duty of every decent citizen to abide by these laws and to see to it that his neighbors observe them. The teacher cannot emphasize enough to the child the moral value of being law-abiding. There should be in every school in the Union children's clubs which should have for their purpose civic honesty and the enforcement of laws which affect the city, village, or township.

Almost any stream with suitable water may be stocked with trout from the national or the state hatcheries, but what is the use of this expense if the game laws are not observed and these fish are caught before they reach maturity, as is so often the case?

LESSON 40

The Brook Trout

Leading Thought — The brook trout have been exterminated in many streams in our country largely because the game laws were passed too late to save them; and because of misuse of our waters. The trout is one of the most cunning and beautiful of our common fishes and the most delicious for food. Many mountain streams in our country could be well stocked with brook trout.

Method — For this lesson secure a trout from a fisherman at the opening of trout season. In some states, a permit is required before a trout may be legally kept in captivity, unless it is a legally captured specimen and is kept only during fishing season.

Observations — 1. In what streams are the brook trout found? Must the water be warm or cold? Can the trout live in impure water? Can it live in salt water?

2. Do the trout swim about in schools or do they live solitary? Where do they like to hide?

3. With what kind of bait are trout caught? Why do they afford such excellent sport for fly-fishing? Can you tell what the food of the trout is?

4. What is the color of the trout above? What colors along its sides? What markings make the fish so beautiful? What is its color below? Has the trout scales? Do you see the lateral line?

5. What is the general shape of the brook trout? Describe the shape, position, and color of the dorsal fin. Describe the little fin behind the dorsal. Why is it unlike the other fins? What is the shape of the tail fin? Is it rounded, square, or crescent-shaped across the end? What is the position and size of the anal fin compared with the dorsal? What are the colors on the ventral fins and where are these fins placed in relation to the dorsal fin? W color are the pectoral fins and how they placed in relation to the gill arc

6. Describe the trout's eyes. Do think the trout is keen-sighted?

7. When and where are the eggs l Describe how the nest is made. How the eggs covered and protected?

8. Could a trout live in the stream your neighborhood? Can you get state in stocking the streams?

9. What are the game laws concer trout fishing? When is the open seas How long must the trout be to be ta legally? If you are a good citizen what you do about the game laws?

10. Write a story telling all you k about the wariness, cunning, and stren of the brook trout.

TROUT

It is well for anglers not to make tr of all fishes, the prime objective of a d sport, as no more uncertain game loves sunlight. Today he is yours for the v asking; tomorrow, the most luscious l will not tempt him. One hour he de you, the next, gazes at you from some sconcement of the fishes, and knows y not, as you pass him, casting, by.

I believe I accumulated some of t angling wisdom years ago, in a cert trout domain in New England, wh there were streams and pools, ripples, c cades and drooping trees; where eve thing was fair and promising to the e for trout; but it required superhuman tience to lure them, and many a day scored a blank. Yet on these very days wh lures were unavailing, the creel emp save for fern leaves, I found they we not for naught; that the real fishing d was a composite of the weather, the win even if it was from the east, the splend colors of forest trees, the blue tourmali of the sky that topped the stream amid t trees, the flecks of cloud mirrored the surface. The delight of anticipatio the casting, the play of the rod, the ex cise of skill, the quick turns in the strea opening up new vistas, the little openin in the forest, through which were seen d neadows and nodding flowers — all went to make up the real trout fish- e actual catch being but an incident g many delights.

t how long one could be content mere scenery in lieu of trout, I am repared to say; if pushed to the wall, fess that when fishing I prefer trout nic effects. Still, it is a very imprac- e and delightful sentiment with truth to it, the moral being that the r should be resourceful, and not be ly cast down on the days when the is in the east.

m aware that this method of angling is not in vogue with some, and would be deemed fanciful, indeed inane, by many more; yet it is based upon a true and homely philosophy, not of today, the philosophy of patience and contentment. "How poor are they that have not patience," said Othello. It is well to be content with things as we find them, and it is well to go a-fishing, and not to catch fish alone, but every offering the day has to give. This should be an easy matter for the angler, as Walton tells us that Angling is somewhat like poetry; men are to be born so.

— "Fish Stories," Holder and Jordan

State of New York Conservation Department

Brook stickleback and nest
Eucalia inconstans

THE STICKLEBACK

'his is certainly the most sagacious of Lilliputian vertebrates; scarcely more n an inch in length when full-grown, azes at you with large, keen, shining- med eyes, takes your measure and darts with a flirt of the tail that says plainly, 'atch me if you can." The sticklebacks are delightful aquarium pets because their natural home is in still water sufficiently stagnant for algæ to grow luxuriously; thus we but seldom need to change the water in the aquarium, which, however, should be well stocked with water plants and have gravel at the bottom.

When the stickleback is not resting, he is always going somewhere and he knows just where he is going and what he is going to do, and earthquakes shall not deter him. He is the most dynamic creature in all creation, I think, except perhaps the dragon fly, and he is so ferocious that if he were as large as a shark he would destroy all other fishes. His ferocity is frightful to behold as he seizes his prey and shakes it as a terrier does a rat.

Well is this fish named stickleback, for along the ridge of its back are sharp, strong spines — five of them in our tiny brook species. These spines may be laid back flat or they may be erected stiffly, making an efficient saw which does great damage to fish many times larger than the stickleback. When we find the minnows in the aquarium losing their scales, we may be sure they are being raked off by this sawback; and if the shiner or sunfish undertakes to make a stickleback meal, there is only one way to do it, and that is to catch the quarry by the tail, since he is too alert to be caught in any other way. But swallowing a stickleback tail first is a dangerous performance, for the sharp spines rip open the throat or stomach of the captor. Dr. Jordan says that the sticklebacks of the Puget Sound region are called "salmon killers" and that they well earn the name; these fierce midgets unhesitatingly attack the salmon, biting off pieces of their fins and also destroying their spawn.

As seen from the side, the stickleback is slender and graceful, pointed like an arrow at the front end, and with the body behind the dorsal fin forming a long and slender pedicel to support the beautifully rounded tail fin. The dorsal fin is placed well back and is triangular in shape; the anal fin makes a similar triangle opposite it below and has a sharp spine at its front edge. The color of the body varies with the light; when the stickleback is floating among the water weeds, the back is greenish mottled with paler green, but when the fish is down on the gravel, it is much darker. The lateral line is marked by a silver stripe.

If large eyes count for beauty, then the stickleback deserves "the apple," f eyes are not only large but gemlike, broad iris of golden brown aroun black pupil. I am convinced tha stickleback has a keener vision than fish; it can move its eyes backwar forward rapidly and alertly. The m opens almost upward and is a w little mouth, in both appearance action.

When swimming, the stickleback about rapidly, its dorsal and anal fir tended, its spines all abristle, and it lashing the water with strong str When the fish wishes to lift itself thr the water, it seems to depend en upon its pectoral fins and these are used for balancing. Its favorite pos is hanging motionless among the weeds, with the tail and the dorsal ventral fins partially closed; it usually upon the pectoral fins which are br against some stem; in one case I saw ventrals and pectorals used togethe clasp a stem and hold the fish in plac moving backward the pectorals do work, with a little beckoning motio the tail occasionally. When resting u the bottom of the aquarium, it close fins and makes itself quite inconspicu It can dig with much power, accomp ing this by a comical auger-like motio plunges head first into the gravel and t by twisting the body and tail around around, it soon forms a hiding place.

But it is as house builder and fa and home protector that the stickle shines. In the early spring he builds hi nest made from the fine green algæ ca frog-spittle. This would seem too deli a material for the house construction, he is a clever builder. He fastens his fi walls to some stems of reed or grass, us as a platform a supporting stem; the o which I have especially studied were tened to grass stems. The stickleback a little cement plant of his own, suppo to be situated in the kidneys, which at t time of year secretes the glue for build purposes. The glue is waterproof. It spun out in fine threads or in filmy mas through an opening near the anal fin. C

es weights his platform with sand
h he scoops up from the bottom, but
not detect that our brook stickleback
this. In his case, home is his sphere
lly, for he builds a spherical house
t the size of a glass marble, three-
ters of an inch in diameter. It is a
w sphere; he cements the inside
so as to hold them back and give
, and he finishes his pretty structure
a circular door at the side. When fin-
l, the nest is like a bubble made of
ds of down, and yet it holds to-
er strongly.

the case of the best-known species,
male, as soon as he has finished his
er to his satisfaction, goes a-wooing;
lects some lady stickleback, and in his
way tells her of the beautiful nest he
made and convinces her of his ability
ke care of a family. He certainly has
ing ways, for he soon conducts her to
ome. She enters the nest through the
circular door, lays her eggs within it,
then, being a flighty creature, she
s responsibilities and flits off carefree.
follows her into the nest, scatters the
ilizing milt over the eggs, and then
ts off again and rolls his golden eyes on
e other lady stickleback and invites
also to his home. She comes without
jealousy because she was not first
ice; she also enters the nest and
her eggs and then swims off uncon-
edly. Again he enters the nest and
ps more milt upon the eggs and then
s forth again, a still energetic wooer.
here was ever a justified polygamist, he
ne, since it is only the cares and respon-
lities of the home that he desires. He
y stops wooing when his nest holds as
ny eggs as he feels equal to caring for.
now stands on guard by the door, and
h his winnowing pectoral fins sets up a
rent of water over the eggs; he drives off
intruders with the most vicious attacks,
keeps off many an enemy simply by a
play of reckless fury; thus he stands
rd until the eggs hatch and the tiny
le sticklebacks come out of the nest
l float off, attaching themselves by
ir mouths to the pond weeds until they become strong enough to scurry around in the water.

Some species arrange two doors in this spherical nest so that a current of water can flow through and over the eggs. Mr. Eugene Barker, who has made a special study of the little brook sticklebacks of the Cayuga Basin, has failed to find more than one door to their nests. Mr. Barker made a most interesting observation on this stickleback's obsession for fatherhood. He placed in the aquarium two nests, one of which was still guarded by its loyal builder, who allowed himself to be caught rather than desert his post; the little guardian soon discovered the unprotected nest and began to move the eggs from it to his own, carrying them carefully in his mouth. This addition made his own nest so full that the eggs persistently crowded out of the door, and he spent much of his time nudging them back with his snout. We saw this stickleback fill his mouth with algæ from the bottom of the aquarium and holding himself steady a short distance away, apparently blow the algæ at the nest from a distance of half an inch; we wondered if this was his method of laying on his building materials before he cemented them.

N. Y. State Conservation Dept.

Horned dace
Semotilus atromaculatus

The eggs of this species are white and shining like minute pearls, and seem to be fastened together in small packages with gelatinous matter. The mating habits of this species have not been thoroughly studied; therefore, here is an opportunity for investigation on the part of the boys and girls. The habits of other species of sticklebacks have been studied more than have those of the brook stickleback.

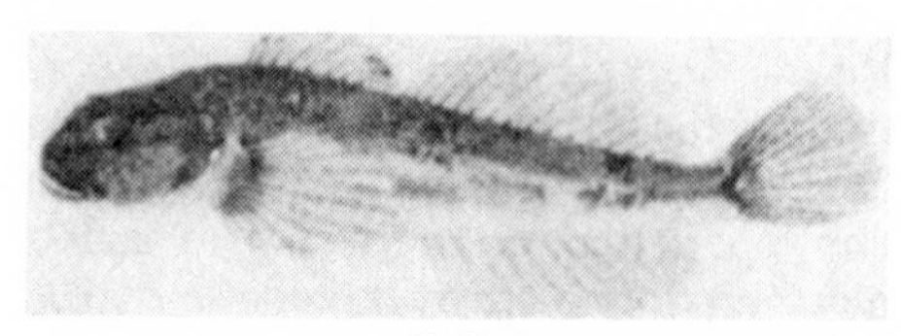

N. Y. State Conservation Dept.

A sculpin
Cottus cognatus

LESSON 41

The Stickleback

Leading Thought — The stickleback is the smallest of our common fish. It lives in stagnant water. The father stickleback builds his pretty nest of algæ and watches it very carefully.

Method — To find sticklebacks go to a pond of stagnant water which does not dry up during the year. If it is partly shaded by bushes, so much the better. Take a dip net and dip deeply; carefully examine all the little fish in the net by putting them in a Mason jar of water so that you can see what they are like. The stickleback is easily distinguished by the five spines along its back. If you collect these fish as early as the first of May and place several of them in the aquarium with plenty of the algæ known as frog-spittle and other water plants they may perhaps build a nest for you. They may be fed upon bits of meat or liver chopped very fine or upon earthworms cut into small sections.

Observations — 1. How did the stickleback gets its name? How many spines has it? Where are they situated? Are they always carried erect? How are these sp used as weapons? How do they act means of protection to the sticklebac

2. Describe or make a sketch sho the shape and position of the dorsal, anal, the ventral, and the pectoral What is the shape of the tail? What i general shape of the fish?

3. What is the color of the sti backs? Is the color always the same? W is the color and position of the lateral l

4. Describe the eyes. Are they larg small? Can they be moved? Do you th they can see far?

5. Describe the mouth. Does it c upward, straight ahead, or downwar

6. When the stickleback is swimm what are the positions and motions of dorsal, anal, tail, and pectoral fins? you see the ventral pair? Are they tended when the fish is swimming?

7. When resting among the pond w of the aquarium what fins does the st leback use for keeping afloat? How are other fins held? What fins does it us move backward? Which ones are u when it lifts itself from the bottom to top of the aquarium? How are its placed when it is at rest on the bottom

8. Drop a piece of earthworm or sc liver or fresh meat cut finely into aquarium and describe the action of sticklebacks as they eat it. How large full-grown stickleback?

9. In what kind of ponds do we f sticklebacks? Do you know how the st leback nest looks? Of what is it bu How is it supported? Is there one doo two? Does the father or mother stic back build the nest? Are the young in nest cared for? At what time is the n built?

THE SUNFISH

This little disc of gay color has won many popular names. It is called pumpkinseed, tobacco box, and sunfish because of its shape, and it is also called bream and pondfish. I have always wondered that it was not called chieftain also, for when it raises its dorsal fin with its saw crest spines, it looks like the headdress of Indian chief; and surely no warrior e had a greater enjoyment in a battle th does this indomitable little fish.

The sunfish lives in the eddies of

State of New York Conservation Department

Sunfish or pumpkinseed
Eupomotis gibbosus

ar brooks and ponds. It is a near rela- : of the rock bass and also of the black s and it has, according to its size, just ;amy qualities as the latter. I once had ınfish on my line which made me think ad caught a bass and I do not know ether I or the mad little pumpkinseed s the more disgusted when I discovered : truth. I threw him back in the water, t his fighting spirit was up and he bbed my hook again within five min- :s, which showed that he had more ırage than wisdom; it would have ved him right if I had fried him in a n, but I never could make up my mind kill a fish for the sake of one mouthful food.

Perhaps of all its names, "pumpkin- ed" is the most graphic, for it resembles is seed in the outlines of its body when en from the side. Looked at from above, has the shape of a powerful craft with ıooth, rounded nose and gently swelling d tapering sides; it is widest at the eyes d this is a canny arrangement, for these eat eyes turn alertly in every direction; and thus placed they are able to discern the enemy or the dinner coming from any quarter.

The dorsal fin is a most militant looking organ. It consists of ten spines, the hind one closely joined to the hind dorsal fin, which is supported by the soft rays. The three front spines rise successively, one above another, and all are united by the membrane, the upper edge of which is deeply toothed. The hind dorsal fin is gracefully rounded and the front and hind fin work independently of each other, the latter often winnowing the water when the former is laid flat. The tail is strong and has a notch in the end; the anal fin has three spines on its front edge and ten soft rays. Each ventral fin also has a spine at the front edge and is placed below and slightly behind the pectorals. The pecto- ral fins, I have often thought, are the most exquisite and gauzelike in texture of all the fins I have ever seen; they are kept al- most constantly in motion and move in such graceful flowing undulations that it is a joy to look at them.

The eye of the sunfish is very large and quite prominent; the large black pupil is surrounded by an iris that has shining lavender and bronze in it, but is more or less clouded above; the young ones have a pale silver iris. The eyes move in every direction and are eager and alert in their expression. The mouth is at the front of the body but it opens upward. The gill opening is prolonged backward at the upper corner, making an earlike flap; this, of course, has nothing to do with the fish's ears, but it is highly ornamental, as it is

Male sunfish guarding his nest

greenish-black in color, bordered by iridescent, pale green, with a brilliant orange spot on its hind edge. The colors of the sunfish are too varied for description and too beautiful to reduce to mere words. There are dark, dull, greenish or purplish cross-bands worked out in patterns of scale-mosaic, and between them are bands of pale, iridescent green, set with black-edged orange spots. But just as we have described his colors our sunfish darts off and all sorts of shimmering, shining blue, green and purple tints play over his body; and as he settles down into another corner of the aquarium, his colors seem much paler and we have to describe him over again. The body below is brassy yellow.

The beautiful colors which the male sunfish dons in spring, he puts at once to practical use. Professor Reighard says that when courting and trying to persuade his chosen one to come to his nest and there deposit her eggs, he faces her, wit gill covers puffed out, the scarlet or or spot on the ear-flap standing out bra and his black ventral fins spread wid show off their patent-leather finish. T does he display himself before her an timidate her; but he is rarely allowe do this in peace. Other males as bril as he arrive on the scene and he r forsooth stop parading before his lady in order to fight his rival, and he fi with as much display of color as he co In the sunfish duel, however, the par pants do not seek to destroy each other to intimidate each other. The vanquis one retires. Professor Gill says: "M while the male has selected a spot in shallow water near the shore, and ge ally in a mass of aquatic vegetation, too large or close together to entirely clude the light and heat of the sun, mostly under an overhanging plant. T choice is apt to be in some general s of shallow water close by the shore wh is favored by many others so that a n ber of similar nests may be found close gether, although never encroaching each other. Each fish slightly excava and makes a saucer-like basin in the cho area which is carefully cleared of all p bles. Such are removed by violent jerks the caudal fin or are taken up by mouth and carried to the circular bou ary of the nest. An area of fine, clean sa or gravel is generally the result, but infrequently, according to Dr. Reigha the nest bottom is composed of the ro lets of water plants. The nest has a dia eter of about twice the length of the fis

On the nest thus formed, the sunf belle is invited to deposit her eggs, wh as soon as laid fall to the bottom and come attached to the gravel at the bott of the nest by the viscid substance wh surrounds them. Her duty is then do and she departs, leaving the master charge of his home and the eggs. If tru be told, he is not a strict monogami Professor Reighard noticed one of the males which reared in one nest two broo laid at quite different times by two males. For about a week, depending up mperature, the male is absorbed in re of the eggs and defends his nest much ferocity; but after the eggs hatched he considers his duty done ets his progeny take care of them- as best they may.

nfish are easily taken care of in an ium, but each should be kept by elf, as they are likely to attack any er fish and are most uncomfortable bors. I have kept one of these beauti- himmering pumpkinseeds for nearly ar by feeding him every alternate with an earthworm; the unfortunate ns are kept stored in damp soil in an kettle during the winter. When I v one of them into the aquarium the sh would seize it and shake it as a er shakes a rat; but this was perhaps make sure of his hold. Once he at- pted to take a second worm directly the first; but it was a doubtful pro- ing, and the worm reappeared as often prima donna, waving each time a fren- farewell to the world.

LESSON 42

The Sunfish

eading Thought — The pumpkin- ls are very gamy little fishes which e the hook with much fierceness. They in the still waters of our streams or ponds and build nests in the spring, which the eggs are laid and which they end valiantly.

Method — The common pumpkinseed the jar aquarium is all that is neces- for this lesson. However, it will add ch to the interest of the lesson if the s who have fished for pumpkinseeds tell of their experiences. The chil- n should acquire from this lesson interest in nesting habits of the sun- es.

Observations — 1. Where are the sun- found? How do they act when they e the hook?

2. What is the general shape of the sunfish's body as seen from above? As seen from the side? Why is it called pump- kinseed?

3. Describe the dorsal fin. How many spines has it? How many soft rays? What is the difference in appearance between the front and hind dorsal fin? Do the two act together or separately? Describe the tail fin. Describe the anal fin. Has it any spines? If so, where are they? Where are the ventral fins in relation to the pec- torals? What is there peculiar about the appearance and movements of the pec- toral fins?

4. Describe the eye of the sunfish. Is it large or small? Is it placed so that the fish can see on each side? Does the eye move in all directions?

5. Describe the position of the mouth. In which direction does it open?

6. What is the color of the upper por- tion of the gill opening or operculum? What is the general color of the sunfish? Above? Below? Along the sides? What markings do you see?

7. Where does the sunfish make its nest? Does the father or mother sunfish make the nest? Does one or both protect it? Describe the nest.

8. How many names do you know for the sunfish? Describe the actions of your sunfish in the aquarium. How does he act when eating an earthworm?

The lamprey is not a fish at all, only a wicked imitation of one which can deceive nobody. But there are fishes which are un- questionably fish — fish from gills to tail, from head to fin, and of these the little sunfish may stand first. He comes up the brook in the spring, fresh as "coin just from the mint," finny arms and legs wide spread, his gills moving, his mouth open- ing and shutting rhythmically, his tail wide spread, and ready for any sudden motion for which his erratic little brain may give the order. The scales of the sun- fish shine with all sorts of scarlet, blue, green, and purple and golden colors. There is a black spot on his head which looks like an ear, and sometimes grows out in a long black flap, which makes the imitation still closer. There are many species of the sunfish, and there may be half a dozen of them in the same brook, but that makes no difference; for our purposes they are all one.

They lie poised in the water, with all fins spread, strutting like turkey-cocks, snapping at worms and little crustaceans and insects whose only business in the brook is that the fishes may eat them. When the time comes, the sunfish makes its nest in the fine gravel, building it with some care — for a fish. When the female has laid her eggs the male stands until the eggs are hatched. His sharp and snappish ways, and the bigness appearance when the fins are all disp keep the little fishes away. Sometim his zeal, he snaps at a hook baited w worm. He then makes a fierce fight the boy who holds the rod is sure th has a real fish this time. But whe sunfish is out of the water, strung willow rod, and dried in the sun, th sees that a very little fish can make a deal of a fuss.

— David Starr Jo

State of New York Conservation Depa

Johnny darter
Boleosoma nigrum

THE JOHNNY DARTER

We never tired of watching the little Johnny, or Tessellated darter (Boleosom grum); although our earliest aquarium friend, (and the very first specimens sho us by a rapid ascent of the river weed how "a Johnny could climb trees,") he has many resources which we have never learned. Whenever we try to catch him with hand we begin with all the uncertainty that characterized our first attempts, even i have him in a two-quart pail. We may know him by his short fins, his first dorsal ha but nine spines, and by the absence of all color save a soft, yellowish brown, whic freckled with darker markings. The dark brown on the sides is arranged in seven or e W-shaped marks, below which are a few flecks of the same color. Covering the side the back are the wavy markings and dark specks which have given the name of "Tessellated Darter"; but Boleosoma is a preferred name, and we even prefer "bo for short. In the spring the males have the head jet black; and this dark color often tends on the back part of the body, so that the fish looks as if he had been taken by tail and dipped into a bottle of ink. But with the end of the nuptial season this c disappears and the fish regains his normal, strawy hue.

His actions are rather bird-like; for he will strike attitudes like a tufted titmouse he flies rather than swims through the water. He will, with much perseverance, p his body between a plant and the sides of the aquarium and balance himself on a s *tem. Crouching catlike before a snail shell, he will snap off a horn which the un- owner pushes timidly out. But he is also less dainty and seizing the animal by the he dashes the shell against the glass or stones until he pulls the body out or s the shell.* — DAVID STARR JORDAN

e johnny darters are, with the stickle- s, the most amusing little fish in quarium. They are well called darters their movements are so rapid when are frightened that the eye can ely follow them; and there is some- g so irresistibly comical in their bright, y eyes, placed almost on top of the , that no one could help calling one em "Johnny." A "johnny" will look ou from one side, and then as quick flash, will flounce around and study with the other eye and then come ard you head-on so that he may take in with both eyes; he seems just as rested in the Johnny out of the jar the latter in the johnny within.

he johnny darter has a queerly shaped y for a fish, for the head and shoulders the larger part of him — not that he denly disappears into nothingness; by means! His body is long and very htly tapering to the tail; along his ral line he has a row of olive-brown s worked out in scale-mosaic; and he some other scale-mosaics also follow- a pattern of angular lines and making tches along his back. The whole upper t of his body is pale olive, which is a d imitation of the color of the brook.

The astonished and anxious look on the nny darter's face comes from the pe- iar position of the eyes, which are set the top of his forehead; they are big, t eyes, with large black pupils, sur- nded by a shining, pale yellow line at inner edge of the green iris; and as the pil is not set in the center of the eye, iris above being wider than below, result is an astonished look, as from sed eyebrows. The eyes move, often swiftly that it gives the impression of nking. The eyes, the short snout, and e wide mouth give johnny a decidedly glike aspect.

Although he is no frog, yet johnny rter seems to be in a fair way to de- velop something to walk upon. His pec- toral fins are large and strong and the ventral pair are situated very close to them; when he rests upon the gravel he supports himself upon one or both of these pairs of fins. He rests with the pec- toral fins outspread, the sharp points of the rays taking hold of the gravel like toenails and thus giving him the appear- ance of walking on his fins; if you poke him gently, you will find that he is very firmly planted on his fins so that you can turn him around as if he were on a pivot. He also uses the pectorals for swimming and jerks himself along with them in a way that makes one wonder if he could not swim well without any tail at all. The tail is large and almost straight across the end and is a most vigorous pusher. There are two dorsal fins. The front one has only spiny rays; when the fin is raised it appears almost semicircular in shape. The second dorsal fin is much longer, and when lifted stands higher than the front fin; its rays are all soft except the front one. As soon as the johnny stops swim- ming he shuts the front dorsal fin so that it can scarcely be detected; when he is frightened, his body lies motionless on the bottom; this act always reminds one of the "freezing" habit of the rabbit. But johnny does not stay scared very long; he lifts his head up inquisitively, stretching up as far as he is able on his feet, that is, his paired fins, in such a comical way that one can hardly realize he is a fish.

The tail and the dorsal fin of the johnny darter are marked with silver dots which give them an exquisite spun-glass look; they are as transparent as gauze.

The johnny darters live in clear, swift streams where they rest on the bottom, with the head upstream. Dr. Jordan has said they can climb up water weed with their paired fins. I have never observed them doing this but I have often seen one walk around the aquarium on his fins as if they were little fan-shaped feet; and when swimming he uses his fins as a bird uses its wings. There are many species of darters, some of them the most brilliantly colored of all our fresh-water fishes. The darters are perchlike in form.

Dr. Jordan says of the breeding habits of the darters: "On the bottom, among the stones, the female casts her spawn. Neither she nor the male pays any further attention to it, but in the breeding season the male is painted in colors as beautiful as those of the wood warblers. When you go to the brook in the spring you will find him there, and if you catch him and turn him over on his side you will see the colors that he shows to his mate, and which observation shows are most useful in frightening away his younger rivals. But do not hurt him. Put him back in the brook and let him paint its bottom with colors of a rainbow, a sunset or a garden of roses. All that can be done with blue, crimson and green pigments, in fish ornamentation, you will find in some brook in which the darters live."

LESSON 43

Johnny Darter

Leading Thought — The johnny darter naturally rests upon the bottom of the stream. It uses its two pairs of paired fins somewhat as feet in a way interesting to observe.

Method — Johnny darters may be caught in nets with other small fish and placed in the aquarium. Place one or two of them in individual aquaria where the pupils may observe them at their leisure. They do best in running water.

Observations — 1. Describe or sketch the johnny darter from above. From the side. Can you see the W-shaped marks
along its side? How is it colored abo
2. How are the pectoral fins pl
Are they large or small? How are
used in swimming? Where are the ve
fins placed? How are the ventrals
dorsals used together? When restin
the bottom how are the pectoral fins
3. What is there peculiar about
dorsal fins of the johnny darter? Whe
is resting, what is the attitude of the d
fins? What is the difference in sha
the rays of the front and hind dorsal
4. When resting on the bottom o
aquarium how is the body held? On
does it rest? In moving about the bo
slowly why does it seem to walk?
does it climb up water weed?
5. When frightened how does it
Why is it called a darter? What is
attitude of all the fins when the fis
moving swiftly?
6. What is the shape of the tail?
7. What is there peculiar about the
of the johnny? Describe the eyes and t
position. What is there in the life of
fish that makes this position of the
advantageous?
8. Where do we find the johnny
ers? In what part of the stream do t
live? Are they usually near the surfac
the water or at the bottom?

To my mind, the best of all subj
for nature-study is a brook. It affords s
ies of many kinds. It is near and dea
every child. It is an epitome of the na
in which we live. In miniature, it i
trates the forces which have shaped m
of the earth's surface. It reflects the
It is kissed by the sun. It is rippled by
wind. The minnows play in the po
The soft weeds grow in the shallows.
grass and the dandelions lie on its su
banks. The moss and the fern are s
tered in the nooks. It comes from
knows not whence; it flows to one kn
not whither. It awakens the desire to
plore. It is fraught with mysteries. It ty
fies the flood of life. It goes on forever

In other words, the reason why
brook is such a perfect nature-study s
ject is the fact that it is the central the

scene of life. Living things appeal to
lren.

ature-study not only educates, but
ucates nature-ward; and nature is ever
companion, whether we will or no.
ı though we are determined to shut
elves in an office, nature sends her
sengers. The light, the dark, the moon,
cloud, the rain, the wind, the falling
the fly, the bouquet, the bird, the
roach — they are all ours.

one is to be happy, he must be in
pathy with common things. He must
in harmony with his environment.
cannot be happy yonder nor to-
row: he is happy here and now, or
er. Our stock of knowledge of com-
ı things should be great. Few of us
travel. We must know the things at
ne.

ature-love tends toward naturalness,
toward simplicity of living. It tends
ntry-ward. One word from the fields
vorth two from the city. "God made
country."

expect, therefore, that much good will come from nature-study. It ought to revolutionize the school life, for it is capable of putting new force and enthusiasm into the school and the child. It is new, and therefore, is called a fad. A movement is a fad until it succeeds. We shall learn much, and shall outgrow some of our present notions, but nature-study has come to stay. It is in much the same stage of development that manual-training and kindergarten work were twenty-five years ago. We must take care that it does not crystallize into science-teaching on the one hand, nor fall into mere sentimentalism on the other.

I would again emphasize the importance of obtaining our fact before we let loose the imagination, for on this point will largely turn the results — the failure or the success of the experiment. We must not allow our fancy to run away with us. If we hitch our wagon to a star, we must ride with mind and soul and body all alert. When we ride in such a wagon, we must not forget to put in the tail-board.

— "The Nature-Study Idea,"
L. H. Bailey

AMPHIBIANS

Especially during early spring, one is likely to see many frogs, toads, and salamanders about ponds and other shallow water. These animals are harmless creatures; they do not bite and their chief method of defense is to escape to some place of concealment.

While there are exceptions to the general rule, and great variations in the life habits of these animals, it may be said that they are fitted to spend certain periods of their lives on land and other periods in water. In general, the immature stages are passed in or quite near water and the young are commonly called tadpoles. Of course, this means that the males and females of most species must return each year to the ponds, streams, or pools for the purpose of mating. Eggs are laid at once and usually hatch within a few days; the length of time varies according to the species and the weather conditions.

To this entire group of cold-bloode imals the term *amphibian* is applied; term was selected because it really m "double life" — these animals live pa their lives on land and part in or q near water. The presence or absence tail, during adult life, divides the amp ans into two more or less natural gro *the tailed and the tailless amphibian*

THE TAILLESS AMPHIBIANS

This group includes the frogs and toads. In attaining the adult stage these animals lose their tadpole tails; but we do not mean that the tail drops from the body; rather let us say that it is absorbed the body before the animal reaches t adult stage.

THE COMMON TOAD

The toad hopped by us with jolting springs. — Akers

Whoever has not had a pet toad has missed a most entertaining experience. Toad actions are surprisingly interesting; one of my safeguards against the blues is the memory of the thoughtful way one of my pet toads rubbed and patted its stomach with its little hands after it had swallowed a June bug. Toads do not make warts upon attacking hands, neither do they rain down nor are they found in the bedrock of quarries; but they do have most interesting history of their ow which is not at all legendary, and whi is very like a life with two incarnatio

Tadpoles

The mother toad lays her eggs in M and June in ponds, or in the still poo along streams; the eggs are laid in lo strings of jelly-like substance, and a ed upon the pond bottom or at- l to water weeds; when first depos- he jelly is transparent and the little eggs can be plainly seen; but after or two, bits of dirt accumulate upon lly, obscuring the eggs. At first the are spherical, like tiny black pills; they begin to develop, they elongate nally the tadpoles may be seen wrig- in the jelly mass, which affords them nt protection. After four or five the tadpoles usually work their way nd swim away; at this stage, the only o detect the head is by the direction e tadpole's progress, since it naturally head first. However, the head soon mes decidedly larger, although at first not provided with a mouth; it has, ad, a V-shaped elevation where the th should be, which forms a sucker ting a sticky substance. By means of substance the tadpole attaches itself ater weeds, resting head up. When adpoles are two or three days old, we detect little tassels on either side of

e toad in various stages of development from the egg to the adult

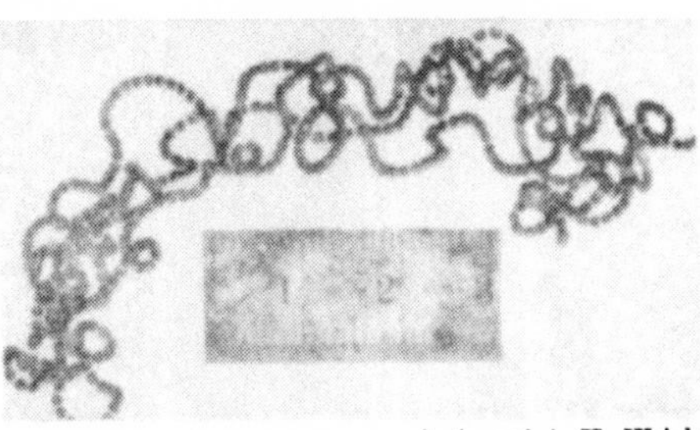

A. A. and A. H. Wright

Eggs of the spadefoot toad, Bufo compactilis. *Some toads lay as many as 8,000 eggs in a season*

the throat, which are the gills by which the little creature breathes; the blood passes through these gills, and is purified by coming in contact with the air which is mixed in the water. About ten days later, these gills disappear beneath a membrane which grows down over them; but they are still used for breathing, simply having changed position from the outside to the inside of the throat. The water enters the nostrils to the mouth, passes through an opening in the throat and flows over the gills and out through a little opening at the left side of the body; this opening or breathing-pore can be easily seen in the larger tadpoles; and when the left arm develops, it is pushed out through this convenient orifice.

When about ten days old, the tadpole has developed a small, round mouth which is constantly in search of something to eat, and at the same time is constantly opening and shutting to take in air for the gills; the mouth is provided with horny jaws for biting off pieces of plants. As the tadpole develops, its mouth gets larger and wider and extends back beneath the eyes, with a truly toadlike expansiveness.

At first, the tadpole's eyes are even with the surface of the head and can scarcely be seen, but later they become more prominent and bulge like the eyes of the adult toad.

The tail of the tadpole is long and flat, surrounded by a fin, and so is an organ for swimming. It strikes the water, first this side and then that, making most graceful curves, which seem to originate

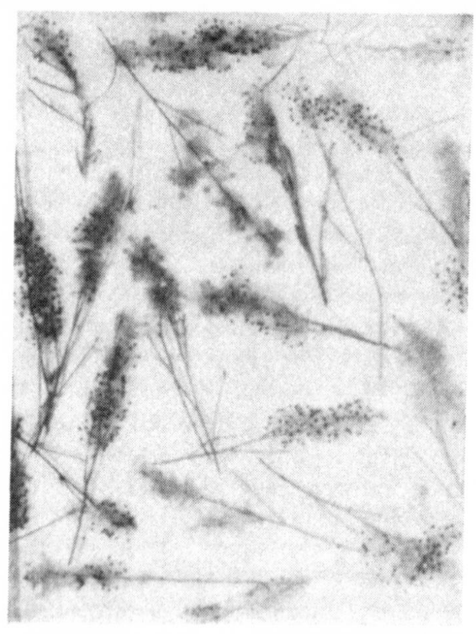

A. A. and A. H. Wright

Eggs of Hammond's spadefoot, Scaphiopus hammondii. *Although it looks so like our common toad, the spadefoot belongs to a different genus; it lays its eggs in cylindrical masses on submerged twigs or grass*

near the body and multiply toward the tip of the tail. This movement propels the tadpole forward, or in any direction. The tail is very thin when seen from above; and it is amusing to look at a tadpole from above, and then at the side; it is like squaring a circle.

There is a superstition that tadpoles eat their tails; and in a sense this is true, because the material that is in the tail is absorbed into the growing body; but the last thing a right-minded tadpole would do would be to bite off its own tail. However, if some other tadpole should bite off the tail or a growing leg, these organs conveniently grow anew.

When the tadpole is a month or two old, depending upon the species, its hind legs begin to show; they first appear as mere buds which finally push out completely. The feet are long and are provided with five toes, of which the fourth is the longest; the toes are webbed so that they may be used to help in swimming. Two weeks later the arms begin to appear, the left one pushing out throug breathing-pore. The " hands " hav fingers and are not webbed; they ar in the water for balancing, while th legs are used for pushing, as the t comes smaller.

As the tadpole grows older, not does its tail become shorter but its a change. It now comes often to the s of the water in order to get more a its gills, although it lacks the frog pole's nice adjustment of the gr lungs and the disappearing gills. At some fine rainy day, the little cre feels that it is finally fitted to live th of a land animal. It may not be a half in length, with big head, attenuated and stumpy tail, but it swims to the s lifts itself on its front legs, whic scarcely larger than pins, and walk toeing in, with a very grownup air; at this moment the tadpole attains ship. Numbers of tadpoles come o the water together, hopping hither thither with all of the eagerness and of untried youth. It is through iss thus in hordes from the water that gain the reputation of being rained d when they really were rained up. It is impossible for a beginner to detect difference between the toad and the tadpole; usually those of the toads black, while those of the frogs are o wise colored, though this is not a variable distinction. The best way to tinguish the two is to get the eggs develop the two families separately.

The Adult Toad

The general color of the com American toad is extremely variable may be yellowish brown, with spot lighter color, and with reddish or ye warts. There are likely to be four irr lar spots of dark color along each sid the middle of the back, and the u parts are light-colored, often somew spotted. The throat of the male toa black and he is not so bright in color a the female. The warts upon the back glands, which secrete a substance d ble for the animal seeking toad din- This is especially true of the glands e elongated swellings above and just of the ear, which are called the pa- glands; these give forth a milky, poi- us substance when the toad is seized a enemy, although the snakes do not to mind it. Some people have an that the toad is slimy, but this is not the skin is perfectly dry. The toad cold to the hand because it is a -blooded animal, which means an ani- with blood the temperature of the ounding atmosphere; the blood of the n-blooded animal has a temperature ts own, which it maintains whether surrounding air is cold or hot.

he toad's face is well worth study; its are elevated and very pretty, the pupil g oval and the surrounding iris shin- like gold. The toad winks in a whole- fashion, the eyes being pulled down the head; the eyes are provided with itating lids, which rise from below, are similar to those found in birds. en a toad is sleeping, its eyes do not ge but are drawn in, so as to lie even h the surface of the head. The two

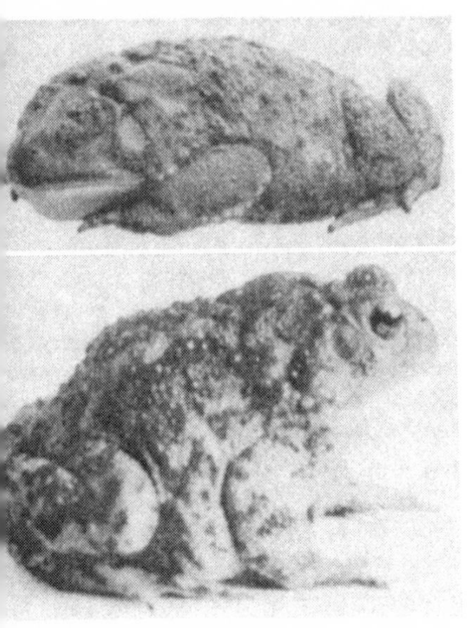

S. H. Gage

A common toad, Bufo americanus, *as he pears in winter sleep and after awakening the spring*

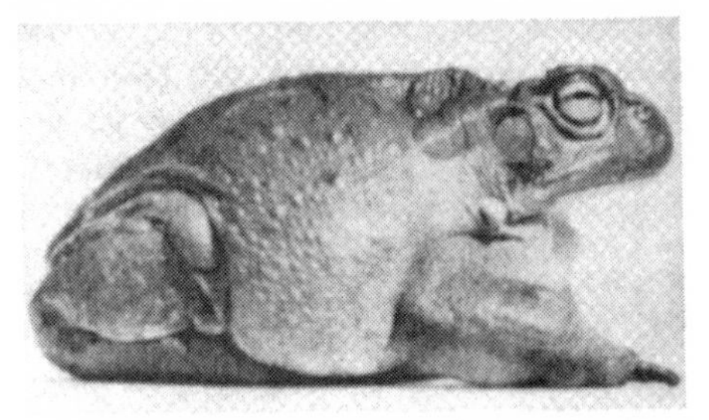

A. A. and A. H. Wright

The giant toad, Bufo alvarius. *This huge toad of the Southwest is from 3¼ to 6½ inches long. If molested it will secrete a fluid which is strong enough to paralyze a dog*

tiny nostrils are black and are easily seen; the ear is a flat, oval spot behind the eye and a little lower down; in the common species it is not quite so large as the eye; this is really the eardrum, since there is no external ear like ours. The toad's mouth is wide and its jaws are horny; it does not need teeth since it swallows its prey whole.

The toad is a jumper, as may be seen from its long, strong hind legs, the feet of which are also long and strong and are armed with five toes that are somewhat webbed. The "arms" are shorter and there are four "fingers" to each "hand"; when the toad is resting, its front feet toe-in in a comical fashion. If a toad is removed from an earth or moss garden, and put into a white wash-bowl, in a few hours it will change to a lighter hue, and vice versa. This is part of its protective color, making it inconspicuous to the eyes of its enemy. It prefers to live in cool, damp places, beneath sidewalks or porches, etc., and its warty upper surface resembles the surrounding earth. If it is disturbed, it will seek to escape by long leaps, and acts frightened; but if very much frightened, it flattens out on the ground, and looks so nearly like a clod of earth that it may escape even the keen eyes of its pursuer. When seized by the enemy, it will sometimes "play possum," acting as if it were dead; but when actually in the mouth of the foe, it emits terrified and heart-rending cries.

The toad's tongue is attached to the

lower jaw, at the front edge of the mouth; it can thus be thrust far out, and since it secretes a sticky substance over its surface, any insects which it touches adhere, and are drawn back into the mouth and swallowed. It takes a quick eye to see this tongue fly out and make its catch. The tadpole feeds mostly upon vegetable matter, but the toad lives entirely upon small animals, usually insects; it is not particular as to what kind of insects, but be-

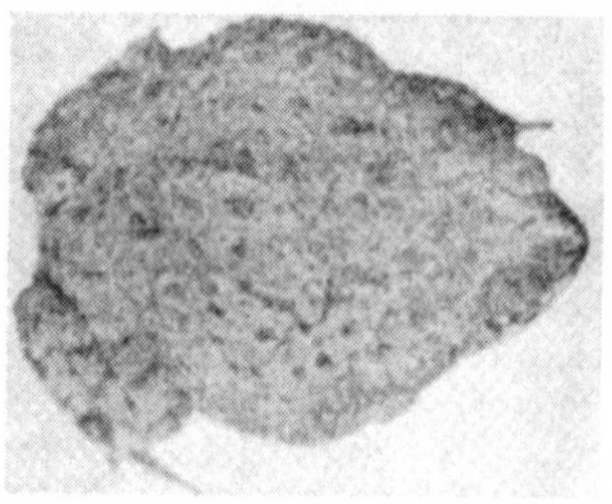

A. A. and A. H. Wright

The little green toad, Bufo debilis. *This small amphibian, resembling a lichen in appearance, is about 1½ inches long. It lives in grassy flat lands from Kansas and Colorado south into northern Mexico*

cause of the situations which it haunts, it usually feeds upon those which are injurious to grass and plants. Indeed, the toad is really the friend of the gardener and the farmer, and has been most ungratefully treated by those whom it has befriended. If you doubt that a toad is an animal of judgment, watch it when it finds an earthworm and set your doubts at rest! It will walk around the squirming worm, until it can seize it by the head, apparently knowing well that the horny hooks extending backward from the segments of the worm are likely to rasp the throat if swallowed the wrong way. If the worm prove too large a mouthful, the toad promptly uses its hands in an amusing fashion to stuff the wriggling morsel down its throat. When swallowing a large mouthful, it closes its eyes; but whether this aids the process, or is merely an expression of bliss, we have not determined. The toad never drinks by taking in water
through the mouth, but absorbs it thr
the skin; when it wishes to drin
stretches itself out in shallow water
thus satisfies its thirst; it will waste
and die in a short time, if kept in a
atmosphere.

The toad burrows in the earth
method of its own, hard to describ
kicks backward with its strong hind
and in some mysterious way, the e
soon covers all excepting its head; t
if an enemy comes along, back goes
head, the earth caves in around it,
where is your toad! It remains in its
row or hiding place usually during the
and comes out at night to feed. This h
is an advantage, because snakes are t
safely at home and, too, there are m
more insects to be found at night.
sagacious toads have discovered that
vicinity of street lights is swarming
insects, and there they gather in numb
In winter they burrow deeply in
ground and go to sleep, remaining
mant until the warmth of spring awak
them; then they come out, and the mo
toads seek their native ponds there to
eggs for the coming generation. They
excellent swimmers; when they are sw
ming rapidly, the front legs are laid b
ward along the sides of the body, so a
offer no resistance to the water; but wh
they are moving slowly, the front
are used for balancing and for keep
afloat.

The song of the toad is a pleasa
crooning sound, a sort of guttural trill
is made when the throat is puffed out
most globular, thus forming a vocal s
the sound is made by the air drawn in
the nostrils and passed back and fo
from the lungs to the mouth over
vocal chords, the puffed-out throat acti
as a resonator.

The toad has no ribs by which to infl
the chest, and thus draw air into the lun
as we do when we breathe; it is oblig
to swallow the air instead and thus fo
it into the lungs. This movement is sho
in the constant pulsation, in and out,
the membrane of the throat.

As the toad grows, it sheds its hor

which it swallows; as this process is lly done strictly in private, the ordi- observer sees it but seldom. One of toad's nice common qualities is its yment in having its back scratched ly.

he toad has many enemies; chief ng these is the snake and only less re crows and also birds of prey.

LESSON 44

The Tadpole Aquarium

Leading Thought — The children uld understand how to make the tad- es comfortable and thus be able to rear m.

Materials — A tin or agate pan, a deep thenware wash-bowl, a glass dish, or a le-mouthed glass jar.

Things to Be Done — 1. Go to some nd where tadpoles live.

2. Take some of the small stones on e bottom and at the sides of the pond, ting them very gently so as not to dis- rb what is growing on their surface. ace these stones on the bottom of the n, building up one side higher than the her, so that the water will be more shal- w on one side than on the other; a one or two should project above the ater.

3. Take some of the mud and leaves om the bottom of the pond, being care- l not to disturb them, and place upon e stones.

4. Take some of the plants found grow- g under water in the pond and plant em among the stones.

5. Carry the pan thus prepared back to the schoolhouse and place it where the sun will not shine directly upon it.

6. Bring a pail of water from the pond and pour it very gently in at one side of the pan, so as not to disarrange the plants; fill the pan nearly to the brim.

7. After the mud has settled and the water is perfectly clear, remove some of the tadpoles which have hatched in the glass aquarium and place them in the "pond." Not more than a dozen should be put in a pan of this size, since the amount of food and microscopic plants which are on the stones in the mud will afford food for only a few tadpoles.

8. Every week add a little more mud from the bottom of the pond or another stone covered with slime, which is probably some plant growth. More water from the pond should be added to replace that evaporated.

9. Care should be taken that the tadpole aquarium be kept where the sun will not shine directly upon it for any length of time, because if the water gets too warm the tadpoles will die.

10. Remove the "skin" from one side of a tulip leaf, so as to expose the pulp of the leaf, and give to the tadpoles every day or two. Bits of hard-boiled egg should be given now and then.

Toads' Eggs and Tadpoles

Leading Thought — The toads' eggs are laid in strings of jelly in ponds. The eggs hatch into tadpoles which are creatures of the water, breathing by gills, and swimming with a long fin. The tadpoles gradually change to toads, which are air-breathing creatures, fitted for life on dry land.

Method — The eggs of toads may be found in almost any pond about the first of May and may be scraped up from the bottom in a scoop-net. They should be placed in the aquarium where the children can watch the stages of development. Soon after they are hatched, a dozen or so should be selected and placed in the tadpole aquarium and the others put back into the stream. The children should ob-

F. Harper and A. A. Wright

Southern toad, Bufo terrestris. *When the male is croaking his throat is puffed out as in the picture. The color of the Southern toads varies from red or gray to black, and in size they range in length from 1½ inches to 3½ inches. They are found from North Carolina to Florida and west to the Mississippi River*

serve the tadpoles every day, watching carefully all the changes of structure and habit which take place. If properly fed, the tadpoles will be ready to leave the water in July as tiny toads.

OBSERVATIONS — 1. Where were the toads' eggs found and on what date? Were they attached to anything in the water or were they floating free? Are the eggs in long strings? Do you find any eggs laid in jelly-like masses? If so, what are they? How can you tell the eggs of toads from those of frogs?

2. Is the jelly-like substance in which the eggs are placed clear or discolored? What is the shape and the size of the eggs? A little later how do they look? Do the young tadpoles move about while they are still in the jelly mass?

3. Describe how the little tadpole works its way out from the jelly covering. Can you distinguish then which is head and which is tail? How does the tadpole act at first? Where and how does it rest?

4. Can you see with the aid of a lens the little fringes on each side of the neck? What are these? Do these fringes disappear a little later? Do they disappear on both sides of the neck at once? What becomes of them? How does the tadpole breathe? Can you see the little hole on the left side, through which the water used for breathing passes?

5. How does the tail look and hc it used? How long is it in proportio the body? Describe the act of swimn

6. Which pair of legs appears How do they look? When they get a larger are they used as a help in sv ming? Describe the hind legs and fe

7. How long after the hind legs ap before the front legs or arms appear? W happens to the breathing-pore when left arm is pushed through?

8. After both pairs of legs are develc what happens to the tail? What beco of it?

9. When the tadpole is very young you see its eyes? How do they look grows older? Do they ever bulge out toads' eyes?

10. As the tadpole gains its legs loses its tail how does it change in actions? How does it swim now? Do come oftener to the surface? Why?

11. Describe the difference betw the front and the hind legs and the fr and the hind feet on the fully grown pole. If the tail or a leg is bitten off some other creature will it grow again

LESSON 45

THE TOAD

LEADING THOUGHT — The toad is ored so that it resembles the soil and t often escapes the observation of its e mies. It lives in damp places and e insects, usually hunting them at night has powerful hind legs and is a vigor jumper.

METHOD — Make a moss garden in glass aquarium jar thus: Place some sto or gravel in the bottom of the jar a cover with moss. Cover the jar with a w screen. The moss should be deluged w water at least once a day and the jar sho be placed where the direct sunlight w not reach it. In this jar, place the toad study.

OBSERVATIONS — 1. Describe the ge eral color of the toad above and belo How does the toad's back look? Of wh use are the warts on its back?

2. Where is the toad usually foun

it feel warm or cold to the hand? Is y or dry? The toad is a cold-blooded l; what does this mean?

Describe the eyes and explain how situation is of special advantage to ad. Do you think it can see in front behind and above all at the same Does the bulge of the eyes help is? Note the shape and color of pupil and iris. How does the toad

Find and describe the nostrils. Find lescribe the ear. Note the swelling and just back of the ear. Do you the use of this?

What is the shape of the toad's h? Has it any teeth? Is the toad's ne attached to the front or the back of the mouth? How is it used to catch ts?

Describe the "arms and hands." many "fingers" on the "hand"? ch way do the fingers point when the is sitting down?

Describe the legs and feet. How y toes are there? What is the relative h of the toes and how are they coned? What is this web between the for? Why are the hind legs so much r than the front legs?

Will a toad change color if placed a different colored objects? How long it take it to do this? Of what advan- is this to the toad?

Where does the toad live? When disturbed how does it act? How far can it jump? If very frightened does it flatten out and lie still? Why is this?

10. At what time does the toad come out to hunt insects? How does it catch the insect? Does it swallow an earthworm head or tail first? When swallowing an earthworm or large insect, how does it use its hands? How does it act when swallowing a large mouthful?

11. How does the toad drink? Where does it remain during the day? Describe how it burrows into the earth.

12. What happens to the toad in the winter? What does it do in the spring? Is it a good swimmer? How does it use its legs in swimming?

13. How does the toad look when croaking? What sort of noise does it make?

14. Describe the action of the toad's throat when breathing. Did you ever see a toad shed its skin?

15. What are the toad's enemies? How does it act when caught by a snake? Does it make any noise? Is it swallowed head or tail first? What means has it of escaping or defending itself from its enemies?

16. How is the toad of great use to the farmer and gardener?

In the early years we are not to teach nature as science, we are not to teach it primarily for method or for drill: we are to teach it for loving — and this is nature-study. On these points I make no compromise.

—L. H. Bailey

THE SPRING PEEPER OR PICKERING'S HYLA

Ere yet the earliest warbler wakes, of coming spring to tell,
From every marsh a chorus breaks, a choir invisible,
As if the blossoms underground, a breath of utterance had found. — Tabb

ssociated with the first songs of robin bluebird, is the equally delightful rus of the spring peepers, yet how inquently do most of us see a member this invisible choir! There are some atures which are the quintessence of slang word "cute," which, interpreted, ans the perfection of Lilliputian proportions, permeated with undaunted spirit. The chickadee is one of these, and the spring peeper is another. I confess to a thrill of delight when the Pickering's hyla lifts itself on its tiny front feet, twists its head knowingly, and turns on me the full gaze of its bronze-rimmed eyes. This is one of the tiniest froglets of them all, being little more than an inch long when fully grown; it wears the Greek cross in

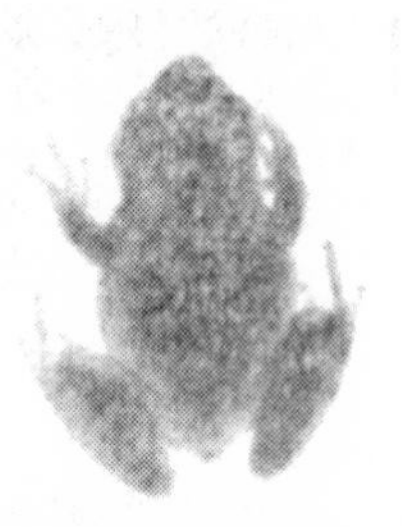

A. A. and A. H. Wright

The spring peeper, Hyla crucifer. *Here is shown the characteristic St. Andrew's cross on the peeper's back. This small frog, measuring ¾ inch to 1½ inches in length will be found from Manitoba to Maine and southward*

darker color upon its back, with some stripes across its long hind legs, which join the pattern on the back when the frog is " shut up," as the boys say.

The reason we see so little of spring peepers is that they are protected from discovery by their color. They have the chameleon power of changing color to match their background. This change can be effected in twenty minutes; the darker lines forming the cross change first, giving a mottled appearance which is at once protective. I have taken three of these peepers, all of them pale yellowish brown with gray markings, and have placed one upon a fern, one on dark soil, and one on the

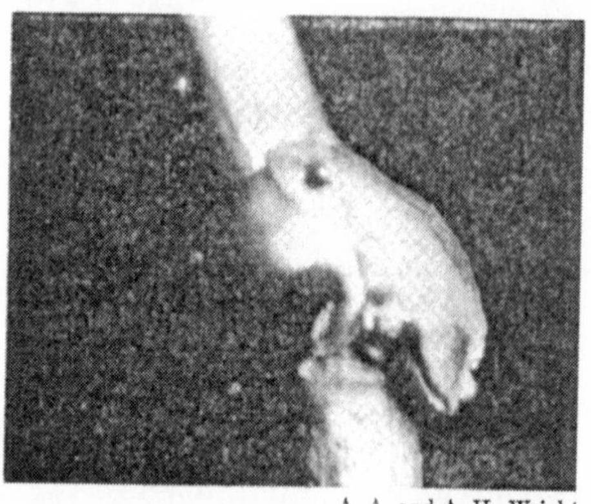

A. A. and A. H. Wright

The note of the male spring peeper is a shrill, clear call and while it is being given his throat expands into a large bubble

purple bud of a flower. Within ha hour, each matched its surroundin closely that the casual eye would detect them. The song of the Picke hyla is a resonant chirp, very stirring heard nearby; it sounds somewhat lik note of a water bird. How such a creature can make such a loud nois mystery. The process, however, ma watched at night by the light of a light or lantern, as none of the pe seem to pay any attention to an arti light; the thin membrane beneath throat swells out until it seems al

A. A. and A. H. W

The green tree frog, Hyla cinerea cin *These frogs, 1½ to 2½ inches long are b green in color with a straw-colored s along each side. On the tips of their toes discs which enable them to cling to ver surfaces. The green tree frogs are found Virginia to Texas and up the Mississ River to Illinois*

large enough to balloon the little c off his perch. No wonder that, with s a sounding-sac, the note is stirring.

The spring peepers have toes and fin ending in little round discs which sec at will a substance by means of wh they can cling to vertical surfaces, e to glass. In fact, the time to study th wonderful feet is when the frog is cli ing up the sides of the glass jar. T fingers are arranged as follows: two sh inside ones, a long one, and another sh outside. The hind feet have three er inside toes quite far apart, a long at the tip of the foot and a shorter outside. When climbing a smooth ce like glass, the toes are spread wide , and there are other little clinging on their lower sides, although not so as those at the tips. It is by means of : sticky, disclike toes that the animals themselves upon the tree trunks or r upright objects.

he whole body of the tree frog, a rela- of the spring peeper, is covered with : tubercles, which give it a roughened earance. The eyes are black with the of reddish color. The tongue is like of other frogs, hinged to the front of lower jaw; it is sticky and can be st far out to capture insects, of which tree frogs eat vast numbers.

he spring peepers breathe by the rapid ation of the membrane of the throat, ch makes the whole body tremble. : nostrils are two tiny holes on either : of the tip of the snout. The ears are tle below and just behind the eyes, and in the form of circular discs.

he eggs of the spring peepers are laid onds during April; each egg has a little be of jelly about it and is fastened to one or a water plant. The tadpoles are all and delicate; the under side of the ly is reddish and shines with metallic luster. These tadpoles differ from those of other frogs in that they often leave the water while the tail is still quite long. In summer, they may be found among the leaves and moss around the banks of ponds. They are indefatigable in hunting for gnats, mosquitoes, and ants; their destruction of mosquitoes, as pollywogs and as grown up frogs, renders them of great use to us. The voice of this peeper may be occasionally heard among the shrubs and vines or in trees during late summer and until November. The little creatures sleep beneath moss and leaves during the winter, waking to give us the earliest news of spring.

A. A. and A. H. Wright

Anderson tree frog, Hyla andersonii. *This is a small, beautiful, green frog with a light-bordered, plum-colored band along each side of its body. It lives chiefly in white cedar swamps from New Jersey to South Carolina*

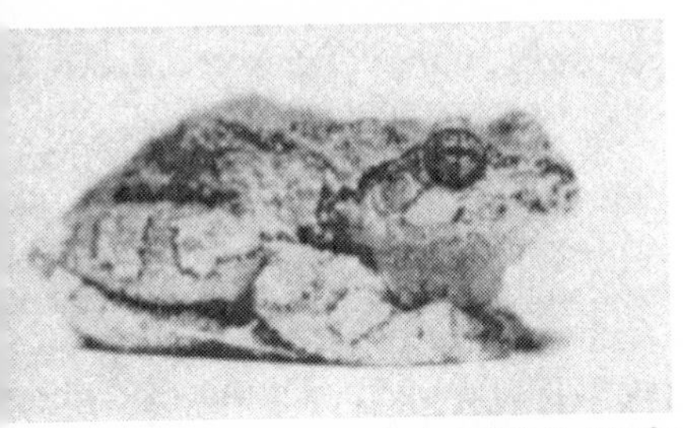

A. A. and A. H. Wright

Common tree toad, Hyla versicolor versi- lor. *From Maine and southern Canada to e Gulf states is the range of these tree ads; their habitat is trees, logs, or stone nces. The color varies from ashy gray to own or green; on the back is an irregular rk star. The eggs, in groups of thirty to rty, are attached to vegetation at the sur- ce of the water*

LESSON 46

Spring Peeper or Pickering's Hyla

Leading Thought — The prettiest part of the spring chorus of the frog ponds is sung by the spring peepers. These little frogs have the tips of their toes specially fitted for climbing up the sides of trees.

Method — Make a moss garden in an aquarium jar or a two-quart can. Place stones in the bottom and moss at one side, leaving a place on the other side for a tiny pond of water. In this garden place a spring peeper, cover the jar with mosquito netting, and place in the shade. The frogs may be found by searching the banks of a pond at night with a lantern. However, this lesson is usually given when by accident the spring peeper is discovered. Any species of tree frog will do; but the Pickering's hyla, known everywhere as the spring peeper, is the most interesting species to study.

Observations — 1. How large is the peeper? What is its color? Describe the markings.

2. Place the peeper on some light-colored surface like a piece of white blotting paper. Note if it changes color after a half hour. Later place it upon some dark surface. Note if it changes color again. How does this power of changing color benefit the animal? Place a peeper on a piece of bark. After a time does it become inconspicuous?

3. Describe the eyes. Note how little the creature turns its head to see anything behind it. Describe its actions if its attention is attracted to anything. What color is the pupil? The iris?

4. Note the movement of breathing. Where does this show the most? Examine the delicate membrane beneath throat. What has this to do with breathing?

5. What is the peeper's note? At time of day does it peep? At what of year? Describe how the frog looks peeping.

6. How does the peeper climb? Wh is climbing up a vertical surface stud toes. How many on the front foot? are they arranged? How many toes on hind foot? Sketch the front and hind How do the toe-discs look when pre against the glass? How does it manag make the discs cling and then let go? there any more discs on the under of the toes? Is there a web between toes of the hind feet? Of the front

7. Look at a peeper very closely describe its nostrils and its ears.

8. Are the peepers good jump What is the size and length of the legs as compared with the body?

9. When and where are the eggs of peeper laid? How do they look?

10. How do the peeper tadpoles d from other tadpoles? Describe then you have ever seen them. In what si tions do they live?

11. Of what use are the peepers to

THE FROG

The stroller along brooksides is likely to be surprised some day at seeing a bit of moss and earth suddenly make a long, high leap, without apparent provocation. An investigation resolves the clump of moss into a brilliantly green-spotted frog with two light-yellow raised stripes down his back; and then the stroller wonders how he could have overlooked such an obvious creature. But the leopard frog is only obvious when it is out of its environment. The common green frog is quite as well protected since its color is exactly that of green pools. Most frogs spend their lives in or about water, and if caught on land they make great leaps to reach their native element; the leopard frog and a few other species, howe sometimes wander far afield.

In form, the frog is more slim than toad, and is not covered with great wa it is cold and slippery to the touch. frog's only chance of escaping its enen is through the slipperiness of its body by making long, rapid leaps. As a jum the frog is much more powerful than toad because its hind legs are so m larger and more muscular, in compari with its size. The first toe in the fr foot of the male leopard frog is m swollen, making a fat thumb; the chanics of the hind legs make it possi for the frog to feather the webbed f as it swims. On the bottom of the toes

A. H. Wright

'he bullfrog, Rana catesbeiana. *This is our largest frog, sometimes attaining a length of* ɪches. *It is widely distributed east of the Rocky Mountains from Canada to Mexico. The* 'frog *has a greenish drab back and a yellowish underside. The eggs are laid in a film,* haps *2 feet square on the surface of still water. Its sonorous bass notes, jug-o'-rum, are* rd *in the evenings of early summer*

dened places at the joints, and some-es others besides, which give the foot rong hold when pushing for the jump. e toe tips, when they are pressed against glass, resemble slightly the peepers' cs. The hind foot is very long, while the front foot the toes radiate almost a circle. The foot and leg are colored the back of the body above, and on under side resemble the under parts.

The frog is likely to be much more ghtly colored than the toad, and usually much of green and yellow in its dress. t the frog lives among green things, ile it is to the toad's advantage to be color of the soil. Frogs also have the ameleon power of changing color to rmonize with their environment. I have n a very green leopard frog change to slate-gray when placed upon slate-col-ed rock. The change took place in the een portions. The common green frog ll likewise change to slate-color, in a nilar situation. A leopard frog changed ickly from dark green to pale olive, hen it was placed in the water after hav-g been on the soil.

The eyes of frogs are very prominent, and are beautiful when observed closely. The green frog has a dark bronze iris with a gleaming gold edge around the pupil, and around the outer margin. The eye of the leopard frog is darker; the iris seems to be black, with specks of ruddy gold scat-tered through it, and there is an outer band of red-gold around the margin. When the frog winks, the nictitating membrane rises from below and covers the whole eye; and when the frog makes a special effort of any sort, it has a comical way of drawing its eyes back into its head. When trying to hide at the bottom of the aquarium, the leopard species lets the eyelids fall over the eyes, so that they do not shine up and attract pursuers.

The ear is in a similar position to that of the toad, and in the bullfrog is larger than the eye. In the green frog, it is a dull grayish disc, almost as large as the eye. In the leopard frog, it is not so large as the eye, and may have a giltish spot at the center.

The nostrils are small and are closed when below the water, as may be easily seen by a lens. The mouth opens widely, the corners extending back under the eye.

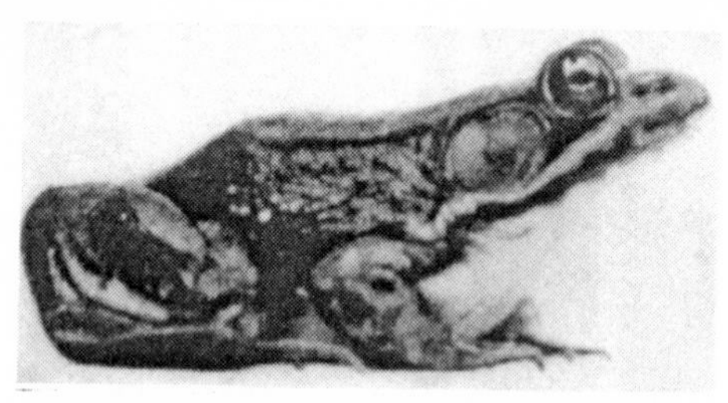

A. A. and A. H. Wright

Male green frog, Rana clamitans. *These inhabitants of deep and shallow ponds are found in eastern North America from Hudson Bay to the Gulf. In the North they are among the largest frogs, ranging from 2 to 4 inches in length. The female is shown in the following picture*

The jaws are horny and are armed with teeth, which are for the purpose of biting off food rather than for chewing it. When above water, the throat keeps up a rhythmic motion which is the process of breathing; but when below water this motion ceases. The food of frogs is largely composed of insects which frequent damp places or live in the water.

The sound-sacs of the leopard frogs, instead of being beneath the throat, as is the case with toads and peepers, are at the side of the throat; and when inflated may extend from just back of the eyes, out above the front legs and part way down the sides. The song is characteristic, and pleasant to listen to, if not too close by. Perhaps exception should be

A. A. and A. H. Wright

Female green frog, Rana clamitans. *The color of these frogs in general is greenish brown with a bright green mark from the eardrum forward along the jaw. Note that the eardrum of the male is larger than that of the female*

made to the lay of the bullfrog, which the song of some noted opera singe more wonderful than musical; the b of the bullfrog makes the earth f quake. If we seize the frog by the l leg, it will usually croak and thus den strate for us the position of its sound-

In addition to the snakes, the frogs l inveterate enemies in the herons, w frequent shallow water and eat then great numbers. The frogs hibernate mud and about ponds, burrowing d enough to escape freezing. In the spr they come up and sing their spring sc and the mother leopard frogs lay t eggs in masses of jelly on the bottom the pond, usually where the wate

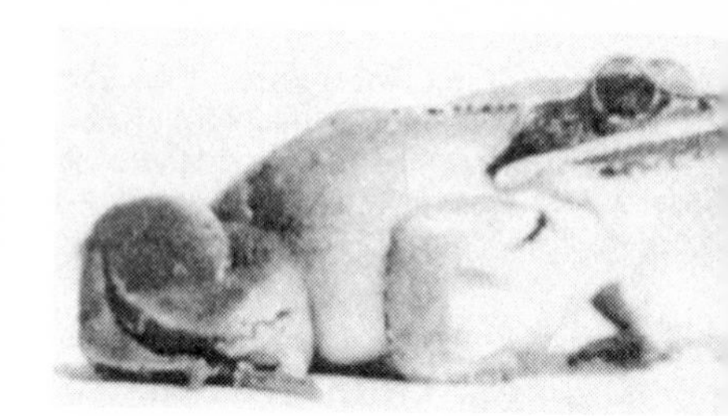

A. A. and A. H. Wr

Wood frog, Rana sylvatica. *In spring th frogs are found about ponds and tempor pools in wooded areas; at other times t are in the woods. They even hibernate un stumps, stones, or logs in or near woo Their color varies from tan to brown, prominent black mask covering the sides the head. They range from Quebec and No Scotia south to the Carolinas and westw to the plains*

deeper than in the situations where t toads' eggs are laid. The eggs of the tv can always be distinguished, since t toads' are laid in strings of jelly, while t leopard frogs' are laid in masses. The bu frog and green frog lay large films of eg on the surface of the water.

It is amusing to watch with a lens t frog tadpoles seeking for their microscop food along the glass of the aquariu There are horny upper and lower jaws, t latter being below and back of the forme The upper jaw moves back and for slightly and rhythmically, but the dro ping of the lower jaw opens the mout There are three rows of tiny black teet

the mouth and one row above; at
des and below these teeth are little,
-like fringes. Fringes, rows of teeth,
ws all work together, up and down,
nd in, in the process of breathing.
ostrils, although minute, are present
tadpole in its early stages. The pupil
e eye is almost circular and the iris
ally yellow or copper-bronze, with
mottling. The eyes do not wink or
draw. The breathing-pore, which is

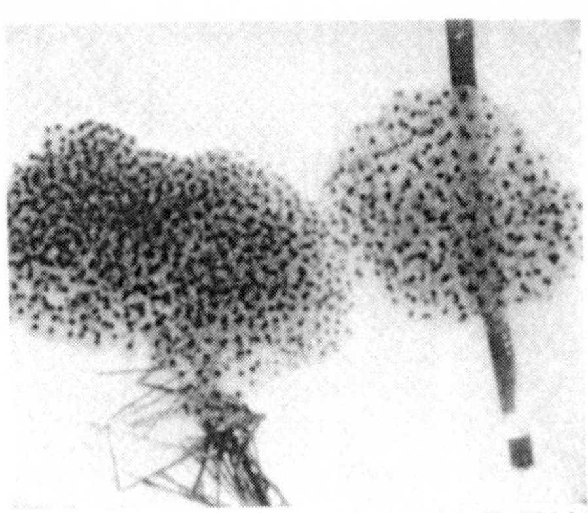

A. A. and A. H. Wright

Eggs of leopard frog, Rana pipiens pipiens, *and wood frog,* Rana sylvatica. *The eggs of the leopard frog are laid in a flattened sphere in waters of swampy marshes, overflows, and ponds. In summer, the adults are found in swampy areas, grassy woodlands, or even hay or grain fields. They range from the Pacific coast states into Mexico. The eggs of the wood frog are laid in round masses*

A. A. and A. H. Wright

uthern leopard frog, Rana sphenoceph- *The home of this frog is in swamps, over- ed areas, or ponds in the southeastern s and northward along the coast to New ey. The pointed snout, glistening white erside, and ridges extending backward each eye are characteristic*

the left side, is a hole in a slight pro-
erance.

t first, the tadpoles of the frogs and
ls are very much alike; but later most
he frog tadpoles are lighter in color,
ally being olive-green, mottled with
cks of black and white. The frog tad-
es usually remain much longer than
toads in the tadpole stage, and when
lly they change to adults, they are far
er in size than the toads are when
y attain their jumping legs.

SUGGESTED READING — Along the Brook,
Raymond T. Fuller; The Frog Book,
Mary C. Dickerson; Handbook of

LESSON 47

The Frog

LEADING THOUGHT — The frog lives near or in ponds or streams. It is a power-

A. A. and A. H. Wright

Wright's bullfrog, Rana heckscheri. *This is a transforming tadpole. Note that the left front leg has not yet pushed through the skin. The range of this frog is from South Carolina to Mississippi*

TAILLESS AMPHIBIANS

1 *and* 2. American Bell Toad, Ascaphus truei, *male and female. The size of this toad is* 1⅛ *to* 2 *inches. Note that the male is tailed.*

Range: *Northern California, Oregon, and Washington, and eastward into Montana.* Habitat: *Usually under rocks in small, cold mountain streams; in rainy seasons they may be found a short distance away from the water. They seem to be rather solitary in habit.*

3 *and* 4. Oak Toad, Bufo quercicus. *The adults of this pigmy toad range in size from* ¾ *to* 1¼ *inches. Its color varies from light brown to almost black. Note the expanded vocal sac of the male (No. 4); when deflated it is an apron fold under the throat. The call is a high whistle, which is more birdlike than froglike. A chorus of calls can be heard for more than an eighth of a mile.*

Range: *North Carolina to Florida, west to Louisiana.* Habitat: *Pine barrens.*

5. Narrow Mouth Toad, Microhyla carolinensis. *The size of these dark, smooth-skinned toads ranges from* ⅘ *to* 1⅖ *inches. The voice of the males resembles the bleating of sheep. The eggs are laid in a surface film, each egg being clearly outlined.*

Range: *From Virginia to Florida, westward to Texas.* Habitat: *In moist places under virtually any kind of cover, even haycocks and decaying logs.*

6. Canyon *or* Spotted Toad, Bufo punctatus. *This toad is* 1⅗ *to* 3 *inches in size; its color varies from greenish tan to red. The call is high pitched and birdlike. The eggs are laid singly in pools of intermittent streams. This toad breeds from April to July.*

Range: *South central Texas to Lower California and California.* Habitat: *Desert canyons.*

7. Great Plains Toad, Bufo cognatus. *These large-bodied, brown, gray, or greenish toads measure from* 1⅞ *to* 4 *inches. Their call is harsh and low pitched. The vocal sac is shaped like a sausage stood on end.*

Range: *Mostly west of the 100th me[...] from North Dakota southwestward to [...] and eastern California.* Habitat: *Grazin[...] in flood plains.*

8. Spadefoot Toad, Bufo compactili[...] *size of this desert toad is* 2 *to* 3½ *inch[...] color is pinkish drab. It breeds in po[...] even in cattle tanks. Note the expanded sa[...] like vocal sac of this male.*

Range: *Utah and Nevada eastward [...] lahoma and southward into Mexico.* Ha[...] *Deserts.*

9. Hammond's Spadefoot, Scaph[...] hammondii. *This toad ranges from* 1½ [...] *inches in size. It breeds in temporary [...] the tadpoles eat many mosquitoes, and the [...] eat many tadpoles. It is seldom seen above g[...] except during rains of long duration. Th[...] usual call is plaintive and catlike.*

Range: *From North Dakota southwa[...] Mexico, and westward to the Pacific* [...] Habitat: *Burrows, which it digs in [...] ground with its strong, spadelike feet, an[...] which it pushes itself by rocking its body.*

10. Canadian *or* Winnipeg Toad, [...] hemiophrys. *In size this toad ranges from [...] to* 3⅕ *inches. It has a very prominent h[...] horny boss between its eyes and on its s[...] It may breed in the shallows at the edges o[...] body of fresh water.*

Range: *North Dakota to Manitoba.* H[...] tat: *Lakes and stream valleys.*

11 *and* 12. Yosemite Toad, Bufo can[...] *male and female. This is the only toad i[...] United States that shows marked diffe[...] between male and female. The male (No. [...] is olive-colored, while the female (No.* 12 [...] *light gray with many black areas. Its si[...] from* 2 *to* 3 *inches.*

Range: *Yosemite National Park and [...] tral Sierra Nevada at altitudes of* 100[...] 1100 *feet.* Habitat: *Wet meadows and [...] gins of streams and lakes.*

Photographs by A. A. and A. H. Wright

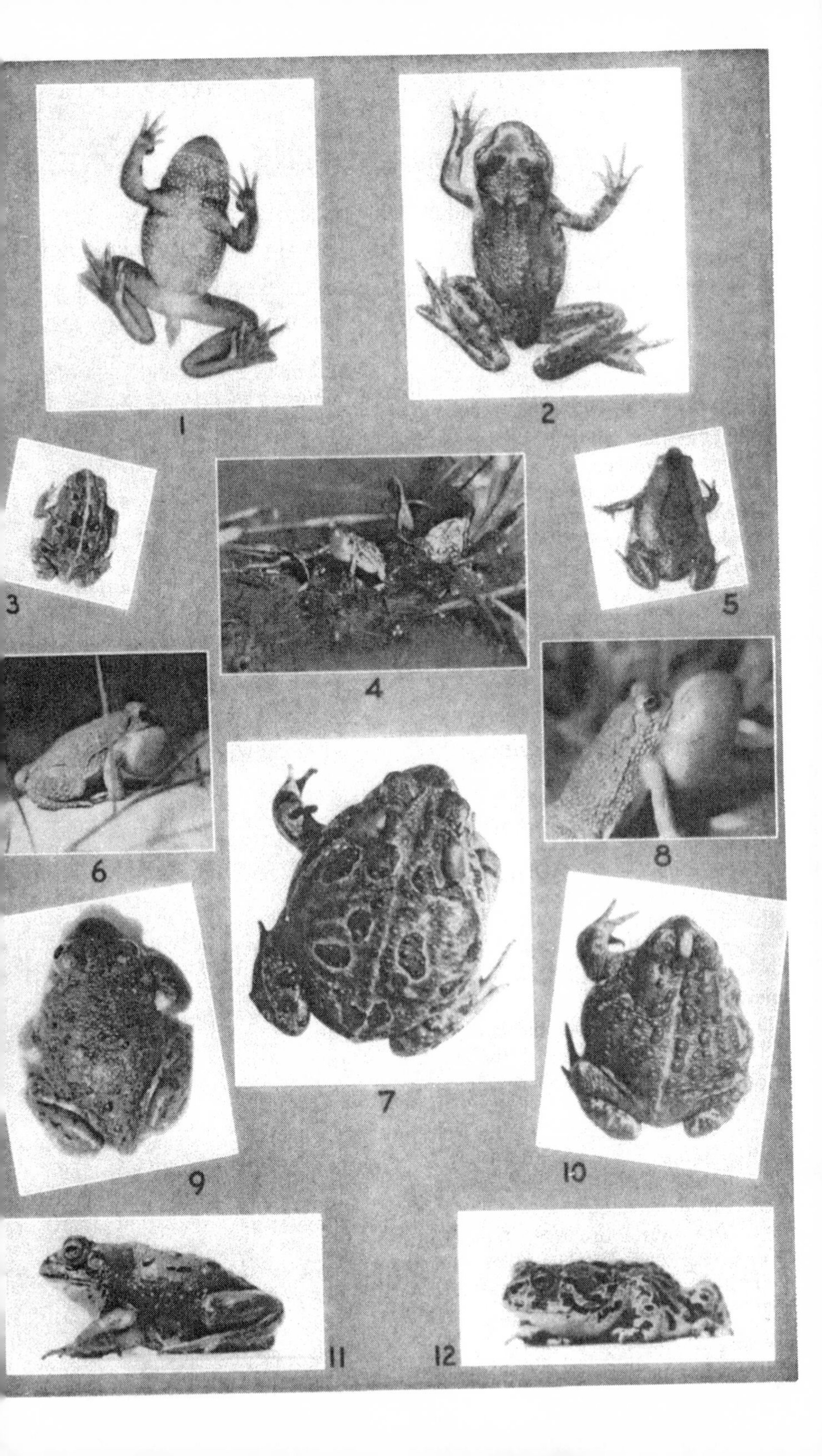
1
2
3
4
5
6
7
8
9
10
11
12

ful jumper and has a slippery body. Its eggs are laid in masses of jelly at the bottom of ponds.

METHOD — The frog may be studied in its native situation by the pupils or it may be brought to the school and placed in an aquarium; however, to make a frog aquarium there needs to be a stick or stone projecting above the water, for the frog likes to spend part of the time entirely out of water or only partially submerged.

OBSERVATIONS — 1. Where is the frog found? Does it live all its life in the water? When found on land how and where does it seek to escape?

2. Compare the form of the frog with that of the toad. Describe the frog's skin, its color and texture. Compare the skins of the two.

3. Describe the colors and markings of the frog on the upper and on the under side. How do these protect it from observation from above? From below? How do we usually discover that we are in the vicinity of a frog?

4. Describe the frog's ears, eyes, nostrils, and mouth.

5. Compare its "hands and feet" with those of the toad. Why the difference in the hind legs and feet?

6. How does the frog feel to your hand? Is it easy to hold him? How does this slipperiness of the frog benefit it?

7. On what does the frog feed? What feeds on it? How does it escape its enemies?

8. What sounds does the frog make? Where are the sound-sacs of the leopard frog located? How do they look when they are inflated?

9. Is the frog a good swimmer? Is it a better jumper than the toad? Why?

10. Where are the leopard frog's eggs laid? How do they look?

11. Can you tell the frog tadpoles from those of the toad? Which remains longer in the tadpole stage? Study the frog tadpoles, following the questions given in Lesson 44.

12. What happens to the frog in winter?

FESTINA LENTE

Once on a time there was a pool
Fringed all about with flag-leaves c
And spotted with cow-lilies garish,
Of frogs and pouts the ancient paris
Alders the creaking redwings sink on,
Tussocks that house blithe Bob o'
coln,
Hedged round the unassailed seclusio
Where muskrats piled their cells Ca
sian;
And many a moss-embroidered log,
The watering-place of summer frog,
Slept and decayed with patient skill,
As watering-places sometimes will.
Now in this Abbey of Theleme,
Which realized the fairest dream
That ever dozing bull-frog had,
Sunned, on a half-sunk lily pad,
There rose a party with a mission
To mend the polliwog's condition,
Who notified the selectmen
To call a meeting there and then.
"Some kind of steps," they said, "
needed;
They don't come on so fast as we did
Let's dock their tails; if that don't m
'em
Frogs by brevet, the Old One take 'em
That boy, that came the other day
To dig some flag-root down this way,
His jack-knife left, and 'tis a sign
That Heaven approves of our design:
'T were wicked not to urge the step
When Providence has sent the weapo
Old croakers, deacons of the mire,
That led the deep batrachian choir,
"Uk! Uk! Caronk!" with bass that m
Have left Lablache's out of sight,
Shook nobby heads, and said "No go!
You'd better let 'em try to grow:
Old Doctor Time is slow, but still
He does know how to make a pill."
But vain was all their hoarsest bass,
Their old experience out of place,
And spite of croaking and entreating
The vote was carried in marsh-meeting.
"Lord knows," protest the polliwogs,
"We're anxious to be grown-up frogs;
But don't push in to do the work
Of Nature till she prove a shirk;
'Tis not by jumps that she advances,

wins her way by circumstances;
wait awhile, until you know
'e so contrived as not to grow;
Nature take her own direction,
she'll absorb our imperfection;
mightn't like 'em to appear with,
we must have the things to steer with."
," piped the party of reform,
great results are ta'en by storm;
holds her best gifts till we show
ve strength to make her let them go;
Providence that works in history,
seems to some folks such a mystery,
s not creep slowly on, incog.,
moves by jumps, a mighty frog;
more reject the Age's chrism,
r queues are an anachronism;
more the future's promise mock,
lay your tails upon the block,
nkful that we the means have voted
To have you thus to frogs promoted."
The thing was done, the tails were cropped,
And home each philotadpole hopped,
In faith rewarded to exult,
And wait the beautiful result.
Too soon it came; our pool, so long
The theme of patriot bull-frog's song,
Next day was reeking, fit to smother,
With heads and tails that missed each other, —
Here snoutless tails, there tailless snouts;
The only gainers were the pouts.

MORAL

From lower to the higher next,
Not to the top is Nature's text;
And embryo Good, to reach full stature,
Absorbs the Evil in its nature.

— Lowell

THE TAILED AMPHIBIANS

'he best-known representatives of this up are the salamanders of various types. ring accidents, a salamander retains its throughout life. Salamanders resem- lizards in shape, and many people e incorrectly called them lizards. It not difficult to distinguish them, if bears in mind that the covering the salamander is rather soft and newhat moist, while that of the rd is rather dry and in the form of les.

The red-backed salamander lacks the phibian habits usual to the group; it es on land during its entire life. The s are laid in a small cluster, in a decay- log or stump; the adult is often to be nd quite near the egg cluster. On the other extreme, the mud puppies and hell-benders spend their entire lives in the water. They are rarely seen, live chiefly under rocks in stream beds, and feed chiefly at night.

The many local forms of amphibians offer excellent opportunities for interesting outdoor studies. Of the tailed amphibians, the newt is considered in detail, and pictures of other representative salamanders are shown.

THE NEWT OR EFT

One of the most commonly seen sala- anders is the newt or eft. After a rain spring or summer, we see these little ange-red creatures sprawling along roads or woodland paths, and since they are rarely seen except after rain, the wise people of old declared they rained down. which was an easy way of explaining their

A spotted salamander in natural surroundings

presence. But the newts do not rain down, they rain up instead, since if they have journeys to make they must needs go forth when the ground is damp; otherwise they would dry up and die. Thus, the newts make a practice of not going out except when the ground is rather moist. A closer view of the eft shows plenty of peculiarities in its appearance to interest us. Its colors are decidedly gay, the body color being orange, ornamented with vermilion dots along each side of the back, each red dot being usually margined with tiny black specks; but the eft is careless about these decorations and may have more spots on one side than on the other. Besides these vermilion dots, it is also adorned with black specks here and there, and especially along its sides looks as if it had been peppered. The newt's greatest beauty lies in its eyes; these are black, with elongated pupils, almost parallel with the length of the head, and bordered above and below with bands of golden, shining iris which give the eyes a fascinating brilliancy. The nostrils are mere pinholes in the end the snout.

The legs and feet look queerly ina quate for such a long body, since tl are short and far apart. There are fc toes on the front feet and five on t hind feet, the latter being decide pudgy. The legs are thinner where tl join the body and wider toward the fe The eft can move very rapidly with scant equipment of legs. It has a m leading way of remaining motionless a long time and then darting forward li a flash, its long body falling into grace curves as it moves. But it can go ve slowly when exploring; it then places little hands cautiously and lifts its head high as its short arms will allow, in ord to take observations. Although it can s quite well, yet on an unusual surface, li glass, it seems to feel the way by touc ing its lower lip to the surface as if to te it. The tail is flattened at the sides a is used to twine around objects in time need; and I am sure it is also used

the eft while crawling, for it curves
ray and that vigorously, as the feet
ess, and obviously pushes against the
d. Then, too, the tail is an aid when,
ne chance, the eft is turned over on
ck, for with its help it can right itself
ily. The eft's method of walking is
sting; it moves forward one front
nd then the hind foot on the other
after a stop for rest, it begins just
it left off when it again starts on.
autiful eyes seem to serve the newt
ndeed, for I find that, when it sees
ace approaching the moss jar, it
s promptly over to the other side.
e are no eyelids for the golden eyes,
he eft can pull them back into its
and close the slit after them, thus
ng them very safe.

e eft with whose acquaintance I was
favored was not yet mature and was
l of earthworms; but he was very fond
lant lice and it was fun to see the
creature stalking them. A big rose
louse would be squirming with satis-
on as it sucked the juice of the leaf,
the eft would catch sight of it and
me greatly excited, evidently holding
breath, since the pulsating throat
d become rigid. There was a particu-
alert attitude of the whole front part
e body and especially of the eyes and
head; then the neck would stretch
long and thin, and the orange snout
oach stealthily to within half an inch
e smug aphid. Then there would be a
as of lightning, something too swift
e coming out of the eft's mouth and
ping up the unsuspecting louse. Then

Red-spotted newt stalking plant lice

S. C. Bishop

Giant or California newt, Triturus torosus. *About ponds and streams from lower California to Alaska this newt may be seen; its body is stout and is about six inches long*

there would be a gulp or two and all would be over. If the aphid happened to be a big one, the eft made visible effort to swallow it. Sometimes his eftship would become greatly excited when he first saw the plant louse, and he would sneeze and snort in a very comical way, like a dog eager for game.

This is the history of this species as summarized from Mrs. S. H. Gage's charming *Story of Little Red Spot*. The egg is laid in some fresh-water pond or the still borders of some stream where there is a growth of water weed. The egg, which is about the size of a sweet pea seed, is fastened to a water plant. It is covered with a tough but translucent envelope, and has at the center a little yellowish globule. In a little less than a month the eft hatches, but it looks very different from the form with which we are most familiar. It has gray stripes upon its sides and three tiny bunches of red gills on each side, just back of its broad head. The keeled tail is long and very thin. The newt is an expert swimmer and breathes water as does a fish. After a time it becomes greenish above and buff below, and by the middle of August it develops legs and has changed its form so that it is able to live upon land; it no longer has gills; soon the coat changes to the bright orange hue which makes the little creature so conspicuous.

The newt usually keeps hidden among moss, or under leaves, or in decaying wood, or in other damp and shady places; but after a rain, when the whole world is damp, it feels confidence enough to go out in the open and hunt for food. For about two and a half years it lives upon land; then it returns to the water. When this

TAILED AMPHIBIANS

1 *and* 2. Spotted Salamander, Ambystoma maculatum. *The adults are 6 inches long or more; the body is glistening black with prominent yellow spots. These, like other salamanders, are entirely harmless; they neither bite nor scratch. Their egg-masses are deposited during early spring, while the water is still very cold, in swampy areas or stagnant pools, and are often attached to sticks or to submerged parts of plants. While the eggs are developing, a greenish color, caused by the presence of numerous algae, appears in the gelatin of the egg-mass. This seems to be peculiar to the egg-mass of this salamander, and biologists are trying to learn the reason for it.*

Range: *Locally in central North America from Wisconsin and Nova Scotia southward.* Habitat: *Damp dark places during most of the year. In spring they migrate to ponds to breed.*

3. Red Salamander, Pseudotriton ruber. *Adults are about 6 inches long; young adults are coral red with irregular black spots; older adults are somewhat purplish brown; the eggs, laid in autumn, are attached to the underside of a stone in a stream.*

Range: *Locally from New York to Georgia, westward to the Mississippi River.* Habitat: *Under flat stones in shallow water.*

4. Marbled Salamander, Ambystoma opacum. *Adults are about 5 inches long, bluish beneath and slaty gray on the back, with about 14 grayish-white bars. The creature is not likely to be mistaken for any other large salamander found within its range, because the others are marked with yellow.*

Range: *Eastern and central North America.* Habitat: *Under flat stones or in burrows in the soil.*

5. Mud Puppy, Necturus maculosus. *This animal, which looks like a huge salamander, has no scales, and its body is shiny. I not come out on land.*

Range: *Eastern and central United* Habitat: *Rivers and lakes.*

6. Tiger Salamander, Ambystom grinum. *This is a large, dark brown, y splotched salamander. The young, whic called Axolotl, may even breed while st taining their external gills and living water.*

Range: *The United States east of the cades.*

7. Slimy Salamander, Plethodon gl sus. *Adults are about 5 to 6 inches long body, which is very sticky, has a ground c black; the speckles vary from white to gr even silver. The belly has a dull lead color may or may not be flecked with white.*

Range: *New York to Wisconsin, sou Florida and Texas.*

8. Slender Salamander, Batrach attenuatus. *The body of this salamand slender, the legs are small and weak, an tail is long. The color in general is brown slightly lighter on the back than on the and sides.*

Range: *The Pacific slope from southwe Oregon to California.*

9. Cave Salamander, Typhlotriton læus. *This inconspicuous salamander h uniformly pale — almost white — body. eyes are rudimentary and are somewhat cealed by the skin.*

Range: *The Ozark plateau region of Ar sas, Kansas, and Missouri.* Habitat: *Cave.*

10. Cave Salamander, Eurycea lucif *The back of this salamander is vermilio orange, with irregular dark brown or b spots.*

Range: *The central portion of the Mi sippi drainage basin.* Habitat: *Caves.*

Photographs, except Figure 2, by S. C. Bishop; Figure 2 by Charles E. Mohr

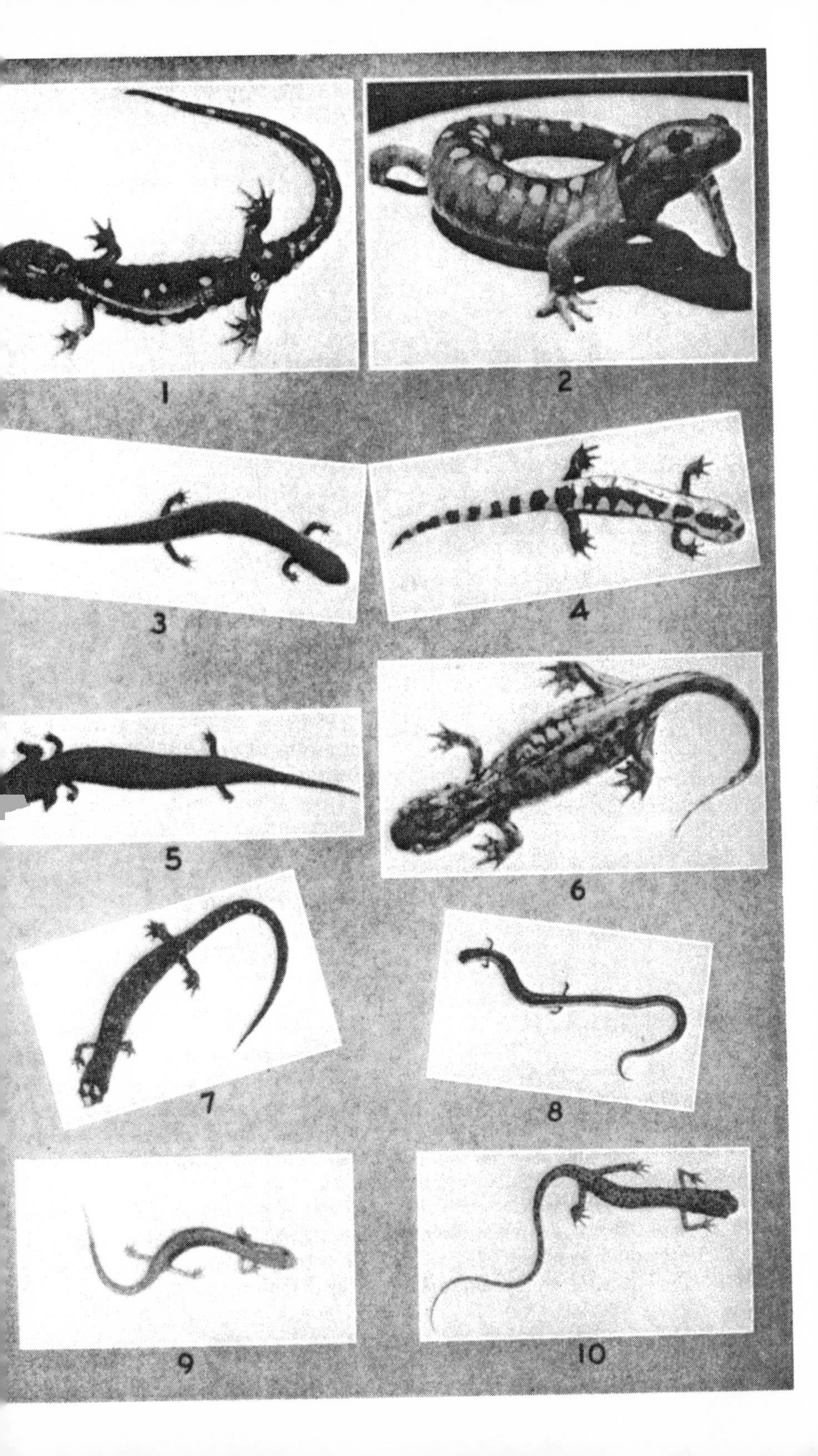
1
2
3
4
5
6
7
8
9
10

impulse comes upon it, it may be far from any stream; but it seems to know instinctively where to go. After it enters the water, it is again transformed in color, becoming olive-green above and buff below, although it still retains the red spots

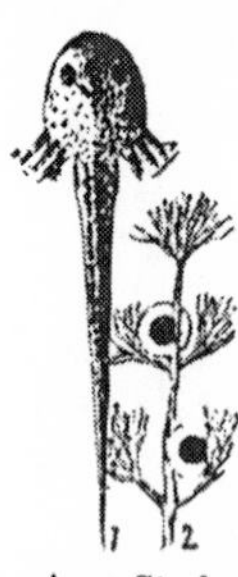

Anna Stryke

Early stage of vermilion-spotted newt. Eggs of newt attached to water plant

along the back; and it also retains its pepper-like dots. Its tail develops a keel which extends along its back and is somewhat ruffled.

The male has the hind legs very large and flat; the lighter-colored female has more delicate and smaller legs. It is here in the water that the efts find their mates and finish careers which must surely have been hazardous. During its long and varied life, the eft often sheds its skin like the snake; it has a strange habit of swallowing its cast-off coat.

LESSON 48

The Newt or Eft

Leading Thought — The newts are born in the water and at first have gills. Later they live on land and have lungs for breathing air; then they go back to the water and again develop the power of breathing the oxygen contained in water; they also develop a keeled tail.

Method — The little, orange eft or red-spotted salamander may be kept i aquarium which has in it an object, as a stone or a clump of moss, which jects above the water. For food it sh be given small earthworms or leaves ered with plant lice. In this way it ma studied at leisure.

Observations — 1. Look at the closely. Is it all the same color? How n spots upon its back and what colors they? Are there the same number of s on both sides? Are there any spots or besides these larger ones? How does eft resemble a toad?

2. Is the head the widest part of body? Describe the eyes, the shape color of the pupil and of the iris. I does the eft wink? Do you think it see well?

3. Can you see the nostrils? How the throat move and why?

4. Are both pairs of legs the same s How many toes on the front feet? H many toes on the hind feet? Does the toe in with its front feet like a toad?

5. Does it move more than one at a time when walking? Does it use feet on the same side in two consecu steps? After it puts forward the ri front foot what foot follows next? Ca move backward?

6. Is the tail as long as the head body together? Is the tail round or fla the sides? How is it used to help the when traveling? Does the tail drag or lifted, or does it push by squirming?

7. How does the eft act when startl Does it examine its surroundings? Do think it can see and is afraid of you?

8. Why do we find more of these c tures during wet weather? Why do peo think they rain down?

9. What does the eft eat? How d it catch its prey? Does it shed its sk How many kinds of efts have you se

10. From what kind of egg does eft hatch? When is this egg laid? H does it look? On what is it fastened?

REPTILES

Yet when a child and barefoot, I more than once, at morn,
Have passed, I thought, a whiplash unbraided in the sun,
When, stooping to secure it, it wrinkled, and was gone.
— Emily Dickinson

he animals in the reptile group have a ring of bony plates or scales. These nals vary greatly in size and shape and ude such forms as snakes, lizards, tur- crocodiles, and alligators. They make r homes in a great variety of places; alligators, the crocodiles, and some of snakes and turtles live in or near water, le many of the snakes and lizards are e at home in desert regions.

the teacher could bring herself to as much interest as did Mother Eve that "subtile animal," as the Bible s the serpent, she might, through such rest, enter the paradise of the boyish rt instead of losing a paradise of her 1. How many teachers, who have an rsion for snakes, are obliged to teach ll boys whose pet diversion is cap- ing these living ribbons and bringing m into the schoolroom stowed away too securely in pockets! In one of the burban Brooklyn schools, boys of this pe sought to frighten their teacher with ir weird prisoners. But she was equal the occasion, and surprised them by de- ring that there were many interesting ngs to be studied about snakes, and thwith sent to the library for books ich discussed these reptiles; and this s the beginning of a nature-study club rare efficiency and enterprise.

There are abroad in the land many roneous beliefs concerning snakes. Most ople believe that they are all venomous, ich is far from true. The rattlesnake ll holds its own in rocky, mountainous aces, and the moccasin haunts the bay- s of the southern coast; however, in ost localities, snakes are not only harm- ss but are beneficial to the farmer. The perstition that if a snake is killed, its tail will live until sundown is general and has but slender foundation in the fact that with snakes, which are lower in their nerve-organization than mammals, the process of death is a slow one. Some people firmly believe that snakes spring or jump from the ground to seize their prey, which is quite false since no snake jumps clear of the ground as it strikes, nor does it spring from a perfect coil. Nor are

F. Harper

Alligator, Alligator mississippiensis. *Alligators may reach a length of twelve feet; they live in or about rivers and swamps of tropical and sub-tropical regions. Their food consists chiefly of fish, mammals, and waterfowl. They are unique among reptiles in being able to produce a loud bellowing noise. In the past, alligators have been ruthlessly slaughtered and even now need more protection*

snakes slimy; on the contrary, they are covered with perfectly dry scales. But the most general superstition of all is that a snake's thrusting out its tongue is an act of animosity; the fact is, the tongue is a sense organ and is used as an insect uses its feelers or antennæ, and the act is also supposed to aid the creature in hearing; thus when a snake thrusts out its tongue, it is simply trying to find out about its surroundings and what is going on.

Snakes are the only creatures able to swallow objects larger than themselves.

F. Harper and A. A. Wright

Alligator eggs. More than 30 eggs may be laid by one female alligator; they are placed above water level in a nest of swamp vegetation. When hatching, the young alligators are about 8 inches long. Turtle eggs, often laid in the same pile of vegetation, are shown in the foreground

This is rendered possible by the elasticity of the body walls, and by the fact that snakes have an extra bone hinging the upper to the lower jaw, allowing them to spread widely; the lower jaw also separates at the middle of its front edge and spreads apart sidewise. In order to force a creature into a " bag " so manifestly too small, a special mechanism is needed; the teeth supply this by pointing backward, and thus assisting in the swallowing. The snake moves by literally walking on the ends of its ribs, which are connected with the crosswise plates on its lower side; each of these crosswise plates has the hind edge projecting down so that it can hold to an object. Thus, the graceful, noiseless progress of the snake is brought about by many of these crosswise plates worked by the movement of the ribs.

Some species of snakes simply c their prey, striking at it and catchi in the open mouth, while others, like pilot black snake, wind themselves a their victims and crush them to de Snakes can live a long time without f many instances on record show that have been able to exist a year or n without anything to eat. In our north climate they hibernate in winter, g to sleep as soon as the weather beco cold and not waking up until spring. snakes grow, they shed their skins; occurs only two or three times a y The crested flycatcher adorns its nest these phantom snakes.

THE GARTER OR GARDEN SNAKE

A chipmunk, or a sudden-whirring quail,
Is startled by my step as on I fare.
A gartersnake across the dusty trail,
Glances and — is not there.

— RILEY

Garter snakes can be easily tamed, and are ready to meet friendly advances half way. A handsome yellow-striped, black garter lived for four years beneath o porch and was very friendly and unafra of the family. The children of the camp

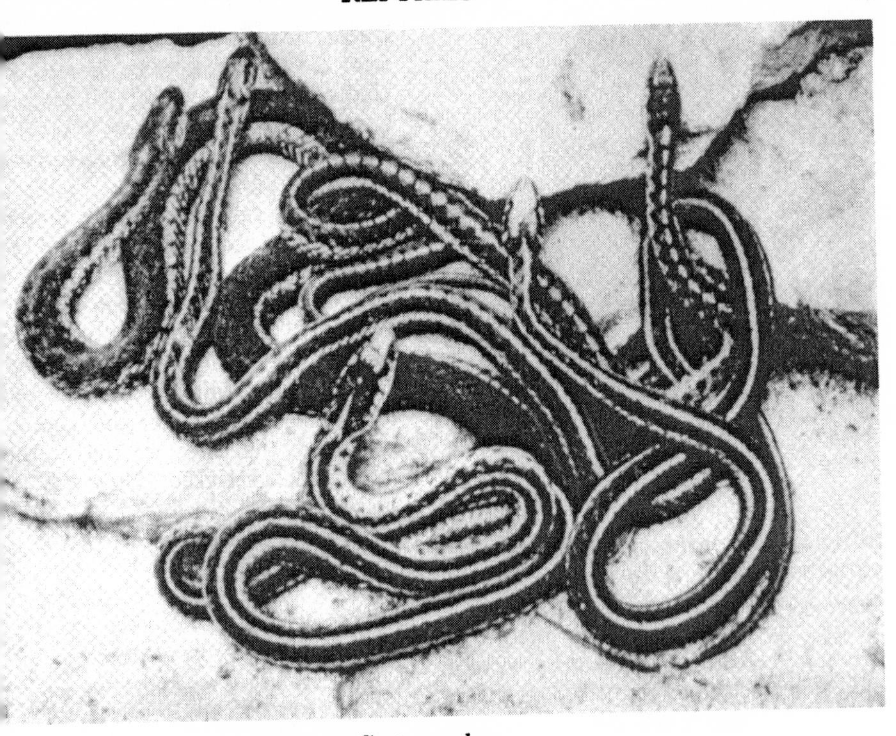

Garter snakes

e it frequent visits, and never seemed e weary of watching it; but the birds cted to it very much, although it r attempted to reach their nests in vine above. The garter snakes are the t common of all, in our northeastern es. They vary much in color; the nd color may be olive, brown, or k, and down the center of the back sually a yellow, green, or whitish stripe, ally bordered by a darker band of und-color. On each side is a similar pe, but not so brightly colored; somees the middle stripe and sometimes side stripes are broken into spots or ent; the lower side is greenish white or ow. When fully grown this snake is to two and one-half feet in length.

The garters are likely to congregate in mbers in places favorable for hibernan, like rocky ledges or stony sidehills. re each snake finds a safe crevice, or kes a burrow which sometimes extends ard or more underground. During the rm days of Indian summer, these winter hermits crawl out in the middle of the day and sun themselves, retiring again to their hermitages when the air grows chilly toward night; and when the cold weather arrives, they go to sleep and do not awaken until the first warm days of spring; then, if the sun shines hot, they crawl out and bask in its welcome rays.

After the warm weather comes, the snakes scatter to other localities more favorable for finding food, and thus these hibernating places are deserted during the summer. The banks of streams and the edges of woods are places which furnish snakes their food, which consists of earthworms, insects, toads, salamanders, frogs, etc. The young are born from late July to mid September and are about six inches long at birth; one mother may have in her brood from eleven to fifty snakelings; she often stays with them only a few hours. There are many stories about the way the young ones run down the mother's throat in case of attack; but as yet no scientist has seen this act or placed it

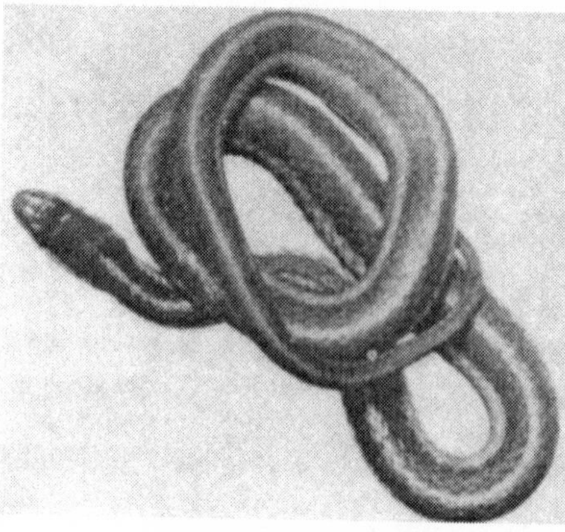

A. A. and A. H. Wright

Common garter snake
Thamnophis sirtalis sirtalis

on record. The little snakes shift for their own food, catching small toads, earthworms, and insects. If it finds food in plenty, the garter snake will mature in one year. Hawks, crows, skunks, weasels, and other predacious animals seem to find the garter snake attractive food.

LESSON 49

The Garter or Garden Snake

Leading Thought — The garter snake is a common and harmless little creature and has many interesting habits which are worth studying.

Method — A garter snake may be captured and placed in a box with a glass cover and thus studied in detail in the schoolroom, but the lesson should begin with observations made by the children on the snakes in their native haunts.

Observations — 1. What are the colors and markings of your garter snake? Do the stripes extend along the head as well as the body? How long is it?

2. Describe its eyes, its ears, its nostrils, and its mouth.

3. If you disturb it how does it act? Why does it thrust its tongue out? What shape is its tongue?

4. In what position is the snake when it rests? Can you see how it moves? upon the lower side. Can you see the plates extending crosswise? Do you it moves by moving these plates? I crawl across your hand, and see if yo tell how it moves.

5. What does the garter snake eat you ever see one swallow a toad? A Did it take it head first or tail first?

6. Where does the garter spend winter? How early does it appear ir spring?

7. At what time of year do you the young snakes? Do the young run down the throat of the mothe safety when attacked? Does the m snake defend her young?

8. What enemies has the garter sn

No life in earth or air or sky;
The sunbeams, broken silently,
On the bared rocks around me lie, —

Cold rocks with half warmed lic
scarred,
And scales of moss; and scarce a yar
Away, one long strip, yellow-barred.

Lost in a cleft! 'Tis but a stride
To reach it, thrust its roots aside,
And lift it on thy stick astride!

Yet stay! That moment is thy grace!
For round thee, thrilling air and spac
A chattering terror fills the place!

A sound as of dry bones that stir,
In the dead valley! By yon fir
The locust stops its noon-day whir!

The wild bird hears; smote with the sou
As if by bullet brought to ground
On broken wing, dips, wheeling round

The hare, transfixed, with trembling
Halts breathless, on pulsating hip,
And palsied tread, and heels that slip

Enough, old friend! — 'tis thou. Forge
My heedless foot, nor longer fret
The peace with thy grim castanet!

From "Crotalus
(The Rattlesnake
Bret Ha

THE MILK SNAKE OR SPOTTED ADDER

The grass divides as with a comb, a spotted shaft is seen,
And then it closes at your feet, and opens farther on.
— EMILY DICKINSON

ıis is the snake which is said to milk ;, a most absurd belief; it would not a cow if it could, and it could not if ould. It has never yet been induced rink milk when in captivity; and if it very thirsty, it could not drink more two teaspoonfuls of milk at most; in any case, its depredations upon the supply need not be feared. Its ob- in frequenting milk houses and sta- is far other than the milking of cows, it is an inveterate hunter of rats and e and is thus of great benefit to the ıer. It is a constrictor, and squeezes orey to death in its coils.

'he ground color of the milk snake is gray, but it is covered with so many wn or dark gray saddle-shaped blotches, they seem rather to form the ground or; the lower side is white, marked h square black spots and blotches. The ke attains a length of two and one-half hree feet when fully grown. Although s commonly called the spotted adder, loes not belong to the adders at all, to the family of the king snakes.

)uring July and August, the mother ke lays from seven to twenty eggs; they deposited in loose soil, in moist rub- h, in compost heaps, etc. The egg is a nmetrical oval in shape and is about e and one-eighth inches long by a half h in diameter. The shell is soft and ite, like kid leather, and the egg resem- s a puffball. The young hatch nearly o months after the eggs are laid; mean- ile the eggs have increased in size so at the snakelings are nearly eight inches ng when they hatch. The saddle-shaped otches on the young have much red them. The milk snake is not venomous; will sometimes, in defense, try to chew e hand of the captor, but the wounds it can inflict are very slight and heal quickly.

LESSON 50

THE MILK SNAKE OR SPOTTED ADDER

LEADING THOUGHT — The milk snake is found around stables where it hunts for rats and mice; it never milks the cows.

METHOD — Although the snake acts fierce, it is perfectly harmless and may be captured in the hands and placed in a glass-covered box for a study in the schoolroom.

OBSERVATIONS — 1. Where is the milk snake found? Why is it called milk snake? Look at its mouth and see if you think it could possibly suck a cow. See if you can get the snake to drink milk.

A. A. and A. H. Wright

Milk snake
Lampropeltis triangulum triangulum

2. What does it live upon? How does it kill its prey? Can the milk snake climb a tree?

3. Where does the mother snake lay her eggs? How do the eggs look? How large are they? How long are the little snakes when they hatch from the egg? Are they the same color as the old ones?

4. Describe carefully the colors and markings of the milk snake and explain how its colors protect it from observation. What are its colors on the underside?

5. Have you ever seen a snake shed its skin? Describe how it was done. How does the sloughed-off skin look? What bird usually puts snake skins around its nest?

I have the same objection to killing a snake that I have to the killing of any other animal, yet the most humane man I know never omits to kill one.

Aug. 5, 1853.

The mower on the river meadows, when he comes to open his hay these days, encounters some overgrown water a full of young (?) and bold in defen its progeny, and tells a tale when he c home at night which causes a shudd run through the village — how it can him and he ran, and it pursued and took him, and he transfixed it with a p fork and laid it on a cock of hay, b revived and came at him again. This i story he tells in the shops at evening. big snake is a sort of fabulous animal. always as big as a man's arm and o definite length. Nobody knows exa how deadly is its bite but nobody is kn to have been bitten and recovered. I men introduced into these meadows the first time, on seeing a snake, a crea which they have seen only in pictures fore, lay down their scythes and run it were the Evil One himself and can be induced to return to their work. T sigh for Ireland, where they say ther no venomous thing that can hurt you.

— Thoreau's Jour

THE WATER SNAKE

Every boy who goes fishing knows the snake found commonly about milldams and wharves or on rocks and bushes near the water. The teacher will have accomplished a great work, if these boys made to realize that this snake is m interesting as a creature for study, th as an object to pelt with stones.

A. A. and A. H. Wright

Common water snake
Natrix sipedon sipedon

The water snake is a dingy brown color, with cross-bands of brown or redd brown which spread out into blotches the side. Its color is very protective it lies on stones or logs in its favorite a tude of sunning itself. It is very local its habits, and generally has a favor place for basking and returns to it ye after year on sunny days.

This snake lives mostly upon frogs a salamanders and fish; however, it pre usually upon fish of small value, so it is little economic importance. It catches victims by chasing and seizing them its jaws. It has a very keen sense of sm and probably traces its prey in this ma ner, something as a hound follows a fo It is an expert swimmer, usually lifti the head a few inches above the wat when swimming, although it is able

and remain below the water for a
t time.
he water snake is a bluffer, and, when
ered, it flattens itself and strikes
ely. But its teeth contain no poison
it can inflict only slight and harmless
nds. When acting as if it would
her fight than eat," if given a slight
ace to escape, it will flee to the water
a "streak of greased lightning," as
boy will assure you.
he water snake may attain a length of
ut four feet; but the usual size is two
one-half to three feet. The young do
hatch from eggs, but are born alive
August and September; they differ
ch in appearance from their parents
hey are pale gray in color, with jet-
k cross-bands. The young often num-
twenty-five to forty and are about eight
hes long.

LESSON 51

The Water Snake

Leading Thought — The water snake
unts the banks of streams because its
d consists of creatures that live in and
out water.

Method — If water snakes are found in
locality, encourage the boys to capture
e without harming it, and bring it to
ool for observation. However, as the
ter snake is very local in its habits, and
unts the same place year after year, it
ll be better nature-study to get the chil-
en to observe it in its native surround-
gs.

Observations — 1. Where is the water
ake found? How large is the largest one
u ever saw?

2. Why does the water snake live near
ter? What is its food? How does it
tch its prey?

3. Describe how the water snake swims. How far does its head project above the water when swimming? How long can it stay completely beneath the water?

4. Describe the markings and colors of the water snake. How do these colors protect it from observation? How do the young look?

5. Does each water snake have a favorite place to which it will usually go to sun itself?

6. Where do the water snakes spend the winter?

May 12, 1858.

Found a large water adder by the edge of Farmer's large mudhole, which abounds with tadpoles and frogs, on which it was probably feeding. It was sunning on the bank and would face me and dart its head toward me when I tried to drive it from the water. It is barred above, but indistinctly when out of the water, so that it appears almost uniformly dark brown, but in the water, broad, reddish brown bars are seen, very distinctly alternating with very dark-brown ones. The head was very flat and suddenly broader than the neck behind. Beneath, it was whitish and reddish flesh-color. It was about two inches in diameter at the thickest part. The inside of its mouth and throat was pink. They are the biggest and most formidable-looking snakes that we have. It was awful to see it wind along the bottom of the ditch at last, raising wreaths of mud amid the tadpoles, to which it must be a very sea-serpent. I afterward saw another, running under Sam Barrett's grist-mill, the same afternoon. He said that he saw a water-snake, which he distinguished from a black snake, in an apple tree near by, last year, with a young robin in its mouth, having taken it from the nest. There was a cleft or fork in the tree which enabled it to ascend.

— Thoreau's Journal

SNAKES

1. Ribbon Snake, Thamnophis sauritus sauritus. *This slender, harmless snake feeds chiefly upon earthworms and young frogs and toads.*

Range: *From Maine, Ontario, and Michigan to Georgia, Alabama, and Mississippi.* Habitat: *Swamps and moist places.*

2. Coral Snake, Micrurus fulvius fulvius. *This beautiful snake is extremely poisonous. Few persons are bitten by it, however, for it is nocturnal in habit and during the day it hides in burrows. Moreover, it does not strike, as most snakes do, but bites into the flesh and chews. It injects so much venom in that way that when it does attack its bite is very dangerous. This dangerous coral snake can be easily distinguished from certain other snakes, which appear to mimic its coloration, by the yellow bands which separate its black from its red bands. Look out for the snake with the yellow bands! Gentle though it may seem, do not play with it.*

3. Rubber Boa, Charina bottæ. *Often spoken of as blind, this boa does have rudimentary eyes, which are, however, almost useless.*

Range: *In humid regions from Utah and Montana to the Pacific coast.*

4. Rough Green Snake, Opheodrys æstivus. *Gentle and harmless, this snake is chiefly insectivorous. It can seldom be induced to bite, and when it does so, its teeth rarely break the skin.*

Range: *From New Jersey south to the Gulf of Mexico and west to Missouri and New Mexico.* Habitat: *Trees and bushy places.*

5. Timber Rattler, Crotalus horridus. *In North America, this rattlesnake is the best known and the most widely distributed. It is more variable in color than is any other rattler. In winter, great numbers hibernate in the same area, and in early spring, when there is a* [illegible] *day, may crawl out into the sunshine.* [illegible] *usually remain near the den and again see*[illegible] *protection if the temperature drops appreci*[illegible] *The food of the timber rattler consists chief*[illegible] *warm-blooded animals such as birds, rats,* [illegible] *and rabbits. It is generally 3 to 5 feet long.*

Range: *Eastern United States to Mississ*[illegible] *Valley states.* Habitat: *More various than* [illegible] *of any other rattler; it is found in both swa*[illegible] *and mountainous regions.*

6. Desert Gopher Snake *or* Bull Sn[illegible] Pituophis catenifer deserticola. *This us*[illegible] *snake, which feeds chiefly on rodents, is* [illegible] *some states protected by law. The length* [illegible] *an adult is usually more than 4 feet.*

Range: *Southern California to Idaho* [illegible] *Washington. Other bull snakes are found f*[illegible] *British Columbia to Mexico.* Habitat: *De*[illegible] *areas.*

7. Ring-necked Snake *or* Eastern Ri[illegible]necked Snake, Diadophis punctatus [illegible]wardsii. *The food of this snake shows g*[illegible] *variety; it includes other small snakes, liza*[illegible] *salamanders, and earthworms.*

Range: *Species are found generally* [illegible] *southern Canada, the United States,* [illegible] *Mexico.* Habitat: *Under old boards, lo*[illegible] *stones, or pieces of bark.*

8. Sidewinder *or* Horned Rattlesna[illegible] Crotalus cerastes. *Its peculiar means of lo*[illegible]*motion gives this snake its name: the body* [illegible] *thrown forward in a series of large loops, a*[illegible] *moves at an angle from the direction in wh*[illegible] *the head is pointed. This way of getting o*[illegible] *the ground seems better adapted than the g*[illegible] *of most snakes would be to life in sandy deser*[illegible] *to which the sidewinder's habitat is virtua*[illegible] *limited. It is known to feed on such animals* [illegible] *pocket mice, kangaroo rats, and lizards.*

Range: *Lower California to southwest Ut*[illegible]

Photographs by A. A. and A. H. Wright

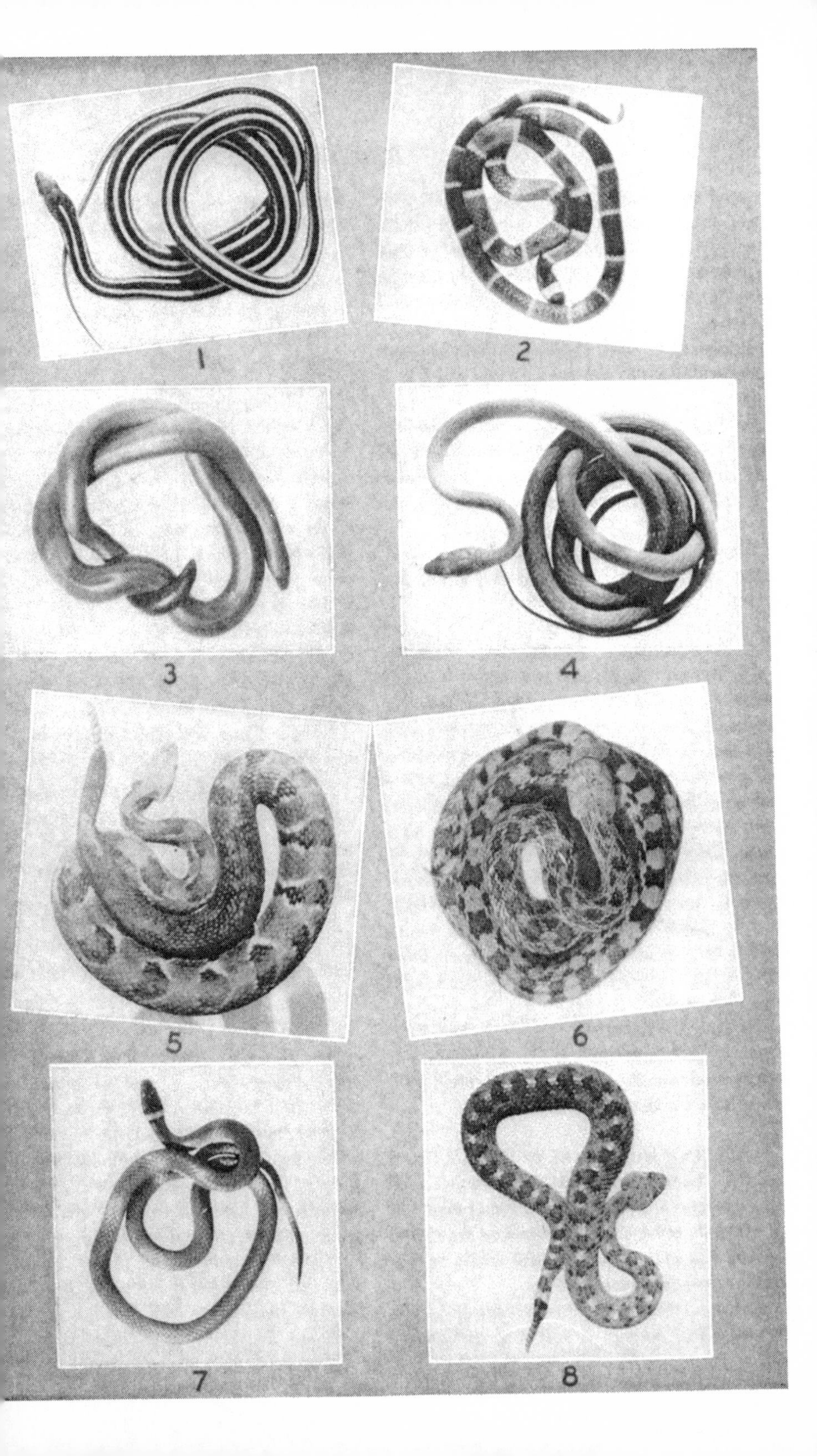
1
2
3
4
5
6
7
8

SNAKES

1. Pike-headed Tree Snake *or* Arizona Long-headed Snake, Oxybelis micropthalamus. *This gentle, slender snake can produce a poisonous bite, which it uses to paralyze its prey. It feeds chiefly on lizards and various small animals.*

Range: *In the United States, southern Arizona.* Habitat: *Trees.*

2. Pilot Black Snake, Elaphe obsoleta obsoleta. *Rats and other small rodents are the food of this useful snake. Adults are usually 5½ feet long, but have reached a length of 7 and 8 feet.*

Range: *From southern New England westward to Michigan, southward to Florida and Texas.*

3. Copperhead, Agkistrodon mokasen mokasen. *The copperhead is common in many parts of the United States, and is probably responsible for more bites than is any other kind of snake. Deaths from its bite have been recorded, but reports from the Antivenin Institute over a period of two years show that although in this time more than three hundred persons were bitten, there were no fatalities, whether or not treatment was given. The food of the copperhead consists mainly of insects, birds, small rodents, and amphibians. It is rather sluggish in habits, and, when molested, usually tries to escape; but if it is taken by surprise or cornered, it defends itself vigorously.*

Range: *Massachusetts to Florida and westward to Arkansas and Texas.* Habitat: *The copperhead usually inhabits drier ground than its relative the moccasin* (*No.* 6).

4. Boyle's King Snake *or* Boyle's Milk Snake, Lampropeltis getulus boylii. *This snake belongs to a great group of king snakes, all of which do much good to farmers by destroying rodents and many other harmful creatures, including even poisonous snakes.*

Range: *Arizona, western Nevada, and California. Other species are widely distributed.* Habitat: *Regions of small streams, esp where chaparral is present.*

5. Gray Pilot Snake, Elaphe ob confinis. *The habits of this snake are s to those of the pilot black snake* (*No.* 2).

Range: *The lower Mississippi Valley, Atlantic, and Gulf states.*

6. Water Moccasin *or* Cottonm Agkistrodon piscivorus. *This poisonous is heavier and larger than the copperhead, it grows from 3 to 5 or even 6 feet in length name of* cottonmouth *has been given it be of the white appearance of the open mou is found in regions of swamps or slow-fl streams, and in sunny hours is often to be at rest on any object that overhangs the w it stays in such a position that if danger ap it can dive into the water. It eats both w and cold-blooded animals, even including snakes. The young are born alive.*

Range: *From southern Virginia to Fl and the Gulf states.* Habitat: *Swampy a*

7. California Lyre Snake, Trimorph vandenburghi. *The bite of this slender, aggressive snake, which it uses to kill or the small animals that are its prey, is pos poisonous to man.*

Range: *California. Other snakes of group are found in the southwestern U States, Mexico, and Central and South A ica.*

8. Southern Hognose Snake, Hete simus. *When threatened, this harmless s may "play possum"; or it may expand its flatten its head, and hiss. It seems to feel all dead snakes should lie on their backs; f turned on its belly when playing dead, it flop over on its back. After a short time, if not disturbed again, it will turn over and c away. Because their threatening actions ferocious appearance have led people to sider them dangerous to man, many of these offensive snakes have been killed.*

Range: *From Florida to Indiana.*

Photographs by A. A. and A. H. Wright

1
2
3
4
5
6
7
8

THE TURTLE

A turtle is at heart a misanthrope; its shell is in itself proof of its owner's distrust of this world. But we need not wonder at this misanthropy, if we think for a moment of the creatures that lived on this earth at the time when turtles first appeared. Almost any of us would have been glad of a shell in which to retire if we had been contemporaries of the smilodon and other monsters of earlier geologic times.

When the turtle feels safe and walks abroad for pleasure, his head projects far from the front end of his shell, and the

A. A. and A. H. Wright

Mud turtle, Kinosternon subrubrum hippocrepis, *viewed from above. Many species of mud turtles are found in the eastern, central, and southern United States. The one pictured is found from Alabama to Texas and north to Kansas. When in captivity, mud turtles will eat lettuce and meat*

legs, so wide and soft that they look they had no bones in them, project o the side, while the little, pointed brings up an undignified rear; but frig

A. A. and A. H. W

Mud turtle viewed from below

him and at once head, legs, and tail disappear, and even if we turn him ov we see nothing but the tip of the no the claws of the feet and the tail turn deftly sidewise. When frightened, hisses threateningly; the noise seems be made while the mouth is shut, a the breath emitted through the nostr

The upper shell of the turtle is call the carapace and the lower shell, t plastron. There is much difference in t different species of turtles in the shape the upper shell and the size and shape the lower one. In most species the ca pace is sub-globular but in some it quite flat. The upper shell is grown fast the backbone of the animal, and t lower shell to the breastbone. The ma

and colors of the shell offer excellent
ects for drawing. The painted terra-
has a red-mottled border to the shell,
ornamental; the wood turtle has a
made up of plates each of which
namented with concentric ridges; and
box turtle has a front and rear trap
, which can be pulled up against the
pace when the turtle wishes to retire,
covering it entirely.

he turtle's head is decidedly snakelike.
olor differs with different species. The
d turtle has a triangular, horny cover-
on the top of the head, in which the
r and beautiful pattern of the shell
repeated; the underparts are brick-red
indistinct yellowish lines under the

A. A. and A. H. Wright

ainted turtle, or terrapin, Chrysemys belli rginata. *The painted turtle pictured is nd from the Mississippi River eastward; species can be found anywhere in the ited States except in deserts and very high untains. This turtle often swims about ks and logs that protrude above the water*

v. The eyes are black with a yellowish s, which somehow gives them a look intelligence. The turtle has no eyelids e our own, but has a nictitating mem- ane which comes up from below and mpletely covers the eye; if we seize e turtle by the head and attempt to uch its eyes, we can see the use of this elid. When the turtle winks, it seems to rn the eyeball down against the lower 1.

The turtle's nostrils are mere pinholes the snout. The mouth is a more or less ooked beak, and is armed with cutting dges instead of teeth. The constant pul- ation in the throat is caused by the tur- e's swallowing air for breathing.

A. A. and A. H. Wright

Chicken turtle, Deirochelys reticularia. *This turtle is at home on the coastal plain from North Carolina to Mississippi. Its high shell may reach a length of eight inches; its neck is long and snakelike*

The turtle's legs, although so large and soft, have bones within them, as the skele- ton shows. The claws are long and strong; there are five claws on the front and four on the hind feet. Some species have a distinct web between the toes; in others it is less marked, depending upon whether the species lives mostly in water or out of it. The color of the turtle's body varies with the species; the body is covered with coarse, rough skin which frequently bears many scales or plates. Thus, large bright- colored scales are conspicuous on the fore legs of the wood turtle, and the tail of the snapping turtle bears a saw-toothed armor of dorsal plates.

The enemies of turtles are the larger fishes and other turtles. Two turtles should never be kept in the same aquar-

A. A. and A. H. Wright

Diamond back terrapin, Malaclemys cen- trata. *The home of the diamond back is in salt marshes from Florida to Massachusetts. In captivity it will eat lettuce, oysters, beef, chopped clams, or fish. Its flesh is used as meat and for making soup*

A. A. and A. H. Wright

Florida snapper, Chelydra osceola, *viewed from above. Snappers live in slow-running streams, ponds, or marshes; the female often goes some distance from her regular home to bury her round, white eggs — usually about two dozen in number.*

ium, since they eat each other's tails and legs with great relish. They feed upon insects, small fish, or almost anything soft-bodied which they can find in the water; they are especially fond of earthworms. The species which frequent the land feed upon tender vegetation and also eat berries. In an aquarium, a turtle should be fed earthworms, chopped fresh beef, lettuce leaves, and berries. The wood turtle is especially fond of cherries.

The aquarium should always have in it a stone or some other object projecting above the water, so that the turtle may

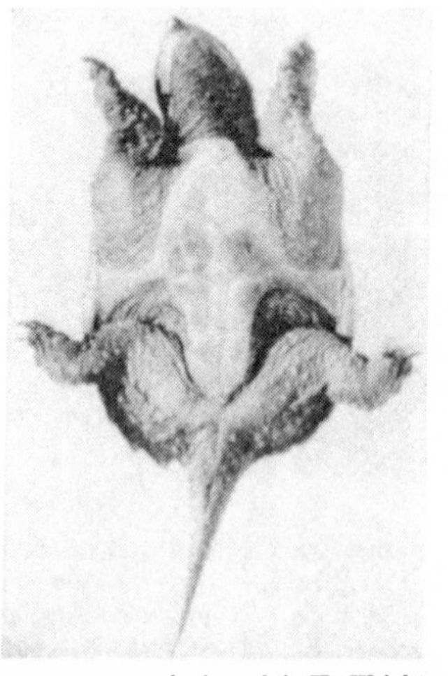

A. A. and A. H. Wright

Florida snapper viewed from below

climb out, if it chooses. In winter, w turtles may bury themselves in the at the bottom of ponds and streams. land turtles dig themselves into the ea Their eggs have white leathery shells, oblong or round, and are buried by mother in the sand or soil near a str or pond. The long life of turtles is a v authenticated fact; dates carved u their shells show them to have attai the age of thirty or forty years.

The following are common kinds:

(a) *The Snapping Turtle* — This so times attains a shell fourteen inches l and a weight of forty pounds. It is a vic

A. A. and A. H. Wr

Gopher turtle, Gopherus berlandieri. *Th turtles are related to the huge turtles of Galapagos Islands. The one pictured is fo in the Rio Grande region; but the range of gopher turtles extends widely through South and the Southwest*

creature and inflicts a severe wound w its sharp, hooked beak; it should not used for a nature-study lesson unless specimen is very young. The large alliga snapper of the South may attain a wei of one hundred pounds.

(b) *The Mud Turtle* — The musk t tle and the common mud turtle both habit slow streams and ponds; they truly aquatic and only come to shore deposit their eggs. They cannot eat unl they are under water, and they seek th food in the muddy bottoms. The mu turtle, when handled, emits a very stro odor; it has on each side of the head tv broad yellow stripes. The mud turtle h no odor. Its head is ornamented wi greenish yellow spots.

(c) *The Painted Terrapin, or Po Turtle* — This can be determined by t ıottled border of its shell. It makes d pet, if kept in an aquarium by it- ɔut will destroy other creatures. It at meat or chopped fish, and is fond ɾthworms and soft insects. It finds od most readily under water.

) *The Spotted Turtle*—This has ıpper shell black with numerous l yellow spots upon it. It is common ınds and marshy streams and its fa- : perch is upon a log with many of ımpanions. It feeds under water, eat- ısect larvæ, dead fish, and vegetation. es fresh lettuce.

A. A. and A. H. Wright

ɔotted turtle, Clemmys guttata. *The e of the spotted turtles extends from higan to Maine and south to Florida. In ivity they often become very tame; they er raw food—earthworms, aquatic in- s, ground beef, or fish*

e) *The Wood Terrapin*—This is our st common turtle; it is found in damp ɔds and wet places, since it lives largely ın the land. Its upper shell often :hes a length of six and one-half inches l is made up of many plates, orna- nted with concentric ridges. This is turtle upon whose shell people carve :ials and dates and then set it free. the fleshy parts of this turtle, except : top of the head and the limbs, are ck-red. It feeds on tender vegetables, ries, and insects, but also enjoys ɔpped meat. It makes an interesting : and will soon learn to eat from the gers of its master.

A. A. and A. H. Wright

Eggs of spotted turtle
Clemmys guttata

(f) *The Box Turtle*—This is easily distinguished from the others, because the front and rear portions of the lower shell are hinged so that they can be pulled up against the upper shell. When this turtle is attacked, it draws into the shell and closes both front and back doors, and is very safe from its enemies. It lives entirely upon land and feeds upon berries, tender vegetation, and insects. It, too, in captivity will eat chopped meat. It lives to a great age.

A. A. and A. H. Wright

A young wood turtle
Clemmys insculpta

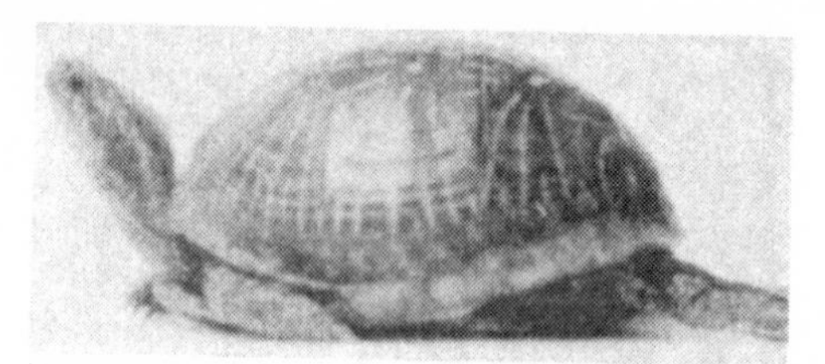

A. A. and A. H. Wright

Box turtle, Terrapene major. *One or more species of box turtle can be found in almost any portion of the United States from the Rocky Mountains eastward*

(g) *The Soft-shelled Turtle* — These are found in streams and canals. The upper shell looks as if it were of one piece of soft leather, and resembles a griddle-cake. The neck is very long and the head particularly snakelike with a piglike snout. Although soft-shelled, these turtles are far from soft-tempered, and must be handled with care. In captivity they must be kept in water.

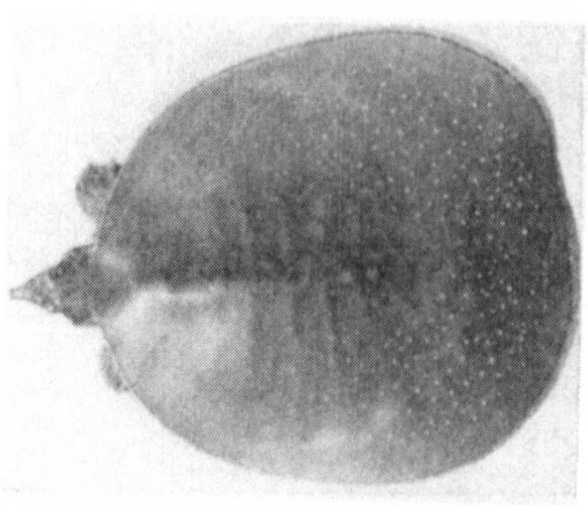

A. A. and A. H. Wright

Soft-shelled turtle, Amyda emoryi. *The species pictured is found in Texas, Oklahoma, and Arkansas; other species may be found from Canada south to the Gulf and as far west as Colorado*

LESSON 52

The Turtle

Leading Thought — The turtle's is for the purpose of protecting its from the attack of enemies. Some t live upon land and others in water.

Method — A turtle of any kind, i schoolroom, is all that is needed to this lesson interesting.

Observations — 1. How much ca see of the turtle when it is walkin

J. T.

A snapping turtle

you disturb it what does it do? How m of it can you see then? Can you see n of it from the lower side than from upper? What is the advantage to the tle of having such a shell?

2. Compare the upper shell with lower as follows: How are they sha differently? What is their difference color? Would it be a disadvantage to turtle if the upper shell were as light ored as the lower? Why? Make a draw of the upper and the lower shell show the shape of the plates of which they composed. Where are the two grown gether?

3. Is the border of the upper shell ferent from the central portion in co and markings? Is the edge smooth or s loped?

How far does the turtle's head pro-
from the front of the shell? What is
hape of the head? With what colors
pattern is it marked? Describe the
How are they protected? How does
turtle wink? Can you discover the
eyelid which comes up from below
over the eye?

Describe the nose and nostrils. Do
think the turtle has a keen sense of
ll?

Describe the mouth. Are there any
h? With what does it bite off its food?
cribe the movement of the throat.
at is the cause of this constant pulsa-
?

What is the shape of the leg? How
t marked? How many claws on the
t feet? Are any of the toes webbed?
which feet are the webbed toes? Why
should they be webbed? Describe the way a turtle swims. Which feet are used for oars?

8. Describe the tail. How much can be seen from above when the turtle is walking? What becomes of it, when the turtle withdraws into its shell?

9. How much of the turtle's body can you see? What is its color? Is it rough or smooth?

10. What are the turtle's enemies? How does it escape from them? What noise does the turtle make when frightened or angry?

11. Do all turtles live for part of the time in water? What is their food and where do they find it? Write an account of all the species of turtles that you know.

12. How do turtle eggs look? Where are they laid? How are they hidden?

LIZARDS

1 *and* 2. BANDED GECKO, Coleonyx brevis. *The gecko, a male, shown in* (1) *has lost the tip of its fragile tail. In* (2) *another gecko, a female, is pictured with a complete tail. An interesting fact about these creatures is that after the tail has been lost another complete tail may later be regenerated. This is characteristic of lizards. The banded gecko is 2 to 3 inches long, and is yellow and brown in color; its small scales give it a very soft, smooth appearance.*

Range: *Found only in Texas.* Habitat: *Under stones; it comes out at night.*

3. CHAMELEON, Anolis carolinensis. *This well-known lizard changes color with temperature conditions: it may fade from dark brown to pale green in three minutes. Often seen in captivity, it can be fed on meal worms and flies; it needs water to drink.*

Range: *North Carolina and Florida to the Rio Grande.*

4. FENCE LIZARD, Sceloporus thayeri. *Like other lizards, this animal eats insects. It is about 5 inches long.*

5. GLASS SNAKE *or* LEGLESS LIZARD, Ophisaurus ventralis. *This long, slender lizard is smooth and glassy. It has a ground color of olive, black, or brown, with greenish to black markings, and a greenish white on the under portions of the body. The long tail makes up about two-thirds of the total length of the animal. An average full-grown specimen is about 24 inches long, but some individuals may attain a length of 3 feet. Like most other lizards, the glass snake, if seized, can shed its tail. While its astonished pursuer gazes at the tail, the body escapes. A new tail begins to grow at once, but it seems never to grow quite as large as the original. The glass snake can be distinguished from a true snake by an ear opening on each side of the head, by numerous rows of small, overlapping scales on its belly, and by movable eyelids.*

Range: *Virginia to Florida westward to Nebraska, Wisconsin, and Mexico.* Ha[illegible] *Chiefly in the ground.*

6. ALLIGATOR LIZARD *or* PLATED LI[illegible] Gerrhonotus infernalis. *Whatever this l[illegible] hears must "go in one ear and out the ot[illegible] for one can look through the ear openings di[illegible] through the head. These lizards, which [illegible] about 18 inches long, make interesting pet[illegible]*

Range: *Southern Texas and northern [illegible]ico.*

7. SONORAN SKINK, Eumeces obsol[illegible] *Skinks are seldom seen in captivity, for [illegible] are hard to capture. They are active in [illegible] light. The females of some skinks stay [illegible] their eggs until they hatch.*

Range: *Utah and Kansas to northern [illegible]ico. Other kinds are widely distributed [illegible] North and Central America; there are [illegible] in the Old World.*

8. GILA MONSTER, Heloderma suspect[illegible] *As far as is known, no two gila monsters [illegible] exactly the same color patterns. Orange, sal[illegible] and brown or black are the chief colors, but [illegible] are variously arranged. This and the closel[illegible]lated Mexican beaded lizard are the only poi[illegible]ous lizards known in the New World. In [illegible] gila monster the poison glands are situate[illegible] the lower jaw and the venom flows out aro[illegible] the teeth and gums. Therefore, since the t[illegible] are above the level of the glands, the poison so[illegible]times does not enter a wound made by the te[illegible] This lizard is rather sluggish and quite [illegible] will not bite even if it is given a good chanc[illegible] do so. When it does bite, it holds on with a str[illegible] grip. In walking it moves slowly and se[illegible] awkward, but it is active enough to climb t[illegible] and bushes, evidently in search of bird's eggs [illegible] which it is very fond. If it is given plenty [illegible] drinking water, it can be kept in captivity [illegible] years on a diet of hen's eggs.*

Range: *Arizona and New Mexico.* Habi[illegible] *Deserts.*

Photographs by A. A. and A. H. Wright

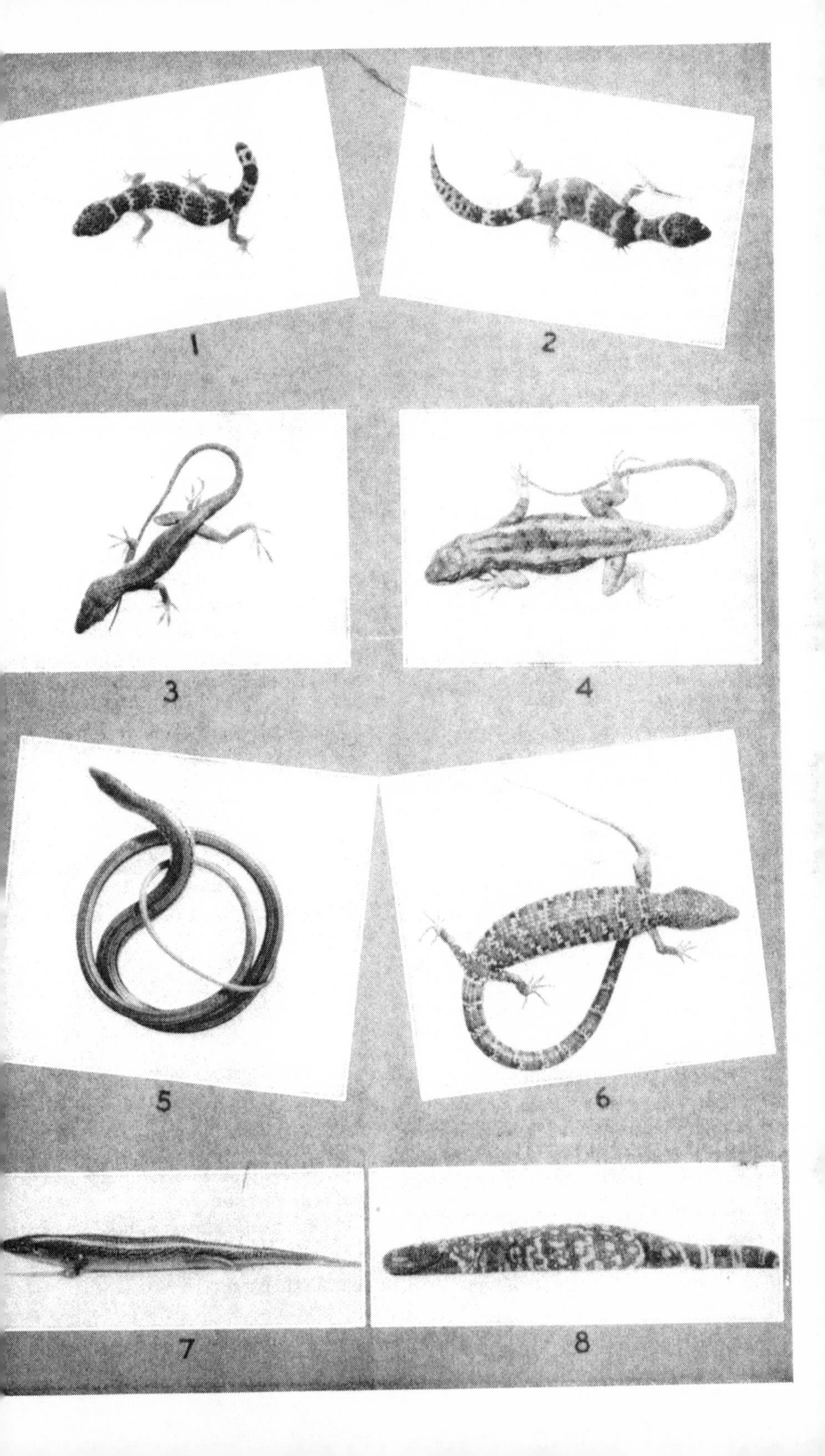
1
2
3
4
5
6
7
8

1 *and* 2. REGAL HORNED TOAD, Phrynosoma solare. *This lizard is called "regal" because the row of spines across the sides and rear of the head gives the effect of a crown. Its color is yellowish, brownish, reddish, or grayish. The eggs are shown in No.* 2.

Range: *Arizona and Lower California. Other kinds are found throughout the western and southwestern states and northern Mexico.*

3. HORNED TOAD, Phrynosoma blainvillii. *These lizards, commonly called "horned toads," are inhabitants of hot, dry regions. In the warmer months they live above ground during the hours of daylight, and are most active when the heat is greatest. Before dark they bury themselves in the sand. They hibernate in winter. In color they often resemble somewhat the ground where they live. A strange habit of the horned toad is that of "squirting blood" from one or both eyes, perhaps as a means of self-defense. The blood has not been found to be poisonous, and must be ejected more to scare than actually to injure the enemy. The horned toad can be tamed, and is often kept for a pet. All too often, however, its owner does not provide enough of the right kind of food—various kinds of small insects—for it, and in such circumstances its ability to live for a long time without food or water serves only to prolong its discomfort. In the Southwest these lizards are sometimes stuffed and sold to tourists as souvenirs, but some states have passed laws prohibiting such sales.*

Range: *San Francisco into Lower California.*

4. HORNED TOADS FEEDING ON ANTS. *In this picture several kinds of horned toads are shown feeding on ants in a pile of sand. They did not dash into the pile, but stood about it in a circle and caught the ants as they came out.*

5. MALE FENCE LIZARD, Sceloporus spinosus. *On either side of the belly the male lizard has a large blue or purple spot margined black. Such marks are used to identify m male lizards.*

Range: *Northern Mexico, New Mexico, Texas to western Florida.* Habitat: *Trunk standing or fallen trees.*

6. MOUNTAIN BOOMER *or* COLLARED ARD, Crotaphytus collaris baileyi. *This usual looking animal makes a good pet if en food can be provided for it. It lives chiefly insects and blossoms of various plants, bu also has cannibalistic habits, and so must be kept in a cage with other lizards of equa smaller size. It is found about rocks at h altitudes. If alarmed or pursued, it runs unt can find a crevice in the rocks. It is a s runner and a high jumper, being able to c an object as much as two feet high. In the ho part of the day its colors seem brighter t during the cooler hours.*

Range: *Southwestern United States Mexico.* Habitat: *Dry, rocky regions.*

7. WHIP-TAIL *or* RACE RUNNER, Cnemidophorus gularis. *These striped lizards active all day under the hottest sun in op areas. In the specimen pictured here, note balls of dirt on its toes from running in s dirt after a rain.*

Range: *Southwestern United States a northern Mexico. A six-line race runner common in the East.*

8. CHUCK-WALLA, Sauromalus obesus. *Th large lizard,* 10 *to* 16 *inches long, is a veg tarian. It protects itself by escaping in crevices. This specimen ran into a crevice an puffed himself up to such a size that it w hard to get him out.*

Range: *Southwestern United States.* Habitat: *Rocky places in desert areas.*

Photographs by A. A. and A. H. Wright

1
2
3
4
5
6
7
8

MAMMALS

Mammals, in contrast to fishes, amphibians, and reptiles, are warm-blooded animals, as are birds. The skin of most mammals is more or less hairy, in contrast to the scale-covered fish and the feathered birds. The young of most mammals are born alive, whereas the young of birds, fish, amphibians, and some species

Marthe Ann, one year old. Human beings are mammals

of reptiles hatch from eggs. After birth young mammals breathe by lungs rather than by gills as do the fish; for a time they are nourished with milk produced by the mother.

Great variations exist in the mammal group. Some of the typical animals in the mammal group which illustrate these variations are opossum, armadillo, whale, deer, buffalo, rabbit, mouse, woodchuck, mole, bat, bear, horse, cat, dog, and man.

Man has always depended a great deal on the lower mammal forms; he uses them for food, clothing, transportation, and numerous other purposes. Many forms are domesticated and have se as man's obedient servants for many turies.

Some of the so-called game ani have suffered wanton destruction at hands of "civilized man," but in recent years many laws and regulat have been passed to give these ani more chances to live. Even more strin laws are needed and rigid enforcer must be exacted if wild animals in eral are to be expected to increas number.

THE COTTON–TAIL RABBIT

The Bunnies are a feeble folk whose weakness is their strength.
To shun a gun a Bun will run to almost any length. — Oliver Herford

is well for Molly Cotton-tail and her ly that they have learned to shun e than guns, for almost every preda- animal and bird makes a dinner of n on every possible occasion. But de- these enemies, moreover, with the tion of guns, men, and dogs, the on-tail lives and flourishes in our st. A "Molly" raised two families last in a briar-patch back of our garden the Cornell campus, where dogs of y breeds abound; and after each fresh of snow this winter we have been able trace our bunny neighbors in their it wanderings around the house, be- th the spruces and in the orchard. track consists of two long splashes, ed, and between and a little behind n, two smaller ones; the rabbit uses ront feet as a boy uses a vaulting pole lands the hind feet on each side and ad of them; because the bottoms of the are hairy the print is not clear-cut. ien the rabbit is not in a hurry it has a uliar lope, but when frightened it kes long jumps. The cotton-tails are ht wanderers and usually remain hid- during the day. In summer, they feed clover or grass or other juicy herbs and w a fondness for sweet apples and fresh bage; in our garden last summer Molly s very considerate. She carefully pulled the grass out of the garden-cress bed, ving the salad for our enjoyment. In ter, the long, gnawing teeth of the ton-tail are sometimes used to the dam- of fruit trees and nursery stock since rabbits are obliged to feed upon bark order to keep alive.

The long, strong hind legs and the long s tell the whole bunny story. Ears to ar the approach of the enemy, and legs propel the listener by long jumps to safe retreat. The attitude of the ears is a good indication of the bunny's state of mind; if they are set back to back and directed backward, they indicate placidity, but a placidity that is always on guard; if lifted straight up they signify attention and anxiety; if one is bent forward and the other backward the meaning is: "Now just where did that sound come from?"

A cotton-tail rabbit

When the rabbit is running or resting in the form, the ears are laid back along the neck. When the cotton-tail stands up on its haunches with both ears erect, it looks very tall indeed.

Not only are the ears always alert, but also the nose; the nostrils are partially covered and in order to be always sure of getting every scent they wabble constantly, the split upper lip aiding in this performance; when the rabbit is trying to get a scent it moves its head up and down in a sagacious, apprehensive manner.

The rabbit has an upper and lower

Verne Morton

The rabbits' ears are ever alert for any sign of danger

pair of incisors like other rodents, but on the upper jaw there is a short incisor behind each of the large teeth; these are of no use now but are inherited from some ancestor which found them useful. There are at the back of each side of the upper jaw six grinding teeth, and five on each side of the lower jaw. The split upper lip allows the free use of the upper incisors. The incisors are not only used for taking the bark from trees, but also for cutting grass and other food. The rabbit has a funny way of taking a stem of grass or clover at the end and with much wabbling of lips finally taking it in, meanwhile chewing it with a sidewise motion of the jaws. The rabbit's whiskers are valuable as feelers, and are always kept on the *qui vive* for impressions; when two cotton-tails meet each other amicably, they rub whiskers together. The eyes are large and dark and placed on the bulge at the side of the head, so as to command the view both ways. Probably a cotton-tail winks, but I never caught one in the act.

The strong hind legs of the rabbit enable it to make prodigious jumps, of eight feet or more; this is a valuable asset to an animal that escapes its enemies by running. The front feet are short and cannot be turned inward like those of the squirrel, to hold food. There are five toes on the front feet, and four on the feet; the hair on the bottom of the is a protection, much needed by an mal which sits for long periods upon snow. When sleeping, the rabbit fold front paws under and rests on the e hind foot, with the knee bent, ready spring at the slightest alarm; when aw it rests on the hind feet and front toes; when it wishes to see if the coast is c it rises on its hind feet, with front drooping.

The cotton-tail has a color well c lated to protect it from observation; brownish-gray on the back and a lighter along the sides, grayish under chin and whitish below; the ears are e with black, and the tail when raised sh a large, white fluff at the rear. The eral color of the rabbit fits in with ural surroundings; since the cotton often escapes its enemies by "freezi this color makes the scheme work I once saw a marsh hare, on a ston a brook, "freezing" most successfull could hardly believe that a living t could seem so much like a stone; onl bright eyes revealed it to us.

The rabbit cleans itself in amu ways. It shakes its feet one at a t with great vigor and rapidity to get the dirt and then licks them clean washes its face with both front paw once. It scratches its ear with the h foot, and pushes it forward so that it be licked; it takes hold of its fur with front feet to pull it around within re of the tongue.

The cotton-tail does not dig a burr

A Dutch rabbit and Belgian hares

sometimes occupies the deserted bur- of a woodchuck or skunk. Its nest lled a "form," which simply means ace beneath a cover of grass or briars, re the grass is beaten down or eaten for a space large enough for the ani- to sit. The mother prepares a shal- excavation in which she makes a soft for the young, using grass and her hair for the purpose; and she con- cts a coarse felted coverlet, under ch she tucks her babies with care y time she leaves them. Young rab- are blind at first, but when about e weeks old are sufficiently grown to quite rapidly. Although there may be or six in a litter, yet there are so many mies that only a few escape.

'ox, mink, weasel, hawk, owl, snake, occasionally red squirrel all relish the ng cotton-tail if they can get it. Noth- but its runways through the briars can e it. These roads wind in and out and oss, twisting and turning perplexingly; y are made by cutting off the grass ms, and are just wide enough for the bit's body. However, a rabbit has apons and can fight if necessary; it leaps r its enemy, kicking it on the back rcely with its great hind feet. Mr. Seton ls of this way of conquering the black ake, and Mr. Sharp saw a cat completely nquished by the same method. Mr. E. . Cleeves told me of a Belgian doe ich showed her enmity to cats in a culiar way. She would run after any cats at came in sight, butting them like a billy at. The cats soon learned her tricks, and ould climb a tree as soon as they caught ht of her. The rabbit can also bite, and ien two males are fighting, they bite ch other savagely. The rabbit's sound of fiance is thumping the ground with the rong hind foot. Some have declared that e front feet are used also for stamping; though I have heard this indignant umping more than once, I could not see e process. The cotton-tail and the com- on domestic rabbit are true rabbits. The ck rabbit is a true hare.

Not the least of tributes to the rabbit's gacity are the Negro folk stories told by Uncle Remus, wherein Brer Rabbit, although often in trouble, is really the most clever of all the animals. I have often thought when I have seen the tactics which rabbits have adopted to escape dogs, that we in the North have underrated the cleverness of this timid animal. In one instance at least that came under our observation, a cotton-tail led a dog to the verge of a precipice, then doubled

Rabbits playing in the moonlight

back to safety, while the dog went over, landing on the rocks nearly three hundred feet below.

An interesting relative of the cotton-tail is the varying hare or snow-shoe rabbit that lives in the wooded regions of north-eastern North America. Of all animals he is one of the most defenseless; foxes, mink, and other flesh-eating inhabitants of the woods find him an easy prey. He has not even a burrow to flee to when pursued by his enemies.

He passes the day half asleep and motionless beneath the sheltering branches of a low fir tree or in a dense thicket. With the coming of night he starts off in search of food.

He has one important advantage over his enemies: twice each year his heavy coat of fur is shed. In the summer the coat is a reddish brown that so blends with his surroundings that he is hardly noticeable; in the winter it is perfectly white so that against a background of snow he is nearly invisible.

LESSON 53

The Cotton-tail Rabbit

Leading Thought — The cotton-tail thrives amid civilization; its color protects it from sight; its long ears give it warning of the approach of danger; and its long legs enable it to run by swift, long leaps. It feeds upon grasses, clover, vegetables, and other herbs.

Method — This study may be begun in the winter, when the rabbit tracks can be observed and the haunts of the cotton-tail discovered. If caught in a box trap, the cotton-tail will become tame if properly fed and cared for, and may thus be studied at close range. The cage I have used for rabbits thus caught is made of wire screen nailed to a frame, making a wire-covered box two feet high and two or three feet square, with a door at one side and no bottom. It should be placed upon oilcloth or linoleum, and thus may be moved to another carpet when the floor needs cleaning. If it is impossible to study the cotton-tail, the domestic rabbit may be used instead.

Observations — 1. What sort of tracks does the cotton-tail make in the snow? Describe and sketch them. Where do you find these tracks? How do you know which way the rabbit was going? Follow the track and see if you can find where the rabbit went. When were these tracks made, by night or by day? What does the rabbit do during the day? What does it find to eat during the winter? How are its feet protected so that they do freeze in the snow?

2. What are the two most notice peculiarities of the rabbit? Of what are such large ears? How are the ears when the rabbit is resting? When tled? When not quite certain about direction of the noise? Explain the sons for these attitudes. When the ra wishes to make an observation to se there is danger coming, what does it How does it hold its ears then? How the ears held when the animal is runn

3. Do you think the rabbit has a k sense of smell? Describe the movem of the nostrils and explain the rea How does it move its head to be sur getting the scent?

4. What peculiarity is there in the per lip? How would this be an aid to rabbit when gnawing? Describe the te how do these differ from those of mouse or squirrel? Of what advantage the gnawing teeth to the rabbit? H does it eat a stem of grass? Note the r bit's whiskers. What do you think t are used for?

5. Describe the eyes. How are t placed so that the rabbit can see forw and backward? Do you think that it sle with its eyes open? Does it wink?

6. Why is it advantageous to the r bit to have such long, strong hind le Compare them in size with the front le Compare the front and hind feet. H many toes on each? How are the botto of the feet protected? Are the front f ever used for holding food like the squ rel's? In what position are the legs wh the rabbit is resting? When it is standin When it is lifted up for observation?

7. How does the cotton-tail escape b ing seen? Describe its coat. Of what use the white fluff beneath the tail? Have y ever seen a wild rabbit "freeze"? Wh is meant by "freezing" and what is t use of it?

8. In making its toilet how does t rabbit clean its face, ears, feet, and fur?

9. What do the cotton-tails feed up during the summer? During the winte Do they ever do much damage?

Describe the cotton-tail's nest. is it called? Does it ever burrow in round? Does it ever use a second-burrow? Describe the nest made for ɔung by the mother. Of what is the :omposed? Of what is the coverlet ? What is the special use of the let? How do the young cotton-tails How old are they before they are to take care of themselves?

What are the cotton-tail's enemies? does it escape them? Have you ever the rabbit roads in a briar-patch? ou think that a dog or fox could fol-hem? Do rabbits ever fight their enemies? If so, how? How do they show anger? Do they stamp with the front or the hind foot?

12. Tell how the cotton-tail differs in looks and habits from the common tame rabbit. How do the latter dig their burrows? How many breeds of tame rabbits do you know?

13. Write or tell stories on the following topics: "A Cotton-tail's Story of Its Own Life until It Is a Year Old"; "The Jack Rabbit of the West"; "The Habits of the White Rabbit or Varying Hare"; "The Rabbit in Uncle Remus' Tales."

Silas Lottridge

Winter lodge of muskrats

THE MUSKRAT

Having finished this first course of big-neck clams, they were joined by a third ɔkrat, and, together, they filed over the bank and down into the meadow. Shortly of them returned with great mouthfuls of the mud-bleached ends of calamus-les. Then followed the washing.

'hey dropped their loads upon the plank, took up the stalks, pulled the blades apart, soused them up and down in the water, rubbing them with their paws until they e as clean and white as the whitest celery one ever ate. What a dainty picture! o little brown creatures, humped on the edge of a plank, washing calamus in onlit water! — DALLAS LORE SHARP

Tracking is a part of the education of ry boy who aspires to a knowledge of od lore; and a boy with this accomplishment is sure to be looked upon with great admiration by other boys less skilled in the interpretation of that writ-

Silas Lottridge

The Muskrat

ing made by small feet on the soft snow or on the mud of stream margins. To such a boy, the track of the muskrat is well known and very easily recognized.

The muskrat is essentially a water animal, and therefore its tracks are to be looked for along the edges of ponds, streams, or in marshes. Whether the tracks are made by walking or jumping depends upon the depth of the snow or mud; if it is deep, the animal jumps, but in shallow snow or mud it simply runs along. The tracks show the front feet to be smaller than the hind ones. The muskrat track is, however, characterized by the tail imprint. When the creature jumps through the snow, the mark of the tail follows the paired imprints of the feet; when it walks, there is a continuous line made by this strong, naked tail. This distinguishes the track of the muskrat from that of the mink, as the bushy tail of the latter does not make so distinct a mark. Furthermore the claws of the feet show distinctly in a muskrat track; those of the mink do not. Measuring the track is a simple device for making the pupils note its size and shape more carefully. The tracks may be looked for during the thaws of March or February, when the muskrats come out of the water tc food.

In appearance the muskrat is pec The body is usually about a foot in l and the tail about eight inches. The is stout and thickset, the head is rou and looks like that of a giant me mouse; the eyes are black and shi the ears are short and close to the the teeth, like those of other rod consist of a pair of front teeth on jaw, then a long, bare space, and then grinders on each side. There are sensitive hairs about the nose and m like the whiskers of mice.

The muskrat's hind legs are n larger and stronger than the front the hind feet are likewise much lo than the front feet and have a wet tween the toes; there are also stiff which fill the space between the outside the web, thus making this hind foot an excellent swimming or The front toes are not webbed and used for digging. The claws are long, st and sharp. The tail is long, stout, and tened at the sides; it has little or no upon it but is covered with scales; used as a scull and also as a rudder w the muskrat is swimming.

The muskrat's outer coat consist long, rather coarse hairs; its under is of fur, very thick and fine, and altho short, it forms a waterproof protection the body of the animal. In color, the is dark brown above with a darker st along the middle of the back; bene the body is grayish, changing to whi on the throat and lips, with a brown s on the chin. In preparing the pelts commercial use, the long hairs are so

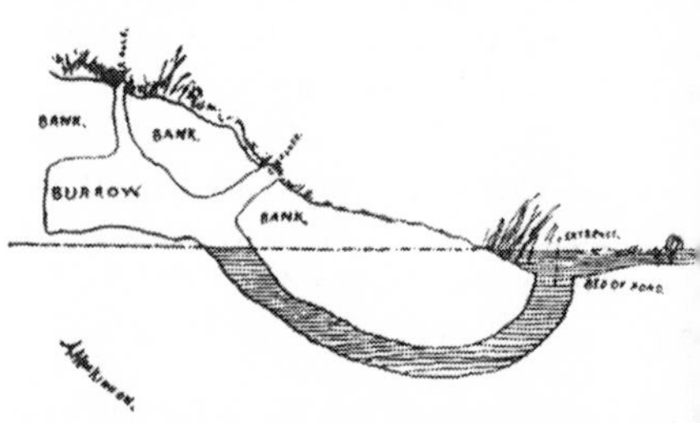

A muskrat's summer home, drawn by MacKinnon, a boy of thirteen years

plucked out leaving the soft, fine r coat, which is often dyed black sold under the name of "Hudson

e muskrat is far better fitted by form fe in the water than upon the land. e it is heavy-bodied and short-legged annot run rapidly, but its strong, oed hind feet are most efficient oars, it swims rapidly and easily; for rud- and propeller the strong, flattened serves admirably, while the fine fur the body is so perfectly waterproof , however much the muskrat swims lives, it is never wet. It is a skillful r and can stay under water for several utes; when swimming, its nose and etimes the head and the tip of the appear on the surface of the water. he food of muskrats is largely roots, ecially those of the sweet flag and the ow lily. Muskrats also feed on other atic plants and are fond of the fresh- er shell-fish. Mr. Sharp tells us, in one his delightful stories, how the musk- wash their food by sousing it up and vn in water many times before eating Often, a muskrat chooses some special ce upon the shore which it uses for a ing room, bringing there and eating ces of lily root or fresh-water clams, l leaving the debris to show where it oitually dines. It does most of its hunt- for food at night, although sometimes

National Parks Bureau, Dominion of Canada

Adult Beaver. The habits of beavers somewhat resemble those of muskrats. Beavers may weigh from 40 to 60 pounds and reach a length of 40 inches. In North America they range from Hudson Bay and Alaska south into Mexico in the West and the southern Alleghenies in the East

it may be seen thus employed during the day.

The winter lodge of the muskrat is a most interesting structure. A foundation of tussocks of rushes, in a stream or shallow pond, is built upon with reeds, making a rather regular dome which may be nearly two or three feet high; or, if many-chambered, it may be a grand affair of four or five feet elevation; but it always looks so much like a natural hummock that the eye of the uninitiated never regards it as a habitation. Beneath this dome and above the water line is a snug, covered chamber carpeted with a soft bed of leaves and moss, which has a passage leading down into the water below, and in some instances an air-hole. In these cabins, closely cuddled together, three or four in a chamber, the muskrats pass the winter. After the pond is frozen they are safe from their enemies except the mink and are always able to go down into the water and feed upon the roots of water plants. These cabins are sometimes built in the low, drooping branches of willows or on other objects.

Whether the muskrat builds itself a winter lodge or not depends upon the nature of the shore which it inhabits; if it is a place particularly fitted for burrows, then a burrow will be used as a winter

Frank H. Steinicke

A beaver lodge in winter. In the foreground the "air hole." In general this home looks ke that of the muskrat, but it is larger and is ade of coarser materials

retreat; but if the banks are shallow, the muskrats unite in building cabins. The main entrance to the muskrat burrow is usually below the surface of the water, the burrow slanting upward and leading to a nest well lined, which is above the reach of high water; there is also often a

National Parks Bureau, Dominion of Canada

Young beavers feeding in the shallow water near the lower edge of a beaver dam

passage, with a hidden entrance, leading out to dry land.

The flesh of the muskrat is delicious, and therefore the animal has many enemies; foxes, weasels, dogs, minks, and also hawks and owls prey upon it. It is, indeed, a good human food. It escapes the sight of its enemies as does the mouse, by having inconspicuous fur; when discovered, it escapes its enemies by swimming, although when cornered it is courageous and fights fiercely, using its strong incisors as weapons. In winter, it dwells in safety when the friendly ice protects it from all its enemies except the mink; but it is exposed to great danger when the streams break up in spring, for it is then often driven from its cabin by floods, and preyed upon while thus helplessly exposed.

It is called muskrat because of the odor, somewhat resembling musk, which it excretes from two glands on the lower side of the body between the hind legs; these glands may be seen when the skin is removed, which is the too common p of this poor creature, since it is hu mercilessly for its pelt.

The little muskrats are born in and there are usually from three to s in a litter. Another litter may be prod in June or July and a third in Augu September. It is only thus, by re large families often, that the muskrat able to hold their own against the hu and trappers and their natural enemi

LESSON 54

THE MUSKRAT

LEADING THOUGHT — The musk while a true rodent, is fitted for lif the water more than for life upon land. Its hind feet are webbed for us oars and its tail is used as a rudder builds lodges of cattails and rushes which it spends the winter.

METHOD — It might be well to b this work by asking for observations the tracks of the muskrat which may found about the edges of almost creek, pond, or marsh. If there are m rat lodges in the region they should visited and described. For studying muskrat's form a live muskrat in capti is almost necessary. The pupils can t study it at leisure although they sho not be allowed to handle the creature it inflicts very severe wounds and is ne willing to be handled. If a live musk cannot be obtained, perhaps some hun in the neighborhood will supply a d one for this observation lesson.

While studying the muskrat the c dren should read all the stories of beav which are available, as the two anim are very much alike in their habits.

OBSERVATIONS — 1. In what local have you discovered the tracks of t muskrat? Describe its general appearan Measure the muskrat's track as follo width and length of the print of one ; (b) the width between the prints ie two hind feet; (c) the length be-n the prints made by the hind feet in ral successive steps or jumps.

Was the muskrat's track made when animal was jumping or walking? Can see in it a difference in the size of front and hind feet? Judging from track, where do you think the musk-came from? What do you think it was ting for?

What mark does the tail make in snow or mud? Judging by its imprint, uld you think the muskrat's tail was g or short, bare or brushy, slender or it?

. How long is the largest muskrat you r saw? How much of the whole length ail? Is the general shape of the body rt and heavy or long and slender?

. Describe the muskrat's eyes, ears, l teeth. For what are the teeth espe-ly fitted? Has the muskrat whiskers mice and rats?

. Compare the front and hind legs as size and shape. Is there a web between toes of the hind feet? What does s indicate? Do you think that the iskrat is a good swimmer?

7. Describe the muskrat fur. Compare outer and under coat. What is its lor above and below? What is the name muskrat fur in the shops?

8. Describe the tail. What is its cover-g? How is it flattened? What do you ink this strong, flattened tail is used r?

9. Do you think the muskrat is better ted to live in the water than on land? ow is it fitted to live in the water in the llowing particulars: Feet? Tail? Fur?

10. How much of the muskrat can you e when it is swimming? How long can stay under water when diving?

11. What is the food of the muskrat? Vhere does it find it? How does it pre-re the food for eating? Does it seek s food during the night or day? Have you ver observed the muskrat's dining room? so, describe it.

12. Describe the structure of the musk-rat's winter lodge, or cabin, in the following particulars: What is its size? Where built? Of what material? How many rooms in it? Are these rooms above or below the water level? Of what is the bed made? How is it arranged so that the entrance is not closed by the ice? Is such a home built by one or more muskrats? How many live within it? Do the muskrats always build these winter cabins? What is the character of the shores where they are built?

13. Describe the muskrat's burrow in the bank in the following particulars: Is the entrance above or below water? Where and how is the nest made? Is it ventilated? Does it have a back door leading out upon the land?

14. What are the muskrat's enemies? How does it escape them? How does it fight? Is it a courageous animal? How does the muskrat give warning to its fellows when it perceives danger? At what time of year is it comparatively safe? At what time is it exposed to greatest danger?

15. Why is this animal called muskrat? Compare the habits of muskrats with those of beavers and write an English theme upon the similarity of the two.

Leonard K. Beyer

Trees felled by beavers. Unlike muskrats, beavers fell trees. They have cut these birches either to use the bark for food or the trunks for reinforcement of a dam. In the background, note the area covered by water held by a beaver dam

16. At what time of year do you find the young muskrats? How many in a litter?

Nature Photography around the Year, Percy A. M
© D. Appleton-Century Co.,

THE HOUSE MOUSE

Somewhere in the darkness a clock strikes two;
And there is no sound in the sad old house,
But the long veranda dripping with dew,
And in the wainscot — a mouse. — BRET HARTE

Were mouse-gray a less inconspicuous color, there would be fewer mice; when a mouse is running along the floor, it is hardly discernible, it looks so like a flitting shadow; if it were black or white or any other color, it would be more often seen and destroyed. It has been very closely associated with man; as a result of this fact the species has been able to spread over the world.

At first glance one wonders what possible use a mouse can make of a tail which is as long as its body, but a little careful observation will reveal the secret. The tail is covered with transverse ridges and is bare save for sparse hairs, except toward the tip. Dr. Ida Reveley first called my attention to the fact that the house mouse uses its tail in climbing. I verified this interesting observation, and found that my mouse used the tail for aid when climbi a string. He would go up the string ha over hand like a sailor, and then in tryi to stretch to the edge of his jar, he variably wound his tail about the stri two or three times, and hanging to t string with the hind feet and tail, wou reach far out with his head and front fe Also, when clinging to the edge of t cover of the jar, he invariably used tail as a brace against the side of the gla so that it pressed hard for more than h its length. Undoubtedly the tail is of gre service in climbing up the sides of wal

The tail is also of some use when t mouse jumps directly upward. The hi legs are very much longer and strong than the front legs. The hind feet are al much longer and larger than the fro feet; and although the mouse, when s its remarkable jumps, depends its strong hind legs, I am sure that the tail is used as a brace to guide ssist the leap. The feet are free from but are downy; the hind foot has front toes, a long toe behind on the de and a short one on the inside. claws are fairly long and very sharp at they are able to cling to almost ing but glass. When exploring, a e stands on its hind feet, folding its front paws under its chin while it es up ready to catch anything in ; it can stretch up to an amazing t. It feeds upon almost anything that le like to eat and, when eating, fre- tly holds its food in its front paws a squirrel.

he thin, velvety ears are flaring cornu- as for taking in sound; the large, ded outer ear can be moved forward ack to test the direction of the noise. eyes are like shining, black beads; if a mouse can wink, it does it so dly as not to be discernible. The nose ong, inquisitive, and always sniffing new impressions. The whiskers are cate and probably sensitive. The th is furnished with two long, curved wing teeth at the front of each jaw, a bare space, and then four grinding h on each side, above and below, like teeth of woodchucks and other ro- ts. The gnawing teeth are very strong enable the mouse to gnaw through rd partitions and other obstacles.

he energy with which the mouse ns itself is inspiring to behold. It bles its fur and licks it with fervor, hing around so as to get at it from ind, and taking hold with its little ds to hold firm while it cleans. When hing its face and head, it uses both nt feet, licking them clean and rub- g them both simultaneously from be- d the ears down over the face. It takes hind foot in both front feet and nib- s and licks it. It scratches the back of head with its hind foot.

Young mice are small, downy, pink, and nd when born. The mother makes for m a nice, soft nest of pieces of cloth, paper, grass, or whatever is at hand; the nest is round like a ball and at its center is nestled the family. Mice living in houses have runways between the plaster and the outside wall, or between ceiling and floor. In winter they live on what food they can find, and upon flies or other insects hibernating in our houses. The house mice sometimes live under stacks of corn or grain in the fields, but usually confine themselves to houses or barns.

Verne Morton

Young field mice, blind, pink, and hairless

They are thirsty little fellows and they like to make their nests within easy reach of water.

Our house mice came from ancestors which lived in Asia originally; they have always been great travelers and they have followed men wherever they have gone, over the world. They came to America on ships with the first explorers and the Pil- grim fathers. They now travel back and forth, crossing the ocean in ships of all sorts. They also travel across the continent on trains. Wherever our food is carried they go; and the mouse which you see in your room one day may be a thousand miles away within a week. They are clever creatures, and learn quickly to connect cause and effect. For two years I was in an office in Washington, and while there I observed that as soon as the bell rang for noon, the mice would appear instantly, hunting wastebaskets for scraps of lunch. They had learned to connect the sound of the bell with food.

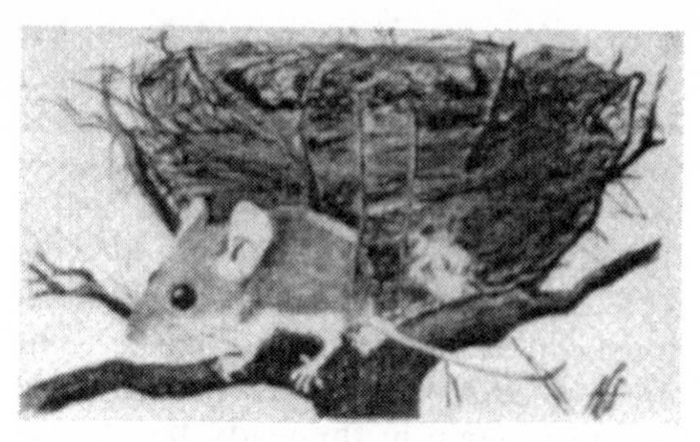

Anna Stryke

A white-footed or deer mouse may use an old bird's nest for its home

Of all our wild mice, the white-footed or deer mouse is the most interesting and attractive. It is found almost exclusively in woods and is quite different in appearance from other mice. Its ears are very large; its fur is fine and beautiful and a most delicate gray in color. It is white beneath the head and under the sides of the body. The feet are pinkish, the front paws have short thumbs, while the hind feet are very much longer and have a long thumb which looks like an elfin hand in a gray-white silk glove. On the bottom of the feet are callous spots which are pink and serve as foot pads. This mouse makes its nest in hollow trees and stores nuts for winter use. We once found two quarts of shelled beechnuts in such a nest. It also likes the hips of the wild rose and many kinds of berries; it sometimes makes its home in a bird's nest, which it roofs over to suit itself. The young mice are usually carried in the mother's mouth, one at a time. As an inhabitant of summer cottages, white-foot is cunning and mischievous; it pulls cotton out of quilts, takes covers from jars, and as an explorer is equal to the squirrel. I once tried to rear some young deer mice by feeding them warm milk with a pipette; although their eyes were not open, they invariably washed their faces after each meal, showing that neatness was bred in the bone. This mouse has a musical voice and often chirps as sweetly as a bird. Like the house mouse it is more active at night.

The meadow mouse is the one that makes its runways under the snow, making strange corrugated patterns over the ground which attract our attentic spring. It has a heavy body, short short ears, and a short tail. It is bro or blackish in color. It sometimes burrows straight into the ground, more often makes its nest in waste r ows. It is the nest of this field n which the bumblebee so often possession of, after it is deserted. meadow mouse is a good fighter, si up like a woodchuck and facing its e bravely. It needs to be courageous it is preyed upon by almost every cre that feeds upon small animals; the h and owls especially are its enemies. well for the farmer that these mice so many enemies, for they multiply idly and would otherwise soon ov and destroy the grain fields. They tremendous damage by girdling valu fruit trees. This mouse is an exce swimmer.

A part of winter work is to make pupils familiar with the tracks of meadow mice and to teach them ho distinguish them from other tracks.

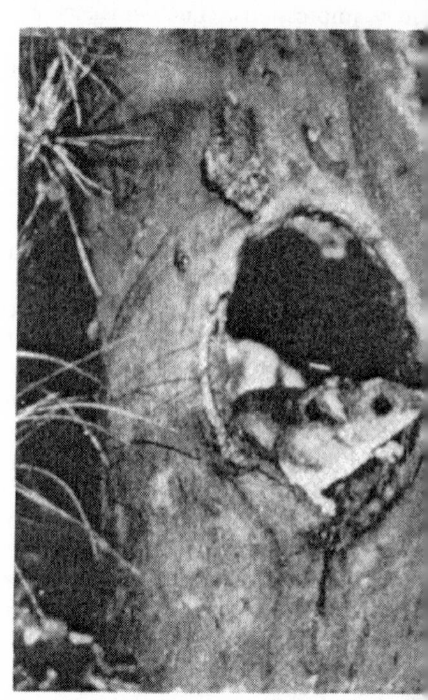

Country Life in An

A white-footed mouse at her own doorwa the woods

apping Field Mice — Probably wild
als have endured more cruelty
ıgh the agency of traps than through
ther form of human persecution. The
e steel traps often catch the animal
ıe leg, holding it until it gnaws off
imprisoned foot, and thus escapes
ned and handicapped for its future
gle for food; or if the trap gets a
g hold, the poor creature may suffer
ıres during a long period, before the
er of the trap appears to put an end
s sufferings by death. If box traps are
, they are often neglected and the
risoned animal is left to languish and
e. The teacher cannot enforce too
ıgly upon the child the ethics of trap-
. Impress upon him that the box traps
ar less cruel; but that if set, they must
xamined regularly and not neglected.
study of mice affords a good oppor-
ity for giving the children a lesson in
ıane trapping. Let them set a tin-can
for meadow mice or deer mice. They
st examine the traps every morning.
little prisoners may be brought to
ool and studied; meanwhile, they
uld be treated kindly and fed bounti-
y. After a mouse has been studied it
uld be set free, even though it be one
the quite pestiferous field mice. The
ral effect of killing an animal after a

Tracks of a white-footed mouse. Note how e long tail has left a print in the snow. As is mouse does not hibernate, its tracks are ten seen on snow

Nature Photography around the Year, Percy A. Morris, © D. Appleton-Century Co., Inc.

A meadow mouse

child has become thoroughly interested in it and its life is always bad.

LESSON 55
The House Mouse

Leading Thought — The mouse is fitted by color, form, agility, and habits to thrive upon the food which it steals from man, and to live in the midst of civilized people.

Method — A mouse cage can be easily made of wire window-screen tacked upon a wooden frame. I have even used aquarium jars with wire screen covers; by placing one jar upon another, opening to opening, and then laying them horizontal, the mouse can be transferred to a fresh cage without trouble, and thus the mousy odor can be obviated while the little creature is being studied. A little water in a wide-necked bottle can be lowered into this glass house by a string, and the food can be given in like manner. Stripped paper should be put into the jar for the comfort of the prisoner; a stiff string hanging down from the middle of the cage will afford him a chance to show his feats as an acrobat.

Observations — 1. Why is the color of the mouse of special benefit to it? Do you think it protects it from the sight of

Robert T. Hatt

A tin-can trap for catching small rodents alive. To a choke trap is wired a tin can with a piece slightly larger than the bait treadle of the trap cut out. To the choke wire of the trap is fastened a square of coarse wire mesh

its enemies? Can you see a mouse easily as it runs across the room? What is the nature of the fur of a mouse?

2. How long is a mouse's tail as compared with its body? What is the covering of the tail? Of what use to the mouse is this long, ridged tail? Watch the mouse carefully and discover, if you can, the use of the tail in climbing.

3. Is the mouse a good jumper? Are the hind legs long and strong when compared with the front legs? How high do you think a mouse can jump? Do you think it uses its tail as an aid in jumping? How much of the legs are covered with hair? Compare the front and hind feet. What sort of claws have they? How does the mouse use its feet when climbing the string? How can it climb up the side of a wall?

4. Describe the eyes. Do you think the mouse can see very well? Does it wink? What is the shape of the ears? Do you think it can hear well? Can it move its ears forward or backward?

5. What is the shape of the snout what advantage is this? Note the whis What is their use? Describe the mc Do you know how the teeth are arran For what other purpose than to bite does the mouse use its teeth? What c animals have their teeth arranged those of the mouse? What food does house mouse live upon? How doe get it?

6. How does the mouse act when reaching up to examine something? I does it hold its front feet? Describe the mouse washes its face; its back; its

7. Where does the house mouse b its nest? Of what material? How do baby mice look? Can they see when are first born?

8. House mice are great travelers. you tell how they manage to get f place to place? Write a story telling you know of their habits.

9. How many kinds of mice do know? Does the house mouse ever in the field? What do you know of habits of the white-footed mouse? Of meadow mice? Of the jumping mice?

American Humane Soc

A woodchuck caught in a humane trap. such traps are visited frequently, anim caught in them do not suffer such agonies in ordinary steel traps. Information ab various types of humane traps can be secu from the American Humane Associati Denver, Colorado.

THE WOODCHUCK

He who knows the ways of the wood-uck can readily guess where it is likely be found; it loves meadows and pastures ere grass or clover lushly grows. It is o fond of garden truck and has a special lectation for melons. The burrow is ely to be situated near a fence or stone ap, which gives easy access to the osen food. The woodchuck makes its ırrow by digging the earth loose with its ont feet, and pushing it backward and t of the entrance with the hind feet. his method leaves the soil in a heap near e entrance, from which paths radiate to the grass in all directions. If one un-rtakes to dig out a woodchuck, one eeds to be not only a husky individual, ıt something of an engineer; the direc-on of the burrow extends downward for little way, and then rises at an easy angle, that the inmate may be in no danger f flood. The nest is merely an enlarge-ıent of the burrow, lined with soft grass hich the woodchucks bring in in their ıouths. During the early part of the sea-on, the father and mother and the litter f young may inhabit the same burrow, though there are likely to be at least two parate nests. There is usually more than ne back door to the woodchuck's dwelling, through which it may escape if pressed too closely by enemies; these back doors differ from the entrance in that they are usually hidden and have no earth heaped near them.

The woodchuck usually feeds in the morning and again in the evening, and is likely to spend the middle of the day resting. It often goes some distance from its burrow to feed, and at short intervals lifts

The woodchuck is at home in grassy meadows

itself upon its hind feet to see if the coast is clear; if assailed, it will seek to escape by running to its burrow; and when running, it has a peculiar gait well described as "pouring itself along." If it reaches its burrow, it at once begins to dig deeply and throw the earth out behind it, thus making a wall to keep out the enemy. When cornered, the woodchuck is a courageous and fierce fighter; its sharp incisors prove a powerful weapon and it will often whip a dog much larger than itself. Every boy knows how to find whether the wood-

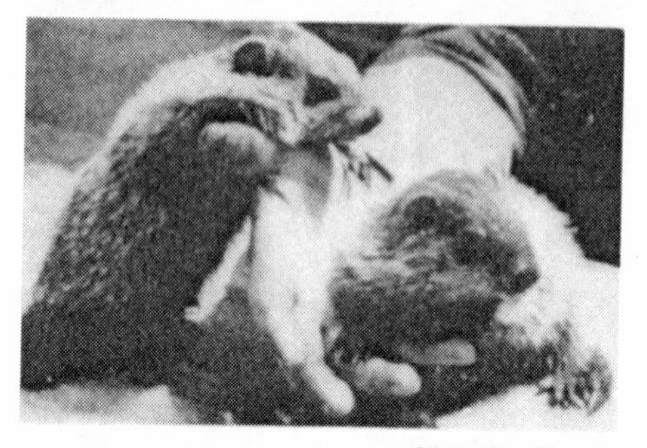

W. J. Hamilton, Jr.

These young woodchucks are as tame as kittens

chuck is in its den or not, by rolling a stone into the burrow, and listening; if the animal is at home, the sound of its digging apprises the listener of the fact. In earlier times, the ground hogs were much preyed upon by wolves, wildcats, and foxes; now only the fox remains and he is fast disappearing, so that at present the farmer and his dog are about the only enemies this burrower has to contend with. In recent years it has been considered a game animal and furnishes much sport for the rifleman. It is an animal of resources and will climb a tree if attacked by a dog; it will also climb trees for fruit, such as wild cherries or peaches. During the late summer, it is the ground hog's business to feed very constantly and become very fat. About the first of October, it retires to its den and sleeps until the end of February or early March, in the eastern United States. During this dormant state, the beating of its heart is so faint as to be scarcely perceptible, and very little nourishment is required to kee alive; this nourishment is supplied by fat stored in its body, which it uses u spring, when it comes out of its bur looking gaunt and lean. The old sa that the ground hog comes out on Can mas Day, and if it sees its shadow, back to sleep for six weeks more, savor of meteorological truth, but it is tainly not true of the ground hog.

The full-grown woodchuck ordina measures about two feet in length. color is grizzly or brownish, someti blackish in places; the under parts are dish and the feet black. The fur is rat coarse, thick, and brown, with longer h which are grayish. The skin is very th and tough and seems to fit loosely, a co tion which gives the peculiar "pour along" appearance when it is runni The hind legs and feet are longer th those in front. Both pairs of feet are fit for digging, the front ones being used loosening the earth and the hind p for kicking it out of the burrow.

The woodchuck's ears are roundish a not prominent; the sense of hearing acute. The teeth consist of two large wh incisors at the front of each jaw, ther bare space, and then four grinders on ea side, above and below; the incisors used for biting food and also for fighti The eyes are full and bright. The tail short and brushy, and it, with the hi legs, forms a tripod which supports t animal as it sits with its forefeet lifted.

When feeding, the woodchuck oft makes a contented grunting noise; wh attacked and fighting, it growls; it al can whistle. I had a woodchuck acquai ance once which always gave a high, shri almost birdlike whistle when I came view. There are plenty of statements books that woodchucks are fond of musi and Mr. Ingersoll states that at Wellesl College a woodchuck on the chapel law was wont to join the morning song exe cises with a "clear soprano." The you woodchucks are born from late March mid May, and the litter usually numbe four or five. In June the "chucklings may be seen following the mother in th with much babyish grunting. If cap- at this period, they make very in- ing pets. By July the young wood- ks leave the home burrow and start ws of their own.

LESSON 56

ie Woodchuck or Ground Hog

ading Thought — The woodchuck thriven with civilization, notwith- ding the farmer's dog, gun, traps, and on. It makes its nest in a burrow in earth and lives upon vegetation; it rnates in winter.

ethod — Within convenient distance observation by the pupils of every ntry schoolhouse and of most village olhouses, may be found a woodchuck its dwelling. The pupils should be n the outline for observations which ıld be made individually through ching the woodchuck for weeks or ıths.

bservations — 1. Where is the wood- ck found? On what does it live? At ıt time of day does it feed? How does ːt when startled?

. Is the woodchuck a good fighter? th what weapons does it fight? What its enemies? How does it escape its mies when in or out of its burrow? w does it look when running?

. What noises does the woodchuck ke? Play a mouth organ near the wood- ck's burrow and note if it likes music.

. How does the woodchuck make its row? Where is it likely to be situated? here is the earth placed which is taken m the burrow? How does the wood- ck bring it out? How is the burrow de so that the woodchuck is not wned in case of heavy rains? In what direction do the underground galleries go? Where is the nest placed in relation to the galleries? Of what is the nest made? How is the bedding carried in? Of what special use is the nest?

5. Do you find paths leading to the entrances of the burrow? If so, describe them. How can you tell whether a woodchuck is at home or not if you do not see it enter? Where is the woodchuck likely to station itself when it sits up to look for intruders?

6. How many woodchucks inhabit the same burrow? Are there likely to be one or more back doors to the burrow? What for? How do the back doors differ from the front doors?

7. How long is the longest woodchuck that you have ever seen? What is the woodchuck's color? Is its fur long or short? Coarse or fine? Thick or sparse? Is the skin thick or thin? Does it seem loose or close fitting?

8. Compare the front and hind feet and describe the difference in size and shape. Are either or both slightly webbed? Explain how both front and hind feet and legs are adapted by their shape to help the woodchuck. Is the tail long or short? How does it assist the animal in sitting up?

9. What is the shape of the woodchuck's ear? Can it hear well? Of what use are the long incisors? Describe the eyes.

10. How does the woodchuck prepare for winter? Where and how does it pass the winter? Did you ever know a woodchuck to come out on Candlemas Day to look for its shadow?

11. When does the woodchuck appear in the spring? Compare its general appearance in the fall and in the spring and explain the reason for the difference.

12. When are the young woodchucks born? What do you know of the way the mother woodchuck cares for her young?

As I turned round the corner of Hubbard's Grove, saw a woodchuck, the first of the season, in the middle of the field six or seven rods from the fence which bounds the wood, and twenty rods distant.

I ran along the fence and cut him off, or rather overtook him, though he started at the same time. When I was only a rod and a half off, he stopped, and I did the same; then he ran again, and I ran up within three feet of him, when he stopped again, the fence being between us. I squatted down and surveyed him at my leisure. His eyes were dull black and rather inobvious, with a faint chestnut iris, with but little expression and that more of resignation than of anger. The general aspect was a coarse grayish brown, a sort of grisel. A lighter brown next the skin, then black or very dark brown and tipped with whitish rather loosely. The head between a squirrel and a bear, flat on the top and dark brown, and darker still or black on the tip of the nose. The whiskers black, two inches long. The ears very small and roundish, set far back and nearly buried in the fur. Black feet, with long and slender claws for digging. It appeared to tremble, or perchance shivered with cold. When I moved, it gritted its teeth quite loud, sometimes striking the under jaw against the other chatteringly, sometimes grinding one jaw on the other, yet as if more from instinct than anger. Whichever way I turned, that way it headed. I took a twig a foot long and touched its snout, at which it started forward and bit the stick, lessening the distance between us to two feet, and still it held all the ground it gained. I played with it tenderly awhile with the stick, trying to open its gritting jaws. Ever its long incisors, two above and two below, were presented. But I thought it would go to sleep if I stayed long enough. It did not sit upright as sometimes, but standing on its fore feet with its head down, i. e., half sitting, half standing. We sat looking at one another about half an hour, till we began to feel mesmeric influences. When I was tired, I moved away, wishing to see him run, but I could not start him. He would not stir as long as I was looking at him or could see him. I walked around him; he turned as fast and fronted me still. I sat down by his side within a foot. I talked to him quasi forest lingo, baby-talk, at any rate in a conciliatory tone, and thought that I some influence on him. He gritte teeth less. I chewed checkerberry l and presented them to his nose a without a grit; though I saw that much gritting of the teeth he had them rapidly and they were covered a fine white powder, which, if you r ured it thus, would have made his a terrible. He did not mind any no might make. With a little stick I l one of his paws to examine it, and it up at pleasure. I turned him over t what color he was beneath (darke most purely brown), though he tu himself back again sooner than I c have wished. His tail was also br though not very dark, rat-tail like, loose hairs standing out on all sides a caterpillar brush. He had a rather look. I spoke kindly to him. I reac checkerberry leaves to his mouth stretched my hands over him, tho he turned up his head and still gritt little. I laid my hand on him, but mediately took it off again, instinct being wholly overcome. If I had ha few fresh bean leaves, thus in advanc the season, I am sure I should have tar him completely. It was a frizzly tail. is a humble, terrestrial color like the tridge's, well concealed where dead v grass rises above darker brown or chest dead leaves — a modest color. If I had some food, I should have ended w stroking him at my leisure. Could ea have wrapped him in my handkerch He was not fat nor particularly lean finally had to leave him without see him move from the place. A large, clum burrowing squirrel. Arctomys, bear-mou I respect him as one of the natives. lies there, by his color and habits so n uralized amid the dry leaves, the wither grass, and the bushes. A sound nap, t he has enjoyed in his native fields, the p winter. I think I might learn some w dom of him. His ancestors have lived h longer than mine. He is more thorough acclimated and naturalized than I. Be leaves the red man raised for him, but can do without them.

— Thoreau's Journ

THE RED SQUIRREL OR CHICKAREE

Just a tawny glimmer, a dash of red and gray,
Was it a flitting shadow, or a sunbeam gone astray!
It glances up a tree trunk, and a pair of bright eyes glow
Where a little spy in ambush is measuring his foe.
I hear a mocking chuckle, then wrathful, he grows bold
And stays his pressing business to scold and scold and scold.

'e ought to yield admiring tribute to e animals which have been able to ish in our midst despite man and his this weapon being the most cowardly unfair invention of the human mind. only time that man has been a fair ter in combating his four-footed hren was when he fought them with eapon which he wielded in his hand. re is nothing in animal comprehen- which can take into account a pro- ile, and much less a shot from a gun; though it does not understand, it ex- ences a deathly fear at the noise. It athetic to note the hush in a forest follows the sound of a gun; every song, y voice, every movement is stilled and y little heart filled with nameless ter- How any man or boy can feel manly en, with this scientific instrument of th in his hands, he takes the life of ttle squirrel, bird, or rabbit, is beyond comprehension. In pioneer days when vas a fight for existence, man against wilderness, the matter was quite dif- ent; but now it seems to me that any- who hunts what few wild creatures have left, and which are in nowise in- ious, is, whatever he may think of him- f, no believer in fair play.

Within my own memory, the beautiful ck squirrel was as common in our ods as was his red cousin; the shotgun exterminated this splendid species lo- ly. Well may we rejoice that the red irrel has, through its lesser size and ater cunning, escaped a like fate; and t, pugnacious and companionable d shy, it lives in our midst and climbs r very roofs to sit there and scold us for ming within its range of vision. It has succeeded not only in living despite man, but because of man, for it rifles our grain bins and corn cribs and waxes opulent by levying tribute upon our stores.

Thoreau describes most graphically the movements of this squirrel. He says: "All day long the red squirrels came and went. One would approach at first warily, warily,

Dorothy M. Compton

Red squirrel at feeding log

through the shrub-oaks, running over the snow crust by fits and starts and like a leaf blown by the wind, now a few paces this way, with wonderful speed and waste of energy, making inconceivable haste with his "trotters," as if it were for a wager, and now as many paces that way, but never getting on more than half a rod at a time; and then suddenly pausing with a ludicrous expression and a gratuitous somersault, as if all the eyes of the universe were fixed on him . . . and then suddenly, before you could say "Jack Robinson" he would be in the top of a

A red squirrel on his vine bridge

young pitch pine, winding up his clock, and chiding all imaginary spectators, soliloquizing and talking to all the universe at the same time."

It is surely one of the most comical of sights to see a squirrel stop running and take observations; he lifts himself on his haunches, and with body bent forward, presses his little paws against his breast as if to say, "Be still, O my beating heart!" which is all pure affectation because he knows he can scurry away in perfect safety. He is likely to take refuge on the far side of a tree, peeping out from this side and that, and whisking back like a flash as he catches our eye; we might never know he was there except that, as Riley puts it, "he lets his own tail tell on him." When climbing up or down a tree, he goes head first and spreads his legs apart to clasp as much of the trunk as possible; meanwhile his sharp little claws cling securely to the bark. He can climb out on the smallest twigs quite as well, when he needs to do so, in passing from tree to tree or when gathering acorns.

A squirrel always establishes certain roads to and from his abiding place and almost invariably follows them. Such a path may be entirely in the tree tops, air bridges from a certain branch of tree to a certain branch of another, may be partially on the ground bet trees. I have made notes of these pat the vicinity of my own home, and noted that if a squirrel leaves them exploring, he goes warily; while, whe lowing them, he is quite reckless i haste. When making a jump from to tree, he flattens himself as wide possible and his tail is held some curved, but on a level with the bod if its wide brush helped to buoy hin and perhaps to steer him also.

During the winter the chickare brightly colored and is a conspicuous ject; his back is bright russet, almost and along his sides, where the red m the grayish white of the underside, t is a dark line which is very ornamen With the coming of summer, however coat becomes quite dingy. In Noven he moults, and his bright color retu When dashing up a tree trunk, his c is never very striking but looks like glimmer of sunlight; this has prob saved many of his kind from the gun whose eyes, being at the front of his h cannot compare in efficiency with th of the squirrel, which, large and full alert, are placed at the sides of the h so as to see equally well in all directions

The squirrel's legs are short because is essentially a climber rather than a r ner; the hips are very strong, which sures his power as a jumper, and his le are truly remarkable. A squirrel uses front paws for hands in a most hun way; with them he washes his face a holds his food up to his mouth wh eating, and it is interesting to note the s of his claws when used as fingers. The tr he makes in the snow is quite charact istic. The tracks are paired and those the large five-toed hind feet are always front.

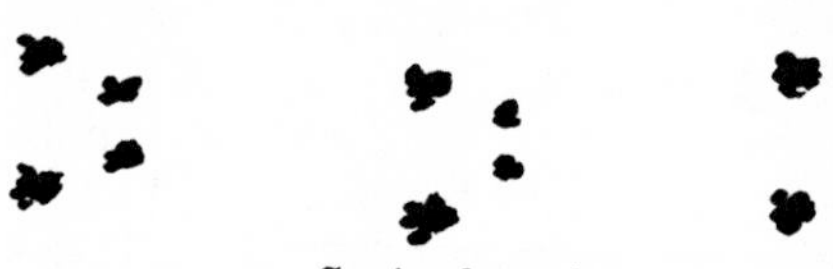

Squirrel tracks

he squirrel has two pairs of gnawing ı which are very long and strong, as l rodents, and he needs to keep busy ving hard things with them, or they grow so long that he cannot use them l and will starve to death. He is very er about opening nuts so as to get all meats. He often opens a hickory nut two holes which tap the places of nut meats squarely; with walnuts butternuts, which have much harder ls, he makes four small holes, one op- te each quarter of the kernel. He has cheek pouches like a chipmunk but can carry corn and other grain. He n fills his mouth so full that his cheeks ge out like those of a boy eating pop- n; but anything as large as a nut he ies in his teeth. His food is far more ed than many suppose and he will almost anything eatable; he is a little te and enjoys stealing from others with nest zest. In spring, he eats leaf buds hunts our orchards for apple seeds. winter, he feeds on nuts, buds, and es; it is marvelous how he will take a e apart, tearing off the scales and leav- them in a heap while searching for ds; he is especially fond of the seeds Norway spruce and hemlock. Of course, is fond of nuts of all kinds and will the chestnut burs from the tree before y are ripe, so that he may get ahead of other harvesters. He stores his food winter in all sorts of odd places and en forgets where he puts it. We often find his winter stores untouched the next summer. He also likes birds' eggs and nestlings, and if it were not for the chastisement he gets from the parent robins, he would work much damage in this way.

Dwight E. Sollberger

Flying squirrel just leaving home

The red squirrels use a great variety of places for nests. In different localities various types of nests are constructed; some individuals prefer hollow trees, some build nests in clumps of vines, such as wild grape vines, and still others make their homes in the ground under or about stumps. During the winter, the red squirrel does not remain at home except in the coldest weather, when he lies cozily with his tail wrapped around him like a fur neck-piece to keep him warm. He is too full of interest in the world to lie quietly long, but comes out, hunts up some of his stores, and finds life worth while despite the cold. One squirrel adopted a birdhouse in one of our trees, and he or his kin have lived there for years; in winter, he takes his share of the suet put on the trees for birds, and because of his greediness we have been compelled to use picture wire for tying on the suet.

A. A. Allen

A gray squirrel with food in its paws

The young are born in a well-protected nest. There are four to six in a litter and they usually appear in April. If it is necessary to move the young the mother grasps the babies by the loose skin of their underparts and carries them to safety.

The squirrel has several ways of expressing his emotions; one is by various curves in his long, beautiful bushy tail. If the creatures of the wood had a stage, the squirrel would be their chief actor. Surprise, incredulousness, indignation, fear, anger, and joy are all perfectly expressed by tail gestures and also by voice. As a vocalist he excels; he chatters with curiosity, "chips" with surprise, scolds by giving a guttural trill, finishing with a falsetto squeal. He is the only singer I know who can carry two parts at a time. Notice him sometimes in the top of a hickory or chestnut tree when nuts are ripe, and you will hear him singing a duet all by himself, a high shrill chatter with a chuckling accompaniment. Long may he abide with us as an uninvited guest at our cribs! For, though he be a freebooter and conscienceless, yet our world would lack its highest example of incarnate grace and activity if he were not in it.

LESSON 57

The Red Squirrel or Chickaree

Leading Thought — The red squirrel by its agility and cleverness has lived on, despite its worst enemy — man. By form and color and activity it is fitted to elude the hunter.

Method — If a pet squirrel in a cage can be procured for observation at the school, the observations on the form and habits of the animal can be best studied thus; but a squirrel in a cage is an anomaly and it is far better to stimulate the p to observe the squirrels out of doors. the following questions, a few at a t and ask the pupils to report the ans to the entire class. Much should be with the supplementary reading, as t are many interesting squirrel stories trating its habits.

Observations — 1. Where have seen a squirrel? Does the squirrel along or leap when running on ground? Does it run straight ahea stop at intervals for observation? H does it look? How does it act when l ing to see if the "coast is clear"?

2. When climbing a tree, does it straight up, or move around the tru How does it hide itself behind a trunk and observe the passer-by? Desc how it manages to climb a tree. Doe go down the tree head first? Is it abl climb out on the smallest branches? what advantage is this to the squirrel?

3. Look closely and see if a squirrel lows the same route always when p ing from one point to another. H does it pass from tree to tree? How d it act when preparing to jump? H does it hold its legs and tail when the air during a jump from branch branch?

4. Describe the colors of the red sq rel above and below. Is there a dark str along its side; if so, what color? How d the color of the squirrel protect it fr its enemies? Is its color brighter in su mer or in winter?

5. How are the squirrel's eyes place Do you think it can see behind as well in front all the time? Are its eyes brig and alert, or soft and tender?

6. Are its legs long or short? Are hind legs stronger and longer than t front legs? Why? Why does it not ne long legs? Do its paws have claws? H does it use its paws when eating and making its toilet?

7. Describe the squirrel's tail. Is it long as the body? Is it used to expre emotion? Of what use is it when the squ rel is jumping? Of what use is it in t winter in the nest?

What is the food of the squirrel dur- ıe autumn? Winter? Spring? Sum- Where does it store food for the r? Does it steal food laid up by jays, nunks, mice, or other squirrels? How it carry nuts? Has it cheek-pouches ıe chipmunk for carrying food? Does in its nest all winter living on stored like a chipmunk?

Where does the red squirrel make me? Of what is it made and where ? In what sort of nest are the young and reared? At what time of the are the young born? How does the mother squirrel carry her little ones if she wishes to move them?

10. How much of squirrel language can you understand? How does it express surprise, excitement, anger, or joy during the nut harvest? Note how many different sounds it makes and try to discover what they mean.

11. Describe or sketch the tracks made by the squirrel in the snow.

12. How does the squirrel get at the meats of the hickory nut and the walnut? How are its teeth arranged to gnaw holes in such hard substances as shells?

FURRY

rry was a baby red squirrel. One day lay his mother was moving him from tree to another. He was clinging with ittle arms around her neck and his clasped tightly against her breast something frightened her, and in sudden movement she dropped her y baby in the grass. Thus, I inherited and entered upon the rather onerous es of caring for a baby of whose needs ew little; but I knew that every well-l-for baby should have a book detailall that happens to it, and therefore ıde a book for Furry, writing in it each the things he did. If the children who pets keep similar books, they will them most interesting reading after-l, and they will surely enjoy the writvery much.

XTRACTS FROM FURRY'S NOTEBOOK

Iay 18, 1902 — The baby squirrel is large enough to cuddle in one hand. cuddles all right when once he is cap-d; but he is a terrible fighter, and when tempt to take him in my hand, he tches and bites and growls so that ave been obliged to name him Fury. old him, however, if he improved in per I would change his name to Furry.

Iay 19 — Fury greets me, when I open box, with the most awe-inspiring little wls, which he calculates will make me turn pale with fear. He has not cut his teeth yet, so he cannot bite very severely, but that isn't his fault, for he tries hard enough. The Naturalist said cold milk would kill him, so I warmed the milk and put it in a teaspoon and placed it in front of his nose; he batted the spoon with both forepaws and tried to bite it, and thus got a taste of the milk, which he drank eagerly, lapping it up like a kitten. When I hold him in one hand and cover him with the other, he turns contented little somersaults over and over.

May 20 — Fury bit me only once today, when I took him out to feed him. He is cutting his teeth on my devoted fingers. I tried giving him grape-nuts soaked in milk, but he spat it out in disgust. Evidently he does not believe he needs a food for brain and nerve. He always washes his face as soon as he is through eating.

May 21 — Fury lies curled up under his blanket all day. Evidently good little squirrels stay quietly in the nest, when the mother is not at home to give them permission to run around. When Fury sleeps, he rolls himself up in a little ball with his tail wrapped closely around him. The squirrel's tail is his "furs," which he wraps around him to keep his back warm when he sleeps in winter.

May 23 — Every time I meet Uncle

John he asks, "Is his name Fury or Furry now?" Uncle John is much interested in the good behavior of even little squirrels. As Fury has not bitten me hard for two days, I think I will call him Furry after this. He ate some bread soaked in milk to-day, holding it in his hands in real squirrel fashion. I let him run around the room and he liked it.

May 25 — Furry got away from me this morning and I did not find him for an hour. Then I discovered him in a pasteboard box of drawing paper with the cover on. How did he squeeze through?

May 26 — He holds the bowl of the spoon with both front paws while he drinks the milk. When I try to draw the spoon away to fill it again after he has emptied it, he objects and hangs on to it with all his little might, and scolds as hard as ever he can. He is such a funny, unreasonable baby.

May 28 — Tonight I gave Furry a walnut meat. As soon as he smelled it he became greatly excited; he grasped the meat in his hands and ran off and hid under my elbow, growling like a kitten with its first mouse.

May 30 — Since he tasted nuts he has lost interest in milk. The nut meats are too hard for his new teeth, so I mash them and soak them in water and now he eats them like a little piggy-wig with no manners at all. He loves to have me stroke his back while he is eating. He uses his thumbs and fingers in such a human way that I always call his front paws *hands*. When his piece of nut is very small he holds it in one hand and clasps the other hand behind the one which holds the dainty morsel, so as to keep it safe.

May 31 — When he is sleepy he scolds if I disturb him and turning over on his back bats my hand with all of his soft little paws and pretends that he is going to bite.

June 4 — Furry ranges around the room now to please himself. He is a little mischief; he tips over his cup of milk and has commenced gnawing off the wall-paper behind the bookshelf to mak a nest. The paper is green and will ably make him sorry.

June 5 — This morning Furry wa den in a roll of paper. I put my hand one end of the roll and then reach with the other hand to get him; b got me instead, because he ran u sleeve and was much more content be there than I was to have him. glad enough when he left his hiding and climbed to the top shelf of the case, far beyond my reach.

June 6 — I have not seen Furr twenty-four hours, but he is here s enough. Last night he tipped over ink bottles and scattered nut shells the floor. He prefers pecans to any nuts.

June 7 — I caught Furry today an bit my finger so that it bled. But a wards, he cuddled in my hand for a time, and then climbed my shoulder went hunting around in my hair wanted to stay there and make a When I took him away, he pulled ou two hands full of my devoted tresses not employ him as a hairdresser.

June 9 — Furry sleeps nights in the drawer of my desk; he crawls in from hind. When I pull out the drawer he out and scares me nearly out of my but he keeps his wits about him and away before I can catch him.

June 20 — I keep the window ope Furry can run out and in and learn take care of himself out-of-doors.

Furry soon learned to take care of h self, though he often returned for n which I kept for him in a bowl. He d not come very near me out-of-doors, he often speaks to me in a friendly man from a certain pitch pine tree near house.

There are many blank leaves in Fur notebook. I wish that he could have w ten on these of the things that he thou about me and my performances. It wo certainly have been the most interest book in the world concerning squirrels.

THE CHIPMUNK

hile the chipmunk is a good runner umper, it is not so able a climber as e red squirrel, and it naturally stays er the ground. One windy day I was k by the peculiar attitude of what t thought was a red squirrel gathergreen acorns from a chestnut oak in of my window. A second glance ed me that it was a chipmunk lying to the branch, hanging on for "dear and with an attitude of extreme cau- quite foreign to the red squirrel in nilar situation. He would creep out, an acorn in his teeth, creep back larger limb, take off the shell, and his little paws stuff the kernel into cheek-pouches; he took hold of one of his mouth with one hand to tch it out, as if opening a bag, and fed the acorn in with the other. I do know whether this process was neces- or not at the beginning, for his cheeks e distended when I first saw him; and kept on stuffing them until he looked f he had a hopeless case of mumps. n with obvious care he descended the

Leonard K. Beyer

is chipmunk has his cheek-pouches well stuffed

e and retreated to his den in the side- l, the door of which I had already dis- vered, although it was well hidden by bunch of orchard grass.

Chipmunks are more easily tamed than red squirrels and soon learn that pockets may contain nuts and other things good to eat. The first tame chipmunk of my

"Chipsie," a chipmunk of the Sierras

acquaintance belonged to a species found in the California mountains. He was a beautiful little creature and loved to play about his mistress' room; she, being a naturalist as well as a poet, was able to understand her little companion, and the relations between them were full of mutual confidence. He was fond of English walnuts and would always hide away all that were placed in a dish on the table. One day his mistress, when taking off her hat after returning from church, discovered several of these nuts tucked safely in the velvet bows; they were invisible from the front but perfectly visible from the side. Even yet, she wonders what the people at church that day thought of her original ideas in millinery; and she wonders still more how "Chipsie" managed to get into the hatbox, the cover of which was always carefully closed.

The chipmunk is a good home builder and carries off, presumably in its cheek-pouches, all of the soil which it removes in making its burrow. The burrow is usu-

Dorothy M. Compton

Peanuts are a favorite food of tame chipmunks

ally made in a dry hillside, the passageway just large enough for its own body, widening to a nest which is well bedded down. There is usually a back door also, so that in case of necessity the inmate can escape. It retires to this nest in late November and does not appear again until March. In mild winters it may be up and about on bright, sunny days. In the nest it stores nuts and other grains so that when it wakens, at long intervals, it can take refreshment.

If you really wish to know whether you see what you look at or not, test yourself by trying to describe the length, position, and number of the chipmunk's stripes. These stripes, like those of the tiger in the jungle, make the creature less conspicuous; when on the ground, where its stripes fall in with the general shape and color of the grass and underbrush, it is quite invisible until it stirs. Its tail is not so long nor nearly so bushy as that of the

Chipmunks sometimes cache their food under stumps

squirrel; it does not need a tail to ba and steer with in the tree tops; and si lives in the ground, a bushy tail w soon be loaded with earth and w be an incubus instead of a thing of be

The chipmunk is not a vocalist lik red squirrel, but he can cluck like a cu and chatter gayly or cogently; and he make himself into a little bunch wit tail curved up his back, while he e nut from both his hands. He is more amusing than the red squirrel in attitude, probably because he is mor nocent and not so much of a *poseur.* food consists of all kinds of nuts, g and fruit, but he does little or no dam

Dorothy M. Con

Common chipmunk, often called gro squirrel

as a rule. He does upon occasion rob flower garden of valued bulbs. He pretty and distinctly companionable, I can rejoice that I have had him his whole family as my near neighbors many years. I always feel especially pr when he shows his confidence by scamp ing around our porch floor and peep in at our windows, as if taking a recipro interest in us.

LESSON 58

The Chipmunk

Leading Thought — The chipmu lives more on the ground than does t rel; its colors are protective and it cheek-pouches in which it carries , and also soil when digging its bur- It stores food for winter in its

ETHOD — The field notebook should ne basis for this work. Give the pupils utline of observations to be made, and for reports now and then. Meanwhile ulate interest in the little creatures by ing aloud from some of the references n.

BSERVATIONS — 1. Do you see the munk climbing around in trees like red squirrel? How high in a tree have ever seen a chipmunk?

. What are the chipmunk's colors ve and below? How many stripes has Where are they and what are their rs? Do you think that these stripes ceal the animal when among grasses bushes?

. Compare the tails of the chipmunk the red squirrel. Which is the longer and bushier? Tell if you can the special advantage to the chipmunk in having this less bushy tail.

4. What does the chipmunk eat? How does it carry its food? How does it differ in this respect from the red squirrel? Does it store its food for winter use? How does it prepare its nuts? How does it hold its food while eating?

5. Where does the chipmunk make its home? How does it carry away soil from its burrow? How many entrances are there? How is the den arranged inside? Does it live in the same den the year round? When does it retire to its den in the fall? When does it come out in the spring?

6. Does the chipmunk do any damage to crops? What seeds does it distribute? At what time do the little chipmunks appear in the spring?

7. Observe carefully the different tones of the chipmunk and compare its chattering with that of the squirrel.

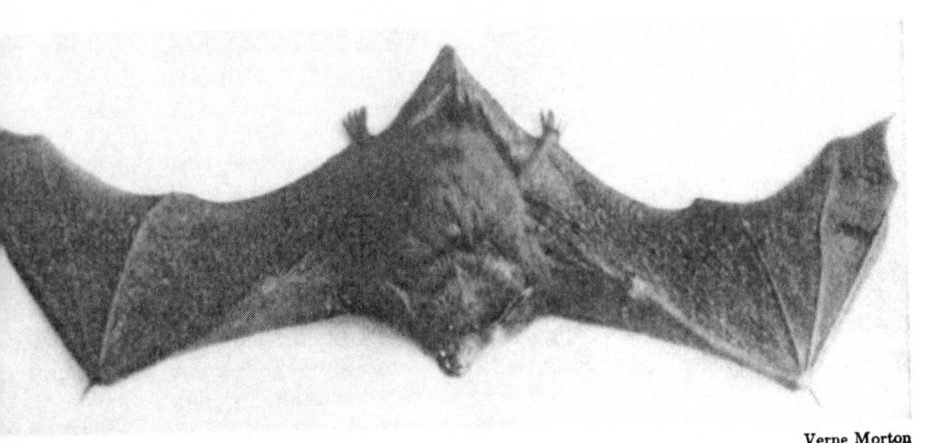

Verne Morton

A bat

THE LITTLE BROWN BAT

His small umbrella, quaintly halved,
Describing in the air an arc alike inscrutable, —
Elate philosopher! — EMILY DICKINSON

Whoever first said "as blind as a bat," rely never looked a bat in the face, or e would not have said it. The deep-set, een, observant eyes are quite in keeping with the alert attitude of the erect, pointed ears; while the pug nose and the wide-open little pink bag of a mouth, set with tiny, sharp teeth, give this anomalous little animal a deliciously impish look. Yet how have those old artists belied the bat, who fashioned their demons after his pattern, ears, eyes, nose, mouth, wings, and all! The superstitions which link the bat with evil malign this bright, engaging little creature. There are no other wings so wonderful as the bat's; the thin mem-

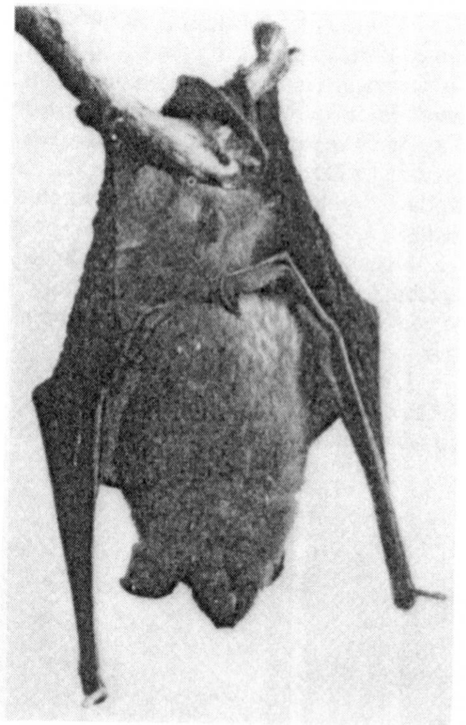

Hung up for his daytime nap

brane is equipped with sensitive nerves which inform the flier of the objects in his path, so that he darts among the branches of trees at terrific speed and never touches a twig; a blinded bat was once set free in a room, across which threads were stretched, and he flew about without ever touching one. After we have tamed one of these little, silky flitter-mice we soon get reconciled to his wings for he proves the cunningest of pets; he soon learns who feeds him, and is a constant source of entertainment.

The flight of the bat consists of darting hither and thither with incredible swiftness, and making sharp turns with no apparent effort. Swifts and swallows cannot compete with the bat in wing celerit agility; it is interesting to note that birds also catch insects on the win food. He makes a collecting net o wing membrane stretched between hind legs and tail, doubling it up lik apron on the unfortunate insects, then reaching down and gobbling t up; and thus he is always doing good ice to us on summer evenings by swal ing a multitude of insects.

The short fur of the bat is as so silk, and covers the body but not wings; the plan of the wing is somet like that of the duck's foot; it consist a web stretched between very much e gated fingers. If a boy's fingers were as in proportion as a bat's, they would m ure four feet. Stretched between the fingers is a thin, rubbery membr which extends back to the ankles thence back to the tip of the bony thus, the bat has a winged margin around his body. Since fingers make framework, it is the thumb that proj from the front angle of the wing, in form of a very serviceable hook, res bling that used by a one-armed man replace the lost member. These hooks bat uses in many ways. He drags him along the floor with their aid, or scratches the back of his head with th if occasion requires. He is essentiall creature of the air and is not at all fit for walking; his knees bend backward an opposite direction from ours. This r ders him unable to walk, and when tempting to do so, he has the appeara of "scrabbling" along on his feet a elbows. When thus moving he keeps wings fluttering rapidly, as if feeling way in the dark, and his movements trembly. He uses his teeth to aid in clin ing.

The little brown bat's wings often me ure nine inches from tip to tip, and y he folds them so that they scarcely sho he does not fold them like a fan, b rather like a pocket-knife. The hind le merely act as a support for the side win and the little hip bones look pitiful sharp. The membrane reaches only to t

; the tiny foot projecting from it is
l with five wirelike toes, tipped with
hooked claws. It is by these claws
he hangs when resting during the
for he is upside-downy in his sleep-
abits, slumbering during the daytime
hanging head downward, without
nconvenience from a rush of blood
e brain; when he is thus suspended,
ail is folded down. Sometimes he
s by one hind foot and a front hook;
ne is a wee thing when all folded to-
r and hung up, with his nose tucked
een his hooked thumbs, in a very
ish fashion.

e bat is very particular about his
nal cleanliness. People who regard
at as a dirty creature might well look
that they be even half as fastidious
e. He washes his face with the front
of his wing, and then licks his wash-
clean; he scratches the back of his
with his hind foot and then licks the
; when hanging head down, he will
h one hind foot down and scratch
nd his ear with an *aplomb* truly comi-
in such a mite; but it is most fun of
o see him clean his wings; he seizes
edges in his mouth and stretches and
s the membrane until we are sure it
nade of silk elastic, for he pulls and
ls it in a way truly amazing.

he bat has a voice which sounds like
squeak of a toy wheelbarrow, and yet
s expressive of emotions. He squeaks
one tone when holding conversation
h other bats, and squeaks quite differ-
ly when seized by the enemy.

he mother bat feeds her little ones
n her breasts as a mouse does its young,
y she cradles them in her soft wings
ile so doing; often she takes them with
when she goes out for insects in the
nings; they cling to her neck during
se exciting rides; but when she wishes
work unencumbered, she hangs her
y youngsters on some twig and goes
k for them later. The little ones are
rn in July and usually occur as twins.
ring the winter, some bats hibernate
e woodchucks or chipmunks. They se-
t for winter quarters some hollow tree or cave or other protected place. They go to sleep when the cold weather comes, and do not awake until the insects are flying; they then come forth in the evenings, or perhaps early in the morning, and do their best to rid the world of insect nuisances. Others migrate to the south with the advent of cold weather.

There are many senseless fears about the bat; for instance, that he likes to get tangled in a lady's tresses, a situation which would frighten him far more than the lady; or that he brings bedbugs into the house when he enters on his quest for insects, which is an ungrateful slander. Some people believe that all bats are vampires, and only await an opportunity to suck blood from their victims. It is true that in South America there are two species which occasionally attack people who are careless enough to sleep with their toes uncovered, but feet thus injured seem to recover speedily. These bats do little damage to people, although they sometimes pester animals; and there are no vampires in the United States. Our bats, on the contrary, are innocent and beneficial to man. There are a few species in our country which have little, leaflike growths on the end of the nose; these growths serve the purpose of sensory organs.

LESSON 59

The Bat

Leading Thought — Although the bat's wings are very different from those of the bird, yet it is a rapid and agile flier. It flies in the dusk and catches great numbers of mosquitoes and other troublesome insects, upon which it feeds.

Method — This lesson should not be given unless there is a live bat to illustrate it; the little creature can be cared for comfortably in a cage in the schoolroom, as it will soon learn to take flies or bits of raw meat when presented on the point of a pencil or toothpick. Any bat will do for this study, although the little brown bat is the one on which my observations were made.

OBSERVATIONS — 1. At what time of day do we see bats flying? Describe how

Charles E. Mohr

Little brown bats hibernating in a Pennsylvania cave

the bat's flight differs from that of birds. Why do bats dart about so rapidly?

2. Look at a captive bat and describe its wings. Can you see what makes the framework of the wings? Do you see the three finger bones extending out into the wings? How do the hind legs support the wing? The tail? Is the wing membrane covered with fur? Is it thick and leathery or thin and silky and elastic? How does the bat fold up its wings?

3. In what position does the bat rest? Does it ever hang by its thumb hooks?

4. Can you see whether the knees of the hind legs bend upward or downward? How does the bat act when trying to walk or crawl? How does it use its thumb h in doing this?

5. What does the bat do dayti Where does it stay during the day many bats congregate together in roosts?

6. Describe the bat's head, inclu the ears, eyes, nose, and mouth. Wh its general expression? Do you thi can see and hear well? How is its m fitted for catching insects? Does it its mouth while chewing or keep it c Do you think that bats can see by light?

7. What noises does a bat make? does it act if you try to touch it? Ca bite severely? Can you understand the Germans call it a flitter-mouse?

8. Do you know how the mother cares for her young? How does she c them? At what time of year may we pect to find them?

9. When making its toilet, how do bat clean its wings? Its face? Its back? feet? Do you know if it is very clean its habits?

10. How and where do the bats the winter? How are they beneficial to Are they ever harmful? What are so superstitions about the bat?

Nature-study should not be unrela to the child's life and circumstances stands for directness and naturalness is astonishing when one comes to th of it, how indirect and how remote fr the lives of pupils much of our educat has been. Geography still often beg with the universe, and finally, perha comes down to some concrete and fami object or scene that the pupil can unc stand. Arithmetic has to do with brok age and partnerships and partial payme and other things that mean nothing the child. Botany begins with cells a protoplasm and cryptogams. History de with political and military affairs, and o rarely comes down to physical facts a to those events that express the real li of the people; and yet political and soc affairs are only the results of expressic of the way in which people live. Read

with mere literature or with stories enes the child will never see. Of e these statements are meant to be general, as illustrating what is even great fault in educational methods. e are many exceptions, and these are becoming commoner. Surely, the best education is that which begins with the materials at hand. A child knows a stone before it knows the earth.

— "THE NATURE-STUDY IDEA,"
L. H. BAILEY

THE SKUNK

nose who have had experience with animal surely are glad that it is small; the wonder always is that so little a ture can make such a large impression n the atmosphere. A fully grown skunk out two feet long; its body is covered long, shining, rather coarse hair, and tail, which is carried like a flag in the is very large and bushy. In color, the is sometimes entirely black, but most n has a white patch on the back of the k, with two stripes extending down back and along the sides to the tail; face, also, has a white stripe.

'he skunk has a long head and a rather nted snout; its front legs are very much rter than its hind legs, which gives it ery peculiar gait. Its forefeet are armed h long, strong claws, with which it digs burrow, which is usually made in light . It also often makes its home in some vice in rocks, or even takes possession of abandoned woodchuck's hole; or trust- to its immunity from danger, makes its me under the barn. In the fall it be- nes very fat, and during the early part winter it hibernates within its den; it nes out during the thaws of winter and ly spring.

The young skunks appear in May; they born in an enlarged portion of the rrow, where a nice bed of grass and ves is made for them; the skunk is scru- lously neat about its own nest. The ung skunks are very active and inter- ting to watch when playing together e kittens.

The skunk belongs to the same family the mink and weasel, which also give a disagreeable odor when angry. The tid material which is the skunk's defense contained in two glands near the base of the tail. These little glands are about the size of marbles, and the quantity of liquid forced from them in a discharge is considerable and it will permeate the atmosphere with its odor for a distance of half a mile down wind. Because this discharge is so disagreeable to all other creatures, the skunk's intelligence has not become so highly developed as has that of some animals. It has not been obliged to rely upon its cunning to escape its enemies, and has therefore never developed

Verne Morton

A skunk. Note the long, pointed head and the bushy tail

either fear or cleverness. It marches abroad without haste, confident that every creature which sees it will give it plenty of room. It is a night prowler, although it is not averse to a daytime promenade. The white upon its fur gives warning at night that here is an animal which had best be left alone. This immunity from attack makes the skunk careless in learning wisdom from experience; it never learns to avoid a trap, or the dangers of a railway or trolley track. It plods deliberately across highways, leaving its protection to the motorist.

The skunk's food consists largely of fruits and berries, insects, mice, snakes, frogs, and other small animals. It also destroys the eggs and young of birds which nest upon the ground. It uses its strong forepaws in securing its prey. Dr. Merriam, who made pets of young skunks after removing their scent capsules, found them very interesting. He says of one which was named "Meph": "We used to walk through the woods to a large

Doubleday, Page & Co.

Pet skunks

meadow that abounded in grasshoppers. Here, Meph would fairly revel in his favorite food, and it was rich sport to watch his manœuvres. When a grasshopper jumped, he jumped, and I have seen him with as many as three in his mouth and two under his forepaws at the same time."

The only injury which the skunk is likely to do farmers is the raiding of hens' nests or the beehives; this can be obviated by properly housing the poultry and bees. On the other hand, the skunk is of great use in destroying injurious insects and mice. Often when skunks burrow beneath barns, they completely rid the place of mice. Skunk fur is very valuable and is sold, surprisingly, under its own name; it is exported in great quantities to Europe.

The skunk takes short steps, and goes so slowly that it makes a double track, the imprints being very close together. The foot makes a longer track than that of the cat, as the skunk is plantigrade; that is, it walks upon its palms and heels as well as its toes.

LESSON 60

The Skunk

Leading Thought — The skunk ha pended so long upon protecting itself its enemies by its disagreeable odor it has become stupid and unadapt and seems never to be able to lear keep off railroad tracks or highways. a very beneficial animal to the farme cause its food consists so largely of in ous insects and rodents.

Method — The questions should given the pupils and they should an them from personal observations or quiries.

Observations — 1. How large i skunk? Describe its fur. Where does black and white occur in the fur? Of w use is the white to the skunk? Is the valuable? What is its commercial na

2. What is the shape of the sku head? The general shape of the body? tail? Are the front legs longer or sho than the hind legs? Describe the fr feet. For what are they used?

3. Where and how does the skunk m its nest? Does it sleep like a woodch during the winter? What is its food? H does it catch its prey? Does it hunt its food during the day or the night? D the skunk ever hurry? Is it afraid? H does it protect itself from its enem Do you think that the skunk's freed from fear has rendered the animal intelligent?

4. At what time do the skunk kitt appear? Have you ever seen little sku playing? If so, describe their antics. H is the nest made soft for the young on

5. How does the skunk benefit farme Does it ever do them any injury? Do y think that it does more good than har

6. Describe the skunk's track as f lows: How many toes show in the trac Does the palm or heel show? Are the trac

together? Do they form a single or ble line?

w animals are so silent as the skunk. ɔgical works contain no information its voice, and the essayists rarely ion it except by implication. Mr. ɔughs says: "The most silent creature n to me, he makes no sound, so far as e observed, save a diffuse, impatient ;, like that produced by beating your with a whisk-broom, when the farm-has discovered his retreat in the stone ." Rowland Robinson tells us that: e voiceless creature sometimes fright-the belated farm-boy, whom he curi- follows with a mysterious hollow ing of his feet upon the ground." reau, as has been mentioned, heard keep up a "fine grunting, like a little ɔr a squirrel"; but he seems to have misunderstood altogether a singular loud patting sound heard repeatedly on the frozen ground under the wall, which he also listened to, for he thought it "had to do with getting its food, patting the earth to get the insects or worms." Probably he would have omitted this guess if he could have edited his diary instead of leaving that to be done after his death. The patting is evidently merely a nervous sign of impatience or apprehension, similar to the well-known stamping with the hind feet indulged in by rabbits, in this case probably a menace like a doubling of the fists, as the hind legs, with which they kick, are their only weapons. The skunk, then, is not voiceless, but its voice is weak and querulous, and it is rarely if ever heard except in the expression of anger.

— "Wild Neighbors,"
Ernest Ingersoll

General Biological Supply House, Chicago

raccoon. In the picture the heavy dark portion over the top of his head is caused by a shadow — but he does have a black mask across his eyes

THE RACCOON

None other of our little brothers of the est has such a mischievous countenance the coon. The black patch across the face and surrounding the eyes like large goggles, and the black line extending from the long, inquisitive nose directly up the

Treed

forehead give the coon's face an anxious expression; and the keenness of the big, beady, black eyes and the alert, "sassy" looking, broadly triangular ears, convince one that the anxiety depicted in the face is anxiety lest something that should *not* be done be left undone; and I am sure that anyone who has had experience with pet coons will aver that their acts do not belie their looks.

What country child, wandering by the brook and watching its turbulence in early spring, has not viewed with awe a footprint on the muddy banks looking as if it were made by the foot of a very little baby? The first one I ever saw I promptly concluded was made by the foot of a brook fairy. However, the coon is no fairy; it is a rather heavy, logy animal and, like the bear and skunk, is plantigrade, walking on the entire foot instead of on the toes, like a cat or dog. The hind foot is long, with a well-marked heel, and five comparatively short toes, giving it a remarkable resemblance to a human foot. The front foot is smaller and looks like a wide, little hand, with four long fingers and a rather short thumb. The claws are strong and sharp. The soles of the feet and the palms of the hands look as if they were covered with black kid, while the feet above and the backs of the hands are covered with short fur. Coon tracks are likely to be found during the thawing days of winter, along s stream or the borders of swamps, o following the path made by cattle. full-length track is about two inches as the coon puts the hind foot in track made by the front foot on the s side, only the print of the hind fe left, showing plainly five toe prints the heel. The tracks may vary from half inch to one foot or more apart pending on how fast the animal is go when it runs it goes on its toes, but w walking it sets the heel down; the tra are not in so straight a line as those m by the cat. Sometimes it goes at a s jump, when the prints of the hind are paired, and between and behind th are the prints of the two front feet.

The coon is covered with long, rat coarse hair, so long as almost to drag wh the animal is walking; it really has different kinds of hair, the long, coa gray hair, blackened at the tips, cover the fine, short, grayish or brownish und coat. The very handsome bushy tail ringed with black and gray.

The raccoon feeds on almost anyth eatable, except herbage. It has a spec predilection for corn in the milk sta and, in attaining this sweet and too some luxury, it strips down the husks a often breaks the plant, doing much da age. It is also fond of poultry and oft raids hen houses; it also destroys bir nests and the young, thus doing harm the farmer by killing both domestic an wild birds. It is especially fond of fish an is an adept at sitting on the shore an catching them with its hands; it likes tu tle eggs, crayfish, and snakes; it haunts th bayous of the Gulf Coast for the oyste which grow there; it is also a skillful fr catcher. Although fond of animal die it is also fond of fruit, especially of berri and wild grapes. It usually chooses for home a hollow tree or a cavern in a ledg near a stream, because of its liking fo water creatures.

Coons when in captivity have bee known to wash their meat before eatin it. I have watched a pet coon perform th act; he would take a piece of meat in h s, dump it into the pan of drinking r and souse it up and down a few s; then he would get into the pan with play feet and roll the meat beneath between them, meanwhile looking : unconcernedly at his surroundings, washing the meat were an act too me- ical to occupy his mind. After the : had been soaked until it was white flabby, he would take it in his hands hang onto it with a tight grip while he :d off pieces with his teeth; or some- s he would hold it with his feet, and hands as well as teeth in tearing it t. The coon's teeth are very much those of the cat, having long, sharp es or canines, and sharp, wedge-shaped ding teeth, which cut as well as grind. :r eating, the pet coon always washed eet by splashing them in the pan.

is an amusing sight to watch a coon nge itself for a nap, on a branch or he fork of a tree; it adapts its fat body he unevenness of the bed with ap- :nt comfort; it then tucks its nose n between its paws and curls its tail ut itself, making a huge, furry ball. all probability, the rings of gray and :k on the tail serve as protective color the animal sleeping in a tree during daytime, when sunshine and shadow ice down between the leaves with ever- nging light. The coon spends much its day asleep in some such situation, l comes forth at night to seek its food. n the fall, the coon lays on fat enough last it during its winter sleep. Usually eral inhabit the same nest in winter, g curled up together in a hollow tree, l remain dormant during the most se- e weeks of winter, coming out during iods of thaw.

The young are born in April; there are m three to six in a litter; they are blind l helpless at first, and are cared for efully by their parents; the family re- ins together until fall. If removed from :ir parents the young ones cry pitifully, nost like babies. The cry or whistle of : fully grown coon is anything but a ppy sound, and is quite impossible to scribe. I have been awakened by it many a night in camp, and it always sounded strange, taking on each time new quavers and whimperings. As a cry, it is first cousin to that of the screech owl.

The stories of pet coons are many. I knew one which, chained in a yard, would lie curled up near its post looking like an innocent stone except for one eye kept watchfully open. Soon a hen filled with curiosity would come warily near, looking longingly at remains of food in the pan; the coon would make no move until the disarmed biddy had come close to the pan. Then there would be a scramble and a squawk and with astonishing celerity he would wring her neck and strip off her feathers. Another pet coon was allowed to range over the house at will, and finally had to be sent away because he had learned to open every door in the house, including cupboard doors, and could also open boxes and drawers left unlocked; and I have always believed he could have learned to unlock drawers if he had been given the key. All coons are very curious, and one way of trapping them is to suspend above the trap a bit of bright tin; in studying this glittering mystery, they forget all about traps.

Marion E. Wesp

This pet raccoon is angry because she has been taken from the shoulder of her mistress and placed on a post to have her picture taken

LESSON 61

The Raccoon

Leading Thought — The raccoon lives in hollow trees or caves along the banks of streams. It sleeps during the day and seeks its food at night. It sleeps during the winter.

Method — If there are raccoons in the vicinity, ask the older boys to look for their tracks near the streams and to describe them very carefully to the class. The ideal method of studying the animal is to have a pet coon where the children may watch at leisure its entertaining and funny performances. If this is impossible, then follow the less desirable method of having the pupils read about the habits of the coon and thus arouse their interest and open their eyes, so that they may make observations of their own when opportunity offers. I would suggest the following topics for oral or written work in English:

"How and Where Coons Live and What They Do"; "The Autobiography of a Coon One Year Old"; "The Queer Antics of Pet Coons"; "Stories of the Coon's Relative, the Bear."

Observations — 1. Where have you found raccoon tracks? How do they differ from those of fox or dog? How far are the foot prints apart? Can you see the heel and toe prints? Do you see the tracks of all four feet? Are the tracks in a straight line like those of the cat? What is the size of the track, the length, the breadth?

2. What do coons eat and how do get their food? Which of our crop they likely to damage? What other age do they do? Have you ever heard c cry or whistle during August nights ir cornfields?

3. Why do raccoons like to live the water? What do they find of int there? How do they prepare their before eating it? How does a coon ha its meat while eating it?

4. What kind of fur has the coon? does it need such a heavy covering? scribe the color of the fur. Describe tail. Of what use is such a large and b tail to this animal?

5. Describe the coon's face. How marked? What is its expression? Desc the eyes, ears, and nose. Has it teeth sembling those of the cat and dog?

6. Describe the coon's feet. How m toes on the front feet? How many on hind feet? How does this differ from cat and dog? How do the front and h feet differ in appearance? Can both used as hands?

7. How do coons arrange themse for a nap in a tree? How do they cc the head? How is the tail used? Do think this bushy tail used in this would help to keep the animal warm winter? Do coons sleep most by day or night?

8. At what time of year are coons test? Leanest? Why? Do they ever co out of their nests in winter? Do they together or singly in winter?

9. At what time of year are the you coons born? Do you know how they lc when they are young? How are they ca for by their parents?

10. Are the coon's movements slow fast? What large animal is a near relat of the coon?

THE WOLF

The study of the wolf should precede the lessons on the fox and the dog. After becoming familiar with the habits of wolves, the pupils will be much better able to understand the nature of the dog and its life as a wild animal. In most calities, the study of the wolf must, course, be a matter of reading, unless t pupils have an opportunity to study t animal in zoological gardens.

might be well to begin this lesson :he wolf with a talk about the gray ·es which our ancestors had to con- (with, and also with stories of the ›te or prairie wolf which has learned lapt itself to civilization and flourishes he regions west of the Rocky Moun- s, despite men and dogs. Literature is in wolf stories. Although Kipling's ous Mowgli Stories belong to the m of fiction, yet they contain inter- ıg accounts of the habits of the wolves ndia, and are based upon the hunter's tracker's knowledge of these animals. have many thrillingly interesting sto- in our own literature which deal with native wolves. Some of the best are ed in the suggested reading at the end his section.

E. H. McCleery

Wolves, seldom seen now, once ranged over many parts of North America

Leonard K. Beyer

A captive wolf

From some or all of these stories, the pupils should get information about the habits of the wolves. This information may be incorporated in an essay or an oral exercise and should cover the following points: Where do the wolves live? On what do they feed? How do they get their prey? How do they call to each other? Description of the den where the young are reared. The wolf's cleverness in eluding hunters and traps.

THE FOX

Do we not always, on a clear morning winter, feel a thrill that must have mething primitive in its quality at see- g certain tracks in the snow that some- ·w suggest wildness and freedom! Such the track of the fox. Although it is mewhat like that of a small dog, yet it very different. The fox has longer legs an most dogs of his weight, and there more of freedom in his track and more strength and agility expressed in it. His ıit is usually an easy lope; this places the imprint of three feet in a line, one ahead of another, but the fourth is off a little at one side, as if to keep the balance.

The fox lives in a den or burrow. The only fox home which I ever saw was a rather deep cave beneath the roots of a stump, and there was no burrow or retreat beyond it. However, foxes often select woodchuck burrows, or make burrows of their own, and if they are caught within, they can dig rapidly, as many a hunter can attest. The mother usually selects an open

Red fox cubs

place as a den for the young foxes; often an open field or sidehill is chosen for this. The den is carpeted with grass and is a very comfortable place for the fox puppies.

The face of the red fox shows plainly why he has been able to cope with man, and thrive despite and because of him. If ever a face showed cunning, it is his. Its pointed, slender nose gives it an expression of extreme cleverness, while the width of the head between the upstanding, triangular ears gives room for a brain of power. In color the fox is russet-red, the hind quarters being grayish. The legs are black outside and white inside; the throat is white, and the broad, triangular ears are tipped with black. The glory of the fox is his " brush," as the beautiful, bushy tail is called. This is red, with black toward the end and is white-tipped. This tail is not merely for beauty, for it affords the fox warmth during the winter, as anyone who has observed the way it is wrapped around the sleeping animal may see. But this bushy tail is a disadvantage, if it becomes bedraggled and heavy with snow and sleet, when the hounds are giving close chase to its owner. The silver fox and the black fox are color phases of the red fox.

The fox is an inveterate hunter of the animals of the field; meadow mice, rabbits, woodchucks, frogs, snakes, and grasshoppers are all acceptable food; he is also destructive of birds. His fondness for the latter has given him a bad reputation with the farmer because of his attacks on poultry. Not only will he raid hen-roosts if he can force entrance, but he catches many fowls in the summer when they are wandering through the fields. The way he carries the heavy burden of his larger prey shows his cleverness: he slings a or a goose over his shoulders, keeping head in his mouth to steady the bur Mr. Cram says, in *American Animal*

" Yet, although the farmer and the are such inveterate enemies, they ma to benefit each other in a great many quite unintentionally. The fox dest numberless field mice and woodch for the farmer and in return the fa supplies him with poultry, and builds venient bridges over streams and places, which the fox crosses oftener t the farmer, for he is as sensitive as a about getting his feet wet. On the wh I am inclined to believe that the fox the best part of the exchange, for, w the farmer shoots at him on every o sion, and hunts him with dogs in the ter, he has cleared the land of wolves panthers, so that foxes are probably s than before any land was ploughed."

The bark of the fox is a high, sh yelp, more like the bark of the co than of the dog. There is no doubt a c siderable range of meaning in the f language, of which we are ignorant. growls when angry, and when pleased smiles like a dog and wags his beaut tail.

Many are the wiles of the fox to misl dogs following his track: he often retra his own steps for a few yards and th makes a long sidewise jump; the dogs on, up to the end of the trail pocket, a try in vain to get the scent from that po Sometimes he walks along the top r of fences or takes the high and dry rid where the scent will not remain; he of

Verne Mo

The attentive ears and bright eyes of th fox cubs show a keen interest in their s roundings

ws roads and beaten paths and also around and around in the midst of rd of cattle or sheep so that his scent dden; he crosses streams on logs and nts various other devices too numer- and intricate to describe. When ed by dogs, he naturally runs in a e, probably so as not to be too far home. If there are young ones in the the father fox leads the hounds far , into the next county if possible. aps one of the most clever tricks of fox is to make friends with the dogs. ve known of two instances where a and fox were daily companions and fellows.

he young foxes are born in the spring. y are gray and woolly at first and are inating little creatures, being exceed- y playful and active. Their parents are devoted to them, and during all their pyhood the mother fox is a menace he poultry of the region, because the essity of feeding her rapidly growing r is upon her.

LESSON 62

The Fox

EADING THOUGHT — The red fox is so ver that it has been able, in many parts our country, to maintain itself despite s and men.

METHOD — This lesson is likely to be en largely from hearsay or reading. wever, if the school is in a rural district, re will be plenty of hunters' stories afloat, from which may be elicited facts concerning the cunning and cleverness of the red fox. In such places there is also the opportunity in winter to study fox tracks upon the snow. The lesson may well be given when there are fox tracks for observation. The close relationship between foxes and dogs should be emphasized.

OBSERVATIONS — 1. Describe the fox's track. How does it differ from the track of a small dog?

2. Where does the fox make its home? Describe the den. Describe the den in which the young foxes live.

3. Describe the red fox, its color and form, as completely as you can. What is the expression of its face? What is there peculiar about its tail? What is the use of this great bushy tail in the winter?

4. What is the food of the fox? How does it get its food? Is it a day or a night hunter? How does the fox benefit the farmer? How does it injure him? How does the fox carry home its heavy game, such as a goose or a hen?

5. Have you ever heard the fox bark? Did it sound like the bark of a dog? How does the fox express anger? Pleasure?

6. When chased by dogs, in what direction does the fox run? Describe all of the tricks which you know by which the fox throws the dog off the scent.

7. When are the young foxes born? How many in a litter? What color are they? How do they play with each other? How do they learn to hunt?

U. S. Bureau of Biol. Survey

Silver fox

National Sport

English setter. This is the famous Brownie's Spot, field trial winner and bench sho champion

DOGS

Not only today but in ancient days, before the dawn of history, the dog was the companion of man. Whether the wild species from whence he sprang was wolf or jackal or some other similar animal, we do not know, but we do know that many types of dogs have been tamed independently by savages, in the region where their untamed relatives run wild. As the whelps of wolves, jackals, and foxes are all easily tamed, and are most interesting little creatures, we can understand how they became companions to the children of the savage and barbarous peoples who hunted them.

In the earliest records of cave dwellers, in the picture writing of the ancient Egyptians and of other ancient peoples, we find record of the presence and value of the dog. But man, in historical times, has been able to evolve breeds that vary more in form than do the wild species of the present. There are 200 distinct breeds of dogs known today, and many of these h been bred for special purposes. The p ontologists, moreover, assure us that th has been a decided advance in the s and quality of the dog's brain since days of his savagery; thus, he has be the companion of man's civilization al It is not, therefore, to be wondered at t the dog is now the most companionab and has the most human qualities a intelligence of all our domesticat animals.

Dogs run down their prey; it is a nec sity, therefore, that they be equipped w legs that are long, strong, and muscul The cat, which jumps for her prey, much more delicate legs but has power hips to enable her to leap. The dog's f are much more heavily padded than th of the cat, because in running he m not stop to save his feet. Hounds oft return from a chase with bleeding fe despite the heavy pads, but the woun usually cuts between the toes. The are heavy and are not retractile; thus, afford a protection to the feet when ing, and they are also used for dig- out game which burrows into the nd. They are not used for grasping like those of the cat and are used incidentally in fighting, while the claws are the most important weap- n her armory. It is an interesting fact Newfoundland dogs, which are such

U. S. Dept. Agriculture

oston terrier. This small popular breed is of the few to originate in America. It is companionable and highly intelligent

ous swimmers, have their toes some- at webbed.

The dog's body is long, lean, and very scular, a fat dog being usually pam- ed and old. The coat is of hair and is of fine fur like that of the cat. It is nterest to note that the Newfoundland has an inner coat of fine hair com- able to that of the mink or muskrat. hen a dog is running, his body is ex- ded to its fullest length; in fact, it ms to "lie flat," the outstretched legs ghtening the effect of extreme muscu- effort of forward movement. A dog naster of several gaits; he can run, walk, t, bound, and crawl.

The iris of the dog's eye is usually of

U. S. Dept. Agriculture

Beagle. These hounds hunt individually, in pairs, or in packs; they are used chiefly for hunting rabbits

a beautiful brown, although this varies with breeds; in puppies, the iris is usually blue. The pupil is round like our own; and although dogs probably cannot see as well in the dark as the cat, they see well at night and in daylight they have keen sight. The nose is so much more efficient than the eyes, that it is on the sense of smell the dog depends for following his prey and for recognizing friend and foe. The damp, soft skin that covers the nose has in its dampness the conditions for carry- ing the scent to the wide nostrils; these are situated at the most forward part of the face, and thus may be lifted in any

U. S. Dept. Agriculture

Greyhound. This swiftest of all large dogs hunts by sight

St. Bernard. These dogs stand about thirty inches high and have an average weight of 175 pounds

direction to receive the marvelous impressions, so completely beyond our comprehension. Think of being able to scent the track of a fox made several hours previously, and not only to scent it, but to follow it by scent for many miles without ever having a glimpse of the fleeing foe! In fact, while running, the dog's attention seems to be focused entirely upon the sense of smell, for I have seen hounds pass within a few rods to the windward of the fox they were chasing, without observing him at all. Furthermore, according to E. H. Baynes, the dog's sense of smell is keen enough to distinguish the scent of the particular creature he is hunting from that of all others, and to distinguish the scent of several animals from that of only one. He knows the difference between foot scent and body scent, and he can immediately tell the scent of a wounded animal from that of a dead one. He can tell, moreover, the direction in which foot scent leads, and some dogs, at least, can follow a particular trail no matter how many other scents have been superimposed upon it. It has been said that the sense of smell in dogs, and especially in hounds, is so acute that the amount of odor required to stimulate the nose is too slight to be expressed. When the nose of a dog becomes dry it is a sign of illness.

A light fall of damp snow gives the dog the best conditions for following a track by scent. A hound, when on the t
will run until exhausted. There are m
authentic observations which show
hounds have followed a fox for twe
four hours without food, and prob
with little rest.

Because the dog's sense of smell i
important to him, he should never
punished by being struck over the n
Nor should he be struck at all about
head and ears, lest his hearing be d
aged. A dog is so sensitive to inflecti
and tones of voice that a severe wor
usually punishment enough; if it se
necessary to strike him, he should
struck only on the foreshoulders
sides. A folded newspaper is good for
purpose.

The dog's weapons for battle, like th
of the wolf, are his tushes; with th
he holds and tears his prey; with them
seizes the woodchuck or other small
mal through the back and shakes its
out. In fighting a larger animal, the
leaps against it and often incident
tears its flesh with his strong claws;
he does not strike a blow with his f
like the cat, nor can he hold his qua
with it.

Dogs' teeth are especially fitted for th
work. The incisors are small and sharp;

H. M. Isenho

*Pointer. These dogs are called pointers b
cause of their habit of pointing at the co
cealed game birds they have scented. This
Isenhower's Flaro, a champion*

e teeth or tushes are very long, but are bare spaces on the jaws so that are able to cross past each other; the r teeth are not adapted for grinding, the teeth of a cow, but are especially l for cutting, as may be noted if we h the way a dog gnaws bones, gnaw-with the back teeth first on one side then on the other. In fact, a dog does seem to need to chew anything, but ly needs to cut his meat in small gh pieces so that he can gulp them n without chewing. His powers of di-ng unchewed food are something that hustling American may well envy.

Helen F. Hill

English springer spaniel. No other family of dogs contains so many recognized breeds as the spaniel family — seven hunting and two toy breeds. Formerly these dogs were trained to flush or "spring" the game so that swifter dogs or falcons could catch it; today they are popular as all-purpose dogs

Mr. and Mrs. J. W. Page

eagle pups. Beagles are small models of ounds; they are not so swift as foxhounds, seem to have a keener sense of smell

Of all domestic animals, the dog is most nanly understandable in expressing otions. If delighted, he leaps about giv- ecstatic little barks and squeals, his in the air and his eyes full of happy icipation. If he wishes to be friendly, looks at us interestedly, comes over to ell of us in order to assure himself ether he has ever met us before, and en wags his tail as a sign of good faith. he wishes to show affection, he leaps on us and licks our face or hands with soft, deft tongue and follows us jeal-sly. When he stands at attention he lds his tail stiff in the air, and looks with one ear lifted as if to say, "Well, hat's doing?" When angry, he growls d shows his teeth and the tail is held gidly out behind, as if to convince us

The Seeing Eye, Inc.

A Seeing Eye dog. The training of dogs to lead the blind began in the United States; the same methods have now become popular in Europe. The Seeing Eye has headquarters in New York City

that it is really a continuation of his backbone. When afraid, he whines and lies flat upon his belly, often looking beseechingly up toward his master as if begging not to

H. M. Isenhower

English pointer pups

be punished; or he crawls away out of sight. When ashamed, he drops his tail between his legs and with drooping head and sidewise glance slinks away. When excited, he barks and every bark expresses high nervous tension.

Almost all dogs that chase their prey bark when so doing. This action would at first sight seem foolish, in that it reveals their whereabouts to their victims and also adds an incentive to flight. These dogs have been trained through many generations and have been selected because of various peculiarities; a good fox hound, coon hound, or rabbit hound barks in order to tell the hunter, not only where it is but what it is doing. A certain kind of bark may indicate to the hunter that the game is " treed " or chased into a hole.

Most breeds of dogs have an acute sense of hearing. When a dog bays at the moon or howls when he hears music, it is simply a reversion to the wild habit of howling to call together the pack or in answer " to the music of the pack." It is interesting that our music, which is the flower of our civilization, should awaken the sleeping ancestral traits in the canine breast. But perhaps that, too, is why we respond to music, because it awakens in us the strong, primitive emotions, and for the time enables us to free ourselves from all conventional shackles and trammels.

LESSON 63
Dogs

Leading Thought — The dog is a mesticated descendant of wolflike mals and has retained certain of the h and characteristics of his ancestors.

Method — For the observation le it would be well to have at hand a disposed dog which would not objec being handled; a collie or a hound w be preferable. Many of the quest should be given to the pupils to ans from observations at home, and the le should be built upon the experienc the pupils with dogs.

Observations — 1. Why are the of the dog compared with those of cat long and strong in proportion to body?

2. Compare the feet of the cat those of the dog and note which has heavier pads. Why is this of use to ea

3. Which has the stronger and hea

Leonard K. B

Collie. This breed of dogs shows great telligence in the herding of various kinds domestic animals; it has long been used Scotland, but its popularity has spread many other countries. The one pictured he is not today the show type

s, the dog or the cat? Can the dog ct his claws so that they are not visi- as does the cat? Of what use is this ıgement to the dog? Are the front just like the hind feet? How many impressions show in the track of the

What is the general characteristic of body of the dog? Is it soft like that he cat, or lean and muscular? What ıe difference between the hair cover- of the dog and the cat? What is the

londike Jack. The dog that pulled four dred fifty pounds five hundred miles ugh the White Horse Pass in the winter he first gold excitement in Alaska

tude of the dog when running fast? w many kinds of gaits has he?

. In general, how do the eyes of the differ from those of the cat? Does he as much upon his eyes for finding his y as does the cat? Can a dog see in dark? What is the color of the dog's s?

. Study the ear of the dog; is it cov- d? Is this outer ear movable, is it a flap, s it cornucopia-shaped? How is this flap d when the dog is listening? Roll a et of paper into a flaring tube and place small end upon your own ear, and te if it helps you to hear better the nds in the direction toward which the be opens. Note how the hound lifts his g earlaps, so as to make a tube for con- ying sounds to his inner ear. Do you ink that dogs can hear well?

7. What is the position of the nose in e dog's face? Of what use is this? De-

Mr. and Mrs. J. W. Page

English setter. This breed originated in England from a cross between a field spaniel and a pointer

scribe the nostrils; are they placed on the foremost point of the face? What is the condition of the skin that surrounds them? How does this condition of the nose aid the dog? What other animals have it? Does the dog recognize his friends or become acquainted with strangers by means of his sight or of his powers of smelling?

8. How long after a fox or rabbit has passed can a hound follow the track? Does

St. Bernard. This breed of huge dogs was developed by monks in the Swiss Alps to aid in the rescue of people lost in the mountains

he follow it by sight or by smell? What are the conditions most favorable for retaining the scent? The most unfavorable? How long will a hound follow a fox trail without stopping for rest or food? Do you think the dog is your superior in ability to smell?

9. How does a dog seize and kill his prey? How does he use his feet and claws when fighting? What are his especially strong weapons? Describe a dog's teeth and explain the reason for the bare spaces on the jaw next to the tushes. Does the dog use his tushes when chewing? What teeth does he use when gnawing a bone? Make a diagram of the arrangement of the dog's teeth.

10. How by action, voice, and especially by the movement of the tail does the dog express the following emotions: delight, friendliness, affection, attention, anger, fear, shame, excitement? How does he act when chasing his prey? Why do wolves and dogs bark when following the t Do you think of a reason why dogs o howl at night or when listening to mu What should we feed to our pet d What should we do to make them c fortable in other ways?

11. Tell or write a story of some of which you know by experience or h say. Of what use was the dog to the neer? How are dogs used in the Ar regions? In Holland?

12. How many breeds of dogs do know? Describe these breeds as follo The length of the legs as compared w the body; the general shape of the bo head, ears, nose; color and character hair on head, body, and tail.

13. Find if you can the reasons wh have led to the developing of the lowing breeds: Newfoundland, St. B nard, mastiffs, hounds, collies, spani setters, pointers, bulldogs, terriers, a pugs.

Verne Mor

A cat family

THE CAT

Of all people, the writer should regard the cat sympathetically, for when she was a baby of five months she was adopted by a cat. My self-elected foster-mother was Jenny, a handsome black and white ca which at that time lost her first litter kittens, through the attack of a savag cat from the woods. She was as Rach g for her children, when she seemed enly to comprehend that I, although r than she, was an infant. She haunted radle, trying to give me milk from her breasts; and later she brought half-d mice and placed them enticingly in radle, coaxing me to play with them, rformance which pleased me much e than it did my real mother. Jenny ys came to comfort me when I cried, bing against me, purring loudly, and ng me with her tongue in a way to e mad the modern mother, wise as to sources of children's internal parasites. s maternal attitude toward me lasted ong as Jenny lived, which was until I nine years old. Never during those s did I lift my voice in wailing, that she not come to comfort me; and even to-I can remember how great that com-was, especially when my naughtiness the cause of my weeping, and when, efore, I felt that the whole world, ex-t Jenny, was against me.

enny was a cat of remarkable intelli-ce and was very obedient and useful. ming down the kitchen stairs one day, played with the latch, and someone o heard her opened the door. She did several times, when one day she nced to push down the latch, and thus

Kittens

opened the door herself. After that, she always opened it herself. A little later, she tried the trick on other doors, and soon succeeded in opening all the latched doors in the house, by thrusting one front leg through the handle, and thus supporting her weight and pressing down with the foot of the other on the thumb-piece of the latch. I remember that guests were greatly astonished to see her coming thus swinging into the sitting room. Later she tried the latches from the other side, jumping up and trying to lift the hook; but now, her weight was thrown against the wrong side of the door for opening, and she soon ceased this futile waste of energy; but for several years, she let herself into all the rooms in this clever manner, and taught a few of her bright kittens to do the same.

"Folks are so tiresome!"

A pet cat enjoys long conversations with favored members of the household. She will sit in front of her mistress and mew, with every appearance of answering the questions addressed her; and since the cat and the mistress each knows her own part of the conversation, it is perhaps more typical of society chatter than we might like to confess. Of our language, the cat learns to understand the call to food, its own name, "Scat," and "No, No," probably inferring the meaning of the latter from the tone of voice. On the other hand, we understand when it asks to go out, and its polite recognition to the one who opens the door. I knew one cat which invariably thanked us when we let him in as well as out. When the cat is hungry, it mews pleadingly; when happy in front of the

Marion E. Wesp

On the doorstep

fire, it looks at us sleepily out of half-closed eyes and gives a short mew expressive of affection and content; or it purrs, a noise which we do not know how to imitate and which expresses perfectly the happiness of intimate companionship. When frightened the cat yowls, and when hurt it squalls shrilly; when fighting, it is like a savage warrior in that it howls a war-song in blood-curdling strains, punctuated with a spitting expressive of fear and contempt; and unfortunately, its love song is scarcely less agonizing to the listener. The cat's whole body enters into the expression of its emotions. When feeling affectionate toward its mistress, it rubs against her gown, with tail erect, and vibrating with a purr which seems fundamental. When angry, it lays its ears back and lashes its tail back and forth, the latter being a sign of excitement; when frightened, its hair stands on end, especially the hair of the tail, making that expressive appendage twice its natural size; when caught in disobedience, the cat lets its tail droop, and when running lifts it in a curve.

While we feed cats milk and scraps from our own table, they have never become entirely civilized in their tastes. They always catch mice and other small animals and prove pestiferous in destroying birds. Jenny was wont to bring her quarry, as an offering, to the front steps of our home every night; one morning we found seven mice, a cotton-tail rabbit and two snakes, which represented night's catch. The cat never chase prey like the dog. It discovers the ha of its victims and then lies in amb flattened out as still as a statue and a feet beneath it, ready to make the sp The weight of the body is a factor w enters into the blow with which the strikes down and stuns its victim, w it later kills by gripping the throat the strong tushes. It carries its victim it does its kittens, by the back.

The cat's legs are not long comp with the body, and it runs with a lea gallop; the upper legs are armed with p erful muscles. It walks on the padded t five on the front feet and four on the h feet. The cat needs its claws to be sh and hooked, in order to seize and hol prey, so they are kept safely sheat when not thus used. If the claws str the earth during walking, as do the do they would soon become dulled. W sharpening its claws it reaches high against a tree or post, and strikes them i the wood with a downward scratch; t act is probably more for exercising muscles which control the claws than sharpening them.

John W. Dec

Anticipation

ıe cat's track is in a single line as if d only two feet, one set directly ahead ıe other. It accomplishes this by set- its hind feet exactly in the tracks e by the front feet. The cat can easily upward, landing on a window-sill five from the ground. The jump is made the hind legs and the alighting is silently on the front feet.

ats' eyes are adapted better than ours eeing in the dim light; in the daytime pupil is simply a narrow, up and down under excitement, and at night, the l covers almost the entire eye. At the of the eye is a reflecting surface, ch catches such light as there is, and eflecting it enables the cat to use it e. It is this reflected light which gives peculiar green glare to the eyes of all cats when seen in the dark. Some t-flying moths have a like arrange- t for utilizing the light, and their eyes like living coals. Of course, since the is a night hunter, this power of multi- ng the rays of light is of great use. iris of the eye is usually yellow, but in ens it may be blue or green.

'he cat's teeth are peculiarly fitted for needs. The six doll-like incisors of the er and lower jaw are merely for scrap- meat from bones. The two great nes, or canines, on each jaw, with a e place behind so that they pass each er freely, are sharp, and are for seizing carrying prey. The cat is able to open mouth as wide as a right angle, in order

Amicable advances

Marion E. Wesp

An aristocrat

better to hold and carry prey. The back teeth, or molars, are four on each side in the upper jaw and three below. They are sharp-edged wedges made for cutting meat fine enough so that it may be swallowed.

The tongue is covered with sharp papillæ directed backwards, also used for rasping juices from meat. The cat's nose is moist, and her sense of smell very keen, as is also her sense of hearing. The ears rise like two hollow half-cones on either side of the head and are filled with sensitive hairs; they ordinarily open forward, but are capable of movement. The cat's whiskers consist of from twenty-five to thirty long hairs set in four lines, above and at the sides of the mouth; they are connected with sensitive nerves and are therefore true feelers. The cat's fur is very fine and thick, and is also sensitive, as can readily be proved, by trying to stroke it the wrong way. While the wild cats have gray or tawny fur, variously mottled or shaded, the more striking colors we see in the domestic cats are the result of man's breeding.

Cats are very cleanly in their habits. Puss always washes her face directly after eating, using one paw for a washcloth and licking it clean after she rubs her face.

She cleans her fur with her rough tongue and also by biting; and she promptly buries objectionable matter. The mother cat is very attentive to the cleanliness of her kittens, licking them clean from nose tip to tail tip. The ways of the mother cat with her kittens do much to sustain the assertions of Mr. Seton and Mr. Long that young animals are trained and educated by their parents. The cat brings half-dazed mice to her kittens, that they may learn to follow and catch them with their own

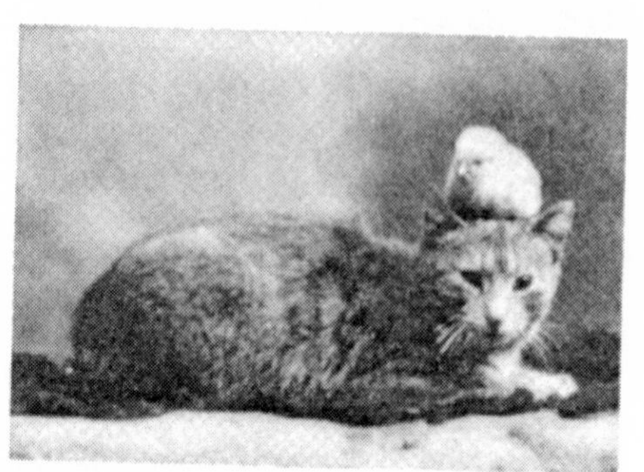

This cat has been trained to be friendly with birds

little claws. When she punishes them, she cuffs the ears by holding one side of the kitten's head firm with the claws of one foot, while she lays on the blows with the other. She carries her kittens by the nape of the neck, never hurting them. She takes them into the field when they are old enough, and shows them the haunts of mice, and does many things for their education and welfare. The kittens meantime train themselves to agility and dexterity, by playing rough and tumble with each other, and by chasing every small moving object, even to their own tails.

The cat loves warmth and finds her place beneath the stove or at the hearthside. She likes some people, and dislikes others, for no reason we can detect. She can be educated to be friendly with dogs and with birds. In feeding her, we should give her plenty of sweet milk, some cooked meat, and fish, of which she is very fond; and we should keep a bundle of catnip to make her happy, for even the larger cats of the wilderness seem to have a pas ate liking for this herb. The cat laps with her rough tongue, and when ea meat, she turns the head this way that, to cut the tough muscle with back teeth.

Cats Should Be Trained to Leave Birds Alone

Every owner of a cat owes it to world to train Puss to leave birds al If this training is begun during kit hood, by switching the culprit every t it even looks at a bird, it will soon le to leave them severely alone. I have t this many times, and I know it is eff cious, if the cat is intelligent. We h never had a cat whose early training controlled, that could ever be indu even to watch birds. If a cat is not t trained as a kitten, it is likely to be alw treacherous in this respect. But in case one has a valuable cat which is given catching birds, I strongly advise the lowing treatment which has been pro practicable by a friend of mine. Whe cat has made the catch, take the b away and sprinkle it with red pepper, a then give it back. One such treatment this resulted in making one cat, wh was an inveterate bird hunter, run a hide every time he saw a bird thereaft Any persons taking cats with them to th summer homes, and abandoning th there to prey upon the birds of the vicini and to become poor, half-starved, w creatures, ought to be arrested and fin It is not only cruelty to the cats, but is positive injury and damage to the co munity, because of the slaughter of ma beneficial and beautiful birds which it e tails.

LESSON 64
The Cat

eading Thought — The cat was made mestic animal before man wrote his- s. It gets prey by springing from am- and is fitted by form of body and h to do this. It naturally hunts at night has eyes fitted to see in the dark.

Method — This lesson may be used in nary grades by asking a few questions time and allowing the children to e their observations on their own kit- at home, or a kitten may be brought chool for this purpose. The upper grade k consists of reading and retelling or ting exciting stories of the great, wild, age cats, like the tiger, lion, leopard, x, and panther.

Observations — 1. How much of Pus- language do you understand? What s she say when she wishes you to n the door for her? How does she for something to eat? What does she when she feels like conversing with ? How does she cry when hurt? When ghtened? What noise does she make en fighting? When calling other cats? hat are her feelings when she purrs? hen she spits? How many things which say does she understand?

2. How else than by voice does she ex- ss affection, pleasure, and anger? When carries her tail straight up in the air she in a pleasant mood? When her tail oristles up" how does she feel? What it a sign of, when she lashes her tail ck and forth?

3. What do you feed to cats? What do ey catch for themselves? What do the ts that are wild live upon? How does e cat help us? How does she injure us?

4. How does a cat catch her prey? Does e track mice by the scent? Does she tch them by running after them as a g does? Describe how she lies in am- sh. How does she hold the mouse as e pounces upon it? How does she carry home to her kittens?

5. Study the cat's paws to see how she lds her prey. Where are the sharp claws? re they always in sight like a dog's? Does she touch them to the ground when she walks? Which walks more silently, a dog or a cat? Why? Describe the cat's foot, including the toe-pads. Are there as many toes on the hind feet as on the front feet? What kind of track does the cat make in the snow? How does she set her feet to make such a track? How does she sharpen her claws? How does she use her claws for climbing? How far have you ever seen a cat jump? Does she use her front or her hind feet in making the jump? On which feet does she alight? Does she make much noise when she alights?

6. What is there peculiar about a cat's eyes? What is their color? What is the color of kittens' eyes? What is the shape of the pupil in daylight? In the dark? Describe the inner lid which comes from the corner of the eye.

7. How many teeth has Puss? What is the use of the long tushes? Why is there a bare space behind these? What does she use her little front teeth for? Does she use her back teeth for chewing or for cutting meat?

8. How many whiskers has she? How long are they? What is their use? Do you think Puss has a keen sense of smell? Why do you think so? Do you think she has a keen sense of hearing? How do the shape and position of the ears help in listening? In what position are the ears when Puss is angry?

9. How many colors do you find in our domestic cats? What is the color of wild cats? Why would it not be beneficial to the wild cat to have as striking colors as our tame cats? Compare the fur of the cat with the hair of the dog. How do they differ? If a cat chased her prey like the dog do you think her fur would be too warm a covering?

10. Describe how the cat washes her face. How does she clean her fur? How does her rough tongue help in this? How does the mother cat wash her kittens?

11. How does a little kitten look when a day or two old? How long before its eyes open? How does the cat carry her kittens? How does a kitten act when it is being carried? How does the mother

cat punish her kittens? How does she teach them to catch mice? How do kittens play? How does the exercise they get in playing fit them to become hunters?

12. How should cats be trained not to touch birds? When must this training begin? Why should a person be punished for injury to the public who takes cats to summer cottages and leaves them there to run wild?

13. Where in the room does Puss best like to lie? How does she sun herself? What herb does she like best? Does she like some people and not others? What strange companions have you known a to have? What is the cat's chief ene How should we care for and make comfortable?

14. Write or tell stories on the foll ing subjects: (1) The Things Which Pet Cat Does; (2) The Wild Cat; The Lion; (4) The Tiger; (5) The pard; (6) The Panther and the Moun Lion; (7) The Lynx; (8) The Histor Domestic Cats; (9) The Different R of Cats, describing the Manx, the Pers and the Angora Cats.

A. A. Wr

A herd of goats by the Nueces River, Texas

THE GOAT

Little do we in America realize the close companionship that has existed in older countries, from time immemorial, between goats and people. This association began when man was a nomad, and took with him in his wanderings his flocks, of which goats formed the larger part. He then drank their milk, ate their flesh, wove their hair into raiment, or made cloth of their pelts, and used their skins for water bags. Among peoples of the East all these uses continue to the present day. In t streets of Cairo, old Arabs may be se with goatskins filled with water upon th backs; and in any city of western Asia southern Europe, flocks of goats are driv along the streets to be milked in sight the consumer.

In order to understand the goat's p culiarities of form and habit, we shou consider it as a wild animal, living up the mountain heights amid rocks and sn scant vegetation. It is marvelously footed, and on its native mountains n climb the sharpest crags and leap ns. This peculiarity has been seized . by showmen who often exhibit ; which walk on the tight rope with and even turn themselves upon it out falling. The instinct for climbing lingers in the domestic breeds, and in country the goat may be seen on top one piles or other objects, while, in suburbs, its form may be discerned he roofs of shanties and other low lings.

U. S. Dept. Agriculture

Saanen goats in Switzerland

t is a common saying that a goat will anything, and much sport is made of peculiarity. This fact has more mean- for us when we realize that wild goats in high altitudes, where there is little nt life, and are, therefore, obliged to l sustenance on lichens, moss, and such nt vegetation as they can find.

The goat is closely allied to the sheep, fering from it in only a few particulars; horns rise from the forehead curving er backward and do not form a spiral e those of the ram; its covering is usu- y of hair, and the male has a beard from hich we get the name goatee; the goat s no gland between the toes, and it does ve a rank and disagreeable odor. In a ld state, it usually lives a little higher the mountains than do the sheep, and is a far more intelligent animal. Mary ıstin says: "Goats lead naturally by ason of a quicker instinct, forage more cely and can find water on their own ac- unt, and give voice in case of alarm.

Saanen doe

Goat leaders exhibit jealousy of their rights to be first over the stepping-stones or to walk the teetering log bridges at the roaring creeks." On the great plains, it is a common usage to place a few goats in a flock of sheep, because of the greater sagacity of these animals as leaders, and also as defenders in case of attack.

Goats' teeth are arranged for cropping herbage and especially for browsing. There are six molar teeth on each side of each jaw; there are eight lower incisors and none above. The goat's sense of smell is very acute; the ears are movable and the sense of hearing is keen; the eyes are full and very intelligent; the horns are somewhat flattened and angular, are often knobbed somewhat in front, and curve backward above the neck; they are, however, very efficient as weapons of defense. The legs are strong, though not large, and are well fitted for leaping and running.

N. Y. Agr. Exp. Station, Geneva

Toggenburg goat. This Swiss breed, developed by a careful selection of animals for many years, has attained a very definite standard of size, color, and conformation

N. Y. Agr. Exp. Station, Geneva

French alpine doe. Alpines are sturdy, and have been bred for high production of fine-flavored milk

The feet have two hoofs, that is, the animal walks upon two toenails. There are two smaller toes behind and above the hoofs. The goat can run with great rapidity. The tail of the goat is short like that of the deer, and does not need to be amputated like that of the sheep. Although the normal covering of the goat is hair, there are some species which have a more or less woolly coat. When angry the goat shakes its head, and defends itself by butting with the head, also by striking with the horns, which are very sharp. Goats are very tractable and make affectionate pets when treated with kindness; they display far more affection for their owner than do sheep.

Our famous Rocky Mountain goat, although it belongs rather to the antelope family, is a large animal, and is the special prize of the hunter; however, it still holds its own in the high mountains of the Rocky and Cascade Ranges. Both sexes have slender black horns, white hair, and black feet, eyes, and nose. Owen Wister says of this animal: "He is white, all white, and shaggy, and twice as large as any goat you ever saw. His white hair hangs long all over him like a Spitz dog's or an Angora cat's; and against its shaggy white mass the blackness of his hoofs and horns, and nose looks particularly black. His are thick, his neck is thick, everyt about him is thick, save only his black horns. They're generally abou (often more than nine) inches long, spread very slightly, and they curve slig backward. At their base they are a rough, but as they rise they become c drically smooth and taper to an ugly p His hoofs are heavy, broad and blunt. female is lighter than the male, and horns more slender, a trifle. And (t turn to the question of diet) we vis the pasture where the herd (of thirty- had been, and found no signs of growing or grass eaten; there was no on that mountain. The only edible stance was a moss, tufted, stiff and dr the touch. I also learned that the is safe from predatory animals. With impenetrable hide and his disembowe horns he is left by the wolves and mo tain lions respectfully alone."

Milch Goats — Many breeds of th have been developed, and the highest t is, perhaps, found in Switzerland. Swiss farmers have found the goat ticularly adapted to their high mount and have used it extensively; thus, g developed in the Saane and Toggenb valleys have a world-wide reputati Above these valleys the high mounta are covered with perpetual snow, and w ter sets in about November 1, last until the last of May. The goats are k with the cows in barns and fed upon h but as soon as the snow is gone from valleys and the lower foothills, the ca and goats are sent with the herders a boy assistants to the grazing grounds. bell is put upon the cow that leads herd so as to keep it together and the b in their gay peasant dresses, are as ha as the playful calves and goats to get in the spring sunshine. The herds foll the receding snows up the mountains til about midsummer, when they rea the high places of scanty vegetation; th they start on the downward journey, turning to the home and stables ab November 1. The milk from goats mixed with that from cows to make chee this cheese has a wide reputation; e of the varieties are Roquefort, veitzer, and Altenburger. Although cheese is excellent, the butter made goat's milk is inferior to that made the cow's. The milk, when the ani- are well taken care of, is exceedingly rishing; it is thought to be the best in the world for children. Usually, trouble with goat's milk is that the nals are not kept clean, nor is care n in milking. Germany has produced ny distinct and excellent breeds of h goats; the Island of Malta, Spain, land, Ireland, Egypt, and Nubia have developed noted breeds. Of all these, Nubias give the most milk, sometimes ding from four to six quarts a day, le an ordinary goat is considered fairly d if it yields two quarts a day.

'HE MOHAIR GOATS — There are two ed breeds of goats whose hair is used ensively for weaving into fabrics; one hese is the Cashmere and the other the gora. The Cashmere goat has long, ight, silky hair for an outside coat and a winter undercoat of very delicate ol. There are not more than two or ee ounces of this wool upon one goat, l this is made into the famous Cash- re shawls; ten goats furnish barely ough of this wool for one shawl. The shmere goats are grown most largely in bet, and the wool is shipped from the h tableland to the Valley of Cashmere, ere it is made into shawls. It requires work of several people for a year to oduce one of these famous shawls.

The Angora goat has a long, silky, and y curly fleece. These goats were first covered in Angora, a city of Asia Minor th of the Black Sea, and some 200 miles theast from Constantinople. The An- ra goat is a beautiful and delicate ani- al, and furnishes most of the mohair ich is made into the cloths known as ohair, alpaca, camel's hair, and many her fabrics. The Angora goat has been troduced into America, in California, exas, Arizona, and to some extent in the iddle West. It promises to be a very ofitable industry. (See Farmers' Bulletin *The Angora Goat*, United States Department of Agriculture.)

The skins of goats are used extensively; morocco, gloves, and many other articles are made from them. In the Orient, the skin of the goat is used as a bag in which to carry water and wine.

LESSON 65
THE GOAT

LEADING THOUGHT — Goats are among our most interesting domesticated animals, and their history is closely interwoven with the history of the development of civilization. In Europe, their milk is made into cheese that has a world-wide fame; and from the hair of some of the species, beautiful fabrics are woven. The goat is naturally an animal of the high mountains.

METHOD — A span of goats harnessed to a cart is second only to ponies, in a child's estimation; therefore, the beginning of this lesson may well be a span of goats thus employed. The lesson should not be given unless the pupils have an opportunity for making direct observations on the animal's appearance and habits. There should be some oral and written work in English done with this lesson.

Bureau of Animal Industry, U. S. D. A.

Angora goat

Following are topics for such work: "The Milch Goat of Switzerland," "How Cashmere Shawls Are Made," "The Angora Goat," "The Chamois."

Observations — 1. Do you think that goats like to climb to high points? Are they fitted to climb steep, inaccessible places? Can they jump off steep places in safety? How does it happen that the goat is sure-footed? How do its legs and feet compare with those of the sheep?

2. What does the goat eat? Where does it find its natural food on mountains? How are the teeth arranged for cutting its food? Does a goat chew its cud like a cow?

3. What is the covering of the goat? Describe a billy goat's beard. Do you suppose this is for ornament? For what is goat's hair used?

4. Do you think the goat has a keen sense of sight, of hearing, and of smell? Why? Why did it need to be alert and keen when it lived wild upon the mountains? Do you think the goat is intelligent? Give instances of this.

5. Describe the horns. Do they differ from the horns of the sheep? How does a goat fight? Does he strike head on, like the sheep, or sidewise? How does he show anger?

6. What noises does a goat make? Do you understand what they mean?

7. Describe the goat, its looks anc tions. Is the goat's tail short at firs does it have to be cut off like the la tail? Where and how is goat's milk u What kinds of cheese are made fron For what is its skin used? Is its flesh eaten?

Everyone knows the gayety of yo kids, which prompts them to cut the r amusing and burlesque capers. The is naturally capricious and inquisitive, one might say crazy for every specie adventure. It positively delights in p ous ascensions. At times it will rear threaten you with its head and ho apparently with the worst intenti whereas it is usually an invitation to p The bucks, however, fight violently v each other; they seem to have no sciousness of the most terrible blows. ewes themselves are not exempt from vice.

They know very well whether or they have deserved punishment. D them out of the garden, where they forbidden to go, with a whip and they flee without uttering a sound; but st them without just cause and they will s forth lamentable cries.

— "Our Domestic Anima
Charles William Burk

THE SHEEP

The earliest important achievement of ovine intelligence is to know whether own notion or another's is most worth while, and if the other's, which one? Indiv ual sheep have certain qualities, instincts, competences, but in the man-herded flo these are superseded by something which I shall call the flock mind, though I can say very well what it is, except that it is less than the sum of all their intelligences. T is why there have never been any notable changes in the management of flocks si the first herder girt himself with a wallet of sheep-skin and went out of his cave-dw ing to the pastures. — "The Flock," Mary Austin

Both sheep and goats are at home on mountains, and sheep especially thrive best in cool, dry locations. As wild animals, they were creatures of the mountain crag and chasm, although they frequented more open places than the mountain goats, and their wool was developed protect them from the bitter cold of h altitudes. They naturally gathered flocks, and sentinels were set to give wa ing of the approach of danger; as soon the signal came, they made their esca

Verne Morton

Sheep at rest

: in the straight away race like the deer, t in following the leader over rock, ge, and precipice to mountain fast-ses where neither wolf nor bear could low. Thus, the instinct of following the der blindly came to be the salvation of : individual sheep.

The teeth of the sheep are like those of : goat, eight incisors below and none the upper row, and six grinding teeth the back of each side of each jaw. This angement of teeth on the small, deli-:e, pointed jaws enables the sheep to p herbage where cattle would starve; can cut the small grass off at its roots, d for this reason, where vast herds of eep range, they leave a desert behind em. This fact brought about a bitter ıd between the cattle and sheep men the far West. In forests, flocks of sheep mpletely kill all underbrush, and now ey are not permitted to run in gov-nment reserves.

The sheep's legs are short and delicate :low the ankle. The upper portion is eatly developed to help the animal in aping, a peculiarity to which we owe e "leg of lamb" as a table delicacy. The oof is cloven, that is, the sheep walks upon two toes; it has two smaller toes above and behind these. There is a little gland between the front toes that secretes an oily substance, which perhaps serves in preventing the hoof from becoming too dry. The ears are large and are moved to catch better the direction of sound. The eyes are peculiar; in the sunlight the pupil is a mere slit, while the iris is yellow or brownish, but in the dark, even of the stable, the pupils enlarge, almost covering the eye. The ewes either lack horns or have small ones, but the

Bureau of Animal Industry, U. S. D. A.

Cheviot sheep

Bureau of Animal Industry, U. S. D. A.

Ewe with her lamb

horns of wild rams are large, placed at the side of the head and curled outward in a spiral. These horns are perhaps not so much for fighting the enemy as rival rams. The ram can strike a hard blow with head or horns, coming at the foe head on, while the goat always strikes sidewise. So fierce is the blow of the angry sheep that an ancient instrument of war, fashioned like a ram's head and used to knock down walls, was called a battering ram. A sheep shows anger by stamping the ground with the front feet. The habit of rumination enables the sheep to feed in a flock and then retire to some place to rest and chew the cud, a performance peculiarly amusing in the sheep.

Sheep under attack and danger are silent; ordinarily they keep up a constant, gentle bleating to keep each other informed of their whereabouts; they also give a peculiar call when water is discovered, and another to inform the flock that there is a stranger in the midst; they also give a peculiar bleat, when a snake or other enemy which they conquer is observed. Their sense of smell is very acute.

Lambs quickly become true members of the herd. Mary Austin says, "Young lambs are principally legs, the connecting body being simply a contrivance for converting milk into more leg, so you understand how it is that they will follow flock in two days and are able to take trail in a fortnight, traveling four and miles a day, falling asleep on their feet tottering forward in the way." The c lambs have games which they play tiringly, and which fit them to becc active members of the flock; one is regular game of "Follow My Lead each lamb striving to push ahead and tain the place of leader. In playing the head lamb leads the chase over n difficult places, such as logs, stones, brooks; thus is a training begun wh later in life may save the flock. other game is peculiar to stony pastu a lamb climbs to the top of a boulder its comrades gather around and try to l it off; the one which succeeds in do this climbs the rock and is "it." This ga leads to agility and sure-footedness. lamb's tail is long and is most express of lambkin bliss, when feeding ti comes; but, alas! it has to be cut off that later it will not become matted w burrs and filth. In southern Russia th is a breed of sheep with large, flat, fat t which are esteemed as a great table d cacy. This tail becomes so cumberso that wheels are placed beneath it, so t it trundles along behind its owner.

In the Rocky Mountains we have noble species of wild sheep which is lik

Mutual contentment

ecome extinct soon. The different ls of domesticated sheep are supd to have been derived from different species. Of the domesticated vari-, we have the Merinos, which origid in Spain and which give beautiful, , fine wool for our fabrics; but their is not very attractive. The Merinos wool on their faces and legs and have kled skins. The English breeds of p have been especially developed for ton, although their wool is valuable. e of these like the Southdown, Shrop-e, and Dorset, give a medium length

Bureau of Animal Industry, U. S. D. A.

Corriedale ram

Verne Morton

Rams in pasture

vool, while the Cotswold has very long ol, the ewes having long strings of wool r their eyes in the fashion of " bangs." The dog is the ancient enemy of sheep; even now, after hundreds of years of nestication, some of our dogs will ret to savagery and chase and kill sheep. is, in fact, has been one of the great wbacks to sheep-raising in the eastern ited States. The collie, or sheep dog, been bred so many years as the special etaker of sheep, that a beautiful relanship has been established between ese dogs and their flocks.

LESSON 66

The Sheep

Leading Thought — Sheep live natully in high altitudes. When attacked by enemies, they follow their leader over difficult and dangerous mountain places.

Method — The questions of this lesson should be given to the pupils and the observations should be made upon the sheep in pasture or stable. Much written work may be done in connection with this lesson. The following topics are suggested for themes: " The Methods by Which Wool Is Made into Cloth," " The Rocky Mountain Sheep," " The Sheep-herders of California and Their Flocks," " The True Story of a Cosset Lamb."

Observations — 1. What is the chief characteristic that separates sheep from other animals? What is the difference between wool and hair? Why is wool of special use to sheep in their native haunts? Is there any hair on sheep?

2. Where do the wild sheep live? What

Bureau of Animal Industry, U. S. D. A.

Hampshire ewe

is the climate in these places? Does wool serve them well on this account? What sort of pasturage do sheep find on mountains? Could cows live where sheep thrive? Describe the sheep's teeth and how they are arranged to enable it to crop vegetation closely. What happens to the vegetation on the range when a great flock of sheep passes over it? Why are sheep not allowed in our forest preserves?

3. What are the chief enemies of sheep in the wilderness? How do the sheep escape them? Describe the foot and leg of the sheep and explain how they help the animal to escape its enemies. We say of certain men that they "follow like a flock of sheep." Why do we make this comparison? What has this habit of following the leader to do with the escape of sheep from wolves and bears?

4. How do sheep fight? Do both rams and ewes have horns? Do they both fight? How does the sheep show anger? Give your experience with a cross cosset lamb.

5. Do you think that sheep can see and hear well? What is the position of the sheep's ears when it is peaceful? When there is danger? How do the sheep's differ from those of the cow?

6. Does the sheep chew its cud lik cow? Describe the action as performe the sheep. How is this habit of cud c ing of use to the wild sheep?

7. Describe a young lamb. Why h such long legs? How does it use its to express joy? What happens to this later? What games have you seen la play? Tell all the stories of lambs that know.

8. How much of sheep language do understand? What is the use to the flock of the constant bleating?

9. For what purposes do we keep sh How many breeds of sheep do you kn What are the chief differences betw the English breeds and the Meri Where and for what purposes is the of sheep used?

10. Have you ever seen a collie loo after a herd of sheep? If so, describe actions. Did you ever know of dogs ing sheep? At what time of day or n was this done? Did you ever know of dog attacking a flock of sheep alone?

THE HORSE

There was once a little animal no bigger than a fox,
And on five toes he scrambled over Tertiary rocks.
They called him Eohippus, and they called him very small,
And they thought him of no value when they thought of him at all.

Said the little Eohippus, I am going to be a horse!
And on my middle finger nails to run my earthly course!
I am going to have a flowing tail! I am going to have a mane!
And I am going to stand fourteen hands high on the Psychozooic plain!

— Mrs. Stetson

It was some millions of years ago that Eohippus lived out in the Rocky Mountain Range; its forefeet had four toes and the splint of the fifth; the hind feet had three toes and the splint of the fourth. Eohippus was followed down the geologic ages by the Orohippus and the Mesohippus and various other hippuses, which showed in each age a successive enlargement and specialization of the middle and the minimizing and final loss of others. This first little horse with m toes lived when the earth was a dar warm place and when animals needed t to spread out to prevent them from mir in the mud. But as the ages went on, earth grew colder and drier, and a lo leg ending in a single hoof was very s

Marion E. Wesp

Mares and colts in shady pasture

ıble in running swiftly over the dry ins. According to the story read in fossils of the rocks, our little American ses migrated to South America, and) trotted dry-shod over to Asia in the d-pliocene age, arriving there sufficntly early to become the companion of historic man. In the meantime, horses re first hunted by savage man for their h, but were later ridden. At present, re are wild horses in herds on the plains Tartary; and there are still sporadic rds of mustangs on the great plains of r own country, although for the most rt they are branded and belong to somee, even though they live like wild horses; ese American wild horses are supposed be descendants of those brought over nturies ago by the Spaniards. The Shetıd ponies are also wild in the islands rth of Scotland, and the zebras, the ost truly wild of all, roam the plains of rica. In a state of wildness, there is alıys a stallion at the head of a herd of ares, and he has to win his position and ep it by superior strength and prowess. ghts between stallions are terrible to witss, and often result in the death of one the participants. The horse is well med for battle; his powerful teeth can flict deep wounds and he can kick and strike hard with the front feet; still more efficient is the kick made with both hind feet while the weight of the body is borne on the front feet, and the head of the horse is turned so as to aim well the terrible blow. There are no wild beasts of prey which will not slink away to avoid a herd of horses. After attaining their growth in the herd with their mothers, the young males are forced by the leader to leave and go off by themselves; in turn, they must by their own strength and attractions win their following of mares. However, there are times and places where many of these herds join, making large bands wandering together.

Field Museum of Natural History

Ancestors of the horse — a restoration

Bureau of Animal Industry, U. S. D. A.

Morgan horse

The length of the horse's leg was evidently evolved to meet the need for flight before fierce and swift enemies, on the great ancient plains. The one toe, with its strong, sharp hoof, makes a fit foot for such a long leg, since it strikes the ground with little waste of energy and is sharp enough not to slip, but it is not a good foot for marshy places; a horse will mire where a cow can pass in safety. The development of the middle toe into a hoof results in lifting the heel and wrist far up the leg, making them appear to be the knee and elbow, when compared with the human body.

The length of neck and head are necessary in order than an animal with such length of leg as the horse may be able to graze. The head of the horse tells much of its disposition; a perfect head should be not too large; it should be broad between the eyes and high between the ears, while below the eyes it should be narrow. The ears, if lopped or turned back, denote a treacherous disposition; they should point upward or forward. If the ears are laid back it is a sign that the horse is angry; sensitive, quick-moving ears indicate a high-strung, sensitive animal. The eyes are placed so that the horse can see in front, at the side, and behind, the last being necessary in order to aim a kick. Hazel eyes are usually preferred to dark ones, and they should be bright and prominent. nostrils should be thin-skinned, wide-ing, and sensitive; in the wild stage, s was one of the horse's chief aids in det ing the enemy. The lips should not be thick and the lower jaw should be nar where it joins the head.

The horse's teeth are peculiar; th are six incisors on each jaw; behind th is a bare space called the bar, of which have made use for placing the bit. B of the bar, there are six molars or grin on each side of each jaw. At the age about three years, canine teeth or tus appear behind the incisors; these are m noticeable in males, and never seem to of much use. Thus, the horse has on e jaw, when full-grown, six incisors, canines, and twelve molars, making fo teeth in all. The incisors are promin and enable the horse to bite the grass m closely than can the cow. The horse wh chewing does not have the sidewise tion of the jaws peculiar to the cow a sheep.

The horse's coat is, when rightly ca for, glossy and beautiful; but if the ho is allowed to run out in the pasture winter, the coat becomes very shaggy, th reverting to the condition of wild hor which stand in need of a warmer coat winter; the hair is shed every year. T mane and the forelock are useful in p tecting the head and neck from flies;

Bureau of Animal Industry, U. S. D.

Percheron draft horse

also is an efficient fly-brush. The mane tail have thus a practical value, and y also add greatly to the animal's uty. To dock a horse's tail for pures of ornament is as absurd as the ed ears and welted cheeks of savages; horses thus mutilated suffer greatly n the attacks of flies.

)wing to the fact that wild horses made ft flight from enemies, the colts could be left behind at the mercy of wolves. us it is that the colt, like the lamb, quipped with long legs from the first, l can run very rapidly; as a runner, it ld not be loaded with a big compound mach full of food, like the calf, and erefore must needs take its nourishment m the mother at frequent intervals. e colt's legs are so long that it must ead the front legs wide apart in order reach the grass with its mouth. When e colt or the horse lies down out of doors d in perfect freedom, it lies flat upon e side. In lying down, the hind quarters first, and in rising, the front legs are rust out first.

The horse has several natural gaits and me that are artificial. Its natural methods progression are the walk, the trot, the nble, and the gallop. When walking ere are always two or more feet on the ound and the movement of the feet conits in placing successively the right hind ot, the right fore foot, left hind foot, left

Bureau of Animal Industry, U. S. D. A.

Man o' War. A famous race horse and the father of famous racers

Bureau of Animal Industry, U. S. D. A.

Carriage stallion

fore foot, right hind foot, etc. In trotting, each diagonal pair of legs is alternately lifted and thrust forward, the horse being unsupported twice during each stride. In ambling, the feet are moved as in the walk, only differing in that a hind foot or a fore foot is lifted from the ground before its fellow fore foot or hind foot is set down. In a canter, the feet are landed on the ground in the same sequence as in a walk but much more rapidly; and in the gallop, the spring is made from the fore foot and the landing is on the diagonal hind foot, and just before landing the body is in the air and the legs are all bent beneath it.

An excellent horseman once said to me, "The whip may teach a horse to obey the voice, but the voice and hand control the well-broken horse," and this epitomizes the best horse training. He also said, "The horse knows a great deal, but he is too nervous to make use of his knowledge when he needs it most. It is the horse's feelings that I rely on. He always has the use of his feelings and the quick use of them." It is a well-known fact that those men who whip and scold and swear at their horses are meantime showing to the world that they are fools in this particu-

A herd of ponies in the Isle of Shetland guarded by a sheep dog

lar business. Many of the qualities which we do not like in our domesticated horses were most excellent and useful when the horses were wild; for instance, the habit of shying was the wild horse's method of escaping the crouching foe in the grass. This habit as well as many others is better controlled by the voice of the driver than by a blow from the whip.

Timothy hay, or hay mixed with clover, form good, bulky food for the horse, and oats and corn are the best concentrated food. Oats are best for driving-horses and corn for the working team. Dusty hay should not be fed to a horse; but if unavoidable, it should always be dampened before feeding. A horse should be fed with regularity, and should not be used for a short time after having eaten. If the horse is not warm, it should be watered before feeding, and in the winter the water should have the chill taken off. The frozen bit should be warmed before being placed in the horse's mouth; if anyone doubts the wisdom of this, let him put a frozen piece of steel in his own mouth. The cruel use of the tight-drawn over checkrein should not be permitted, although a moderate check is often needed and is not cruel. When the horse is sweating, it should be blanketed immediately if hitched outside in cold weather; but in the barn the blanket should not be put on until the perspiration has stopped steaming. The grooming of a horse is a part of its rights, and its legs should receive more attention during this process than its body, a fact not always well understood.

The breeds of horses may always classified more or less distinctly as follc racers or thoroughbreds; the saddle hc or hunter; the coach horse; the draft hc and the pony. For a description of bre see dictionaries or cyclopedias. Of draft horses, the Percherons, Shires, Clydesdales are most common; of carriage and coach horses, the Eng hackney and the French and Gerr coach horses are famed examples. Of roadster breeds, the American trotter, American saddle horse and the Eng thoroughbred are most famous.

LESSON 67

The Horse

Leading Thought — The horse a wild animal depended largely upon

Marion E. W

Percheron colt

gth and fleetness to escape its ene-, and these two qualities have made greatest use to man.

ETHOD — Begin this study of the horse the stories of wild horses. "The Pac-Mustang" in *Wild Animals I Have wn* is an excellent story to show the ts of the herds of wild horses. Before nning actual study of the domestic es, ask for oral or written English exer-; descriptive of the lives of the wild es. After the interest has been thus sed the following observations may be gested, a few at a time, to be made in-ntally in the street or in the stable.

BSERVATIONS — 1. Compare the length he legs of the horse with its height. any other domestic animal legs as g in proportion? What habits of the estral wild horses led to the develop-nt of such long legs? Do you think length of the horse's neck and head responds to the length of its legs? y?

. Study the horse's leg and foot. The se walks on one toe. Which toe do you nk it is? What do we call the toenail the horse? What advantage is this sort foot to the horse? Is it best fitted for nning on dry plains or for marshy land? es the hoof grow as our nails do? Do know whether there were ever any rses with three toes or four toes on each t? Make a sketch of the horse's front l hind leg and label those places which respond to our wrist, elbow, shoulder, nd, heel, knee, and hip.

3. Where are the horse's ears placed on e head? How do they move? Do they p back and forth like the cow's ears en they are moved, or do they turn if on a pivot? What do the following fferent positions of the horse's ears in-cate: When lifted and pointing forward? hen thrown back? Can you tell by the tion of the ears whether a horse is nerv-s and high-strung or not?

4. What is the color of the horse's eyes? he shape of the pupil? What advantage es the position of the eyes on the head ve to the wild horse? Why do we put inders on a horse? Can you tell by the expression of the eye the temper of the horse?

5. Look at the mouth and nose. Are the nostrils large and flaring? Has the horse a keen sense of smell? Are the lips thick or thin? When taking sugar from the hand, does the horse use teeth or lips?

6. Describe the horse's teeth. How many front teeth? How many back teeth? Describe the bar where the bit is placed. Are there any canine teeth? If so, where? Do you know how to tell a horse's age by its teeth? Can a horse graze the grass more closely than a cow? Why? When it chews does it move the jaws sidewise like the cow? Why? Why did the wild horses not need to develop a cud-chewing habit?

7. What is the nature of the horse's coat in summer? If the horse runs in the pasture all winter, how does its coat change? When does the horse shed its coat? What is the use of the horse's mane, forelock, and tail? Do you think it is treating the horse well to dock its tail?

8. Why do colts need to be so long-legged? How does a colt have to place its front legs in order to reach down and eat the grass? Does the colt need to take its food from the mother often? How does it differ from the calf in this respect? How has this difference of habit resulted in a difference of form in the calf and colt?

9. When the horse lies down which part goes down first? When getting up which rises first? How does this differ from the method of the cow? When the horse lies down to sleep does it have its legs partially under it like the cow?

10. In walking which leg moves first? Second? Third? Fourth? How many gaits has the horse? Describe as well as you can all of these gaits.

11. Make a sketch of a horse showing the parts. (See Webster's Unabridged.) When we say a horse is fourteen hands high what do we mean?

12. In fighting, what weapons does the horse use and how?

13. In training a horse, should the voice or the whip be used more? What qualities should a man have to be a good horse trainer? Why is shying a good quality in

wild horses? How should it be dealt with in the domestic horse?

14. What sort of feed is best for the horse? How and when should the horse be watered? Should the water be warmed in cold weather? Why? Should the bit be warmed in winter before putting it in a horse's mouth? Why? Should a tight over checkrein be used when driving? Why not? When the horse has been driven until it is sweating what are the rules for blanketing it when hitched out of doors and when hitched in the barn? What is your opinion of a man who lets his horse stand waiting in the cold, unblanketed? If horses were kept out of doors all the time would this treatment be so cruel and dangerous? Why not? Why should dusty hay be dampened before it is fed to a horse? Why should a horse be groomed? Which should receive more attention, the legs or the body?

15. How many breeds of horses do you know? What is the use of each? Describe as well as you can the characteristics of the following: the thoroughbred, the hackney, and other coach horses; the American trotter, the Percheron, the Clydesdale.

16. Write English themes on the lowing subjects: "The Prehistoric H of America," "The Arabian Horse an Life with Its Master," "The Bron and Mustangs of the West," "The Horses of Tartary," "The Zebras Africa," "The Shetland Ponies and Islands on Which They Run Wild."

Many horses shy a good deal at ob they meet on the road. This mostly a from nervousness, because the objects not familiar to them. Therefore, to the habit, you must get your horse ac tomed to what he sees, and so give confidence. . . . Be careful never to a horse that is drawing a vehicle or in the middle of a hill, except for a and if for a rest, draw him across the and place a big stone behind the whee that the strain on the shoulder may eased. Unless absolutely necessary n stop a horse on a hill or in a rut, so when he starts again it means a heavy Many a horse has been made a jibber his temper spoilt by not observing rule. — "A COUNTRY READER," H. B. BUCHANAN.

CATTLE

That in numbers there is safety is a basic principle in the lives of wild cattle, probably because their chief enemies, the wolves, hunted in packs. It has often been related that, when the herd is attacked by wolves, the calves are placed at the center of the circle made by the cattle, standing with heads out and horns ready for attack from every quarter. But when a single animal, like a bear or tiger, attacks any of the herd, they all gather around it in a narrowing circle of clashing horns, and many of these great beasts of prey have thus met their death. The cow is as formidable as the bull to the enemy, since her horns are strong and sharp and she tosses her victim, unless it is too large. The heavy head, strong neck, and short massive horns of the bull are not so much for defense against enemies as against r bulls. The bull not only tosses and gc his victim, but kneels or tramples upon Both bull and cow have effective we ons of defense in the hind feet, wh kick powerfully. The buffalo bull of In will attack a tiger single-handed, and u ally successfully. It is a strange thing t all cattle are driven mad by the smell blood, and weird stories are told of stampeding of herds from this cause, the plains of our great West.

Cattle are essentially grass and herba eaters, and their teeth are peculiarly ranged for this. There are eight front tee on the lower jaw, and a horny pad oppos them on the upper jaw. Back of the on each jaw there is a bare place and grinding teeth on each side. As a cow cro

John L. Rich

Bison or American buffalo. The original wild cattle of America

: herbage, her head is moved up and wn to aid in severing the leaves, and : peculiar sound of the tearing of the ves thus made is not soon forgotten by ose who have heard it. In the wild or mesticated state the habit of cud-chew- ; is this: The cattle graze mornings d evenings, swallowing the food as fast cropped, and storing it in their ruminat- ; stomachs. During the heat of the day, ey move to the shade, preferably to the ady banks of streams, and there in quiet e food is brought up, a small portion at time, and chewed with a peculiar side- se movement of the jaws and again allowed. There is probably no more per- ct picture of utter contentment than herd of cows chewing their cuds in the ade, or standing knee-deep in the cool ream on a summer's day. The cattle in herd keep abreast and move along when azing, heads in the same direction.

Connected with the grazing habit, is at of the hiding of the newborn calf its mother; the young calf is a wabbly eature and ill-fitted for a long journey; the mother hides it, and there it stays frozen" and will never stir unless ac- ally touched. As the mother is obliged to be absent for some time grazing with the herd, the calf is obliged to go without nourishment for a number of hours, and so it is provided with a large compound stomach which, if filled twice a day, suffices to insure health and growth. The cow, on the other hand, giving her milk out only twice a day, needs a large udder in which to store it. The size of the udder is what has made the cow useful to us as a milch animal.

A fine cow is a beautiful creature, her soft yellow skin beneath the sleek coat of

Cows in pasture. A Jersey and a Holstein

Marion E. Wesp

A very young Jersey calf gets its breakfast

short hair, the well-proportioned body, the mild face, crowned with spreading, polished horns and illuminated with large gentle eyes, are all elements of beauty which artists have recognized, especially those of the Dutch school. The ancients also admired bovine eyes, and called their most beautiful goddess the ox-eyed Juno.

The cow's ears can be turned in any direction, and her sense of hearing is keen; so is her sense of smell, aided by the moist, sensitive skin of the nose; she always sniffs danger and also thus tests her food. Although a cow if well kept has a sleek coat, when she is allowed to run out of doors during the winter her hair grows long and shaggy as a protection. The cow walks on two toes, or as we say has a split hoof. She has two lesser toes above and behind the hoofs which we call dewclaws. The part of her leg which seems at first glance to be her knee is really her wrist or ankle. Although short-legged, the cow is a good runner, as those who have chased her can bear witness. She can walk and gallop, and has a pacing trot; she is a remarkable jumper, often taking a fence like a deer; she also has marvelous powers as a swimmer, a case being on record where a cow swam five miles. But a cow would be ill-equipped for comfort if it were not for her peculiar tail, which is made after the most approved pattern of fly-brushes, and is thus used. Woe betide the fly she hits with it, if the blow is as efficient as that which she incidentally bestows on the head of the milker. It is to get rid of flies that the cattle, and especially the buffaloes, wallow in the mud and thus [cover] themselves with a fly-proof armor.

There is a fairly extensive range of e[mo]tions expressed in cattle language, f[rom] the sullen bellow of the angry anima[l], the lowing which is the call of the h[erd], and the mooing which is meant for [the] calf; and there are many other bellow[ings] and mutterings which we can partially [un]derstand.

Every herd of cows has its leader, [who] has won the position by fair fight. A[dd a] new cow to the herd, and there is at o[nce] a trial of strength, to adjust her to [her] proper place; and in a herd of cows, [the] leader leads; she goes first and no one [may] say her nay. In fact, each member of [the] herd has her place in it; and that is wh[y it] is so easy to teach each cow in a her[d to] take her own stanchion in the stable[. In] a herd of forty cows which I knew, e[ach] cow took her stanchion, no matte[r in] what order she happened to enter [the] stable.

A cow at play is a funny sight; her [tail] is lifted aloft like a pennant and she k[icks] as lightly as if she were made of rub[ber]. She is also a sure-footed beast, as any[one] can attest who has seen her running d[own] the rocky mountainsides of the Alps [at] a headlong pace and never making a [mis]take. In lying down, the cow first kn[eels] with the front legs, or rather drops on [her] wrists, then the hindquarters go do[wn] and the front follow. She does not lie

E. S. Har[t]

Cornell Ormsby Esteem. Holstein heifer, all-American yearling

her side when resting, like the horse
n at ease, but with her legs partially
er her. In getting up, she rests upon
wrists and then lifts the hindquarters.

The Usefulness of Cattle

When man emerged from the savage e, his first step toward civilization was nesticating wild animals and training m for his own use. During the nomad e, when tribes wandered over the face he earth, they took their cattle along. m the first, these animals have been d in three capacities: first, for carry- burdens and as draft animals; second, neat; third, as givers of milk. They were used in the earlier ages as sacrifices to various deities, and in Egypt, some re held sacred.

As beasts of burden and draft animals, en are still used in many parts of the ited States. For logging, especially in neer days, oxen were far more valuable n horses. They are patient and will pull ew inches at a time, if necessary, a tedi- s work which the nervous horse refuses endure. Cows, too, have been used as ft animals, and are so used in China lay, where they do most of the plowing; these Oriental countries milk is not con- med to any extent, so the cow is kept the work she can do. In ancient times the East, white oxen formed a part of yal processions.

Because of two main uses of cattle by vilized man, he has bred them in two directions; for producing beef, and for milk. The beef cattle are chiefly Aberdeen-Angus, Galloway, Shorthorn or Durham, and Hereford; the dairy breeds are the Jersey, Guernsey, Ayrshire, Holstein-Friesian and Brown Swiss. The beef animal is, in cross section, approximately like a brick set sidewise. It should be big and full across the loins and back, the shoulders and hips covered heavily with flesh, the legs stout, the neck thick and short, and the face short; the line of the back is straight, and the stomach line parallel with it. Very different is the appearance of the milch cow. Her body is oval, instead of being approximately square in cross-section. The outline of her back is not straight, but sags in front of the hips, which are prominent and bony. The shoulders have little flesh on them; and if looked at from above, her body is wedge-shaped, widening from shoulders backward. The stomach line is not parallel with the back bone, but slants downward from the shoulder to the udder. The following are the points that indicate a good milch cow: Head high between the eyes, showing large air passages and indicating strong lungs. Eyes clear large, and placid, indicating good disposition. Mouth large, with a muscular lower jaw, showing ability to chew efficiently and rapidly. Neck thin and fine, showing veins through the skin. Chest deep and wide, showing plenty of room for heart and lungs. Abdomen large but well supported, and increasing in size toward the

Eugene J. Hall

Lady Fairfax. A prize winning Hereford cow. Herefords are one of the leading breeds of beef cattle

Animal Husbandry Dept., Cornell U.

len Carnock's Jessie 9th. Angus heifer ready for the show ring

rear. Ribs well spread, not meeting the spine like the peak of a roof, but the spine must be prominent, revealing to the touch the separate vertebræ. Hips much broader than the shoulders. Udder large, the four quarters of equal size, and not fat; the "milk veins" which carry the blood from the udder should be large and crooked, passing into the abdomen through large openings. Skin soft, pliable, and covered with fine, oily hair. She should have good digestion and great powers of assimilation. The milch cow is a milk-making machine, and the more fuel (food) she can use, the greater her production.

E. S. Harrison

Cornell Ollie Catherine. A prize-winning Holstein cow

The physiological habits of the beef and milch cattle have been changed as much as their structure. The food given to the beef cow goes to make flesh; while that given to the milch cow goes to make milk, however abundant her food. Of course, there are all grades between the beef and the milch types, for many farmers use dual herds for both. However, if a farmer is producing milk it pays him well to get the best possible machine to make it, and that is always a cow of the milch type.

A Geography Lesson

All the best breeds of cattle have been evolved in the British Isles and in Europe north of Italy and west of Russia. All our domesticated cattle were developed from wild cattle of Europe and Asia. The cattle which roam in our rapidly narrowing grazing lands of the far West are Europ cattle. America had no wild cattle exc the bison. In geography supplemen readers, read about Scotland, England, Channel Islands, the Netherlands, Fra and Switzerland and the different kind cattle developed in these countries.

How to Produce Good Milk

There are four main ingredients of m — fat, protein, sugar, and ash. The fa for the purpose of supplying the ani with fat, which may be used as such, which may be converted into energy. T protein supplies the material from wh muscle tissue is built. The sugar provi a source of energy. The protein and su considered together form what we kn as curd, which is the main ingredient cheese; however, cheese, to be go should contain a full amount of butter The ash which may be seen as resid when milk is burned, builds up the bo of the animal.

Jersey cows produce a milk contain a higher per cent of fat than any otl common dairy breed in the United Stat The Holstein cows produce a large fl of milk with a low per cent of fat. T quantity of sugar is relatively consta while the protein increases with the but not in direct proportion.

The dairy barn should have concre floors and metal equipment to aid in kee ing the surroundings clean. The produ tion of clean milk requires that the co be brushed or groomed each day; th their udders be washed before each mil ing, preferably with individual washclot saturated in a mild chlorine solution. soon as the milk is drawn from the udde it should be taken to a dairy house whe it should be strained into sterilized can The milk should then be cooled immed ately, and kept at a low temperature unt it is ready to be used. Milk absorbs odo or flavors very readily, and therefore shou never be kept in the dairy barn itself. pure quality of milk that may be safe consumed raw must be produced b healthy cows, cared for by healthy atten under sanitary conditions. Pasteuriza- of milk destroys bacteria and makes it ble to keep the milk sweet for several if stored in a refrigerator.

ilk to be legally sold in New York must possess three per cent of butter 'or upper grades or first-year work in high school, there could not be a : profitable exercise than teaching the ls the use of the Babcock milk tester.

THE CARE OF THE MILCH COW

is impossible to overestimate the im- ance of teaching the pupils in rural icts the proper care of milch cattle he production of milk. The milch cow perfect machine, and should be re- ed as such in producing milk. First, should have plenty of food of the right , that is, well-balanced ration. Second, should have a warm, clean stable and upplied with plenty of good fresh air. old stable makes it necessary to pro- : much more food for the cow; a case record shows that when a barn was ned up in cold weather for necessary airing, the amount of milk from the s stabled in it decreased ten per cent wenty-four hours. There should be a tected place for drinking, if the cattle st be turned out of the barn for water vinter; it is far better to have the water ed into the barn, although the herd uld be given a few hours each day in open air. A dog should never be used driving cows. To be profitable, a cow uld give milk ten months of the year east. Calves should be dehorned when y are a few days old by putting caustic ash on the budding horns, thus obvi- ng the danger of damaging the cow dehorning.

n a properly run dairy, a pair of scales nds near the can for receiving the milk; l as the milk from each cow is brought it is weighed and the amount set down posite the cow's name on a "milk et" that is tacked on the wall nearby. the end of each week the figures on the lk sheet are added, and the farmer ows just how much milk each cow is giving him, and whether there are any in the herd that are not paying their board.

LESSON 68

THE COW

LEADING THOUGHT — Certain characteristics which enable the cow to live successfully as a wild animal have rendered her of great use to us as a domestic animal.

METHOD — Begin the lesson by leading the pupils to understand the peculiar adaptation of cattle for success as wild animals. This will have to be done largely by reading and asking for oral or written work on the following topics: "The Aurochs," "Wild Cattle of the Scottish Highlands," "The Buffaloes of the Orient," "The American Bison," "The Cowboys of the West and Their Work with Their Herds," "The Breeds of Beef Cattle, Where They Came From, and Where Developed," "The Breeds of Milch Cattle, Their Origin and Names." The following questions may be given out a few at a time and answered as the pupils have opportunity for observation.

OBSERVATIONS — 1. What are the characteristics of a fine cow? Describe her horns, ears, eyes, nose, and mouth. Do you think she can hear well? What is the attitude of her ears when she is listening? Do you think she has a keen sense of smell? Is her nose moist? Is her hair long or short? Smooth or rough?

2. The cow walks on two toes. Can you see any other toes which she does not walk on? Why is the cow's foot better adapted than that of the horse for walking in mud and marshes? What do we call the two hind toes which she does not walk on? Can you point out on the cow's leg those parts which correspond with our elbow, wrist, knee, and ankle? Is the cow

a good runner? Is she a good jumper? Can she swim?

3. For what use was the cow's tail evidently intended? How do the wild buffaloes and bison get rid of attacks of flies?

4. How much of cattle language do you understand? How does the cow express pleasure? Lonesomeness? Anger? How does the bull express anger? What does the calf express with the voice?

5. Is there always a leader in a herd of cows? Do certain cows of the herd always go first and others last? Do the cows readily learn to take each her own place in the stable? How is leadership of the herd attained? Describe cattle at play.

6. At what time of day do cattle feed in the pasture? When and where do they chew the cud? Do they stand or lie to do this? Describe how a cow lies down and gets up.

7. How do wild cattle defend themselves from wolves? From bears or other solitary animals?

8. For what purposes were cattle first domesticated? For how many purposes do we rear cattle today?

9. Name and give brief descriptions of the different breeds of cattle with which you are familiar. Which of these are beef and which milch types?

10. What are the distinguishing points of a good milch cow? Of a good beef animal? What does the food do for each of these? Which part of the United States produces most beef cattle? Which the most milch cattle?

11. What do we mean by a bala ration? Do you know how to com one? What is the advantage of fee cattle a balanced ration?

12. What must be the per cent of ter fat in milk to make it legally salab your state? How many months of the should a good cow give milk?

13. Should a dog be used in dri dairy cows? Why not?

14. Why is a cool draughty barn a pensive place in which to keep ca Why is a barn not well ventilate danger?

15. Why is the dehorning of c practiced? When and how should a be dehorned?

16. Why should milk not be stra in the barn? Why is it profitable for dairy farmer to keep his stable clean to be cleanly in the care of milk? How the food of cows affect the flavor of milk? Why should a farmer keep a ord of the number of pounds of r which each cow in his dairy gives e day?

17. For what are oxen used? Whe are they superior to horses as draft mals? Do you know of any place wh oxen are used as riding animals?

18. How many industries are depe ent upon cattle?

19. Give oral or written exercises on following themes: "How the Best Bu Is Made"; "The Use of Bacteria in B ter"; "How Dairy Cheese Is Mad "How Fancy Cheeses Are Made."

THE PIG

I wander through the underbresh,
Where pig tracks pintin' to'rds the crick,
Is picked and printed in the fresh
Black bottom-lands, like wimmern prick
Their pie-crusts with a fork. — Riley

By a forest law of William the First of England in the eleventh century, it was ordained that any that were found guilty of killing the stag or the roebuck or the wild boar should have their eyes put o This shows that the hunting of the w boar in England was considered a sp of gentlemen in an age when nothing w

Bureau of Animal Industry, U. S. D. A.

Breakfast, cafeteria style

sidered sport unless it was dangerous. : wild hog of Europe is the ancestor our common domesticated breeds, al- ugh the Chinese domesticated their wild species, even before the dawn of ory.

'he wild hog likes damp situations ere it may wallow in the water and d; but it also likes to have, close by, ods, thicket, or underbrush, to which can retire for rest and also when in ger. The stiff, bristling hairs which er its thick skin are a great protection en it is pushing through thorny thick- When excited or angry, these bristles : and add to the fury of its appearance. en in our own country the wild hogs of : South whose ancestors escaped from mestication have reverted to their origi- savagery, and are dangerous when in- iated. The only recorded instance when r great national hunter, Theodore osevelt, was forced ignominiously to mb a tree, was after he had emptied rifle into a herd of " javelins," as the ld pigs of Texas are called; the javelins : the peccaries, which are the American presentatives of the wild hog.

That the hog has become synonymous th filth is the result of the influence of n upon this animal, for of all animals, e pig is naturally the neatest, keeping bed clean, often in the most discouraging and ill-kept pens. The pig is sparsely clothed with bristles and hairs, which yield it no protection from the attacks of flies and other insects. Thus it is that the pig, in order to rid itself of these pests, has learned to wallow in the mud. However, this is in the nature of a mud bath, and is for the purpose of keeping the body free from vermin. The wild hogs of India make for themselves grass huts, thatched above and with doors at the sides, which shows that the pig, if allowed to care for itself, understands well the art of nest building.

One of the most interesting things about a pig is its nose; this is a fleshy disc with nostrils in it and is a most sensitive organ of feeling; it can select grain from

Bureau of Animal Industry, U. S. D. A.

Razorback. A hog of no definite breed, which is allowed to roam at will in some of the southern states

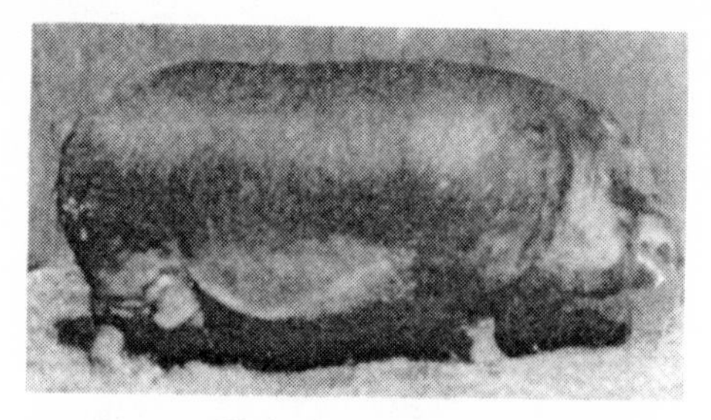

Duroc-Jersey boar

chaff, and yet it is so strong that it can root up the ground in search for food. " Root " is a pig word, and was evidently coined to describe the act of the pig when digging for roots; the pig's nose is almost as remarkable as the elephant's trunk, and the pig's sense of smell is very keen; it will follow a track almost as well as a dog. There are more instances than one of a pig being trained as a pointer for hunting birds, and showing a keener sense of smell and keener intelligence in this capacity than do dogs. French pigs are taught to hunt for truffles, which are fungi growing on tree roots, a long way below the surface of the ground; the pig detects their presence through the sense of smell.

The pig has a full set of teeth, having six incisors, two canines, and seven grinding teeth on each jaw; although in some cases there are only four incisors on the upper jaw. A strange thing about a pig's teeth is the action of the upper canines, or tushes, which curve upward instead of downward; the lower canines grind up against them, and are thus sharpened. The females have no such development of upper tushes as do the males; these

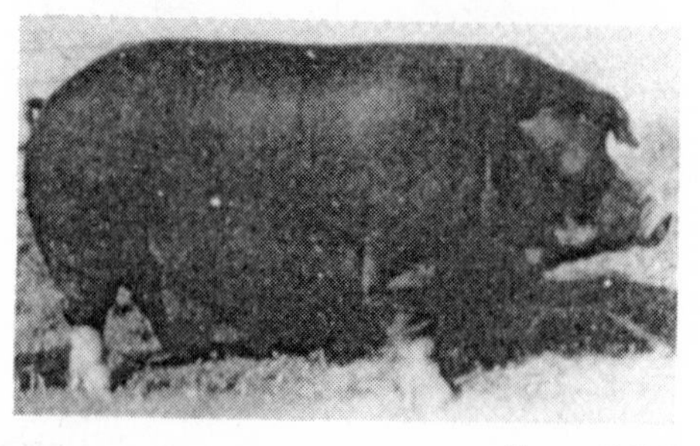

Eugene J. Hall

Poland China hog

tushes, especially the upper ones, are u as weapons; with them, the wild b slashes out and upward, inflicting terr wounds, often disabling horses and ing men. Professor H. F. Button descr the fighting of hogs thus: " To oppose terrible weapons of his rival, the boar a shield of skin over his neck and sh ders, which may become two inches th and so hard as to defy a knife. When of these animals fight, each tries to k the tushes of his opponent against shield, and to get his own tushes un the belly or flank of the other. Thus, e

Bureau of Animal Industry, U. S. D

Hampshire boar

goes sidewise or in circles, which has giv rise to the expression, ' to go sidewise l a hog to war.' "

When, as a small girl, I essayed difficult task of working buttonholes, was told if I did not set my stitches m closely together, my buttonhole wou look like a pig's eye, a remark which ma me observant of that organ ever aft But though the pig's eyes are small, th certainly gleam with intelligence, and th take in all that is going on which may any way affect his pigship.

The pig is the most intelligent of the farm animals, if it is only given chance; it has excellent memory and c be taught tricks readily; it is affectiona and will follow its master around like dog. Anyone who has seen a trained p at a show picking out cards and counti must grant that it has brains. We stuff so with fattening food, however, that does not have a chance to use its brain, e cept now and then when it breaks out the sty and we try to drive it back. Und e circumstances, we grant the pig all sagacity usually imputed to the one once possessed swine and drove them the sea. Hunters of wild hogs prom that they are full of strategy and ning, and are exceedingly fierce.

'he head of the wild hog is wedgeped with pointed snout, and this form bles the animal to push into the thick erbrush along the river banks whenr it is attacked. But civilization has nged this bold profile of the head, so t now in many breeds there is a hollow ween the snout and eyes, giving the m which we call "dished." Some eds have sharp, forward-opening ears, ile others have ears that lop. The wild of Europe and Asia has large, open s extending out wide and alert on each e of the head.

The covering of the pig is a thick skin et with bristling hairs; when the hog excited, the bristles rise and add to the y of its appearance. The bristles aid in tecting the animal when it is pushing ough thorny thickets. The pig's querly l is merely an ornament, although the l of the wart hog of Africa, if pictures y be relied upon, might be used in a nited fashion as a fly-brush.

When the pig is allowed to roam in e woods, it lives on roots, nuts, and escially acorns and beech nuts; in the tumn it becomes very fat through feedg upon the latter. The mast-fed bacon the semi-wild hogs of the southern tes is considered the best of all. But

Bureau of Animal Industry, U. S. D. A.

Tamworth barrow, a bacon type

Bureau of Animal Industry, U. S. D. A.

A champion Berkshire sow

almost anything, animal or vegetable, that comes in its way is eaten by the hog, and it has been long noted that the hog has done good service on our frontier as a killer of rattlesnakes. The pig is well fitted for locomotion on either wet or dry soil, for the two large hoofed toes enable it to walk well on dry ground and the two hind toes, smaller and higher up, help to sustain it on marshy soil. Although the pig's legs are short, it is a swift runner unless it is too fat. The razor-backs of the South are noted for their fleetness.

We understand somewhat the pig's language: the constant grunting, which is a sound that keeps the pig herd together, the complaining squeal of hunger, the satisfied grunt signifying enjoyment of food, the squeal of terror when seized, and the nasal growl when fighting. But there is much more to the pig's conversation than this; I knew a certain lady, a lover of animals, who once undertook to talk pig language as best she could imitate it, to two of her sows when they were engaged in eating. They stopped eating, looked at each other a moment and forthwith began fighting, each evidently attributing the lady's remark to the other, and obviously it was of an uncomplimentary character.

The pig's ability to take on fat was evidently a provision, in the wild state, for storing up from mast fat that should help sustain the animal during the hardships of winter; and this characteristic is what makes swine useful for our own food. Pigs, to do best, should be allowed to have pasture and plenty of fresh green food. Their troughs should be kept clean and they

ANIMALS OF ZOOS AND PARKS

1. Rhinoceros. *From two Greek words which mean " nose " and " horn " we have the word " rhinoceros." Note the hornlike projection on the nose of this African animal which is shown in the picture; a form in Asia differs slightly in appearance.* Range: *Tropical portions of Asia and Africa.*

2. Hippopotamus. *This thick-skinned, short-legged, four-toed animal is at home in the rivers of Africa. It feeds chiefly on grass and aquatic plants. The word " hippopotamus " is derived from two Greek words which mean " river " and " horse."*

3. Kangaroo. *The short forelegs and the powerful hind legs which it uses for jumping give the kangaroo a unique appearance. By means of great leaps, this animal travels rapidly. The immature young are carried in an external pouch. There are various kinds of kangaroos; the red kangaroo is shown in the picture.*

4. Zebra. *These swift, wild animals of Africa are members of the horse family; their unique color arrangement, of dark stripes on a tawny background, is definitely characteristic of them. The colt in the picture is one week old.*

5. Malay Tiger. *The range of this large member of the cat family extends throughout most of Asia from southern Siberia south to Java and Sumatra. In color, a Malay tiger is tawny with black cross stripes. The male, much larger than the female, may reach a length of ten feet including the tail.*

6. Polar Bear. *Found in Arctic regions, this white bear is to be seen on ice floes as well as swimming about in the water; it may weigh as much as* 1500 *pounds and reach a length of* 9 *feet.*

7. Nubian Giraffe. *This uniquely spot African mammal may reach a height twenty feet. By means of a very long n and a grasping tongue, it can easily sec for its food leaves from trees. While it n remind one somewhat of a horse, it is rea to some extent, like a cow; it chews a cud*

8. Bactrian or Two-humped Cam *Long ago the camel was domesticated by n and is to this day an important beast of b den in northern Africa and western Asia. is especially adapted to withstand the ha ships of the deserts; it can go without dri ing water for several days because cert portions of its stomach serve as water res voirs. Water can be taken in large quantit and then used as needed. There is a o humped camel known as the Arabian dromedary camel.*

9. Wapiti or American " Elk." T *American mammal is incorrectly cal " elk "; that title really belongs to our moo which is a true elk. The wapiti's range is n restricted chiefly to more remote regions the western United States and Canada; b formerly the animal was found also in t central and eastern United States. In color, is chestnut red in summer and rather grayi in winter.*

10. Virginia or White-tailed Dee *Only the males possess antlers; these are solid bone, are directed forwards with t prongs upward, and are shed every sprin These deer were formerly very common the plains and forests of the central an southern United States; but now they a abundant in only certain of the wilder po tions of their former range. Their food co sists of buds, leaves, tender bark, and vario other forms of plant life.*

Photographs by New York Zoological Society

should have access to ashes, and above all, they should have plenty of pure water; and as the pig does not perspire freely, access to water where it can take its natural mud baths helps to keep the body cool and the pig healthy in hot weather.

The breeds of hogs most common in America are the Berkshire, which are black and white markings, and have ears extending erect; the Poland China, which are black and white with drooping ears;

Marion E. Wesp

A family meal

the Duroc-Jersey, which are red or chestnut with drooping ears; the Yorkshire and Cheshire, which are white with erect ears; and the Chester White, which are white with drooping ears. The Poland China and Duroc-Jersey are both pure American breeds.

LESSON 69
The Pig

Leading Thought — The pig is something more than a source of pork. It is a sagacious animal and naturally cleanly in its habits when not made prisoner by man.

Method — The questions in this lesson may be given to the pupils a few time, and those who have access to f or other places where pigs are kept make the observations, which shou discussed when they are given to the Supplementary reading should be g the pupils, which may inform them the habits and peculiarities of the hogs. Theodore Roosevelt's experien hunting the wart hog in Africa will p interesting reading.

Observations — 1. How does the nose differ from that of other anir What is it used for besides for smel Do you think the pig's sense of sm very keen? Why do pigs root?

2. Describe the pig's teeth. For are they fitted? What are the tushes Which way do the upper tushes t How do wild hogs use their tushes?

3. Do you think that a pig's eyes intelligent? What color are they? Do think the pig can see well?

4. Is the pig's head straight in fron is it dished? Is this dished appearance found in wild hogs? Do the ears stand straight or are they lopped? What vantage is the wedge-shaped head to wild hogs?

5. How is the pig covered? Do think the hair is thick enough to keep flies? Why does the pig wallow in mud? Is it because the animal is dirty nature or because it is trying to keep cle Do the hog's bristles stand up if i angry?

6. If the pig could have its natural f what would it be and where would it found? Why and on what should pigs pastured? What do pigs find in the fo to eat? What kind of bacon is conside the best?

7. On how many toes does the pig wa Are there other toes on which it does walk? If wading in the mud are the hind toes of use? Do wild pigs run rapic Do tame pigs run rapidly if they are too fat? Do you think the pig can swi Do you think that the pig's tail is of use or merely an ornament?

8. What cries and noises do the p make which we can understand?

How do hogs fight each other? When boars fight, how do they attack or l off the enemy? Where do we get expression "going sidewise like a hog var"?

). How many breeds of pigs do you w? Describe them.

. What instances have you heard that v the hog's intelligence?

2. Give an oral or written English ex-se on one of the following topics: he Antiquity of Swine; How They Were Regarded by the Ancient Egyptians, Greeks, and Romans" (see encyclopedia); "The Story of Hunting Wild Hogs in India"; "The Razor-Back Hogs of the South"; "The Wart Hog of Africa"; "Popular Breeds of Hogs."

The nice little pig with a querly tail,
All soft as satin and pinky pale
Is a very different thing by far
Than the lumps of iniquity big pigs are.

— Nonsense Rhyme

INSECTS

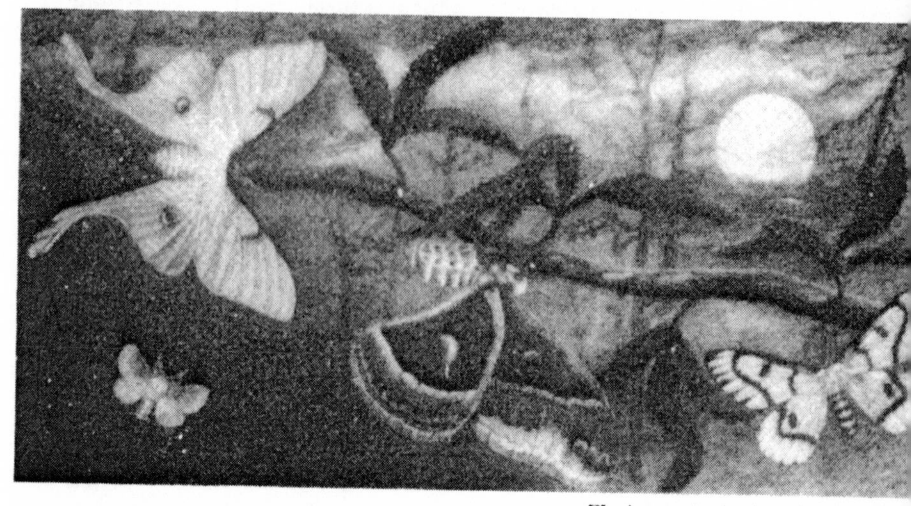

Wood engraving by Anna Botsford Com

Luna moth
Flannel-moth

Cecropia moth

Juno moth

Insects are among the most interesting and available of all living creatures for nature-study. The lives of many of them afford more interesting stories than are found in fairy lore; many of them show exquisite colors; and, most important of all, they are small and are, therefore, easily confined for observation.

About us on every side are myriads of tiny creatures that commonly pass unnoticed, and even when we observe them, we usually think them unworthy of serious consideration. But all life is linked together in such a way that no part of the chain is unimportant. Frequently the action of some of these minute beings seriously affects the material success or failure of a great commonwealth. The introduction and spread of a single species of insect (the cottony-cushion scale) in California threatened the destruction of the extensive orchards of that state; thousands of trees perished. The introduction of a few individuals of a particular kind of lady-bug (*Rodōlia cardinālis*), which feeds u this pest and multiplies rapidly, s checked the pest, and averted the disas

But insects are of interest to us other reasons than the influence they have upon our material welfare; the st of them is a fruitful field for intellect growth. It is not a small matter to be a to view intelligently the facts presented the insect world, to know something what is going on around us. And so ext sive and complex is this field that no gains more than a mere smattering c cerning it.

We know as yet comparatively li about the minute structure of insects; transformations and habits of the grea number of species have not been studi and the blood-relationship of the vari groups of insects is very imperfectly und stood. If, therefore, one would learn sor thing of the action of the laws that gov the life and development of organi beings, and at the same time experie

pleasure derived from original investi-
on, he cannot find a better field than
fered by the study of insects.

ut it is not necessary that one should
the tastes and leisure required for
ful scientific investigation in order to
it by this study. It can be made a
eation, a source of entertainment
n we are tired, a pleasant occupation
our thoughts when we walk. Any one
find out something new regarding in-
architecture — the ways in which
e creatures build nests for themes or for their young. It is easy to ob-
e remarkable feats of engineering,
derful industry, unremitting care of
ng, tragedies, and even war and slav-

he abundance of insects makes it easy
tudy them. They can be found where-
r man can live, and at all seasons. This
ndance is even greater than is com-
nly supposed. The number of individ-
s in a single species is beyond compu-
on: who can count the aphids or the
e-insects in a single orchard, or the
s in a single meadow?

Not only are insects numerous when we
ard individuals, but the number of
cies is far greater than that of all other
mals taken together. The number of
cies in a single family is greater in sev-
cases than the number of stars visible
a clear night.

The word *insect* is often applied incor-
tly to any minute animal; but the term
uld be restricted to those forms possess-
six legs and belonging to the class,
xapoda. The name *Hexapoda* is from
o Greek words: *hex*, six; and *pous*, foot.
refers to the fact that the members of
s order differ from other arthropods in
possession of only six feet. Thus
iders, which have eight legs, are not in-
ts.

Insects breathe by means of a system
air-tubes (tracheæ) which extend
rough the body. This is true even in the
se of those that live in water and are
pplied with gill-like organs (the tracheal
ls). The head is distinct from the tho-
x, and bears a single pair of antennæ; in these respects they are allied to the millipedes and centipedes although they are ap-
parently more closely related to a small
group of animals known as symphylids.

Insects can be easily distinguished by
the number of their feet, and usually, also
by the presence of wings.

While the young pupils should not be
drilled in insect anatomy as if they were
embryo zoologists, yet it is necessary for
the teacher who would teach intelligently to know something of the life
stories, habits, and structure of the common insects.

Nearly all insects in the course of their
lives undergo remarkable changes in form.
Thus the butterfly, which delights us with
its airy flight, was at one time a caterpillar;
and the busy bee lived first the life of a
clumsy grub. Generally speaking, insects
develop from eggs. The word egg brings
before most of us the picture of the egg of
the hen or of some other bird. But insect
eggs are often far more beautiful than those
of any bird; they are of widely differing
forms and are often exquisitely colored;
the shells may be ornately ribbed and pitted, are sometimes adorned with spines, and
are as beautiful to look at through a microscope as the most artistic piece of mosaic.

From the eggs, larvæ (*singular, larva*)
issue. These larvæ may be caterpillars, or
the creatures commonly called worms, or
perhaps maggots or grubs. The larval stage
is devoted to feeding and to growth.
It is the chief business of the larva to
eat diligently and to attain maturity as
soon as possible; for often the length
of the larval period depends more upon
food than upon lapse of time. All insects have their skeletons on the outside
of the body; that is, the outer covering of
the body is chitinous, and the soft and
inner parts are attached to it and supported by it. This skin is so firm that it cannot stretch to accommodate the increasing size of the growing insect, so from
time to time it is shed. But before this is
done, a new skin is formed beneath the
old one. After the old skin bursts open
and the insect crawls forth, the new skin
is sufficiently soft and elastic to allow for

the increase in the size of the insect. Soon the new skin becomes hardened like the old one, and after a time is shed. This shedding of the skin is called molting.

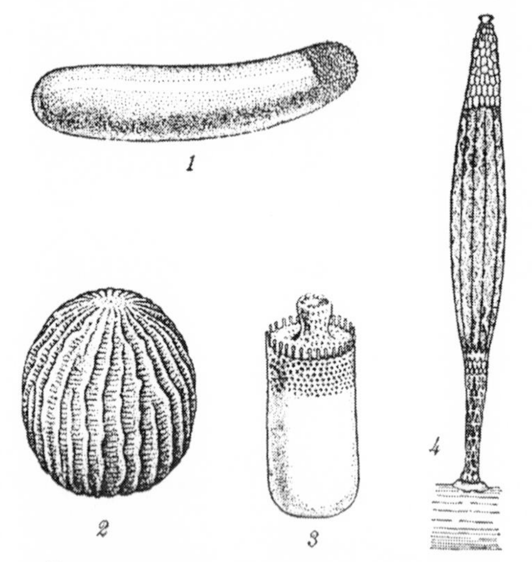

Eggs of insects: 1, *the tree-cricket,* Œcanthus nigricornus; 2, *the White Mountain butterfly,* Œnis semidea; 3, *stinkbug,* Piezosterum subulatum; 4, *water-measurer,* Hydrometra martini

Some insects shed their skins only four or five times during the period of attaining their growth, while other species may molt twenty times or more.

After the larva has attained its full growth it changes its skin and its form, and becomes a pupa. The pupa stage is ordinarily one of inaction, except that very wonderful changes take place within the body itself. Usually the pupa has no power of moving around, but in many cases it can squirm somewhat, if disturbed. The pupa of the mosquito is active and is an exception to the rule. The pupa is usually an oblong object and seems to be without head, feet, or wings;

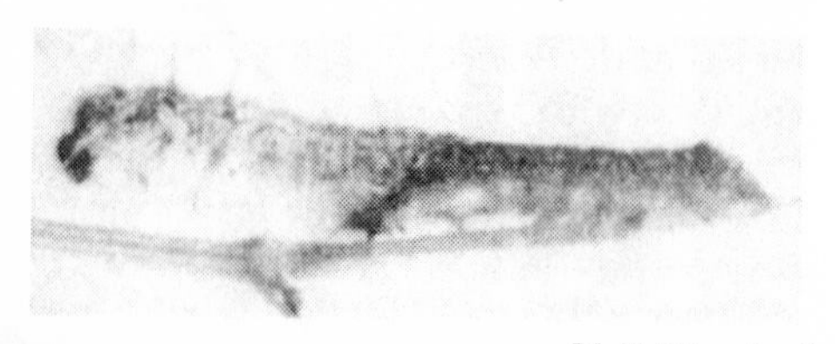

M. V. Slingerland

The forest tent-caterpillar shedding its skin

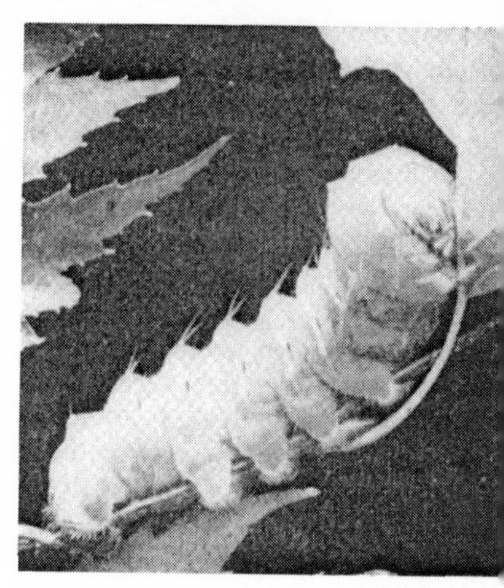

M. V. Slinge

Full-grown caterpillar of the luna mot

but if it is examined closely, especially the case of butterflies and moths, the tennæ, wings, and legs may be seen, fol down beneath the pupa skin.

Many larvæ, especially among moths, weave about themselves a cover of silk which serves to protect them fr their enemies and the weather during helpless pupa period. This silken cover is called a cocoon. The larvæ of butterf do not make a silken cocoon, but pupa is suspended to some object b silken knob, sometimes by a halter of s and remains entirely naked. The pupa c butterfly is called a chrysalis. Care sho be taken to have the children use words pupa, chrysalis, and cocoon und standingly.

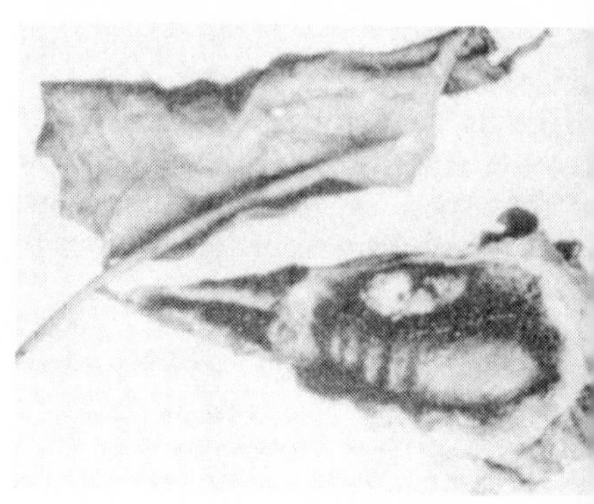

M. V. Slingerl

A luna cocoon cut open, showing the pu

A butterfly chrysalis

ter a period varying from days to
:hs, depending upon the species of
t and the climate, the pupa skin
s open and from it emerges the adult
t, often equipped with large and
tiful wings and always provided with
:gs and a far more complex structure
e body than characterized it as a larva.
insect never grows after it reaches this
t stage and therefore never molts.
e people seem to believe that a small
vill grow into a large fly, and a small
le into a large beetle; but after an in-
attains its perfect wings it does not
larger. Many adult insects take very
: food, although some continue to eat
rder to support life. The adult stage is
narily shorter than the larval stage; it
ıs a part of nature's economic plan
the grown-up insects should live only
; enough to lay eggs, and thus secure

A luna moth

he delicate, exquisite green of the luna's wings is set y the rose-purple, velvet border of the front wings, the white fur on the body and inner edge of the hind s. Little wonder that it has been called the "Empress ne Night." The long swallow tail of the hind wings s the moth a most graceful shape, and at the same probably affords it protection from observation. ing the daytime the moth hangs, wings down, be- h the green leaves, and these long projections of the wings folded together resemble a petiole, making the :t look very much like a large leaf

the continuation of the species. Insects having the four distinct stages in their growth, egg, larva, pupa, and adult, are said to undergo *complete metamorphosis*.

But not all insects pass through an inactive pupa stage. With some insects, like the grasshoppers, the young, as soon as they are hatched, resemble the adult forms

Insect brownies; tree hoppers as seen through a lens

in appearance. These insects, like the larvæ, shed their skins to accommodate their growth, but they continue to feed and move about actively until the final molt when the perfect insect appears. Such insects are said to have *incomplete metamorphosis*, which simply means that the form of the body of the adult insect is not greatly different from that of the young; the dragonflies, crickets, grasshoppers, and bugs are of this type. It must be remembered that while many people refer to all insects as bugs, the term bug is cor-

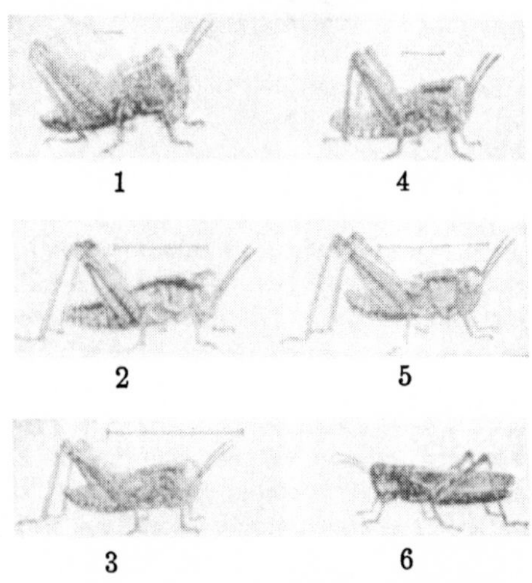

The grasshopper is an example of incomplete metamorphosis

1, nymph, first stage; 2, nymph, second stage; 3, nymph, third stage; 4, nymph, fourth stage; 5, nymph, fifth stage; 6, adult

rectly applied only to one group of insects. This group includes such forms as stinkbugs, squash bugs, plant lice, and tree hoppers. The young of insects an incomplete metamorphosis are c nymphs instead of larvæ.

Summary of the Metamorphoses of Insects

Kinds of Metamorphosis	*Names of Stages*
I. Complete metamorphosis (example, butterfly)	Egg. Larva. Pupa. (Among the moths the pu sometimes enclosed in a cocoon.) Adult or winged insect.
II. Incomplete metamorphosis (example, grasshopper)	Egg. Nymph (several stages). Adult, or imago.

The Structure of Insects

The insect body is made up of ringlike segments which are grown together. These segments are divided into groups according to their use and the organs which they bear. Thus the segments of an insect's body are grouped into three regions: the head, the thorax, and the abdomen. The head bears the eyes, the antennæ, and the mouth-parts. On each side

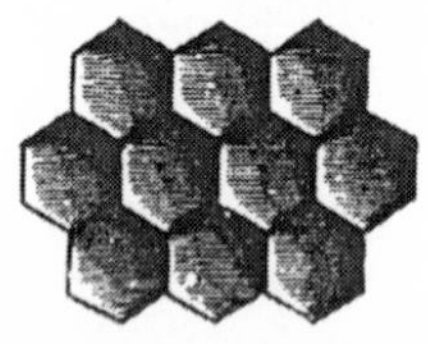

A part of the compound eye, enlarged, of an insect

of the head of the adult insect may be seen the compound eyes; these are so called because they are made up of many small eyes set together, much like the cells of the honeycomb. These compound eyes are not found in larvæ of insects with complete metamorphosis, such as caterpillars, maggots, and beetle grubs. In addition to the compound eyes, many adult insects possess simple eyes; these are placed between the compound eyes and are usually three in number. Often they cannot be seen without the aid of a lens.

The antennæ or feelers are composed of many segments and are inserted in front of the eyes or between them. They vary greatly in form. In some insects are mere threads; in others, like the worm moths, they are large, feathe organs.

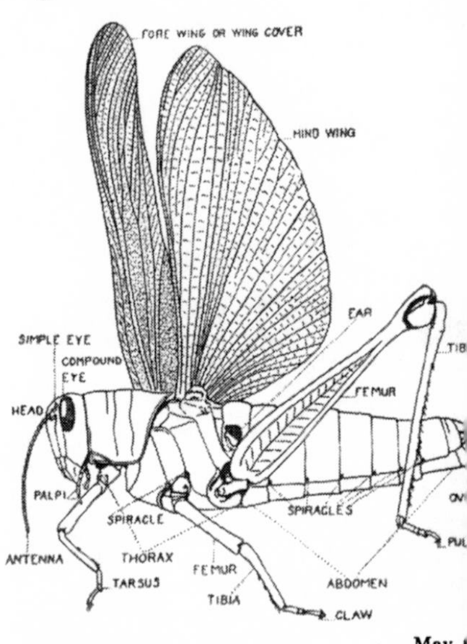

Grasshopper, with the parts of the exte anatomy named

The mouth-parts of insects vary gre in structure and in form, being adapte the life of the insect species to which t minister. Some insects have jaws fitted seizing their prey, others for chew leaves; others have a sucking tube for ting the juices from plants or the bl from animals, and others long delic tubes for sipping the nectar from flow

M. V. Slingerland

phinx moth with the sucking tongue unrolled

n the biting insects, the mouth-parts sist of an upper lip, the labrum, under lip, the labium, and two pairs jaws between them. The upper pair jaws is called the mandibles and the ver pair, the maxillæ (*singular maxilla*). ere may be also within the mouth one two tonguelike organs. Upon the xillæ and upon the lower lip there may o be feelers, which are called palpi *ngular palpus*). The jaws of insects, en working, do not move up and down, do ours, but move sidewise like shears. many of the insects, children can obve the mandibles and the palpi without e aid of a lens.

The thorax is the middle region of the sect body. It is composed of three of the dy segments more or less firmly joined gether. The segment next the head is lled the prothorax, the middle one, the mesothorax, and the hind one, the metathorax. Each of these segments bears a pair of legs and, in the winged insects, the second and third segments bear the wings.

The mouth of the tree hopper, shown here tending beneath the body, is a long, threeinted sucking tube

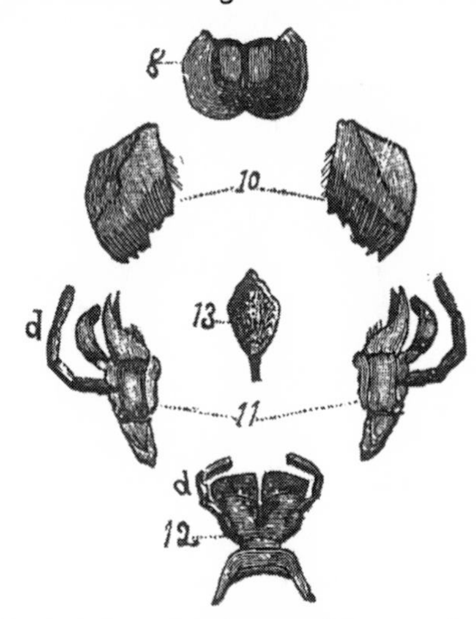

The mouth-parts of a grasshopper, enlarged and named

8, upper lip or labrum; 10, mandibles or upper jaws; 11, maxillae or lower jaws; 12, under lip or labium; 13, tongue; d, palpi

Each leg consists of two small segments next to the body, next to them a longer segment, called the femur, beyond this a segment called the tibia, and beyond this the tarsus or foot. The tarsus is made up of a number of segments, varying from one to six, the most common number being five. The last segment of the tarsus usually bears one or two claws.

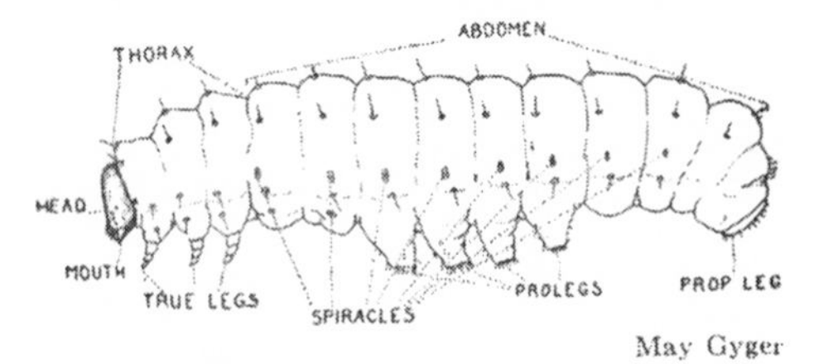

May Gyger

A caterpillar, with the parts of the external anatomy named

While we have little to do with the internal anatomy of inscets in elementary nature-study, the children should be taught something of the way that insects breathe. The child naturally believes that the insect, like himself, breathes through the mouth, but as a matter of fact insects breathe through their sides. If we examine almost any insect carefully, we can find along the sides of the body a series of openings. These are called the spiracles, and through them the air passes into the insect's body. The number of spiracles varies greatly in different insects. There is, however, never more than one pair on a single segment of the body, and they do not occur on the head. The spiracles, or breathing pores, lead into a system of air tubes which are called tracheæ (*tra'-ke-ee*), which permeate the insect's body and thus carry the air to every smallest part of its anatomy. The blood of the insect bathes these thin-walled air tubes and thus becomes purified, just as our blood becomes purified by bathing the air tubes of our lungs. Thus, although the insects do not have localized breathing organs, like our lungs, they have, if the expression may be permitted, lungs in every part of their little bodies.

Summary of Structure of an Insect

- Head
  - Antennæ.
  - Compound eyes.
  - Simple eyes or ocelli.
  - Mouth-parts
    - Labrum, or upper lip.
    - Mandibles, or upper jaws.
    - Maxillæ, or lower jaws, and maxillary palpi.
    - Labium and labial palpi.
- Thorax
  - Prothorax and first pair of legs.
  - Mesothorax and
    - second pair of legs.
    - first pair of wings.
  - Metathorax and
    - third pair of legs.
    - second pair of wings.
  - Wing
    - veins.
    - cells.
  - Leg
    - Two small segments called coxa and tr chanter.
    - Femur.
    - Tibia.
    - Tarsus and claws.
- Abdomen
  - The abdomen bears
    - ears (in locusts only)
    - spiracles.
    - ovipositor.

INSECTS OF THE FIELDS AND WOODS

ʒome insects go through all the stages their development on land; these are : insects of fields and woods. This ·up includes some of the most interest- ; and beautiful of insects. They are especially well adapted for nature-study because specimens are constantly available. The insects presented from page 301 to page 400 are common examples of this group.

THE BLACK SWALLOWTAIL BUTTERFLY

Γhis graceful butterfly is a very good ɛnd to the flowers, being a most effint pollen-carrier. It haunts the gardens d sips nectar from all the blossom cups ld out for its refreshment; and it is ınd throughout almost all parts of the ıited States. The grace of its appearance much enhanced by the " swallowtails," o projections from the hind margins of e hind wings. The wings are velvety ıck with three rows of yellow spots ross them, the outer row being little ·scents set in the margin of the wing; d each triplet of yellow spots is in the me cell of the wing between the same o veins. The hind wings are more elabote, for between the two inside rows of llow spots, there are exquisite metallic ue splashes, more vivid and more sharply ıtlined toward the inside of the wing ıd shading off to black at the outside. nd just above the inner angle of the nd wing is an orange eyespot with a ack center. On the lower surface of the wings, most of the yellow spots are replaced with orange.

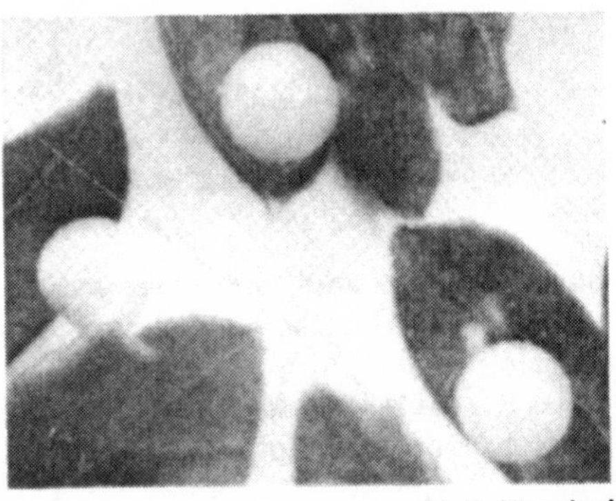

M. V. Slingerland

The eggs of the black swallowtail butterfly, enlarged

M. V. Slingerland

Black swallowtail butterfly

The mother butterfly is larger than her mate and has more blue on her wings, while he has the yellow markings of the hind wings much more conspicuous. She lays her eggs, which are just the color of a drop of honey, on the under surface of the leaves of the food plant. After about ten days there hatch from these eggs spiny little fellows, black and angular, each with a saddle-shaped, whitish blotch in the middle of the back. But it would take an elfin rider to sit in this warty, spiny saddle. The caterpillar has six spines on each segment, making six rows of spines the whole length of the body; the spines on the black portions are black and those on the saddle white, but they all have orange-colored bases.

When little, spiny saddle-back gets ready to change its skin to one more commodious for its increased size, it seeks

Dept. of Entomology, Cornell U.

Black swallowtail caterpillars, showing two stages of growth

some convenient spot on the leaf or stem and spins a little silken carpet from the silk gland opening in its under lip; on this carpet it rests quietly for some time, and then the old tight skin splits down the back, the head portion coming off separately. Swelling out to fill its new skin to the utmost, the caterpillar leaves its cast-off clothes clinging to the silken carpet and marches back to its supper.

But after one of these changes of skin it becomes a very different looking caterpillar, for now it is as smooth as it was formerly spiny; it is now brilliant caraway green, ornamented with roundwise stripes of velvety black; and set in the front margin of each of these stripes are six yellow spots. In shape, the caterpillar is larger toward the head; its true feet have little, sharp claws and look very different from the four pairs of prolegs and the hind prop-leg, all of which enable him to hold fast to the stem or the leaf; these fat legs are green, each ornamented with a black, velvety polka dot.

When we were children we spent hours poking these interesting creatures with straws to see them push forth their brilliant orange horns. We knew this was an act of resentment, but we did not realize that from these horns was exhaled the nauseating odor of caraway which greeted our nostrils. We incidentally discove that they did not waste this odor u each other, for once we saw two of full-grown caterpillars meet on a cara stem. Neither seemed to know that other was there until they touched; t both drew back the head and butted e other like billy goats, whack! whack! T both turned laboriously around and h ried off in a panic.

The scent organs of these caterpil are really little Y-shaped pockets in segment back of the head, pockets full this peculiar caterpillar perfume. Un the stimulus of attack, the pocket turned wrong side out and pushed far making the "horns," and at the sa time throwing the strong odor upon air. This spoils the flavor of these cat pillars as bird food, so they live on serene peace, never hiding under t leaves but trusting, like the skunk, to peculiar power of repelling the enemy.

We must admire this caterpillar for t methodical way in which it eats the le beginning near the base, it does not bu its bridges behind it by eating through t midrib, but eats everything down to t midrib; after it arrives at the tip of the le it finishes midrib and all on its retu journey, doing a clean job, and finishi everything as it moves along.

When the caterpillar has completed growth, it is two inches long; it then see some sheltered spot, the lower edge of clapboard or fence rail being a favori place; it there spins a button of silk whic it grasps firmly with its hind prop-leg, an then, with head up, or perhaps horizonta it spins a strong loop or halter of silk, fa tening each end of it firmly to the obje on which it rests. It thrusts its hea through, so that the halter acts as a slin holding the insect from falling. There sheds its last caterpillar skin, whic shrinks back around the button, revealin the chrysalis, which is angular with earlik projections in front. Then comes the crit cal moment, for the chrysalis lets go the button with its caterpillar feet, and trusting to the sling for support, pushe off the shrunken skin just shed and i the hooks with which it is furnished y in the button of silk. Sometimes ıg this process, the chrysalis loses its entirely and falls to the ground, ·h is a fatal disaster. The chrysalis is wish brown and usually looks very h like the object to which it is at-ed, and is thus undoubtedly protected the sight of possible enemies. Then e day it breaks open, and from it issues ımpled mass of very damp insect vel- which soon expands into a beautiful erfly.

LESSON 70

ᴉE Black Swallowtail Butterfly

ᴌEADING THOUGHT — The caterpillars of swallowtail butterflies have scent ans near the head which they thrust :h when attacked, thus giving off a dis-eeable odor which is nauseating to ls.

ᴍETHOD — In September, bring into schoolroom and place in the terrarium, breeding cage, a caraway or parsley nt on which these caterpillars are feed-, giving them fresh food day by day, l allow the pupils to observe them at ess and thus complete the lesson.

The Caterpillar and Chrysalis

OBSERVATIONS — 1. Touch the caterpil- on the head with a bit of grass. What es it do? What color are the horns? here do they come from? Are there two ›arate horns or two branches of one ›rn? What odor comes from these ›rns? How does this protect the caterpillar? Does the caterpillar try to hide under the leaves when feeding?

2. Describe the caterpillar as follows What is its shape? Is it larger toward the head or the rear end? What is its ground color? How is it striped? How many black stripes? How many yellow spots in each black stripe? Are the yellow spots in the middle, or at each edge of the stripe?

3. How do the front three pairs of legs look? How do they compare with the pro-legs? How many prop-legs are there? What is the color of the prolegs? How are they marked? Describe the prop-leg. What is its use?

4. Observe the caterpillar eating a leaf. How does it manage so as not to waste any?

5. Have you found the egg from which the caterpillar came? What color is it? Where is it laid?

6. How does the young caterpillar look? What are its colors? How many fleshy spines has it on each segment? Are these white on the white segments and black on the black segments? What is the color of the spines at their base?

7. Watch one of these caterpillars shed its skin. How does it prepare for this? How does it spin its carpet? Where does the silk come from? Describe how it acts when shedding its skin.

At the top is a caterpillar of the black swallowtail butterfly ready to change to the chrysalis form. Below is shown a chrysalis of the black swallowtail butterfly

A tiger swallowtail butterfly visiting a lily

8. When a caterpillar is full grown, how does it hang itself up to change to a chrysalis? How does it make the silk button? How does it weave the loop or halter? How does it fasten it? When the halter is woven what does the caterpillar do with it? Describe how the last caterpillar skin is shed. How does the insect use its loop or halter while getting free from the molted skin?

9. Describe the chrysalis. What is its general shape? What is its color? Is it easily seen? Can you see where the wings are, within the chrysalis? How is the chrysalis supported?

10. How does the chrysalis look when the butterfly is about to emerge? Where does it break open? How does the butterfly look at first?

The Butterfly

1. Why is this butterfly called the b
swallowtail? What is the ground colo
the wings? How many rows of ye
spots on the front wings? Are they all
same shape? How are they arranged
tween each two veins? Describe the
wings. What colors are on them that
not on the front wings? Describe wh
this color is placed. Describe the eye
on the hind wing. Where is it? How
the markings on the lower side of the w
differ from those above? How does
ground color differ from the upper s

2. What is the color of the body of
butterfly? Has it any marks? Has it
same number of legs as the monarch
terfly? Describe its antennæ. Watch
butterfly getting nectar from the pet
blossom and describe the tongue. Wh
is the tongue when not in use?

3. How does the mother butterfly
fer in size and in markings from her m

The "caraway worms" were the c
that revealed to us the mystery of the p
and butterfly. We saw one climb up
side of a house, and watched it as w
many slow, graceful movements of
head it wove for itself the loop of
which we called the "swing" and wh
held it in place after it changed to a ch
alis. We wondered why such a brilli
caterpillar should change to such a d
colored object, almost the color of
clapboard against which it hung. Th
one day, we found a damp, crumpl
black butterfly hanging to the em
chrysalis skin, its wings "all mussed"
we termed it; and we gazed at it pityin
but even as we gazed, the crumpled wi
expanded and then there came to
childish minds a dim realization of
miracle wrought within that little, din
empty shell.

— "How to Know the Butterflie
Comsto

American Museum of Natural History

Migrating monarch butterflies

THE MONARCH BUTTERFLY

t is a great advantage to an insect to 'e the bird problem eliminated, and monarch butterfly enjoys this ad- ıtage to the utmost. Its method of ht proclaims it, for it drifts about in a y, leisurely manner, its glowing red king it like a gleaming jewel in the air, ery different flight indeed from the zig- ; dodging movements of other butter- s. The monarch has an interesting race tory. It is a native of tropic America, d has probably learned through some e instinct that by following its food ınt north with the opening season, it ins immunity from special enemies ıer than birds, which attack it in some ge in its native haunts. Each mother tterfly follows the spring northward as it advances, as far as she finds the milk-weed sprouted. There she deposits her eggs, from which hatch individuals that carry on the migration as far to the north as possible. It usually arrives in New York State early in July. As cold weather approaches, the monarchs often gather in large flocks and move back to the South. How they find their way we cannot understand, since there are among them none of the individuals which pressed northward early in the season.

The very brilliant copper-red color of the upper sides of the wings of the monarch is made even more brilliant by the contrasting black markings which outline the veins and border the wings, and also cover the tips of the front wings with a

The monarch butterfly

triangular patch; this latter seems to be an especially planned background for showing off the pale orange and white dots set within it. There are white dots set, two pairs in two rows, between each two veins in the black margin of the wings; and the fringe at the edge of the wings shows corresponding white markings. The hind wings and the front portions of the front wings have, on their lower sides, a ground color of pale yellow, which makes the insect less conspicuous when it alights and folds its wings above its back, upper surfaces together. The black veins, on the lower surface of the hind wings, are outlined with white, and the white spots are much larger than on the upper surface. The body is black, ornamented with a few pairs of white spots above and with many large white dots below. The chief distinguishing characteristic of insects is the presence of six legs; but in this butterfly the front legs are so small that they scarcely look like legs.

It is easy to observe the long, coiled tongue of the butterfly. If the act is done gently, the tongue may be uncoiled lifting it out with a pin. It is very in esting to see a butterfly feeding u nectar; this may be observed in garden almost any day. I have also served it indoors, by bringing in p nias and nasturtiums for my impriso butterflies, but they are not so likely eat when in confinement. The ante are about two-thirds as long as the b and each ends in a long knob; this kn in some form, is what distinguishes antennæ of the butterflies from thos moths. The male monarch has a bl spot upon one of the veins of the h wing; this is a perfume pocket and is fi with what are called scent scales. Th are scales of peculiar shape which co the wing at this place and give forth odor which we with our coarse sense smell cannot perceive; but the lady m arch is attracted by this odor. The m monarch may be described to the child as a dandy carrying a perfume pocket attract his sweetheart.

It is very interesting to the pupils if th are able to see a bit of the butterfly's w through a lens or microscope; the cov ing of scales, arranged in such perf

The viceroy butterfly. Note the black ba on the hind wings. This band distinguish the viceroy from the monarch, which it sembles in color and markings

s, is very beautiful and also very won-'ul. The children know that they get t upon their fingers from butterflies' gs, and they should know that each n of this dust is an exquisite scale with ched edges and a ribbed surface.

'he monarch is, for some reason un-wn to us, distasteful to birds, and its liant colors are an advertisement to all ls of discretion that here is an insect ich tastes most disagreeable and which, refore, should be left severely alone. ere is another butterfly called the vice-which has taken advantage of this im-nity from bird attack on the part of monarch and has imitated its colors truly remarkable way, differing from it y in being smaller in size and having a ck band across the middle of the hind ıg.

The milkweed caterpillar, which is the ung of the monarch butterfly, is a strik-; object, and when fully grown is about) inches long. The milkweed is a suc-ent food and the caterpillar may ma-e in eleven days; it is a gay creature, :h ground color of green and cross ipes of yellow and black. On top of the :ond segment, back of the head, are two ıg, slender, whiplash-like organs, and on : seventh segment of the abdomen is a nilar pair. When the caterpillar is fright-ed, the whiplashes at the front of the dy twitch excitedly; when it walks, they

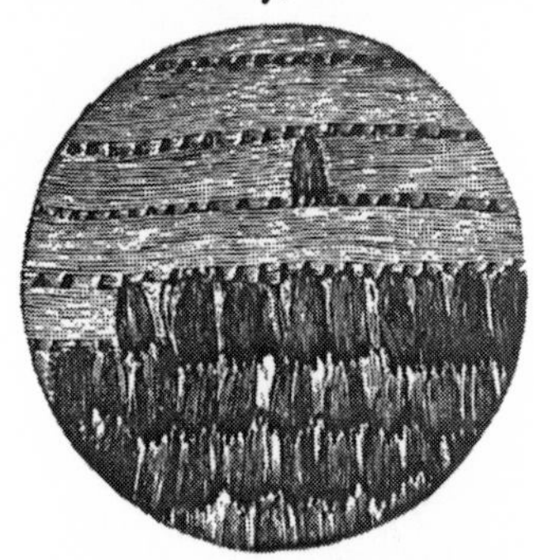

he scales on a butterfly's wing as seen through a microscope

ove back and forth. Those at the rear of e body are more quiet and not so expres-ve of caterpillar emotions. These fila-ents are undoubtedly of use in frighten-

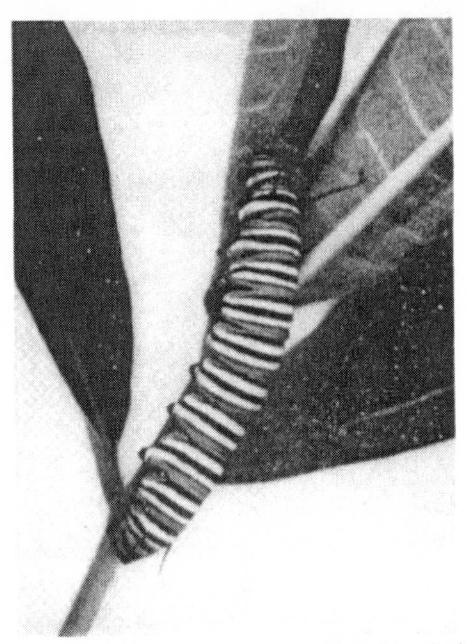

M. V. Slingerland

The monarch caterpillar

ing away the little parasitic flies that lay their eggs upon the backs of caterpillars; these eggs hatch into little grubs that feed upon the internal fatty portions of the caterpillar and bring about its death through weakness. I remember well when I was a child, the creepy feeling with which I beheld these black- and yellow-ringed caterpillars waving and lashing their whips back and forth after I had disturbed them; if the ichneumon flies were as frightened as I, the caterpillars were surely safe.

The caterpillar will feed upon no plant except milkweed; it feeds both day and night, with intervals of rest, and when resting hides beneath the leaf. Its striking colors undoubtedly defend it from birds, because it is as distasteful to them as is the butterfly. However, when frightened, these caterpillars fall to the ground where their stripes make them very inconspicuous among the grass and thus perhaps save them from the attack of some animals other than birds. These caterpillars, like all others, grow by shedding the skeleton skin as often as it becomes too tight.

The monarch chrysalis is, I maintain,

the most beautiful gem in Nature's jewel casket; it is an oblong jewel of jade, darker at the upper end and shading to

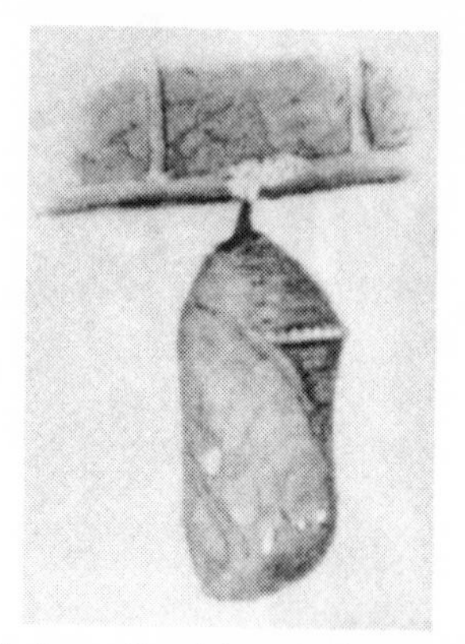

Monarch chrysalis. A jewel of living jade and gold

the most exquisite whitish green below; outlining this lower paler portion are shining flecks of gold. If we look at these gold flecks with a lens, we cannot but believe that they are bits of polished gold foil. There may be other gold dots also, and outlining the apex of the jewel is a band of gold with a dotted lower edge of jet; and the knob at the top, to which the silk which suspends the chrysalis is fastened, is also jet. The chrysalis changes to a darker blue-green after two days, and black dots appear in the gold garniture. As this chrysalis is usually hung to the underside of a fence rail or overhanging rock, or to a leaf, it is usually surrounded by green vegetation, so that its green color protects it from prying eyes. Yet it is hardly from birds that it hides; perhaps its little gilt buttons are a hint to birds that this jewel is not palatable. As it nears the time for the butterfly to emerge, the chrysalis changes to a duller and darker hue. The butterfly emerges about twelve days after the change to a chrysalis.

LESSON 71

The Monarch Butterfly

Leading Thought — The monarch b
terfly migrates northward in spring
summer, moving up as the milkweed
pears, so as to give food to its caterpi
and it has often been noticed migrat
back southward in the autumn in la
swarms. This insect is distasteful to bi
in all its stages. Its chrysalis is one of
most beautiful objects in all nature.

Method — This lesson may be given
September, while yet the caterpillars
the monarch may be found feeding u
milkweed, and while there are yet m
specimens of this gorgeous butterfly to
seen. The caterpillars may be brought
on the food plant, and their habits a
performances studied in the schoolroo
but care should be taken not to have
atmosphere too dry.

L. W. Brown

Monarch butterfly emerging from the chrys lis

The Butterfly

ʙservations — 1. How can you tell monarch butterfly from all others? ɪt part of the wings is red? What por- ; are black? What portions are white? ɪt are the colors and markings on the r side of the wings? What is the color ɪe body and how is it ornamented?

Is the flight of the monarch rapid, ow and leisurely? Is it a very showy ct when flying? Are its colors more iant in the sunshine when it is flying at any other time? Why is it not id of birds?

When the butterfly alights, how s it hold its wings? Do you think it is onspicuous when its wings are folded when they are open?

. Can you see the butterfly's tongue? cribe the antennæ. How do they differ n the antennæ of moths? How many has this butterfly? How does this dif- from other insects? Note if you can see indications of front legs.

. Is there on the butterfly you are dying a black spot near one of the ɪs on each hind wing? Do you know ɪt this is? What is it for?

. Why are the striking colors of this terfly a great advantage to it? Do you ɔw of any other butterfly which imi- es it and thus gains an advantage?

The Caterpillar

. Where did you find the monarch erpillar? Was it feeding below or above the leaves? Describe how it eats the kweed leaf.

2. What are the colors and the mark- ;s of the caterpillar? Do you think these ɪke it conspicuous?

3. How many whiplash-shaped fila- ɪnts do you find on the caterpillar? On ɪich segments are they situated? Do ese move when the caterpillar walks or ɪen it is disturbed? Of what use are they the caterpillar?

4. Do you think this caterpillar would d upon anything except milkweed? ɔes it rest, when not feeding, upon the per or the lower surface of the leaves?

Above, a monarch butterfly; below, a viceroy. In color and markings, except for the black bands on the hind wings of the viceroy, they are similar

Does it feed during the night as well as the day?

5. If disturbed, what does the caterpillar do? When it falls down among the grass, how do its cross stripes protect it from observation?

6. Tell all the interesting things which you have seen this caterpillar do.

The Chrysalis

1. When the caterpillar gets ready to change to a chrysalis what does it do? How does it hang up? Describe how it sheds its skin.

2. Describe the chrysalis. What is its color? How and where is it ornamented? Can you see, in the chrysalis, those parts which cover the wings of the future butterfly?

3. To what is the chrysalis attached? Is it in a position where it does not attract attention? How is it attached to the object?

4. After three or four days, how does the chrysalis change in color? Observe, if you can, the butterfly come out from the chrysalis, noting the following points: Where does the chrysalis skin open? How does the butterfly look when it first comes out?

How does it act for the first two or three hours? How does the empty chrysalis skin look?

A BUTTERFLY AT SEA

Far out at sea — the sun was high,
While veered the wind and flapped the sail;
We saw a snow-white butterfly
Dancing before the fitful gale
Far out at sea.

The little wanderer, who had lost
His way, of danger nothing knew;
Settled a while upon the mast;
Then fluttered o'er the waters blue
Far out at sea.

Above, there gleamed the boundless sky;
Beneath, the boundless ocean sheen;
Between them danced the butterfly,
The spirit-life of this fair scene,
Far out at sea.

The tiny soul that soared away,
Seeking the clouds on fragile wings
Lured by the brighter, purer ray
Which hope's ecstatic mor
brings —
Far out at sea.

Away he sped, with shimmering glee,
Scarce seen, now lost, yet onward b
Night comes with wind and rain, an
No more will dance before the mo
Far out at sea.

He dies, unlike his mates, I ween,
Perhaps not sooner or worse crosse
And he hath felt and known and seen
A larger life and hope, though lost
Far out at sea.

— R. H. Hc

THE ISABELLA TIGER MOTH OR WOOLLY BEAR

Brown and furry,
Caterpillar in a hurry,
Take your walk
To the shady leaf or stalk,
Or what not,
Which may be the chosen spot;
No toad spy you,
Hovering bird of prey pass by you;
Spin and die,
To live again a butterfly.

— Christina Rossi

Many times during autumn, the children find and bring in the very noticeable caterpillar which they call the "woolly bear." It seems to them a companion of the road and the sunshine; it usually seems in a hurry, and if the children know that it is hastening to secure some safe place in which to hide during the season of cold and snow, they are far more interested in its future fate. If the caterpillar is already curled up for the winter, it will "come to" if warmed in the hand or in the sunshine.

The woolly bear is variable in appearance; sometimes five of the front segments are black, four of the middle reddish brown, and three of the hind segments black. In others only four front segments are black, six are reddish, and two are black at the end of the body; there are still other variations, so that each individ will tell its own story of color. There really thirteen segments in this caterpi not counting the head; but the last are so joined that probably the child will count only twelve. There are a reg number of tubercles on each side of e segment, and from each of these arise little rosette of hairs; but the tuber are packed so closely together, that i difficult for the children to see how ma rosettes there are on each side. While body of the caterpillar looks as if it w covered with evenly clipped fur, there usually a few longer hairs on the r segment.

There is a pair of true legs on each the three front segments which fo the thorax, and there are four pairs prolegs. All of the segments behind t three belong to the abdomen, and prolegs are on the 3rd, 4th, 5th, and abdominal segments; the prop-leg is ıe rear end of the body. The true legs his caterpillar have little claws, and as shining as if encased in patent her; but the prolegs and prop-leg are ely prolongations of the sides of the y to assist the insect in holding to leaf. The yellow spot on either side he first segment is a spiracle; this is opening leading into the air tubes hin the body, around which the blood vs and is thus purified. There are no acles on the second and third segnts of the thorax, but eight of the lominal segments have a spiracle on ıer side.

The woolly bear's head is polished black; antennæ are two tiny, yellow projecns which can easily be seen with the ed eye. The eyes are too small to be s seen; because of its minute eyes, the olly bear cannot see very far and, theree, it is obliged to feel its way. It does s by stretching out the front end of body and reaching in every direction, observe if there is anything to cling to its neighborhood. When we try to ze the woolly bear it rolls up in a little ll, and the hairs are so elastic that we e it up with great difficulty. These

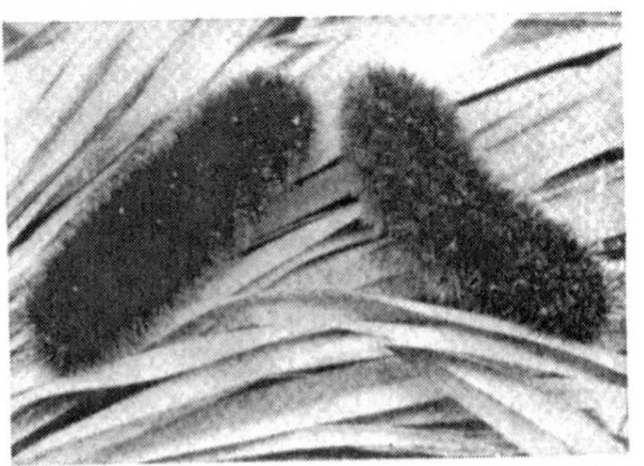

M. V. Slingerland

Woolly bears

hairs are a protection from the attacks of birds which do not like bristles for food; and when the caterpillar is safely rolled up, the bird sees only a little bundle of bristles and lets it alone. The woolly bear feeds upon many plants: grass, clover, dandelion, and others. It does not eat very much after we find it in autumn, because its growth is completed. The woolly bear should be kept in a box which should be placed out of doors, so that it may be protected from storms but have the ordinary winter temperature. Keeping it in a warm room during the winter often proves fatal.

M. V. Slingerland

he Isabella tiger moths, the adults of the woolly bear. The larger is the female

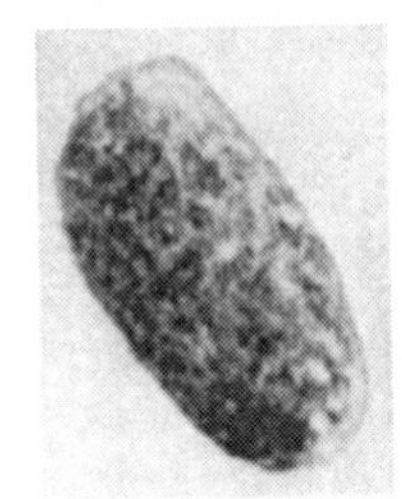

M. V. Slingerland

The cocoon of the woolly bear

Normally, the woolly bear does not make its cocoon until April or May. It finds some secluded spot in the fall, and there curls up in safety for the long winter nap; when the warm weather comes in the spring, it makes its cocoon by spinning silk about itself; in this silk are woven the hairs which it sheds easily at that time, and the whole cocoon seems made of felt. It seems amazing that such a large caterpillar can spin about itself and squeeze itself into such a small cocoon; and it is quite as amazing to see within the cocoon the smooth little pupa, in which is condensed all that was essential of the caterpillar. Sometimes when the caterpillars are kept in a warm room they make their cocoons in the fall, but this is not natural.

The issuing of the moth from the cocoon is an interesting lesson for the last of May. The size of the moth which comes from the cocoon seems quite miraculous compared with the size of the caterpillar that went into it. The moth is in color dull, grayish, tawny yellow with a few black dots on the wings; sometimes the hind wings are tinted with dull orange-red. On the middle of the back of the moth's body there is a row of six black dots; and on each side of the body is a similar row. The legs are reddish above and tipped with black. The antennæ are small and inconspicuous. The moths are night fliers, and the mother moth seeks some plant that will be suitable food for the little caterpillar as soon as it is hatched; here she lays her eggs.

LESSON 72

The Isabella Tiger Moth or Woolly Bear

Leading Thought — When we see the woolly bear hurrying along in the fall, it is hunting for some cozy place in which to pass the winter. It makes its cocoon, usually in early spring, of silk woven with its own hair. In late spring, it comes forth a yellowish moth with black dots on its wings.

Method — Have the children bring in woolly bears as they find them; place them in boxes or breeding jars which have grass or clover growing in them. The children can handle the caterpillars while they are studying them, and then they should be put back into the breeding jars and be set out of doors where they can have nat conditions; thus the entire history be studied.

The Caterpillar

Observations — 1. How can you the woolly bear from all other caterpil Are they all colored alike? How many ments of the body are black at the fi end? How many are red? How many ments are black at the rear end of body? How many segments does this m in all?

2. Look closely at the hairs of woolly bear. Are they set separately o rosettes? Are any of the hairs of the b longer than others or are they all even

3. Can you see, just back of the he the true legs with their little sharp cla How many are there?

4. Can you see the fleshy legs along sides of the body? How many are th of these?

5. Can you see the prop-leg, or hindmost leg of all? Of what use to caterpillar are these fleshy legs?

6. Describe the woolly bear's he How does it act when eating?

7. Can you see a small, bright yell spot on each side of the segment j behind the head? What do you supp this is? Can you see little openings alo each side of all the segments of the bo except the second and third? What a they? Describe how the woolly b breathes.

8. On what does the woolly bear fee If you can find a little woolly bear, gi it fresh grass to eat and see how it grov Why does it shed its skin?

9. When the woolly bear is hurryi along, does it lift its head and the fro end of its body now and then? Why do it do this? Do you think it can see far?

10. What does the woolly bear when you try to pick it up? Do you fir you can pick it up easily? Do you thir that these stiff hairs protect the wool bear from its enemies? What are its en mies?

11. Where should the woolly bear kept in winter to make it comfortable?

THE COCOON

. When does the woolly bear usually :e its cocoon?

. Of what material is it made? How s the woolly bear get into its cocoon?

. What happens to it inside the co- n?

. Cut open a cocoon and describe how woolly bear looks now.

THE MOTH

. Where did the moth come from?

.. How did it come out of the cocoon? : if you can find the empty pupa case the cocoon.

:. What is the color of the moth and v is it marked? Are the front and hind gs the same color?

4. What are the markings and colors of the body? Of the legs?

5. What do you think that the mother Isabella will do, if you give her liberty?

The mute insect, fix't upon the plant
On whose soft leaves it hangs, and from
whose cup
Drains imperceptibly its nourishment,
Endear'd my wanderings.

— WORDSWORTH

Before your sight,
Mounts on the breeze the butterfly, and
soars,
Small creature as she is, from earth's bright
flowers
Into the dewy clouds.

— WORDSWORTH

THE CECROPIA

The silkworm which gives us the silk commerce has been domesticated for ituries in China. Because of this do- :stication, it is willing to be handled d is reared successfully in captivity, and s thus come to be the source of most of r silken fabrics. However, we have in nerica native silkworms which produce strong and lustrous silk; but the cater- llars have proved difficult to rear in large imbers. Moreover, it would take years to mesticate them, and the amount of la- r involved in the production of their k would be so great that they are un- :ely, for many years at least, to be of mmercial importance.

The names of our common native silk- orms are cecropia, promethea, polyphe- us, and luna. In all of these species the oths are large and beautiful, attracting e attention of everyone who sees them. he caterpillars are rarely found, since eir varied green colors render them in- nspicuous among the leaves on which ey feed. None of the caterpillars of the ant silkworms occur in sufficient num- ers to injure the foliage of our trees to ny extent; they simply help Nature to do a little needful pruning. All of the moths are night flyers and are, therefore, seldom seen except by those who are in-

M. V. Slingerland

The cecropia moth

terested in the visitors to our street lights.

The cecropia is the largest of our giant silkworms, the wings of the moth expand- ing sometimes six and one-half inches. It occurs from the Atlantic Coast to the Rocky Mountains.

The cecropia cocoon is found most abundantly on our orchard and shade trees; it is called by the children the

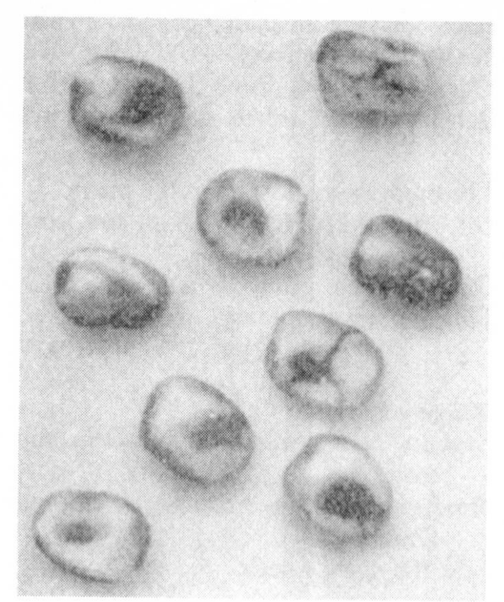

M. V. Slingerland

The eggs of the cecropia moth, enlarged

"cradle cocoon," since it is shaped like a hammock and hung close below a branch, and it is a very safe shelter for the helpless creature within it. It is made of two walls of silk, the outer one being thick and paper-like and the inner one thin and firm; between these walls is a matting of loose silk, showing that the insect knows how to make a home that will protect it from winter weather. It is a clever builder in another respect, since at one end of the cocoon it spins the silk lengthwise instead of crosswise, thus making a valve through which the moth can push, when it issues in the spring. It is very interesting to watch one of these caterpillars spin its cocoon. It first makes a framework by stretching a few strands of silk, which it spins from a gland open-

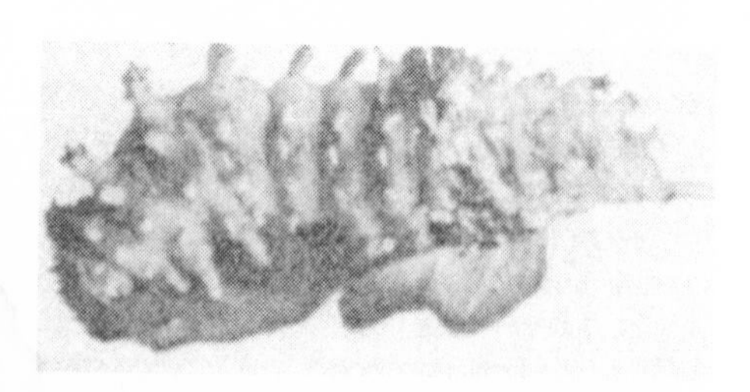

M. V. Slingerland

The cecropia caterpillar molting

ing in the lower lip; it then makes a lc
network upon the supporting strands,
then begins laying on the silk by mov
its head back and forth, leaving the sti
thread in the shape of connecting M's
of figure 8's. Very industriously does
work, and after a short time it is
screened by the silk that the rest of
performance remains to us a mystery. I
especially mysterious, since the inner v
of the cocoon encloses so small a cell t
the caterpillar is obliged to compress it
in order to fit within it. This achievem
would be something like that of a m
who should build around himself a l

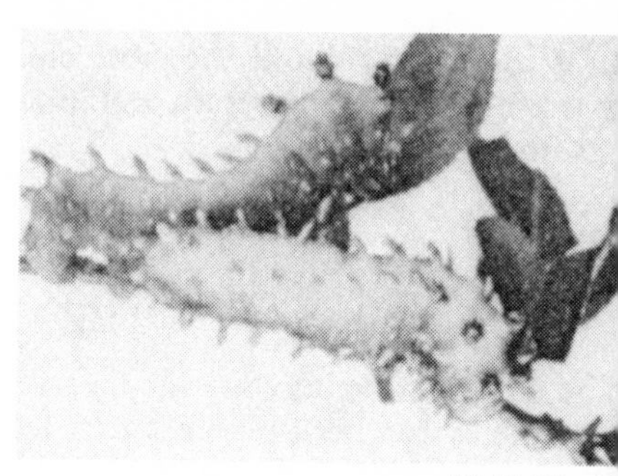

M. V. Slingerl

Full-grown cecropia caterpillars

only a few inches longer, wider, a
thicker than himself. After the cocoon
entirely finished, the caterpillar sheds
skin for the last time and changes in
a pupa.

Very different, indeed, does the pu
look from the brilliantly colored, wa
caterpillar. It is compact, brown, oval, a
smooth, with ability to move but ve
little when disturbed. The cases whi
contain the wings, which are later to
the objects of our admiration, are nc
folded down like a tight cape around t
body; and the antennæ, like great feat
ers, are outlined just in front of the wi
cases. There is nothing more wonderful
all nature than the changes which a
worked within one of these little, brow
pupa cases; for within it, processes go
which change the creature from a crawl
among the leaves to a winged inhabitar
of the air. When we see how helpless th
pupa is, we can understand better ho

h the strong silken cocoon is needed protection from enemies, as well as inclement weather.

spring, usually in May, after the es are well out on the trees, the pupa is shed in its turn, and out of it comes wet and wrinkled moth, its wings all npled, its furry, soft body very untidy; it is only because of this soft and npled state that it is able to push its out through the narrow door into outer world. It has, on each side of its y just back of the head, two little horny ks that help it to work its way out. It ertainly a sorry object as it issues, look-

M. V. Slingerland

Cecropia caterpillar weaving its cocoon

; as if it had been dipped in water and d been squeezed in an inconsiderate nd. But the wet wings soon spread, the ght antennæ stretch out, the furry cov- ng of the body becomes dry and fluffy, d the large moth appears in all its per- tion. The ground color of the wings is dusky, grayish brown while the outer argins are clay-colored; the wings are ossed, beyond the middle, by a white nd which has a broad outside margin red. There is a red spot near the apex the front wing, just outside of the zig- g white line; each wing bears, near its nter, a crescent-shaped white spot bor- red with red. But though it is so large, it oes not need to eat; the caterpillar did l the eating that was necessary for the hole life of the insect; the mouth of the oth is not sufficiently perfected to take od.

M. V. Slingerland

A cecropia cocoon

When the cecropia caterpillar hatches from the egg it is about a quarter of an inch long and is black; each segment is ornamented with six spiny tubercles. Like all other caterpillars, it has to grow by shedding its horny, skeleton skin, the soft skin beneath stretching to give more room at first, then finally hardening and being shed in its turn. This first molt of the cecropia caterpillar occurs about four days after it is hatched, and the caterpillar which issues looks quite different than it did before; it is now dull orange or yellow with black tubercles. After six or seven days more of feeding, the skin is again shed and now the caterpillar appears with a yellow body; the two tubercles on the

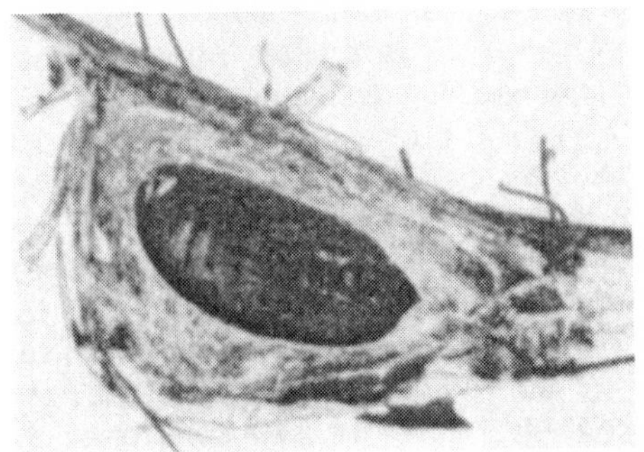

M. V. Slingerland

A cecropia cocoon cut open, showing the pupa within it

top of each segment are now larger and more noticeable. They are blue on the first segment, large and orange-red on the second and third segments, and greenish blue with blackish spots and spines on all the other segments except the eleventh, which has on top, instead of a pair of tubercles, one large, yellow tubercle, ringed with black. The tubercles along the side of the insect are blue during this stage. The next molt occurs five or six days later; this time the caterpillar is blu-

M. V. Slingerland

Just out of the cocoon

ish green in color, the large tubercles on the second and third segments being deep orange, and those on the upper part of the other segments yellow, except those on the first and last segments, which are blue. All the other tubercles along the sides are blue. After the fourth molt it appears as an enormous caterpillar, often attaining the length of three inches, and is as large through as a man's thumb; its colors are the same as in the preceding stage. There is some variation in the colors of the tubercles on the caterpillars during these different molts; in the third stage, it has been observed that the tubercles usually blue are sometimes black. After the last molt the caterpillar eats voraciously for perhaps two weeks or longer and then begins to spin its cocoon.

LESSON 73

The Cecropia

Leading Thought — The cecr moth passes the winter as a pupa in a coon which the caterpillar builds ou silk for the purpose. In the spring moth issues and lays her eggs on some t the leaves of which the caterpillar relis The full-grown caterpillars are large green with beautiful blue and orange bercles.

Method — It is best to begin with cocoons, for these are easily found a the leaves have fallen. These cocoons kept in the schoolroom, should be th oughly wet at least once a week. H ever, it is better to keep them in a box of doors where they can have the adv tage of natural moisture and temperat and from those that are kept outside moths will not issue until the leaves op upon the trees and provide food for young caterpillars to eat when the e hatch.

The Cocoon

Observations — 1. How does the coon look on the outside? What is general shape? To what is it fastened? it fastened to the lower or the upper si of a twig? Are there any dried leaves tached to it?

2. Where do you find cecropia coons? How do they look on the tre Are they conspicuous?

3. Cut open the cocoon, being care not to hurt the inmate. Can you see th it has an outer wall which is firm? Wh lies next to this? Describe the wall next the pupa. How does this structure p

the pupa from changes of tempera- and dampness?
Is the outside covering easy to tear? at birds have been known to tear this on apart?
Are both ends of the cocoon alike? you find one end where the silk is not en across but is placed lengthwise? y is this so? Do you think that the h can push out at this end better than he other?

The Pupa

. Take a pupa out of a cocoon care- y and place it on cotton in a wide- ithed fruit jar where it may be ob- ed. Can the pupa move at all? Is it ble to defend itself? Why does it not d to defend itself?
. Can you see in the pupa the parts t will be the antennæ and the mouth?
. Describe how the wing coverings k. Count the rings in the abdomen.
. Why does the pupa need to be pro- ted by a cocoon?

The Moth

. What is the first sign that the moth coming out of the cocoon? Can you r the little scratching noise? What do suppose makes it? How does the moth k when it first comes out? If it were t all soft and wet, how could it come from so small an opening?
. Describe how the crumpled wings spread out and dry. How does the covering of the wings change in appearance?
3. Make a water-color drawing or describe in detail the fully expanded moth, showing the color and markings of wings, body, and antennæ.
4. Do the moths eat anything?
5. If one of the moths lays eggs, describe the eggs, noting color, size, and the way they are placed.

The Caterpillar

1. On what do you find the cecropia caterpillar feeding? Describe its actions while feeding.
2. What is the color of the caterpillar? Describe how it is ornamented.
3. Can you see the breathing pores, or spiracles, along the sides of the body? How many of these on each segment? How do they help the caterpillar to breathe?
4. Describe the three pairs of true legs on the three segments just back of the head. Do these differ in form from the prolegs along the sides of the body? What is the special use of the prolegs? Describe the prop-leg, which is the hindmost leg of all.
5. Do you know how many times the cecropia caterpillar sheds its skin while it is growing? Is it always the same color?
6. Watch the caterpillar spin its cocoon; describe how it begins and how it acts as long as you can see it. Where does the silk come from?

THE PROMETHEA

The promethea is not so large as the cropia, although the female resembles e latter somewhat. It is the most com- on of all our giant silkworms. Its cater- lars feed upon wild cherry, lilac, ash, ssafras, buttonwood, and many other es.

During the winter, leaves may often seen hanging straight down from the anches of wild cherry, lilac, and ash. If ese leaves are examined, they will be found to be wrapped around a silken case containing the pupa of the promethea. It is certainly a canny insect which hides itself during the winter in so good a disguise that only the very wisest of birds ever suspect its presence. When the promethea caterpillar begins to spin, it selects a leaf and covers the upper side with silk, then it covers the petiole with silk, fastening it with a strong band to the twig, so that not even most violent winter winds

The female promethea

will be able to tear it off. Then it draws the two edges of the leaf about itself like a cloak as far as it will reach, and inside this folded leaf it makes its cocoon, which always has an opening in the shape of a conical valve at the upper end, through which the moth may emerge in the spring. This caterpillar knows more botany than some people do, for it makes no mistake in distinguishing a compound leaf from a simple one. When it uses a leaflet of hickory for its cocoon, it fastens the leaflet to the mid stem of the leaf and then fastens the stem to the twig. The male pupa is much more slender than that of the female. The moths do not issue until May or June.

The moth works its way out through the valve at the top of the cocoon. The female is a large, reddish brown moth with markings resembling somewhat those of the cecropia. The male is very different in appearance; its front wings have very graceful, prolonged tips, and both wings are almost black, bordered with ash color. The promethea moths differ somewhat in habit from the other silkworms, in that they fly during the late afternoon as well as at night. The eggs are whitish with brown stain, and are laid in rows, a good many on the same leaf.

The caterpillars, as they hatch from the eggs, have bodies ringed with black and yellow. They are sociable little fellows and live together side by side amicably, not exactly " toeing the mark " like a spelling class, but all heads in a row at the edge of the leaf where each is eating as fast as possible. When they are small, caterpillars remain on the undersid the leaves out of sight. In about five d the first skin is shed and the color of caterpillar remains about the same. F or five days later the second molt occ and then the caterpillar appears in a b tiful bluish green costume, with b tubercles, except four large ones on second and third segments, and one l one on the eleventh segment, which yellow. This caterpillar has an interes habit of weaving a carpet of silk on wh to change its skin; it seems to be be able to hold on while pushing off the skin, if it has the silken rug to cling After the third molt, the color is a dee greenish blue and the black tubercles smaller, and the five big ones are la and bright orange in color. After fourth molt, which occurs after a pe of about five or six days, the caterpi appears in its last stage. It is now over inches long, quite smooth and most p perous looking. Its color is a beauti light, greenish blue, and its head is yell It has six rows of short, round, black tu cles. The four large tubercles at the fr end of the body are red, and the la tubercle on the rear end of the body yellow.

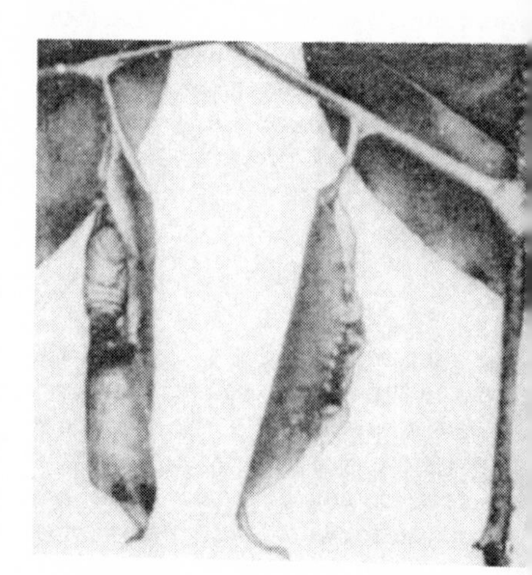

M. V. Slingerl

Promethea cocoons; the one on the l has been cut away to show the pupa. N how the leaves are fastened by silk to twigs

The Cynthia

'he cynthia is a beautiful moth which come to us from Asia; it is very large h a ground color of olive green, with :nder tints and white markings; there white tufts of hairs on the abdomen. ›uilds its cocoon like the promethea, ening the petiole to the twig; there- : the lesson indicated for the prome- a will serve as well for the cynthia. e cynthia caterpillars live upon the ınthus tree and are found only in the ions where this tree has been intro- :ed.

LESSON 74

The Promethea

Leading Thought — The promethea ɛerpillar fastens a leaf to a twig with silk d then makes its cocoon within this ɩf. The male and female moths are very fferent in appearance.

Method — This work should begin in e late fall, when the children bring in ese cocoons which they find dangling ı the lilac bushes or wild cherry trees. uch attention should be paid to the ay the leaf is fastened to the twig so it ill not fall. The cocoons should be kept ıt of doors, so that the moths will issue

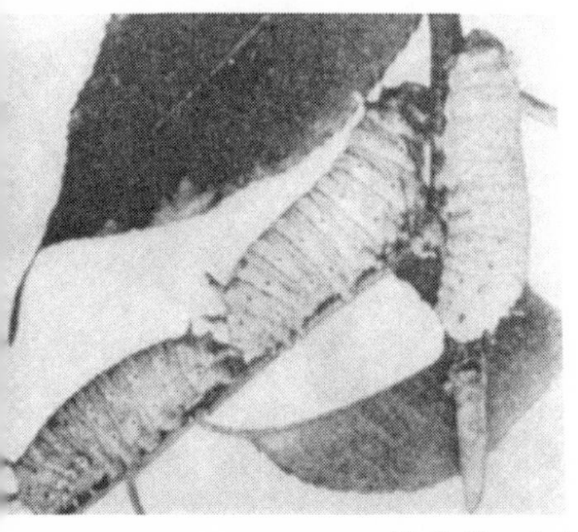

M. V. Slingerland

Promethea caterpillars

A polyphemus moth and cocoon. This is a yellowish or brownish moth with a window-like spot in each wing

late in the spring when they can have natural conditions for laying their eggs, and the young caterpillars will be supplied with plenty of food consisting of new and tender leaves.

The Cocoon

Observations — 1. On what tree did you find it? Does it look like a cocoon? Does it not look like a dried leaf still clinging to the tree? Do you think that this disguise keeps the birds from attacking it? Do you know which birds are clever enough to see through this disguise?

2. How is the leaf fastened to the twig? Could you pull it off readily? What fastened the leaf to the twig?

3. Tear off the leaf and study the cocoon. Is there an opening to it? At which end? What is this for?

4. Cut open a cocoon. Is it as thick as that of the cecropia?

5. Study the pupa. Is it as large as that of the cecropia?

6. Can you see where the antennæ of the moth are? Can you see the wing covers? Can the pupa move?

The Moth

1. Are there two kinds of moths that come from the promethea cocoons? Does one of them look something like the cecropia? This is the mother promethea.

2. Are any of the moths almost black in color with wings bordered with gray and

with graceful prolonged tips to the front wings? This is the father moth.

3. Make water-color drawings of promethea moths, male and female.

4. If the promethea mother lays eggs, describe them.

The Caterpillar

1. How do the promethea caterpillars look when they first hatch from the eggs? Do they stay together when they are very young? How do they act? Where do they hide?

2. How do they change color as they grow older? Do they remain togethe scatter? Do they continue to hide on lower sides of leaves?

3. What preparation does a promet caterpillar make before changing its sk Why does it shed its skin? Does the c of the caterpillar change with every cha of skin?

4. Describe the caterpillar when i full grown. What is its ground co What are the colors of its ornamen tubercles? The color of its head?

5. Describe how a promethea cater lar makes its cocoon.

THE HUMMINGBIRD OR SPHINX MOTHS

M. V. Slingerland

The moth of the sphinx caterpillar which feeds on tomato plants

If during the early evening, when all the swift hummingbirds are abed, we hear the whirr of rapidly moving wings and detect the blur of them in the twilight, as if the creature carried by them hung entranced before some deep-throated flower, and then whizzed away like a bullet, we know that it is a hummingbird or sphinx moth. And when we see a caterpillar with a horn on the wrong end of the body, a caterpillar which, when disturbed, rears threateningly, then we may know it is the sphinx larva. And when we find a strange, brown, segmented shell, with a long jug handle at one side, buried in the earth as we spade up the garden in the spring, then we know we have the sphinx pupa.

The sphinx was a vaudeville person of ancient mythology, who went about boring people by asking them riddles, and they could not give the right answers, v promptly ate them up. Although Linna gave the name of sphinx to these mot because he fancied he saw a resembla in the resting or threatening attitude the larvæ to the Egyptian Sphinx, th are still other resemblances. These inse present three riddles: The first one "Am I a hummingbird?" the seco "Why do I wear a horn or an eyespot the rear end of my body where horns a eyes are surely useless?" and the thi "Why do I look like a jug with a han and no spout?"

The sphinx moths are beautiful a

Sphinx larva in sphinx attitude

nt creatures. They have a distinctly :-made appearance, their colors are so el and "the cut" so perfect. They long, rather narrow, strong wings h enable them to fly with extraordi- rapidity. The hind wings are shorter, act as one with the front wings. The is stout and spindle-shaped. The nnæ are thickened in the middle or rd the tip, and in many species have tip recurved into a hook. Their colors most harmonious combinations and t exquisite contrasts; the pattern, al- gh often complex, shows perfect re- nent. Olive, tan, brown and ochre, k and yellow, and the whole gamut of s, with eyespots or bands athwart the l wings of rose color or crimson, are e of the sphinx color schemes.

Iost of the sphinx moths have re- kably long tongues, which are some- es twice the length of the body. When in use, the tongue is curled like a ch spring in front of and beneath the d; but of what possible use is such a g tongue? That is a story for certain vers to tell, the flowers which have the tar-wells far down at the base of tubu- corollas, like the petunia, the morn-

M. V. Slingerland

e tobacco sphinx moth with tongue extended

ing glory, or the nasturtium. Some of these flowers, like jimson weed and flowering tobacco, open late in the day when these evening visitors are flying about. In some cases, especially among the or-

The pupa of the common tomato sphinx caterpillar. Note that the part encasing the long tongue is free and looks like the handle of a jug

chids, there is a special partnership established between one species of flower and one species of sphinx moth. The tobacco sphinx is an instance of such partnership; this moth visits tobacco flowers and helps develop the seeds by carrying pollen from flower to flower; and in turn it lays its eggs upon the leaves of this plant, on which its great caterpillar feeds and waxes fat, and in high dudgeon often disputes the smoker's sole right to the "weed." Tobacco probably receives enough benefit from the ministrations of the moth to compensate for the injury it suffers from the caterpillars; but the owner of the tobacco field, not being a plant, does not look at it in this equitable manner.

The sphinx caterpillars are leaf-eaters, and each species feeds upon a limited number of plants which are usually related; for instance, one feeds upon both the potato and tomato; another upon the Virginia creeper and grapes. In color these caterpillars so resemble the leaves that they are discovered with difficulty. Those on the Virginia creeper, which shades porches, may be located by the black pellets of waste material which fall from them to the ground; but even after this unmistakable hint I have searched a long time to find the caterpillar in the leaves above; its color serves to hide the insect from birds which feed upon it eagerly. In some species, the caterpillars are ornamented with oblique stripes along the sides, and in others the stripes are length-

Dept. of Entomology, Cornell U.
Adults of the Myron sphinx

wise. There is often a great variation in color between the caterpillars of the same species; the tomato worm is sometimes green and sometimes black.

In the young larva the horn on the rear end is often of different color from the body; in some species it stands straight up and in some it is curled toward the back. It is an absolutely harmless projection and does not sting nor is it poisonous. However, it looks awe-inspiring and perhaps protects its owner in that way. The *Pandora* sphinx has its horn curled over its back in the young stage but when fully grown the horn is shed; in its place is an eyespot which, if seen between the leaves, is enough to frighten away any cautious bird fearing the evil eye of serpents. The sphinx caterpillars have a habit, when disturbed or when resting, of rearing up the front part of the body, telescoping the head back into the thoracic segments, which in most species are enlarged, and assuming a most threatening and ferocious aspect. If attacked they will swing sidewise, this way and then that, ma a fierce crackling sound meanwhile, calculated to fill the trespasser with te When resting they often remai this lifted attitude for hours, absol rigid.

The six true legs are short with s little claws. There are four pairs of fl prolegs, each foot being armed with h for holding on to leaf or twig; and large, fleshy prop-leg on the rear segr is able to clasp a twig like a vise. All t fleshy legs are used for holding on, w the true legs are used for holding edges of the leaf where the sidewise w ing jaws can cut it freely. These cate lars do clean work, leaving only the ha and more woody ribs of the leaves. Myron caterpillar seems to go out o way to cut off the stems of both the g and Virginia creeper.

There are nine pairs of spiracles, a on each segment of the abdomen an the first thoracic segment. The edge these air openings are often strikingly ored. Through the spiracles the air is mitted into all the breathing tubes of body around which the blood flows is purified; no insect breathes throug mouth. These caterpillars, like all ot grow by shedding the skeleton skin, w splits down the back.

Often one of these caterpillars is covered with white objects which the informed, who do not know that cate lars never lay eggs, have called eggs. the sphinx moths at any stage would h horror of such eggs as these! They are

M. V. Slingerland
Eggs of the Myron sphinx

eggs but are little silken cocoons spun the larvæ of a hymenopterous paras It is a tiny four-winged "fly" which

ggs within the caterpillar. The little
s which hatch from these eggs feed
1 the fleshy portions of the caterpillar
they get their growth, at which time
poor caterpillar is almost exhausted;
then they have the impudence to
e out and spin their silken cocoons
fasten them to the back of their vic-
Later, they cut a little lid to their
n cells which they lift up as they come
into the world to search for more
rpillars.

s soon as the sphinx larva has obtained
growth, it descends and burrows into
earth. It does not spin any cocoon
packs the soil into a smooth-walled
in which it changes to a pupa. In the
ng the pupa works its way to the sur-
of the ground and the moth issues.
the case of the tomato and tobacco
inx pupa, the enormously long tongue
its case separate from the body of the
a, which makes the "jug handle." The
g cases and the antennæ cases can be
inctly seen. In other species the pupæ
e the tongue case fast to the body. The
a of the Myron sphinx does not enter
ground, but draws a few leaves about
n the surface of the ground, fastens
m with silk, and there changes to a
pa.

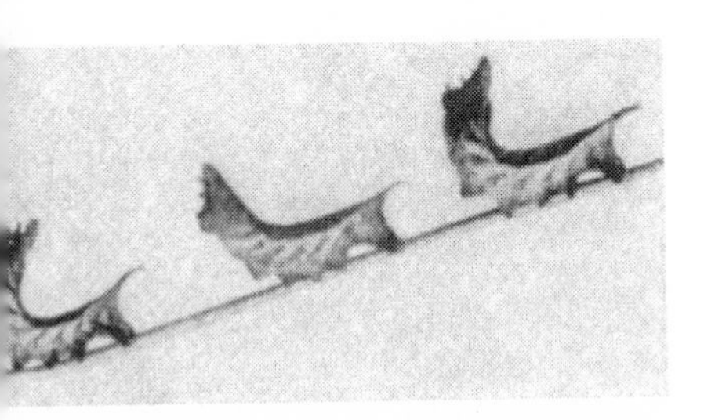
M. V. Slingerland

" cake walk." Caterpillars of the Myron sphinx in an attitude of defense

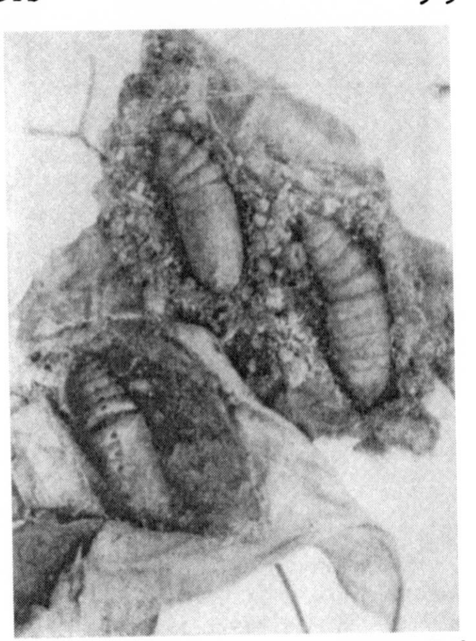
M. V. Slingerland

Pupæ of the Myron sphinx within the cocoons

LESSON 75

The Hummingbird or Sphinx Moths

Leading Thought — The sphinx caterpillars have a slender horn or eyespot on the last segment of the body. When disturbed or when resting they rear the front part of the body in a threatening attitude. They spin no cocoons but change to pupæ in the ground. The adults are called hummingbird moths, because of their swift and purring flight. The sphinx moths carry pollen for many flowers.

Method—The sphinx caterpillar found on the potato or tobacco, or one of the species feeding upon the Virginia creeper, is in autumn available in almost any locality for this lesson. The caterpillars should be placed in a breeding cage in the schoolroom. Fresh food should be given them every day and moist earth be placed in the bottom of the cages. It is useless for the amateur to try to rear the adults from the pupæ in breeding cages. The moths may be caught in nets during the evening when they are hovering over the petunia beds. These may be placed on leaves in a tumbler or jar for observation.

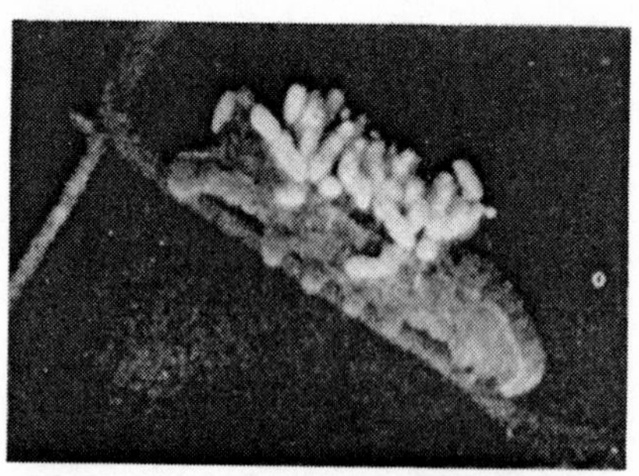

M. V. Slingerland

A Myron caterpillar that has been parasitized. The white objects upon it are the cocoons of the little grubs which feed upon the fatty parts of the caterpillar

The Caterpillar

Observations — 1. On what plant is it feeding? What is its general color? Is it striped? What colors in the stripes? Are they oblique or lengthwise stripes? Are all the caterpillars the same color?

2. Can you find the caterpillar easily when feeding? Why is it not conspicuous when on the plant? Of what use is this to the caterpillar?

3. Note the horn on the end of the caterpillar. Is it straight or curled? Is it on the head end? What color is it? Do you think it is of any use to the caterpillar? Do you think it is a sting? If there is no horn, is there an eyespot on the last segment? What color is it? Can you think of any way in which this eyespot protects the caterpillar?

4. Which segments of the caterpillar are the largest? When the creature is disturbed what position does it assume? How does it move? What noise does it make? Do you think this attitude scares away enemies? What position does it assume when resting? Do you think that it resembles the Egyptian Sphinx when resting?

5. How many true legs has this caterpillar? How does it use them when feeding? How many prolegs has it? How are these fleshy legs used? How are they armed to hold fast to the leaf or twig? Describe the hind or prop-leg. How is it used?

6. Do you see the breathing pores or spiracles along the sides of the body? H many are there? How are they colc How does the caterpillar breathe? Do think it can breathe through its mo

7. How does the sphinx caterp grow? Watch your caterpillar and s shed its skin. Where does the old break open? How does the new, soft look? Do the young caterpillars reser the full-grown ones?

8. Describe how the caterpillar Can you see the jaws move? Does it up the plant clean as it goes?

9. Have you ever found the spl caterpillar covered with whitish, oval jects? What are these? Does the ca pillar look plump or emaciated? Exp what these objects are and how they c to be there.

10. Where does the caterpillar gc change to a pupa? Does it make cococ How does the pupa look? Can you the long tongue case, the wing cases, antennæ cases?

The Moth

1. Where did you find this moth? V it flying by daylight or in the dusk? H did its rapidly moving wings sound? V it visiting flowers? What flowers? Wh is the nectar in these flowers?

The white-lined sphinx moth

. What is the shape of the moth's y? Is it stout or slender? What colors it? How is it marked?

. The wings of which pair are longer? tch or describe the form of the front the hind wings. Are the outer edges loped, notched, or even? What colors on the front wing? On the hind one? these colors harmonious and beauti- Make a sketch of the moth in water or.

. What is the shape of the antennæ? scribe the eyes. Can you see the coiled gue? Uncoil it with a pin and note v long it is. Why does this moth need h a long tongue?

5. From what flowers do the sphinx moths get nectar? How does the moth support itself when probing for nectar? Do you know any flowers which are dependent on the sphinx moths for carrying their pollen? How many kinds of sphinx moths do you know?

Hurt no living thing:
Ladybird, nor butterfly,
Nor moth with dusty wing,
Nor cricket chirping cheerily,
Nor grasshopper so light of leap,
Nor dancing gnat, nor beetle fat,
Nor harmless worms that creep.

— Christina Rossetti

THE CODLING MOTH

t is difficult to decide which seems the re disturbed, the person who bites into apple and uncovers a worm, or the rm which is uncovered. From our ndpoint, there is nothing attractive out the worm which destroys the beauty l appetizing qualities of our fruit, but m the insect standpoint the codling erpillar (which is not a worm at all) is t at all bad. When full grown, it is out three-fourths of an inch long, and likely to be flesh color, or even rose or, with brownish head; as a young va, it has a number of darker rose spots each segment and is whitish in color; shield on the first segment behind the ad, and that on the last segment of the dy, are black. When full grown, the ap- worm is plump and lively; and while is jerking angrily at being disturbed, we n see its true legs, one pair to each of e three segments of the body behind the ad. These true legs have sharp, single ws. Behind these the third, fourth, th, and sixth segments of the abdomen each furnished with a pair of fleshy olegs and the hind segment has a op-leg. These fleshy legs are mere makeifts on the part of the caterpillar for caring the long body, since the three pairs front legs are the ones from which delop the legs of the moth. The noticing of the legs of the codling moth is an important observation on the part of the pupils, since, by their presence, this insect may be distinguished from the young of the plum curculio, which is also found in apples but which is legless. The codling moth has twelve segments in the body, back of the head.

The codling larva usually enters the apple at the blossom end and tunnels down by the side of the core until it reaches the middle, before making its way out into the pulp. The larva weaves a web

M. V. Slingerland

The adult of the codling moth, showing the variations of its markings. The two larger ones are about three times natural size

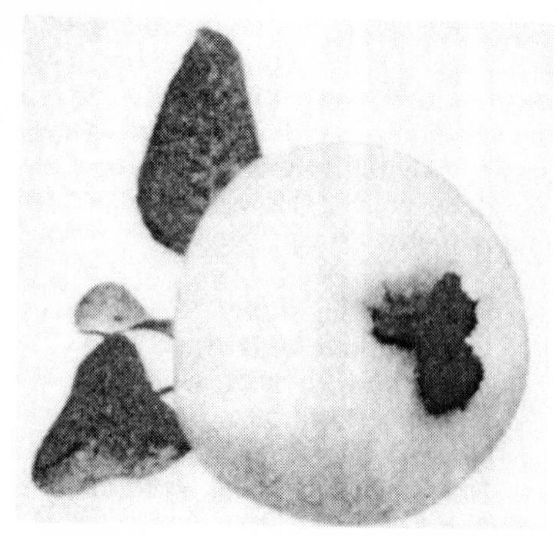

M. V. Slingerland

A wormy apple

as it goes, but this is probably incidental, since many caterpillars spin silk as they go, "street yarn" our grandmothers might have called it. In this web are entangled the pellets of indigestible matter, making a very unsavory looking mass. The place of exit is usually circular, large enough to accommodate the body of the larva, and it leads out from a tunnel which may be a half inch or more in diameter beneath the rind. Often the larva makes the door some time before it is ready to leave the apple, and plugs it with a mass of debris, fastened together with the silk. As it leaves the apple, the remnants of this plug may be seen streaming out of the opening. Often also, there is a mass of waste pellets pushed out by the young larva from its burrow, as it enters the apple; thus it injures the appearance of the apple at both entrance and exit. If the apple has not received infection by lying next to another rotting apple, it first begins to rot around the burrow of the worm, especially near the place of exit.

The codling caterpillar injures the fruit in the following ways: The apples are likely to be stunted and fall early; the apples rot about the injured places and thus cannot be stored successfully; the apples thus injured look unattractive, and therefore their market value is lessened; wormy apples, packed in barrels with others, rot and contaminate all the neighboring apples. This insect also attacks pears sometimes peaches.

The larvæ usually leave the apples fore winter. If the apples have fallen, t crawl up the tree and there make t cocoons beneath the loose bark; bu they leave the apples while they are on trees, they spin silk and swing down carried into the storeroom or placed barrels, they seek quarters in protec crevices. In fact, while they particul like the loose bark of the apple trees, t are likely to build their cocoons on nea fences or on brush, wherever they can f the needed protection. The cocoon made of fine but rather rough silk wh is spun from a gland opening near mouth of the caterpillar; the cocoon is beautiful, although it is smooth inside is usually spun between a loose bit of b and the body of the tree; but after mak it, the insect seems in no hurry to char its condition and remains a quite liv caterpillar until spring. It is while the c ling larvæ are in their winter quarters t our bird friends of the winter, the n hatches, woodpeckers, and chickadees, stroy them in great numbers, hunti eagerly for them in every crevice of t trees. It is therefore good policy for us

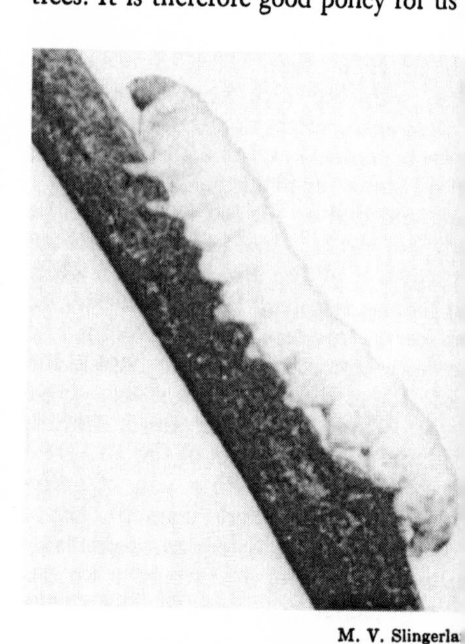

M. V. Slingerla

Larva of the codling moth, greatly enlarg

these birds to our orchards by plac- beef fat on the branches and thus e these little caterpillar hunters to the trees every day.

is an interesting fact that the codling pillars, which make cocoons before ıst first, change immediately to pupæ h soon change to moths, and thus her generation gets in its work before ıpples are harvested.

he codling moth is a beautiful little ture with delicate antennæ and a vn, mottled and banded body; its ;s are graced by wavy bands of ashy brown lines, and the tips of the front ;s are dark brown with a pattern of bronze wrought into them; the hind ;s are shiny brown with darker edges little fringes. The moths which have ered in cocoons issue in the spring lay their eggs on the young apples just the petals fall. The egg looks like a ute drop of dried milk and is laid on side of the bud; but the little larva, after it is hatched, crawls to the blos- and finds entrance there; and it is efore important that its first lunch ıld include a bit of arsenic and thus

M. V. Slingerland

Just ready to spray. A pear and two apples with the petals recently fallen and with the calyx lobes widely spread

end its career before it fairly begins. The trees should be sprayed with some stom- ach poison directly after the petals fall, and before the five lobes of the calyx close up around the stamens. If the trees are sprayed while blossoming, the pollen is washed away and the apples do not set; moreover, the bees which help us much in carrying pollen are killed. If the trees are sprayed directly after the calyx closes up around the stamens the poison does not lodge at the base of the stamens and the little rascals get into the apples without getting a dose. (See the lesson on the apple.)

M. V. Slingerland

The pupæ and cocoons of codling moths

LESSON 76

The Codling Moth

Leading Thought — The codling moth is a tiny brown moth with bronze mark- ings, which lays its egg on the apple. The larva hatching from the egg enters the blossom end and feeds upon the pulp of the apple, injuring it greatly. After attain- ing its growth it leaves the apple and hides beneath the bark of the tree or in some other protected place, and in the spring

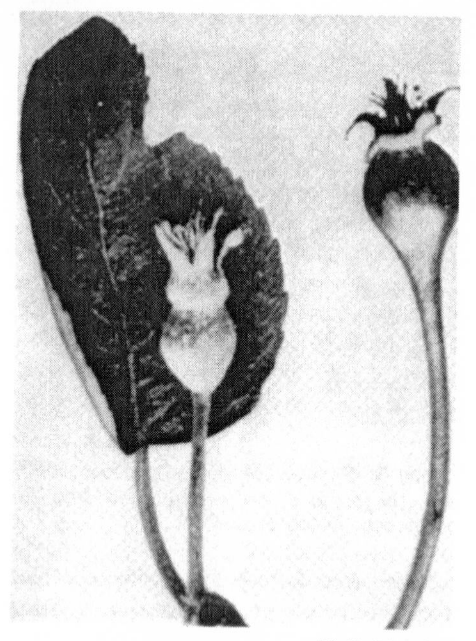

M. V. Slingerland

Almost too late to spray. The apple on the left has the calyx lobes nearly drawn together. The pear on the right still has the calyx cavity open

makes the cocoon from which the moth issues in time to lay eggs upon the young apples.

METHOD — The lesson should begin with a study of wormy apples, preferably in the fall when the worms are still within their burrows. After the pupils become familiar with the appearance of the insect and its methods of work, a prize of some sort might be offered for the one who will bring to school the greatest number of hibernating larvæ found in their winter quarters. Place these larvæ in a box with cheesecloth tacked over its open side; place this box out of doors in a protected position. Examine the cocoons to find the pupæ about the last of April; after the pupæ appear, look for the moths in about five days.

It would be a very good idea for the pupils to prepare a Riker mount showing specimens of the moths, of the cocoons showing the cast pupa skin, and of the caterpillar in a homeopathic vial of alcohol; pictures illustrating the work of the insect may be added. The pictures sh be drawn by the pupils, showing wormy apple, both the outside and in tion. The pupils can also sketch, fron pictures here given, the young apple v just in the right condition to spray, a note explaining why.

OBSERVATIONS — 1. Find an apple a codling moth larva in it. How lar the worm? How does it act when turbed?

2. What is the color of the caterpil body? Its head?

3. How many segments are there in body? How many of these bear What is the difference in form betw the three front pairs of legs and others?

4. Look at a wormy apple. How can tell it is wormy from the outside? Can see where the worm entered the ap Was the burrow large or small at f Can you find an apple with a worm which has the door for exit made, closed with waste matter? How is matter fastened together? If the apple no worm in it, can you see where it the apple? Make a sketch or describe evidence of the caterpillar's prog through the apple. Do you find a wel silk in the wormy part? Why is this? D the worm eat the seeds as well as the p of the apple?

5. Take a dozen rotting apples; h many of them are wormy? Do the part the apple injured by the worm begin rot first? In how many ways does the c ling moth injure the apple? Does it inj other fruits than apples?

6. How late in the fall do you find codling larvæ in the apple? Wh do these larvæ go when they leave apple?

Work to be done in March or ea April — Visit an orchard and look un the loose bark on old trees, or along p tected sections of fences or brush pi and bring in all the cocoons you can fi Do not injure the cocoons by tearing th from the places where they are wov but bring them in on bits of the bark other material to which they are attach

How does the cocoon look outside inside? What is in the cocoon? Why the cocoon made? When was it le?

. Place the cocoons in a box covered cheesecloth and place the box out loors where the contents can be fre-ntly observed and make the following es:

a) When does the larva change to pupa? Describe the pupa. How does cocoon look after the moth issues n it?

b) Describe the moth, noting color ead, thorax, body, and front and hind gs.

3. If these moths were free to fly around the orchard, when and where would they lay their eggs?

4. When should the trees be sprayed to kill the young codling moth? With what should they be sprayed? Why should they not be sprayed during the blossoming period? Why not after the calyx closes?

5. How do the nuthatches, downy woodpeckers, and chickadees help us get rid of the codling moth?

6. Write an essay on the life history of the codling moth, the damage done by it, and the best methods of keeping it in check.

LEAF-MINERS

And there's never a leaf nor a blade too mean
To be some happy creature's palace.

— Lowell

May not Lowell have had in mind, en he wrote these lines, the canny le creatures which find sustenance for ir complete growth between the upper d lower surfaces of a leaf which seems us as thin as a sheet of paper? To most ildren, it seems quite incredible that ere is anything between the upper and ver surfaces of a leaf, and this lesson ould hinge on the fact that in every f, however thin, there are rows of cells ntaining the living substance of the leaf, th a wall above and a wall below to otect them. Some of the smaller in-ts have discovered this hidden treasure, ich they mine while safely protected m sight, and thus make strange figures on the leaves.

Among the most familiar of these are e serpentine mines, so called because e figure formed by the eating out of the een pulp of the leaf curves like a ser-nt. Some serpentine mines are made by e caterpillars of certain tiny moths, hich have long fringes upon the hind ings. The life story of such a moth is as llows: The little moth, whose expanded wings measure scarcely a quarter of an inch across, lays an egg on the leaf; from this, there hatches a tiny caterpillar that soon eats its way into the midst of the leaf. In shape, the caterpillar is somewhat "square built," being rather stocky and wide for its length; it feeds upon the juicy tissues of the leaf and divides, as it goes, the upper from the lower surface of the leaf; and it teaches us, if we choose to look, that these outer walls of the leaf are thin, colorless, and paper-like. We can trace the

Grace H. Griswold

Serpentine mines in a columbine leaf

Grace H. Griswold

A verbena leaf, showing mines that are mere blotches

whole life history and wanderings of the little creature, from the time when, as small as a pinpoint, it began to feed, until it attained its full growth. As it increased in size, its appetite grew larger also, and these two forces working together naturally enlarged its house. When finally the little miner got its growth, it made a rather larger and more commodious room at the end of its mine, which to us looks like the head of the serpent; here it changed to a pupa, perhaps after nibbling a hole with its sharp little jaws, so that when it changed to a soft, fluffy little moth with mouth unfitted for biting, it was able to escape. In some species, the caterpillar comes out of the mine and goes into the ground to change to a pupa. By holding up to the light a leaf thus mined, we can see why this little chap was never obliged to clean house; it mined out a new room every day, and left the sweepings in the abandoned mine behind. Mines of this sort are often seen on the leaves of the nasturtium, the smooth pigweed, the columbine, and many o plants. There are mines of many sha each form being made by a different cies of insect. Some flare suddenly fro point and are trumpet-shaped, while s are mere blotches. The blotch mines made through the habits of the in within them; it feeds around and arou instead of forging ahead as the serpen miners do. The larvæ of beetles, flies, moths may mine leaves, each species ing its own special food plant. Most the smaller leaf mines are made by caterpillars of the moths which are called the Tineina or Tineids. Most these barely have a wing expanse that reach a quarter of an inch, and many smaller; they all have narrow wings, hind wings being mere threads borde with beautiful fringes. The specific na of these moths usually end in "ell thus, the one that mines in apple is *m foliella*, the one in grain is *granella*. C of these little moths, *Gelechia pinifolie* lives the whole of its growing life in of a pine needle. The moth lays the eg

Dept. of Entomology, Cornell

Mines of the trumpet leaf-miner

ut the middle of the needle, and the e caterpillar that hatches from it ws its way directly into the heart of needle; and there, as snug as snug can it lives and feeds until it is almost a rter of an inch long; think of it! Many me I have held up to the light a pine dle thus inhabited, and have seen the le miner race up and down its abode f it knew that something was happen-. When it finally attains its growth it kes wider the little door through which ntered; it does this very neatly; the door n even oval, and looks as if it were made h the use of dividers. After thus open- the door, the caterpillar changes to a le, long pupa, very close to its exit; and er it emerges, an exquisite little moth h silvery bands on its narrow, brown ngs, and a luxurious fringe on the edges its narrow hind wings and also on the ter hind edges of the front wings.

The gross mines in the leaves of dock d beet are not pretty. The leaves are tted, sometimes for their whole length, d soon turn brown and lie prone on the ound, or dangle pathetically from the lk. These mines are made by the larvæ a fly, and a whole family live in the me habitation. If we hold a leaf thus ined up to the light, while it is still een, we can see several of the larvæ orking, each making a bag in the life bstance of the leaf, and yet all joining gether to make a great blister. The flies at do this mischief belong to the family *nthomyidæ*; and there are several species hich have the perturbing habit of min- g the leaves of beets and spinach. It be- ooves those of us who are fond of these greens," as our New England ances- rs called them, to hold every leaf up to e light before we put it into the skillet, st we get more meat than vegetable in ese viands. The flies who thus take our eens ahead of us are perhaps a little rger than houseflies, and are generally ay in color with the front of the head lver white. These insects ought to teach s the value of clean culture in our gar- ens, since they also mine in the smooth igweed.

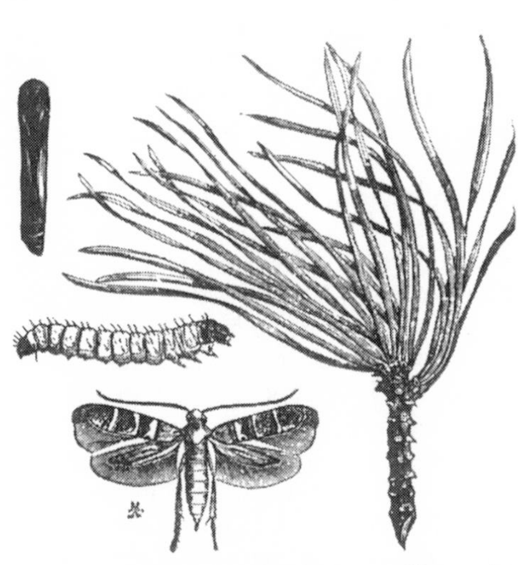

The pine-needle leaf-miner. Note the mined leaves. The pupa, the caterpillar, and the moth of the leaf-miner are much enlarged

LESSON 77

Leaf-Miners

Leading Thought — The serpent-like markings and the blister-like blotches which we often see on leaves are made by the larvæ of insects which complete their growth by feeding upon the inner living substance of the leaf.

Method — The nasturtium leaf-miner is perhaps the most available for this lesson since it may be found in its mine in early autumn. However, the pupils should bring to the schoolroom all the leaves with mines in them that they can find and study the different forms.

Observations — 1. Sketch the leaf with the mine in it, showing the shape of the mine. What is the name of the plant on which the leaf grew?

2. Hold the leaf up to the light; can you see the insect within the mine? What is it doing? Is there more than one insect in the mine? Open the mine and see how the miner looks.

3. There are three general types of mines: those that are long, curving lines, called serpentine mines; those that begin small and flare out, called trumpet mines; and those that are blister-like, called blotch mines. Which of these is the mine you are studying?

4. Study a serpentine mine. Note that where the little insect began to eat, the mine is small. Why does it widen from this point? What happened in the part which we call the serpent's head?

5. Look closely with a lens and find if there is a break above the mine in the upper surface of the leaf or below the mine in the lower surface of the leaf. If the insect is no longer in the mine can you find where it escaped? Can you find a shed pupa skin in the "serpent's head"?

6. Why does an insect mine in a leaf? What does it find to eat? How is it protected from the birds or insects of prey while it is getting its growth?

7. Look on leaves of nasturtium, columbine, lamb's quarters, dock, and burdo for serpentine mines. Are the mines these different plants alike? Do you s pose the mines are made by the sa insect?

8. Look on leaves of dock, burdo beet, and spinach for blotch mines. there more than one insect in th mines? If the insects are present, hold t leaf up to the light and watch them eat

9. Look in the leaves of pitch or ot thick-leaved pines (not white pine) pine needles which are yellow at the t Examine these for miners. If the mine not within, can you find the little cir lar door by which it escaped? Would y think there was enough substance half a pine needle to support a little cr ture while it grew up?

10. If you find leaf-miners at work, not pluck off the leaves being mined b cover each with a little bag of Swiss m lin tied close about the petiole and th capture the winged insect.

THE LEAF–ROLLERS

If we look closely at sumac leaves before they are aflame from autumn's torch, we find many of the leaflets rolled into little cornucopias fastened with silk. The silk is not in a web, like that of the spider, but the strands are twisted together, hundreds of threads combined in one strong cable, and these are fastened from roll to leaf, like tent ropes. If we look at the young basswoods, we find perhaps many of their leaves cut across, and the flap made into a roll and likewise fastened with silken ropes. The witch hazel, which is a veritable insect tenement, also shows these rolls. In fact, we may find them upon the leaves of almost any species of tree or shrub, and each of these rolls has its own special maker or indweller. Each species of insect which rolls the leaves is limited to the species of plant on which it is found; and one of these caterpillars would sooner starve than take a mouthful from a leaf of any other plant. Some pe ple think that insects will eat anythi that comes in their way; but of all anima insects are the most fastidious as to th food.

Some species of leaf-rollers unite seve leaflets together, while others use a sing leaf. The sumac leaf-roller begins in a si gle leaf; but in its later stages, it faste together two or three of the terminal le lets in order to gain more pasturage. T little silken tent ropes which hold t folded leaves are well worth study with lens. They are made of hundreds threads of the finest silk, woven from gland opening near the lower lip of t caterpillar. The rope is always larger whe it is attached to the leaf than at the ce ter, because the caterpillar crisscrosses t threads in order to make the attachme to the leaf larger and firmer. Unroll a te carefully, and you may see the fastenin

in an earlier stage, and may even
the first turned-down edge of the leaf.
ever, the center of a leaf-roller's habi-
n is usually very much eaten, for the
e reason for making its little house is
the soft-bodied caterpillar may eat its
ompletely hidden from the eyes of
or other animals. When it first
nes from the egg, it feeds for a short
, usually on the underside of the leaf;
when still so small that we can barely
it with the naked eye, it somehow
ages to fold over itself one edge of
eaf and peg it down. The problem of
so small a creature is able to pull over

Leaf-rollers in sumac, showing the fastening of the silk stay-ropes

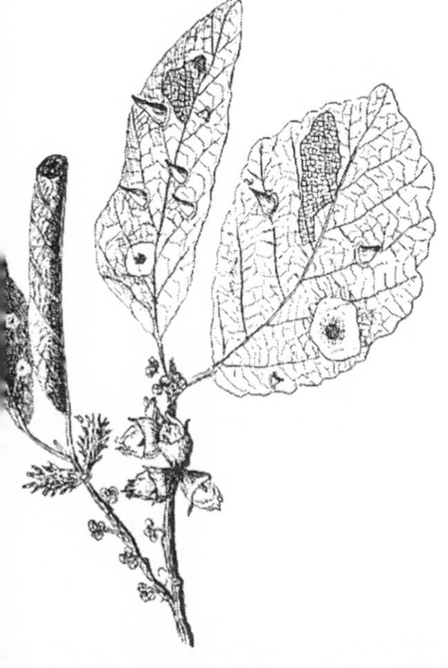

ch hazel, showing work of leaf-rollers, leaf-miners, and gall-makers

fold down or to make in a roll a stiff
is hard to solve. I myself believe it is
e by making many threads, each a lit-
more taut than the last. I have watched
eral species working, and the leaf comes
wly together as the caterpillar stretches
head and sways back and forth hun-
ds of times, fastening the silk first to
side and then to the other. Some ob-
vers believe that the caterpillar throws
weight upon the silk, in order to pull
the leaf together; but in the case of the sumac leaf-roller, I am sure this is not true, as I have watched the process again and again under a lens, and could detect no signs of this method. Many of the caterpillars which make rolls change to small moths known as Tortricids. This is a very large family, containing a vast number of species, and not all of the members are leaf-rollers. These little moths have the

Grace H. Griswold

A leaf of hollyhock rolled by a leaf-roller

front wings rather wide and more or less rectangular in outline. The entomologists have a pleasing fashion of ending the names of all of these moths with "ana"; the one that rolls the currant leaves is *Rosana*, the one on juniper is *Rutilana*, etc. Since many of the caterpillars of this family seek the ground to pupate and do not appear as moths until the following spring, it is somewhat difficult to study

Leaflets fastened together by the skipper caterpillar to make a nest. The adult skipper is shown

their complete life histories, unless one has well-made breeding cages with earth at the bottom; and even then it is difficult to keep them under natural conditions, since in an ordinary living room the insects dry up and do not mature.

LESSON 78

THE LEAF-ROLLERS

LEADING THOUGHT — Many kinds of insects roll the leaves of trees and plants into tents, in which they dwell and feed during their early stages.

METHOD — This is an excellent lesson for early autumn when the pupils may find many of these rolled leaves, which they may bring to the schoolroom, and
which will give material for the le
The rolls are found plentifully on su
basswood, and witch hazel.

OBSERVATIONS — 1. What is the r
of the trees and shrubs from which t
rolled leaves that you have collected
taken?

2. Is more than one leaf or le
used in making the roll?

3. Is the leaf rolled crosswise or le
wise? How large is the tube thus mad

4. Is the nest in the shape of a tub
are several leaves fastened together,
ing a box-shaped nest?

5. How is the roll made fast? Exa
the little silken ropes with a lens an
scribe one of them. Is it wider where
attached to the leaf than at the mi
Why?

6. How many of these tent ropes
there which make fast the roll? Unr
leaf carefully and see if you can find
of the tent ropes that fastened the rol
gether when it was smaller. Can you
where it began?

7. As you unroll the leaves what do
see at the center? Has the leaf been ea
Can you discover the reason why the c
pillar made this roll?

8. How do you think a caterpillar
ages to roll a leaf so successfully? W
is the spinning gland of a caterpillar?
does the insect act when spinning thr
back and forth when rolling the
What sort of insect does the caterp
which rolls the leaf change into? Do
suppose that the same kind of caterpi
make the rolls on two different specie
trees?

9. In July or early August get som
the rolls with the caterpillars in them,
roll a nest, take the caterpillar out and
it on a fresh leaf of the same kind of
or shrub on which you found it,
watch it make its roll.

L. H. Weld

…e spiny oak-gall *The pointed bullet-gall on oak twigs* *A cluster of galls on mid-rib of an oak leaf* *The acorn plum-gall*

THE GALL DWELLERS

He retired to his chamber, took his lamp, and summoned the genius as usual. " Gen-…," said he, " build me a palace near the sultan's, fit for the reception of my spouse, …e princess; but instead of stone, let the walls be formed of massy gold and silver, …d in alternate rows; and let the interstices be enriched with diamonds and emeralds. …he palace must have a delightful garden, planted with aromatic shrubs and plants, …aring the most delicious fruits and beautiful flowers. But, in particular, let there be … immense treasure of gold and silver coin. The palace, moreover, must be well pro-…ded with offices, storehouses, and stables full of the finest horses, and attended by …uerries, grooms, and hunting equipage." By the dawn of the ensuing morning, the …nius presented himself to Aladdin, and said, " Sir, your palace is finished; come and …e if it accords with your wishes." — ARABIAN NIGHTS' ENTERTAINMENTS

Although Aladdin is out of fashion, we …ll have houses of magic that are even …ore wonderful than that produced by his …sourceful lamp. These houses are built …rough an occult partnership between in-…cts and plant tissues; we do not under-…and exactly how they are made, although …e are beginning to understand a little …ncerning the reasons for the growth. …hese houses are called galls and are thus …ell named, since they grow because of an …ritation to the plant caused by the insect.

There are many forms of these gall …wellings, and they may grow upon the …ot, branch, leaf, blossom, or fruit. The …iraculous thing about them is that each kind of insect builds its magical house on a certain part of a certain species of tree or plant; and the house is always of a certain definite form on the outside and of a certain particular pattern within. Many widely differing species of insects are gall-makers; and he who is skilled in gall lore knows, when he looks at the outside of the house, just what insect dwells within it.

We may take the history of the common oak apple as an example. A little, four-winged, flylike creature, a wasp, lays its eggs, early in the season, on the leaf of the scarlet oak. As soon as the larva hatches, it begins to eat into the substance of one of the leaf veins. As it eats, it discharges through its mouth into the tissues of the leaf a substance which is secreted from glands within its body. Immediately

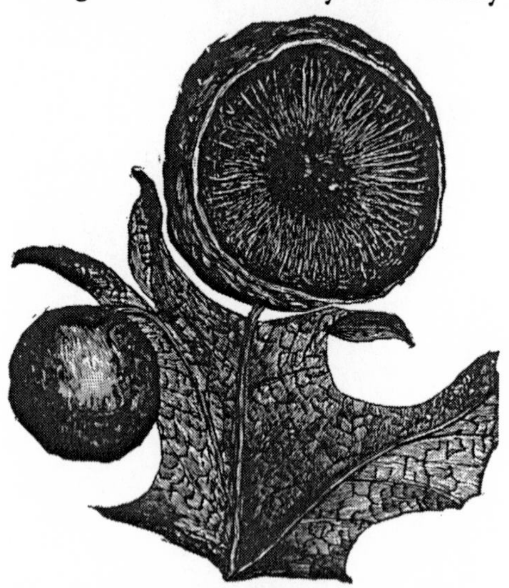

Oak apple, showing the larva of the gall insect

the building of the house commences; out around the little creature grow radiating vegetable fibers, showing by their position plainly that the grub is the center of all of this new growth; meanwhile, a smooth, thin covering completely encloses the globular house; larger and larger grows the house until we have what we are accustomed to call an oak apple, so large is it. The little chap inside is surely content and happy, for it is protected from the sight of all of its enemies, and it finds the walls of its house the best of food. It is comparable to a boy living in the middle of a giant sponge cake, who when hungry would naturally eat out a larger cave in the heart of the cake. After the inmate of the oak apple completes its growth, it changes to a pupa and finally comes out into the world a tiny wasp, scarcely a quarter of an inch in length.

The story of the willow cone-gall is quite different. A little gnat lays her eggs on the tip of the bud of a twig; as soon as the grub hatches and begins to eat, the growth of the twig is arrested, the leaves are stunted until they are mere scales and are obliged to overlap in rows around little inmate, thus making for it a c shaped house which is very thorou shingled. The inhabitant of this gall hospitable little fellow, and his house s ters and feeds many other insect gu He does not pay any attention to th being a recluse in his own cell, but civilly allows them to take care of th selves in his domain, and to feed u the walls of his house. He stays in his s home all winter and comes out in spring a tiny, two-winged fly.

There are two galls common on stems of goldenrod. The more nume is spherical in form and is made by a and prosperous looking little grub wh later develops into a fly. But althoug is a fly that makes the globular gall in stem of goldenrod, the spindle-shaped often seen on the same stem has quite other story. A little brown and gray m tled moth, about three-fourths of an i long, lays her egg on the stem of the yo goldenrod. The caterpillar, when hatches, lives inside the stem, which commodatingly enlarges into an obl room. The caterpillar feeds upon the s stance of the stem until it attains growth, and then it cuts, with sharp ja

Verne Mo

Willow cone-galls

ttle oval door at the upper end of its
ıse and makes an even bevel by widen-
the opening toward the outside. It
n makes a little plug of debris which
npletely fills the door; but because of
bevel, no intrusive beetle or ant can
h it in. Thus the caterpillar changes
a helpless pupa in entire safety; and
en the little moth issues from the pupa
n, all it has to do is to push its head
inst the door, and out it falls, and the
luse is now a creature of the outside
rld.

Many galls are compound, that is, they
made up of a community of larvæ,
h in its own cell. The mossy rose gall
an instance of this. The galls made by
tes and aphids are open either below or
ove the surface of the leaf; the little
nical galls on witch hazel are examples
these. In fact, each gall has its own par-
ular history, which proves a most inter-
ing story if we seek to read it with our
n eyes.

LESSON 79

The Gall Dwellers

Leading Thought — The galls are pro-
ctive habitations for the little insects which dwell within them. Each kind of insect makes its own peculiar gall on a certain species of plant.

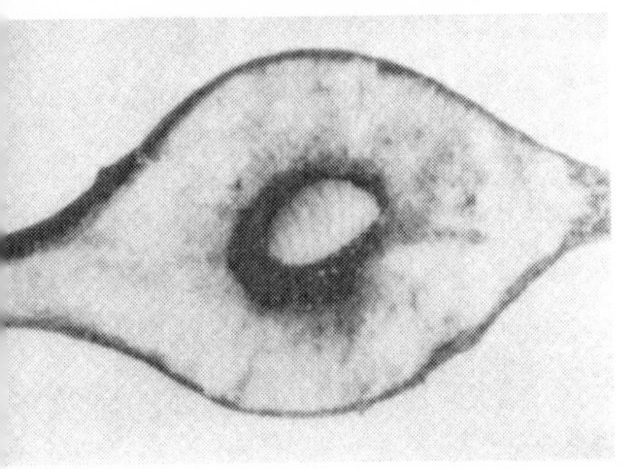

M. V. Slingerland

pherical gall of the goldenrod, opened, showing its prosperous looking owner

Glenn W. Herrick

The vagabond gall of the cottonwood

Method — Ask the pupils to bring in as many of these galls as possible. Note that some have open doors and some are entirely closed. Cut open a gall and see what sorts of insects are found within it. Place each kind of gall in a tumbler or jar covered with cheesecloth and put them where they may be under observation for perhaps several months; note what sort of winged insect comes from each.

Observations — 1. On what plant or tree did this gall grow? Were there many like it? Did they grow upon the root, stem, leaf, flower, or fruit? If on the leaf, did they grow upon the petiole or the blade?

2. What is the shape of the little house? What is its color? Its size? Is it smooth or wrinkled on the outside? Is it covered with fuzz or with spines?

3. Open the gall; is there an insect within it? If so, where is it and how does it look? What is the appearance of the inside of the gall?

4. Is there a cell for the insect at the very center of the gall, or are there many such cells?

5. Has the house an open door? If so, does the door open above or below? Is there more than one insect in the galls

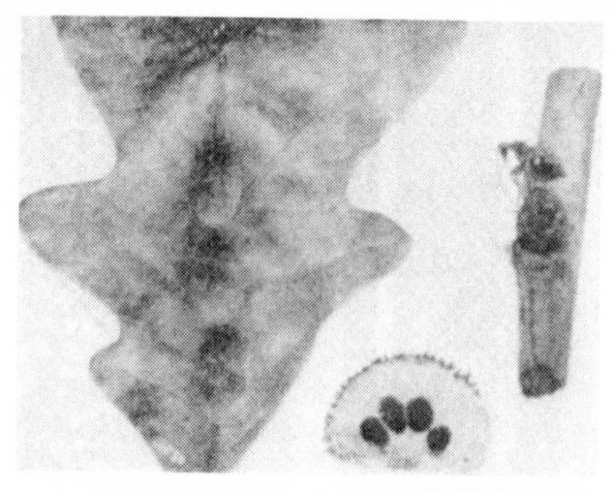

C. V. Triggerson

Pictured here are porcupine galls on the leaf of a white oak; a section of a porcupine gall showing the cells; and a female gall-fly laying eggs in an oak bud

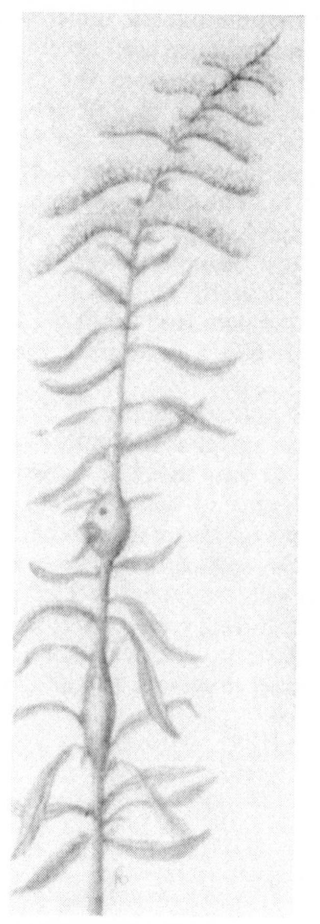

Stem of goldenrod showing the spheri gall above, made by the larva of a fly; a the spindle-shaped gall below, made by t caterpillar of a moth

with open doors? What sort of insect makes this kind of house?

6. Do you find any insects besides the original gall-maker within it? If so, what are they doing?

7. Of what use are these houses to their little inmates? How do they protect them from enemies? How do they furnish them with food?

8. Do the gall insects live all their lives within the galls or do they change to winged insects and come out into the world? If so, how do they get out?

9. How many kinds of galls can you find upon oaks? Upon goldenrod? Upon witch hazel? Upon willow?

A green little world
With me at its heart!
A house grown by magic,
Of a green stem, a part.

My walls give me food
And protect me from foes,
I eat at my leisure,
In safety repose.

My house hath no window,
'Tis dark as the night!
But I make me a door
And batten it tight.

And when my wings grow
I throw wide my door;
And to my green castle
I return nevermore.

THE GRASSHOPPER

Because the grasshopper affords special facilities for the study of insect structure, it has indeed become a burden to the students in the laboratories of American universities. But in nature-study we mu not make anything a burden, least all the grasshopper, which, bei such a famous jumper as well as fli

not long voluntarily burden any
ct.
nce we naturally select the most sali-
characteristic of a creature to present
to young pupils, we naturally begin
lesson with the peculiarity which
es this insect a "grasshopper." When
creature has unusually strong hind
we may be sure it is a jumper, and the
hopper shows this peculiarity at first
ce. The front legs are short, the mid-
egs a trifle longer, but the femur of the
l leg is nearly as long as the entire
y, and contains many powerful mus-
which have the appearance of being

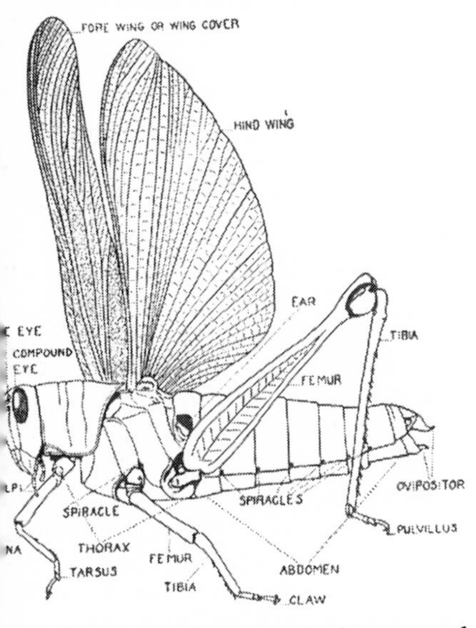

grasshopper with parts of the external anatomy named

ided, because of the way they are at-
hed to the skeleton of the leg; the tibia
the hind leg is long and as stiff as if
de of steel. When getting ready to
np the grasshopper lowers the great
nur below the level of the closed wings
d until the tibia is parallel with it and
entire foot is pressed against the
und. The pair of double spines at the
d of the tibia, just back of the foot, are
essed against the ground like a spiked
el, and the whole attitude of the insect
tense. Then, like a steel spring, the long
s straighten and the insect is propelled
gh into the air and far away. This is a
remarkable example of insect dynamics. Since so many species of birds feed upon the grasshopper, its leaping power is much

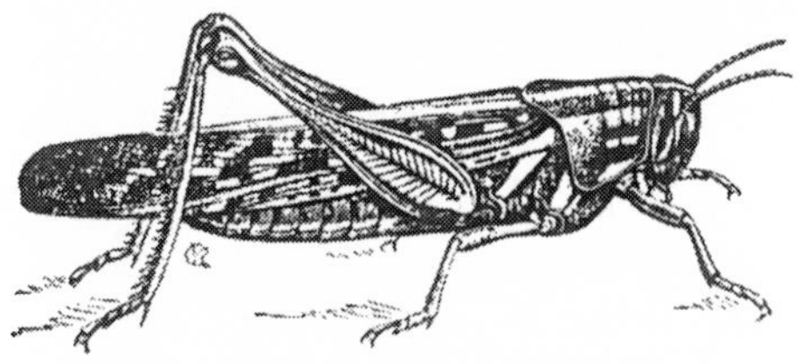

The American bird grasshopper

needed to escape them. However, when the grasshopper makes a journey it uses its wings.

As we watch a grasshopper crawling up the side of a vial or tumbler we can examine its feet with a lens. Between and in front of the claws is an oval pad which clings to the glass, not by air pressure as was once supposed, but by means of microscopic hairs, called tenent hairs, which secrete a sticky fluid. Each foot consists of three segments and a claw; when the insect is quiet, the entire foot rests upon the ground; but when it is climbing on glass, the toe pads are used.

The grasshopper's face has a droll expression; would that some caricaturist could analyze it! It is a long face, and the compound eyes placed high upon it give a look of solemnity. The simple eyes can

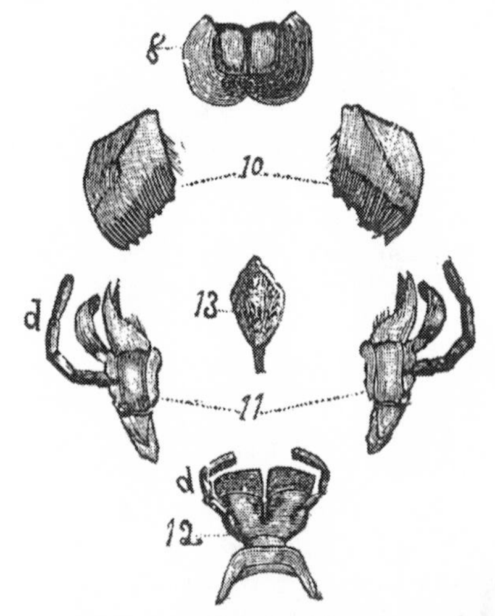

The mouth-parts of a grasshopper, enlarged

8, upper lip or *labrum*. 10, upper jaws or *mandibles*. 11, lower jaws or *maxillae*. 12, under lip or *labium*. 13, tongue. *d*, palpi

be made out with a lens. There is one just in front of each big eye, and another, like the naughty little girl's curl, is "right in the middle of the forehead." The antennæ are short but alert. The two pairs or palpi connected with the mouth-parts are easily seen, likewise the two pairs of jaws, the notched mandibles looking like a pair of nippers. We can see these jaws much better when the insect is eating, which act is done methodically. First, it begins at one edge of a leaf, which it seizes between the front feet so as to hold it firm; it eats by reaching up and cutting downwards, making an even-edged, long hole on the leaf margin; it makes the hole deeper by repeating the process. It some-

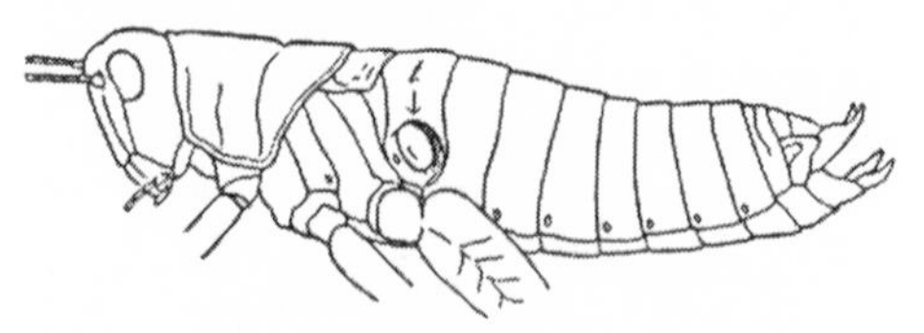

A drawing of a grasshopper without the wings to show an ear, labeled t

times makes a hole in the middle of a leaf and bites in any direction, but it prefers to move the jaws downward. While it is feeding, its palpi tap the leaf continually and its whole attitude is one of deep satisfaction. There is an up-rolled expression to the compound eyes which reminds us of the way a child looks over the upper edge of its cup while drinking milk. The grasshopper has a preference for tender herbage, but in time of drought will eat almost any living plant.

Back of the head is a sunbonnet-shaped piece, bent down at the sides, forming a cover for the thorax. The grasshopper has excellent wings, as efficient as its legs; the upper pair are merely strong, thick, membranous covers, bending down at the sides so as to protect the under wings; these wing covers are not meant for flying and are held stiff and straight up in the air during flight. The true wings, when the grasshopper is at rest, are folded lengthwise like a fan beneath the wing covers; they are strongly veined and circular in shape, giving much surface for beating the

air. The grasshopper's flight is us swift and short; but in years of fa some kinds of grasshoppers fly high ir air and for long distances, a fact reco in the Bible regarding the plague o custs. When they thus appear in hordes, they destroy all the vegetatio the region where they settle.

The wings of grasshoppers vary in c those of the red-legged species being while those of the Carolina locusts black with yellow edges. The abdom segmented, as in all insects, and along lower side there are two lengthwise sut or creases which open and shut bell like when the grasshopper breathes. spiracles or breathing pores can be s on each segment, just above this sutur

The grasshopper has its ears well tected; to find them, we must lift wings in order to see the two large sou ing discs, one on each side of the first ment of the abdomen. These are la and much more like ears than are the li ears in the elbows of the katydids.

The singing of the short-horned gr hoppers is a varied performance, each cies doing it in its own way. One spe makes a most seductive little note by p ing the femur and tibia of the hind together; with the hind feet complet off the ground, the legs are moved up down with great rapidity, giving off a li purr. The wings in this case do not lif all. There are other species that make sound by rubbing the legs against wing covers.

The grasshopper makes its toilet th It cleans first the hind feet by rubb them together and also by reaching b and scrubbing them with the middle fe the big hind femur it polishes with bent elbow of the second pair of legs. cleans the middle feet by nibbling a licking them, bending the head far neath the body in order to do it. It p ishes its eyes and face with the front fe stopping to lick them clean betwe whiles, and it has a most comical man of cleaning its antennæ; this is acco plished by tipping the head sidewise, a bending it down so that the antenna side rests upon the floor; it then plants front foot of that side firmly upon the nna and pulls it slowly backward be- en the foot and floor.

he grasshopper has some means of de- e as well as of escape; it can give a ful nip with its mandibles; and when ed, it emits copiously from the mouth rownish liquid which is acrid and ill lling. This performance interests chil- n who are wont to seize the insect by umping legs and hold it up, command- it to "chew tobacco."

rasshoppers are insects with incom- e metamorphosis, which merely means the baby grasshopper, as soon as it erges from the egg, is similar in form ts parent except that it has a very large d and a funny little body, and that it no quiet pupal stage during life. When nature, the under wings or true wings e a position outside of the wing covers look like little fans.

he short-horned grasshoppers lay their s in oval masses protected by a tough rcoat. The ovipositor of the mother sshopper is a very efficient tool, and h it she makes a deep hole in the und, or sometimes in fence rails or er decaying wood; after placing her s in such a cavity, she covers the hid- place with a gummy substance so that intruders or robbers may work harm

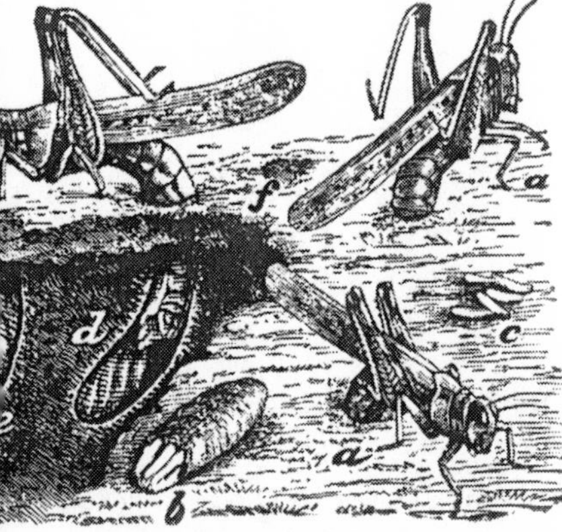

Grasshoppers laying eggs

, a, a, females, ovipositing; b, egg-pod extracted n the ground with the end broken open; c, a few s lying loose upon the ground; d, e, show the earth tially removed to illustrate an egg-mass already in ce and one being placed; f shows where such a mass been covered up

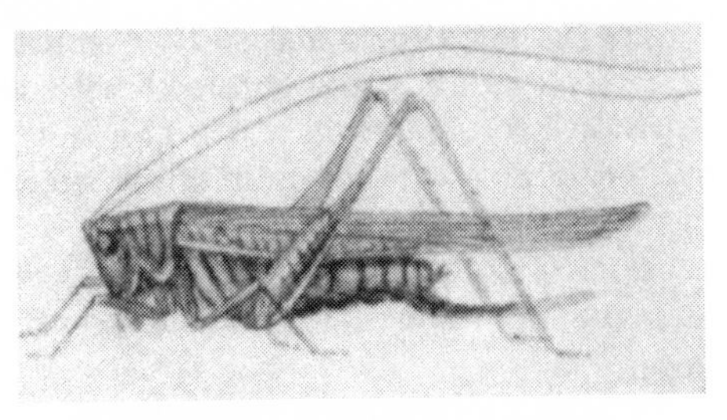

Long-horned or meadow grasshopper

to her progeny. Most species of grasshoppers pass the winter in the egg stage; but sometimes we find in early spring the young ones which hatched in the fall, and they seem as spry as if they had not been frozen stiff.

LESSON 80

The Red-legged Grasshopper

Leading Thought — The grasshopper feeds upon grass and other herbage and is especially fitted for living in grassy fields. Its color protects it from being seen by its enemies, the birds. If attacked, it escapes by long jumps and by flight. It can make long journeys on the wing.

Method — The red-legged grasshopper (*M. femur-rubrum*) has been selected for this lesson because it is the most common of all grasshoppers in many parts of our country, though other species may be used as well. The red-legged locust or grasshopper has, as is indicated by its name, the large femur of the hind legs reddish in color. Place the grasshopper under a tumbler and upon a spray of fresh herbage, and allow the pupils to observe it at leisure. It might be well to keep some of the grasshoppers in a cage similar to that described for crickets. When one is studying the feet, or other parts of the insect requiring close scrutiny, the grasshopper should be placed in a vial so that it may be passed around and observed with a lens. Give the questions a few at a time, and encourage the pupils to study these insects in the field.

OBSERVATIONS — 1. Since a grasshopper is such a high jumper, discover if you can

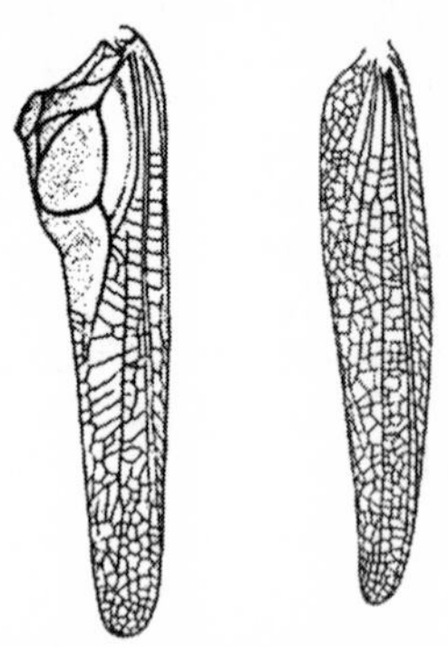

Left, *wing of male; and* right, *wing of female meadow grasshoppers*

how he does this "event." Which pair of legs is the longest? Which the shortest? How long are the femur and tibia of the hind leg compared with the body? What do you think gives the braided appearance to the surface of the hind femur? What is there peculiar about the hind femur? Note the spines at the end of the tibia just behind the foot.

2. Watch the grasshopper prepare to jump and describe the process. How do you think it manages to throw itself so far? If a man were as good a jumper as a grasshopper in comparison to his size, he could jump 300 feet high or 500 feet in distance. Why do you think the grasshopper needs to jump so far?

3. As the grasshopper climbs up the side of a tumbler or vial, look at its feet through a lens and describe them. How many segments are there? Describe the claws. How does it cling to the glass? Describe the little pad between the claws.

4. Look the grasshopper in the face. Where are the compound eyes situated? Can you see the tiny simple eyes like mere dots? How many are there? Where are they? How long are the antennæ? For what are they used?

5. How does a grasshopper eat? Do the jaws move up and down or sidewise? W does the grasshopper eat? How many p of palpi can you see connected with mouth-parts? How are these used w the insect is eating? When there are m grasshoppers, what may happen to crops?

6. What do you see just back of grasshopper's head, when looked at fr above?

7. Can the grasshopper fly as wel jump? How many pairs of wings has Does it use the first pair of wings to with? How does it hold them when ing? Where is the lower or hind pai wings when the grasshopper is walki How do they differ in shape from front wings?

8. Note the abdomen. It is made many rings or segments. Are these ri continuous around the entire bo Where do their breaks occur? Descr the movement of the abdomen as the sect breathes. Can you see the spiracles breathing pores? Lift the wings, and f the ear on the first segment of the domen.

9. If you seize the grasshopper h may it show that it is offended?

10. How does the grasshopper perfo

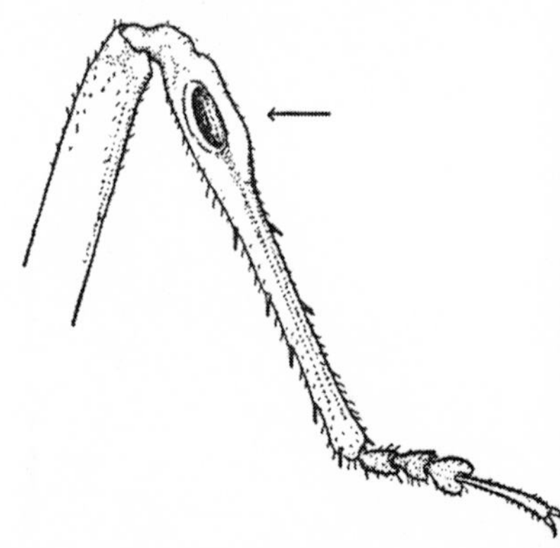

Front leg of a katydid showing the ear n the elbow

its toilet? Describe how it cleans its a tennæ, face, and legs.

11. What becomes of the grasshopp in the winter? Where are the eggs la How can you tell a young from a fu grown grasshopper?

2. Do all grasshoppers have antennæ rter than half the length of their bod- Do some have antennæ longer than ir bodies? Where are the long-horned grasshoppers found? Describe how they resemble the katydids in the way they make music and in the position of their ears.

THE KATYDID

I love to hear thine earnest voice
Wherever thou art hid,
Thou testy little dogmatist,
Thou pretty katydid,

Thou mindest me of gentle folks,
Old gentle folks are they,
Thou say'st an undisputed thing
In such a solemn way.

— Holmes

)istance, however, lends enchantment he song of the katydid, for it grates on nerves as well as on our ears, when at se quarters. The katydid makes his sic in a manner similar to that of the ket but is not, however, so well uipped, since he has only one file and y one scraper for playing. As with the adow grasshoppers and crickets, only males make the music, the wings of females being delicate and normally ned at the base. The ears, too, are in same position as those of the cricket, d may be seen as a black spot in the nt elbow. The song is persistent and y last the night long: "Katy did, she n't, she did." James Whitcomb Riley s, "The katydid is rasping at the si- ce," and the word "rasping" well de- ibes the note.

The katydids are beautiful insects, with en, finely veined, leaflike wing covers der which is a pair of well developed ngs, folded like fans; they resemble in m the long-horned grasshoppers. The mmon northern species (*Cyrtophyllus*) all green above except for the long, deli- te fawn-colored antennæ, and the ownish fiddle of the male, which con- ts of a flat triangle just back of the orax where the wing covers overlap. metimes this region is pale brown and metimes green, and with the unaided e we can plainly see the strong cross- in, bearing the file. The green eyes have rker centers and are not so large as the es of the grasshopper. The body is green ith white lines below on either side. There is a suture the length of the ab- domen in which are placed the spiracles. The insect breathes by sidewise expansion and contraction, and the sutures rhythmi- cally open and shut; when they are open, the spiracles can be seen as black dots.

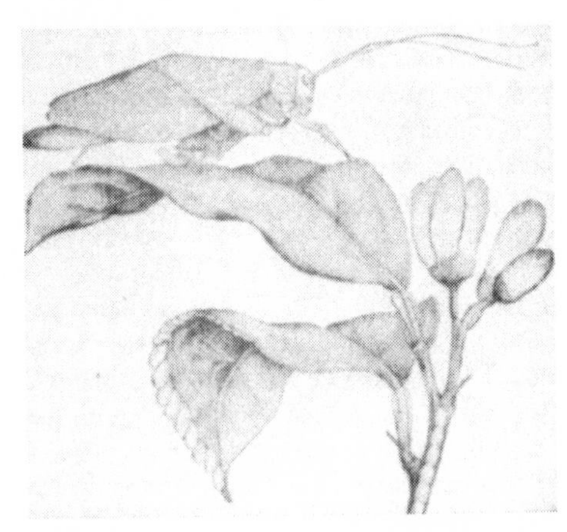

The angular-winged katydid and her eggs

The legs are slender and the hind pair very long. The feet are provided with two little pads, one on each side of the base of the claw. In the grasshopper there is only one pad, which is placed between the two hooks of the claw. The female has a green, sickle-shaped ovipositor at the end of the body. With this she lays her flat, oval eggs, slightly overlapping in a neat row.

The katydids are almost all dwellers in trees and shrubs; although I have often found our common species upon asters and similar high weeds. The leaflike wings of these insects are, in form and color, so similar to the leaves that they are very completely hidden. The katydid is rarely discovered except by accident; although when one is singing, it may be approached and ferreted out with the aid of a lantern.

The katydid, when feeding, often holds the leaf or the flower firmly with the front

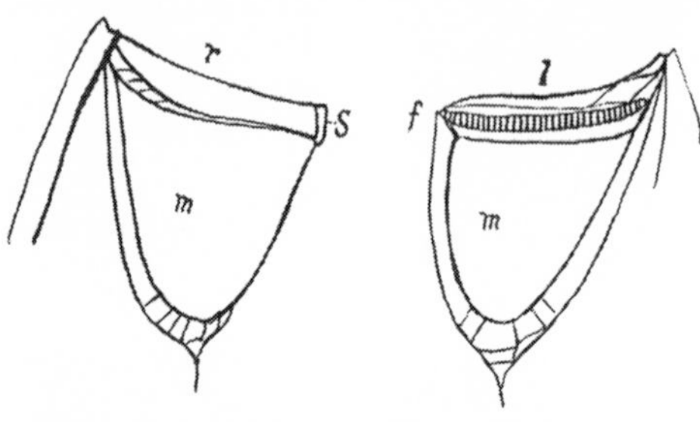

The front portions of the wings of a male katydid showing the file, f, *on one wing and the scraper,* s, *on the other*

feet, while biting it off like a grazing cow, and if it is tough, chews it industriously with the sidewise-working jaws. A katydid will often remain quiet a long time with one long antenna directed forward and the other backward, as if on the lookout for news from the front and the rear. But when the katydid "cleans up," it does a thorough job. It nibbles its front feet, paying special attention to the pads, meanwhile holding the foot to its mandibles with the aid of the palpi. But one washing is not enough; I have seen a katydid go over the same foot a dozen times in succession, beginning always with the hind spurs of the tibia and nibbling along the tarsus to the claws. It cleans face with its front foot, drawing it do ward over the eye and then licking clean. It cleans its antenna with its m dibles by beginning at the base and dr ing it up in a loop as fast as finish After watching the process of th lengthy ablutions, we must conclude t the katydid is among the most fastidi members of the insect "four hundre

LESSON 81

The Katydid

Leading Thought — The katydids semble the long-horned grasshoppers a the crickets. They live in trees, and t male sings "katy did" by means of musical instrument similar to that of t cricket.

Method — Place a katydid in a crick cage in the schoolroom, giving it fre leaves or flowers each day, and encoura ing the pupils to watch it at recess. It m be placed in a vial and passed around f close observation. In studying this inse use the lesson on the red-legged grassho per and also that on the cricket. These le sons will serve to call the attention of t pupils to the differences and resen blances between the katydid and the two related insects.

THE BLACK CRICKET

If we wish to become acquainted with these charming little troubadours of the field, we should have a cricket cage with a pair of them within it. They are most companionable, and it is interesting to note how quickly they respond to a musical sound. I had a pair in my room at one time, when I lived very near a cathedral. Almost every time that the bells rang during the night, my cricket would respond with a most vivacious and sympathetic chirping.

The patent-leather finish to this cric et's clothes is of great use; for, althoug the cricket is an efficient jumper, it is, aft

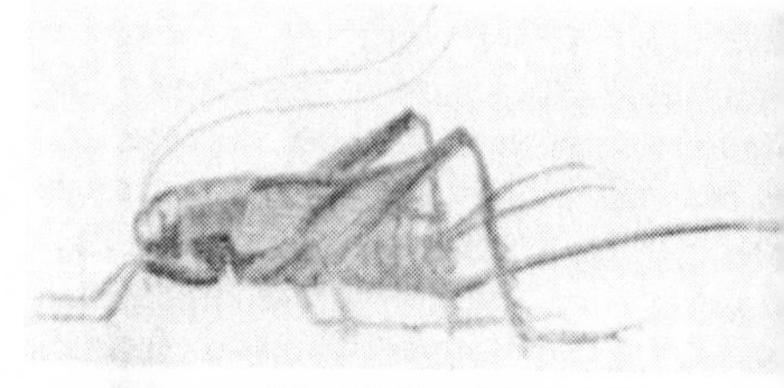

The field cricket

mostly by running between grass es that it escapes its enemies. If we o catch one, we realize how slippery and how efficiently it is thus able to through the fingers.

he haunts of the cricket are usually y; it digs a little cave beneath a stone od in some field, where it can have vhole benefit of all the sunshine when ues from its door. These crickets can- fly, since they have no wings under

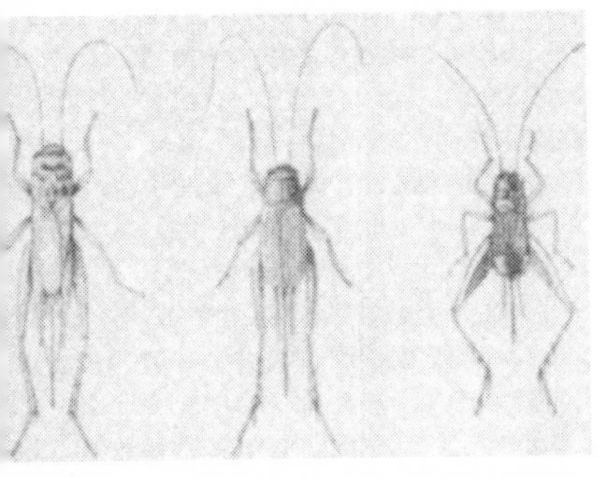

, *a house cricket;* center, *field cricket;* right, *another species of field cricket*

r wing covers, as do the grasshoppers. hind legs have a strong femur, and ort but strong tibia with downward- ting spines along the hind edge, which oubtedly help the insect in scrambling ough the grass. At the end of the tibia, t to the foot, is a rosette of five spines, two longer ones slanting to meet foot; these spines give the insect a n hold, when making ready for its ing. When walking, the cricket places whole hind foot flat on the ground, rests only upon the claw and the ad- ning segment of the front pairs of feet. e claws have no pads like those of the ydid or grasshopper; the segment of tarsus next the claw has long spines the hind feet and shorter spines on middle and front feet, thus showing t the feet are not made for climbing, t for scrambling along the ground. hen getting ready to jump, the cricket uches so that the tibia and femur of hind legs are shut together and almost the ground. The dynamics of the cket's leap are well worth studying.

The cricket's features are not so easily made out, because the head is polished and black; the eyes are not so polished as the head; the simple eyes are present, but are discerned with difficulty. The antennæ

The wing of a male cricket enlarged. a, *file;* b, *scraper*

are longer than the body and very active; there is a globular segment where they join the face. I have not discovered that the crickets are so fastidious about keep- ing generally clean as are some other in-

A section of the file enlarged

sects, but they are always cleaning their antennæ. I have seen a cricket play his wing mandolin lustily and at the same time carefully clean his antennæ; he pol- ished these by putting up a foot and bend-

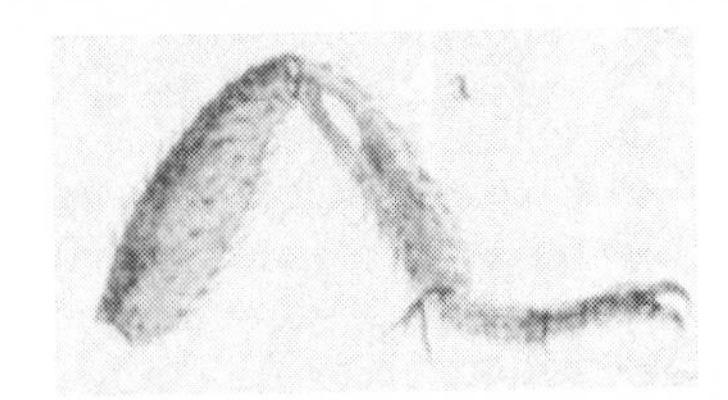

The front leg of a cricket enlarged to show the ear at a

ing the antenna down so that his mouth reached it near the base; he then pulled the antenna through his jaws with great deliberation, nibbling it clean to the very end. The lens reveals to us that the flexi- bility of the antennæ is due to the fact that they are many jointed. The palpi are easily seen, a large pair above and a smaller pair beneath the "chin." The palpi are used to test food and prove if it be palatable. The crickets are fond of melon or other sweet, juicy fruits, and by putting such food into the cage we can see them bite out pieces with their sidewise-working jaws, chewing the toothsome morsel with gusto. They take hold of the sub-

A cricket cage

stance they are eating with the front feet as if to make sure of it.

The wing covers of the cricket are bent down at the sides at right angles, like a box cover. The wing covers are much shorter than the abdomen and beneath them are vestiges of wings, which are never used. The male has larger wing covers than the female, and they are veined in a peculiar scroll pattern. This veining seems to be a framework for the purpose of making a sounding board of the wing membrane, by stretching it out as a drumhead is stretched. Near the base of the wing cover there is a heavy cross-vein covered with transverse ridges, which is called the file; on the inner edge of the same wing, near the base, is a hardened portion called the scraper. When he makes his cry, the cricket lifts his wing covers at an angle of forty-five degrees and draws the scraper of the under wing against the file of the overlapping lest his musical apparatus become out, he can change by putting the wing cover above. The wing cover excellent sounding boards and they q as the note is made, setting the air bration, and sending the sound a lon tance. The female cricket's wing c are more normal in venation; and she always be distinguished from her sp by the long swordlike ovipositor at th of her body; this she thrusts into ground when she lays her eggs, thus ing them where they will remain s protected during the winter. Both have a pair of "tail feathers," as the dren call them, which are known as cerci (*sing. cercus*) and are fleshy pr at the end of the abdomen.

There would be no use of the cric playing his mandolin if there were no appreciative ear to listen to his m This ear is placed most convenientl the tibia of the front leg, so that the c ets literally hear with their elbows, a the katydids and the meadow grass pers. The ear is easily seen with the na eye as a little white, disclike spot.

The chirp of the cricket is, in literat usually associated with the coming of tumn; but the careful listener may it in early summer, although the son not then so insistent as later in the sea He usually commences singing in the ternoon and keeps it up periodically night. I have always been an admire the manly, dignified methods of this li "minnesinger," who does not wan abroad to seek his ladylove but sta sturdily at his own gate, playing his m dolin the best he is able; he has faith his sable sweetheart is not far away, that if she likes his song she will com him of her own free will. The cricke ever a lover of warmth and his mand gets out of tune soon after the eveni become frosty. He is a jealous music When he hears the note of a rival he once "bristles up," lifting his wings a higher angle and giving off a sharp m tant note. If the two rivals come in si of each other there is a fierce duel. T at each other with wide-open jaws, fight until one is conquered and re-ts, often minus an antenna, cercus, or n a leg. The cricket's note has a wide ge of expression. When waiting for his love, he keeps up a constant droning; e hears his rival, the tone is sharp and ant; but as the object of his affection roaches, the music changes to a seduc- whispering, even having in it an un-tain quiver, as if his feelings were too ng for utterance.

THE BLACK CRICKETS

Of the insect musicians the cricket is ily the most popular. Long associated h man, as a companion of the hearth d the field, his song touches ever the ords of human experience. Although , in America, do not have the house-cket which English poets praise, yet our ld-crickets have a liking for warm cor-rs, and will, if encouraged, take up their ode among our hearthstones. The great- tribute to the music of the cricket is e wide range of human emotion which expresses. "As merry as a cricket" is a ry old saying and is evidence that the icket's fiddling has ever chimed with the y moods of dancers and merrymakers. gain, the cricket's song is made an em-em of peace; and again we hear that the icket's "plaintive cry" is taken as the rbinger of the sere and dying year. From ppiness to utter loneliness is the gamut vered by this sympathetic song. Leigh unt found him glad and thus addresses m:

nd you, little housekeeper who class
Vith those who think the candles come too soon,
oving the fire, and with your tricksome tune
Nick the glad, silent moments as they pass.

— "WAYS OF THE SIX-FOOTED,"
COMSTOCK

LESSON 82

THE BLACK CRICKET

LEADING THOUGHT — The crickets are among the most famous of the insect musicians. They live in the fields under stones and in burrows, and feed upon grass and clover. As with most birds, only the male makes music; he has his wing covers developed into a mandolin or violin, which he plays to attract his mate and also for his own pleasure.

METHOD — Make some cricket cages as follows: Take a small flowerpot and plant in it a root of fresh grass or clover. Place over this and press well into the soil a glass chimney, or a small piece of fine mesh screen rolled into the shape of a cylinder and fastened securely with string or fine wire. Cover the top with mosquito netting. Place the pot in its saucer, so that it may be watered by keeping the saucer filled. Ask the pupils to collect some crickets. In each cage, place a male and one or more females, the latter being readily distinguished by the long ovipositors. Place the cages in a sunny window, where the pupils may observe them, and ask for the following observations. In studying the cricket closely, it may be well to put one in a vial and pass it around. In observing the crickets eat, it is well to give them a piece of sweet apple or melon rind, as they are very fond of pulpy fruits.

OBSERVATIONS — 1. Is the covering of the cricket shining, like black patent leather, or is it dull? What portions are dull? Of what use do you think it is to the cricket to be so smoothly polished?

2. Where did you find the crickets? When you tried to catch them, how did they act? Did they fly like grasshoppers or did they run and leap?

3. Look carefully at the cricket's legs. Which is the largest of the three pairs? Of what use are these strong legs? Look

carefully at the tibia of the hind leg. Can you see the strong spines at the end, just behind the foot or tarsus? Watch the cricket jump and see if you can discover the use of these spines. How many joints in the tarsus? Has the cricket a pad like the grasshopper's between its claws? When the cricket walks or jumps does it walk on all the *tarsi* of each pair of legs?

4. Study the cricket's head. Can you see the eyes? Describe the antennæ—their color, length, and the way they are used. Watch the cricket clean its antennæ and describe the process. Can you see the little feelers, or palpi, connected with the mouth? How many are there? How does it use these feelers in tasting food before it eats? Watch the cricket eat, and see whether you can tell whether its mouth is made for biting or sucking.

5. Study the wings. Are the wings of the mother cricket the same size and shape as those of her mate? How do they differ? Does the cricket have any wings under these front wings, as the grasshopper does? Note the cricket when he is playing his wing mandolin to attract his mate. How does he make the noise? Can you see the wings vibrate? Ask your teacher to show you a picture of the musical wings of the cricket, or to show you the wings themselves under the microscope, so t you may see how the music is made.

6. Why does the mother cricket n such a long ovipositor? Where does put her eggs in the fall to keep them until spring?

7. Look in the tibia, or elbow, of front leg for a little white spot. What you suppose this is? Are there any wh spots like it on the other legs? Ask y teacher to tell you what this is.

8. Can you find the homes of the cr ets in the fields? Do the black crick chirp in the daytime or after dark? they chirp in cold or windy weather, only when the sun shines?

CRICKET SONG

Welcome with thy clicking, cricket!
Clicking songs of sober mirth;
Autumn, stripping field and thicket,
Brings thee to my hearth,
Where thy clicking shrills and quicke
While the mist of twilight thickens.

. . . .

No annoy, good-humored cricket,
With thy trills is ever blent;
Spleen of mine, how dost thou trick i
To a calm content?
So, by thicket, hearth, or wicket,
Click thy little lifetime, cricket!

—Bayard Tayl

THE SNOWY TREE CRICKET

This is a slim, ghostlike cricket. It is pale green, almost white in color, and about three-fourths of an inch long. Its long, slender hind legs show that it is a good jumper. Its long antennæ, living threads, pale gray in color, join the head with amber globelike segments. The pale eyes have a darker center and the palpi are very long. The male has the wing covers shaped and veined like those of the black cricket, but they are not so broad and are whitish and very delicate. The wings beneath are wide, for these crickets can fly. The female has a long, swordlike ovipositor.

The snowy tree cricket, like its re tives, spends much time at its toilet. whips the front foot over an antenna a brings the base of the latter to the man bles with the palpi and then cleans it ca fully to the very tip. It washes its face wi the front foot, always with a downwa movement. If the hind foot becomes e tangled in anything it first tries to kick clean, and then, drawing it beneath t body, bends the head so as to reach with the mandibles and nibbles it clea The middle foot it also thrusts benea the body, bringing it forward between t front legs for cleaning. But when clea

its front feet, the snowy tree cricket
s on airs; it lifts the elbow high and
ws the foot through the mouth with
esture very like that of a young lady
h a seal ring on her little finger, hold-
the ornate member out from its com-
ions as if it were stiff with a conscious-
s of its own importance.

There are two common species of the
wy tree crickets which can hardly be
arated except by specialists or by watch-
their habits. One is called "the whis-
" and lives on low shrubs or grass;
ives a clear, soft, prolonged, unbroken
te. The other is called "the fiddler"
l lives on shrubs and in trees and vines.
note is a pianissimo performance of
katydid's song; it is delightful, rhyth-
c, and sleep-inspiring; it begins in the
e afternoon and continues all night un-
the early, cold hours of the approach-
dawn. The vivacity of the music de-
nds upon the temperature, as the notes
given much more rapidly during the
t nights.

"So far as we know, this snowy tree
cket is the only one of the insect mu-
ians that seems conscious of the fact
at he belongs to an orchestra. If you
ten on a September evening, you will
ar the first player begin; soon another

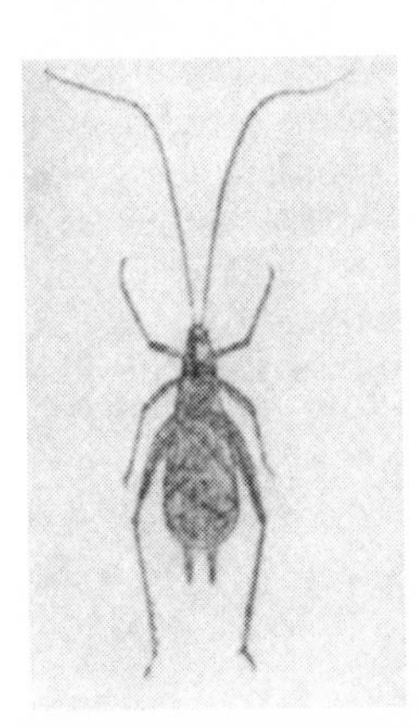

The snowy tree cricket

ill join, but not in harmony at first. For
me time there may be a seesaw of ac-
nted and unaccented notes; but after a
hile the two will be in unison; perhaps
not, however, until many more players
have joined the concert. When the rhyth-
mical beat is once established, it is in as
perfect time as if governed by the baton of

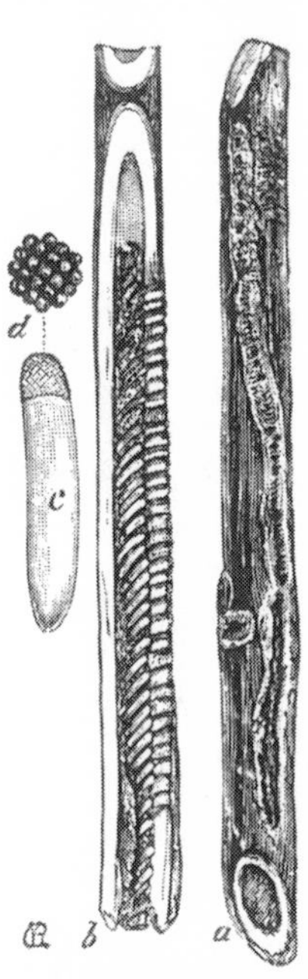

Eggs of the snowy tree cricket, laid in rasp-berry cane; c *and* d, *egg enlarged*

a Damrosch or a Thomas. The throbbing
of the cricket heart of September, it has
been fitly named. Sometimes an injudi-
cious player joins the chorus at the wrong
beat, but he soon discovers his error and
rectifies it. Sometimes, also, late at night,
one part of the orchestra in an orchard gets
out of time with the majority, and discord
may continue for some moments, as if
the players were too cold and too sleepy
to pay good attention. This delectable
concert begins usually in the late after-
noons and continues without ceasing until just before dawn the next morning.
Many times I have heard the close of the
concert; with the 'wee sma' hours the
rhythmic beat becomes slower; toward
dawn there is a falling off in the number
of players; the beat is still slower, and the
notes are hoarse, as if the fiddlers were
tired and cold; finally, when only two or
three are left the music stops abruptly."
(*Ways of the Six-footed*, Comstock.)

The lesson on this cricket may be adapted from that on the black cricket.

THE COCKROACH

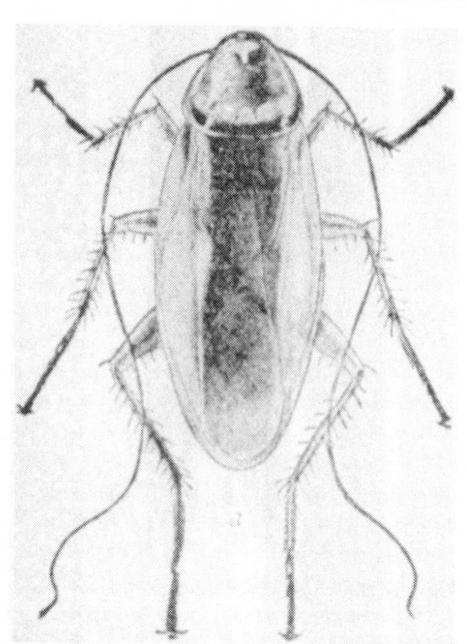

The American cockroach

Cockroaches in our kitchens are undoubtedly an unmitigated nuisance, and yet, as in many other instances, when we come to consider the individual cockroach, we find him an interesting fellow and exceedingly well adapted for living in our kitchens despite us.

In shape, the cockroach is flat, and is thus well adapted to slide beneath utensils and into crevices and corners. Its covering is smooth and polished like patent leather, and this makes it slippery and enables it to get into food without becoming clogged by the adherence of any sticky substance. The antennæ are very long and flexible and can be bent in any direction. They may be placed far forward to touch things which the insect is approaching, or may be placed over the back in order to be out of the way. They are like graceful, living threads, and the cockroach tests its whole environment with their aid. T
mouth has two pairs of palpi or feel
one of which is very long and noticeab
these are kept in constant motion as
to test the appetizing qualities of fo
The mouth-parts are provided with ja
for biting and, like all insect jaws, th
work sidewise instead of up and dov
The eyes are black but not prominent
large, and seem to be merely a part of t
sleek, polished head-covering.

Some species of cockroaches have win
and some do not. Those which have wi
have the upper pair thickened and us
for wing covers. The under pair are th
ner and are laid in plaits like a fan. T
wing covers are as polished as the bo
and quite as successful in shedding dir

The legs are armed with long spi
which are very noticeable and might pro
to be a disadvantage in accumulating fil
but they are polished also; and too, t
insect spends much time at its toilet.

Cockroaches run "like a streak," ch
dren say; so speedily, indeed, do they
that they escape our notice, although
may be looking directly at them. Th
celerity in vanishing, saves many a coc
roach from being crushed by an avengi
foot.

When making its toilet, the cockroa
draws its long antenna through its ja

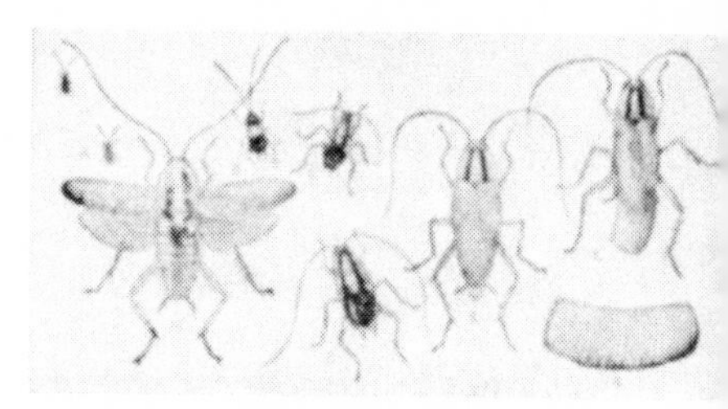

The Croton bug

a, b, c, d, successive stages of development; *e,* adu
f, adult female laying her case of eggs; *g,* the egg ca
h, adult with the wings spread

it were a whiplash, beginning at the and finishing at the tip. It cleans leg by beginning near the body and troking downward the long spines h seem to shut against the leg. It les its feet clean to the very claws, scrubs its head vigorously with the t femur.

he cockroach's eggs are laid in a mass osed in a pod-shaped covering, which aterproof and polished and protects ontents from dampness.

ke the grasshopper, the cockroach has complete metamorphosis; that is, the g insect when hatched from the egg mbles the adults in shape and general arance, but is of course quite small.

Vhen the cockroaches, or the Croton s, as the small introduced species of roach is called, once become estab- d in a house, one way to get rid of n is to fumigate the kitchen; this is a gerous performance and should be e only by an expert. In storerooms and lling houses sprinkling the runways rally with sodium fluoride has proved successful; this can be done by any- although great care should be exer- d, for sodium fluoride is somewhat onous to man.

LESSON 83

The Cockroach

EADING THOUGHT — The cockroach is pted for living in crevices, and although its haunts may be anything but clean, the cockroach keeps itself quite clean. The American species live in fields and woods and under stones and sticks and only occasionally venture into dwellings. The species that infest our kitchens and water-pipes are European.

METHOD — Place a cockroach in a vial with bread, potato, or some other food, cork the vial, and pass it around so that the children may observe the prisoner at their leisure.

OBSERVATIONS — 1. What is the general shape of the cockroach? Why is this an advantage? What is the texture of its covering? Why is this an advantage?

2. Describe the antennæ and the way they are used. Note the two little pairs of feelers at the mouth. If possible, see how they are used when the cockroach is inspecting something to eat. Can you see whether its mouth is fitted for biting, lapping, or sucking its food?

3. Note the eyes. Are they as large and prominent as those of the bees or butterflies?

4. Has this cockroach wings? If so, how many and what are they like? Note two little organs at the end of the body. These are the cerci, like those of the crickets.

5. Describe the general appearance of the cockroach's legs, and tell what you think about its ability as a runner.

6. Note how the cockroach cleans itself and how completely and carefully this act is performed. Have you ever seen a cockroach's eggs? If so, describe them.

7. How can you get rid of cockroaches if they invade your kitchen?

THE APHIDS OR PLANT LICE

know of no more diverting occupation n watching a colony of aphids through ns. These insects are the most help- and amiable little ninnies in the whole ct world; and they look the part, prob- y because their eyes, so large and wide rt, seem so innocent and wondering. e usual color of aphids is green; but re are many species which are otherwise colored, and some have most bizarre and striking ornamentations. In looking along an infested leafstalk, we see them in all stages and positions. One may have thrust its beak to the hilt in a plant stem, and be so satisfied and absorbed in sucking the juice that its hind feet are lifted high in the air and its antennæ curved backward, making all together a gesture which

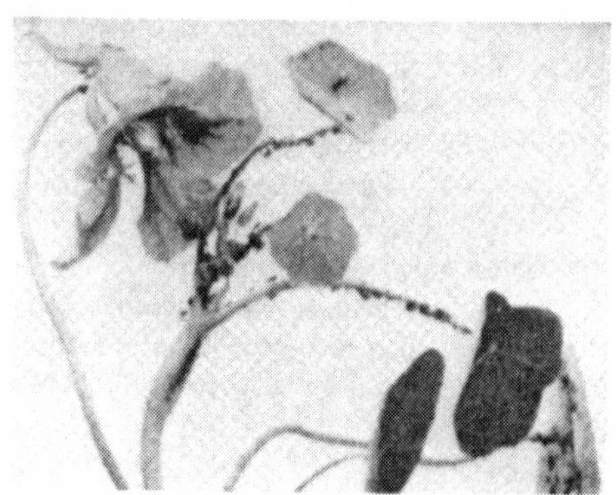

Grace H. Griswold

Aphids on a nasturtium

seems an adequate expression of bliss; another may conclude to seek a new well, and pull up its sucking tube, folding it back underneath the body so it will be out of the way, and walk off slowly on its six rather stiff legs; when thus moving, it thrusts the antennæ forward, patting its pathway to insure safety. Perhaps this pathway may lead over other aphids which are feeding, but this does not deter the traveler or turn it aside; over the backs of the obstructionists it crawls, at which the disturbed ones kick the intruder with both hind legs; it is not a vicious kick but a push rather, which says, "This seat reserved, please!" It is comical to see a row of them sucking a plant stem "for dear life," the heads all in the same direction, and they packed in and around each other as if there were no other plants in the world to give them room, the little ones wedged in between the big ones, until sometimes some of them are obliged to rest their hind legs on the antennæ of the neighbors next behind.

Aphids seem to be born to serve as food for other creatures — they are simply little machines for making sap into honeydew, which they produce from the alimentary canal for the delectation of ants; they are, in fact, merely little animated drops of sap on legs. How helpless they are when attacked by any one of their many enemies! All they do, when they are seized, is to claw the air with their six impotent legs and two antennæ, keeping up this performance as long as there is a leg left, and apparently to the very never realizing "what is doing." But are not without means of defense; t two little tubes at the end of the body not for ornament or for producing ho dew for the ants, but for secreting at tips a globule of waxy substance w smears the eyes of the attacking in I once saw an aphid perform this when confronted by a baby spider; a of yellow liquid oozed out of one t and the aphid almost stood on its in order to thrust this offensive glo directly into the face of the spider — whole performance reminding me boy who shakes his clenched fist in opponent's face and says, "Smell of th The spider beat a hasty retreat.

A German scientist, Mr. Busgen, covered that a plant louse smeared eyes and jaws of its enemy the aphis with this wax, which dried as soon as plied. In action it was something throwing a basin of paste at the hea the attacking party; the aphis lion treated was obliged to stop and clean i before it could go on with its hunt, the aphid walked off in safety. The ap surely need this protection because have two fierce enemies, the larvæ of aphis lions and the larvæ of the ladyb They are also the victims of parasitic sects; a tiny four-winged "fly" lays an within an aphid; the larva hatching f it feeds upon the inner portions of aphid, causing it to swell as if afflic with dropsy. Later the aphid dies, and interloper with malicious impertine cuts a neat circular door in the aphid's skeleton skin and issues from full-fledged insect.

Bureau of Ent., U. S.

An aphid parasite laying its eggs withi aphid, enlarged

he aphids are not without their re-
ces to meet the exigencies of their
in colonies. There are several dis-
t forms in each species, and they seem
pe needed for the general good. Dur-
the summer, we find most of the
ids on plants are without wings; these
females which give birth to living
ng and do not lay eggs. They do this
il the plant is overstocked and the food
ply seems to be giving out; then an-
er form which has four wings is pro-
ed. These fly away to some other
t and start a colony there; but at the
roach of cold weather, or if the food
nts give out, male and female individu-
are developed, the females being al-
s wingless, and it is their office to lay
eggs which shall last during the long
ter months, when the living aphids
st die for lack of food plants. The next
ing each winter egg hatches into a fe-
le which we call the "stem mother"
ce she with her descendants will popu-
the entire plant.

Plant lice vary in their habits. Some live
the ground on the roots of plants and
very destructive; but the greater num-
of species live on the foliage of plants
d are very fond of the young, tender
ves and thus do great damage. Some
hids have their bodies covered with
ite powder or with tiny fringes, which
e them the appearance of being cov-
d with cotton; these are called "woolly
hids."

The aphids injuring our flowers and
nts may, in general, be killed by spray-
them with nicotine sulfate in the pro-
rtion of one teaspoonful to one gallon
water in which three or four ounces
soap have been dissolved. The spraying
ust be done very thoroughly so as to
ch all the aphids hidden on the stems
d beneath the leaves. A second applica-
n may be necessary in three or four
ys.

LESSON 84

The Aphids or Plant Lice

Leading Thought — Aphids have the mouth in the form of a sucking-tube which is thrust into the stems and leaves of plants; through it the plant juices are drawn for nourishment. Aphids are the source of honeydew of which ants are fond.

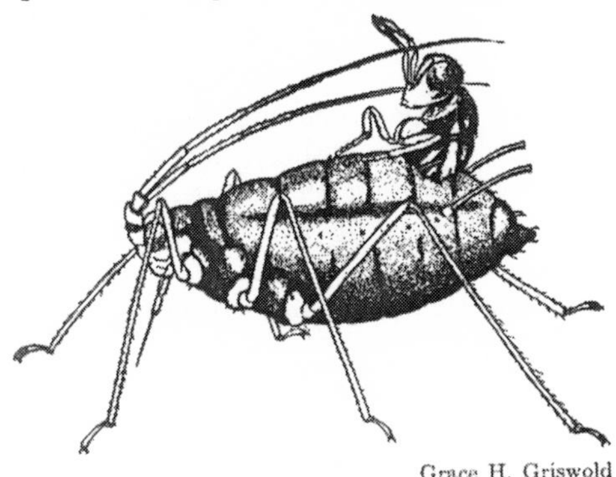
Grace H. Griswold

A parasite emerging from a geranium aphid, enlarged

Method — Bring into the schoolroom a plant infested with aphids, place the stem in water, and let the pupils examine the insects through the lens.

Observations — 1. How are the aphids settled on the leaf? Are their heads in the same direction? What are they doing?

2. Touch one and make it move along. What does it do in order to leave its place? What does it do with its sucking-tube as it walks off? On what part of the plant was it feeding? Why does not Paris green when applied to the leaves of plants kill aphids?

3. Describe an aphid, including its eyes, antennæ, legs, and tubes upon the back. Does its color protect it from observation?

4. Can you see cast skins of aphids on the plant? Why does an aphid have to shed its skin?

5. Are all the aphids on a plant wingless? When a plant becomes dry, are there, after several days, more winged aphids? Why do the aphids need wings?

6. Do you know what honeydew is? Have you ever seen it upon the leaf? How is honeydew made by the aphids? Does it

come from the tubes on their back? What insects feed upon this honeydew?

7. What insect enemies have the aphids?

8. What damage do aphids do to plants? How can you clean plants of plant lice?

I saw it [an ant], at first, pass, without stopping, some aphids which it did not, however, disturb. It shortly after stationed itself near one of the smallest, and appeared to caress it, by touching the extremity of its body, alternately with its antennæ, with an extremely rapid movement. I saw, with much surprise, the f proceed from the body of the aphid, the ant take it in its mouth. Its anten were afterwards directed to a much la aphid than the first, which, on being ressed after the same manner, dischar the nourishing fluid in greater quant which the ant immediately swallowed then passed to a third which it caress like the preceding, by giving it several g tle blows, with the antennæ, on the p terior extremity of the body; and the liq was ejected at the same moment, and ant lapped it up.

— Pierre Huber, 1

THE ANT LION

A child is thrilled with fairy stories of ogres in their dens, with the bones of their victims strewn around. The ants have real

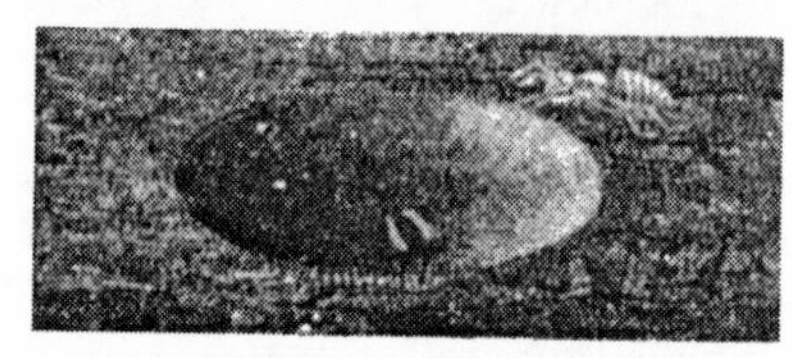

Pitfall of an ant lion

ogres, but luckily they do not know about it and so cannot suffer from agonizing fears. The ant ogres seem to have depended upon the fact that the ant is so absorbed in her work that she carries her booty up hill and down dale with small regard for the topography of the country. By instinct they build pits which will someday be entered by ants obsessed by industry and careless of what lies in the path. The pits vary with the size of the ogre at the bottom; there are as many sized pits as there are beds in the story of Golden Locks and the bears; often the pits are not more than an inch across, or even less, while others are two inches in diameter. They are always made in sandy or crumbly soil and in a place protected from wind and rain; they vary in depth in proportion to their width, for the slope is always as steep as the soil will stand without slipping.

All that can be seen of the ogre at t bottom is a pair of long, curved jaws, lo ing innocent enough at the very center the pit. If we dig the creature out, we fi it a comical looking insect. It is hun backed, with a big, spindle-shaped ab men; from its great awkward body p jects a flat, sneaking looking head, arm in front with the sickle jaws, which spiny and bristly near the base, a smooth, sharp, and curved at the tip. T strange thing about these jaws is that th lead directly to the throat, since the a lion has no mouth. Each jaw is made of two pieces which are grooved whe they join and thus form a tube with a ho in the tip through which the industrio blood of the ants can be sucked; not on do the sharp sickle points hold the victi but there are three teeth along the side each jaw to help with this. The two fro pairs of legs are small and spiny; the hi

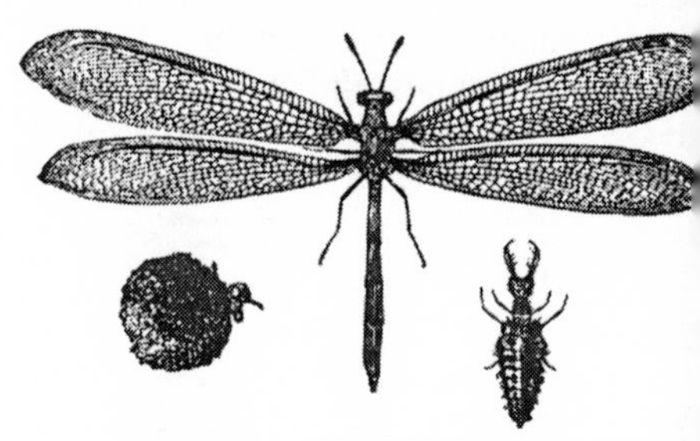

Ant lion with its cocoon and larva

are strong and peculiarly twisted, and e a sharp spikelike claw at the end, ch is so arranged as to push the insect kward vigorously if occasion requires; fact, the ant lion in walking about ves more naturally backward than ford because of the peculiar structure of legs.

Having studied the ogre, we can see bethow he manages to trap his victim. As ant goes scurrying along, she rushes r the edge of the pit and at once bes to slide downward; she is frightened l struggles to get back; just then a jet sand, aimed well from the bottom of pit, hits her and knocks her back. She l struggles, and there follows a fusillade sand jets, each hitting her from above l knocking her down to the fatal center ere the sickle jaws await her and are omptly thrust into her; if she is large l still struggles, the big, unwieldy body the ogre, buried in the sand, anchors n fast and his peculiar, crooked hind s push his body backward in this ange tug of war; thus, the ant ogre is t dragged out of his den by the struggles the ant, and soon the loss of blood akens her and she shrivels up.

The secret of the jets of sand lies in the t head of the ogre; if we look at it rerding it as a shovel, we can see that it is ll fitted for its purpose; for it is a shovel th a strong mechanism working it. In t, the whole pit is dug with this shovel ad. Wonderful stories are told about e way that ant lions dig their pits, markg out the outer margin in a circle, and rking inward. However, our common t lion of the East simply digs down into e sand and flips the sand out until it akes a pit. If an ant lion can be caught d put in a jar of sand it will soon make pit, and the process may be noted carelly.

There is one quality in the ogre which erits praise, and that is his patience. here he lies in his hole for days or perps weeks, with nothing to eat and no t coming that way; so when we see an sent-minded ant scrambling over into e pit, let us think of the empty stomach of this patient little engineer who has constructed his pit with such accuracy and so much labor. So precarious is the living picked up by the ant lions, that it may require one, two, or three years to bring one to maturity. At that time it makes a perfectly globular cocoon of silk and sand, the size of a large pea, and within it changes to a pupa; and when finally ready to emerge, the pupa pushes itself part way out of the cocoon, and the skin is shed and left at the cocoon door. The adult resembles a small dragonfly; it has large netveined wings and is a most graceful insect, as different as can be from the humpbacked ogre which it once was — a transformation quite as marvelous as that which occurred in Beauty and the Beast. Throughout the Middle West, the ant lion in its pit is called the "doodlebug."

LESSON 85

The Ant Lion

Leading Thought — The ant lion or "doodlebug" makes a little pit in the sand with very steep sides, and hidden at the bottom of it, waits for ants to tumble in to be seized by its waiting jaws. Later the ant lion changes to a beautiful insect with gauzy wings, resembling a small dragonfly.

Method — The pupils should see the ant lion pits in their natural situations, but the insects may be studied in the schoolroom. Some of the ant lions may be dug out of their pits and placed in a dish of sand. They will soon make their pits, and may be watched during this interesting process. It is hardly advisable to try to rear these insects, as they may require two or three years for development.

Observations — 1. Where were the ant lion pits out of doors? Were they in a windy place? Were they in a place protected from storms? In what kind of soil were they made?

2. Measure one of the pits. How broad

across was it and how deep? Are all the pits of the same size? Why not?

3. What can you see as you look down into the ant lion's pit? Roll a tiny pebble in and see what happens. Watch until an ant comes hurrying along and slips into the pit. What happens then? As she struggles to get out how is she knocked back in? What happens to her if she falls to the bottom?

4. Take a trowel and dig out the doodlebug. What is the shape of its body? What part of the insect did you see at the bottom of the pit? Do you know that these great sickle-shaped jaws are hollow tubes for sucking blood? Does the ant lion eat anything except the blood of its victim?

5. Can you see that the ant lion moves backward more easily than forward? H are its hind legs formed to help pus backward? How does this help the ant l in holding its prey? How does the big a ward body of the ant lion help to hol in place at the bottom of the pit whe seizes an ant in its jaws?

6. What shape is the ant lion's he How does it use this head in taking prey? In digging its pit?

7. Take a doodlebug to the sch room, place it in a dish of sand cove with glass, and watch it build its pit the sand.

8. Read in the entomological bo about the cocoon of the ant lion and w the adult looks like, and then write ant lion autobiography.

THE MOTHER LACEWING AND THE APHIS LION

Flitting leisurely through the air on her green gauze wings, the lacewing seems like a filmy leaf, broken loose and drifting on the breeze. But there is purpose in her

Aphis lion. Larva, eggs, cocoon, and the adult lacewing

flight, and through some instinct she is enabled to seek out an aphis-ridden plant or tree, to which she comes as a friend in need. As she alights upon a leaf, she is scarcely discernible because of the pale green of her delicate body and wings; however, her great globular eyes that shine like gold attract the attention of the careful observer. But though she is fairy-like in appearance, if you pick l up, you will be sorry if your sense of sm is keen, for she exhales a most disagr able odor when disturbed — a habit whi probably protects her from birds or oth creatures which might otherwise eat h

However, if we watch her we shall s that she is a canny creature despite h frivolous appearance; her actions a surely peculiar. A drop of sticky fluid sues from the tip of her body, and s presses it down on the surface of the le then lifting up her slender abdomen li a distaff, she spins the drop into a thre a half inch long or more, which t air soon dries; and this silken thread stiff enough to sustain an oblong egg, large as the point of a pin, which she la at the very tip of it. This done she la another egg in a like manner, and wh she is through, the leaf looks as if it we covered with spore cases of a glitteri white mold. This done she flies off an disports herself in the sunshine, care fre knowing that she has done all she can f her family.

After a few days the eggs begin to loo dark, and then if we examine them wit a lens, we may detect that they conta

doubled-up creatures. The first we
of the egg inmate as it hatches is a
of jaws thrust through the shell, open-
it for a peephole; a little later the
er of the jaws, after resting a while
an eye on the world which he is so
to enter, pushes out his head and
and drags out a tiny, long body, very
w looking and clothed in long, soft
. At first the little creature crawls
it his eggshell, clinging tightly with
is six claws, as if fearful of such a dizzy
ht above his green floor; then he
rms around a little and thrusts out a
l inquiringly while still hanging on
r dear life." Finally he gains courage
prospects around until he discovers
gg stalk, and then begins a rope climb-
performance, rather difficult for a little
not more than ten minutes old. He
s a careful hold with his front claws,
two other pairs of legs carefully bal-
ing for a second, and then desperately
ing the stalk with all his clasping
vs, and with many new grips and
ics, he finally achieves the bottom in
ty. As if dazed by his good luck, he
ds still for a time, trying to make up
mind what has happened and what
do next; he settles the matter by trot-
g off to make his first breakfast of
ids; and now we can see that it is a
ky thing for his brothers and sisters,
unhatched, that they are high above
head and out of reach, for he might
be discriminating in the matter of
breakfast food, never having met any
his family before. He is a queer looking
le insect, spindle-shaped and with pe-
iarly long, sickle-shaped jaws project-
from his head. Each of these jaws is
de up of two pieces joined lengthwise
as to make a hollow tube, which has
opening at the tip of the jaw, and anner one at the base which leads directly
the little lion's throat. Watch him as
catches an aphid; seizing the stupid
tle bag of sap in his great pincers, he
ts it high in the air, as if drinking a
mper, and sucks its green blood until it
rivels up, kicking a remonstrating leg
the last. It is my conviction that aphids never realize when they are being eaten; they simply dimly wonder what is happening.

It takes a great many aphids to keep an aphis lion nourished until he gets his growth; he grows like any other insect by shedding his skeleton skin when it becomes too tight. Finally he doubles up and spins around himself a cocoon of glistening white silk, leaving it fastened to the leaf; when it is finished, it looks like a seed pearl, round and polished. I wish some child would watch an aphis lion weave its cocoon and tell us how it is done! After a time, a week or two perhaps, a round little hole is cut in the cocoon, and there issues from it a lively little green pupa, with wing pads on its back; but it very soon sheds its pupa skin and issues as a beautiful lacewing fly with golden eyes and large, filmy, iridescent, pale green wings.

LESSON 86

The Mother Lacewing and the Aphis Lion

Leading Thought — The lacewing fly or goldeneyes, as she is called, is the mother of the aphis lion. She lays her eggs on the top of stiff, silken stalks. The young aphis lions when hatched, clamber down upon the leaf and feed upon plant lice, sucking their blood through their tubular jaws.

Method — Through July and until frost, the aphis lions may be found on almost any plant infested with plant lice; and the lacewing's eggs or eggshells on the long stalks are also readily found. All these may be brought to the schoolroom. Place the stem of a plant infested with aphids in a jar of water, and the acts of the aphis lions as well as the habits of the aphids may be observed at convenient times by all the pupils.

Observations — 1. When you see a leaf with some white mold upon it, examine it with a lens; the mold is likely to be the eggs of the lacewing. Is the egg as large as a pinhead? What is its shape? What is its color? How long is the stalk on which it is placed? Of what material do you think the stalk is made? Why do you suppose the lacewing mother lays her eggs on the tips of stalks? Are there any of these eggs near each other on the leaf?

2. If the egg is not empty, observe through a lens how the young aphis lion breaks its eggshell and climbs down.

3. Watch an aphis lion among the plant lice. How does it act? Do the aphids seem afraid? Does the aphis lion move rapidly? How does it act when eating an aphid?

4. What is the general shape of the aphis lion? Describe the jaws. Do you think these jaws are used for chewin merely as tubes through which the g blood of the aphids is sucked? Do aphis lions ever attack each other or c insects? How does the aphis lion c in appearance from the ladybird larva

5. What happens to the aphis lion it gets its growth? Describe its cocoo you can find one.

6. Describe the little lacewing fly comes from the cocoon. Why is she ca goldeneyes? Why lacewing? Does sh rapidly? Do you suppose that if she sh lay her eggs flat on a leaf, the first a lion that hatched would run about and all its little brothers and sisters which still in their eggshells? How do the a lions benefit our rose bushes and o cultivated plants?

THE HOUSEFLY

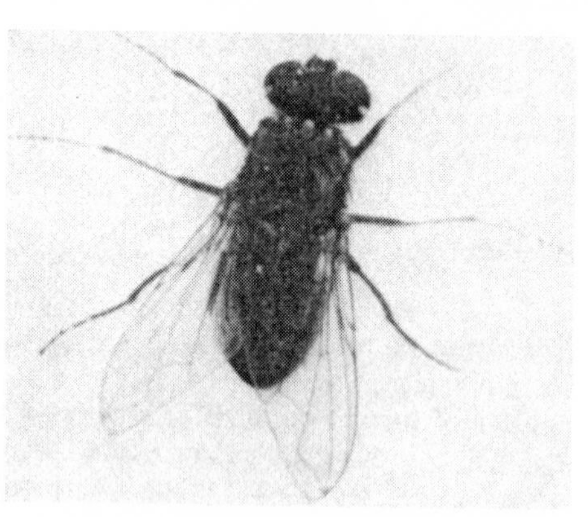

Dept. of Entomology, Cornell U.

The housefly

The housefly is one of the most cosmopolitan members of the animal kingdom. It flourishes in every land, plumping itself down in front of us at table, whether we be eating rice in Hong Kong, dhura in Egypt, macaroni in Italy, pie in America, or tamales in Mexico. There it sits, impertinent and imperturbable, taking its toll, letting down its long elephant-trunk tongue, rasping and sucking up such of our meal as fits its needs. As long as we simply knew it as a thief we, during untold ages, merely slapped it and shooed it, which effort on our part apparently g it exhilarating exercise. But during rec years we have begun trapping and poi ing, trying to match our brains agains agility; although we slay it by thousa we seem only to make more room fo well-fed progeny of the future, and in end we seem to have gained nothing. the most recent discoveries of science h revealed to us that what the housefly ta of our food is of little consequence c pared with what it leaves behind. Beca of this we have girded up our loins gone into battle in earnest.

I have always held that nature-st should follow its own peaceful path not be the slave of economic science. occasionally it seems necessary, when a question of creating public sentim and of cultivating public intelligence combating a great peril, to make nat study a handmaiden, if not a slave, in work. If our woods were filled with wol and bears, as they were in the days of grandfather, I should give nature-st lessons on these animals which would l to their subjugation. Bears and wol trouble us no more; but now we h enemies far more subtle, in the ever p

microbes, which we may never hope to
quer but which, with proper precau-
s, we may render comparatively harm-
Thus, our nature-study with insects
ch carry disease, like the mosquitoes,
, and fleas, must be a reconnaissance
a war of extermination; the fighting
ics may be given in lessons on health
hygiene.

erhaps if a fly were less wonderfully
le, it would be a less convenient vehi-
for microbes. Its eyes are two great,
wn spheres on either side of the head,
are composed of thousands of tiny
sided eyes that give information of
at is coming in any direction; in addi-
, it has on top of the head, looking
ight up, three tiny, shining, simple
s, which cannot be seen without a lens.
antennæ are peculiar in shape, and are
se organs; it is attracted from afar by
tain odors, and so far as we can dis-
er, its antennæ are all the nose it has.
mouth-parts are all combined to make
ost amazing and efficient organ for get-
g food; at the tip are two flaps, which
rasp a substance so as to set free the
ces, and above this is a tube, through
ich the juices may be drawn to the
mach. This tube is extensible, being
veniently jointed so that it can be
ded under the "chin" when not in use.
is is usually called the fly's tongue, but
s really all the mouth-parts combined,
if a boy had his lips, teeth, and tongue,
nding out from his face, at the end of
ube a foot long.

The thorax can be easily studied; it
striped black and white above and
ars the two wings, and the two little
ps called balancers, which are probably
nnants of hind wings with which the
mote ancestors of flies flew. The fly's
ng is a transparent but strong mem-
ane strengthened by veins, and is pret-
y iridescent. The thorax bears on its
wer side the three pairs of legs. The ab-
men consists of five segments and is
vered with stiff hairs. The parts of the
g seen when the fly is walking consist of
ree segments, the last segment or tarsus
ing more slender; if looked at with a
lens the tarsus is seen to be composed of five segments, the last of which bears the claws; it is with these claws that the fly walks, although all of the five segments really form the foot; in other words, it walks on its tiptoes. But it clings to ceilings by means of the two little pads below the claws, which are covered with hairs that excrete at the tips a sticky fluid. Chiefly because of the hairs on its feet, the fly becomes a carrier of microbes and a menace to health.

The greatest grudge I have against this little, persistent companion of our household is the way it has misled us by appearing to be so fastidious in its personal habits. We have all of us seen, with curiosity and admiration, its complex ablutions and brushings. It usually begins, logically, with its front feet, the hands; these it cleans by rubbing them against each other lengthwise. The hairs and spines on

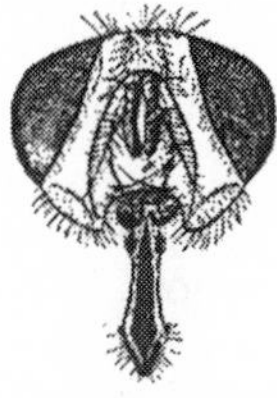

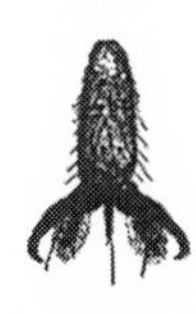

At the left is the head of a housefly showing eyes, antennæ, and mouth parts. At the right is a much enlarged foot of the fly

one leg act as a brush for the other, and then, lest they be not clean, it nibbles them with its rasping disc, which is all the teeth it has. It then cleans its head with these clean hands, rubbing them over its big eyes with a vigor that makes us wink simply to contemplate; then bobbing its head down so as to reach what is literally its back hair, it brushes valiantly. After this is done, it reaches forward first one and then the other foot of the middle pair of legs, and taking each in turn between the front feet, brushes it vigorously, and maybe nibbles it. But as a pair of military brushes, its hind feet are conspicuously efficient; they clean each other by being rubbed together and then they work simultaneously on each side in cleaning the wings, first the under side and then the upper side. Then over they come and comb the top of the thorax; then they brush the sides, top, and under sides of the abdomen, cleaning each other between the acts. Who, after witnessing all this, could believe that the fly could leave any tracks on our food which would lead to our undoing! But the housefly, like many housekeepers with the best intentions in the matter of keeping clean, has not mastered the art of getting rid of the microbes. Although it has so many little eyes, none of them can magnify a germ so as to make it visible; and thus it is that,

Cornell Extension Bulletin

The larva, or maggot, and the pupa of a housefly, much enlarged

when feeding around where there have been cases of typhoid and other diseases, the housefly's little claws become infested with disease germs; and when it stops some day to clean up on our table, it leaves the germs with us. In recent years the fly has been conspicuous in spreading amœbic dysentery. Our only safety lies in the final extermination of this little nuisance.

It is astonishing how few people know about the growth of flies. People of the highest intelligence in other matters, think that a small fly can grow into a large one. A fly when it comes from the pupa stage is as large as it will ever be, the young stages of flies being maggots. The housefly's eggs are little, white, elongated bodies about as large as the point of a pin. These are laid preferably in horse manure. After a few hours, they hatch into slender, pointed, white maggots which feed upon the excrement. After five or six days, the larval skin thickens and turns brown, making the insect look like a small grain of wheat. This is the pupal stage, which lasts about five days, and then the skin bursts open and the full-grown fly appears. Of course, not all the flies multiply according to the example given to the children in the following lesson. The housefly has many enemies and, therefore, probably one hibernating mother fly is the an tress of billions by September; howe despite enemies, flies multiply with g rapidity.

I know of no more convincing exp ment as an example of the dangerous t of the fly, than that of letting a hous walk over a saucer of nutrient gela After three or four days, each track plainly visible as a little white growth bacteria.

Much is being done now to eradic the housefly, and undoubtedly there be new methods of fighting it devi every year. The teacher should keep touch with the bulletins on this subj published by the United States Dep ment of Agriculture, and should give pupils instructions according to the lat ideas. At present the following are methods of fighting this pest: Keep pre ises clean and place food and waste ma rials under cover. All of the windows the house should be well screened. All flies which get into the house should killed by using commercial flypap sprays, or swatters.

LESSON 87

The Housefly

Leading Thought — The housefly h conquered the world and is found pra tically everywhere. It breeds in filth a especially in horse manure. It is very p lific; the few flies that manage to pass t winter in this northern climate are t ancestors of the millions which attack and our food later in the season. These a a menace to health because they ca germs of disease from sputa and exc ment to our tables, leaving them upon o food.

Method — Give out the questions

ervation and let the pupils answer m either orally or in their notebooks. If sible, every pupil should look at a sefly through a lens or microscope. If is not possible, pictures should be wn to demonstrate its appearance.

OBSERVATIONS — 1. Look at a fly, using ns if you have one. Describe its eyes. you see that they have a honeycomb ngement of little eyes? Can you see, top of the head between the big eyes, ot? A microscope reveals this dot to be de of three tiny eyes, huddled together. er seeing a fly's eyes, do you wonder t you have so much difficulty in hitting or catching it?

2. Can you see the fly's antennæ? Do u think that it has a keen sense of smell? hy?

3. How many wings has the fly? How es it differ from the bee in this respect? n you see two little white objects, one t behind the base of each wing? These called poisers, or balancers, and all s have them in some form. What is e color of the wings? Are they trans- rent? Can you see the veins in them? n what part of the body do the wings ow?

4. Look at the fly from below. How any legs has it? From what part of the dy do the legs come? What is that part the insect's body called to which the gs and wings are attached?

5. How does the fly's abdomen look? hat is its color and its covering?

6. Look at the fly's legs. How many gments can you see in a leg? Can you e that the segment on which the fly alks has several joints? Does it walk on of these segments or on the one at the p?

7. When the fly eats, can you see its ngue? Can you feel its tongue when it sps your hand? Where does it keep its ngue usually?

8. Describe how a fly makes its toilet follows: How does it clean its front feet? Its head? Its middle feet? Its hind feet? Its wings?

9. Do you know how flies carry disease? Did you ever see them making their toilet on your food at the table? Do you know what diseases are carried by flies? What must you do to prevent flies from bringing disease to your family?

10. Do you think that a small fly ever grows to be a large fly? How do the young of all kinds of flies look? Do you know where the housefly lays its eggs? On what do the maggots feed? How long before they change to pupæ? How long does it take them to grow from eggs to flies? How do the houseflies in our northern climate pass the winter?

11. *Lesson in Arithmetic* — It requires perhaps twenty days to span the time from the eggs of one generation of the housefly to the eggs of the next, and thus there might easily be five generations in one summer. Supposing the fly which wintered behind the window curtain in your home last winter, flew out to the stables about May 1 and laid 120 eggs in the sweepings from the horse stable, all of which hatched and matured. Supposing one-half of these were mother flies and each of them, in turn, laid 120 eggs, and so on for five generations, all eggs laid developing into flies, and one-half of the flies of each generation being mother flies. How many flies would the fly that wintered behind your curtain have produced by September?

12. Pour some gelatin, unsweetened, on a clean plate. Let a housefly walk around on the gelatin as soon as it is cool; cover the plate to keep out the dust and leave it for two or three days. Examine it then and see if you can tell where the fly walked. What did it leave in its tracks?

13. Write an essay on the housefly, its dangers and how to combat it, basing the essay on bulletins of the United States Department of Agriculture.

THE COLORADO POTATO BEETLE

The potato beetle is not a very attractive insect, but it has many interesting peculiarities. No other common insect so clearly illustrates the advantage of warning colors. If we take a beetle in the hand, it

M. V. Slingerland

Adult Colorado potato beetles

at first promptly falls upon its back, folds its legs and antennæ down close to its body, and " plays possum " in a very canny manner. But if we squeeze it a little, immediately an orange-red liquid is ejected on the hand, and a very ill-smelling liquid it is. If we press lightly, only a little of the secretion is thrown off; but if we squeeze harder it flows copiously. Thus a bird trying to swallow one of these beetles would surely get a large dose. The liquid is very distasteful to birds, and it is indeed a stupid bird that does not soon learn to let severely alone orange and yellow beetles striped with black. The source of this offensive and defensive juice is at first a mystery, but if we observe closely we can see it issuing along the hind edge of the thorax and the front portion of the wing covers; the glands in these situations secrete the protective juice as it is needed. The larvæ are also equipped with similar glands and, therefore, have the brazen habit of eating the leaves of our precious potatoes without attempting to hide.

The life history of the potato beetle is briefly as follows: Some of the adult b
tles or pupæ winter beneath the surf
of the soil, burrowing down a foot or m
to escape freezing. As soon as the pot
plants appear above ground the mot
beetle comes out and lays her eggs up
the undersides of the leaves. Th
orange-yellow eggs are usually laid in cl
ters. In about a week there hatch fr
the eggs little yellow or orange hun
backed larvæ, which begin at once to fe
upon the leaves. These larvæ grow, as
other insects, by shedding their ski
They do this four times, and during t
last stages are very conspicuous insects
the green leaves; they are orange or y
low with black dots along the sides, a
so humpbacked are they that they seem
be " gathered with a puckering strin
along the lower side. It requires from s
teen days to three weeks for a larva to co
plete its growth. It then descends in
the earth and forms a little cell in whi
it changes to a pupa. It remains in th
condition for one or two weeks, accordi
to the temperature, and then the fu
fledged beetle appears. The entire l
cycle from egg to adult beetle may
passed in about a month, although if t
weather is cold, this period will be long
The beetles are very prolific, a moth
beetle having been known to produce fi
hundred eggs, and there are two gene

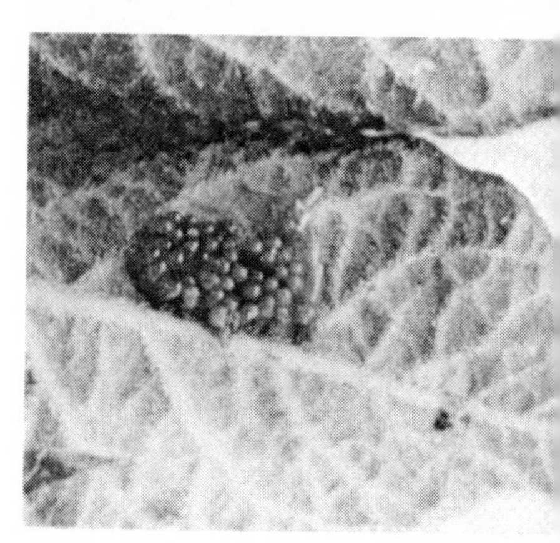

M. V. Slingerla

Eggs of the Colorado potato beetle

s each year. These beetles damage the
to crop by stopping the growth
ugh destroying the leaves, thus caus-
the potatoes to be of inferior quality.
he adult beetle is an excellent object
on in the study of beetle form. Atten-
should be called to the three regions
he body: a head, which is bright
ge; the compound eyes, which are
k; and three simple eyes on the top
he head, which are difficult to see
out a lens. The antennæ are short,
r joints easily noted, and special atten-
should be paid to their use, for they
constantly moving to feel approaching
ects. The two pairs of mouth palpi may
seen, and the beetle will eagerly eat
potatoes, so that the pupils may see
t it has biting mouth-parts. The tho-
c shield is orange, ornamented with
k. The three pairs of legs are short,
ch is a proof that these beetles do not
rate on foot. The claws and the pads
eath can be seen with the naked eye.
h wing cover bears five yellow stripes,
five black ones, although the outside
ck stripe is rather narrow. These beetles
very successful flyers. During flight,
wing covers are raised and held mo-
nless while the gauzy wings beneath
unfolded and do the work. Children
always interested in seeing the way
beetles fold their wings beneath the
ng covers.

One of the most remarkable things
out the Colorado potato beetle is its his-
y. It is one of the few insect pests which
native to America. It formerly fed upon
dbur, a wild plant allied to the potato,
ich grows in the region of Colorado,
izona, and Mexico, and was a well-be-
ved, harmless insect. With the advance
civilization westward, the potato came
o, and proved to be an acceptable plant
this insect; and here we have an exam-
e of what an unlimited food supply will
for an insect species. The beetles mul-
lied so much faster than their parasites,
at it seemed at one time as if they would
nquer the earth by moving on from po-
to field to potato field. They started on
eir march to the Atlantic seaboard in
59; in 1874 they reached the coast, and
judging by the numbers washed ashore,
they sought to fly or swim across the
Atlantic.

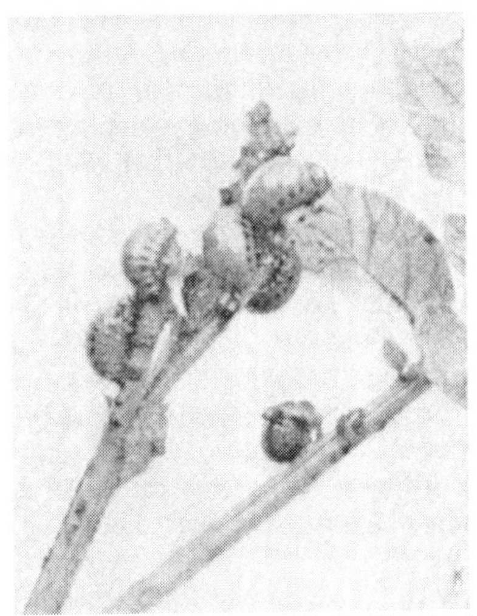

M. V. Slingerland

Larvæ of Colorado potato beetle

LESSON 88

The Colorado Potato Beetle

Leading Thought — The Colorado potato beetle is a very important insect, since it affects to some extent the price of potatoes each year. It is disagreeable as a food for birds, because of an acrid juice which it secretes. We should learn its life history and thus be able to deal with it intelligently in preventing its ravages.

Method — The study of the potato beetle naturally follows and belongs to gardening. The larvæ should be brought into the schoolroom and placed in a breeding cage on leaves of the potato vine. Other plants may be put into the cage to prove that these insects prefer to eat the potato. The children should observe how the larvæ eat and how many leaves a full-grown larva will destroy in a day. Earth should be put in the bottom of the breeding cage so that the children may see the larvæ descend and burrow into it. The adult beetles should be studied carefully, and the children should see the excretion of the acrid juice.

Observations — 1. At what time do you see the potato beetles? Why are they more numerous in the fall than in the spring? Where do those which we find in the spring come from? What will they do if they are allowed to live?

2. What is the shape of the potato beetle? Describe the markings on its head. What color are its eyes? Describe its antennæ. How are they constantly used? Can you see the palpi of the mouth? Give the beetle a bit of potato and note how it eats.

3. What is the color of the shield of the thorax? Describe the legs. Do you think the beetle can run fast? Why not? How many segments has the foot? Describe the claws. Describe how it clings to the sides of a tumbler or bottle.

4. If the beetle cannot walk rapidly, how does it travel? Describe the wing covers. Why is this insect called the ten-lined potato beetle?

5. Describe the wings. How are they folded when at rest? How are the wing covers carried when the beetle is flying?

6. Take a beetle in your hand. What does it do? Of what advantage is it to the insect to pretend that it is dead? If you squeeze the beetle, what happens? I does the fluid which it ejects look smell? Try to discover where this comes from. Of what use is it to the tle? Why will birds not eat the po beetle?

7. Where does the mother be lay her eggs? Are they laid singly o clusters? What color are the eggs? I long is it after they are laid before hatch?

8. Describe the young larva whe first hatches. What color is it at first? D it change color later? Describe the co and markings of a full-grown larva.

9. How does this larva injure the tato vines? Does it remain in sight w it is feeding? Does it act as if it were af of birds? Why is it not eaten by bird

10. Where does the larva go when full grown? How many times does it s its skin during its growth? Does it ma little cell in the ground? How does pupa look? Can you see in it the e antennæ, legs, and wings of the beetle

11. Write an English theme giving history of the Colorado potato beetle, the reasons for its migration from its tive place.

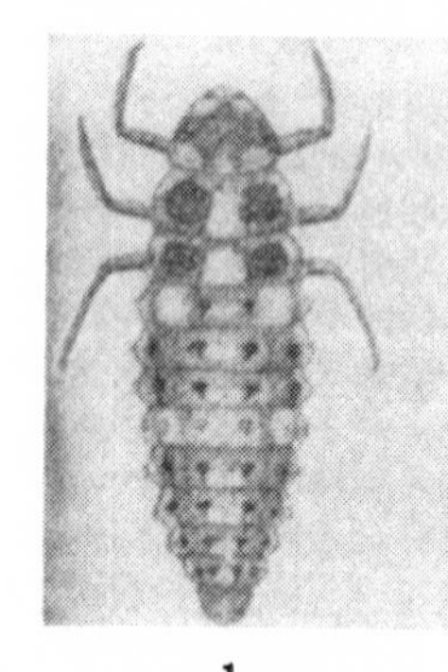
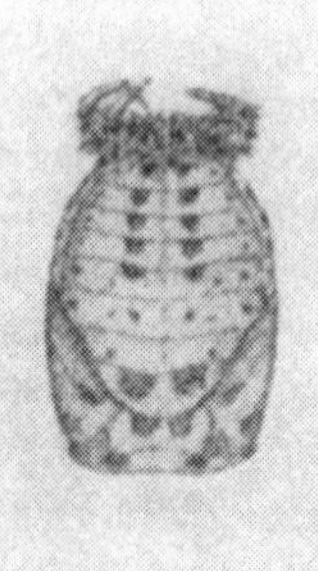

1 2 3

The ladybird. 1, larva. 2, pupa. 3, adult. The small beetle represents actual size

THE LADYBIRD

Ladybird, Ladybird, fly away home!
Your house is on fire, your children will burn.

This incantation we, as children, repeated to this unhearing little beetle, probably because she is, and ever has been, the incarnation of energetic indecisi She runs as fast as her short legs can ca her in one direction, as if her life

ded on getting there, then she turns
it and goes with quite as much vim in
her direction. Thus, it is no wonder
children think that when she hears
news of her domestic disasters, she
els about and starts for home; but she
not any home now nor did she ever
e a home, and she does not carry even
unk. Perhaps it would be truer to say
she has a home everywhere, whether
is cuddled under a leaf for a night's
ging or industriously climbing out on
s, only to scramble back again, or per-
nce to take flight from their tips.

here are many species of ladybirds, but
eneral they all resemble a tiny pill cut
alf, with legs attached to the flat side.
netimes it may be a round and some-
es an oval pill, but it is always shining,
the colors are always dull dark red, or
ow, or whitish, and black. Sometimes
is black with red or yellow spots, some-
es red or yellow with black spots and
spots are usually on either side of the
rax and one on each snug little wing
er. But if we look at the ladybird carely we can see the head and the short,
blike antennæ. Behind the head is the
rax with its shield, broadening toward
rear, spotted and ornamented in vari-
ways; the head and thorax together oc-
y scarcely a fourth of the length of the
ect, and the remainder consists of the
nispherical body, encased with pol-
ed wing covers. The little black legs,
ile quite efficient because they can be
ved so rapidly, are not the ladybird's
y means of locomotion; she is a good
r and has a long pair of dark wings
ich she folds crosswise under her wing
vers. It is comical to see her pull up her
ngs, as a lady tucks up a long petticoat;
d sometimes ladybird is rather slovenly
out it and runs around with the tips of
r wings hanging out behind, quite un-
lily.

But any untidiness must be inadvertent,
cause the ladybird takes very good care
herself and spends much time in "wash-
g up." She begins with her front legs,
eaning them with her mandibles, indus-
iously nibbling off every grain of dust; she then cleans her middle and hind legs by rubbing the two on the same side back and forth against each other, each acting as a whisk broom for the other; she cleans her wings by brushing them between the edges of the wing cover above and the tarsus of her hind leg below.

The ladybird is a clever little creature, even if it does look like a pill, and if you disturb it, it will fold up its legs and drop as if dead, playing possum in a most deceptive manner. It will remain in this attitude of rigid death for at least a minute or two and then will begin to claw the air with all its six legs in its effort to turn right side up.

From our standpoint the ladybird is of great value, for during the larval as well as adult stages, all species except one feed upon those insects which we are glad to be rid of. They are especially fond of aphids and scale insects. One of the greatest achievements of economic entomology was the introduction on the Pacific Coast of a ladybird from Australia which preys upon the cottony cushion scale insect, a species very dangerous to orange and lemon trees. Within a few years the introduced ladybirds had exterminated this pest.

The ladybird's history is as follows: The mother beetle, in the spring, lays her eggs here and there on plants; as soon as the larva hatches, it starts out to hunt for aphids and other insects. It is safe to say that no ladybird would recognize her own children in time to save them, even if the house were burning, for they do not in the least resemble her; they are neither rolypoly nor shiny, but are long and segmented and velvety, with six queer, short legs that look and act as if they were whittled out of wood; they seem only efficient for clinging around a stem. The larvæ are usually black, spotted with orange or yellow; there are six warts on each segment, which make the creature's back look quite rough. The absorbing business of the larva is to crawl around on plants and chew up the foolish aphids or the scale insects. I have seen one use its front foot to push an aphid, which it was eating, closer to its jaws; but when one green leg of its victim still clung to its head, it did not try to rub it off as its mother would have done, but twisted its head over this way and that, wiping off the fragment on a plant stem and then gobbling it up.

After the larva has shed its skeleton skin several times, and destroyed many times its own bulk of insects, it hunts for some quiet corner, hangs itself up by the rear end, and condenses itself into a sub-globular form; it sheds its spiny skin, pushing it up around the point of attachment, and there lets it stay like the lion's skin of Hercules. As a pupa, it is more nearly rectangular than round, and if we look closely we can see the wing cases, the spotted segments of the abdomen, and the eyes, all encased in the pupa skin; the latter bursts open after a few days and the shining little half-globe emerges a full-grown ladybird, ready for hiding through the winter in some cozy spot from which she will emerge in the spring, to stock our trees and vines, next year, with her busy little progeny.

LESSON 89

The Ladybird

Leading Thought — The ladybird is a beetle. Its young are very different from the adult in appearance, and feed upon plant lice.

Method — These little beetles are very common in autumn and may be brought to the schoolroom and passed around in vials for the children to observe. Their larvæ may be found on almost any p
infested with plant lice. Plant and all
be brought into the schoolroom and
actions of the larvæ noted by the pu
during recess.

Observations — 1. How large is
ladybird? What is its shape? Would
of them make a little globe if they
put flat sides together?

2. What colors do you find on
ladybird?

3. Do you see the ladybird's head
antennæ? What is the broad shield
rectly back of the head called? How
marked, and with what colors? What c
are the wing covers? Are there any s
upon them? How many? Does the la
bird use its wing covers when it flies?
scribe her true wings. Does she fold th
beneath the wing covers?

4. Note the legs and feet. Are the
long? Are they fitted for running?
which part of the body are they attach

5. If you disturb the ladybird how d
she "play possum"? Describe how
makes her toilet.

The Larva

1. Describe the ladybird larva. Doe
look like its mother? What is its for
Is it warty and velvety or shiny?

2. Describe its head and jaws as far
you can see. How does it act when eati
Can you see its little stiff legs? Is ther
claw at the end of each?

3. Describe the actions of the ladyb
larva in attacking and eating the plant li
Does it shed its skin as it grows?

4. Watch a larva until it changes to
pupa. How does the pupa look? Can y
see the shed skin? Where is it? To wh
is the pupa attached? When the pupa sk
breaks open what comes out of it?

5. Why is the ladybird of great use
us? Write a story about the ladybird whi
saved the orange orchards of California

THE FIREFLY

And lavishly to left and right,
The fireflies, like golden seeds,
Are sown upon the night.
— Riley

The time of the sowing of these seeds is ing warm, damp nights in July and Au- t, and even in September, although y are sown less lavishly then. How lit- most of us know of the harvest, al- ugh we see the sowing, which begins the early twilight against the back- und of tree shadows, and lasts until the d atmosphere of the later night damp- the firefly ardor! The flight of various cies differs in the height from the und; some species hover next to the ss, others fly above our heads, but ely as high as the tree tops in northern itudes. Some species give a short flash t might be called a refulgent blinking; ers give a longer flash so that we get an a of the direction of their flight; and re is a common species in the Gulf tes which gives such long flashes that y mark the night with gleaming curli- es.

It is likely to be an exciting chase be- e we are able to capture a few of these ects for closer inspection; but when ce captured, they do not sulk but will ep on with their flashing and give us a st edifying display. The portion of the efly which gives the light is in the abdo- en, and it glows steadily like phosphor- ent wood; then suddenly it gleams th a green light that is strong enough reveal all its surroundings; and it is evidently an act of will on the part the beetle that it is startling to mem- rs of our race, who cannot even blush turn pale voluntarily. The fireflies may truly said to be socially brilliant, for the shing of their lights is for the attraction their mates.

The fireflies are beetles, and there are any species which are luminous. A com- on one is here figured (*Photinus pyralis*). It is pale gray above and the head is completely hidden by the big shield of the thorax. The legs are short; thus this beetle trusts mostly to its wings as a means of locomotion. The antennæ are rather long and are kept in constant motion, evidently conveying intelligence of surroundings to the insect. Beneath the gray elytra, or wing covers, is a pair of large, dark-veined membranous wings which are folded in a very neat manner crosswise and lengthwise, when not in use. When in use, the wing covers are lifted stiffly and the flying is done wholly with the membranous wings. Looked at from beneath, we can at once see that some of the segments of the abdomen are partly or entirely sulphur yellow, and we recognize them as the lamp. If the specimen is a male, the yellow area covers all of the end of the abdomen up to the fourth or fifth segment; but if it is a female, only the middle portion of the abdomen, especially the fifth segment, is converted into a lamp. These yellow areas, when dissected under the microscope, prove to be filled with fine tracheæ, or air-tubes; and we know very little about the way the light is made.

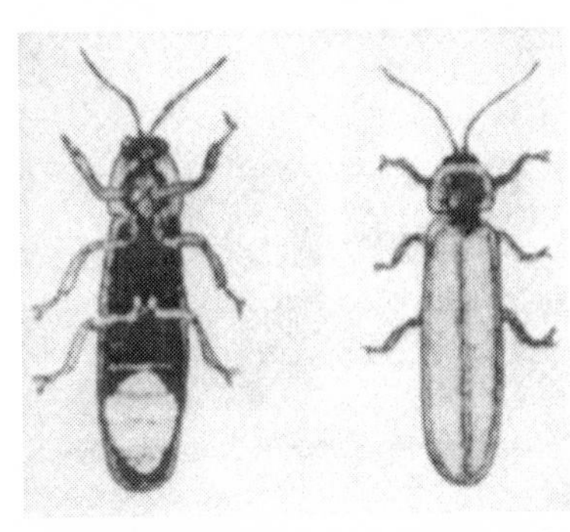

A common firefly. The view of the underside shows the " lamp "

In some species, the female is wingless and has very short wing covers, and a portion of her body emits a steady, greenish light which tells her lord and master where to find her. These wingless females are called glowworms.

Fireflies during their larval stages are popularly called wireworms, although there are many other beetle larvæ thus called. In many of the species, the firefly eggs, larvæ, and pupæ are all luminescent, but not so brilliant as when adults. The

Larva and pupa of a common firefly

larva of the species here figured was studied by C. V. Riley, who gave us an interesting account of its habits. It lives in the ground and feeds on soft-bodied insects and earthworms. Each segment of this wireworm has a horny, brown plate above, with a straight white line running through the middle and a slightly curved white line on each side; the sides of the larva are soft and rose-colored; the white spiracles show against little, oval, brown patches. Beneath, the larva is cream color with two brown comma-like dots at the center of each segment. The head can be pulled back completely beneath the first segment. The most interesting thing about this larva is the prop-leg at the end of its body, which naturally aids it in locomotion; but this prop-leg also functions as a brush; after the larva has become soiled with too eager delving into the tissues of some earthworm, it curls its body over, and with this fan-shaped hind foot scrubs its head and face very clean. This is a rare instance of a larva paying any attention to its toilet.

When full grown, the larva makes a little oval cell within the earth and char to a pupa; after about ten days, the p skin is shed and the full-fledged be comes forth. The larva and pupa of species give off light, but are not so liant as the adult. The pupils should encouraged to study the early stages the fireflies, because very little is knc concerning them.

In Cuba a large beetle called the cuc has two great oval spots on its thorax, sembling eyes, which give off light. Cuban ladies wear cucujos at the op in nets in the hair. I once had a pair wh I tethered with gold chains to the bod of my ball gown. The eyespots glov steadily, but with the movement of da ing, they grew more brilliant until no g tering diamonds could compete with th glow.

LESSON 90

The Firefly

Leading Thought — When the fir wishes to make a light, it can produce c which, if we knew how to make it, wo greatly reduce the price of artificial lig for the light made by fireflies and ot creatures requires less energy than a other light known.

Method — After the outdoor obser tions have been made, collect some these beetles in the evening with a swe net; place them under a glass jar or tu bler, so that their light can be studied close range. The next day give the obs vation lesson on the insects.

Observations — 1. At what time year do you see fireflies? Do they beg to lighten before it is dark? Do you s them high in the air or near the groun Is the flash they give short, or long enou to make a streak of light? Do you see the on cold and windy nights or on warm, st damp evenings? Make a note of the ho when you see the first one flash in evening.

2. Catch a few fireflies in the night; p them under a glass jar. Can you see t

when they are not flashing? What is it? When they make the flash can ee the outline of the "firefly lamp"? ch closely and see if you think the ing is a matter of will on the part of irefly. Do you think the firefly is sig- g to his mate when he flashes?

Study the firefly in daylight. Is it a r is it a beetle? What color is it above? en you look squarely down upon it, you see its head and eyes?

Are the firefly's legs long or short? en a beetle has short legs is it a sign it usually walks or runs, or flies?

Describe the antennæ. Are they in tant motion? What service do you k the firefly's antennæ perform for it?

Lift one of the wing covers carefully. at do you find beneath it? Does the le use its wing covers to beat the air help it during flight? How does the le hold its wing covers when flying?

Turn the beetle on its back. Can you he part of the body that flashes? What r is it?

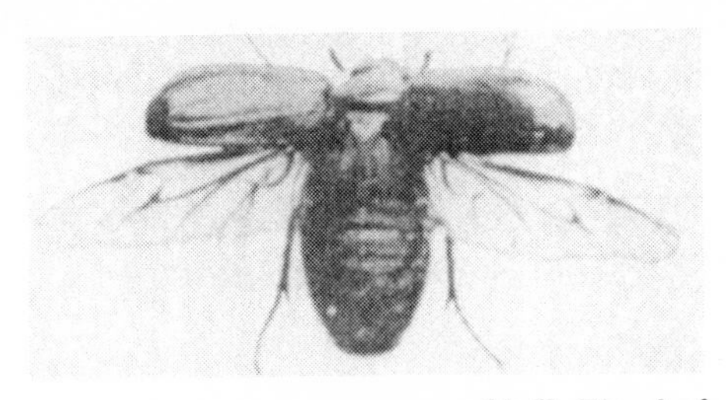

M. V. Slingerland

A maybeetle flying, showing that the beetles hold the wing covers rigid and still in flight, the hind wings doing the work

8. Do you know the life history of the firefly? What is it like in its earlier stages? Where does it live? Does it have the power of making light when it is in the larval stage?

There, in warm August gloaming,
With quick, silent brightenings,
From meadow-lands roaming,
The firefly twinkles
His fitful heat-lightnings.

— LOWELL

THE WAYS OF THE ANT

My child, behold the cheerful ant,
How hard she works, each day;
She works as hard as adamant
Which is very hard, they say.

— OLIVER HERFORD

ery many performances on the part he ant seem to us without reason; un- btedly many of our performances n likewise to her. But the more un- tandingly we study her and her ways, more we are inclined to believe that knows what she is about; I am sure none of us can sit down by an ant- and watch its citizens come and go, out discovering things to make us vel.

y far the greater number of species of find exit from their underground rows beneath stones in fields. They the stone for more reasons than one: ecomes hot under the noon sun and ains warm during the night, thus giv- ing them a cozy nursery in the evening for their young. Some species make mounds, and often several neighboring mounds be- long to the same colony, and are con- nected by underground galleries. There are usually several openings into these mounds. In the case of some of the west- ern species which make galleries beneath the ground there is but one opening to the nest, and Dr. McCook says that this gate is closed at night; at every gate in any ants' nest, there are likely to be sentinels stationed, to give warning of intruders.

As soon as a nest is disturbed, the scared little citizens run helter-skelter to get out of the way; but if there are any larvæ or pupæ about, they take them up and make off with them; when too hard pressed, however, they will in most cases drop the precious burden, although I have several

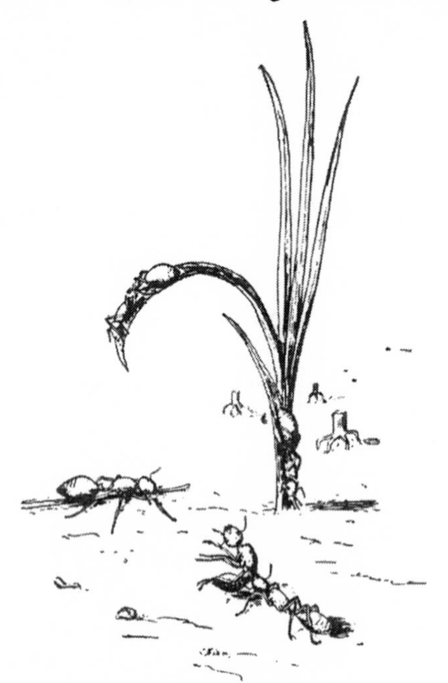

Agricultural ants. Note that one ant is carrying a sister

times seen an ant, when she dropped a pupa, stand guard over it and refuse to budge without it. The ant's eggs are very small objects, being oblong and about the size of a pinpoint. The larvæ are translucent creatures, like rice grains with one end pointed. The pupæ are yellowish, covered with a parchment-like sac, and resemble grains of wheat. When we lift stones in a field, we usually find, directly beneath, the young of a certain size.

There are often, in the same species of ants, two sizes; the large ones are called majors and the smaller minors; sometimes there is a smaller size yet, called minims. The smaller sizes are probably the result of lack of nutrition. But whatever their size, they all work together in bringing food for the young and in caring for the nest. We often see an ant carrying a dead insect or some other object larger than herself. If she cannot lift it or shove it, she turns around, and going backwards pulls it along. It is rarely that we see carrying the same load, although we observed this several times. In one or cases, the two seemed not to be in pe accord as to which path to take. If ants find some large supply of food, n of them will form a procession to bri into the nest bit by bit; such process go back by making a little detour so as to meet and interfere with those com During most of the year, an ant co consists only of workers and laying que but in early summer the nest may found swarming with winged fo which are the kings and queens. S warm day these will issue from the and take their marriage flight, the time in their lives when they use t wings; for ants, like seeds, seem to provided with wings simply for the of scattering wide the species. It strange fact that often on the same swarms will issue from all the nests of species in the whole region; by what n terious messenger word is sent that bri about this unanimous exodus is sti mystery to us. This seems to be a provis for crossbreeding; and as bearing u this, Miss Fielde discovered that an a king is not only made welcome in a n but is sometimes seized by workers pulled into a nest; this is most significa since no worker of any other colony of same species is permitted to live in but its own nest.

After the marriage flight, the ants to the ground and undoubtedly a la number perish; however, just here knowledge is lamentably lacking, and servations on the part of pupils as to w happens to these winged forms will valuable. In the case of most species, know that a queen finds refuge in so shelter and there lays eggs. Mr. Comst once studied a queen of the big, bla carpenter ant which lives under the ba of trees. This queen, without taking a food herself, was able to lay her eggs a rear her first brood to maturity; she gurgitated food for this first brood, a then they went out foraging for the c ony. However, Miss Fielde found that species she studied the queen could do this; a question most interesting to e is whether any of the young queens, r the marriage flight, are adopted into er colonies of the same species. As n as a queen begins laying eggs, she ds her then useless wings, laying them le as a bride does her veil.

Vhen we are looking for ants' nests ieath stones, we often stumble upon a ony consisting of citizens differing in or. One has the head and thorax rust- with the abdomen and legs brown; ociated with this brown ant is a black ash-colored species. These black ants the slaves of the brown species; but very in the ant world has its ameliora- ns. When the slave-makers attack the ve nest, they do not fight the inmates less they are obliged to. They simply t the nest of the larvæ or pupæ, which y carry off to their own nests; and there y are fed and reared, as carefully as are eir own young. The slaves seem to be rfectly contented, and conduct the usehold affairs of their masters with ap- rent cheerfulness. They do all the tasks volved in taking care of the nest and ding the young, but they are never per- tted to go out with war parties; thus ey never fight, unless their colony is acked by marauders.

If one chances upon an ant battle, one ust needs compare it to a battle of men fore the invention of gunpowder; for those days fighting was more gory and eadful than now, since man fought man til one of the two was slain. There is great variation in military skill as well in courage shown by different species of ts; the species most skilled in warfare arch to battle in a solid column and hen they meet the enemy the battle re- lves itself into duels, although there is code of ant honor which declares that e must fight the enemy singlehanded. lthough some ants are provided with nomous stings, our common species use eir jaws for weapons; they also eject pon each other a very acid liquid which e know as formic acid. Two enemies ap- roach each other, rear on their hind legs, throw this ant vitriol at each other, then close in deadly combat, each trying to cut the other in two. Woe to the one on which the jaws of her enemy are once set! For the ant has bulldog qualities, and if she once gets hold, she never lets go even though she be rent in pieces herself. At night the ant armies retreat to their citadels, but in the morning fare forth again to battle; and thus the war may be waged for days, and the battlefield be strewn with the remains of the dead and dying. So far as we are able to observe, there are two chief causes for ant wars; one is when two colonies desire the same ground, and the other is for the purpose of making slaves.

M. V. Slingerland

An aphid stable on a dogwood twig, built by ants to protect their herds

Perhaps the most interesting as well as most easily observed of all ant practices are those that have to do with plant lice, or aphids. If we find an ant climbing a plant of any sort, it is very likely that we shall find she is doing it for the purpose of tending her aphid herds. The aphid is a stupid little creature which lives by thrust- ing its bill or sucking tube into a stem or leaf of a plant, and thus settles down for life, nourished by the sap which it sucks up; it has a peculiar habit of exud- ing from its alimentary canal drops of honeydew when it feels the caress of the ant's antennæ upon its back. I had one year under observation a nest of elegant little ants with shining triangular abdomens which they waved in the air like pennants when excited. These ants were most devoted attendants on the plant lice infesting an evening primrose; if I jarred the primrose stem, the ants had a panic, and often one would seize an aphid in her jaws and dash about madly, as if to rescue it at all hazards. When the ant wishes honeydew, she approaches the aphid, stroking it or patting it gently with her antennæ, and if a drop of the sweet fluid is not at once forthcoming, it is probably because other ants have previously exhausted its individual supply; if the ant gets no response, she hurries on to some other aphid not yet milked dry.

This devotion of ants to aphids has been known for a hundred years, but only recently has it been discovered to be of economic importance. Professor Forbes, in studying the corn root-louse, discovered that the ants care for the eggs of this aphid in their own nests during the winter, and take the young aphids out early in the spring, placing them on the roots of smartweed; later, after the corn is planted, the ants move their charges to the roots of the corn. Ants have been seen to give battle to the enemies of the aphid. The aphids of one species living on dogwood are protected while feeding by stables, which a certain species of ant builds around them, from a mortar made of earth and vegetable matter.

LESSON 91

FIELD OBSERVATIONS ON ANTS

LEADING THOUGHT — However aim to us may seem the course of the as we see her running about, undoubte if we understood her well enough, should find that there is rational ant se in her performances. Therefore, wh ever we are walking and have time, let make careful observations as to the acti of the ants which we may see.

METHOD — The following questi should be written on the blackboard a copied by the pupils in their notebo This should be done in May or June, a the answers to the questions worked by observations made during the sumn vacation.

OBSERVATIONS — 1. Where do you fi ants' nests? Describe all the differ kinds you have found. In what sort of s do they make their nests? Describe entrance to the nest. If the nest is mound, is there more than one entran Are there many mounds near each oth If so, do you think they all belong to same colony?

2. When the nest is disturbed, how the ants act? Do they usually try to sa themselves alone? Do they seek to sa their young at the risk of their own liv If an ant carrying a young one is ha pressed, will she drop it?

3. Make notes on the difference in a pearance of eggs, larvæ, and pupæ in a ants' nest.

4. In nests under stones, can you fi larvæ and pupæ assorted according sizes?

5. How many sizes of ants do you fi living in the same nest?

6. What objects do you find ants ca rying to their nests? Are these for foo How does an ant manage to carry an o ject larger than herself? Do you ever s two ants working together carrying t same load?

7. If you find a procession of ants ca

ng food to their nest, note if they fol-
the same path coming and going.

. If you find winged ants in a nest, ch a few in a vial with a few of the rkers, and compare the two. The nged ants are kings and queens, kings being much smaller than the eens.

). If you chance to encounter a swarm winged ants taking flight, make ob- vations as to the size of swarm, the ght above the ground, and whether any falling to the earth.

10. Look under the loose bark of trees nests of the big black carpenter ant. u may find in such situations a queen t starting a colony, which will prove st desirable for stocking an artificial ts' nest.

11. If you find ants climbing shrubs, es, or other plants, look upon the leaves r aphids and note the following points:

(a) How does an ant act as she ap- oaches an aphid?

(b) If the aphids are crowded on the af, does she step on them?

(c) Watch carefully to see how the t touches the aphid when she wishes e honeydew.

(d) Watch how the aphid excretes the neydew, and note if the ant eats it.

(e) If you disturb aphids which have ts tending them, note whether the ants tempt to defend or rescue their herds.

(f) If there are aphis lions or ladybird rvæ eating the aphids, note if the ants tack them.

12. If you find a colony of ants under ones where there are brown and black nts living together, the black members re the slaves of the brown. Observe as arefully as possible the actions of both ne black and the brown inhabitants of ne nest.

13. If you chance to see ants fighting, ote how they make the attack. With hat weapons do they fight? How do they y to get at the adversary?

14. Write a story covering the follow- ng points: How ants take their slaves; the ttitude of masters and slaves toward each ther; the work which the slaves do; the story of the ant battle; and how ants care for and use their herds.

A Lubbock ant-nest

LESSON 92

How to Make the Lubbock Ant-Nest

Material — Two pieces of window glass, 10 inches square; a sheet of tin, 11 inches square; a piece of plank, 1¼ inches thick, 20 inches long, and at least 16 inches wide; a sheet of tin or a thin, flat board, 10 inches square.

To Make the Nest — Take the plank and on the upper side, a short distance from the edge, cut a deep furrow. This furrow is to be filled with water, as a moat, to keep the ants imprisoned. It is necessary, therefore, that the plank should have no knotholes, and that it be painted thoroughly to keep it from checking. Take the sheet of tin 11 inches square, and make it into a tray by turning up the edges three-eighths of an inch. Place this tray in the middle of the plank. Place within the tray one pane of glass. Lay around the edges of this glass four strips of wood about half an inch wide and a little thicker than the height of the ants which are to live in the nest. Cover the glass with a thin layer of fine earth. Take the remain- ing pane of glass and cut a triangular piece from one corner, then place the pane on top of the other, resting upon the pieces of wood around the sides. The cover of the nest may be a piece of tin, with a han- dle soldered to the center, or a board with a screw eye in the center with which to lift it. There should be a piece of blotter or of very thin sponge introduced into the nest between the two panes of glass, in a position where it may be reached with a pipette, without removing the upper glass, for it must be kept always damp.

To establish a colony in this nest proceed as follows: Take a two-quart glass fruit jar and a garden trowel. Armed with these, visit some pasture or meadow near by, and find under some stone a small colony of ants which have plenty of eggs and larvæ. Scoop up carefully eggs, ants, dirt, and all, and place them in the jar, being as careful as possible not to injure the specimens. While digging, search carefully for the queen, which is a larger ant and is sometimes found. But if you have plenty of eggs, larvæ, and pupæ, the ants will become very contented in their new nest while taking care of them. After you have taken all the ants desirable, place the cover on the jar, carry them to the Lubbock nest and carefully empty the c
tents of the fruit jar on top of
board which covers the nest. Of course
furrow around the plank has been fi
with water, so the stragglers cannot
cape. The ants will soon find the way i
the nest through the cut corner of
upper pane of glass, and will transfer t
larvæ to it because it is dark. After t
are in the nest, which should be wit
two or three hours, remove the dirt
the cover, and the nest is ready for ob
vation. But, since light disturbs the li
prisoners, the cover should be remo
only for short periods.

The Fielde nest is better adapted fo
serious study of ants, but it is not so w
adapted for the schoolroom as is the L
bock nest.

THE ANT–NEST AND WHAT MAY BE SEEN WITHIN IT

Ant anatomy becomes a very interesting study when we note the vigorous way the ant uses it — even to the least part. The slender waist characterizes the ant as well

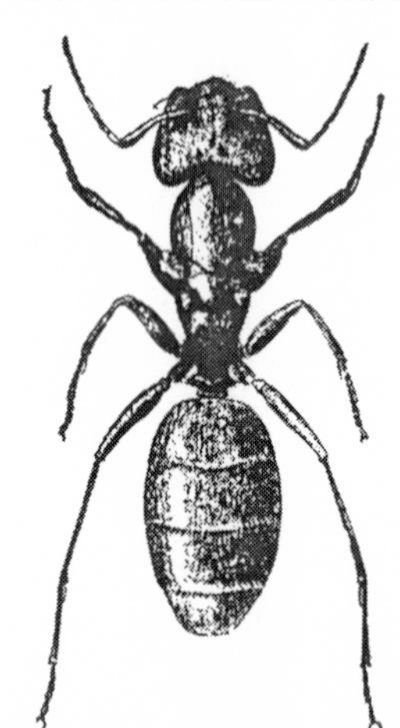

The black carpenter ant, much enlarged

as the wasp; the three regions of the body are easily seen, the head with its ever moving antennæ, the slender thorax with its three pairs of most efficient legs, and the long abdomen. The ant's legs are fairly long as compared with the size of the body and the ant can run with a rapid
that, comparatively, would soon out
tance any Marathon runner, howe
famed. I timed an ant one day when s
was taking a constitutional on my f
rule. She was in no hurry, and yet s
made time that if translated into hum
terms would mean sixteen yards per s
ond. In addition to running, many a
when frightened will make leaps with
credible swiftness.

The ant does not show her cleverness
her physiognomy, probably because l
eyes seem small and dull and she has a
cidedly "retreating forehead"; but t
brain behind this unpromising appearan
is far more active and efficient than th
behind the gorgeous great eyes of t
dragonfly or behind the "high brow"
the grasshopper. The ant's jaws are ve
large compared with her head; they wo
sidewise like a pair of shears and are arm
with triangular teeth along the biti
edges; these are not teeth in a vertebra
sense, but are like the teeth of a sa
These jaws are the ant's chief utens
and weapons; with them she seizes t
burdens of food which she carries hom

them she gently lifts her infant ges; with them she crushes and breaks ıard food; with them she carries out from her tunnel, and with them she .s her enemies. She also has a pair of palpi, or feelers.

lthough her eyes are so small and ished with coarse facets, as compared other insects, this fact need not count nst her, for she has little need of eyes. home life is passed in dark burrows re her antennæ give her information er surroundings. Note how these an- æ are always moving, seeming to be mble in eagerness to receive sensa- s. But aside from their powers of tell- things by the touch, wherein they are e delicate than the fingers of the blind, have other sense organs which are parable to our sense of smell. Miss de has shown that each of the five end nents of the antennæ has its own pow- in detecting odor. The end segment ects the odor of the ant's own nest enables her to distinguish this from er nests. The next, or eleventh seg- nt, detects the odor of any descendant he same queen; by this, she recognizes sisters wherever she finds them. rough the next, or tenth segment, she

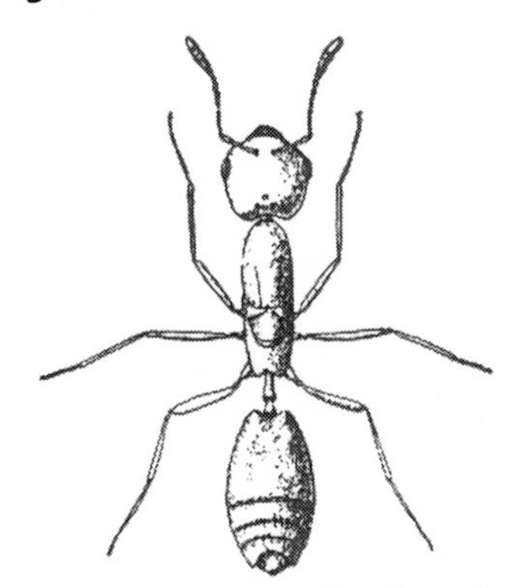

The red ant, much enlarged

ognizes the odor of her own feet on trail, and thus can retrace her own ps. The eighth and ninth segments con- to her the intelligence and means of ing for the young. If an ant is deprived these five end-joints of the antennæ, loses all power as a social ant and becomes completely disfranchised. Miss Fielde gives her most interesting experi- ments in detail in the Proceedings of the

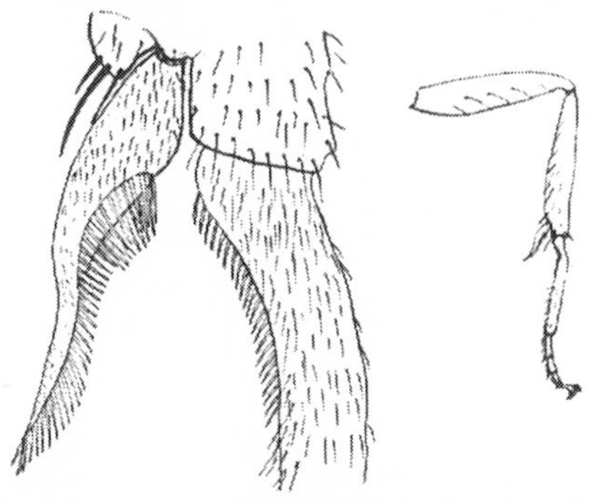

The antenna-comb on the front leg of an ant

Academy of Natural Sciences of Philadel- phia, July and October, 1901.

It is natural enough that the ant, de- pending so much on her antennæ for impressions and stimuli, should be very particular to keep them clean and in good order. She is well equipped to do this, for she has a most efficient antennæ brush on her wrist; it is practically a circular comb, which just fits over the antenna; and to see the ants using these brushes is one of the most common sights in the ant-nest and one of the most amusing. The ant usually commences by lifting her leg over one antenna and deftly passing it through the brush, and then licks the brush clean by passing it through her mouth, as a cat washes her face; then she cleans the other in a similar manner and possibly finishes by doing both alternately, winding up with a flourish, like a European gentleman curling his mustaches. Her antennæ cleaned, she starts promptly to do some- thing, for she is a little six-footed Martha, always weighed down or buoyed up by many duties and cares. Keeping her an- tennæ on the qui vive, she assures herself, by touch, of the nature of any obstacle in her path. If she meets another ant, their antennæ cross and pat each other, and thus they learn whether they are sisters or aliens; if they are sisters, they may stand for some time with their antennæ flutter- ing. One who has watched ants carefully, is compelled to believe that they thus convey intelligence of some sort, one to the other. The ant is a good sister "according to her lights"; if her sister is hungry, she will give to her, even from her own par-

Ants making their toilets

tially digested food; the two will often stand mouth to mouth for some minutes during this process; if she feels inclined, she will also help a sister at her toilet, and lick her with her tongue as one cow licks another. The tongue of the ant is very useful in several ways; with it she takes up liquids, and also uses it with much vigor as a washcloth. Sometimes an ant will spend a half hour or more at her own toilet, licking every part of her own body that her tongue can reach, meanwhile going through all sorts of contortions to accomplish it; she uses her feet to scrub portions of her body not to be reached by her tongue.

But it is as infant nurse that the ant is a shining example. No mother instinct is hers, for she has yielded the power of motherhood to the exigencies of business life, since all workers are females but are undeveloped sexually. She shows far more sense in the care of her infant sisters than the mother instinct often supplies to human mothers. The ant nurse takes the eggs as soon as laid, and whether her care retards or hastens hatching we know not; but we do know that although the queen ant may not lay more than two eggs a day, a goodly number of these seem to hatch at the same time. The eggs are massed in bundles and are sticky on the outside; so they are held together in a bundle. Miss Fielde says that as the eggs are hatching, one ant will hold up the bundle, while another feeds those which have br
the shell. The larvæ, when young,
hang together by means of tiny hook
their bodies. This habit of the eggs
young larvæ is a convenient one, sinc
ant is thus able to carry many at a tim

The larvæ are odd looking little
tures, shaped like crookneck squashes,
small end being the head and neck
the latter being very extensible. The
nurses, by feeding some more than ot
are able to keep a brood at the same s
of development; and in a well-ord
ant-nest, we find those of the same siz
one nursery. I have often thought
graded school as I have noted in
nests the youngsters assorted accordin
size.

The ants seem to realize the cost
care of rearing their young; and whe
nest is attacked, the oldest, which are
ally in the pupa stage, are saved
When the larvæ are young, they are
on regurgitated food; but as they g
older, the food is brought to them, or
to the food, and they do their own eat
In one of my nests, I placed part of
yolk of an egg hard boiled, and the
nurses dumped the larvæ down around
edges of it; there they munched indu
ously, until through their transparent
ies I could see the yellow of the egg
whole length of the alimentary canal.
ant nurses are very particular about t
peratures for their young, and Miss Fi
says they are even more careful ab
draughts. Thus they are obliged to m
them about in the ground nests, carry
them down to the lower nurseries in
heat of the day, and bringing them
nearer to the warm stones, during the
nings. This moving is always done c
fully, and though the ant's jaws are s
formidable nippers, she carries her b
sisters with gentleness; and if they
pupæ, she holds them by the loose pu
skin, like carrying a baby by its clot
The pupæ look like plump little gr
bags, tied at one end with a black stri
They are the size of small grains of wh
and are often called ants' eggs, whic
absurd, since they are almost as larg

ant Ants' eggs are not larger than oints.

he ant nurses keep the larvæ and pupæ clean by licking them; and when a ngster issues from the pupa skin, it is atter of much interest to the nurses. ve often seen two or three of them straighten out the cramped legs and nnæ of the young one, and hasten eed her with regurgitated food. When first issue from the pupa skin they pale in color, their eyes being very k in contrast; they are usually helpless stupid, although they often try to n their antennæ and make a toilet; they do not know enough to follow r elders from one room to another, they are a source of much care to the ses. In case of moving, a nurse will lock s with a "callow," as a freshly hatched lt ant is called, and drag her along, the of the callow sprawling helplessly anwhile. If in haste, the nurse takes d anywhere, by the neck or the leg, and tles her charge along; if she takes her the waist the callow curls up like a kit- and is thus more easily moved. After nurses have moved them from one mber to the next, I have noticed that callows are herded together, their at- dants ranged in a circle about them. ten we see one ant carrying another ich is not a callow, and this means that rtain number of the colony have made their minds to move, while the others not awake to this necessity. In such ase, one of these energetic sisters will ze another by the waist, and carry her with an air that says plainly, "Come ng, you stupid!"

Ants are very cleanly in their nests, and find the refuse piled in a heap at one ner, or as far as possible from the brood. If we are fortunate enough to find a een for the nest, then we may observe attention she gets; she is always kept a special compartment, and is sur- nded by ladies in waiting, who feed r and lick her clean and show solicitude her welfare; although I have never ob- ved in an ant-nest that devotion to roy- y which we see in a beehive.

Not the least interesting scene in an ant-nest is when all, or some, are asleep and are as motionless as if dead.

LESSON 93

Observations of Ants in an Artificial Nest

Leading Thought — The ants are very devoted to their young and perhaps the care of them is the most interesting feature in the study of the artificial nest.

Method — Have in the schoolroom a Lubbock's nest with a colony of ants within it, with their larvæ in all stages, and if possible, their queen. For observing the form of the ant, pass one or two around in a vial.

Observations — 1. What is there peculiar about the shape of the ant's body? Can you see which section bears the legs? Are the ant's legs long compared with her body? Can she run rapidly?

2. Look at the ant's head through a lens, and describe the antennæ, the jaws, and the eyes.

3. Note how the ant keeps her antennæ in motion. Note how she gropes with them as a blind person with his hands. Note how she uses them in conversing with her companions.

4. How does the ant clean her antennæ? Does she clean them more often than any other part of her body? How does she make her toilet?

5. See how an ant eats syrup. How do ants feed each other?

6. How does the ant carry an object? How does she carry a larva or a pupa? Have you ever seen one ant carry another? If so, describe it.

7. Note the way the ants feed their young. How do they keep them clean? Does an ant carry one egg or one small larva at a time or a bundle of them? How do you suppose the bundle is fastened together?

8. Describe an egg, a larva, and a pupa of the ant and tell how they differ. Do you know which ant is the mother of the larvæ in the nest?

9. Do you find larvæ of different sizes

all together in your nest? Do you find larvæ and pupæ in the same group? Do the ants move the young often from one nest to another? Why do you suppose they do this?

10. Note how the ant nurses take care of the callow ant when it is coming out from the pupa skin. How do they a her and care for her? How do they her around? How do ants look when ing?

11. Note where the ants throw the use from the nest. Do they ever cha the position of this dump heap?

THE MUD-DAUBER

This little cement worker is a nervous and fidgety creature, jerking her wings constantly as she walks around in the sunshine; but perhaps this is not nervousness,

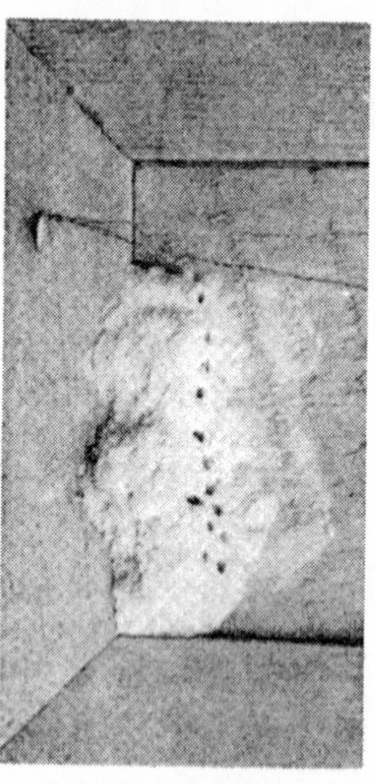

Nest of a mud-dauber on the back of a picture frame

but rather to show off the rainbow iridescence of her black wings. Her waist is a mere pedicel and the abdomen is only a knob at the end of it. The latter, seen from the outside, would seem of little use as an abdomen; but if we watch the insect flying, we can see plainly that the abdomen is an aid to steering.

In early summer, we find this black wasp at her trade as a mason. She seeks the edges of pools or puddles where she works industriously, leaving many little holes whence she takes mud to mix with the saliva which she secretes from mouth to make firm her cement. cement she plasters on the undersid some roof or rafter or other protec place, going back and forth until she built a suitable foundation. She w methodically, making a tube about inch long, smooth inside but rough side, the walls about one-eighth of an i thick. She does all the plastering with jaws, which she uses as a trowel. W the tube is completed except that the is left open, she starts off in quest spiders, and very earnestly does she s them. I have seen her hunt every nook corner of a porch for this prey. When finds a spider, she pounces upon it stings it until it is helpless, and carrie to her cement tube, which is indee spider sarcophagus, and thrusts it with She brings more spiders until her t is nearly full; she then lays an egg wit it and makes more cement and nea closes the door of the tube. She pla another tube by the side of this, wh she provisions and closes in the same w she may make another and another tu often a half dozen, under one adobe ro

The wasp in some mysterious w knows how to thrust her sting into spider's nervous system in a peculiar w which renders her victim unable to mo although it yet lives. The wasp is vegetarian like the bee, and she m supply her young with wasp-meat stead of beebread. Since it is duri the summer and hot weather when t young wasps are hatched and begin th growth, their meat must be kept fre for a period of two or three wee these paralyzed spiders do not die, al-ugh they are helpless. It is certainly ractical joke with justice in it, that se ferocious creatures lie helpless while ng eaten by a fat little grub which they ld gladly devour, if they could move. he wasp larva is a whitish, plump grub it eats industriously until the spider at is exhausted. It then weaves a co-n of silk about itself which just covers walls of its home tube, like a silken estry; within this cocoon the grub nges to a pupa. When it finally erges, it is a full-grown wasp with jaws ich are able to cut a door in the end its tube, through which it comes out o the world, a free and accepted mason. e females, which issue late in the sea-, hide in warm or protected places dur-the winter; they particularly like the ds of lace window curtains for hiber-ing quarters. There they remain until ing comes, when they go off to build ir plaster houses.

There are about seventy species of mud sps in our country. Some provision their sts with caterpillars instead of spiders. is is true of the jug-builder, which kes her nest jug-shaped and places two three of them side by side upon a twig. e uses hair in her mortar, which makes tronger. This is necessary, since the jug saddled upon twigs and is more exposed the rain than is the nest of the most mmon mud-dauber. The jug-builder is wn in color and has yellow markings the abdomen; but she does not resem-ble the yellow jackets, because she has a threadlike waist. There are other species of mud wasps which use any small cavity

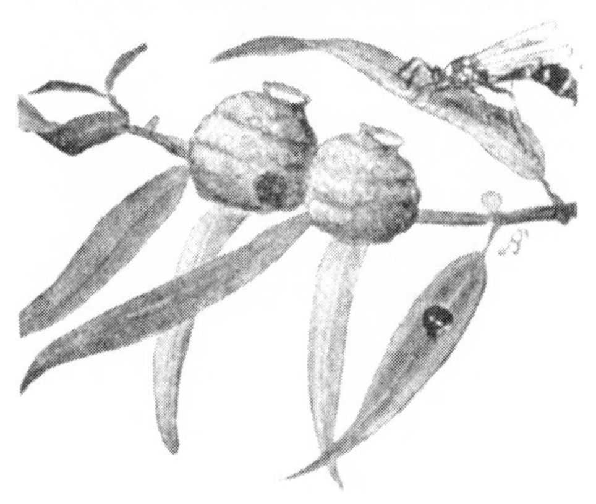

The jug-builder and her nests

they can find for the nest, plastering up the opening after the nest has been pro-visioned and the egg laid. We often find keyholes, knotholes, and even the cavity in the telephone transmitter plastered up by these small opportunists.

The mud-dauber, which is the most common and most likely to be selected for this lesson, is a slender creature and looks as if she were made of black tinsel; her body gives off glints of steel and blue; her abdomen constantly vibrates with the movement of breathing. Her eyes are large and like black beads; her black antennæ curve gracefully outward, and her wings, corrugated with veins, shimmer with a smoky blue, green, and purple. She stands on her black tiptoes when she walks, and she has a way of turning around constantly as if she expected an attack from the rear. Her wings, like those of other mud wasps, are not folded fanwise like those of the yellow jacket, but are folded beside each other over her back.

A mud-dauber and her nests, with cells cut en to show from left to right, larva full own, cocoon, young larva feeding on its ider-meat, and an empty cell

LESSON 94

The Mud-Dauber

Leading Thought — There are certain wasps which gather mud and mix it into mortar with which to build nests for their young. Within these nests, the mother wasp places spiders or insects which are disabled by her sting, to serve as the food of the young wasps.

Method — Have the pupils bring the homes of the mud wasps to school for observation. The wasps themselves are very common in spring and also in autumn, and they may be studied at school and may be passed around in vials for closer observation; they do not sting severely when handled, the sting being a mere prick. The purpose of the lesson should be to stimulate the pupils to watch the mud-daubers while building their nests and capturing their prey.

Observations — 1. Where did you find the mud-dauber's nest? How was it protected from the rain? Was it easily removed? Could you remove it all, or did some of it remain stuck fast?

2. What is the shape of the nest? How does it look inside? Of how many tubes does it consist? How long is each tube? Were the tubes laid side by side?

3. Of what material was the nest made? Is it not much harder than mud? How did the wasp change the mud to cement? Where did she get the mud? How did she carry it? With what tools did she pla[illegible] it?

4. For what purpose was the nest ma[illegible] Is the inside of the tubes smooth as c[illegible]pared with the outside of the nest?

5. Write a little story about all t[illegible] happens in one of these tubes, includ[illegible] the following points: What did [illegible] mother wasp place in the tube? How [illegible] why did she close it? What hatched fr[illegible] the egg she placed within it? How d[illegible] the young wasp look? On what doe[illegible] feed? What sort of cocoon does it sp[illegible] How does it get out of the nest wh[illegible] full grown?

6. Describe the mud-dauber wasp. H[illegible] large is she? What is the color of her bo[illegible] Of her wings? How many wings has s[illegible] How are her wings folded differently fr[illegible] those of the yellow jacket? Describe [illegible] eyes; her antennæ; her legs; her waist; [illegible] abdomen.

7. Where did you find the wasp? H[illegible] did she act? Do you think that she c[illegible] sting? How does this wasp pass the w[illegible]ter?

8. Do you know the mud wasps wh[illegible] build the little jug-shaped nests for th[illegible] young? Do you know the mud wa[illegible] which utilize crevices and keyholes [illegible] their nests and plaster up the opening[illegible]

9. Do you know about the digger wa[illegible] which pack away grasshoppers or cater[illegible]lars in a hole in the ground, in which th[illegible] lay their egg and then cover it?

THE YELLOW JACKET

Many wasps are not so waspish after all when we understand one important fact about them; i.e., although they are very nervous themselves, they detest that quality in others. For years the yellow jackets have shared with us our meals at our summer camp on the lake shore. They make inquisitive tours of inspection over the food on the table, often seeming to include ourselves, and coming so near that they fan our faces with their wings. They usually end by selecting the sweetened fruits, but they also carry off bits [illegible] roast beef, pouncing down upon the m[illegible] platter and seizing a tidbit as a hawk d[illegible] a chicken. We always remain calm duri[illegible] these visitations, for we know that unl[illegible] we inadvertently pinch one we shall n[illegible] be harmed; and it is grea[illegible] fun to wat[illegible] one of these graceful creatures poisi[illegible] daintily on the side of the dish lappi[illegible] up the fruit juice as a cat does milk, t[illegible] slender, yellow-banded abdomen palpit[illegible]ing as she breathes. Occasionally, two [illegible]

:he same place, and a wrestling match .es which is fierce while it lasts, but participants always come back to the unharmed. They are extra polite in r manners, for after one has delved rly into the fruit syrup, she proceeds lean her front feet by passing them ugh her jaws, which is a wasp's way of g a finger bowl.

oth yellow jackets and the white-faced k hornets build in trees, and their s are much alike, although the paper le by the yellow jackets is finer in tex- :. However, some species of yellow ets build their nests in the ground, of similar form. The nest is of paper le of bits of wood which the wasps off with their jaws from weather-worn :es or boards. This wood is reduced to ulp by saliva which is secreted from wasp's mouth, and is laid on in little :rs which can be easily seen by exam- ıg the outside of the nest. These layers y be of different colors. A wasp will ıe with her load of paper pulp, and ıg her jaws and front feet for tools she join a strip to the edge of the paper pat it into shape. The paper tears re readily along the lines of the joining n across. The cover of the nest is made nany layers of shell-like pieces fastened ether, and the outer layers are water- of; the opening of a nest is at the bot- n. Mr. Lubbock has shown that cer- n wasps are stationed at the door, as

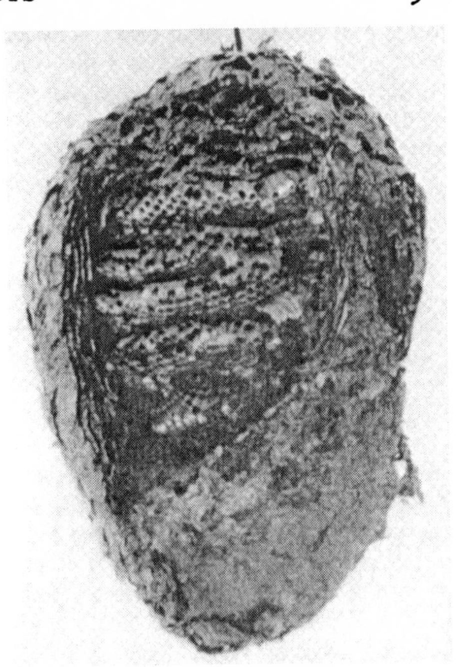

A wasps' nest with the side walls removed

sentinels, to give warning on the approach of an enemy. The number of stories of combs in a nest depends upon the age and size of the colony. They are fastened together firmly near the center by a central core or axis of very strong, firm paper, which at the top is attached to a branch or whatever supports the nest. The cells all open downward, in this respect differing from those of the honeybee, which are usually placed horizontally. The wasp comb differs from the honeycomb in that it is made of paper instead of wax, and that the rows of cells are single instead of double. The cells in the wasp comb are not for storing honey, but are simply the cradles for the young wasps. (See figure above.)

Sometimes a wasp family disaster makes it possible for us to examine one of these nests with its inmates. Here we find, in some of the cells, the long white eggs fastened to the very bottom of the cell, in an inner angle, as if a larva when hatched needed to have a cozy corner. These wasp larvæ are the chubbiest little grubs imaginable and are very soft bodied. It was once a mystery to me how they were able

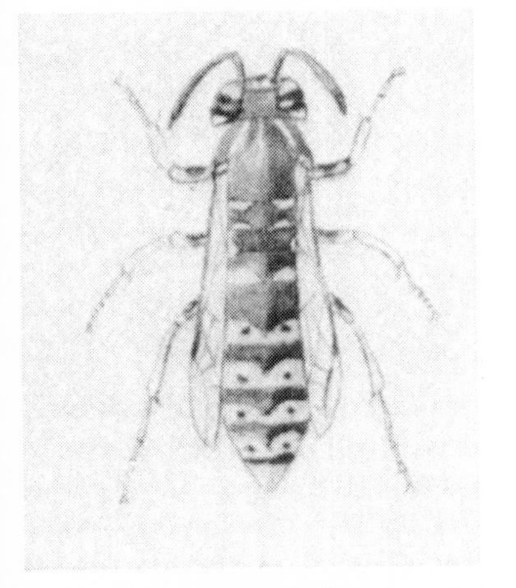

A yellow jacket

to hang in the cells, head down, without getting " black in the face " or falling out; but this was made plain by studying the little disc at the rear end of the larva's body, which is decidedly sticky; after a larva is

Looking a wasp in the face

dead, its heavy body can be lifted by pressing a match against this disc; thus it evidently suffices to keep the baby wasp stuck fast to its cradle. The larva's body is mostly covered with a white, papery, soft " skeleton skin "; the head is yellowish and highly polished, looking like a drop of honey. At one side may be seen a pair of toothed jaws, showing that it is able to take and chew the food brought by the nurses. They seem to be well-trained youngsters, for they all face toward the center of the nest, so that a nurse, when feeding them, can move from one to another without having to pass to the other side of the cell. It is a funny sight to behold a combful of well-grown larvæ, each fitting in its cell like meal in a bag and with head and several segments projecting out as if the bag were overflowing. It behooves the wasp larva to get its head as far out of the cell as possible, so that it will not be overlooked by the nurses; the little ones do this by holding themselves at the angle of the cell; this they accomplish by wedging the back into the corner. These young larvæ do not face inwards like the older ones, but rest in an inner angle of the cell.

After a larva has reached the limit of its cell room, it spins a veil around itself and fastens it at the sides, so that it forms a lining to the upper part of the cell and makes a bag over the " head and shoulders " of the insect. This cocoon is very tough, and beneath its loose dome the larva skin is shed; the pupa takes on a decidedly waspish form, except that the

color is all black; the legs and the w
are folded piously down the breast
the antennæ lie meekly each side of
face, with the " hands " folded outsid
them; the strong toothed jaws are re
so that when the pupa skin is molted,
insect can cut its silken curtain and c
out into its little nest world as a
fledged yellow jacket.

What a harlequin the wasp is, in
costume of yellow and black! Ofte
the invertebrate world these colors m
" sit up and take notice," and the wa
costume is no exception. Whoever
had any experience in meddling with
low jackets avoids acquaintance with
yellow and black insects. Yet we m
confess that the lady wasp has good t
in dress. The yellow crossbands on
black skirt are scalloped, and, in fact,
her yellow is put on in a most *chic* m
ner; she, being slender, can well affor
dress in roundwise stripes; and she f
her wings prettily like a fan, not over
back like the mud wasp, which wo
cover her decorations. There is a se
tion coming to the one who, armed w
a lens, looks a wasp in the face; she alw
does her hair pompadour, and the yel
is here put on with a most bizarre eff
in points and arabesques. Even her j
are yellow with black borders and bl
notches. Her antennæ are velvety bl

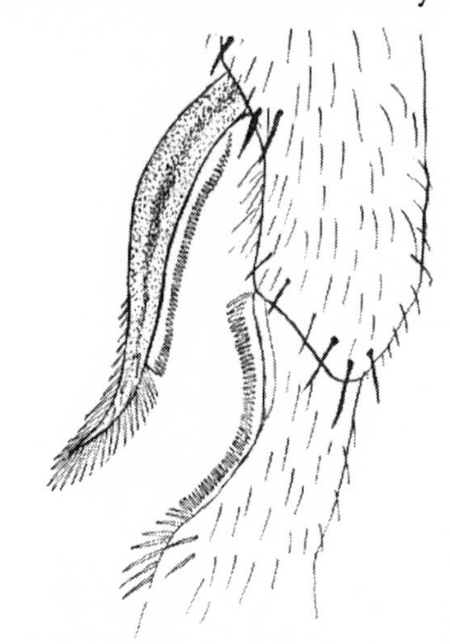

The antenna-comb or cleaner on the for of a wasp

legs are yellow, and her antennæ
ıb, on her wrist, is a real comb and
te ornate.

n the nest which we studied in late
;ust, the queen cells were just being
eloped. They were placed in a story
by themselves, and they were a third
er than the cells of the workers. The
en of this nest was a most majestic
p, fully twice as large as any of her sub-
s; her face was entirely black, and the
ow bands on her long abdomen were
uite a different pattern from those on
workers; her sting was not so long in
portion, but I must confess it looked
cient. In fact, a yellow jacket's sting is
ormidable looking spear when seen
ough a microscope, since it has on
side some backward projecting barbs,
ant to hold it firm when driving home
thrust.

While wasps are fond of honey and
er sweets, they are also fond of animal
d and eat a great many insects, benefit-
us greatly by destroying mosquitoes
flies. As no food is stored for their win-
use, all wasps excepting the queens die
the cold. The queens crawl away to
tected places and seem to be able to
hstand the rigors of winter; each queen,
the spring, makes a little comb of a few
ls, covering it with a thin layer of paper.
e then lays eggs in these cells and gath-
food for the young; but when these
t members of the family, which are
ays workers, come to maturity, they
e upon themselves the work of enlarg-
the nest and caring for the young.
ter that, the queen devotes her energies
laying eggs.

Wasps enlarge their houses by cutting
ay the paper from the inside of the cov-
ng, to give more room for building the
mbs wider; to compensate for this, they
ild additional layers on the outside of
nest. Thus it is that every wasp's nest,
wever large, began as a little comb of a
v cells and was enlarged to meet the
eds of the rapidly growing family. Or-
narily the nest made one year is not
d again.

LESSON 95

The Yellow Jacket

Leading Thought — The wasps were the original paper makers, using wood pulp for the purpose. Some species construct their houses of paper in the trees or bushes while others build in the ground.

Method — Take a deserted wasp-nest, the larger the better, and with sharp scissors remove one side of the covering of the nest, leaving the combs exposed and follow with the questions and suggestions indicated. From this study of the nest encourage the children to observe more closely the wasps and their habits, which they can do in safety if they learn to move quietly while observing. (See Fig. p. 381.)

Observations — 1. Which kind of wasp do you think made this nest? Of what is the nest made? Where did the wasp get the material? How do the wasps make wood into paper?

2. What is the general shape of the nest? Is the nest well covered to protect it from rain? Where is the door where the wasps went in and out? Is the covering of the nest all of the same color? Do these differences in color give you any idea of how the wasps build the paper into the nest? Does the paper tear more easily one way than another? Is the covering of the nest solid or in layers?

3. How many combs or stories are there in the wasp house? How are they fastened together and how suspended?

4. Compare the combs of the wasp-nest with those of the honeybee. How do they resemble each other and how differ? Do the cells open upward or downward? For what purpose are the combs in the wasp-nest used? Are all the cells of the same size? Do you know the reason for this difference in size?

5. How do the young wasp grubs manage to cling to the cells head downward? Are the cells lined with a different color and does this lining extend out over the opening in some cases? Is this lining of the cells made of paper also? Do you know how a young wasp looks and how the white lining of the cells is made?

6. Do you believe that some wasps of the colony are always posted as sentinels at the door to give warning if the colony is attacked?

7. Do wasps store food to sustain them during the winter? What happens to them during the winter? Is the same nest used year after year?

8. Can you describe the beginning of this wasp-nest? When was it made? Tell the story of the wasp that made it. How large was the nest at first? How was nest enlarged?

9. What is the food of wasps? How these insects benefit us?

10. Write a story giving the life his of a wasp.

11. In the summer watch a yel jacket eat from a dish of sweetened f which you may place out of doors to c her to come where you can carefully serve her. What are the colors of the low jacket? Where is the yellow? How the yellow bands made ornamental? H does she fold her wings? How many wi has she? What is the color of her le Describe her antennæ and eyes. How d she eat the fruit juice? Can you obse the motion of her body when breathes?

THE LEAF-CUTTER BEE

One beautiful day in late June when I was picking some roses, I saw a bee, almost as large as a honeybee but different in shape and darker in color, alight on a leaf

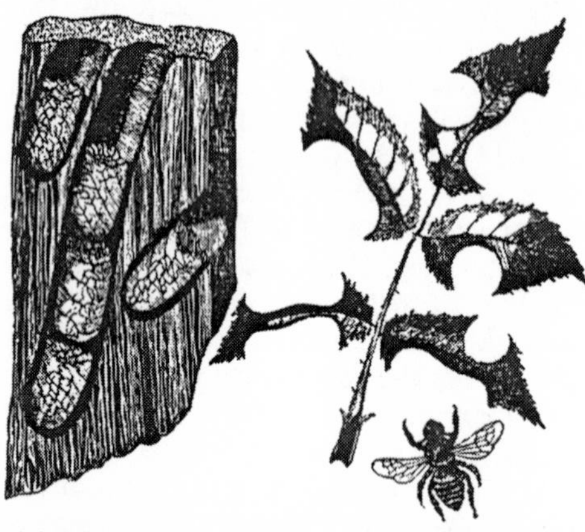

A leaf-cutter bee, its nest, and rose leaves cut by the bee

and, moving with nervous rapidity, cut a circle out of a leaf with her jaws "quicker'n a wink"; then, taking the piece between her forefeet and perhaps holding it also with her jaws, she flew away, the green disc looking as large in proportion to her size as a big bass drum hung to the neck of a small drummer. I wai long for her to come back, but she ca not; meanwhile I examined the leaves the rose bush and found many circl and also many oblong holes with the e deeply rounded, cut from the leaflets.

I knew the story of the little bee a was glad I had seen her cut a leaflet w her jaw shears, which work sidewise l real shears. I knew that somewhere she h found a cavity big enough for her nee perhaps she had tunneled it herself in dead wood of some post or stump, us her jaws to cut away the chips; maybe had found a crevice beneath the shing of a roof or beneath a stone in the fie or she may have rolled a leaf; anyway, little cave was several inches long, circu in outline and large enough to admit body. She first cut a long piece from rose leaf and folded it at the end of tunnel; and then she brought another a another long piece and bent and shap them into a little thimble-like cup, fast ing them together with some saliva gl from her mouth. After the cup was ma to her liking, she went in search of foo which she found in the pollen of so

vers. This pollen was carried not as the
ıeybees do, because she has no pollen
kets on her legs; but it was dusted into
fur on the lower side of her body; as
: scraped the pollen off, she mixed it
h some nectar which she had also
ınd in the flowers, and made it into a
ıty mass and heaped it at the bottom of
: cup; she probably made many visits to
wers before she had a sufficient amount
this bee pastry, and then she laid an egg
on it; after this, she immediately flew
ck to the rose bush to cut a lid for her
). She is a nice mathematician and she
:s the lid just a little larger than the rim
the cup, so that it may be pushed down
making it fit very closely around the
ges; she then cuts another and perhaps
other of the same size and puts them
er and fastened to the first cover. When
ished, it is surely the prettiest baby bas-
t ever made by a mother, all safely en-
ısed to keep out enemies. But her work
then only begun. She has other baby
skets to make and she perhaps makes
ı or more, placing one cup just ahead of
other in the little tunnel.

But what is happening meanwhile to
e bee babies in the baskets? The egg
tches into a little white bee grub which
ls to and eats the pollen and nectar
ste with great eagerness. As it eats, it
ows and sheds its skeleton skin as often
it becomes too tight, and then eats and
ows some more. How many mothers
uld know just how much food it would
quire to develop a child from infancy
ıtil it grows up! This bee mother knows
ll this amount, and when the food is all
ne, the little bee grub is old enough to
ange to a pupa; it looks very different
w, and although it is mummy-shaped,
: can see its folded wings and antennæ.
ter remaining a motionless pupa for a
w days, it sheds its pupa skin and now
is a bee just like its mother; but as the
dest bee is at the bottom of the tunnel,
en after it gets its wings and gnaws its
ıy out of its basket, it very likely cannot
cape and find its way out into the sun-
iny world, until its younger brothers and
ters have gone out before it.

Anna C. Stryke

A pansy cut by a leaf-cutter bee

There are many species of these leaf-cutter bees and each species makes its own kind of nest, always cutting the same size of circlets and usually choosing its own special kind of leaf to make this cradle. Some are daintier in their tastes and use rolled petals instead of leaves; and we have found some tiny cups made of gorgeous peony petals, and some of pansy petals, a most exquisite material.

At Chautauqua, New York, we found a species which rolled maple leaves into a tube that held three or four cups, and we also found there a bee stowing her cups in the open end of a tubular rod used to hold up an awning. There are other species which make short tunnels in the ground for their nests; perhaps the most common of all wedge their cups between or beneath the shingles on the roofs of summer cottages. But, however or wherever the leaf-cutter works, she is a master mechanic and does her work with niceness and daintiness.

LESSON 96

The Leaf-cutter Bee

Leading Thought — When we see the edges of rose leaves with holes of regular pattern in them, some of the holes being oblong and some circular, we know the leaf-cutter bee has cut them to make her cradle cups.

Method — It is very easy to find in June or autumn the leaves from which the leaf-cutter bee has cut the bedding for her young. Encourage the pupils to look for the nest during the summer and to bring some of the cups to school when they return, so that they may be studied in detail; meanwhile the teacher may tell the story of the nest. This is rather difficult for the pupils to work out.

Observations — 1. Do you find rose leaves with round holes cut in their edges? Do you find on the same bush some leaflets with oblong holes in them? Sketch or describe the rose leaf thus cut, noting exactly the shape of the holes. Are the circular holes of the same size? Are the long holes about equal in size and shape? Do you find any other plants with holes like these cut in them? Do you find any petals of flowers thus cut?

2. What do you think made these holes? If an insect were taking a leaf for food would the holes be as regular? Watch the rose bush carefully and see if you can discover the insect which cuts the leaf.

3. Have you ever seen the little black bee carrying pieces of rose leaves between her front feet? With what instrument do you suppose she cut the leaves? Where
you think she was going?

4. Have you ever found the nest of
leaf-cutter bee? Was it in a tunnel ma
in dead wood or in some crack or cran
How many of the little rose-leaf cups
there in it? How are the cups placed? A
the little bees still in the cups or can y
see the holes through which they crawl
out?

5. Take one cup and study it careful
How are the pieces of leaves folded
make the cups? How is the lid put o
Soak the cup in water until it comes ap
easily. Describe how many of the lo
pieces were used and how they were be
to make a cup. Of how many thicknes
is the cover made? Are the covers just t
same size as the top of the cup or a lit
larger? How does the cover fit so tight

6. If you find the nest in July or ea
August, examine one of the cups carefu
and see what there is in it. Take off t
cover without injuring it. What is at t
bottom of the nest? Is there an ins
within it? How does it look? What is
doing? Of what do you think its food w
made? How and by whom was the fo
placed in the cup? Place the nest in a b
or jar with mosquito netting over the to
and put it out of doors in a safe a
shaded place. Look at it often and s
what this insect changes into.

7. If the mother bee made each lit
nest cup and put in the beebread a
honey for her young, which cup contai
the oldest of the family? Which t
youngest? How do you think the fu
grown bees get out of the cup?

8. Do you think that the same speci
of bee always cuts the same sized holes
a leaf? Is it the same species which c
the rose leaves and the pansy petals?

THE LITTLE CARPENTER BEE

Take a dozen dead twigs from almost any sumac or elder, split them lengthwise, and you will find in at least one or two of them a little tunnel down the center where there was once pith. In the month of June or July, this narrow tunnel is ma
into an insect apartment house, one lit
creature in each apartment, partitioned
from the one above and the one belo
The nature of this partition reveals to

her the occupants are bees or wasps; is made of tiny chips, like fine saw- glued together, a bee made it and : are little bees in the cells; if it is e of bits of sand or mud glued to- er, a wasp was the architect and young s are the inhabitants. Also, if the food e cells is pollen paste, it was placed e by a bee; if paralyzed insects or ers are in the cells, a wasp made the

he little carpenter bee (*Ceratina a*) is a beautiful creature, scarcely one ter of an inch in length, with metallic body and rainbow tinted wings. In ng, she selects some twig of sumac, r, or raspberry which has been broken, thus gives her access to the pith; this at once begins to dig out, mouthful by thful, until she has made a smooth nel several inches long; she gathers pol- and packs beebread in the bottom of cell to the depth of a quarter-inch, then lays upon it a tiny white egg. She gs back some of her chips of pith and s them together, making a partition at one-tenth of an inch thick, which fastens firmly to the sides of the tun- this is the roof for the first cell and floor of the next one; she then gathers

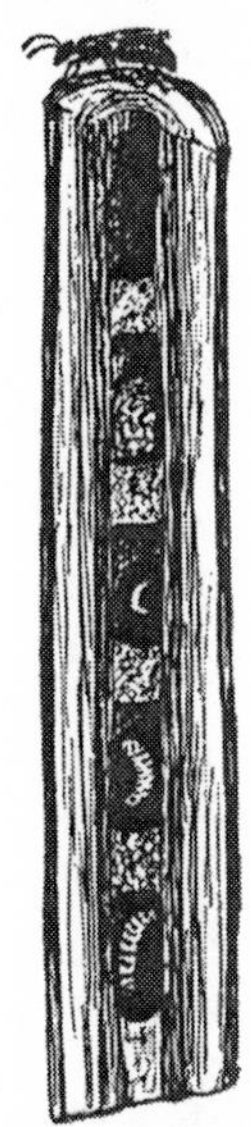

he little carpenter bee; her nest, cut open how the eldest larva at the bottom and youngest nearest the entrance

more pollen, lays another egg, and builds another partition.

Thus she fills the tunnel, almost to the opening, with cells, sometimes as many as fourteen; but she always leaves a space for

Nest of the carpenter wasp

a vestibule near the door, and in this she makes her home while her family below her are growing up.

The egg in the lowest cell of course hatches first; a little bee grub issues from it and eats the beebread industriously. This grub grows by shedding its skin when it becomes too tight, then changes to a pupa, and later to a bee resembling its mother. But, though fully grown, it cannot get out into the sunshine, for all its younger brothers and sisters are blocking the tunnel ahead of it; so it simply tears down the partition above it and kicks away the little pieces. The little grub bides its time until the next youngest brother or sister tears down the partition above its head and pushes the fragments into the very face of the elder, which, in turn, pushes them away. Thus, while the young bees are waiting, they are kept more or less busy pushing behind them the broken bits of all the partitions above them. Finally, the youngest gets its growth, and there they all are in the tunnel, the broken partitions behind the hindmost at the bottom of the nest, and the young bees packed closely together in a row with heads toward the door. When we find the nest at this period, we know the mother because her head is toward her young ones and her back to the door. A little later, on some bright morning, they all come out into the sunshine and flit about on gauzy, rainbow wings, a very happy family, out of prison.

But if the brood is a late one, the home must be cleaned out and used as a winter nest, and still the loyal little mother bee stays true to her post; she is the last one to enter the nest; and not until they are all housed within, does she enter. It is easy to distinguish her, for her wings are torn and frayed with her long labor of

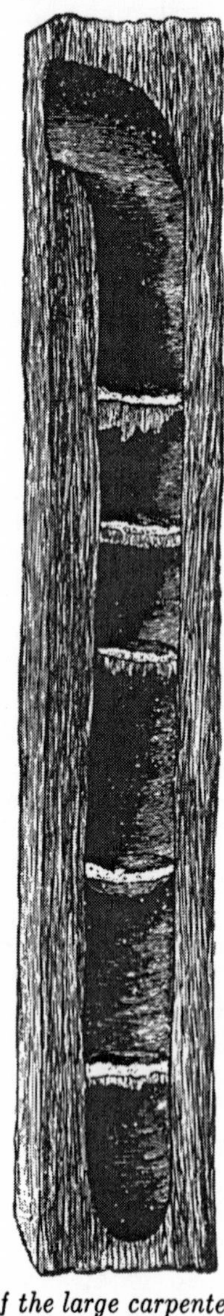

Nest of the large carpenter wasp

building the nest, until they scarcely serve to carry her afield; but she remains on guard over her brood.

The story of the little carpenter wasps is similar to that of the bee, except that we have reason to believe they often use her abandoned tunnels instead of making new ones. They make their little partitions out of mud; their pupæ are always in long, slender, silken cocoons, and we have no evidence that the mother remains in att ance.

LESSON 97

The Little Carpenter Bee

Leading Thought — Not all bees in colonies like the honeybees and b blebees. One tiny bee rears her b within a tunnel which she makes in pith of sumac, elder, or raspberry.

Method — This lesson may be give early summer or in autumn. In sprin early summer, the whole family of in their apartments may be observe autumn, the empty tenement with fragments of the partitions still clin may be readily found and examined; sometimes a whole family may be fo stowed away in the home tunnel, for winter.

Observations — 1. Collect dead t of sumac or elder and cut them in lengthwise. Do you find any with the tunneled out?

2. How long is the tunnel? Are its s smooth? Can you see the partitions w divide the long narrow tunnel into c Look at the partitions with a lens, if ne sary, to determine whether they are m of tiny bits of wood or of mud. If m of mud, what insect made them? If o tle chips, how and by what were constructed?

3. Are there any insects in the cell so, describe them. Is there beebread in cells?

4. For what was the tunnel m With what tools was it made? How the partitions fastened together? H does a young bee look?

5. Write the story of the oldest of bee family which lived in this tun Why did it hatch first? On what di feed? When it became a full-fledged what did it do? How did it finally get

Take a glass tube, the hollow at the er being about one-eighth of an inch ss, a tube which you can get in any store. Break this tube into sections r seven inches long, wrap around each a black paper or cloth made fast with rubber bands, and suspend them in a hedge or among thick bushes in May. Examine these tubes each week to see if the wasps or bees are using them.

THE BUMBLEBEE

Thou, in sunny solitudes,
Rover of the underwoods,
The green silence dost replace
With thy mellow, breezy bass.
— EMERSON

here seems to have been a hereditary between the farm boy and the bumbee, the hostilities usually initiated by boy. Like many wars, it is very foolish wicked, and has resulted in great harm oth parties. Luckily, the boys of today more enlightened; and it is to be hoped t they will learn to endure a bee sting two for the sake of protecting these inishing hosts, upon which so many vers depend for carrying their pollen; of all the insects of the field, the bumbees are the best and most needed nds of flowers.

The bumblebees are not so thrifty and ehanded as are the honeybees, and do provide enough honey to sustain the ole colony during the winter. Only the ther bees, or queens as they are called, vive the cold season. Just how they do ve do not know, but probably they are ter nourished and therefore have more durance than the workers. In early May, e of the most delightful of spring visitts is one of these great buzzing queens, ing low over the freshening meadows, ing to find a suitable place for her nest; d the farmer or fruit grower who knows business is as anxious as she that she d suitable quarters, knowing well that e and her children will render him most icient aid in growing his fruit and seed. e finally selects some cosy place, very ely a deserted nest of the field mouse, d there begins to build her home. She ils early and late, gathering pollen and nectar from the blossoms of the orchard and other flowers which she mixes into a loaf as large as a bean upon which she lays a few tiny eggs and then covers them with wax. She then makes a honey-pot of wax

A bumblebee

as large as a small thimble and fills it with honey; thus provided with food she broods over her eggs, keeping them warm until they hatch. Each little bee grub then burrows into the beebread, making for itself a cave while satisfying its hunger. When fully grown, it spins about itself a cocoon, changes to a pupa, and then comes out a true bumblebee but smaller than her queen mother. These workers are daughters and are happy in caring for the growing family; they gather pollen and nectar and add to the mass of beebread for the young to burrow in; meanwhile the queen remains at home and devotes her energies

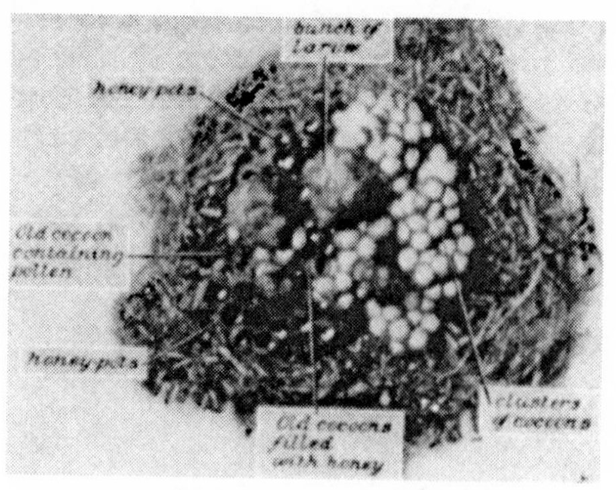

A bumblebee nest in midsummer

to laying eggs. The workers not only care for the young, but later they strengthen the silken pupa cradles with wax, and thus make them into cells for storing honey. When we understand that the cells in the bumblebee's nest are simply made by the young bees burrowing in any direction, we can understand why the bumblebee comb is so disorderly in the arrangement of its cells. Perhaps the boy of the farm would find the rank bumblebee honey less like the ambrosia of the gods if he knew that it was stored in the deserted cradles and swaddling clothes of the bumblebee grubs.

All of the eggs in the bumblebee nest in the spring and early summer develop into workers which do incidentally the vast labor of carrying pollen for thousands of flowers; to these only is granted the privilege of carrying the pollen for the red clover, since the tongues of the other bees are not sufficiently long to reach the nectar. The red clover does not produce seed in sufficient quantity to be a profitable crop unless there are bumblebees to pollinate its blossoms. Late in the summer, queens and drones are developed in the bumblebee nest, the drones, as with the honeybees, being mates for the queens. But of all the numerous population of the bumblebee nest, only the queens survive the rigors of winter, and on them and their success depends the future of the bumblebee species.

There are many species of bumblebees, some much smaller than others, but they all have the thorax covered with plush above and the abdomen hairy, and their fur is usually marked in various patt of pale yellow and black. The bumble of whatever species, has short but active antennæ and a mouth fitted for ing as well as for sucking. Between large compound eyes are three simple The wings are four in number and str the front legs are very short; all the have hairs over them and end in a th jointed foot, tipped by a claw. On hind leg, the tibia and the first tarsal j are enlarged, making the pollen basket which the pollen is heaped in gol masses. One of the most interesting ob vations possible to make is to note the bumblebee brushes the pollen f her fur and packs it into her pollen kets.

LESSON 98
The Bumblebee

Leading Thought — The bumbleb are the chief pollen carriers for most our wild flowers as well as for the clov and other farm plants. They should, the fore, be kindly treated everywhere; and should be careful not to hurt the big que bumblebee, which we see often in May

Method — Ask the questions and courage the pupils to answer them as th have opportunity to observe the bumb bees working in the flowers. A bumbleb may be imprisoned in a tumbler for a sh period for observation, and then allow to go unharmed. It is not advisable study the nest, which is not only a dang ous proceeding for the pupil, but al means the destruction of a colony of the very useful insects. However, if the lo of a nest is discovered, it may be dug and studied after the first heavy frost. cial stress should be laid upon the ob- ations of the actions of the bees when ting flowers.

)BSERVATIONS — 1. In how many flow- do you find the bumblebee? Watch closely and see how she gets the nec- Notice how she "bumbles around" in ower and becomes dusted with pollen. atch her and note how she gets the len off her fur and packs it in her pollen kets. On which legs are her pollen bas- s? How does the pollen look when ked in them? What does she do with llen and nectar?

2. Catch a bumblebee in a jelly glass d look at her closely. Can you see three le eyes between the big compound es? Describe her antennæ. Are they active? How many pairs of wings has she? Do you think they are strong? Which pair of legs is the shortest? How many segments are there in the leg? Do you see the claws on the foot?

3. What is the bumblebee's covering? What is the color of her plush? Is she furry above and below?

4. Can you see that she can bite as well as suck with her mouth-parts? Will a bumblebee sting a person unless she is first attacked?

5. Have you seen the very large queen bumblebee in the spring, flying near the ground hunting for a place to build a nest? Why must you be very careful not to hurt her? How does she pass the winter? What does she do first, in starting the nest?

6. In how many ways does the bumblebee benefit us?

THE HONEYBEE

During many years naturalists have en studying the habits and adaptations the honeybees, and, as yet, the story of eir wonderful ways is not half told. Al- ough we know fairly well what the bees , yet we have no inkling of the processes hich lead to a perfect government and anagement of the bee community; and en the beginner may discover things ver known before about these fascinat- g little workers. In beginning this work might be well to ask the pupils if they ave ever heard of a republic that has any kings and only one queen; and here the citizens do all the governing ithout voting, and where the kings are owerless and the queen works as hard as nd longer than any of her subjects; and hen tell them that the pages of history ontain no account of a republic so won- erful as this; yet the nearest beehive is he home of just this sort of government.

In addition to the interest of the bee olony from a nature-study standpoint, it s well to get the children interested in bee- eeping as a commercial enterprise. A mall apiary well managed may bring in n acceptable income; and it should be the source of a regular revenue to the boys and girls of the farm, for one hive should net the young beekeeper from three to five dollars a year and prove a business education to him in the meantime.

Bees are perfect socialists. They have noncompetitive labor, united capital, communal habitations, and unity of interests. The bee commune is composed of castes as immutable as those of the Brahmins, but these castes exist for the benefit of the whole society instead of for the individuals belonging to them. These castes we have named queens, drones, and workers, and perhaps we should first of all

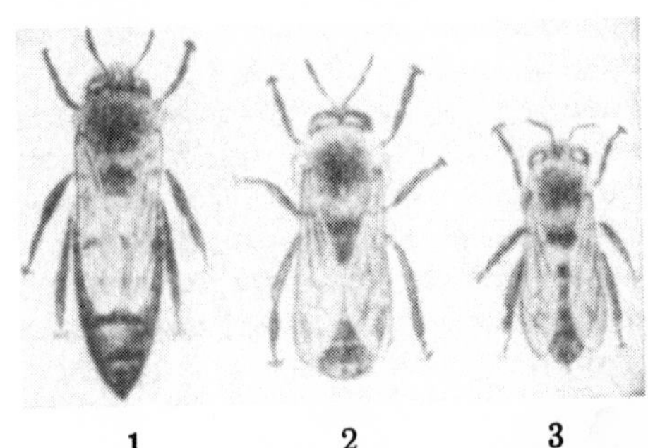

1 2 3

1, *queen bee*. 2, *drone*. 3, *worker*

study the physical adaptations of the members of these castes for their special work in the community.

The Worker

There are three divisions to the body of the bee, as in all insects — head, thorax,

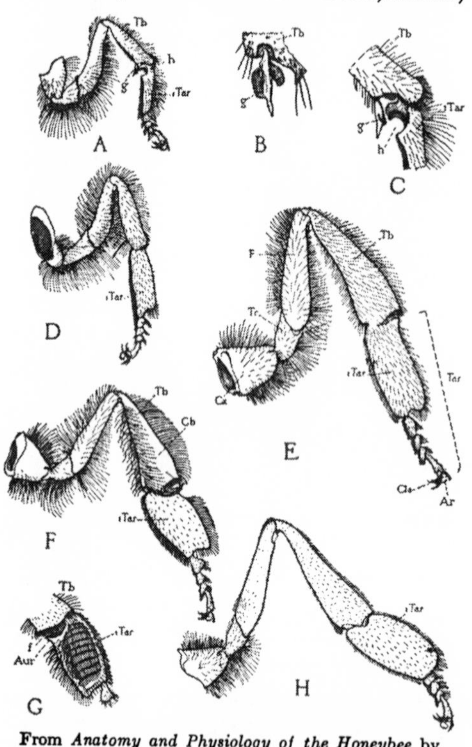

From *Anatomy and Physiology of the Honeybee* by Snodgrass. McGraw-Hill Book Company, Inc.

Legs of the honeybee

A, front leg of worker showing antenna cleaner (g, h); B, spine of antenna cleaner; C, antenna cleaner enlarged; D, middle leg of worker, anterior surface; E, left hind leg of queen, outer surface; F, left hind leg of worker, outer surface, showing pollen basket (Cb); G, first tarsal joint (Tar.), inner surface, of leg of worker, showing pollen comb (transverse rows of spines); H, left hind leg of drone, outer surface

and abdomen. The head bears the eyes, antennæ, and mouth-parts (p. 393, W). There are two large compound eyes on either side of the head and three simple eyes between them. The antennæ arise from the face, each consisting of two parts, one straight segment at the base, and the end portion which is curved and made up of many segments. There is also a short, beadlike segment where the antenna joins the face. A lens is needed to see the jaws of the bee, folded across, much like a pair of hooks, and below them the tong which is a sucking tube; the length of tongue is very important, for upon depends the ability of the bee to get ne from the flowers.

The thorax bears three pairs of legs low and two pairs of wings above. Each consists of six segments, and the foot tarsus has four segments and a pair claws. The front leg has an antenna co between the tibia and tarsus, A(g,h) a C; the hind leg has a pollen bask which is a long cavity bordered by ha wherein the pollen is packed and carri F(Cb).

On the other side of the large joint yond the pollen basket are rows of spi which serve to collect pollen grains fr other parts of the body, G, and tween these two large segments is a cl through which pollen is forced in loadi the baskets. This loading must occur wh the bee is on the wing, so that the le may be free for the peculiar actions which the loading is brought about.

The front pair of wings is larger th the hind pair. The wings of the old be that have done much work are alwa frayed at the edges.

There are six segments or rings to t abdomen, plainly visible from above. the three to five segments next the thor are marked above with yellow bands their front edges, the bee is an Italia On the lower side of the abdomen, fo of the segments are composed of a ce tral part with an overlapping plate each side. These flaps cover the eig areas through which wax is secreted; b without dissection this cannot be see except when the wax plates are abnormal large, in which event they may protru and be visible. The flecks of wax the formed are used by the bees to buil their combs.

The Queen

The queen bee is a truly royal insec She is much larger than the worker, h body being long and pointed, and exten ing far beyond the tips of her closed wing giving her a graceful form. She has no po baskets or pollen comb upon her legs, use it is not a part of her work to er pollen or honey. The queen bee s life as an ordinary worker egg, which lected for special development. The cers tear down the partitions of the around the chosen egg and build a ection over the top, making an apart-t. The little white bee grub, as as it hatches, is fed for five days the same food that is given to the ker grubs in the earliest part of their ling period; it is a special substance, eted by the worker bees, called royal . This food is very nourishing, and af-being reared upon it, the princess larva ves around herself a silken cocoon and nges to a pupa. Meanwhile the workers e sealed her cell with wax.

Vhen the princess pupa changes to the -grown queen she cuts a circular door he cover of the cell and pushes through nto the world. Her first real work is to nt for other queen cells, and if she finds she will, if not hindered, make a hole its side and destroy the poor princess hin. If she finds another full-grown zen, the two fight until one succumbs. e queen rarely uses her sting upon any-ng or anyone except a rival queen.

After a few days she takes her marriage ht in the air, where she mates with ne drone, and then returns to her hive 1 begins her great work as mother of the colony. She runs about on the comb, pokes her head into a cell to see if it is ready, then turning about thrusts her abdomen in and neatly glues an egg fast to the bottom.

When the honey season is at its height she works with great rapidity, sometimes laying at the rate of six eggs a minute, often producing two thousand eggs during a day, which would equal in weight her own body. If the workers do not allow her to destroy the other developing queens, she departs from the hive with a major portion of the worker bees in what is known as a swarm, seeking a home elsewhere.

The Drone

The drone differs much from the queen and the worker. He is broad and blunt, being very different in shape from the queen, and larger than the worker (p. 391, Fig. 2). He has no pollen baskets on his legs and has no sting. His eyes are very much larger than those of the queen or the worker and unite at the top of the head (D, below). His wings are larger and stronger than those of the worker or

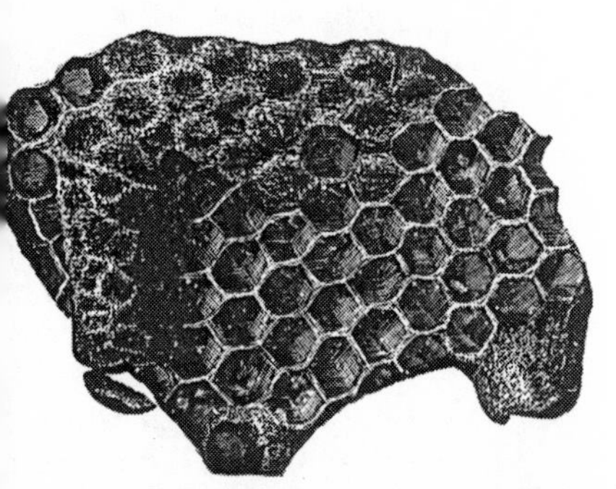

Comb of the honeybee. The beginnings of o queen cells are represented on the lower ge of the comb, and a completed queen cell tends over the face of the comb near the left le. From the lower end of it hangs a lid hich was cut away by the workers to allow e queen to emerge

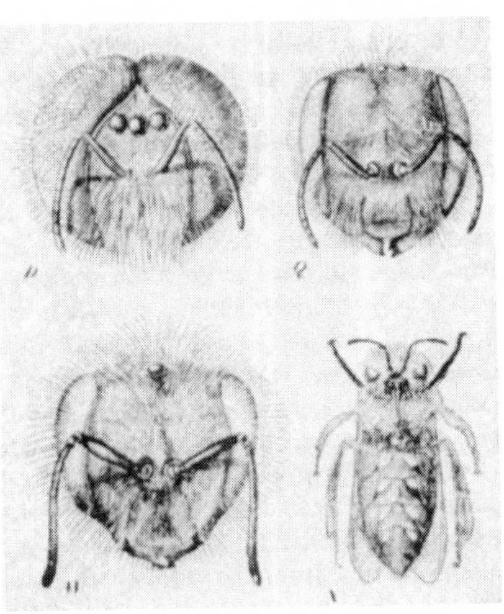

A. J. Hammar

D, head of drone. Q, head of queen bee. W, head of worker. X, worker bee seen from underneath, showing plates of wax secreted from the wax pockets

queen. It is not his business to go out and gather honey or to help in the work of the hive. His tongue is not long enough to get honey from the flowers; he has no pollen basket in which to carry pollen; he has no sting to fight enemies and no pockets for secreting wax; he is fed by his sister workers until the latter part of the season when the honey supply runs low, and then he is driven from the hive to die of starvation. The drone should be called a prince or king, since his particular office in the hive is to mate with the queen.

LESSON 99

The Honeybee

Leading Thought — In a colony of honeybees there are three different forms of bees, the queens, the drones, and the workers. All of these have their own special work to do for the community.

Method — In almost every country or village community there is an apiary, or at least someone who keeps a few colonies of bees; to such the teacher may turn for material for this lesson. If this is not practical the teacher may purchase specimens from any bee dealer; she may, for instance, get an untested queen with attendant workers in a queen cage sent by mail for a small sum. These could be kept alive for some time by feeding them with honey, during which time the pupils can study the forms of the two castes. Any apiary during September will give enough dead drones for a class to observe. Although ordinarily we do not advocate the study of dead specimens, yet common sense surely has its place in nature-study; and in case of the honeybee, a closer study o form of the insect than the living might see fit to permit is desirable. T are no more wonderful instances of a tation of form to life than is foun the anatomy of the workers, queens, drones; moreover, it is highly desir if the pupils are ever to become beek ers, that they know these adaptations.

A lens is almost necessary for these sons and a compound microscope with a low power would be a very desir adjunct. This lesson should not be g below the fifth grade; and it is be adapted to eighth-grade work.

The Worker

Observations — 1. How many sions of the body are there?

2. What organs are borne on the h

3. Are there small, simple eyes betw the large compound ones?

4. What is the difference between large eyes and the small?

5. Describe the antennæ.

6. What can you see of the mouth? scribe it.

7. Look at the tongue under the mi scope and see how it is fitted for gett nectar from flowers.

8. What organs are borne on the t rax?

9. Study the front or middle leg. H many joints has it?

10. With a lens find the anten cleaner on the front leg. Describe it.

11. Describe the feet and claws.

12. Compare the third segment of hind leg with that of the front leg.

13. Note that this segment of the h leg is much wider. Note its form and scribe how it forms the pollen basket.

14. Note the cleft through which pollen is forced in loading the pollen b kets and the pollen combs just below it

15. Compare the front and hind w as to shape and size.

16. How many rings are there on abdomen and how are the rings colo above?

·. Study the lower side of the body; ou know where the wax comes from?

. Write an account of the develop- t of the larva of the worker bee; the es of a worker bee from the time it s from its cocoon until it dies work- for the colony.

The Queen Bee

. How does the queen differ in size shape from the worker?

. Has she pollen baskets or pollen bs on her hind legs?

. How does the shape of the abdomen er from that of the worker?

. Write a story of the life of a queen . This should cover the following nts: The kind of cell in which the en is developed; the kind of food on ch she is reared; the fact that she rarely gs people, but reserves her sting for er queens; why she does not go out to her honey; how and by whom and on at she is fed; she would not use pollen kets if she had them; the work she does the colony; the length of her life com- ed with that of a worker; the time of year when new queens are developed, and what becomes of the old queen when a new one takes her place; why she is called a queen.

The Drone

1. How does the drone differ in size and form of body from the worker?
2. How does he differ in these respects from the queen?
3. Has he pollen baskets on his legs?
4. Has he a sting?
5. Compare his eyes with those of the queen and the worker.
6. Compare the size of his wings with those of the queen and the worker.
7. Write a composition on the drone. This should cover the following points: In what sort of cell the drone is developed; whether he goes out to gather honey or help in the work of the hive; how he is fed; how he is unfitted for work for the colony in the following particulars: tongue, lack of pollen baskets, lack of sting, and of wax pockets; why the drone should be called a prince or king; the death of the drones; when and by what means it occurs.

HONEYCOMB

The structure of honeycomb has been ages admired by mathematicians, who ve measured the angles of the cells and monstrated the accurate manner in ich the rhomb-shaped cell changes at base to a three-faced pyramid; and have oved that, considering the material of nstruction, honeycomb exemplifies the ongest and most economic structure ssible for the storing of liquid contents. hile recent instruments of greater pre- ion in measuring angles show less per- ction in honeycomb than the ancients lieved, yet the fact still stands that the neral plan of it is mathematically ex- llent.

Some have tried to detract from bee ill, by stating that the six-sided cell is nply the result of crowding cells to- ther. Perhaps this was the remote origin of the hexagonal cell; but if we watch a bee build her comb, we find that she begins with a base laid out in triangular pyramids, on either side of which she builds out six-sided cells. A cell just begun, is as distinctly six-sided as when completed.

The cell of a honeycomb is six-sided in cross section. The bottom is a three-sided pyramid and its sides help form pyramids at the bottom of the cells opposite, thus economizing every particle of space. In the hive, the cells usually lie horizontal, although sometimes the combs are twisted. The honey is retained in the cell by a cap of wax which is made in a very cunning fashion; it consists of a circular disc at the middle supported from the six angles of the cell by six tiny girders. The comb is made fast to the section of the hive by being plastered upon it. The comb foundation sold to apiarists is quite thick, so that the edges of the cell may be drawn out and almost complete the sides of the cell. This comb foundation is beautifully constructed in imitation of the base of the normal cells, and some-

A section of honey. Each cell is capped and supported by six girders

times has some surplus wax in it which can be used to draw out the first part of the sides of the cell. In order to make a fine section of comb honey, the apiarist uses a full sheet of this material, which guides the bees in the direction of their comb and gives uniform cells throughout. The cells of honeycomb are used also for the storing of beebread and also as cradles for the young bees.

LESSON 100

The Honeycomb

Leading Thought — The cells of honeycomb are six-sided and in double rows and are very perfectly arranged the storing of honey, so as to save roor

Materials — A section filled honey and also a bit of empty comb a bit of commercial comb founda which may be obtained in any apiary.

Observations — 1. Look at a bi empty honeycomb; what is the shap the cell as you look down into it?

2. What is the shape of the botton the cell?

3. How does the bottom of the cell the bottom of the cell opposite? Exp how honeycomb economizes space as s age for honey, and why an economy space is of use to bees in the wild stat

4. In the hive is the honeycomb pla so that the length of the cells is horizo or up and down?

5. Observe honeycomb contair honey; how is the honey retained in cells?

6. Carefully take off a cap from honey cell and see if you can find the girders that extend inward from the an of the cell to support the circular port in the center.

7. By what means is the honeyco made fast to the sides of the section or hive?

8. Study a bit of comb foundation note where the bees will pull out the to form the cell.

9. Why and how is comb foundat used by the bee-keeper?

10. For what purpose besides stor honey are the cells of honeycomb used the bees?

INDUSTRIES OF THE HIVE AND THE OBSERVATION HIVE

Beehives are the houses which man furnishes for the bee colonies, the wild bees ordinarily living in hollow trees or in caves. The usual hive consists of a box which is the lower story and of one or more upper stories, called "supers." In the lower story are placed frames for the brood and for storing the honey for the winter use of the bees. In the supers are placed either large frames containing comb for the storage of honey which is to be thrown and sold as liquid honey by means o honey extractor, or smaller sections wh contain about one pound of honey a which are sold as made by the bees. I the habit of the bees to place their br in the lower part of their nests and st honey in the upper portions. The b keepers have taken advantage of this ha of the bees and remove the supers w r filled combs and replace them with ers to be filled, and thus get a large of honey. The number of bees in a ny varies; there should be at least y thousand in a healthy colony. Of se a large proportion are workers; there be a few hundred drones the latter t of the season, but only one queen.

Honeycomb is built of wax and is hung n the frame so that the cells are hori-tal; its purpose is to cradle the young for the storage of pollen and honey. e wax used for building the comb is a retion of the bees; when comb is ded, a number of self-elected bee citi-s gorge themselves with honey and g themselves up in a curtain, each bee ching up with her fore feet and taking d of the hind feet of the one above her. er remaining thus for some time the x appears in little plates, one on each e of the second, third, fourth, and fifth ments of the abdomen. This wax is ewed by the bees and made into comb.

Honey is made from the nectar of flow- which the bee takes into her honey mach. This, by the way, is not the true mach of the bee and has nothing to do th digestion. It is simply a receptacle storing the nectar, which is mixed with me secretion from the glands of the bee at brings about chemical changes, the ief of which is changing the cane sugar the nectar into the more easily digested ape sugar and fruit sugar of the honey. ter the honey is emptied from the ney stomach into the cell, it remains posed to the air for some time before e cell is capped, and thus ripens. It is interesting fact that up to the seven-enth century honey was the only means eople had for sweetening their food, as gar was unknown.

Beebread is made from the pollen of owers which is mixed with nectar or oney so as to hold it together; it is car-ed from the field on the pollen baskets of e hind legs of the workers; it is packed to the cell by the bees and is used for od for the developing brood. Propolis is ee glue; it is used as a cement and varnish; is gathered by the bees from the leaf

A. I. Root Co.

An observation hive

buds of certain trees and plants, although when they can get it, the bees will take fresh varnish. It is used as a filler to make smooth the rough places of the hive; it often helps hold the combs in place; it calks every crack; it is applied as a varnish to the cells of the honeycomb if they remain unused for a time, and if the door of the observation hive be left open, the bees will cover the inside of the glass with this glue, and thus make the interior of the hive dark.

The young bees are footless, white grubs. Each one lives in its own little cell and is fed by the nurse bees, which give it food already largely digested; this food the nurse bees secrete from glands in their heads.

The removal of honey from the supers does not do any harm to the bee colony if there is enough honey left in the brood chambers to support the bees during the winter. There should be forty pounds of honey left in the brood chamber for winter use. In winter, the hives should be protected from the cold by being placed in special houses or by being encased in larger boxes, an opening being left so that the bees may come out in good weather. The chaff hive is best for both winter and summer, as it surrounds the hive with a space which is filled with chaff, and keeps the hive warm in winter and cool in summer. Many beekeepers put their bees in cellars during the winter, but this method is not as safe as the packed hive. Care should be taken in summer to place the hives so that they are shaded at least part of the day. The grass should be mown around the hives so that the bees will not become entangled in it as they return from the fields laden with honey.

What may be seen in the observation hive — First of all, it is very interesting to watch the bees build their comb. When more comb is needed, certain members of the colony gorge themselves with honey and remain suspended while it oozes out of the wax pockets on the lower side of the abdomen. This wax is collected and chewed to make it less brittle and then is carried to the place where the comb is being built and is molded into shape by the jaws of the workers. However, the bee that puts the wax in place is not always the one that molds it into comb.

A bee comes into the hive with her honey stomach filled with nectar and disgorges this into a cell. When a bee comes in loaded with pollen, she first brushes it from the pollen baskets on her hind legs into the cell; later another worker comes along and packs the pollen grains into the cell.

The bee nurses run about on the comb feeding the young bee grubs partially digested honey and pollen. Whenever the queen moves about the comb she is followed by a retinue of devoted attendants which feed her on the rich and perfectly digested royal jelly and also take care of her royal person and give her every attention possible. The queen, when laying, thrusts her abdomen into the cell and glues a little white egg to the bottom. The specially interesting thing about this is that the queen always lays an egg which will produce a female or worker in the smaller cells, and will always lay an egg to produce a drone or male in the larger cells.

If there is any foreign substance in the observation hive, it is interesting to see the bees go to work at once to remove it. They dump all of the debris out in front of the hive. They close all crevices in the hive; and they will always curtain the glass, if the door is kept open too much, with propolis or bee glue, the sticky substance which they get from leaf buds and other vegetable sources. When bees fan to up a current of air in the hive, they g back and forth, moving the wings so idly that we can only see a blur about t bodies.

If drones are developed in the hiv is interesting to see how tenderly t are fed by their sister workers, altho they do not hesitate to help themsel to the honey stored in the cells; and if observation hive is working during S tember, undoubtedly the pupils may able to see the murder of the drones their sisters. But the children should derstand that this killing of the dro is necessary for the preservation of colony, as the workers could not st enough honey to keep the colony a during the winter if the drones were lowed to go on feeding.

If you see the worker bees fighting means that robbers are attempting to at the stores of the observation hive. T entrance to the hive should at once contracted by placing a block of wood front, so that there is room for only c bee at a time to pass in and out.

LESSON 101

The Industries of the Hive

Leading Thought — In the hive carried on the industries of wax-maki building of honeycomb, storing of hon and beebread, caring for the young, ke ing the hive clean and ventilated, and ca ing all crevices with bee glue.

Method — This lesson should be the nature of a demonstration. If th is an apiary in the neighborhood, it quite possible that the teacher may sh the pupils a hive ready for occupancy the bees; in any case she will have difficulty in borrowing a frame of bro comb, and this with a section of hon which can be bought at the grocery, sufficient if there is no observation hi This lesson may be an informal talk b tween teacher and pupils.

An observation hive in the schoolroo is an object of greatest interest to t pupils, as through its glass sides they m

ble to verify for themselves the won-
l tales concerning the lives and do-
of the bees which have been told us
aturalists. Moreover, the study thus
e of the habits of the bees is an ex-
nt preparation for the practical apia-
and we sincerely believe that bee-
ing is one of the ways by which the
and girls of the farm may obtain
ey for their own use.

he observation hive is very simply con-
cted and can be made by anyone who
ws how to use ordinary carpenter tools.
simply a small, ordinary hive with a
of glass on each side which is cov-
by a hinged door. A hive thus made
aced so that the front end rests upon
indow-sill; the sash is lifted an inch
o, a strip of wood or a piece of wire
ing being inserted underneath the
except in front of the entrance of the
, to hinder the bees from coming
into the room. A covered passageway
uld extend from the entrance of the
to the outside of the window-sill.
s window should be one which opens
y from the playground so that the bees
ing and going will not come into col-
n with the pupils. The observation
dow should be kept carefully shut,
ept when the pupils are using it,
e the bees object to light in their
nes.

isted in the Source of Materials at
back of this book is an observation hive
ich we have used by stocking it afresh
h season, it being too small for a
-sustaining colony. But it has the ad-
tage of smallness which enables us to
all that is going on within it, which
uld be impossible in a larger hive. This
e comes in several sizes, and will be
pped from the makers stocked with
s.

OBSERVATIONS — *Industries and Care the Hive* — 1. What is the hive, and
at do wild bees use instead of the hive?
scribe as follows:

2. A brood chamber and a super and
uses of each.

3. How many and what bees live in a
e.

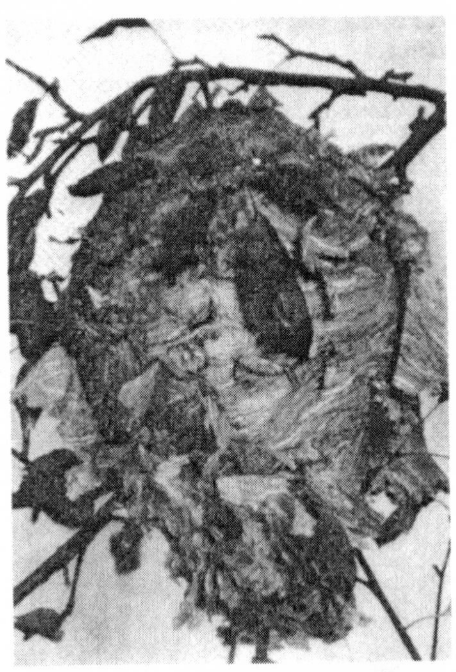

Verne Morton

A hornets' nest. The entrance, near the bottom, may be seen

4. How the honeycomb is made and placed and the purpose of it.

5. How the wax is produced and built into the comb.

6. How honey is made.

7. What beebread is and its uses.

8. What propolis is and what it is used for.

9. How young bees look and how they are cradled and fed.

10. Does the removal of the honey from the supers in the fall do any harm to the bee colony?

11. How much honey should a good-sized colony have in the fall to winter well?

12. How should the hives be protected in the winter and summer?

What may be seen in the observation hive — 13. Describe how a bee works when building honeycomb.

14. How does the bee act when storing honey in a cell?

15. How does a bee place pollen in a cell and pack it into beebread?

16. Describe how the nurse bees feed

the young, and how the young look when eating.

17. Describe how the "ladies in waiting" feed and care for the queen.

18. Try to observe the queen when she is laying eggs and describe her actions.

19. How do the bee workers keep their house clean?

20. How do they stop all crevices in the hive? If you keep the hive uncov too long, how will they curtain the dow?

21. Describe the actions of the when they are ventilating the hive.

22. If there are any drones in the describe how they are fed.

23. How can you tell queens, dr and workers apart?

INSECTS OF THE BROOK AND POND

The insects considered on pages 400 to 415 spend a part or all of their lives in brooks and ponds. These insects may be

From *Elementary Lessons on Insects*, Needham

A tray of water with floating algae, weeds, etc., dipped from a pond

A, a may-fly nymph. B, a damsel-fly nymph. C, a midge larva. D, the flocculent dwelling tube of a smaller midge larva. E, a water strider. F, a small dragonfly nymph. G, a tadpole

easily studied in the schoolroom, if an aquarium is available. The aquarium may be quite simple, or it may be a more elaborate one. See Fig. page 5.

LESSON 102

How to Make an Aquarium

The schoolroom aquarium may b very simple affair and still be effective. most any glass receptacle will do, glass ing chosen because of its transparency that the life within may be obser Tumblers, jelly tumblers, fruit jars, bu jars, candy jars, and battery jars are available for aquaria. The tumblers especially recommended for observing habits of aquatic insects.

To Make an Aquarium: 1. Place the jar a layer of sand an inch or more depth.

2. In this sand plant the water pla which you find growing under water a pond or stream; the plants most av able are waterweed, bladderwort, w starwort, water cress, stoneworts, frog s tle, or water silk.

3. Place on top a layer of small sto or gravel; this is to hold the plants place.

4. Tip the jar a little and pour in v gently at one side water taken from a po or stream. Fill the jar to within two three inches of the top; if it be a je tumbler, fill to within an inch of the t

5. Let it settle.

6. Place it in a window which does get too direct sunlight. A north wind is the best place; if there is no north w dow in the schoolroom, place it far enou at one side of some other window so th it will not receive too much sunlight.

To get living creatures for the aquar- use a dip net, which is made like a ow insect net.

Dip deep into the edges of the pond be sure to bring up some of the leaves mud, for it is in these that the little r animals live.

As fast as dipped up, these should be ed in a pail of water, so that they may arried to the schoolroom.

). After the material has been brought the schoolroom, it should be poured into a shallow pan so that it can be ed into other containers for further y. A little experience will soon show t kinds of creatures are likely to eat rs in an aquarium. By putting similar s together in one container, it will be e possible to distribute them in such anner that there will not be many fa- ies. It is well to put only a few crea- s in each container.

HE CARE OF THE AQUARIUM — Care ıld be taken to preserve the plant life he aquarium, as the plants are neces- to the life of the animals. They not / supply the food, but they give off oxygen which the animals need for ıthing, and they also take up from the er the poisonous carbonic acid gas n off from the bodies of the animals.

. The aquarium should be kept where re is a free circulation of air.

. If necessary to cover the aquarium revent the insects, like the water boat- n and water beetles, from escaping, tie r it a bit of mosquito netting, or lay n the top a little square of the wire ting used for window screens.

. The temperature should be kept ıer cool; it is better that the water of aquarium should not be warmer than 50 degrees Fahrenheit, but this is not al- ways possible in the schoolroom.

An inexpensive and durable aquarium

4. If any insects or animals die in the aquarium, they should be removed at once, as the decomposing bodies render the water foul.

5. To feed the animals that live upon other animals, take a bit of raw beef, tie a string to it and drop it in, leaving the free end of the string outside of the jar. After it has been in one day, pull it out; for if it remains longer it will make the water foul.

6. As the water evaporates it should be replaced with water from the pond.

THE DRAGONFLIES AND DAMSEL FLIES

A pond without dragonflies darting ve it, or without the exquisitely irides- t damsel flies clinging to the leaves of border would be a lonely place indeed. one watches these beautiful insects, e wonders at the absurd errors which ve crept into popular credence about them. Who could be so silly as to believe that they could sew up ears or that they could bring dead snakes to life! The queer names of these insects illustrate the prej- udices of the ignorant — devil's-darning- needles, snake doctors, snake feeders, etc. Despite all this slander, the dragonflies re-

AQUATIC INSECTS

1. Stone Fly, Plecoptera. *Left, adult; right, nymph. The adults are most commonly seen in numbers about street lights. The nymphs swim or crawl; they are found on the underside of rocks in swiftly-flowing permanent streams.*

2. May Fly, Ephemerida. *Left, nymph; right, adult. At most, the adults live only one or two days. The nymphs live in all sorts of aquatic situations.*

3. Back Swimmer, Notonecta. *The back of this insect is shaped like the bottom of a boat, so that by using the hind legs for oars it swims on its back with great ease.*

4. Water Boatman, Corixa. *Although this insect swims, it spends much of the time anchored on the bottom of the stream or pond. Even during the winter months the water boatman is active beneath the ice.*

5. Water Walking Stick, Ranatra. *A long breathing-tube is to be found at the end of this insect's abdomen; by means of this tube, the insect can rest at the bottom of a very shallow pond and still breathe by projecting the tube to the surface of the water.*

6. Water Scorpion, Nepa. *This insect looks quite lifeless as it waits quietly in the trash of a shallow pond for its prey. With its sharp sucking beak and its strong front legs it attacks many aquatic animals which are larger than itself.*

7. Water Bug, Belostoma. *After the female has glued her numerous eggs to the back of the male, he very obligingly stays near the surface of the water and elevates the eggs into the air where they hatch.*

8. Giant Water Bug or Electric-Light Bug, Benacus. *The striped eggs of this insect are large and are laid in clusters on some piece of vegetation which projects from water.*

9. Water Strider, Gerris. *These predacious insects move at a rapid but somewhat uncertain rate over the surface of more quiet waters.*

10. Dobson, Corydalis. *Larva at left; female adult in center; head of male at right. The larvæ, known as hellgrammites, are found under stones in the beds of swiftly-flowing streams.*

11. Predacious Diving Beetle, Dyti *Larva at left; adult at right. The larvæ aquatic creatures much larger than thems and suck the softer portions from their ies. The brownish-black adults are see quiet water.*

12. Diving Beetle, Acilius. *Larva at adult at right. It is a common sight to se slender larvæ hanging head down with air-breathing "tails" projecting into th through the upper surface of the water.*

13. Water Scavenger Beetle, Hydr ilus. *Larva at left; adult at right. In pools, these black beetles may be found sı ming through the water or hanging hea at the surface.*

14. Whirligig Beetle, Gyrinus. *One see companies of these bluish-black, flatt beetles whirling about over the surfac brooks or ponds. Their eyes are divide such a manner as to appear as four eye two looking into the water and two loo into the air.*

15. Water Penny or Riffle Bee Psephenus. *Larva, left, shows back vi right shows side view. During any se these larvæ can be found clinging tightl the lower surface of stones in rapid stre In general appearance they resemble a cru cean more than an insect.*

16. Black Fly, Simulium. *The larvæ aquatic and are able to maintain their posi in rapid water by means of a sucking dis the tail end of the body. Great number these larvæ form the so-called "black mo which is so evident in some streams in e summer.*

17. Crane Fly, Tipula. *Larva at l adult at right. Great variations as to ha and habitats exist among the larvæ of various kinds of crane flies; some are aqu some live on plants, and still others live soil.*

18. Drone Fly, Eristalis. *Left, larva, o called the rat-tailed maggot; right, ad The drone fly resembles so closely a m honey bee that as it hovers about flowers often mistaken for a drone bee. The larva rat-tailed maggot, lives about decaying pl and animal material in foul water. The t like appendage at the rear end of the bod a breathing-tube.*

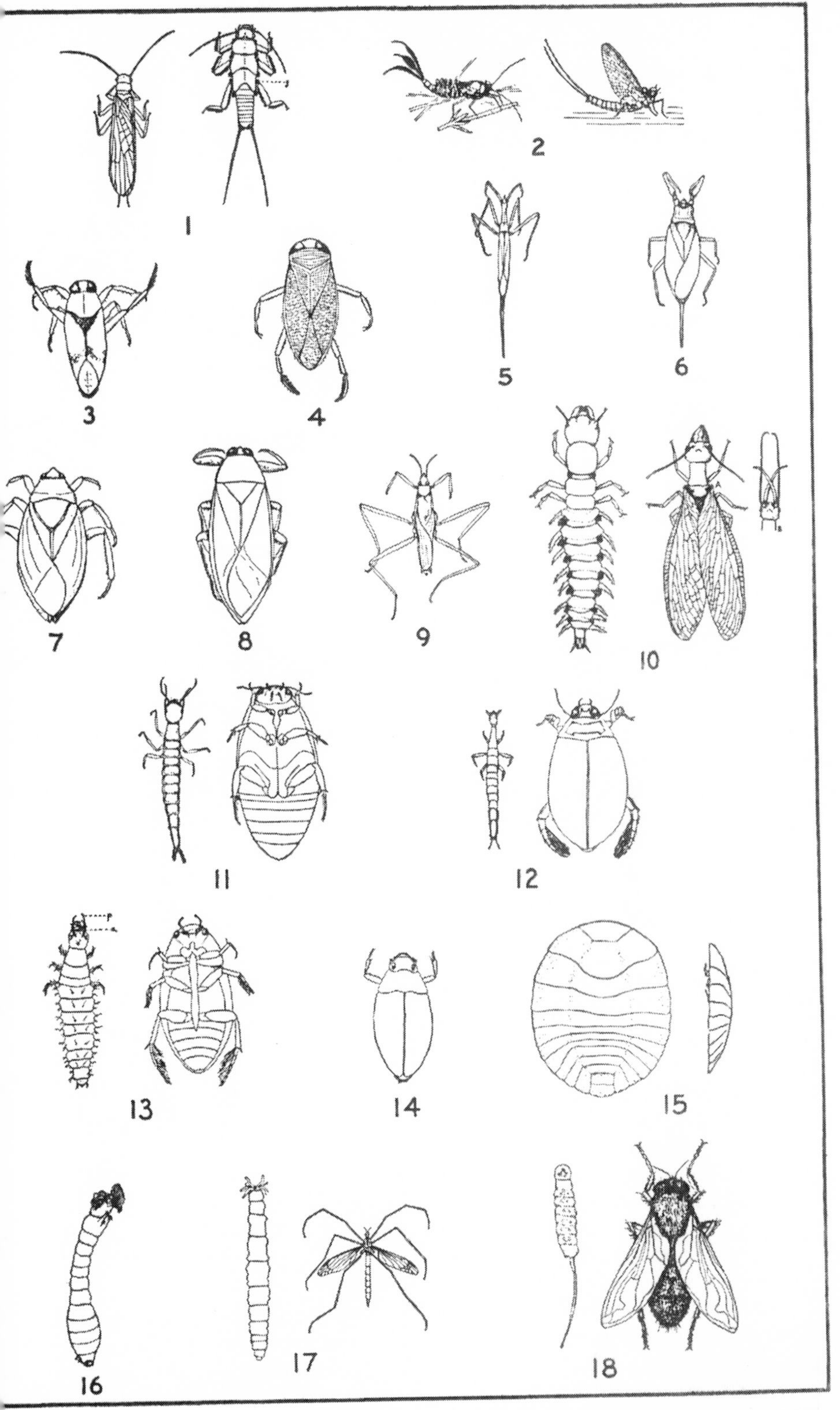

From *Common Water Insects*, by Ann Morgan

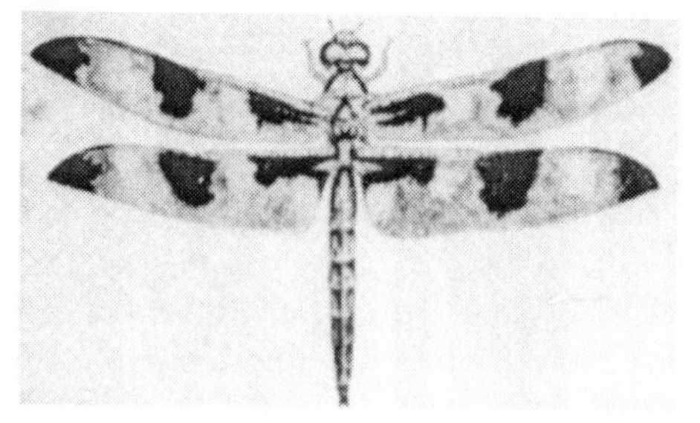

From *Outdoor Studies*, Needham

The ten-spot, a common dragonfly

main not only entirely harmless to man, but in reality his friends and allies in waging war against flies and mosquitoes; they are especially valuable in battling mosquitoes since the nymphs, or young, of the dragonfly take the wrigglers in the water, and the adults, on swiftest wings, take the mosquitoes that are hovering over ponds laying their eggs.

The poets have been lavish in their attention to these interesting insects and have paid them delightful tributes. Riley says:

Till the dragon fly, in light gauzy armor
burnished bright,
Came tilting down the waters in a wild,
bewildered flight.

Tennyson drew inspiration for one of his most beautiful poems from the two stages of dragonfly life. But perhaps Lowell in that exquisite poem, *The Fountain of Youth*, gives us the perfect description of these insects:

In summer-noon flushes
When all the wood hushes,
Blue dragon-flies knitting
To and fro in the sun,
With sidelong jerk flitting,
Sink down on the rushes.
And, motionless sitting,
Hear it bubble and run,
Hear its low inward singing
With level wings swinging
On green tasselled rushes,
To dream in the sun.

It is while we, ourselves, are dreamin
the sun by the margin of some pond,
these swift children of the air seem b
natural part of the dream. Yet if we wa
to note them more closely, we find m
things very real to interest us. First, t
are truly children of the sun, and if s
cloud throws its shadow on the waters
some moments, the dragonflies disap
as if they wore the invisible cloak of
fairy tale. Only a few of the comn
species fly alike in shade and sunshine,
early and late. The best known of thes
the big, green skimmer which does
care so much for ponds, but darts
fields and even dashes into our ho
now and then. Probably it is this spe
which has started all of the drago
slander, for it is full of curiosity, and
hold itself on wings whirring too rap
to make even a blur, while it exami
our faces or inspects the pictures or
niture or other objects which attract i

Another thing we may note w
dreaming by the pond is that the la
species of dragonflies keep to the hig
regions above the water, while the sma
species and the damsel flies flit near its
face. Well may the smaller species k

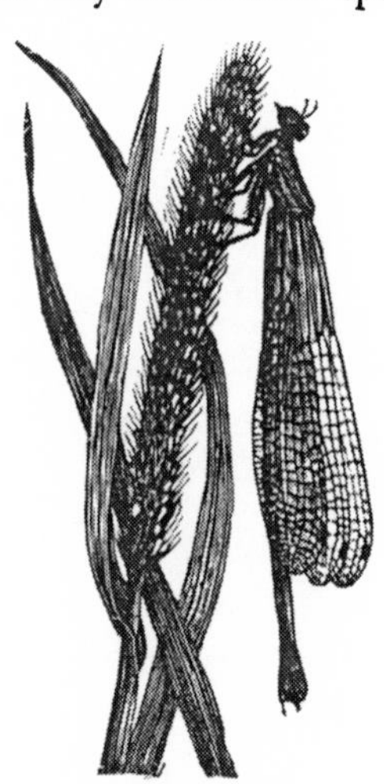

A damsel fly

below their fierce kindred; otherwise t
would surely be utilized to sate their h
ger, for these insects are well named d
ons, and dragons do not stop to inqu

her their victims are relatives or not. when they are resting that the dragon- and damsel flies reveal their most no- ble differences. The dragonfly ex- s both wings as if in flight while it ; in the sun or rests in the shadow. e is a big, white-bodied species called whitetail which slants its wings for- and down when it rests; but the sel flies fold their wings together over back when resting. The damsel flies

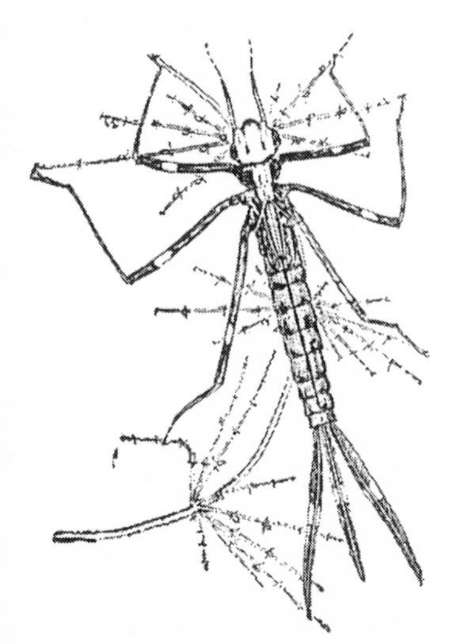

nph of a damsel fly on an aquatic plant

e more brilliantly colored bodies than the dragonflies, many of them being escent green or coppery; they are more nder and delicate in form. The damsel has eyes which are so placed on the es of the head as to make it look like ross on the front of the body fastened the slender neck, and with an eye at tip of each arm. There are very many cies of dragonflies and damsel flies, but y all have the same general habits.

The dragonfly nymphs are the ogres of pond or stream. To anyone unused to m and their ways in the aquarium, re is a surprise in store, so ferocious are ey in their attacks upon creatures twice eir size. The dragonfly's eggs are laid in e water; in some instances they are sim- y dropped and sink to the bottom; but the case of damsel flies, the mother nctures the stems of aquatic plants and places the eggs within them. The nymph in no wise resembles the parent dragon- fly. It is a dingy little creature, with six

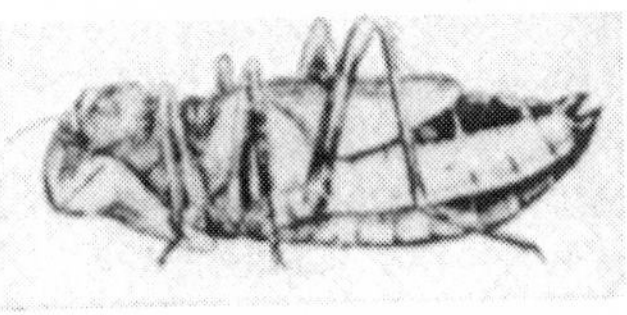

From *Outdoor Studies*, Needham

Nymph of a dragonfly, showing the position of the large lower lip folded beneath the head

queer, spider-like legs and no wings, although there are four little wing pads extending down its back, which encase the growing wings. It may remain hidden in the rubbish at the bottom of the pond or may cling to water weeds at the sides, for different species have different habits. But in them all we find a most amazing lower lip. This is so large that it covers the lower part of the face like a mask, and when folded back it reaches down between the front legs. It is in reality a grappling organ with hooks and spines for holding prey; it is hinged in such a manner that it can be thrust out far beyond the head to seize some insect, unsuspecting of danger. These nymphs move so slowly and look so much like their background, that they are

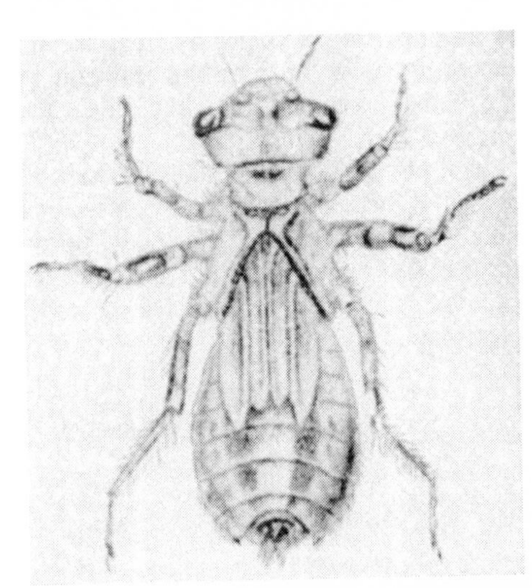

From *Outdoor Studies*, Needham

The same dragonfly nymph seen from above

always practically in ambush awaiting their victims.

The breathing of the dragonfly nymphs is peculiar; there is an enlargement of the rear end of the alimentary canal, in the walls of which tracheæ or breathing-tubes extend in all directions. The nymph draws water into this cavity and then expels it, thus bathing the tracheæ with the air mixed with water and purifying the air within them. Expelling the water so forcibly propels the nymph ahead, so this act

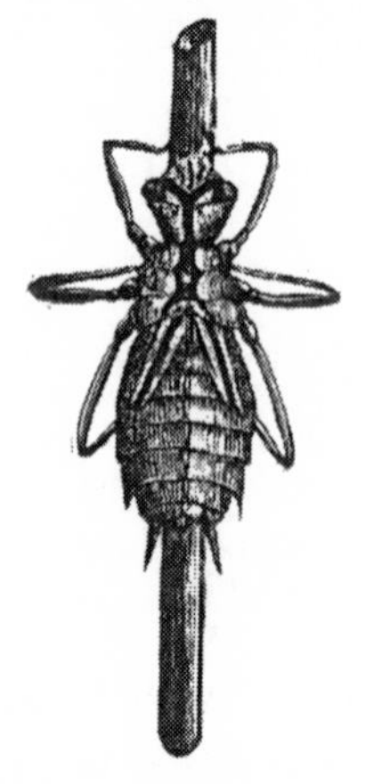

The cast skin of a dragonfly nymph. The skin splits along the back and the adult emerges, leaving the empty skin attached to the object upon which the transformation took place

serves as a method of swimming as well as of breathing. Damsel fly nymphs, on the other hand, have at the rear end of the body three long, platelike gills, each ramified with tracheæ.

Nymphs grow by shedding the skin as fast as it becomes too small; and when finally ready to emerge, they crawl up on some object out of the water and molt for the last time, and are thereafter swift creatures of the air.

LESSON 103

THE DRAGONFLIES AND DAMSEL F

LEADING THOUGHT — The drago are among the swiftest of all winged tures and their rapid, darting flight ables them to hawk their prey, which sists of other flying insects. Their stages are passed in the bottoms of p where they feed voraciously on aq creatures. The dragonflies are bene to us because, when very young and w full grown, they feed largely upon quitoes.

METHOD — The work of observing habits of adult dragonflies should largely done in the field during late s mer and early autumn. The points for servation should be given the pupils summer vacation use, and the res placed in the field notebook.

The nymphs may be studied in spring, when getting material for aquarium. April and May are good mo for securing them. They are collected using a dip net, and are found in the toms of reedy ponds or along the e of slow-flowing streams. These nym are so voracious that they cannot trusted in the aquarium with other sects; each must be kept by itself. T may be fed by placing other water ins in the aquarium with them or by giv them pieces of fresh meat. In the la case, tie the meat to a thread so tha may be removed after a few hours, if eaten, since it soon renders the water f

The dragonfly aquarium should h sand at the bottom and some water we planted in it, and there should be so object in it which extends above the face of the water which the nymphs, w ready to change to adults, can climb u while they are shedding the last nymp skin and spreading their new wings.

OBSERVATIONS ON THE YOUNG OF DR ONFLIES AND DAMSEL FLIES — 1. Wh did you find these insects? Were they the bottom of the pond or along the ed among the water weeds?

2. Are there any plumelike gills at

of the body? If so, how many? Are e platelike gills used for swimming? here are three of these, which is the ger? Do you know whether the nymphs these long gills develop into dragon- or into damsel flies?

. If there are no plumelike gills at the of the body, how do the insects move? they swim? What is the general color he body? Explain how this color pro- s them from observation. What ene- s does it protect them from?

. Are the eyes large? Can you see the e wing pads on the back in which the gs are developing? Are the antennæ g?

. Observe how the nymphs of both gonflies and damsel flies seize their y. Describe the great lower lip when ended for prey. How does it look when led up?

. Can you see how a nymph without plumelike gills breathes? Notice if the er is drawn into the rear end of the ly and then expelled. Does this proc- help the insect in swimming?

. When the dragonfly or damsel fly nph has reached its full growth, where es it go to change to the winged form? w does this change take place? Look the rushes and reeds along the pond rgin, and see if you can find the empty mph skins from which the adults erged. Where is the opening in them?

OBSERVATIONS ON THE ADULT DRAGON-ES — 1. Catch a dragonfly, place it un- a tumbler, and see how it is fitted for e in the air. Which is the widest part its body? Note the size of the eyes com- red with the remainder of the head. Do ey almost meet at the top of the head? ow far do they extend down the sides the head? Why does the dragonfly need ch large eyes? Why does a creature with ch eyes not need long antennæ? Can you e the dragonfly's antennæ? Look with a ns at the little, swollen triangle between e place where the two eyes join and the rehead; can you see the little, simple es? Can you see the mouth-parts?

2. Next to the head, which is the widest and strongest part of the body? Why does the thorax need to be so big and strong? Study the wings. How do the hind wings differ in shape from the front wings? How is the thin membrane of the wings made strong? Are the wings spotted or colored? If so, how? Can you see if the wings are folded along the front edges? Does this give strength to the part of the wing which cuts the air? Take a piece of writing paper and see how easily it bends; fold it two or three times like a fan and note how much stiffer it is. Is it this principle which strengthens the dragonfly's wings? Why do these wings need to be strong?

3. Is the dragonfly's abdomen as wide as the front part of the body? What help is it to the insect when flying to have such a long abdomen?

OUTLINE FOR FIELD NOTES — Go to a pond or sluggish stream when the sun is shining, preferably at midday, and note as far as possible the following things:

1. Do you see dragonflies darting over the pond? Describe their flight. They are hunting flies and mosquitoes and other insects on the wing; note how they do it. If the sky becomes cloudy, can you see the dragonflies hunting? In looking over a pond where there are many dragonflies darting about, do the larger species fly higher than the smaller ones?

2. Note the way the dragonflies hold their wings when they are resting. Do they rest with their wings folded together over the abdomen or are they extended out at an angle to the abdomen? Do you know how this difference in attitude of resting determines one difference between the damsel flies and the dragonflies?

3. The damsel flies are those which hold their wings folded above the back when resting. Are these as large and strong-bodied as the dragonflies? Are their bodies more brilliantly colored? How does the shape of the head and eyes differ from those of the dragonflies? How many different-colored damsel flies can you find?

4. Do you see some dragonflies dipping down in the water as they fly? If so, they

are laying their eggs. Note if you find others clinging to reeds or other plants with the abdomen thrust below the surface of the water. If so, these are da flies inserting their eggs into the ste the plant.

THE CADDIS WORMS AND THE CADDIS FLIES

People who have never tried to fathom the mysteries of the bottom of brook or pond are to be pitied. Just to lie flat, face downward, and watch for a time all that happens down there in that water world

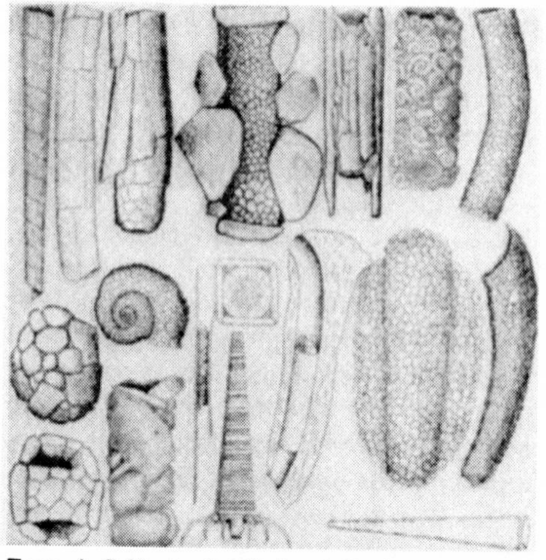

From *A Guide to the Study of Freshwater Biology*, Needham

Types of caddis-worm cases

is far more interesting than witnessing any play ever given at matinee. At first one sees nothing, since all the swift-moving creatures have whisked out of sight, because they have learned to be shy of moving shadows; but soon the crayfish thrusts out his boxing gloves from some crevice, then a school of tiny minnows " stay their wavy bodies 'gainst the stream "; and then something strange happens. A bit of rubbish on the bottom of the brook walks off. Perhaps it is a dream, or we are under the enchantment of the water witches! But no, there goes another, and now a little bundle of sand and pebbles takes unto itself legs. These mysteries can only be solved with a dip net and a pail half filled with water, in which we may carry home the treasure trove.

When we finally lodge our catch in the aquarium jar, our mysterious mo sticks and stones resolve themselves little houses built in various fashions, each containing one inmate. Some of houses are made of sticks fastened gether lengthwise; some are built like cabins, crosswise; some consist simply hollow stem cut a convenient len some are made of sand and pebbles; one, the liveliest of all, is a little tube m of bits of rubbish and silk spun in a sp making a little cornucopia.

On the whole, the species which liv the log cabins are the most convenien study. Whatever the shape of the cas house, it has a very tough lining of s which is smooth within, and forms framework to which the sticks and sto are fastened. These little dwellings alw have a front door and a back door. Ou the front door may protrude the d colored head followed by two dark s ments and six perfectly active legs, front pair being so much shorter than other two pairs that they look almost l mouth palpi. In time of utter peace, m of the little hermit is thrust out, and see the hind segment of the thorax, wh is whitish, and behind this the abdom of nine segments. At the sides of the domen, and apparently between the s ments, are little tassels of short, wh threadlike gills. These are filled with impure from contact with the blo which exchanges its impurities speed for the oxygen from the air that is mix with the water. Water is kept flowing

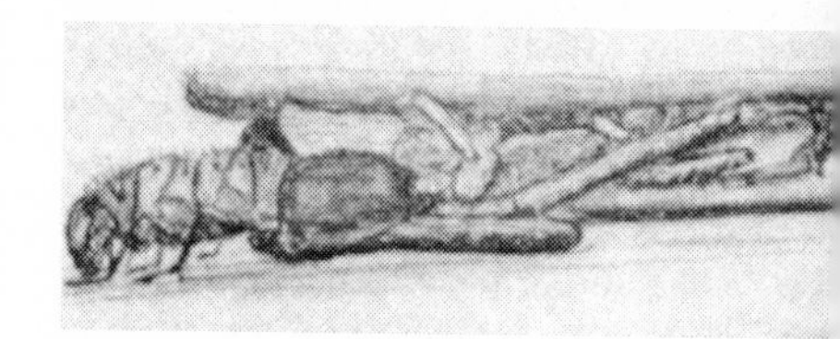

A caddis worm in its case

he front door of the cabin, over the s and out at the back door, by the rhyth- movement of the body of the little mit, and thus a supply of oxygen is adily maintained.

The caddis worm is not grown fast to case as is the snail to its shell. If we d down with forceps a case in which occupant is wrong side up, after a few ggles to turn itself over, case and all, it l turn over within the case. It keeps its d upon the case by two forward-curv- hooks, one on each side of the tip of rear segment. These hooks are inserted the tough silk and hold fast. It also has top of the first segment of the abdo- n a tubercle, which may be extended at ll; this helps to brace the larva in its onghold, and also permits the water to w freely around the insect. So the little rmit is entrenched in its cell at both ds. When the log-cabin species wishes swim, it pushes almost its entire body t of the case, thrusts back the head, reads the legs wide apart, and then ubles up, thus moving through the ater spasmodically, in a manner that re- inds us of the crayfish's swimming ex- pt that the caddis worm goes head first. his log-cabin species can turn its case er dexterously by movements of its legs.

The front legs of the caddis worm are much shorter than the other two pairs at they look like palpi, and their use is hold close to the jaws bits of food which e being eaten. The other legs are used r this too if the little legs cannot manage ; perhaps also these short front legs help old the bits of building material in place hile the web is woven to hold it there.

upa of caddis fly removed from its case. Note the threadlike gills

The caddis worm, like the true caterpillars, has the opening of the silk gland near the lower lip. The food of most caddis worms is vegetable, usually the various

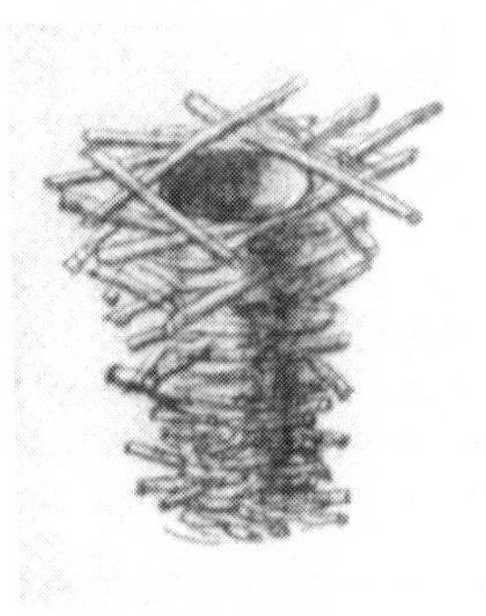

Log-cabin type of case

species of water plants; but there are some species which are carnivorous, like the net-builder, which is a fisherman.

The caddis-worm case protects its inmate in two ways, first, from the sight of the enemy, and second, from its jaws. A fish comes along and sees a nice white worm and darts after it, only to find a bundle of unappetizing sticks where the worm was. All of the hungry predatory creatures of the pond and stream would be glad to get the caddis worm, if they knew where it went. Sometimes caddis-worm cases have been found in the stomachs of fishes.

While it is difficult to see the exact operation of building the caddis-worm house, the general proceeding may be readily observed. Take a vigorous half-grown larva, tear off part of the sticks and bits of leaves that make the log cabin, and then place the little builder in a tumbler with half an inch of water at the bottom, in which are many bright flower petals cut into strips, fit for caddis lumber. In a few hours the little house will look like a blossom with several rows of bright petals set around its doorway.

When the caddis worm gets ready to pupate, it fastens its case to some object In the water and then closes its front and back doors. Different species accomplish this in different ways; some spin and fasten a silken covering over the doors, which is often in the form of a pretty grating; others simply fasten the material of which the case is made across the door. But though the door be shut, it is so arranged as to allow the water to flow through and to bring oxygen to the threadlike gills, which are on the pupæ as well as on the larvæ. When ready to emerge, the pupa crawls out of its case and climbs to some object above the water and sheds its pupa skin, and the adult insect flies off. In some species, living in swift water, the adult issues directly from the water, its wings expanding as soon as touched by the air.

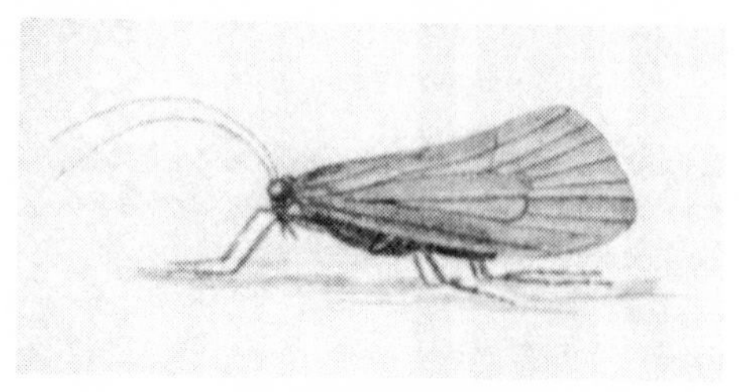

A caddis fly

Caddis flies are familiar to us all even if we do not know them by name. They are night fliers and flame worshipers. Their parchment-like or leathery wings are folded like a roof over the back, and from the side the caddis fly appears as an elongated triangle with unequal sides. The front wings are long and the hind ones shorter and wider; the antennæ are long and threadlike and always waving about for impressions; the eyes are round and beadlike; the tarsi, or feet, are long and these insects have an awkward way of walking on the entire tarsus which gives them an appearance of kneeling. Most of the species are dull-colored, brownish or gray, the entire insect often being of one color. The mother caddis flies lay their eggs in the water. Perhaps some species drop the eggs in when hovering above, but in some cases the insect must make a diving bell of her wings and go down into the water to place her eggs securely. The wings are covered with hairs and not with scales, and therefore they are better fitted for diving than would be those of the moth. I have seen caddis flies swim vigorously.

LESSON 104

The Caddis Worms and Caddis Fl

Leading Thought — The caddis wo build around themselves little houses of bits of sticks, leaves, or stones. T crawl about on the bottom of the pon stream, protected from sight, and able withdraw into their houses when attack The adults of the caddis worm are win mothlike creatures which come in n bers to the light at night.

Method — With a dip net the cad worms may be captured and then may placed in the school aquarium. Duckw and other water plants should be k growing in the aquarium. The log-ca species is best for this study, because lives in stagnant water and will theref thrive in an aquarium.

Observations — 1. Where do you fi the caddis worms? Can you see th easily on the bottom of the stream pond? Why?

2. Of what are the caddis-worm hou made? How many kinds have you e found? How many kinds of material c you find on one case? Describe one as actly as possible. Find an empty case a describe it inside. Why is it so smooth side? How is it made so smooth? Are the cases the same size?

3. What does the caddis worm when it wishes to walk around? What the color of the head and the two s ments back of it? What is the color of t body? Why is this difference of color tween the head and body protective? the caddis worm grown fast to its case, the turtle is to its shell?

4. Note the legs. Which is the short pair? How many pairs? What is the use the legs so much shorter than the other If the caddis-worm case happens to wrong side up, how does it turn over?

5. When it wishes to come to the su face or swim, what does the caddis wor do? When reaching far out of its case do

er lose its hold? How does it hold on? the caddis worm out of its case and he hooks at the end of the body with :h it holds fast.

How does the caddis worm breathe? en it reaches far out of its case, note breathing gills. Describe them. Can see how many there are on the seg-ts? How is the blood purified through e gills?

What are the caddis worm's ene-s? How does it escape them? Touch when it is walking; what does it do?

On top of the first segment of the omen is a tubercle. Do you suppose this helps to hold the caddis worm in ase?

What does the caddis worm eat? De-be how it acts when eating.

o. How does the caddis worm build its ? Watch one when it makes an addi- to its case, and describe all that you see.

1. Can you find any of the cases with front and back doors closed? How are y closed? Open one and see if there is a a within it. Can you see the growing gs, antennæ, and legs? Has it breathing ments like the larva? Cover the aquar- with mosquito netting so as to get all moths which emerge. See if you can cover how the pupa changes into a dis fly.

2. How does the caddis fly fold its gs? What is the general shape of the ect when seen from the side with wings sed? What is the texture of the wings? w many wings are there? Which pair is longer?

3. Describe the eyes; the antennæ. Does the caddis fly walk on its toes, or on its complete foot?

14. Examine the insects which come around the lights at night in the spring and summer. Can you tell the caddis flies from the other insects? Do they dash into the light? Do they seem anxious to burn themselves?

Little brook, so simple, so unassuming — and yet how many things love thee!

Lo! Sun and Moon look down and glass themselves in thy waters.

And the trout balances itself hour-long against the stream, watching for its prey; or retires under a stone to rest.

And the water-rats nibble off the willow leaves and carry them below the wave to their nests — or sit on a dry stone to trim their whiskers.

And the May-fly practices for the millionth time the miracle of the resurrection, floating up an ungainly grub from the mud below, and in an instant, in the twinkling of an eye (even from the jaws of the baffled trout) emerging, an aerial fairy with pearl-green wings.

And the caddis-fly from its quaint disguise likewise emerges.

And the prick-eared earth-people, the rabbits, in the stillness of early morning play beside thee undisturbed, while the level sunbeams yet grope through the dewy grass.

And the squirrel on a tree-root — its tail stretched far behind — leans forward to kiss thee,

Little brook, for so many things love thee.

— Edward Carpenter

THE MOSQUITO

In defiance of the adage, the mother of most common mosquitoes does not itate to put her eggs all in one basket, t perhaps she knows it is about the saf- little basket for eggs in this world of certainties. If it were possible to begin s lesson with the little boat-shaped egg skets, I should advise it. They may be found in almost any rain barrel, and the eggs look like a lot of tiny cartridges set side by side, points up, and lashed or glued together, so there shall be no spilling. Like a certain famous soap, they "float," coming up as dry as varnished corks when water is poured upon them.

The young mosquito, or wriggler, breaks

through the shell of the lower end of the egg and passes down into the water, and from the first, it is a most interesting creature to view through a hand lens. The

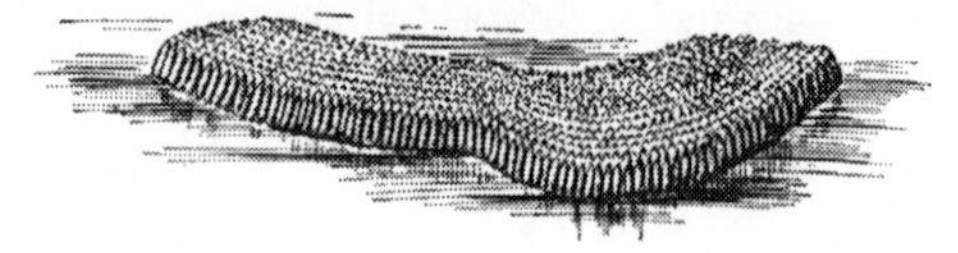

The egg-raft of a mosquito

head and the thorax are rather large while the body is tapering and armed with bunches of hairs. At the rear of the body are two tubes very different in shape; one is long, straight, and unadorned; this is the breathing-tube through which air passes to the tracheæ of the body. This tube has at the tip a star-shaped valve, which can be opened and shut; when it is opened at the surface of the water, it keeps the little creature afloat and meanwhile allows air to pass into the body. When the wriggler is thus hanging to the surface of the water, it feeds upon small particles of decaying vegetation; it has a remarkable pair of jaws armed with brushes, which in our common species, by moving rapidly, set up currents and bring the food to the mouth. This process can be seen plainly with a lens. When disturbed, the wriggler shuts the valve to its breathing-tube, and sinks. However, it is not much heavier than the

A mosquito aquarium. Note the egg-raft, larvæ, pupæ, and the adult emerging

water; I have often seen one rise for s distance without apparent effort. other tube at the end of the body sup the swimming organs, which consis four finger-like processes and va bunches of hairs. When swimming, wriggler goes tail first, the swimmin gans seeming to take hold of the water to pull the creature backward, in a s of spasmodic jerks; in fact, the ir seems simply to " throw somersaults," an acrobat. I have often observed glers standing on their heads in the

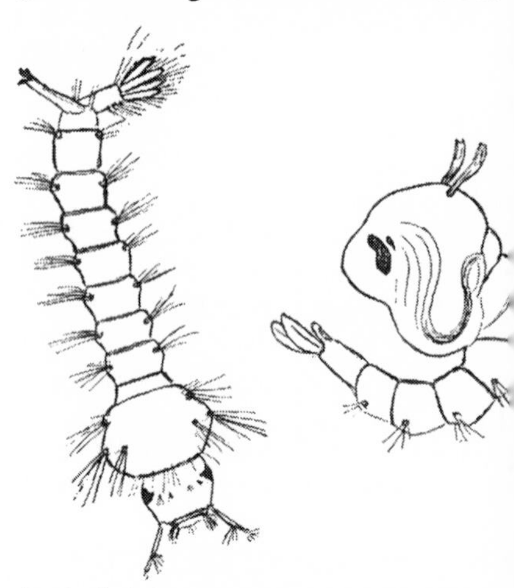

At the left is a larva and at the right a p of the mosquito

tom of the aquarium, with their jaws b under, revolving their brushes briskly; they never remain very long below surface, as it is necessary for them to t in fresh air often.

The pupa has the head and thoracic ments much enlarged making it all " h and shoulders " with a quite insignific body attached. Upon the thorax are t breathing-tubes, which look like two e and therefore when the pupa rests at surface of the water, it remains head up that these tubes may take in the air; the end of the body are two swimming gans which are little, leaflike projectio At this stage the insect is getting ready live its life in the air, and for this reas probably, the pupa rests for long peri at the surface of the water and does swim about much, unless disturbed. H

, it is a very strange habit for a pupa nove about at all. In the case of other , butterflies, and moths, the pupa e is quiet.

Vhen fully mature, the pupa rises to surface of the water, the skeleton skin aks open down its back and the mos-to carefully works itself out; until its gs are free and dry, it rests upon the ting pupa skin. This is indeed a frail k, and if the slightest breeze ruffles the er, the insect is likely to drown before wings are hard enough for flight.

The reason that kerosene oil put upon surface of the water where mosquitoes ed kills the insects is that both the væ and pupæ of mosquitoes are obliged rise to the surface and push their breath--tubes through the surface film so that y will open to the air; a coating of oil the water prevents this, and they are focated. Also when the mosquito erges from the pupa skin, if it is even ched by the oil, it is unable to fly and n dies.

The male mosquitoes have bushy or thery antennæ. These antennæ are aring organs of very remarkable con-uction. The Anopheles may be distin-ished from the Culex by the following

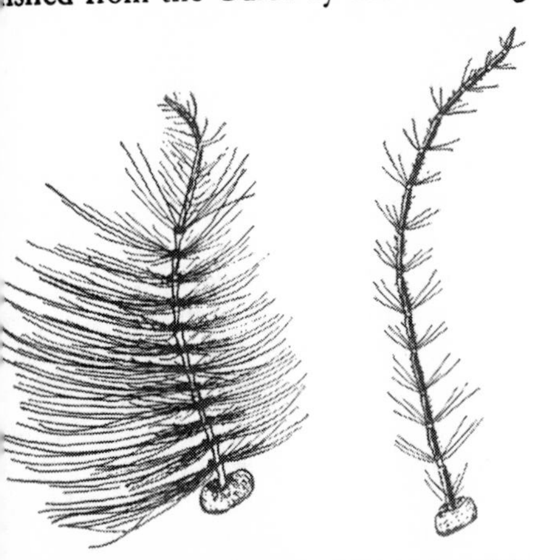

ntennæ of the mosquito; left, male; right, female

haracteristics: Its wings are spotted in-tead of plain. When at rest it is perfectly traight, and is likely to have the hind legs n the air. It may also rest at an angle to the surface to which it clings. The Culex is not spotted on the wings and is likely to be humped up when at rest. In our climate the Anopheles is more dan-

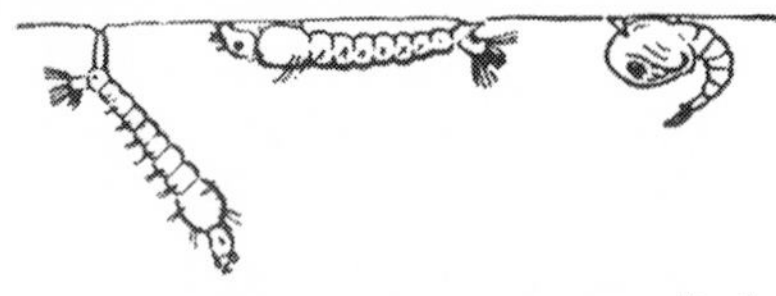

Normal resting position of mosquito larvæ and pupæ. Left; larva of Culex, *the common house mosquito. Middle; larva of* Anopheles, *the carrier of malaria. Right; pupa of* Culex. *Note the breathing-tubes*

gerous than the Culex because it carries the germs of malaria. A mosquito's wing under a microscope is a most beautiful object, as it is " trimmed " with ornamen-tal scales about the edges and along the veins. The male mosquitoes neither sing nor bite; the song of the female mosquito is supposed to be made by the rapid vibra-tion of the wings, and her musical per-formances are for the purpose of attract-ing her mate, as it has been shown that he can hear through his antennæ a range of notes covering the middle and next higher octaves of the piano.

Science has shown us that the mos-quitoes are in a very strange way a menace to health. Through a heroism, as great as ever shown on field of battle, men have imperiled their lives to prove that the germs of the terrible yellow fever are transmitted by the biting mosquito, and with almost equal bravery other men have demonstrated that the germs of malaria are also thus carried.

In the North, our greatest danger is from the mosquitoes which carry the ma-larial germs. These are the mosquitoes with spotted wings and belong to the genus *Anopheles*. This mosquito in order to be of danger to us must first feed upon the blood of some person suffering from malaria and thus take the germ of the disease into its stomach. Here the germ develops and multiplies into many minute germs, which pass through another stage and finally get into the blood of the mos-quito and accumulate in the salivary glands. The reason any mosquito bite or insect bite swells and itches is that, as the insect's beak is inserted into the flesh, it carries with it some of the saliva from the insect's mouth. In the case of Anopheles these malarial germs are carried with the saliva into the blood of the victim. It has been proved that in the most malarial

Normal position of Culex *and* Anopheles *on a wall;* Culex *above and* Anopheles *below*

countries, like Italy and India, people are entirely free from malaria if they are not bitten by mosquitoes. Thus the mosquito is the sole carrier of the malaria germs.

After this explanation has been made, it would be well for the teacher to take the pupils on a tour of inspection through the neighborhood to see if there are any mosquito larvæ in rain barrels, ponds, or pools of stagnant water. If such places are found, let the pupils themselves apply the following remedies.

1. Rain barrels should be securely covered.

2. Stagnant pools should be drained and filled up if possible.

3. Wherever there are ponds or pools where mosquitoes breed that cannot be filled or drained, the surface of the water should be covered with a spray of kerosene oil. This may be applied with a spray pump or from a watering can.

4. If it is impracticable to cover such places with oil, introduce into such pools the following fish: minnows, sticklebacks, sunfish, and goldfish.

The effect of this lesson upon the children should be to impress them with danger to life and health from mosqui and to implant in them a determina to rid the premises about their home these pests.

LESSON 105
The Mosquito

Leading Thought — The wrigglers wigglers, which we find in rain barrels stagnant water are the larvæ of n quitoes. We should study their life hist carefully if we would know how to get of mosquitoes.

Method — There is no better way interest the pupils in mosquitoes than place in an aquarium jar in the sch room a family of wrigglers from so pond or rain barrel. For the pupils' sonal observation, take some of the w glers from the aquarium with a pipe and place them in a homeopathic vial; the vial three-fourths full of water cork it. Pass it around with a hand l and give each pupil the opportunity to serve it for five or ten minutes. It wo be well if this vial could be left on e desk for an hour or so during study p ods, so that the observations may be m casually and leisurely. While the pu are studying the wrigglers, the follow questions should be placed upon blackboard, and each pupil should m notes which may finally be given at a son period. This is particularly availa work for September.

In studying the adult mosquito, a l or microscope is necessary. But it is great importance that the pupils be taug to discriminate between the compa ly harmless species of Culex and the gerous Anopheles; and therefore they uld be taught to be observant of the mosquitoes rest upon the walls, and ether they have mottled or clear wings.

The Larva

)bservations — 1. Note if all the wrig-s are of the same general shape, or if ne of them have a very large head; these er are the pupæ and the former are the æ. We will study the larvæ first. iere do they rest when undisturbed? they rest head up or down? Is there part of their body that comes to the face of the water?

. When disturbed what do they do? hen they swim, do they go head or tail t?

,. Observe one resting at the top. At at angle does it hold itself to the sur-e of the water? Observe its head. Can see the jaw brushes revolving rapidly? hat is the purpose of this? Describe its s. Can you see its antennæ?

1. Note the two peculiar tubes at the l of the body and see if you can make their use.

;. Note especially the tube that is ust up to the surface of the water when creatures are resting. Can you see how opening of this tube helps to keep the iggler afloat? What do you think is the rpose of this tube? Why does it not be-ne filled with water when the wriggler wimming? Can you see the two air ves-s, or tracheæ, extending from this tube ng the back the whole length of the dy?

5. Note the peculiarities of the other e at the rear end of the body. Do you nk the little finger-like projections are aid in swimming? How many are there?

7. Can you see the long hairs along the e of the body?

8. Does the mosquito rest at the bot-m of the bottle or aquarium?

The Pupa

1. What is the most noticeable difference in appearance between the larva and pupa?

2. When the pupa rests at the surface of the water, is it the same end up as the wriggler?

3. Note on the "head" of the pupa two little tubes extending up like ears. These are the breathing-tubes. Note if these open to the air when the pupa rests at the surface of the water.

4. Can you see the swimming organs at the rear of the body of the pupa? Does the pupa spend a longer time resting at the surface than the larva? How does it act differently from the pupæ of other flies and moths and butterflies?

5. How does the mosquito emerge from the pupa skin? Why does kerosene oil poured on the surface of the water kill mosquitoes?

The Adult Mosquito

1. Has the mosquito feathery antennæ extending out in front? If so, what kind of mosquito is it?

2. Do the mosquitoes with bushy antennæ bite? Do they sing?

3. Are the wings of the mosquito spotted or plain? How many has it?

4. When at rest, is it shortened and humpbacked or does it stand straight out with perhaps its hind legs in the air?

5. What are the characteristics by which you can tell the dangerous Anopheles?

6. Why is the Anopheles more dangerous than the Culex?

7. Examine a mosquito's wing under a microscope and describe it.

8. Examine the antennæ of a male and of a female mosquito under a microscope, and describe the difference.

9. Which sex of the mosquito does the biting and the singing?

10. How is the singing done?

INVERTEBRATE ANIMALS OTHER THAN INSECTS

This group includes backboneless animals other than insects. Among these are spiders and their relatives, centipedes and millipedes, crustaceans, mollusks or shelled animals, worms, and seashore creatures representing several other groups.

THE GARDEN SNAIL

Perchance if those who speak so glibly of a "snail's pace" should study it, they would not sneer at it, for, carefully observed, it seems to be one of the most wonderful methods of locomotion ever devised by animal. Naturally enough, the snail cannot gallop, since it has but one foot; but it is safe to assert that this foot, which is the entire lower side of the body, is a remarkable organ of locomotion. Let a snail crawl up the side of a tumbler and note how this foot stretches out and holds on. It has flanges along the sides, which secrete an adhesive substance that enables the snail to cling, and yet it also has the power of letting go at will. The slow, even, pushing forward of the whole body, weighted by the unbalanced shell, is as mysterious, and seemingly as inevitable, as the march of fate, so little is the motion connected with any apparent muscular effort. But when his snailship wishes to let go and retire from the world, this foot performs a feat which is certainly wor of a juggler; it folds itself lengthwise, a the end on which the head is retires f into the shell, the tail end of the f being the last to disappear. And now fi your snail!

Never was an animal so capable stretching out and then folding up all organs, as is this little tramp who carr his house with him. Turn one on his ba when he has withdrawn into his lit hermitage, and watch what happens. So he concludes he will find out where he and why he is bottomside up; as the fi evidence of this, the hind end of the fo which was folded together, pushes for then the head and horns come bubbli out. The horns are not horns at all, b each is a stalk bearing an eye on the t This is arranged conveniently, like a m ble fastened to the tip of a glove fing When a snail wishes to see, it stretch forth the stalk as if it were made of rubb f danger is perceived, the eye is pulled exactly as if the marble were pulled through the middle of the glove fin- or as a boy would say, "it goes into ıole and pulls the hole in after it." Just w the stalked eyes is another pair of ter horns, which are feelers, and :h may be drawn back in the same ner; they are used constantly for test- he nature of the surface on which the is crawling. It is an interesting ex- nent to see how near to the eyes and feelers we can place an object, before ng them back in. With these two of sense organs pushed out in front

Snail sketches. 1, *The thorny path to bliss.* *Snail showing the breathing pore.* 3, *Pros- :ting*

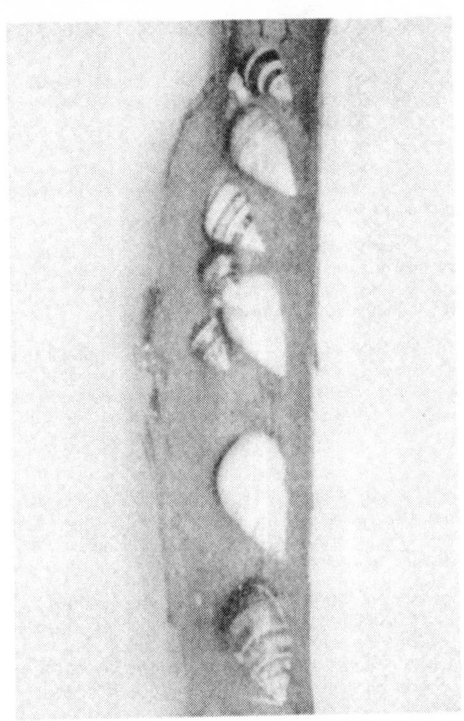

E. Morton Miller

Tree snails æstivating on the under side of a piece of bark

of him, the snail is well equipped to observe the topography of his immediate vicinity; if he wishes to explore above, he can stand on the tip of his tail and reach far up; and if there is anything to take hold of, he can glue his foot fast to it and pull himself up. Moreover, I am convinced that snails have decided views about where they wish to go, for I have tried by the hour to keep them marching lengthwise on the porch railing, so as to study them; and every snail was determined to go crosswise and crawl under the edge, where it was nice and dark.

It is interesting to observe through a lens the way a snail takes his dinner; place before him a piece of sweet apple or other soft fruit, and he will lift himself on his foot and begin to work his way into the fruit. He has an efficient set of upper teeth, which look like a saw and are colored as if he chewed tobacco; with these teeth and with his round tongue, which we can see popping out, he soon makes

1. Crown Melongena, Melongena corona, *Gmelin. Reported from Florida and West Indies. The species lives in brackish water and is fond of the razor-back clam and oyster. It obtains its name from the crownlike appearance of the projections on the shell whorls. Length, 2 to 5 inches.*

2. Brown-mouth Cymatium, Cymatium chlorostomum, *Lamarck. This species is commonly found in the West Indies and Florida Keys. Length, 3 inches.*

3. White-mouth Cymatium, Cymatium tuberosum, *Lamarck. The illustration shows an immature individual. A mature specimen is similar to Figure 2. The species is distributed in Florida Keys and West Indies. Length, 2 to 3 inches.*

4. Lined Murex, Murex cobritti, *Bernardii. Collected at depths of from 10 to 150 fathoms from Cedar Keys to Texas and the West Indies. Length, 3 inches.*

5. Mossy Ark, Arca umbonata, *Lamarck. These bivalves are distributed from North Carolina to the West Indies and Gulf of Mexico. They are often cast up on the Florida beaches by storms. Length, 2 to 3 inches.*

6. Black Lace Murex, Murex rufus, *Lamarck. This species is gathered in water from 1 to 30 fathoms in depth from North Carolina to the West Indies. Length, 2 inches.*

7. Apple Murex, Murex pomum, *Gmelin. Abundant in West Indies; also reported from North Carolina and the Gulf of Mexico. The shell mouth is lined with bright yellow. Length, 2 to 5 inches.*

8. White-spike Murex, Murex fulvescens, *Sowerby. The color varies from white to pink. The shell is found from North Carolina to Florida and the West Indies. Length, 6 inches.*

9. Moon Shell, Polinicies duplicata, *Say. The species ranges from New England to the Gulf of California. It possesses a chitinous operculum, and is either bluish or brownish tinged on the upper surface. Diameter, 3 inches.*

10. Rock Worm Shell, Vermetus nigricans, *Dall. This mollusk forms a much-coiled and cylindrical shell. It is commonly found attached to rocks, and even contributes to reef building. Distributed from West Florida to Florida Keys.*

11. Mouse Cone, Conus mus, *Hwass. This mottled chestnut-colored cone possesses a striated body-whorl. It is commonly found in shallow bays in Florida and the West Indies. Length, 1 to 2 inches.*

12. Florida Cone, Conus floridanus, *Gabb. This species is referred to as "chinese tops," since the surface markings resemble the characters of the Chinese alphabet. It ranges from North Carolina to the Gulf of California. Length, 1¼ inches.*

13. Giant Band Shell, Fasciolaria giga *Kiener. The illustration shows a young 2 to 3 inches long; a mature one may rea length of 24 inches. The surface is yello and the aperture orange-red. It is foun North Carolina, West Indies, and Brazil.*

14. Lettered Olive, Oliva litterata, *marck. These polished shells with hierogly markings are fairly common from North C lina to Texas and the West Indies. They in colonies and are sand burrowers. Len 1½ to 2½ inches.*

15. Netted Olive, Oliva reticularis, *marck. This shell possesses a woven patter fine brown lines on a white background. It curs in the West Indies and Florida. Len 1 to 2½ inches.*

16. Mottled Top Shell, Calliostoma ju num, *Gmelin. The shell is conical, and is pe within the mouth. Distributed from N Carolina to the West Indies. Length, 1 inc*

17. Ridged Chione, Chione cancellata, *næus. This shell occurs in abundance in Gulf of California. It is distributed from N Carolina to Brazil. Diameter, 1 inch.*

18. Beaming Scallop, Pecten irradians, *marck. This common edible species ra from New England to Cape Hatteras. The terior is brown marked by bars of red, pur or orange. Diameter, 2 to 3 inches.*

19. Vase Shell, Vasum muricatum, *B. The shell color is white lined with pink. I found in the Florida Keys, West Indies, Panama. Length, 3 inches.*

20. Ponderous Ark, Arca ponderosa, *S This species is distributed from Cape Coa Texas and the West Indies. In the fossil st it is found in New Jersey. Diameter, 2 inc*

21. Spiny Pearl Shell, Margitifera radi *Leach. Distributed from Georgia to the W Indies. They are found associated with sponges in Florida. Diameter, 1½ inches.*

22. Little Red Murex, Murex messor *Reeve. This shell is found in the Florida K and West Indies. Length, about 1 inch.*

23. Rose Euglandina, Euglandina ros *Ferussac. This rose-colored land shell is fou in Western Florida. It conceals itself in br during the rainy season.*

24. Calico Scallop, Pecten gibbus, *L næus. The shell is mottled with red, bro and orange. Distributed in North Carolina a the West Indies. Diameter, 1½ to 2 inches.*

25. Volcano Shell, Fissurella fasicula *Lamarck. The common name is based on t resemblance of the shell to a volcano; it found in the Florida Keys and West Indi Diameter, 1 inch.*

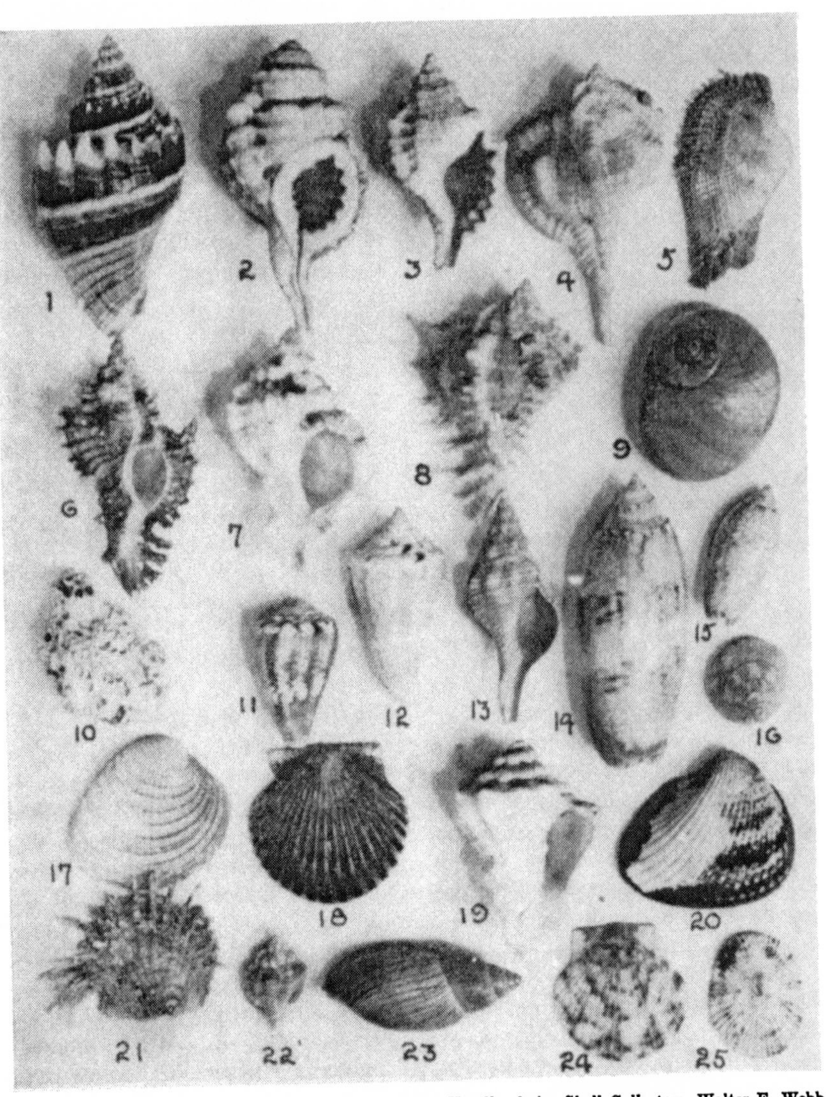

From *Handbook for Shell Collectors*, Walter F. Webb

an appreciable hole in the pulp; but his table manners are not nice, since he is a hopeless slobberer.

There are right and left spiraled snails. All those observed for this lesson show the spiral wound about the center from left over to right, or in the direction of the movement of the hands of a clock, and this is usually the case. With the spiral like this, the breathing pore is on the right side of the snail and may be

Hugh Spencer

Slugs with eye-stalks extended. Slugs are relatives of land snails but they have no shells

seen as an opening where the snail joins the shell. This pore may be seen to open and contract slowly; by this motion, the air is sucked into the shell where it bathes the snail's lung, and is then forced out — a process very similar to our own breathing.

The snail acts quickly when attacked; at the first scare, he simply draws in his eyes and feelers and withdraws his head, so that nothing can be seen of him from above except a hard shell which would not attract the passing bird. But if the attack continues, he lets go all hold on the world, and nothing can be seen of him but a little mass which blocks the door to his house; and if he is obliged to experience a drought, he makes a pane of glass out of mucus across his door, and thus stops evaporation. This is a very wise precaution, because the snail is made up largely of moisture and much water is needed to keep his mucilage factory running.

The way the snail uses his eyes is co cal; he goes to the edge of a leaf and p one eye over to see what the new t tory is like; but if his eye strikes an ject, he pulls that one back, and prosp for a time with the other. He can lengt the eyestalk amazingly if he has n How convenient for us if we could t see around a corner! If a small boy v as well off as a snail, he could see the tire ball game through a knothole in fence. In fact, the more we study the sn the more we admire, first his powers ascertaining what there is in the wo and then his power of getting around the world by climbing recklessly relentlessly over obstacles, not car whether he is right side up on the fl or hanging wrong side up from the c ing; and, finally, we admire his utter r cence when things do not go to suit h I think the reason I always call a sn "he" is that he seems such a philosop — a Diogenes in his tub. However, si the snail combines both sexes in individual the pronoun is surely appli ble.

When observed through a lens, snail's skin looks like that of the alliga rough and divided into plates, with a s face like pebbled leather; and no insect truder can crawl up his foot and get i the shell "unbeknownst," for the shel grown fast to the flange that grows of the middle of the snail's back. T smoother the surface the snail is cra ing upon, the harder to make him let The reason for this lies in the mu which he secretes as he goes, and whi enables him to fasten himself anywhe he can crawl up walls or beneath a horizontal surface, shell downward, a he leaves a shining trail behind him wh ever he goes.

Snail eggs are as large as small peas, most transparent, covered with very s shells, and fastened together by muc They are laid under stones and decayi leaves. As soon as the baby snail hatches has a shell with only one spiral turn in it; it grows, it adds layer after layer to the sh on the rim about the opening — which ed the lip; these layers we can see as ges on the shell. If we open an empty ll, we can see the progress of growth in size of the spirals. Snails eat succulent ves and other soft vegetable matter. ring the winter, they bury themselves neath objects or retire into soft humus. preparing for the winter, the snail kes a door of mucus and lime, or some- nes three doors, one behind another, oss the entrance to his shell, leaving a y hole to admit the air. There are vari- es of snails which are eaten as dainties in rope and are grown on snail farms for e markets. The species most commonly ed is the same as that which was re- rded as a table luxury by the ancient omans.

LESSON 106

The Garden Snail

Leading Thought — The snail carries s dwelling with him, and retires within in time of danger. He can climb on any nooth surface.

Method — The pupils should make a ailery, which may consist of any glass jar, ith a little soil and some moss or leaves at e bottom, and a shallow dish of water at ne side. The moss and soil should be kept oist. Place the snails in this and give nem fresh leaves or pulpy fruit, and they ill live comfortably in confinement. A it of cheesecloth fastened with a rubber and should be placed over the top of the r. A tumbler inverted over a dish, on hich is a leaf or two, makes a good obser- ation cage to pass around the room for loser examination. An empty shell should e at hand, which may be opened and xamined.

Observations — 1. Where do you find nails? Why do they like to live in such laces?

2. How does a snail walk? Describe its foot." How can it move with only one oot? Describe how it climbs the side of he glass jar. How does it cling?

3. What sort of track does a snail leave behind it? What is the use of this mucus?

4. Where are the snail's eyes? Why is this arrangement convenient? If we touch one of the eyes what happens? What advantage is this to the snail? Can it pull in one eye and leave the other out?

5. Look below the eyes for a pair of feelers. What happens to these if you touch them?

6. What is the use of its shell to a snail? What does the snail do if startled? If attacked? When a snail is withdrawn into its shell can you see any part of the body? Is the shell attached to the middle of the foot? How did the shell grow on the snail's back? How many spiral turns are there in the full-grown shell? Are there as many in the shell of a young snail? Can you see the little ridges on the shell? Do you think that these show the way the shell grew?

7. Can you find the opening through which the snail draws its breath? Where is this opening? Describe its action.

8. Put the snail in a dry place for two or three days, and see what happens. Do you think this is for the purpose of keeping in moisture? What does the snail do during the winter?

9. Place a snail on its back and see how it rights itself. Describe the way it eats. Can you see the horny upper jaw? Can you see the rasping tongue? What do snails live on?

10. Do you know how the snail eggs look and where they are laid? How large is the shell of the smallest garden snail you ever saw? How many spiral turns were there in it? Open an empty snail shell and see how the spirals widened as the snail grew. Do you think the shell grew by layers added to the lip?

11. Do all snails have shells? Describe all the kinds of snails you know. What people consider snails a table delicacy?

TO A SNAIL

Little Diogenes bearing your tub, whither
away so gay,
With your eyes on stalks, and a foot that
walks, tell me this I pray!

Is it an honest snail you seek that makes
you go so slow,
And over the edges of all things peek?
Have you found him, I want to know,

Or do you go slow because you know, y
house is near and tight?
And there is no hurry and surely no w
lest you stay out late at night.

THE EARTHWORM

Although not generally considered attractive, for two reasons the earthworm has an important place in nature-study: it furnishes an interesting example of lowly organized creatures, and it is of great economic importance to the agriculturist. The lesson should have special reference to the work done by earthworms and to the simplicity of the tools with which the work is done.

Hugh Spencer

Earthworms

The earthworm is, among lower animals, essentially the farmer. Long before men conceived the idea of tilling the soil, this seemingly insignificant creature was busily at work plowing, harrowing, and fertilizing the land. Nor did it overlook the importance of drainage and the addition of amendments — factors of comparatively recent development in the management of the soil by man.

Down into the depths, sometimes as far as seven or eight feet, but usually from twelve to eighteen inches, goes the little plowman, bringing to the surface the subsoil, which is exactly what we do when we plow deeply. To break up the soil as our harrows do, the earthworm grinds it in a gizzard stocked with grains of sand or fine gravel, which act as millstones. Thu turns out soil of much finer texture t we, by harrowing or raking, can prod In its stomach it adds the lime ame ment, so much used by the mod farmer. The earthworm is apparently adept in the use of fertilizers; it even sh discrimination in keeping the organic n ter near the surface, where it may be corporated into the soil of the root zo It drags into its burrows dead leaves, fl ers, and grasses, with which to line upper part. Bones of dead animals, she and twigs are buried by it, and, being m or less decayed, furnish food for pla These minute agriculturists have ne studied any system of drainage, but t bore holes to some depth which carry surplus water. They plant seeds by co ing those that lie on the ground with s from below the surface — good, enrich well granulated soil it is, too. They f ther care for the growing plants by cu vating, that is keeping fine and granu the soil about the roots.

It was estimated by Darwin that, garden soil in England, there are m than fifty thousand earthworms in an ac and that the whole superficial layer vegetable mold passes through their b ies in the course of every few years, at t rate of eighteen tons an acre yearly.

This agricultural work of the earthwo has been going on for ages. Wild la owes much of its beauty to this dimin tive creature which keeps the soil in go condition. The earthworm has und mined and buried rocks, changing grea the aspect of the landscape. In this w it even has preserved ruins and ancie works of art. Several Roman villas in En land owe their preservation to the ear worm. All this work is accomplished wi the most primitive tools: a tiny probosc stensible pharynx, a rather indeter-ate tail, a gizzard, and the calcar-; glands peculiar to this lowly crea-.

n earthworm has a peculiar, crawling ement. Unlike the snake, which also es without legs, it has no scales to ction in part as legs; but it has a very ial provision for locomotion. On the er side of a worm are found numerous — tiny, bristle-like projections. These be seen to be in double rows on each nent, excepting the first three and the . The setæ turn so that they point in opposite direction from that in which worm is moving. It is this use of these ging bristles, together with strong scles, which enables a worm to hold tly to its burrow when bird or man mpts its removal. A piece of round tic furnishes an excellent example of traction and extension, such as the thworm exhibits. Under the skin of the rms are two sets of muscles; the outer sing in circular direction around the ly, the inner running lengthwise. The vement of these may be easily seen in ood-sized living specimen. The body engthened by the contraction of circu-and the extension of longitudinal mus-s, and shortened by the opposite move-nt.

The number of segments may vary with age of the worm. In the immature thworm, the *clitellum*, a thick, whitish g near the end, is absent. The laying of earthworm's eggs is an interesting per-mance. A saclike ring is formed about body in the region of the clitellum. is girdle is gradually worked forward d, as it is cast over the head, the sac-ds snap together enclosing the eggs. ese capsules, yellowish-brown, football-ped, about the size of a grain of wheat, y be found in May or June about ma-re piles or under stones.

Earthworms are completely deaf, al-ough sensitive to vibration. They have eyes, but can distinguish between light d darkness. The power of smell is feeble. e sense of taste is well developed; the se of touch is very acute; and we are not so sure as is Dr. Jordan that the angle-worm is at ease on the hook.

Any garden furnishes good examples of the home of the earthworm. The burrows are made straight down at first, then wind about irregularly. Usually they are about one or two feet deep, but may reach even eight feet. The burrow terminates generally in an enlargement where one or several worms pass the winter. Toward the surface, the burrow is lined with a thin layer of fine, dark colored earth, voided by the worm. This creature is an excavator and builder of no mean ability. The tower-like "castings" so characteristic of the earthworm are formed with excreted earth. Using the tail as a trowel, it places earth now on one side and now on the other. In this work, of course, the tail protrudes; in the search for food, the head is out. A worm, then, must make its home, narrow as it is, with a view to being able to turn in it.

An earthworm will bury itself in loose earth in two or three minutes and in compact soil, in fifteen minutes. Pupils should be able to make these observations easily either in the terrarium or in the garden.

In plugging the mouths of their burrows, earthworms show something that seems like intelligence. Triangular leaves are invariably drawn in by the apex, pine needles by the common base, the manner varying with the shape of the leaf. They do not drag in a leaf by the footstalk, unless its basal part is as narrow as the apex. The mouth of the burrow may be lined with leaves for several inches.

The burrows are not found in dry ground or in loose sand. The earthworm lives in the finer, moderately wet soils. It must have moisture, since it breathes through the skin, and it has sufficient knowledge of soil texture and plasticity to recognize the futility of attempts at burrow building with unmanageable, large grains of sand.

These creatures are nocturnal, rarely appearing by day unless "drowned out" of the burrows. During the day they lie near the surface extended at full length, the head uppermost. Here they are discovered by keen-eyed birds and sacrificed by thousands, notwithstanding the strong muscular protest of which they are capable.

Seemingly conscious of its inability to find the way back to its home, an earthworm anchors tight by its tail while stretching its elastic length in a foraging expedition. It is an omnivorous creature, including in its diet earth, leaves, flowers, raw meat, fat, and even showing cannibalistic designs on fellow earthworms. In the schoolroom, earthworms may be fed on pieces of lettuce or cabbage leaves. A feeding worm will show the proboscis, an extension of the upper lip used to push food into the mouth. The earthworm has no hard jaws or teeth, yet it eats through the hardest soil. Inside the mouth opening is a very muscular pharynx, which can be extended or withdrawn. Applied to the surface of any small object it acts as a suction pump, drawing food into the food tube. The earth taken in furnishes some organic matter for food; calcareous matter is added to the remainder before being voided. This process is unique among animals. The calcareous matter is supposed to be derived from leaves which the worms eat. Generally the earth is swallowed at some distance below the surface and finally ejected in characteristic "castings." Thus, the soil is slowly worked over and kept in good condition by earthworms, of which Darwin says: "It may be doubted whether there are many other animals which have played so important a part in the history of the world as have these lowly organized creatures."

Fly fishing is an art, a fine art beyond a doubt, but it is an art and, like all art, it is artificial. Fishing with an angleworm is natural. It fits into the need of the occasion. It fits in with the spirit of the It is not by chance that the anglew earthworm, fishworm, is found in e damp bank, in every handy bit of sod, green earth over, where there are whose boys are real boys with en enough to catch a fish. It is not by ch that the angleworm makes a perfec on a hook, with no anatomy with whic feel pains, and no arms or legs to be brc off or to be waved helplessly in the Its skin is tough enough so as no tear, not so tough as to receive unsee bruises, when the boy is placing it on hook. The angleworm is perfectly at h on the hook. It is not quite comfort anywhere else. It crawls about on sidew after rain, bleached and emaciated. never quite at ease even in the grou but on the hook it rests peacefully, the apparent feeling that its natural sion is performed.

— "Boys' Fish and Boys' Fishi
David Starr Jor

LESSON 107

The Earthworm

Leading Thought — The earthwor a creature of the soil and is of great nomic importance.

Method — Any garden furnishes ab dant material for the study of earthwor They are nocturnal workers and may observed by lantern or flashlight. To fo some estimate of the work done i single night, remove the "casts" fro square yard of earth one day, and exam that piece of earth the next. It is well have a terrarium in the schoolroom frequent observation. Scatter grass or d leaves on top of the soil, and note w happens. For the study of the individ worm and its movements, each pu should have a worm with some earth u his desk.

Observations — 1. How does earthworm crawl? How does it turn ov Has it legs? Compare its movement w that of a snake, another legless anim What special provision for locomoti has the earthworm?

. Compare the lengths of the conted and extended body. How can the nge be accounted for?

. Describe the body — its shape and r — above and below. Examine the nents. Do all the worms have the e number? Compare the head end n the tail end of the body. Has every m a "saddle," or clitellum?

. Does the earthworm hear easily? Has yes? Is it sensible to smell or to touch? nat sense is most strongly developed?

. Describe the home of the earthworm. t occupied by more than one worm? w long does it take a worm to make urrow? How does it protect its home? How does it make a burrow? In what kind of soil do you find earthworms at work?

6. Is the earthworm seen most often at night or by day? Where is it the rest of the time? How does it hold to its burrow? When is the tail end at the top? When the head end?

7. What is the food of the earthworm? How does it get its food?

8. Look for the eggs of the earthworm about manure piles or under stones.

9. What are the enemies of the earthworm? Is it a friend or an enemy to us? Why?

10. The earthworm is a good agriculturist. Why?

THE CRAYFISH

Vhen I look at a crayfish I envy it, so is it in organs with which to do all t it has to do. From the head to the , it is crowded with a large assortment executive appendages. In this day of ltiplicity of duties, if we poor human atures only had the crayfish's capabils, then might we hope to achieve what before us.

The most striking thing in the appeare of the crayfish is the great pair of pers on each of the front legs. Wondery are its "thumb and finger" put toher; the "thumb" is jointed so that it move back and forth freely; and both armed, along the inside edge, with teeth and with a sharp claw at the tip that they can get a firm grip upon an ect. Five segments in these great legs be easily seen; that joining the body mall, but each successive one is wider d larger, to the great forceps at the end. e two stout segments behind the nips give strength, and also a suppleness t enables the claws to be bent in any ection.

The legs of the pair behind the big pers have five segments readily visible; t these legs are slender and the nippers the end are small; the third pair of legs armed like the second pair; but the fourth and fifth pairs lack the pincers, and end in a single claw.

But the tale of the crayfish's legs is by no means told; for between and above the great pincers is a pair of short, small legs

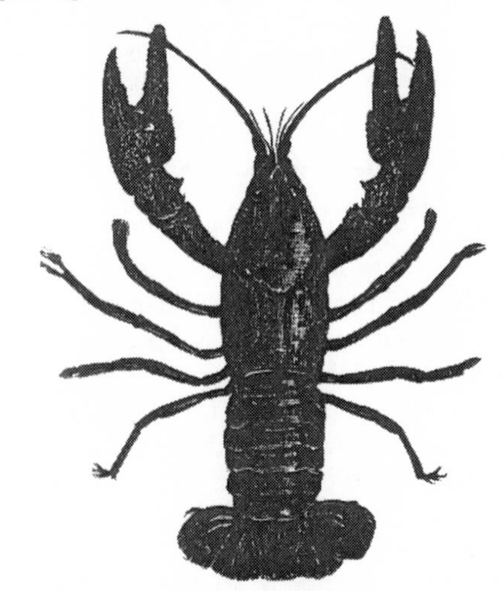

The crayfish

tipped with single claws, and fringed on their inner edges. These are the maxillipeds, or jaw-feet; and behind them, but too close to be seen easily, are two more pairs of jaw-feet. As all of these jaw-feet assist at meals, the crayfish apparently always has a "three-fork" dinner; and as

Charles E. Mohr

A blind white crayfish found living in the darkness of a cave in Kentucky

if to provide accommodations for so many eating utensils, it has three pairs of jaws all working sidewise, one behind the other. Two of these pairs are maxillæ and one, mandibles. The mandibles are the only ones we see as we look in between the jaw-feet; they are notched along the biting edge. Connected with the maxillæ, on each side, are two pairs of threadlike flappers that wave back and forth vigorously and have to do with setting up currents of water over the gills.

Thus we see that, in all, the crayfish has three pairs of jaw-feet, one pair of great nippers, and four pairs of walking feet, two of which also have nippers and are used for digging and carrying.

When we look upon the crayfish from above, we see that the head and thorax are fastened solidly together, making what is called a cephalothorax. The cephalothorax is covered with a shell called the carapace, which is the name given alsc the upper part of thc turtle's shell. T suture where the head joins the thora quite evident. In looking at the head, eyes first attract our attention; each black and oval and placed on the tip a stalk, so it can be extended or retrac or pushed in any direction, to look danger. These eyes are like the compou eyes of insects, in that they are made of many small eyes, set together i honeycomb pattern.

The long antennæ are as flexible braided whiplashes, large at the base a ending in a threadlike tip. They are cc posed of many segments, the basal o being quite large. Above the antennæ each side is a pair of shorter ones cal antennules, which come from the sa basal segment; the lower one is the m slender and is usually directed forwa the upper one is stouter, curves upwa and is kept always moving, as if it w constantly on the alert for impressio The antennæ are used for exploring ahead or behind the creature, and often thrust down into the mud a gravel at the bottom of the aquarium, if probing for treasure. The antennu seem to give warning of things closer hand. Between the antennæ and ant nules is a pair of finger-like organs that hinged at the outer ends and can be lif back, if we do it carefully.

In looking down upon a crayfish, can see six abdominal segments and flaring tail at the end, which is really other segment greatly modified. The fi segment, or that next to the cepha thorax, is narrow; the others are abc equal in size, each graceful in shape, w a widened part at each side which exter down along the sides of the creatu These segments are well hinged togetl so that the abdomen may be complet curled beneath the cephalothorax. T plates along the sides are edged w fringe. The tail consists of five parts, o semicircular in the center, and two f shaped pieces at each side, and all margined with fringe. This tail is a rema able organ. It can be closed or extend ewise like a fan; it can be lifted up or led beneath.

Looking at the crayfish from below, we on the abdomen some very beautiful ther-like organs called swimmerets. ch swimmeret consists of a basal seg- nt with twin paddles joined to its tip, h paddle being narrow and long and ged with hairs. The mother crayfish four pairs of these, one pair on each the second, third, fourth, and fifth seg- nts; her mate has an additional larger r on the first segment. These swim- rets, when at rest, lie close to the ab- men and are directed forward and htly inward. When in motion, they ldle with a backward, rhythmic motion, first pair setting the stroke and the er pairs following in succession. This tion sends the body forward, and the immerets are chiefly used to aid the s in forward locomotion. A crayfish, the bottom of a pond, seems to glide out with great ease; but place it on d, and it is an awkward walker. The son for this difference lies, I believe, in e aid given by the swimmerets when the ature is in water. Latter says: " In walk- , the first three pairs of legs pull and e fourth pair pushes. Their order of ovement is as follows: The first on the ht and the third on the left side move gether, next the third right and the first t, then the second right and fourth left, d lastly the fourth right and second t."

When the crayfish really wishes to im, the tail is suddenly brought into e; it is thrust out backward, lays hold of e water by spreading out widely, and en doubles under with a spasmodic k which pulls the creature swiftly back- rd.

The crayfish's appearance is magically ansformed when it begins to swim; it is longer a creature of sprawling awkward gs and great clumsy nippers; now, its any legs lie side by side supinely and e great claws are limp and flow along in aceful lines after the body, all obedient the force which sends the creature fly- g through the water. I cannot discover that the swimmerets help in this movement.

The mother crayfish has another use for her swimmerets; in the spring, when she is ready to lay eggs, she cleans off her paddles with her hind legs, covers them with waterproof glue, and then plasters her eggs on them in grapelike clusters of little dark globules. What a nice way to look after her family! The little ones hatch, but remain clinging to the maternal swimmerets until they are large enough to scuttle around on the brook bottom and look out for themselves.

The breathing apparatus of the crayfish cannot be seen without dissection. All the walking legs, except the last pair, have gills attached to that portion of them which joins the body, and which lies hidden underneath the sides of the carapace or shell. The blood is forced into these gills, sends off its impurities through their thin walls, and takes in the oxygen from the water, currents of which are kept steadily flowing forward.

Crayfishes haunt still pools along brooksides and river margins and the shallow ponds of our fresh waters. There they hide beneath sticks and stones, or in caves of their own making, the doors of which they guard with the big and threatening nippers, which stand ready to grapple with anybody that comes to inquire if the folks are at home. The upper surface of the crayfish's body is always so nearly the color of the brook bottom that the eye seldom detects the creature until it moves;

E. Morton Miller

A land crab, a relative of the crayfish. Note the eye-stalks

and if some enemy surprises one, it swims off with terrific jerks which roil all the water around; thus it covers its retreat. In the winter, our brook forms hibernate in the muddy bottoms of their summer haunts. There are many species; some in our southern states, when the dry season comes on, live in little wells which they dig deep enough to reach water. They heap up the soil which they excavate

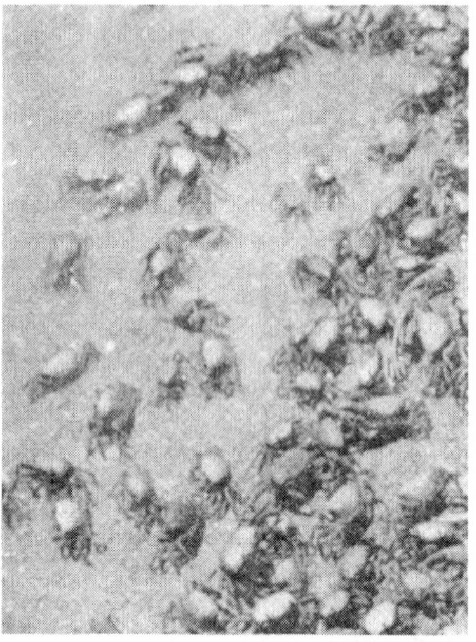

E. Morton Miller

Fiddler crabs, so called from the position in which the male often holds the enlarged claw, are burrowing crabs of the Atlantic coast

around the mouth of the well, making well-curbs of mud; these are ordinarily called "crawfish chimneys." The crayfishes find their food in the flotsam and jetsam of the pool. They seem fond of the flesh of dead fishes and are often trapped by its use as bait.

The growth of the crayfish is like that of insects; as its outer covering is a hard skeleton that will not stretch, it is shed as often as necessary; it breaks open down the middle of the back of the carapace, and the soft-bodied creature pulls itself out, even to the last one of its claws. While its new skin is yet elastic, it stretches to its utmost; but this skin also hardens after a time and is, in its t shed. Woe to the crayfish caught in helpless, soft condition after molting! it then has no way to protect itself. sometimes find the old skin floating, fect in every detail, and so transparent t it seems the ghost of a crayfish.

Not only is the crayfish armed in beginning with a great number of l antennæ, etc., but if it happens to any of these organs they will grow ag We have often found one of these c tures with one of the front claws m larger than the other; it had probably its big claw in a fight, and the new grov was not yet completed.

I have been greatly entertained watching a female crayfish make her n in my aquarium, which has, for her c fort, a bottom of three inches of cl gravel. She always commences at one s by thrusting down her antennæ and n pers between the glass and stones; seizes a pebble in each claw and pull up and in this way starts her excavati but when she gets ready to carry off load, she comes to the task with her tucked under her body, as a lady tucks her skirts when she has something to that requires freedom of movement. Th with her great nippers and the two p of walking feet, also armed with nipp she loads up as much as she can carry tween her great claws and her breast. S keeps her load from overflowing by ho ing it down with her first pair of jaw-fe just as I have seen a schoolboy use chin, when carrying a too large load books; and she keeps the load from fall out by supporting it from beneath w her first pair of walking legs. Thus, starts off with her "apron" full, walki on three pairs of feet, until she gets to dumping place; then she suddenly lets and at the same time her tail straight out with a gesture which says plain "There!" Sometimes when she gets very large load, she uses her second p of walking legs to hold up the burd and crawls off successfully, if not w ease, on two pairs of legs, — a most natural quadruped.

had two crayfishes in a cage in an rium, and each made a nest in the el at opposite ends of the cage, heap- up the debris into a partition between n. I gave one an earthworm, which she nptly seized with her nippers; she then up a good-sized pebble in the nippers er front pair of walking legs, glided to the other nest, spitefully threw n both worm and pebble on top of fellow prisoner, and then sped home- l. Her victim responded to the act by g up and expressing perfectly, in his ude and the gestures of his great claws, most eloquent of crayfish profanity. watching crayfishes carry pebbles, I been astonished to see how con- tly the larger pair of jaw-feet are used elp pick up and carry the loads.

LESSON 108

The Crayfish

EADING THOUGHT — The crayfish, or wfish, as it is sometimes called, has pair of legs developed into great pin- s for seizing and tearing its food and defending itself from enemies. It can in mud or water. It belongs to the ne animal group as do the insects, and s a near cousin of the lobster.

METHOD — Place a crayfish in an aquar- n (a battery jar or a two-quart Mason in the schoolroom, keeping it in clear ter until the pupils have studied its m. It will rise to explore the sides of aquarium at first, and thus show its uth-parts, legs, and swimmerets. After- rds, place gravel and stone in the bot- n of the aquarium, so that it can hide elf in a little cavity which it will make carrying pebbles from one side. Wash gravel well before it is put in, so that water will be unclouded and the chil- n can watch the process of excavation.

OBSERVATIONS — 1. What is there peculiar about the crayfish which makes it difficult to pick it up? Examine one of these great front legs carefully and see how wonderfully it is made. How many parts are there to it? Note how each succeeding part is larger from the body to the claws. Note the tips which form the nippers, or chelæ, as they are called. How are they armed? How are the gripping edges formed to take hold of an object? How wide can the nippers be opened, and how is this done? Note the two segments behind the great claw and describe how they help the work of the nippers.

2. Study the pair of legs behind the great claws or chelæ, and compare the two pairs, segment by segment. How do they differ except as to size? How do the nippers at the end compare with the big ones? Look at the next pair of legs behind these; are they similar? How do the two pairs of hind legs differ in shape from the two pairs in front of them?

3. Look between the great front claws and see if you can find another pair of small legs. Can you see anything more behind or above these little legs?

4. When the crayfish lifts itself up against the side of the jar, study its mouth. Can you see a pair of notched jaws that work sidewise? Can you see two or three pairs of threadlike organs that wave back and forth in and out of the mouth?

5. How many legs, in all, has the crayfish? What are the short legs near the mouth used for? What are the great nippers used for? How many legs does the crayfish use when walking? In what order are they moved? Is the hind pair used for pushing? What use does it make of the pincers on the first and second pairs of walking legs?

6. Look at the crayfish from above; the head and the covering of the thorax are soldered together into one piece. When this occurs, the whole is called a cephalothorax; and the cover is called by the same name as the upper shell of the turtle, the carapace. Can you see where the head is joined to the thorax?

7. Look carefully at the eyes. Describe how they are set. Can they be pushed out

SEASHORE CREATURES

1. Sea Urchin, Strongylocentrotus. *The sea urchin is found along the Atlantic, Arctic, and Pacific coasts. Its habitat varies from tide pools and shallow waters to very deep water. The body of the living animal is a flattened hemisphere covered with short spines. What is usually described as a sea urchin is really the skeleton or "test."*

2. Fiddler Crab, Uca. *Fiddler crabs are common along the Atlantic coast of the United States. Above high tide great numbers of these crustaceans are found rushing into their burrows for shelter as one approaches.*

3. Common Starfish, Asterias. *One or the other of the two common varieties may be expected along the coast from the Gulf of Mexico to Labrador. The mouth is at the center of the lower surface of the animal and through the mouth the stomach is turned inside out to engulf and digest food. Its ability to devour mollusks makes the starfish a great enemy of oyster beds; it can force open the shells of an oyster, mussel, or other mollusk by pulling steadily with its strong arms and tubular feet. For that reason starfish caught at oyster beds are destroyed by plunging them into boiling water. Oystermen used to chop them in two and throw the pieces back into the water until they learned that each of the pieces could become a new starfish.*

4. Egg Cases *or* Fisherman's Purses, Elasmobranch. *These queer egg cases of sharks and skates are found empty along the shore.*

5. The Notch-side Shell, Pleurotoma nana. *This species belongs to a large family of shells with a world-wide distribution.*

6. Sand Dollar, Echinarachnius. *While sand dollars are capable of moving about over the ocean bottom by means of suckerlike feet, they do not seem able to right themselves if they are turned on their backs; and so thousands are cast helpless upon the shores by storms. are found mostly from New Jersey north to the Arctic Ocean. The specimens picke on the shore are usually only the skeleto "test." The skeleton shows a design in five branching from a common center, a clear cation that sand dollars and starfish are rel*

7. Giant Whelk, Busycon *or* Fu *These whelks may be found from Cape to the Gulf of Mexico, being most abun along the coast of New Jersey and Long Is Sound on sandy or gravelly beaches nea low-tide level. The strings of egg cap which are often found on the beach are un each tough capsule may contain about dozen eggs or young whelks.*

8. Great Ark Shell, Arca. *The ark s are cosmopolitan in their distribution; in dition to being distributed in both the Atl and Pacific oceans they are found in the M terranean.*

9. Star Coral, Astrangia danæ. *Wha usually knows as coral is only the stony sk remains of coral animals; but pictured are the living animal forms, known as po They are glassy in appearance and each p has eighteen to twenty-four tentacles or stin organs by means of which it captures its*

10. Sand Crab, Hippa. *This very com yellowish-white crustacean, sometimes for fish bait, lives in shallow water along sandy beaches from New Jersey to Island. With its pointed abdomen as a to sand crab digs a burrow very quickly.*

11. Jellyfish. *The jellyfish is shaped an umbrella and has its mouth and stom in the position occupied by the handle of a umbrella; the tentacles and other sense or are attached to the outer edge of the umbr By means of its tentacles the jellyfish cap the small animals upon which it feeds.*

Courtesy of American Museum of Natural History

or pulled in? Can they be moved in all directions? Of what advantage is this to the crayfish?

8. How many antennæ has the crayfish? Describe the long ones and tell how they are used. Do the two short ones on each side come from the same basal segment? These little ones are called the antennules. Describe the antennules of each side and tell how they differ. Can you see the little finger-like organs which clasp above the antennæ and below the antennules on each side of the head? Can these be moved?

9. Look at the crayfish from above. How many segments are there in the abdomen? Note how graceful is the shape of each segment. Note that each has a fan-shaped piece down the side. Describe how the edges of the segments along the sides are margined.

10. Of how many pieces is the tail made? Make a sketch of it. How are the pieces bordered? Can the pieces shut and spread out sidewise? Is the tail hinged so it can be lifted up against the back or curled under the body?

11. Look underneath the abdomen and describe the little fringed organs called the swimmerets. How many are there?

12. How does the crayfish swim? With what does it make the stroke? Describe carefully this action of the tail. When it is swimming, does it use its swimmerets? Why do not the many legs and big nippers obstruct the progress of the crayfish when it is swimming?

13. When does the crayfish use its swimmerets? Do they work so as to p the body backward or forward? Do know to what use the mother crayfish p her swimmerets?

14. Do you know how crayfis breathe? Do you know what they eat where they find it?

15. Where do you find crayfisl Where do they like to hide? Do they headfirst into their hiding place, or they back in? Do they stand ready to fend their retreat? When you look do into the brook, are the crayfishes usu seen until they move? Why is this? Wh do the crayfishes pass the winter? Did ever see the crayfish burrows or m chimneys?

16. If the crayfish loses one of its l or antennæ, does it grow out again? H does the crayfish grow?

17. Put a crayfish in an aquarium wh has three inches of coarse gravel on bottom, and watch it make its den. H does it loosen up a stone? With how ma legs does it carry its burden of pebb when digging its cave? How does it use jaw-feet, its nippers, and its first and s ond pairs of walking legs in this work?

A rock-lined, wood-embosomed nook,
Dim cloister of the chanting brook!
A chamber within the channeled hills,
Where the cold crystal brims and spills,
By dark-browed caverns blackly flows,
Falls from the cleft like crumbling sno
And purls and splashes, breathing roun
A soft, suffusing mist of sound.

— J. T. Trowbrid

DADDY LONGLEGS OR GRANDFATHER GREYBEARD

I wonder if there ever was a country child who has not grasped firmly the leg of one of these little sprawling creatures and demanded: "Grandfather Greybeard, tell me where the cows are or I'll kill you," and Grandfather Greybeard, striving to get away, puts out one of his long legs this way, and another that way, and points in so many directions that he usually saves his life, since the cows must be som where. It would be more interesting the children and less embarrassing to t "daddy" if they were taught to look mo closely at those slender, hairlike legs.

"Daddy's" long legs are seven-jointe The first segment is seemingly solder fast to the lower side of his body, and called the coxa. The next segment is re knob, usually black and ornamental, is called the trochanter. Then comes femur, a rather long segment directed vard; next is a short swollen segment he "knee joint" or patella; next the a, which is also rather long. Then ne the metatarsus and tarsus, which mingly make one long downward-di- ted segment, outcurving at the tips, on ich the "daddy" tiptoes along.

have seen a "daddy" walk into a drop water, and his foot was never wetted, light was his touch on the water sur- e film. The second pair of legs is the gest; the fourth pair next, and the first r usually the shortest. The legs of the ond pair are ordinarily used in explor- the surroundings. Notice that, when "daddy" is running, these two legs spread wide apart and keep in rapid tion; their tips, far more sensitive than y nerves of our own, tell him the na- e of his surroundings, by a touch so ht that we cannot feel it on the hand. e have more respect for one of these irlike legs, when we know it is capable transmitting intelligence from its tip.

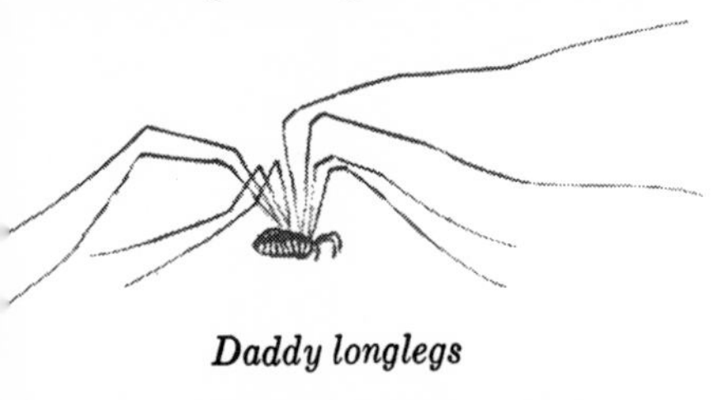

Daddy longlegs

The "daddy" is a good traveler and oves with remarkable rapidity. And why ot? If our legs were as long in comparison his, they would be about forty feet in ngth. When the "daddy" is running, e body is always held a little distance ove the ground; but when the second ir of legs suggests to him that there may something good to eat in the neighbor- ood, he commences a peculiar teetering otion of the body, apparently touching to the ground at every step; as the body carried tilted with the head down, this ovement enables the creature to explore e surface below him with his palpi, hich he ordinarily carries bent beneath his face, with the ends curled up under his "chin." The palpi have four segments that are easily seen, and although they are ordinarily carried bent up beneath the

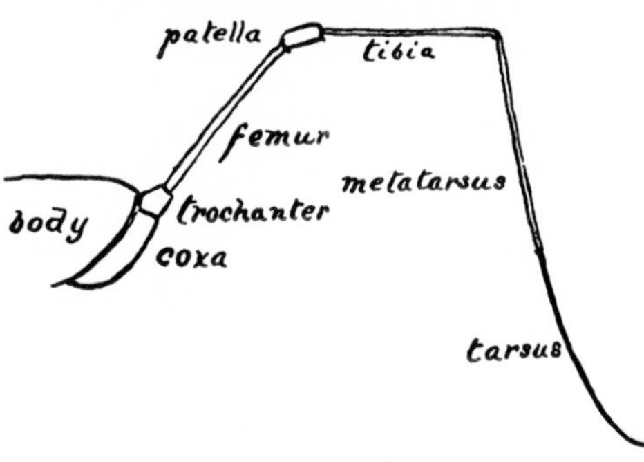

One of "daddy's" long legs with segments named

head, they can be extended quite a distance if "daddy" wishes to test a substance. The end segment of the palpus is tipped with a single claw.

Beneath the palpi is a pair of jaws; these, in some species, extend beyond the palpi. I have seen a daddy longlegs hold food to his jaws with his palpi and he seemed also to use them for stuffing it into his mouth.

The body of the daddy longlegs is a little oblong object, looking more like a big grain of wheat than anything else, because in these creatures the head, thorax, and abdomen are all grown together compactly. On top of the body, between the feeler-legs, is a little black dot, and to the naked eye it would seem that if this were an organ of sight the creature must be a Cyclops with only one eye. But under the lens this is seen to be a raised knob, and there is on each side of it a little shining black eye. We hardly see the use of two eyes set so closely together, but probably the "daddy" does.

The most entertaining thing which a "daddy" in captivity is likely to do is to clean his legs; he is very particular about his legs, and he will grasp one close to the basal joint in his jaws and slowly pull it through, meanwhile holding the leg up to the jaws with the palpi, while he industriously nibbles it clean for the whole length to the very toe. Owing to the likelihood of his losing one of his legs, he has the power of growing a new one; so we often see a "daddy" with one or more legs only half grown.

There are many species of daddy long-legs in the United States, and some of them do not have the characteristic long legs. In the North, all except one species die at the approach of winter; but not until after the female, which, by the way, ought to be called "granny longlegs," has laid her eggs in the ground, or under some protecting stone, or in some safe crevice of wood or bark. In the spring the eggs hatch into tiny creatures which look just like the old daddy longlegs, except for their size. They get their growth like insects, by shedding their skins as fast as they outgrow them. It is interesting to study one of these cast skins with a lens. There it stands with a slit down its back, and with the skin of each leg absolutely perfect to the tiny claw! Again we marvel at these legs that seem so threadlike, and which have an outer covering that can be shed. Some say that the daddy longlegs live on small insects which they straddle over and pounce down upon, and some say they feed upon decaying matter and vegetable juices. This would be an interesting line of investigation for pupils, since they might be able to give many new facts about the food of these creatures. The "daddies" are night prowlers, and like to hide in crevices by day, waiting for the dark to hunt for their food. They have several common names. Besides the two given they are called "harvestmen" and the French call them "haymakers." Both of these names were very probably given because the creatures appear in greater numbers at the time of haying and harvesting.

LESSON 109

The Daddy Longlegs

Leading Thought — These long-legged creatures have one pair of legs too many to allow them to be classed with the insects. They are more nearly related to spiders, which also have eight legs. Th are pretty creatures when exami closely, and they do many interest things.

Method — Put a grandfather greybe in a breeding cage or under a large tu bler, and let the pupils observe him leisure. If you place a few drops of swe ened water at one side of the cage, children will surely have an opportun to see this amusing creature clean his le

Observations — 1. Where did you fi the harvestman? What did it do as soon it was disturbed? How many names do y know for this little creature?

2. A "daddy" with such long legs c tainly ought to have them studied. H many segments in each leg? How do t segments look? How do the legs lo where they are fastened to the bod Which is the longest pair of legs? T next? The next? The shortest?

3. If you had such long stilts as he h they would be about forty feet lor Would you lift yourself that high in t air? Does the "daddy" lift his body hi or swing it near the ground? What sha is the body? Can you see if there is a d tinct head? Can you see a black dot on t of the front end of the body? If you shou see this dot through a microscope would prove to be two bright black ey Why should the daddy's eyes be on to

4. Do you see a pair of organs that lo like feelers at the front end of the bod These are called palpi. How does he u his palpi? Give him a little bruised or d caying fruit, and see him eat. Where you think his mouth is? Where does keep his palpi when he is not using the for eating?

5. Note what care he takes of his leg How does he clean them? Which does l clean the oftenest? Do you think the ve long second pair of legs is used as muc for feeling as for walking? Put some obje in front of the "daddy" and see him e plore it with his legs. How much of th leg is used as a foot when the "daddy stands or runs?

6. When running fast, how does th

ddy" carry his body? When explor- how does he carry it? Do you ever find "daddy" with his body resting on the ace on which he is standing? When ing, are all eight of his legs on the und? Which are in the air? Is the head usually tilted up or down?

. Do you see the daddy longlegs early in the spring? When do you find him most often? How do you suppose he passes the winter in our climate? Have you ever seen a "daddy" with one leg much shorter than the other? How could you explain this?

8. Try to discover what the daddy long-legs eats, and where he finds his food.

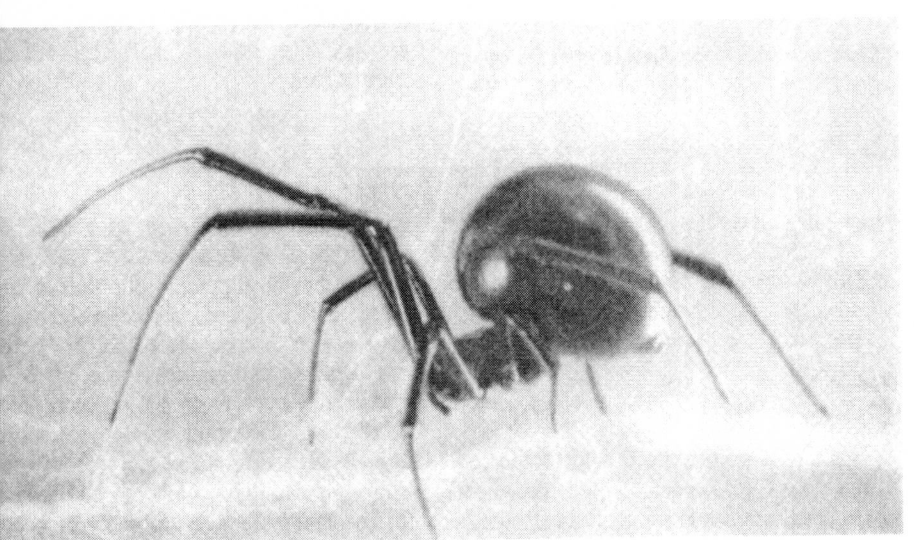

G. E. Jenks

'he poisonous black widow or hourglass spider. It and the tarantula are the only dangerous spiders in the United States

SPIDERS

The spiders are the civil engineers iong the small inhabitants of our fields d woods. They build strong suspension idges, from which they hang nets made th exquisite precision; and they build planes and balloons, which are more icient than any that we have yet con- ucted; for although they are not exactly rigible, yet they carry the little balloon- s where they wish to go, and there are w fatal accidents. Moreover, the spiders e of much economic importance, since ey destroy countless millions of insects ery year, most of which are noxious — e flies, mosquitoes, bugs, and grasshop- rs.

There is an impression abroad that all iders are dangerous to handle. This is a mistake; the bite of any of our common spiders is not nearly so dangerous as the bite of a malaria-laden mosquito. Al-

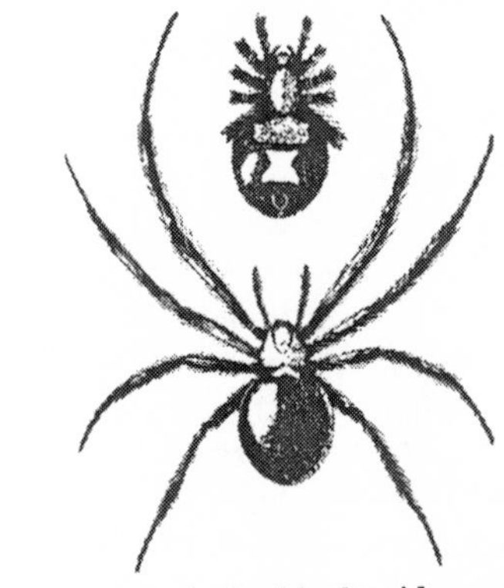

Another view of the black widow. Above is the underside to show the red hourglass

though there is a little venom injected into the wound by the bite of any spider, yet there are few species found in the United States whose bite is sufficiently venomous to be feared. With the ex tion of the tarantulas of the Southv and the hourglass or black widow, wl seems now to be extending its range f the South, the spiders of the United S are really as harmless to handle as are r of our common insects.

The tarantula, a large, dark-colored, hairy spider found in the Southwest. It is poisonous

There is no need for studying the a omy of the spider closely in nature-st Our interest lies much more in the v derful structures made by the spiders, t in a detailed study of the little creat themselves.

COBWEBS

Here shy Arachne winds her endless thread,
And weaves her silken tapestry unseen,
Veiling the rough-hewn timbers overhead,
And looping gossamer festoons between.

— ELIZABETH AKERS

Our house spiders are indefatigable curtain-weavers. We never suspect their presence, until suddenly their curtains appear before our eyes, in the angles of the ceilings — invisible until laden with dust. The cobwebs are made of crisscrossed lines, which are so placed as to entangle any fly that comes near. The lines are stayed to the sides of the wall and to each other quite firmly, and thus they are able to hold a fly that touches them. The spider is likely to be in its little den at the side of the web; this den may be in a crevice in the corner or in a tunnel made of the silk. As soon as a fly becomes entangled in the web, the spider runs to it, seizes it in its jaws, sucks its blood, and then throws away the shell, the wings, and the legs. If a spider is frightened, it at first tries to hide and then may drop by a thread to the floor. If we catch the little acrobat it will usually "play possum" and we may examine it more closely through a lens. We shall find it is quite different in form from an insect. First to be noted, it has eight legs; but most important of all, it has only two parts to the body. The head and thorax are consolidated into one pi which is called the cephalothorax. The domen has no segments like that of the sects, and is joined to the cephalotho by a short, narrow stalk. At the fron the head is the mouth, guarded by mandibles, each ending in a sharp claw the tip of which the poison gland op It is by thrusting these mandibles into prey that it kills its victims. On each of the mandible is a palpus, which in males is of very strange shape. The are situated on the top of the head. Th are usually four pairs of these eyes, each looks as beady and alert as if it w the only one.

The spinning organs of the spider situated near the tip of the abdom while the spinning organ of the caterpi

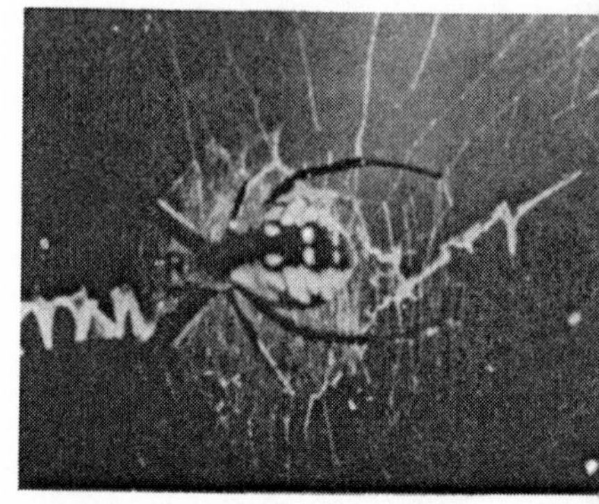

An orange garden spider and web. This der is common in the United States; its is spun in fields and gardens.

ituated near its lower lip. The spider's comes from two or three pairs of spin-ets which are finger-like in form, and on the end of each are many small tubes m which the silk is spun. The silk is in uid state as it issues from the spinner- but it hardens immediately on contact h the air. In making their webs, spiders duce two kinds of silk: one is dry and lastic, making the framework of the b; the other is sticky and elastic, cling- to anything that it touches. The ly and the legs of spiders are usually ry.

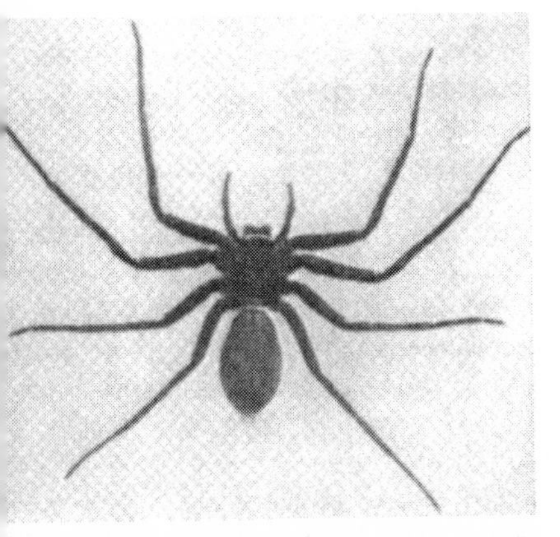

The banana spider. These spiders differ m other Arachnida by having the abdo-n unsegmented and joined to the thorax by short, narrow stalk as shown here

LESSON 110

Cobwebs

Leading Thought — The cobwebs which are found in the corners of ceilings and in other dark places in our houses are

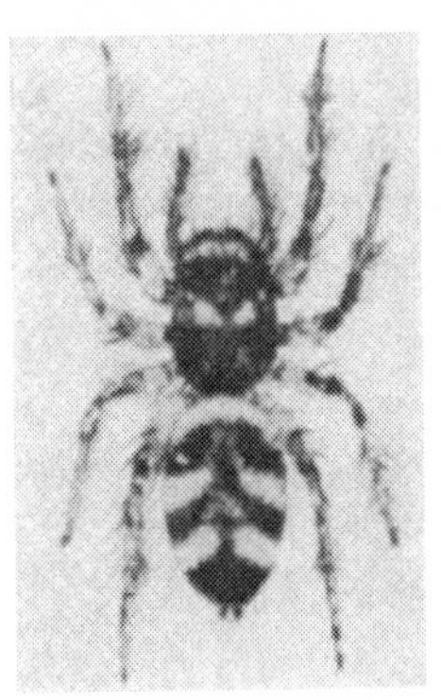

A jumping spider

made by the house spider, which spins its web in these situations for the purpose of catching insects.

Method — The pupils should have under observation a cobweb in a corner of a room, preferably with a spider in it.

Observations — 1. Is the web in a sheet or is it a mass of crisscrossed, tangled threads? How are the threads held in place?

2. What is the purpose of this web? Where does the spider hide? Describe its den.

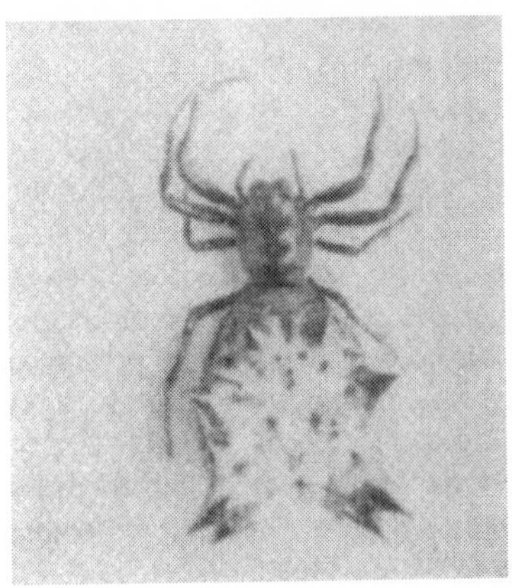

The spiny-bellied spider

3. If a fly becomes tangled in a web, describe the action of the spider. Does the spider eat all of the fly? What does it do with the remains?

4. If the spider is frightened, what does it do? Where does the silken thread come from, and how does its source differ from the source of the silken thread spun by caterpillars?

5. Imprison a spider under a tumbler or in a vial, and look at it very carefully. How many legs has it? How does the spider differ from insects in this respect? How many sections are there to the body? How does the spider differ from insects in t respect?

6. Look closely at the head. Can you the hooked jaws, or fangs? Can you see palpi on each side of the jaws? Where the spider's eyes? How many pairs of e does it have?

When the tangled cobweb pulls
The cornflower's cap awry,
And the lilies tall lean over the wal
To bow to the butterfly,
It is July.

— Susan Hartley Sw

THE FUNNEL WEB OF A GRASS SPIDER

And dew-bright webs festoon the grass
In roadside fields at morning.

— Elizabeth Akers

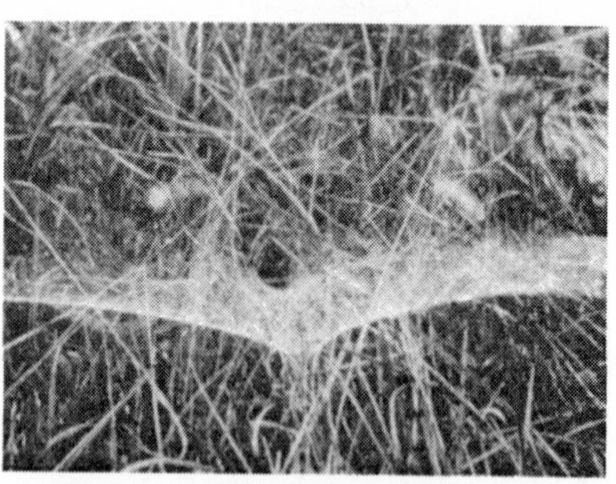

The funnel web of a grass spider

Sometimes, on a dewy morning, a field will seem carpeted with these webs, each with its opening stretched wide, and each with its narrow hallway of retreat. The general shape of the web is like that of a broad funnel with a tube leading down at one side. This tube is used as a hiding place by the architect, the grass spider, which thus escapes the eyes of its enemies, and also keeps out of sight of any insects that might be frightened at seeing it, and so avoid the web. But the tube is no cul-de-sac; quite to the contrary, it has a rear exit, through which the spider, if frightened, escapes from attack.

The web is formed of many lines of silk crossing each other irregularly, formin firm sheet. This sheet is held in place many guy-lines, which fasten it to s rounding objects. If the web is touch lightly, the spider rushes forth from lair to seize its prey; but if the w be jarred roughly, the spider spe out through its back door and can found only with difficulty. The smal insects of the field, such as flies a bugs, are the chief food of this spider rarely attempts to seize a grown grassh per.

The funnel-shaped webs in dark c ners of cellars are made by a species whi is closely related to the grass spider a has the same general habits, but whi builds in these locations instead of the grass.

LESSON 111

The Funnel Web

Leading Thought — The grass spic spins funnel-shaped webs in the grass entrap the insects of the field. This w has a back door.

METHOD — Ask the pupils to observe a b on the grass with a spider within it.

OBSERVATIONS — 1. What is the gen- shape of the web? Is there a tunnel ding down from it? Why is it called unnel web?

2. Of what use is the funnel tube, and at is its shape? Where does it lead, and what use is it to the spider? Can you ner a spider in its funnel tube? Why t?

3. How is the web made? Is there any regularity in the position of the threads that make it? By what means is it stayed in place?

4. Touch the web lightly, and note how the spider acts. Jar the web roughly, and what does the spider do?

5. What insects become entangled in this web?

6. Compare this web with similar funnel webs found in corners of cellars, sheds, or porches, and see if you think the same kind of spider made both.

THE ORB WEB

Of all the structures made by the lower eatures, the orb web of the spider is, be- nd question, the most intricate and autiful in design, and the most exquisite workmanship. The watching of the con- uction of one of these webs is an expe- ence that brings us close to those mys- ries which seem to be as fundamental they are inexplicable in the plan of the iverse. It is akin to watching the growth a crystal, or the stars wheeling across e heavens in their appointed courses.

The orb web of the large black and yel- w garden spider is, perhaps, the best sub- ct for this study, although many of the naller orbs are far more delicate in struc- re. These orb webs are most often aced vertically, since they are thus more kely to be in the path of flying insects. he number of radii, or spokes, differs ith the different species of spiders, and ey are usually fastened to a silken frame- ork, which in turn is fastened by guy- nes to surrounding objects. These radii spokes are connected by a continuous iral line, spaced regularly except at the enter or hub; this hub or center is of more lid silk, and is usually surrounded by an pen space; and it may be merely an ir- egular network, or it may have wide ands of silk laid across it.

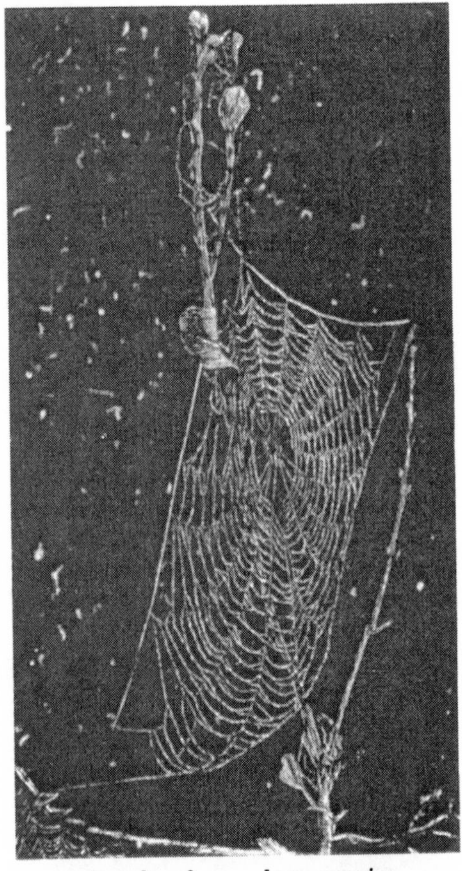

An orb web on a dewy morning

The radii or spokes, the guy-lines, the amework, and the center of the web are ll made of inelastic silk, which does not dhere to an object that touches it. The piral line, on the contrary, is very elastic, and adheres to any object brought in contact with it. An insect which touches one of these spirals and tries to escape becomes entangled in the neighboring lines and is thus held fast until the spider can

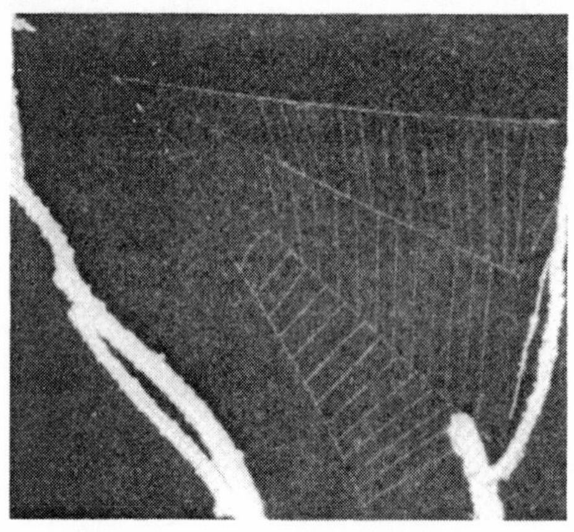

The finished web of a triangle spider

reach it. If one of these elastic lines be examined with a microscope, it is a most beautiful object. There are strung upon it, like pearls, little drops of sticky fluid which render it not only elastic but adhesive.

Some species of orb weavers remain at the center of the web, while others hide in some little retreat near at hand. If in the middle, the spider always keeps watchful claws upon the radii of the web so that if there is any jarring of the structure by an entrapped insect, it is at once apprised of the fact; if the spider is in a den at one side, it keeps a claw upon a trap line which is stretched tightly from the hub of the web to the den, and thus communicates any vibration of the web to the hidden sentinel. When the insect becomes entangled, the spider rushes out and envelops it in a band of silk, which feat it acco plishes by turning the insect over and o rapidly, meanwhile spinning a bro silken band which swathes it. It may b the insect before it begins to swathe it silk or afterwards. It usually hangs swathed insect to the web near where was caught, until ready to eat it; it th takes the prey to the center of the web the spider usually sits there, or to its c at one side, if it is a den-making spec and there sucks the insect's blood, ca fully throwing away the hard parts.

The spider does not became entang in the web, because when it runs it st upon the dry radii and not upon the stic spiral lines. During the busy season, spider is likely to make a new web ev

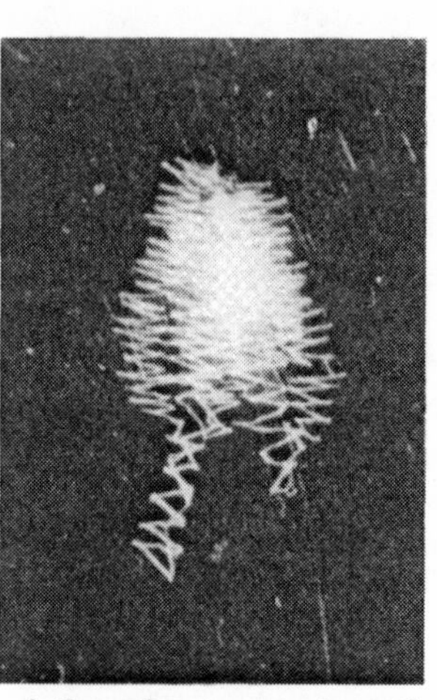

Some of the orb weavers strengthen th orb webs by spinning a zigzag ribbon, as p tured above, across the center

twenty-four hours, but this depe largely upon whether the web has mea while been destroyed by large insects.

The spider's method of making its fi bridge is to place itself upon some hi point and, lifting its abdomen in the to spin out on the breeze a thread of si When this touches any object, it adhe and the spider draws in the slack until linc is " taut "; it then travels across t bridge, which is to support its web, a makes it stronger by doubling the li From this line, it stretches other lines fastening a thread to one point, and th

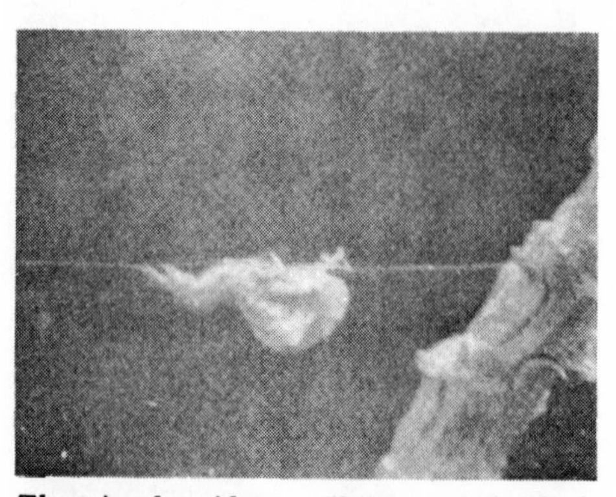

The triangle spider usually rests on the single line of the web

‹ing along to some other point, spin- ; the thread as it goes and holding the clear of the object on which it is walk- by means of one of its hind legs. When right point is reached, it pulls the line .t, fastens it, and then, in a similar fash- proceeds to make another. It may :e its first radius by dropping from its lge to some point below; then climbing k to the center, it fastens the line for ther radius, and spinning as it goes, ks down and out to some other point, ding the thread clear and then pulling ight before fastening it. Having thus cted the center of the web, it goes back forth to and from it, spinning lines il all of the radii are completed and ened at one center. It then starts at center and spins a spiral, laying it on to radii to hold them firm. However, the :s of this spiral are farther apart and ch more irregular than the final spiral. us far, all of the threads the spider has n are inelastic and not sticky; and this t or temporary spiral is used by the ler to walk upon when spinning the l spiral. It begins the latter at the outer ;e instead of at the center, and works 'ard the middle. As the second spiral gresses, the spider with its jaws cuts y the spiral which it first made, and which it has used as a scaffolding. A careful observer may often see remnants of this first spiral on the radii between the lines of the permanent spiral. The spider works very rapidly and will complete a web in a very short time. The final spiral is made of the elastic and adhesive silk.

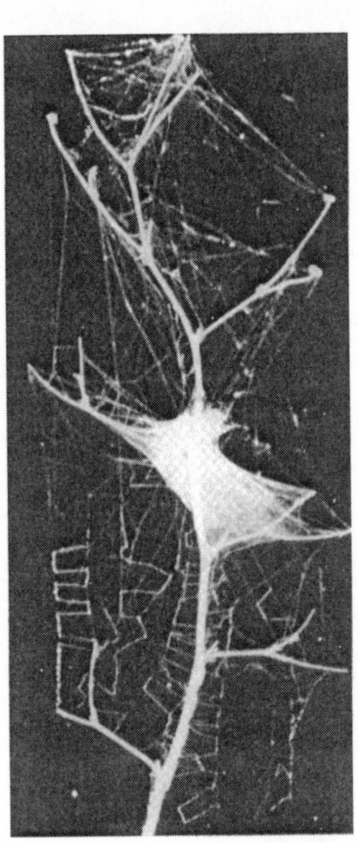

The irregular web of a dictynid

ɾhe spinner of this web, Amaurobius, *lives a crevice in the cliff. The web was spun ›ut the entrance*

LESSON 112

The Orb Web

Leading Thought — Perhaps no structure made by a creature lower than man is so exquisitely perfect as the orb web of the spider.

Method — There should be an orb web where the pupils can observe it, preferably with the spider in attendance.

Observations — 1. Is the orb web usually hung horizontally or vertically?

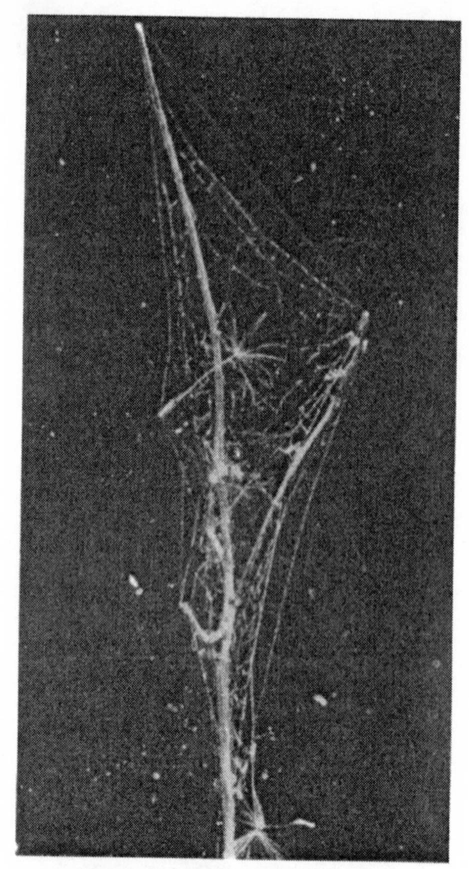

Web of a hackled-band spider

2. Observe the radii, or "spokes," of the web. How many are there? How are they fastened to surrounding objects? Is each spoke fastened to some object or to a framework of silken lines?

3. Observe the silken thread laid around the spokes. Is it a spiral line or is each circle complete? Are the lines the same distance apart on the outer part of the web as at the center? How many of the circling lines are there?

4. Is the center of the web merely an irregular net, or are there bands of silk put on in zigzag shape?

5. Touch any of the "spokes" lightly with the point of a pencil. Does it adhere to the pencil and stretch out as you pull the pencil away? Touch one of the circling lines with a pencil point, and see if it adheres to the point and is elastic. What is the reason for this difference in the stickiness and elasticity of the different k of silk in the orb web?

6. If an insect touches the web, does it become more entangled by see to get away?

7. Where does the spider stay, at center of the web or in a little retrea one side?

8. If an insect becomes entangle the web, how does the spider discover fact and act?

9. If the spider sits at the middle of orb, it has a different method for dis ering when an insect strikes the web t does the spider that hides in a den at side. Describe the method of each.

10. How does the spider make fast insect? Does it bite the insect befor envelops it in silk? Where does it c the insect to feed upon it?

11. How does the spider manage to about its web without becoming tangled in the sticky thread? How of does the orb weaver make a new web?

How an Orb Web is Made

Spiders may be seen making their w in the early morning or in the eveni Find an orb web with a spider in atte ance; break the web without frighten the spider and see it replace it in the e evening, or in the morning about c break. An orb weaver may be brought i

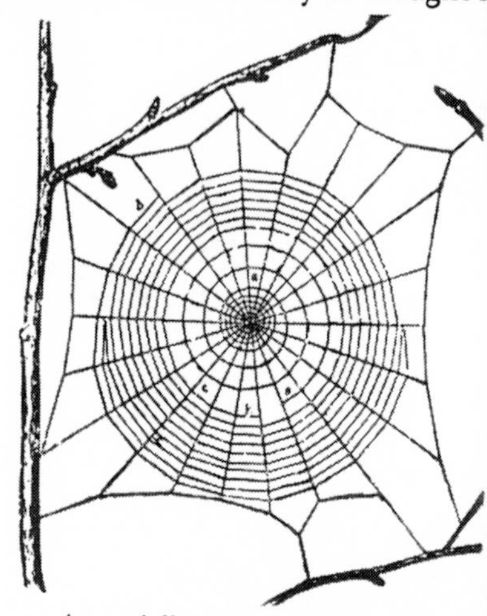

A partially completed orb web

a, the temporary spiral stay-line, b, the sticky sp line, c, the fragments of the temporary spiral hangin a radius

house on its web, when the web is on ranch, and placed where it will not be urbed, and thus be watched at leisure.

)BSERVATIONS — 1. How does the spider nage to place the supporting line be-en two points?

. How does it make the framework for ding the web in place?

. How does it make the first radius?

.. How does it make the other radii and ct the point which is to be the center he web?

. How does it keep the line which it is nning clear of the line it walks upon?

. After the radii are all made, are they tened at the center?

7. How and where does the spider first begin to spin a spiral? Are the lines of this spiral close together or far apart? For what is the first spiral that the spider spins used?

8. Where does it begin to spin the permanent spiral? Where does it walk when spinning it? By the way it walks on the first spiral, do you think it is sticky and elastic? What does it do with the first spiral while the second one is being finished?

9. If the center of the web has a zigzag ribbon of silk, when was it put on?

10. How many minutes did it take the spider to complete the web?

THE FILMY DOME

Like bubbles cut in half, these delicate mes catch the light rays and separate m like a prism into waves of rainbow ors. One of these domes is usually out the size of an ordinary bowl, and is pended with the opening on the lower e. It is held in place by many guy-lines ich attach it to surrounding objects. ove a filmy dome are always stretched ny crisscrossed threads for some disce up. These are for the purpose of ndering the flight of insects, so that they l fall into the web. The little spider, ich always hangs, back downward, just low the center of the dome, rushes to prey from the lower side, pulls it rough the meshes of the web, and feeds on it. But any remains of the insect or eces of sticks or leaves which may drop on the web, it carefully cuts out and ops to the ground, mending the hole ry neatly.

LESSON 113
The Filmy Dome

Leading Thought — One little spider ins a filmy dome, beneath the apex of hich it hangs, back downward, awaiting prey.

Method — On a sunny day in late sumer or early autumn, while walking along woodland paths, the careful observer is sure to see suspended among the bushes or in the tops of weeds, or among dead branches of young hemlocks, the filmy dome webs. They are about as large as a small bowl, and usually so delicate that they cannot be seen unless the sun shines upon them; they are likely to be exquisitely iridescent under the sun's rays. Such a dome may be studied by a class or by the pupils individually.

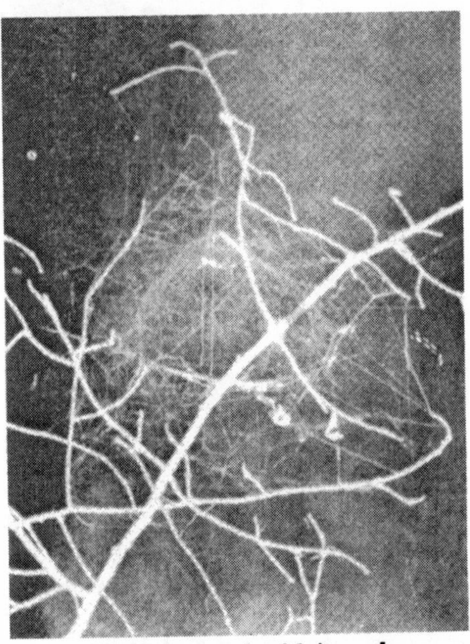

A filmy dome web with its maker

Observations — 1. Where did you discover the filmy dome? What is the size of the dome? Does it open above or below? How is it held in place?

2. Are there many crisscrossed threads extending above the dome? If so, what do you think they are for?

3. Where does the spider stay? Is the spider large and heavy, or small and delicate?

4. What does the spider do if an in becomes entangled in its web?

5. Throw a bit of stick or leaf upo filmy dome web, and note what beco of it.

With spiders I had friendship made,
And watch'd them in their sullen tra

— "The Prisoner of Chillo
Lord By

BALLOONING SPIDERS

If we look across the grass some warm sunny morning or evening of early fall, we see threads of spider silk clinging every-

A sea of gossamer. The webs of ballooning spiders

where; these are not regular webs for trapping insects, but are single threads spun from grass stalk to grass stalk until the fields are carpeted with glistening silk. We have a photograph of a plowed field, taken in autumn, which looks like the waves of a lake; so completely is the ground covered with spider threads that it shows the " path of the sun " like water.

When we see so many of these random threads, it is a sign that the young spiders have started on their travels, and it is not difficult then to find one in the act. The spiderling climbs up some tall object, like a twig or a blade of grass, and sends out its thread of silk upon the air. If the thread becomes entangled, the spiderling sometimes walks off on it, using it as a bric or sometimes it begins again. If the thr does not become entangled with any ject, there is soon enough given off for friction of the air current upon it to s port the weight of the body of the li creature, which promptly lets go its h of earth as soon as it feels safely buoyed and off it floats to lands unknown. Spid thus sailing through the air have been covered in midocean.

Thus we see that the spiders have same way of distributing their species o the globe as have the thistles and dan lions. It has been asked what the spid live upon while they are making these lc journeys, especially those that have drif out to sea. The spider has very convenie habits of eating. When it finds plenty food it eats a great deal; but in time famine it lives on, apparently comfortab without eating. One of our captive spid was mislaid for six months and when found her she was as full of " grit " as ev and she did not seem to be abnorma hungry when food was offered her.

A noiseless, patient spider,
I mark'd where on a little promontory
stood isolated,
Mark'd how to explore the vacant vast s
rounding,
It launch'd forth filament, filament, fi
ment out of itself:
Ever unreeling them, ever tirelessly spe
ing them.

d you O my soul where you stand,
rounded, detached, in measureless oceans of space,
aselessly, musing, venturing, throwing, seeking the spheres to connect them,
l the bridge you will need be form'd, till the ductile anchor hold;
l the gossamer thread you fling catch somewhere, O my soul.

— WALT WHITMAN

LESSON 114

BALLOONING SPIDERS

LEADING THOUGHT — The young of ny species of spiders scatter themselves e thistle seeds in balloons which they ke of silk.

METHOD — These observations should made out-of-doors during some warm nny day in October.

OBSERVATIONS — 1. Look across the grass some warm sunny morning or evening of early fall, and note the threads of spider silk gleaming everywhere, not regular webs, but single threads spun from grass stalk to grass stalk, or from one object to another, until the ground seems glistening with silk threads.

2. Find a small spider on a bush, fence post, or at the top of some tall grass stalk; watch it until it begins to spin out its thread.

3. What happens to the thread as it is spun out?

4. If the thread does not become entangled in any surrounding object what happens? If the thread does become entangled, what happens?

5. How far do you suppose a spider can travel on this silken airplane? Why should the young spider wish to travel?

THE WHITE CRAB SPIDER

There are certain spiders which are crab-e in form, and their legs are so arranged at they can walk more easily sidewise or ckward than forward. These spiders spin webs, but lie in wait for their prey. any of them live upon plants and fences d, in winter, hide in protected places.

The white crab spider is a little rascal at has discovered the advantage of protective coloring as a means of hiding itself om the view of its victims, until it is too

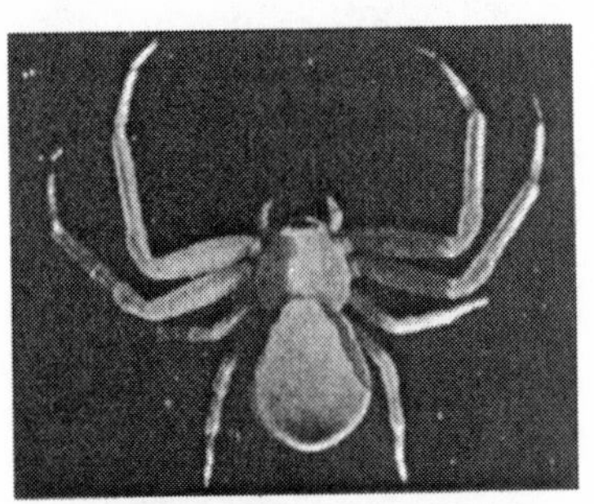

The white crab spider

te for them to save themselves; the small ssassin always takes on the color of the ower in which it lies concealed. In the

A white crab spider with a bee it has captured

white trillium, it is greenish white; while in the goldenrod its decorations are yellow. It waits in the heart of the flower, or in the flower clusters, until the visiting insect alights and seeks to probe for the nectar; it then leaps forward and fastens its fangs into its struggling victim. I have seen a crab spider in a milkweed attack a bee three times its size. This spider was white with lilac or purple markings. If disturbed, the crab spider can walk off awkwardly or it may drop by a silken thread. It is especially interesting, since it illustrates another use for protective coloring; and also because this species seems to be able to change its colors to suit its surroundings.

LESSON 115

The White Crab Spider

Leading Thought — The white crab spider has markings upon its body of the same color as the flower in which it rests and is thus enabled to hide in ambush out of the sight of its victims — the insects which come to the flower for nectar.

Method — Ask the children to bring one of these spiders to school in the flo in which it was found; note how inc spicuous it is, and arouse an interest in different colors which these spiders sume in different flowers.

Observations — 1. What is the sh of the body of the crab spider? Which the legs are the longest? Are these l directed forward or backward?

2. How is the body marked? W colors do you find upon it? Are the col the same in the spiders found in the t liums as those in other flowers? Why this? Do you think that the color of spider keeps it from being seen?

3. Place the white spider which y may find in a trillium or in a daffodil, a note if the color changes.

4. Do the crab spiders make webs? H do they trap their prey?

HOW THE SPIDER MOTHERS TAKE CARE OF THEIR EGGS

M. V. Slingerland

A crab spider on a goldenrod, upper right. The spider is white when lurking in the white trillium and yellow when among the flowers of the goldenrod

Protecting her eggs from the vicissitudes of the weather seems to be the spider mother's chief care; though at the same time and by the same means she protects them from the attacks of predacious insects. Many of the species make silken egg-sacs, which are often elaborate in construction, and are carefully placed in protected situations.

Often a little silvery disc may be seen attached to a stone in a field. It resembles a circular lichen on the stone, but if it examined it is found to consist of an u per, very smooth, waterproof coat, wh below is a soft, downy nest, complet enfolding the spider's eggs.

The egg-sacs of the cobweb weavers a often found suspended in their webs. O of the large orb weavers makes a very markable nest, which it attaches to t branches of weeds or shrubs. This sac about as large as a hickory nut, and ope like a vase at the top. It is very secur suspended by many strong threads of si

Entrance to the underground nest of a turr spider

ıat the blasts of winter cannot tear it ɛ. The outside is shining and water-ɸf, while inside it has a fit lining for ıderling cradle.

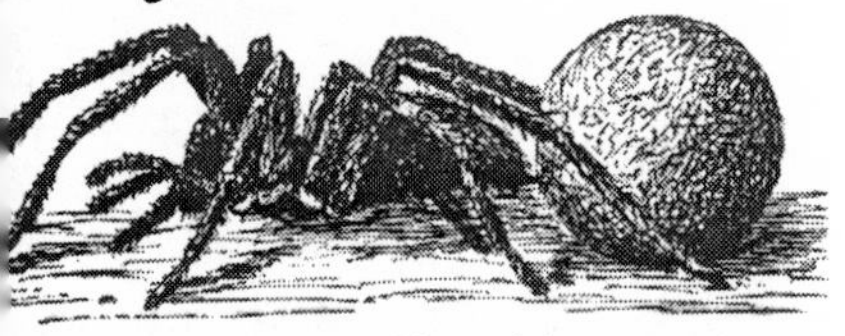

A female turret spider with egg-sac

ɸr. Burt G. Wilder studied the devel-ıent of the inmates of one of these :s by cutting open different nests at ɛrent periods of the winter. In the ımn, the nest contained five hundred nore eggs. These eggs hatched in early ter but it seemed foreordained that ıe of the little spiders were born to ʼe as food for their stronger brethren. ɛy seemed resigned to their fate, for ɛn one of these victims was seized by cannibalistic brother, it curled up its ; and submitted meekly. The result of , process was that, out of the five hun-d little spiders hatched from the eggs, y a few healthy and apparently happy ıng spiders emerged from the nest in : spring, sustained by the nourishment ɔrded them by their own family, and ed for their life in the outside world. ɔome spiders make a nest for their eggs hin folded leaves, and some build them :revices of rocks and boards.

Ɩhe running spiders, which are the large ones found under stones, make globular egg-sacs; the mother spider drags after her this egg-sac attached to her spinnerets; the young, when they hatch, climb upon their mother's back, and there remain for a time.

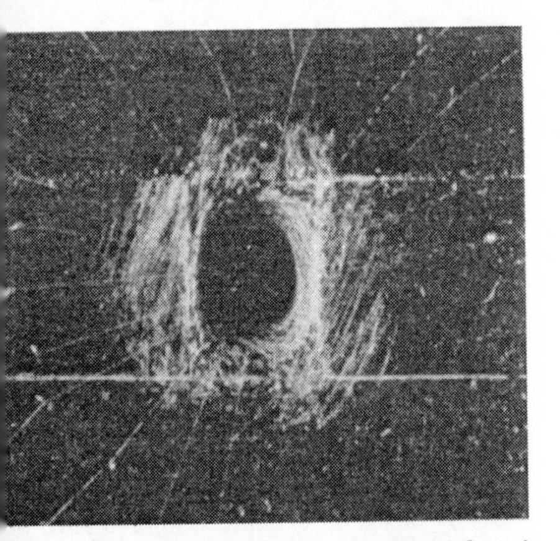

e nest of Ariadna, *a long, slender tube of silk in a crack in a wood block*

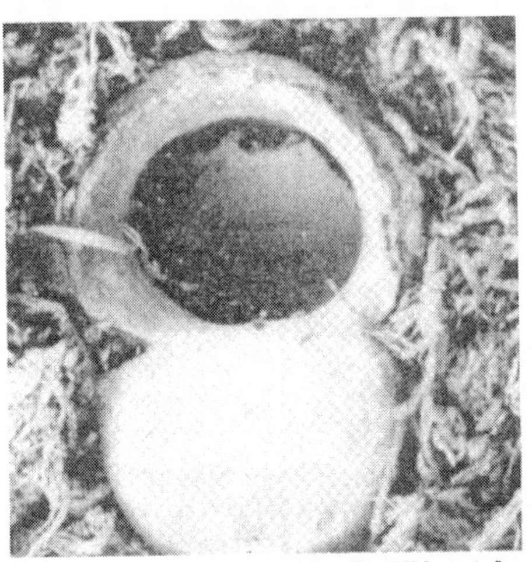

Ward's Natural Science Establishment, Inc.

The trap-door spider and her nest. The spider digs a tunnel in the ground, coats the walls with earth and saliva, and then spins a lining of silk. The hinged door is a continuation of the walls with the outer surface covered with earth

LESSON 116

The Nests of Spiders

Leading Thought — The spider mothers have many interesting ways of protect-

INVERTEBRATES

1. Water Spider, Lycosa. *This is one of very few spiders that frequent the water.*

2. House Centipede, Scutigera. *Each segment of the flattened body of this centipede bears a single pair of very long legs.*

3. Scorpion, Scorpionida. *A scorpion is characterized by a long, slender, flattened body which ends in a curved, venomous stinger. The sting causes much pain but is seldom if ever fatal to man.*

4. Millipede, Spirobolus. *These animals live in damp places and feed chiefly on decaying matter.*

5. Water Sow Bug, Asellus. *In the decaying vegetation and bottom trash of stagnant, shallow water, one often finds these flattened crustaceans.*

6 and 8. Fairy Shrimps, Eubranchipus. *These crustaceans always swim on their backs; they are about one inch long.*

7. Tadpole Shrimp, Apus. *This near relative of the fairy shrimp is an aquatic animal. It is shield-shaped like the horseshoe crab.*

9. Dog Louse, Linognathus piliferus. *This is the common louse of dogs; to the casual observer it could not be distinguished from the lice which infest other animals.*

10. Scud, Gammarus. *In the eastern United States, these may be found the year round in streams or ponds.*

11. Water Flea, Daphnia. *Daphnia is one of the many kinds of crustaceans called water fleas. They are usually found in quiet water where they feed on algæ. Water fleas are an important source of food for fish and aquatic insects.*

12. Pleurocera. *This mollusk is found in great variety and abundance in rivers in the eastern United States from the Great Lakes south.*

13. Copepod, Cyclops. *Cyclops represents a group of tiny crustaceans known as copepods.*

14. Fresh-water Limpet, Ancylus. *These snails are generally distributed; they live in streams as well as in quiet water.*

15. Goniobasis. *Full grown fresh-w snails of this species are 1¼ inches long; t are found in rapid currents as well as am plant growth of quiet waters.*

16. Vivipara. *These snails may reac length of 2 inches; they are found on muddy bottoms of streams and lakes.*

17. Wheel Snails, Helisoma (Planorl *The shell is coiled in a flat spiral wit sunken center.*

18. Campeloma. *This snail is found f the St. Lawrence River to the Gulf of Me The young are born alive.*

19. Valvata. *This small, widely distribu snail exists in great numbers in both deep shallow water.*

20. Bythinia. *This European snail has b introduced, by the operations of comme into the Hudson River and the Great La region.*

21. Amnicola. *On sandy bottoms among water vegetation, these snails widely distributed in shallow water.*

22. Paludestrina. *These tiny snails, ab one-sixth of an inch long, are distributed fr the Atlantic to the Pacific in fresh water.*

23. Common Pond Snail, Lymnæa. T *snail represents a widely distributed grou common snails which differ greatly in si they form an important item in the food water birds, fishes, and frogs.*

24. Pouch Snail, Physa. *Pouch snails remarkably active. In color and shape t vary so greatly that it is often difficult identify them. They are interesting to obse and may be kept easily in an aquarium.*

25. Fingernail Clam, Sphærium. *Th small white mussels are about half an i long; they are widely distributed, be found in the fresh water of almost any po stream, or lake.*

26. Paper-shell Mussel, Anodonta. T *shell of this snail is thin, usually smooth, o often marked by concentric rings. They found from the Atlantic to the Pacific Oce*

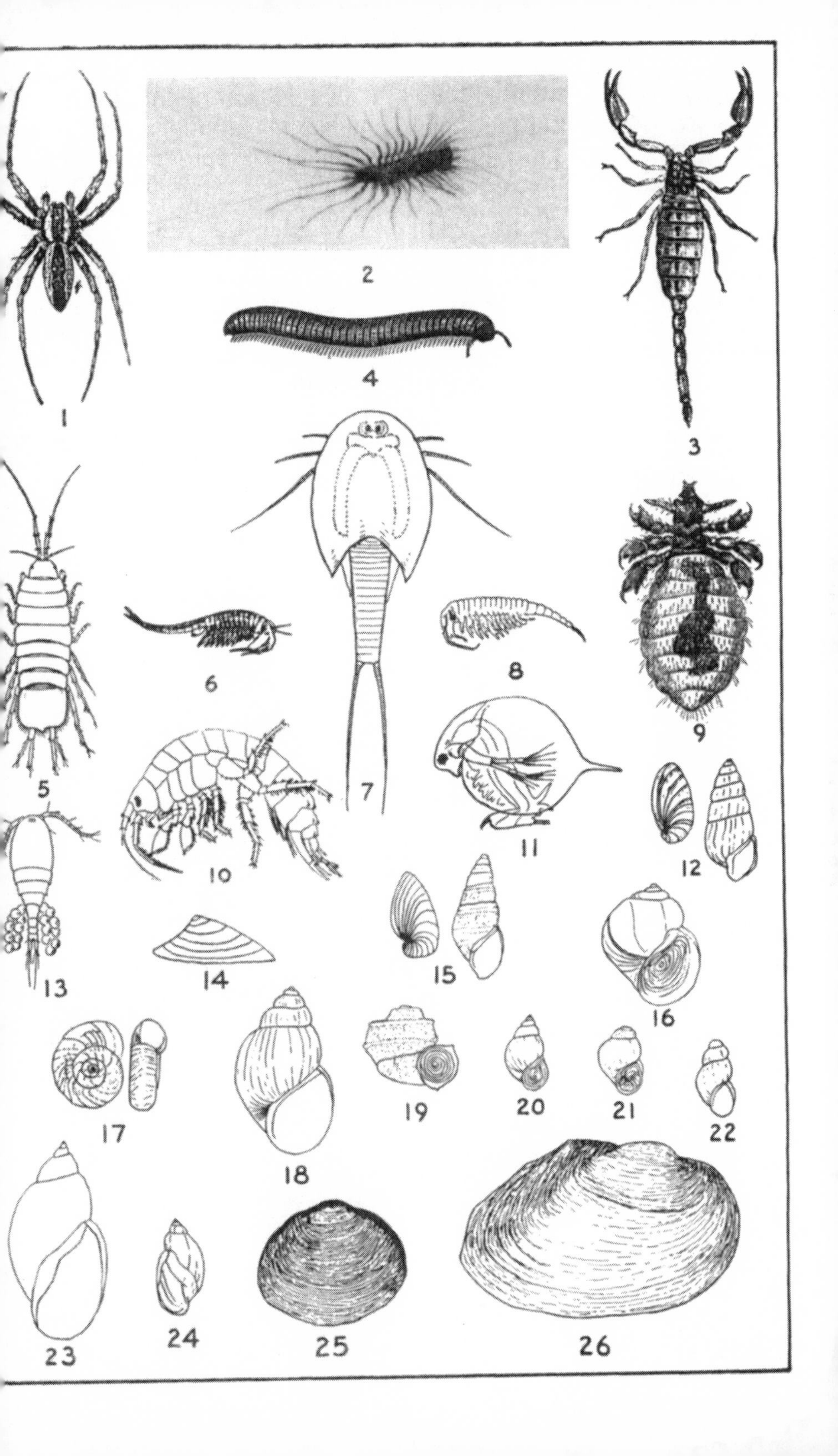
1
2
3
4
5
6
7
8
9
10
11
12
13
14
15
16
17
18
19
20
21
22
23
24
25
26

The egg-sac of one of the orb weavers. It is made in the autumn and contains 500 or more eggs. The eggs hatch early in the winter but no spiders emerge until spring. During the winter the stronger spiders calmly devour their weaker brothers, and in the spring those which survive emerge well nourished to fight their battles in the outside world

ing their eggs, which they envelop silken sacs and place in safety.

Method — Ask the pupils to bring all the spider egg-sacs that they can fi Keep some of them unopened, and o others of the same kind, and thus disco how many eggs are in the sac and h many spiderlings come out. This is a go lesson for September and October.

Observations — 1. In what situat did you find the nest? How was it p tected from rain and snow? To what it attached?

2. Of what texture is the outside of sac? Is the outside made of waterpr silk? What is the texture of the lining?

3. How many eggs in this sac? Wha the color of the eggs? When do the s derlings hatch? Do as many spiders co out of the sac as there were eggs? Wh this?

PART III
PLANTS

HOW TO BEGIN THE STUDY OF PLANTS AND THEIR FLOWERS

W. C. Muenscher

Mountain laurel

'he only right way to begin plant study 1 young children is through awaken- their interest in and love for flowers. st children love flowers naturally; they by bringing flowers to school, and here, teaching the recognition of flowers by ne, may be begun this delightful study. is should be done naturally and in- mally. The teacher may say: "Thank , John, for this bouquet. Why, here pansy, a bachelor's button, a larkspur, a poppy." Or, "Julia has brought me beautiful flower. What is its name, I nder?" Then may follow a little dis- sion, which the teacher leads to the per conclusion. If this course is con- tently followed, the children will learn names of the common flowers of od, field, and garden, and never realize it they are studying anything.

The next step is to inspire the child with a desire to care for and preserve his bouquet. The posies brought in the perspiring little hand may be wilted and look dejected; ask their owner to place the stems in water, and call attention to the way they lift their drooping heads. Parents and teachers should very early inculcate in children this respect for the flowers which they gather; no matter how tired the child or how disinclined to further effort, when he returns from the woods or fields or garden with plucked flowers, he should be made to place their stems in water immediately. This is a lesson in duty as well as in plant study. Attention to the behavior of the thirsty flowers may be gained by asking the following questions:

1. When a plant is wilted how does it look? How does its stem act? Do its

Columbine, Aquilegia

leaves stand up? What happens to the flowers?

2. Place the cut end of the stem in water and look at it occasionally during an hour; describe what happens to the stem, the leaves, the blossom.

Some Needs of Plants

Another step in plant study comes naturally from planting the seeds in window-boxes or garden. This may be done in the kindergarten or in the primary grades. As soon as the children have had some experience in the growing of flowers, they should conduct some experiments which will teach them about the needs of plants. These experiments are fit for the work of the second or third grade. Uncle John says, "All plants want to grow; all they ask is that they shall be made comfortable." The following experiments should be made vital and full of interest, by impressing upon the children that through them they will learn to give their plants what they need for growth.

Experiment 1. *To find out in what kind of soil plants grow best* — Have the children of a class, or individuals representing a class, prepare four little pots or boxes, as follows: Fill one with rich, woods humus, or with potting earth from a florist's; another with poor, hard soil, which may be found near excavations; another with clean sand; another with sawdust. Plant the same kind of seeds in all four, and place them where they will get plenty of light. It is best to select s[illegible] that germinate quickly, such as be[illegible] radishes, lettuce, or calendula. W[illegible] them as often as needful. Note w[illegible] plants grow the best. This trial sh[illegible] cover six weeks at least and atten[illegible] should now and then be called to the [illegible] tive growth of the plants.

Experiment 2. *To prove that pl[illegible] need light in order to grow* — Fill two [illegible] with the same rich soil; plant in these [illegible] same kind of seeds. Keep the soil m[illegible] place one pot in the window and place [illegible] other in a dark closet or under a box, [illegible] note what happens; in which pot do [illegible] plants have the more normal grow[illegible] Or take two potted geraniums which l[illegible] equally thrifty; keep one in the light [illegible] the other in darkness. What happens?

Experiment 3. *To show that the le[illegible] turn toward light* — Place a geranium [illegible] a window and let it remain in the sa[illegible] position for two weeks. Which way [illegible] all the leaves face? Turn it around, [illegible] note what the leaves have done after a [illegible] days.

Experiment 4. *To show that pla[illegible] need water* — Fill three pots with [illegible] earth, plant the same kind of seeds [illegible] each, and place them all in the sa[illegible] window. Give one water sufficient to k[illegible]

Eva L. Go[illegible]

A terrarium. This glass box with a hin[illegible] lid is easily made of six pieces of glass [illegible] some adhesive tape. When some soil is pla[illegible] in such a box, various plants can be gro[illegible] and many kinds of insects, reptiles, or [illegible] phibians will be perfectly at home there

soil moist, keep another flooded with
er, and give the other none at all.
at happens to the seeds in the three
?

he success of these four experiments
ends chiefly upon the genius of the
her. The interest in the results should
een; every child should feel that every
l planted is a living thing and that it
ruggling to grow; every look at the ex-
iments should be like another chapter
continued story.

he explanations of these experiments
uld be simple, with no attempt to
h the details of plant physiology. The
d of plants for rich, loose earth and
water is easily understood by the chil-
n; but the need for light is not so ap-
ent, and Uncle John's story of the
ch factory is the most simple and
phic way of making known to the chil-
n the processes of plant nourishment.
tells us that plants are like us; they
e to have food to make them grow;
ere is the food and how do they find
Every green leaf is a factory to make
d for the plant; the green pulp in the
leaf is the machinery; the leaves get the raw materials from the sap and from the air, and the machinery unites them and makes them into plant food. This is mostly starch, for this is the chief food of plants, although they require some other kinds of food also. The machinery is run by sunshine-power, so the leaf-factory can make nothing without the aid of light; the leaf-factories begin to work as the sun rises, and stop working when it sets. But the starch has to be changed to sugar before the plant can use it for nourishment and growth; and so the leaves, after making the starch from the sap and the air, are obliged to digest it, changing the starch to sugar; for the growing parts of the plant feed upon sweet sap. Although the starch-factory in the leaves can work only during the daytime, the leaves can change the starch to sugar during the night. So far as we know, there is no starch

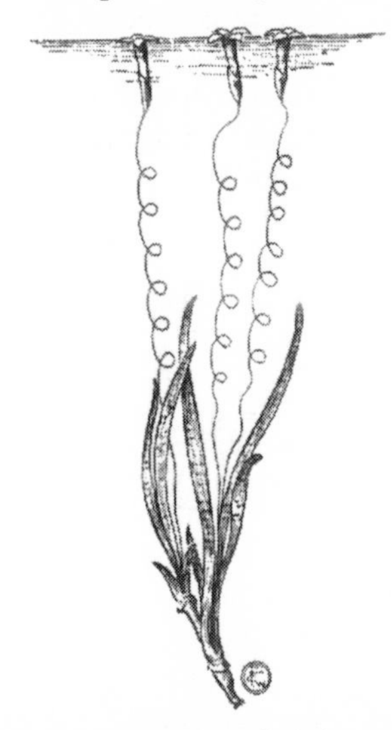

Eel grass, Vallisneria. *A quiet-water plant,*
l grass produces its male flowers under
ıter; its female flowers bloom at the top.
hen mature, the male flowers float to the
rface, where pollination occurs; the female
wers are then retracted to mature the
uits under water. This plant is the favorite
od of canvas-back ducks

W. C. Muenscher

Gray or old field birch. Although these birches grow in clumps, several trunks from a common root, observe that the trunks soon separate widely, thus providing abundant light for the leaves

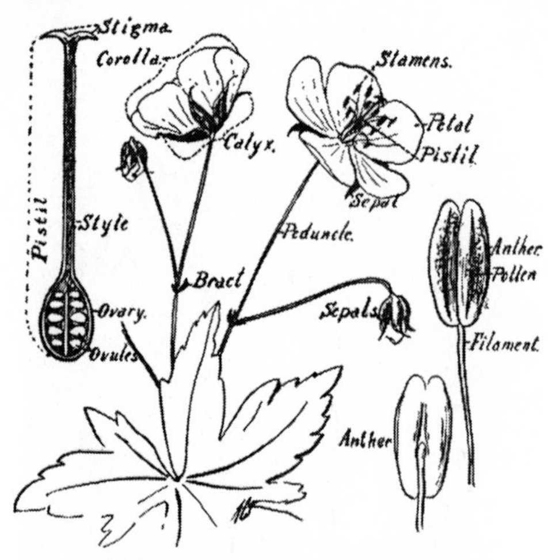

A flower with the parts named

in the whole world which is not made in the leaf-factories.

This story should be told and repeated often, until the children realize the work done by leaves for the plants and their need of light.

The clouds are at play in the azure space
And their shadows at play on the bright
green vale.
And here they stretch to the frolic chase;
And there they roll on the easy gale.

There's a dance of leaves in that aspen
bower,
There's a titter of winds in that beechen
tree,
There's a smile on the fruit and a smile
on the flower,
And a laugh from the brook that runs to
the sea.

— Bryant

How to Teach the Names of the Parts of a Flower and of the Plant

The scientific names given to the parts of plants have been the stumbling block to many teachers, and yet this part of plant study should be easily accomplished. First of all, the teacher should have in mind clearly the names of the parts which she wishes to teach; the illustrations here given are for her convenience. When talking with the pupils about flowers let use these names naturally:

"See how many geraniums we h the corolla of this one is red and of t one is pink. The red corolla has fourt petals and the pink one only five," et

"This arbutus which James brought a pretty little pink bell for a corolla."

"The purple trillium has a purple rolla, the white trillium a white corc and both have green sepals."

The points to be borne in mind that children like to call things by th names because they are *real* names, a they also like to use "grownup" nar for things; but they do not like to com to memory names which to them meaningless. Circumlocution is a waste breath; calling a petal a "leaf of a flowe or the petiole "the stem of a leaf," is l calling a boy's arm "the projecting p of James's body" or Molly's golden h "the yellow top" to her head. All t names should be taught gradually by co stant unemphasized use on the part the teacher; and if the child does r learn the names naturally then do r make him do it unnaturally.

The lesson on the garden or horsesh geranium with single flowers may be giv first in teaching the structure of a flow since the geranium blossom is simple a easily understood.

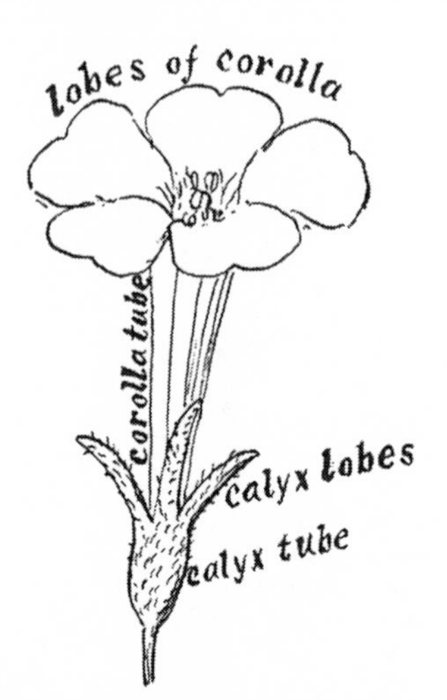

A flower with petals united forming a tub and with sepals likewise united

TEACH THE USE OF THE FLOWER

rom first to last the children should aught that the object of the flower is levelop seed. They should look eagerly the maturing flower for the growing t. Poetry is full of the sadness of the ng flower, whereas rightly it should the gladness of the flower that fades, ause its work is done for the precious at its heart. The whole attention of child should be fixed upon the de- ping fruit instead of the fading and ng petals.

ll places then and in all seasons,
'lowers expand their light and soul-like wings,
ching us by most persuasive reasons,
ow akin they are to human things.

— LONGFELLOW

FLOWER AND INSECT PARTNERS

t is undoubtedly true that the proc- s of cross-pollination and the compli- d devices of flowers for insuring it only be well taught to older pupils only fully understood in the college oratory; yet there are a few simple facts ich even the young child may know, as ows:

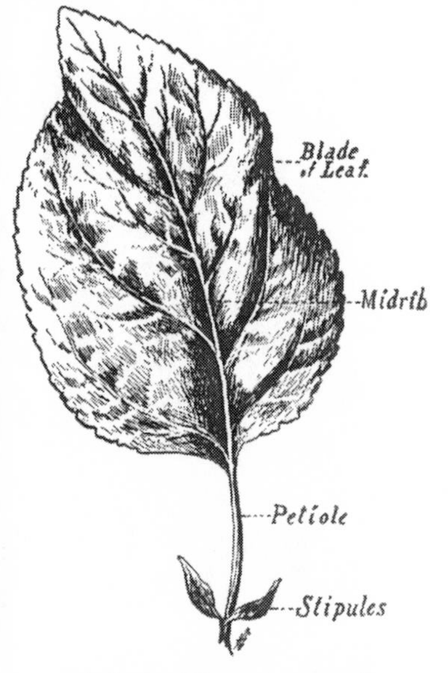

A leaf with parts named

A hawk moth or sphinx. A moth may carry pollen from one flower to another

1. Pollen is needed to make most seeds form; some flowers need the pollen from other flowers of the same kind to produce their seeds; but many flowers use the pollen from their own flowers to pollinate their ovules, which grow into seeds.

2. Flowers have neither legs like some animals, nor have they wings like butterflies, therefore they cannot go after pollen; in seeking food and drink from flowers insects carry pollen from one flower to another.

I taught this to a four-year-old once in the following manner: A pine tree in the yard was sifting its pollen over us and little Jack asked what the yellow dust was; we went to the tree and saw where it came from, and then I found a tiny young cone and explained to him that this was a pine blossom, and that in order to become a cone with seeds, it must have some pollen fall upon it. We saw that the wind sifted the pollen over it and then we examined a ripe cone and found the seeds. Then we looked at the clovers in the lawn. They did not have so much pollen and they were so low in the grass that the wind could not carry it for them; but right there was a bee. What was she doing? She was getting honey for her hive or pollen

A skipper visiting flowering heads of English plantain

for her brood, and she went from one clover head to another; we caught her in a glass fruit jar, and found she was dusted with pollen and that she had pollen packed in the baskets on her hind legs; and we concluded that she carried plenty of pollen on her clothes for the clovers, and that the pollen in her baskets was for her own use. After that he was always watching the bees at work. We found afterwards, that insects seem to be called to the flowers by color or by fragrance, or by both of these means. The dandelion we watched was very bright and the insects were busy there; then we found bees working on mignonette whose blossoms were so small that Jack did not think they were blossoms at all, and we concluded that in this case the bees were attracted by fragrance. We found other flowers which attracted bees by both color and fragrance; and this insect-flower partnership remained a factor of great interest in the child's mind.

The Relation of Plants to Geography

There should be from first to last steady growth in the intelligence of t child as to the places where certain pla grow. He finds hepaticas and trilliums the woods, daisies and buttercups in t sunny fields, mullein on the dry hillsid cattails in the swamp, and water lil floating on the pond. This may all taught by simply asking the pupils qu tions relating to the soil and the spec conditions of the locality where th found the flowers they bring to school.

Seed Germination

In the early days of nature-study, t one feature of plant life came near "g bling up" all of nature-study, and yet is merely an incident in the growth of t plant. To sprout seeds is absurd as object in itself; it is incidental as is t breaking of the egg-shell to the study the chicken. The peeping into a seed li a bean or a pea to see that the plant really there, with food material for

W. P. Alexar

Early saxifrage. This spring flower is home on exposed rocks and dry hillsides

:e growth packed all around it, is in-
ting to the child. To watch the little
: develop, to study its seed leaves and
becomes of them, to know that they
the plant its first food and to know
a young plant looks and acts, are all
s of legitimate interest in the study
e life of a plant; in fact the struggle
e little plant to get free from its seed
; may be a truly dramatic story. But
gard this feature as the chief object
anting seed is manifestly absurd.
he object of planting any seed should
o rear a plant which shall fulfill its
le duty and produce other seed. The
wing observations regarding the ger-
ation of seeds should be made while
children are eagerly watching the com-
of the plants in their gardens or win-
-boxes:

Which comes out of the seed first,
root or the shoot and leaves? Which
does the root grow, up or down?
ich way do the leaves grow, no matter
h side up the seed is planted?

How do the seed leaves get out of
seed coat, or shell? How do the seed
es differ in form from the leaves which
e later? What becomes of the seed
es after the young plant begins to
v?

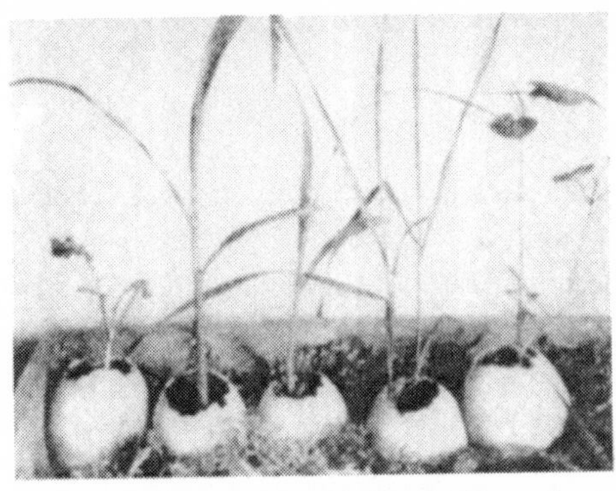

Egg-shell experiment farm. The plants from left to right: cabbage, field corn, pop-corn, wheat, buckwheat

WILD FLOWERS

Because of their beauty and scientific value, special need exists for the protection of our native wild flowers and shrubs. It is understandable that these uncultivated

O. L. Foster

Spring beauty

plants should attract the visitor, but in too many instances he is not satisfied to enjoy their beauty as they exist in their natural habitats. All too frequently he picks flowers in large numbers, only to discard them faded and wilted a few hours later. Often valuable plants are dug out or pulled up by their roots, probably with the idea that these flowers or shrubs would have the same beauty in a garden as in the woods or fields where they grow naturally. Such practices are to be discouraged. In the first place, wild flowers are almost always most attractive in their natural surroundings. Furthermore the transplanting of flowers and shrubs from woods or swamps to a cultivated garden is a delicate operation, and there is very little likelihood of its being accomplished successfully.

Extensive removal of these plants whether from field, marsh, or wooc likely to bring about the extinctio certain species and from both scien and æsthetic standpoints this is hi unfortunate.

The malicious destruction of flowe plants should, of course, not be allo Some plants are so rare, or otherwis danger of extinction, that state laws h been enacted which protect them. example in New York State, trailing butus, flowering dogwood, fringed tian, pink lady's-slipper, yellow lady's- per, and mountain laurel are protecte law.

Some flowers are so abundant that t can be picked in moderation if the r are not disturbed, if plenty of flowers left for seed, and if the plant itself is taken with the flower. Trilliums, for ample, cannot be picked without serio harming the plant, for the food-produ leaves and stem are taken with the flo Everyone should have the privilege of

O. L. F

Cut-leaved toothwort or pepper-root

g the natural beauty of the country-
Such enjoyment is impossible if a
ively small number of people insist
picking and destroying native plants
heir own selfish interests.

THE HEPATICA

he wise men say the hepatica flower has no petals but has pink, white or pur- sepals instead: and they say, too, that the three leaflets of the cup which holds flower are not sepals but are bracts; and they offer as proof the fact that they do not w close to the blossom, but are placed a little way down the stem. But the hepatica s not care what names the wise men give to the parts of its blossom: it says as plainly f it could talk: "The bees do not care whether they are sepals or petals since they are tty in color, and show where the pollen is to be found. I will teach the world that cts are just as good to wrap around flower-buds as are sepals, and that sepals may be as beautiful as petals. Since my petticoat is pretty enough for a dress why should I wear it thus?" — "THE CHILD'S OWN BOOK OF WILD FLOWERS"

Ve seek the hepatica in its own haunts, ause there is a longing for spring in hearts that awakens with the first m sunshine. As we thread our way into den woods, avoiding the streams and dles which are little glacial rivers and s, having their sources in the snow- ts still heaped on the north side of ngs, we look eagerly for signs of return- life. Our eyes slowly distinguish among various shades of brown in the floor the forest, a bit of pale blue or pink- ple that at first seems like an optical il- ion; but as we look again to make sure, it is the hepatica, lifting its delicate ssoms above its mass of purple-brown ves. These leaves, moreover, are always utiful in shape and color and suggest terns for sculpture like the acanthus, for rich tapestries like the palm leaf in Orient. It warms the heart to see these ve little flowers stand with their faces to the sun and their backs to the snow-drifts, looking out on a gray-brown world, nodding to it and calling it good.

In the spring, new leaves may appear very soon after the flowers; these leaves are present until the following spring. The hepatica flowers are white, pink, and lavender; the latter are sometimes called "blue." The colored floral parts, so-called petals, are in reality all sepals and often vary in number, from six to twelve. On dark days and during the night, the young blossoms close; but when they become old and faded they remain open all the time.

There are many stamens with greenish white anthers and pollen. They stand erect around the many pistils at the center of the flower. The number of pistils varies from six to twenty-four. Each holds aloft the little horseshoe-shaped, whitish stigma, which if pollinated usually develops into a fruit. The hepatica is a per-

Hepaticas

ennial, and its natural habitat is rich, moist woods. While it is adapted to the shade of woods, it can be successfully transplanted to suitable situations in lawns and gardens. The leaves which have passed the winter under the snow are rich purple be-

Robert Conn

Hepaticas in natural surroundings

neath and mottled green and pu above, making beautiful subjects for w color drawings. The new hepatica lea are put forth in the spring, before leaves appear on the trees. The roots quite numerous and fine.

LESSON 117

The Hepatica

Leading Thought — The hepa flower buds are developed in the fall, are ready to blossom early in the spri This plant thrives best in moist and sh woods.

Method — The pupils should have questions before they go into the wo

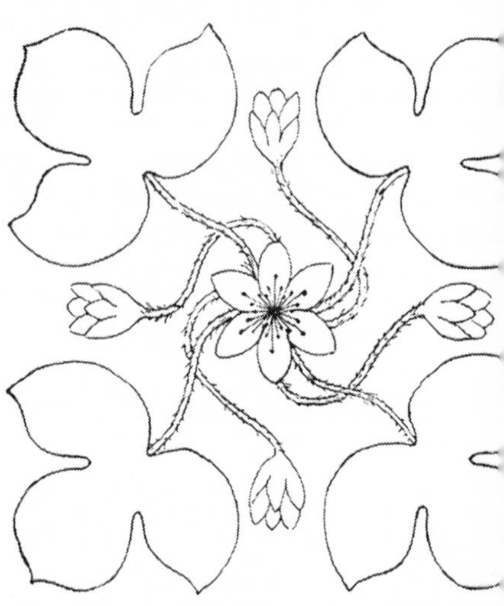

Evelyn Mitc

Embroidery design from the hepatica

to observe spring flowers, and should swer them individually.

Observations — 1. Where do you fi the hepaticas? Do you ever find them the open fields? Do you ever find the in the pine woods?

2. How do the leaves look in ea spring? Sketch in color one of these c leaves. How do the young leaves look? A the leaves that come up late in the spri as fuzzy as those that appear early? Wh is the difference in texture and color b tween the leaves that remained over wi ter and those that appear in the spring?

3. Find a hepatica plant before it

; to blossom. Look, if possible, at its / center. Describe these little flower ls.

. How does the bud look when it be- ; to lift up? Describe the stems and little bracts that hold the bud.

. Are the hepaticas in your woods all k, or blue, or white? Do those which at first pink or blue fade to white later? the blossoms keep open during the ht and in stormy weather? Are they all grant?

. How many sepals has your hepatica? Turn back the three sepal-like bracts and see that they are not a part of the flower at all but join the stem below the flower.

7. Describe the stamens in the hepatica. How many pistils are there? Does each pistil develop into a fruit?

8. What insects do you find visiting the hepaticas?

9. Describe a hepatica plant in the woods; mark it so that you will know it, and visit it occasionally during the summer and autumn, noting what happens to it.

THE YELLOW ADDER'S-TONGUE

Once a prize was offered to a child if she would find two leaves of the adder's tongue at were marked exactly alike: and she sought long and faithfully, but the only prize she on was a lesson in Nature's book of variations, where no two leaves of any plant, shrub tree are exactly alike: for even if they seemed so to our eyes, yet there would exist in em differences of strength and growth too subtle for us to detect. But this child was ow in learning this great fact, and, until she was a woman, the adder's-tongue leaves, beautifully embroidered with purple and green, were to her a miracle, revealing the finite diversity of Nature's patterns.

— "THE CHILD'S OWN BOOK OF WILD FLOWERS"

This little lily of the woods is a fasci- ting plant. Its leaves of pale green mot- d with brownish purple often cover osely large irregular areas in the rich soil our woodlands and it is sometimes und in open fields; yet I doubt if the derground story of these forest rugs is ten thought of. The leaves are twins, d to the one who plucks them care- sly they seem to come from one slen- r stem. It requires muscle as well as cision of character to follow this weak em down several inches, by digging ound it, until we find the corm at base. A corm is the swollen base of a em and is bulblike in form; but it is not ade up of layers, as is a bulb. It is a store- ouse for food and also a means of spread- g the species; for from the corms there ow little corms called cormels, and each rmel develops a separate plant. This un- erground method of reproduction is the cret of why the leaves of the adder's- ngue appear in patches, closely crowded gether.

Only a few of the plants in a "patch" produce flowers, and it is interesting to see how cleverly these lily bells hide from the casual eye. Like many of the lilies, the three sepals are petal-like and are identi- fied as sepals only by their outside posi- tion, although they are thicker in texture.

W. P. Alexander

Adder's-tongue or dog's-tooth violet

They are purplish brown outside, which makes the flower inconspicuous as we look down upon it; on the inner side, they are a pure yellow, spotted with darker yellow near where they join the stem. The three petals are pure yellow, paler outside than in, and they have dark spots like the tiger lilies near the heart of the flower; and

Verne Morton

Adder's-tongue going to seed

where they join the stem, each has on each side an ear-shaped lobe.

The open flower is bell-shaped; and like other bells it has a clapper, or tongue. This is formed by six downward-hanging stamens, the yellow filaments of which have broad bases and taper to points where the oblong anthers join them. The anthers are red or yellow. It is this stamen clapper that the visiting insects cling to when probing upward for nectar from this flower at the upper end of the bell. The pale green pistil is somewhat three-sided, and the long style remains attached long after the flower disappears. The flower is slightly fragrant, and it is visited by the queen bumblebees and the solitary bees, of which there are many species. The flower closes nights and during clou stormy days. The seed capsule is plu and rather triangular, and splits into th sections when ripe. The seeds are nun ous and are fleshy and crescent-shaped

But the adder's-tongue, like many ot early blooming flowers, is a child of spring. The leaves, at first so prettily m tled, fade out to plain green; and by m summer they have entirely disappear the place where they were being cover with other foliage of far different patte But down in the rich woods soil are t plump globular corms filled with the fo manufactured by the spotted leaves duri their brief stay, and next spring two pairs spotted leaves may appear where there w but one pair this year.

LESSON 118

Adder's-Tongue or Dog's-Tooth Violet

Leading Thought — The adder tongue is a lily, and its mottled leav appear in the spring, each pair comi n a corm deep in the soil below. It has ways of spreading, one underground neans of new corms growing from the er ones, and the other by means of ls, many of which are probably per-ed through the pollen carried by in-s.

IETHOD — This plant should be studied he woods, notes being made on it there. a plant showing corm, roots, leaves, blossom may be brought to the oolhouse for detailed study, and then nted in a shady place in the school den.

)BSERVATIONS — 1. Where does the ad-'s-tongue grow? Do you ever find it in n fields? How early do you find its ves above ground? At what time do its ssoms appear?

. How many leaves has each plant? hat colors do you find in them? What he color of their petioles? Do the leaves nain mottled later in the season?

3. Do the adder's-tongue plants occur gly or in patches?

4. Is the flower lifted up, or is it droop-? What is its general shape? How many als? How would you know they were als? How do they differ in color, out-e and in, from the petals? How are the tals marked? Can you see the lobes at base of each petal? When sepals and tals are so much alike the botanists call m all together the perianth.

5. If the perianth, or the sepals and tals together, makes a bell-shaped flower, at makes the clapper to the bell? How the insects use this clapper when they visit the flower? Do the flowers stay open nights and dark days?

Adder's-tongue in natural surroundings

6. How many stamens are there? Describe or sketch one, noting its peculiar shape. Are the stamens all the same length? Can you see the pistil and its stigma? Where is it situated in relation to the stamens? Do you think the stigma is ready for pollen at the time the anthers are shedding it?

7. After the petals and sepals fall what remains? How does the ripe seed capsule look? How does it open to let out the seeds? Are there many seeds in a capsule? What is the shape of the seeds of this plant?

Verne M

BLOODROOT

What time the earliest ferns unfold,
And meadow cowslips count their gold;
A countless multitude they stood,
A Milky Way within the wood. — Danske Dandridge

Only a few generations ago, this land of ours was peopled by those who found it fitting to paint their bodies to represent their mental or spiritual conditions or intentions. For this purpose they had studied the plants of our forests to learn the secrets of the dyes which they yielded, and a dye that would remain on the flesh permanently, or until it wore off, was highly prized. Such a dye was found in the bloodroot, a dye appropriate in its color to represent a thirst for blood; with it they made their war paint, and with it they ornamented their tomahawks to symbolize their sanguinary purpose.

The Indian warriors have passed away from our forests, and the forests themselves are passing away, but the bloodroot still lingers, growing abundantly in rich moist woods or in shaded areas in glades, borders of meadows, and fence corners. Its beautiful white flowers, open to the morning sun in early April, attract hungry bees which come for pollen; like many other early flowers, it offers nectar. Probably many of the little bees prefer pollen to nectar at this tim year, for it is an important element in food of all kinds of bee brood. But bloodroot's fragile blossoms are elu and do not remain long; like their relati the poppies, their petals soon fall, their white masses disappear like the sn drifts which so recently occupied the sa nooks.

The way the bloodroot leaf enfolds flower bud seems like such an obvi plan for protection, that we are unthi ingly prone to attribute consciousness the little plants.

Not only does the leaf enfold the b but it continues to enfold the flowerst after the blossom opens. There are t sepals which enclose the bud, but fall ıe flower opens. There are ordinarily t white petals, although there may be ve; usually every other one of the eight ls is longer than its neighbors, and makes the blossom rather square than ılar in outline. There are many sta- ıs, often twenty-four, and the anthers brilliant yellow with whitish filaments. two-lobed stigma opens to receive en before the pollen of its own flower pe. The stigma is large, yellow, and set ctly on the ovary, and is quite notice- in the freshly opened blossoms. It is ly to shrivel before its homegrown pol- is ripe. The blossoms open wide on ny mornings; the petals rise up in the rnoon and close at night, and also re- n closed during dark, stormy days un- hey are quite old, when they remain n; they are now ready to fall to the ınd at the slightest jar, leaving the ob- g, green seed pod set on the stem at a t bevel, and perhaps still crowned with yellowish stigma. The seed pod is ob- g and pointed and remains below the tecting leaf. There are many yellowish brownish seeds.

When the plant appears above ground, leaf is wrapped in a cylinder about the l, and it is a very pretty leaf, especially "wrong side," which forms the out- of the roll; it is pale green with a net- k of pinkish veins, and its edges are actively lobed; the petiole is fleshy, ıt, and reddish amber in color. The verstalk is likewise fleshy and is tinged h raw sienna; the stalks of both leaf flower stand side by side. After the als of the flower have fallen, the leaf ws much larger, often measuring six hes across and having a petiole ten hes long. It is then one of the most utiful leaves in the forest carpet, its cular form and deeply lobed edges dering it a fit subject for decorative ign.

The rootstock is large and fleshy, and in is stored the food which enables the wer to blossom early, before any food been made by the new leaves. There many stout and rather short roots ıt fringe the rootstock. Once in clear-

O. L. Foster

Bloodroot, showing leaf not yet unrolled

ing a path through a woodland, we hap- pened to hack off a mass of these root- stocks, and we stood aghast at the gory results. We had admired the bloodroot flowers in this place in the spring, and we felt as guilty as if we had inadvertently hacked into a friend.

LESSON 119

Bloodroot

Leading Thought — The bloodroot has a fleshy rootstock, in which is stored food for the nourishment of the plant in early spring. The flower bud is at first pro- tected by the folded leaf. The juice of the rootstock is a vivid light crimson, and was used by Indians as a war paint. The juice is acrid, and the bloodroot is not relished as food by grazing animals, but it is used by us as a medicine.

Method — The bloodroot should be

studied in the woods where it is to be found growing.

Observations — 1. At what time of year does bloodroot blossom? In what situations does it thrive?

2. What do we see first when the bloodroot puts its head above the soil? Where is the flower bud? How is it protected by the leaf?

3. Study the flower. How many sepals has it? What is their color? What is the position of the sepals when the flower is in bud? What is their position when the flower opens? How many petals? What is their color and texture? Describe the position of the petals in the bud and in the open flower. Look straight into the flower; is its shape circular or square?

4. Do the flowers close nights and during dark days? Do the flowers longest open do this? Describe how the petals and sepals fall.

5. Describe the stamens. What is color of the anthers? Of the pollen? scribe the pistil. Does the two-groc stigma open before the pollen is shed after? What insects do you find visi the bloodroot?

6. Sketch or describe a bloodroot as it is wrapped around the stalk of flower. How are both flowerstalk and petiole protected at the base? Desc or sketch a leaf after it is unfolded open. Describe the difference betw the upper and lower surfaces of the l What sort of petiole has it? Break petiole; what sort of juice comes from Describe and measure the leaf later in season; do they all have the same num of lobes?

7. Compare the bloodroot with poppies; do you find any resemblance tween the habits of these two kinds flowers?

THE TRILLIUM

Buffalo Museum of Science

The white trillium

It would be well for the designer of tapestries to study the carpets of our forests for his patterns, for he would find there a new carpet every month, quite different in plan and design from the c spread there earlier or later. One of most beautiful designs from Natu looms is a trillium carpet, which is at best when the white trilliums are in bl som. It is a fine study of the artistic p sibilities of the triangle when reduced terms of leaves, petals, and sepals.

The trillium season is a long one; begins in April with the purple wa robin or birthroot, the species with pur red, or sometimes yellowish flowers. T season ends in June with the last of great white trilliums, which flush pink stead of fading, when old age comes up them.

The color of the trillium flower depe upon the species studied; there are th petals, and the white and painted trilliu have the edges of the petals ruffled; t red and nodding trilliums have petals a sepals nearly the same size, but in t white trillium the sepals are narrower a shorter than the petals. The sepals alternate to the petals, so that when look straight into the flower we see it x-pointed star, three of the points be- green sepals. The pistil of the trillium ix-lobed. It is dark red in the purple lium and very large; in the white spe- , it is pale green and smaller; it opens the top with three flaring stigmas. ere are six stamens with long anthers, l they stand between the lobes of the til. The flowerstalk rises from the cen- where three large leaves join. The werstalk has a tendency to bend a little, l is rather delicate. The three leaves ve an interesting venation, and make good subject for careful drawing. The werstalk varies with different species, d so does the length of the stem of the nt, the latter being fleshy and green to- rd the top and reddish toward the root. e trilliums have a thick, fleshy, and ich scarred rootstock from which ex- d rootlets which are often corrugated. e trilliums are perennial, and grow ostly in damp, rich woods. The painted llium is found in cold, damp woods ong the banks of brooks; the white llium is likely to be found in large num- rs in the same locality, while the purple llium is found only here and there. Flies d beetles carry the pollen for the red llium, apparently attracted to it by its nk odor, which is very disagreeable to

The stemless trillium

Buffalo Museum of Science

Red trilliums

us. The large white trillium is visited by bees and butterflies. The fruit of the trillium is a berry; that of the purple species is somewhat six-lobed and reddish. In late July the fruit of the white trillium is a cone with six sharp wings, or ridges, from apex to base, the latter being three-quarters of an inch across. These vertical ridges are not evenly spaced, and beneath them are packed as closely as possible the yellow-green seeds, which are as large as homeopathic pills. In cross section, it can be seen that the trillium berry is star-shaped with three compartments, the seeds growing on the partitions. This trillium fruit is very rough outside, but smooth inside, and the dried stamens often still cling to it.

The trilliums are so called from the word *triplum*, meaning threefold, as there are three leaves, three petals, and three sepals.

White trilliums in natural surroundings

LESSON 120

The Trillium

Leading Thought — The trilliums are lilies, and are often called wood lilies, because of their favorite haunts. There are several species, but they are all alike in that they have three sepals, three petals, and three leaves.

Method — This lesson may be given from trilliums observed in the woods by the pupils, who should be encouraged to watch the development of the berry and also to learn all the different species common to a locality.

Observations — 1. How many leaves has the trillium? How are they arranged? Draw a leaf, showing its shape and veins. Describe the stem of the plant below the leaves, giving the length and color.

2. How far above the leaves does the flowerstalk or peduncle extend? Does the flower stand upright or droop? Describe or sketch the colors, shape, and arrangem of the petals and sepals. Do the pe have ruffled margins?

3. Describe the pistil and the stigm Describe the stamens and how they placed in relation to the pistil.

4. Do the flowers remain open dur cloudy days and nights?

5. What insects do you find visiting trilliums? Do the same insects visit purple and the white trilliums? What the difference in odor between the pur and the white trillium? Does this se to bring different kinds of insects to ea

6. How does the color of the white t lium change as the blossom matur What is the color and shape of the fr of each different species of trillium? Wh is the fruit ripe?

Trillium and adder's-tongue. These plan may often be found growing in the san habitat

O. L. Foster

Dutchman's-breeches

DUTCHMAN'S–BREECHES AND SQUIRREL CORN

In a gymnasium where things grow,
Jolly boys and girls in a row,
Hanging down from cross-bar stem
Builded purposely for them.
Stout little legs up in the air,
Kick at the breeze as it passes there;
Dizzy heads in collars wide
Look at the world from the underside;
Happy acrobats a-swing,
At the woodside show in early spring.
— A. B. C.

And toward the sun, which kindlier burns,
The earth awaking, looks and yearns,
And still, as in all other Aprils,
The annual miracle returns.
— Elizabeth Akers

There are many beautiful carpets spread ore the feet of advancing spring, but haps none of them are so delicate in tern as those woven by these two plants t spread their fernlike leaves in April l May. There is little difference in the iage of the two; both are delicate green l lacelike above, and pale, bluish green the under side. And each leaf, although finely divided, is, after all, quite simple; for it has three chief divisions, and these in turn are divided into three, and all the leaves come directly from a stem under the ground. These plants grow in the woodlands, and by spreading their green leaves early, before the trees are in foliage, they have the advantage of the spring sunshine. Thus they make their food for maturing their seeds, and also store some of it in their underground parts for use early the following spring. By midsummer the leaves have entirely disappeared, and another carpet is spread in the place which they once covered.

Dutchman's-breeches and squirrel corn resemble each other so closely that they are often confused; however, they are quite different in form; the "legs" of

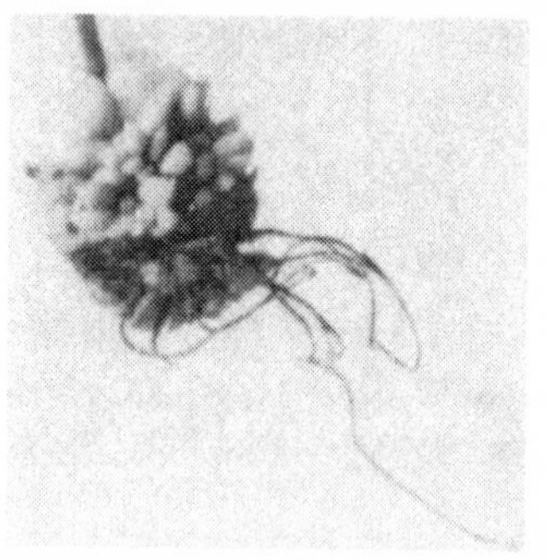

The underground storehouse of Dutchman's-breeches

the Dutchman's-breeches are quite long and spread wide apart, while the blossoms of the squirrel corn are rounded bags instead of "legs." The underground parts of the two are quite different. The Dutchman's-breeches grows from a little bulb made up of grayish scales, while the squirrel corn develops from a round, yellow tuber; these yellow, kernel-like tubers are scattered among the roots, each capable of developing a plant next year. The Dutchman's-breeches grow in thin woodlands and on rocky hillsides, but the squirrel corn is found more often in rich, moist woods. The blossom of the Dutchman's-breeches comes the earlier of the two. These flowers are white with yellow tips, and are not fragrant. The flowers of the squirrel corn are grayish with a tinge of magenta and are fragrant.

The legs of the Dutchman's-breeches are nectar pockets with tubes leading to them, and are formed by two petals. Opposite these two petals are two others more or less spoon-shaped, with the spoon bowls united to protect the anthers and stigma. There are two little sepals which are scalelike.

The seed capsule of the Dutchman's-breeches is a long pod with a slender, pointed end, and it opens lengthwise. ' seed capsules of the squirrel corn are s lar and I have found in one cap twelve seeds, which were shaped like l

Seed capsule of squirrel corn

kernels of corn, black in color, and s like patent leather.

LESSON 121

Dutchman's-Breeches and Squirrel Corn

Leading Thought — The Dutchma breeches, or "boys and girls," as it is o called, is one of the earliest flowers of woodlands. There are interesting di ences between this flower and its c relative, squirrel corn. The flowers of b of these resemble in structure the flo of the bleeding heart.

Dutchman's-breeches in rich woodland vironment

Verne Morton

Squirrel corn

METHOD — As the Dutchman's-breeches blossoms in April and May, usually earlier than squirrel corn, we naturally study the former first and compare the latter with it in form and in habits. The questions should be given the pupils for them to answer for themselves during their spring walks in the parks or woodlands.

OBSERVATIONS — 1. Where do you find Dutchman's-breeches? Which do you prefer to call these flowers, Dutchman's-breeches or boys and girls? Are there leaves on the trees when these flowers are in bloom?

2. Which blossoms earlier in the season, Dutchman's-breeches or squirrel corn? How do the flowers of the two differ in shape? In odor?

3. In the flower of the Dutchman's-breeches find two petals which protect the nectar. How do they look? What part of the breeches do they form? Find two other petals which protect the pollen and stigma.

4. Find the two sepals. How many bracts do you find on the flower stalk?

5. What insects visit these flowers? Describe how they get the nectar.

6. Have you ever seen squirrels harvesting squirrel corn? What is the purpose of the kernels of the squirrel corn?

7. Study the leaf. How many main parts are there to it? How are these parts divided? What is the color of the leaf above? Below? Can you distinguish the leaves of the Dutchman's-breeches from those of the squirrel corn?

8. Describe the seed capsule of Dutchman's-breeches. How does it open? How many seeds has it? Compare this with the fruit of squirrel corn and describe the difference.

9. What happens to the leaves of these two plants late in summer? How do the plants get enough sunlight to make food to mature their seed? What preparations have they made for early blossoming the next spring?

JACK-IN-THE-PULPIT

With hooded heads and shields of green,
Monks of the wooded glen,
I know you well; you are, I ween,
Robin Hood's merry men.

— "CHILD'S OWN BOOK OF FLOWERS"

This little preacher is a prime favorite with all children, its very shape, like that of the pitcher plant, suggesting mystery; and what child could fail to lift the striped hood to discover what might be hidden beneath! And the interest is enhanced when it is discovered that the hood is but a protection for the true flowers, standing

Verne Morton

Jack-in-the-pulpit or Indian turnip

upon a club-shaped stem, which has been made through imagination into "Jack," the little preacher.

Jack-in-the-pulpit prefers wet locations but is sometimes found on dry, wooded hillsides; an abundance of blossoms occurs in late May. This plant has another name, which it has earned by being interesting below ground as well as above. It has a solid, flattened, food-storehouse called a corm with a fringe of coarse rootlets encircling its upper portion. This corm was used as a food by the Indians, which fact gave the plant the name of Indian turnip. I think all children test the corm as a food for curiosity, and retire from the field with a new respect for the stoicism of the Indian when enduring torture; but this is an undeserved tribute. When raw, these corms are peppery because they are filled with minute, needle-like crystals which, however, soften with boiling, and the Indians boiled them before eating them.

Jack-in-the-pulpit is a near cousin to the calla lily; the white part of the calla and the striped hood over "Jack" are both spathes, and a spathe is a leaf modified for the protection of a flower or flowers. "Jack" has but one leg and his flowers are set around it, all safely enfolded in the lower part of the spathe. The pistillate flowers which make the berries are round and greenish, and are packed like berries on the stalk; they have purple stigmas with whitish centers. The pollen-bearing fl ers are mere little projections, aln white in color, each usually bearing purplish, cuplike anthers filled with w pollen. Occasionally both kinds of flov may be found on one spadix (as "Ja is called in the botanies), the pollen-b ing flowers being set above the others; usually they are on separate plants. fessor Atkinson has demonstrated when a plant becomes very strong thrifty, its spadix will be set with the pi late flowers and its berries will be m but if the same plant becomes weal produces the pollen-bearing flowers next year.

When "Jack" first appears in spring it looks like a mottled, pointed for it is well sheathed. Within this sh the leaves are rolled lengthwise to a pc and at the very center of the rolled le is a spathe, also rolled lengthwise, wl enfolds the developing flower buds. a most interesting process to watch unfolding of one of these plants. On older plants there are two, or someti three leaves, each with three large leaf on the younger plants there may be one of these compound leaves, but leaflets are so large that they seem three entire leaves.

The spathes, or pulpits, vary in co

Leonard K. B

Calla lily or water arum

ıe being maroon and white or greenish, ɫ some greenish and white. They are y pretty objects for water-color draw- s.

Small flies and some beetles seem to be : pollen carriers for this plant. Various enious theories have been suggested to ›ve that our Jack-in-the-pulpit acts as a p to imprison visiting insects, as does : English species; but I have studied : flowers in every stage, and have :n the insects crawl out of the hoods as ily as they crawled in, and by the same en, though somewhat narrow passage tween the spadix and the spathe.

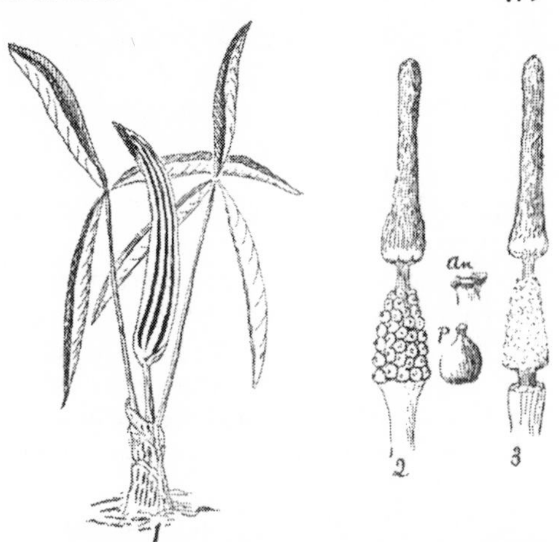

1, *Jack-in-the-pulpit unfolding*. 2, *Spadix with pistillate flowers*. 3, *Spadix with staminate flowers*

P, Pistillate flower, enlarged. An, a staminate flower enlarged, showing four anthers

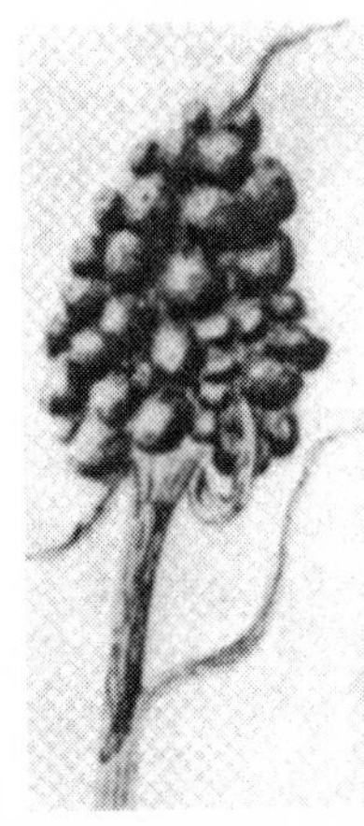

The berries of Jack-in-the-pulpit

After a time the spathe falls away, show- g the globular, green, shining berries. In ugust even the leaves may wither away, which time the berries are brilliant scar- t. Jack-in-the-pulpit is a perennial. It ›es not blossom the first year after it is seedling. I have known at least one case here blossoms were not produced until ıe third year. Below ground, the main ›rm gives off smaller corms and thus the ant spreads by this means as well as by eds.

LESSON 122

Jack-in-the-pulpit

Leading Thought — The real flowers ɛ Jack-in-the-pulpit are hidden by the riped spathe which is usually spoken of : the flower. This plant has a peppery ›ot which the Indians used for food.

Method — The questions should be answered from observation in the woods; a single plant may be dug up and brought to school for study, and later planted in some shady spot in the school garden.

Observations — 1. Where do you find Jack-in-the-pulpit? Is the soil dry or damp? Do you ever find it in the fields?

2. How early in the season does this plant blossom? How late?

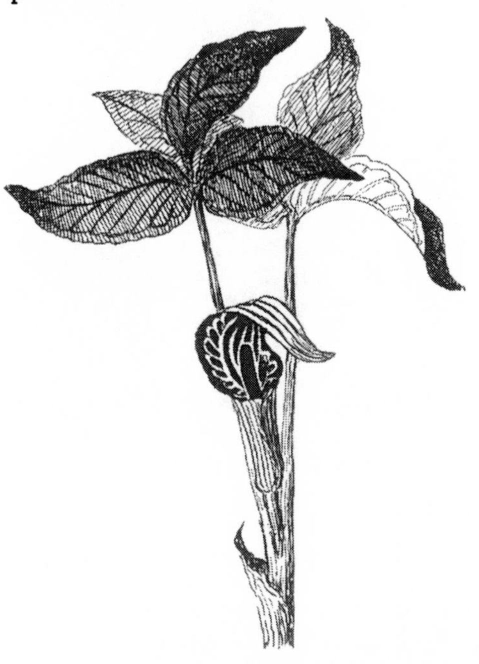

3. How does the Jack-in-the-pulpit look when it first pushes out from the ground? How are its leaves rolled in its spring overcoat?

4. How does the pulpit, or spathe, look when the plant first unfolds? Is its tip bent over or is it straight?

5. Describe or sketch the leaves of Jack-in-the-pulpit. Note how they rise above the flower. How many leaflets has each leaf? Sketch the leaflets to show the venation. How do these stand above the flower? Can you find any of the plants with only one leaf?

6. Why is the spathe called a pulpit? What are the colors of the spathe? Are all the spathes of the same colors?

7. Open up the spathe and see the rows of blossoms around the base of the spadix; if you call the spadix "Jack," t the flowers clothe his one leg. Are all blossoms alike? Describe, if you can, th flowers which will produce the seed a those which produce the pollen. Do y find the two on the same spadix or different plants?

8. What insects do you find carry the pollen for "Jack"? Do you know h its berries look in June? How do they l in August? Do the leaves last as long the berries?

9. What other name has "Jack"? H does the plant multiply below the grou

10. Compare the Jack-in-the-pulpit w the calla lily.

11. Write an English theme on "T Sermon That Jack Preached from His P pit."

THE VIOLET

It is interesting to note the flowers which have impinged upon the imagination of the poets; the violet more than most flowers has been loved by them, and they have sung in varied strains of its fragrance and loveliness.

Verne Morton

Round-leaved yellow violet

Browning says:

Such a starved bank of moss,
'Till that May morn,
Blue ran the flash across;
Violets were born.

And Wordsworth sings:

A violet by a mossy stone,
Half hidden from the eye;
Fair as a star, when only one
Is shining in the sky.

And Barry Cornwall declares that the olet

Stands first with most, but always with t lover.

But Shakespeare's tribute is the m glowing of all, since the charms of bo the goddesses of beauty and of love a made to pay tribute to it:

violets dim
But sweeter than the lids
of Juno's eyes
Or Cytherea's breath.

However, the violets go on living th own lives, in their own way, quite u mindful of the poets. There are many d ferent species, and they frequent quite d ferent locations. Some live in the wood others in meadows, and others in dam

:hy ground. They are divided into two
nct groups — those where the leaf-
:s come directly from the underground
stocks, and those where the leaves
e from a common stem, the latter be-
:alled the leafy-stemmed violets. Much
ntion should be given to sketching
studying accurately the leaves of the
:imens under observation, for the dif-
nces in the shapes of the leaves, in
y instances, determine the species; in
e cases the size and shape of the stip-
determine the species; and whether
leaves and stems are downy or smooth
nother important characteristic. In the
: of those species where the leaves
ng from the rootstock, the flower stems
from the same situation; but in the
y-stemmed violets the flower stems
e off at the axils of the leaves. In some
:ies the flower stems are long enough
ift the flowers far above the foliage,
le in others they are so short that the
vers are hidden.

'he violet has five sepals and their shape
length is a distinguishing mark. There
five petals, one pair above, a pair one
each side, and a broad lower petal,
ich gives the bees and butterflies a rest-

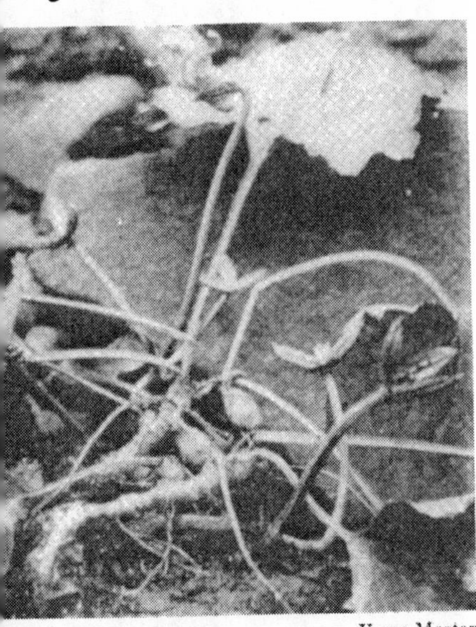

Verne Morton

Common blue violet, showing two of the tle flowers which never open lying beneath e bare rootstocks. Note the three-valved ed capsules

Verne Morton

Dog violet

ing place when they are seeking nectar. This lower petal is prolonged backward into a spur which holds the nectar.

The spur forms the nectary of the violet, and in order to reach the sweet treasure, which is at the rearmost point of the nectary, the insect must thrust its tongue through a little door guarded by both anthers and pistil; the insect thus becomes laden with pollen, and carries it from flower to flower. In many of the species, the side petals have at their bases a little fringe which forms an arch over the door or throat leading to the nectary. While this is considered a guard to keep out undesirable insects like ants, I am convinced that it is also useful in brushing the pollen from the tongues of the insect visitors.

Some species of violets are very fragrant, while others have little odor. The color of the anthers also differs with different species. The children should be interested in watching the development of the seeds from the flower. The seed pods are three-lobed, each one of these lobes dividing lengthwise, with a double row of seeds within. Each lobe curls back and thus scatters the seed.

At the base of most of the species of violets can be found the small flowers which never open; they have no petals, but

Verne Morton

Long-spurred violet

within them the pollen and the pistil are fully developed. These flowers seem to be developed only for self-pollination, and in the botanies they are called cleistogamous flowers; in some species they are on upright stems, in others they lie flat. There is much difference in the shape of the rootstock in the different species of violet; some are delicate and others are strong, and some are creeping.

LESSON 123

The Violet

Leading Thought — Each violet flower has a well of nectar, with lines pointing to it. Violets have also down near their roots flowers that never open, which are self-pollinated and develop seeds.

Method — To make this work of the greatest use and interest, each pupil should make a portfolio of the violets of the locality. This may be in the form of pressed and mounted specimens, or of water-color drawings. In either case, the leaf, leafstalk, flower, flowerstalk, and rootstock should be shown, and each blossom should be neatly labeled with name, locality, and date. From the nature-study standpoint, a portfolio of drawings is the more desirable, since from making the drawings the pupils become more observant of the differences in structure and color which distinguish the species. Such a portfolio may be a most beautiful object; the cover of thick cardboard may have an original, ventionalized design made from the f ers and leaves of the violets. Each drav may be followed by a page contai notes by the pupil and some approp quotation from botany, poetry, or o literature.

Verne Mo

The Canada white violet, a leafy-stemm species

Leonard K. Beyer

Bird's-foot violet

)BSERVATIONS — 1. Describe the local- and general nature of the soil where violet was found. That is, was it in the ›ds, dry fields, or near a stream?

. Sketch or describe the shape of the , paying particular attention to its mar- and noting whether it is rolled toward stem at its base. Is the petiole longer horter than the leaf? Are there stipules :re the leaf joins the main stem? If so, they toothed on the edge?

. What is the color of the leaf above? the leaves and stems downy and vel- ,, or smooth and glossy?

. Does the flowerstalk come from the tstock of the plant, or does it grow n the main stem at the axil of the ? Are the flowerstalks long enough to the flowers above the foliage of the ıt?

. How many sepals has the violet? Are y long or short, pointed or rounded? w many petals has the violet? How are y arranged? Is the lower petal shaped the others? What is the use of this ad lower petal? Are there any marks n it? If you should follow one of these s, where would it lead?

. Look at the spur at the back of the ver. Of which petal is it a part? How g is it, compared with the whole ver? What is the use of this spur?

. Find the opening that leads to the tar-spur and note what the tongue of bee or butterfly would brush against en reaching for the nectar. Are the side als which form the arch over the open- that leads to the nectar fringed at their es?

. What colors are the petals? Are they same on both sides? How are they marked and veined? Are the flowers fragrant?

9. What color are the anthers? What color is the stigma? Examine a fading violet, and describe how the seed is developed from the flower.

10. Find the seed-pods of the violet. How are the seeds arranged within them? How do the pods open? How are the seeds scattered?

11. Look at the base of the violet and find the little flowers there which never open. Examine one of these flowers and find if it has sepals, petals, anthers, and pistil. Are these closed flowers on upright stems or do the stems lie flat on the earth? Of what use to the plant are these little closed flowers?

THE MAY APPLE OR MANDRAKE

This is a study of parasols and, therefore, perennial interest to the little girls who the small ones for their dolls, and with ny airs and graces hold the large ones ›ve their own heads. And when this ersion palls, they make mandarin dolls these fascinating plants. This is easily done by taking one of the small plant umbrellas and tying with a grass sash all but two of the lobes closely around the stem, thus making a dress, the lobes left out being cut in proper shape for flowing sleeves; then for a head some other flower is robbed of its flower bud, which is put

Verne Morton

May apple or mandrake

into place and surmounted with a clover leaflet hat. Then a pin is thrust through hat, head, and neck into the stem of the dressed plant, and the whole is properly finished by placing a small umbrella above the little green mandarin.

The mandrakes grow in open places where there is sun, and yet not too much of it; they like plenty of moisture, and grow luxuriantly in open glades or in meadows or pastures bordering woodlands, and in the fence-corners, along roadsides. The first lesson of all should be to notice how nature has folded these little umbrellas. Study the plants when they first put their heads above ground, each like a parasol wrapped in its case, and note how similarly to a real umbrella it is folded around its stem. Later, after the umbrellas are fairly spread, they afford a most interesting study in varieties of form and size. Some of the leaves have only four lobes while others have many more. I have found them with as many as nine, although the botanies declare seven to be the normal number. One of the special joys afforded by nature-study is finding things different from the descriptions of them in the books.

One of these little parasols is a worthy object for careful observation. Its stem is stout and solid, and at its base may be seen the umbrella-case, now discarded like other umbrella-cases; the stem is pink wherever the sun touches it, but close under the leaves it is likely to be gree ends at the middle of the parasol by se ing out strong, pale green, fuzzy ribs i each lobe. The lobes are narrow tow the stem but broad at the outer edge, e lobe being sparsely toothed on its o margins and with a deep, smooth nc at the center. From the ribs of each l extend other ribs, an arrangement q different from that which we find in c umbrellas. The lobes of the mandrake or parasol are divided almost to the cen The parasol is a beautiful shining g on the upper side, and has a pale gr lining that feels somewhat woolly.

In examining any patch of May app we find that many of the leaves are dou one of these twin leaves is always la than the other and evidently belongs the main stem, since its stem is stou and it is likely to have seven lobes w the smaller one may have but five. H ever, the number of lobes varies. Neit of these double leaf-parasols has its extending out toward the other; inst they are at the side next each other, actly as if the original single stem been split and the whole parasol had b torn in twain.

Brooklyn Botanic Ga

May apple, showing flowers and leaves

But of greatest interest is the bud car-d under this double parasol. At first it is ttle, elongate, green ball on a rather stiff le stalk, which arises just where the two inches fork. One of the strange things out this bud is, that when the plant is t coming from the ground, the bud shes its head out from between the two ded parasols, and takes a look at the orld before it is covered by its green nshade. As the bud unfolds, it looks as it had three green sepals, each keeping cup form and soon falling off, as a tle girl drops her hood on a warm day; t each of these sepals, if examined, will found to be two instead of one; the ter is the outside of the green hood ile the inner is a soft, whitish mem-ane. As the greenish white petals spread t, they disclose a triangular mass of yel-w stamens grouped about the big seed x, each side of the triangle being op-site one of the inner petals. After the wer is fully open, the stamens spread d each anther is easily seen to be ooved, and each edge of the groove ens for the whole of its length; but ecause of its shape and position, it lets e pollen fall away from the pistil instead toward it; nor do the tips of the anthers ach the waxy, white, ruffled stigma. here is no nectar in this flower; but the g queen bumblebee collects the pollen r her new nest, and " bumbles " around the flower while getting her load, so at she becomes well dusted with the pol-n, and thus carries it from flower to ower. But the whole story of the pollen arriers of the May apple is, as yet, untold; nd any child who is willing to give time nd attention to discovering the different sects which visit this flower may give the world valuable and as yet unknown cts. It is said that a white moth is often ound hanging to the flowers, but it is ifficult to understand why the moth hould be there if the flower does not have ny nectar.

The seed vessel at the center of the lower is large and chunky, and, although rowned with its ruffled stigma, looks as f it were surely going to " grow up " into a May apple. There are usually six wide, white, rounded petals, three on the out-side and three on the inside; but some-times there are as many as nine. There are usually twice as many stamens as petals, but I have often found thirteen stamens, which is not twice any possible number of petals. The petals soon fall, and the green fruit — which is a berry instead of an apple — has nothing to do but grow, until in July it is as juicy and luscious to the thirsty child as if it were the fruit of the gods. It is about two inches long, a rich yellow in color, and is sometimes called the " wild lemon," although it is not sour. It is also called the hog-apple because the clever swine of the South know how to find it. Riley thus celebrates this fruit:

May apples as they grow

And will any poet sing of a lusher, richer thing,
Than a ripe May apple, rolled like a pulpy lump of gold
Under thumb and finger tips; and poured molten through the lips?

While the May apple itself is edible, certainly its root is not, except when given by physicians as a medicine, for it is quite poisonous when eaten. When we see plants growing in colonies or patches, it usually means that very interesting things are going on underground beneath them, and the mandrake is no exception to this. Each plant has a running underground stem, straight and brown and fairly smooth; at intervals of a few inches, there are attached to it rosettes of stout, white roots, which divide into tiny, crooked rootlets. There is a large rosette of these roots under the plant we are studying, and we can always find a rosette of them under the place where the plant stood last year. Beneath the present plant we can find the bud from which will grow the rootstock for the coming year. The working out of the branching and the peculiarities of these rootstocks is an excellent lesson in this peculiar and interesting kind of plant reproduction.

LESSON 124

The Mandrake

Leading Thought — These interesting plants grow in colonies because of the spreading of their underground stems. Their fruit is well hidden by its green parasol until it is ripe.

Method — Begin the study just as the mandrakes are thrusting their heads up through the soil in April, and continue the work at intervals until the fruit is ripe.

Observations — 1. How do the mandrakes look when they first appear above the ground? How are the little umbrellas folded in their cases? What do the cases look like? How can you tell from the first the plants which are to bear the flowers and fruit?

2. Study a patch of mandrakes, and see how many varieties of leaves or parasols you can find. Do they all have the same number of main ribs and lobes? How many lobes do most of them have? Are there more single or double leaves in the patch?

3. Take a simple plant and study it carefully. What sort of stem has it? Can you find at its base the old umbrella case? How high is the stem? What is its co
at the bottom and at the top? How ma
ribs does it divide into at the top? A
these ribs as smooth as the stem? H
does the parasol lining differ from its o
side in color and feeling?

4. Study the leaf lobes. What is th
general shape? Are they all notch
at the wide end? How close to t
stem does the division between the
extend?

5. Take a plant with two leaves. Wh
is the flower bud to be found? How is
protected from the sun? Does the ste
divide equally on each side of it or is c
part larger than the other? Are the tw
leaves of the same size? How many lo
has each? What are the chief differenc
in shape between one of these twin lea
and one which has no flower bud?

6. How does the flower bud look? Wh
happens to the green hood or sepals wh
the flower opens? Can you find six sep
in the hood?

7. Does the open flower bow dow
ward? As the flower opens, what is t
shape of the group of stamens at the ce
ter? Are there the same number of whi
waxy petals in all the flowers? Are the
always about twice as many stamens
petals? How do the anthers open to sh
the pollen? Do they let the pollen f
away from the ruffled stigma of the " fa
little seed box at the center of the flowe

8. Does the flower have a strong odo
Does not the plant itself give off th
odor? Do you think it is pleasant? Do t
cattle eat the mandrake when it is
pastures?

9. What insects do you find visiti
the mandrake flowers?

10. Do you like the May apple? Wh
is it ripe? Cut a fruit across and see h
the seeds are arranged.

11. Where are mandrakes found? I
they always grow in patches?

12. Why must we not taste of the ma
drake root?

13. In late July, visit the mandra
patch again. Are there any leaves no
What is left of the plants?

THE BLUETS

uring April, great patches of blue ap- in certain meadows, seeming almost reflections from the sky; and yet when look closely at the flowers which give azure hue to the fields, we find that are more lavender than blue. The lla of the bluet is a tube, spreading into four long, lavender, petal-like s; each lobe is paler toward its base the opening of the tube has a ring vivid yellow about it, the tube itself g yellow even to its very base, where four delicate sepals clasp it fast to the ry. After the corolla has fallen the ls remain.

f we look carefully at the bluets we two forms of flowers: a, those with a -lobed stigma protuding from the ning of the flower-tube; *b*, those where throat of the tube seems closed by anthers which join like four finger- pressed together. In opening the ver, we observe that those which have stigmas protruding from the tube have anthers fastened to the sides of the e about half-way down; while those t have the four anthers near the open- of the tube have a pistil with a short e which brings the stigmas about half- up the tube. An insect visiting the ver a gets her tongue dusted with pol- from the anthers at the middle of the e; and this pollen is ready to be brushed against the stigmas of a flower of the orm. A bee visiting a bluet of the *b* m receives the pollen at the base of her gue; from here it can be brushed off the protruding stigmas of the flowers the a form.

This arrangement in flowers for the iprocal exchange of pollen also char- erizes members of the primrose family; s certainly a very clever arrangement for uring cross-pollination.

LESSON 125

The Bluets

Leading Thought — The bluets have o forms of flowers, the anthers and stigmas being placed in different positions in the two.

Method — Ask the children to bring in several bits of sod covered with bluets. Let the pupils, with the aid of a lens if necessary, find the two different forms of flowers. Later, let each see a flower of each form with the tube opened lengthwise.

Cyrus Crosby

Bluets

Observations — 1. Where do the bluets grow? Do they grow singly or in masses? On what kind of soil do they grow, in woods or meadows? At what time of year do they bloom?

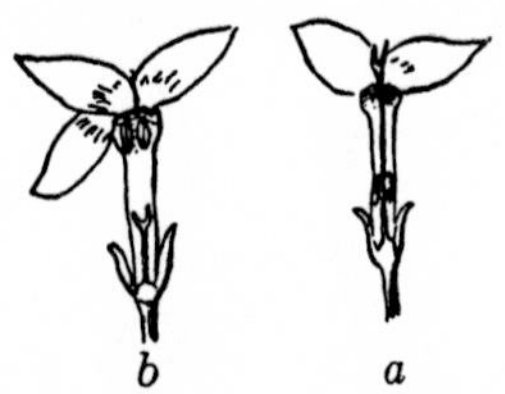

b, Section of a bluet blossom that has the anthers at the throat of the tube and the stigmas below. a, Section of a bluet with the stigmas protruding and the anthers below

2. Describe the bluet flower, its color, the shape of its sepals, the form of the corolla, the color of the corolla-tube and lobes.

3. Where is the nectar in the bluet?

4. Look directly into the flowers. Do you see any with the stigmas thrust out of the corolla-tube? Is there more than one style? Has it one or two stigmas?

Open this flower-tube and describe where the anthers are situated in it. How many anthers are there?

5. Look for a flower where the stigmas do not protrude and the anthers close the throat of the tube. Where are the stigmas in this flower, below or above the anthers? Where are the anthers attached?

6. Work out this problem: How do insects gathering nectar from one forn the bluets become dusted with polle such a way as to leave it upon the stig of the other form of the bluet flov

7. How many sepals has the flowe the bluet? Do these sepals fall off wl the corolla falls?

THE YELLOW LADY'S-SLIPPER

Graceful and tall the slender drooping stem,
With two broad leaves below,
Shapely the flower so lightly poised between,
And warm its rosy glow. — Elaine Goodale

Brooklyn Botanic Garden

Showy lady's-slipper

These showy flowers look so strange in our woodlands that we gaze at them as curiously as we might upon a veiled lady from the Orient who had settled in our midst. There is something abnormal and mysterious in the shape of this flower, and though it be called the lady's-slipper, yet it would be a strange foot that could fit such a slipper; and if it is strange at the first glance, it is still more so as try to compare it with other flowers. Th are two long sepals that extend up down, the lower one being made up two grown together. The sepals are low, and are wider than the two l streamers that extend out at right an to them, which are petals; the brigh color of the latter, their markings of dish dots, the hairs near their bases, go to show that these petals, although different in shape, belong to the sa series as the big lower petal which is pu out into a sac, shaped like a deep, l bowl, with its upper edges incurved. If look carefully at this bowl, we find t openings besides the main one; these t are near the stem, and their edges not incurved. Extending out into each these openings is a strange little rou object, which is an anther; but if we to get pollen from this anther wit pencil or a knife we get, instead of powd pollen, a smear that sticks to what touches, like melted rubber or gum. T secret of this is that the lower side of anther is gummy, and, adhering to wl ever touches it, brings with it, wl pulled away, the mealy pollen which loose above it. Another strange thing that, if this lower part of the anthe not carried away, it seems to parti harden and opens downward, letting pollen escape in a way usual with ot ers. We have to remove a side of the l to see the stigma; it is fan-shaped, is bent at right angles to the flower n; and above it, as if to protect it, is iff triangular piece which is really a ngely modified stamen. I think one on why the lady's-slipper always is ed " she " is because of this tendency her part to divert an object from its ural use. Surely a hairpin used for a er knife or a monkey wrench for a nmer is not nearly so feminine a diver- n as a stamen grown wide and long to ke an awning above a stigma.

'he general color of the flower is yellow, there are some dark red spots on the nen-awning and along the folded-in face of the petal sac. The little bee hts on the flower and crawls into the l at the center, the recurved edges pre- ting it from returning by the same ning. At the bottom of the sac there vegetable hairs to be browsed upon;

Buffalo Museum of Science

Smaller yellow lady's-slipper

Leonard K. Beyer

Lesser purple-fringed orchis

if there is nectar, I have never been able to detect it with my coarse organs of taste; and Mr. Eugene Barker, who has examined hundreds of the flowers, has not been able to detect the presence of nectar in them at any stage; but he made no histological study of the glands.

After a satisfying meal the bee, which is a lively crawler, seeks to get out to the light again through one of the openings near to the stem. In doing this, she presses her head and back, first against the projecting stigma and then against the sticky anther, which smears her with a queer kind of plaster; and it sticks there until she brushes it off on the stigma of another flower, when crowding past it; and there she again becomes smeared with pollen plaster from this flower's anthers. Mr. Barker, who has especially studied these flowers, has found that the little mining bees of the genus *Andrena* were the most frequent visitors; he also found honeybees and one stray young grasshopper in the sacs. The mining bees which he sent to me had their backs plastered with the pollen. Mr. Barker states that the flowers are not visited frequently by insects, and adds feelingly: " My long waiting was rewarded with little insect activity aside from the mosquitoes which furnished plenty of entertainment."

Brooklyn Botanic Garden

Larger yellow lady's-slipper

The ovary looks like a widened and ribbed portion of the flowerstalk, and is hairy outside; its walls are thick and obscurely three-angled; seen in cross section

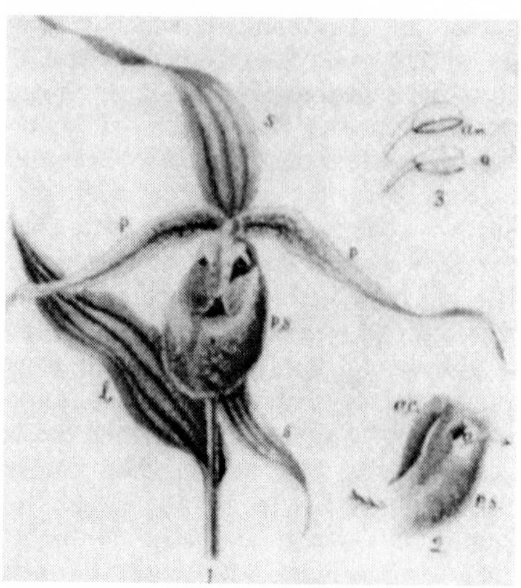

Detail of yellow lady's-slipper

1, l, leaf; s,s, sepals; p,p, petals; p.,s., petal sac. 2, Side view: a.c., anther cover; p.s., petal sac; a, anther. 3, an, anther closed; o, anther open

the seeds are arranged in a triangular f ion which is very pretty.

The leaves of the yellow lady's-slip are oval or elliptic, with smooth ed and parallel veins; they often have row veins between each two heavier o The leaves are of vivid yellowish green are scattered, in a picturesque man alternately along the stem, which tl bases completely clasp. The stem is so what rough and ribbed and is likely

Brooklyn Botanic Ga

Pink moccasin flower or stemless lad slipper

grow crooked; it grows from one to t feet in height. The roots are a mass small rootlets. This species is found woods and in thickets.

The pink moccasin flower, also call the stemless lady's-slipper (*C. acaule*), perhaps prettier than the yellow spec and differs from it in several particula The sac opens by the merest crevice, a there are dark-pink lines which lead the little opening of the well. The dow ward-folded edges prevent the visiti insect from getting out by this openi even more surely than in the other speci The side petals are not so long as in t ow species, and they extend forward if to guide the insect to the well in lower petal. The sepals are greenish ple, and are likewise shorter; and the er one is wide, indicating that it is de up of two grown together. At the e of the ovary there is a pointed green ct or leaf, which lifts up and bends ve the flower. There are but two leaves the stemless lady's-slipper; they arise m the base of the flowerstalk. They are adly ovate, and from six to seven inches g. This species grows in sandy or rocky ods.

Another species more beautiful than se is the showy lady's-slipper, which white with a pink entrance to the petal . This grows by preference in peaty gs, and is not so common as the others.

The interesting points for observation these flowers are the careful noting of e kinds of insects which visit them, and w they enter and leave the " slipper," or .

Buffalo Museum of Science

Showy lady's-slipper. In this native habitat it is surrounded by such plants as horse-tail, Indian cucumber, and ferns

LESSON 126

The Yellow Lady's-Slipper

Leading Thought — The moccasin wer belongs to that family of flowers own as orchids which especially depend on insects for bringing and carrying llen, and which have developed many ange devices to secure insect aid in llination.

Method — A trip may be taken to see ese plants where they grow.

Observations — 1. Where does the llow lady's-slipper grow? Look carefully its leaves and describe them. How they join the stem? Are they opposite alternate?

2. What is there peculiar about the pals? How many are there?

3. Describe the three petals and the fference and likeness in their form and lor. What is the shape of the lower etal? Is there a hole in this sac? What is the color of the sac? Is there anything about it to attract insects? If an insect should enter the mouth of the well in the lower petal could it easily come out by the same opening? Why not? Where do you think it would emerge?

4. Note the two roundish objects projecting into the two openings of the sac near the stem. Thrust a pencil against the under side of one of these. What happens? How does this pollen differ from the pollen of ordinary flowers?

5. Explain how a bee visiting these flowers, one after another, must carry the pollen from one to another and deposit it on the waiting stigmas.

6. How is the insect attracted? How is it trapped?

7. Look at the seed capsule and describe it from the outside.

8. How many species of lady's-slippers do you know? Do you know the pink, or stemless species? How does it differ from the yellow species?

THE EVENING PRIMROSE

Children came
To watch the primrose blow. Silent they stood,
Hand clasped in hand, in breathless hush around,
And saw her shyly doff her soft green hood
And blossom — with a silken burst of sound.

— MARGARET DELANI

To the one who has seen the evening primrose unfold, life is richer by a beautiful, mysterious experience. Although it may be no more wonderful than the unfolding of any other flower, yet the sud-

Brooklyn Botanic Garden
Evening primrose in flower

denness of it makes it seem more marvelous. For two or three days it may have been getting ready; the long tube which looks like the flowerstalk has been turning yellow; pushing up between two of the sepals, which clasp tips beyond it, there appears a row of petals. Then some warm evening, usually about sunset, but varying from four o'clock in the aft noon to nine or ten in the evening, t petals begin to unfurl; they are wrapp around each other in the bud as an u brella is folded, and thus one edge each petal becomes free first. The pe first in freeing its edge seems to be doi all the work, but we may be sure that the others are opening too; little by lit the sepals are pushed downward, un their tips, still clasped, are left beneat and the petals now free suddenly fla open before our delighted eyes, with movement so rapid that it is difficult f us not to attribute to them consciousn of action. Three or four of these flow may open on a plant the same evenin and they, with their fellows on the neig boring plants, form constellations of sta bloom that invite attention, and night-fl ing insects are often seen on them. The is a difference in the time required f a primrose flower to unfold, probably c pending upon its vigor; once I watch for half an hour to see it accomplishe and again I have seen it done in two three minutes. The garden species see to unfold more rapidly than the wi species, and is much more fragrant. T rapidity of the opening of the blosso depends upon the petals getting free fro the sepals, which seem to try to repre them. The bud is long, conical, obscure four-sided, and is completely covered the four sepals, the tips of which a cylindrical and twisted together; this an interesting habit, and one wonders they hold the petals back until the latt are obliged to burst out with the for of repressed energy; after they let go of t petals, they drop below the flower ang larly, and finally their tips open and ea ıl turns back lengthwise along the l-tube.

'he four lemon-yellow petals are broad, ı the outer margin notched. The eight nens are stout, and set one at the ldle of each petal and one between ı two petals. The long, pale yellow hers discharge their pollen in cob-by strings. When the flower first ns, the stigma is egg-shaped and lies ɔw the anthers; later, it opens into a ss and usually hangs off at one side he anthers. If we try to trace the style k to the ovary, we find that it extends ɣn into what seems to be the very base he flowerstalk, where it joins the main n. This base is enlarged and ribbed is the seed box, or ovary. The tube is in nectar, but only the long sucking-es of moths can reach it, although I e sometimes seen the ubiquitous bees empting it. The butterflies may take nectar in the daytime, for the blossoms the wild species remain open, or par-ly open, for a day or two. But the night-ng moths which gather nectar have

, Evening primrose showing buds, one ly to open, a flower just opened above at left, an older flower at the right, a fading ver and seed capsules below. 2, seed cap-s. Cross section of seed capsule with seeds ve

Verne Morton

Winter rosette of evening primrose

the first chance, and it is they who carry the flower's pollen.

There are times when we may find the primrose blossoms with holes in the petals, which make them look very ragged. If we look at such plants carefully, we may find the culprit in the form of a green caterpillar very much resembling the green tube of the bud; and we may conclude, as Dr. Asa Fitch did, that this caterpillar is a rascal, because it crawls out on the bud-ends and nibbles into them, thus damaging several flowers. But this is only half the story. Later this caterpillar descends to the ground, digs down into it and there changes to a pupa; it remains there until the next summer and then, from this winter cell, emerges an exquisitely beautiful moth called the *Alaria florida*; its wings expand about an inch, and all except the outer edges of the front wings are rose-pink, slightly mottled with lemon-yellow, which latter color decorates the outer margins for about one-quarter of their length; the body and hind-wings are whitish and silky, the face and antennæ are pinkish. Coiled up beneath the head is a long sucking-tube which may be unfolded. This moth is the special pollen-carrier of the evening primrose; it flies about during the evening, and thrusts its long, tubular mouth into the flower to suck the nectar, meanwhile gathering strings of pollen upon the front part of its body. During the day, it hides within the partially closed flower, thus carrying the pollen to the ripened stigmas, its colors meanwhile protecting it almost completely from observation. The fading petals of the primrose turn pinkish, and the pink color of the moth renders it invisible when in the old flowers, while the lemon-yellow tips of its wings, protruding from a flower still fresh and yellow, form an equally perfect protection from observation.

The evening primrose is an ornamental plant in both summer and winter. It is straight, and is ordinarily three or four feet tall, although it sometimes reaches twice that height. It is branched somewhat, the lower portion being covered with leaves and the upper portion bearing the flowers. The leaves are pointed and lanceolate, with few whitish veins. The leaf edges are somewhat ruffled and obscurely toothed, especially in the lower leaves. The leaves stand up in a peculiar way, having a short, pink petiole, which is swollen and joins the stalk like a bracket. The upper leaves are narrower; the leafy bracts at the base of the flower grow from the merest slender leaflet at the base of the bud to a leaf as long as the seed pod, when the flower blooms.

The seed capsules are four-sided, long, and dark green. In winter they are crowded in purplish-brown masses on the dry stalks, each one a graceful vase with four flaring tips. At the center of each there projects a needlelike point; and within the flaring, pale, satin-lined divisions of these urns we may see the brown seeds, which are tossed by the winter winds far and near. The young plants develop into vigorous rosettes during the late summer and autumn, and thus pass the winter under the snow coverlet.

LESSON 127

The Evening Primrose

Leading Thought — The pollen of the evening primrose is carried by night-flying insects. The evening primrose's flowers open in the evening; their pale yel color makes them noticeable objects the twilight, and even in the dark.

Method — The form of the even primrose may be studied from pla brought to the schoolroom; but its spe interest lies in the way its petals exp in the evening, so the study should continued by the pupils individually the field. This is one of the plants wh is an especially fit subject for the sum notebook; but since it blossoms v late and the plants are available even October, it is also a convenient plant study during the school year. The gar species is well adapted for this lesson.

Observations — 1. Look at the pl as a whole. How tall is it? Is the st stiff and straight? Where do you fin growing? Does it grow in the woods?

2. Are the leaves near the base the sa shape as those at the top of the pla What is their shape? Are the ed toothed? What is there peculiar about veins? How do the leaves join the ste How do the leaves which are at the b of the flowerstalk look? Those at base of the buds?

3. Where on the plant do the flow grow? Which flowers blossom first, th above or below? Take a bud nearly re to open; what is there peculiar in the pearance of the budstalk? What is general shape of the bud? Describe sepals. Look at their tips carefully, see how they hold together. Cut a b across and see how the petals are fol within it.

4. Take an open flower; where are sepals now? Describe the open pet their shape and color.

5. How many stamens are there? H are they placed? What is the shape the anthers? How does the pollen look

6. What is the shape and the posit of the stigma in the freshly opened flow Later? Open the flower-tube and find h far down the style extends. Where is ovary? How does the ovary look on outside? Taste the opened tube; can detect the nectar? What sort of ton must an insect have to reach this nec

w do the fading flowers look and act?

. Describe the seed pod. Cut it across, see how many compartments there within it. How are the seeds arranged t? How do the pods open and how are seeds scattered?

. Watch the flower of the evening nrose open, and describe the process fully. At what hour did it open? What the movement of the petals? Can you how they unfold in relation one to another? How do they get free from the sepals? How many minutes are required for the whole process of the opening of the flower? How many flowers on a plant expand during the same evening? Look at the open blossoms in the dark; can you see them? How do they look? What insects do you find visiting these flowers?

9. How long does the primrose blossom remain open? How do the young plants of the evening primrose pass the winter?

THE MILKWEED

Little weavers of the summer, with sunbeam shuttle bright,
And loom unseen by mortals, you are busy day and night,
Weaving fairy threads as filmy, and soft as cloud swans, seen
In broad blue sky-land rivers, above earth's fields of green.

— RAY LAURANCE

s there any other young plant that ws off its baby clothes as does the ng milkweed! When it comes up ough the soil, each leaf is folded length- e around the stem, flannel side out, l it is entirely soft and white and in- tile. The most striking peculiarity of milkweed plant is its white juice, ich is a kind of rubber. Let a drop of lry on the back of the hand, and when try to remove it we find it quite elastic l possessed of all the qualities of crude ober. At the first trial it seems quite possible to tell from which part of the m this white juice comes, but by blot- g the cut end once or twice, we can see t the hollow of the center of the stem s around it a dark green ring, and that tside this is a light green ring. It is m the dark green ring encircling the m cavity that the milk exudes. This lk is not the sap of the plant any more an resin is the sap of the pine; it is a ecial secretion, and is very acrid to the te. Milkweed is seldom eaten by graz- g animals. If a milkweed stem be broken gashed, this juice soon heals the wound. cut across, every vein in every leaf pro- ces "milk," and so does every small wer pedicel. When the "milk" is by ance smeared on cloth and allowed to y, soap and water will not remove it, but it yields readily to chloroform, which is a solvent of rubber.

Verne Morton

Milkweed in blossom

The milkweed leaves are in stately conventional pairs; if one pair points east and west, the pair above and the pair below point north and south. The leaf is beautiful in every particular; it has a dark green upper surface, with veins that join in scallops near the border; it is soft to the touch on the upper surface, and is velvety below.

Leonard K. Beyer

Milkweed in natural surroundings

The lens reveals that the white under surface, or the nap of the velvet, is a cover of fine white hairs.

The flower of the milkweed is too complicated for little folks even to try to understand; but for the pupils of the seventh and eighth grades it will prove an interesting subject for investigation, if they study it with the help of a lens. In examining the globular bud, we see the five hairy sepals, which are later hidden by the five long, pinkish green petals that bend back around the stem. When we look into the flower, we see five little cornucopias — which are really horns of plenty, since they are filled with nectar; in the center of each is a little, fleshy tongue, with its curved point resting on the disk at the center of the flower. Between each two of these nectar-horns can be seen the white bordered opening of a long pocket — like a dress pocket — at the upper end of the opening of which is a black dot. Slip a needle into the pocket opening until it pushes against the black dot, and out pops a pair of yellow saddle-bags, each attached to the black dot which joins them. These are the pollen-bags, and each was borne in a sac, shaped like a vest-pocket, one lying either side of upper end of the long pocket. These len-bags are sticky, and they contract as to close over the feet of the visiting

Since the stalk of the flower clu droops and each flower pedicel dro the bee is obliged to cling, hanging b down, while getting the nectar, and to turn about as if on a pivot in orde thrust her tongue into the five cornuco in succession; she is then certain to th her claws into a long pocket, and it ceeds to close upon them, its edges ing like the jaws of a trap. The bee trying to extricate her feet, leaves whate pollen-bags she had inadvertently gathe in this trap-pocket, which gives them sage to the stigma. But the milkw flower, like some folks, is likely to ove matters, and sometimes these poc grasp too firmly the legs of the bee hold her a prisoner. We often find ins thus caught and dead. Sometimes bees come so covered with these pollen-b which they are unable to scrape off, they die because of the clogging. But one bee that suffers there are thousa that carry off the nectar.

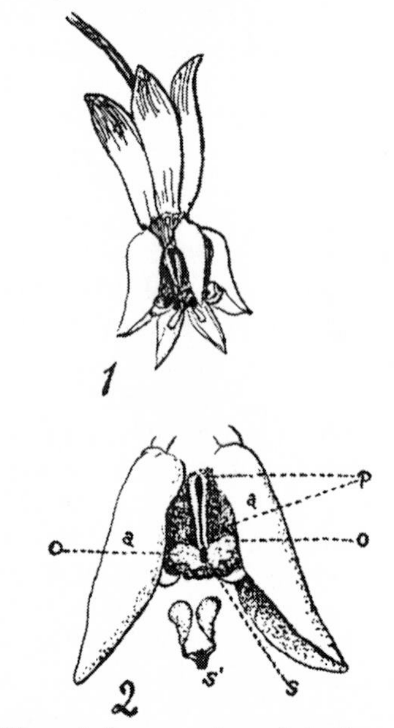

1, *Milkweed flower, enlarged*. 2, *Same, m enlarged*

a, a, nectar-horns; p, pocket; o, openings to poc s, pollen-bags in place; s' pollen-bags removed

The milkweed pod has been the ad-
ration of nature students from the be-
ning, and surely there are few plant
uctures that so interest the child as this
use in which the milkweed carries its
ds. When we look at a green pod, we
t admire its beautiful shape; on either
e of the seam, which will sometime
en, are three or four rows of projecting
ints rising from the felty surface of the
d in a way that suggests embossed
broidery. We open the pod by pulling
apart along the seam; and this is not
seam with a raw edge but is finished
th a most perfect selvage. When we
re children we were wont to dispossess
ese large green pods of their natural
ntents, and because they snapped shut
easily, we imprisoned therein bumble-
es "to hear them sing," but we always
t them go again. We now know that
ere is nothing so interesting as to study
e contents of the pod just as it is. Be-
w the opening is a line of white velvet;
one end, and with their "heads all in
ne direction," are the beautiful, pale-
mmed, brown, overlapping seeds; and at
e other end we see the exquisite milk-
eed silk with the skein so polished that
o human reel could give us a skein of
ich luster. If we remove the contents of
e pod as a whole, we see that the velvety
ortion is really the seed-support and that
joins the pod at either end. It is like
hammock full of babies, except that the
ilkweed babies are fastened on the out-
de of the hammock.

No sooner is our treasure open to the
ir than the shining silk begins to separate
nto floss of fairy texture. But before one
eed comes off, let us look at the beautiful
attern formed by the seeds overlapping
- such patterns we may see in the mosaics
f mosques.

Pull off a seed, and with it comes its
wn skein of floss, shining like a pearl;
ut if we hold the seed in the hand a
noment the skein unwinds itself into a
luff of shining threads as fine as spiders'
ilk, and each individual thread thrusts
tself out and rests upon the air; and al-
ogether there are enough of the threads

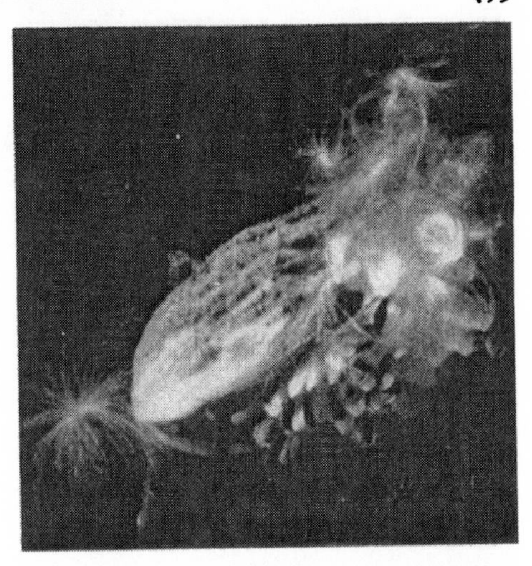

Milkweed seed balloons just leaving the sheltering pod

to float the seed, a balloon of the safest sort. If we wreck the balloon by rubbing the floss through our fingers, we shall feel one of the softest textile fibers spun by Mother Nature.

If we look closely at our seed we see a margin all around it. Well, what if the balloon should be driven over a stream or lake, and the seed dropped upon the water? It must then sink unless it has a life preserver; this margin that we have noted keeps it afloat; if you do not believe it, try it.

If we pull off all the seeds, we can see that the velvety support is flat and that all of the seeds are attached to it, but before we stop our admiring study we should look carefully again at the inside of the pod, for never was there a seed cradle with a lining more soft and satiny.

LESSON 128

The Milkweed

Leading Thought — The milkweed when wounded secretes a milky juice which is of a rubber-like composition; it flows out of the wounded plant and soon hardens, thus protecting the wound. Milkweed flowers depend entirely upon insects for pollination; the pollen is not a free, yellow powder, but it is contained in paired sacs, which are joined in V-shape.

Milkweeds sending forth their seed balloons

The seeds are carried by balloons, and they can float on water as well.

METHOD — Begin the study of the plant when it first appears above ground in April or May. Give the pupils the questions about the blossom for a vacation study, and ask that their observations be kept in their notebooks. The study of the pods and seeds may be made in September or October. When studying the milky juice, add a geography lesson on rubber trees and the way that rubber is made.

OBSERVATIONS — 1. *The plant.* How does the milkweed look as it appears above ground in the spring? How are its leaves folded when it first puts its head up? Cut off a fully expanded plant a few inches above the ground. What flows out of the stem? Blot off the " milk " and study the cross-section of the stem. What is at the center? How many layers do you see around this center? Can you see from which the milkweed juice comes? How does the juice feel as it dries on your fingers? How does it look when dry? Place a few drops on a piece of paper and wh it is dry pull it off and see if it is elast Break the edge of the leaf. Does the mil juice flow from it? Does it come from t veins? Do you think that this is the s of the milkweed? Cut a gash in the mi weed stem and see how the " milk " fl the wound. How does this help the plan Do cattle feed upon the milkweed wh it grows in pastures? Why not?

2. How are the leaves arranged on t stem? How do the upper and under sid of the leaves differ? Examine with a le and see what makes the nap of the velv What gives the light color to the und side? Sketch a leaf showing its shape a venation, noting especially the directi of the veins as they approach the edge the leaf.

3. *The flower.* Where do the flow clusters come off the stems in relati to the leaves? Does the stalk of the flow cluster stand stiff or droop? Take a goo sized flower cluster and count the flowe in it. What would happen if all the flowers should develop into pods? Ho many flower clusters do you find in o plant? Which of these clusters ope first? Last?

4. Take off a single bud with its sta or pedicel. Does the milky juice come the break? Is the bud stalk stiff or droo ing? What is its color and how does it fee What is the shape of the bud? How ma sepals has it? Look at the stalk, sepals, a bud with a lens and describe their cove ing. Look for a flower just opening whe the petals stand out around it like a fiv

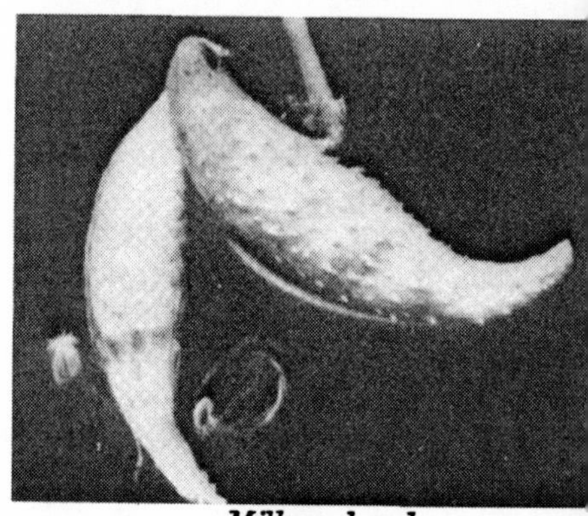

Milkweed pod

ıted star. What is their color? What pens to the petals when the flower is y expanded? Can you see the sepals n? Look straight into the flower. Do see the five nectar-horns? Look at m with a lens and describe them. ıere does the tip of the tongue rest? th a lens, look between two of the tar-horns; can you see a little slit or ket, with white protruding edges? te just above the pocket a black dot; ust a needle into this pocket near its e and lift it toward the crown of the ver, touching the black dot. What pens?

. Describe the little branched object t came out when you touched it with eedle. These are the pollen saddlebags l each bag comes from a pocket at one e of, and above the long pocket. Do se saddlebags cling to the needle? ok with a lens at some of the older wers, and see if you can find the pollen-s protruding from the long pocket. See ou can find how the long pocket is a sageway to the stigma. To see how the little saddlebags were transported, watch a bee gathering nectar. Describe what happens.

6. Since the flowers bend over, how must the bee hold on to the flower while she gathers nectar from the horns? As she turns around, would she naturally pull out some of the saddlebags? Catch a bee in a collecting tube and see if her feet have upon them these pollen-sacs. After these pollen-sacs have been gathered upon her feet, what happens to them when she visits the next flower? Is the opening of the long pocket like a trap? Can you find on milkweed flowers any bees or other insects that have been entangled in these little traps and have thus perished? Try the experiment of drawing a thread into one of these traps and with your lens see if the opening closes over it.

7. How many kinds of insects do you find visiting the milkweed flowers? Can you detect the strong odor of the flowers? How does the milkweed benefit by having so many flowers and by offering such an abundance of nectar?

Verne Morton

THE WHITE WATER LILY

Whence O fragrant form of light,
Hast thou drifted through the night
Swanlike, to a leafy nest,
On the restless waves at rest.

Thus asks Father Tabb, and if the lily uld answer it would have to say: Through ages untold have the waves upheld me until my leaves and my flowers have changed into boats, my root to an anchor, and my stems to anchor-ropes."

There is no better example for teaching the relation between geography and plant life than the water lily. Here is a plant that has dwelt so long in a certain

Egyptian lotus flower and seed vessel

situation that it cannot live elsewhere. The conditions which it demands are quiet water, not too deep, and with silt bottom. Every part of the plant relies upon these conditions. The rootstock has but few rootlets; and it lies buried in the silt, where it acts as an anchor. Rising from the rootstock is a stalk as pliable as if made of rubber, and yet it is strong; its strength and flexibility are gained by having at its center four hollow tubular channels, and smaller channels near the outside. These tubes extend the whole length of the stem, making it light so that it will float, and at the same time giving it strength as well as flexibility. At the upper end of the stalk is a leaf or flower, which is fashioned as a boat. The circular leaf is leathery and often bronze-red below, with prominent veins, making an excellent bottom to the boat; above, it is green with a polished surface, and here are situated its air-pores, although the leaves of most plants have these stomata in the lower surface. But how could the water lily leaf secure air, if its stomata opened in the water? The leaf is large, circular, and quite heavy; it would require a very strong, stiff stem to hold it aloft, but by its form and structure it is fitted to float upon the water, a little green dory, varnished inside, and waterproof outside.

The bud is a little, egg-shaped buoy protected by its four pinkish brown, leathery sepals; as it opens, we can see four rows of petals, each overlapping the sp between the next inner ones; at the cen there is a fine display of brilliant yell anthers. Those hanging over the green yellow pit, which has the stigma at center, are merely golden hooks. Wh the flower is quite open, the four sep each a canoe in form, lie under the and float it; although the sepals are brow ish outside, they are soft white on inside next the flower. Between each t sepals stands a large petal, also can shaped, and perhaps pinkish on the o side; these help the sepals in floating flower. Inside of these there is a row large creamy white petals which sta upright; the succeeding rows of petals smaller toward the center and grade in the outer rows of stamens, which are pe like at the base and pointed at the The inner rows of stamens make a f golden fringe around the cup-shap pistil.

It has been stated that pond lilies, the state of nature, have an interest way of opening in the early morning, c ing at noon, and opening again tow evening. If we knew better the habits

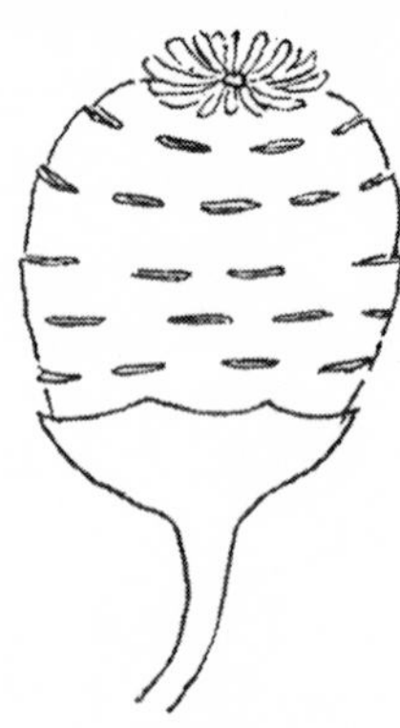

Seed vessel of white pond lily

the insects which pollinate these flow we should possibly have the key to t action. In our ponds in parks and grou we find that each species of pond opens and closes at its own partic time each day. Each flower opens usu for several consecutive days, and the f day of its blooming it opens about an h later and closes an hour earlier than days following. After the lilies have ssomed, the flower stem coils in a spiral brings the ripening seeds below the face of the water. After about two nths the pod bursts, letting the seeds in the water. Each seed is in a little , which the botanists call an aril, and ich serves to float the seed off for some ance from the parent plant. The aril lly decays and the seed falls to the tom where, if the conditions are fa-able, it develops into a new plant.

To emphasize the fact that the water is dependent upon certain geographi- conditions, ask the pupils to imagine vater lily planted upon a hillside. How ld its roots, furnished with such in-ficient rootlets, get nourishment there? w could its soft, flexible stems hold ft the heavy leaves and blossoms to the light? In such a situation it would be nere drooping mass. Moreover, if the pils understand the conditions in which water lilies grow in their own neigh-hood, they can understand the condi-ns under which the plant grows in er countries. Thus, when they read out the great *Victoria regia* of the Ama-n — that water lily whose leaves are ge enough to support a man — they uld have visions of broad stretches of l water and they should realize that the ttom must be silt. If they read about lotus of Egypt, then they should see Nile as a river with borders of still ter and with bottom of silt. Thus, from conditions near at hand, we may culti-e in the child an intelligent geographi- imagination.

LESSON 129

The Water Lily

Leading Thought — The water lily has come dependent upon certain condi-ns in pond or stream, and has become fitted in form to live elsewhere. It must ve quiet waters, not too deep, and with t bottom.

Method — The study should be made first with the water lilies in a stream or pond, to discover just how they grow. For the special structure, the leaves and flowers may be brought to the school-room and floated in a pan of water. The lesson may easily be modified to fit the yellow water lily, which is in many ways even more interesting, since in shallow water it holds its leaves erect while in deeper water its leaves float.

Observations — 1. Where is the water lily found? If in a pond, how deep is the water? If in a stream, is it in the current? What kind of bottom is there to the stream or pond? Do you find lilies in the water of a limestone region? Why?

2. What is the shape of the leaf? What is the color above and below? What is the texture? How is it especially fitted to float? How does it look when very young?

3. Examine the petiole. How long is it? Is it stiff enough to hold up the leaf? Why does it not need to hold up the leaf? How does it serve as an anchor? Cut a petiole across and describe its inside structure. How does this structure help it float?

4. Examine the open flower. How many sepals? How many rows of petals? How do the stamens resemble the petals? How are the sepals fitted to keep the flower afloat? At what times of the day does the lily open? At what hours does it close?

5. Describe the pistil. When the lily first opens, how are the stamens placed around the pistil? What happens to the seed box after the blossoms have faded? Does the seed pod float upon the water as did the flower? What sort of stalk has the flower? How does this stalk hold the seed pod below the water?

6. What sort of seed has the water lily? Sketch the seed pod. How does the seed escape from it? How is it scattered and planted?

7. What sort of root has the water lily? Are there many fine rootlets upon it? Why? How does this rootstock serve the plant aside from getting moisture?

8. Imagine a water lily set on a dry hillside. Could the stalks uphold the

flowers or leaves? Is the petiole large enough to hold out such a thick, heavy leaf?

9. Judging from what you know of the places where water lilies grow and the c dition of the water there, describe Nile where the lotus grows. Describe Amazon where the *Victoria regia* gro

PONDWEED

The study of any plant which has obvious limitations as to where it may grow should be made a help in the study of

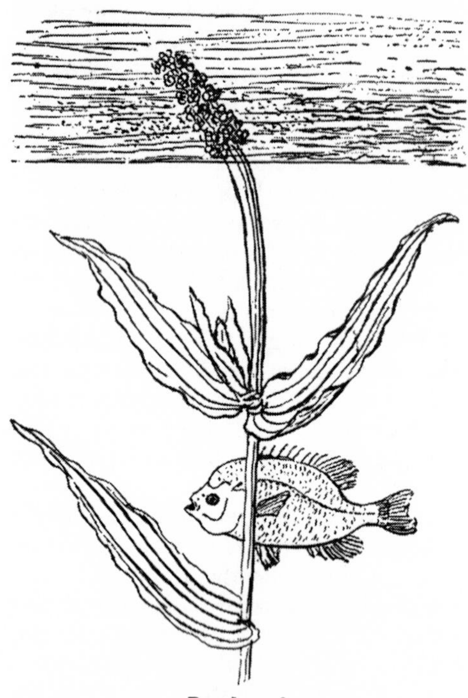

Pondweed

geography. Pondweed is an excellent subject to illustrate this principle; it grows only in quiet beds of sluggish streams or in ponds, or in the shallow protected portions of lakes. It has tremendous powers of stretching up, which render it able to grow at greater depth than one would suppose possible; it often flourishes where the water is from ten to twenty feet deep. Often, when the sun is shining, it may be seen like a bed of seaweed on the bottom. Its roots, like those of most water plants, have less to do with the matter of absorbing water than do the roots of land plants, one of their chief functions being to anchor the plant fast; they have a firm grip on the bottom, and if pondwee cut loose, it at once comes to the surfa floats, and soon dies.

The stem is very soft and pliable a the plant is supported and held upri by the water. A cross-section of the st shows that its substance is spongy, larger open cells being near the outer ed and thus helping it to float. The leaves two or three inches long, their broad ba encircling the stem, their tips tapering slender points. They have parallel ve and ruffled edges. They are dull ol green in color, much darker than stems; in texture they are very thin, pery, and so shining as to give the pression of being varnished. No la plants have such leaves; they remind us once of kelp or other seaweeds. The lea are scattered along the stems, by no me thickly, for water plants do not seem need profuse foliage.

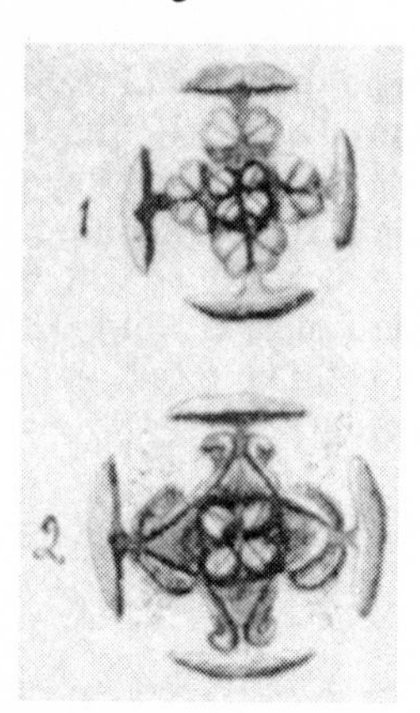

1, *Flower of a pondweed enlarged, ea stage.* 2, *Same at later stage*

In blossom time the pondweed sho its real beauty. The stems grow and gro like Jack's beanstalk, and what was bed of leaves on the pond bottom su denly changes into a forest of high plan h one standing tall and straight and h every leaf extended, as if its stems re as strong and stiff as ironwood; but wave disturbs the water the graceful dulations of the plant tell the true story the pliant stems. There is something at arouses our admiration when we see e of these pondweeds grown so straight d tall, often three or four yards high, h its little, greenish brown flower-head ove the water's surface. We have spent urs looking down into such a submerged est, dreaming and wondering about the l meaning of such adaptations.

Although the stem is flexible, the some-at curved, enlarged stalk just below the wer-head is rigid; it is also more spongy an the lower part of the stem and is us fitted to float the flower. The flower elf is one of the prettiest sights that ture has to show us through a lens. It a Maltese cross, the four reddish stig-as arranged in a solid square at the cen-r; at each side of this central square a double-barrelled anther, and outside each anther is a queer, little, dipper-aped, green flap. When the anthers en, they push away from the stigmas d throw their pollen toward the out-le. There may be thirty or more of these ny, cross-shaped flowers in one flower-ad. In the bud, the cup-shaped flaps ut down closely, exposing the stigmas st, which would indicate that they ripen fore the pollen is shed. The pollen is hite, and is floated from plant to plant the surface of the water; often the ater for yards will be covered with this ving dust.

LESSON 130

Pondweed

Leading Thought — The pondweed ves entirely below the water; at blossom me, however, it sends up its flowerstalks the surface of the water, and there heds its pollen, thus securing cross-pol-ation.

Method — As this is primarily a lesson that relates to geography, the pondweed should be studied where it is growing. It may be studied in the spring or fall, and the pupils asked to observe the blossom-ing, which occurs in late July. After the pupils have seen where it grows, the plants themselves may be studied in an aquar-ium, or by placing them in a pail or basin of water. There are confusing num-bers of pondweeds but any of them will do for this lesson. The one described above is *P. perfoliatus*.

Observations — 1. Where is the pond-weed found? Does it ever grow out of water? Does it ever grow in very deep water? Does it ever grow in swiftly flow-ing water?

2. Has the pondweed a root? Does the pondweed need to have water carried to its leaves, as it would if it were living in the air? What is one of the chief uses of the roots to the pondweed? Break off a plant; does it float? Do you think it would float off and die, if it were not held by its root?

3. Compare the stem of pondweed with that of any land plant standing straight. What is the chief difference? Why does the pondweed not need a stiff stem to hold it up? Cut the stem across, and see if you can observe why it floats.

4. Examine the leaves. Are all of them below the surface of the water? If some float, how do they differ in texture and form from those submerged? How are they arranged on the stem? Are they set close together? What is the difference in texture between its leaves and those of the jewelweed, dock, or any other land plant? If any leaves project out of the water are they different in form and tex-ture from those submerged? Sketch the leaf, showing its shape, its edges, and the way it joins the stem.

5. How far below the surface of the water does the pondweed usually lie? Does it ever rise up to the water's surface? When? Have you ever noticed the pond-weed in blossom? How does the blossom look on the water? Can you see the white pollen floating on the surface of the water? Look down into the water and see

the way the pondweed stands when in blossom.

6. Study the blossom. Note the stalk that bears it. Is the part that bears the flower enlarged and stiffer than the stem below? Do you think that this enlarged part of the stalk acts like the bob on a fish-line? Examine a flower cluster with a lens. How many flowers upon it? Study one flower carefully. Describe the four stigmas at the center. Describe the anthers arranged around them. Describe the flap which protects each anther. When the anthers open do they discharge the pollen toward or away from the stigmas?

7. What happens after the flowers are pollinated? Do they still float? What sort of seed capsule has the pondweed? D the seeds break away and float?

Again the wild cow-lily floats
Her golden-freighted, tented boats,
In thy cool caves of softened gloom,
O'ershadowed by the whispering reed,
And purple plumes of pickerel weed,
And meadow-sweet in tangled bloo

The startled minnows dart in flocks,
Beneath thy glimmering amber rocks,
If but a zephyr stirs the brake;
The silent swallow swoops, a flash
Of light, and leaves with dainty plash,
A ring of ripples in her wake.

— "BIRCH STREAM
ANNA BOYNTON AVER

Verne Mor

Cattails sending off their seed and balloons

THE CATTAIL

In June and early July, if the cattail be closely observed, it will be seen to have the upper half of the cat's tail much narrower and different in shape from the lower half — as if it were covered with a quite different fur. It seems to be clothed with a fine drooping fringe of olive ye low. With the aid of a lens, we can see th this fringe is a mass of crowded anther two or three of them being attached t the same stalk by a short filament. The anthers are packed full of pollen, which ed down upon the pistillate flowers be- by every breeze; and with every puff stronger wind, the pollen is showered r all neighboring flowers to the lee- rd. There is not much use in trying to d the pistillate flowers in the plush of cattail. They have no sepals or pet- and are so imbedded in thick plush t the search is hardly worth while for ure-study, unless a microscope is used. e ovary is rather long, the style slender, l the stigma reaches out to the cut- sh surface of the cattail. The pupils find what these flowers are by study- the fruit; in fact, the fruit does not fer very much from the flower, except t it is mature and is browner in color.

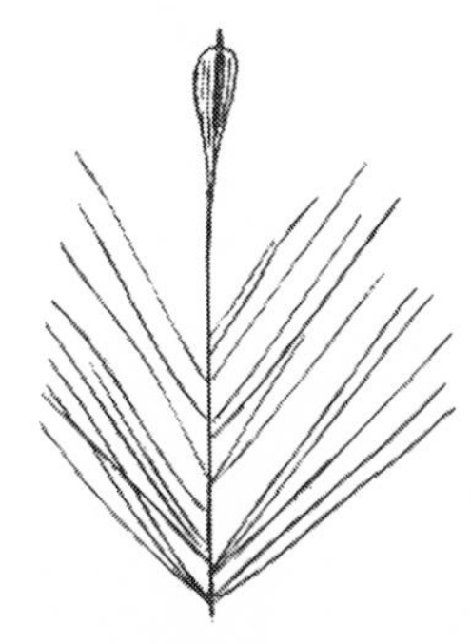

A cattail fruit with its balloon

Brooklyn Botanic Garden

Cattail in blossom. The staminate flowers are massed at the tip, and the pistillate flowers which form the "cattail" are massed lower on the stalk

It is an interesting process to take apart cattail plant; the lower, shorter leaves rround the base of the plant, giving it e and strength. All the leaves have the ne general shape, but vary in length. ch leaf consists of two parts: the free rtion, which is long and narrow and flat ward its tapering tip but is bent into a ugh as it nears the plant, and the lower rtion, which clasps the plant entirely partially, depending upon whether it an outer or inner leaf. This clasping of e stalk by the leaf adds to its strength. 'e almost feel as if these alternate leaves re consciously doing their best to pro- ct the slender flower stem. The free part the leaves is strengthened by lengthwise ins, and they form edges that never tear break. They are very flexible, and there- re yield to the wind rather than defy If we look at a leaf in cross section, we can see the two thick walls strengthened by the framework of stiff veins which divide the interior into long cells. If we cut the leaf lengthwise we can see that these long cells are supported by stiff, coarse partitions.

Where the leaf clasps the stem, it is very stiff and will break rather than bend. The texture of the leaf is soft and smooth, and its shade of green is attractive. The length of the leaves is often greater than that of the blossom stalk, and their graceful curves contrast pleasantly with its ramrod-like stiffness. It is no wonder that artists and decorators have used the cattail lavishly as a model. It is interesting to note that the only portion of the leaves injured by the wind is the extreme tip.

The cattail is adapted for living in swamps where the soil is wet but not under water all the time. When the land is drained, or when it is flooded for a considerable time, the cattails die out and disappear. They usually occur in marshy zones along lakes or streams; and such a zone is always sharply defined by dry land on one side and water on the other. The cattail roots are fine and fibrous and are especially fitted, like the roots of the tamarack, to thread the mud of marshy ground and thus gain a foothold. The cattails form one of the cohorts in the phalanx of encroaching plants, like the reeds and rushes, which surround and, by a slow march of years, finally conquer and dry up ponds. But in this they overdo the matter, since after a time the soil becomes too dry for them and they disappear, giving place to other plants which find there a congenial environment. The place where I studied the cattails as a child is now a garden of joe pye weed and wild sunflowers.

The Cattail

Leading Thought — The cattail is adapted to places where the soil is wet but not under water; its pollen is scattered by the wind, and its seeds are scattered by wind and water. Its leaves and stalks are not injured or broken by the wind.

Method — As this is primarily a geography lesson, it should be given in the field if possible; otherwise the pupils must explore for themselves to discover the facts. The plant itself can be brought into the schoolroom for study. When studying the seeds, it is well to be careful, or the schoolroom and the pupils will be clothed with the " down " for weeks.

Observations — 1. Where are the cattails found? Is the land on which they grow under water all the year? At any part of the year? Is it dry land all the y What happens to the cattails if the l on which they grow is flooded for a son? What happens to them if the l is drained?

2. How wide a strip do the catt cover, where you have found them? they near a pond or brook or stream? they grow out in the stream? Why do t not extend further inland? What is character of the soil on which they gr

3. What sort of root has the catt Why is this root especially adapted the soil where cattails grow? Describe rootstock.

4. *The cattail plant.* Are the leaves ranged opposite or alternate? Tear o few of the leaves and describe the dif ence between the lower and the upper of a leaf as follows: How do they diffe shape? Texture? Pliability? Color? Wid Does each leaf completely encircle stalk at its base? Of what use is this the plant? Of what use is it to have plant stiffer where the leaves clasp stalk? What would happen in a w storm if this top-heavy, slender seeds were bare and not supported by the leav

5. Take a single leaf, cut it across n where it joins the main stalk and also n its tip. Look at the cross section and how the leaf is veined. What do its l veins or ribs do for the leaf? Split leaf lengthwise and see what other s ports it has. Does the cattail leaf br or tear along its edges easily? Does wind injure any part of the leaf?

6. Study the cattail flowers the half of June. Note the part that will velop into the cat's tail. Describe the p above it. Can you see where the pol comes from? The pistillate flowers wh are in the plush of the cattail have sepals, petals, odor, or nectar. Do y think that their pollen is carried to th by the bees? How is it carried?

7. Examine the cattail in fall or w ter. What has happened to that part the stalk above the cattail where anthers grew? Study two or three of fruits, and see how they are provided traveling. What scatters them? Will

ail balloons float? Would the wind he water be more likely to carry the ail seeds to a place where they would v? Describe the difference between cattail balloon and the thistle balloon.

How crowded do the cattail plants v? How are they arranged to keep n shading each other? In how many s is the wind a friend of the cattails?

. How do the cattails help to build land and make narrower ponds and ams?

LESSON 131

A Type Lesson for a Composite Flower

eading Thought — Many plants have ir flowers set close together and thus ke a mass of color, like the geraniums he clovers. But there are other plants ere there are different kinds of flowers one head, those at the center doing a tain kind of work for the production eed, and those around the edges doing other kind of work. The sunflower, denrod, asters, daisies, coneflower, this- dandelion, burdock, everlasting, and ny other common flowers have their ssoms arranged in this way. Before any he wild flower members of this family studied, the lesson on the garden sun- ver should be given. (See Lesson 159.)

Method — These flowers may be stud- in the schoolroom with suggestions field observations. A lens is almost essary for the study of most of these wers.

Observations — 1. Can you see that at you call the flower consists of many wers set together like a beautiful mo- c? Those at the center are called disc wers; those around the edges ray wers.

2. Note that the flowers around the ges have differently shaped corollas than

Daisies and grasses

those at the center. How do they differ? Why could these be called the banner flowers? Why are they called the ray flowers? How many ray flowers are there in the flower-head you are studying? Cut off or pull out all the ray flowers and see how the flower-head looks. Why do you think the ray flowers hold out their banners? Has the ray flower any stigma or stamens?

3. Study the flowers at the center. Are they open, or are they unfolded buds? Can you make a sketch of how they are arranged? Are any of the florets open? What is the shape and the color of the corolla? Can you see the stamen-tubes pushing out from some? What color are the stamen-tubes? Can you see the two-parted stigmas in others? What color is the pollen? Do the florets at the center or at the outside of the disc open first? When they first open, do you see the stamen-tube or the stigma?

4. The flower-heads are protected before they open with overlapping bracts. As the flower-head opens, these bracts are pushed back beneath it. Describe the shape of these bracts. Are they set in regular, overlapping rows? Are they rough or smooth? Do they end bluntly, with a short point, with a long point, with a spine, or with a hook? How do the bracts act when the flower-head goes to sleep? Do they remain after the seeds are ripened?

5. Study the ripe fruits. How are they scattered? Do they have balloons? Is the balloon close to the seed? Is it fastened to all parts of it?

THE GOLDENROD

Once I was called upon to take some ildren into the field to study autumn wers. The day we studied goldenrod, I told them the following story on the way, and I found that they were pleased with the fancy and through it were led

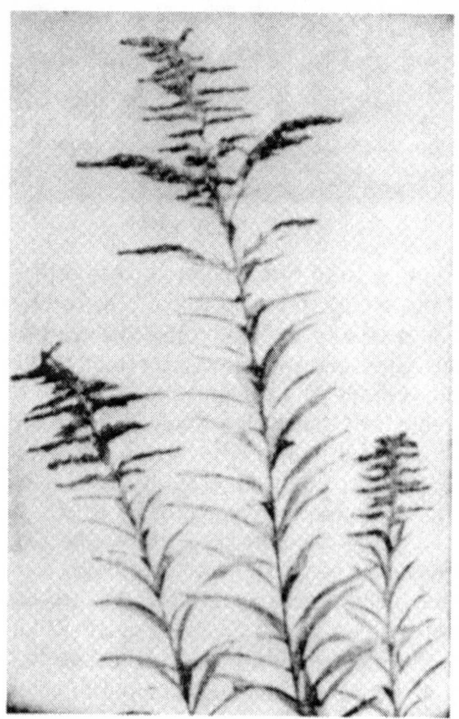

Goldenrod

to see the true purpose of the goldenrod's blossoming:

"There are flowers which live in villages and cities, but people who also live in villages and cities are so stupid that they hardly know a flower city when they see it. This morning we are going to visit a golden city where the people are all dressed in yellow, and where they live together in families; and the families all live on top of their little, green, shingled houses, which are set in even rows along the street. In each of these families, there are some flowers whose business it is to furnish nectar and pollen and to produce fruits which have fuzzy balloons; while there are other flowers in each family which wave yellow banners to all the insects that pass by and signal them with a code of their own, thus: 'Here, right this way is a flower family that needs a bee or a beetle or an insect of some sort to bring it pollen from abroad, so that it can ripen its seed; and it will give nectar and plenty of pollen in exchange.' Of course, if the flowers could walk around people, or fly like insects, they could fe and carry their own pollen, but as i they have to depend upon insect mes gers to do this for them. Let us see v of us will be the first to guess what name of this golden city is, and who be the first to find it."

The children were delighted with riddle and soon found the goldenrod c We examined each little house with ornate, green "shingles." These li houses, looking like cups, were arran on the street stem, right side up, in orderly manner and very close togetl and where each joined the stem, th was a little green bract for a doorst Living on these houses we found flower families, each consisting of a tubular disc flowers opening out like be and coming from their centers were long pollen-tubes or the yellow, t parted stigmas. The ray flowers had sh but brilliant banners; and they, as v as the disc flowers, had young fruits w pretty fringed pappus developing up them. The ray flowers were not set regularly around the edges as in the ast but the families were such close neighb that the banners reached from one ho to another. And all of the families on of the little, green streets were signali to insects, and one boy said, "They m be making a very loud yellow noise." V found that very many insects had sponded to this call — honeybees, bumt bees, mining and carpenter bees, bl black blister beetles with short wings a awkward bodies, beautiful golden-gre chalcid flies, soldier beetles, and ma

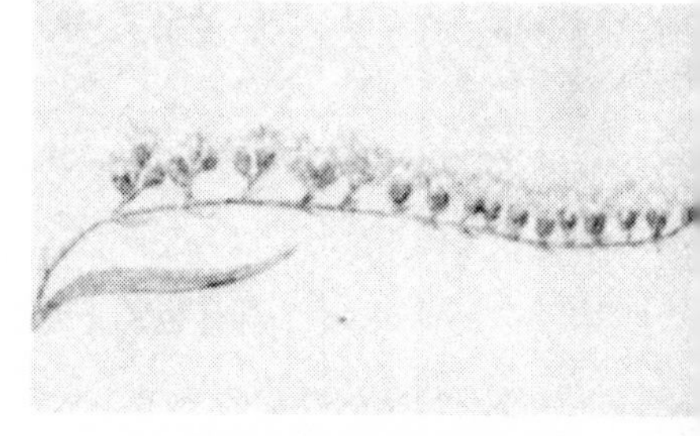

A street in goldenrod city

ers; and we found the spherical gall
the spindle-shaped gall in the stems,
the strange gall up near the top which
w among the leaves.

Jnless one is a trained botanist it is
sted energy to try to distinguish any
the well-marked species of goldenrod;
according to Gray, we have fifty-six
cies, the account of which makes
lve pages of most uninteresting readin the *Manual*. The goldenrod family
not in the least cliquish; the species
ve a habit of interbreeding, to the con-
ion of the systematic botanist.

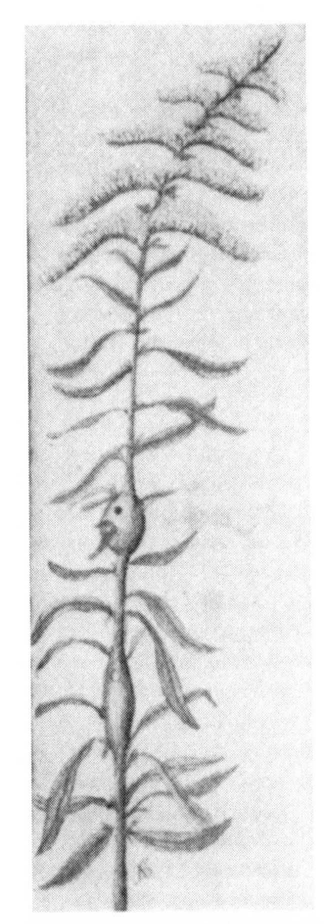

sect galls on goldenrod. The upper one is deserted, the lower inhabited

LESSON 132

The Goldenrod

Leading Thought — In the goldenrod the flower-heads are very small. They attract the attention of the insects because

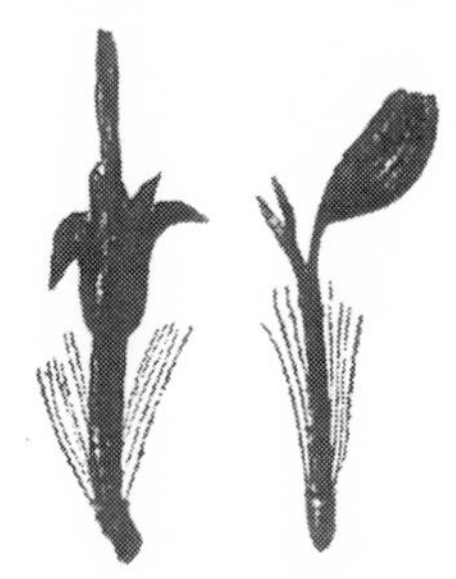

Disc flower and ray flower of goldenrod

they are set closely together along the stem, thus producing a mass of color.

Method — There should be a field excursion to get as many kinds of goldenrod as possible. Bring to the schoolroom any kind of goldenrod, and give further lessons on the flowers there. The following observations will bring out differences in well-marked species.

Observations — 1. Use Lesson 131 to study the flower. How many ray flowers in the head? How many disc flowers? Are the rays arranged as regularly around the edges as in the asters and daisies? How are the flower-heads set upon the stems? Which flower-heads open first — those at the base or at the tip of the stem? Do the upper stems of the plant blossom before those lower down?

2. Do the stems bearing flowers come from the axils of the leaves? What is the general shape of the flower branches? Do they come off evenly at each side, or more at one side? Are the flower branches long or short? Make a sketch of the general shape of the goldenrod you are studying.

3. Is the stem smooth, downy, or covered with bloom? What is its color? In cross-section, is it circular or angular?

4. What is the shape and form of the edges of the lower leaves? The upper ones? Are they set with or without petioles on the stem? Do they have a

Goldenrod in bloom

heart-shaped base? Are the leaves smooth or downy? Are they light or dark green?

5. *Field notes.* Where do you find the goldenrod growing? Do you find one k growing alone or several kinds grow together? Do you find any growing in woods? If so, how do they differ in sh from those in the field?

6. How many kinds of insects do find visiting goldenrod flowers? H many kinds of galls do you find on goldenrod stems and leaves?

7. Study the goldenrods in Novem Describe their fruits and how they scattered.

I am alone with nature,
With the soft September day;
The lifting hills above me,
With goldenrod are gay.
Across the fields of ether
Flit butterflies at play;
And cones of garnet sumac
Glow down the country way.

The autumn dandelion
Beside the roadway burns;
Above the lichened boulders
Quiver the plumèd ferns.
The cream-white silk of the milkweed
Floats from its sea-green pod;
From out the mossy rock-seams
Flashes the goldenrod.

— MARY CLEMMER AM

THE ASTERS

Let us believe that the scientist who gave to the asters their Latin name was inspired. Aster means *star*, and these, of all flowers, are most starlike; and in beautiful constellations they border our fields and woodsides. The aster combination of colors is often exquisite. The ray flowers of many asters are lavender, oar-shaped, and are set like the rays of a star around the yellow disc flowers; these latter send out long, yellow anther tubes, overflowing with yellow pollen, and add to the stellar appearance of the flower-head.

And asters by the brookside make asters
in the brook.

Thus sang H. H. of these beautiful masses of autumn flowers. But if H. H. had attempted to distinguish the speci she would have said rather that asters the brookside make more asters in t book; for Gray's *Manual* assures us th we have 77 species including widely d ferent forms, varying in size, color, a also as to the environment in which th will grow. They range from woodla species, which have a few whitish r flowers hanging shabbily about the yell disc and great, coarse leaves on lon gawky petioles along the zigzag stem, the beautiful and dignified New Engla aster, which brings the glorious purple a orange of its great flower-heads to decora our hills in September and October.

Luckily, there are a few species whi are fairly well marked, and still mo kily, it is not of any consequence ether we know the species or not, so as our enjoyment of the flowers them- 'es is concerned. The outline of this on will call the attention of the pupils the chief points of difference and like- s in the aster species, and they will s learn to discriminate in a general y. The asters, like the goldenrods, be- to bloom at the tip of the branches, flower-heads nearest the central stem oming last. All of the asters are very sitive, and the flower-heads usually se as soon as they are gathered. The flowers are pistillate, and therefore de-

Asters

velop akenes. The akene has attached to its rim a ring of pappus, and is ballooned to its final destination. In late autumn the

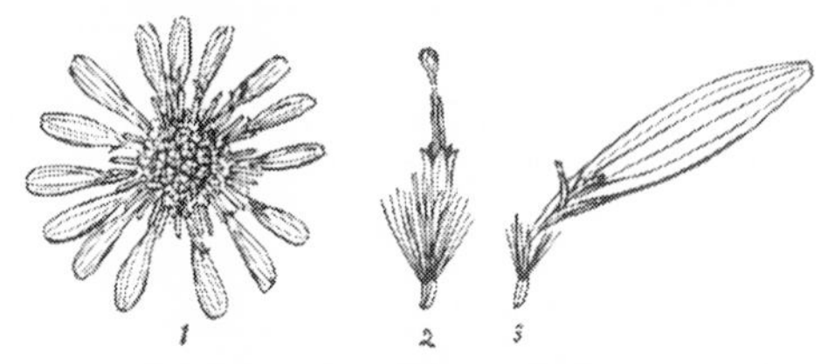

1, *An aster flower-head enlarged;* 2, *a disc flower;* 3, *a ray flower*

matured flower-heads are fuzzy, with seeds ready for invitations from any passing wind to fly whither it listeth.

LESSON 133

THE ASTERS

LEADING THOUGHT — There are very many different kinds of asters, and they all have their flowers arranged similarly to those of the sunflower.

METHOD — Have the pupils collect as many kinds of asters as possible, being careful to get the basal leaves and to take notes on where each kind was found — that is, whether in the woodlands, by the brooksides or in the open fields. This lesson should follow that on the sunflower.

OBSERVATIONS — 1. What was the character of the soil and surroundings where this aster grew? Were there large numbers of this kind growing together? Were the flowers wide open when you gathered them? How soon did they close?

2. How high did the plants stand when growing? Were there many flowers, or few, on each plant?

3. Study the lower and the upper leaves. Describe each as follows: the shape, the size, the edges, the way it was joined to the stem.

4. Is the stem many-branched or few? Do the branches bearing flowers extend in all directions? Are the stems hairy or smooth, and what is their color?

5. What is the diameter of the single flower-head? What is the color of the ray flowers? How many ray flowers are there? What is the shape of a single ray as com-

pared with that of a sunflower? What are the colors of the disc flowers? Of the pollen? Do the disc flowers change color after blossoming?

6. Look at the bracts below the flower-head. Are they all the same shape? What is their color? Do they have recurved tips or do they overlap closely? Are they sticky?

7. Take the aster flower-head apart and look at it with a lens. In a disc flower, note the young fruit, the pappus, the tu lar five-parted corolla, the anther-tu and the stigmas. In the ray flower, f the young fruit, the pappus, and stigma.

8. Watch the bees working on ast and find where they thrust their tong to reach the nectar.

9. Study an aster plant in late autun describe the akenes and how they scattered.

THE JEWELWEED OR TOUCH–ME–NOT

Jewels for the asking at the brookside, pendant jewels of pale gold or red-gold and of strange design! And the pale and the red are different in design, although of the same general pattern. The pale ones seem more simple and open, and we may study

Jewelweed

them first. If the flowers of the jewelweed have been likened to ladies' earrings, then the bud must be likened to the old-fashioned earbob; for it is done up in the neatest little triangular knob imaginable, with a little curly pigtail appendage at one side, and protected above by two cup-shaped sepals, their pale green seeming like enamel on the pale gold of bud. It is worth while to give a glance the stalk from which this jewel han it is so delicate and so gracefully curv and just above the twin sepals is a t green bract, elongate, and following t curve of the stem as if it were just a l artistic touch; and though the flow fall, this little bract remains.

It would take a Yankee, very good guessing, to make out the parts of t flower, so strange are they in form. V had best begin by looking at the bloss from the back. The two little, green sepals are lifted back like butterfly win and we may guess from their position th there are two more sepals, making four all. These latter are yellow; one is notch at the tip and is lifted above the flow the other is below and is made into wide-mouthed triangular sac, ending a quirl at the bottom, which, if we t it, we shall find is the nectary, very f of sweetness. Now, if we look the flow in the face, perhaps we can find t petals; there are two of them "holdi arms" around the mouth of the nec sac. And stiff arms they are too, two a side, for each petal is two-lobed, t front lobe being very short and the p terior lobe widening out below into long frill, very convenient for the bee cling to, if she has learned the tric when prospecting the nectar sac behi for its treasure. The way this treasure s swings backward from its point of attac ment above when the insect is probing st make the bee feel that the joys of are elusive. Meanwhile, what is the b projecting down above the entrance he nectar sac, as if it were a chandelier a vestibule? If we look at it with a , we can see that it is made up of chubby anthers, two in front, one at h side and one behind; their short, at little filaments are crooked, bringing anthers together like five closed fingers ding a fist full of pollen-dust, just ready sift it on the first one that chances to s below. Thus it is that the bumble- gets its back well dusted with the amy-white pollen and does a great busi- s for the jewelweed in transferring it. t after the pollen is shed, some day bumblebee pushes up too hard against anthers and they break loose, all in a nch, looking like a crook-legged table; l there in their stead, thus left bare l ready for pollen, is the long green til with its pointed stigma ready to e the pollen out of the fur of any mblebee that calls.

The red-gold jewelweed is quite dif- ent in shape from the pale species. The al sac is not nearly so flaring at the uth, and the nectar-spur is half as long the sac and curves and curls beneath flower. The shape of the nectar-spur ggests that an insect with a long, flexible king-tube that could curl around and be it to the bottom would be most cessful in securing the nectar; and ne butterflies do avail themselves of contents of this bronze pitcher. Mr. athews mentions the black swallowtail tterfly and I have seen the yellow road- e butterfly partaking of the nectar. But am sure that the flowers which I have d under observation are the special rtners of a small species of bumblebee, hich visits these flowers with avidity, lerity, and certainty, plunging into the ctar sac "like a shot," and out again d in again so rapidly that the eye can rdly follow. One day, one of them ac- mmodatingly alighted on a leaf near e, while she combed from her fur a eamy-white mass of pollen, which atched in color the fuzz on her back, heaping it on her leg baskets. It was comical to see her contortions to get the pollen off her back. The action of these bumblebees in these flowers is in marked contrast to that of the large bumblebees and the honeybees. One medium-sized species of bumblebee has learned the trick of embracing with the front legs the narrow, stiff portion of the petals which encircles the opening to the sac, thus holding the flower firm while thrusting the head into the sac. The huge species — black with very yellow plush — does not attempt to get the nectar in a legitimate manner, but systematically alights, back downward, below the sac of the flower, with head toward the curved spur, and cuts open the sac for the nectar. A nectar-robber of the most pronounced type! The honeybees, Italian hybrids, are the most awkward in their attempts to get nectar from these flowers; they attempt to alight on the expanded portion of the petals and almost invariably slide off between the two petals. They then circle around and finally succeed, as a rule, in gaining a foothold and securing the nectar. But the midget bumblebees in probing the orange jewelweed show a *savoir faire* that is convincing; they are so small that they are quite out of sight when in the nectar sacs.

The jewelweed flowers of the pale species and the pale flowers of the orange species — for this latter has sometimes pale yellow flowers — are not invariably marked with freckles in the nectar sac. But the most common forms are thus speckled. The orange jewelweed flower is a model for an artist in its strange, graceful form and its color combination of yellow spotted and marbled with red.

Gray's *Manual* states that in the jewelweeds are often flowers of two sorts: "The large ones which seldom ripen seeds, and very small ones which are fertilized early in the bud, their floral envelopes never expanding but forced off by the growing pod and carried upward on its apex." My jewelweed patch has not given me the pleasure of observing these two kinds of flowers; my plants blossom luxuriously and profusely, and a large proportion

Jewelweed, showing blossoms, and pods discharging seeds

of the flowers develop seed. The little, straight, elongated seed pods are striped prettily and become quite plump from the large seeds within them. Impatient? We should say so! This pod which looks so smug and straight-laced that we should never suspect it of being so touchy, at the slightest jar when it is ripe, splits lengthwise into five ribbon-like parts, all of which tear loose at the lower end and fly up in spirals around what was once the tip of the pod, but which now looks like a crazy little turbine wheel with five arms. And meanwhile, through this act the fat, wrinkled seeds have been flung, maybe several feet away from the parent plant, and perhaps to some congenial place for growth the following spring. This surprising method of throwing its seeds is the origin of the popular name touch-me-not, and the scientific name *Impatiens* by which these plants are known.

The jewelweed has other names — celandine and silver-leaf, and ladies'-eardrop. It is an annual with a slight and surface-spreading growth of roots, seeming scarcely strong enough to anchor the branching stems, did not the plants h the habit of growing in a community, e helping to support its neighbor. The st is round, hollow, and much swollen at joint; it is translucent, filled with m ture, and its outer covering is a smo silken skin, which may be readily strip off. Both species of jewelweed vary in color of their stems, some being gre others red, and some dark purple; and the differing colors may be found wit a few yards of each other.

The leaves are alternate, dark gr above and a lighter shade below, ovat form with scalloped edges, with mi and veins very prominent beneath depressed on the upper side; they smooth on both sides to the unaided but with a lens a film of fine, short h may be seen, particularly on the un side. When plunged beneath clear wa they immediately take on the appeara of burnished silver; when removed, drop remains on their surface.

The flowerstalks spring from the a of the leaves and are very slender threadlike, and the flowers nod and sw with every breeze. They grow in op drooping clusters, few blossoms open time, and with buds and seed caps present in various stages of growth.

The jewelweed is involuntarily m hospitable, and always houses m guests. Galls are formed on the lea and flowers; the hollow stems are habited by stalk-borers; leaf-miners between the upper and under surfa of the leaves, making curious arabes patterns and initials as if embroider milady's green gown.

LESSON 134

The Jewelweed or Touch-me-no

Leading Thought — The jewelw may be found by the brookside, swamps, or in any damp and well-sha area. It is provided with a remarka contrivance for scattering its seeds afield. It has no liking for open su places, unless they are very damp. Th are two kinds, often found growing her, though the spotted touch-me-not npatiens *biflora*) is said to be more lely distributed than its relative — the den or pale touch-me-not (*Impatiens lida*).

METHOD — The jewelweed should be died where they are growing; but if s is impracticable, a bouquet of both ıds (if possible), bearing buds, blosns, and seed capsules, and one or two nts with roots, may be brought to the oolroom.

In the fields the children may see how ll the plant is provided with means to tain itself in its chosen ground, and ıs be prompted to look with keener es at other common weeds.

OBSERVATIONS — 1. Do you think the velweed is an annual, sustaining life in seeds during winter, or do its roots vive?

2. Do the roots strike deeply into the l, or spread near the surface?

3. Study the stem; is it hard and woody juicy and translucent, rough or smooth, id or hollow?

4. Note the shape and position of the ves; do they grow opposite or altertely on the stalk? Are their edges ene, toothed, or scalloped? Do they vary color on upper and lower surfaces? Are ey smooth or in the least degree rough hairy? Plunge a plant under clear water a good light and observe the beautiful nsformation. Does the water cling to e leaves?

5. Where do the flowerstalks spring m the main stalk? Do the flowers grow ıgly or in clusters? Do the blossoms all en at nearly the same time or form a succession of bud, flower, and seed on the same stem?

6. Study the parts of the flower. Find the four sepals and describe the shape and position of each. Describe the nectar sac in the nectar horn. Can you find the two petals? Can you see that each petal has a lobe near where it joins the stem? Find the little knob hanging down above the entrance of the nectar sac; of what is it composed? Look at it with a lens, and tell how many stamens unite to make the knob. Where is the pollen and what is its color? What insect do you think could reach the nectar at the bottom of the spurred sac? Could any insect get at the nectar without rubbing its back against the flat surface of the pollen boxes? What remains after the stamens fall off? Describe how the bees do the work of pollination of the jewelweeds. Write or tell as a story your own observations on the actions of the different bees visiting these flowers.

7. Carefully observe a seed capsule without touching it; can you see the lines of separation between its sections? How many are there? What happens when the pod is touched? Are the loosened sections attached at the stalk, or at the apex of the pod? Hold a pod at arm's length when it is discharging its contents and measure the distance to which the seeds are thrown. Of what use is this habit of seed-throwing to the plant?

8. Describe the difference in shape and color between the pale yellow and the orange jewelweeds. Watch to see if the same insects visit both of these kinds of jewelweed.

WEEDS

W. C. Muen

Chicory enough to make anyone see blue

The worst weed in corn may be — corn.

— Professor I. P. Roberts

Nature is the great farmer. Continually she sows and reaps, making all the forces of the universe her tools and helpers; the sun's rays, wind, rain and snow, insects and birds, animals small and great, even to the humble burrowing worms of the earth — all work mightily for her, and a harvest of some kind is absolutely sure. But if man interferes and insists that the crops shall be only such as may benefit and enrich himself, she seems to yield a willing obedience, and under his control does immensely better work than when unguided. But Dame Nature is an "eye-servant." Let the master relax his vigilance for ever so short a time, and among the crops of his desire will come stealing in the hardy, aggressive, and to him useless plants t seem to be her favorites.

A weed is a plant growing where wish something else to grow, and a pl may, therefore, be a weed in some lo tions and not in others. Our grandmoth considered "butter-and-eggs" a pre posy, and planted it in their garde wherefrom it escaped, and it is now a b weed wherever it grows. A weed m crowd out our cultivated plants, by ste ing the moisture and nourishment in t soil which they should have; or it m shade them out by putting out bro leaves and shutting off their sunlig When harvested with a crop, weeds m be unpalatable to the stock which fe

n it; or in some cases, as with the
1 parsnip, the plant may be poisonous.
Each weed has its own way of winning
the struggle with our crops, and it be-
oves us to find that way as soon as pos-
e in order to circumvent it. This we
do only by a careful study of the pe-
iarities of the species. To do this we
st know the plant's life history;
ether it is an annual, surviving the
iter only in its seeds; or a biennial,
ing in fleshy root or in broad, green,
y rosette the food drawn from the soil
l air during the first season, to perfect
fruitage in the second year; or a peren-
l, surviving and springing up to spread
kind and pester the farmer year after
r, unless he can destroy it "root and
nch." Purslane is an example of the
t class, burdock or mullein of the sec-
1, and the field sorrel or Canada thistle
the third. According to their nature the
mer must use different means of ex-
mination; he must strive to hinder the
nuals and biennials from forming any
d whatever; and where perennials have
de themselves a pest, he must put in a
oed crop," requiring such constant
1 thorough tillage that the weed roots
l be deprived of all starchy food manu-
tured by green leaves and be starved
t. Especially, every one who plants a
den should know how the weeds look
en young, for seedlings of all kinds
delicate and easy to kill before their
ots are well established.

LESSON 135

Outline for the Study of a Weed

1. Why do we call a plant a weed? Is a
ed a weed wherever it grows? How
out "butter-and-eggs" when it grew in
randmother's garden? Why do we call
at a weed now? What did Grandmother
ll it?

2. Why must we study the habits of a weed before we know how to fight it?

We should ask of every weed in our garden or on our land the following questions, and let it answer them through our observations in order to know why the weed grows where it chooses, despite our efforts.

3. How did this weed plant itself where I find it growing? By what agency was its seed brought and dropped?

4. What kind of root has it? If it has a taproot like the mullein, what advantage does it derive from it? If it has a spreading shallow-growing root like the purslane, what advantage does it gain? If it has a creeping root with underground buds like the Canada thistle, how is it thereby helped?

5. Is the stem woody or fleshy? Is it erect or reclining or climbing? Does it gain any advantage through the character of its stem?

6. Note carefully the leaves. Are they eaten by grazing animals? Are they covered with prickles like the teasel or fuzz like the mullein, or are they bitter and acrid like the wild carrot?

7. Study the blossoms. How early does the weed bloom? How long does it remain in bloom? How are the flower buds and the ripening seeds protected?

8. Does it ripen many seeds? Are these ripened at the same time or are they ripened during a long period? Of what advantage is this? How are the seeds scattered, carried, and planted? Compute how many seeds one plant of this weed matures in one year.

9. What are some ways in which a weed may do harm to our cultivated crops?

That which ye sow ye reap. See yonder
fields!
The sesamum was sesamum, the corn
Was corn. The Silence and the Darkness
know!

— Edwin Arnold

POISON IVY [1]

Poison ivy may be found creeping over the ground, climbing as a vine, attached by aerial rootlets to trees, walls, or fences, or growing erect as a shrub. The alternate,

W. C. Muenscher

Poison ivy climbing on the trunk of a tree

compound leaves are made up of three leaflets; and this has given rise to the line often quoted:

Leaflets three, let it be.

During the fall and winter, the plant can usually be identified by the presence of clusters of small, white, berry-like fruit is thus easily distinguished from the w bines or Virginia creepers which l leaves made up of five or more leaflets in late summer have clusters of blue ber

[1]Because each year thousands of people throughout the country suffer from the effects of poison ivy, the editors have included here recommendations for treatment. Thanks are due to Jane Brown, R.N., of the Gannett Health Center, Cornell University, for providing this information.

Prevention of Ivy Poisoning Afte Contact With the Plant

Wash your hands, face, or affected p as soon as possible, preferably with st kitchen soap. Wash your clothes, too you are sensitive to the plants, washing n than five minutes after exposure may be late. If you are not particularly sensit you may be able to prevent some skin r tion by washing within half an hour of posure.

Curative Treatment When Poisoning Has Begun

To soothe the pain and prevent the gen spread of the inflammation, the best tr ment is something simple. Soaking in water usually gives relief. The applica of baking soda, one or two teaspoons t cup of water, is often effective in reliev the pain caused by the inflammation. C mine lotion helps relieve the itch whil dries the rash. Be careful about treating vere rashes with over-the-counter med tions that contain anesthetics or anthihi mines.

If the case of poisoning is a severe one is best to consult a physician before atten ing to use any remedy.

W. C. Muenscher, *Poison Ivy and Poison Sumac*, Cornell Extension Bulletin 191

Leaves of poison ivy and poison sumac and some harmless plants with which they are often confused

1, Poison ivy, *Rhus toxicodendron*. Leafstalk bearing three leaflets; buds visible. 2, Virginia creeper, *Parthenocissus quinquefolia*. Leafstalk bearing five leaflets. 3, Silky dogwood, *Cornus amomum*. Leafstalk with one blade, leaves opposite. 4, Fragrant sumac, *Rhus canadensis*. Leafstalk bearing three leaflets; buds hidden under base of leafstalk. 5, Poison sumac, *Rhus vernix*. Leaves alternate; leafstalk bearing several leaflets with smooth margins; buds visible. 6, Dwarf sumac, *Rhus copallina*. Margin of leaflets smooth or toothed, leaf axis winged. 7, Smooth sumac, *Rhus glabra*. Margin of leaflet toothed, buds hidden under base of leafstalks. 8, Staghorn sumac, *Rhus typhina*. Like 7, but leaves and twigs are hairy. 9, Mountain ash, *Sorbus americana*. Margin of leaflets toothed; buds visible. 10, Black ash, *Fraxinus nigra Marsh*. Leaves and buds opposite. 11, Elderberry, *Sambucus canadensis*. Leaves and buds opposite

THE COMMON OR FIELD BUTTERCUP

The buttercups, bright-eyed and bold,
Held up their chalices of gold
To catch the sunshine and the dew.

Leonard K. Beyer

Swamp buttercup

There are many widely varying species of buttercups. Some of them grow in woods, others in swamps, and some even in water. The blossoms of most buttercups are yellow but a few kinds have white blossoms. On some plants the blossoms are very showy and on others they are very inconspicuous. The common or field buttercup, which is widely distributed, is the one considered here.

Common buttercups and daisies are always associated in the minds of the children, because they grow in the same fields; yet the two are so widely different in structure that they may reveal to the child something of the marvelous differences between common flowers; for the buttercup is a single flower, while the single daisy is a large group of tiny flowers.

The buttercup sepals are five elongated cups, about one-half as long as the petals; they are pale yellow with brownish tips, but in the globular buds they are gre The petals are normally five in numb but often there are six or more; the pet are pale beneath, but on the inside th are a most brilliant yellow, and shine a: varnished. Probably it is due to t luminous color that one child is able determine whether another likes but or not, by noting when the flower is h beneath the chin, if it makes a yell reflection; it would be a sodden co plexion indeed that would not reflect y low under this provocation. Each pe is wedge-shaped, and its broad outer ec is curved so as to help make a cupl flower; if a fallen petal be examined tiny scale will be found at its base, as its point had been folded back a tri However, this is not a mere fold, bu little scale growing there; beneath it developed the nectar.

When the buttercup first opens, all the anthers are huddled in the cent so that it looks like a golden nest full

W. C. Muens

Marsh marigold or cowslip

den eggs. Later the filaments stretch lifting the anthers into a loose, nded tuft, almost concealing the ıch of pistils, which are packed close ether beneath every stigma. Later, the ments straighten back, throwing the hers in a fringy ring about the pale en pistils; and each pistil sends up a rt, yellowish stigma. The anthers open ay from the pistils and thus prevent -pollination to some degree; they also m to shed much of their pollen before stigmas are ready to receive it.

Sometimes petals and sepals fall simul-eously and sometimes first one or the er; but they always leave the green ıch of pistils with a ragged fringe of stamens clinging to them. Later the tils mature, making a globular head. ch fruit is a true akene; it is flattened

Buttercup

and has at its upper end a short, recurved hook which may serve to help it to catch a ride on passers-by. However, the akenes, containing the seeds, are largely scattered by the winds.

The buttercup grows in sunny situations, in fields and along roadsides, but

Buttercup flower slightly enlarged. Note the scale covering the nectar at the base of the falling petal

it cannot stand the shade of the woods. It is a pretty plant; its long stems are downy near the bottom, but smooth near the flower; the leaves show a variety of forms on the same plant; the lower ones have many (often seven) deeply cut divisions, while the upper ones may have three irregular lobes, the middle one being the longest. Beetles gather the nectar and pollen of buttercups, and therefore are its chief pollen carriers; but flies and small bees and other insects may also find their food in these brilliant colored cups.

LESSON 136

The Buttercup

Leading Thought — The buttercup may grow with the white daisies, in sunny places, but each buttercup is a single flower, while each daisy is a flower cluster.

Method — Buttercups brought by the pupils to school may serve for this lesson.

Observations — 1. Look at the back of a flower of the buttercup. What is there peculiar about the sepals? How do the sepals look on the buttercup bud? How do they look later?

2. Look into the flower. How many petals are there? Are there the same number of petals in all the flowers of the same plant? What is the shape of a petal? Compare its upper and lower sides. Take a

fallen petal, and look at its pointed base with a lens and note what is there.

3. How do the stamens look? Do you think you can count them? When the flower first opens how are the stamens arranged? How later? Do the anthers open towards or away from the pistils?

4. Note the bunch of pistils at the center of the flower. How do they look when the flower first opens? How later?

5. When the petals fall, what is left? Can you see now how each little pi will develop into an akene?

6. Describe the globular head of aker

7. Look at the buttercups' stems. A they as smooth near the base as near t flower? Compare the upper leaf with t lower leaf, and note the difference shape and size.

8. Where do the buttercups grow? we find them in the woods? What inse do you find visiting the flowers?

THE HEDGE BINDWEED

I once saw by the roadside a beautiful pyramid, covered completely with green leaves and beset with pink flowers. I stopped to examine this bit of landscape

Brooklyn Botanic Garden

Bindweed

gardening, and for the first time in my life I felt sorry for a burdock; for this burdock had met its match and more in standing up against a weakling plant which it must have scorned at first, had it been capable of this sensation. Its mighty leaves had withered, its flower-stalks showed no burs, for the bindweed had caught the burdock in its hund embraces and had squeezed the life o of it. Once in northern Florida our e were delighted with the most beauti garden we had ever seen, which resolv itself later into a field of corn, in whi every plant had been made a trellis for t bindweed; there it flaunted its pink a white flowers in the sunshine with a gra and charm that suggested nothing of t oppressor.

Sometimes the bindweed fails to fi support to lift it into the air. Then readily mats itself over the grass, maki a carpet of exquisite pattern. This vi has quite an efficient way of taking ho It lifts its growing tips into the air, swa ing them with every breeze; and the w each extreme tip is bent into a hook see just a matter of grace and beauty, as the two or three loose quirls below it; b when during its graceful swaying t hook catches to some object, it makes fa with amazing rapidity; later the you arrow-shaped leaves get an ear over t support, and in a very short time the vi makes its first loop, and the deed is dor It twines and winds in one way, followi the direction of the hands of the clock from the right, under, and from the le over the object to which it clings. If t support is firm, it makes only enou turns around it to hold itself firmly; but it catches to something as unstable as i own stems, the stems twist until they b come so hard-twisted that they form support in themselves.

t is rather difficult to perceive the al-
ıate arrangement of the leaves on the
dweed stem, since they twist under or
r so that they spread their whole grace-
length and breadth to the sun; to the
:less observer they seem only to grow
the upper or outer side of the vine.
e leaves are arrow-shaped, with two
g backward and outward projecting
nts, or "ears," which are often grace-
y lobed. Early in the year the leaves
glossy and perfect; but many insects
ble them, so that by September they
usually riddled with holes.

The flower bud is twisted as if the bind-
:d were so in the habit of twisting that
:arried the matter farther than neces-
/. Enveloping the base of the flower
l are two large sepal-like bracts, each
:led like a duck's breast down the cen-
; if these are pulled back, it is seen that
:y are not part of the flower, because
:y join the stem below it. There are
: pale green sepals of unequal sizes, so
.t some look like fragments of sepals.
e corolla is long, bell-shaped, opening
h five starlike lobes; each lobe has a
ckened white center; and while its
rgins are usually pink, they are some-
ıes a vivid pink-purple and sometimes
:irely white. Looking down into this
wer-bell, we find five little nectar wells;
l each two of these wells are separated
a stamen which is joined to the co-
la at its base and at its anther-end
:sses close about the style of the pistil.
hen the flower first opens it shows the
oon-shaped stigmas close together, push-
: up through the anther cluster; later,
: style elongates, bringing the stigmas
beyond the anthers. The pollen is
ıite, and through the lens looks like tiny
arls.

When we study the maturing seed
osule, we can understand the uneven
e of the sepals better; for after the co-
la with the attached stamens falls, the
oals close up around the pistil; the small-
sepal wraps it first, and the larger ones
order of size enfold the matured seed
d; and outside of all, the great, leafy
ıcts with their strong keels provide protection. The pod has two cells and two seeds in each cell. But it is not by seeds alone that the bindweed spreads; it is the running rootstock which, when the plant once gets a start, helps it to cover a large area. The bindweed is a relative of the morning-glory and it will prove an interesting study to compare the two in methods of twining, in the time of day of the opening of the flowers, the shape of the leaves, etc. So far as my own observations go, the bindweed flowers seem to remain open only during the middle of the day, but Müller says the flowers stay open on moonlight nights and may attract hawk moths. This is an interesting question for investigation, and it may be settled by a child old enough to make and record truthful observations.

There are several species of bindweed, but all agree in general habits. The field bindweed lacks the bracts at the base of the flower.

LESSON 137

The Hedge Bindweed

Leading Thought — There are some plants which have weak stems and cling to objects for support. The bindweed is one of these, and the way that it takes hold of objects and grows upon them is an interesting story.

Method — It is better to study this plant where it grows; but if this is not practical, the vine with its support should be brought into the schoolroom, the two being carefully kept in their natural relative positions. Several of the questions should be given to the pupils for their personal observation upon this vine in the field. It is an excellent study for pencil or water-color drawing.

Observations — 1. How does the bindweed get support, so that its leaves and its flowers may spread out in the sunshine? Why does its own stem not support it? What would happen to a plant with such a weak stem if it did not twine upon other objects?

2. How does it climb upon other plants? Does its stem always wind or twist

in the same direction? How does it first catch hold of the other plant? If the supporting object is firm, does it wind as often for a given space as when it has a frail support? Can you see the reason for this?

3. Look at the leaves. Sketch one, to be sure that you see its beautiful form and veins. Note if the leaves are arranged alternately on the stem, and then observe how and why they seem to come from one side of the stem. Why do they do this?

4. What is there peculiar about the flower bud? Look at its stalk carefully and describe it. Cut it across and look at the end with a lens and describe it. Turn back two sepal-like bracts at the base of the flower or bud. Are they a part of the flower, or are they below it? Find the true sepals. How many are there? Are they all the same size?

5. Examine the flower in blossom. What is its shape? Describe its colors. Look down into it. How many stamens are there, and how are they set in flower? How does the pistil look when flower first opens? Later? Can you the color of the pollen? Can you where the nectar is borne? How m nectar wells are there?

6. What insects do you find visit bindweed flowers? Do the flowers rem open at night or on dark days?

7. Study the seed capsule. How is it tected on the outside? What next folds it? Cut a seed capsule across with its coverings, and see how it is protect How many seeds are there in the capsu

8. Has the bindweed other methods spreading than by seeds? Look at the ro and tell what you observe about them

9. Make a study of the plant on wh the bindweed is climbing, and tell w has happened to it.

10. Compare the bindweed with morning-glory, and notice the differen and resemblances.

THE DODDER

Brooklyn Botanic Garden

Dodder or love vine

The dodder, which is also known by names as diverse as "strangle-weed" and "love vine," is a good example of the changes that take place in a plant whi has become a parasite. When a pla ceases to be self-supporting, when its g its living from the food made by oth plants for their own sustenance, it lo its own power of food-making. The do der has no leaves of its own, for it do not manufacture or digest its own foo Its dull yellow stems reach out in lo tendrils swayed by every breeze until th come in contact with some other pla The tendrils wind about the victim pla always under from the right side and ov from the left. They get their hold means of suckers which develop on t coiled stem; so firmly are these suck attached that the yellowish stem w break before they can be torn from th hold. The devilfish uses the suckers its tentacles only to hold fast its prey; b the suckers of the dodder penetrate t bark of the victim to the sap channe where they suck the matured sap which necessary to the life of the host plant.

The development of the dodder fro

s point is an example of the further his-y of a parasite. No sooner has it tapped ucculent victim than its now useless ›t and lower portions wither away and ve the dodder wholly deprived of con-t with the earth.

Ihe stems of the dodder are plentifully dded with small, dull-white flowers htly bunched. The calyx has five lobes; : corolla is globular, with five little lobes ›und its margin and a stamen set in :h notch. A few of the species have a ır-lobed calyx and corolla; but however ıny the lobes, the flowers are shiftless ›king and are yellowish or greenish ıite; despite its shiftless appearance, wever, each flower usually matures four rfectly good, plump seeds. The seed ;sels are globular capsules and develop ›idly while the blossoming continues

Brooklyn Botanic Garden

Dodder in flower on stem of goldenrod

Brooklyn Botanic Garden

Dodder. This plant has severed all connection with the soil and secures its sustenance from a host plant, the jewelweed

unabated. They drop their cargo of seeds, which perpetuate the existence of this parasitic plant.

Kinds of dodder which attack clover and other farm crops get their seeds harvested with the rest; and the farmer who does not know how to test his clover seed for impurities, sows with it the seeds of its enemy.

There are nine species of dodder more or less common in America. Some of the species, among which is the flax dodder, live only upon certain other species of plants, while others take almost any plant that comes within reach. Where it flourishes, it grows so abundantly that it makes large yellow patches in fields, completely choking out the leaves of its victims.

LESSON 138

The Dodder

Leading Thought — There are some plants which not only depend upon other plants to hold them up, but even steal

Leonard K. Beyer

White or oxeye daisy

their living by drawing the vital sap from the host plants.

METHOD — Bring in dodder with the host plant for the pupils to study in the schoolroom, and ask them to observe afterwards the deadly work of this parasite in the field.

OBSERVATIONS — 1. What is the color of the stem? In which direction does it wind?

2. How is the stem fastened to the host plant? Tear off these suckers and examine with a lens the place where they were attached, and note if they enter into the stem of the host plant.

3. How does the dodder get hold of its host? Has the dodder any leaves of own? How can it get along and gı without leaves?

4. How do the flowers look throug lens? Are there many flowers? Can see the petal lobes and the stamens?

5. How many seeds does each flo develop? How do the seeds look? In w way are they a danger to our agricultu

I should also avoid the informat method. It does a child little good mer to tell him matters of fact. The facts not central to him and he must ret them by a process of sheer memory; in order that the teacher may kn whether he remembers, the recitation employed, — re-cite, to tell over again. T educational processes of my younger d were mostly of this order, — the book the teacher told, I re-told, but the res were always modified by an unpredicta coefficient of evaporation. Good teach now question the child to discover w he has found out or what he feels, or suggest what further steps may be tak and not to mark him on what he reme bers. In other words, the present-day pr ess is to set the pupil independently work, whether he is young or old, and information-leaflet or lesson does not this. Of course, it is necessary to give so information, but chiefly for the purp of putting the pupil in the way of acq ing for himself and to answer his natu inquiries; but information-giving ab nature subjects is not nature-study.

"THE OUTLOOK TO NATUR

L. H. BAI

THE WHITE DAISY

Every child loves this flower, and yet it is not well understood. It is always at hand for study from June until the frosts have laid waste the fields. However much enjoyment we get from the study of this beautiful flower-head, we should study the plant as a weed also, for it is indeed a pest to those farmers who do not practice a rotation of crops. Its root is long and tenacious of the soil, and it ripens ma seeds which mingle with the grass se and thus the farmer sows it to his o undoing. The bracts of the involucre, the shingles of the daisy-house, are ratl long, and have parchment-like margi They overlap in two or three rows. the daisy flower-head, the ray flowers white; there may be twenty or thirty

Verne Morton

A daisy meadow

ese, making a beautiful frame for the olden-yellow disc flowers. The ray is ther broad, veined, and toothed at the p. The ray flower has a pistil which ows its two-parted stigma at the base of e banner. The disc flowers are brilliant llow, tubular, rather short, with the five oints of the corolla curling back. The ther-tubes and the pollen are yellow; are the stigmas. The arrangement of e buds at the center is exceedingly pretty. The flowers develop no pappus, and therefore the akenes have no balloons. They seem in the present day to depend upon the ignorance and helplessness of man to scatter their akenes far and wide with the grass and clover seed which he sows for his own crops. It was thus that the daisy came to America, and in this manner it still continues to flaunt its banners in our meadows and pastures. The white daisy is not a daisy, but a chrysanthemum. It has never been called by this name popularly, but has at least twenty other common names, among them the oxeye daisy, moonpenny, and herb Margaret.

LESSON 139

The White Daisy

Leading Thought — The white daisy is not a single flower but is made up of many little flowers and should be studied by the outline given in Lesson 131.

Yellow daisy or black-eyed Susan

THE YELLOW DAISY OR BLACK-EYED SUSAN

These beautiful, showy flowers have ch contrasts in their color scheme. The n to twenty ray flowers wave rich, orange banners around the cone of purple-brown disc flowers. The rays are notched and bent downward at their tips; each ray flower has a pistil, and develops a seed. The disc flowers are arranged in a conical, button-like center; the corollas are pink-purple at the base of the tube, but their five recurved, pointed lobes are purple-brown. The anther-tube is purple-brown and the stigmas show the same color; but the pollen is brilliant orange, and adds much to the beauty of the rich, dark florets when it is pushed from the anther-tubes. There is no pappus developed, and therefore the seeds are not carried far by the wind.

The stem is strong and erect; the bracts of the involucre are long, narrow, and hairy, the lower ones being longer and wider than those above; they all spr out flat, or recurve below the open flow head. In blossoming, first the ray flow spread wide their banners; then the flor around the base of the cone open a push out their yellow pollen through brown tubes; then day by day the bl soming circle climbs toward the apex a beautiful way of blossoming upward

LESSON 140

The Black-Eyed Susan

Leading Thought—This flower sho be studied by the outline given in L son 131.

THE THISTLE

O. L. Foster

Bull or common thistle

On looking at the thistle from its own standpoint, we must acknowledge it to be a beautiful and wonderful plant. It is like a knight of old encased in armor and with lance set, ready for the fray. The most impressive species is the great pasture or bull thistle (*Cirsium pumilum*). It ha blossom-head three inches across. This not so common as the lance-leaved thist which ornaments roadsides and fe corners, where it may remain undisturb for the necessary second year of grow before it can mature its seeds. The m pernicious species, from the farme standpoint, is the Canada thistle. Its ro stocks are perennial, and they inva garden, grain-field, and meadow. Th creep for yards in all directions, just de enough to be sure of moisture, and se up new plants here and there, especia if the main stalk is cut off. Rootstoc severed by the plow send up sho from both of the broken parts. Not with the common thistle, which has single main root, with many fibro and clustered branches but with no si shoots.

The stem of the lance-leaved thistle strong and woody, and is closely hugg by pricky leaf stems, except for a f inches above the root. The leaves placed alternately on the stem; they deep green, covered above with rough a bristling hairs, and when young are co ered on the under side with soft, gr wool which falls away later. The spi grow on the edges of the leaves, which ply lobed and are also somewhat wavy ruffled, thus causing the savage spears meet the enemy in any direction. The is are without spines. Small buds or aches may be found at the axils of the es; and if a plant is beheaded, those lary buds nearest the top of the stem grow vigorously.

The thistle flowers are purple in color very fragrant; they grow in single ls at the summit of the stem, and from axils of the upper leaves. The top-st heads open first. Of the individual vers in the head, those of the outer s first mature and their pistils protrude; pollen grains are white. In each flower, corolla is tube-shaped and purple, part-into five fringelike lobes at the top, fading to white at its nectar-filled e.

The stamens have dark purple anthers, ted in a tube in which their pollen is harged. The pistil, ripening later, ves out the pollen with its stigma, ch at first is blunt at the end, its two-ted lips so tightly held together that a grain of its own flower's pollen can taken. But when thrust far out beyond anther-tube, the two-parted stigma ns to receive the pollen which is ught by the many winged visitors; for ll flowers, the thistles with their abun-

A floret from a thistle flower-head

t nectar are the favorites of insects. terflies of many species, moths, bee-, and bees — especially the bumble-bees — are the happy guests of the thistle blooms.

The thistles believe in large families; a single head of the lance-leaved thistle

W. C. Baker

Canada thistle

has been known to have 116 seeds. Each seed is covered by a tight hard shell and the whole fruit is called an akene. Very beautiful and wonderful is the pappus of the thistle; it is really the calyx of the flower, its tube being a narrow collar, and the lobes being split into the silken floss. At the larger end of the akene is a circular depression with a tiny hub at its center; into this ring, and around the knob, is fitted the collar which attaches the down to the akene. Hold the balloon between the eye and the light, and it is easy to see that the down is made of many-branched plumes which interlace and make it more buoyant. When first taken from its crowded position on the flower-head, the pappus surrounds the corolla in a straight, close tube; but if placed for just a few moments in the sun, the threads spread, the filmy branchlets open out, and a fairy parachute is formed, with the seed hanging beneath; if no breath of air touches it while spreading, it will sometimes form a perfect funnel; when blown upon, some of the silken threads lose their places on the rim and rise to the center. When driven before the breeze, this balloon will float for a long distance. When it falls, it lets go of the akene as the wind moves it along the rough surface of the ground, and when it is thus unburdened the down fluffs out in every direction, making a perfect globe.

For the first season after the seed has rooted, the thistle develops only a rosette, meanwhile putting down roots and becoming permanently established. The next season, the flowers and akenes are developed, and then the plant dies. Would that this fact were true of the Canada thistle; but that, unfortunately, is perennial, and its persistent rootstocks can only be starved out by keeping the stalks cut to the ground for the entire season. This thistle trusts to its extensively creeping rootstocks more than to its seeds for retaining its foothold and for spreading. While it develops many akenes, a large number of its seeds are infertile and will not grow.

LESSON 141

The Common or Lance-leaved Thistle

Leading Thought — The thistle is covered with sharp spines, and these serve to protect it from grazing animals. It beautiful purple flowers, arranged in he similar to those of the sunflower.

Method — A thistle plant brought i the schoolroom — root and all — placed in water will serve well for lesson. The pupils should first be q tioned about where the thistles are fou Any thistle will do for the lesson.

Observations — 1. Where do you the thistles growing? Do you find m than one species growing thickly togeth Do you find any of the common this growing in soil which has been cultiva this season?

2. Describe the stalk; is it smooth? I weak, or strong and woody? What sor root has it?

3. Do the leaves grow alternately opposite? Are they smooth or downy one or both sides? Do the spines gr around the margins, or on the leaves veins? Are the leaf edges flat, or wavy ruffled?

4. How does this affect the direct in which the spines point? Are the lea entire or deeply lobed? Have they petio or are they attached directly to the sta

5. Note if any buds or small branc are in the axils of the lower leaves. W effect does cutting the main stalk se to have on each side shoot?

6. Do the flower-heads of the this grow singly or in clusters? Do they co from the summit of the stalk, or do th branch from its sides? Which blosso heads open first — the topmost or th lowest on the stalk? Are the flow fragrant? What insects do you m often see visiting thistle blosson Study the thistle flower according Lesson 131.

7. Carefully study a thistle ballo How is the floss attached to the aken Is it attached to the smaller or the lar end? Hold the thistle balloon betwe your eye and the light. Does the do consist of single separate hairs, or ha they many fine branches? How is the do arranged when all the flowers are pack together in the thistle-head? Take akene from among its closely packed f

·s in the thistle-head, and put it in the or in a warm, dry place where it can- blow away. How long does it take for balloon to open out? What is its pe? Is there any down at the center of balloon or is it arranged in a funnel- ped ring? Can you find a perfectly bular thistle balloon with the akenes attached to it? How far do you think thistle balloons might travel?

8. If a thistle akene finds a place for planting during the autumn, how does the young plant look the next season? Describe the thistle rosette. What growth does it make the second summer? What happens to it then?

9. Why can you not cultivate out the Canada thistle as you can the other species? Why is it less dependent on its akenes for propagation than the others?

THE BURDOCK

Psychologists say that all young things selfish, and the young burdock is a ning example of this principle. Its first ves are broad and long, with long peti- s by means of which they sprawl out m the growing stem in every direction, vering up and choking out all the lesser nts near them. In fact, the burdock re- ins selfish in this respect always, for great basal leaves prevent other plants m getting much sunlight when they ow near its own roots. One wonders at t how a plant with such large leaves can oid shading itself; we must study care- ly the arrangement of its leaves in order understand this. The long basal leaves stretched out flat; the next higher, newhat smaller ones are lifted at an gle so as not to stand in their light. is arrangement characterizes in gen- l the leaves of the plant, for each higher f is smaller and has a shorter petiole, ich is lifted at a narrower angle from e stalk; and all the leaves are so adjusted to form a pyramid, allowing the sun- ht to sift down to each part. While me of the uppermost leaves may be rcely more than an inch long, the lower es are very large. They are pointed at e tip and wide at the base; where the f joins the petiole it is irregular, bor- red for a short distance on each side th a vein, and then finished with a lounce," which is so full that it even aches around the main stem — another culiarity of structure which shuts off nlight from plants below. On the lower side, the leaf is whitish and feltlike to the touch; above, it is a raw green, often somewhat smooth and shiny. The leaf is in quality poor, coarse, and flimsy, and it hangs — a web of shoddy — on its strong supporting ribs; its edges are slightly notched and much ruffled. The petiole and stems are felty in texture; the petiole is grooved, and expands at its base to grasp the stems on both sides with a certain vicious pertinacity which characterizes the whole plant.

D. L. Foster

Common burdock, showing blossom and buds

The flower-heads come off at the axils

of the upper leaves, and are often so crowded that the leaf is almost lost to sight. It is amazing to behold the number of flower-heads which develop on one thrifty plant. The main stem and the pyramid of lower branching stems are often crowded with the green balls beset with bracts which are hooked, spiny, and which hold safe the flowers. This composite flower-head is a fortress bristling with spears which are not changed to peaceful pruning-hooks, although they are hooked at the sharp end, every hook turning toward the flowers at the center; the lower bracts are shorter and stand out at right angles, while the others come off at lesser angles, graded so as to form a globular

A burdock floret with hooked bract

involucre — a veritable blockhouse. The flower might be a tidbit for the grazing animal; but if so, he has never discovered it, for these hooks may have kept him from ever enjoying a taste. The bracts, not only by hooks at the tip, but also by spreading out at the bases, make a thickly battened covering for the flower-cluster.

But if we tear open one of these little heads, we are well repaid in seeing the quite pretty florets. The corollas are long, slender, pink tubes, with five, pointed lobes. The anther-tubes are purple, the pistils and the stigmas white; the stigmas are broad and feathery when they are dusting out the pollen from the anther-tubes, but later they change to very delicate pairs of curly Y's. The young akene is shining white, and the pappus forms a short, white fluff at the upper margin; but this is simply a family trait, for the burdock akenes never need to be ballooned to their d tination; they have a surer method travel. When in full bloom, the burd flower-heads are very pretty and the sl ful child weaver makes them into bea ful baskets. When I was a small gir made whole sets of furniture from th flowers; and then, becoming more am tious, wove some into a coronet whic wore proudly for a few short hours, o to discover later, from my own experien that great truth which Shakespeare voi — "uneasy lies the head that wears crown."

In winter, the tough, gray stalks the burdock still stand; although they n partially break. They insert the hooks their seed storehouses into the clothing covering of the passer-by; and when gets a hold, mayhap a dozen others hold hands and follow. If they catch tail of horse or cow, then indeed they m feel their destiny fulfilled; for the anim switching about with its uneasy appe age, threshes out the seeds, and unhe ingly plants them by trampling them i the ground. Probably some of the li stock of our Pilgrim Fathers came America thus burdened; for the burd is a European weed, although now it flo ishes too successfully in America. T leaves of the burdock are bitter, and avoided by grazing animals. Fortunat for us, certain flies and other insects parently like their bitter taste, and eggs upon them, which hatch into lar that live all their lives between the upp and lower surfaces of the leaf. Often leaves are entirely destroyed by the nute larvæ of a fly, which live togetl cozily between these leaf blankets, givi the leaves the appearance of being flicted with large blisters. A small mo caterpillar finds both food and shelter the ripe fruiting heads.

The burdocks have long vigorous t roots, and it is therefore difficult to era cate them without much labor. But p sistently cutting off the plant at the ro will, if the cut be deep, finally discoura this determined weed.

LESSON 142

The Burdock

ꞁEADING THOUGHT — The burdock ives because its great leaves shade down nts in its vicinity, and also because it ; taproots. It scatters its seed by hooking seed-heads fast to the passer-by.

METHOD — Study a healthy burdock nt in the field, to show how it shades wn other plants and does not shade elf. The flowers may be brought into e schoolroom for detailed study.

OBSERVATIONS — 1. Note a young plant. ow much space do its leaves cover? Is ything growing beneath them? How e its leaves arranged to cover so much ace? Of what advantage is this to the nt?

2. Study the full-grown plant. How are e lower leaves arranged? At what angles the stalks do the petioles lie? Are the per leaves as large as the lower ones? o they stand at different angles to the lk?

3. Study the arrangement of leaves on ɔurdock plant, to discover how it manes to shade down other plants with its leaves and yet does not let its own upper leaves shade those below.

4. Study a lower and an upper leaf. What is the general shape? What peculiarity where it joins the petiole? What is the texture of the leaf above and below? The color? Describe the petiole and how it joins the stem.

5. Where do the flowers appear on the stem? Are there many flowers developed? Count all the flower-heads on a thrifty burdock.

6. The burdock has its flowers gathered into heads, like the sunflower and thistle. Describe the burdock flower-head according to Lesson 131.

7. What insects visit the burdock flowers? Can you make baskets from the flower-heads?

8. Study the burdock again in winter, and see what has happened to it. Describe the fruit. How are the fruits carried far away from the parent plant? How many akenes in a single " house "? How do they escape?

9. Write the biography of a burdock plant which came to America as a fruit, attached to the tail of a Shetland pony.

PRICKLY LETTUCE, A COMPASS PLANT

The more we know of plants, the more e admire their ways of attaining success a world where a species attains success ly after a long struggle. The success of ickly lettuce depends much upon its beg able to live in dry situations and withnd the long droughts of late summer. he pale green stems grow up slim and ll, bearing leaves arranged alternately d from all sides, since between two, e of which is exactly above the other, o other leaves are borne. Thus, if the aves stood out naturally, the shape of e whole plant would be a somewhat unt pyramid. But during the hot, dry eather, the leaves do not stand out raight from the stem; instead, they twist out so that they are practically all in e plane, and usually point north and south, although this is not invariably the case. The way this twisting is accomplished is what interests us in this plant. The long spatulate leaf has a thick, fleshy midrib, and at the base are developed two pointed lobes which clasp the stalk. The leaf is soft and leathery and always seems succulent, because it retains its moisture; it has a ruffled edge near its base, which gives it room for turning without tearing its margin. Each leaf tips over sidewise toward the stem. The ruffled margin of the upper edge is pulled out straight when the leaf stands in this position, while the lower margin is more ruffled than ever. Thus, it stands, turning edgewise to the sun, retaining its moisture and thriving when cultivated plants are dry and dying.

Brooklyn Botanic Garden

Wild lettuce

It also has another "anchor to the windward." A plant so full of juice would prove attractive food for cattle when pastures are dry. The leaves of prickly lettuce perhaps escape because each has a row of very sharp spines on the lower side of the midrib. If we watch a grazing animal, such as a cow, reach out her tongue to pull the herbage into her mouth, we see that these spines repel her. The teasel has the same means of warning off meddlesome tongues. The prickly lettuce also has spines on its stem, and the leaves are toothed with spines at their points.

LESSON 143

Prickly Lettuce

Leading Thought — The sunshine s the machinery in the leaf-factories goi and incidentally increases evaporat from the plant, as it does from any m surface. The wild lettuce plant has edges of its leaves turned to the sun; t they stand in one plane and have surface exposed directly to the sun. T leaves are usually directed north a south. The lettuce also has spines wh protect it from grazing animals.

Method — The lettuce should be st ied in the field, and is a good subject a lesson in late summer or Septemb This lesson should supplement the one transpiration. The young plants show t arrangement of the leaves best. The flo ers may be studied by the outline giv in Lesson 131.

Observations — 1. Where does t prickly lettuce grow? What sort of st has it? How are the leaves arranged the stem?

2. If the leaves stood straight out fr

Cyrus Cros

A common compass plant; note the prick

stem, what would be the shape of plant? How do the leaves stand? Is ir upper surface exposed to the rays of sun? Which portion of the leaf is ned toward the sun?

. If the leaves turn sidewise and stand one plane, do they stand north and th or east and west? How does the ewise position of the leaf protect the nt during drought? Why does any nt wither during drought? If the leaves the lettuce should extend east and west tead of north and south, would they get more sun? (See Lesson on the Sun, page 833.)

4. What is the shape of the lettuce leaf? How does it clasp the stalk? Does the leaf turn toward the stem or away from it?

5. How are the leaves protected against grazing cattle? How does the cow use her tongue to help bring herbage to her mouth? How are the prickly spines arranged on the lettuce leaf, and in what way may these spines protect the lettuce from grazing animals? Sketch a leaf showing its shape, its venation, and its spines.

Hugh Spencer

Common dandelion

THE DANDELION

This is the most persistent and indomible of weeds, yet I think the world ould be very lonesome without its lden flower-heads and fluffy seedheres. Professor Bailey once said that ndelions in his lawn were a great troue to him until he learned to love them, d then the sight of them gave him enest pleasure. And Lowell says of this lear common flower" —

'Tis the Spring's largess, which she scatters now
To rich and poor alike, with lavish hand,
Though most hearts never understand
To take it at God's value, but pass by
The offered wealth with unrewarded eye.

It is very difficult for us, when we watch the behavior of the dandelions, not to attribute to them thinking power, they have so many ways of getting ahead of us. I always look at a dandelion and talk to it as if it were a real person. One spring when all the vegetables in my garden were callow weaklings, I found there, in their midst, a dandelion rosette with ten great leaves spreading out and completely shading a circle ten inches in diameter; I said, "Look here, Madam, this is my garden!" and I pulled up the squatter. But I could not help paying admiring tribute to the taproot, which lacked only an inch of being a foot in length. It was smooth, whitish, and fleshy, and, when cut, bled a milky juice; it was as strong from the end-pull as a whipcord; it also had a bunch of rather fine rootlets about an inch below the surface of the soil and an occasional rootlet farther down; and then I said, "Madam, I beg your pardon; I think this was your garden and not mine."

Dandelion leaves afford an excellent study in variation of form. The edges of the leaf are notched in a peculiar way, so that the lobes were, by some one, supposed to look like lions' teeth in profile; thus the plant was called in France "dents-de-lion" (teeth of the lion), and we have made from this the name dandelion. The leaves are bitter, and grazing animals do not like to eat them.

The hollow stalk of the blossom-head from time immemorial has been a joy to children. It may be made into a trombone, which will give to the enterprising teacher an opportunity for a lesson in the physics of sound, since by varying its length the pitch is varied. The dandelion-curls, which the little girls enjoy making, offer another lesson in physics — that of surface tension, too difficult for little girls to understand. If the plant is in a lawn, the stem is short, indeed so short that the lawn mower cannot cut off the flower-head. In this situation it will blossom and seed within two inches of the ground; but if the plant is in a meadow or in other high grass, the stalk lifts up sometimes two feet or more. We once found two such stems each measuring over thirty inches in height.

Before a dandelion head opens, the stem, unless very short, is likely to b
down, but the night before it is to blo
it straightens up; after the blossoms h
matured it may again bend over,
straightens up when the seeds are to
cast off.

It often requires an hour for a dandel
head to open in the morning and it ra
stays open longer than five or six ho
it may require another hour to close. U
ally not more than half the flowers of
head open the first day, and it may req
several days for them all to blossom. A
they have all bloomed and retired i
their green house and put up the shutt
it may take them from one to two we
to perfect their akenes.

In the life of the flower-head the
volucre, or the house in which the flo
family lives, plays an important part. T
involucral bracts, in the row set next
the flowers, are sufficiently long to co
the unopened flowers; the bracts near
stem are shorter and curl back, mak
a frill. In the freshly opened flower-he
the buds at the middle all curve sligh
toward the center, each bud showin
blunt, five-lobed tip which looks like
tips of five fingers held tightly togeth
The flowers in the outer row blossom fi
straightening back and pushing the
rolla outward; and now we can see t
the five lobes in the bud are the
notches at the end of the corolla. All
flowers in the dandelion head have b
ners, but those at the center, belong
to the younger flowers, have shorter a
darker yellow corollas. After a corolla
opened, there pushes out from its tub
base a darker yellow anther-tube; the
filaments below the tube are visible w
a lens. A little later, the stigma-ram
pushes forth from the tube, its fuzzy si
acting like a brush to bring out all
pollen; later it rises far above the anth
tube and quirls back its stigma-lobes,
if every floret were making a dandel
curl of its own. The lens shows us, bel
the corolla, the akene. The pappus is
set in a collar upon the dandelion se
as it is in the aster seed; there is a sh
stem above the seed which is called

eak" and the pappus is attached to s.

Every day more blossoms may open; t on dark, rainy days and during the ght the little green house puts up its itters around the flower family, and if bracts are not wide enough to cover growing family, the banners of the ter flowers have along their lower sides ck or brownish portions which serve calk the chinks. It is interesting to tch the dandelion stars close as the ght falls, and still more interesting to tch the sleepy-heads awaken long after sun is up in the morning; they often not open until eight o'clock.

After all the florets of a dandelion head ve blossomed, they may stay in retire- ent for several days, and during this riod the flowerstalk often grows in- striously; and when the shutters of the le green house are again let down, what a different appearance has the dandelion head! The akenes with their balloons are set so as to make an exquisite, filmy globe;

1, *Floret of dandelion;* 2, *akene of dandelion. Both enlarged*

now they are ready to coquette with the wind, and one after another all the bal- loons go sailing off. One of these akenes is well worth careful observation through a lens. The balloon is attached to the top of the beak as an umbrella frame is at- tached to the handle, except that the "ribs" are many and fluffy; while the dandelion youngster, hanging below, has an overcoat armed with grappling hooks, which enable it to cling fast when the balloon chances to settle to the ground.

Father Tabb says of the dandelion — "With locks of gold today; tomorrow silver gray; then blossom bald." But not the least beautiful part of the dandelion is this blossom-bald head after all the akenes are gone; it is like a mosaic, with a pit at the center of each figure where the akene was attached. There is an in- teresting mechanism connected with this receptacle. Before the akenes are fully out this soon-to-be-bald head is concave at the center; later it becomes convex, and the mechanism of this movement lib- erates the akenes which are embedded in it.

Each freshly opened corolla-tube is full to overflowing with nectar, and much pol- len is developed; therefore, the dandelion has many kinds of insect visitors. But per- haps the bee shows us best where the nectar is found; she thrusts her tongue

Brooklyn Botanic Garden

andelion, showing stages from bud to "old man"

down into the little tubes below the rays, working very rapidly from floret to floret. The dandelion stigmas have a special provision for securing cross-pollination; if that fails, they may be self-pollinated; and now the savants have found that the pistils can also grow seeds without any pollen from anywhere. It surely is a resourceful plant!

The following are the tactics by which the dandelion conquers us and takes possession of our lands: (a) It blossoms early in the spring and until snow falls, producing seed for a long season. (b) It is broadminded as to its location, and flourishes on all sorts of soils. (c) It thrusts its long taproots down into the soil, and thus gets moisture and food not reached by other plants. (d) Its leaves spread out from the base, and crowd and shade many neighboring plants out of existence. (e) Many insects visit it, and so it has plenty of pollen carriers to insure strong seeds; it can also develop seeds from its own pollen, or it even can develop seeds without any pollen. (f) It develops almost numberless akenes, and the wind scatters them far and wide and they thus take possession of new territory. (g) It forms vigorous leaf-rosettes in the fall, and thus is able to begin growth early in the spring.

LESSON 144

The Dandelion

Leading Thought — The dandelions flourish despite our determined efforts to exterminate them. Let us study the way in which they conquer.

Method — The study should be made with the dandelions on the school grounds. Questions should be given, a few at a time, and then let the pupils consult the dandelions as to the answers.

The dandelion is a composite flower and may be studied according to Lesson 131. All the florets are ray flowers.

Observations — 1. Where do you find dandelions growing? If they are on the lawn, how long are their blossom- or seed-stalks? If in a meadow or among high grass, how long is the blossom-stalk? the blossom-stalk solid or hollow? D it break easily?

2. Dig up a dandelion root a then explain why this weed withsta drought, and why it remains, when o planted.

3. Sketch or describe a dandelion le Why was the plant named "lio teeth"? How are the leaves arrang about the root? How does this help dandelion and hinder other plants? what condition do the leaves pass winter under the snow?

4. Take a blossom not yet open. N the bracts that cover the unopened flow head. Note the ones below and descri them.

5. Note the dandelion flower-head j open. Which flowers open first? How the buds look at the center? Do all t florets have rays? Are the ray flowers the center of the head the same color a length as those outside? Examine a flo and note the young akene. Is the papp attached to it or above it?

6. What happens to the dandeli blossom on rainy or dark days? Do y think that this has anything to do wi the insect visitors? Do bees and other sects gather nectar during dark or rai days?

7. Note at what hour the dandelio on the lawn close and at what hour th open on pleasant days.

8. Make notes on a certain dandeli plant three times a day: How long do it take the dandelion head to open ful on a sunny morning? How long does remain open? How long does it take t flower-head to close? What proportion the flowers in the head blossoms duri the first day? What proportion of t flowers in the head blossoms during t second day? How long before they a blossom? Does the flower-head rema open longer in the afternoon on som days than on others, equally sunny? Do the stem bend over before the blossor head opens?

9. After all the little flowers of a dand lion head have blossomed, what happe

it? Measure the stem, and see if it etches up during the time. How does dandelion look when it opens again? ok at a dandelion head full of seed, and how the seeds are arranged to make a fect globe. Shake the seeds off and ex- ine the "bald head" with a lens. Can see where the seeds were set?

10. Examine a dandelion akene with a s. Describe the balloon, the beak or m of the balloon, and the akene.

11. How early in the spring, and how late in the fall, do dandelions blossom?

12. Watch a bee when she is working on a dandelion flower, and see where she thrusts her tongue and which flowers she probes.

13. Tell all the things that you can remember about the dandelion which help it to live and thrive.

14. What use do we make of the dandelion?

THE PEARLY EVERLASTING

These wraithlike flowers seem never to ve been alive, rather than to have been dowed with everlasting life. Cattle do t often eat them. The stems are cov- d with white felt; the long narrow ves are very pale green, and when ex- ined with a lens, look as if they were vered with a layer of cotton which dis- ises all venation except the thick mid- . The leaves are set alternate, and be- me shorter and narrower and whiter toward the top of the plant. All this cottony covering tends to prevent the evaporation of water from the plant during the long droughts. The everlasting never has much juice in its leaves, but what it has, it keeps.

The flowerstalks are rather stout, wooly, soft, and pliable. They come off at the axils of the threadlike whitish leaves. The pistillate and the staminate flowers are borne on separate plants, and usually in separate patches. The pistillate or seed-developing plants have globular flower buds, almost egg-shaped, with a fluffy lemon-yellow knob at the tip; this fluff is made up of stigmas split at the end.

Verne Morton

he pistillate flower-heads of pearly everlasting

1, *Pistillate floret;* 2, *pappus;* 3, *staminate floret. All enlarged*

At the center of this tassel of lemon-yellow stigma-plush, may often be seen a depression; at the bottom of this well, there are three or four perfect flowers. One of the secrets of the everlasting is, evidently, that it does not put all of its eggs in one basket; it has a few perfect flowers for insurance. This pistillate or

Leonard K. Beyer

A good stand of pearly everlasting

seed-bearing flower has a long, delicate tube, ending in five needle-like points and surrounded by a pretty pappus. The bracts of the flower-cluster seem to cling around the base of the beautiful yellow tassel of fertile flowers, as if to emphasize it. They look as if they were made of white Japanese paper, and when looked at through a lens, they resemble the petals of a water lily. They are dry to begin with, so they cannot wither.

The staminate or pollen-bearing flower-heads are like white birds' nests, the white bracts forming the nest and the little yellow flowers the eggs. The flower has a tubular, five-pointed, starlike corolla, with five stamens joined in a tube at the middle, standing up like a barrel from the corolla. The anther-tube is ocher-yellow with brown stripes, and is closed at first with five little flaps, making a cone at t top. Later, the orange-yellow poll bulges out as if it were boiling over. T flowers around the edges of the flower d open first.

LESSON 145

The Pearly Everlasting

Leading Thought — There are oft found growing on the poor soil in c pastures, clumps of soft, whitish pla which are seldom eaten by cattle. Th is so little juice in them that they ret their form when dried and thus have w their name.

Method — The pupils should see th plants growing, so that they may obse the staminate and pistillate flowers, wh are on separate plants and often in se rate clumps. If this is not practicab bring both kinds of flowers into t schoolroom for study.

Observations — 1. Where does t pearly everlasting grow? Do cattle eat What is the general color of the pla With what is the stem covered?

2. What is the shape of the leav How are they veined? With what are th covered? How are they placed on t stem? What is the relative size of t lower and upper leaves?

3. Do you see some plants which ha egg-shaped flower-heads, each with a y low knob at the tip? Take one apart a look at it with a lens, and see what for the white part and what forms the yell knob. Do you see other flower-heads th look like little white birds' nests fill with yellow eggs? Look at one of the with a lens, and tell what kind of flow head it is.

4. Except that the pistillate and stan nate flowers are on different plants, t flowers of the pearly everlasting should studied according to Lesson 131.

5. What do you know of the edelwe of the Alps? How does it resemble t pearly everlasting? Do you know anoth common kind of everlasting called puss toes?

MULLEIN

I like the plants that you call weeds, —
Sedge, hardhack, mullein, yarrow, —
Which knit their leaves and sift their seeds
Where any grassy wheel-track leads
Through country by-ways narrow.

— LUCY LARCOM

Ve take much pride unto ourselves bese we belong to the chosen few of the ttest," which have survived in the ggle for existence. But, if we look und upon other members of this select d, we shall find many lowly beings ich we do not ordinarily recognize as peers. Mullein is one of them, and r we study its many ways of "winning " then we may bow to it and call it rother."

was wandering one day in a sheep ture and looking curiously at the few nts left uneaten. There was a great stle with its sharp spines and the pearly rlasting — too woolly and anæmic to appetizing even to a sheep; and besides se, there was an army of mullein stalks tall, slim, and stiff-necked, or branching great candelabra, their upper leaves ering alternately to the stalks for half ir length. I stopped before one of them l mentally asked, "Why do the sheep relish you? Are you bitter?" I took a e, Nebuchadnezzar-like, and to my unined taste it seemed as good fodder any; but my tongue smarted and rned for some time after, from being cked by the felt which covered the f. I recalled the practical joke of which grandmother once made me the vic1; she told me that to be beautiful, I eded only to rub my cheeks with muln leaves, an experience which convinced that there were other things far more sirable than beauty — comfort, for innce. This felt on the mullein is beautiwhen looked at through a microscope; consists of a fretwork of little, white, arp spikes. No wonder my cheeks were l one day and purple the next. and no

Verne Morton

Mullein. Note the stone fence in the background. Mullein often grows in such places

wonder the sheep will not eat mullein unless starved! This frostlike felt covering not only may make the plant unpalatable to grazing animals but may also help to keep the water in the leaves from evaporating. I soon discovered another means by which the mullein resists drought, when I tried to dig up the plant with a stick; I followed its taproot down far enough to understand that it was a subsoiler and reached below most other plants for moisture and food. Although it was late autumn, the mullein was still in blossom; there were flowers near the tip and also one here and there on the capsule-crowded stem. I estimated there were hundreds of seed capsules on that one

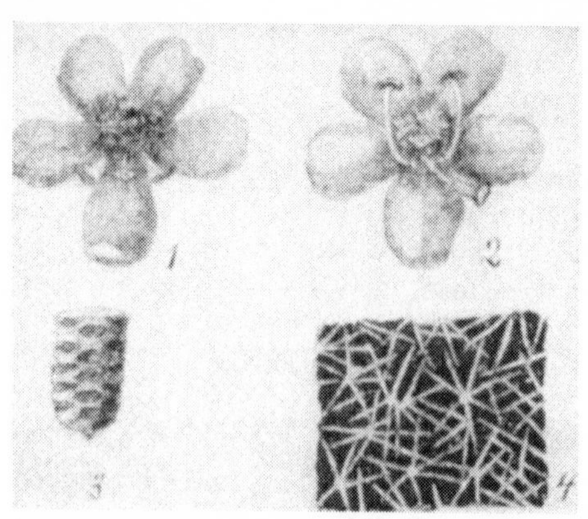

1, 2, *Mullein flowers in different stages.* 3, *Mullein seed enlarged.* 4, *A bit of mullein leaf enlarged*

plant; I opened one, still covered with the calyx-lobes, and found that the mullein was still battling for survival; for I found this capsule and many others inhabited by little brown-headed white grubs, which gave an exhibition of St. Vitus dance as I laid open their home. They were the young of a snout beetle, which is a far more dangerous enemy of the mullein than is the sheep.

The mullein plant is like the old woman who lived in a shoe in the matter of blossom-children; she has so many that they are unkempt and irregular, but there are normally four yellow or white petals and a five-lobed calyx. I have never been able to solve the problem of the five stamens which, when the flower opens, are folded together in a knock-kneed fashion. The upper three are bearded below the anthers, the middle being the shortest. The lower two are much longer and have no fuzz on their filaments; they at first stand straight out, with the stigma between them; but after the upper anthers have shed their pollen, these stamens curve up like boars' teeth and splash their pollen on the upper petals, the stigma protruding one-sidedly below. Later the corolla, with the stamens which are attached to it, falls off, leaving the stigma and style attached to the seed capsule.

The color of the mullein flowers varies from lemon-yellow to white. The filaments are pale yellow; the anthers and pollen, orange. The seed capsule is cased in the long calyx-lobes, and is sha like a blunt egg. By cutting it in two c wise, the central core, tough and flatte and almost filling the capsule, is revea and growing upon its surface are num less tiny brown seeds, as fine as powder. Later the capsule divides tially in quarters, opening wide enc to shake out the tiny seeds with e wandering blast. The seed, when through a lens, is very pretty; it lc like a section of a corncob, pitted ribbed. A nice point of investigation some junior naturalist is to work out fertilization of the mullein flower, note what insects assist. The mullein another spoke in the wheel of its succ The seed, scattered from the sere dried plants, settles in any place whe can reach the soil, and during the season grows a beautiful velvety rosett fuzzy leaves. These rosettes lie flat u the snow, with their taproots strong already deep in the soil, and are re to begin their work of food-making soon as the spring sun gives them po

LESSON 146

Mullein

Leading Thought — The mullein its leaves covered with felt, which

A typical winter rosette of mullein

ɔ to retard evaporation. The plant is lom eaten by grazing animals. It has eep root, and thus gets moisture be- d the reach of most other plants. It ssoms all summer and until the snow ıes in the autumn, and thus forms ny, many seeds, which the wind plants it; and here in our midst it lives and ives despite us.

ΛETHOD — The pupils should have a d trip to see what plants are left un- ɜn in pastures, and thus learn where llein grows best. The flower- or seed- k, with basal leaves and root, may be ught to the schoolroom for the les- .

)BSERVATIONS — 1. Where does the llein grow? Do you ever see it in ımps or woodlands? Do cattle or sheep it? Does it flourish during the summer ught? Look at a mullein leaf with a s and describe its appearance.

:. What sort of root has the mullein? w is its root adapted to get moisture ich other plants cannot reach? De- ibe the flowering stalk. How are the ves arranged on it and attached to it? : there several branching flowerstalks or ingle one?

;. Describe the flower bud. Do the llein flowers nearest the base or the begin to blossom first? Is this invari- e, or do flowers open here and there gularly on the stem during the season?

ɩ. Describe the mullein flower. How many lobes has the calyx? Are these cov- ered with felt? How many petals? Are there always this number? Are the petals of the same size? Are they always regular in shape?

5. How many stamens? How do the upper three differ from the lower two? Describe the style and stigma. What are the colors of petals, anthers, and stigma? What insects do you find visiting the flowers?

6. Describe the seed capsule, its shape and covering. Cut it across and describe the inside. Where are the seeds borne? Are there many? Look at the seed with a lens and describe it. How does the cap- sule open and by what means are the seeds scattered?

7. Does the mullein grow from the seed to maturity in one year? How does it look at the end of the first season? Describe the winter rosette, telling how it is fitted to live beneath the snows of winter. What is the advantage of this habit?

8. Write a theme telling some ways the mullein has of flourishing and of com- bating other plants.

The mullein's pillar, tipped with golden
flowers,
Slim rises upward, and yon yellow bird
Shoots to its top.

— "THE HILL HOLLOW,"
A. B. STREET

THE TEASEL

Γhe old teasel stalks standing gaunt and y in the fields, braving the blasts of ıter, seem like old suits of armor, which cit admiration from us for the strength d beauty of the protecting visor, breast- ıte, and gauntlets, and at the same time r our thoughts to the knights of old ıo once wore them in the fray. Thus, th the teasel, we admire this panoply of ars, which recall the purple flowers and ribbed akenes.

Let us study this plant in armor: First, its stem is tough, woody, hollow, with ridges extending its full length and each ridge armed with spines which are quite wide at the base and very sharp. It is im- possible to take hold anywhere without being pricked by either large or small spines. The leaves are long, lanceolate, set opposite in pairs, rather coarse in texture, with a stiff, whitish midrib; the bases of the two leaves closely clasp the stem; the midrib is armed below with a row of long, white, recurved prickers, and woe unto

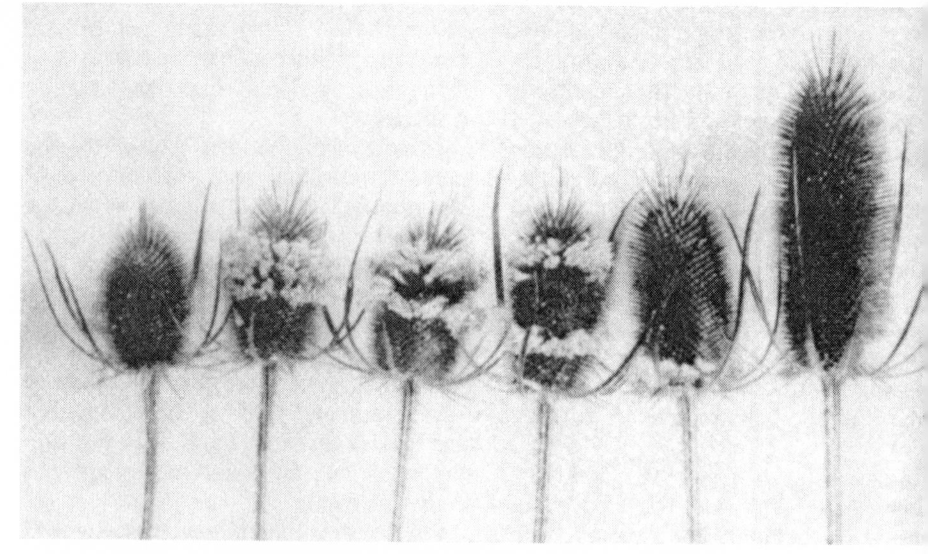

The teasel begins at the middle and blossoms both ways

the tongue of grazing beast that tries to lift this leaf into the mouth. If one pair of clasping leaves points east and west, the next pairs above and below point north and south.

The flowerstalks come off at the axils of the leaves and therefore each pair stands at right angles to the ones above and below. The flowers are set in dense heads armed with spines, and the head is set in an involucre of long, upcurving spiny prongs. If we look at it carefully, the teasel flower-head wins our admiration, because of the exquisite geometrical design made by the folded bases of the spines, set in diagonal rows. If we pull out a spine, we find that it enlarges toward the base to a triangular piece that is folded at right angles, holding the flower. Note that the spiny bracts at the tip of the flower-head are longer and more awesome than those at the sides; if we pass our hands down over the flower-head we feel how stiff the spines or bracts are, and can hear them crackle as they spring back.

The teasel has a quite original method of blossoming. The goldenrod begins to blossom at the tip of the flowering branches and the blossom-tide runs inward and downward toward the base. The clover begins at the base and blossoms ward the tip, or the center. But the tea begins at the middle and blossoms bo ways. Some summer morning we will fi its flower-head girt about its middle w a wide band of purple blossoms; afte few days these fade and drop off, and th there are two bands, sometimes four ro of flowers in each, and sometimes on two. Below the lower band and above t upper band, the enfolding bracts are fill with little round-headed lilac buds, wh between the two rows of blossoms t protecting bracts hold the precious gro ing seed. Away from each other this do ble procession moves, until the lower ba reaches the pronged involucre and t upper one forms a solid patch on the ap of the flower-head. Since the seconda blossom-heads starting from the leaf ax are younger, we may find all stages of th blossoming in the flower-heads of o plant.

No small flower better repays close e amination than does that of the teas If we do not pull the flower-head apa what we see is a little purple flower co sisting of a white tube with four purp lobes at the end, the lower lobe being little longer than the others and turni

slightly at its tip; projecting from be-
een each of the lobes, and fastened to
: tube, are four stamens with long white
ments and beautiful purple anthers
ed with large, pearly white pollen
ins; at the very heart of the flower,
: white stigma may be seen far down
: tube. But a little later, after the an-
ers have fallen or shriveled, the white
gma extends out of the blossom like a
ig white tongue and is crowded with
ite pollen grains.

But to see the flower completely we
ed to break or cut a flower-head in two.
en we see that the long white tube is
ped at one end with purple lobes and
inge of anthers, and at the other is set
on a little green, fluffy cushion which
ps the ovary; the shape of the ovary in
e flower tells us by its form how the
uit will look later. Enfolding ovary and
be is the bract with its spiny edges,
shing its spear outward, but not so far
t as the opening of the flower. The pol-
n of the teasel is white and globular,
ith three little rosettes arranged at equal
stances upon it like a bomb with three

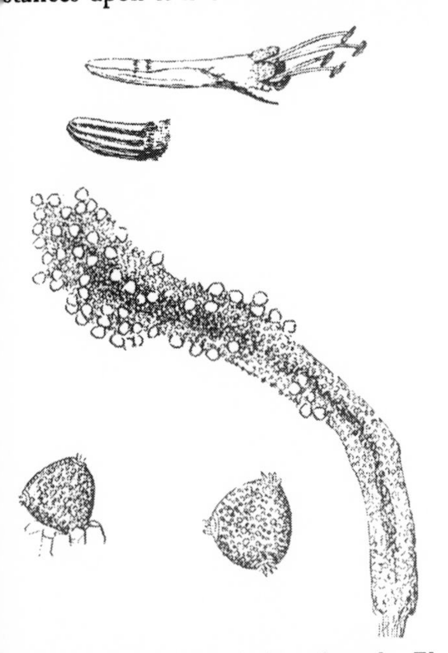

Teasel flower and fruit enlarged. The tigma of a teasel floret much magnified to how the pollen adhering to it. Below are ollen grains greatly magnified

Verne Morton

A winter rosette of teasel

fuses. These little rosettes are the growing points of the pollen grains and from any of them may emerge the pollen tube to push down into the stigma. The teasel pollen is an excellent subject for the children to study, since it is so very large; and if examined with a microscope with a three-fourths objective, the tubes running from the pollen grains into the stigma may be easily seen.

In blossoming, the teasel is not always uniform in the matter of rows of flowers. There may be more rows in the upper band than in the lower, or *vice versa*; this is especially true of the smaller secondary blossoms. But though the teasel flowers fade and the leaves fall off, still the spiny skeleton stands, the thorny stalks holding up the empty flower-heads like candelabra, from which the seeds are tossed far and wide, shaken out by the winds of autumn. But though battered by wintry blasts, the teasel staunchly stands; it will often stand even until the ensuing summer, its heads empty where once were blossom and seed. Alas, because of this emptiness, it has been debased by practical New England housewives into a utensil for sprinkling clothes for ironing.

The spines of one species of teasel were in earlier times used for raising the nap on woolen cloth, and the plant was grown extensively for that purpose. The bees are fond of the teasel blossoms and teasel honey has an especially fine flavor.

The teasels are biennial, and during the first season develop a rosette of crinkled leaves which have upon them short spines.

LESSON 147
The Teasel

Leading Thought — The teasel is a plant in armor. It has a peculiar method of beginning to blossom in the middle of the flower-head and then blossoming upward and downward from this point.

Method — In September, bring in a teasel plant which shows all stages of blossoming, and let the pupils make observations in the schoolroom.

Observations — 1. Where does the teasel grow? Is it eaten by cattle? How is it protected?

2. What sort of stem has it? Is it hollow or solid? Where upon it are the spines situated? Are the spines all of the same size? Can you take hold of the stem anywhere without being pricked?

3. What is the shape of the leaves? How do they join the stem? Are the leaves set opposite or alternate? If one pair points east and west in which direction will the pairs above and below point? How and where are the leaves armed? How does the cow or sheep draw leaves into the mouth with the tongue? If either should try to do this with the teasel, how would the tongue be injured?

4. Where do the flowerstems come off? Do they come off in pairs? How are the pairs set in relation to each other?

5. What is the general appearance of the teasel flower-head? Describe the lo involucre prongs at the base. If the tea is in blossom, where do you find the fl ers? How many girdles of flowers are th around the flower-head? How many ro in one girdle? Where did the first flow blossom in the teasel flower-head? Wh on the head will the last blossoms appe Where are the buds just ready to op Where are the ripened akenes?

6. Examine a single flower. How is protected? Cut out a flower and bract a see how the long-spined bract enfo it. Would the bract spear deter ca from grazing on the blossom? Wh are the longest spines on the tea head?

7. Study a single flower. What is shape of its corolla? How is it color What color are the stamens? How ma Describe the pollen. If the pollen is ing shed where is the stigma? After pollen is shed, what happens to stigma?

8. What do you find at the base of flower? How does the young seed lo Later in the season take a teasel head a describe how it scatters its seed. How the ripe seeds look?

9. For what were teasels once us How many years does a teasel plant li How does it look at the end of its fi season? How is this an advantage as method of passing the winter?

QUEEN ANNE'S LACE OR WILD CARROT

Queen Anne was apparently given to wearing lace made in medallion patterns; and even though we grant that her lace is most exquisite in design as well as in execution, we wish most sincerely that there had been established in America such a high tariff on this royal fabric as to have prohibited its importation. It has for decades held us and our lands prisoners in its delicate meshes, it being one of the most stubborn and persistent weeds that ever came to us from over the seas.

But for those people who admire lace of intricate pattern, and beautiful blossoms, whether they grow on scalaw plants or not, this medallion flower attr uted to Queen Anne is well worth stu ing. It belongs to the family *Umbellife* which one of my small pupils always cal "umbrelliferæ" because, he averred, th have umbrella blossoms. In the case Queen Anne's lace the flower-cluster, umbel, is made up of many smaller u bels, each a most perfect flower-cluster itself. Each tiny white floret has five pet and should have five stamens with crea anthers, but often has only two. Howev it has always at its center a pistil co

:d of two parts set snugly together, :h rests in a solid, bristly, green, cup-calyx. Twenty or thirty of these little soms are set in a rosette, the stalks raded length; and where the bases of stalks meet are some long, pointed, ow bracts. Each of these little flower-ters, or umbels, has a long stalk, its :th being just fit to bring it to its t place in the medallion pattern of

An inner and a border floret and a bract of Queen Anne's lace, enlarged

Verne Morton

Queen Anne's lace or wild carrot

this royal lace. And these stalks also have set at their bases some bracts with long, threadlike lobes, which make a delicate, green background for the opening blossoms; these bracts curl up about the buds and the seeds. If we look straight into the large flower-cluster, we can see that each component cluster, or umbellet, seems to have its own share in making the larger pattern; the outside blossoms of the outside clusters have the outside petals larger, thus forming a beautiful border. At the very center of this flower medallion, there is often a larger floret with delicate, wine-colored petals; this striking floret is not a part of a smaller flower-cluster, but stands in stately solitude upon its own isolated stalk. The reason for this giant floret at the center of the wide, circular flower-cluster is a mystery; and so far as I know, the botanists have not yet explained the reason for its presence. May we not, then, be at liberty to explain its origin on the supposition that her Royal Highness, Queen Anne, was wont to fasten her lace medallions upon her royal person with garnet-headed pins?

When the flowers wither and the fruits begin to form, every one of the little umbels turns toward the center, its stalk curving over so that the outside umbels reach over and close over the whole flower-head; and the threadlike bracts at the base reach up as if they, too, were in the family councils, and must do their slender duty in helping to make the fading flowers into a little, tightfisted clump. Such little porcupines as the fruits are! Each fruit is clothed with long spines set in bristling rows, and is a most forbidding-looking youngster when examined through a lens; and yet there is method in its spininess, and we must grudgingly grant that it is not only beautiful in its ornamentation but is also well fitted to take hold with a will when wandering winds sift it down to the soil.

The wild carrot is known in some localities as the " bird's-nest weed," because the maturing fruit-clusters, their edges curving inward, look like little birds' nests.

Charles F. Fudge

Fruiting cluster or " bird's nest " of wild carrot

But no bird's nest ever contained so many eggs as does this imitation one. In one we counted 34 tiny umbels on which ripened 782 fruits; and the plant from which this " bird's nest " was taken developed nine more quite as large.

Altogether the wild carrot is well fitted to maintain itself in the struggle for existence, and is most successful in crowding out its betters in pasture and meadow. Birds do not like its spiny seeds; the stem of the plant is tough and its leaves are rough and have an unpleasant odor and an acrid taste. Winter's cold cannot harm it, for it is a biennial; its seeds often germinate in the fall, sending down long, slender taproots crowned with tufts of inconspicuous leaves; it thus stores u supply of starchy food which enable to start early the next season with g vigor. The root, when the plant is f grown, is six or eight inches long, as th as a finger and yellowish white in colo is very acrid and somewhat poisonou

The surest way of exterminating Queen Anne's lace is to prevent its lific seed production by cutting or upr ing the plants as soon as the first b soms open.

LESSON 148

Queen Anne's Lace or Wild Car

Leading Thought — Queen An lace is a weed which came to us fr Europe and flourishes better here t on its native soil. It has beautiful bloss set in clusters, and it matures m seeds which it manages to plant succ fully.

Method — The object of this les should be to show the pupils how weed survives the winter and how i able to grow where it is not wanted. weed is very common along most coun roadsides, and in many pastures and me ows. It blossoms very late in the autu and is available for lessons often as as November. Its fruit-clusters may used for a lesson at almost any time dur the winter.

Observations — 1. Look at a wild rot plant; how are its blossoms arrang Take a flower-cluster; what is its sha How many small flower-clusters make large one? How are these arranged to m the large cluster symmetrical?

2. Take one of the little flower-clust from near the center, and one from outside of the large cluster; how ma little flowers, or florets, make up smaller cluster? Look at one of the flo through a lens; can you see the cup-sha calyx? How many petals has it? Can y see its five anthers and its two-par white pistil?

;. Take one of the outer florets of the side cluster; are all its flowers the same pe? How do they differ? Where are florets with the large petals placed in big flower-cluster? How does this help make "the pattern"?

4. Do the outside or the central flowers the large clusters open first? Can you d a cluster with an almost black or very k red floret at its center? Is this dark wer a part of one of the little clusters does it stand alone, its stalk reaching ectly to the main stem? Do you think nakes the flowers of the Queen Anne's e prettier to have this dark red floret at center?

5. Take a flower-cluster with the flow- not yet open. Can you see the thread- e green bracts that close up around each d? Can you see finely divided, thread- e bracts that stand out around the ole cluster? What position do these acts assume when the flowers are open? hat do they do after the flowers fade d the fruits are being matured?

6. What is the general shape of the uit-cluster of the wild carrot? Have u ever found such a cluster broken and blowing across the snow? Do u think this is one way the seed is anted?

7. Examine a fruit of the wild carrot th a lens. Is it round or oblong? Thin flat? Is it ridged or grooved? Has it any oks or spines by which it might cling the clothing of passers-by, or to the ir or fleece of animals, and thus be attered more widely? Does the fruit cling its stem or break away when it is uched?

8. Take one fruit-cluster and count e number of seeds within it. How many uit-clusters do you find on a single plant? How many fruits do you therefore think a single plant produces?

9. What would you consider the best means of destroying this prolific weed?

10. What do you think is the reason that the wild carrot remains untouched, so that it grows vigorously and matures its seeds in lanes and pastures where cattle graze?

11. Have you noticed any birds feeding on the fruits of the wild carrot?

I do not want change: I want the same old and loved things, the same wild flowers, the same trees and soft ash-green; the turtle-doves, the blackbirds, the coloured yellow-hammer sing, sing, singing so long as there is light to cast a shadow on the dial, for such is the measure of his song, and I want them in the same place. Let me find them morning after morning, the starry-white petals radiating, striving upwards to their ideal. Let me see the idle shadows resting on the white dust; let me hear the bumble-bees, and stay to look down on the rich dandelion disc. Let me see the very thistles opening their great crowns — I should miss the thistles; the reed-grasses hiding the moor-hen; the bryony bine, at first crudely ambitious and lifted by force of youthful sap straight above the hedgerow to sink of its own weight presently and progress with crafty tendrils; swifts shot through the air with outstretched wings like crescent-headed shaftless arrows darted from the clouds; the chaffinch with a feather in her bill; all the living staircase of the spring, step by step, upwards to the great gallery of the summer — let me watch the same succession year by year.

— "THE PAGEANT OF SUMMER,"
RICHARD JEFFERIES

GARDEN FLOWERS

People have always admired the wild flowers that grow in the woods and meadows, and have wanted to be able to bring them near their homes. And so when someone would see a flower that he thought especially beautiful growing in field or forest, he would take it from its native home and plant it in a garden. If others admired it, they would ask for seeds, roots, or cuttings; and thus the plant would come to many gardens. As many wild flowers are beautiful, and as various people have varying tastes, in this way many kinds of flowering plants have come into cultivation.

But man is seldom content to leave a thing as he finds it; and so after a time people set about improving upon nature. Plant breeders have tried in many ways to add to the attractiveness of flowers in size, color, and shape. Often they have succeeded, and some strains have been greatly improved: thus, the aster that grows in our gardens is much more elaborate than the natural plant; Shakespeare's modest wild eglantine has yielded sixteen varieties of sweetbriar. Others, like the morning-glory and the calla lily, have changed very little during years of cultivation. Perhaps it is well that we have not always succeeded in improving upon nature; it is pleasant to have cultivated elegance and natural simplicity side by side.

Some garden flowers have been popular for many years; these we call "old-fashioned." They are still popular today; they are not out-of-date, like old-fashioned clothes; in their case, "old-fashione means rather that they have stood test of time. Everyone knows the cha of an old-fashioned garden. But there also new strains that bid fair to win t way and to stand the test as well as older flowers, from which some of th have been developed, as the delphini from the old-fashioned larkspur. All them, however, even the most eleg newcomer, came originally from so wild flower.

Wild or cultivated, simple or orna flowers are among the most import means of decorating our homes and g dens. Every dooryard throughout the la is a picture that is viewed by the passer- Whether the picture is attractive or n may depend very largely upon the pr ence or absence of attractive flowers in t yard or about the door.

Ferry-Morse Seed Co.

THE CROCUS

The crocus, like the snowdrop, cannot ait for the snow to be off the ground fore it pushes up its gay blossoms, and has thus earned the gratitude of those 1o are winter weary.

The crocus has a corm instead of a bulb :e the snowdrop or daffodil. A corm is solid, thickened, underground stem, and not in layers, like the onion. The roots me off the lower side of the corm. The rm of the crocus is well wrapped in veral, usually five, white coats with pa- ry tips. When the plant begins to grow e leaves push up through the coats. The aves are grasslike and may be in number om two to eight, depending on the riety. Each leaf has its edge folded, and e white midrib has a plait on either de, giving it the appearance of being box- aited on the under side. The bases of e leaves enclosed in the corm coats are llow, since they have had no sunlight start their starch factories and the een within their cells. At the center of e leaves appear the blossom buds, each iclosed in a sheath.

The petals and sepals are similar in olor, but the three sepals are on the out- de, and their texture, especially on the outer side, is coarser than that of the three protected petals. But sepals and pet- als unite into a long tube at the base. At the very base of this corolla-tube; away down out of sight, even below the surface of the ground, is the seed box, or ovary. From the tip of the ovary the style extends up through the corolla-tube and is tipped with a ruffled three-lobed stigma.

The three stamens are set at the throat of the corolla-tube. The anthers are very long and open along the sides. The an- thers mature first, and shed their pollen in the cup of the blossom where any in- sect, seeking the nectar in the tube of the corolla, must become dusted with it. However, if the stigma lobes fail to get pollen from other flowers, they later spread

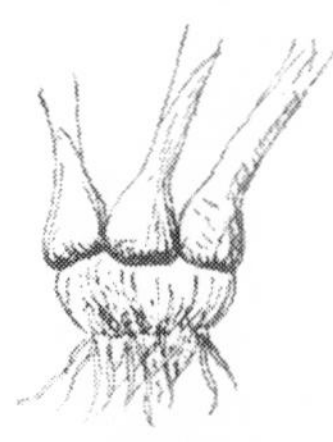

The old and young corms of the crocus

apart and curl over until they reach some of the pollen of their own flower.

Crocus blossoms have varied colors: white, yellow, orange, purple, the latter often striped or feather-veined. And while many seeds like tiny pearls are developed in the oblong capsule, yet it is chiefly by its corms that the crocus multiplies. On top of the mother corm of this year

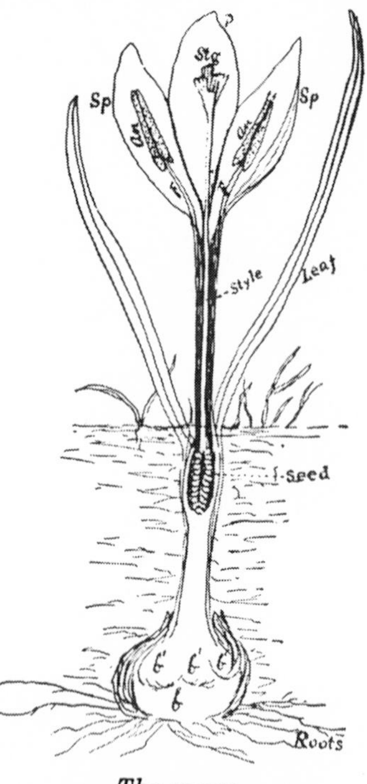

The crocus

p, petal; sp, sepal; an, anther; f, filament; stg, stigma; b, mother corm; b′ b′ b′ young corms

develop several small corms, each capable of growing a plant next year. But after two years of this second-story sort of multiplication the young crocuses are pushed above the surface of the ground. Thus, they need to be replanted every two or three years. Crocuses may be planted from the first of October until the ground freezes. They make pretty borders to garden beds and paths. Or they may be planted in lawns without disturbing the grass, by punching a hole with a stick or dibble and dropping in a corm and then pressing back the soil in place above it. The plants will mature before the grass needs to be mowed.

LESSON 149

The Crocus

Leading Thought — The crocus bl soms appear very early in the spring, cause the plants have food stored in und ground storehouses. Crocuses multiply seeds and by corms.

Method — If it is possible to have c cuses in boxes in the schoolroom windo the flowers may thus best be studi Otherwise, when crocuses are in blo bring them into the schoolroom, cor and all, and place them where the cl dren may study them at leisure.

Observations — 1. At what date the spring have you found crocuses blossom? Why are they able to bloss so much earlier than other flowers?

2. Take a crocus just pushing up of its corm. How many overcoats prot its leaves? What is at the very center the corm? Has the flower bud a spec overcoat?

3. Describe the leaves. How are th folded in their overcoats? What color they where they have pushed out abc their overcoats? What color are th within the overcoats?

4. Do the flowers or the leaves ha stems, or do they arise directly from t corm?

5. What is the shape of the open croc flower? Can you tell the difference tween sepals and petals in color? Can y tell the difference by their position? by their texture above or below? As y look into the flower, which makes t points of the triangle, the sepals or t petals?

6. Describe the anthers. How long they? How many are there? How do th open? What is the color of the polle Describe how a bee becomes dusted wi pollen. Why does the bee visit the croc blossom? If she finds nectar there, whe is it?

7. Describe the stigma. Open a flow and see how long the style is. How do t sepals and petals unite to protect t style? Where is the seed box? Is it so down that it is below ground? How ma seeds are developed from a single blosso

How many colors do you find in the
us flowers? Which are the prettiest
he lawn? Which in the flower beds?
How do the crocus blossoms act in
. and stormy weather? When do they
n? How does this benefit them?
). How do the crocus corms multiply?
y do they often need resetting?
. Describe how to raise crocuses best:
kind of soil, the time of planting, and
best situations.

of the frozen earth below,
of the melting of the snow,
'o flower, but a film, I push to light;
stem, no bud — yet I have burst
: bars of winter, I am the first
Sun, to greet thee out of the night!

p in the warm sleep underground
is still, and the peace profound:

Yet a beam that pierced, and a thrill
that smote
Call'd me and drew me from far away;
I rose, I came, to the open day
I have won, unshelter'd, alone, remote.

— "THE CROCUS,"
HARRIET E. H. KING

When first the crocus thrusts its point of
gold,
Up through the still snow-drifted garden-
mould,
And folded green things in dim woods un-
close
Their crinkled spears, a sudden tremor
goes
Into my veins and makes me kith and kin
To every wild-born thing that thrills and
blows.

— "A TOUCH OF NATURE,"
T. B. ALDRICH

THE DAFFODILS AND THEIR RELATIVES

Daffydown Dilly came up in the cold from the brown mold,
Although the March breezes blew keen in her face,
Although the white snow lay on many a place.

'hus it is that Miss Warner's stanzas us the special reason we so love the odils. They bring the sunshine color he sodden earth, when the sun is chary is favors in our northern latitude; and sight of the daffodils floods the spirit a sense of sunlight.

'he daffodils and their relatives, the quils and narcissuses, are interesting n we stop to read their story in their n. The six segments of the perianth, as we would say, the three bright-col- l sepals and the three inner petals of flower, are different in shape; but they ook like petals and stand out in star- e around the flaring end of the flower- e, which, because of its shape, is called corona, or crown; however, it looks e like a stiff little petticoat extending in the middle of the flower than it s like a crown. When we look down the crown of one of these flowers, see the long style with its three-lobed stigma pushing out beyond the anthers, which are pressed close about it at the throat of the tube; between each two an- thers may be seen a little deep passage, through which the tongues of the moth or butterfly can be thrust to reach the nectar. In a tube, slit open, we can see the nectar at the very bottom; it is sweet to the taste and has a decided flavor. In this open tube we may see that the filaments of the stamens are grown fast to the sides of the tube for much of their length, enough remaining free to press the anthers close to the style. The ovary of the pistil is a green swelling at the base of the tube; by cutting it across we can see that it is triangular in outline, and has a little cavity in each angle large enough to hold two rows of the little, white, shining, unripe seeds. Each of these cavities is partitioned from the others by a green wall.

When the flowerstalk first appears, it

Daffodil

comes up like a sheathed sword, pointing toward the zenith, green, veined lengthwise, and with a noticeable thickening at each edge. As the petals grow, the sheath begins to round out; the stiff stem at the base of the sheath bends at right angles. This brings a strain upon the sheath which bursts it, usually along the upper side, although sometimes it tears it off completely at the base. The slitted sheath, or spathe, hangs around the stem, wrinkled and parchment-like, very like the loose wrist of a suede glove. The stalk is a strong green tube; the leaves are fleshy and are grooved on the inner side. At the base the groove extends part way around the flowerstalk. The number of leaves varies with the variety, and they are usually as tall as the flowerstalk. There is one flower on a stalk in the daffodils and the poet's narcissus, but the jonquils and paper-white narcissus have two or more flowers on the same stalk.

A bed should be prepared by digging deep and fertilizing with stable mar The bulbs should be planted in Sep ber or early October, and should be four to six inches apart, the upper en the bulbs at least four inches below surface of the soil. They should no disturbed but allowed to occupy the for a number of years, or as long as give plenty of flowers. As soon as surface of the ground is frozen in winter, the beds should be covered four to six inches in depth with st mixed stable manure, which can be r off very early in the spring.

The new bulbs are formed at the s of the old one; for this reason the d dils will remain permanently planted, do not lift themselves out of the gro like the crocuses. The leaves of the p should be allowed to stand as long as will after the flowers have disappea so that they may furnish the bulbs plenty of food for storing. The s should not be allowed to ripen, as it the plant too much energy and thus the bulbs. The flowers should be cut as they are opening. Of the white varie the poet's narcissus is the most sati tory, as it is very hardy and very pretty corona being a shallow, flaring, gree yellow rosette with orange-red bor the anthers of its three longest stam making a pretty center. No wonder

Leonard K.

Poet's narcissus

ɜus bent over the pool in joy at view-
; himself, if he was as beautiful a man
the poet's narcissus is a flower.

LESSON 150

ᴀffodils, Jonquils, and Narcissuses

Leading Thought — The daffodil, jon-
il, and narcissus are very closely related,
d quite similar. They all come from
lbs which should be planted in Sep-
nber; but after the first planting, they
ll flower on year after year, bringing
ıch brightness to the gardens in the
ly spring.

Method — The flowers brought to
ıool may be studied for form, and there
ould be a special study of the way the
wer develops its seed, and how it is
opagated by bulbs. The work should
ıd directly to an interest in the cultiva-
ɔn of the plants. In seedsmen's cata-
gues or other books, the children will
ıd methods of planting and cultivating
ese flowers in cities. Daffodils are espe-

Jonquil showing detail of flower

a, corona or crown; b, sepals and petals forming rianth; c, corolla tube; d, ovary or seedcase; e, sheath spathe

W. Atlee Burpee Co.

Paper-white narcissus

cially adapted for both window gardens
and school gardens.

Observations — 1. Note the shape of
the flower. Has it any sepals? Can you see
any difference in color, position, and texture between the petals and sepals?

2. How do the petal-like parts of these
flowers look? How many of them are
there? Do they make the most showy
part of the flower?

3. What does the central part of the
flower look like? Why is it called the
corona, or crown? Peel the sepals and petals off one flower, and see that the tube
is shaped like a trumpet.

4. Look down into the crown of the
flower and tell what you see. Can you see
where the insect's tongue must go to
reach the nectar?

5. Cut open a trumpet lengthwise to
find where the nectar is. How far is it

from the mouth of the tube? How long would the insect's tongue have to be to reach it? What insects have tongues as long as this?

6. In order to reach the nectar how would an insect become dusted with pollen? Are the stamens loose in the flower-tube? Is the pistil longer than the stamens? How many parts to the stigma? Can you see how the flowers are arranged so that insects can carry pollen from flower to flower?

7. What is the green swelling in the stem at the base of the trumpet? Is it connected with the style? Cut it across and describe what you see. How do the young seeds look and how are they arranged?

8. Where the flowerstalk joins the stem, what do you see? Are there one or more flowerstalks coming from this spathe?

9. Describe the flowerstalk. Are the leaves wide or narrow? Are they as long as the flowerstalk, are they flat, or are they grooved?

10. What are the differences between daffodils, jonquils, and poet's narcissus? When should the bulbs for these flowers be planted? Will there be more bulbs formed around the one you plant? Will the same bulb ever send up flowers and leaves again? How do the bulbs divide to make new bulbs?

11. How should the bed for the bulbs be prepared? How near together should the bulbs be planted? How deep in the earth? How can they be protected dur the winter?

12. Why should you not cut the lea off after the flowers have died? W should you not let the seeds ripen? Wh should the flowers be cut for bouque Who was Narcissus, and why should th early spring flowers be named after hi

I emphatically deny the common tion that the farm boy's life is drudge Much of the work is laborious, and thi shares with all work that is producti for the easier the job the less it is wo doing. But every piece of farm work is a an attempt to solve a problem, and the fore it should have its intellectual inter and the problems are as many as the ho of the day and as varied as the face nature. It needs but the informing of t mind and the quickening of the imagi tion to raise any constructive work abo the level of drudgery. It is not mere d work to follow the plow — I have follow it day after day — if one is conscious of the myriad forces that are set at work the breaking of the furrow; and there always the landscape, the free fields, t clean soil, the rain, the promise of t crops. Of all men's labor, the farmer's the most creative. I cannot help wond ing why it is that men will eagerly se work in the grease and grime of a noi factory, but will recoil at what they c the dirty work of the farm. So much we yet bound by tradition!

— L. H. Bail

THE TULIP

We might expect that the Lady Tulip would be a stately flower, if we should consider her history. She made her way into Europe from the Orient during the sixteenth century, bringing with her the honor of being the chosen flower of Persia, where her colors and form were reproduced in priceless webs from looms of the most skilled weavers. No sooner was she seen than worshiped, and shortly all Europe was at her feet.

A hundred years later, the Netherlan was possessed with the tulip mania. Gro ers of bulbs and brokers who bought a sold them indulged in wild speculatio Rare varieties of the bulbs became mo costly than jewels, one of the famous bla tulips being sold for about $1800. Sin then, the growing of tulips has been o of the important industries of the Neth lands.

There are a great many varieties of t

Marion E. Wesp

Tulips in a border

›s, and their brilliant colors make our rdens gorgeous in early spring. Although ιis flower is so prim, yet it bears well ose observation. The three petals, or in-er segments of the perianth, are more quisite in texture and in satiny gloss ι their inner surface than are the three ıter segments or sepals; each petal is like osgrain silk, the fine ridges uniting at ιe central thicker portion. In the red ırieties, there is a six-pointed star at the eart of the flower, usually yellow or yel-w-margined, each point of the star be-g at the middle of a petal or sepal; the ıree points on the petals are longer than ιose on the sepals.

When the flower bud first appears, it nestled down in the center of the plant, arcely above the ground. It is protected y three green sepals. As it stretches up, ιe bud becomes larger and the green of ιe sepals takes on the color of the tulip ower, until when it opens there is little n the outside of the sepals to indicate that they once were green. But they still show that they are sepals, for they surround the petals, each standing out and making the flower triangular in shape as we look into it. During storms and dark days, the sepals again partially close about the rest of the flower.

The seed vessel stands up, a stout, three-sided, pale green column at the center of the flower; in some varieties, its three lobed yellowish stigma makes a Doric capital; in others, the divisions are so curled as to make the capital almost Ionic. The six stout, paddle-shaped stamens have their bases expanded so as to encircle completely the base of the pistil column; these wide filaments are narrower just below the point where the large anthers join. The anther opens along each side to discharge the pollen; however, the anthers flare out around the seed vessel and do not reach half way to the stigma, a position which probably insures cross-pollination by insects, since the bees cannot reach the nectar at the base of the pistil without dusting themselves with pollen.

The flower stem is stout, pale green, covered with a whitish bloom. The leaves

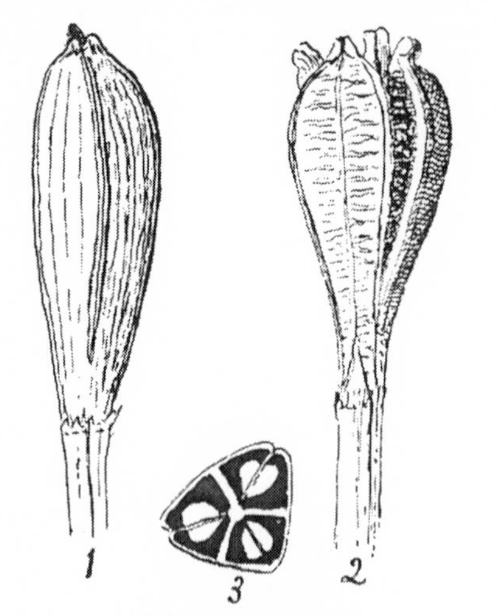

Tulip seed capsule

1, Tulip seed capsule; 2, same opened; 3, cross section of same

are long, trough-shaped, and narrow with parallel veins; the bases of the lower ones encircle the flower stem and have their edges more or less ruffled and their tips recurved; the upper leaves do not completely encircle the flower stem at their bases. The texture of the leaves is somewhat softer on the inside than on the outside, and both sides are grayish green.

After the petals and stamens are dropped the seed vessel looks like an ornamental tip to the flowerstalk; it is three-sided, and has within double rows of seeds along each angle.

The bulb is formed of several coats, or layers, each of which extends upward and may grow into a leaf; this shows that the bulb is made up of leaves which are thickened with the food stored up in them during one season, so as to start the plant growing early the next spring. In the heart of each bulb is a flower bud, sheltered by the fleshy leaf-layers around it, which furnish it food in the spring. This structure of the bulb shows how the leaves clasp the flower stem at their bases. The true roots are below the bulb, making a thick tassel of white rootlets, which reach deep i the soil for minerals and water.

Tulips are very accommodating; t will grow in almost any soil, if it is drained so that excessive moisture may rot the bulbs. In preparing a bed, it sho be rounded up so as to shed water; should also be worked deep and made ri If the soil is stiff and clayey, set bulbs o three inches deep, with a handful of sa beneath each. If the soil is mellow loa set the bulbs four inches deep and fr four to six inches apart each way, depe ing on the size of the bulbs. They sho be near enough so that when they bloss the bed will be covered and show no ga Take care that the pointed tip of bulb is upward and that it does not to one side as it is covered. October is usual time for planting, as the beds often used for other flowers during summer. However, September is not early for the planting, as the more r growth made before the ground freez the better; moreover, the early buyers h best choice of bulbs. The beds should protected by a mulch of straw or lea during the winter, which should be rak off as soon as the ground is thawed the spring. The blossoms should be as soon as they wither, in order that t new bulbs which form within and at t sides of the parent bulb may have all the plant food, which would otherwise to form seed. Tulips may be grown fr seed, but it takes from five to seven ye to obtain blossoms, which may be qu unlike the parent. Most of these seedli will be worthless; a few may develop i desirable new tulips. The bulblets gr to a size for blooming in two or th years; the large one which forms in t center of the plant will bloom the n season.

LESSON 151

The Tulip

Leading Thought — The tulips bl som early, because they have food stor in the bulbs the year before, ready to early in the spring. There are many va ties; each is worth studying carefully, a should all know how to grow these ıtiful flowers.

(ETHOD — These observations may be e upon tulips in school gardens or quets. The best methods of cultivat- hould be a part of the garden training. this, consult the seed catalogues; also he pupils form some idea of the num- of varieties from the seed catalogues. ter-color drawings may be used as s in studying the tulip. The red va- es are best for beginning the study, then follow with the other colors; differences.

)BSERVATIONS — 1. What is the color our tulip? Is it all the same color? Is bottom of the flower different in color? ıat does the pretty shape of these dif- nt colors at the heart of the flower mble?

. Look at a tulip just opening. What ses it to appear so triangular? Can you that the three sepals are placed out- the petals? Is there any difference in or between the sepals and petals on inside? On the outside? Are the sepals petals the same in length and shape? the three petals more satiny on the de than the sepals? Is the center part he petal as soft as the edges?

. When the tulip flower bud first be- s to show, where is it? What color are sepals which cover it? Describe the ning of the flower. Do the green sepals off? What becomes of them?

4. In the open flower, where is the seed pod, and how does it look? How do the anthers surround the seed pod, or ovary? Describe the anthers, or pollen boxes. What color are they? What color is the pollen? Do the anthers reach up to the stigma, or tip of the seed pod? Where is the nectar in tulips? How do the insects become covered with the pollen in reaching it? Do the flowers remain open during dark and stormy days?

5. Describe the tulip stem and the leaves. Do the leaves completely encircle the flower stem at the base? Are their edges ruffled? In the sprouting plant, do these outer basal leaves enfold the leaves which grow higher on the stem? Are the leaves the same color above and below? What shade of green are they?

6. After the petals have dropped, study the seed pod. Cut it crosswise and note how many angles it has. How are these angles filled? Should tulips be allowed to ripen seeds? Why not?

7. Study a bulb of a tulip. There are outer and inner layers and a heart. What part of the plant do the outer layers make? What part does the center make? Where are the true roots of the tulip?

8. When should tulip bulbs be planted? How should you prepare the soil? How protect the bed during the winter? How long would it take to grow the flowers from the seed? Do you know the history of tulips?

THE PANSY

Some people are pansy-faced and some ısies are human-faced, and for some oc- t reason this puts people and pansies on istinctly chummy basis. When we ana- e the pansy face, we find that the dark ots at the bases of the side petals make eyes, the lines radiating from them king quite eyelashy. The opening of nectar-tube makes the nose, while the ot near the base of the lower petal has do for a mouth, the nectar guiding-lines ing not unlike whiskers. Meanwhile, the two upper petals give a "high-browed" look to the pansy countenance, and make it a wise and knowing little face.

The pansy nectar is hidden in the spur made by the lower petal extending behind the flower. The lines on the lower and side petals all converge, pointing directly to the opening which leads to this nectar-well. Moreover, the broad lower petal serves as a platform for the bee to alight upon, while she probes the nectar-well with her tongue.

Verne M

But at the door leading to the nectar-well sits a little man; his head is green, he wears a white cape with a scalloped, reddish brown collar, and he sits with his bandy legs pushed back into the spur as if he were taking a foot bath in nectar. This little pansy man has plenty of work to do; for his mouth, which is large and at the top of his green head, is the stigma. The cape is made of five overlapping stamens, the brown, scalloped collar being the anthers; his legs consist of prolongations of the two lower stamens. And when the bee probes the nectar-well with her tongue, she tickles the little man's feet so that his head and shoulders wriggle; and thus she brushes the pollen dust from his collar against her fuzzy face, and at the same time his mouth receives the pollen from her dusty coat.

As the pansy matures, the little man grows still more manlike; after a time he sheds his anther cape, and we can see that his body is the ribbed seed pod. did not eat pollen for nothing, for h full of growing seeds. Sometimes plush brushes, which are above his h in the pansy flower, become filled w pollen, and perhaps he gets a mouth of it.

The pansy sepals, five in number, fastened at about one-third of their leng their heart-shaped bases making a li green ruffle around the stem where joins the flower. There is one sepal ab and two at each side, but none below nectar-spur. The flowerstalk is quite sh and bends so that the pansy seems to lo sidewise at us instead of staring strai upward. The plant stem is angled a crooked and stout. In form, the leaves most varied; some are long and point others wide and rounded. The edges slightly scalloped and the leaf may h at its base a pair of large, deeply lob stipules. In a whole pansy bed it wo

improbable that one would find two
'es just alike.

'he pansy ripens many seeds. The
ed seed capsule, with its base set in
sepals, finally opens in three valves
the many seeds are scattered. To send
m as far afield as possible, the edges
ach valve of the pod curl inward, and
p the seeds out as boys snap apple seeds
n the thumb and finger.

'ansies like deep, rich, cool, moist soil.
ey are best suited to a northern climate,
prefer the shady side of a garden to
full sunshine. The choice varieties
perpetuated through cuttings. They
y be stuck in the open ground in sum-
r in a half-shady place and should be
l watered in dry weather. All sorts of
sies are readily raised from seed sown
pring or early summer, and seedlings,
en well established, do not suffer, as a
:, from winter frosts.

'he general sowing for the production
arly spring bloom is made out of doors
August, while seeds sown indoors from
ruary to June will produce plants to
ver intermittently during the late sum-
r and fall months. When sowing pansy
d in August, sow the seed broadcast
a seed-bed out-of-doors, cover it very
tly with fine soil or well-rotted ma-
e, and press the seed in with a small
rd; then mulch the seed-bed to the
ckness of one inch with long, strawy
se manure from which the small par-
es have been shaken off, so as to have
soil well and evenly covered. At the
of two weeks the plants will be up.
en remove the straw gradually, a little
a time, selecting a dull day if possible.
ep the bed moist.

f the pansies are allowed to ripen seeds
season of bloom will be short, for
en its seeds are scattered the object of
plant's life is accomplished. Flowers
ne with the forming seeds are smaller
n the earlier ones. But if the flowers
kept plucked as they open, the plants
sistently put forth new buds. The
cked flowers will remain in good con-
ion longer if picked in the early morn-
before the bees begin paying calls, for
a fertilized flower fades more quickly than
one which has received no pollen.

LESSON 152
The Pansy

Leading Thought — The pansy is a member of the violet family. The flower often resembles a face; the colors, markings, and fragrance all attract the bees, who visit it for the nectar hidden in the spur of the lower petal.

Method — The children naturally love pansies because of the resemblance of these flowers to quaint little faces. They become still more interested after they see the little man with the green head who appears in the flower as it fades. A more practical interest may be cultivated by studying the great numbers of varieties in the seed catalogues and learning their names. This is one of the studies which leads directly to gardening. There are many beautiful pansy poems which should be read in connection with the lesson.

Observations — 1. How does the pansy flower resemble a face? Where are the eyes? The nose? The mouth? How many petals make the pansy forehead? The cheeks? The chin?

2. Where is the nectar in the pansy? Which petal forms the nectar-tube?

3. Describe how a bee gets the nectar. Where does she stand while probing with her tongue?

4. Where is the pollen in the pansy? What is the peculiar shape of the anthers? How do the two lower stamens differ in form from the three upper ones?

5. Where is the stigma? Does the bee's tongue go over it or under it to reach the nectar? Describe the pansy arrangement for dusting the bee with pollen and for getting pollen from her tongue.

6. Observe the soft little brushes at the base of the two side petals.

7. Take a fading flower; remove the petals, and see the little man sitting with his crooked legs in the nectar-tube. What part of the flower makes the man's head? What parts form his cape? Of what is his pointed, scalloped collar formed?

8. How many sepals has the pansy? Describe them. How are they attached? When the flower fades and the petals fall, do the sepals also fall?

9. Where in the flower is the young seed pod? Describe how this looks after the petals have fallen.

10. Describe how the seed pod opens. How many seeds are there in it? How are they scattered?

11. Study the pansy stem. Is it solid? Is it smooth or rough? Is it curved? Does it stand up straight or partially recline on the ground?

12. Take a pansy leaf and sketch it with the stipules at its base. Can you find two pansy leaves exactly alike in shape, color, and size?

13. At what time should the pansy seed be planted? How should the soil be prepared?

I dropped a seed into the earth. It grew, and the plant was mine.

It was a wonderful thing, this plant mine. I did not know its name, and plant did not bloom. All I know is th planted something apparently as life as a grain of sand and there came fo a green and living thing unlike the se unlike the soil in which it stood, un the air into which it grew. No one co tell me why it grew, nor how. It had crets all its own, secrets that baffle wisest men; yet this plant was my frie It faded when I withheld the light, wilted when I neglected to give it wa it flourished when I supplied its sim needs. One week I went away on a va tion, and when I returned the plant dead; and I missed it.

Although my little plant had died soon, it had taught me a lesson; and lesson is that it is worth while to hav plant.

— "The Nature-Study Ide
L. H. Bai

Verne Mo

THE BLEEDING HEART

The summer's flower is to the summer sweet,
Though to itself it only live and die.

— Shakespeare

For the intricate structure of this type of flower, the bleeding heart is much more easily studied than its smaller wild sisters, the Dutchman's-breeches or squirrel corn; therefore it is well to study these flowers when we find them in profusion in gardens, and the next spring we may stu the wildwood species more understa ingly.

The flowers of the bleeding heart

utiful jewel-like pendants arranged ng the stem according to their age; the ture flower, ready to shed its petals, is r the main stem, while the tiny un- ned bud is hung at the very tip, where buds are constantly being formed ing a long season of bloom. This flower a strange modification of its petals; the pink outer ones, which make the rt, are really little pitchers with nectar their bottoms, and although they hang uth downward the nectar does not v out. When these outer petals are re- ved, we can see the inner pair placed posite to them, the two of them close gether and facing each other like two oved ladles. Just at the mouth of the chers these inner petals are almost di- led crosswise; and the parts that extend yond are spoon-shaped, like the bowls of o spoons which have been pinched out as to make a wide, flat ridge along their nters. These spoon-bowls unite at the , and between them they clasp the an- ers and stigma. Special attention should given to the division between the two rtions of these inner petals; for it is a nge, the workings of which are of much portance to the flower. On removing e outer petals, we find a strange frame- ork around which the heart-shaped part the flower seems to be modeled. These e filaments of the stamens grouped in rees on each side; the two outer ones each group are widened into frills on e outer edge, while the central one is iffer and narrower. At the mouth of the tchers all these filaments unite in a tube ound the style; near the stigma they split part into six short, white, threadlike fila- ents, each bearing a small, brilliant yel- w anther. So close together are these thers that they are completely covered y the spoon-bowls made by the inner etals, the pollen mass being flat and disc- ke. During the period when the pollen produced, the stigma is flat and imma- re; but after the pollen is shed, it be- omes rounded into lobes ready to receive ollen from other flowers.

Although the description of the plant f this flower is most complex and elabo- ate, the workings of the flower are most

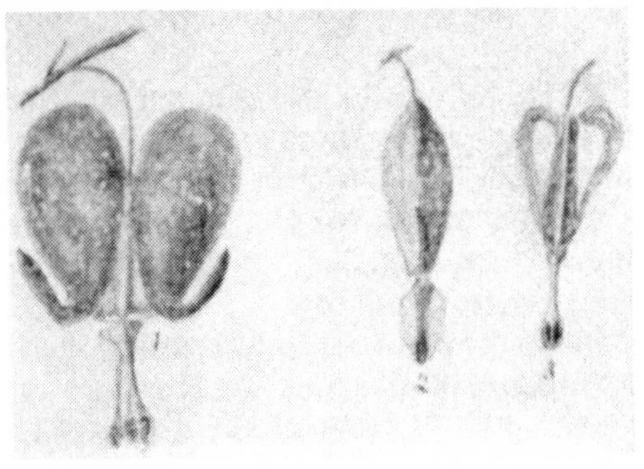

1, *Flower of bleeding heart with swing door ajar.* 2, *Side view of flower showing the broad tips of the inner petals.* 3, *Flower with outer petals removed showing inner petals — and the heart-shaped bases of the stamens*

simple. As the nectar pitchers hang mouth down, the bee must cling to the flower while probing upward. In doing this she invariably pushes against the outside of the spoon-bowls, and the hinge at their base allows her to push them back while the mass of pollen is thrust against her body; as this hinge works both ways, she receives the pollen first on one side and then on the other, as she probes the nec- tar pitchers. And perhaps the next flower she visits may have shed its pollen, and the swing door will uncover the ripe stigma ready to receive the pollen she brings.

The sepals are two little scales opposite the bases of the outer petals. Before the flower opens, the spouts of the nectar pitchers are clamped up on either side of the spoon-bowls; at first they simply spread apart, but later they curve back- ward. The seed pod is long and narrow, and in cross section is seen to contain two compartments with seeds growing on every side of the partition.

The bleeding heart is a native of China, and was introduced into Europe about the middle of the last century.

LESSON 153

The Bleeding Heart

Leading Thought — The bleeding heart flower has its pollen and stigma cov- ered by a double swing door, which the bees push back and forth when they gather the nectar.

METHOD — Bring a bouquet of the bleeding heart to the schoolroom, and let each pupil have a stem with its flowers in all stages. From this study, encourage them to watch these flowers when the insects are visiting them.

OBSERVATIONS — 1. How are these flowers supported? Do they open upward or downward? Can you see the tiny sepals?

2. How many petals can you see in this flower? What is the shape of the two outer petals? How do they open? Where is the nectar developed in these petals?

3. Take off the two outer petals and study the two inner ones. What is their shape near the base? How are their parts which project beyond the outer petals shaped? What does the spoon-end of these petals cover? Can you find the hi in these petals?

4. Where are the stamens? How ma are there? Describe the shape of the mens near the base. How are they uni at the tip?

5. Where is the stigma? The style? T ovary?

6. Supposing a bee is after the nec where must she rest while probing for Can she get the nectar without pushi against the flat projecting portion of inner petals? When she pushes th spoon-bowls back, what happens? D she get dusted with pollen? After s leaves, does the door swing back? Supp she visits another flower which has sh its pollen, will she carry pollen to stigma? Does she have to work the hing door to do this?

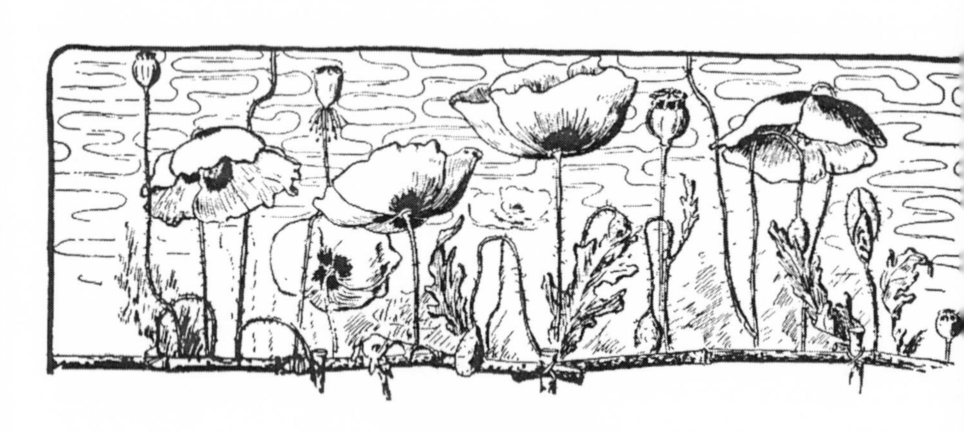

THE POPPIES

Perhaps we might expect that a plant which gives strange dreams to those who eat of its juices should not be what it seems in appearance. I know of nothing so deceptive as the appearance of the poppy buds, which, rough and hairy, droop so naturally that it seems as if their weight must compel the stem to bend; and yet, if we test it, we find the stem is as stiff as if made of steel wire. Moreover, the flower and the ripened seed capsule must be far heavier than the bud; and yet, as soon as the flower is ready to open, the stem straightens up, although it does not always remove the traces of the crook; and after the capsule is full of ripened see the stem holds it up particularly stiff as if inviting the wind to shake out t seeds.

The rough covering of the bud co sists of two sepals, as can be easily see but if we wish to see the poppy shed sepals, we must get up in the morning, f the deed is usually done as soon as the fi rays of the early sun bring their messa of a fair day. The sepals break off at th base and fall to the ground. The two o posite outer petals unfold, leaving the tw inner petals standing erect, until the su shine folds them back. An open popp n looked at below, shows two petals, semicircular, and overlapping each r slightly; looked at from above, we two petals, also half circles, set at : angles to the lower two, and divided ı each other by the pistil.

he pistil of the poppy is, from the nning, a fascinating box. At first, it vase with a circular cover, upon which ridges, placed like the spokes of a el. If these ridges are looked at with ns, particles of pollen may be seen ring to them; this fact reveals the t that each ridge is a stigma, and of these radiating stigmas are joined s better to catch the pollen. In a e of fringe about the pistil are the ens. In the study of the stamens, we ld note whether their filaments ex- l or dilate near the anthers, and we ld also note the color of the masses pollen which crowd out from the ers.

espite the many varieties of poppies, e are only four species commonly cul- ed. The opium poppy has upon its ge a white bloom, the filaments of its ens are dilated at the top, and its seed ule is smooth. The Oriental poppy all of these characteristics, except that oliage is green and not covered with m. Its blossom is scarlet and very

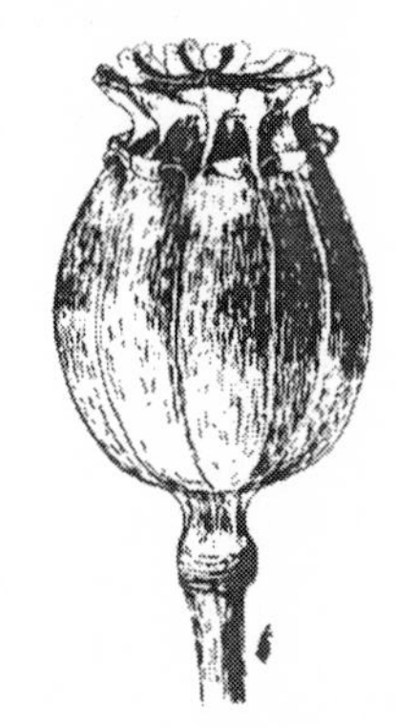

Anna C. Stryke

The poppy seed-shaker

e, and has a purple center in the petals purple stamens; it has three sepals. Its erstalks are stout and leafy. The corn py, which grows in the fields of Eu-

F. A. Southard, Jr.

Oriental poppies, showing buds and blossom

rope, is a weed we gladly cultivate. This naturally has red petals and is dark at the center of the flower; but it has been changed by breeding until now we have many varieties. Its foliage is finely cut and very bristly or hairy. Its seed capsule is not bristly. To see this poppy at its best, we should visit northern Italy or southern France in late May, where it makes the grain fields gorgeous. This is the original parent of all the Shirley poppies. The Arc- tic, or Iceland poppy, has flowers of satiny texture and finely crumpled; its colors are yellow, orange, or white, but never scarlet like the corn poppy; it has no leaves on its flower stem, and its seed capsule is hairy. Of these four species, the opium poppy and the corn poppy are annuals, while the Arctic and the Oriental species are peren- nials.

The bees are overfond of the poppy pollen and it is a delight to watch the fervor with which they simply wallow in it, brushing off all of the grains possible onto their hairy bodies. I have often seen a honeybee seize a bunch of the anthers and rub them against the underside of her body, meanwhile standing on her head in an attitude of delirious joy. As an indica- tion of the honeybee's eye for color, I have several times seen a bee drop to the ground to examine a red petal which had fallen. This was evidence that she trusted, at least in part, to the color to guide her to the pollen.

But perhaps it is the development of the poppy seed capsule which we find the most interesting of the poppy performances. After fertilization, the stigma disc develops a scalloped edge, a stigma rounding out the point of each scallop; and a sharp ridge, which continues the length of the globular capsule, runs from the center of each scallop. If examined on the inside, it will be seen that the ridge on the capsule is the edge of a partition which extends only part way toward the center of the capsule. On these partitions, the little seeds are grown in great profusion, and when they ripen, they fall together in the hollow center of the seed box. But how are they to get out? This is a point of interest for the children to observe, and they should watch the whole process. Just beneath the stigma disc, and between each two of the sharp ridges, the point loosens; later, it turns outward and back, leaving a hole which leads directly into the central hollow portion of the capsule. The way these points open is as pretty a story as I know in flower history. This beautiful globular capsule, with its graceful pedestal where it joins the stem, is a seed-shaker instead of a salt- or pepper-shaker. Passing people and animals push against it and the stiff stem bends and then springs back, sending a little shower of seeds this way and that; or a wind sways the stalk, and the seeds are sown, a few at a time, and in different conditions of season and weather. Thus, although the poppy puts all her eggs in one basket, she sends them to market a few at a time. The poppy seed is a pretty object, as seen through the lens. It is shaped like a round bean, and is covered with a honeycomb network.

LESSON 154

The Poppies

Leading Thought — The poppies shed their sepals when the flowers expand; they offer quantities of pollen to the bees, which are very fond of it. The seed capsule develops holes around the top, through which the seeds are shaken few at a time.

Method — It is best to study these fl ers in the garden, but the lesson may given if some of the plants with the b are brought to the schoolroom, care ing taken that they do not droop.

Observations — 1. Look at the bu the poppy. How is it covered? How m sepals? Can you see where they unite the stem bent because the bud is hea What happens to this crook in the st when the flower opens? Does the cr always straighten out completely?

2. Describe how the poppy sheds sepals. At what time of day do the popp usually open?

3. Look at the back of, or beneath, open flower. How many petals do see? How are they arranged? Look at base of the flower. How many pe do you see? How are they arranged relation to the lower petals and to pistil?

4. Look at the globular pistil. Desc the disc which covers it. How many rid on this disc? How are they arranged? L at the ridges with a lens and tell w they are.

5. Look at the stamens. How are t arranged? Describe the anthers — th color, and the color of the pollen. Wa the bees working on the poppies, and n if they are after nectar or pollen.

6. Find all the varieties of poppies p sible, and note the colors of the petals the outside, the inside, and at the ba of the stamens, including filaments, thers, and pollen; of the pistil disc a ovary. Sketch the poppy opened, and a in the bud. Sketch a petal, a stamen, a the pistil, in separate studies.

7. Study the poppy seed box as it ens. How does the stigma disc look? W is the shape of the capsule below the d Is it ridged? What relation do its rid bear to the stigma ridges on the d Cut a capsule open, and note what th ridges on the outside have to do with partitions inside. Where are the se borne?

8. Note the development of the ho

ıeath the edge of the disc of the poppy sule. How are they made? What are y for? How are the seeds shaken from se holes? What shakes the poppy seed and helps sow the seeds? Look at a d through a lens, and describe its form l decoration.

). Notice the form of the poppy leaf, l note whether it is hairy or covered with bloom. What is there peculiar about the smell of the poppy plant? Where do poppies grow wild?

10. Is the slender stem smooth, or grooved and hairy? Is it solid or hollow?

11. When a stem or leaf is pierced or broken off, what is the color of the juice which exudes? Does this juice taste sweet, or bitter and unpleasant?

THE CALIFORNIA POPPY

California poppies

Although this brilliant flower blossoms ːerfully for us in our Eastern gardens, can never understand its beauty until see it glowing in masses on the Calinia foothills. We can easily understand y it was selected as the flower of that at state, since it burnished with gold hills, above the gold buried below; and that land that prides itself upon its ıshine, these poppies seem to shine up the sun shines down. The literature of lifornia, and it has a noble literature its own, is rich in tributes to this faed flower. There is a peculiar beauty the contrast between the shining flower d its pale blue-green, delicate masses of iage. Although it is called a poppy and ongs to the poppy family, yet it is not true poppy, but belongs to a genus med after a German who visited Calinia early in the nineteenth century, acnpanying a Russian scientific expedin; this German's name was Eschscholtz, l he, like all visitors, fell in love with s brilliant flower, and in his honor it s named *Eschscholtzia* (es-sholts-ia) *ifornica*. This is not nearly so pretty so descriptive as the name given to s poppy by the Spanish settlers on the cific Coast, for they called it *Copa-de-*), cups of gold.

The bud of the *Eschscholtzia* is a pretty ng; it stands erect on the slender, rather g stem, which flares near the bud to urnlike pedestal with a slightly ruffled ı, on which the bud is set. This rim is en pink above, and remains as a pretty se for the seed pod. But in some garden ieties, the rim is lacking. The bud itself is covered with a peaked cap, like a Brownie's toboggan cap stuffed full to the tip. It is the shape of an old-fashioned candle extinguisher; it is pale green, somewhat ribbed, and has a rosy tip; it consists of two sepals, which have been sewed together by Mother Nature so skillfully that we cannot see the seams. One of the most interesting performances to watch that I know is the way this poppy takes off its cap before it bows to the world. Like magic the cap loosens around the base; it is then pushed off by the swelling, expanding petals until completely loosened, and finally it drops.

The petals are folded under the cap in

an interesting manner. The outer petal enfolds all the others as closely as it can, and its mate within it enfolds the other two, and the inner two enfold the stamens

Anna C. Stryke

California poppies

with their precious gold dust. When only partially opened, the petals cling protectingly about the many long stamens; but when completely opened, the four petals flare wide, making a flower with a golden rim and an orange center, although among our cultivated varieties they range from orange to an anæmic white. To one who loves them in their glorious native hues, the white varieties seem almost repulsive. Compare one of these small, pale flowers with the great, rich, orange ones that glorify some favored regions in the Mojave Desert, and we feel the enervating and decadent influence of civilization.

The anthers are many and long, and are likely to have a black dot on the short filament; at first, the anthers stand i close cluster at the center of the flov but later they flare out in a many-poin star. Often, when the flowers first op especially the earlier ones, the stig cannot be seen at all; but after a time three, or even six stigmas, spread w athwart the flower and above the stam star, where they may receive pollen fr the visiting insects. The anthers give ab dance of pollen, but there is said to no nectary present. This flower is a g guardian of its pollen, for it closes dur the nights and also on dark and rainy d only exposing its riches when the s shine insures insect visitors. In our E ern gardens it closes its petals in the sa order in which they were opened, though there are statements that in C fornia each petal folds singly around own quota of anthers. The insects in C fornia take advantage of the closing pe and often get a night's lodging wit them, where they are cozily housed w plenty of pollen for supper and breakf and they pay their bill in a strange way carrying off as much of the golden m as adheres to them, just as the man w weighs gold dust gets his pay from w adheres to the pan of his scales.

After the petals fall, the little pod very small, but its growth is as astonish as that of Jack's beanstalk; it finally atta a slim length of three inches, and of more. It is grooved, the groove runn straight from its rimmed base to its r tip; but later a strange twisting takes pla If we open one of these capsules leng wise, we must admire the orderly way which the little green seeds are faster by delicate white threads, in two crowc rows, the whole length of the pod.

The leaf is delicately cut and makes foliage a fine mass, but each leaf is qu regular in its form. It has a long, flatter petiole, which broadens and clasps stem somewhat at its base. Its blade five main divisions, each of which is dee cut into finger-like lobes. The color of t foliage and its form show adaptations desert conditions.

This plant has a long, smooth tapro

ecially adapted for storing food and isture needed during the long, dry Cali-nia summers; for it is perennial in its tive state, although in the wintry East, plant it as an annual.

LESSON 155

The California Poppy

Leading Thought — The California ppy is a native of California; there it ssoms during the months of February, arch, and April in greatest abundance. is found in the desert as well as among foothills.

Method — If possible, the students uld study this flower in the garden. In East, it flowers until frost comes, and ords a delightful subject for a Septem- lesson. In California it should be stud- in the spring, when the hills are vered with it. But the plant may be ought into the schoolroom, root and and placed in a jar, under which con-tions it will continue to blossom.

Observations — 1. Look at the Cali-rnia poppy as a whole and tell, if you n, why it is so beautiful when in blos-m.

2. Look at the flower bud. What sort stalk has it? What is the shape of the lk just below the bud? What is the lor of the little rim on which the bud sts? What peculiarity has this bud? De-ribe the little cap.

3. Watch a flower unfold. What hap-ns to the "toboggan cap"? How does e bud look after the cap is gone? What its appearance when the petals first en? When they are completely open?

4. Describe the anthers. How do they and when the flower first opens? How ter? Can you see the stigmas at first? escribe them as they look later.

5. Does the poppy remain open at ght? Does it remain open during cloudy rainy weather?

6. Do the petals have the same position at they did in the bud? As the flower atures, note how each petal curls. Do ey all fall at once? Are there any anthers ft after the petals fall?

7. How does the little pod look when the petals first fall? What happens to it later? Note the little rim at its base. Cut the seed pod open lengthwise, examine the seeds with a lens, and describe how they are fastened to the sides of the pod. Are the ribs straight from end to end in the pod at first? Do they remain in this position? How does the pod open and scatter its seeds?

8. Study the leaf of this California poppy. Describe how it joins the stem. Sketch a leaf showing its chief divisions into leaflets and how each leaflet is divided. Note that the juice of the stem has the peculiar odor of muriatic acid.

9. Look at the root. Do you think it is fitted to sustain the plant through a long, dry summer? What kind of summers do they have in California? Where does the poppy grow wild?

10. Read all the accounts you can find of the California poppy, and write a story describing why it was chosen as the flower of that great state, and how it came by its name.

In a low brown meadow on a day
Down by the autumn sea,
I saw a flash of sudden light
In a sweep of lonely gray;
As if a star in a clouded night
One moment had looked on me
And then withdrawn; as if the spring
Had sent an oriole back to sing
A silent song in color, where
Other silence was too bad to bear.

I found it and left it in its place,
The sun-born flower in cloth of gold
That April owns, but cannot hold
From spending its glory and its grace
On months that always love it less,
But take its splendid alms in their distress.
Back I went through the gray and the brown,
Through the weed-woven trail to the distant town;
The flower went with me, fairly wrought
Into the finest fiber of my thought.

— "A California Poppy in November,"
Irene Hardy

THE NASTURTIUM

It is quite fitting that the nasturtium leaves should be shaped like shields, for that is one of their uses; they are shields which protect the young nasturtium seeds from the hot sun and from the view of devouring enemies. The nasturtiums are natives of Peru and Chili, and it is fitting that the leaves should develop in shield-shape, and the shields overlap until they form a tent which shades the tender developing seed from the burning sun. But they do not shield the flower, which thrusts its brilliant petals out between the shields, and calls loudly to the world to admire it. It would indeed be a pity for such a remarkable flower to remain hidden; its five sepals are united at their base,

W. Atlee Burpee Co.

Single nasturtium

and the posterior one is extended into long spur, a tube with a delectable nect well at its tip. The five petals are around the mouth of this tube, the tw upper ones differing in appearance a office from those below; these two sta up like a pair of fans, and on them a lines which converge; on the upper sep are similar lines pointing toward the sar interesting spot. And what do all the lines lead to, except a veritable treasu cave filled with nectar! The lower pet tell another story; they stand out, ma ing a platform or doorstep, on which t visiting bee alights. But it requires a b insect to pollinate this flower, and wh if some inefficient little bee or fly shou alight on the petal-doorstep and steal in the cave surreptitiously? This contingen is guarded against thus: each of the lower petals narrows to a mere insect fo bridge at their inner end; and this fo bridge is quite impassable, because it beset with irregular little spikes and pr jecting fringes, sufficient to perplex discourage any small insect from crawli that way.

But why all these guiding lines a guarded bridges? If you watch the sa blossom for several successive days, it w reveal this secret. When a flower fi opens, the stamens are all bent downwar but when an anther is ready to open en doors, the filament lifts it up and es it like a sentinel blocking the door- to the nectar treasure. Then when the ɔer comes, whether it be butterfly, or hummingbird, it gets a round of en ammunition for its daring. Perhaps e may be two or three anthers stand- guard at the same time, but, as soon heir pollen is exhausted, they shrivel give room for fresh anthers. Mean- le, the stigma has its three lobes ed and lying idly behind and below anthers; after all the pollen is shed, style rises and takes its position at the : entrance and opens up its stigmas, a three-tined fork, to rake the pollen n any visiting insect, thus robbing the ɔer of precious gold dust which shall ilize the seeds in its three-lobed ovary. ıough the flower flares its colors wide, s attracting the bees and humming- ls, yet the growing seeds are protected. : stalk which held the flower up ight now twists around in a spiral and ws the triplet seeds down behind the :n shields.

Nasturtium leaves are very pretty, and often used as subjects for decorative er-color drawings. The almost circu- leaf has its stalk attached below and tle at one side of the center; the leaves brilliant green above but quite pale eath, and are silvery when placed be- th the water. The succulent stems e a way of twisting half around the es of the trellis and thus holding the nt secure to its support. But if there is

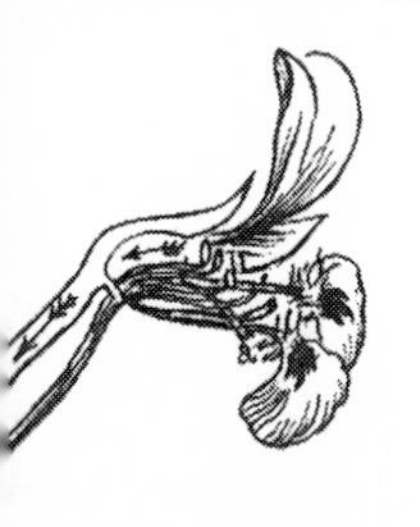

, Nasturtium flower in early stage of blos- ıing. Note the anthers lifted in the path he nectar which is indicated by the arrow. : closed stigma is shown deflected at a. 'he same flower in later stage; the anthers empty and deflected. The stigma is raised in the nectar path*

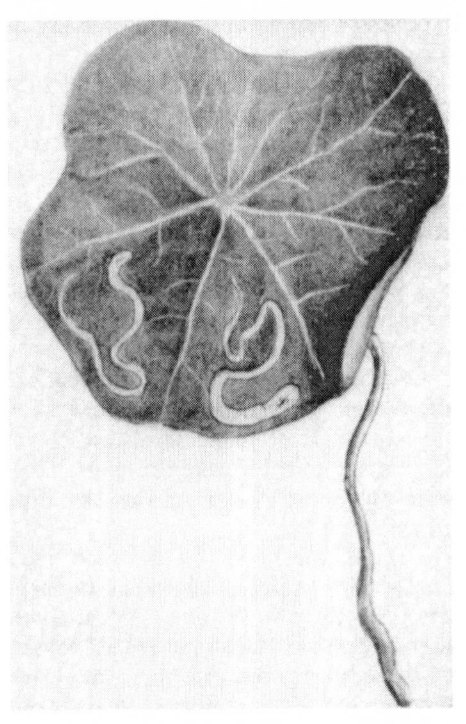

Nasturtium leaf showing the work of serpentine miners

no trellis, the main stem grows quite stocky, often lifting the plant a foot or two in height, and from its summit sending out a fountain of leaf- and flower-stalks. Some nasturtiums are dwarf and need no support.

The nasturtium is among the most interesting and beautiful of our garden flowers, and will thrive in any warm, sunny, fairly moist place. Its combinations of color are exceedingly rich and brilliant. H. H. says of it:

How carelessly it wears the velvet of the
same
Unfathomed red, which ceased when Ti-
tian ceased
To paint it in the robes of doge and priest.

LESSON 156

The Nasturtium

Leading Thought — The nasturtium has a special arrangement by which it sends its own pollen to other flowers and receives pollen from other flowers by insect messengers.

Method — The nasturtiums and their foliage should be brought into the schoolroom in sufficient quantity so that each child may have a leaf and a flower for study. The object of the lesson is to interest the pupils in studying, in their gardens, one flower from the bud until the petals wither, taking note of what happens each day and keeping a list of the insect visitors.

Observations — 1. Look at the back of the flower. What is there peculiar about the sepals? How many sepals are there? How many join to make the spur? What is in this spur? Taste of the tip. Find where the nectar is.

2. Look the flower in the face. How do the two upper petals differ in shape from the three lower ones? What markings are there on the upper petals? Where do these lines point? Are there any markings on the sepals pointing in the same direction? If an insect visiting a flower should follow these lines, where would it go?

3. Describe the shape of the lower petals. Suppose a little ant were on one of these petals and she tried to pass over to the nectar-tube or spur, would the fringes hinder her?

4. Look down the throat of the spur, and tell what a bee or other insect would have to crawl over before it could get at the nectar.

5. In your garden, or in the bouquet in the window if you cannot visit a garden, select a nasturtium that is just o
ing and watch it every day, making
following notes: When the blossom
opens, where are the eight stamens?
the unripe, closed anthers lifted so a
be in the path of the bee which is gat
ing nectar? How do the anthers o
How is the pollen held up in the
to the nectar? Can you see the sti
of this flower? Where is it? *Note the s*
flower on successive days: How m
anthers are open and shedding pollen
day? Are they all in the same positio
yesterday? What happens to the ant
which have shed their pollen?

6. When the stigma rises in the ne
path, how does it look? Where are all
anthers when the stigma raises its t
tines which rake the pollen off the visi
insect? Do you know why it is an ad
tage to the nasturtium to develop its
by the aid of the pollen from ano
plant?

7. Can you see the beginning of
seedcase when the stigma arises to rec
the pollen?

8. The flowers project beyond
leaves. Do the ripening seedcases do t
What happens to their stems to withd
them behind the leaf?

9. Sketch a nasturtium leaf, and
plain in what way it is like a shield. H
does the leaf look when under water?

10. What sort of stem has the
turtium? How does it manage to cl
the trellis? If it has no trellis upon wh
it can climb, does it lie flat upon
ground?

THE BEE-LARKSPUR

This common flower of our gardens, sending up from a mass of dark, deeply cut leaves tall racemes of purple or blue flowers, has a very interesting story to tell those who watch it day by day and get acquainted with it and its insect guests. The brilliant color of the flowers is due to the sepals, which are purple or blue, in varying shades; each has on the back side near its tip, a green thickened spot. If we glance up the flowerstalk, we
see that, in the upper buds, the se
are green, but in the lower buds t
begin to show the blue color; and i
bud just ready to open, we can see t
the blue sepals are each tipped wit
green knob, and this remains green a
the sepals expand. The upper and r
most sepal is prolonged into a spur, wh
forms the outside covering of the nec

r; it is greenish, and is wrinkled like ong-wristed suede glove; two sepals ead wide at the sides and two more be-. All this expanse of blue sepals is a kground for the petals, which, by their trasting color, attract the bees looking nectar. Such inconsequential petals hey are! Two of them " hold hands " nake an arch over the entrance to the tar tube; and just below these on each are two more tiny, fuzzy, spreading als, often notched at the tip and al-/s hinged in a peculiar way about the er petal; they stand at the door to nectar storehouse. If we peel off the nkled sepal-covering of the spur, we see the upper petals extending back o it, making a somewhat double-reled nectary.

f we look into a larkspur flower just opened, we see below the petals a bunch of green anthers, hanging by white thread-like filaments to the center of the flower and looking like a bunch of lilliputian

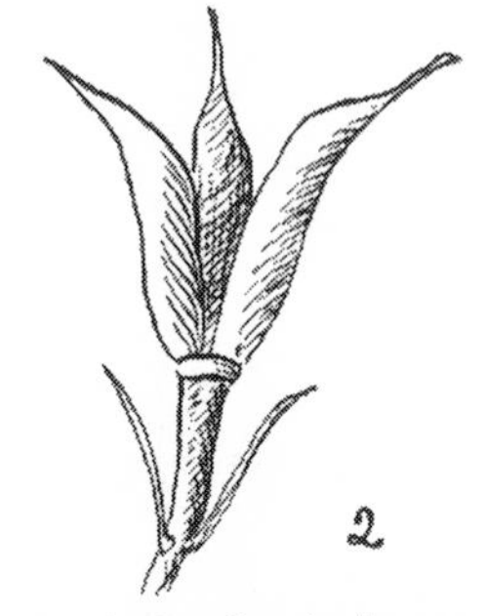

1, *Drawing of the bee-larkspur flower enlarged.* 2, *The seed capsule of the bee-larkspur*

bananas. Behind these anthers is an undeveloped stigma, not visible as yet. After the flower has been open for a short time, three or four of the anthers rise up and stand within the lower petals; while in this position, their white pollen bursts from them, and no bee may then thrust her tongue into the nectar-spur without being powdered with pollen. As soon as the anthers have discharged their pollen, they shrivel, and their places are taken by fresh ones. It may require two or three days for all the anthers to lift up and get rid of their pollen. After this has been accomplished, the three white, closely adhering pistils lift up their three stigmas into the path to the nectar; and now they are ready to receive the pollen which the blundering bee brings from other flowers. Since we cannot always study the same flower for several consecutive days, we can read the whole story by studying the flowers freshly opened on the upper por-

Cyrus Crosby

Bee-larkspur

tion of the stalk, and those below them that are in more advanced stages.

The bees, especially the bumblebee, will tell the pollination story to us in the garden. A visiting bee alights on the lower petals; grasping these firmly she thrusts her head into the opening between them and probes the spur twice, once in each nectar-well. It is a fascinating pastime to follow the bee as it goes from flower to flower like a Madam Pompadour, powdered with white pollen. The tips of the tall flower-

The larkspur

1, Early stage with stigma deflected. 2, Advanced stage with stigma raised

stalks are likely to bend or curl over; but no matter what the direction the broken or bent stem takes, the flowers will twist around on their pedicels until they face the world and the bee, exactly as if they were on a normally erect stem.

All the larkspurs have essentially the same pollen story, although some have only two petals; in every case the anthers at first hang down, and later rise up in the path to the nectar. Thus they discharge their pollen; after they wither, the stigmas arise in a similar position.

The bee-larkspur has a very beautiful fruit. It consists of three graceful capsules rising from the same base and flaring out into pointed tips. The seeds are fastened to the curved side of each capsule, which, when ripe, opens; and then they may be shaken out by the winds. When study the bud, we notice two little bracts se its base and these remain with the fr

LESSON 157

The Bee-Larkspur

Leading Thought — The bee-larks begins blossoming early in the season, blossom stalk elongating and develop new buds at its tip until late in autur The flower has a very interesting way inducing the bees to carry its pollen.

Method — Bring to the schoolroo flowerstalk of the bee-larkspur, and th study the structure and mechanism of flower. This lesson should inspire pupils to observe for themselves the vi ing bees and the maturing seeds. them to write an account of a bumble making morning calls on the larkspurs

Observations — 1. Which flowers the larkspur open first — those near tip of the stem or those below?

2. Examine the buds toward the tip the flowerstalk. What color are the se in these buds? Do the sepals change co as the flower opens? Note the little gr knobs which tip the closed sepals t clasp the bud. What color are the se on the open flower? Is there any gr upon them when open?

3. Where is the nectar-spur? Wh sepal forms this? How are the other se arranged?

4. Now that we know the flower its brilliant color from its sepals, let find the petals. Look straight into flower, and note what forms the contr ing color of the heart of the flower; th are the petals. Can you see that two joined above the opening into the nec tube? How many are at the lower part the entrance? How are these lower pe hinged about the upper one? Peel a se cover from the nectar-spur, and see if upper petals extend back within the sp forming nectar-tubes.

5. Take a flower just opened, and scribe what you see below the pet What is the color of the anthers? Of filaments? Can you see the stigma?

6. Take a flower farther down the stalk, ich has therefore been open longer, d describe the position of the anthers this. Are any of them standing up- ht? Are they discharging their pollen? hat color is the pollen? Are these up- ht anthers in the way of the bee when e thrusts her tongue into the nectar- be?

7. Take the oldest flower you can find. hat has happened to the anthers? Can u see the pistils in this flower? In what sition now are the stigmas?

8. Push aside the anthers in a freshly ened flower and see if you can find the gmas. What is their position? How do ey change in form and position after the llen is shed? Do they arise in the th of the bee before all the pollen from the anthers of their own flower is shed?

9. SUGGESTIONS FOR OBSERVATION IN THE GARDEN — Watch a bumblebee working on the larkspur and answer the following questions: How does she hold on to the flower? Where does she thrust her tongue? Can she get the nectar without brushing the pollen from the anthers which are lifting up at the opening of the nectar-tube? In probing the older flowers, how would she come in contact with the lifted stigmas? How do the petals contrast in color with the sepals? Compare the common larkspur with the bee-larkspur, and notice the likeness and difference. What kind of fruit capsules has the bee-larkspur? Describe the seeds, and how they are scattered.

THE BLUE FLAG OR IRIS

Beautiful lily, dwelling by still rivers
Or solitary mere,
Or where the sluggish meadow brook delivers
Its waters to the weir!

The burnished dragon fly is thine attendant,
And tilts against the field,
And down the listed sunbeams rides resplendent
With steel-blue mail and shield.

— "FLOWER-DE-LUCE," HENRY W. LONGFELLOW

The iris blossom has a strange appear- ce, and this is because nothing in it what it seems. The style of the pistil divided into three broad branches which ok like petals. These, with the sepals, rm a tunnel through which bees may ss. The true petals, marked with beau- ful purple lines, stand between these nnels. It has been said that such lines flowers guide insects to the nectar- ells. This belief is open to question; for rtainly these lines on the iris leading to e center of the flower do not lead to e nectar-wells. If we look directly down to the flower of the blue flag, we see dges on the broad styles and purple ins on the petals, all leading to the center of the flower. If an insect alighting there should seek for nectar-wells at the point where all these lines meet, it would find no nectar.

Dr. Needham, in an admirable study of this flower and its visitors, tells us that he has seen the little butterflies called "skippers," the flag weevils, and other flower beetles apparently made victims of this deceptive appearance; this is some evidence that the guiding lines on flowers are noted and followed by insects.

The blue flag seems to be specially designed for bees; even the large showy blossom is, according to Sir John Lubbock, the favorite color of the bee. The bees seem to have no difficulty in finding

Leonard K. B

Larger blue flag

the nectar. The sepal with its purple and yellow tip and its dark veining and golden guiding lines marks the path to the nectar, which is far from the center of the flower. The bee alights on the lip of the sepal, presses forward scraping her back against the down-hanging stigma, then scrapes along the open anther which lies along the roof of the tunnel. The tunnel leads to the nectar-wells at the very base of the sepal.

The bees which Dr. Needham found doing the greatest work as pollen carriers were small solitary bees (*Clisodon terminalis* and *Osmia destructa*); each of these alighted with precision on the lip which forms the landing-platform of each tunnel, pushed its way in, got the nectar from both wells, came out, and immediately went to another tunnel. One might ask why the bee in coming out did not deposit the pollen from the anther on the stigma of the same flower. The stigma prevents this by hanging down, like a flap to a tent, above the entrance; its surface is so directed that it gathers pollen from the entering bee and turns its blank side to the bee making an exit. Th ingenious arrangement insures the cro fertilization which Darwin has show us is so necessary for the most vigoro and beautiful offspring.

The arrangement of the flower parts the iris may be described briefly thu three petals, three sepals, and a style wi three branches; the latter are broad a flat and cover the bases of the three s pals, forming tubes which lead to t nectar; three anthers lie along the unde side of the styles. The wild yellow i is especially fitted for welcoming the bu blebee as a pollen-carrier, since the e trance between the style and the sep is large enough to admit this larger inse The bumblebees and the honeybees wo in different varieties of iris in garde

In some varieties of iris there is a co ing resembling plush on the style whi forms the floor of the tunnel. Throu a lens this plush is exquisite — the n of white filaments standing up tipp with brilliant yellow. Various theories to the use of this plush have been a vanced, a plausible one being that it

:eep the ants out; but the ants could ly pass along either side of it. One in the garden while I was holding an in my hand, a bumblebee visited it :rly, never noting me: after she had)ed the nectar-wells, she probed or)led among the plush, working it oughly on her way out. Did she possi- find something there to eat?

LESSON 158

The Blue Flag or Iris

eading Thought — Each iris flower three side doors leading to the nectar- s; and the bees, in order to get the :ar, must brush off the pollen dust :heir backs.

/Iethod — While the blue flag is the :t interesting of our wild species of yet the flower-de-luce, or the garden is quite as valuable for this lesson. : form of the flowers may be studied he schoolroom, but the pupils should ch the visiting insects in the garden ield.

Wild iris in natural surroundings

Observations — 1. Look for the side doors of the iris blossom. Which part of the flower forms the doorstep? How is it

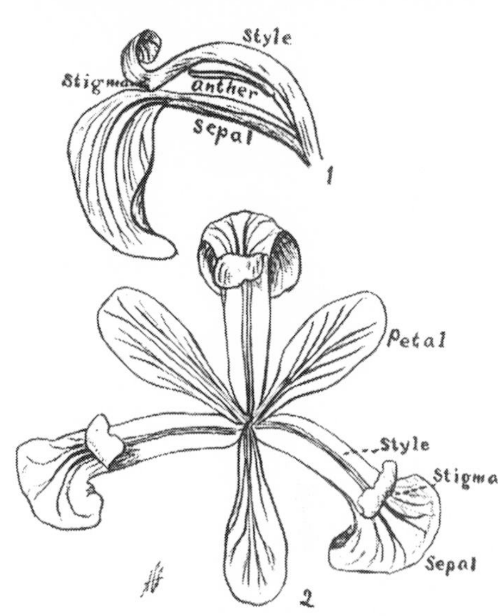

Detail of the blossoms of the blue flag flower

1, Side view of the passage to the nectar. 2, Looking directly into the iris flowers. Note the deceiving guide-lines in the petals

marked? Which part of the flower makes the arch above the entrance?

2. Find the anther, and describe how it is placed. Can you see two nectar-wells? Explain how a bee will become dusted with pollen while getting the nectar.

3. Where is the stigma? What is there very peculiar about the styles of the iris? Can a bee, when backing out from the side door, dust the stigma with the pollen

Cyrus Crosby

Fleur-de-lis

she has just swept off? Why not? How does the stigma of the next flower that the bee visits get some of the pollen from her back?

4. Look straight down into an iris flower. Can you see the three petals? How are they marked?

5. Watch the insects visiting the iris. Do you know what they are? What do the different insects do?

6. Describe the way the iris flower bud is enfolded in bracts. What is there peculiar about the way the iris leaves join the stem?

7. How many kinds of flag, or iris, you know?

8. Describe the seed vessel and seed the iris.

The fleur-de-lis is the national flowe France.

It is said that the Franks of old ha custom, at the proclamation of a king elevating him upon a shield or *target,* *placing in his hand a reed,* or *flag in b som, instead of a sceptre.*

— "Among the Flowers and Tr with the Poets," Wait and Leon

THE SUNFLOWER

Anna C. Stryke

The sunflower. Next to the ray flowers are the florets in the last stages of blossoming with stigmas protruding; next within are rows in the earlier stage with pollen bursting from the anther-tubes, while at the center are unopened buds

Many of the most beautiful of the autumn flowers belong to the *Compositæ,* a family of such complicated flower arrangement that it is very difficult for the child or the beginner in botany to comprehend it; and yet, when once understood, the composite scheme is very simple and beautiful, and is repeated over and over in flowers of very different pearance. It is a plan of flower cooperati there are many flowers associated to for single flower-head. Some of these, "ray" or "banner" flowers, hold bright pennants which attract inse while the disc flowers, which they round, attend to the matter of the poll tion and production of seed.

The large garden sunflower is the tea er's ally to illustrate to the children story of the composites. Its florets are large that it is like a great wax model. A what could be more interesting than watch its beautiful inflorescence — t orderly march toward the center in dou lines of anther columns, with phalan bearing the stigmas surrounding th and outside all, the ranks of ray flov flaunting their flags to herald to the wo this peaceful conquest of the sleepi tented buds at the center?

Ordinarily, in nature-study we do pull the flowers apart, as is necessary botany; in nature-study, all that we c to know of the flower is what it does, we can see that without dissection. with the *compositæ* the situation is qu different. Here we have an assemblage flowers, each individual doing its c work for the community; and in orde make the pupils understand this fac is necessary to study the individual flor

We begin with the study of one

buds at the center of the flower-
d; this shows the white, immature seed
ow, and the closed, yellow corolla-tube
ve. Within the corolla may be seen
brown anther-tube, and on the upper
t of the seed are two little, white, ear-
scales, to which especial notice should
directed, since in other composites
re are many of these scales and they
m the pappus — the balloon to carry
seed. The bud shows best the pro-
ting chaffy scale which enfolds the
d, its pointed, spine-edged tip being
ded over the young bud, as may be
n by examining carefully the center
a freshly opened sunflower. In this
ular bud (shown in figure) there is a
escopic arrangement of the organs, and
after another is pushed out. First,
corolla-tube opens, starlike, with five
nted lobes, very pretty and graceful,
h a bulblike base; from this corolla
shes out the dark-brown tube, made up
five anthers grown together. By open-
the corolla, we see the filaments of
stamens below the joined anthers.
is anther-tube, if examined through a
s, shows rows of tiny points above and
ow, two to each anther, as if they had
n opened like a book to join edges with
ir neighbors. The anther-tube is closed
the tip, making a five-sided cone; and
the seams, the yellow pollen bulges
t, in starlike rays. The pollen bulges
t for good reason, for behind it is the
gma, like a ramrod, pushing all before
in the tube, for it is its turn next to
et the outer world. The two stigma-
es are pressed together like the halves
a sharpened pencil, and they protrude
ough the anther-tube as soon as all the
llen is safely pushed out; then the
gma-lobes separate, each curling back-
rds so as to offer a receptive surface to
llen grains from other florets, or even
er sunflowers. In the process of curl-
back, they press the anther-tube down
o the corolla, and thus make the floret
orter than when in the pollen stage.
e ray flower differs in many essentials
m the perfect florets of the disc. If
remove one from the flower-head, we

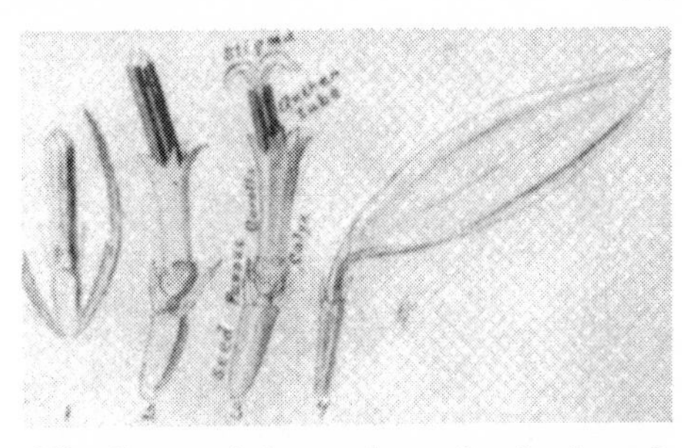

The flower of the sunflower head enlarged

1, A floret of the sunflower in the bud stage as it appears at the center of the sunflower. Note the protecting bract at the right. 2, A floret in earliest stage of blossoming. 3, A floret in the latest stage of bloom with the parts named. 4, A ray or banner flower

find at its base a seedlike portion, which is a mere pretense; it is shrunken, and never can be a seed because it has connected with it no stigma to bring to it the pollen. Nor does this flower have stamens nor a tubular corolla; instead it has one great, petal-like banner, many times longer and wider than the corollas of the other flowers. All this flower has to do is to hold its banner aloft as a sign to the world, especially the insect world, that here is to be found pollen in plenty, and nectar for the probing.

But more wonderful than the perfection of each floret is their arrangement in the flower-head. Around the edge of the disc the ray flowers, in double or treble rank, flare wide their long petals like the rays of the sun, making the sunflower a most striking object in the landscape. If the sunflower has been open for several days, next to the ray flowers will be seen a circle of star-mouthed corollas from which both ripened pollen and stigmas have disappeared, and the fertilized seeds below them are attaining their growth. Next comes a two- or three-ranked circle, where the split, coiled-back stigma-lobes protrude from the anther-tubes; within this circle may be two or three rows of florets, where pollen is being pushed out in starry radiance; and within this ring there may be a circle where the anther-tubes are still closed; while at the center lie the buds, arranged in an exquisite pattern of circling radii, cut by radii circling in the opposite direc-

Agronomy Dept. Cornell U.

A field of sunflowers

tion; and at the very center the buds are covered with the green spear-points of their bracts. I never look at the buds in the sunflower without wondering if the study of their arrangement is not the basis of much of the most exquisite decoration in Moorish architecture. To appreciate fully this procession of the bloom of the sunflower from its rim to its center, we need to watch it day by day — then only can its beauty become a part of us.

The great green bracts, with their long pointed tips, which overlap each other around the base of the sunflower head, should be noted with care, because these bracts have manifold forms in the great *Compositæ* family; and the pupil should learn to recognize this part of the flower-head, merely from its position. In the burdocks, these tracts form the hooks which fasten to the passer-by; in the thistle, they form the prickly vase about the blossom; while in the pearly everlasting, they make the beautiful, white, shell-like mass of the flower which we treasure as immortal. In the sunflower these bracts are very ornamental, being feltlike outside and very smooth inside, bordered with fringes of pretty hairs, which may be seen best through a lens. They overlap each other regularly in circular rows, and each bract is bent so as to fit around the disc.

In looking at a mass of garden sun-flowers, we are convinced that the he heads bend the stems, and this is pr ably true, in a measure. But the ste are very solid and firm, and the bend as stiff as the elbow of a stovepipe; a after examining it, we are sure that t bend is made with the connivance of stem, rather than despite it. Proba most people, the world over, believe t sunflowers twist their stems so that th blossoms face the sun all day. This lief shows the utter contentment of m people with a pretty theory. If you lieve it, you had best ask the first s flower you see if it is true, and she answer you if you will ask the quest morning, noon, and night. My own servations make me believe that the s flower, during the later weeks of bloom, is like the Mohammedan, keep its face toward the east. True, I h found many exceptions to this rule, though I have seen whole fields of s flowers facing eastward, when the sett sun was gilding the backs of their gr heads. If they do turn with the sun must be in the period of earliest blosso ing before they become heavy with rip ing seeds.

The sunflower seed is eagerly sought many birds, and it is raised extensiv for chicken-feed. The inadequate li pappus falls off, and the seeds are large end up, in the very ornamen diamond-shaped sockets. They finally come loosened, and as the great sten assaulted by the winds of autumn, bended heads shake out their seed scatter them far afield.

LESSON 159

The Sunflower

Leading Thought — The sunflowe not a single flower, but is a large num of flowers living together; and each li flower, or floret, as it is called, has own work to do.

Method — Early in September, w school first opens, is the time for lesson. If sunflowers are growing near they should be studied where they sta

l their story may thus be more com-
tely told. Otherwise, a sunflower
ould be brought to the schoolroom and
ced in water. If one is selected which
just begun to blossom, it will show,
by day, the advance of the blossoming
eks. I have kept such a flower fourteen
s, and it blossomed cheerfully from
rim to its very center. A large sun-
wer that has only partially blossomed
also needed for taking apart to show
arrangement of this big flower-cluster.
ke a bud from the center, a floret show-
anther-tube and another showing the
led pair of stigmas, and a ray or banner
wer. (See Fig. p. 575.) Each pupil
ould be furnished with these four
rets; and after he has studied them,
ow him the other half of the sun-
wer, with each floret in place. After
s preliminary study, let the pupils obve the blossoming sunflower for sev-
l consecutive days.

Observations — 1. A little flower
ich is part of a big flower-cluster is
led a floret. You have before you three
rets of a sunflower and a ray floret.
udy first the bud. Of how many parts
it composed? What will the lower,
ite part develop into? Can you see two
tle white points standing up from it
each side of the bud? Note the shape
d color of the unopened floret. Note
at there is a narrow, stiff, leaflike bract,
ich at its base clasps the young seed,
ile its pointed tip bends protectingly
er the top of the bud.

2. Take an open floret with the long,
rk brown tube projecting from it.
ote that the young seed is somewhat
ger than in the bud, and that it still
s its earlike projections at the top. De-
ribe the shape of the open corolla. Look
the brown tube with a lens. How many
les has it? How many little points pro-
cting at the top and bottom on each
le of the tube? How does the tube look
the tip, through a lens? Can you see
e pollen bursting out? If so, how does
look? Do you think that there is just
e tubular anther, or do you think sev-
al anthers are joined together to make
this tube? Open the corolla-tube carefully,
and see if you can answer this last question. Open the anther-tube, and see if
you can find the pistil with its stigmas.

3. Take a floret with the two yellow
horns of the stigma projecting. Where is
the brown anther-tube now? Is it as long
as in the floret you have just studied?
What has happened to it? What did the
stigmas do to the pollen in the anther-
tube? How do the two parts or lobes of
the stigma look when they first project?
How later?

4. Take a ray flower. How many parts
are there to it? How does the seedlike
portion of the blossom look? Do you
think it will ever be a good seed? Describe
the corolla of this flower. How much
larger is it than the corolla of the florets?
Has the ray flower any pistil or stamens?
Of what use is the ray flower to the sunflower cluster? Do you think that we
would plant sunflowers in our gardens
for their beauty if they had no ray flowers?

5. After studying the separate flowers,
study a sunflower in blossom, and note
the following: Where are the ray flowers
placed? How many rows are there? How
are they set so that their rays make the
sunflower look like the sun? Do you see
why the central portion of the sunflower
is called the disc, and the outer flowers
are called the rays — in imitation of the
sun?

6. Next to the ray flowers, what sort
of florets appear? How many rows are
there? What kind form the next circle,
and in how many rows? What stages of
the florets do you find forming the inner
circle, and how many rows? What do you
find at the center of the flower-head?
Note the beautiful pattern in which the
buds are arranged. Can you see the separate buds at the very center of the sunflower? If not, why?

7. Make notes on a sunflower that has
just opened, describing the stages of the
florets that are in blossom; continue these
notes every day for a week, describing
each day what has happened. If the sunflower you are observing is in garden or
field, note how many days elapse between

the opening of the outer row of flowers and the opening of the central buds.

8. Look below or behind the sunflower, and note the way it is attached to the stem. What covers the disc? These green, overlapping, leaflike structures are called bracts. What is the shape of one of these bracts? What is its texture, outside and inside? Look at it with a lens, along the edges, and note what you see. How are the bracts arranged? Do they not "shingle" the house of the sunflower cluster? This covering of the disc, or the house of the sunflower cluster, is called the involucre.

9. Does the stem of the sunflower hold it upright? Some people declare that it twists its stem so as to face the sun all day. Do you think this is true?

10. Study a sunflower head after the seeds are ripe. Do the little ears which you saw at the top of the seeds still remain? How does the sunflower scatter the seeds? Note how the disc looks after t seeds are all gone. What birds are es cially fond of sunflower seeds? Of wh use are the seeds commercially?

Flowers have an expression of coun nance as much as men or animals. So seem to smile; some have a sad expressic some are pensive and diffident; oth again are plain, honest, and upright, li the broad-faced Sunflower, and the hol hock.

— HENRY WARD BEECH

Eagle of flowers! I see thee stand,
And on the sun's noon-glory gaze;
With eye like his thy lids expand
And fringe their disk with golden ra
Though fixed on earth, in darkness root
there,
Light is thy element, thy dwelling air,
Thy prospect heaven.

— "THE SUNFLOWER," MONTGOME

THE BACHELOR'S–BUTTON

Stumpp and Walter Co.

Bachelor's-button

This beautiful garden flower gives variation in form from other composit when studied according to Lesson 13 This valued flower came to us from E rope and it sometimes escapes cultiv tion and runs wild in a gentle way. W call it bachelor's-button; but in Euro it is called the cornflower, and und this name it found its way into literatu None of the flowers that live in cluste repays close study better than does th bachelor's-button. The flowers are all t bular, but they do not have banne Their tubes flare open like trumpets, a they are indeed color trumpets heraldi to the insect world that there is nect for the probing and pollen for exchang Looked at from above, the marginal flo ers do not seem tubular; from the sid they show as uneven-mouthed trumpe with lobed edges; but though we sear each trumpet to its slender depths w can find no pistils. These marginal flowe have no duty in the way of maturing seec ome varieties the marginal flowers are :e, and in others they are blue, pink, ›urple. They vary in number from n to fourteen or more.

he disc flowers have a long corolla-:, which is white and delicately lobed is enlarged toward the upper end to ırple bulb with five long slender lobes. anther-tube is purplish black, and is t into almost a hook, the tip opening ard the middle of the flower-head. pollen is glistening white tinged with ›w, and looks very pretty as it bursts from the dark tubes. The purple na first appears with its tips close to-er, but with a pollen brush just below ter it opens into a short Y. The buds ıe center of the flower are bent hook-›ed over the center of the flower-head. involucral bracts or "shingles" are pretty, each one ornamented with a y fringe; they form a long, elegantly ›ed base to the flower-head. After the 'ers have gone and the seeds, which are ly akenes, have ripened, these bracts : open, making a wide-mouthed urn n which the ripened seeds are shaken he winds; and after the seeds are gone, white fuzz of their empty cases re-ns at the bottom of the urn. The seed lump and shining, with a short fringe appus around the top and a contracted e at one side near the base where it v fast to the receptacle; for these seeds not set on end, as are those of the

Stumpp and Walter Co.

Sweet sultan. This flower comes in many shades

sunflower. The short pappus is hardly sufficient to buoy up the seed, and yet undoubtedly aids it to make a flying jump with the passing breeze.

LESSON 160

The Bachelor's-Button

Leading Thought — Each bachelor's-button is made up of many little flowers, which may be studied by the outline given in Lesson 131.

THE SALVIA OR SCARLET SAGE

'he flower story of the sage is so pe-ar that Darwin has used it to illustrate mechanisms which the visiting in-s must work in some flowers in or-to get the nectar. The scarlet sage, ch gladdens our flower-beds during summer and autumn with its bril-ce, has as interesting a story as has of its family. Looking at it from the side, we should say that its nectar-wells oo deep to be reached by any creature ept a moth or butterfly, or a humming-l; there is no platform for a bee to alight upon, and the tube is too long to be fathomed by a bee's tongue; but the bees are very good business folk; they adapt themselves to flowers of various types, and in autumn the glow of the salvia attracts the eye scarcely more than the hum of the visiting bees attracts the ear.

The calyx of the salvia is as red as the corolla, and is somewhat fuzzy while the corolla is smooth. The calyx is a three-lobed bulging tube held stiff by rather strong veins; there is one large lobe above

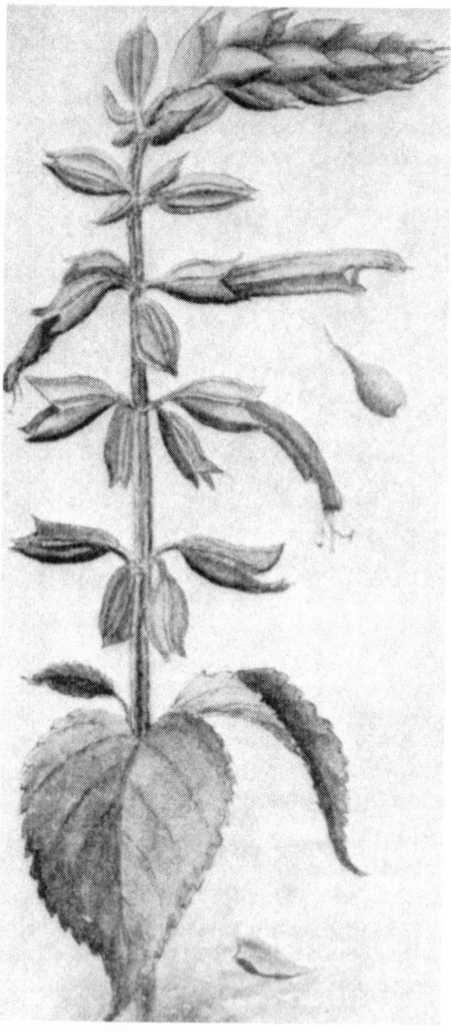

The salvia or scarlet sage, showing the bracts still present above and falling as the flowers open

and two small ones below the corolla. The corolla is a tube which is more than twice the length of the calyx; it is prolonged above into a projecting hood, which holds the anthers and the stigma; it has a short, cuplike lower lip and two little turned-back, earlike lobes at the side.

The special mechanism of the salvia is shown in the stamens; there are two of these lying flat along the floor of the corolla-tube and grown fast to it. Near the mouth of the tube, each of these lifts up at a broad angle to the roof, and is more or less T-shaped; at the tip of one of the a of the T is an anther while the other is longer and slants down and inwar the floor of the tube, as shown at 2 in figure.

The bee visiting the flower and ente the corolla-tube pushes her head aga the inner arms of the stamens, lif them, and in so doing causes the ant on the front arms of the T to lower leave streaks of pollen along her f sides. The stigma is at first concealed in hood; but, when ripe, it projects hangs down in front of the opening of corolla-tube, where it may be brus along one side or the other by the ing insect, which has been dusted the pollen of some other flower. stigma lobes open in such a manner they do not catch the pollen from insect backing out of their own cor As the nectar is at the base of the cor tube, the bees, in order to get it, craw almost out of sight. Late in the se they seem to "go crazy" when gathe this nectar; I have often seen them sea ing the bases of the corolla-tubes w have fallen to the ground, in order to what is left of the sweet treasure.

But the pollen story is not all tha of interest in the salvia. Some of the p of the flower which are green in n blossoms are scarlet as a cardinal's

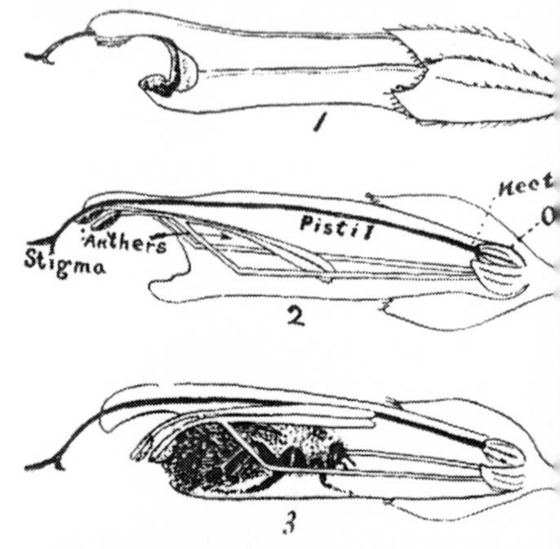

1, *Blossom of scarlet sage as seen from side.* 2, *The same flower with side rem showing the arrangement of its parts. bee working the stamen's mechanism as seeks the nectar*

this. If we glance at a flowerstalk, we that at its tip it looks like a braided, :tened cone; this appearance is caused the scarlet, long-pointed bracts, each which covers with its bulging base the rlet calyx, which in turn enfolds the rlet flower bud. These bracts fall as : flowers are ready to open, making a lliant carpet about the plant. Each werstalk continues to develop buds at tip for a long season; and this, taken gether with its scarlet bracts and flow- , renders the salvia a thing of beauty in r gardens, and makes it cry aloud to llen-carriers that here, even in late au- nn, there is plenty of nectar.

LESSON 161

Salvia or Scarlet Sage

Leading Thought — This flower has : bracts and calyx scarlet instead of en, and this makes it a brilliant mass color which pleases our eyes and at- cts the pollen-carrying insects. Its an- :rs are placed at the tip of two levers, ich the insects push up and down as ey enter the flower, thus becoming sted with pollen.

Method — The structure of this flower y be studied in the schoolroom and its chanism there understood; but the st important part of the lesson is the ob- vation out-of-doors upon the way the es work the stamen levers when seeking : nectar. This is best observed during e September or October, after other wers are mostly gone, and when the es are working with frantic haste to get the honey possible.

Observations — 1. How does the calyx of the salvia differ from that of other flowers in color? How does it differ from the corolla in texture? How many lobes has it? How are they placed about the corolla?

2. What is the shape of the corolla? How does it make a hood over the entrance to the tube? What does the hood hold? Is there any platform made by the lower lip of the corolla for a visiting insect to alight upon?

3. Cut open one side of the corolla and describe how the stamens are arranged. Thrust your pencil into an uninjured flower and see if the anthers in the hood are moved by it. How? Describe how a bee in visiting this flower moves the anthers and becomes dusted with pollen.

4. Where is the stigma? How does it receive pollen from visiting insects? Would it be likely to get the pollen which has just been scraped off from its own anthers by the bee? Why?

5. Experiment to find where the nectar is. Do you ever see bees getting the nectar from fallen flowers? Do they get the nectar from the "front" or the "back door"?

6. What other parts of this flower are red, which in other flowers are green? How does this make the budding portions of the flower stem look? Why does this make the salvia a more beautiful plant for our gardens?

7. Compare the mechanism of the stamens of the scarlet sage with the mechanism of the stamens of the common garden sage.

PETUNIAS

These red-purple and white flowers, ich, massed in borders and beds, make y our gardens and grounds in late sum- :r and early autumn, have an interest- ; history. Professor L. H. Bailey uses it an illustration in his thought-inspiring ok, *The Survival of the Unlike*; he says it our modern petunias are a strange compound of two original species; the first one was found on the shores of the La Plata in South America and was introduced into Europe in 1823. "It is a plant of upright habit, thick sticky leaves and sticky stems, and very long-tubed white flowers which exhale a strong perfume at nightfall." The second species of

W. Atlee Burpee

petunia came from seeds sent from Argentina to the Glasgow Botanical Gardens in 1831. "This is a more compact plant than the other, with a decumbent base, narrower leaves and small, red-purple flowers which have a very broad or ventricose tube, scarcely twice longer than the slender calyx lobes." This plant was called *Petunia violacea* and it was easily hybridized with the white species; it is now, strangely enough, lost to cultivation, although the white species is found in some old gardens. The hybrids of these two species are the ancestors of our garden petunias, which show the purple-red and white of their progenitors. The petunias are of the Nightshade family and are kin to the potato, tomato, eggplant, tobacco, and Jimson weed; and the long-tongued sphinx or hummingbird moths secure much nectar from their blossoms.

The petunia corolla is tubular, and the five lobes open out in salver-shape; each lobe is slightly notched at its middle, from which point a marked midrib extends to the base of the tube. In some varieties t edges of the lobes are ruffled. Within t throat of the tube may be seen a netw of darker veins, and in some varieties t network spreads out over the corolla lob Although many colors have been dev oped in petunias, the red-purple and wh still predominate; when the two col combine in one flower, the pattern m be symmetrical, but is often broken a blotchy.

When a flower bud is nearly ready open, the long, bristly tube of the coro lies with its narrow base set in the cal the long, fuzzy lobes of which flare c in bell-shape; the tube is marked by leng wise lines made by the five midribs; t lobes of the corolla are folded along t outer portions of these midribs, and th folded tips are twisted together much if some one had given them a half tu with the thumb and finger. It is a pleasi experience to watch one of these flow unfold. When a flower first opens, th lies near the bottom of the throat of t be the green stigma, with two anthers uggled up in front of it and two behind the latter being not quite so advanced age as the former. As the filaments of e front pair of anthers are longer than ose of the rear pair, the little group lies a low angle offering a dusty doormat r entering insects. If we open a flower this stage we find another anther, as t unopened, which is on the shortest amen of the five. This seems to be ittle pollen reserve, perhaps for its own e later in the season. There is an in-resting mechanism connected with these amens; each is attached to the corolla-be at the base for about half its length, d at the point of attachment curves sud-nly inward so as to " cuddle up " to the stil, the base of which is set in the nec-r-well at the bottom of the flower. If e introduce a slender pencil or a tooth-ck into the flower-tube along the path hich the moth's tongue must follow to ach the nectar, we can see that the amens, pressing against it at the point here they curve inward, cause the an-ers to move about so as to discharge eir pollen upon it; and as the toothpick withdrawn they close upon it cogently so at it carries off all the pollen with which is brought in contact.

If we look at the stigma at the center its anther guard, it has a certain close-ted appearance, although its outer lges may be dusted with the pollen; as e flower grows older, the stigma stands ove the empty anthers at the throat the flower-tube and opens out into two stinct lobes. Even though it may have cepted some of its own pollen, it ap-rently opens up a new stigmatic surface r the pollen brought from other flowers visiting insects.

Dr. James G. Needham says that at ake Forest he has been attracted to the etunia beds in the twilight by the whir-ng of the wings of countless numbers of hinx, or hummingbird moths, which ere visiting these flowers. We also may nd these moths hovering over petunia eds in almost any region if we visit them n the warmer evenings. And it is a safe guess that the remote white ancestor of our petunias had some special species of sphinx moth which it depended upon for carrying its pollen; and the strong perfume it exhaled at nightfall was an odor signal to its moth friends to come and feast.

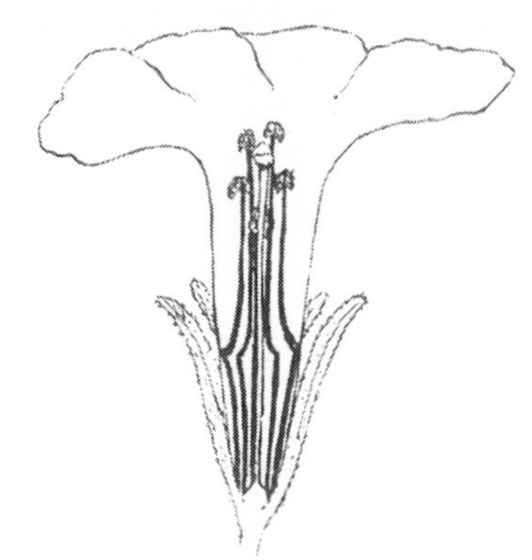

A petunia blossom cut open on the upper side, showing the pistil surrounded by the in-curved stamens and the partially opened stigma surrounded by the anthers. Note the short stamen below the pistil

With their long feeding tubes the hum-mingbird moths have little difficulty in securing the nectar, but bees also will work industriously in the petunias. They will scramble into the blossoms and, ap-parently complaining with high-pitched buzzing because of the tight fit, rifle the nectar-wells that seem to be better adapted to insects of quite different build.

The leaves of the petunia are so broadly ovate as to be almost lozenge-shape, es-pecially the lower ones; they are soft, and have prominent veins on the lower side; they are without stipules, and have short flat petioles. The stems are soft and fuzzy and are usually decumbent at the base, except the central stems of a stool or clump, which stand up straight.

The flower stems come off at the axils of the leaves; the lower flowers open first. The blossoms remain open about two days; at the first sign of fading, the lobes of the corolla droop dejectedly like a frill that has lost its starch, and finally the co-rolla — tube and all — drops off, leaving a little conical seed capsule nestled snugly in the heart of the bell-shaped calyx. At this time, if this peaked cap of the seed capsule be removed, the many seeds look like tiny white pearls set upon the fleshy, conical placenta. As the capsule ripens, it grows brown and glossy like glazed manila paper and it is nearly as thin; then it cracks precisely down its middle, and the seeds are spilled out at any stirring of the stems. The ripe seeds are dark brown, almost as fine as dust, and yet, when examined with a lens, they are seen to be exquisitely netted and pitted.

LESSON 162

The Petunia

Leading Thought — The petunias are native to South America; they have an interesting history. Such insects as hummingbird moths are attacted to their flowers, and from them easily secure pollen and nectar.

Method — The petunias are such determined bloomers that they give us flowers up to the time of killing frosts, and they are therefore good material for nature lessons. Each pupil should have a flower in hand to observe during the lesson, and should also have access to a petunia bed for observations on the habits of the plant.

Observations — 1. What colors do you find in the petunia flowers? If they are striped or otherwise marked, what are the colors? Are the markings symmetrical and regular?

2. Sketch or describe a flower, looking into it. What is the shape of the corolla lobes? How many lobes are there? How are they veined? What peculiar markings are at the throat of the flower?

3. What are the color and position of the stigma? How are the stamens arranged? How many anthers do you see? What is the color of the anthers? Of the pollen?

4. Sketch or describe the flower from the side. What is the shape of the rolla-tube? Is it smooth or fuzzy? How it marked? What are the number a shape of the sepals, or lobes, of the caly

5. Study a freshly opened flower, a describe the position and appearance the anthers and stigma. Do they rem in these relative positions after the flov is old?

6. Cut open a flower, slitting it alo the upper side. Describe the stamens a how they are attached. Is the pistil tached in the same manner? Where the nectar? Thrust a slender pencil o toothpick into the tube of a fresh flow Does this spread the anthers apart a move them around? When it is wi drawn, is there pollen on it? Can you s in your open flower the mechanism which the pollen is dusted on the obj thrust into the flower?

7. What insects have tongues su ciently long to reach the nectar-well at t bottom of the petunia flower? At wh time do these insects fly? At what ti of day do most of the petunia flow open? Visit the petunia beds in the tw light, and note whether there are any i sects visiting them. What insects do y find visiting these flowers during the da

8. Sketch or describe the leaves of t petunia. How do the leaves feel? Look a leaf with a lens and note the fringe hair along its edges. Describe the veini of the leaf.

9. Describe the petunia stems. Are th stout or slender? How do they feel? Wi what are they covered? Where do th flowerstalks come off the main stem?

10. Describe or sketch a flower b just ready to open. How are the tips of th lobes folded? How long does the flow remain in bloom? What is the first si of its fading?

11. Describe the seed capsule. Whe does it open? Are the seeds many or fe large or small? What is their color wh ripe? When examined with a lens, ha the seeds any noticeable pits or mar ings?

THE GARDEN OR HORSESHOE GERANIUM

'he geraniums perhaps do more to ;hten the world than almost any other tivated flowers. They will grow for ryone, whether for the gardener in the servatory of the rich, or in a tin can the window sill of the crowded tene- nt of the poor. And it is interesting know that this common plant has a tivated ancestry of two hundred years' iding. These geraniums, which are ly not geraniums botanically but are argoniums, originally came from south- Africa, and the two ancestors of our nmon bedding geraniums were intro- ed into England in 1710 and 1714.

'he geranium is of special value to the cher, since it is available for study at season of the year, and has a most resting blossom. The single-flowered ieties should be used for this lesson, ce the blossoms that are double have their original form. Moreover, the anium's blossom is so simple that it is pecial value as a subject for a beginning on in teaching the parts of a flower; l its leaves and stems may likewise be d for the first lessons in plant structure. The stem is thick and fleshy, and is

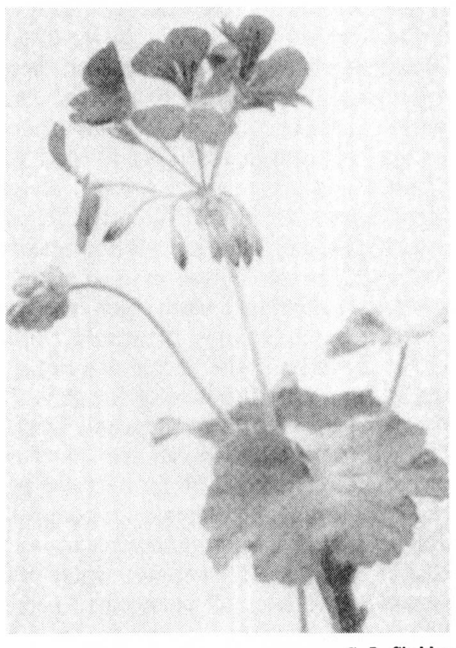

S. L. Sheldon

Horseshoe geranium. Note the positions of the opened flowers and the buds. Note the shape of the two upper petals with their colored lines. The flower at the left, seen in profile, shows that these upper petals project farther forward than those below. Note the cluster of young buds set in a circlet of bracts just below this flower

Swallowtail butterfly on a geranium

downy on the new growth; there is much food stored in these stems, which accounts for the readiness with which cuttings from them will grow. Two stipules are found on the stem at the base of each petiole. These stipules often remain after the leaves have fallen, thus giving the stem an unkempt look. The leaves are of various shapes, although of one general pattern; they are circular and beautifully scalloped and lobed, with veins for every lobe radiating from the petiole; they are velvety above and of quite different texture beneath, and many show the dark horseshoe which gives the name to this variety. The petiole is usually long and stiff and the leaves are set alternately upon the stem.

The flower has five petals, and at first glance they seem of much the same shape and position; but if we look at them carefully, we see that the upper two are much narrower at the base and project farther forward than do the lower three. Moreover, there are certain lines on these upper petals all pointing toward the center of the flower; and if we follow them we find a deep nectar-well just at the base of these upper petals and situated above the ovary of the flower. No other flower shows a prettier plan for guiding insects to the hidden sweets, and in none is there a more obvious and easily seen well of nectar. It extends almost the whole length of the flowerstalk, the nectar-gland forming a hump near the base of the stalk. If we thrust a needle down the whole length of this nectar-tube we can see that this bright flower developed its nectar especially for some long-tongued insect, probably a butterfly. It is interesting to note that in the double geranium where the stamens have been all changed to petals and where, therefore, no seeds are formed, this nectar-well has been lost.

There are five sepals, the lower one being the largest. But the geranium is careless about the number of its stamens; most flowers are very good mathematicians, and if they have five sepals and five petals they are likely to have five or ten stamens. The geranium often shows seven anthers, but if we look carefully we may find ten stamens, three of them without anthers. But this is not always true; there are sometimes five anthers and two or three filaments without anthers. The color of the anthers differs with the variety of the flower. The stamens broaden below, and their bases are joined, making a cup around the lower part of the ovary. The pistil is at the center of the flower and has no style, but at the summit divides into five long, curving stigmas; but again the geranium cannot be trusted to count, for sometimes there are seven or eight stigmas. Although many of our common varieties of geraniums have been bred so long that they have almost lost the habit of producing seed, yet we may often find in these single blossoms the ovary chan into the peculiar long beaklike p which shows the relationship of this p to the crane's-bill or wild geranium.

When the buds of the geranium appear, all of them are nestled in a of protecting bracts, each bud being closed in its own protecting sepals. soon each flowerstalk grows longer droops and often the bracts at its fall off; from this mass of drooping b the ones at the center of the cluster up and open their blossoms first. Of when the outside flowers are in blo those at the center have withered pe

It would be well to say somethin the pupils about those plants which h depended upon man so long for t planting that they do not develop more seed for themselves. In connect with the geraniums, there should be a son on how to make cuttings and s their growth. The small side branche the tips of the main stems may be as cuttings. With a sharp knife mak cut straight across. Fill shallow boxes sand, and plant the cuttings in th boxes, putting the stems for one-thir their length in the sand; place them cool room and keep them consta moist. After about a month the pl may be repotted in fertile soil. The fa the best time to make cuttings.

LESSON 163

The Garden or Horseshoe Geranium

Leading Thought — The gerani are very much prized as flowers for o mental beds. Let us see why they ar valued.

Method — A variety of geranium w single flowers should be chosen for purpose, and it may be studied in schoolhouse window or in the gard As the parts of this flower are of a general type, it is an excellent one w which to teach the names and purp of the flower parts. Each child can m a little drawing of the sepals, petals, mens, and pistil, and label them with proper names.

)BSERVATIONS — 1. What sort of stem the geranium? Is it smooth or downy? ıat makes the geranium stem look so gh and untidy?

. Study the leaf. Show, by description y drawing, its shape, its wings, and its ıs. What are its colors and texture ve? Beneath? Is the petiole long or rt? What grows at the base of the peti- where it joins the stem? What mark- is there on the leaf, which makes us this a "horseshoe geranium"? Are re other geraniums with leaves of simi- shape that have no horseshoe mark?

. Study the flower. Are the petals all same size and shape? How many of m are broad? How many narrow? Do narrow ones project in front of the ers? Do these have lines upon them? ıere do these lines point? Find the tar-well; how deep is it? Does it extend ost the entire length of the flower- k? For what insects is it fitted? Are re nectar-tubes in the stems of the ıniums with double flowers?

. How many sepals are there? Are they the same size? Where is the largest?

. How many stamens can you see? ıat is the color of the filaments and the anthers? How are the stamens ıed at their bases? Can you find any nens without anthers?

. Where is the pistil situated? Can you the ovary, or seed box? How many mas? Describe their color and shape?

. In what part of the flower will the ls be developed? How does the gera- m fruit look? Sketch the pod. Do the niums develop many seeds? Why not? you know the seed pod of the wild ınium? If so, compare it with the pod his plant.

. Take a flower-cluster when the flow- are all in the bud, and note the follow- ing: When the buds first appear, what protects them? What becomes of these bracts later? How do the sepals protect the bud? Are the bud stems upright and

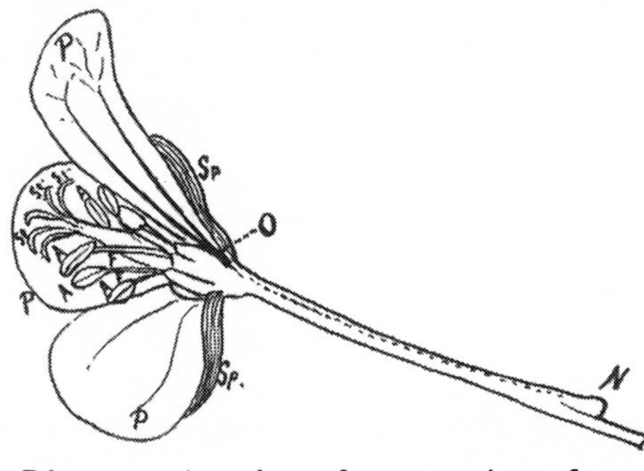

Diagram of a horseshoe geranium flower

Sp, sepals; P, petals; A, anthers; F, filament; St, stigma; O, opening to nectar gland; N, nectar gland

stiff or drooping? How many buds are there in a cluster?

9. Take notes on successive days as follows: What happens to the stalk as the bud gets ready to bloom? Is it a central or an outside blossom that opens first? How many new blossoms are there each day? How long is it from the time that the first bud opens until the last bud of the cluster blossoms? What has this to do with making the geranium a valuable ornamental plant?

10. Make some geranium cuttings, and note how they develop into new plants. Place one of the cuttings in a bottle of water and describe how its roots appear and grow.

God made the flowers to beautify
The earth, and cheer man's careful mood;
And he is happiest who hath power
To gather wisdom from a flower,
And wake his heart in every hour
To pleasant gratitude.

— MARY HOWITT

THE SWEET PEA

Here are sweet peas on tiptoe for a flight,
With wings of delicate flush o'er delicate white,
And taper fingers catching at all things,
To bind them all about with tiny rings.

— KEATS.

Among the most attractive of the seeds which make up the treasure of the children's seed packets are the sweet peas. They are smooth little white or brown globules, marked with a scar on the side showing where they were attached to the

Sweet pea, blossoms and seed pods

pod. One of these peas divides readily into two sections; and after it has been soaked in water for twenty-four hours, the embryo of the future plant may, with the aid of a lens, be seen within it. After planting, the sprout pushes through the seed coat at a point very near the scar, and a leaf shoot emerges from the same place; but the two act very differently. The shoot lifts upward toward the light, and the root plunges down into the soil. As the plant grows, it absorbs the food stored in the seed; but the seed rema below ground and does not lift itself i the air, as happens with the bean. T root forms many slender branches, n the tips of which may be seen the fri of roots, which take up the minerals a water from the soil. The first leaves the pea seedling put forth no tendr but otherwise look like the later on The leaves grow alternately on the sta and they are compound, each having fr three to seven leaflets. The petiole winged, as is also the stem of the pla There is a pair of large, clasping stipu at the base of each leaf. If we comp one of these leaves with a spray of tendr we can see that they resemble each ot in the following points: The basal leaf of the petiole are similar and the stipu are present in each case; but the leafl nearest the tip are marvelously changed little stiff stalks with a quirl at the of each, ready to reach out and hook up any object that offers surface to cling Sometimes we find a leaflet paired w a tendril. The sweet pea could not gr vigorously without a support outside itself.

Of course, the great upper petal of t sweet-pea blossom is called the bann It stands aloft and proclaims the sw pea as open; but before this occurs, tenderly enfolds all the inner part of t flower in the unopened bud, and wh the flower fades it again performs t duty. The wings are also well named; these two petals which hang like a peak roof above the keel seem like wings j ready to open in flight. The two lov petals are sewed together in one of N ture's invisible seams, making a lo curved treasure-chest resembling the k of a boat, and it has thus been call Within the keel are hidden the pistil a

.mens. The ovary is long, pod-shaped d downy; from its tip the style projects, strong as a wire, curving upwards, and vered with a brush of fine, white hairs; the very tip of the style, and often pro- :ting slightly from the keel, is the stigma. ound the sides and below the ovary and le are nine stamens, their filaments oadening and uniting to make a white, ken tube about the ovary, or young pod. om the tip of this stamen-tube, each the nine filaments disengages itself, d lying close to the style thrusts its ther up into the point of the keel, be- v the stigma. But strange to say, one ie, lorn stamen " flocks by itself " above e pistil, curving its anther up stigma- rd. If we touch the point of the keel th the finger, up fly — like a jack-in-the- x — the anthers splashing the finger th pollen; and if a bee, in her search : nectar, alights on the wings at the ry base of the petals, up flies the pollen ush and daubs her with the yellow dust, ich she may deposit on another stigma. ie interesting part of this mechanism the brush near the tip of the style be- v the stigma — a veritable broom, with ints all directed upward. As the pollen discharged around it, the brush lifts it when the keel is pressed down, and the

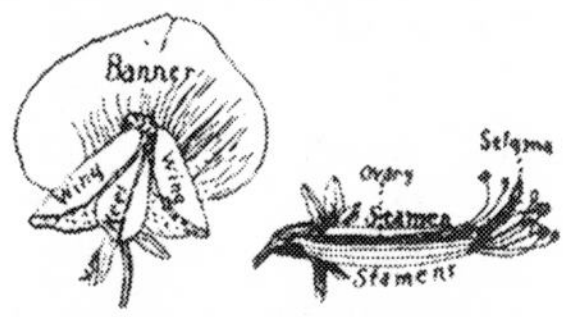

Blossom of sweet pea with parts labelled

ff petals forming the keel, in springing ck to place, scrape off the pollen and aster it upon the visitor. But for all this borate structure, sweet peas, of all flow- ;, are the most difficult to cross-pollinate, ice they are so likely to receive some of eir own pollen during this process.

The sweet-pea bud droops, a tubular lyx with its five-pointed lobes forming bell to protect it. Within the bud the nner petal clasps all in its protecting nbrace.

After the petals fall, the young pod ands out from the calyx, the five lobes which are recurved and remain until

W. Atlee Burpee Co.

Sweet pea in flower

the pod is well grown. As the sweet pea ripens, all the moisture is lost and the pod becomes dry and hard; through the dampness of dews at night and the sun's heat which warps it by day, finally each side of the pod suddenly coils into a spiral, flinging the seed many feet distant in different directions.

LESSON 164

The Sweet Pea

Leading Thought — The sweet pea has some of its leaflets changed to tendrils which hold it to the trellis. Its flower is like that of the clover, the upper petal forming the banner, the two side petals the wings, and the two united lower petals the keel which protects the stamens and pistil.

Method — This should be a garden lesson. A study should be made of the peas before they are planted, and their germination carefully watched. Later, the method of climbing, the flower and the fruit should each be the subject of a lesson.

Sweet-pea pod bursting in spiral

OBSERVATIONS ON GERMINATION — 1. Soak some sweet peas over night; split them the next morning. Can you see the little plant within?

2. Plant some of the soaked peas in cotton batting, which may be kept moist. At what point does the sprout break through the seed covering? Do the root and leaf shoot emerge at the same place, or at different points? Which is the first to appear?

3. Plant some of the soaked peas in the garden. How do the young plants look when they first appear? Does the fleshy part of the seed remain a part of the plant and appear above the ground, as is the case with the bean? What becomes of the meat of the seed after growth has started?

4. Do the first leaves which unfold from the seed pea look like the later ones? Are the leaves simple or compound? Do they grow opposite each other or alternately?

5. Take a leaf and also a spray of the tendrils. How many leaflets are there in the compound leaf? Describe the petiole and the basal leaves. How far apart are the leaflets on the mid-stalk? Compare the stalk on which the tendrils grow with this leaf. Are the basal leaflets like those of the leaf? Is the petiole like that of the leaf? Do you think that the leaflets toward the tip of the stalk often change to tendrils? Why do you think so? How do tendrils aid the sweet pea? Do you see the earlike stipules at the base of the leaf? Are there similar stipules at the base the tendril stem?

OBSERVATIONS ON THE FLOWER A FRUIT — 1. Take the sweet pea in bl som. Why is the large upper petal call the banner? How does it compare in si with the other petals? What is its purp when the flower is open? Why do y think the side petals are called wing What is their position when the flower open?

2. Describe that part of the flower low the wings. Do you think that it made of two petals grown together? W is it called the keel of the flower? Pr down with your finger on the tip of t keel. What happens? Is your fing splashed with pollen? Where is the nect in the sweet pea? Would an insect g ting the nectar press down upon the k and receive a splash of pollen?

3. Open the keel. How many stame do you find within it? How many ha their filaments joined together? Is th one separate from the others? Agai what are the anthers pressed by t keel?

4. Remove the stamens and descri the pistil. Which part of this will ma the pod in which the new peas will velop? Describe how the style is curve How is the style covered near its ti What is this brush for? Can you find t stigma with the help of the lens? Wh the bee is seeking for nectar and push down on the keel, does the stigma pu out at the same point as the pollen? Do this enable the stigma sometimes to ceive pollen which the bees bring fro other flowers?

5. Describe an unopened flower bu What is its position? How many lobes the calyx? What is their shape, and ho do they protect the bud? Which pe is folded over all the others? How do the position of the open flower differ fro that of the bud?

6. How does the young pod look wh the petals fall? How does it look wh ripe? How does it open to scatter litt ripe sweet peas? Do the lobes of the sep still remain with the pod?

CULTIVATED CROP PLANTS

THE CLOVERS

Sweet by the roadside, sweet by the rills,
Sweet in the meadows, sweet on the hills,
Sweet in its wine, sweet in its red,
Oh, half of its sweetness cannot be said;
Sweet in its every living breath,
Sweetest, perhaps, at last, in death.

— "A SONG OF CLOVER," HELEN HUNT JACKSON

Ida Baker

lover has for centuries been a most
able forage crop; and for eons it has
n the special partner of the bees, giving
m honey for their service in carrying
pollen; and it has been discovered that
as also a mysterious and undoubtedly
ancient partnership with bacteria be-
ground, which, moreover, brings fer-
y to the soil. The making of a collec-
of the clovers of a region is a sure
of enlisting the pupils' interest in
se valuable plants. The species have
e similarities and differences, which
opportunity for much observation in
nparing them. There may be found in
st localities the white and yellow sweet
vers, the black and spotted medics, and their relative the alfalfa; while of the true clovers there are the red, the zigzag, the buffalo, the rabbit's-foot, the white, the alsike, the crimson, and two yellow or hop clovers.

In all the clovers, those blossoms which are lowest, or on the outside of the head, blossom first, and all of them have upon their roots the little swellings, or nodules, which are the houses in which the benef-icent bacteria grow.

If we pull up or dig out the roots of

Heads of crimson clover

Farmers' Bulletin 455, U. S. D. A.

A young clover plant showing nodules or root tubercles

alfalfa, or of the true clovers or vetches, we find upon the rootlets little swellings which are called nodules, or root tubercles. Although these tubercles look so uninteresting, no fairy story was ever more wonderful than is theirs. They are, in fact, the home of the clover brownies, which help the plants to do their work. Each nodule is a nestful of living beings so small that it would take twenty-five thousand of them end to end to reach an inch; therefore, even a little swelling can hold many of these minute organisms, which are called bacteria. For many years people thought that these swellings were injurious to the roots of the clover, but now we know that the bacteria which live in them are simply underground partners of these plants. The clover roots give the bacteria homes and places to grow, and in return these are able to extract a very valuable chemical fertilizer from the air, and to change its form so that the clo can absorb it. The name of this substa is nitrogen, and it makes up more t three-fourths of the air we breathe. O plants are unable to take the nitro from the air and use it in making fo but these little bacteria extract it f the air which fills every little space tween every two grains of soil and t change it to a form which the clovers use. After the clover crop is harvested roots remain in the ground, their li storehouses filled with this precious s stance, and the soil falls heir to it.

Nitrogen in the form of commercial tilizer is very expensive when the far has to buy it. So when he plants clo or alfalfa on his land, he is bringing to soil this expensive element of pl growth, and it costs him nothing. Th why a good farmer practices the rotat of crops and puts clover upon his l every three or four years.

Alfalfa is so dependent on its little derground partners, that it cannot gr

Stumpp and Walter

Red clover

ll without them; and so the farmer nts, with the alfalfa seed, some of the l from an old alfalfa field, which is rich these bacteria, or better still, he inocues the clover seed with a culture of the cteria. On a farm I know, the bacterial l gave out before all of the seed was nted; and when the crop was ready to t it was easy to see just where the seed thout the inoculated soil had been nted, for the plants that grew there re small and poor, while the remainder the field showed a luxurious growth.

It is because of the great quantity of rogen absorbed from the air through e bacteria on its roots that the alfalfa such a valuable fodder; for it contains e protein which otherwise would have be furnished to cattle in expensive grain cottonseed meal. The farmer who gives s stock alfalfa does not need to pay such ge bills for grain. Other plants belonging to the same family as the clovers — like the vetches and cowpeas — also have bacteria on their roots. But each species of legume has its own species of bacteria, although in some cases soil inoculated with bacteria from one species of legume will grow them on roots of another species.

Stumpp and Walter Co.

Alfalfa blossoms

Farmers' Bulletin 1722, U. S. D. A.

ngle plant of alfalfa showing a portion of the root system

In addition to the enriching of the soil, clover roots, which penetrate very deeply, protect land from being washed away by freshets and heavy rains; and since clover foliage makes a thick carpet over the surface of the soil, it prevents evaporation and thus keeps the soil moist. Crimson clover is used extensively as a cover crop; it is sowed in the fall, especially where clean culture is practiced in orchards, and spreads its leaves above and its roots within the soil, keeping out weeds and protecting the land. (See also pages 770–75.) In the spring it may be plowed under, and thus it will add again to the fertility. This is also an æsthetically pleasing crop, for a field of crimson clover in bloom is one of the most beautiful sights in our rural landscape.

Red clover has such deep florets that, of all our bees, only the bumblebees have sufficiently long tongues to reach the nectar. It is, therefore, dependent upon this bee for developing its seed, and the enlightened farmer of today looks upon the

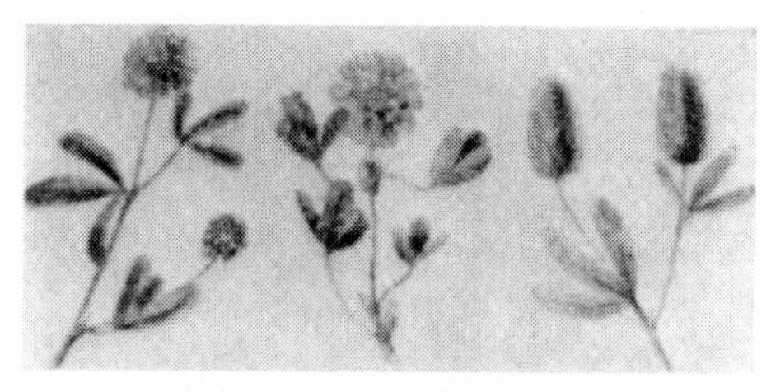

Yellow or hop clover, buffalo clover, and rabbit-foot or pussy clover

bumblebees as his good friends. The export of clover seed from the United States has sometimes reached the value of two million dollars a year, and this great industry can only be carried on with the aid of the bumblebee. There are sections of New York State where the growing of clover seed was once a most profitable business, but where now, owing to the dearth of bumblebees, no clover seed whatever is produced.

LESSON 165
The Clovers

Leading Thought — The clovers enrich with nitrogen the soil in which they are planted. They are very valuable food for stock. Their flowers are pollina by bees.

Method — Each pupil should dig u root of red clover or alfalfa to use for lesson on the nodules. The flowers sho be studied in the field, and also in de in the schoolroom.

Observations — 1. How many kind clover do you know? How many of medics?

2. In all clovers, which flowers of head blossom first, those on the lo or outside, or those on the upper inside?

3. Take up a root of red clover or alfa noting how deep it grows. Wash the r free from soil, and find the little swelli on it. Write the story of what these sw ings do for the clover, and incident for the soil.

4. How must the soil be prepared that alfalfa may grow successfully? W does the farmer gain by feeding alfal and why?

5. How do clover roots help to prot the land from being washed away heavy rains?

6. How do clovers keep the soil moi How does this aid the farmer?

7. What is a cover crop, and what its uses?

8. Upon what insects does the clover depend for carrying pollen? Can produce seed without the aid of th valuable bees? Why not?

SWEET CLOVER

In passing along the country roads, especially those which have suffered upheaval from the road machines, suddenly we are conscious of a perfume so sweet, so suggestive of honey and other delicate things, that we involuntarily stop to find its source. Close at hand we find this perfume laboratory in the blossoms of the sweet clover. It may be the species with white blossoms, or the one with yellow flowers, but the fragrance is the same. There stands the plant, lifting high its beautiful blue-green foliage and its spik of flowers for the enjoyment of t passer-by, while its roots are feeling th way down deep in the poor, hard soil, t ing air and drainage with them and buil ing, with the aid of their undergrou partners, nitrogen factories which will e rich the poverty-stricken earth, so th other plants may find nourishment in

Never was there such another bene cent weed as the sweet clover — bene cent alike to man, bee, and soil. Usual see it growing on soil so poor that it only attain a height of from two to r feet; but if it once gets foothold on a erous soil, it rises majestically ten feet

.ike the true clover, its leaf has three lets, the middle one being longer and er than the other two and separated n them by a naked midrib; the leaflets long, oval in shape, with narrow, thed edges, and they are dull, velvety en; the two stipules at the base of the are little and pointed.

'he blossoming of the sweet clover is a tty story. The blossom stalk, which nes from the axil of the leaf, is at first inch or so long, packed closely with e green buds having pointed tips. But oon as the blossoming begins, the stalk ngates, bringing the flowers farther rt — just as if the buds had been fased to a rubber cord which had been tched. The buds lower down open t; each day some of the flowers bloom, ile those of the day before linger, and s the blossom tide rises, little by little, the stalk. But the growing tip develops re and more buds, and thus the blossom story continues until long after the frosts have killed most other plants; finally the tip is white with blossoms, while the seeds developed from the first flowers on the plant have been perfected and scattered.

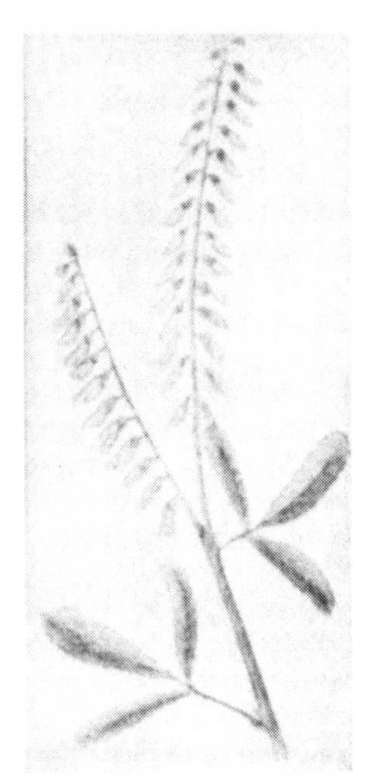

Yellow sweet clover

White sweet clover

The blossom is very much like a diminutive sweet pea; the calyx is like a cup with five points to its rim, and is attached to the stalk by a short stem. The banner petal is larger than the wings and the keel. A lens shows the stamens united into two groups, with a threadlike pistil pushing out between; both stamens and pistil are covered by the keel, as in the pea blossom.

The flowers are visited by bees and many other insects, which are attracted to them by their fragrance as well as by the white radiance of their blossoms. The ripened pod is well encased in the calyx at its base. The foliage of the sweet clover is fragrant, especially so when drying; it is to some extent used for fodder. The sweet clovers came to us from Europe and are, in a measure, compensation for some of the other emigrant weeds which we wish had remained at home.

LESSON 166

Sweet Clover

Leading Thought — This beneficent plant grows in soil that is often too poor for other plants to thrive in. It brings available nitrogen into the soil, and thus makes it fertile so that other plants soon find in its vicinity nourishment for growth.

Method — Plants of the sweet clover with their roots may be brought to the schoolroom for study. The children should observe sweet clover in the field; its method of inflorescence, and the insects which visit it, should be noted.

Observations — 1. What first makes you aware that you are near sweet clover? On what kinds of soil, and in what localities, does sweet clover abound?

2. Do you know how sweet clover growing in poor soils and waste places acts as a pioneer for other plants?

3. Dig up a sweet-clover plant, and see how far its stems go into the soil.

4. How high does the plant grow? What is the color of its foliage?

5. Compare one of the leaves with the leaf of a red clover, and describe the likeness and the difference. Note especially the edges of the upper and the lov leaves, and also the stipules.

6. Describe the way the sweet clo blossoms. Do the lower or upper flow open first? How does the flowerstalk lc before it begins to blossom? What ha pens to it after the blossoming begi How long will it continue to blossom?

7. Take a blossom and compare it wi that of a sweet pea. Can you see the ba ner? The wings? The keel? Can you see the stamens are united into two sets? C you see the pistil? Note the shape of t calyx.

8. How many flowers are in blossom a time? Does it make a mass of white attract insects? In what other way does attract insects? What insects do you fi visiting it?

9. How do the ripened pods look?

The blooming wilds His gardens are; so
cheering
Earth's ugliest waste has felt that flo
ers bequeath,
And all the winds o'er summer hills care
ing
Sound softer for the sweetness that th
breathe.

— Theron Bro

THE WHITE CLOVER

The sweet clover should be studied first, for after making this study it is easier to understand the blossoming of the white and the red clover. In the sweet clovers the flowers are strung along the stalk, but in the red, the white, and many others, it is as if the blossom stalk were telescoped, so that the flowers are all in one bunch, the tip of the stalk making the center of the clover head. We sometimes use the white clover in our lawns because of a peculiarity of its stem which, instead of standing erect, lies flat on the ground, sending leaves and blossoms upward and thus making a thick carpet over the ground. The leaves are very pretty; and although they grow upon the stems alternately, they always manage to twist around so as to lift their three leaflets upward to the lig The three leaflets are nearly equal in si with fine, even veins and toothed edg and each has upon it, near the middle pale, angular spot. The white clover, common with other clovers, has the pre habit of going to sleep at night. Botani may object to this human term, but t great Linnæus first called it sleep, and may be permitted to follow his examp Certainly the way the clover leaves fo at the middle, the three drawing ne each other, looks like going to sleep, a is one of the things which even the lit child will enjoy observing.

The clover head is made up of ma little flowers; each one has a tubular cal with five delicate points and a little st old it up into the world. In shape, the lla is much like that of the sweet pea, each secretes nectar at its base. The ide blossoms open first; and as soon ney are open, the honey bees, which rly visit white clover wherever it is ving, begin at once their work of gath- g nectar and carrying pollen; as soon he florets are pollinated they wither droop below the flower-head.

ere *I made One, turn down an empty Glass,*

s old Omar, and I always think of it n I see the turned-down florets of the e-clover blossom. In this case, how- , the glass is not empty, but holds the uring seed. This habit of the white- er flowers saves the bees much time, e only those which need pollinating lifted upward to receive their visits. length of time the little clover head ires for the maturing of its blossoms ends much upon the weather and n the insect visitors.

/hite-clover honey is in the opinion nany the most delicious honey made n any flowers except, perhaps, orange soms. So valuable is the white clover honey plant that apiarists often grow s of it for their bees.

LESSON 167

The White Clover

EADING THOUGHT — The white clover creeping stems. Its flowers depend n the bees for their pollination, and bees depend upon the white-clover soms for honey.

IETHOD — The plant may be brought the schoolroom while in blossom, and orm be studied there. Observations as he fertilization of the flowers should nade out-of-doors.

BSERVATIONS — 1. Where does the te clover grow? Why is it sometimes l in lawns?

Note carefully the clover leaf, the oe of the three leaflets, stalks, and

Stumpp and Walter Co.

White clover

edges. Is part of the leaflet lighter colored than the rest? If so, describe the shape. Are the leaflets unequal or equal in size? Does each leaf come directly from the root? Are they alternately arranged? Why do they seem to come from the upper side of the stem?

3. Note the behavior of the clover

leaves at night. How do the two side leaflets act? The central leaflet?

4. Take a white-clover head, and note that it is made up of many little flowers. How many? Study one of the little flowers with a lens. Can you see its calyx? Its petals? Its stalk? In what way is it similar to the blossom of the sweet pea?

5. Take a head of white clover which has not yet blossomed. Tie a string about its stalk so that you may be sure you are observing the same flower and make the following observations during several days: Which blossoms begin to open first — those outside or inside? How many buds open each day? What happens to the blossoms as they fade? How many days pass from the time the flowers begin to blossom until the last flower at the center opens?

6. What insects do you see working on the white-clover blossoms? How does the bee act when collecting nectar? Can you see where she thrusts her tongue? W does the bee do for the clover bloss

7. Tie little bags of cheesecloth o two or three heads of white clover and if they produce any seed.

Little flower; but if I could understan
What you are, root and all, and all in
I should know what God and man is.

— Tenny

To me the meanest flower that blows
give
Thoughts that do often lie too deep
tears.

— Wordswo

I know a place where the sun is like g
And the cherry blooms burst with sn
And down underneath is the lovel
nook
Where the four leaf clovers grow.

— Ella Higgin

THE MAIZE OR INDIAN CORN

Hail! Ha-wen-ni-yu! Listen with open cars to the words of thy people. Continue listen. We thank our mother earth which sustains us. We thank the winds wh have banished disease. We thank He-no for rain. We thank the moon and st which give us light when the sun has gone to rest. We thank the sun for warmth a light by day. Keep us from evil ways that the sun may never hide his face from us shame and leave us in darkness. We thank thee that thou hast made our corn to gr Thou art our creator and our good ruler, thou canst do no evil. Everything thou dc is for our happiness.

Thus prayed the Iroquois Indians when the corn had ripened on the hills and valleys of New York State long before it was a state, and even before Columbus had turned his ambitious prows westward in quest of the Indies. Had he found the Indies with their wealth of fabrics and spices, he would have found there nothing so valuable to the world as has proved this golden treasure of ripened corn.

The origin of Indian corn, or maize, is shrouded in mystery. There is a plant which grows on the tablelands of Mexico which is possibly the original species; but so long had maize been cultivated by the American Indians that it was thoroug domesticated when America was first covered. In those early days of Ameri colonization, it is doubtful, says Profe John Fiske, if our forefathers could h remained here had it not been for Ind corn. No plowing, or even clearing, necessary for the successful raising of t grain. The trees were girdled, thus kil their tops to let in the sunlight, the r earth was scratched a little with a pri tive tool, and the seed put in and cove and the plants that grew therefrom t care of themselves. If the pioneers been obliged to depend alone upon

Arthur C. Parker

Seneca Indian women husking corn for braiding

eat and rye of Europe, which only ws under good tillage, they might have ved before they gained a foothold on forest-covered shores.

THE CORN PLANT

n studying the maize it is well to keep mind that a heavy wind is a serious my to it; such a wind will lay it low, from such an injury it is difficult for corn to recover and perfect its seed. us, the mechanism of the corn stalk leaf is adapted for prevention of this ster. The corn stalk is, practically, a ng cylinder with a pithy center; the rs of the stalks are very strong, and at rt intervals the stalk is strengthened hard nodes, or joints; if the whole stalk e as hard as the nodes, it would be astic and would break instead of bend; it is, the stalk is very elastic and will d far over before it breaks. The nodes nearer each other at the bottom, thus ing strength to the base; they are farther apart at the top, where the wind strikes, and where the bending and bowing of the stalk is necessary.

The leaf comes off at a node and clasps the stalk for a considerable distance, thus making it stronger, especially toward the base. Just where the leaf starts away from the stalk there is a little growth which serves as a rain guard; if water should seep between the stalk and the clasping leaf, it would afford harbor for destructive fungi. The structure of the corn leaf enables it to escape injury from the wind; the strong veins are parallel with a strong but flexible midrib at the center; often, after the wind has whipped the leaves severely, only the tips are split and injured. The edges of the corn leaf are ruffled, and where the leaf leaves the stalk there is a wide fold in the edge at either side; this arrangement gives play for a sidewise movement without breaking the leaf margins. The leaf is thus protected from the wind, whether it is struck from above or horizontally. The

Farmers' Bulletin 537, U. S. D. A.

A good hill of corn. The hills of corn about this one have been removed so that it may stand out more clearly. Note the tassels and the drooping position of the well-filled ears

true roots of the corn plant go quite deep into the soil, but are hardly adequate to the holding of such a tall, slender stalk upright in a wind storm; therefore, all about the base of the plant are brace-roots, which serve to hold the stalk erect — l the stay-ropes about a flagpole.

The Ear of Corn

The ears of corn are borne at the joi or nodes; and the stalk, where the presses against it, is hollowed out so as hold it snugly. In the following ways, husks show plainly that they are modif leaves: the husk has the same structure the leaf, having parallel veins; it com off the stem like a leaf; it is often gre and therefore does the work of a leaf; changes to leaf shape at the tip of the e thus showing that the husk is really th part of the leaf which usually clasps t stem. If a husk tipped with a leaf is exa ined, the part serving as a rain guard w be found at the place where the two jo As a matter of fact, the ear of corn is or branch stalk which has been very mu shortened, so that the nodes are very clo together, and therefore the leaves co off close together. By stripping the hu back one by one, the change from the o side, stiff, green leaf structure to the inn delicate, papery wrapping for the see may be seen in all its stages. This is beautiful lesson in showing how the mai protects its seed, and the husk may w be compared to the clothing of a ba The pistillate flowers of the corn, whi finally develop into the kernels, grow pairs along the sides of the end portion the shortened stalk, which is what we c the " cob." Therefore, the ear will sho an even number of rows, and the c shows distinctly that the rows are pair The corn silk is the style and stigma the pistillate flowers; and therefore, order to secure pollen, it must exte from the ovule, which later develops in a kernel, to the tip of the ear, where protrudes from the end of the husk. computation of the number of kernels a row and on the ear makes a very go arithmetic lesson for the primary pupi especially as the kernels occur in pairs.

The Growth of the Corn

If we cut a kernel of corn crosswise v can see, near the point where it joins th

ɔ, the little plant. Corn should be ger-
nated between wet blotters in a seed-
ting experiment before observations are
de on the growing corn of the fields.
hen the corn first appears, the corn
ves are in a pointed roll within a color-
s sheath which pierces the soil. Soon
y spread apart, but it may be some
ne before the corn stalk proper appears.
en it stretches up rapidly, and very
n will be tipped with beautiful pale
wn tassels. These tassels merit careful
dy, for they are the staminate flowers.
ch floret has two anthers hanging down
m it, and each half of each anther is a
le bag of pollen grains; and in order
t they shall be shaken down upon the
iting corn silk below, the bottom of
h bag opens wide when the pollen
ripe. The corn silk, at this stage, is
nched at the tip and clothed with fine
irs, so that it may catch a grain of the
cious pollen. Then occurs one of the
st wonderful pollen stories in all na-
e, for the pollen-tube must push down
ough the center of the corn silk for its
ole length, in order to reach the wait-
g ovule and thus make possible the development of a kernel of corn. These young, unfertilized kernels are pretty objects, looking like seed pearls, each wrapped in furry bracts. If the silk from one of these young flowers does not receive its grain of pollen, then the kernel

tassel of corn, showing the pollen-bearing flowers

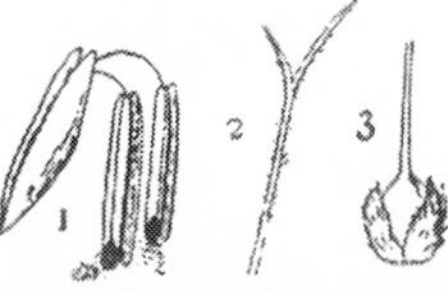

1, *The anthers of corn.* 2, *The tip of the corn silk showing the stigma.* 3, *The pistillate flower, which will develop into the kernel*

will not develop and the ear will be imperfect. On the other hand if the pollen from another variety of corn falls upon the waiting stigmas of the silk, we shall find the ear will have upon it a mixture of the two varieties. This is best exemplified when we have the black and white varieties of sweet corn growing near each other.

One reason why corn is such a valuable plant to us is that its growth is so rapid. It is usually not planted until late spring, yet, with some varieties, by September the stalks may be as much as twenty feet in height. The secret of this is that the corn, unlike many other plants, has many points of growth. While young, the part of the stalk just above each node is a growing center and the tip of the stalk also grows; the first two experiments suggested below will demonstrate this. In most plants, the tip of the stem is the only center of growth. When blown down by the wind, the corn has a wonderful way of lifting itself, by inserting growing wedges in the lower sides of the nodes. A corn stalk blown down by the wind will often show this wedge shape at every joint, and the result will be an upward curve of the whole stalk. Of course, this cannot be seen unless the stalk is cut lengthwise through the center. Experiment 3 is suggested to demonstrate this.

During drought the corn leaves check the transpiration of water by rolling together lengthwise in tubes, thus offering less surface to the sun and air. The farmer calls this the curling of the corn, and it is

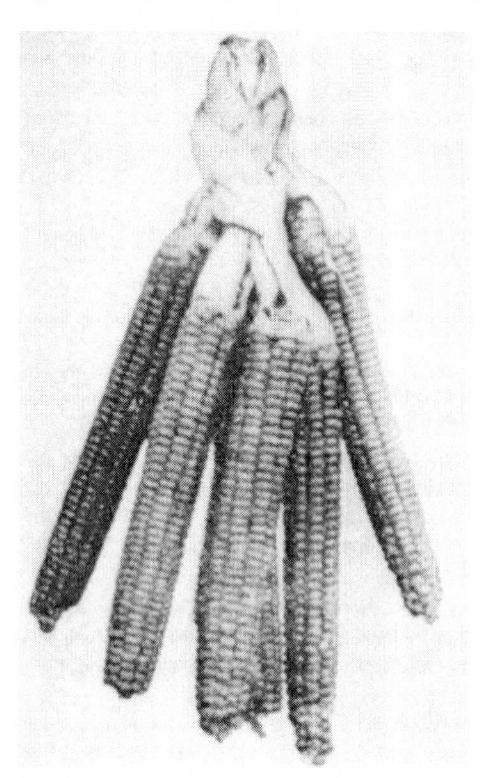

Ears of corn with braided husks as the Indians used to carry them

always a sign of lack of moisture. If a corn plant with leaves thus curled be given plenty of water, the leaves will soon straighten out again into their normal shape.

LESSON 168

The Maize

Leading Thought — The Indian corn, or maize, is a plant of much beauty and dignity. It has wonderful adaptations for the development of its seed and for resisting the wind.

Method — The study may begin spring when the corn is planted, givi the pupils the outline for observations be filled out in their notebooks during t summer, when they have opportunit for observing the plant; or it may studied in the autumn as a matured pla It may be studied in the schoolroom or the field, or both.

Observations on the Corn Plant

1. Describe the central stem. How ma joints or nodes has it? Of what use to t plant are these nodes? Are the joi nearer each other at the bottom or t top of the plant?

2. Where do the leaves come off t stem? Describe the relation of the ba of the leaves to the stem. Of what use this to the plant?

3. Note the little growth on the l where it leaves the stalk. Describe h this prevents the rain from seeping do between the stalk and the clasping le What danger would there be to the pla if the water could get into this narr space?

4. What is the shape of the leaf? D scribe the veins. Does the leaf tear eas across? Does it tear easily lengthwi

Agronomy Dept., State College of Agr., Cornell

A corn shock. In regions where corn is harvested by machinery, and where it is used for silage, it is often shocked to permi to mature

what use to the plant is this condi-
ı?
. Are the edges of the corn leaf straight
uffled? How does this ruffled edge per-
the leaf to turn without breaking?
scribe at length the benefit the corn
nt derives from having leaves which are
easily broken across and which can
ıd readily sidewise as well as up and
vn.
. Describe the roots of the corn plant.
scribe the brace-roots. Explain their

. Describe all the ways in which the
n plant is strengthened against the
ıd.
)BSERVATIONS ON THE EAR OF CORN —
Vhere on the corn plant are the ears
ne? Are two ears borne on the same
of the stalk? Remove an ear, and see
v it was fitted against the side of the
k.
. Where do the ears come off the stalk
elation to the leaves?
o. Examine the outside husks, and
ıpare them with the green leaves.
ıat is there to suggest that the corn
k is a leaf changed to protect the seed?
you think that the husk represents
t portion of the leaf which clasps the
k? Why? Describe how the inner husk
ers from the outer in color and texture.
scribe how this is a special protection
he growing kernels.
1. After carefully removing the husk,
mine the silk and see if there is a
ead for every kernel. Is there an equal
ount of silk lying between every two
s? Do you know what part of the corn
ver is the corn silk? What part is the
nel?
2. How many rows of kernels are there
an ear? How many kernels in a row?
w many on the whole ear? Do any of
rows disappear toward the tip of the
? If so, do they disappear in pairs? Do
know why? Are the kernels on the
of the ear and near the base as perfect
hose along the middle? Do you know
ether they will germinate as quickly and
orously as the middle ones?
3. Study a cob with no corn on it and

Extension Chart, U. S. D. A.

Sugar cane, a near relative of corn, is a crop of tropical and subtropical regions. The stalks in the foreground have been stripped of leaves and are ready for the mill

note if the rows of kernel-sockets are in distinct pairs. This will, perhaps, show best if you break the cob across.

14. Break an ear of corn in two, and sketch the broken end showing the relation of the cob to the kernels.

15. Are there any places on the ear you are studying where the kernels did not grow or are blasted? What may have caused this?

16. Describe the requisite for a perfect ear of seed corn. Why should the plant from which the seed ear is taken be vigorous and perfect?

OBSERVATIONS ON THE GROWTH OF CORN — WORK FOR THE SUMMER VACATION — 17. How does the corn look when it first comes up? How many leaves are there in the pointed roll which first appears above the ground? How long before the central stalk appears?

18. When do the tassels first appear? What kind of flowers are the corn tassels? Describe the anthers. How many on each flower? Where do the anthers open to discharge their pollen?

19. Note that the kernel is the ovary. The silk is the style; it is attached to the ovary and is long enough to extend out beyond the husks; at its tip is the branched stigma.

20. What carries the pollen for the

corn plant? If you have rows of popcorn and sweet corn or of sweet corn and field corn next to each other why is it that the ears will show a mixture of both kinds?

EXPERIMENT 1

Compare the growth of the corn plant with that of the pigweed. When the corn stalk first appears above ground, tie two strings upon it, one just above a joint and one below it. Tie two strings the same distance apart on the stem of a pigweed. Measure carefully the distance between these two strings on the two plants. Two weeks later measure the distance between the strings again. What is the result?

EXPERIMENT 2

Measure the distance between two of the nodes or joints near the tip of a certain corn stalk. Two weeks later measure distance again and compare the two.

EXPERIMENT 3

When a stalk of corn is still greer August, bend it down and place a s across it at about half its length. Desc how it differs in position after two three weeks. Cut lengthwise across on the nodes, beyond the point held d by the stick, and see the we shaped growth within the joint wl helps to raise the stalk to an upright p tion.

EXPERIMENT 4

During the August drought, note t the corn leaves are rolled. Give a c plant with rolled leaves plenty of w and note what happens. Why?

U. S. Department of Agricu

Cotton pickers at work

THE COTTON PLANT

There are some plants which have made great chapters in the histories of nations, and cotton is one of them. The fiber of cotton was used for making clothing so long ago that its discovery is shrouded in the myths of prehistoric times. But we believe it first came into use in India, for in this land we find certain laws conce ing cotton which were codified about B.C.; and allusions to the fine, white ment on the peoples of India are frequ in ancient history. Cotton was introdu into Egypt from India at an early date was in common use there about 150 not until our Civil War laid fallow cotton fields of the United States, did ·pt realize the value of its crop; and al- ugh much money was lost there in ag- ıltural speculation after our own prod- was again put on the market, still ton has remained since that time one Egypt's most valuable exports.

Vhen Columbus discovered America found cotton growing in the West ies, and the chief articles of clothing he native Mexicans were made of cot- . Cloths of cotton were also found in ient tombs of Peru, proving it was used re long before the white man set his t upon those shores. When Magellan de his famous voyage around the world 500, he found the cotton fiber in use Brazil.

t is a strange fact that the only region he world between the parallels of 40° th and 40° south latitude where cot- did not grow as a native or cultivated nt when America was discovered was region of our Gulf states, which now duces more cotton than any other. first mention of cotton as a crop in American colonies is in the report lished in 1666. At the time of the olutionary War the cotton industry thoroughly established. It is one of significant facts of history that the in- tion of the cotton gin by Eli Whitney 793, which revolutionized the cotton ustry and brought it to a much more fitable basis, wrought great evil in the ited States, since it revived the profits slave-holding. The institution of slav- was sinking out of sight by its own ght; Washington showed that it was most expensive way to work land, and erson failed to liberate his own slaves ply because he believed that liberty ıld come to all slaves inevitably, since e-holding was such an expense to the ntation owners. But the cotton gin, ich removed the seeds rapidly — a cess theretofore done slowly and labo- sly by hand — suddenly made the rais- of cotton so profitable that slaves were in employed in its production with at financial benefits. And thus it came about that the cotton plant innocently wielded a great influence in the political as well as the industrial life of our country.

Cotton in blossom

The cotton plant has a taproot, with branches which go deep into the soil. The stem is nearly cylindrical, the branches often spreading and sometimes irregular; the bark is dark and reddish; the wood is white. In Egypt, and probably in other arid countries, the stalks are gathered for fuel in winter.

The leaves are alternate, with long petioles. The upper leaves are deeply cut, some having five, some seven, some three, and some even nine lobes; strong veins extend from the petiole along the center of each lobe; the leaves near the ground may not be lobed at all. Where the petiole joins the stem, there is a pair of long, slender, pointed stipules, but they often fall off early. A strange characteristic of the cotton leaves is that they bear nectar-glands; these may be seen on the underside and along the main ribs of the leaf; they appear as little pits in the rib; some leaves may have none, while others may have from one to five.

The flower bud is partially hidden beneath the clasping bracts of the involucre. These bracts are three or four in number, and they have the edges so deeply lobed

U. S. Department of Agriculture

A single cotton plant loaded with maturing bolls

that they seem branched. By pushing back the bracts we can find the calyx, which is a shallow cup with five shallow notches in its rim. The petals are rolled in the bud like a shut umbrella. The open flower has five broadly spreading petals; when the bud first opens in the morning, the petals are whitish or pale yellow with a purplish spot at the base, by noon they are pale pink, by the next day they are a deep purplish red and they fall at the end of the second day. There are nectar-glands also in the flower at the base of the calyx, and the insects are obliged to thrust their tongues between the bases of the petals to reach the nectar; only long-tongued bees, moths, and butterflies are able to attain it.

There are many stamens which have their filaments united in a tube extending up into the middle of the flower and enlarging a little at the tip; below the enlarged base of this tube is the ovary which later develops into the cotton boll; within the stamen-tube extends the long st and from its tip are thrust out three five stigmas like little pennants from top of a chimney; and sometimes they more or less twisted together. The you boll is covered and protected by fringed bracts, which cover the bud a remain attached to the ripened boll. T calyx, looking like a little saucer, also mains at the base of the boll. The b soon assumes an elongated, oval sha with long, pointed tip; it is green outs and covered with little pits, as large pinpoints. There are, extending b from the pointed tip, three to five crea or sutures, which show where the boll open. If we open a nearly ripened boll, find that halfway between each two tures where the boll will open there i partition extending into the boll divid it into compartments. These are rea carpels, as in the core of an apple, a their leaf origin may be plainly seen in venation. The seeds are fastened by th pointed ends along each side of the c tral edge of the partition, from which th break away very easily. The number seeds varies, usually two or three alo each side; the young seeds are wrapp in the young cotton, which is a strin soft white mass. The cotton fibers are tached to the covering of the seed arou the blunt end, and usually the poin end is bare. When the boll opens, the c ton becomes very fluffy and if not pick will blow away. The wild cotton disse nates its seeds by sending them off the wings of the wind. Heavy winds at cotton-picking time are a menace to crop and often occasion serious loss.

The mechanism of the opening of cotton boll is very interesting; along central edge of each partition and exte ing up like beaks into the point of the b is a stiff ridge, about the basal portion which the seeds are attached; as the b becomes dry, this ridged margin becom as stiff as wire and warps outward; at same time, the outside of the boll is shr eling. This action tears the boll ap along the sutures and exposes the see with their fluffy balloons to the action

: wind. The ripe, open, empty boll is rth looking at; the sections are wide ırt and each white, delicate, parch- :nt-like partition or carpel, has its wire ʒe curved back gracefully. The outside the boll is brown and shriveled, but ide it is still white and shows that it had oft lining for its seeds.

The amount of the cotton crop per acre ies with the soil and climate; the ıount that can be picked per day also ɔends upon the cotton as well as upon : picker. Children have been known to :k one hundred pounds per day, and a t-class picker from five hundred to six ndred pounds, or even eight hundred; e man has made a record of picking ty pounds in an hour. Cotton is one of : most important crops grown in Amer- , and there are listed more than one ndred and thirty varieties which have ginated in our country.

LESSON 169

Cotton

LEADING THOUGHT — Cotton has had a at influence upon our country politi- ly as well as industrially. Its fiber was :d by the ancients, and it is today one the most important crops in the regions ere it is grown.

METHOD — A cotton plant with blos- ns and ripe bolls upon it may be ught into the schoolroom or studied in : field.

OBSERVATIONS — 1. How many varieties cotton do you know? Which kind is it ı are studying?

2. What sort of root has the cotton nt? Does it go deep into the soil?

3. How high does the plant grow? Are : stems tough or brittle? What is the or of the bark? Of the wood? Do you ɔw of a country where cotton stalks are d for fuel? Do the stem and branches w erect or very spreading?

4. Are the leaves opposite or alternate? Are the petioles as long as the leaves? Are there any stipules where the petioles join the main stem? How many forms of leaves can you find on the same stem? How do the upper differ from the lower leaves? Describe or sketch one of the large upper leaves, paying especial attention to the veins and the shape of the lobes.

5. Look at the lower side of a leaf and find, if you can, a little pit on the midrib near its base. How many of these pits can you find on the veins of one leaf? What is

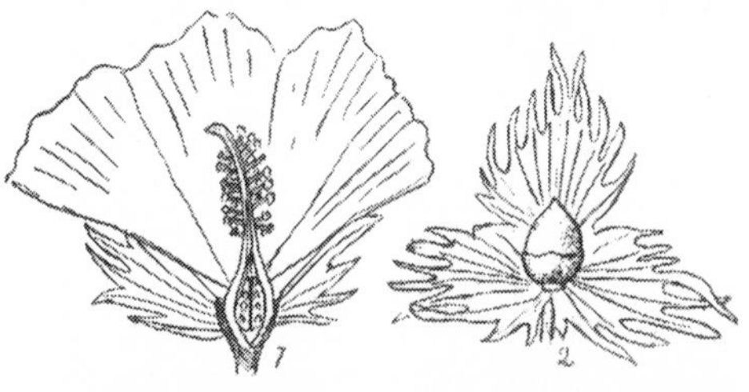

1, *The cotton flower cut in half, showing the stamen-tube at the center, up through which extends the style of the pistil. Note the bracts and calyx.* 2, *A young boll, with calyx at its base, set in the involucral bracts*

the fluid in these pits? Taste it and see if it is sweet. Watch carefully a growing plant and describe what insects you find feeding on this nectar. Note if the wasps and ants feeding on this nectar attack the caterpillars of the cotton worm which destroy the leaf. Where are the nectar-glands of plants usually situated?

6. Study the flower bud; what covers it? How many of these bracts cover the flower bud? What is their shape and how do their edges look? Push back the bracts and find and describe the calyx. How are the petals folded in the bud?

7. Take the open flower; how many petals are there, and what is their shape? At what time of day do the flowers open? What color are the petals when the flowers first open? What is their color later in the day? What is their color the next day? When do the petals fall?

8. Describe the stamens; how are they joined? How are the anthers situated on the stamen-tube? Is the stamen-tube perfectly straight or does it bend at the tip?

9. Peel off carefully the stamen-tube and describe what you find within it. How many stigmas come out of the tip of the tube? Find the ovary below the stamen-tube. Which part of the flower grows into the cotton boll?

10. Take a boll nearly ripe; what covers it? Push away the bracts; can you find the calyx still present? What is the shape of the boll? What is its color and texture? Can you see the creases where it will open? How many are there of them?

11. Open a nearly ripe boll very carefully. How many partitions are there in it? Where are they in relation to the openings? Gently push back the cotton from the seeds without loosening them, and describe how the seeds are connected with the partitions. Is the seed attached by its pointed or by its blunt end?

12. How many seeds in each chamber in the cotton boll? Where on the seed does the cotton grow? How does the cotton blanket wrap about the seed? If t cotton is not picked what happens to Of what use to the wild cotton plant a seeds covered with cotton?

13. What makes the cotton boll ope Describe an open and empty boll outsi and inside.

14. How much cotton is considered good crop per acre in your vicinity? H much cotton can a good picker gather a day?

15. Write English themes on the f lowing topics: "The History of the C ton Plant from Ancient Times until T day," "How the Cotton Plant Has A fected American History."

Queen-consort of the kingly maize,
The fair white cotton shares his thro
And o'er the Southland's realm she clai
A just allegiance, all her own.

— Minnie Curtis Wa

Verne Mor

THE STRAWBERRY

Of all the blossoms that clothe our open fields, one of the prettiest is that of the wild strawberry. And yet so influenced is man by his stomach that he seldo heeds this flower except as a promise a crop of strawberries. It is comforting

w that the flowers of the field " do not
: a rap " whether man notices them or
; insect attentions are what they need,
they are surely as indifferent to our
ifference as we are to theirs.

'he field strawberry's five petals are lit-
cups of white held up protectingly
ınd anthers and pistils; each petal has
ɔase narrowed into a little stalk which
botanists call a claw. When the blos-
first opens, the anthers are little, flat,
dly lemon-yellow discs, each disc con-
ng of two clamped together sternly and
erminedly as if they meant never to
n and yield their gold dust. At the very
ter of the flower is a little, greenish-yel-
cone, which, if we examine with a lens,
can see is made up of many pistils set
ether, each lifting up a little circular
ma. Whether all the stigmas receive
len or not determines the formation of
od strawberry.

'he sepals are slender and pointed and
n to be ten in number, every other one
ng smaller and shorter than its neigh-
s; the five shorter ones, however, are
sepals but are bracts below the calyx.

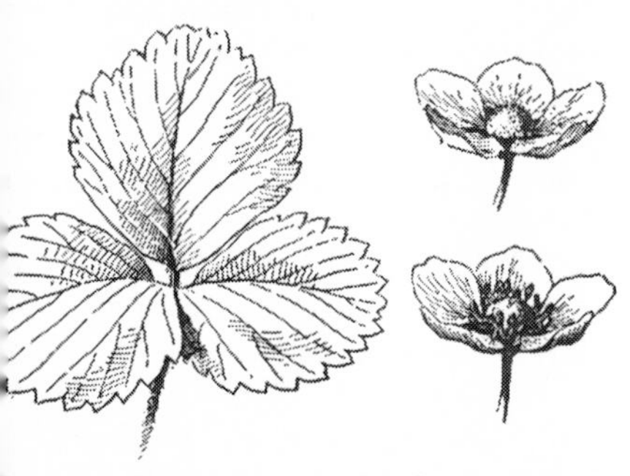

*t, a strawberry leaf. Right above, a pistil-
ıte flower. Right below, a perfect flower*

e sepals unite at their bases so that the
wberry has really a lobed calyx instead
separate sepals. The blossom stalk is
., pinkish, and silky; it wilts easily.
ere are several blossoms borne upon
stalk and the central one opens first.

'he strawberry leaf is beautiful; each of
three leaflets is oval, deeply toothed,
has strong regular veins extending
n the midrib to the tip of each tooth.
color it is rich, dark green and turns to

Strawberry fruit

wine color in autumn. It has a very pretty way of coming out of its hairy bud scales, each leaflet folded lengthwise and the three pressed together. Its whole appearance then is infantile in the extreme, it is so soft and helpless looking. But it soon opens out on its pink, downy stem and shows the world how beautiful a leaf can be.

If a comparison of the wild and cultivated strawberries is practicable, it makes this lesson more interesting. While the wild flowers are usually perfect, many cultivated varieties have the pollen and pistils borne in different flowers, and they depend upon the bees to carry their pollen. The blossom stalk of the garden strawberry is round, smooth, and quite strong, holding its branching panicle of flowers erect, and it is usually shorter than the leafstalks among which it nestles. The flowers open in a series, so that ripe and green fruit, flowers, and buds may often be found on the same stem. As the strawberry ripens, the petals and stamens wither and fall away; the green calyx remains as the hull, which holds in its cup the pyramid of pistils which swell and ripen into the juicy fruit. To the botanists the strawberry is not a berry, that definition being limited to fruits having a juicy pulp and containing many seeds, like the currant or grape. The strawberry is a fleshy fruit bearing its akenes, the hard parts which we have always called seeds, in shallow pits on its surface. These akenes are so small that we do not notice them when eating the fruit, but each one is a tiny nut, almond-shaped, and containing within its tough little shell a starchy meat to sustain the future plant which may grow from it. It is by planting these akenes that growers obtain new varieties.

The root of the strawberry is fibrous and threadlike. When growers desire plants for setting new strawberry beds they are careful to take only such as have light colored and fresh-looking roots. On old plants the roots are rather black and woody and are not so vigorous.

The stem of the strawberry is partially underground and so short as to be unnoticeable. However, the leaves grow upon it alternately one above another, so that the crown rises as it grows. The base of each leaf has a broad, clasping sheath which partly encircles the plant and extends upward in a pair of earlike stipules.

The runners begin to grow after the fruiting season has closed; they originate from the upper part of the crown; they are strong, fibrous, and hairy when young. Some are short between joints, others seem to reach far out as if seeking for the best location before striking root; a young plant will often have several leaves before putting forth roots. Each runner may start one or more new strawberry plants. After the young plant has considerable root growth, the runner ceases to carry sap from the main stem and withers to a mere dry fiber. The parent plant continues to live and bear fruit, for the strawberry is a perennial, but the later crops are of less value. Gardeners usually renew their plots each year, but if intending to harvest a second year's crop, they cut off the runners as they form.

LESSON 170

The Strawberry

Leading Thought — The strawberry plant has two methods of perpetuating itself, one by the akenes which are gr on the outside of the strawberry fruits, one by means of runners which start plants wherever they find place to root. Cultivated plants are grown f runners, but new species must be gr from seed.

Method — It would be well to ha strawberry plant, with roots and run attached, for an observation lesson by class. Each pupil should have a leaf, cluding the clasping stipules and sh at its base. Each one should also ha strawberry blossom and bud, and if sible a green or ripe fruit.

Observations — 1. What kind of has the strawberry? What is its color?

2. How are the leaves of the strawb plant arranged? Describe the base of leaf and the way it is attached to the st How many leaflets are there? Sketc strawberry leaf, showing the edges form of the leaflets, and the veins.

3. From what part of the plant do runners spring? When do the runners gin to grow? Does the runner strike before forming a new plant or does the tle plant grow on the runner and draw tenance from the parent plant?

4. What happens to the runners a the new plants have become establish Does the parent plant survive or die a it sends out many runners?

5. Describe the strawberry blosso How many parts are there to the hul calyx? Can you see that five of these set below the other five?

6. How many petals are there? D the number differ in different flow Has the wild strawberry as many pe as the cultivated ones?

7. Study with a lens the small g button at the center of the flower. T is made up of pistils so closely set only their stigmas may be seen. Do find this button of pistils in the same b som with the stamens? Does the wild b som have both stamens and pistils in same flower?

8. Describe the stamens. What ins carry pollen for the strawberry plants?

9. Are the blossoms arranged in c

s? Do the flowers all open at the same ıe? What parts of the blossom fall ay and what parts remain when the it begins to form?

ıo. Are the fruits all of the same shape d color? Is the pulp of the same color thin as on the surface? Has the fruit an outer coat or skin? What are the specks on its surface?

11. How many kinds of wild strawberries do you know? How many kinds of cultivated strawberries do you know?

12. Describe how you should prepare, plant, and care for a strawberry bed.

Verne Morton

"When the frost is on the punkin and the fodder's in the shock"

THE PUMPKIN

If the pumpkin were as rare as some ːhids, people would make long pilmages to look upon so magnificent a ınt. Although it trails along the ground, ting Mother Earth help it support its ;antic fruit, yet there is no sign of weakss in its appearance; the vine stem is ong, ridged, and spiny. The spines upon are surely a protection under some cirınstances, for I remember distinctly ıt when, as children, barefooted and ning the world, we "played Indian" d found our ambush in the long rows ripening corn, we skipped over the pumpkin vines, knowing well the punishment they inflicted on the unwary feet.

From the hollow, strongly angled stem arise in majesty the pumpkin leaves, of variously lobed patterns, but all formed on the same decorative plan. The pumpkin leaf is as worthy of the sculptor's chisel as is that of the classic acanthus; it is palmately veined, having from three to five lobes, and its broad base is supported for a distance on each side of the angled petiole by the two basal veins. The leaves are deep green above and paler below; they are covered on both sides with minute bristles, and their edges are finely toothed. The bristly, angled stalk which lifts it aloft is a quite worthy support for so beautiful a leaf. And, during our child-

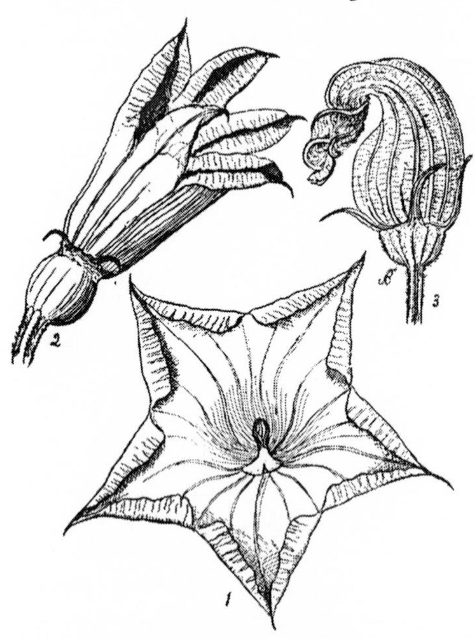

The closing of a pumpkin flower

1, Staminate flower beginning to close; note the folded edges of the lobes. 2, Pistillate flower nearly closed. 3, Staminate flower closed and in its last stage

hood, it was also highly esteemed as a trombone, for it added great richness of quality to our orchestral performances, balancing the shrillness of the basswood whistle and the sharp buzzing of the dandelion-stem pipe.

Growing from a point nearly opposite a leaf may be seen the pumpkin's elaborate tendril. It has a stalk like that of the leaf, but instead of the leaf blade it seems to have the three to five naked ribs curled in long, small coils very even and exact. Perhaps, at some period in the past, the pumpkin vines lifted themselves by clinging to trees, as do the gourd vines of today. But the pumpkin was cultivated in fields with the maize by the North American Indians, long before the Pilgrim Fathers came to America and made its fruit into pies. Since the pumpkin cannot sustain itself in our northern climate without the help of man, it was evidently a native of a warmer land. With cultivation it now sends its long stems out for many feet, r ing entirely upon the ground. But, lik conservative, elderly maiden lady, it s wears corkscrew curls in memory of a fa ion long since obsolete. Occasionally, see the pumpkin vines at the edge of field pushing out and clambering o stone piles, and often attempting to cli the rail fences, as if there still remair within them the old instinct to climb.

But though its foliage is beautiful, glory of the pumpkin is its vivid yell blossom and, later, its orange fruit. Wh the blossom first starts on its career a bud, it is enfolded in a bristly, ribbed ca with five stiff, narrow lobes, which cl up protectingly about the green, co shaped bud, a rib of the cone appeari between each two lobes of the calyx. we watch one of these buds day after d we find that the green cone changes to yellow color and a softer texture as t bud unfolds, and then we discover that is the corolla itself; however, these r which extend out to the tip of t corolla lobes remain greenish below, p manently. The expanding of the flow bud is a pretty process; each lobe, su ported by a strong midrib, spreads c into one of the points of a five-point star; each point is very sharp and angu because, folded in along these edges one of the prettiest of Nature's hems the ruffled margin of the flower. Not un the sun has shone upon the star for so little time of a summer morning do the

Verne Mor

The staminate blossom of the pumpk showing the anther knob at the left. A bud the staminate flower at the center and closed blossom at the right

ned-in margins open out; and, late in : afternoon or during a storm, they fold wn again neatly before the lobes close ; if a bee is not lively in escaping she y, willy-nilly, get a night's lodging, for se folded edges literally hem her in.

The story of the treasure at the heart this starry, bell-shaped flower is a uble one, and we had best begin it by ecting a flower that has below it a little en globe — the ovary — which will er develop into a pumpkin. At the heart such a flower there stand three stigmas, t look like lilliputian boxing gloves; h is set on a stout, postlike style, which ; its base in a great nectar-cup, the ges of which are slightly incurved over welling sweetness. In order to reach s nectar, the bee must stand on her d and brush her pollen-dusted side inst the stigmas. Professor Duggar has ted that in dry weather the margins of s nectar-cup contract noticeably, and t in wet weather the stigmas close wn as if the boxing gloves were on sed fists.

The other half of the pumpkin-blossom ry is to be found in the flowers which ve no green globes below them, for se produce the pollen. Such a flower ; at its center a graceful pedestal with a ad base and a slender stem, which up-lds a curiously folded, elongate knob, t looks like some ancient or primitive vel wrought in gold. The corrugations its surface are the anther cells, which curiously joined and curved around a tral oblong support; by cutting one oss, we can see plainly the central core, rdered by cells filled with pollen. But ere is the nectar-well in the smooth cup this flower? Some have maintained that : bees visit this flower for the sake of : pollen, but I am convinced that this not all of the story. In the base of the destal which supports the anther knob re appear, after a time, three incon-cuous openings; and if we watch a bee, shall see that she knows these openings there and eagerly thrusts her tongue wn through them. If we remove the hers and the pedestal, we shall find nectar below the latter; the nectar-cup is carpeted with the softest of buff velvet, and while it does not reek with nectar, as does the cup which encompasses the

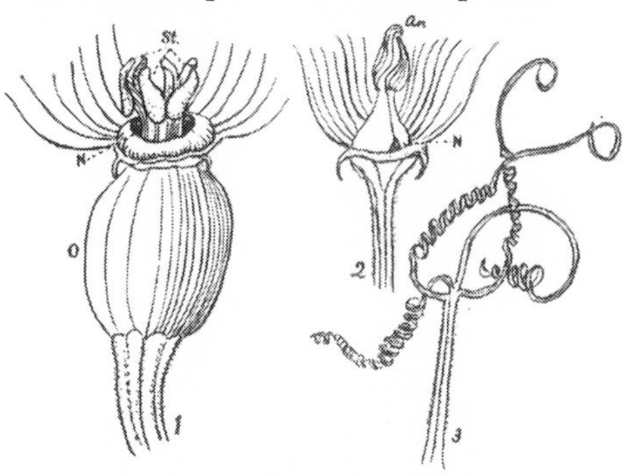

1, *Base of pistillate blossom; O, ovary which develops into the pumpkin; N, nectar cup; St, stigmas.* 2, *Base of a staminate blossom; N, opening into the nectar cup; An, anthers joined, forming a knob.* 3, *Pumpkin tendril*

styles of the pistil, yet it secretes enough of the sweet fluid so that we can taste it distinctly. Thus, although the bees find pollen in this flower they also find nectar there. The pumpkin is absolutely dependent upon the work of bees and other insects for carrying its pollen from the blossom that bears it to the one which lacks it, as this is the only way that the fruit may be developed.

And after the pollen has been shed and delivered, the flower closes, to open no more. The fading corolla looks as if its lobes had been twisted about by the thumb and finger to secure tightness; and woe betide the bee caught in one of these prisons, unless she knows how to cut through its walls or can find within sustenance to last until the withered flower falls. The young pumpkin is at first held up by its stiff stem but later rests upon the ground.

The ripe pumpkin is not only a colossal but also a beautiful fruit. The glossy rind is brilliant orange and makes a very efficient protection for the treasures within it. The stem is strong, five-angled, and stubborn, and will not let go its hold until the fruit is over-ripe. It then leaves a star-shaped scar to match the one at the other end of the fruit, where once the blossom sat enthroned. The pumpkin in shape is like a little world flattened at the poles, and with the lines of longitude creased

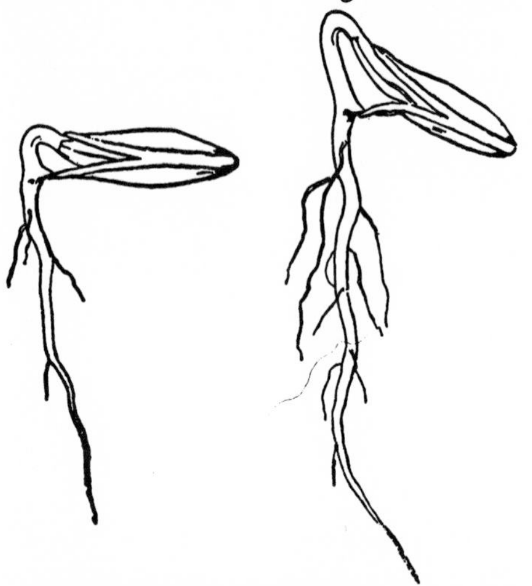

A squash plant breaking out of the seed coats at the left. The operation further progressed is shown at the right

into its surface. But the number of these longitudinal creases varies with individual pumpkins, and seems to have no relation to the angles of the stem or the three chambers within.

If we cut a small green pumpkin across, we find the entire inside solid. There are three fibrous partitions extending from the center, dividing the pulp into thirds; at its outer end each partition divides, and the two ends curve in opposite directions. Within these curves the seeds are borne. A similar arrangement is seen in the sliced cucumber. As the pumpkin ripens, the partitions surrounding the seeds become stringy and very different from the "meat" next to the rind, which makes a thick, solid outer wall about the central chamber, where are contained six rows of crowded seeds, attached by their pointed tips and supported by a network of yellow, coarse fibers — like babies supported in hammocks. All this network, making a loose and fibrous core, allows the seeds to fall out in a mass when the pumpkin is broken. If we observe where the cattle have been eating pumpkins we find these masses of seeds left and trampled into t mud, where, if our winter climate p mitted, they could grow into plants ne year.

The pumpkin seed is attached by pointed end; it is flat, oblong, and has rounded ridge at its edge, within which a delicate "beading." The outside is ve mucilaginous; but when it is wiped d we can see that it has an outer, very thi transparent coat; a thicker white, midd coat; while the meat of the seed is cover with a greenish, membranous coat. T meat falls apart lengthwise and flatwis the two halves forming later the se leaves and containing the food laid up the "pumpkin mother" for the nouris ment of the young plant. Between the two halves, at the pointed end, is the e bryo, which will develop into a new pla

When sprouting, the root pushes o through the pointed end of the seed a grows downward. The shell of the seed forced open by a little wedge-shaped p jection, while the seed leaves are pull from their snug quarters. If the se leaves are not released, the seed sh clamps them together like a vise, and t little plant is crippled.

Both squashes and pumpkins figure the spicy Thanksgiving pies, but the ch value of the pumpkin crop in America as food for milch cows; it causes a yield milk so rich that the butter made fro it is as golden as its flesh. But the H

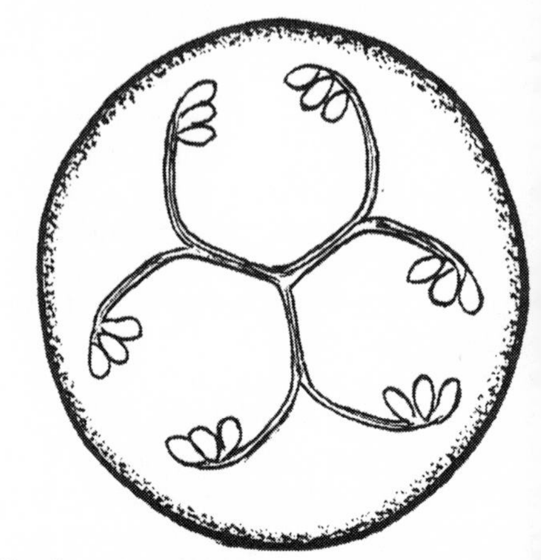

Section of a pumpkin just after the blosso has fallen. Note how the seeds are borne

e'en jack-o'-lantern appeals to the chil-
. In this connection, a study of ex-
sion might be made interesting; the
ing of the corners of the mouth up or
n, and the angles of the eyebrows,
ing all the difference between a jolly
and an "awful face."

LESSON 171
The Pumpkin

eading Thought — The pumpkin and
sh were cultivated by the American
ans in their cornfields long before Co-
bus discovered the new world. Insects
y the pollen for the flowers of these
ts, which must be cross-pollinated in
r to develop their fruit.

Iethod — This work may be done in
garden or field in September or early
ober; or a vine bearing both kinds
lowers, leaves, and tendrils may be
ght to the schoolroom for observa-
. The lesson on the pumpkin fruit
be given later. A small green pump-
should be studied with the ripe one,
also with the blossoms, so as to show
position of the seeds during develop-
t. This lesson can be modified to fit
cucumber, the melon, the squash, and
gourd.

The Pumpkin Vine and Flowers

bservations — 1. How many differ-
forms of flowers do you find on a
pkin vine? What are the chief dif-
nces in their shape?

Look first at the flowers with the
slender stalks. What is the shape and
r of the blossom? How many lobes
it? Is each lobe distinctly ribbed or
ed? Is the flower smooth on the inner
the outer surface? Are the edges of
lobes scalloped or ruffled?

What do you see at the bottom of
golden vase of this flower? This yellow
, or knob, is formed by the joining
hree anthers, one of which is smaller

Verne Morton

Pistillate flowers of the squash in various stages

than the others. Do all the pumpkin flowers have this knob at the center? Look at the base of the standard which bears the anther-knob, and note if there are some openings; how many? Cut off the anther pedestal, and describe what is hidden beneath it. Note if the bees find the openings to the nectar-well and probe there for the nectar. Do they become dusted with pollen while seeking the nectar?

4. What color is the pollen which is clinging to the anther? Is it soft and light, or moist and sticky? Do you think that the wind would be able to lift it from its deep cup and carry it to the cup of another flower?

5. Describe the calyx behind this pollen-bearing flower. How many lobes has it? Are the lobes slender and pointed?

6. Find one of the flowers which has below it a little green globe, which will later develop into a pumpkin. How does this flower differ from the one that bears the pollen?

7. Describe or sketch the pistil which is at the bottom of this flower vase. Into how many lobes does it divide? Do these three stigmas face outward or toward each other? Are the styles which uphold the stigmas short or long? Describe the cup

Verne Morton

A pumpkin vine showing tendrils, a flower, and an immature pumpkin

in which they stand. Break away a bit of this little yellow cup and taste it. Why do you think the pumpkin flowers need such a large and well-filled nectary? Could insects get the nectar from the cup without rubbing against the stigmas the pollen with which they became so thoroughly dusted when they visited the staminate flowers?

8. Cut through the center of one of the small green pumpkins. Can you see into how many sections it is divided? Does the number of seed-clusters correspond with the number of stigmas in the flower? Make a sketch of a cross section, showing where the seeds are placed.

9. What insects do you find visiting the pumpkin flowers?

10. Carefully unfold a flower bud which is nearly ready to open, and note how it is folded. Then notice late in the afternoon how the flower closes. What part is folded over first? What next? How does it look when closed?

11. Describe the stems of the pumpkin vine; how are they protected? Sketch or describe a pumpkin leaf.

12. Describe one of the tendrils of the pumpkin vine. Do you think that these tendrils could help the vine in climbing? Have you ever found a pumpkin vine climbing up any object?

The Pumpkin Fruit

Observations — 1. Do you think the pumpkin is a beautiful fruit? Why? Describe its shape and the way it is creased. Describe the rind, its color and its text Describe the stalk; does it cling to pumpkin? How many ridges in the s where it joins the vine? How many wl it joins the pumpkin? Which part of stalk is larger? Does this give it a fir hold?

2. Cut in halves crosswise a small g pumpkin and a ripe one. Which is m solid? Can you see how the seeds are bo in the green pumpkin? How do they l in the ripe pumpkin? What is next to rind in the ripe fruit? What part of pumpkin do we use for pies?

3. Can you see in the ripe pump where the seeds are borne? How are t suspended? How many rows of s lengthwise of the pumpkin? What is of a pumpkin after the cattle have ea it? Might the seeds thus left plant th selves?

4. Is the pumpkin seed attached at round end or at the pointed end? Desc the pumpkin seed, its shape, and its ed How does it feel when first taken from pumpkin? How many coats has the se

5. Describe the meat of the seed. D it divide naturally into two parts? Can see the little embryo plant? Have you tried roasting and salting pumpkin squash seeds, to prepare them for foo almonds and peanuts are prepared?

6. Plant a pumpkin seed in damp s and give it warmth and light. From wh end does it sprout? What comes first,

Bodger Seeds,

Gourds of many varieties are grown ornamental purposes. Pumpkins, squas and cucumbers are all members of the go family

t or the shoot? What part of the seed ms the seed leaves?

. Describe how the pumpkin sprout es open the shell to its seed in order to its seed leaves out. What happens if it s not pull them out? Which part of seedling pumpkin appears above und first?

. How do the true leaves differ in pe from the seed leaves? In what ways the seed leaves useful to the plant?

! on Thanksgiving day, when from East and from West,
m North and from South come the pilgrim and guest,
hen the gray-haired New-Englander sees round his board
e old broken lines of affection restored,
hen the care-wearied man seeks his mother once more,
d the worn matron smiles where the girl smiled before,
What moistens the lip and brightens the eye?
What calls back the past, like the rich Pumpkin pie?

Oh, fruit loved of boyhood! the old days recalling,
When wood-grapes were purpling and brown nuts were falling,
When wild, ugly faces we carved in its skin,
Glaring out through the dark with a candle within!
When we laughed round the corn-heap, with hearts all in tune,
Our chair a broad pumpkin — our lantern the moon,
Telling tales of the fairy who travelled like steam,
In a pumpkin-shell coach, with two rats for her team!

— J. G. WHITTIER

TREES

W. C. Muens

Douglas firs, San Juan Island, Washington. These trees show the effects of strong prevailin winds from one direction

I wonder if they like it — being trees?
I suppose they do.
It must feel so good to have the ground so flat,
And feel yourself stand straight up like that.
So stiff in the middle, and then branch at ease,
Big boughs that arch, small ones that bend and blow,
And all those fringy leaves that flutter so.
You'd think they'd break off at the lower end
When the wind fills them, and their great heads bend.
But when you think of all the roots they drop,
As much at bottom as there is on top,
A double tree, widespread in earth and air,
Like a reflection in the water there.

— "Tree Feelings," Charlotte Perkins Stetson

Natural is our love for trees! A tree is a living being, with a life comparable to our own. In one way it differs from us greatly: it is stationary, and it has roots and trunk instead of legs and body; it is obliged to wait to have what it needs come to it, instead of being able to search the wide world over to satisfy its wants.

The Parts of the Tree

The head, or crown, is composed of the branches as a whole, which in turn are composed of the larger and smal branches and twigs. The *spray* is the te given to the outer twigs, the finest di sions of the trunk, which bear the leav and fruit. The branches are divisions the *bole* or *trunk*, which is the body stem of the tree. The bole, at the ba divides into roots, and the roots into ro lets, which are covered with root hai It is important to understand what ea of the parts of a tree's anatomy does help carry on the life of the tree.

The roots, which extend out in every
ection beneath the surface of the
ound, have two quite different offices to
rform: first, they absorb the water and
nerals from the soil; second, they hold
e tree in place against the onslaught of
e winds. If we could see a tree standing
its head with its roots spread in the air
the same manner as they are in the
ound, we could then better understand
at there is as much of the tree hidden
low ground as there is in sight above
ound; although the part beneath the
ound is of quite different shape, being
tter and in a more dense mass. The
ots seem to know in which direction
grow to reach water; thus, the larger
mber of the roots of a tree are often
und to extend out toward a stream
wing perhaps some distance from the
ee; when they find plenty of food and
ater the rootlets interlace forming a solid
at. On the Cornell University campus
e certain elms whose roots, every six or
ven years, fill and clog the nearby sewers;
ese trees send most of their roots in the
rection of the sewer pipe. The fine root-
ts upon the tree roots are covered with
ot hairs, which really form the mouths

Avenue of trees

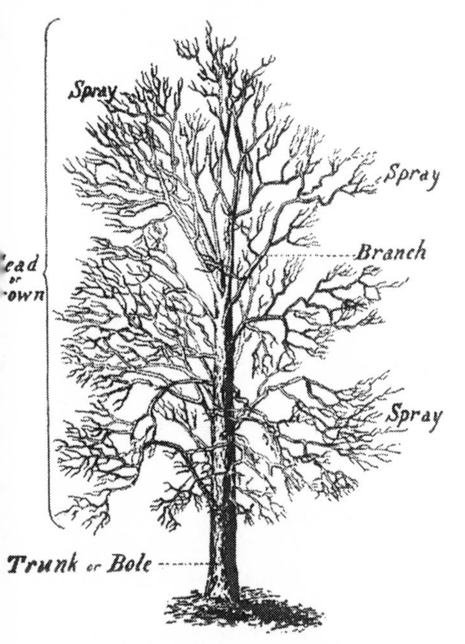

A tree with parts named

by which liquids are taken into the tree.

To understand how firm a base the roots form to hold up the tall trunk, we need to see an uprooted tree. The great roots seem to be molded to take firm grasp upon the soil. It is interesting to study some of the " stump fences " which were made by our forefathers, who uprooted the white pines when the land was cleared of the primeval forest, and made fences of their widespreading but rather shallow extending roots. Many of these fences stand today with branching, out-reaching roots, white and weather-worn, but still staunch and massive as if in memory of their strong grasp upon the soil of the wilderness.

The trunk, or bole, or stem of the tree has also two chief offices: it holds the branches aloft, rising to a sufficient height in the forest so that its head shall push

U. S. Forest Service

Loblolly pine. Annual rings near the center are narrow, but they become much wider. This increased rate of growth was due to thinning of the stand

through the leaf canopy and expose the leaves to the sunlight. It also is a channel by which the sap surges from root to leaf and back again through each growing part. The branches are divisions of the trunk, and have the same work to do.

In cross section, the tree trunk shows on the outside the layer of protective bark; next to this comes the cambium layer, which is the vital part of the trunk; it builds on its outside a layer of bark, and on its inside a layer of wood around the trunk. Just within the cambium layer is a lighter colored portion of the trunk, which is called the sapwood because it is filled with sap, which moves up and down its cells in a mysterious manner; the sapwood consists of the more recent annual rings of growth. Within the sapwood are concentric rings to the very center or pith; this portion is usually darker in color and is called the heartwood; it no longer has anything to do with the life of the tree, but simply gives to it strength and staunchness. The larger branches, if cut across, show a structure similar to that of the trunk — the bark on the outside, the cambium layer next, and within this the rings of annual growth. Even the smaller branches and twigs show similar structure, but they are young and have not attain many annual rings.

The leaves are borne usually on the o ermost parts of the tree. A leaf would of little use, unless it could be reached the sunlight. Therefore the trunk lifts t branches aloft, and the branches hold t twigs far out, and the twigs divide in the fine spray, so as to spread the lea and hold them out into the sunshine. structure, the leaf is made up of the pe ole and the blade, or widened portion the leaf, which is sustained usually wit framework of many ribs or veins. T petioles and the veins are sap channels li the branches and twigs.

WOOD–GRAIN

This is the way that the sap-river ran
From the root to the top of the tree
Silent and dark,
Under the bark,
Working a wonderful plan
That the leaves never know,
And the branches that grow
On the brink of the tide never see.

— John B. Ta

The Way a Tree Grows

The places of growth on a tree may found at the tips of the twigs and the ti of the rootlets; each year through th growth the tree pushes up higher, dow deeper, and out farther at the sides. But addition to all of these growing tips, the is a layer of growth over the entire tree over every root, over the trunk, over th limbs, and over each least twig, just as if thick coat of paint had been put over th complete tree. It is a coat of growth i stead, and these coats of growth make th concentric rings which we see when th trunks or branches are cut across. Su growth as this cannot be made witho food; but the tree can take only water a minerals from the soil; the root hairs ta up the water in which the "fertilizer" dissolved, and it is carried up through th larger roots, up through the sapwood the trunk, out through the branches the leaves, where in the leaf-factories th er and free oxygen are given off to the and the nourishing elements retained mixed with certain chemical ele- ıts of the air, thus becoming tree food. : leaf is a factory; the green pulp in the cells is part of the machinery; the :hinery is set in motion by sunshine ver; the raw materials are taken from air and from the sap containing min- s from the soil; the finished product is ely starch. Thus, it is well, when we in the study of the tree, to notice that leaves are so arranged as to gain all the light possible, for without sunlight starch factories would be obliged to ıut down." It has been estimated that a mature maple of vigorous growth re is exposed to the sun nearly a half e of leaf surface. Our tree appears to us new phase when we think of it as a ch factory covering half an acre.

tarch is plant food in a convenient m for storage, and it is stored in sap- od of the limbs, the branches, and nk, to be used for the growth of the t year's leaves. But starch cannot be as- ıilated by plants in this form; it must changed to sugar before it may be used build up the plant tissues. So the leaves obliged to perform the office of stom- and digest the food they have made the tree's use. In the mysterious labo- ory of the leaf cells, the starch is ınged to sugar; and nitrogen, sulphur, osphorus, and other substances are en from the sap and starch added to m, and thus are made the proteids ich form another part of the tree's diet. is interesting to note that while the rch factories can operate only in the ılight, the leaves can digest the food it can be transported and used in the wing tissues in the *dark*. The leaves also an aid to the tree in breathing, but y are not especially the lungs of the e. The tree breathes in certain respects we do; it takes in oxygen and gives off bon dioxide; but the air containing the gen is taken in through the numerous res in the leaves called stomata, and also ough lenticels in the bark; so the tree lly breathes all over its active surface.

A big tulip poplar, in Jackson County, North Carolina

The tree is a rapid worker and achieves most of its growth and does most of its work by midsummer. The autumn leaf which is so beautiful has completed its work. The green starch-machinery or chlo- rophyl, the living protoplasm in the leaf cells, has been withdrawn and is safely se- cluded in the woody part of the tree. The

Verne Morton

Trees in winter

autumn leaf which glows gold or red, has in it only the material which the tree can no longer use. It is a mistake to believe that the frost causes the brilliant colors of autumn foliage; they are caused by the natural old age and death of the leave and where is there to be found old and death more beautiful? When the assumes its bright colors, it is mak ready to depart from the tree; a thin, co layer is being developed between petiole and the twig, and when is perfected, the leaf drops from own weight or the touch of the slight breeze.

A tree, growing in open ground, recc in its shape the direction of the prevail winds. It grows more luxuriantly on leeward side. It touches the heart of one who loves trees to note their stu endurance of the onslaughts of this, th most ancient enemy.

HOW TO BEGIN TREE STUDY

During autumn the attention of the children should be attracted to the leaves by their gorgeous colors. It is well to use this interest to cultivate their knowledge of the forms of leaves of trees; but the teaching of the tree species to the young child should be done quite incidentally and guardedly. If the teacher says to child bringing a leaf, "This is a white-c leaf," the child will soon quite unc sciously learn that leaf by name. Th tree study may be begun in the kinderg ten or the primary grades.

1. Let the pupils use their leaves a

U. S. Forest Service

Mt. Baker from Table Mountain, Washington. Trees growing near timberline are stunted

lor lesson by classifying them according color, and thus train the eye to disminate tints and color values.

2. Let them classify the leaves accord- to form, selecting those which resem- each other.

3. Let each child select a leaf of his own oosing and draw it. This may be done placing the leaf flat on paper and out- ing it with pencil or with colored ayon.

4. Let the pupils select paper of a color nilar to the chosen leaf and cut a paper af like it.

5. Let each pupil select four leaves aich are similar and arrange them on a rd in a symmetrical design. This may done while the leaves are fresh, and the rd with leaves may be pressed and thus eserved.

In the fourth grade, begin with the udy of a tree which grows near the hoolhouse. In selecting this tree and in eaking of it, impress upon the children at it is a living being, with a life and th needs of its own. I believe so much making this tree seem an individual, at I would if necessary name it Pocahontas or Martha Washington. First, try to ascertain the age of the tree. Tell an interesting story of who planted it and who were children and attended school in the school building when the tree was planted. To begin the pupils' work, let each have a little notebook in which shall be written, sketched, or described all that happens to this particular tree for a year. The following words with their meanings should be given in the reading and spelling lessons: *Head, bole, trunk, branches, twigs, spray, roots, bark, leaf, petiole, foliage, sap.*

LESSON 172

Tree Study

Autumn Work — 1. What is the color of the tree in its autumn foliage? Sketch it in water colors or crayons, showing the shape of the head, the relative proportions of head and trunk.

2. Describe what you can see of the tree's roots. How far do you suppose the roots reach down? How far out at the sides? In how many ways are the roots useful to the tree? Do you suppose, if the tree were turned bottomside up, that it

Mountain maple, sugar maple, and red maple

would show as many roots as it now shows branches?

3. How high on the trunk from the ground do the lower branches come off? How large around is the trunk three feet from the ground? If you know how large around it is, how can you get the distance through? What is the color of the bark? Is the bark smooth or rough? Are the ridges fine or coarse? Are the furrows between the ridges deep or shallow? Of what use is the bark to the tree?

4. Describe the leaf from your tree, paying special attention to its shape, its edges, its color above and below, its veins or ribs, and the relative length and thickness of its petiole. Are the leaves set opposite or alternate upon the twigs? As the leaves begin to fall, can you find two which are exactly the same in size and shape? Draw in your notebook the two leaves which differ most from each other of any that grew on your tree. At what date do the leaves begin to fall from your tree? At what date are they all off the tree?

5. Do you find any fruit or seed upon your tree? If so describe and sketch it, and tell how you think it is scattered and planted.

Winter Work — 1. Make a sketch of the tree in your notebook, showing its shape as it stands bare. Does the trunk divide into branches, or does it extend through the center of the tree and the branches come off from its sides? Of what use are the branches to a tree? Is the spray — the twigs at the end of the branche coarse or fine? Does it lift up or droop the bark on the branches like that on trunk? Is the color of the spray the sa as that of the large branches?

2. Study the cut end of a log or stu and also study a slab. Which is the he wood and which is the sapwood? Can y see the rings of growth? Can you co these rings and tell the age of the t from which this log came? Describe, you can, how a tree trunk grows lar each year. What is it that makes the gr in the wood which we use for furniture we girdle a tree why may it die? If place a nail in a tree three feet from ground this winter, will it be any hig from the ground ten years from now? H does the tree grow tall?

3. Take a twig of a tree in February a look carefully at the buds. What is th color? Are they shiny, rough, sticky, downy? Are they arranged on the tw opposite or alternate? Can you see t scar below the buds where the last yea leaf was borne? Place the twig in wa and put it in a light, warm place, and what happens to the buds. As the lea push out, what happens to the sca which protected the buds?

4. What birds do you find visiting yo tree during winter? Tie some strips

Trunks of young birches, black on left, yell on right

ef fat upon its branches, and note all the kinds of birds which come to feast on it.

Spring Work — 1. At what date do the ung leaves appear upon your tree? hat color are they? Look carefully to e how each leaf was folded in the bud. ere all the leaves folded in the same y? Are the young leaves thin, downy, d tender? Do they stand out straight as d the old leaves last autumn, or do they oop? Why? Will they change position d stand out as they grow stronger? Why the leaves stand out from the twigs in

Ralph W. Curtis

Tamarack

Verne Morton

Sycamore in winter

order to get sunshine? What would happen to a tree if it lost all its leaves in spring and summer? Tell all of the things you know which the leaves do for the tree?

2. Are there any blossoms on your tree in the spring? If so, how do they look?

Hemlocks under a load of snow

Are the blossoms which bear the fruit on different trees from those that bear the pollen, or are these flowers placed separately on the same tree? Or does the same flower which produces the pollen also produce the seed? Do the insects carry the pollen from flower to flower, or does the wind do this for the tree? What sort of fruits are formed by these flowers? How are the fruits scattered and planted?

3. At what date does your tree stand in full leaf? What color is it now? What birds do you find visiting it? What insects? What animals seek its shade? Do the squirrels live in it?

4. Measure the height of your tree as follows: Choose a bright, sunny morning for this. Take a stick $3\frac{1}{2}$ feet long and thrust it in the ground so that three feet will project above the soil. Immediately measure the length of its shadow and the length of the shadow which your tree makes from its base to the shadow of its topmost twigs. Supposing that the shad from the stick is 4 feet long and t shadow from your tree is 80 feet lo then your example will be: 4 ft.: ft.:: 80 ft.:(?), which will make the t 60 feet high.

To measure the circumference of t tree, take the trunk three feet from t ground and measure it exactly with tape measure. To find the thickness of t trunk, divide the circumference just fou by 3.14.

LESSON 173

How to Make Leaf Prints

A very practical help in interesting ch dren in trees is to encourage them make portfolios of leaf prints of all t trees of the region. Although the proc is mechanical, yet the fact that eve print must be correctly labeled makes f useful knowledge. One of my treasur possessions is such a portfolio made by t lads of St. Andrews School of Richmon Virginia, who were guided and inspired this work by their teacher, Profess W. W. Gillette. The impressions we made in green ink and the results are beautiful as works of art. Professor G lette gave me my first lesson in maki leaf prints.

Material — 1. A smooth surface su as a slate, a thick plate of glass, or a m ble slab about 12×15 inches.

2. A tube of printer's ink, either gre or black; one tube contains a sufficie supply of ink for making several hundr

Alder showing staminate catkins of curre year, and fruits matured from pistillate ca kins of preceding year

ts. Or a small quantity of printer's ink be purchased at any printing office.
. Two six-inch rubber rollers, such as tographers use in mounting prints. A er press may be used instead of one er.
. A small bottle of kerosene to dilute ink, and a bottle of gasoline for cleanthe outfit after using, care being taken tore them safe from fire.
. Sheets of paper; $8\frac{1}{2} \times 11$ inches is a d size. The paper should be of good lity, with smooth surface, in order that nay take and hold a clear outline. The inary paper used in printers' offices for nting newspapers works fairly well.
'o make a print, place a few drops of upon the glass or slate, and spread it ut with the roller until there is a thin t of ink upon the roller and a smooth ch in the center of the glass or slate. hould never be so liquid as to "run,"

Linden in blossom

Leaf print of a sycamore maple

for then the outlines will be blurred. Ink the leaf by placing it on the inky surface of the glass and passing the inked roller over it once or twice until the veins show that they are smoothly filled. Now place the inked leaf between two sheets of paper and roll *once* with the *clean* roller, bearing down with all the strength possible; a second passage of the roller blurs the print. Two prints are made at each rolling, one of the upper, and one of the under side of the leaf. Dry and wrinkled leaves may be made pliant by soaking in water, and drying between blotters before they are inked.

Prints may also be made a number at a time by pressing them under weights, being careful to put the sheets of paper with the leaves between the pages of old magazines or folded newspapers, in order that the impression of one set of leaves may not mar the others. If a letter press is available for this purpose, it does the work quickly and well.

SAP

Strong as the sea and silent as the grave,
It flows and ebbs unseen,
Flooding the earth, a fragrant tidal wave,
With mists of deepening green.

— John B. Tabb

Troy St

Sugar bush in spring

THE MAPLES

The sugar maple, combining beauty with many kinds of utility, is dear to the American heart. Its habits of growth are very accommodating; when planted where it has plenty of room, it shows a short trunk and oval head, which, like a dark green period, prettily punctuates the summer landscape; but when it occurs in the forest, its noble bole, a pillar of granite gray, rises to uphold the arches of the forest canopy; and it often attains there the height of one hundred feet. It grows rapidly and is a favorite shade tree, twenty years being long enough to make it thus useful. The foliage is deep green in the summer, the leaf being a glossy, dark green above and paler beneath. It has five main lobes, the two nearest the petiole being smaller; the curved edges between the lobes are marked with a few smoothly c large teeth; the main veins extend direc from the petiole to the sharp tips of lobes; the petiole is long, slender, and casionally red. The leaves are placed op site. The shade made by the foliage of maple is so dense that it shades down plants beneath it; even grass grows sparsely there. If a shade tree stands in exposed position, it grows luxuriously the leeward of the prevailing winds, a thus makes a one-sided record of th general direction.

It is its autumn transfiguration whi has made people observant of the mapl beauty; yellow, orange, crimson, and sc let foliage makes these trees gorge when October comes. Nor do the trees their color uniformly; even in Septemb maple may show a scarlet branch in midst of its green foliage. I believe this a hectic flush and a premonition of ath to the branch which, less vigorous an its neighbors, is being pruned out by ture's slow but sure method. After the id color is on the maple, it begins to d its leaves. This is by no means the sad which the poets would have us believe; brilliant colors are an evidence that e trees have withdrawn from the leaves ich of the manufactured food and have red it snugly in trunk and branch for nter keeping. Thus, only the mineral bstances and waste materials are left in e leaf, and they give the vivid hues. It is mistake to think that frost causes this illiance; it is caused by the natural, beau- ul, old age of the leaf. When the leaves ally fall, they form a mulch-carpet, and d their substance to the humus from ich trees and other plants draw new wers for growth.

After every leaf has fallen, the maple ows why its shade is dense. It has many anches set close and at sharp angles to e trunk, dividing into fine, erect spray, ing the tree a resemblance to a giant isk broom. Its dark, deep-furrowed rk smoothes out and becomes light gray the larger limbs, while the spray is pur- ish, a color given it by the winter buds. ese buds are sharp-pointed and long. late winter, their covering of scales ows premonitions of spring by enlarg-

Sugar maple leaves

Sugar maple blossoms

ing, and as if due to the soft influence, they become downy, and take on a sun- shine color before they are pushed off by the leaves. The leaves and the blossoms appear together. The leaves are at first yellowish, downy, and drooping. The flowers appear in tassel-like clusters, each downy, drooping thread of the tassel bear- ing at its tip a five-lobed calyx, which may hold seven or eight long, drooping stamens or a pistil with long, double stig- mas. The flowers are greenish yellow, and those that bear pollen and those that bear the seeds may be borne on separate trees or on the same tree, but they are always in different clusters. If on the same tree, the seed-bearing tassels are at the tips of the twigs, and those bearing pollen are along the sides.

The ovary is two-celled, but there is usually only one seed developed in the pair which forms a " key "; to observe this, however, we have to dissect the fruits; they have the appearance of two seeds joined together, each provided with a thin, closely veined wing and the two attached to the tree by a single long, drooping stem. This twin-winged form is well fitted to be whirled off by the autumn winds, for the

Sugar maple growing in the open

seeds ripen in September. I have seen seedlings growing thickly for rods to the leeward of their parent tree, which stood in an open field. The maples may bear blossoms and produce seeds every year.

There are six species of native maples which are readily distinguishable. The silver and the red maples and the box elder are rather large trees; the mountain and the striped (or goosefoot) maples are scarcely more than shrubs, and mostly grow in woods along streams. The Norway and the sycamore maples have been introduced from Europe for ornamental planting. The cut-leaf silver maple comes from Japan.

The maple wood is hard, heavy, strong, tough, and fine-grained; it is cream color, the heartwood showing shades of brown; it takes a fine polish and is used as a finishing timber for houses and furniture. It is used in construction of ships, cars, piano action, and tool handles; its fine-grained quality makes it good for wood carving; it is an excellent fuel and has many other uses.

Maple Sugar Making

Although we have tapped the trees America for many hundred years, we not as yet understand perfectly the m teries of the sap flow. In 1903, the sci tists at the Vermont Experiment Stat did some very remarkable work in clear up the mysteries of sap movement. Th results were published in their Bullet 103 and 105, which are very interest and instructive.

The starch which is changed to su in the sap of early spring was made previous season and stored within tree. If the foliage of the tree is injured caterpillars one year, very little sugar be made from that tree the next spri because it has been unable to store enou starch in its sapwood and in the ou ray cells of its smaller branches to mak good supply of sugar. During the lat part of winter, the stored starch dis pears, being converted into tree-food the sap, and then begins that wonder surging up and down of the sap tide. D ing the first part of a typical sugar seas more sap comes from above down th from below up; toward the end of the s son, during poor sap days, there is m sap coming up from below than do from above. The ideal sugar weather c sists of warm days and freezing nigh This change of temperature between c and night acts as a pump. During the c when the branches of the tree are warm the pressure forces into the hole bo into the trunk all the sap located in adjacent cells of the wood. Then the s tion which follows a freezing night dri more sap into those cells, which is in t forced out when the top of the tree again warmed. The tree is usually tapp on the south side, because the action the sun and the consequent temperatu pump more readily affects that side.

"Tapping the sugar bush" are magi words to the country boy and girl. W do we older folk remember those days March when the south wind settled t snow into hard, marble-like drifts, and t father would say, "We will get the s ckets down from the stable loft and sh them, for we shall tap the sugar bush on." In those days the buckets were ade of staves and were by no means so sily washed as are the metal buckets of day. Well do we recall the sickish smell musty sap that greeted our nostrils en we poured in the boiling water to an those old brown buckets. Previously ring the winter evenings, we all had lped fashion sap spiles from stems of mac. With buckets and spiles ready en the momentous day came, the large, n caldron kettle was loaded on a stone-at together with a sap cask, log chain, and various other utensils, and as many ildren as could find standing room; en the oxen were hitched on and the ocession started across the rough pas-re to the woods, where it eventually ar-ed after numerous stops for reloading most everything but the kettle.

When we came to the boiling place, we ted the kettle into position and flanked with two great logs against which the e was to be kindled. Meanwhile the en and stoneboat returned to the house r a load of buckets. The oxen, blinking, ith bowed heads, or with noses lifted oft to keep the underbrush from strik-g their faces, "gee'd and haw'd" up hill d down dale through the woods, stop-ng here and there while the men with gers bored holes in certain trees near her holes which had been made in years gone by. When the auger was withdrawn, the sap followed it, and enthusiastic young tongues met it half way, though

Maple seedling

they received more chips than sweetness therefrom; then the spiles were driven in with a wooden mallet.

The next day after "tapping," those of us large enough to wear the neck yoke donned cheerfully this badge of servitude and with its help brought pails of sap to the kettle, and the "boiling" began. As the evening shades gathered, how delicious was the odor of the sap steam, permeating the woods farther than the shafts of firelight pierced the gloom! How weird and delightful was this night experience in the woods! And how cheerfully we swallowed the smoke which the contrary wind seemed ever to turn toward us! We poked the fire to send the sparks upward, and now and then added more sap from a barrel, and removed the scum from the boiling liquid with a skimmer thrust into the cleft of a long stick for a handle. As the evening wore on, we drew closer to each other as we told stories of the Indians, bears, panthers, and wolves which had roamed these woods when our father was a little boy; and came to each of us a

Verne Morton

A foretaste. An old-fashioned sugar bush

M. V. Slingerland

Leaves of silver maple

disquieting suspicion that perhaps they were not all gone yet, for everything seemed possible in those night-shrouded woods; and our hearts suddenly "jumped into our throats" when nearby there sounded the tremulous, blood-curdling cry of the screech owl.

After about three days of gathering and boiling sap, came the "siruping down." During all that afternoon we added no more sap and we watched carefully the tawny, steaming mass in the kettle; when it threatened to boil over, we threw in a thin slice of fat pork which seemed to have some mysterious calming influence. The odor grew more and more delicious and presently the sirup was pronounced sufficiently thick. The kettle was swung off the logs and the sirup dipped through a cloth strainer into a carrying-pail. Oh, the blackness of the residue left on that strainer! But it was clean woods-dirt and never destroyed our faith in the maple sugar, any more than did the belief that our friends were made of dirt destroy our friendship for them. The next day our interests were transferred to the house, where we "sugared off." There we boiled the sirup to sugar on the stove and pouring it thick and hot upon snow made that most delicious of all sweets — t maple wax; or we stirred it until "grained," before we poured it into t tins to make the "cakes" of maple sug

Now the old stave bucket and the sum spile are gone; in their place the pate galvanized spile not only conducts t sap but holds in place a tin bucket ca fully covered. The old caldron kettle broken, or lies rusting in the shed. In place, in the newfangled sugar-hous are evaporating vats, set over furnac with chimneys. But we may as well co fess that the maple sirup of today seen to us a pale and anemic liquid, lacking t delicious flavor of the rich, dark nect which we, with the help of cinders, smok and various other things, brewed of yo in the open woods.

LESSON 174

The Sugar Maple

Leading Thought — The sugar map grows very rapidly, and is therefore a us ful shade tree. Its wood is used for ma purposes, and from its sap is made a d licious sugar.

Method — This study of the map should be done by the pupils out of doo with a tree to answer the questions. T

Ralph W. Cur

Blossoms of the silver maple

ly of the leaves, blossoms, and fruit be made in the schoolroom. The le is an excellent subject for Lesson The observations should begin in fall and continue at intervals until e.

BSERVATIONS. FALL WORK — 1. Where ne maple you are studying? Is it near er trees? What is the shape of the d? What is the height of the trunk w the branches? What is the height he tree? How large around is the trunk e feet from the ground? Can you when the tree was planted? Can tell by the shape of the tree from ch direction the wind blows most n?

. Can you find fruits on your tree? h fruit is called a key. Sketch a key, wing the way the seeds are joined and direction of the wings. Sketch the n which holds the key to the twig. Are h halves of the key good or is one pty? How are the fruits scattered and nted? How far will a maple key fly on wings? Plant a maple fruit where you watch it grow next year.

. Make leaf prints and describe a leaf he maple, showing its shape, its veins,

Ralph W. Curtis

Blossoms of mountain maple

Ralph W. Curtis

Blossoms of striped maple

and petiole. Are the leaves arranged opposite or alternate on the twig? Make leaf prints or sketches of the leaves of all the other kinds of maples which you can find. How can you tell the different kinds of maples by their leaves?

4. If your tree stands alone, measure the ground covered by its shadow from morning until evening. Mark the space by stakes. What grows beneath the tree? Do grass and other plants grow thriftily beneath the tree? Do the same plants grow there as in the open field?

5. Does your maple get its autumn colors all at once, or on one or two branches first? At what time do you see the first autumn colors on your tree? When is it completely clothed in its autumn dress? Is it all red or all yellow, or mixed? If it is yellow this year do you think it will be red next year? Watch and see. Sketch your maple in water colors.

6. At what time do the leaves begin to fall? Do those branches which first colored brightly shed their leaves before the

Ralph W. Curtis

Leaves and fruit of striped maple

others? At what date does your tree stand bare?

7. Find a maple tree in the forest and compare it with one that grows as a shade tree in a field. Why this difference?

WINTER WORK — 8. Make a sketch of your maple with the leaves off. What sort of bark has it? Is the bark on the branches like that on the trunk? Are the main branches large? At what angle do they come off the trunk? Does the trunk extend up through the entire tree? Is the spray fine or coarse? Is it straight or crooked?

9. Study the winter buds. Are they alternate or opposite on the twigs? Are they shining or dull?

SPRING WORK — 10. At what time do we tap maple trees for sap? On which side of the tree do we make the hole? If we tapped the tree earlier would we get any sap? What kind of weather is the best causing sap flow? Do you suppose th is the sap going up from the root to tree and the branches, or that com down from the branches to the root wh flows into the bucket? Why do we make maple sugar all summer? Do suppose the sap ceases to run beca there is no more sap in the tree?

11. Write a story telling all you find in books or that you know from y own experience about the making maple sugar.

12. When do the leaves of your ma first appear? How do they then look? they stand out or droop?

13. Do the blossoms appear with leaves or after them? How do the bloss look? Can you tell the blossoms w stamens from those with pistils? Do find them in the same cluster? Do find them on the same tree?

14. What uses do we find for ma wood? What is the character of the wo

Ralph W. C

Blossoms of red maple

THE AMERICAN ELM

Although the American elm loves moist woods, it is one of those trees that enjoy gadding; and without knowing just how it has managed to do it, we can see plainly that it has planted its seeds along fence corners, and many elms now grace our fields on sites of fences long ago laid low. Because of its beautiful form and its rapid growth, the elm has been from earliest times a favorite shade tree in the East and Middle States. Thirty years after ing planted, the elms on the Cornell U versity campus clasped branches acr the avenues; and the beauty of man village and city is due chiefly to these gra ful trees of bounteous shade. Moreov the elm is at no time more beautiful th when it traces its flowing lines agai background of snow and gray horizon. hether the tree be shaped like a vase fountain, the trunk divides into great ifting branches, which in turn divide o spray that oftentimes droops gracely, as if it were made purposely to susn from its fine tips the woven pockett of the oriole. No wonder this bird so en chooses the elm for its rooftree!

n winter, the dark, coarsely-ridged bark l the peculiar, wiry, thick spray, as well the characteristic shape of the tree re-l to us its identity; some elms have a culiar habit of growing their short nches all the way down their trunk, king them look as if they were en-ned with a vine. The elm leaf, although ribs are straight and simple, shows a le quirk of its own in the uneven sides its base where it joins the petiole; it is k green and rough above, light green l somewhat rough below; but this leaf ough only when stroked in certain di-tions, while the leaf of the slippery is rough whichever way it may be oked. The edges of the leaf have saw th, which are in turn toothed; the iole is short. The leaf comes out of the bud in the spring folded like a little fan; but before the fans are opened to the spring breezes, the elm twigs are furry with reddish green blossoms. The blossom consists of a calyx with an irregular number of lobes, and for every lobe, a stamen which consists of a threadlike filament from which hangs a bright red anther; at the center is a two-celled pistil with two light green styles. These blossoms appear in March or early April, before the leaves.

Ralph W. Curtis

Blossoms of slippery elm

Ralph W. Curtis

American elm, vase type

When full-grown the fruit hangs like beaded fringe from the twigs. The fruit is flat and has a wide, much-veined margin or wing, notched at the tip and edged with a white silken fringe; the seed is at the center, wrinkled and flat. Each fruit shows at its base the old calyx and is attached by a slender threadlike stem to the twig at the axils of last year's leaves. A little later the lusty breezes of spring break the frail threads and release the fruits, al-

G. F. Morgan

Elm fruit

though few of them find places fit for growth of seeds.

The elm roots are water hunters and extend deep into the earth; most of them grow toward water, seeming to know the way. The elm heartwood is reddish, the sapwood being broad and whitish in color; the wood is very tough because of the interlaced fibers, and therefore very hard to split. It is used for cooperage, wheel hubs, saddlery, and is now used more extensively for furniture; its grain is most ornamental. It is fairly durable as posts, but perhaps the greatest use of all for the tree is for shade. The slippery elm is much like the white elm, except that its inner bark is very mucilaginous, and children love to chew it. The cork elm has a peculiar corky growth on its branches, giving it a very unkempt look. The wahoo, or winged elm, is a small tree, and its twigs are ornamented on each side by a corky layer. The English elm has a solid, round head, very different from that of our graceful species. The elms are long-lived, unless attacked by insects or disease; some elm trees have lived for centuries. The Washington elm in Cambridge, and the William Penn elm in Philadelphia, which now has a monument to mark its place, were famous trees.

Lovers of the elm are at present much alarmed at the inroads made by the Dutch elm disease, so called because it was first discovered in Holland. It first appeared in this country in Ohio in 1930, and many hundreds of infected trees have since b discovered in northern New Jersey a southeastern New York. The disease caused by a fungus, *Ceratostomella ul* the spores of which are carried from t to tree by the European elm bark-bee

A tree seriously infected is doomed, a will serve as a focal point of infection healthy trees. Dead branches should cut and burned as soon as seen, and th oughly infected trees should be complet destroyed, root and branch. Any com tent plant pathologist can from a study infected twigs identify the disease, wh in its early stages resembles less seri disorders.

Unless the disease is eradicated by th drastic measures, our beautiful elms m succumb as did the chestnuts to the ch nut blight. No one would willingly pict America without its elms.

LESSON 175

The Elm

Leading Thought — The elm has a culiarly graceful form, which makes it

Ralph W. C

American elm

lue as a shade tree. It grows best in moist :ations. Its wood is very tough.

METHOD — This work should be begun the fall with the study of the shape of e tree and its foliage. Sketches should made when the tree is clothed in au- mn tints, and later it should be sketched ain when its branches are naked. Its ossoms should be studied in March and oril and its fruits in May.

OBSERVATIONS — 1. Where does the n grow? Does it thrive where there is tle water? What is the usual shape of e elm? How does the trunk divide into anches to make this shape possible? 'hat is the shape of the larger elms? De- ribe the spray. Describe the elm bark. ow can you tell the elm from other trees winter?

2. Study the elm leaf. What is its form? 'hat kind of edges has it? How large is ' What is the difference in appearance d feeling between the upper and lower les? Are the leaves rough above which- er way you stroke them? If a leaf is lded lengthwise are the two halves ex- tly alike? How are the leaves arranged the twig? What is their color above and :low? Describe the leafy growth along e trunk.

3. What is the color of the elm tree in tumn? Make a sketch of the elm tree u are studying.

American elm

Verne Morton

American elm in winter

4. What sort of roots has the elm? Do they grow deep into the earth? What is the character of its wood? Is it easy to split? Why? What are the chief uses of the elm?

5. Do you know what distinguishes the slippery elm, the cork elm, the winged elm or wahoo, and the English elm from the common American or white elm which you have been studying?

6. Write an essay on two famous American elms.

7. What birds love to build in the elm trees?

8. What disease threatens our elms? What steps should be taken to save the elms?

Spring Work—9. Which appear first, the blossoms or the leaves? Describe the elm blossom. How long before the frui ripen? How are the fruits attached to th twig? Describe an elm fruit. How are th fruits scattered? How are the young leav folded as they come out of the bud?

THE OAKS

The symbol of rugged strength since man first gazed upon its noble proportions, the oak more than other trees has

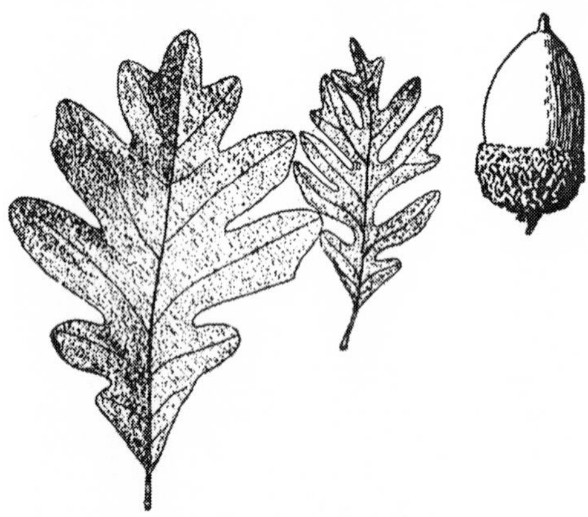

White-oak leaves and acorn

been entangled in human myth, legend, and imagination. It was regarded as the special tree of Zeus by the Greeks; while in primitive England the strange worship of the Druids centered on it. Virgil sang of it thus:

Full in the midst of his own strength he stands
Stretching his brawny arms and leafy hands,
His shade protects the plains, his head the hills commands.

Although the oak is a tree of grandeur when its broad branches are covered with leafage, yet it is only in winter when it stands stripped like an athlete that we realize wherein its supremacy lies. Then only can we appreciate the massive trunk and the strong limbs bent and gnarled with combating the blasts of centuries. But there are oaks and oaks, and each species fights time and tempest in his ow peculiar armor and in his own way. Ma of the oaks achieve the height of eigh to one hundred feet. The great branch come off the sturdy trunk at wide angle branches crooked or gnarled but whic may be long and strong; the small

Lewis W. Hendersh

White oak in winter

ches also come off at wide angles, in turn bear angular individual spray ll of which, when covered with leaves, e the broad, rounded head which racterizes this tree. The oaks are did into two classes which the children learn to distinguish, as follows:

. *The white oak group*, the leaves which have rounded lobes and are h and light-colored below; the wood ight-colored, the acorns have sweet els and mature in one year, so that e are no acorns on the branches in ter. To this class belong the white, stnut, bur, post, and chinquapin oaks.

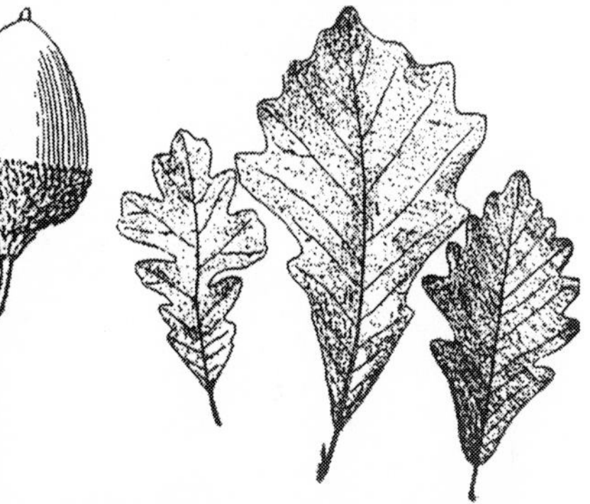

eaves and acorn of swamp white oak

A. *The black oak group*, the leaves of ch are nearly as smooth below as ve, and have angular lobes ending in p points. The bark is dark in color, acorns have bitter kernels and require years for maturing, so that they may seen on the branches in winter. To group belong the black, red, scarlet, nish, pin, scrub, blackjack, laurel, and ow oaks.

here is a great variation in the shape he leaves on the same tree, and while black, the red, and the scarlet oaks are l-marked species, it is possible to find es on these three different trees which similar in shape. Oaks also hybridize, thus their leaves are a puzzle to the anist; but in general, the species can determined by tree books, and the ils may learn to distinguish some of m.

he acorns and their scaly saucers are varied in shape, and are a delight to children as well as to pigs. The great acorns of the red oak are made into cups and saucers

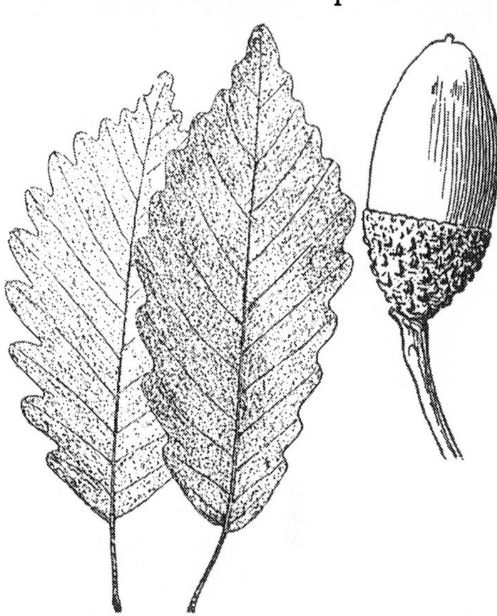

Leaves and acorn of chestnut oak

by the girls, and those of the scarlet oak into tops by the boys. The white oaks turn

Ralph W. Curtis

Blossoms of chestnut oak. Compare with chestnut blossoms, p. 646

a rich wine-color in the autumn, while the bur and the chestnut are yellow. The red oak is a dark, wine-red; the black oak russet, and the scarlet a deep and brilliant red.

White oak in winter

When the oak leaves first come from the buds in the spring, they are soft and downy and drooping, those of the red and scarlet being reddish, and those of the white, pale green with red tints. Thoreau says of them, " They hang loosely, flacidly down at the mercy of the wind, like a newborn butterfly or dragonfly."

The pollen-bearing flowers are like beads on a string, several strings hanging down from the same point on the twig, making a fringe, and they are attractive to the eye that sees. The pistillate flowers are inconspicuous, at the axils of the lea and have irregular or curved stigmas; t are on the same branch as the pollen-b ing flowers.

Cup and saucer made from the acorn of red oak

The oak is long-lived; it does not duce acorns until about twenty year age and requires about a century to ture. Although from two to three hund years is the average age of most oaks, y scarlet oak of my acquaintance is ab four hundred years old, and there are c still living in England which were th when William the Conqueror came. T famous Wadsworth Oak at Gene

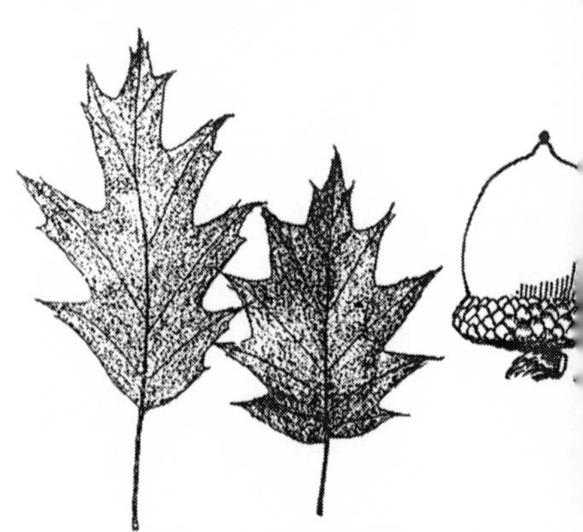

Leaves and acorn of red oak

New York, had a circumference twenty-seven feet. This was a swa white oak. One reason for their attain great age is long, strong, taproots wh plant them deep; doubtless the g number of roots near the surface wh act as braces, and their large and luxuri heads, also help the oaks to survive.

Oak wood is usually heavy, very stro tough, and coarse. The heart is brown, sapwood whitish. It is used for m purposes — ships, furniture, wagc cars, cooperage, farm implements, pi wharves, railway ties, etc. The white a live oaks give the best wood. Oak bar used extensively for tanning.

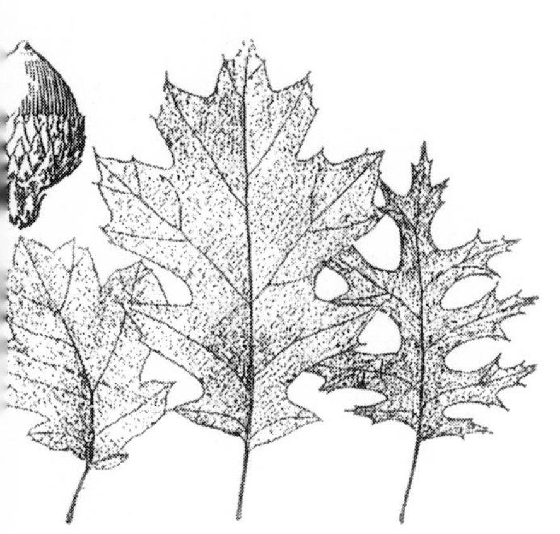

Leaves and acorn of black oak

LESSON 176
The Oaks

Leading Thought — The oak tree is : symbol of strength and loyalty. Let study it and see what qualities in it have ıs distinguished it.

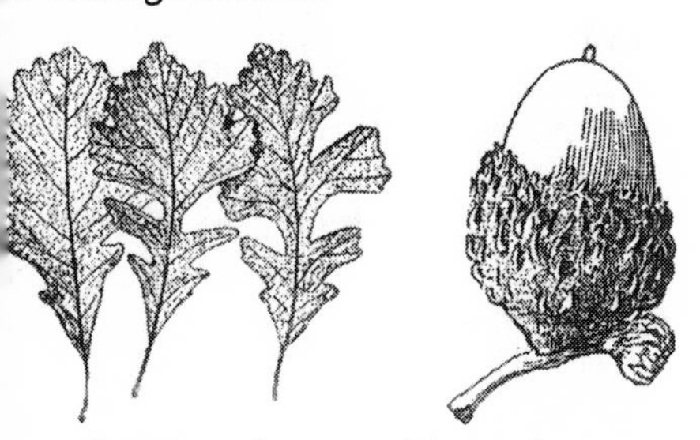

Leaves and acorn of bur oak

Method — Any oak tree may be used this lesson; but whatever species is ɜd, the lesson should lead to the knowledge of all the species of oaks in the neighborhood. The tree may be sketched, essays concerning the connection of the oak with human history may be written, while the leaves and acorns may be brought into the

Verne Morton

Beech, a near relative of the oak

schoolroom for study. Use Lesson 173 for a study of leaves of all the oaks of the neighborhood.

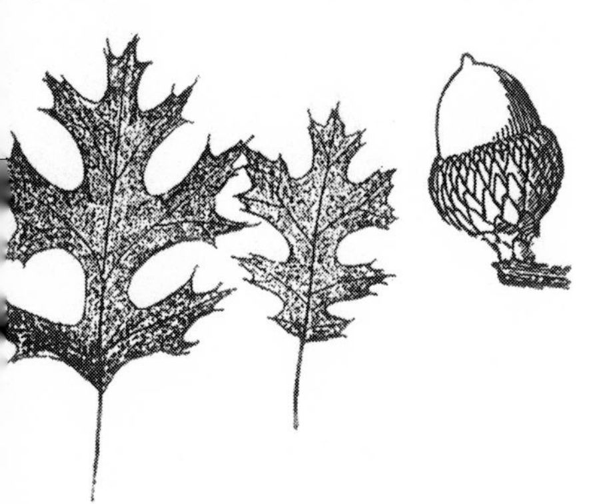

Leaves and acorn of scarlet oak

Observations — 1. Describe the oak tree which you are studying. Where is it

Verne Morton

Beech photographed in April showing last year's leaves

growing? What shape is its head? How high in proportion to the head is the trunk? What is the color and character of its bark? Describe its roots as far as you can see. Are the branches straight or crooked? Delicate or strong? Is the spray graceful or angular?

2. What is the name of your oak tree? What is the color of its foliage in autumn? Find three leaves from your tree which differ most widely in form, and sketch them or make leaf prints of them for your notebook. Does the leaf have the lobes rounded, or angular and tipped with sharp points? Is the leaf smooth on the lower side or rough? Is there much difference in color between the upper and the lower side?

3. Describe the acorns which grow on your oak. Has the acorn a stem, or is it set directly on the twig? How much of the acorn does the cup cover? Are the scales on the cup fine or coarse? Is the cup rounded inwards at its rim? What is the length of the acorn including the cup? The diameter? Are there acorns on your oak in winter? Is the kernel of the acorn sweet or bitter? Plant an acorn and watch it sprout.

4. Read stories about oak trees, a write them in your notebook in your o words.

5. How great an age may the oak tain? Describe how the country rou about looked when the oak you are stu ing was planted.

6. How many kinds of oaks do y know? What is the difference in lea between the white and the black groups? What is the difference in length of time required for the acorns mature in these two groups? The dif ence in taste of the acorns? The differe in the general color of the bark?

7. How do the oak leaves look wh they first come out of the bud in spri What is the color of the tree covered w new leaves? When does your oak blosso Find the pollen-bearing blossoms, wh are hung in long, fuzzy, beady stri Find the pistillate flower which is to fo the acorn. Where is the pistillate flo situated in relation to the pollen-bear flower?

8. Make a sketch of your oak tree the fall, and another in the winter. W the autobiography of some old oak t in your neighborhood.

9. For what is the oak wood used? H is the bark used?

Beech nuts and "husks"

THE SHAGBARK HICKORY

How pathetically the untidy bark of is dignified tree suggests the careless iment of a great man! The shagbark is busy being something worth while that does not seem to have time or energy to othe itself in tailor-made bark, like the ech, the white ash, and the basswood. nd just as we may like a great man more cause of his negligence of fashion's de-ands, so do we esteem this noble tree, d involuntarily pay it admiring tribute we note its trunk with the bark scaling f in long, thin plates that curve outward the top and bottom and seem to be ly slightly attached at the middle.

In general shape, the shagbark resembles e oak; the lower branches are large and, though rising as they leave the bole, eir tips are deflected; and, for their whole ngth, they are gnarled and knotted as if show their strength. The bark on the rger branches may be scaly toward their ses but above is remarkably smooth. he spray is angular and extends in al-ost every direction. The leaves, like those other hickories, are compound. There e generally five leaflets, but sometimes ly three and sometimes seven. The sal pair is smaller than the others. he hickory leaves are borne alternately the twig, and from this character the ckory may be distinguished from the hes, which have leaves of similar type, it which are placed opposite on the vigs. The shagbark usually has an un-mmetrical oblong head; the lower anches are usually shorter than the up-er ones, and the latter are irregularly aced, causing gaps in the foliage.

The nut is large, with a thick, smooth iter husk channeled at the seams and parating readily into sections; the inner ell is sharply angled and pointed and ightly flattened at the sides; the kernel sweet. The winter buds of the shagbark e large, light brown, egg-shaped, and owny; they swell greatly before they ex-and. There are from eight to ten bud scales; the inner ones, which are red, in-crease to two or three inches in length before the leaves unfold, after which they fall away. The young branches are smooth, soft, delicate in color, and with conspicu-ous leaf scars.

The hickory bears its staminate and pis-tillate flowers on the same tree. The pol-len-bearing flowers grow at the base of the season's shoots in slender, pendulous

Ralph W. Curtis

Shagbark hickory. Note loose strips of bark

Ralph W. Curtis

Shellbark or shagbark hickory

green catkins, which occur usually in clusters of three swinging from a common stem. The pistillate flowers grow at the tips of the season's shoots singly or perhaps two or three on a common stem. In the shagbark the middle lobe of the staminate calyx is nearly twice as long as the other two, and is tipped with long bristles; it usually has four stamens with yellow anthers; its pistillate calyx is four-toothed and hairy, and has two large, fringed stigmas.

The big shagbark, or king nut, is similar to the shagbark in height, manner of growth, and bark. However, its leaves have from seven to nine leaflets, which are more oblong and wedgelike than are those of the shagbark; they are also more downy when young and remain slightly downy beneath. The nut is very large, thick-shelled, oblong, angled, and pointed at both ends. The kernel is large and sweet but inferior in flavor to that of the smaller shagbark. The big shagbark has larger buds than has the other. Their fringy, reddish purple inner scales grow so large that they appear tulip-like before they fall away at the unfolding of the leaves.

Hickory wood ranks high in value; is light-colored, close-grained, heavy, a very durable when not exposed to mo ture. It is capable of resisting immen strain, and therefore it is used for t handles of spades, plows, and other too As a fuel, it is superior to most woo making a glowing, hot, and quite lasti fire.

LESSON 177

The Shagbark

Leading Thought — The hickories a important trees commercially. They ha compound leaves which are set alternat upon the twig. The shagbark can be to from the other hickories by its ragg scaling bark.

Method — This lesson may be beg in the winter when the tree can be studi carefully as to its shape and method branching. Later, the unfolding of t leaves from the large buds should watched, as this is a most interesting pr ess; and a little later the blossoms may studied. The work should be taken again in the fall, when the fruit is ripe.

Opening leaf bud of shagbark hickory

)BSERVATIONS. WINTER WORK — 1. ıat is the general shape of the whole :? Are the lower branches very large? what angle do the branches, in general, w from the trunk? Are there many e branches?

. Where is the spray borne? What is character — that is, is it fine and ooth, or knotted and angled? What is color?

. Describe the bark. Is the bark on limbs like that on the trunk?

. What is the size and shape of the ls? Are the buds greenish-yellow, yel- ish-brown, or do they have a reddish ge?

. Count the bud scales. Are they vny or smooth?

PRING WORK — 6. Describe how the kory leaf unfolds from its bud. How is h leaflet folded within the bud?

. Describe the long greenish catkins ich bear the pollen. On what part of twigs do they grow? Do they grow gly or in clusters?

. Take one of the tiny, pollen-bearing vers and hold it under a lens on the nt of a pin. How many lobes has the calyx? Count the stamens, and note the color of the anthers.

9. Upon what part of the twigs do the pistillate flowers grow? How many points or lobes has the pistillate calyx? Describe the growth of the nut from the flower.

AUTUMN WORK — 10. Does the hickory you are studying grow in an open field or in a wood?

11. Are the trunk and branches slender and lofty, or sturdy and wide spreading?

12. Note the number and shape of the leaflets. Are they slim and tapering, or do they swell to the width of half their length? Are they set directly upon or are they attached by tiny petioles or petiolules to the mid-stem or petiole? Are they smooth or downy on the underside? Are the leaves set upon the twigs alternately or opposite each other? How are the leaflets set upon the mid-stem?

13. Describe the outer husk of the nut. Into how many sections does it open? Does it cling to the nut and fall with it to the ground? Is the nut angled and pointed, or is it roundish and without angles? Is the taste of the kernel sweet or bitter?

THE CHESTNUT

Che chestnut, formerly one of the most ful and valuable trees in the eastern ited States, has been eliminated over st of its natural range by the deadly stnut bark disease. In the Southeast chestnut trees are still to be found, over most of the land, where they w originally, growing chestnut exists y as small sprouts. These sprouts are ıost always badly diseased and able to for only a few years. It is almost cer- ı that within a short span of time, all ture chestnut trees will disappear.

Che interest in native chestnut, even ugh most of it is gone, and not likely reappear, is still so great that a discus- n of it is included here.

This splendid tree, sometimes reaching height of one hundred feet, seldom eives the admiration due to it, simply because humanity is so much more interested in food than in beauty. The fact that the chestnuts are sought so eagerly has taken away from interest in the appearance of the tree. The chestnut has a great round head set firmly on a handsome bole, which is covered with grayish brown bark divided into rather broad, flat, irregular ridges. The foliage is superb; the long, slender, graceful leaves, tapering at both ends, are glossy, brilliant green above and paler below; and they are placed near the ends of the twigs, those of the fruiting twigs seeming to be arranged in rosettes to make a background for blossom or fruit. The leaves are placed alternately and have deeply notched edges, the veins extending straight and unbroken from midrib to margin; the petiole is short. The leaf is like that of the beech, except

Verne Morton

Not long ago these chestnuts were living and flourishing. Now, as is true of most of the other chestnuts in the United States, only their gaunt skeletons remain

that it is much longer and more pointed; it resembles in general shape the leaf of the chestnut oak, except that the edges of the latter have rounded scallops instead of being sharply toothed. The burs appear at the axils of the leaves near the end of the twig. Thoreau has given us an admirable description of the chestnut fruit:

" What a perfect chest the chestnut is packed in! With such wonderful care Nature has secluded and defended these nuts as if they were her most precious fruits, while diamonds are left to take care of themselves. First, it bristles all over with sharp, green prickles, some nearly a half inch long, like a hedgehog rolled into a ball; these rest on a thick, stiff, barklike rind one-sixteenth to one-eighth of an inch thick, which again is most daintily lined with a kind of silvery fur or velvet plush one-sixteenth of an inch thick, even rising into a ridge between the nuts, like the lining of a casket in which the most precious commodities are kept. At last frost comes to unlock this ch it alone holds the true key; and then ture drops to the rustling leaves a ' do nut, prepared to begin a chestnut's co again. Within itself again each indivic nut is lined with a reddish velvet, a to preserve the seed from jar and injur falling, and perchance from sudden da and cold; and within that a thin, wh skin envelops the germ. Thus, it has lin within lining and unwearied care, no count closely, six coverings at least be you reach the contents."

The red squirrels, as if to show th spite because of the protection of treasure chest, have the reprehens habit of cutting off the young burs thus robbing themselves of a rich l harvest — which serves them right. Th are usually two nuts in each bur, set v flat sides together; but sometimes th are three and then the middle on squeezed so that it has two flat sides. casionally there is only one nut develo in a bur, and it grows to be almost glc lar. The color we call chestnut is deri from the beautiful red-brown of the ished shell of the nut, polished exc where the base joins the bur, and at apex, which is gray and downy.

The chestnut is a beautiful t

Leaves and blossoms of the chestnu

ther green in summer or glowing len yellow in autumn, or bare in win- but it is most beautiful during late e and July, when covered with constel- ons of pale yellow stars. Each of these s is a rosette of the pollen-bearing blos- s; each ray consists of a catkin often six eight inches in length, looking like a ead of yellowish chenille fringe; cloth- this thread in tufts for its whole length the stamens, standing out like minute eads tipped with tiny anther balls. If observe the blossom early enough, we

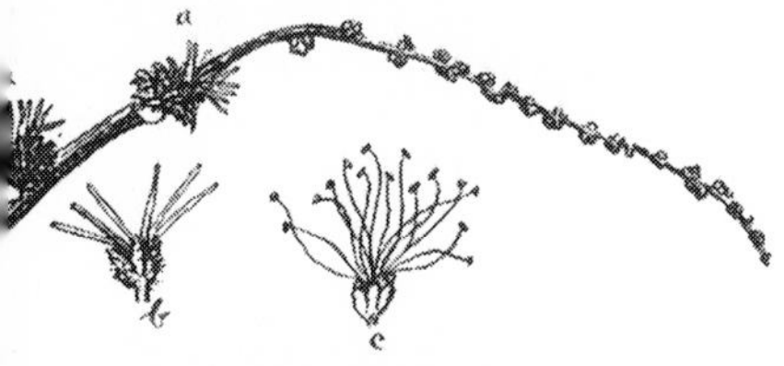

Detail of chestnut blossoms

, a, pistillate flowers set in a base of scales; b, pistil- flower enlarged; c, staminate flower enlarged

see these stamens curled up as they ne forth from the tiny, pale yellow, six- ed calyx. One calyx, although scarcely -sixteenth of an inch across, develops m ten to twenty of these stamens; these y flowers are arranged in knots along central thread of the catkin. No won- it looks like chenille! There are often many as thirty of these catkin rays in star rosette; the lower ones come from axils of the leaves; but toward the tips the twig, the leaves are ignored and the kins have possession. In one catkin I imated that there were approximately oo stamens developed, each anther ked with pollen. When we think that re may be thirty of the catkins in a ssom star, we get a glimmering of the ount of pollen produced.

And what is all this pollen for? Can it simply to fertilize the three or four onspicuous flowers at the tip of the g beyond and at the center of the star? ese pistillate flowers are little bunches green scales with some short, white eads projecting from their centers; and ond them a skimpy continuation of stalk with more little green bunches scattered along it, which are undeveloped pistillate blossoms. The one or two flow-

Verne Morton

Chestnuts in burs

ers at the base of the stalk seem to get all the nourishment and the others do not develop. If we examine one of these nests of green scales, we find that there are six threads belonging to one tiny, green flower with a six-lobed calyx; the six threads are the stigmas, each one reaching out and asking for no more than one grain of the rich shower of pollen.

Whereas the chestnut blooms in the summer, the blossoms of the other members of its family appear earlier; and their fruit has formed when the chestnut comes into bloom.

Chestnut wood is light, rather soft, stiff, coarse, and not strong. It is used in cabinet work, cooperage, and for telegraph poles and railway ties. When burned as fuel, it snaps and crackles almost as much as hemlock.

O. L. Foster

Chestnuts

THE HORSE CHESTNUT

Ralph W. Curtis

Horse chestnut in blossom

The wealth of children is, after all, the truest wealth in this world; and the horse chestnuts, brown and smooth, looking so appetizing and so belying their looks, have been used from time immemorial by boys as legal tender — a fit use, for these handsome nuts seem coined purposely for boys' pockets.

The horse chestnut is a native of Asia Minor. It has also a home in the high mountains of Greece. In America, it is essentially a shade tree. Its head is a broad cone, its dark green foliage is dense, and, when in blossom, the flower clusters stand out like little white pyramids against the rich background in a most striking fashion. "A pyramid of green supporting a thousand pyramids of white" is a clever description of this tree's blossoming. The brown bark of the trunk has a tendency to break into plates, and the trunk is just high enough to make a fitting base for the handsome head.

The blossom panicle is at the tip end of the twig and stops its growth at that point; the side buds continue to grow th making a forking branch. Each bloss panicle stands erect like a candle flan and the flowers are arranged spira around the central stem, each pedicel c rying from four to six flowers. The ca has five unequal lobes, and it and t stem are downy. Five spreading and equal petals with ruffled margins raised on short claws, to form the corol seven stamens with orange colored thers are thrust far out and up from t flower. The blossoms are creamy or pi ish white and have purple or yell blotches in their throats. Not all the flo ers have perfect pistils. The stigmas rip before the pollen, and are often thr forth from the unopened flower. The flo ers are fragrant and are eagerly visited bumblebees, honeybees, and wasps.

Very soon after the blossom falls, th may be seen one or two green, pric balls, which contain the fruits. By Oc ber the green, spherical husk breaks op in three parts, showing its white satin ing and the roundish, shining, smooth at its center. At first there were six lit nuts in this husk, but all except one g up to the single burly occupant that there when the husk opens. The gr round, pale scar on the nut is where

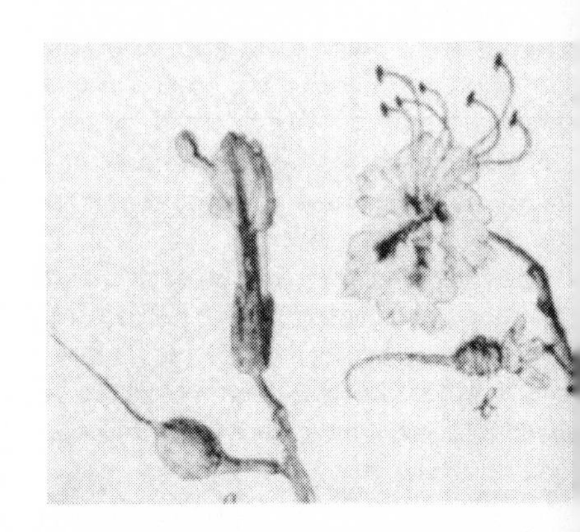

a, Blossom of the sweet buckeye and yo fruit; b, Blossom and young fruit of ho chestnut

ned the husk. Very few American animals will eat the nut; the squirrels scorn and horses surely disown it.

In winter, the horse chestnut twig has its tip a large bud and looks like a obbed antenna thrust forth to test the ety of the neighborhood. There are, beles the great varnished buds at the ends the twigs, smaller buds opposite to each her along the sides of the twig, standing t stiffly. On each side of the end bud, d below each of the others, is a horseoe-shaped scar left by the falling leaf of t year. The "nails" in this horseshoe formed by the leafy fibers which ined the petiole to the twig. The great rminal buds hold both leaves and flows. The buds in winter are brown and ining as if varnished; when they begin swell, they open, displaying the silky ay floss which swaddles the tiny leaves. he leaves unfold rapidly and lift up their een leaflets, looking like partly opened nbrellas, and giving the tree a very owny appearance, which Lowell so well scribes:

nd gray hoss-chestnut's leetle hands unfold
fter'n a baby's be at three days old.

The leaf, when fully developed, has ven leaflets, of which the central ones

Ralph W. Curtis

Horse chestnut blossoms and leaves

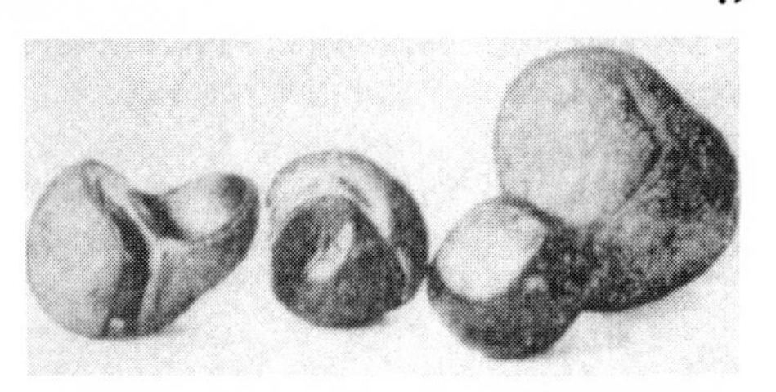

O. L. Foster

Buckeyes. They resemble horse chestnuts

are the larger. They are all attached around the tip of the petiole. The number of leaflets may vary from three to nine, but is usually seven. The leaflets are oval in shape, being attached to the petiole at the smaller end; their edges are irregularly toothed. The veins are large, straight, and lighter in color; the upper surface is smooth and dark green, the underside is lighter in color and slightly rough. The petiole is long and shining and enlarges at both ends; when cut across, it shows a woody outer part encasing a bundle of fibers, one fiber to each leaflet. The places where these fibers were attached to the twig make the nails in the horseshoe scar. The leaves are placed opposite on the twigs.

Very different from that of the horse chestnut is the flower of the yellow or sweet buckeye; the calyx is tubular, long, and five-lobed; the two side petals are on long stalks and are closed like spoons over the stamens and anthers; the two upper petals are also on long stalks, lifting themselves up and showing on their inner surfaces a bit of color to tell the wandering bee that here is a tube to be explored. The flowers are greenish yellow. The flowers of the Ohio buckeye show a stage between the sweet buckeye and the horse chestnut. The Ohio buckeye is our most common native relative of the horse chestnut. Its leaves have five leaflets instead of seven. The sweet buckeye is also an American species and grows in the Allegheny Mountains.

LESSON 178

The Horse Chestnut

Leading Thought — The horse chestnut has been introduced in America as a shade tree from Asia Minor and southern Europe. Its foliage and its flowers are both beautiful.

METHOD — This tree is almost always at hand for the village teacher, since it is so often used as a shade tree. Watching the leaves develop from the buds is one of the most common of the nature-study lessons. The study of the buds, leaves, and fruits may be made in school; but the children should observe the tree where it grows and pay special attention to its insect visitors when it is in bloom.

OBSERVATIONS — 1. Describe the horse chestnut tree when in blossom. At what time does this occur? What is there in its shape and foliage and flowers which makes it a favorite shade tree? Where did it grow naturally? What relatives of the horse chestnut are native to America?

2. Study the blossom cluster; are the flowers borne on the ends or on the sides of the twig? Describe the shape of the cluster. How are the flowers arranged on the main flowerstalk to produce this form? Do the flowers open all at once from top to bottom of the cluster? Are all the flowers in the cluster the same color? Are they fragrant? What insects visit them?

3. Take a single flower; describe the form of the calyx. Is it smooth or downy? Are the lobes all the same size? Are the petals all alike in size and shape? What gives them the appearance of Japanese paper? Are any connected together? Are they all splashed with color alike?

4. How many stamens are there? Where do you see them? What color are the anthers? Search the center of a flower for a pistil with its green style. Do you find one in every flower? Could a bee reach the nectar at the base of the blossom without touching the stigma? Could she withdraw without dusting herself with pollen?

5. How long after the blossom does the young fruit appear? How does it loo How many nuts are developed from ea cluster of blossoms? What is the shape the bur? Into how many parts does open? Describe the outside; the insi Describe the shape of the nuts, their co and markings. Open a nut. Can you fi any division in the kernel? Is it good eat?

HORSE-CHESTNUT TWIGS AND LEAVES SPRING — 6. Are the buds on the tw nearly all the same size? Where are t larger ones situated? What is the color the buds? How are the scales arranged them? Are they shiny or dull? What the scales enfold? Can you tell withc opening them which buds contain flow and which ones leaves?

7. Describe the scars below the bu What caused them? What marks are them? What made the "nails" in t horseshoe? Has the twig other scars? H do the ring-marks show the age of t twig? Do you see the little, light colo dots scattered over the bark of the tw What are they?

8. Describe how the leaf unfolds fr the bud. What is the shape of the le Do all the leaves have the same num of leaflets? Do any of them have an ev number? How are the leaflets set upon t petiole? Describe the leaflets, includi shape, veins, edges, color above and low. Is the petiole pliant, or stiff a strong? Is it the same shape and s throughout its length? Break a petiole; it green throughout? What can you see its center? Are the leaves opposite or ternate? When they fall, do they drop tire or do the leaflets fall apart from t petiole?

9. Make a sketch of the horse-chestn tree.

10. How do the flowers and leaves the horse chestnut differ from those the sweet buckeye and of the Ohio bu eye?

Ralph W. Curtis

Yellow-twigged and weeping willows

THE WILLOWS

They shall spring up among the grass, as willows by the water courses.
— ISAIAH

Vhen I cross opposite the end of Willow Row the sun comes out and the trees very handsome, like a rosette, pale, tawny or fawn color at base and red-yellow or ıge-yellow for the upper three or four feet. This is, methinks, the brightest object he landscape these days. Nothing so betrays the spring sun. I am aware that the has come out of the cloud just by seeing it light up the osiers. — THOREAU

'he willow Thoreau noted, is the len osier, a colonial dame, a descendant n the white willow of Europe. It is the ;t common tree planted along streams onfine them to their channels, and af-ls an excellent subject for a nature-ly lesson. The golden osier has a short ugh magnificent trunk, giving off tre-ıdous branches, which in turn branch uphold a mass of golden terminal ɔts. But there are many willows besides , and the one who tries to determine he species and hybrids must conclude of making willows there is no end. : species most beloved by children he pussy willow, which is often a shrub, rarely reaching twenty feet in height. It loves moist localities, and on its branches in early spring are developed the silky, furry pussies, larger than the pussies of other willows. These are favorite objects for a nature-study lesson, and yet how little have the teachers or pupils known about these flowers!

The showy willow pussies are the pollen-bearing flowers; they are covered in winter by a brown, varnished, double, tentlike bract. The pussy in full bloom shows beneath each fur-bordered scale two stamens with long filaments and plump anthers; but there are no pistils in this blossom. The flowers which produce seed are borne on another tree entirely and in similar greenish gray catkins, but not so soft and furry. In the pistillate cat-

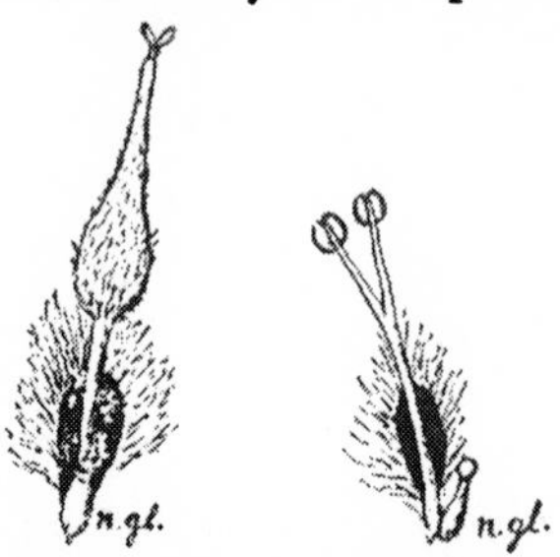

Enlarged willow blossoms

Pistillate blossom showing nectar-gland, (n.gl.) Staminate flower showing the nectar-gland (n.gl.)

kin each fringed scale has at its base a pistil which thrusts out a Y-shaped stigma. The question of how the pollen from one gets to the pistils of another is a story which the bees and the wind can best tell. The willow flowers give the bees almost their earliest spring feast, and when they are in blossom, the happy hum of the bees working in them can be heard for some distance from the trees. The pollen gives them bee bread for their early brood, and they get their honey supply from the nectar which is produced in little jug-shaped glands, at the base of each pollen-bearing flower on the " pussy " catkin, and in a long pocket at the base of each flower on the pistillate catkin. So they pass back and forth, carrying their pollen loads, which fertilize the stigmas on trees where there is no pollen.

In June the willow seed is ripe. The catkin then is made up of tiny pods, which open like milkweed pods and are filled with seed equipped with balloons. When these fuzzy seeds are being set free people say that the willows " shed cotton."

Although the seed of the willow is produced in abundance, it is hardly needed for preserving the species. Twigs which we place in water to develop flowers will also put forth roots; even if the twigs are placed in water wrong side up, rootlets will form. A twig lying flat on moist soil will push out rootlets along its entire length as though it were a root; and shoots will grow from the buds on its upper side. This habit of the willows and the fact that the roots are long, strong, and fibrous, m
these trees of great use as soil bin
There are few things better than a t
hedge of willows to hold streams to
proper channels during floods; the

Verne M

Willow pussies; the staminate blossom the willow

:h out in all directions, interlacing mselves in great masses, and thus hold soil of the banks in place. The twigs everal of the species, notably the crack l sand-bar willows, are broken off ily by the wind and carried off down am, and where they lodge, they take t; thus, many streams are bordered by -planted willow hedges.

The willow foliage is fine and makes a utiful, soft mass with delicate shadows. e leaf is long, narrow, pointed, and ıder, with finely toothed edges and rt petiole; the exact shape of the leaf, course, depends upon the species, but of them are much lighter in color be- than above. The willows are, as a ole, water lovers and quick growers.

Although willow wood is soft and ex- dingly light, it is very tough when sea- ed and is used for many things. The oden shoes of the European peasant, ficial limbs, willowware, and charcoal he finest grain used in the manufacture gunpowder are all made from the wil- wood. The toughness and flexibility the willow twigs have given rise to ny industries; baskets, hampers, and niture are made of them. To get these gs the willow trees are pollarded, or back every year between the fall of leaves and the flow of the sap in the ing. This pruning results in many gs. The use of willow twigs in basketry ncient. The Britons fought the Roman diers from behind shields of basket rk; and the wattled huts in which they :d were woven of willow saplings eared with clay. Salicylic acid, used lely in medicine, is made from willow k, which produces also tannin and ne unfading dyes.

There are many insect inhabitants of willow, but perhaps the most interest- is the little chap who makes a conelike ect on the twig of certain species of low growing along our streams. This ie is naturally considered a fruit by the orant, but we know that the willow ds are grown in catkins instead of ies. This willow cone is made by a small ıt which lays its egg in the tip of the twig; as soon as the little grub hatches, it begins to gnaw the twig, and this irrita- tion for some reason stops the growth. The leaves instead of developing along the stem are dwarfed and overlap each other. Just in the center of the cone at the tip of the twig the little larva lives its whole life surrounded by food and protected from enemies; it remains in the cone all winter, in the spring changes to a pupa, and after a time comes forth — a very deli- cate little fly. The larva in this gall does not live alone. It has its own little apart- ment at the center, but other gall gnats live in outer chambers and breed there in great numbers. It is well to gather these cones in winter; examine one by cutting it open to find the larva, and place others in a fruit jar with a cover so as to see the

Verne Morton

Pistillate blossoms of the pussy willow

Verne Morton

Seeds of the willow

little flies when they shall issue in the spring. (See p. 337.)

There is another interesting winter tenant of willow leaves, but it is rather difficult to find. On the lower branches may be discovered, during winter and spring, leaves rolled lengthwise and fastened, making elongated cups. Each little cup is very full of a caterpillar which just fits it, the caterpillar's head forming the plug of the opening. This is the partially grown larva of the viceroy butterfly. A larva of the autumn brood of this butterfly eats off the tip of the leaf each side of the midrib for about half its length, fastens the petiole fast to the twig with silk, then rolls the base of the leaf into a cup, lines it w silk and backs into it, there to remain til fresh leaves on the willow in spring ford it new food.

LESSON 179

The Willows

Leading Thought — The willows h their pollen-bearing flowers and their se bearing flowers on separate trees; the p len is distributed by bees and by the wi The willow pussies are the pollen-bear flowers.

Method — As early in March as is p ticable, have the pupils gather twigs of many different kinds of willows as can found; these should be put in jars water and placed in a warm, sunny w dow. The catkins will soon begin to pu out from the bud scales, and the wh process of flowering may be watched.

Observations — 1. How can you t the common willow tree from afar? what localities do these trees grow? W is the general shape of the big willo How high is the trunk, or bole? W sort of bark has it? Are the main branc large or small? Do they stand out at a wi angle or lift up sharply? What color the terminal shoots, or spray?

2. Are the buds opposite or altern on the twigs? Is there a bud at exactly t end of any twig? How many bracts there covering the bud?

3. Which appear first, the leaves or t blossoms? Study the pussies on your tw and see if they are all alike. Is one ki more soft and furry than the other? they of different colors?

4. Take one of the furry pussies. D scribe the little bract, which is like a p tecting hood at its base. What color is t fur? After a few days, what color is t pussy? Why does it change from silv color to yellow? Pick one of the catki apart and see how the fur protects the s mens.

5. Take one of the pussies which is n so furry. Can you see the little pistils wi the Y-shaped stigmas set in it? Is ea

tle pistil set at the base of a little scale th fringed edges?

6. Since the pollen-bearing catkins are one tree and the seed-bearing catkins on the other, and since the seeds can- t be developed without the pollen, how the pollen carried to the pistils? For this swer, visit the willows when the pussies all in bloom and listen. Tell what u hear. What insects do you see work- on the willow blossoms? What are ey after?

7. What sort of seed has the willow? ow is it scattered? Do you think the nd or water has most to do with plant- willow seed?

Work for May or September — 8. scribe willow foliage and leaves. How n you tell willow foliage at a distance?

9. What sort of roots has the willow? Why are the willows planted along the banks of streams? If you wished to plant some willow trees how would you do it? Would you plant seeds or twigs?

10. For what purposes is willow wood used? How are the twigs used? Why are they specially fitted for this use? What is pollarding a tree? What chemicals do we get from willow bark?

11. Do you find willow cones on your willows? Cut one of these cones through and see if you can find any seeds. What is in the middle of it? What do you think made the scales of the cone? Do you think this little gall insect remains in here all winter?

12. In winter, hunt the lower branches of willows for leaves rolled lengthwise, making a winter cradle for the young cat- erpillars of the viceroy butterfly.

THE COTTONWOOD OR CAROLINA POPLAR

The sojourner on our western plains ere streams are few and sluggish, dis- pearing entirely in summer, soon learns love the cottonwoods, for they will ow and cast their shade for men and cat- where few other trees could endure. e cottonwood may be unkempt and ged, but it is a tree, and we are grateful it for its ability to grow in unfavorable uations. In the Middle West it attains perfection, although in New York we ve some superb specimens — trees ich are more than one hundred feet in ight and with majestic trunks, perhaps e or six feet through. The deep-fur- wed, pale gray bark makes a handsome vering. The trunk divides into great out- inging, widely spaced branches, which ar a fine spray on their drooping ends. rgent declares that at its best the cot- nwood is one of the stateliest inhabit- ts of our eastern forests. The variety we nt in cities we call the Carolina poplar, t it is a cottonwood. It is a rapid grower, d therefore a great help to the " boom vns " of the West and to the boom burbs in the East; although for a city e its weak branches break too readily in wind storms in old age. However, it keeps its foliage clean, the varnished leaves shedding the dust and smoke; because of this latter quality it is of special use in towns that burn soft coal.

The cottonwood twigs which we gather for study in the spring are yellowish or

Cottonwood

reddish, those of last year's growth being smooth and round, while those showing previous growth are angular. The buds are

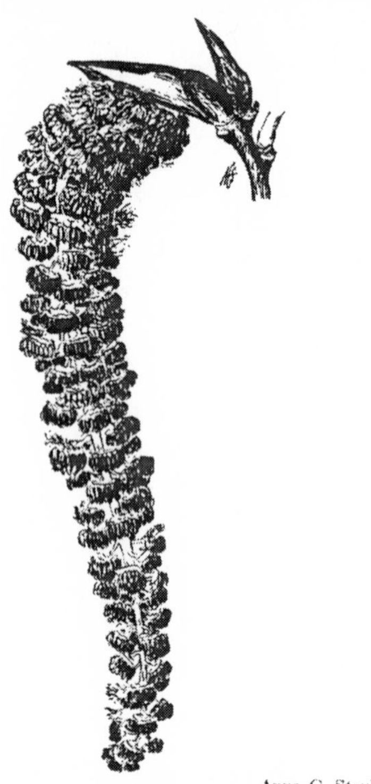

Anna C. Stryke

Staminate catkin of cottonwood

red-brown and shining, and covered with resin which the bees like to collect for their glue. The leaf buds are slender and sharp-pointed; the flower buds are wider and plumper.

The two sexes of the flowers are borne on separate trees. The trees bearing pollen catkins are so completely covered with them that they take on a very furry, purplish appearance when in blossom. These catkins are from three to five inches long and half an inch thick, looking fat and pendulous; each fringed scale of the catkin has at its base a disc looking like a white bracket, from which hang the reddish-purple anthers; these catkins fall after the pollen is shed and look like red caterpillars upon the ground.

The seed-bearing flowers are very different; they look like a string of little, greenish beads loosely strung. Each pistil is globular and set in a tiny cup, and it h three or four stigmas which are widen or lobed; as it matures, it becomes larg and darker green, and the string elonga to six or even ten inches. The lit pointed pods open into two or mo valves and set free the seeds, which a provided with a fluff of pappus to s them off on the breeze; so many of t seeds develop that every object in t neighborhood is covered with their fu and thus the tree has gained its na "cottonwood."

The foliage of the cottonwood is li that of other poplars, trembling with t breeze. The heavy, subcircular leaf is su ported on the sidewise flattened petic so that the slightest breath of air sets quaking; a gentle breeze sets the wh tree twinkling and gives the eye a fas nating impression as of leaves beckoni The leaf is in itself pretty. It is from th to five inches long, broad, slightly an lar at the base, and has a long, taperi pointed tip. The edge is saw-toothed, a also slightly ruffled except near the st where it is smooth; it is thick and shini green above and paler beneath. The lo slender petiole is red or yellowish, and t leaves are placed alternately on the twi

In the autumn the leaves are brillia yellow. The wood is soft, weak, fi grained, whitish or yellowish, and ha

Cyrus Cr

The growing fruits of the cottonwood

ny luster; it is not durable. It is used ıewhat for building and for furniture, ome kinds of cooperage, and also for es and woodenware; but its greatest is for making the pulp for paper. ny newspapers and books are printed cottonwood paper. It is common from Middle States to the Rocky Moun- ıs and from Manitoba to Texas.

LESSON 180

The Cottonwood

ɹeading Thought — The cottonwood ı poplar. It grows rapidly and flour- :s on the dry western plains where er trees fail to gain a foothold. It grows l in the dusty city, its shining leaves dding the smoke and dirt.

Лethod — Begin this study in spring ore the cottonwoods bloom. Bring in gs in February, give them water and mth, and watch the development of catkins. Afterwards watch the unfold- of the leaves and study the tree. Twigs the aspen, if brought indoors in early ing, provide a very interesting study.

Observations — 1. What is the color he bark on the cottonwood? Is it ridged :ply? What is the color of the twigs?

Ralph W. Curtis

wers of trembling aspen, sometimes called opple tree," a near relative of the cotton- od

Ralph W. Curtis

Lombardy poplar, another relative of the cottonwood

Are they round or angular, or both? Describe the winter buds and bud scales. Can you tell which bud will produce leaves and which flowers?

2. Describe the catkin as it comes out. Has this catkin anthers and pollen, or will it produce seed? Do you think the seeds are produced on the same trees as the pollen?

3. Find a pollen-bearing catkin. Describe the stamens. Can you see anything but the anthers? On what are they set?

What color are they? What color do they give to the tree when they are in blossom? What happens to the catkins after their pollen is shed?

4. Find a seed-bearing catkin. How long is it? Do you see why this tree is called the necklace poplar? Describe the pistils which make the beads on the necklace.

5. When do the seeds ripen? If you have lived near the tree, how do you know when they are ripe? How long is the catkin with the ripened seeds? How many balls on the necklace now? What is the color? How many seeds come out of each little pod? How are the seeds floated on the air? Why do we call this tree " cottonwood "?

Seed pod of poplar, shut and open

6. How large is the largest cottonwood that you know? Sketch it to show the shape of the tree. Are the main branc large? Do they droop at the tips?

7. How does the foliage of the cott wood look? Does it twinkle with wind? Examine the leaves upon a bra and tell why you think they twinkle. the petioles round or flat? Are they tened sidewise or up and down? Are t stiff or slender? Describe the leaves, giv their shape, veins, edges, color, and ture above and below. Are the edges fled as well as toothed? Is the leaf hea If a breeze comes along how would affect such a heavy, broad leaf on suc slender, thin petiole? Blow against leaves and see how they move. Do understand, now, why they tremble in slightest breeze? Can you see why leaves shed smoke and dust, when used shading city streets?

8. Why is the cottonwood used a shade tree? Do you think it makes a be tiful shade tree? How long does it tak to grow? What kind of wood does it p duce? For what is the wood of the cott wood used?

THE WHITE ASH

Myths and legends cluster about the ash tree. It was, in the Norse mythology, the tree " Igdrasil," the tree of the universe, which was the origin of all things. " As straight as a white ash tree " was the highest compliment that could be paid to the young pioneer; so straight is its fiber and so strong its quality that the American Indians made their canoe paddles from it.

The bark of the ashes is very beautiful. It is divided into fine, vertical ridges, giving the trunks the look of being shaded with pencil lines; the bark smooths out on the lower branches. But even more characteristic than the bark are the ash branches and twigs; the latter are sparse, coarse, and clumsy, those of the white ash being pale orange or gray, and seemingly warped into curves at the ends; they are covered with whitish gray dots, which reveal themselves under the lens to be breathing-pores.

The white ash loves to grow in woods or in rich soil anywhere, e though it be shallow; at its best, it reac the height of 130 feet, with a trunk feet through. Its foliage is peculiarly gra ful; the leaves are from eight to twe inches long and are composed of fr five to nine leaflets. The leaflets have lit petiolules connecting them with the p ole; in shape they are ovate with edges scurely toothed or entire; the two ba leaflets are smaller than the others a the end one largest; in texture, they satiny, dark green above, whitish benea with feather-like veins, often hairy on lower side. The petioles are swollen at base. The leaves are set opposite up the twig; except for the horse chestn the ashes are our only common trees w compound leaves which have the lea opposite. This characteristic distinguis the ashes from the hickories. The autu foliage has a very peculiar color; the lea

dull purple above and pale yellow be-
; this brings the sunshine color into
shadowy parts of the tree, and gives
urious effect of no perspective. Not-
standing this, the autumn coloring is
y to the artistic eye and is very char-
ristic.

'he fruits of the ash are borne in
wded clusters; the delicate stalk, from
e to five inches long, is branched into
ller stalks to which are joined two or
e keys. Often several of these main
ks come from the same bud at the tip
last year's wood, so that they seem
wded. The fruit is winged, the wing
ng almost twice as long as the seed set
ts base. Thoreau says: "The keys of
white ash cover the trees profusely, a sort of mulberry brown, an inch and a half long, and handsome." The fruits cling persistently to the tree, and I have often observed them being blown over the surface of the snow as if they were skating to a planting place.

G. F. Morgan

Pistillate blossoms of white ash

Ralph W. Curtis

White ash

The flowers appear in April or May, before the leaves. The pistillate flowers make an untidy fringe, curling in every direction around the twigs. The chief flowerstalk is three to four inches long, quite stout, pale green, and from this arise short, fringed stalks, each carrying along its sides the knobs on little stalks — which are the pistillate flowers. Each tiny flower seems to be bristling with individuality, standing off at an angle to get a share of the pollen. The flower has the calyx four-lobed; the style is long and slender and is divided into a V-shaped purple stigma.

The staminate flowers appear early in the spring, and look like knobs on the tips of the coarse, sparse twigs; they consist of masses of thick, green anthers with very short, stout filaments; each calyx is four-lobed. These flowers are attached to a five-branching stem; but the stem and its branches cannot be seen unless the anthers are plucked off, because they hang

Verne Morton

Young ash

in such a crowded mass. Later the leaves come out beyond them.

The leaf buds in winter are very pretty; they are white, bluntly pointed, with a pale gray half-circle below, on which was set last year's leaf. Another one of Nature's miracles is the bouquet of leaves coming from one of the big four-parted terminal buds, which is made up of four scales, two of which are longer and narrower than the others. Within the bud each little leaflet is folded like a sheet of paper lengthwise, and folded with the other leaflets like the leaves of a book; and when they first appear they look like tiny, scrawny birds' claws. But it is not merely one pair of leaves that comes from this bud, but many, each pair being set on a twig opposite a at right angles to the next pair on eit side. Even as many as five pairs of th splendid compound leaves may co from this one prolific bud. As they p out, the green stem of the new w grows, thus spacing the pairs properly the making of beautiful foliage.

LESSON 181

Ash Trees

Leading Thought — The ashes among our most valuable timber trees; white ash is one of the most beautiful a useful of them all. It does not make ests, but it grows in them, and its w is of great value for many things.

Method — The pupils should all the tree where it grows. The questi may be given to them for their field n books. The lesson may begin in the and be continued in the spring.

Observations — 1. What is there ab the bark of the ash tree which dis guishes it from other trees? Where d the white ash grow? What is the hei and thickness of the ash tree you studying?

2. The ash leaf is a compound leaf how many leaflets is it composed? W is the texture and shape of the leafl Describe the veins. Do the leaflets h petioles (petiolules)? Are the edges of leaflets toothed? Which of the leaflet largest? Which smallest? Is the pet swollen at the base? How are the le arranged on the twigs? How does this tinguish the ashes from all our other t which have compound leaves? How the hickories have their leaves arrang What color is the ash foliage in autu

3. Describe the seeds of the ash and way they are arranged on their ste Where are they placed on the tree? H long do they cling? How does the sn help to scatter them?

4. When does the white ash blosso Are the pistillate and staminate flov together or separate? Find and desc them.

5. What are our uses for ash timb

or what are the saplings used? How did e Indians use the white ash? Write a eme on all the interesting things you n find about the ash trees.

6. How many species of the ash trees you know?

care not how men trace their ancestry,
o ape or Adam; let them please their
whim;
ut I in June am midway to believe
tree among my far progenitors,
Such sympathy is mine with all the race,
Such mutual recognition vaguely sweet
There is between us. Surely there are
times
When they consent to own me of their
kin,
And condescend to me and call me cousin,
Murmuring faint lullabies of eldest time,
Forgotten, and yet dumbly felt with thrills
Moving the lips, though fruitless of the
words.

— "UNDER THE WILLOWS," LOWELL

THE APPLE TREE

As the apple tree among the trees of the wood, so is my beloved among the sons.
sat down under his shadow with great delight, and his fruit was sweet to my taste.

— "THE SONG OF SOLOMON"

An old-fashioned orchard is always a deght to those of us who love the picturque. The venerable apple tree with its eat twisted and gnarled branches, rearg aloft its rounded head, and casting its adow on the green turf below, is a picre well worthy of the artist's brush. And at is the kind of orchard I should always ave, because it suits me, just as it does uebirds, downies, and chickadees, as a ace to live in. However, if I wished to ake money by selling apples, I should eed to have an orchard of comparatively oung trees, which should be straight and ell pruned, and the ground beneath them ell cultivated; for there are few plants at respond more generously to cultivaon than does the apple tree. In such an rchard, a few annual crops might be rown while the trees were young, and ach year there should be planted in Auust or September the seed of crimson over or of some other good cover crop. This would grow so as to protect the round from washing during the heavy ains and thaws of fall and winter, and in e spring it would be plowed under to dd more humus to the soil.

The apple originally came from southestern Asia and the neighboring parts of urope, but it has been cultivated so long that we have no accounts of how it began. The prehistoric lake dwellers of Switzerland ate this fruit. In this country the apple thrives best on clay loam, although it grows on a great variety of soils; where wheat and corn grow, there will the apple grow also. In general, the shape of the apple-tree head is rounded or broadly pyramidal; however, this differs somewhat with varieties. The trunk is short and rather stocky, the bark is a beautiful soft

A Baldwin apple tree

Verne Morton

Apple tree in winter

gray and is decidedly scaly, flaking off in pieces which are more or less quadrangular. The wood is very fine-grained and heavy. On this account for many years it was used for wood engraving and is also a favorite wood for wood carving; it makes an excellent fuel. The spray is fine, and while at the tips of the limbs it may be drooping or horizontal, it often grows erect along the upper sides of the limbs, each shoot looking as if it were determined to be a tree in itself. The leaves are oval, with toothed edges and long petioles. When the leaves first appear each has two stipules at its base. The shape of the apple leaves depends to some extent upon the variety of the apple.

It has long been the practice not to depend upon the seeds for reproducing a variety; for, since insects do such a large work in pollinating the apple flowers, would be quite difficult to be sure that seed would not be a result of a cross b tween two varieties. Therefore, the ma ter is made certain by the process of gra ing or budding. There are several mod of grafting; one in common use is t cleft-graft. A scion, which is a twig be ing several buds, is cut from a tree of t desired variety, and its lower end is c wedge-shaped. The branch of the tree be grafted is cut off across and split dov through the end to the depth of abo two inches; the wedge-shaped end of t scion is pressed into this cleft, so that bark will come in contact with the inn edge of the bark on one side of the cl branch. The reason for this is that t growing part of the tree is the cambiu layer, which is just inside of the bark, a if the cambium of the scion does not co in contact with the cambium of the bran they will not grow together. After t graft becomes well established, the oth branches of the tree are cut off and t tree produces apples only from that pa of it which grows from the graft. Aft the scion has been set in the stock, all

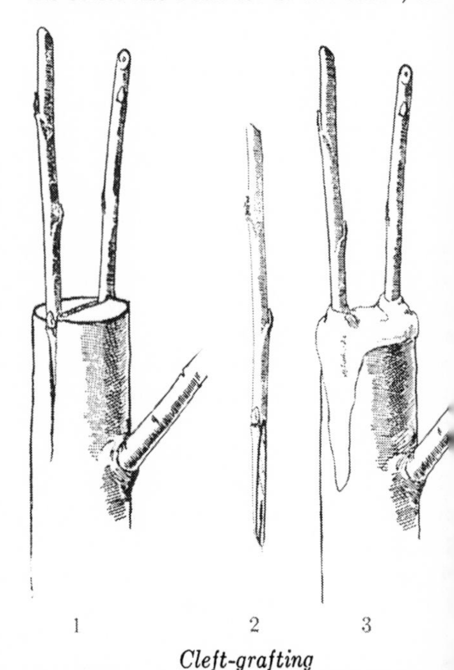

Cleft-grafting

1, Cleft-graft; 2, scion for cleft-grafting; 3, the gr waxed wounded parts are covered with graft-wax, which keeps in the moisture and ›s out disease germs.

udding is done on a similar principle, in a different fashion. A seedling apple about a year and a half old has a aped slit cut into its bark; into this re a bud cut from a tree of the desired ety is inserted, and is bound in with . The next spring this tree is cut back ust above the place where the bud set in, and this bud shoot grows several feet; the next year the tree may be to the orchardist. Budding is done a large scale in the nurseries, for it is his method that the different varieties placed on the market.

Most varieties of apple trees should be forty feet apart each way. It is possible, done judiciously, to raise some small ›s on the land with the young orchard, care should be taken that they do not the trees of their share of the water minerals in the soil. Some varieties in to bear much sooner than others, n at seven years; but an orchard does come into full bearing until after it been planted fifteen or twenty years. e present practice is to prune a tree that the trunk shall be short. This kes the picking of the fruit much easier also exposes the tree less to wind and -scald.

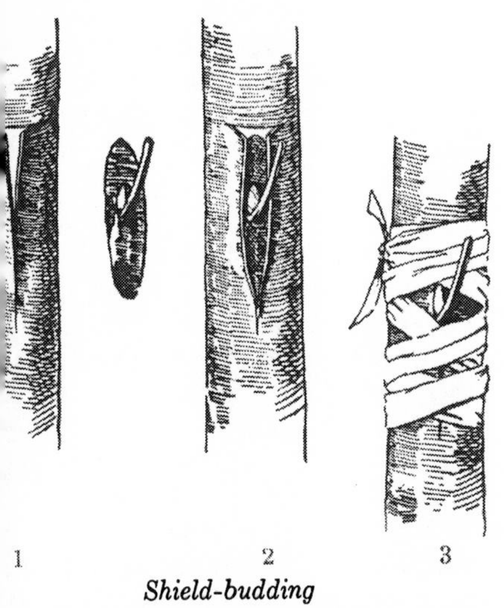

Shield-budding

he T-shaped slit and the bud; 2, the bud set in the slit; 3, the bud tied

Verne Morton

Thorn apple. In winter, the low broad form of this tree is quite evident

There are certain underlying principles of pruning that every child should know: The pruning of the root cuts down the amount of moisture which the tree is able to get from the soil. The pruning of the top throws the food into the branches which are left and makes them more vigorous. If the buds at the tips of the twigs are pruned off, the food is forced into the side buds and into the fruit, which make greater growth. Thinning the branches allows more light to reach down into the tree, and gives greater vigor to the branches which are left. A limb should be pruned off smoothly where it joins the larger limb, and no stump should be left projecting; the wound should be painted so as not to allow fungus spores to enter.

We should not forget that we have a native apple, which we know as the thorn apple. Its low, broad head in winter makes a picturesque point along the fences; its fine, thick spray, spread horizontally, makes a fit framework for the bridal bouquet which will grow upon it in June; and it is scarcely less beautiful in autumn, when covered with the little, red apples called "haws." Though we may refrain from eating these native apples, which consist of a bit of sweet pulp around large seeds, the codling moth finds them most acceptable.

LESSON 182

The Apple Tree

Leading Thought — The tree of each variety of apple has its own characteristic shape, although all apple trees belong to one general type. If apple trees of a certain variety are desired, they can be produced by budding or grafting; trees grown from apple seeds do not produce apples of the same variety as those of the parent tree.

Method — A visit to a large, well-grown orchard in spring or autumn will aid in making this work interesting. Any apple tree near at hand may be used for the lesson.

Observations — 1. How tall is the largest apple tree you know? What variety is it? How old is it? How can you distinguish old apple trees from young ones at a glance?

2. Choose a tree for study: How thick is its trunk? What is the shape of its head? Does the trunk divide into large branches or does it extend up through the center of the head?

3. What sort of bark has it? What is the color of the bark?

4. Does the spray stand erect or is it gnarled and querly? Does the spray grow simply at the ends of the branches or along the sides of the branches?

5. Are the leaves borne at the tip the spray? Are the leaves opposite o ternate? Describe or sketch an apple Does it have stipules at its base whe first appears?

6. What is the character of apple wood? What is it used for?

7. Did this tree come from a seed b in an apple of the same variety whic produces? What is the purpose of graf a tree? What is a scion? How and do we choose a scion? How do we pre a branch to receive the scion? If should place the scion at the cente the branch would it grow? Where n it be placed in order to grow? How we protect the cut end of the branch a it is grafted? Why?

8. What is meant by the term "b ding"? What is the difference betw grafting and budding? Describe the p ess of budding.

9. Where is budding done on a l scale? How do nurserymen know w special varieties of apples their nur stock will bear? How old is a tree w it is budded? How old when it is sol the orchardist?

10. Why should the soil around a trees be tilled? Is this the practice in best-paying orchards?

11. What is often used as a cover in orchards? When is this planted? what purpose?

12. How far apart should apple t be set? How may the land be util while the trees are growing? At what does an apple tree usually come into b ing?

13. Is the practice now to allow apple tree to grow tall? Why is an a tree with a short trunk better than with a long trunk?

14. What does it do to a tree to pr its roots? What does it do to a tre prune its branches?

15. How does it affect a tree to pr the buds at the tips of the twigs?

16. How does it affect a tree to the branches? Describe how a limb sh be pruned and how the wound thus m should be treated. Why?

HOW AN APPLE GROWS

An apple tree in full blossom is a beau-ul sight. If we try to analyze its beauty find that on the tip of each twig there a cluster of blossoms, and set around em, as in a conventional bouquet, are e pale, soft, downy leaves. These leaves d blossoms come from the terminal win- buds, which are protected during win- by little scales which are more or less wny. With the bursting of the bud, ese scales fall off, each one leaving its ark crosswise on the twig, marking the d of the year's growth; these little ges close together and in groups mark e winters which the twig has experi-ced, and thus reveal its age.

Varieties of apples differ in whether the ossoms or the leaves push out first; the ason may cause a like difference. The ite, downy leaves at first have two nar-w stipules at the base of their petioles. ey are soft, whitish, and fuzzy, as are o the flower stem and the calyx, which lds fast in its slender, pointed lobes the obular flower bud. We speak of the lobes the calyx because they are joined at the se, and are not entirely separate as are pals. The basal part of the calyx is cup-aped, and upon its rim are set the large, al petals, each narrowing to a slender m at its base. The petals are set be-een the sepals or lobes of the calyx, the ter appearing as a beautiful, pale green, e-pointed star at the bottom of the wer. The petals are pink on the outside d white on the inside, and are veined om base to edge like a leaf; they are mpled more than are the cherry petals.

The many pale, greenish white stamens different lengths and heights stand up e a column at the center of the flower. ey are tipped with pale yellow anthers, d are attached to the rim of the calyx-p. They are really attached in ten differ-t groups, but this is not easy to see.

The five pale green styles are very silky d downy and are tipped with green gmas. The pistils all unite at their bases making a five-lobed, compound ovary. The upper part of this ovary may be seen above the calyx-cup, but the lower por-tion is grown fast to it and is hidden within it. The calyx-cup is what develops into the pulp of the apple, and each of these pistils becomes one of the five cells in the apple core. If one of the stigmas does not receive pollen, its ovary will de-velop no seed; this often makes the apple lopsided. When the petals first fall, the calyx lobes are spread wide apart; later they close in toward the center, making a tube. To note exactly the time of this change is important, since the time of spraying for the codling moth is before the calyx lobes close. These lobes may be seen in any ripe apple as five little, wrin-kled scales at the blossom end; within them may be seen the dried and wrin-kled stamens, and within the circle of stamens, the sere and blackened styles.

Verne Morton

Apple blossoms

There may be five or six, or even more

Verne Morton

Peach blossoms

blossoms developed from one winter bud, and there may be as many leaves encircling them, forming a bouquet at the tip of the twig. However, rarely more than two of these blossoms develop into fruit, and the fruit is much better when only one blossom of the bouquet produces an apple; if a tree bears too many apples it cannot perfect them.

The blossoms and fruit are usually at the end of the twigs and spurs of the apple tree; and only rarely do they grow along the sides of the branches as do those of the cherry and the peach. However, there are many buds which produce only leaves; and just at the side and below the spur, where the apple is borne, a bud is developed, which pushes on and continues the growth of the twig, and will in turn be a spur and bear blossoms the following year.

LESSON 183

How an Apple Grows

Leading Thought — The purpose of the apple blossom is to produce apples which shall contain seeds to grow into more apple trees.

Method — This lesson should begin with the apple blossoms in the spring and should continue, with occasional obse tions, until the apples are well grown this is not possible, the blossom may studied, and directly afterward the ap may be observed carefully, noting its lation to the blossom.

Observations — 1. How are the ap buds protected in the winter? As the b open what becomes of the protect scales? Can you see the scars left by scales after they have fallen? How d this help us to tell the age of a twig branch?

2. As the winter buds open, which pear first — the flowers or the leaves? they both come from the same bud? all the buds produce both flowers leaves?

3. Study the bud of the apple bloss Describe its stem; its stipules; its ca What is the shape and position of lobes or sepals of the calyx? Why do usually call them the " lobes of the caly instead of sepals?

4. Sketch or describe an open ap blossom. How many petals? What is th shape and arrangement? Can you see calyx lobes between the petals as you l down into the blossom? What of figure do they make? Are the pe usually cup-shaped? What is th color outside and inside? Why do buds seem so pink and the blossoms white?

5. How many stamens are there? they all of the same length? What is

Verne M

Pear blossoms

lor of the filaments and anthers? On
at are they set?

6. How many pistils do you see? How any stigmas are there? Are the ovaries ited? Are they attached to the calyx?

7. Describe the young leaves as they pear around the blossoms. What is their lor? Have they any stipules? Why do ey make the flowers look like a bouquet?

8. After the petals fall, what parts of e blossom remain? What part develops to the apple? Does this part enclose the aries of the pistils? How can you tell the ripe apple if any stigma failed to ceive pollen?

9. What is the position of the calyx bes directly after the petals fall? Do they ange later? How does this affect spraying for the codling moth?

10. Watch an apple develop; look at once a week and tell what parts of the ossom remain with the apple.

11. How many blossoms come from e winter bud? How many leaves? Do e blossoms ever appear along the sides the branches, as in the cherries? How any blossoms from a single bud develop to apples?

M. V. Slingerland

Just ready to spray. A pear and two apples from which the petals have already fallen, with calyx lobes widely spread

12. Since the apple is developed on the tip of the twig, how does the twig keep on growing?

13. Compare the apple with the pear, the plum, the cherry, and the peach in the following particulars: position on the twigs; number of petals; number and color of stamens; number of pistils; whether the pistils are attached to the calyx-cup at the base.

THE APPLE

Man fell with apples and with apples rose,
If this be true; for we must deem the mode
In which Sir Isaac Newton could disclose,
Through the then unpaved stars, the turnpike road,
A thing to counterbalance human woes.

— Byron

Apples seem to have played a very imrtant part in human history, and from e first had much effect upon human stiny, judging from the trouble that ened both to Adam and to Helen of Troy om meddling, even though indirectly, th this much esteemed fruit. It is surely more than just to humanity — shut t from the Garden of Eden — that the ple should have led Sir Isaac Newton discover the law which holds us in e universe; and that, in these later cenries, apples have been developed so beautiful and so luscious as almost to reconcile us to the closing of the gates of Paradise.

While it is true that no two apples were ever exactly alike, any more than any two leaves, yet their shapes are often very characteristic of the varieties. From the big, round Baldwin to the cone-shaped gillyflower, each has its own peculiar form, and also its own colors and markings and its own texture and flavor. Some have tough skins, others bruise readily even with careful handling; but to all kinds, the skin is an armor against those ever-present foes, the fungus spores, myriads of which are floating in the air ready to enter the smallest breach, and by their growth bring about decay. Even the tip of a branch or twig swayed by the wind may bruise an apple and cause it to rot;

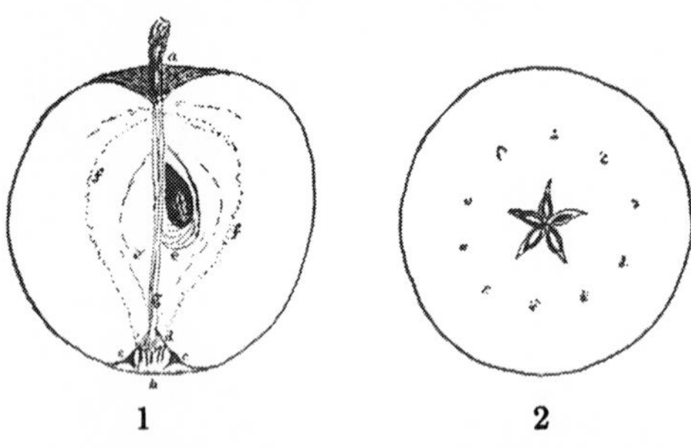

1, *Apple cut along core; 2, Apple cut across core showing the five carpels and the ten outer core lines*

a, cavity; b, basin; c, calyx lobes; d, calyx-tube with withered stamens attached; e, carpels; f, outer core lines, terminating at a point where stamens are attached; g, fibers extending from stem to basin

windfalls are always bruised and will not keep. Greater care in packing, wrapping, picking, and storing, so as to avoid contact with other apples, is a paying investment of labor to the apple grower.

The cavities at the stem and basin ends of the fruit are also likely to have, in the same variety, a likeness in their depth or shallowness, and thus prove a help in identifying an apple. At the blossom or basin end of the fruit may be seen five scales, which are all that remain of the calyx lobes which enclosed the blossom; and within them are the withered and shrunken stamens and styles.

When the fruit is cut, we see that the inner parts differ as much in the different varieties as do the outer parts. Some have large cores, others small. The carpels, or seed cells, are five in number, and when the fruit is cut across through the center these carpels show as a pretty, five-pointed star; in them the seeds lie, all pointing toward the stem. Some apples have both seeds and carpels smooth and shining, while in others they are tufted with a soft, fuzzy outgrowth. The number of seeds in each cell varies; quite often the number is two. If a carpel is emp the apple is often lopsided, and this s nifies that the stigma of that ovary ceived no pollen. The apple seed is ov plump, and pointed, with an outer she and a delicate inner skin covering t white meat; this separates readily into tw parts, between which, at the point, m be seen the germ. The entire core, wi the pulp immediately surrounding t seed cells, is marked off from the rest the pulp by the core lines, faint in so varieties but distinct in others. In o native crab apples this separation so complete that, when the fruit is ri the core may be plucked out leaving globular cavity at the center of t apple.

Extending from the stem to the bas through the center of the apple, is bundle of fibers, five in number, each tached to the inner edge of a carpel, seed box. Other bundles of fibers p through the flesh about halfway betwe the core and the skin. Delicate as th are, so that no one observes them in e ing the fruit, they show clearly as a seco core line, and each terminates at a poi in the calyx-tube where the stamens we attached — as can be easily seen by disse ing an apple. In transverse section, the show as ten faint dots placed oppos each outer point and inner angle of t star at the center formed by the carpe Sometimes the seeds are very close to t stem, and the apple is said to have a sess core; if at the center of the fruit, it ha medium core; if nearest to the blossc end, it has a distant core. This position the core marks different varieties.

Apples, even of the same variety, dif much in yield and quality according the soil and climate in which they gro Varieties of apples are constantly char ing; new varieties are introduced and old varieties are discarded. The Baldwin still the leading variety in New York St but it has a distinct downward trend the newer plantings. Northern Spy a Rhode Island Greening are still holdi their own. In the plantings of recent yea McIntosh and Cortland have been m ular; it is only a question of time un- the McIntosh will lead in New York e.

'oo often in passing through the coun- we see neglected and unprofitable hards, with soil untilled, the trees un- ned and scale-infested, yielding scanty t, fit only for the cider mill and the egar barrel. This kind of orchard must s away and give place to the new horti- ture.

LESSON 184

The Apple

EADING THOUGHT — The apple is a ritious fruit, wholesome and easily di- ted. The varieties of apple differ in pe, size, color, texture, and flavor. A fect apple has no bruise upon it and wormholes in it.

METHOD — Typical blossoms of differ- varieties of apples should be brought the schoolroom, where the pupils y closely observe and make notes about ir appearance. Each pupil should have or two apples that may be cut in tical and transverse sections, so that the p, core lines, carpels, and seeds may be erved. After this lesson there should an apple exhibit, and the pupils should taught how to score the apples accord- to size, shape, color, flavor, and tex- e.

OBSERVATIONS — 1. Sketch the shape of r apple. Is it almost spherical, or flat- ed, or long and egg-shaped, or with equal tapering sides? How does the pe of the apple help in determining its iety?

2. What is the color of the skin? Is it ied by streaks, freckles, or blotches? s it one blushing cheek, the rest being a different color?

3. Is the stem thick and fleshy, or short l knobby, or slender and woody and g? Does each variety have a character- c stem?

4. Is the cavity or depression where the m grew narrow and deep like a tunnel, shallow like a saucer?

5. Examine the blossom end, or basin. What is its shape? Can you find within it the remnants of the calyx lobes, the stamens, and the pistils of the flower?

6. What is the texture of the skin of the apple? Is it thin, tough, waxy, or oily? Has it a bloom that may be rubbed off? From what sort of injury does the skin protect the apple?

EXPERIMENT 1. Take three apples of equal soundness and peel one of them; place them on a shelf. Place one of the un- peeled apples against the peeled one, and the other a little distance from it. Does the peeled apple begin to rot before the other two? Does the unpeeled apple touching the peeled one begin to decay first at the point of contact?

EXPERIMENT 2. Take an apple with a smooth, unblemished skin and vaccinate it with some juice from an apple that has begun to decay; perform the operation with a pin or needle, pricking first the unsound fruit and then the sound one; this may be done in patterns around the apple or with the initials of the operator's name. Where does this apple begin to de- cay? What should these two experiments teach us about the care and storage of fruit?

7. Cut an apple through its center from stem to blossom end. Describe the color, texture, and taste of the pulp. Is it coarse or fine-grained? Crisp or smooth? Juicy or dry and mealy? Sweet or sour? Does it exhale a fragrance or have a spicy flavor?

8. Is the flesh immediately surrounding the core separated from the rest of the pulp by a line more or less distinct? This is called the core line and differs in size and outline in different varieties. Can you find any connection between the stem and blossom ends and the core? Can you see the fibrous threads which connect them?

9. Cut an apple transversely across the middle. In what shape are the seed cells arranged in the center? Do the carpels or seed cells, vary in shape in different varieties? Are they closed, or do they all open into a common cavity? Can you see, between the core lines and the skin, faint

little dots? Count, and tell how they are arranged in relation to the star formed by the core.

10. The stiff, parchment-like walls of the seed cells are called carpels. How many of these does the apple contain? Do all apples have the same number of carpels? Are the carpels of all varieties smooth and glossy, or velvety? How many seeds do you find in a carpel? Do they lie with the points toward the stem end or the blossom end of the apple? Where are they attached to the apple? Describe the apple seed — its outer and inner coat its "meat."

11. Is the core at the center of apple, or is it nearer to the stem end to the blossom end of the fruit? Are apples alike in this particular?

12. Describe fully all the varieties apples which you know, giving the a age size, texture, and color of the skin, shape of the cavities at the stem and b som ends, the color, texture, and flavo the pulp, and the position within the ap of the core.

THE PINE

None other of our native trees is more beautiful than the pine. In the East, we have the white pine with its fine-tasselled foliage, growing often one hundred and fifty to two hundred feet in height and reaching an age of from two to three hundred years. On the Pacific coast, the splendid sugar pine lifts its straight trunk from two to three hundred feet in height; and although the trunk may be from six to ten feet in diameter yet it looks slender, so tall is the tree. A sugar pine cone on my desk measures twenty-two inches in length and weighs almost one pound, although it is dried and emptied of seed.

There is something majestic about the pines, which even the most unimpressionable feel. Their dark foliage outlined against wintry skies appeals to the imagination, and well it may, for it represents an ancient tree costume. The pines are among the most ancient of trees, and were the contemporaries of those plants which were put to sleep, during the Devonian age, in the coal beds. It is because the pines and the other evergreens belong essentially to earlier ages, when the climate was far different from that of today, that they do not shed their leaves like the more recent, deciduous trees. They stand among us, representatives of an ancient race, and wrap their green foliage about them as an Indian sachem does his blanket, in calm disregard of modern fashion of attire.

All cone-bearing trees have typicall central stem from which the branc come off in whorls, but so many thi have happened to the old pine trees t the evidence of the whorls is not v plain; the young trees show this meth of growth clearly, the white pine hav five branches in each whorl. Someti pines are seen which have two or th stems near the top; but this is a story injury to the tree and its later victory.

The very tip of the central stem in evergreens is called "the leader," beca it leads the growth of the tree upward stretches up from the center of the wh of last year's young branches, and there its tip are the buds which produce t year's branches. There is a little bee which appears to be possessed of evil, it seems to like best of all to lay its e in the very tip of this leader; the grub, af hatching, feeds upon the bud and bo down into the shoot, killing it. Th comes the question of which branch the upper whorl shall rise up and take place of the dead leader; but this is election which we know less about th we do of those resulting from our blan ballots. We do know that one branch this upper whorl arises and continues growth of the tree. Sometimes there two candidates for this position, and th each make such a good struggle for place that the tree grows on with t stems instead of one — and sometin

th even three. This evil insect injures
the leaders of other conifers also, but
these are less likely to allow two competitors to take the place of the dead leader.
The lower branches of many of the pines come off almost at right angles from
the bole; the foliage is borne above the
branches, which gives the pines a very
different appearance from that of other
trees. The foliage of most of the pines is
dark green, looking almost black in winter; the pitch pine has the foliage yellowish green, and the white pine, bluish green; each species has its own peculiar shade. There is great variation in the color and form of the bark of different species. The white pine has nearly smooth bark on the young trees, but on the older ones it has ridges that are rather broad, flat, and scaly, separated by shallow sutures, while the pitch pine has its bark in scales like the covering of a giant alligator.

G. F. Morgan

Austrian pine in blossom showing staminate flowers

Verne Morton

White pine in Winter

The foliage of the pine consists of pine needles set in little bundles on raised points which look like little brackets along the twigs. When the pine needles are young, the bundle is enclosed in a sheath making the twig look as if it were covered with pinfeathers. In many of the species this sheath remains, encasing the base of the bundle of needles; but in the white pine it is shed early. The number of leaves in the bundle helps us to determine the species of the tree; the white pine has five needles in each bunch, the pitch pine has three, while the Austrian pine has two. There is a great difference in the length and the color of the needles of different species of pine. Those of the white pine are soft, delicate, and pliable, and from three to four inches in length; the needles of the pitch pine are stiff and coarse and about the same length; the

G. F. Morgan

Austrian pine, staminate blossoms and empty cones

white pine needles are triangular in section, and are set so as to form distinct tassels, while those of the Austrian pine simply clothe the ends of the twigs. The needles of the pine act like the strings of an æolian harp; and the wind, in passing through the tree, sets them into vibration, making a sighing sound which seems to the listener like the voice of the tree. Therefore, the pine is the most companionable of all our trees and, to one who observes them closely, each tree has its own tones and whispers a different story.

The appearance of the unripe cone is another convincing evidence that mathematics is the basis of the beautiful. The pattern of the overlapping scales is intricate and yet regular — to appreciate it one needs to try to sketch it. Beneath each scale, when it opens wide, we find nestled at its base two little seeds; each provided with a little wing so that it can sail off with the wind to find a place to grow. The shape of the scales of the cone is another distinguishing characteristic of the pine, and sketching the outside of scales from several different species of pine cones will develop the pupils' powers of observation; the tip of the scale may be thickened or armed with a spine.

The pine cone requires two years for maturing; the pistillate flower from wh it is developed is a tiny cone with e scale spread wide and standing upright catch the pollen for the tiny ovule nest within it. The pistillate flower of the wh pine grows near the tip of the new tw and is pinkish in color. In the Austr pine it is the merest pink dot at first, after a little shows itself to be a true c with pink-purple scales, which stand very erect and make a pretty object wh viewed through a lens; each scale is p at its three-pointed tip, with pink wi just below, the inner portions being p green. The cone is set just beside the gr ing tip of the twig, is pointed upward, a its sheath scales are turned back like ch around its base.

In June when the new shoots of pine twigs stand up like pale green cand on a Christmas tree, at their bases m be found the staminate catkins set radiating whorls, making galaxies golden stars against the dark green ba ground of foliage. In the Austrian pi

Ralph W. C

Young and mature cones of white pin

ıe of these pollen catkins may be an ch or two long and a half-inch in width; ch little scale of this cone is an anther c, filled to bursting with yellow pollen. :om these starry pollen cones there de-ends a yellow shower when a breeze ısses; for the pine trees depend upon .e wind to sift their pollen dust into the ted cups of the cone scales, which will ose upon the treasure soon. The pollen ains of pine are very beautiful when en through a microscope; and it seems most incredible that the masses of yel-w dust sifted in showers from the pines hen in blossom should be composed of ıese beautiful structures. When the pine rests on the shores of the Great Lakes e in bloom, the pollen covers the waves r miles out from the shores.

If we examine the growing tips of the ine branches, we find the leaves look llow and pinfeathery. The entire leaf wrapped in a smooth, shining, silken ıeath, at the tip of which its green point rotrudes. The sheath is tough like parch-ıent and is cylindrical, because the pine eedles within it are perfectly adjusted

R. E. Horse

White pines

one to another in cylindrical form. The sheath is made up of several layers, one over the other, and may be pulled apart. The new leaves are borne on the new, pale green wood.

The uses of pines are many. The lumber of many of the species, especially that of the white pine, is free from knots and is used for almost everything from house-building to masts for ships. In the southern states, the long-leaved pines are tapped for resin, which is not the sap of the tree, as is generally supposed. Pine sap is like other sap; the resin is a product of certain glands of the tree, and is of great use to it in closing wounds and thus keeping out the spores of destructive fungi. It is this effort of the tree to heal its wounds that makes it pour resin into the cuts made

Ralph W. Curtis

Vhite pine *Norway spruce*
'itch pine *Hemlock*

U. S. Geological Survey, G. K. Gilbert

Yellow pine on the brink of the Little Yosemite Valley

Leonard K. Beyer

Seminole boy in a cypress dugout, Everglades, Florida

by the turpentine gatherers. This resin is taken to a distillery, where the turpentine is given off as a vapor and condensed in a coiled tube which is kept cold. What is left is known as "rosin."

LESSON 185

THE PINE

LEADING THOUGHT — The pines are among our most ancient trees. Their foliage is evergreen but is shed gradually. The pollen-bearing and the seed-producing flowers are separate on the tree. The seeds are winged and are developed in cones.

METHOD — At least one pine tree should be studied in the field. Any species will do, but the white pine is the most interesting. The Austrian pine which is commonly planted in parks is a good subject. The leaves and cones may be studied in the schoolroom, each pupil having a specimen.

OBSERVATIONS — 1. What is the general shape of the pine tree? Is there one central stem running straight up through the center of the tree to the top? Do you find any trees where this stem is divided into two or three near the top? Descri how the pine tree grows. What is t "leader"? What happens if the leader injured? How do the topmost branch of the young pine look? Do they all co off from the same part of the stem? H many are there in a whorl?

2. What color is the bark? Is it ridg or in scales?

3. Do the branches come off the ma stem at right angles or do they lift or droop down? Where is the foliage bor on the branches? What is the color of t foliage? Is the pine foliage ever shed, does the pine leaf, when it comes, stay as long as the tree lives?

4. Study the pine leaves. Why are th called needles? Note that they grow s eral together in what we call a bund How many in one bundle? Is the bun enclosed in a little sheath at the bas Are the bundles grouped to make distin tassels? Study one of the needles. H long is it? Is it straight or curved? Flexib or coarse and stiff? Cut it across and exa ine it with a lens. What is the outline cross section? Why does the wind ma a moaning sound in the pines?

White-pine cone

. Study a pine cone. Does it grow r the tip of the branch or along the s? Does it hang down or stand out ly? What is its length? Sketch or de- be its general shape. Note that it is le up of short, overlapping scales. at pattern do the scales make as they set together? Describe or sketch one e; has it a thickened tip? Is there a e at the tip of the scale?

. Where in the cone are the seeds? cribe or sketch a pine seed. How long s wing? How is it carried and planted? en the cone opens, how are the seeds tered? What creatures feed upon the e seed?

. Study the pine when in blossom, ch is likely to be in June. This time asily determined because the air around tree is then filled with the yellow pol- dust. Study the pollen-bearing flower. t conelike in form? Does it produce a at deal of pollen? If you have a micro- pe, look at the pollen through a high ective and describe it. How many of pollen catkins are clustered together? what part of the twigs are they borne? Where are the pistillate flowers which are to form the young cones? How large are they and how do they look at the time the pollen is flying? Do they point upward or droop downward? Look beneath the scales of a little cone with a lens and see if you can find the flowers. What is it that carries the pine pollen to the flowers in the cone?

8. Name all the uses for pine lumber that you know. Write an English theme on how turpentine is produced from pines and the effect of this industry upon pine forests. Where does resin appear on the pine? Of what use is it to the tree? Do you think it is pine sap? What is the difference between resin and rosin?

9. How long do the pine trees live? Write a story of some of the changes that have taken place in your neighborhood since the pine tree which you have been studying was planted.

10. Make the following drawings: a bundle of pine needles showing the sheath and its attachment to the twig; the cone; the cone scale; the seed. Sketch a pine tree.

THE NORWAY SPRUCE

he Norway spruce is a native of Eu- e, and we find it in America one of most satisfactory of all spruces for amental planting; it lifts its slender e from almost every park and private te in our country, and is easily distin- shed from all other evergreens by the oping, pendant habit of its twigs, ch seem to hang down from the ight, uplifted branches. We have ices of our own — the black, the white, the red spruces; and it will add much he interest of this lesson for the pupils ead in the tree and forestry books con- ning these American species. Chewing n and spruce beer are the products of black and red spruce of our eastern sts. The Douglas spruce, which is a and not a spruce, is also commonly nted as an ornamental tree, but it is only at its best on the Pacific Coast, where it is one of the most magnificent of trees.

The Norway spruce tree is in form a beautiful cone, slanting from its slender tip to the ground, on which its lower drooping branches rest; the upper branches come off at a narrower angle from the sturdy central stem than do the widespreading lower branches. On the older trees, the twigs hang like pendulous fringes from the branches, enabling them to shed the snow more readily — a peculiarity which is of much use to the tree, because it is a native of the snowy northern countries of Europe and also grows successfully in the high altitudes of the Alps and other mountains. If we stroke a spruce branch toward the tip, the hand slides smoothly over it; but brush backward from the tip, and the hand is pricked

Norway spruce

by hundreds of the sharp, bayonet-pointed leaves; this arrangement permits the snow to slide off.

If we examine a twig of the present year's growth, we can see on every side of its brown stem the pointed leaves, each growing from a short ridge; but the leaves on the lower side stretch out sidewise, and those above lift up angularly. Perhaps the twig of last year's growth has shed its leaves which grew on the underside and thus failed to reach the sun. The leaf of the spruce is curved, stiff, and four-sided, and ends in a sharp point. It is dark yellowish above and lighter beneath and is set stiffly on the twig. The winter buds for next year's growth may be seen at the tips of the twigs, covered with little, recurved, brown scales quite flower-like in form. In the balsam fir, which is often planted with the Norway spruce, these buds are varnished.

The cones are borne on the tips of the branches and hang down. In color they are pale wood-brown; they are from four to six inches long, and are very conspicuous. They are made up of broad sc that are thin toward the notched tips; t are set around the central stem in spi of five rows. If we follow one spiral aro marking it with a winding string, it prove to be the fifth row above the pl where we started. These manifold spi can be seen sometimes by looking into tip end of a cone. The cone has m resin on it, and is a very safe place seeds; but when it begins to open, sq rels impatiently tear it to pieces, harv ing the seeds and leaving a pile of c scales beneath the tree to tell of th piracy.

A Norway spruce in blossom is a bea ful sight; the little, wine-red pistil cones are lifted upwards from the tips the twigs, while short terminal branc are laden with the pollen-bearing catk which are soft and caterpillarish, grow on soft, white stems from the base scales which enclosed and protected th during the winter; these catkins are fil with the yellow dust. The young co continue to stand upright after the sc have closed on the pollen which has b sifted by the wind to the ovules at base of the scales; and for some time t remain most ornamentally purplish Before the cone is heavy enough to b from its own weight, it turns around downward, and then changes its colo green, ripening into brown in the fal

The Norway spruce grows on the A

G. F. M

Staminate blossoms and young cone of a N way spruce

ındantly, and like the youth with the ıner, "excelsior" is its motto; this apırs even in its scientific name (*Picea ex-sa*). Here it grows to the height of one ndred to one hundred and fifty feet. Its od is valuable and its pitch is marketed. this country, it is used chiefly for ornantal planting and for windbreaks.

LESSON 186

The Norway Spruce

Leading Thought — The Norway uce is one of the most valuable of the es which have come to America from rope. It grows naturally in high places d in northern countries where there is ıch snow; its drooping twigs cannot ld a great burden of snow, and thus it apes being crushed.

Method — This lesson should begin in autumn when the cones are ripe. The

L. H. Bailey

Cones of Norway spruce

tree should be observed by all of the pupils, and they should bring in twigs and cones for study in the schoolroom. The

Cyrus Crosby

A cone of Norway spruce, showing that the spiral of the scales is in rows of five

lesson should be taken up again in May when the trees are in blossom.

Observations — 1. What is the general shape of the tree? Do the lower branches come off at the same angle as the upper? If untrimmed, what can you see of the trunk? Do the lower branches rest upon the ground? What advantage would this be to the tree in winter? Do the twigs stand out, or droop from the branches? Of what advantage is this in case of heavy snow? What is the color of the foliage? Where did the Norway spruce come from?

2. What is the color of the twig? How are the leaves set upon it? Are there more leaves on the upper than on the under side of the twigs of this year's growth? Of last year's growth? Brush your hand along a branch toward the tip. Do the leaves prick? Brush from the tip backward. Is the result the same? Why is this angle of the leaves to the twig a benefit during snowstorms?

3. Take a single leaf. What is its shape? How many sides has it? Is it soft or stiff? Is it sharp at the tip? Describe the buds which are forming for next year's growth. Look along the twigs and see if you can

discover the scales of the bud which produced last year's growth.

4. Where are the cones borne? How long does it take a cone to grow? Is it heavy? Is there resin on it? Note that the scales are set in a spiral around the center of the cone. Wind a string around a cone following the same row of scales. How many rows between those marked with a string? Look into the tip of a cone and see the spiral arrangement. Sketch and describe a cone scale, paying special attention to the shape of the tip. Try to tear a cone apart. Is this easily done? Hang a closed cone in a dry place and note what happens.

5. Describe the seed, its wings, and where it is placed at the base of the scale. How many seeds under each scale? When do the cones open of themselves to scatter the seed? Do you observe squirrels tearing these apart to get the seed?

6. The Norway spruce blossoms in May. Find the little flower which will produce the cone, and describe it. What color is it? Is it upright or hanging down? Do the scales turn toward the tip or backward? Why is this? Where are the pollen catkins borne? How many of them arise from the same place on the twig? Can you see the little scales at the base of each pistillate catkin? What are they? Are they very full of pollen? Do the insects carry the pollen for the Norway spruce, or does the wind sift it over the pistillate blossoms? After the pollen is shed, note if the scales of the young cones close up. How long before the cones begin to droop?

7. What use do we make of the Norway spruce? What is it used for in Europe?

All outward wisdom yields to that with
Whereof nor creed nor canon holds t
key;
We only feel that we have ever been
And evermore shall be.

And thus I know, by memories unfur
In rarer moods, and many a name
sign,
That once in Time, and somewhere in t
world,
I was a towering pine.

Rooted upon a cape that overhung
The entrance to a mountain go
whereon
The wintry shade of a peak was flung,
Long after rise of sun.

There did I clutch the granite with fi
feet,
There shake my boughs above the ro
ing gulf,
When mountain whirlwinds through t
passes beat,
And howled the mountain wolf.

There did I louder sing than all the flo
Whirled in white foam adown the pr
ipice,
And the sharp sleet that stung the na
woods,
Answer with sullen hiss.

I held the eagle till the mountain mist
Rolled from the azure paths he came
soar,
And like a hunter, on my gnarled wris
The dappled falcon bore.

— From "THE SPIRIT OF THE PIN
BAYARD TAYL

THE HEMLOCK

O'er lonely lakes that wild and nameless lie,
Black, shaggy, vast and still as Barca's sands
A hemlock forest stands. Oh forest like a pall!
Oh hemlock of the wild, Oh brother of my soul,
I love thy mantle black, thy shaggy bole,
Thy form grotesque, thy spreading arms of steel.

— PATTEE

In its prime, the hemlock is a magnifi-nt tree. It reaches the height of from ty to one hundred feet and is cone-aped. Its fine, dense foliage and droop-g branches give it an appearance of ex-uisite delicacy; and I have yet to see else-here such graceful tree-spires as are the mlocks of the Sierras, albeit they have nding tips. However, an old hemlock comes very ragged and rugged in appear-ce; and dying, it rears its wind-broken anches against the sky, a gaunt figure of ark loneliness.

The hemlock branches are seldom oken by snow; they droop to let the bur-n slide off. The bark is reddish, or some-nes gray, and is furrowed into wide, aly ridges. The foliage is a rich dark een, but whitish when seen from below. he leaves of the hemlock are really ar-nged in a spiral, but this is hard to dem-nstrate. They look as though they were ranged in double rows along each side of e little twig; but they are not in the me plane and there is usually a row of ort leaves on the upper side of the twig. he leaf is blunt at the tip and has a lit-: petiole of its own which distinguishes from the leaves of any other species of nifer; it is dark, glossy green above, pale een beneath, marked with two white, ngthwise lines. In June, the tip of every rig grows and puts forth new leaves hich are greenish yellow in color, mak-g the tree very beautiful and giving it e appearance of blossoming. The leaves e shed during the third year. The hem-ck cones are small and are borne on the s of the twigs. The seeds are borne o beneath each scale, and they have wings nearly as large as the scale itself. Squirrels are so fond of them that proba-bly but few have an opportunity to try their wings. The cones mature in one year, and usually fall in the spring. The hem-lock blossoms in May; the pistillate flow-ers are very difficult to observe, as they are tiny and greenish and are placed at the tip of the twig. The pollen-bearing flowers are little, yellowish balls on delicate, short stems, borne along the sides of the twig.

Ralph W. Curtis

Hemlock branch showing young and mature cones

Hemlock bark is rich in tannin and is used in great quantities for the tanning of leather. The timber, which is coarse-grained, is stiff and is used in framing buildings and for railroad ties; nails and spikes driven into it cling with great te-nacity and the wood does not split in nail-ing. Oil distilled from the leaves of hem-lock is used as an antiseptic.

The dense foliage of the hemlock offers a shelter to birds of all kinds in winter; even the partridges roost in the young trees. These young trees often have branches drooping to the ground, making an evergreen tent which forms a winter harbor for mice and other beasties. The seed-eating birds which remain with us during the winter feed upon the seeds; and as the cones grow on the tips of the delicate twigs, the red squirrels display their utmost powers as acrobats when gathering this, their favorite food.

LESSON 187

The Hemlock

Leading Thought — This is one of the most common and useful and beautiful of our evergreen trees. Its fine foliage makes it an efficient winter shelter for birds.

Method — Ask the children the questions and have them make notes on the hemlock trees of the neighborhood. The study of the leaves and the cones may be made in the schoolroom.

Observations — 1. Where does the hemlock tree grow in your neighborhood? What is the general shape of the tree? What sort of bark has it? How tall does it grow? How are its branches arranged to shed the snow?

2. What is the color of the foliage? How are the leaves arranged on the twigs? Are all the leaves of about the same siz What is the position of the smaller leav

3. Break off a leaf and describe shape; its petiole. Does the leaf of a other evergreen have a petiole? What the color and marking of the hemlock le above? Below? At what time of year a the new leaves developed? How does t hemlock tree look at this time? Does t hemlock ever shed its leaves?

4. Are the hemlock cones borne on t tip of the twigs or along the side? Ho long does it take a cone to mature? Wh does it fall? How many scales has Where are the seeds borne? How ma seeds beneath each scale? Describe a sketch a hemlock seed. How are the see scattered? Study the tree in May, and s if you can find the blossom.

5. Make drawings of the following: t hemlock twig, showing the arrangeme of the leaves; single leaf, enlarged; co cone scale; seed.

6. What creatures feed upon the he lock seed? What birds find protection the hemlock foliage in winter?

7. For what purposes is hemlock ba used? What is the timber good for? Is nail easily pulled out from a hemlo board?

THE DOGWOOD

Through cloud rifts the sunlight is streaming in floods to far depths of the wood
Retouching the velvet-leafed dogwood to crimson as vital as blood.

There is no prettier story among the flowers than that of the bracts of the dogwood, and it is a subject for investigation which any child can work out for himself. I shall never forget the thrill of triumph I experienced when I discovered for myself the cause of the mysterious dark notch at the tip of each great white bract, which I had for years idly noticed. One day my curiosity mastered my inertia, and I hunted a tree over for a flower bud, for it was rather late in the season; finally I was rewarded by finding the bracts in all stages of development.

The flowering dogwood forms its bu during the summer, and of course th must have winter protection. They a wrapped in four close-clasping, purpli brown scales, one pair inside and one pa outside, both thick and well fitted to pr tect the bunch of tiny flower buds at the center. But when spring comes, these b scales change their duties, and by rap growth become four beautiful white pinkish bracts which we call the dogwoo flower. For months these bracts cover th true flowers which are at their center a then display them to an admiring worl artistic eye loves the little notch at tip of the bracts, even before it has in it the story of winter protection, hich it is an evidence.

he study of the flowers at the center is e interesting if aided by a lens. Within blossom can be seen its tube, set in four-lobed calyx. It has four slender ls curled back, its four chubby, green-yellow anthers set on filaments which them up between the petals; and at center of all is the tiny green pistil.

Hugh Spencer

Dogwood

re may be twenty, more or less, of se perfect flowers in this tiny, greenish ow bunch at the center of the four t, flaring bracts. These flowers do not n simultaneously, and the yellow buds open flowers are mingled together in rosette. The calyx shows better on the than on the open flower. It might be l to explain to the pupils that a bract is ply a leaf in some other business than t ordinarily performed by leaves.

he twigs have a beautiful, smooth k, purplish brown above and greenish ow. The flowers grow at the tips of the gs; and the young leaves are just below flowers and also at the tips of the gs. These twigs are spread and bent in eculiar way, so that each white flower-d may be seen by the admiring world and not be hidden behind any of its neighbors. This habit makes this tree a favorite for planting, since it forms a mass of white bloom.

The dogwood banners unfurl before the flowers at their hearts open, and they re-

Blossom and bud of dogwood enlarged

main after the last flower has received within itself the vital pollen which will enable it to mature into a beautiful berry. This long period of bloom is another quality which adds to the value of the dogwood as an ornamental tree. At the time the bracts fall, the curly petals also fall out leaving the little calyx-tubes standing with style and stigma projecting from their centers, making them look like a bunch of lilliputian churns with dashers. In autumn, the foliage turns to a rich, purplish crimson — a most satisfying color.

Charles E. Mohr

Dwarf cornel or bunchberry — a dwarf dogwood

During the winter, the flowering dogwood, which renders our forests so beautiful in early spring, may be readily recognized by its bark, which is broken up into small scales and mottled like the skin of

The flower buds of the dogwood are formed during the previous season

a serpent; and on the tips of its branches are the beautiful clusters of red berries, or speaking more exactly, drupes. This fruit is oval, with a brilliant, shining, red, pulpy covering which must be attractive to birds. At its tip it has a little purple crown, in the center of which may be seen the remnant of the style, but this attractive outside covers a seed with a very thick, hard shell, which is quite indigestible and fully able to protect, even from the attack of the digestive juices of the bird's stomach, the tender white kernel within it, which includes the stored food and the embryo. There are in the North other common species of dogwood which have dark blue fruit.

LESSON 188

The Dogwood

Leading Thought — The real petals of the dogwood are not the chief means of attracting insects to its flowers. The showy portions are really bracts and the true flower.

Method — Observe a branch of dogwood when it is in flower. The bra should have upon it some flowers that unopened. Study the flower first, and the pupils to discover for themselves the great white bracts have a notch in tip. A lens is a great help to the inte in studying these tiny flowers.

Observations — 1. What is there the center of the group of bracts? How the parts at the center look? Are t of the same shape? Are some opened others not? Can you see how many pe this tiny flower has? Describe its ca How many stamens has it? Can you the pistil? If a flower has a calyx and mens and a pistil, has it not all tha flower needs?

2. How many of these flowers are th at the center of what is often called dogwood "blossom"? What color they? Would they show off much i were not for the great white bann around them? Do we not think of th great white bracts as the dogwood flow

3. Study one of these banners. Wha its shape? Are the four white bracts same shape and size? Make a sketch these four bracts with the bunch of fl ers at the center. What is there pecu about each one of these white bra Find one of the flower-heads which is yet opened and watch it develop int small flower.

4. Sketch the bracts from below. Is pair wider than the other? Is the wi pair inside or outside?

5. Where are the flowers of the d wood borne? How are the twigs arrang so as to unfurl all the banners and hide one behind another, so that whole tree is a mass of white?

6. While studying the flowers, stu where the young leaves come from. C you still see the scales which protected t leaf buds?

7. What kind of fruit develops fr the dogwood blossoms? What colors its leaves in autumn?

THE VELVET OR STAGHORN SUMAC

The sumacs with flame leaves at half-mast, like wildfire spread over the glade;
Above them, the crows on frayed pinions move northward in ragged parade.

Ihe sumacs, in early autumn, form a ring line" along the borders of wood-ids and fences, before any other plant t the Virginia creeper has taken on ghter colors. No other leaves can emu-e the burning scarlet of their hues. The nacs are a glory to our hills; and some-ie, when Americans have time to culti-e a true artistic sense, these shrubs will y an important part in landscape gar-ning. They are beautiful in summer, en each crimson "bob" (a homely w England name for the fruit panicle) set at the center of the bouquet of eading, fernlike leaves. In winter naked-ss they are most picturesque, with their adly branching twigs bearing aloft the ne-colored pompons against the back-und of snow; at this time and in early ing when more desirable food is lack-, the birds eat the pleasantly acid ipes. In spring, they put out their soft ves in exquisite shades of pale pinkish en, and when in blossom their stami-e panicles of greenish white cover m with loose pyramids of delicate om.

Well may it be called velvet sumac, for s year's growth of wood and the leaf ms are covered with fine hairs, pinkish first, but soon white; if we slip our fin-s down a branch, we can tell, even with-looking, where last year's growth be- and ended, because of the velvety feel. e name staghorn sumac is just as fit-g, for its upper branches spread widely a stag's horns and, like them, the new wth is covered with velvet.

The leaves are borne on the new wood, l therefore at the ends of branches; y are alternate; the petiole broadens ere it clasps the branch, making a sery for the next year's bud, which is tled below it. The leaves are com-ind and the number of leaflets varies from eleven to thirty-one. Each leaflet is set close to the midrib, with a base that is not symmetrical; the leaflets have their edges toothed, and are long and narrow; they do not spread out on either side of the midrib like a fern, but naturally droop somewhat, and thus conceal their under-sides, which are much lighter in color. The leaflets are not always set exactly

Verne Morton

Sumac

Verne Morton

The staghorn sumac

opposite; the basal ones are bent back toward the main stem, making a fold in the base of each. The end leaflets are not always three, symmetrically set, but sometimes are two and sometimes one, with two basal lobes.

The wine-colored "bob" is cone-shaped, but with a bunchy surface. Remove all the drupes from it and note its framework of tiny branches, and again pay admiring tribute to nature's way of doing up compact packages. Each fruit is a drupe, as is also the cherry. A drupe is merely a seed within a fleshy layer, all being enclosed in a firmer outside covering; here, the outside case is covered with dark red fuzz, a clothing of furs for winter, the fur standing out in all directions. The fleshy part around the seed has a pleasantly acid taste, and one of my childhood diversions was to share these fruits in winter with the birds. I probably inadvertently ate also many a little six-footed brother hidden away for winter safe-keeping, for every sumac panicle is a crowded insect tenement.

It is only in its winter aspect that we can see the peculiar way of the sumac's branching, which is in picturesque zigzags, ending with coarse, widespreading twigs. Each terminal twig was a stem for the bouquet of blossom and fruit set about with graceful leaves, but in the winter, after the leaves have fallen, the coarse branching is very noticeable. The wood of the sumac has a pith, and is coarse in texture.

During late May the new growth st near the end of last year's twig; the b are yellowish and show off against dark gray twigs. From the center of th buds comes the fuzzy new growth, wh is usually reddish purple; the tiny lea are folded, each leaflet creased at its m rib and folded tightly against itself; as leaves unfold, they are olive-green tin with red, and look like tassels coming around the old dark red "bob." When sumacs are in blossom, we see in ev group of them two kinds; one with py mids of white flowers, and the other w pinkish callow bobs. The structure these two different flower-clusters is re the same, except that the white ones looser and more widely spread. E flower of the white panicle is stamin and has five greenish, somewhat h sepals and five yellowish white petals, the center of which are five large anth

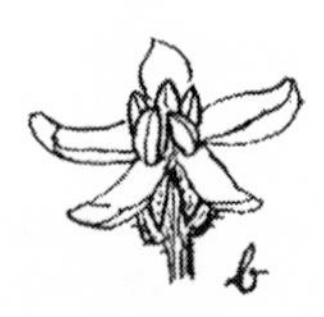

a, Pistillate flower from a "bob." b, Staminate flower from the greenish panicle

A flower from the bob is quite differe it has the five hairy sepals alternating w five narrow, yellowish white petals, b clasping the globular base or ovary, wh is now quite covered with pinkish plu and bears at its tip the three styles flar into stigmas.

The velvet sumac is larger than smooth species (*Rhus glabra*), and easily distinguished from it, since the r wood of the latter is smooth and cove with bloom but is not at all velvety. T poison sumac (page 514), dangerous many people when handled, is a swa species and its fruit is a loose, droop panicle of whitish berries, very much that of poison ivy; therefore, any su at has the red bob is not dangerous. The ison species has the edges of its leaflets tire and each leaflet has a distinct peti- of its own where it joins the midrib.

There is much tannin in sumac and it is ed extensively to tan leather. The bobs used for coloring a certain shade of own. The famous Japanese lacquer is ade from the juice of a species of mac.

LESSON 189

The Velvet or Staghorn Sumac

Leading Thought — The sumac is a autiful shrub in summer because of its nlike leaves; it is picturesque in winter, d its colors in autumn are most bril- nt. Its dark red fruit clusters remain on it during the entire winter. In June shows two kinds of blossoms on dif- ent shrubs; one is whitish and bears the llen, the other is reddish and is a pistil- e flower, later developing into the seed the "bob," or fruit cluster.

Method — Begin this study in Octo- r when the beautiful autumn color of e leaves attracts the eye. Observations to made in the field should be outlined d should be answered in the field note- oks. The study of the fruit and leaf may made in the schoolroom, and an in- rest should be developed which will lead the study of the interesting flowers the llowing spring. The sumacs in autumn ake a beautiful subject for water-color etches, and their peculiar method of anching with their dark red seed clus- rs or bobs makes them excellent subjects r winter sketching.

Observations — 1. Why is this called e velvet sumac? Why is it called the aghorn sumac? Look at the stems with lens and describe the velvet. Can you ll this year's wood by the velvet? Is there ny velvet on last year's wood? Is there any the wood below? What is there pecul- in the appearance of last year's wood? hat are the colors of the hairs that ake the velvet on this year's growth? On st year's growth? What is the color of is year's growth under the velvet? here are the leaves borne?

2. Look at the leaves. How many come off the stem between two, one of which is above the other? Is the midrib velvety? What is its color at base and at tip? What is the shape of the petiole where it joins the stem? Remove the leaf. What do you find hidden and protected by its broad base?

3. How many leaflets are there on the longest leaf which you can find? How many on the shortest? Do the leaflets have little petioles, or are they set close to the midrib? How does the basal pair differ from the others? Are the leaflets the same color above as below? Are the pairs set exactly opposite each other? Look at the three leaflets at the tips of several leaves and see if they are all regular in form. Draw a leaflet, showing its base, its veins, and its margin. Draw an entire leaf, and color it as accurately as possible.

4. Study the fruit. Pick one of the bobs and note its general shape. Is it smooth or bunchy? Sketch it. Remove one of the little bunches and find out why it is of that shape. Remove all of the seeds from one of last year's bobs and see how the fruit is borne. Sketch a part of such a bare stem.

5. Take a single fruit; look at it through a lens and describe it. What are the colors? Cut or pare away the flesh, and describe the seed. What birds live on the sumac seeds in winter? How many kinds of insects can you find wintering in the bob? Find a fruit free from insects and taste it.

Winter Work — 6. Study the sumac after the leaves have fallen and sketch it. What is there peculiar in its branching? Of what use to the plant is its method of branching? Break a branch and look at the end. Is there a pith? What color are the wood and pith?

May or June Work — 7. Where on the branch does the new growth start? How are the tiny leaves folded? Look over a group of sumacs and see if their blossoms all look alike. Are the different kinds of blossoms found on the same tree or on different trees? Take one of the white pyramidal blossom clusters; look at one of these flowers with a lens and describe its sepals and petals. How many anthers has it and where are they? This is a pollen-bearing flower and has no pistil. How are its tiny staminate flowers arranged on the stem to give the beautiful pyramid shape? This kind of flower cluster is called a panicle.

8. Take one of the green bobs and see if it is made up of little round flowers. Through a lens study one of these. How many sepals? How many petals? Describe the middle of the flower around which the petals and sepals clasp. Is this the ovary, or seed box? Can you see the stigmas protruding beyond it? What inse visit these flowers?

9. How can you tell the velvet or st horn sumac from the smooth sumac? H can you tell both of these from the pois sumac?

10. To what uses are the sumacs put

I see the partridges feed quite exte sively upon the sumach berries, at my c house. They come to them after eve snow, making fresh tracks, and have n stripped many bushes quite bare.

— THOREAU'S JOURNAL, Feb. 4, 18

THE WITCH HAZEL

In the dusky, somber woodland, thwarting vistas dull and cold,
Thrown in vivid constellations, gleam the hazel stars of gold,
Gracious gift of wealth untold.

Hazel blossoms brightly glowing through the forests dark and drear,
Work sweet miracles, bestowing gladness on the dying year,
Joy of life in woods grown sere.

Witch hazel is not only a most interesting shrub in itself, but it has connected with it many legends. From its forked twigs were made the divining rods by which hidden springs of water or mines of precious metals were found, as it was firmly believed that the twig would turn in the hand when the one who held it passed over the spring or mine. At the present day, its fresh leaves and twigs are used in large quantities for the distilling of the healing extract so much in demand as a remedy for cuts and bruises and for chapped or sunburned skins. It is said that the Oneida Indians first taught the white people concerning its medicinal qualities.

The witch hazel is a large shrub, usually from six to twelve feet high, although under very advantageous circumstances it has been known to take a treelike form and attain a height of more than twenty feet. Its bark is very dark grayish brown, smooth, specked with little dots, which are the lenticels, or breathing-pores. If the season's growth has been rapid, the new twigs are lighter in color, but wh stunted by drouth or poor soil, the ne growth has a tint similar to the old. T wood is white, very tough and fibro with a pith or heartwood of softer su stance and yellow in color. The leav are alternate, and the leaf buds appe at the tips of the season's twigs, while t blossoms grow at the axils of the leav

The witch-hazel leaf is nearly as bro as it is long, bluntly pointed at its ti with a stem generally less than one-ha inch in length. The sides are unequal size and shape, and the edges are rough scalloped. The veins are almost straigh and are depressed on the upper side b very prominent beneath, and they a lighter in color than the rest of the lea Witch-hazel leaves are likely to be apa ment houses for insects, especially t insects that make galls. Of these the are many species, each making a diff ently shaped gall. One of the most co mon is a gall shaped like a little horn spur on the upper side of the leaf and ha a tiny door opening on the underside :he leaf. If one of these snug little ıes is torn open, it will be found occu- l by a community of little aphids, or ıt lice.

'he witch-hazel blossoms appear at the of a leaf or immediately above the scar ı which a leaf has fallen, the season loom being so late that often the bush are of leaves and is clothed only with yellow, fringelike flowers. Usually the ers are in clusters of three, but occa- ally four or five can be found on the e very short stem. The calyx is four- :d, the petals are four in number, ›ed like tiny, yellow ribbons, about -half inch long and not much wider ı a coarse thread. In the bud, these ıls are rolled inward in a close spiral, a watch spring, and are coiled so tly that each bud is a solid little ball arger than a bird shot. There are four ıens lying between the petals, and be- ɜn each two of these stamens is a lit- cale just opposite the petal. The an- s are most interesting. Each has two little doors which fly open, as if by magic springs, and throw out the pollen which clings to them. The pistil has two stig-

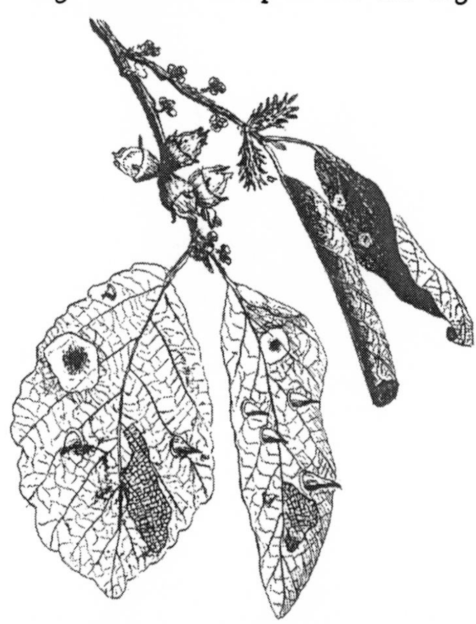

Leaves, fruits, and blossoms of witch hazel. Various types of galls, and the work of miners can be seen on the leaves

Ralph W. Curtis

soms, leaves, and fruits of witch hazel

mas, which are joined above the two-celled ovary within which the seeds develop. The blossoms sometimes open in late September, but the greater number appear in October and November. They are more beautiful in November after the leaves have fallen, since these yellow, starry flowers seem to bring light and warmth into the landscape. After the petals fall, the calyx forms a beautiful little urn, holding the growing fruit.

The nuts seem to require a sharp frost to separate the closely joined parts; it requires a complete year to mature them. One of these nuts is about half an inch long and is covered with a velvety green outer husk, until it turns brown; cutting through it discloses a yellowish white inner shell, which is as hard as bone; within this are the two brown seeds each ornamented with a white dot; note particularly that these seeds lie in close-fitting cells. The fruit, if looked at when the husk is opening, bears an odd resemblance to a grotesque monkey-like face with staring eyes. Frosty nights will open the husks, and the dry warmth of sunny days or of the heated schoolroom will cause the edges of the cups which hold the seeds to

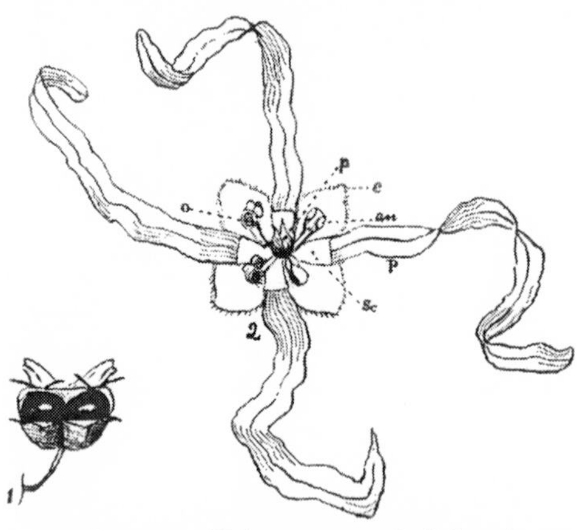

1, *A queer little face — witch-hazel nut ready to shoot its seeds.* 2, *Enlarged flower of witch hazel showing the long petals*

p, with dotted line, the pistil; an, anther; o, anther with doors open; c, lobes of calyx; sc, scale opposite the base of petal

curve inward with such force as to send the seeds many feet away; ordinarily they are thrown from ten to twenty feet, but Hamilton Gibson records one actual measurement of forty-five feet. The children should note that the surface of the seeds is very polished and smooth, and the way they are discharged may be likened to that by which an orange seed is shot from between the fingers.

LESSON 190

The Witch Hazel

Leading Thought — The witch hazel blossoms during the autumn, and thus adds beauty to the landscape. It has an interesting mechanism by which its seeds are shot for a distance of many feet.

Method — This lesson divides naturally into two parts; a study of the way the seeds are distributed is fitted for the primary grades, and a study of the flower for more advanced grades. For the primary grades the lesson should begin by the gathering of the twigs which bear the fruit. These should be brought to the schoolroom — there to await results. Soon seeds will be popping all over the sch room, and then the question as to h this is done, and why, may be made topic of the lesson. For the study of flower and the shrub itself, the w should begin in October when the b soms are still in bud. As they expand t may be studied, a lens being necessary observing the interesting little doors the anthers.

Observations — 1. Is the witch ha a shrub or a tree?

2. What is the color of the bark? I thick or thin, rough or smooth, dark light, or marked with dots or lines? there any difference in color between older wood and the young twigs? Is wood tough or brittle? Dark or light color?

3. Do the leaves grow opposite e other or alternate? On what part of plant do the leaf buds grow?

4. What is the general shape of

G. F. Mc

Flowers and fruit of witch hazel

f? Is it more pointed at the base or at tip? Are the leaves regular in form, or ger on one side than the other? Are the ges entire, toothed, or wavy? Are the ioles short or long? Are the veins aight or branching? Are they promi-nt? Are the leaves of the same color on th sides?

;. Are there many queerly shaped little ellings on the leaf above and below? how many of these you can find. Tell at you think they are.

. Do the flowers grow singly or in clus-s? What is the shape and color of the als, and how many of them are there each blossom? Describe the calyx. If re are any flower buds just opening, serve and describe the way the petals folded within them.

7. How many stamens? With a lens serve the way the two little doors to the ther fly open; how is the pollen thrown out? What is the shape of the pistil? How many stigmas?

8. Does each individual flower have a stem or is there a common stem for a cluster of blossoms? Do the flowers grow at the tips or along the sides of the twigs? When do the witch-hazel flowers appear and how long do they last?

9. Make a drawing of a witch-hazel nut before it opens. What is the color of the outer husk when ripe? Cut into a closed nut and observe the extreme hardness and strength of the inner shell.

10. Where are the seeds situated? Can you see that the shell, when partially open, ready to throw out the seeds, resembles a queer little face? Describe the color and marking of the seeds; are they rough or smooth? How far have you known the witch hazel to throw its seeds? Study the nut and try to discover how it throws the seeds so far.

THE MOUNTAIN LAUREL

As a child I never doubted that the rel wreaths of Grecian heroes were de from mountain laurel, and I sup-sed, of course, that the flowers were d also. My vision was of a hero crowned h huge wreaths of laurel bouquets, ich I thought so beautiful. It was a ock to exchange this sumptuous head-r of my dreams for a plain wreath of ves from the green bay tree.

However, the mountain laurel leaf is rgreen and beautiful enough to crown ictor; in color it is a rich, lustrous green ove, with a yellow midrib, the lower e being of a much lighter color. In pe, the leaf is long, narrow, pointed at h end and smooth-edged, with a rather ort petiole. The leaves each year grow the new wood, which is greenish and igh, in contrast with the old wood, ich is rich brownish red. The leaves are anged below the flower-cluster, so that y make a shining green base for this tural bouquet.

The flowers grow on the tips of the

W. C. Muenscher

Mountain laurel

branching twigs, which are huddled together in a manner that brings into a mass many flowers. I have counted seventy-five of them in a single bunch; the youngest

Diagram of flower of laurel
p, pocket; st, stamen

flowers grow nearest the tip of the twig. The blossom stems are pink, and afford a rich background for the starry open flowers and knobby closed buds. The bud of the laurel blossom is very pretty and resembles a bit of rose-colored pottery; it has a five-sided, pyramidal top, and at the base of the pyramid are ten little buttresses which flare out from the calyx. The calyx is five-lobed, each lobe being green at the base and pink at the point. Each one of the ten little buttresses or ridges is a groove in which a stamen is growing, as we may see by looking into an opening flower; each anther is " headed " toward the pocket which ends the groove. The filament lengthens and shoves the anther into the pocket, and then keeps on growing until it forms a bow-shaped spring, like a sapling with the top bent to the ground. The opening flower is saucer-like, pinkish white, and in form is a five-pointed star. At the bottom of the saucer a ten-pointed star is outlined in crimson; and bowed above this crimson ring are the ten white filaments with their red-brown anthers stuffed cozily into the pockets, one pocket at the center of each lobe, and one halfway between; each pocket is marked with a splash of crimson with spotty edges. From the center of the flower projects the stigma, far from and above the pollen pockets.

Each laurel flower is thus set with ten spring-traps; a moth or bee, seeking t nectar at the center of the flower, is s to touch one or all of these bent filamen As soon as one is touched, up it spri and slings its pollen hard at the intrud The pollen is not simply a shower powder, but is in the form of a stic string, as if the grains were strung on c web silk. When liberating these spri with a pencil point, I have seen the poll thrown a distance of thirteen inches; th if the pollen ammunition does not str the bee, it may fall upon some open flov in the neighborhood. The anthers spri back after this performance and the f ments curl over each other at the cen of the flower below the pink stigma; l after a few hours they straighten out a each empty anther is suspended above own pocket. The anthers open while the pocket; each one is slit open at its so that it is like the leather pocket o sling.

After the corollas fall, the long stig still projects from the tip of the ripen ovary, and there it stays, until the caps is ripe and open. The five-pointed ca remains as an ornamental cup for fruit. The capsule opens along five val and each section is stuffed with little, most globular seeds.

The mountain laurel grows in woo and shows a preference for rocky mo tain sides or sandy soil.

Another of the common species is sheep laurel, which grows in swam places, especially on hillsides. The flow of this are smaller and pinker than mountain laurel, and are set below leaves on the twig. Another species, cal the pale or swamp laurel, has very sn flowers, not more than half an inch breadth and its leaves have rolled-b edges and are whitish green benea This species is found only in cold p bogs and swamps.

LESSON 191

The Mountain Laurel

Leading Thought — The laurel bl som is set with ten springs, and e ring acts as a sling in throwing pollen pon visiting insects, thus sprinkling the sitor with pollen which it carries to her flowers.

METHOD — Have the pupils bring to e schoolroom a branch of laurel which ows blossoms in all stages from the bud. lthough this lesson is on the mountain urel, any of the other species will do as ell.

OBSERVATIONS — 1. How are the laurel aves set about the blossom clusters? hat is the shape of the laurel leaf? hat are its colors above and below? ow do the leaves grow with reference the flowers? Do they grow on last year's this year's wood? How can you tell the w wood from the old?

2. Take a blossom bud. What is its ape? How many sides to the pyramid- e tip? How many little flaring ridges the base of the pyramid? Describe the lyx.

3. What is the shape of the flower hen open? How many lobes has it? hat is its color? Where is it marked ith red?

4. In the open blossom, what do you e of the ten ridges, or keels, which you ticed in the bud? How does each one these grooves end? What does the urel blossom keep in these ten pockets? ouch one of the ten filaments with a ncil and note what happens.

5. Take a bud scarcely open. Where e the stamens? Can you see the anthers? ake a blossom somewhat more open. here are the anthers now? From these servations explain how the stamens ace their anthers in the pockets. How the filaments grow into bent springs?

6. Are the anthers open when they are ill in the pocket? Look at an anther with lens and tell how many slits it has. How they open? Are the pollen grains ose when they are thrown from the ther? How are they fastened together? oes this pollen mass stick to whatever touches?

7. What is the use to the flower of this rangement for throwing pollen? What sects set free the stamen springs? Where is the nectar which the bee or moth is after? Can it get this nectar without setting free the springs? Touch the filaments with a pencil and see how far they will sling the pollen.

8. Describe the pistil in the open flower. Is the stigma near the anthers? Would they be likely to throw their pollen on the stigma of their own flower? Could they throw it on the stigmas of neighboring flowers?

9. How does the fruit of the laurel look? Does the style still cling after the corolla falls? Describe the fruit capsule. How does it open? How do the seeds look? Are there many of them?

10. Where does the mountain laurel grow? What kind of soil does it like? Do you know any other species of laurel? If so, are they found in the same situations as the mountain laurel?

A childish gladness stays my feet,
As through the winter woods I go,
Behind some frozen ledge to meet
A kalmia shining through the snow.

I see it, beauteous as it stood
Ere autumn's glories paled and fled,
And sigh no more in pensive mood,
" My leafy oreads are all dead."

I hear its foliage move, like bells
On rosaries strung, and listening there,

Verne Morton

Spray of mountain laurel

Forget the icy wind that tells
Of turfless fields, and forests bare.

All gently with th' inclement scene
I feel its glossy verdure blend;—
I bless that lovely evergreen
As heart in exile hails a friend.

Its boughs, by tempest scarcely stirred,
Are tents beneath whose emerald fold
The rabbit and the snowbound bird
Forget the world is white and cold.

And still, 'mid ruin undestroyed,
Queen arbor with the fadeless crown
Its brightness warms the frosty void,
And softens winter's surliest frown.

—From "THE MOUNTAIN LAUREL"
THERON BROWN

FLOWERLESS PLANTS

FERNS

Iany interesting things about ferns · be taught to the young child, but the re careful study of these plants is bet- adapted to the pupils in the higher les, and is one of the wide-open doors t lead directly from nature-study to ematic science. While the pupils are lying the different forms in which s bear their fruit, they can make col- ions of all the ferns of the locality. ce ferns are easily pressed and are beau- l objects when mounted on white pa- , the making of a fern herbarium is a ghtful pastime; or leaf prints may be de which give beautiful results (see p.); but better, perhaps, than either col- ions or prints, are pencil or water-color wings with details of the fruiting organs enlarged. Not only is such a portfolio a thing of beauty, but the close observation needed for drawing brings much knowledge to the artist.

THE CHRISTMAS FERN

No shivering frond that shuns the blast sways on its slender chaffy stem;
Full veined and lusty green it stands, of all the wintry woods the gem.

— W. N. Clute

The rootstock of the fern is a humble mple of "rising on stepping stones of dead selves," this being almost literally e of the tree ferns. The rootstock, ich is a stem and not a root, has, like er stems, a growing tip from which, h year, grow beautiful green fronds and nerous rootlets. These graceful fronds oice the world and our eyes for the sum- r, and make glad the one who, in win- , loves to wander often in the woods to uire after the welfare of his many nds during their period of sleeping and king. These fronds, after giving their ssage of winter cheer, and after the fol- ving summer has made the whole wood- d green and the young fronds are growing thriftily from the tip of the rootstock, die down; and in midsummer we can find the old fronds lying sere and brown, with broken stipes, just back of the new fern clump; if we examine the rootstock we can detect, behind those fronds, remains of the stems of the fronds of year before last; and still farther behind we may trace all the stems of fronds which gladdened the world three years ago. Thus we learn that this rootstock may have been creeping on an inch or so each season for many years. One of the chief differences between our ferns and the tree ferns of the tropics, which we often see in greenhouses, is that in the tree fern the rootstock rises in the air instead of creeping on, or below, the

Verne Morton

The Christmas fern. The contracted tips of some of the fronds consist of fruiting pinnæ

surface of the ground. This upright rootstock of the tree fern also bears fronds at its tip, and its old fronds gradually die down, leaving it rough below its crown of green plumes.

The Christmas fern has its green stipe, or petiole, and its rachis, or midrib, more or less covered with ragged, brownish scales, which give it an unkempt appearance. Its pinnæ, or leaflets, are individually very pretty; in color they are dark, shining green, lance-shaped, with a pointed lobe or ear at the base projecting upward. The edges of the pinnæ are delicately toothed; each point is armed with a little spine, and the veins are fine, straight, and free to the margin; the lower pinnæ often have the earlike lobe completely severed.

In studying a fertile fern from above, we notice that about a dozen pairs of the pinnæ near the tip are narrowed and roughened and are more distinctly toothed on the margins. Examining them underneath, we find on each a double row of circular raised dots, which are the fruit dots, or sori; there is a row between the midrib and margin on each side, and also a double row extending up into the point at the base. Early in the season these spots look like pale blisters; later they turn pale brown, each blister having a depression at its center; by the middle of J masses of tiny globules, not larger t pinpoints, push out from beneath margin of these dots. The blister-membrane is simply a cover for the g ing spores, and is called the indusium July it shrivels into an irregular scroll, clinging to the pinnule by its depre center; and by this time the profusio tiny globules covers the entire under of the pinna like a brown fuzz. If we sc off some of this fuzz and examine it a lens, we can see that it consists of n berless little globules, each with a s to attach it to the leaf; these are the sp cases, or sporangia, each globule b packed full of spores which, even thro the lens, look like yellowish powder. each particle of this dust has its own st ture and contains in its heart the li fern-substance.

Not all the fronds of the fern cl bear these fruit dots. The ones we se for decoration are usually the st fronds, for the fertile ones are no graceful, and many uninformed pe think the brown spore cases are a fun The Christmas fern, being evergreen very firm in texture, is much used holiday decoration, whence its com name, which is more easily rememb than *Polystichum acrostichoides*, its name. It grows best in well-shaded w lands, liking better the trees which s their leaves than the evergreens. It is deed well adapted to thrive in damp, shade; it is rarely found on slopes wh

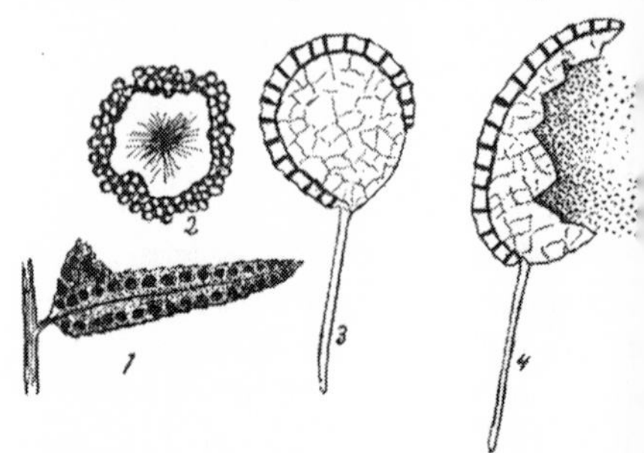

1, *Fertile leaflet of Christmas fern sho indusia and spore cases.* 2, *An indusium spore cases, enlarged.* 3, *A spore case, larged.* 4, *A spore case discharging sp enlarged*

e the south, and full sunshine may kill

LESSON 192

The Christmas Fern

Leading Thought — The fern has a eping underground stem called the otstock, which pushes forward and ıds up fresh fronds each year. Some of : fronds of the Christmas fern bear ɔres on the lower surface of the ter- nal pinnæ.

Method — This lesson should be given ring the latter part of May, when the ıit dots are still green. Take up a fern d transplant it in a dish of moss in the ıoolroom, and later plant it in some nvenient shady place. The pupils ɔuld sketch the fertile frond from the per side so as to fix in their minds the ntracted pinnæ of the tip; one of : lower pinnæ should be drawn in detail, ɔwing the serrate edge, the ear, and the nation. The teacher should use the fol- wing terms constantly and insistently, as to make the fern nomenclature a part the school vocabulary, and thus fit the pils for using fern manuals.

A *frond* is all of the fern which grows one stem from the rootstock; the *blade* that portion which bears leaflets; the

Buffalo Museum of Science

ɔmmon polypody, often mistaken for the Christmas fern

stipe is the stem or petiole; the *rachis* is the midrib and is a continuation of the

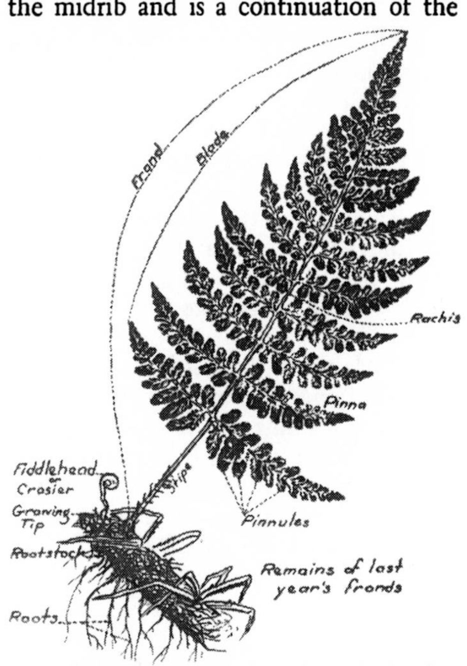

Leaf print of a fern with the parts named. This fern is twice pinnate

stipe; the *pinna* is a chief division of the midrib or rachis, when the fern is compound; the *pinnule* is a leaflet of the last division; the *sori* are the fruit dots; the *indusium* is the membrane covering the fruiting organs; the *sporangia* are the tiny brown globules, and are the spore cases; the *spores* make up the fine dust which comes from the spore cases. It would be well to make on the blackboard a diagram of the fern with its parts named, so that the pupils may consult it while studying ferns.

Observations — 1. Study a stump of the Christmas ferns. Are there any withered fronds? Where do they join the rootstock? Do the green fronds come from the same place on the rootstock as the withered ones?

2. Take a frond of the Christmas fern. Are the stem, or stipe, and the midrib, or rachis, smooth or rough? What color are the scales of the stalk? Do you think that these scales once wrapped the fern bud?

3. Does each frond of a clump have the

same number of pinnæ on each side? Can you find fronds where the pinnæ near the tip are narrower than those below? Take a lower pinna and draw it carefully, showing its shape, its edges, and its veins. Is there a point, or ear, at the base of every pinna? Is it a separate lobe or a mere point of the pinna?

4. Take one of the narrow pinnæ near the tip of the frond, and examine it beneath. Can you see some circular, roundish, blister-like dots? Are they dented at the center? How many of these dots on a pinna? Make a little sketch showing how they are arranged on the pinna and on the little earlike point. Look at the fruiting pinnæ of a fern during July, and describe how they look then.

5. Do all the fronds of a fern clump have these narrowed spore-bearing pinnæ? Do you know what those fronds are cal that bear the fruit dots?

6. Where do you find the Christm fern growing? Do you ever find it in sunny place? Why is it called the Chr mas fern?

FERN SONG

Dance to the beat of the rain, little Fe
And spread out your palms again,
And say, "Tho' the sun
Hath my vesture spun,
He had labored, alas, in vain,
But for the shade
That the Cloud hath made,
And the gift of the Dew and the Rain."
Then laugh and upturn
All your fronds, little Fern,
And rejoice in the beat of the rain!

— John B. Ta

THE BRACKEN

It is well for the children to study the animals and plants which have a world-wide distribution. There is something comforting in finding a familiar plant in strange countries; and when I have found the bracken on the coast ranges of California, on the rugged sides of the Alps, and in many other far places, I have always experienced a thrill of delightful memories of the fence-corners of the homestead farm. Since the bracken is so widespread, it is natural that it should find a place in literature and popular legend. As it clothes the mountains of Scotland, it is much sung of in Scottish poetry. Many superstitions cluster around it — its spores, if caught at midnight on a white napkin, are supposed to render the possessor invisible. Professor Clute, in *Our Ferns in Their Haunts*, gives a delightful chapter about the relation of the bracken to people.

For nature-study purposes, the bracken is valuable as a lesson on the intricate patterns of the fern leaf; it is in fact a lesson in pinnateness. The two lower branches are large and spreading, and are in themselves often three times pinnate; the branches higher up are twice pinnate; while the main branch near the tip is or pinnate, and at the tip is merely lob The lesson, as illustrated in the diagr of the fern, should be well learned for ture study, because this nomenclature used in all the fern manuals. The fact th a pinnule is merely the last division o frond, whether it be twice or thrice p nate, should also be understood.

The bracken does not grow best in co plete shade, but it becomes established waste places which are not too shad it thrives especially in woodsides, and fence-corners on high and cold land. Professor Clute says, " It is found both woodland and in the open field; its fav ite haunt is neither, but is that halfw ground where man leaves off and natu begins, the copse or the thicket." With it usually grows about three feet high, b varies much in this respect. The great angular fronds often measure two or thr feet across, and are supposed to bear a lil ness to an eagle with spread wings. Its ro stock is usually too deeply embedded earth for the study of any except the m energetic; it is about the size of a lead pe cil and is black and smooth; in its way is a great traveler, sending up fronds en or twenty feet from its starting place. also sends off branching rootstocks.

The fruiting pinnules look as if they ere hemmed and the edges of the hems nbroidered with brown wool; but the nbroidery is simply the spore cases pushg out from under the folded margin hich protected them while developing. Much on which to base necromancy s been found in the figure shown in e cross section of the stem or stipe. The ter C thus made, supposed to stand for hrist, is a potent protection from tches. But this figure has also been comred to the devil's hoof, an oak tree, or the initial of one's sweetheart, and all these imaginings have played their part in the lives of the people of past ages. It was

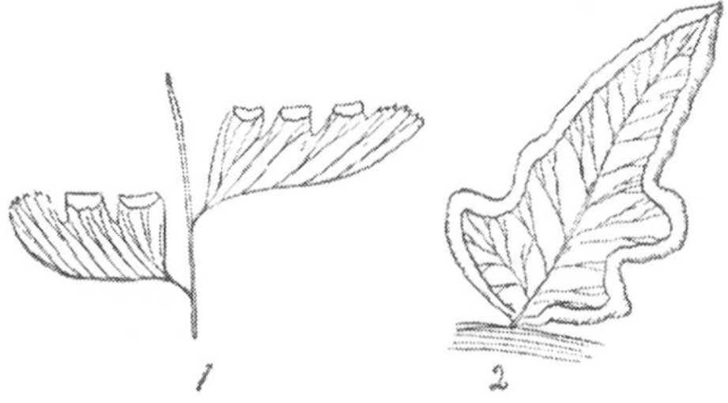

1, *Fruiting pinnules of the maidenhair fern, enlarged.* 2, *Fruiting pinnule of the bracken, enlarged. In both these species the spores are borne under the recurved edges of the pinnules*

Verne Morton

Bracken

believed in England that burning the bracken from the fields brought rain; the roots in time of scarcity have been ground and mixed with flour to make bread. The young ferns, or crosiers, are sometimes cooked and eaten like asparagus. The fronds have been used extensively for tanning leather and for packing fish and fruit, and when burned their ashes are used instead of soap.

In Europe, bracken grows so rankly that it is used for roof-thatching and for the bedding of cattle. The name "brake," which is loosely used for all ferns, comes from the word "bracken"; some people think that brakes are different from ferns, whereas this is simply a name which has strayed from the bracken to other species.

LESSON 193

The Bracken

Leading Thought — The bracken is a fern which is found in many parts of the world. It is much branched and divided, and it covers the ground in masses where it grows. The edges of its pinnules are folded under to protect the spores.

Method — Bring to the schoolroom large and small specimens of the bracken, and after a study is made tell about the superstitions connected with this fern and as far as possible interest the pupils in its literature.

Observations — 1. Do you find the bracken growing in the woods or open places? Do you find it in the cultivated fields? How high does it stand? Could you find the rootstock?

2. Take a bracken frond. What is its general shape? Does it remind you of an eagle with spread wings? Look at its very tip. Is it pinnate or merely lobed? Can you find a place farther down where the leaflets, or pinnules, are not joined at their bases? This is once pinnate. Look farther down and find a pinna that is lobed at the tip; at the base it has distinct pinnules. This is twice pinnate. Look at the lowest divisions of all. Can you find any part of this which is three times pinnate? Four times pinnate? Pinna means feather, pinnate therefore means feathered. If thing is once pinnate, it means that has along each side divisions similar to feather; twice pinnate means that ea feather has little feathers along each si thrice pinnate means that the little feat ers have similar feathers along each sic and so on.

3. Can you see whether the edges the pinnules are folded under? Lift up o of these edges and see if you can find wh is growing beneath it. How do the folded margins look during August a September?

4. Cut the stem, or stipe, of a brack across and see the figure in it. Does it lo like the initial C? Or a hoof, or an o tree, or another initial?

5. Discover, if you can, the differe uses which people of other countries fi for this fern.

HOW A FERN BUD UNFOLDS

Verne Morton

Fiddle heads or crosiers. Young ferns unfolding

All of the parts of the frond of a fe are tightly folded spirally within the b and every fold of every leaflet is al folded in a spiral. But the first glance one of these little woolly spirals gives but small conception of its marvelous e folding. Every part of the frond is prese in that bud, even to the fruiting orga all the pinnæ and the pinnules are pack in the smallest compass — each divisio even to the smallest pinnule, coiled in spiral toward its base. These coiled fe buds are called crosiers; they are wooll with scales instead of hairs. When t fern commences to grow, it stretches u and seems to lean over backward in i effort to be bigger. First the main stem, rachis, loosens its coil; but before this completed, the pinnæ, which are coile at right angles to the main stem, begin unfold; a little later the pinnules, whic are folded at right angles to the pinn loosen and seem to stretch and yawn b fore taking a look at the world which th have just entered; it may be several da before all signs of the complex coiling d appear. The crosiers of the bracken a er-looking creatures, soon developing :e claws which some people say look the talons of an eagle; and so intricate e action of their multitudinous spirals, to watch them unfolding impresses as would a miracle.

LESSON 194

How a Fern Bud Unfolds

EADING THOUGHT — All of the parts he frond of a fern are tightly folded ally within the bud, and every lobe very leaflet is also folded in a spiral.

IETHOD — The bracken crosier is a st illuminating object for this lesson, ause it has so many divisions and is so e; it is also convenient, because it may found in September. However, any bud will do. The lesson may be best n in May when the woodland ferns starting. A root of a common fern with its buds may be brought to the schoolroom, where the process of unfolding may be watched at leisure. Later, the plant may be set out in a suitable place.

OBSERVATIONS — 1. Take a very young bud. How does it look? Do you see any reason why uninformed people call these buds caterpillars? Can you see why they are popularly called "fiddle heads"? What is their true name? How many turns of the coil can you count? What is the covering of the crosier? How is the stem grooved to make the spiral compact?

2. Take a crosier a little further advanced. How are its pinnæ folded? How is each pinnule of each pinna folded? How is each lobe of a pinnule folded? Is each smaller part coiled toward each larger part?

3. Write in your notebook the story of the unfolding fern, and sketch its stages each day from the time it is cuddled down in a spiral until it is a fully expanded frond.

THE FRUITING OF THE FERN

f we were required to know the position of the fruit dots or the character of the in-ium, nothing could be easier than to ascertain it; but if it is required that you be cted by ferns, that they amount to anything, signify anything to you, that they be ther sacred scripture and revelation to you, help to redeem your life, this end is not asily accomplished. — THOREAU.

'he fern, like the butterfly, seems to e several this-world incarnations; and haps the most wonderful of these is the re. Shake the dust out of the ripened and each particle, although too small the naked eye to see, has within it the sibilities of developing a mass of grace-ferns. Each spore has an outside hard r, and within this an atom of fern-stance; but it cannot be developed un-it falls into some warm, damp place orable for its growth; it may have to t many years before chance gives it this orable condition, but it is strong and ins its vital power for years. There are es known where spores grew after nty years of waiting. But what does s microscopic atom grow into? It develops into a tiny heart-shaped, leaflike structure which botanists call the prothallium; this has on its lower side little roots which reach down into the soil for nourishment; and on its lower surface are two kinds of pockets, one round and the other long. In the round pockets are developed bodies which may be compared to the pollen; and in the long pockets, bodies which may be compared to the ovules of flowering plants. In the case of ferns, water is necessary to float the pollen from the round pockets to the ovules in the long pockets. From a germ thus fertilized in one of the long pockets, a little green fern starts to grow, although it may be several years before it becomes a plant strong enough to send up fronds with spore dots on them. To study the structure of the spore requires the highest powers of the microscope; and even the prothallium in most species is very small, varying from the size of a pinhead to that of a small pea, and it is therefore quite

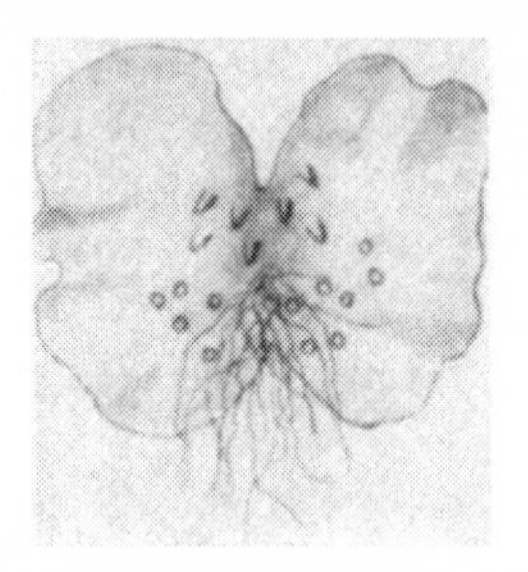

Prothallium, greatly enlarged, showing the two kinds of pockets and the rootlets

difficult to find. I found some once on a mossy log that bridged a stream, and I was never so triumphant over any other outdoor achievement. They may be found in damp places or in greenhouses, but the teacher who is able to show her pupils this stage of the fern will be very fortunate. The prothallium is a stage of the fern to be compared to the flower and seed combined in the higher plants; but this is difficult for young minds to comprehend. I like to tell the children that the fern, like a butterfly, has several stages: Beginning with the spore-bearing fern, we next have the spores, next the prothallium stage, and then the young fern. In the other case we have first the egg, then the caterpillar, then the chrysalis, and then the butterfly. Looking at the ripe fruit dots on the lower side of the fern leaf, we can easily see with a lens a mass of tiny globules; each one of these is a spore case, or sporangium (plural *sporangia*), and is fastened to the leaf by a stalk and has, almost encircling it, a jointed ring. (See figure on p. 694.)

When the spores are ripe, this ring straightens out and ruptures the globule, and out fly the spores. By scraping a little of the brown fuzz from a fruiting pinna of the Christmas fern upon a glass slide and placing a cover glass upon it, we find it very easy to examine through the microscope, and we are able thus to find the spore cases in all stages, and to see the spores tinctly. The spore cases may also be s

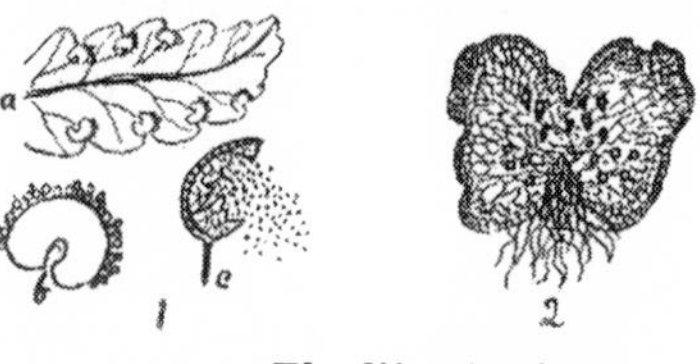

The life of a fern

1, a, pinna bearing fruit; b, a fruit dot, enla showing spore cases pushing out around the edges c indusium; c, spore case, enlarged, showing how it charges the spores. 2, Prothallium, enlarged. 3, Y fern growing from the prothallium

with a hand lens, the spores seeming t to be mere dust.

The different ways the ferns blar their spore cases is a delightful study, one which the pupils enjoy very mu All of our common ferns except the tle polypody thus protect their spo Whether this blanket be circular, or ho shoe-shaped, or oblong, or in the forn pocket or cup, depends upon the ge to which the fern belongs. The little p tecting blanket membrane is called indusium, and while its shape dis guishes the genus, the position in wh it grows determines the species. I sh never forget my surprise and delight wh as a young girl, I visited the Philadelp Centennial Exposition, and there in

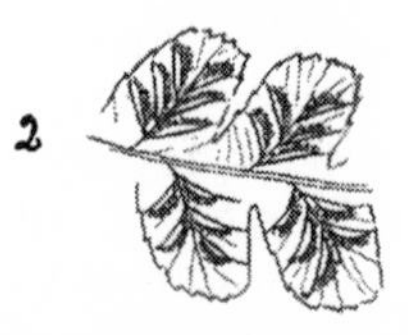

1, *Fruiting pinnule of the boulder fern,* *larged.* 2, *Fruiting pinnules of spleenw* *enlarged*

great conservatories saw for the first ti the tree ferns of the tropics. One of th was labeled *Dicksonia*, and, mystifie asked the privilege of examining fronds for fruiting organs; when lo!

usium proved to be a little cup, borne the base of the tooth of the pinnule, ctly like that of our boulder fern, which lso a *Dicksonia*. I had a sudden feeling t I must have fern friends all over the rld.

The children are always interested in way the maidenhair folds over the tips

ruiting pinnules of evergreen wood fern

her scallops to protect her spore nury. While many of our ferns have their tile fronds very similar in form to the rile ones, yet there are many common ns with fertile fronds that look so difent from the others that one would not nk they were originally of the same patn; but although their pinnules are nged into cups, or spore pockets, of ious shapes, if they be examined carely they will be seen to have the same eral structure and the same divisions, wever much contracted, as have the ge sterile fronds. The *Osmundas*, which lude the interrupted, the cinnamon, d the flowering ferns, are especially od for this part of the lesson. The sensie fern, so common in damp places in

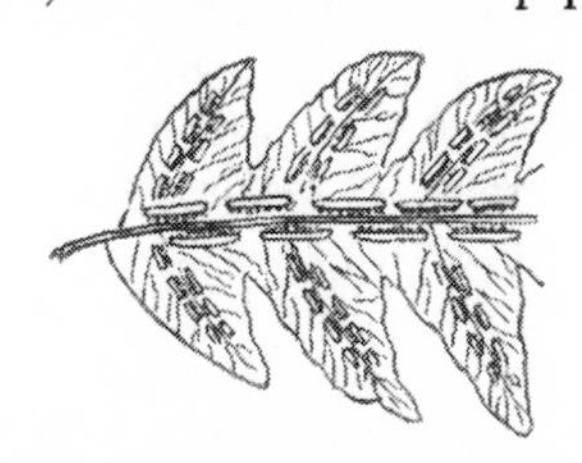

Fruiting pinnules of the chain fern

en fields, is also an excellent illustran of this method of fruiting. While dying the ferns, the teacher should lay ess upon the fact that they represent an ly and simple form of plant, that they reached the zenith of their growth in the Carboniferous age, and that, to a large extent, our coal is composed of them. It is interesting to think that the exquisite and intricate leaf patterns of the ferns should belong to a primitive type.

LESSON 195

The Fruiting of the Fern

Leading Thought — Ferns do not have flowers, but they produce spores. Spores are not seeds; but fern spores grow into a tiny prothallium, and this in turn produces a young fern. Each genus of ferns has its own peculiar way of protecting its spores; and if we learn these different ways, we can recognize most ferns without effort.

Method — July is the best time for this lesson, which is well adapted for summer schools or camping trips. However, if it is desired to use it as a school lesson, it should be begun in June, when the fruiting organs are green, and it may be finished in September after the spores are discharged. Begin with the Christmas

A sensitive fern, showing sterile and fertile fronds

fern, which ripens in June, and make the fruiting of this species a basis for comparison. Follow this with other wood ferns which bear fruit dots on the back of

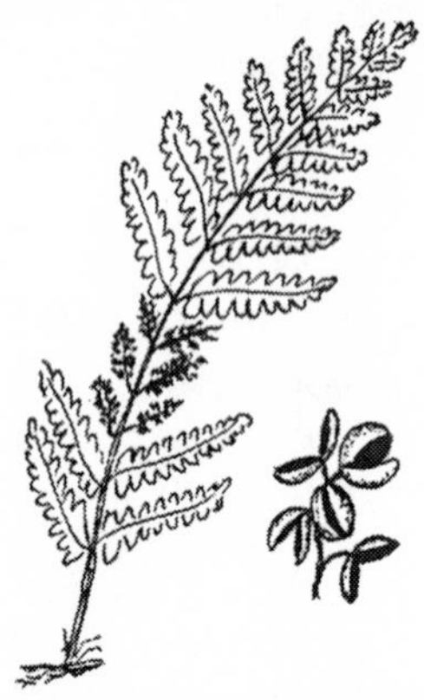

Diagram of the interrupted fern, showing the three pairs of fruiting pinnæ, and a part of one of these enlarged. This fern often has fronds four or five feet high

the fronds. Then study the ferns which live in more open places, and which have fronds changed in form to bear the spores — like the sensitive, the ostrich, the royal, and the flowering ferns. A study of the interrupted fern is a desirable preparation for the further study of those which have special fruiting fronds; the interrupted fern has, at about the middle of its frond, one to five pinnæ on each side, fitted for spore-bearing, the pinnules being changed into globular cups filled with spore cases.

While not absolutely necessary, it is highly desirable that each member of the class should look at a fruit dot of some fern through a three-quarters objective of a compound microscope, and then examine the spore cases and the spores through a one-sixth objective. It must be remembered that this lesson is for advanced grades, and is a preparation for systematic scientific work. If a microscope is not available, the work may be done with a hand lens aided by pictures.

Observations — 1. Take a fern that is in fruit; lay it on a sheet of white paper and leave it thus for a day or two, where it will not be disturbed and where there is no draught; then take it up carefully; the form of the fern will be outlined in d
What is this dust?

2. What conditions must the spo
have in order to grow? What do they gr
into?

3. Look at a ripe fruit dot on the b
of a fern leaf and see where the spo
come from. Can you see with a lens ma
little, brown globules? Can you see t
some of them are torn open? These are
spore cases, called *sporangia*, each glob
being packed with spores. Can you
how the sporangia are fastened to the l
by little stems?

4. Almost all our common wood fe
have the spore cases protected by a t
membrane when very young; this li
membrane is called the indusium, an
is of different shape in those ferns wh
do not have the same surname, or gene
name. Study as many kinds of wood fe
as you can find. If the blanket, or in
sium, is circular with a dent at the cen
where it is fastened to the leaf, and
spore cases push out around the mar
it is a *Christmas fern*; if horseshoe-shap
it is one of the *wood ferns*; if oblong,
rows on each side of the midrib, it
chain fern; but if oblong and at an an
to the midrib, it is a *spleenwort*; if i
pocket-shaped and opening at one side
is a *bladder fern*; if it is cup-shaped, it
boulder fern; if it breaks open and
back in star shape, it is a *Woodsia*; if
edge of the fern leaf is folded over all al
its margin to protect the spore cases, i
a *bracken*; if the tips of the scallops of
leaf are delicately folded over to mak
spore blanket, it is the *maidenhair*.

5. If you know of swampy land wh
there are many tall brakes, look for a k
that has some of its pinnæ withered
brown. Examine these withered pin
and you will see that they are not withe
at all but are changed into little cups
hold spore cases. This is the *interrup*
fern. The *flowering fern* has the pinna
its tip changed into cups for spore ca
The *cinnamon fern*, which grows
swampy places, has whole fronds which
cinnamon-colored and look withered,
which bear the spores. The *ostrich fe*

Bulletin 119, Agricultural Extension Service, Ohio State U.

Important characteristics which distinguish fern groups

1, Sensitive fern: a, frond; b, spore-bearing frond
2, Hartford or climbing fern
3, Grape fern: a, frond; b, modified frond producing ›re cases; c, detail of spore cases
4, Spleenwort: a, frond; b, spore cases like pockets ›ve veins
5, Common polypody: a, frond; b, large fruiting dots
6, Interrupted fern: a, spore-bearing pinnæ (leaflets) ›fined to a few in middle of frond; b and c, details ›wing arrangement of spore cases
7, Cinnamon fern: a, fertile frond
8, Royal fern: a, frond; b, the modified pinnæ at the tip producing spores
9, Maidenhair fern; a, pinnæ; b, detail showing spores beneath the folded margins of pinnæ
10, Bracken: a, entire frond; b, detail of pinnæ; c, spore-bearing folded margins of a pinnule
11, Chain fern: a, pinnæ; b, spore areas showing chain formation
12, Christmas fern: a, frond; b, spore cluster; c, detail of spore clusters
13, Bladder fern: a, frond; b, fruit dots

1. Purple Cliff Brake, Pellaea atropurpurea. *Sometimes called "winter brake" because in its southern range the fronds remain green all winter, this fern usually grows in situations which can be reached only with difficulty. It can be grown in cultivation if it is always kept in the same position. Should the plant be moved, the change in relation to the light will retard its growth.*
Range: *New England and British Columbia south and west to California and northern Mexico.* Habitat: *Crevices in dry rocks. (Photo by Dr. and Mrs. John Small)*

2. Climbing Fern, Lygodium palmatum. *Seventy-five years ago this fern was common, but in many places reckless picking has almost exterminated it. The fronds, 1 to 3 feet long, twine or climb about other plants or trail on the ground. The specimen shown in the picture is a typical young plant, which differs somewhat in general appearance from a mature plant.*
Range: *Massachusetts south to Florida.* Habitat: *Banks of streams. (Photo by Charles E. Mohr)*

3. Grape Fern. *This is one of the many variations of* Botrychium dissectum.
Range: *Nova Scotia and New Brunswick west to Wisconsin and Iowa, south to South Carolina, Georgia, and Florida.* Habitat: *A great variety of habitats: sterile hilltops, dry pastures, meadows, thickets, rich swampy woods and sandy banks in pine barrens. (Photo by Dr. and Mrs. John Small)*

4. Hart's-Tongue, Phyllitis Scolopendrium. *Although very rare in this country, this fern is common in parts of Eurasia, especially Great Britain.*
Range: *Locally in New Brunswick, Ontario, central New York, Tennessee, and North Carolina.* Habitat: *Shaded ravines in regions where there are limestone cliffs. (Photo by Dr. and Mrs. John Small)*

5. Hay-scented Fern, Dennstædtia punctilobula. *The fronds of this slender, tapering, pale green fern are 1½ to 2 feet long. They are very fragrant when dried.*
Range: *Nova Scotia to Minnesota south to Georgia.* Habitat: *Rocky pastures, meadows, thickets, and near swamps. (Photo Leonard K. Beyer)*

6. Maidenhair Fern, Adiantum pedatu *Except on the Atlantic coastal plains, t fern is rather widely distributed througho North America. It is found in sheltered, sha places rather than in open areas of the woo*
Range: *Nova Scotia to British Colum south to Georgia and Arkansas.* Habit *Rich, moist woodlands. (Photo by Dr. a Mrs. John Small)*

7. Interrupted Fern, Osmunda Clayt iana. *This fern, the cinnamon fern (No. and the royal fern (No. 10), are similar many ways; they are tall, showy, and bea ful. They can be transplanted if shade, ple of water, and good soil are provided.*
Range: *Minnesota to Newfoundland so to Missouri, Kentucky, and North Caroli* Habitat: *Swampy areas. (Photo by Bro lyn Botanic Garden)*

8. Walking Leaf Fern, Camptosorus r zophyllus. *The end frond of this fern see capable of taking root at its tip; thus n plants are started. Sometimes the third g eration maintains a connection with original plant.*
Range: *From Maine and southern Can to Georgia and westward.* Habitat: *Loc in shady ravines on cliffs or decaying stum (Photo by Verne Morton)*

9. Cinnamon Fern, Osmunda cin momea. *The sterile fronds of this fern quite similar in appearance to those of interrupted fern; they can be distinguis by a tuft of wool at the base of the pinna*
Range: *Canada to Florida west to N Mexico and into South America.* Habit *Swampy areas. (Photo by Brooklyn Bota Garden)*

10. Royal or Flowering Fern, Osmu regalis. *In North America the usual height this fern is 2 to 5 feet, but in Europe i said to reach a height of 10 feet.*
Range: *Southern and eastern Canada Nebraska, Mississippi, and Florida into tro cal America.* Habitat: *Swampy areas. (Ph by Buffalo Museum of Science)*

1 2 3 4

5 6

7 8

9 10

which has fronds that look like magnificent ostrich feathers, has stiff little stalks of fruiting fronds very unlike the magnificent sterile fronds. The *sensitive fern*, which grows in damp meadows and along roadsides, also has contracted fruiting fronds. If you find any of these, compare carefully the fruiting with the sterile fronds, and note in each case the resemblance in branching and in pinnules and also the shape of the openings through which the spores are sifted out.

6. Gather and press specimens of many ferns in the fruiting stage as can find, taking both sterile and frui fronds in those species which have specialization.

7. Read in the geologies about the fe which helped in the making of our beds.

Nature made ferns for pure leaves to what she could do in that line.

— THOR

THE FIELD HORSETAIL

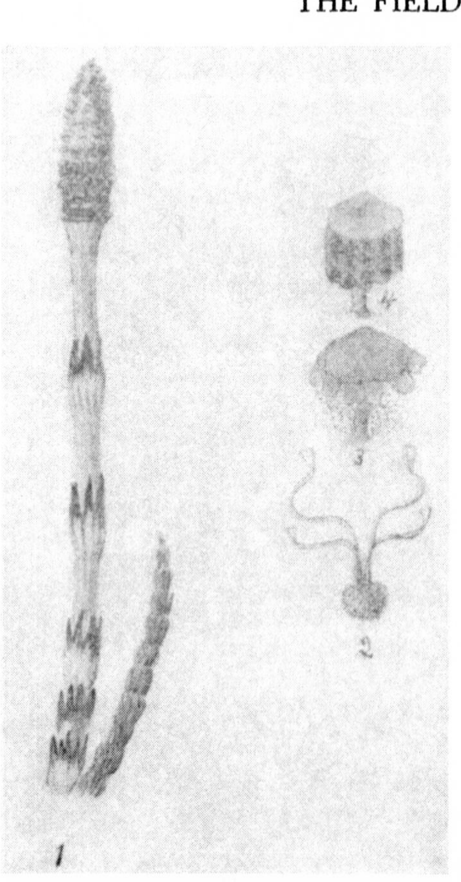

1, *Fertile plant of the field horsetail;* 2, *spores;* 3, *disc discharging spores;* 4, *disc with spore sacs*

These queer, pale plants grow in sandy or gravelly soil, and since they appear so early in the spring they are objects of curiosity to children. The stalk is pale and uncanny looking; the pinkish stem, all same size from bottom to top, is or mented at intervals with upward-po ing, slender, black, sharp-pointed sca which unite at the bottom and enci the stalk in a slightly bulging ring, a which shows a ridge for every scale, tending down the stem. These bl scales are really leaves springing fror joint in the stem, but they forgot l ago how to do a leaf's work of getting f from the air. The " blossom," which is a real blossom in the eye of the botar is made up of rows of tiny discs which set like miniature toadstools around central stalk. Before it is ripe, there tends back from the edge of each dis row of little sacs stuffed so full of gr spores that they look united like a row tiny green ridges. The discs at the top the fertile spike discharge their sp first, as can be seen by shaking the pl over white paper, the falling spores lo ing like pale green powder. The burst empty sacs are whitish, and hang aro the discs in torn scallops, after the sp are shed. The spores, when seen under microscope, are wonderful objects, e a little green ball with four spiral ba wound about it. These spirals uncoil throw the spore, giving it a movemen of something alive. The motor powe these living springs is the absorbing moisture.

The beginning of the sterile shoot be seen like a green bit of the bloss

W. C. Muenscher

ld horsetail, Equisetum arvense, *though t a flowering plant "blooms" with spores*

ike of the plantain; but later, after the tile stalks have died down, these cover e ground with their strange fringes. Not kinds of horsetails have separate fruitg and vegetative shoots; in some the ne stalk bears both fruiting and vegetae parts.

The person who first called these sterile ants "horsetails" had an overworked agination, or none at all; for the only ality the two have in common is brushiss. A horse which had the hair of its il set in whorls with the same precision this plant has its branches would be one the world's wonders. The *Equisetum* is e of the plants which give evidence of ture's resourcefulness; its remote ancesrs probably had a whorl of leaves at each int or node of the main stem and anches; but the plant now has so many een branches that it does not really need e leaves, and thus they have been reiced to mere points, and look like nothg but "trimming," they are so purely namental. Each little cup or socket, of e joint or node, in branch or stem, has row of points around its margin, and ese points are terminals of the angles in the branch. If a branch is triangular in cross section, it will have three points at its socket, if quadrangular it will have four points, and the main stem may have six or a dozen, or even more points. The main stem and branches are made up entirely of these segments, each set at its lower end in the socket of the segment behind or below it. These green branches, rich in chlorophyl, manufacture for the plant all the food that it needs. Late in the season this food is stored in the rootstocks, so that early next spring the fertile plants, nourished by this stored material, are able to push forth before most other plants, and thus develop their spores early in the season. There is a prothallium stage as in the ferns.

Above where the whorl of stems comes from the main branch, may be seen a row of upward-standing points which are the remnants of leaves; each branch as it leaves the stem is set in a little dark cup with a toothed rim. There is a nice gradation from the stout lower part of the stem to the tip, which is as delicate as one of the side branches.

The rootstock dies out behind the plant and pushes on ahead like the rootstock of ferns. The true roots may be seen attached on the underside. The food made in the summer is stored in little tubers, which may be seen in the rootstocks.

Dr. and Mrs. John Small

Club moss or ground pine, Lycopodium alopecuroides. *This plant is common in the bogs of pine-barrens*

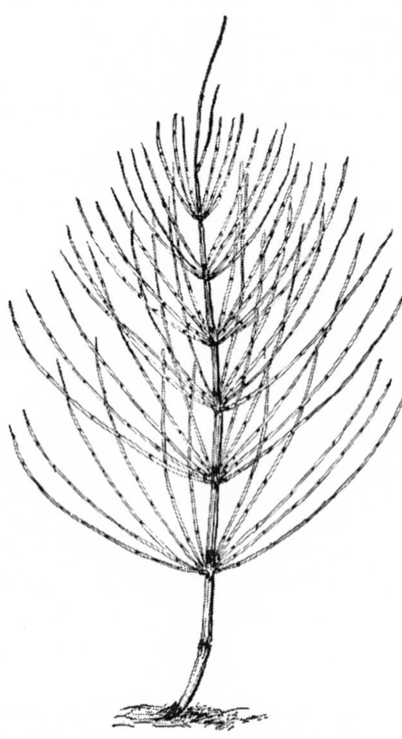

Sterile plant of the field horsetail

LESSON 196

THE FIELD HORSETAIL

THE FERTILE PLANT

LEADING THOUGHT — The horsetail is a plant that develops spores instead of seeds, and has green stems instead of leaves.

METHOD — In April and May, when the children are looking for flowers, they will find some of these weird-looking plants. These may be brought to the schoolroom and the observation lesson given there.

OBSERVATIONS — 1. Where are these plants found? On what kind of soil?

2. In what respect does this plant differ from other plants in appearance? Can you find any green part to it?

3. What color is the stem? Is it the same size its whole length? Is it smoc or rough?

4. Do you see any leaves on the sten Do you see the black-pointed scales? which direction do these scales point? A they united at the bottom? What sort ring do they make around the stem? Sp a stem lengthwise and see if there joints, or nodes, where the ring joins t stalk.

5. How does the "blossom" loc What color are the little discs that ma up the blossom? How are the discs set?

6. Take one of the plants which has t discs surrounded by green ridges. Sha it over a white paper. What comes frc it? Where does it come from? Whi discs on the stalk shed the green spo first?

THE STERILE PLANT

LEADING THOUGHT — The horsetail *Equisetum* is nourished by very differe looking stems from those which bore t spores. It lacks leaves, but its branches green and do the work of making food the plant.

METHOD — The sterile plants of t horsetail do not appear for several wee after the fertile ones; they are much mc

Dorothy M. Comp

Ground pine, Lycopodium complanatu *is widely distributed in dry coniferous woo throughout North America, Europe, and A*

merous, and do not resemble the fer- plants in form or color. These sterile nts may be used for a lesson in Septem- or October. Some of these plants with ir roots may be brought into the oolroom for study.

OBSERVATIONS — 1. Has this plant any ves? How does it make and digest its d without leaves? What part of it is en? Wherever there is green in a plant, re is the chlorophyl-factory for making d. In the horsetail, then, what part of plant does the work of leaves?

. Take off one little branch and study vith the lens. How does it look? Pull it rt. Where does it break easily? How ny joints, or nodes, are there in the nch?

. Study the socket from which one of segments was pulled off. What do you around its edge? How many of these nts? Look at the branch in cross sec- n. How many angles has it? What rela- n do the points bear to the angles? Do you think these points are all there are left of true leaves?

4. How do the little green branches come off the main stem? How many in a place? How many whorls of branches on the main stem?

5. Study the bases of the branches. What do you see? Look directly above where the whorl of branches comes off the main stem. What do you see? Cut the main stem in cross section just below this place, and see if there are as many little points as there are angles or ridges in the stem. Do you suppose these little points are the remnants of leaves on the main stem?

6. What kind of root has the horsetail? Do you think this long running root is the true root or an underground stem? Where are the true roots? Do you think the root-stock dies off at the oldest end each year, like the fern? Can you find the little tubers in the rootstock which contain nourish- ment for next year's spore-bearing stalks?

THE HAIR-CAP MOSS OR PIGEON WHEAT

The mosses are a special delight to chil- n because they are green and beautiful ore other plants have gained their enness in the spring and after they have t it in the fall; to the discerning eye, ossy bank or a mossy log is a thing of uty always. When we were children we arded moss as a forest for fairy folk, h moss stem being a tree, and we natu- y concluded that fairy forests were rgreen. We also had other diversions h pigeon wheat, for we took the fruit- stem, pulled the cap off the spore cap- e, and tucked the other end of the red m into the middle of the capsule, mak- a beautiful coral ring with an emerald t." To be sure, these rings were rather delicate to last long, but there were nty more to be had for nothing; so we de these rings into long chains which wore as necklaces for brief and happy ments, their evanescence being one of ir charms.

Pigeon wheat is a rather large moss which grows on dry knolls, usually near the margins of damp woodlands in just those places where wintergreens love to grow. In fall or winter it forms a greenish brown mass of bristling stems; in the early summer the stems are tipped with the vivid green of the new growth. The bristling appearance comes from the long sharp leaves set thickly upon the ruddy brown stems; each leaf is pretty to look at with a lens, which reveals it as thick though narrow, grooved along the middle, the edges usually armed with sharp teeth, and the base clasping the stem. These leaves, although so small, are wonderfully made; during the hot, dry weather they shut up lengthwise and twist into the merest threads; thus their soft, green sur- faces do not lose as much moisture by ex- posure to the air. More than this, they huddle close to the stem and in this posi- tion they are less likely to suffer from the effect of drought. But as soon as the rains come, they straighten back at right angles

George E. Nichols

Hair-cap moss or pigeon wheat

to the stem, and curve their tips downward. Bring in some of this moss and let it dry, and then drop it into a glass of water and watch this miracle of leaf movement! And yet it is no miracle but a mechanism quite automatic, and therefore like other miracles, when once they are understood.

In early June the mossy knoll shows us the origin of the name pigeon grass or pigeon wheat, for it is then covered with a forest of shining, ruddy, stiff, little stalks, each stalk bearing on its tip a woolly object about the size of a grain of wheat. But it is safe to say that the pigeons and other birds enjoy our own kind of wheat better than this, which is attributed to them.

A study of one of these wheat grains reveals it as covered with a yellowish mohair cap, ending in a golden-brown peak at its tip, as if it were the original patt of the toboggan cap; it closes loos around the stem below. This "grain" the spore capsule of the moss; the ha cap pulls off easily when seized by its t This cap is present at the very beginni even before the stalk lengthens; it p tects the delicate tissues of the growi spore case. It is only through a lens t we can see it in all its silky softness. T capsule revealed by the removal of cap is a beautiful green object, usu four-sided, set upon an elegant little ped tal where it joins the coral stalk, and w a lid on its top like a sugar-bowl co with a point instead of a knob at its cen

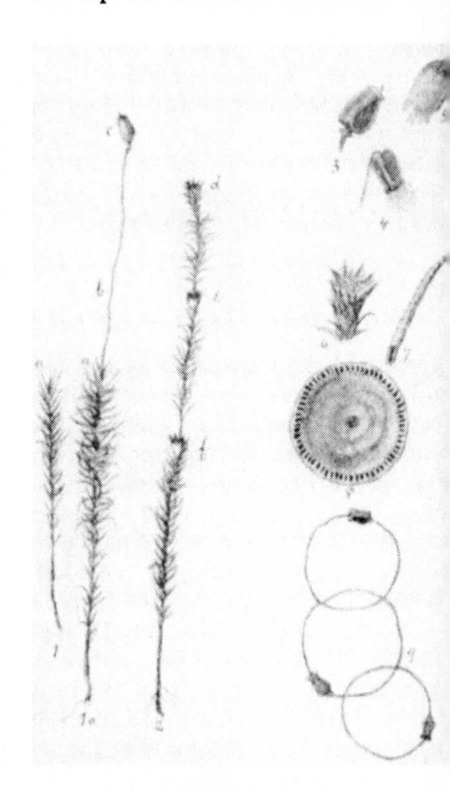

Hair-cap moss

1, Moss stem carrying the egg before fertilize 1a, The same stem carrying the spore capsule an stalk; b, stalk; c, spore capsule, with cap coverin 2, Stalk showing the starlike cups; d, the cup in w was developed the sperms which fertilized the egg this year; e, last year's cup; f, the cup of year b last; only the upper leaves of the stem are aliv Spore capsule with the cap removed, showing the 4, Spore capsule with lid off and shaking out the sp 5, The cap. 6, Starlike cup in which the sperm is veloped. 7, Leaf of moss. 8, The top of the capsule showing the teeth around the edge between w the spores sift out. 9, A part of a necklace chain of the spore capsules and their stems

ıen the spores are ripe, this lid falls and then if we have a lens we may another instance of moss mechanism. ɔking at the uncovered end of the cap-;, we see a row of tiny teeth around margin, which seem to hold down an er cover with a little raised rim. The anists have counted these teeth and l there are 64. The teeth themselves not important, but the openings be-en them are, since only through these nings can the spores escape. In fact, capsule is a pepper box with a grating und its upper edge instead of holes in cover; and when it is fully ripe, instead tanding right side up, it tips over; thus spores are shaken out more easily. ese teeth are like the moss leaves; they ll with moisture, and thus in rainy ther they, with the inner cover, swell that not a single spore can be shaken . If spores should come out during the ı, they would fall among the parent nts where there is no room for growth. t when they emerge in dry weather, wind scatters them far and wide where re is room for development.

Vhen seen with the naked eye, the res seem to be simply fine dust, but h dust grain is able to produce moss nts. However, the spore does not grow into a plant like a seed; it grows into ;, green, branching threads which push ng the surface of damp soil; and on se threads little buds appear, each of ich grows up into a moss stem.

f we examine some other plants of eon wheat moss, we find that some ms end in yellowish cups which look ıost like blossoms; on closer examina-n, we find that there are several of se cups, one below the other, with the m extending up through the middle. e upper cup matured this year, the one ow it last year, and so on. These cups star-pointed, and inside, at the bottom, starlike cluster of leaves. Among the ves of this star-rosette are borne the *theridia*, too small for us to see without igh-power microscope. The sperm cells m these *antheridia* are carried to other nts, some of which produce egg cells at their very tips, although the egg cell has no leaf rosette to show where it is. This egg cell, after receiving the sperm cells, grows into the spore capsule supported on its coral stem. These — stalk, capsule, and all — grow up out of the mother plant; the red stalk is enlarged at its base, and fits into the moss stem like a flagstaff in the socket. After the star-shaped cup has shed its sperm cells, the stem grows up from its center for an inch or so in height and bears new leaves, and next year will bear another starry cup. This condition is true of pigeon wheat and some others; but many other mosses have sperm and egg cells on the same plant.

The brown leaves on the lower part of the moss stem are dead, and only the green leaves on the upper part are living.

And this is the story of the moss cycle:

1. A plant with an egg cell at its tip; another plant with a star-cup holding the moss sperm cell which is splashed by a raindrop over to the waiting egg.

2. The egg cell as soon as fertilized develops into a spore capsule, which is lifted up into the world on a beautiful shining stem and is protected by a silky cap.

3. The cap comes off; the lid of the spore case falls off, the spores are shaken out and scattered by the wind.

4. Those spores that find fitting places grow into a net of green threads.

5. These green threads send up moss stems which repeat the story.

LESSON 197

The Hair-Cap Moss

Leading Thought — The mosses, like the butterfly and the fern, have several stages in their development. The butterfly stages are the egg, the caterpillar, the chrysalis, the butterfly. The moss stages are the egg (or ovule), the spores, the branching green threads, and moss plants with their green foliage. In June we can

MOSSES AND HEPATICS

These plants make up that portion of the plant kingdom known as Bryophytes *or* Bryophyta. *Although they number more than* 16,000 *species, most of which are mosses, comparatively few are of any economic importance. The group is, however, of great interest to scientists because its members seem to represent a step in development between the algae and higher plants.*

Bryophytes do not have true roots as do the higher plants; instead of roots they have numerous hairlike growths called rhizoids. The rhizoids of the hepatics are only one-celled; those of the mosses are much more highly developed.

Hepatics grow in such a way as to make a flat covering over the ground. Mosses, because of their vegetative multiplication, usually grow in compact clusters. Their compactness enables mosses to store up for a long time any moisture that they collect. When dry they are dormant.

Mosses and liverworts are best distinguished by the way in which their spores are released. In mosses, with a few exceptions, the end of the capsule forms a lid which falls off, thus releasing the spores. In liverworts, the end of the capsule splits lengthwise and the segments bend apart, allowing the spores to fall out. Although not all liverworts have flat thalli, like those shown in Figures 6 *and* 8, *and many are surprisingly mosslike in appearance, these can be distinguished from mosses because the leaves of liverworts are only a single cell in thickness throughout all their area.*

1. Broom Moss, Dicranum scoparium. *This moss has its name from the resemblance it bears to a hair broom or long brush. It is so abundant throughout most parts of the Northern Hemisphere that it is often used by florists to produce the effect of green banks in exhibits. (Photo by E. B. Mains)*

2. Common Hair-Cap, Bird Wheat, *or* Pigeon Wheat Moss, Polytrichium commune. *This very common moss, found not only in all parts of North America but in Europe and Asia as well, is the plant most people have in mind when they speak of moss. It grows, not only in woods but also in old fields and meadows, where it must withstand great variations in temperature and moisture. When dry, the leaves fold up against the stem and dry plants present a very different aspect from damp ones. In Europe the hair-cap mosses are for small brooms and bed fillings. This is largest moss; its stems may reach a foo length but they are usually much sm (Photo by E. B. Mains)*

3. Common Fern Moss, Thuidium catulum. *In distribution this moss is gen it grows on soil, stones, and logs in d places. In appearance, it resembles a del fern; its general form reminds one of son the types of frost pictures seen on window winter. (Photo by W. C. Steere)*

4. Awned Hair-Cap Moss, Polytrich piliferum. *This moss is much smaller than pigeon wheat; it may be found growin very dry situations, at high altitudes, even on bare rock ledges. (Photo by Bu Museum of Science)*

5. Plume Moss, Hypnum Crista-castre *In moist cool forests, one may expect to see yellow-green moss at its best; there it ma found even completely covering old logs stumps. (Photo by Buffalo Museum of Sci*

6. Purple-fringed Riccia, Ricciocar natans. *This near relative of the mosses longs to a group of plants known as hep It is shown in the picture as it grows in nant pools; it may also grow on land, in w case it has a very different appearance. (P by W. C. Steere)*

7 *and* 8. True Liverwort, Marcha polymorpha. *These plants, known as worts or hepatics, are close allies of the mo The plant body is in the form of a thallu shown in the pictures) rather than a main and leaves. The thallus creeps on the grou which it is attached by large hairs called rhiz these rhizoids perform the functions of r In No. 7, the portions of the plants whic semble the ribs of an umbrella bear the fe reproductive bodies. The little cups on surface of the thallus in No. 8 produce tative reproductive bodies called gemma brood bodies. A slightly different umbr shaped growth on another plant (not figu produces the male reproductive bodies. (P by E. B. Mains)*

1
2
3
4
5
6
7
8

find all these stages, except perhaps the branching-thread stage.

METHOD — The children should bring to the schoolroom a basin of moss in its fruiting stage; or still better, go with them to a knoll covered with moss. Incidentally tell them that this moss, when dried, is used by the Laplanders for stuffing their pillows, and that the bears use it for their beds. Once, a long time ago, people believed that a plant, by the shape of its leaf or flower, indicated its nature as a medicine, and as this moss looked like hair, the water in which it was steeped was used as a hair tonic.

OBSERVATIONS — 1. Take a moss stem with a grain of pigeon wheat at the end. Examine the lower part of the stalk. How are the leaves arranged on it? Examine one of the little leaves through a lens and describe its shape, its edges, and the way it joins the stem. Are the lower leaves the same color as the upper ones? Why?

2. Describe the pretty shining stem of the fruit, which is called the pedicel. Is it the same color for its entire length? Can you pull it easily from the main plant? Describe how its base is embedded in the tip of the plant.

3. Note the silken cap on a grain of the pigeon wheat. This is called the veil. Is it all the same color? Is it grown fast to the plant at its lower margin? Take it by the tip, and pull it off. Is this done easily? Describe what it covers. This elegant little green vase is called a spore capsule. How many sides has it? Describe its base which stands upon the stem. Describe the little lid. Pull off the lid; is there another lid below it? Can you see around the e
the tiny teeth which hold this lid in pla

4. Do all the spore vases stand strai
up, or do some bend over?

5. Do you think the silken cap falls
of itself after a while? Can you find a
capsules where the cap or veil and the
have fallen off? See if you can shake a
dust out of such a spore vase. What
you think this dust is? Ask your teach
or read in the books, about moss spo
and what happens if they find a da
place in which to grow.

6. Hunt among the moss for so
stems that have pretty, yellowish, star
cups at their tips. How does the inside
one of these cups look? Ask the teache
tell you what grows in this cup. Lo
down the stem and see if you can find
year's cup; the cup of two years ago. Me
ured by these cups how old do you th
this moss stem is?

7. Select some stems of moss, b
those that bear the fruit and those t
bear the cups. After they are dried,
scribe how the leaves look. Examine
plant with a lens and note how th
leaves are folded and twisted. Do
leaves stand out from the stem or lie cl
to it? Is this position of the leaves of
use to the plant in keeping the water fr
evaporating? How do the star-cups l
when dry?

8. Place these dried stems in a glass
water and describe what happens to
cup. Examine some of the dried moss a
the wet moss with a lens, and describe
difference. Of what use to the moss is t
power of changing form when damp?

MUSHROOMS AND OTHER FUNGI

There is something uncanny about plants which have no green parts; indeed, many people find it difficult to think of them as plants. It is, therefore, no wonder that many superstitions cluster about toadstools. In times of old, not only was it believed that toads sat on them, but that fairies danced upon them and used them for umbrellas. The poisonous qualities
some species made them also a nat
ingredient of the witch's cauldron.
science, in these days, brings revelati
concerning these mysterious plants wh
are far more wonderful than the
which superstition wove about them
days of yore.

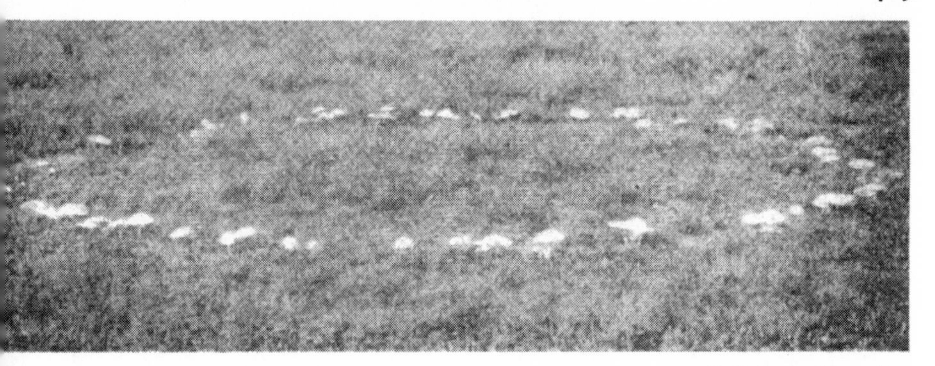

U. S. Dept. of Agriculture

A fairy ring

When we find plants with no green rts which grow and thrive, though unle to manufacture their own organic od through the alchemy of chlorophyll, nlight, and air, we may safely infer that one way or another they gain the prodts of this alchemy at second hand. Such ants are either parasites or saprophytes; parasites, they steal the food from the lls of living plants; if saprophytes, they e on such of this food material as reains in dead wood, withered leaves, or ils enriched by their remains.

Thus, we find mushrooms and other ngus fruiting bodies, pallid, brownve, yellow, or red in color, but with no ns of the living green of other plants; d this fact reveals their history. Some of em are parasites, as certain species of acket fungi which are the deadly eneies of living trees; but most of the fungus species that we ordinarily see are saprophytes, and live on dead vegetation. Fungi, as a whole, are a great boon to the world. Without them our forests would be choked out with dead wood. Decay is simply the process by which fungi and other organisms break down dead material, so that the major part of it returns to the air in gaseous form, and the remainder, now mostly humus, mingles with the soil.

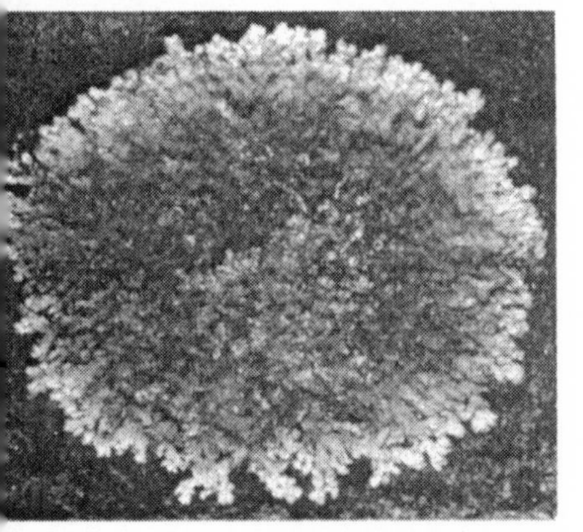

Brooklyn Botanic Garden

A lichen

As a table delicacy, mushrooms are highly prized. A very large number of

George F. Atkinson

Meadow mushroom. A common edible mushroom

species are edible. But every year the newspapers report deaths resulting from eating the poisonous kinds — the price of an ignorance which comes from a lack of the powers of observation developed in nature-study. It would be very unwise for any teacher to give rules to guide her pupils in separating edible from poisonous mushrooms, since the most careful directions may be disregarded or misunderstood. She should emphasize the danger incurred by mistaking a poisonous for an edible species. One small button of the deadly kind, if eaten, may cause death. A few warning rules may be given, which, if firmly impressed on the pupils, may result in saving human life.

First and most important, avoid all mushrooms that are covered with scales, or that have the base of the stem included in a sac, for two of the poisonous species, often mistaken for the common edible mushroom, have these distinguishing characteristics. Care should be taken that every specimen be collected in a way to show the base of the stem, since in some poisonous species this sac is hidden beneath the soil.

Second, avoid the young, or button, stages, since they are similar in appearance in species that are edible and in those that are poisonous.

Third, avoid those that have milky juices; unless the juices are reddish in color, the mushrooms should not be eaten.

Fourth, avoid those with shiny, thin, or brightly colored caps, and those with whitish or clay-colored spores.

Fifth, no mushroom or puffball should be eaten after its meat has begun to turn

George F. Atkinson

The deadly amanita, Amanita phalloides. *Note the form of the ring, and the cup at the base of the stem*

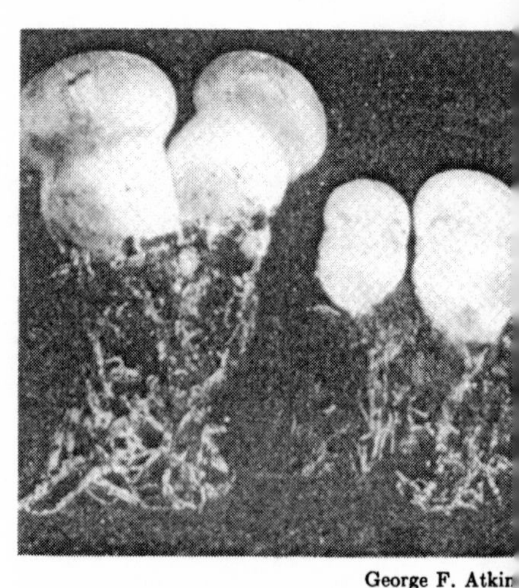

George F. Atkin

Young stages of cultivated edible mushroo showing spawn

brown or has become infested with larvæ.

How Mushrooms Look and How Th Live

There are many kinds of mushroo varying greatly in form, color, and size, b wherever they appear it means that son time previous the mushroom spores ha been planted there. There they threw o threads which have penetrated the fo substance and gained a successful growt which finally resulted in sending up in the world the fruiting organs. In gene shape these consist of a stem with a c upon it, making it usually somewhat u brella-shaped. Attached to the cap, a usually under it, are platelike growt called gills, or a fleshy surface which is fu of pores. In the gills, each side of ea plate develops spores. These, as fine dust, are capable of producing other mus rooms.

In the common edible species of mus room (*Agaricus campestris*), the stem white and almost cylindrical, taperi slightly toward the base; it is solid, though the core is not so firm as the ou side. When it first pushes above t ground, it is in what is called the " butt stage " and consists of a little, round cap covered with a membrane which is a ned to the stem. Later the cap spreads e, for it is naturally umbrella-shaped, it tears loose this membrane, leaving ece of it attached to the stem; this remt is called the ring or collar. The cols very noticeable in many species, but he common mushroom it soon shrivels disappears. The cap is at first rounded then convex; its surface is at first ooth, looking soft and silky; but as the it becomes old, it is often broken up triangular scales which are often dark wn, although the color of the cap is ally white or pale brown. The gills beth the cap are at first white, but later, he spores mature, they become brownblack because of the ripened spores.

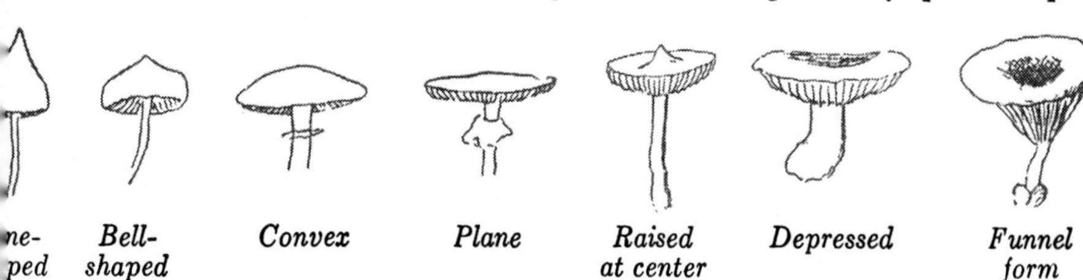

LESSON 198

Mushrooms

Leading Thought — Mushrooms are fruiting organs of the fungi which w in the form of threads, spreading in ry direction through the food material. dust which falls from ripe mushms is made up of spores, which are true seeds, but which will start a new wth of the fungus.

Method — The ideal method would be to study the mushrooms in the field and forest, making an excursion for the purpose of collecting as many species as possible. But the lesson may be given from specimens brought into the schoolroom by pupils, care being taken to bring with them the soil, dead wood, or leaves on which they were found growing. After studying one species thus, encourage the pupils to bring in as many others as possible. There are a few terms which the pupils should learn to use, and one method of teaching them is to place the diagrams shown above and on page 719, on the blackboard, and leave them there for a time.

Since mushrooms are especially good subjects for water-color and pencil studies, it would add much to the interest of the work if each pupil, or the school as a whole, should make a portfolio of sketches of all the species found. With each drawing there should be made on a supple-

George F. Atkinson

Inky-cap mushroom

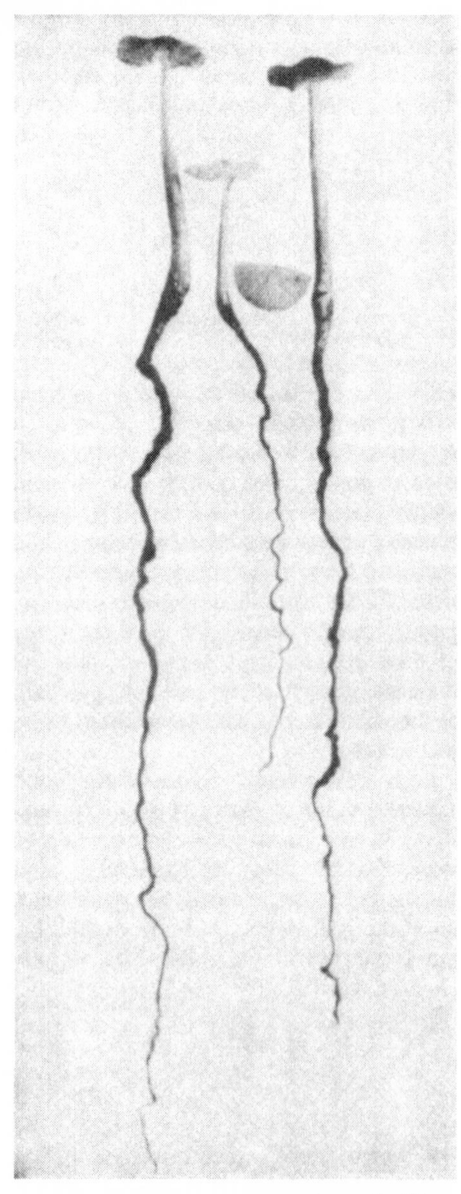

George F. Atkinson

Rooted Collybia, Collybia radicata. *In woods, during summer, in groups or singly this mushroom is common. The stem extends into the ground for some distance, giving the appearance of a " tap root "*

mentary sheet a spore print of the species. White paper should be covered very thinly with white of egg or mucilage, so as to hold fast the discharged spores when making these prints for portfolio or herbarium.

OBSERVATIONS — 1. Where was mushroom found? If on the ground, the soil wet or dry? Was it in open fi or in woods? Or was it found on rot wood, fallen leaves, old trees or stumps roots? Were there many or few sp mens?

2. Is the cap cone-shaped, bell-sha convex, plane, concave, or funnel-fo Has it a raised point at the center? H wide is it?

3. What is the color of the upper face of the cap when young? When Has it any spots of different colors on Has it any striate markings, dots, or grains on its surface? Is its texture smo or scaly? Is its surface dull, or polish or slimy? Break the cap and note the c of the juice. Is it milky?

4. Look beneath the cap. Is the ur surface divided into plates like the le of a book, or is it porous?

5. The plates which may be compa to the leaves of a book are called gills though they are not for the purpose breathing, as are the gills of a fish. there more gills near the edge of the than near the stem? How does this oc What are the colors of the gills? Are gills the same color when young as w old? Are the lower edges of the gills sh blunt, or saw-toothed?

George F. At

A spore print from the common edible m room

5. Break off a cap and note the relation the gills to the stem. If they do not join stem at all they are termed "free." If ey end by being joined to the stem, ey are called "adnate" or "adnexed." they extend down the stem they are led "decurrent."

7. Take a freshly opened mushroom, off the stem even with the cap, and the cap, gills down, on white paper; er with a tumbler, or other dish to exde draught; leave it for twenty-four urs and then remove the cover, lift the carefully and examine the paper. hat color is the imprint? What is its

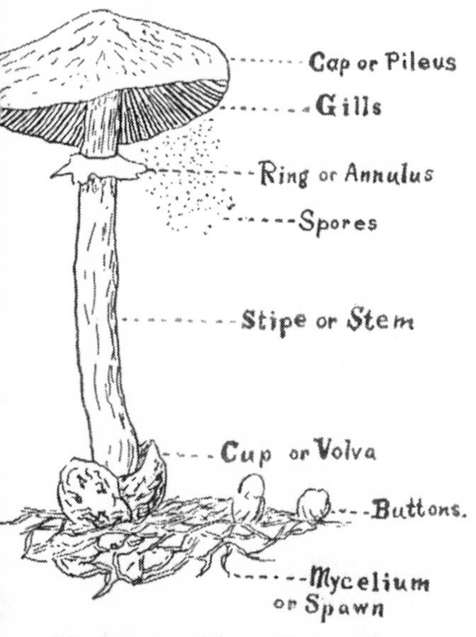

Mushroom with parts named

ape? Touch it gently with a pencil and what makes the imprint. Can you tell the pattern where this fine dust came m? Examine the dust with a lens. This st is made up of mushroom spores, ich are not true seeds, but which do for shrooms what seeds do for plants. How you think the spores are scattered? Do u know that one little grain of this spore st would start a new growth of mushoms?

8. Look at the stem. What is its length? Its color? Is it slender or stocky? Is its surface shiny, smooth, scaly, striate, or dotted? Has it a collar or ring around it near the top? What is the appearance of

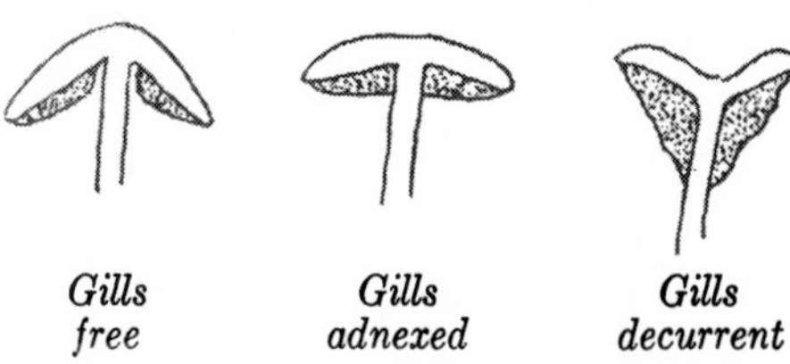

Gills free *Gills adnexed* *Gills decurrent*

this ring? Is it fastened to the stem, or will it slide up and down? Is the stem solid or hollow? Is it swollen at its base? Is its base set in a sac or cup, or is it covered with a membrane which scales off? Do you know that the most poisonous of mushrooms have the sac or the scaly covering at the base of the stem?

9. Examine with a lens the material on which the mushroom was growing; do you see any threads in it that look like mold? Find if you can what these threads do for the mushroom. If you were to go into the mushroom business what would you buy to start your beds? What is mushroom "spawn"?

10. If you can find where the common edible mushrooms grow plentifully, or if you know of any place where they are grown for the market, get some of the young mushrooms when they are not larger than a pea and others that are larger and older. These young mushrooms are called "buttons." Find by your own investigation the relation between the buttons and the threads. Can you see the gills in the button? Why? What becomes of the veil over the gills as the mushrooms grow large?

11. Do you know the common edible mushroom when you see it? What characteristics separate this from the poisonous species? What is the "death cup," as it is called, which covers the base of the stem of the most common poisonous species?

PUFFBALLS

The puffballs are always interesting to children, because of the " smoke " which issues from them in clouds when they are pressed between thumb and finger. The common species are white or creamy when young; and some of the species are warty or roughened, so that as children we

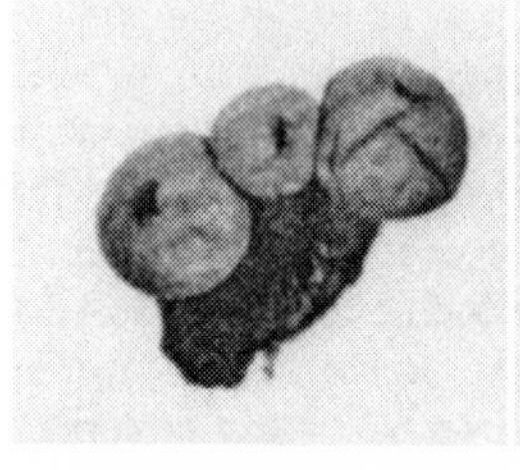
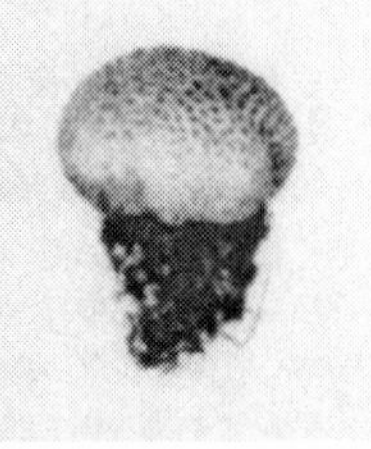

Dr. and Mrs. John Small

Puffballs. On the left is shown Lycoperdon, *on the right,* Scleroderma. *The mature spores escape through the openings to be seen on* Lycoperdon

called them " little lambs." They grow on the ground usually, some in wet, shady places, and others, as the giant species, in grassy fields in late summer. This giant puffball always excites interest when found. It is a smoothish, white, rounded mass, apparently resting on the grass as if thrown there; when lifted it is seen that it has a connection below at its center, through its mycelium threads, which form a network in the soil. It is often a foot in diameter, and specimens four feet through have been recorded. When its meat is solid and white to the very center, it makes very good food. The skin should be pared off, the meat sliced and sprinkled with salt and pepper and fried in hot fat until browned. All the puffballs are edible, but uninformed persons might mistake the button stages of some of the poisonous mushrooms for little puffballs, and it is not well to encourage the use of small puffballs for the table.

A common species — " the beaker puffball " — is pear-shaped, with its small end made fast to the ground, which is permeated with its vegetative threads.

The interior of a puffball, " the mea is made up of the threads and spores. they ripen, the threads break up so t with the spores they make the " smok as can be seen if the dust is examin through a microscope. The outer wall m become dry and brittle and break open allow the spores to escape, or one or m openings may appear in it as spore do The spores of puffballs were used ext sively in pioneer days to stop the ble ing of wounds and especially for no bleed.

In one genus of the puffball family, t outer coat splits off in points on maturi like an orange peel cut lengthwise in or seven sections but still remaining tached to the base. There is an inner c that remains as a protection to the spor so that these little balls are set each i little star-shaped saucer. These star poi straighten out flat or even curl under dry weather, but when damp they lift and again envelop the ball to a greater less extent.

Wm. P. Alexa

Cup-shaped puffball, Calvatia cyathi mis. *This edible puffball may reach 6 inc in diameter; it is found on open gra ground in early autumn*

LESSON 199

Puffballs

Leading Thought — The puffballs are ngi that grow from the threads or mylia which permeate the ground or other atter on which the puffballs grow. The ıffballs are the fruiting organs, and smoke" which issues from them is rgely made up of spores, which are car:d off by the wind and thus sown and anted.

Method — Ask the pupils to bring to hool any of the globular or pear-shaped ngi in the early stages when they are hite, taking pains to bring them on the il or wood on which they are growing.

Observations — 1. Where did you find .e puffball? On what was it growing? ʼere there many growing in company? emove the puffball, and examine the ace where it stood with a lens to find the atted and crisscrossed fungus threads.

Verne Morton

An earthstar

2. What is the size and shape of the uffball? Is its surface smooth or warty? ʼhat is its color inside and outside?

3. Have you ever found the giant puff-

George F. Atkinson

Giant puffball, Calvatia gigantea. *It is not unusual to find these puffballs* 10 *to* 20 *inches in diameter. This is the largest puffball and is a great favorite among the edible varieties. In prime condition the flesh is white; it is edible as long as it remains white*

ball, which may become four inches to four feet through? Where was it growing? Have you ever eaten this puffball sliced and fried? Do you know by the looks of the meat when it is fit to eat?

4. If the puffball is ripe, what is its color outside and in? What is the color of its "smoke"? Does the smoke come out through the broken covering of the puffball, or are there one or more special openings to allow it to escape?

5. Puff some of the "smoke" on white paper and examine it with a lens. What do you think this dust is? Of what use is it to the puffball?

6. Have you ever found what are called earthstars, which look like little puffballs set in star-shaped cups? If you find these note the following things:

(a) Of what is the star-shaped base made?

(b) Let this star saucer become very dry; how does it act?

(c) Wet it; how does it behave then?

(d) Where and how does the spore dust escape from the earthstars?

7. For what medicinal purpose is the "smoke" of the puffball sometimes used?

THE BRACKET FUNGI

There are some naturalists who think ıat one kind of life is as good as another nd therefore call all things good. Perhaps this is the only true attitude for the nature lover. To such the bracket-like fungi which appear upon the sides of our

Verne Morton

A bracket fungus

forest and shade trees are simply an additional beauty, a bountiful ornamentation. But some of us have become special pleaders in our attitude toward life, and those of us who have come to feel the grandeur of tree life can but look with sorrow upon these fungus outgrowths, for they mean that the doom of the tree is sealed.

There are many species of bracket fungi. Three of these are very common. The gray bracket, gray above and with creamy surface below (*Polyporus applanatus*), is a favorite for amateur etchers, who with a sharp point make interesting sketches upon this naturally prepared plate; this species often grows to great size and is frequently very old. Another species (*P. lucidus*) is in color a beautiful mahogany or coral-red above and has a peculiar stem from which it depends; the stem and upper surface are polished as if burnished and the lower surface is yellowish white. Another species (*P. sulphureus*) is sulphur yellow above and below; usually many of these yellow brackets are grouped together, their fan-shaped caps overlapping. Many of the shelf fungi live only on dead wood, and those are an aid in reducing dead branches and stumps until they crumble and become again a part of the soil. However, several of the species attack living trees and do great damage. They can gain access to the living tree only through an injured place in the bark, a break caused perhaps by the wind, by a bruise from a falling tree, or more oft from the hack of the careless wood-chc per; often they gain entrance through unhealed knot hole. To one who und stands trees and loves them, these woun inflicted by forces they cannot withsta are truly pathetic. After the wound made and before the healing is accom lished, the wind may sift into the wou the almost omnipresent spores of the fungi and the work of destruction begi From the spores grows the mycelium, t fungus threads which push into the he of the wood, getting nourishment fro it as they go. When we see wood th diseased we say that it is rotting, b rotting merely means the yielding up the body substance of the tree to the voracious fungus threads. They push radially and then grow upward and dov ward, weakening the tree where it m needs strength to withstand the onslaug of the wind. Later these parasitic threa may reach the cambium layer, the livi ring of the tree trunk, and kill the tr entirely; but many a tree has lived lo with the fungus attacking its heartwo

George F. Atkir

A bracket fungus, Polyporus versico *This is a common form of* Polyporus *found dead wood. When wet this fungus is flexi but when dry it is woody, and almost bri*

bracket fungus found by Professor
inson was eighty years old; however,
may have shortened the life of the
a century or more.

After these fungus threads are thor-
hly established in the tree, they again
a wound in the protecting bark where
y may push out and build the fruiting
an, which we call the bracket. This
be at the same place where the fatal
ry was made, or it may be far from it.
e bracket is at first very small and is
nposed of a layer of honeycomb cells,
ed and hard above and opening below
ells so small that we can see the cell
nings only with a lens. These cells are
hexagonal like the honeycomb, but
tubes packed together. Spores are de-
oped in each tube. Next year another
r of cells grows beneath this first
cket and extends out beyond it; each
it is thus added to, making it thicker
marking its upper surface with con-
tric rings around the point of attach-

George F. Atkinson

Oyster mushroom, Pleurotus ostreatus

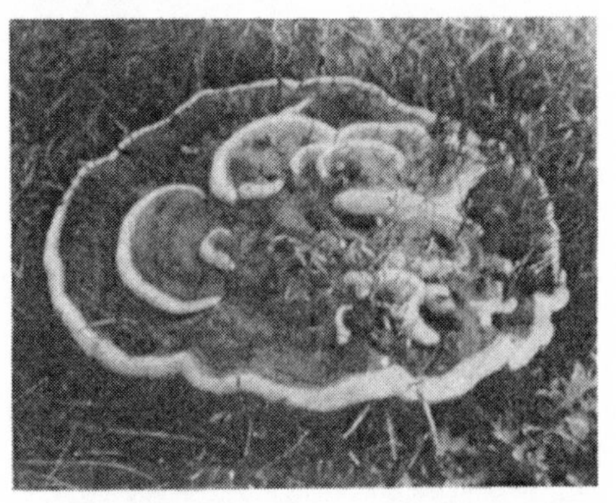

George F. Atkinson

This woody type of pore fungi, Gandoderma, *usually found growing on old wood has a brittle polished crust*

ment. The creamy surface of the great bracket fungus on which etchings are made is composed of a layer of these minute spore-bearing tubes. Not all bracket fungi show their age by these annual growths, for some species form new shelves every year, which decay after the spores are ripened and shed.

When once the mycelium of such a fungus becomes established, the tree is probably doomed and its lumber made worthless even though, as sometimes happens, the tree heals its wounds so that the fungus is imprisoned and can never send out fruiting brackets. Thus it is most important to teach the pupils how to protect trees from the attacks of these enemies, which are devastating our forests, and which sometimes attack our orchards and shade trees.

As soon as a tree is bruised, the wound should be painted or covered with a coat of tar. If the wind breaks a branch, the splinters left hanging should be sawed off, leaving a smooth stump, and this should be painted. While ordinary paint if renewed each year will suffice, experiment has shown that the coat of tar is better and should be used.

Especially should teachers impress on pupils the harm done by careless hacking with axe or hatchet. We shall do an invaluable service in the protection of our forests if we teach the rising generation the respectful treatment of trees — which

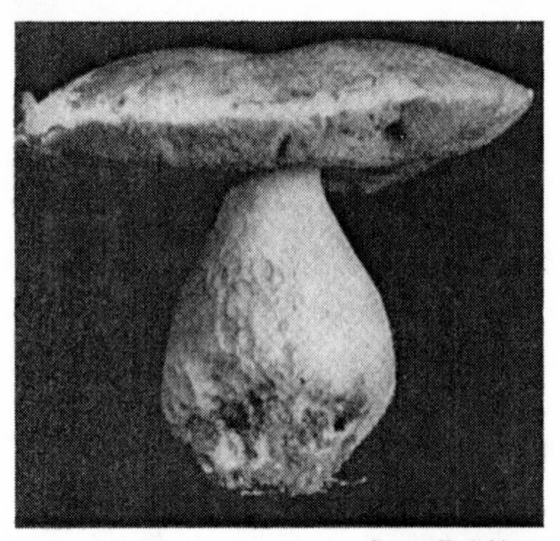

George F. Atkinson

Edible boletus, Boletus edulis. *This is a common plant in woods and open places during July and August. It has tubes instead of gills below the cap. The spores are developed within the tubes, as in the bracket fungi*

is due living conditions whose span of life may cover centuries.

LESSON 200

Bracket Fungi

Leading Thought — The fungi which we see growing shelflike from trees are deadily enemies to the trees. Their spores germinate and penetrate at some open wound, and the growing fungus weakens the wood.

Method — It is desirable that a tree on which shelf fungus grows should be studied by the class, for this is a lesson on the care of trees. After this lesson the fungus itself may be studied at leisure in the schoolroom.

Observations — 1. On what kind of tree is the bracket fungus growing? Is it alive or dead? If living, does it look vigorous or is it decaying?

2. Is the fungus bracket growing against the side of the tree, or does it stand out on a stem?

3. Look at the place where the bracket joined the tree. Does it seem to be a part of the wood?

4. What color is the fungus on its upper surface? How large is it? How thick near the tree? How thick at the ec
Can you detect concentric layers or ri
If it is the large species used for etch
cut down through it with a knife
hatchet and count the layers; this sho
show its age.

5. Look at the lower surface. How
it appear to the naked eye? If you scra
it with a pin or knife does the br
show? Examine the surface with a
and describe what you see. Cut or b
the fungus and note that each of th
holes is an opening to a little tube.
each of these tubes spores are borne.

6. Have you ever seen toadstools t

William P. Alex

Bear's head fungus, Hydnum caput-
This beautiful fungus grows in white clu
of irregular shape and hangs from deca
trees or logs like clumps of icicles; the sp
are produced on teeth instead of in pore
gills as in many other mushrooms

ıtead of having the leaflike gills, have neath the cap a porous surface like a le honeycomb or like the under side of shelf fungi?

7. How many kinds of shelf fungi can u find? Which of them is on living es, and which on stumps or dead wood?

8. If the fungus is on a living tree, then tree is ruined, for the fungus threads ve worked through it and weakened it that it will break easily and is of no use lumber. There must have been an open und in the tree where the fungus en- ed; see whether you can find this und. There must also have been a und where the shelf grew out; see ether you can detect it. If the tree ould heal all its wounds after the fun- entered, what would become of the igus?

9. What does the shelf fungus feed on? hat part of it corresponds to the roots d leaves of other plants? What part may compared to the flowering and fruiting rts of other plants?

10. What treatment must we give trees keep them free from this enemy?

LESSON 201

Hedgehog Fungi

There is something mysterious about fungi, but perhaps none of these won- rful organisms so strangely impresses observer as the fountain-like masses creamy white or the branching white al that we see growing on a dead tree nk. The writer remembers as a child t the finding of these woodland treas- s made her feel as if she were in the sence of the supernatural, as if she had covered a fairy grotto or a kobold cave. e prosaic name of hedgehog fungi has n applied to these exquisite growths. eir life story is simple enough. The res falling upon dead wood start eads which ramify within it and feed its substance, until strong enough to d out a fruiting organ. This consists of tem, dividing into ascending branches; m these branches, depending like the lactites in a cave, are masses of drooping spines, the surface of each bearing the spores. And it is so natural for these spines to hang earthward that they are invariably so placed, unless the position of the tree has changed since they grew. There is one species called the "satyr's beard," sometimes found on living trees, which is a mere bunch of downward-hanging spines; the coral-like species is called *Hydnum coraloides*, and the one that looks like an exquisite white frozen fountain, and may be seen in late summer or early autumn growing from dead limbs or branches, is the bear's-head fungus; it is often eight inches across.

Suggested Reading — Readings on page 717.

Observations — 1. These fungi come from a stem which extends into the wood.

2. This stem divides into many branchlets.

3. From these branchlets there hang long fleshy fringes like miniature icicles.

4. These fringes always hang downward when the fungus is in natural position.

5. These fringes bear the spores.

LESSON 202

The Scarlet Saucer

(*Sarcocypha coccinea*)

The heart of the child, searching the woods for hepaticas — woods where snow banks still hold their ground on north slopes — is filled with delight at finding these exquisite saucer-like fungi. They are most often found on fallen rotting branches which are more or less buried in leaves, and there are likely to be several of different sizes on the same stick. When they grow unhindered, and while

George F. Atkinson

The scarlet saucer, Sarcocypha coccinea

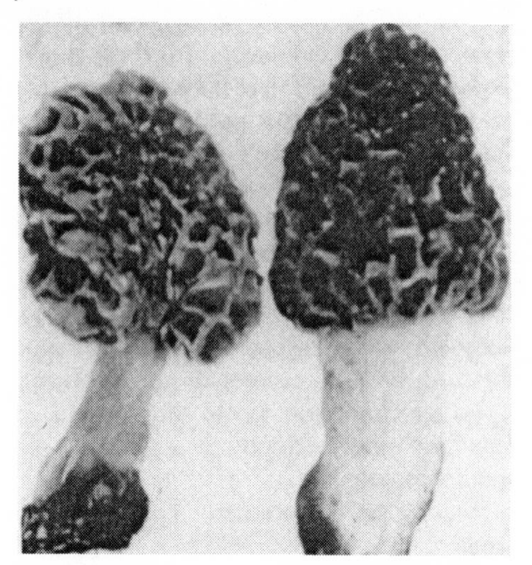

George F. Atkinson

An edible morel, Morchella esculenta

they are young, they are very perfectly saucer-shaped and range from the size of a pea to an inch or two across. But the larger they are the more likely are they to be distorted, either by environment or by the bulging of rapid growth. The underside of the saucer is beautifully fleshlike in color and feeling, and is attached at the middle to the stick. The inside of the saucer is the most exquisite scarlet, shading to crimson. This crimson lining bears the spores in little sacs all over its surface.

Observations — 1. Where did you find the fungus?

2. What is the shape of the saucer? How large is it? Is it regular and beautiful or irregular and distorted?

3. What is the color inside?

4. What is the color outside?

5. Turn the saucer bottom side up — that is, scarlet side down — on a piece of white paper, and see whether you can get a spore harvest.

LESSON 203

The Morels

In May or June in open, damp places, such as orchards or the moist fence-corners of meadows, the morels may be found. This mushroom family contains member that is poisonous, and the me bers are very unlike any other family appearance. They are very pretty w their creamy white, thick, swollen ste and a cap more or less conical, made up the deep-celled meshes of an unequal n work. The outside edges of the n work are yellowish or brownish when morel is young and edible, but later t dark as the spores develop. In some s cies the stems are comparatively smo and in others their surface is more or wrinkled. The spores are borne in the pressions of the network. These mu rooms should not be eaten after the c change from creamy white to brownish

Observations — 1. Where did you f the morels?

2. Describe the stem. Is it solid or h low? Is it smooth or rough?

3. What is the shape of the cap? H does it look? What color is the outer e of the network? What is the color wit the meshes?

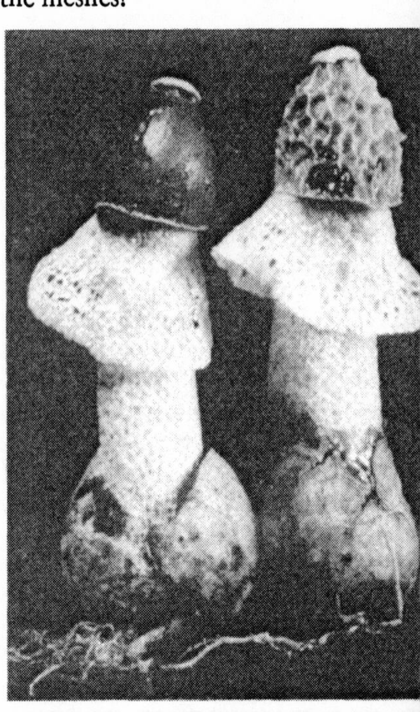

George F. Atk

Stinkhorns

4. Take one of these fungi, lay it on a eet of white paper, and note the color the spores.

LESSON 204

The Stinkhorns

To give a nature-study lesson on the nkhorn is quite out of the question, for e odor of these strange growths is so useating that even to come near to one them in the garden is a disagreeable perience. The reason for mentioning em at all is because of the impression ade by them that most mushrooms are smelling, which is a slander.

It is a pity that these fungi are so offen- e that we do not care to come near ough to them to admire them, for they e most interesting in appearance. The ientific name of our commonest genus hen translated means "the net bear- s," and it is a most appropriate name. he stout, white stem is composed of net- ork without and within. The outer cov- ing of the stem seems to tear loose from e lower portion as the stem elongates, d is lifted so that it hangs as a veil ound the bottom of the bell-shaped cap, hich is always covered with a pitted net- ork. The mycelium, or spawn, of the inkhorn consists of strands which push eir way through the ground or through e decaying vegetable matter on which they feed. On these strands are produced the stinkhorns, which at first look like eggs; but later the top of the egg is broken, and the strange horn-shaped fungus pushes up through it. The spores are borne in the chambers of the cap, and when ripe the substance of these chambers dissolves into a thick liquid in which the spores float. The flies are attracted by the fetid odor and come to feast upon these fungi and to lay their eggs within them, and incidentally they carry the spores away on their brushy feet, and thus help to spread the species.

George F. Atkinson

Bird's nest fungi

MOLDS

It is lucky for our peace of mind that r eyes are not provided with micro- opic lenses, for then we should know at the dust, which seems to foregather on our furniture from nowhere, is com- sed of all sorts of germs, many of them the deadly kind. The spores of mold e very minute objects, the spore cases ing the little white globes, not larger an the head of a small pin, which we see on mold; yet each of these spore cases eaks and lets out into the world thou- nds of spores, each one ready to start a owth of mold and perfectly able to do it under the right conditions; almost any substance which we use for food, if placed in a damp and rather dark place, will prove a favorable situation for the development of the spore, which swells, bursts its wall, and sends out a short thread. This gains nourishment, grows longer, and branches, sending out many threads, some of which go down into the nutritive material and are called the mycelium. While these threads, in a way, act like roots, they are not true roots. Presently the tip ends of the threads, which are spread out in the air above the bread or other material, begin to enlarge, forming little gobules; the substance (protoplasm) within them breaks up into little round bodies, and each develops a cell wall and thus becomes a spore. When these are unripe they are white, but later they become almost black. In the blue mold the spores are borne in clusters of chains, and resemble tiny tas-

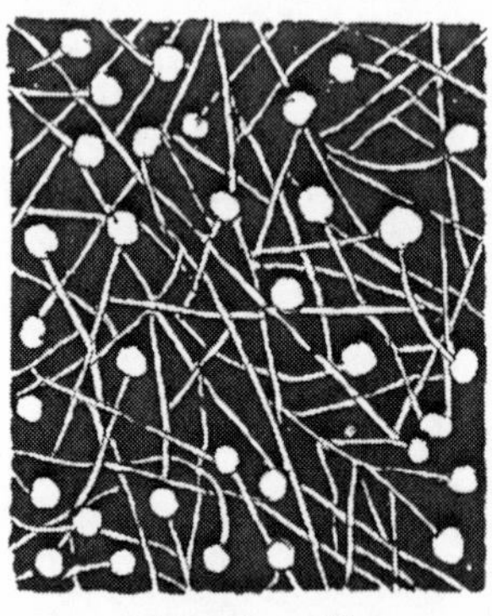

Bread mold

sels instead of growing within little globular sacs.

Molds, mildews, blights, rusts, and smuts are all flowerless plants and, with the mushrooms, belong to the great group of fungi. Molds and mildews will grow upon almost any organic substance, if the right conditions of moisture are present, and the temperature is not too cold.

Molds of several kinds may appear upon the bread used in the experiments for this lesson. Those most likely to appear are the bread mold — consisting of long, white threads tipped with white, globular spore cases, and the green cheese mold — which looks like thick patches of blue-green powder. Two others may appear, one a smaller white mold with smaller spore cases, and the other a black mold. However, the bread mold is the one most desirable for this lesson, because of its comparatively large size. When examined with a lens, it is a most exquisite plant. The long threads are fringed at the sides, and they pass over and through each other, making a web fit for fairies — a web all beset with the spore cases, like fairy pearls. However, as the spores ripen, these spore cases turn black, and after a time so many of them are developed and ripen that the whole mass of mold is black. T time required for the development mold varies with the temperature. For tv or three days nothing may seem to be ha pening upon the moist bread; then queer, soft whiteness appears in patch In a few hours or perhaps during t night, these white patches send up wh fuzz which is soon dotted with tiny pea like spore cases. At first there is no od when the glass is lifted from the sauc but after the spores ripen, the odor quite disagreeable.

The special point to teach the childr in this lesson is that dryness and sunlig are unfavorable to the development mold; and it might be well to take one the luxuriant growths of mold develop in the dark, uncover it and place it the sunlight, and see how soon it withe The lesson should also impress upon the that dust is composed, in part, of livi germs waiting for a chance to grow.

LESSON 205

Molds

Leading Thought — The spores mold are everywhere and help to ma what we call dust. These spores will gr on any substance which gives them no ishment, if the temperature is warm, t air is moist, and the sunlight is exclud

Method — Take bread in slices t inches square, and also the juice of ap sauce or other stewed fruit. Have ea pupil, or the one who does the work the class, provided with tumblers a saucers. Use four pieces of bread cut about two-inch squares, each placed o saucer; moisten two and leave the ot two dry. With a feather or the fin take some dust from the woodwork of room or the furniture and with it ligh ch each piece of bread. Cover each h a tumbler. Set one of the moistened ces in a warm, dark place and the other dry, sunny place. Place a dry piece in ilar situations. Let the pupils examine se every two or three days.

'ut fruit juice in a saucer, scatter a little t over it and set it in a warm, dark ce. Take some of the same, do not scat- any dust upon it, cover it safely with a ibler, and put it in the same place as other. A lens is necessary for this les- , and it is much more interesting for pupils if they can see the mold under icroscope with a three-fourths objec- :.

)BSERVATIONS — 1. When does the ld begin to appear? Which piece of ad showed it first? Describe the first nges you noticed. What is the color of mold at first? Has it any odor?

. At what date did the little branching ld threads with round dots appear? Is re an odor when these appear? What the colors of the dots, or spore cases, first? When do these begin to change or? How does the bread smell then? hat caused the musty odor?

;. Did the mold fail to appear on any the pieces of bread? If so, where were these placed? Were they moist? Were they exposed to the sunlight?

4. Did more than one kind of mold appear on the bread? If so, how do you know that they are different kinds? Are there any pink or yellow patches on the bread? If so, these are made by bacteria and not by mold.

5. From the results of the experiments, describe in what temperature mold grows best; in what conditions of dryness or moisture? Does it flourish in the sunlight or in the dark?

6. Where does the mold come from? What harm does it do? What should we do to prevent the growth of mold? Name all of the things on which you have seen mold or mildew growing.

7. Examine the mold through a microscope or a lens. Describe the threads. Describe the little round spore cases. Look at some of the threads that have grown down into the fruit juice. Are they like the ones which grow in the air?

8. If you have a microscope cut a bit of the mold off, place it in a drop of water on a glass slide, and put on a cover glass. Examine the mold with a three-fourths objective, and describe the spores and spore cases.

BACTERIA

The yellow, pink, or purple spots devel- ed upon the moist and moldy bread y be caused by bacteria and yeast. Bac- ia are one-celled organisms; they are smallest known living things, and can seen only through a high-power micro- pe.

3acteria are found almost everywhere in the soil, on foods and fruits, in the ter of ponds, streams, and wells, in the uths and stomachs of all animals, and fact in almost all possible places. They ur also in the air. Most of them are rmless, some of them are useful, and ny produce disease in both plants and mals, including man.

What bacteria do would require many ge volumes to enumerate. Some of them develop colors or pigments; some produce gases, often ill-smelling; some are phosphorescent; some take nitrogen from the air and fix it in the soil; some produce putrefaction; and some produce disease. Nearly all of the contagious diseases are produced by bacteria. Diphtheria, scarlet fever, typhoid fever, tuberculosis, influenza, grippe, colds, cholera, lockjaw, leprosy, blood poisoning, and many other diseases are thought to be the result of bacteria. On the other hand, many of the bacteria are beneficial to man. Some forms ripen the cream before churning, others give flavor to butter; some are an absolute necessity in making cheese. The making of cider into vinegar is the work of bacteria; some help to decompose the dead bodies of animals, so that they return to the dust whence they came.

We have in our blood little cells whose business it is to destroy the harmful bacteria which get into the blood. These little fighting cells move everywhere with our blood, and if we keep healthy and vigorous by right living, right food, and exercise, these cells may prove strong enough to kill the disease germs before they harm us. Direct sunlight also kills some of the bacteria. Exposure to the air is also a help in subduing disease germs. Bichloride of mercury, carbolic acid, formaldehyde, and burning sulphur also kill germs. We can do much to protect ourselves from harmful bacteria by being very clean in our persons and in our homes, by bathing frequently, and washing often with soap. We should eat only pure and freshly cooked food, we should get plenty of sleep and admit the sunlight to our homes; we should spend all the time possible in the open air and be careful to drink pure water. If we are not sure that the water is pure, it should be boiled for twenty minutes and then cooled for drinking.

In Experiment A the milk vials and the corks are all boiled, so that we may be sure that no other bacteria than the ones we chose are present, since boiling kills these germs. As soon as the milk becomes discolored we know that it is full of bacteria.

Experiment B shows that bacteria can be transplanted to gelatin, which is a material favorable for their growth. But the point of this experiment is to show the child that a soiled finger will have upon it germs which, by growing, cloud the gelatin. They should thus learn the value of washing their hands often or of keeping their fingers out of their mouths.

Experiment C shows the way the de-structive bacteria attack the potato. ' discolored spots show where the de begins, and the odor is suggestive of de If a potato thus attacked is put in bright sunlight the bacteria are destroy and this indicates a value of sunshine

LESSON 206

Bacteria

Leading Thought — Bacteria are s small plants that we cannot see th without the aid of a microscope, but t can be planted and will grow. The ob of this lesson is to enforce cleanliness.

1, *A bacillus which causes cholera.* 2, *A bacillus which causes typhoid.* 3, *A bacillus found in sewage*
All these are much enlarged

4, *Bacteria from tubercle on white s clover, much enlarged.* 5 and 6, *Bacteri lactic acid ferments in ripening of ch much enlarged*

Method — Experiment A — The br used for the mold experiment is likel develop spots of yellow, red, or pu upon it, and cultures from these sp may be used in this lesson as follows: T some vials, boil them and their corks, nearly fill them with milk that has b boiled. Take the head of a pin or hair sterilize the point by holding in a fla let it cool, touch one of the yellow sp on the bread with the point, being car to touch nothing else, and thrust the p with the bacteria on it into the milk; t cork the vials.

Experiment B — Prepare gelatin a the table but do not sweeten. Pour s of this gelatin on clean plates or sau After it has cooled let one of the chil touch lightly the gelatin in one saucer a few seconds with his soiled finger. M the place. Ask him to wash his ha thoroughly with soap and then app

ger to the surface of the gelatin in the
ıer plate. Cover both plates to keep out
: dust and leave them for two or three
ys in a dark place. The plates touched
the soiled finger will show a clouded
owth in the gelatin; the other plate will
ɔw a few irregular, scattered growths or
ne.

EXPERIMENT C — Take a slice of boiled tato, place it in a saucer, leave it uncov- :d for a time or blow dust upon it, label with the date, then cover it with a tum- :r to keep it from drying and place it in :ool, somewhat dark place.

The pupils should examine all these ltures every day and make the following ›tes:

EXPERIMENT A — How soon did you serve a change in the color of the milk? ɔw can you tell when the milk is full of e bacteria? How do you know that the cteria in the milk were transplanted by e pin?

EXPERIMENT B — Can you see that the latin is becoming clouded where the iled finger touched it? This is a growth the bacteria which were on the soiled ıger.

EXPERIMENT C — What change has ken place in the appearance of the slice potato? Are there any spots growing oon it? What is the odor? What makes e spots? Describe the shape of the spots. he color. Are any of them pimple- aped? Make a drawing of the slice of po- to showing the bacteria spots. What e the bacteria doing to the potato? Take part of the slice of potato with the bac- ria spots upon it, and put it in the sun- ine. What happens? Compare this with e part kept in the dark.

After this lesson the children should : asked the following questions:

1. Why should the hands always be ashed before eating?
2. Why should the fingernails be kept ean?
3. Why should we never bite the ngernails or put the fingers in the outh?
4. Why should we never put coins in e mouth?
5. Why should wounds be carefully cleansed and dressed at once?
6. Why should clothing, furniture, and the house be kept free from dust?
7. Why should house cleaning be done as far as possible without raising dust?
8. Why are hardwood floors more healthful than carpets?
9. Why is a damp cloth better than a dry duster for removing dust?
10. Why should the prohibition against spitting in public places be strictly enforced?
11. Why should the dishes, clothes, and other articles used by sick persons be kept distinctly separate from those used by well members of the family?
12. Why should food not be exposed for sale on the street?
13. Why, during an epidemic of such a disease as typhoid fever, should water be boiled before drinking?

This habit of looking first at what we call the beauty of objects is closely associated with the old conceit that everything is made to please man: man is only demanding his own. It is true that everything is man's because he may use it or enjoy it, but not because it was designed and " made " for " him " in the beginning. This notion that all things were made for man's special pleasure is colossal self-assurance. It has none of the humility of the psalmist, who exclaimed, " What is man, that thou art mindful of him? "

" What were these things made for, then? " asked my friend. Just for themselves! Each thing lives for itself and its kind, and to live is worth the effort of living for man or bug. But there are more homely reasons for believing that things were not made for man alone. There was logic in the farmer's retort to the good man who told him that roses were made to make man happy. " No, they wa'n't," said the farmer, " or they wouldn't a had prickers." A teacher asked me what snakes are " good for." Of course there is but one answer: they are good to be snakes.

— " THE NATURE-STUDY IDEA,"
L. H. BAILEY

PART IV
EARTH AND SKY

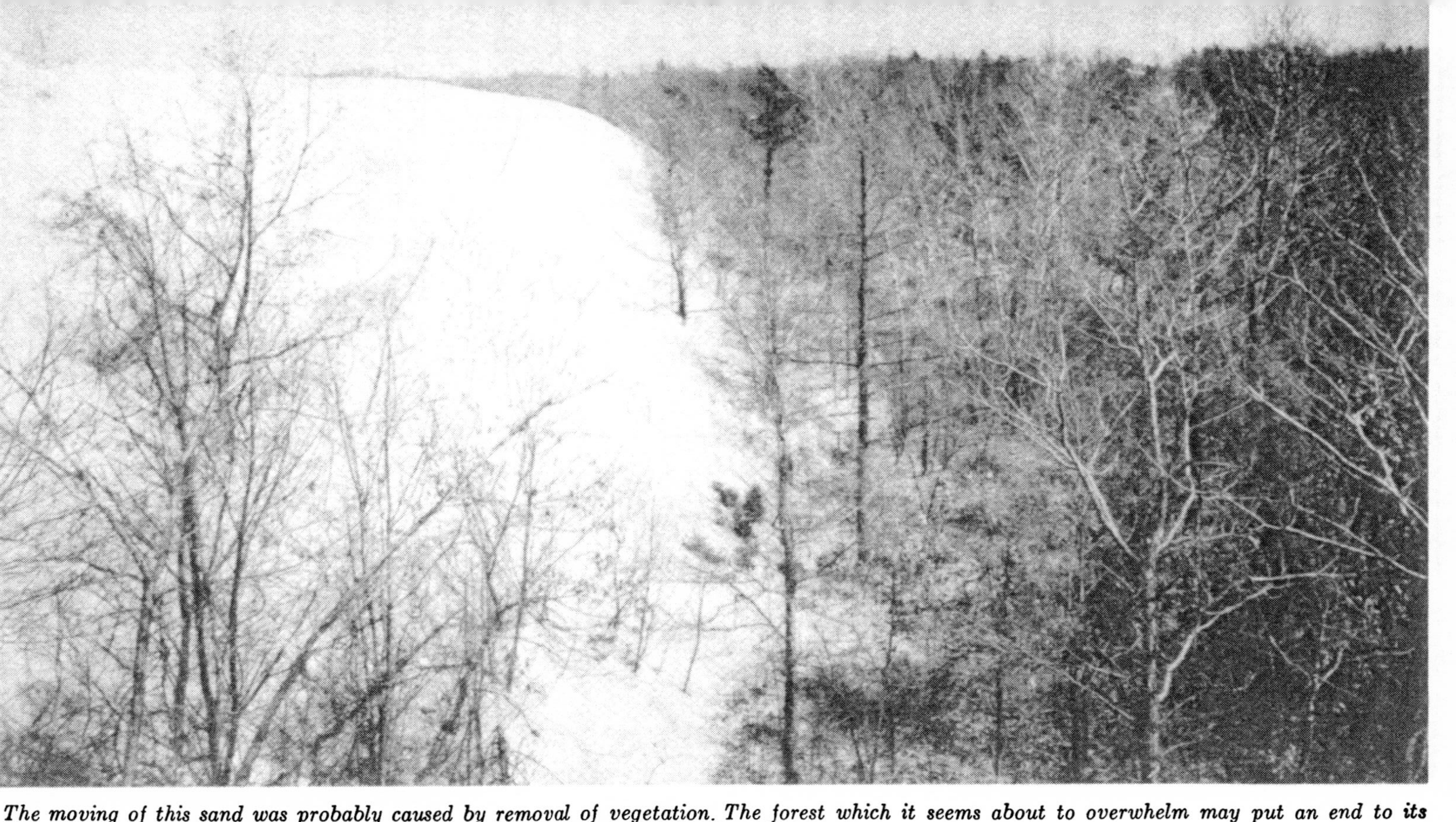

The moving of this sand was probably caused by removal of vegetation. The forest which it seems about to overwhelm may put an end to its further progress in this direction

THE BROOK

Little brook, sing a song of a leaf that sailed along,
Down the golden braided center of your current swift and strong.

— J. W. Rili

A brook is undoubtedly the most fascinating bit of geography which the child encounters; and yet how few children who happily play in the brook — wading, making dams, drawing out the crayfish by his own grip from his lurking place under the log, or watching schools of tiny minnows — ever dream that they are dealing with real geography. The geography lesson on the brook should not be given for the purpose of making work out of play, but to conserve all the natural interest in the brook, and to add to it by revealing other and more interesting facts concerning the brook. A child who thus studies it will master some of the fundamental facts of physical geography, so that ever after he will know and understand all streams, whether they are brooks or rivers. An interesting time to study a brook is after a rain; and May or October gives attractive surroundings for the study. However, the work should be continued now and then during the entire year, for each season gives it some new features of interest.

Each brook has its own history, which can be revealed only to the eyes of those who follow it from its beginning to where it empties its water into a larger stream or pond. At its source the brook usually is a small stream with narrow banks; not until it receives water from surrounding slopes does it gain enough power to cut its bed deep in the earth. Where it flows with swift current down a hillside, it cuts its bed deeper, because swift-moving water has great power for cutting and carrying away the soil. However, if the hillside happens to be in the woods, the roots of trees or bushes will help to keep the soil from being washed away. Unless there obstacles, the course of the brook is lik to be more direct in flowing down a h side than when crossing level fields. T delightful way in which brooks mean across level areas is due to some obst tion, such as a tree, a stone, or a bu of grass or shrubs, which interferes m with the movement of water on a pl than on a hillside. Gravity, which fore pulls water down a steep slope, acts u it less forcibly on gently sloping or ne level lands. After a stream has thus sta its crooked course, in time of flood current strikes with great force on outside of the curves, thereby cut them back and making the stream co still more crooked. The places on banks where the soil is bare and expo to the force of the current are the po where the banks are cut most rapidly flood time.

But the brook is not simply an objec look at and admire; it is a very busy wor its chief labor being that of a digger carrier. When it is not carrying anyth — that is, when its waters are perfe clear — the stream is doing the least w The poets, as well as common peo speak of the playing of the brook w its limpid waters catch the sunbeams their dimpling surface; but when waters are roily, the brook is working hard. This usually occurs after a r which adds much more water to the ume of the brook; the action of gra upon this larger volume forces it to more swiftly, and every drop in the str that touches the bank or bottom snatc up a tiny load of earth and carries it al And every drop thus laden, when it str inst a corner of the bank, tears more loose through the impact, and other ps snatch it up and carry it on down stream. Thus, after a time there are many drops carrying loads and bump- along, knocking loose more earth, that whole brook, which is made up of ps, looks muddy. In its work as a dig-, every drop of water that touches the at the bottom or on the banks of brook uses its own little load of earth gravel as a crowbar or pickaxe to loosen er bits of dirt and gravel; and all the ps hastening on, working hard to-her, cut the channel of the brook wider deeper. In some steep places, so many the drops are working together that y are able to pick up pebbles or stones, h which they batter and tear down ger pieces of the bank and scrape out ater holes in the bottom of the stream. d when the drops have torn loose a k or a pebble, they do not merely carry they pound and grind it with other ks and pebbles, wearing away its sharp es and breaking it into smaller and aller pieces, until it may finally be a k or a pebble no longer, but only a wder as fine as flour. On and on the ok flows, a gang of workers each of ich is using its own load as a tool, in close procession, and working dou- quick. But as soon as the brook ches a plain or level, it slows down the drops act tired; they have no bition to pick up more soil, and each s fall its own load as soon as possible, pping the larger pieces of gravel and k first, carrying the finer soil farther, finally letting that down also. If we mine the sediment of a flooded brook, find that the gravel is always dropped t, and that the fine mud is carried thest before it is deposited.

The roar of a flooded stream is very ferent from the murmur of its waters en they are low. It is not to be won-red at, when we once think of all that going on in the brook during periods flood. There are some simple experi-nts to show what the force of water n do when turned against the soil. Pour water from a pitcher into a bed of soft soil, and note how quickly a hole will be made; if the pitcher is held near the soil, a smaller hole will be formed than if the pitcher is held high up; this shows that

The brook. Its snowbank source, its tools, and its workshop

the farther the water falls, the greater is its force. This explains why the banks of streams are undermined when a strong current is driven against them. The swift current, of course, tears away more earth at bends and curves than when it is flow-ing in a straight line; for ordinarily, when flowing straight, the current is swiftest in the bed of the stream, and is therefore only digging at the bottom; but when it flows around curves, it is directed against the banks, and therefore has much more surface to work upon. Thus it is that bends are cut deeper and deeper. If the bare arm is thrust into a flooded brook, we find

Even the large stones along the bank were probably brought there by the brook when it was working hard. When it works hard again, it may carry them somewhere else

that many pieces of gravel strike against it; and if we reach the bottom, we can feel the pebbles being moved along over the brook bed.

LESSON 207

The Brook

Leading Thought — The water from the little brook near our school is flowing toward the ocean, and is meanwhile digging out and carrying along with it the soil through which it flows.

Method — The best time to study a brook is after a rain, and October or May is an interesting time for beginning this lesson. The work should be continued during the entire year. It may be done at noon or recess, if the brook is near at hand; or there may be excursions after school, if the brook is at some distance. The observations should be made by the class as a whole.

Observations — 1. Does the brook have its source in a spring or a swamp, or does it receive its water as drainage from surrounding hills? Follow it back to its very beginning. Do you find this in open fields or in woods? Is the land about level, or does it slope?

2. Are its banks deeper at the beg ning, or is the brook at first almost a level with the surrounding fields? the banks become deeper farther from source? Are the banks higher where brook flows down hill, or where it is a level?

3. Is the course of the brook m crooked on a hillside or when it is flow through a level area? Are the banks m worn away and steep where the brc flows through woods or bushes th where it is flowing through the op fields?

4. Can you find the places where t water is cutting the banks most, wh the brook is flooded? Why does it cut banks at these particular points?

5. Into what stream, pond, or lake d the brook flow? If you should launch toy boat upon the waters of this bro and it should keep afloat, through wh streams would it pass to reach the oce Through what townships, counties, stat or countries would it pass?

6. When is the brook working a when is it playing? What is the differe between the color of the water ordina and when the brook is flooded? W causes this difference?

7. Make the following experiment show what the brook is carrying afte storm when the water is roily. Dip fr the swift portion of the stream a gl fruit jar full of water. Place it on a w dow sill and do not disturb it until water is clear. How much sediment settled at the bottom of the jar? Wh was this sediment when you dipped the water? If this quart of water co carry so much soil or sediment, how mu do you think, would the whole brc carry?

8. Where did the brook get the to make the water roily? Study its ba in order to answer this question. Do think the soil in the water came from banks that are covered by vegetation from those which are bare?

9. How did the brook pick up the

at it carried when it was flooded? Do u think that one of the tools the brook ;s with is the current? Try to find place where the swift current strikes the nk, and note if the latter is being worn ay.

10. Does the swift current take more l where it is flowing straight, or where ere are sharp bends? How are the bends the brook or creek made?

11. Thrust your bare hand or arm into e swift current of the brook when it is oded. Do you feel the gravel strike ainst your arm or hand? Wade in the ter. As the pebbles are being rolled ng the bed of the stream, do you feel em strike against your feet or legs?

12. Does the water, loaded with soil d pebbles, dig into the banks more orously than just the water alone could ? Which washes away more earth and ries it downstream — a fast or a slow rent?

13. Does the brook flow fastest when waters are low or high? When the ook is at its highest flood, do you nk it is working the hardest? If so, plain why. When it is working the rdest and carrying most soil and gravel, es it make a different sound than when is flowing slowly and its waters are ar?

14. How does the brook look when it is doing the least amount of work possible?

15. Make a map of your brook showing every pool, indicating the places where the current is swiftest, and showing the bends in its course. To test the rapidity of the current, put something afloat on it and measure how far it will go in a minute.

16. How many kinds of trees, bushes, and plants grow along the banks of your brook? How many kinds of fish and insects do you find living in it? How many kinds of birds do you see frequently near it?

A Brook Puzzle for Pupils to Solve — When we have a load to carry we go slowly because we are obliged to; and the heavier the load, the slower we go. On the other hand, when we wish to run very swiftly we drop the load so as not to be weighted down; when college or high school boys run races in athletic games, they do not wear even their ordinary clothing, but dress as lightly as possible; they also train severely so that they do not have to carry any more flesh on their bones than is necessary. How is it that in the case of a brook just the opposite is true? The faster the brook runs, the more it can carry; and the heavier it becomes the faster it runs; and the faster it runs the more work it can do.

LIFE IN THE BROOK

By any body of water, whether brook, er, pond, lake, canal, or sea, there will found many kinds of plant and animal :, which constitute a wealth of nature aterial. The plant life is somewhat different from that which grows far away m bodies of water; the forms of animal : vary with the quantity and condition the water.

All bodies of water serve as highways, er which not only man but other anils travel from one region to another. en many birds follow watercourses in eir migrations. Plants growing along watercourse often have their seeds card by the water and dropped at points many miles downstream. The seeds of plants growing near large bodies of water may be carried by waves to distant shores.

Information about many forms of life occurring in or near water may be found in the parts of this book dealing with plants and animals.

In the bottom of the valley is a brook that saunters between oozing banks. It falls over stones and dips under fences. It marks an open place on the face of the

In such a situation may be found many kinds of plant and animal life

earth, and the trees and soft herbs bend their branches into the sunlight. The hangbird swings her nest over it. Mossy logs are crumbling into it. There are still pools where the minnows play. The brook runs away and away into the forest. As a boy I explored it but never found its source. It came somewhere from the Beyond and its name was Mystery.

The mystery of this brook was its char ing moods. It had its own way of recordi the passing of the weeks and months. remember never to have seen it twice the same mood, nor to have got the sar lesson from it on two successive days: y with all its variety, it always left that sar feeling of mystery and that same vag longing to follow to its source and to kn the great world that I was sure must beyond. I felt that the brook was grea and wiser than I. It became my teach I wondered how it knew when Ma came, and why its round of life recur so regularly with the returning seasons remember that I was anxious for the spri to come, that I might see it again. longed for the earthy smell when the sn settled away and left bare brown marg along its banks. I watched for the suck that came up from the river to spawn made a note when the first frog peep I waited for the unfolding spray to soft the bare trunks. I watched the greeni of the banks and looked eagerly for t bluebird when I heard his curling n somewhere high in the air.

— "The Nature-Study Ide
L. H. Bai

HOW A BROOK DROPS ITS LOAD

The brook is most discriminating in the way it takes up its burdens, and also in the way it lays them down. With quite superhuman wisdom, it selects the lightest material first, leaving the heaviest to the last; and when depositing the load, it promptly drops the heaviest part first. And thus the flowing waters of the earth are eternally lifting, selecting, and sifting the soils on its surface.

The action of rain upon the surface of the ground is in itself an excellent lesson in erosion. If there is on a hillside a bit of bare ground which has been recently cultivated or graded, we can plainly see, after a heavy rain, where the finer material has been sorted out and carried away, leaving the larger gravel and stones. And if we examine the pools in the brook, we sh find deltas as well as many examples of t way the soil is sifted as it is dropped. T water of a rill flowing through past and meadow is clear, even after a ha rain. This is owing, not so much to t fact that the roots hold the banks the brook firmly, as that the grass on t surface of the ground acts as a mulch a protects the soil from the erosive imp of the raindrops. On the other hand, a for a reverse reason, a rill through plow ground is muddy. On a hillside, therefo contour plowing is practiced — that plowing crosswise the hillside instead up and down. When the furrow is carri crosswise, the water after showers c not dash away, carrying off in it all t r and more fertile portions of the soil. ere are many instances in our southern es where this difference in the direc- of plowing has saved or destroyed the ility of hillside farms.

'he little experiment suggested at the inning of the following lesson should w the pupils clearly the following nts: It is through motion that water es up soil and holds it in suspension. e tendency of still water is to drop all load which it is carrying, and it drops heaviest part first. We find the peb- at the bottom of the jar, the sand gravel next, and the fine mud on top. e water may become perfectly clear the jar and yet, when stirred a little, vill become roily again because of the vement. Every child who wades in a ok knows that the edges and the still ls are more comfortable for the feet n is the center of the stream under swift current. This is because, where water is less swift at the sides, it de- its its mud and makes a soft bottom; ile under the swifter part of the cur- t, mud is washed away leaving the er stones bare. For the same reason, bottom of a stream crossing a level d is soft, because the silt, washed down m the hills by the swift current, is pped when the waters come to a more et place. If the pupils can build across tony brook a dam that will hold for or three months in the fall or spring en the brook is flooded, they will be e to note that the stones will soon be re or less covered with soft mud; for dam, stopping the current, causes the er to drop its load of silt. It would have be a very recently made pool in a am which would not have a soft mud tom. The water at times of flood is ced to the side of the streams in eddies; current is thus checked, and its load nud dropped.

t should be noted that at points where brook is narrowest the current is swift- and where the current is swiftest the tom is more stony. Also, where there bend in the stream, the brook digs per into the bank where it strikes the curve, and much of the soil thus washed out is removed to the other side of the stream where the current is very slow, and there is dropped. If possible, note that where a muddy stream empties into a pond or lake, the waters of the latter are made roily for some distance out, but beyond this the water remains clear. The pupils should be made to see that the swift current of the brook is checked when its waters empty into a pond or lake, and because of this they drop their load. This happens year after year, and a point extending out into the lake or pond is thus built up. In this manner the great river deltas are formed.

W. G. Pierce, U. S. Geological Survey

A meandering stream. In time of flood, this stream brought down much of the soil of the flood plain through which it wanders

LESSON 208

How a Brook Drops Its Load

Leading Thought — The brook carries its load only when it is flowing rapidly. As soon as the current is checked, it drops the larger stones and gravel first and then the finer sediment. It is thus that deltas are built up where streams empty into lakes and ponds.

Method — Study the rills made in freshly graded soil directly after a heavy rain. Ask the pupils individually to make observations on the flooded brook.

Experiment — Take a glass fruit jar nearly full of water from the brook, add gravel and small stones from the bed of the brook, sand from its borders, and mud from its quiet pools. Have it brought into the schoolroom, and shake it thoroughly. Then place it in a window and ask the pupils to observe the following things:

(a) Does the mud begin to settle while the water is in motion; that is, while it is being shaken?

(b) As soon as it is quiet, does the settling process begin?

(c) Which settles first — the pebbles, the sand, or the mud? Which settles on top — that is, which settles last?

(d) Notice that as long as the water is in the least roily, it means that the soil in it has not all settled; if the water is disturbed even a little it becomes roily again, which means that as soon as the water is in motion it takes up its load.

OBSERVATIONS — 1. Where is the current swiftest, in the middle or at the side of the stream?

2. What is the difference, in the bottom of the brook, between the place below the swift current and the edges? That is, if you were wading in the brook, where would it be more comfortable for your feet — at the sides or in the swiftest part of the current? Why?

3. Does the brook have a more stony bed where it flows down a hillside than where it flows through a level place?

4. Place a dam across your brook wh
the bottom is stony, and note how soo
will have a soft mud bottom. Why is t

5. Can you find a still pool in y
brook that has not a soft, muddy bott
Why is this?

6. Does the brook flow more swiftly
the steep and narrow places than in
wide portions and where it is damme

7. Do you think if water, flow
swiftly and carrying a load of mud, w
to come to a wider or more level pl
like a pool or millpond dam, that it wo
drop some of its load? Why?

8. If the water flows less swiftly al
the edges than in the middle, would
make the bottom below softer and m
comfortable to the feet than where
current is swiftest? If so, why?

9. If you can see the place wher
brook empties into a pond or lake, h
does it make the waters of the latter l
after a storm? What is the water of
brook doing to give this appearance, a
why?

10. What becomes of the soil drop
by the brook as it enters a pond or la
Do you know of any points of land
tending out into a lake or pond where
stream enters it? What is the delta o
stream?

ROCKS AND MINERALS

REVISED BY H. RIES
Professor of Geology in Cornell University

J. K. Hillers, U. S. Geological Survey

: showing glacial striæ — the scratches made by other rocks and gravel carried in the ice

ny brook or stream which you may : observed has doubtless been rolling ts way for countless ages; and, however ll and insignificant its appearance, it probably caused great changes in the ıtryside through which it flows. Some- re along its course it may have cut) gorges; and where it empties into ke or into another stream, it may : built out great points or sandbars. ough all these years, it has been carry- with it great masses of the materials ch it excavates, transports, and rede- ts, and it will probably continue to do or centuries to come.

ı a general way, the materials that it carries are of two types, coarse and fine, the first consisting of rocks, pebbles, and sand, and the second of silty and clayey substances. Both of these types, and the brook's way with them, are of great importance to human life.

As we have seen, the brook both picks things up and lays them down. Both these acts are of benefit to man; they have given us, for instance, the rich bottom lands of the Mississippi and Missouri valleys. But they can also be of great harm, for water may carry off the soil which it or some other agency long ago deposited; and when it has done so, centuries will pass before that soil can be replaced.

Many children are naturally interested in stones. The peculiar shapes, odd markings, and colorings of stones attract a child's attention and arouse in him a desire to know more about them.

I once knew two children, aged seven and five, who could almost invariably recognize the different boulders and pebbles of rock which they found scattered over the surface in the region about Ithaca, New York. They also could tell, when the pebbles were broken, which parts were quartz and which mica. They had incidentally asked about one of these stones, and I had told them the story of the glacial period and how these stones were torn away from the mountains in Canada and brought down by glaciers and dropped in Ithaca. It was a story they liked, and their interest in these granite voyagers was always one of the many elements that helped to make our walks in the field delightful.

Photomicrograph by W. A. Bentley

Snow crystal. Many minerals are crystalline in form

The term *mineral* is not generally u in its broadest sense; it really means a substance which is neither plant nor a mal. To be specific we will restrict its to the more limited meaning — " an in ganic substance occurring in nature, h ing a definite chemical composition, a usually a distinct crystalline form." *mineral* may also be defined as a sin chemical element, or two or more ments chemically combined, formin part of the earth's crust.

Some eight or ten chemical eleme in various combinations, make up m of our common rock-forming miner they are oxygen, silicon, aluminum, ir calcium, magnesium, sodium, potassiu sulphur, and carbon.

A rock is an aggregation of minerals may be made up entirely of a single m eral, as is rock salt; but more often a r contains two or more minerals. Gran for example, is composed of felds quartz, and mica, and may contain ho blende.

I. ROCKS

Perhaps you have heard someone use the expression " rock bottom " to mean the foundation, base, or beginning of something. The words are very expressive, and have their meaning buried in the earth. When we look out over a lawn, park, field, or even a large body of water, we see the surface and do not stop to think that far beneath are beds of solid rock. We can see exposures of various types of rock in such places as cuts made for highways or railroads, along deeply cut stream banks, in quarries, or sometimes outcropping in the slope of a mo tain.

To understand what is meant by term *rock*, we need to recall what was s in the discussion of minerals. *Rocks aggregations of minerals, and mine are composed of elements or chem combinations of elements.*

The study of rocks is treated under t branch of science called geology. In a t book on that subject will be found m interesting information on rocks; but our purposes it seems best to consider o w of the more common rocks, or stones, they are sometimes called.

It will be well to mention, however, at rocks are divided into three main oups; these divisions are determined by eir origin, their position in the earth's ıst, and their location in respect to each ıer. The three groups are *sedimentary rocks*, formed from sediments deposited chiefly by water, sometimes by wind or glaciers; *igneous rocks*, formed by the solidification of molten rock; and *metamorphic rocks*, formed from the other two groups, by processes which produced such changes in them as to warrant placing them in a separate group.

SEDIMENTARY ROCKS

The materials in these rocks are in lay-; they were deposited by the water or nd assisted by the force of gravity; they re laid down according to size or weight individual particles. The materials vary cording to the places where they were d down, such as deserts, river beds, del-, beaches, or ocean bottoms. (See also e Brook, p. 735.)

If you will put some muddy water in a ss tumbler and watch the mud settle, u will notice that some of the larger par-les settle very soon, while some of the er particles will be held for hours before ey are dropped. This simple experiment ows, in a general way, what takes place a muddy stream. During and after a rd rain, a stream carries much more liment than at any other time. Some of e finer particles are not dropped until ey reach the body of water into which e stream is flowing, and there in quiet ter they settle down. Thus sediment y build up deltas or settle on the ocean lake bottoms.

In the ocean there live many animals ich secrete shells, and there are fishes th bony skeletons. When these animals e, their hard parts settle to the bottom, ere they are covered with sediment ich preserves them. Later, when this liment has hardened to rock, we find ese animal remains preserved as fossils. plant organism may leave an impression its form in the sediment, even though e vegetable matter has decayed. The various kinds of fossils serve as a sort of key or index to aid scientists in determining at what time, in the history of the earth, the particular rock-forming materials were laid down. In shale, a rock formed from old clay beds, we sometimes find footprints of prehistoric animals and impressions of raindrops that fell many ages ago.

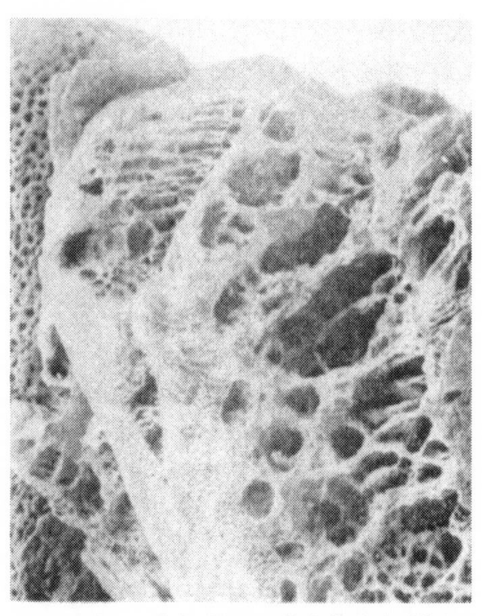

C. D. Walcott, U. S. Geological Survey

A large piece of sandstone showing the uneven effects of weathering

Some common examples of sedimentary rock are limestone, shale, and sandstone; even iron ore beds, coal, and rock salt are included in this group.

IGNEOUS ROCKS

These rocks have been formed, by cooling, from materials that have been forced up from the interior of the earth. These materials are in the form of molten lava, which does not always reach the surface before it cools. They do not show assortment and stratification as do sedimentary rocks, but have, instead, a crystalline texture. The size of the grains is determined

Photomicrograph by W. A. Bentley

Snow crystal

by the position in which the molten material cooled; the portions cooling at or near the surface of the earth contain smaller crystals than do the materials which cooled more slowly at points far below the surface of the earth. Granite is one of the most common igneous rocks.

GRANITE

In granite, the quartz may be detected by its fracture, which is always conchoidal and never flat; that is, it has no cleavage planes. It is usually white or smoky, and is glassy in luster. It cannot be scratched with a knife. The feldspar is usually whitish or flesh-colored and the smooth surface of its cleavage planes shines brilliantly as the light strikes upon it; it can be scratched with a knife but this requires effort. The mica is in pearly scales, sometimes whitish and sometimes black. The scales of these mica particles may be lifted off with a knife, and it may thus be distinguished. If there are black particles in t granite which do not separate, like t mica, into thin layers, they probably cc sist of hornblende.

Granite is used extensively for buildi purposes and for monuments. It is a ve durable stone; when polished it endu better than when rough-finished, since t polished surface gives less opportun for water to lodge and freeze. If grani by prolonged weathering, is broken do to grains of sand and clay, these may washed away and carried into lakes or t ocean where they settle down in mo or less sorted forms. If the sand gra form a deposit of appreciable size a extent, this becomes a sandstone rock

Cleopatra's Needle, which stood thousands of years in the dry climate Egypt, soon commenced to weather a crumble when placed in Central Pa New York. The Department of Parks the City of New York has furnished t following information concerning t treatment given to the monument:

After the obelisk had been standing its new home about two years, the surf began to chip. In 1885, four years after erection, a coating of paraffin wax w recommended for its surface. It was fou that there were many loose flakes of la size; all such flakes that could be sa were left in place. To have removed th flakes would have damaged the hie glyphics to a serious degree. Within a f months the entire monument was coa with wax.

In 1893 these areas were treated by p sure to insure that a solid body of para wax would fill all openings and prev any accidental movement. The flakes mapped and numbered for the purpose ascertaining at any time whether t have increased in area or whether n ones have developed. A thorough exa ination in 1913 showed no further terioration; but as a precautionary meas a coating of liquid veneer was applied

surface; this veneer did not penetrate
l was soon washed away by the rain.
n 1930, forty-five years after the appli-
ion of the preservative, no indication
ld be found of any need for renewing
treatment. The preservative had ab-
itely stopped and prevented what
ild have been the rapid disintegration
the oldest and perhaps the most inter-
ng monument in America.

'his shaft has a most interesting his-
. It was quarried near Assuan, in the
st famous of all the granite quarries of
ient Egypt. It was cut as a solid shaft
the quarry and carried down the Nile
er for 500 miles — an engineering feat
ich would be hard to accomplish to-
, with all our modern appliances. It
one of the obelisks that graced the an-
it city of On, later called Heliopolis,
ated on a plateau near the present city
Cairo; On was the city where Moses
born and reared. One of these obe-
s still stands where it was first placed
part of a magnificent temple, the tem-
a part of a magnificent city. It now
ids alone in the middle of a great fer-
plain, which is vividly green with
wing crops; a road shaded by tamarisk
lebbakh trees leads to it; nearby is a
iah, creaking as the blindfolded bul-
walks around and around, turning the
eel that lifts the chain of buckets from
well to irrigate the crops; and a hooded
w, whose ancestors were contempo-
es of its erection, caws hoarsely as it
hts on the beautiful apex of this an-
it shaft, which has stood there nearly
r thousand years and has seen a great
go down to dust to fertilize a grassy
n.

J. H. Comstock

The granite obelisk still standing on the site of the ancient city of On

LESSON 209

Igneous Rocks: Granite

Leading Thought — Granite is com-
ed of feldspar, quartz, and mica, and
en contains hornblende.

Method — Specimens of coarse granite and a pocket knife are needed.

Observations — 1. What minerals do you find in granite? How can you tell what these minerals are? Look at the granite with a lens. How can you tell the quartz from feldspar? Take a knife and scratch the two. Can you tell them apart in that way? How can you tell the mica? How can you tell the hornblende?

2. What buildings made of granite have you seen? What monuments made from it have you seen?

3. What is weathering? Mention some of the characteristics of weathering. Why does the rough-finished granite weather sooner than that which is polished?

4. Examine some sand with a lens. What mineral do you find present in it in the greatest quantity?

5. Write the story of the Cleopatra's Needle in Central Park, New York City.

METAMORPHIC ROCKS

Metamorphic means changed; it is therefore to be understood that metamorphic rocks are those rocks whose texture and mineral composition have been changed since they were originally formed as igneous or sedimentary rocks. They may have decayed or they may have been made stronger by the process; but in general the term is used to mean changes that have occurred in rocks as a result of great weight, heat, movement, and pressure the presence of water and gases. depth within the earth's crust at wh the process took place determines chi which factor was most important in br ing about the change.

Some common metamorphic rocks slate, formed from clay; marble, fr limestone; quartzite, from sandstone; anthracite coal from soft coal.

CALCITE, LIMESTONE, AND MARBLE

Calcite or calc spar, which is calcium carbonate, is a mineral and is the material of which marble, limestone, and chalk are

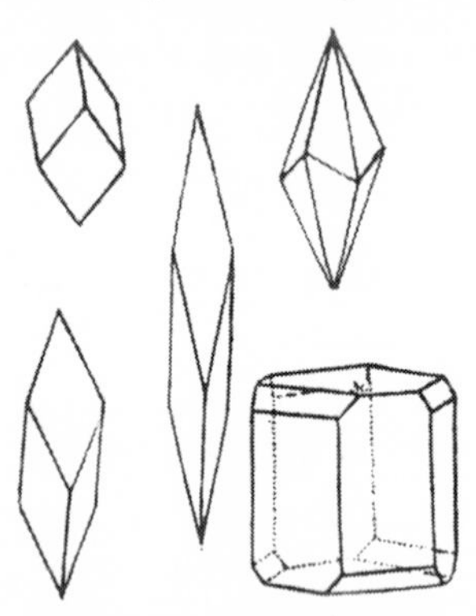

Forms of calcite crystals

made. The faces of the calcite crystal are always arranged in groups of three or multiples of three — a three-sided pyramid or two pyramids joined base to base. When acute and formed of three pairs of faces, the crystals are called dogtooth spar. The crystals appear in a great variety of forms, but they all have the common quality of splitting readily in three directions, the fragments forming rhombs. When these cleaved or split pieces are transparent, they are called Iceland spar. When an object is viewed through Iceland spar at least one-quarter inch thick, it appears double. The calcite crystal is often transparent with a slight yellowish tinge, but it shows other colors; and it has a sligl cloudy or slightly pearly or almost gl luster like feldspar. It is easily scratc with a knife and will not scratch g If a drop of strong vinegar or weak hy chloric (muriatic) acid is dropped o it will effervesce.

Limestone has often been formed the bottoms of oceans; its substance c chiefly from the skeletons of corals the shells of other sea creatures, since shells and coral skeletons are calc carbonate in composition. In the wa the shells and corals were broken do

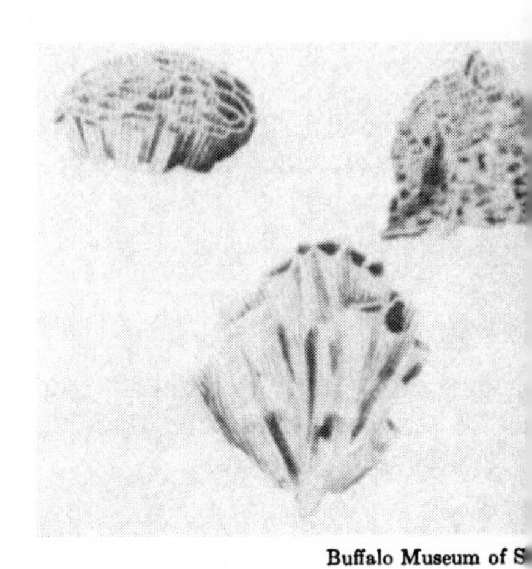

Buffalo Museum of S

Fossilized Coral: the skeletons of coral mals

l then deposited in layers on the bot-
n of the sea; in addition some car-
nate of lime has been precipitated from
water. Layers of limestone are now be-
deposited off the shores of Florida,
ere corals grow in great abundance.
nestone is used extensively for building
poses, and in many climates is very
able. The great pyramids of Egypt are
limestone. It is slowly dissolved in
ter, especially if the water be acid; thus,
imestone regions, there are caves where
water has dissolved out the rock; and
ched to their roofs and piled upon
ir floors may be large icicle-shaped sta-
ites and stalagmites, which were made
the lime-bearing water dripping down
l evaporating, leaving its burden in crys-
behind it. When the roof of a cave
s in, the cavity thus made is called a
k hole and is often dangerous. The fa-
us Natural Bridge in Virginia is all
t is left of what was once the roof of
h a cavern. Other famous caves are
ay and Mammoth in the East and
lsbad Caverns of New Mexico. The
er of limestone regions is always hard,
ause of the lime which it holds in solu-
n; and in such regions the streams usu-
have no silt, but have clean bottoms;
reover, the springs are likely to bene contaminated because the water has
through long caves instead of filtering
ough sand.

Chalk is similar in origin to limestone;
s made up of the shells of minute sea
atures, so small that we can only see
m with the aid of a microscope. Try to
nk how many years it must have re-
red for the shells of such tiny beings to
ld up the beds which make the great
lk cliffs of England!

Marble is formed inside the earth from
estone, under the influence of heat and
ssure; it differs from limestone chiefly
that the grains are of crystalline struc-
e, and are larger; it is usually white or
y in color, and sometimes is found in
ering colors. The most famous marbles
the Carrara of Italy, the Parian from
Island of Paros, and the Pentelican
m the mountain of that name near
Athens. The reason why these marbles are so famous is that in ancient times sculptors carved beautiful statues from them, and the architects used them for building magnificent temples. The principal marble deposits in the United States are in Vermont, Georgia, and Tennessee. Some marbles do not last well when exposed to severe climatic conditions. Marble is also used to make lime. When either marble or limestone is heated very hot, it separates into two parts, one of which is lime, and the other carbonic acid gas — the same that is used for charging soda water.

LESSON 210

Calcite, Limestone, and Marble

Leading Thought — Calcite or calc spar is lime carbonate. The best known forms of its crystals are rhombic; but instead of having twelve right-angled edges, the sides are lozenge-shaped, and are set together with six obtuse angles and six acute. Dogtooth spar is one form of calcite crystal. Limestone is a solid form of calcite. Marble is granular limestone, which is made of crystalline grains of calcite. Chalk is soft, fine-grained limestone.

Method — Specimens of dogtooth spar, limestone, marble, shells of oysters or other sea creatures, and coral should be provided for this lesson; also a bottle of dilute hydrochloric acid, and a piece of glass tubing about six inches long with which to drop the acid on the stones. Some strong vinegar will do instead of the acid.

Marbles which are composed of mineral dolomite, a carbonate of lime and magnesia, will not effervesce with cold acid.

Observations — 1. What is the form of the calcite crystal? What is the luster of the crystal? Is it the same as the inside of sea-shells? Will calcite scratch glass? Can you scratch it with a knife? What happens to calcite if you put a drop of weak hydrochloric acid upon it?

2. Is marble made up of crystals? Examine it with a lens to see. What is its color? Have you seen marble of other col-

ors than white? Do you know the reason why marble is sometimes clouded and streaked?

3. What are the uses of marble? What have you ever seen made from marble? Why is it used for sculpture? What famous statues which were made of marble have you seen? Name some of the famous ancient marble buildings.

4. Test a piece of limestone for hardness. Can you scratch it with a knife? Is it as soft as marble? Put a drop of acid on it. Does it effervesce? If there are any fossils in your piece of limestone, test them with acid and see if they will effervesce. Any other mineral that you have which will effervesce when touched with acid is probably some form of calcite.

5. Are there any buildings in your town made of limestone? How do you know the stone is limestone? Where was it obtained? Is it affected by the weather?

6. Why is water in limestone regions hard? Why are limestone regions likely to have caves within the rocks? How are stalactites and stalagmites formed in caves? What are sink holes? How are they formed? In what county of your state is limestone found?

7. How is the lime which is used for plastering houses made?

8. Write a theme on how the chalk rocks are made.

9. Test a shell with acid; test a piece of coral with acid. How does it happen that these, which were once a part of living creatures, are now limestone? Of what material are the bones of our own bodies made?

A great chapter in the history of world is written in the chalk. Few passa in the history of man can be supported such an overwhelming mass of direct indirect evidence as that which testifie the truth of the fragment of the history the globe, which I hope to enable you read, with your own eyes, to-night. Let add, that few chapters of human hist have a more profound significance for selves. I weigh my words well when I sert, that the man who should know true history of the bit of chalk which ev carpenter carries about in his breec pocket, though ignorant of all other tory, is likely, if he will think his kno edge out to its ultimate results, to hav truer, and therefore a better, concept of this wonderful universe, and of ma relation to it, than the most learned dent who is deep-read in the records humanity and ignorant of those of ture.

During the chalk period, or "Cr ceous epoch," not one of the present gr physical features of the globe was in ex ence. Our great mountain ranges, P enees, Alps, Himalayas, Andes, have been upheaved since the chalk was posited, and the Cretaceous sea flov over the sites of Sinai and Ararat. this is certain, because rocks of Cretace or still later date, have shared in the ele tory movements which give rise to th mountain chains; and may be fou perched up, in some cases, many thous feet high upon their flanks.

— THOMAS HUX

II. MINERALS

For the pupils in the elementary grades it seems best to limit the study of minerals to those which make up our common rocks. In order to teach about these minerals well, the teacher should have at least one set of labeled specimens. Such a collection may be obtained from a supply house. These collections vary in number of specimens and price. The teacher should have one or two perfect crystals of quartz, feldspar, and calcite. An ex lent practice for a boy is to copy these c tals in wood for the use of the teacher.

The physical characteristics used identifying minerals are briefly as lows:

1. *Form*. This may be crystalli which shows the shape of the crystals d nitely; granular, as in marble, the gra having the internal structure, but not

ternal form, of crystals; compact, which without crystalline form, as in limestone flint.

2. *Color*.

3. *Luster* or *shine*, which may be glossy .e quartz, pearly like the inside of a shell, ky like asbestos, dull, or metallic like ld.

4. *Hardness* or resistance to scratching, thus: easily scratched with the fingernail; cannot be scratched by the fingernail; easily scratched with steel; with difficulty scratched with steel; not to be scratched by steel. A pocket knife is usually the implement used for scratching.

CRYSTAL GROWTH

To watch the growth of a crystal is to tness a miracle; involuntarily we stand awe before it, as a proof that of all ıths mathematics is the most divine and e most inherent in the universe. The acher will fail to make the best use of is lesson if she does not reveal to the ild through it something of the marvel crystal growth.

That a substance which has been dislved in water should, when the water aporates, assemble its particles in solid rm of a certain shape, with its plane sur;es set exactly at certain angles one to other, always the same whether the ystal be large or small, is quite beyond r understanding. Perhaps it is no more iraculous than the growth of living bezs, but it seems so. The fact that when imperfect crystal, unfinished or broken, placed in water which is saturated with

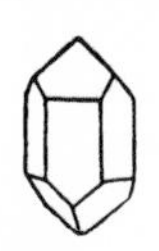

Forms of quartz crystals

e same substance, it will be built out d made perfect, shows a law of growth exquisitely exemplified as to again make glad to be part of a universe so perfectly verned. Moreover, when crystals show variation in numbers of angles and ines it is merely a matter of division multiplication. A snow crystal is a sixed star, yet sometimes it has three rays.

The window sill of a schoolroom may a place for the working of greater wonrs than those claimed by the alchemists old, when they transmuted baser metals to gold and worthless stones to diamonds. It may be a place where strings of gems are made before the wondering eyes of the children; gems fit to make necklaces for any naiad of the brook or oread of the caves.

Photomicrograph by W. A. Bentley

Snow crystal

It adds much to the interest of this lesson if different colored substances are used for the forming of the crystals. Blue vitriol, potassium bichromate, and alum give beautiful crystals, contrasting in shape as well as in colors.

Copper sulphate and blue vitriol are two names for one substance; it is a poison when taken internally and, therefore, it is best for the teacher to carry on the experiment before the pupils instead of trusting the substance to them indiscriminately. Blue vitriol forms an exquisitely beautiful blue crystal, which is lozenge-shaped with oblique edges. Often, as purchased from the drug store, we find it in the form of rather large, broken, or imperfect crystals. One of the pretty experiments is to place some of these broken crystals in a saucer containing a saturated solution of the vitriol, and note that they straightway assert crystal nature by building out the broken places, and growing into perfect crystals. Blue vitriol is used much in the dyeing and in the printing of cotton and linen cloths. It has quite wonderful preservative qualities; if either animal or vegetable tissues are permeated by it, they will remain dry and unchanged.

Copper sulphate solutions have also been effectively used in treating seeds of some farm crops to kill spores which might be present and would later cause smuts or other fungus diseases to develop on the growing crops.

Potassium bichromate is also a poison, and therefore the teacher should make the solution in the presence of the class. It forms orange-red crystals, more or less needle-shaped. It crystallizes so readily that if one drop of the solution be placed on a saucer the pupils may see the formation of the crystals by watching it for a few moments through a lens.

The common alum we buy in crystal form; however, it is very much broken. Its crystals are eight-sided and pretty. Alum is widely used in dyes, in medicines, and in many other ways. It is very astringent, as every child knows who has tried to eat it, and has found the lips and tongue much puckered thereby.

Although we are more familiar with crystals formed from substances dissolved in water, yet there are some minerals, like iron, which crystallize only when they are melted by heat; and there are other crystals, like the snow, which are formed from vapor. Thus, substances must be molten hot, or dissolved in a liquid, or in the form of gas, in order to grow into crystals.

LESSON 211

Crystal Growth

Leading Thought — Different substances when dissolved in water will reform as crystals; each substance forms crystals of its own peculiar color and shape.

Method — Take three test tubes, long vials, or clear bottles. Fill one with a solution made by dissolving one part of blue vitriol in three parts of water; fill another by dissolving one part of bichromate potash with twenty-five parts of wat fill another with one part of alum in th parts of water. Suspend from the mou of each test tube or vial a piece of wh twine, the upper end tied to a toothp which is placed across the mouth of t vial; the other end should reach the b tom of the vial. If necessary tie a pebl to the lower end so that it will ha straight. Place the bottles on the wind sill of the schoolroom, where the childr may observe what is happening. All them to stand for a time, until the str in each case is encrusted with cryst then pull out the string and the cryst Dry them with a blotter, and let the c dren observe them closely. Care sho be taken to prevent the children from t ing to eat these beautiful crystals, by t ing them that the red and blue cryst are poisonous.

Observations — 1. In which bottle the crystals form first? Which string is heaviest with the crystals?

2. What was the color of the water which the blue vitriol was dissolved? I as brilliant in color now as it was wh it was first made? Do you think that growth of the crystals took away fr the blue material of the water? Look at t blue vitriol crystals with a lens, and scribe their shape. Are the shapes of t large crystals of the vitriol the same those of the small ones?

3. What is the shape of the crystals the potassium bichromate? What is color? Are these crystals as large as th of the blue vitriol or of the alum?

4. What shapes do you find among crystals of alum?

5. Do you think that vitriol and pot sium bichromate and alum will, un favorable circumstances, always form e its own shape of crystal wherever it occ in the world? Do you think crystals co be formed without the aid of water?

6. How many kinds of crystals do y know? What is rock candy? Do you thi you could make a string of rock car if you dissolved sugar in water and plac a string in it?

SALT

Saturated solution" is an uninspiring n to one not chemically trained; and it merely means water which holds as ch as it can take of the dissolved sub- ıce; if the water is hot, it dissolves re of most substances. To make a sat- ted solution of salt we need two parts alt or a little more, for good measure, ive parts of water; the water should be red until it will take up no more salt.

Form of a salt crystal

slip of paper placed in a saucer of solution will prove a resting place for crystals as they form. In about two s the miracle will be working, and the ils should now and then observe its gress. Those saucers set in a draft or warm place will show crystals sooner n others, but the crystals will be ller; for the faster a crystal grows, the ller is its stature. If the water evapo- s rapidly, the crystals are smaller, be- se so many crystals which do not have erial for large growth are started. en the water is evaporated, to appreci- the beauty of the crystals we should at them with a lens or microscope. h crystal is a beautiful little cube, often a pyramid-shaped depression in each or side. After the pupils have seen se crystals, the story of where salt is nd should be told them.

alt is obtained by several methods. more common ones include mining arge deposits of rock salt, evaporation ake or ocean water which is salty; and pumping of water down to a salt de- it and thereby dissolving the salt. In the latter case, the salt solution is brought to the surface and evaporated. The oldest salt works in this country were in Syracuse, New York, where the salt was obtained from salt springs which were famous among the American Indians. At Ithaca, New York, the salt deposits are about 2,000 feet below the surface of the earth. Salt is obtained in a number of states, either from wells or through mines. Salt is ob- tained by evaporating sea water on San Francisco Bay, and from lake water at Salt Lake City. The largest salt mines in the world include those of Germany and Poland; these have been worked for many, many years.

When the United States was first set- tled, salt was brought over from England; but this was so expensive that people could not afford it, and so they soon be- gan to make their own salt by evaporating sea water in kettles on the beach. In those countries where it is scarce, salt is said to be literally worth its weight in gold. The necessity for salt to preserve the health of both people and animals has tempted the governments of some countries to place a special tax upon it.

Salt lakes are found in natural basins of arid lands, and are always without outlets. The water which runs in escapes by evapo- ration, but the salt it brings cannot escape, and accumulates. A salt lick is a place where salt is found on the surface of the earth, usually near a salt spring. Animals will travel a long distance to visit a salt lick, which gained its name through their attentions.

LESSON 212

Salt

Leading Thought — Salt dissolves in water, and as the water evaporates the salt appears in beautiful crystals.

Method — Let each pupil, if possible, have a cup and saucer, a square of paper small enough to go into the saucer, and some salt and water. Let each pupil take five teaspoonfuls of water and add to this two spoonfuls of salt, stirring the mixture until it is dissolved. When the water will take no more salt, let each pupil write his name and the date on the square of paper and lay it in the saucer, pressing it down beneath the surface. Let some place their saucers in a warm place, others where they may be kept cool, and others in a draft. If it is impossible for each pupil to have a saucer, two or three pupils may be selected to perform the experiments.

OBSERVATIONS — 1. When you pour the salt into the water, what becomes of it? How do you know when the water will hold no more salt?

2. After a saucer filled with the salt water has stood exposed to the air for several days, what becomes of the water? From which saucers did the water evaporate fastest — those in the warm pla or those in the cold? In which did crystals form first?

3. Which saucers contained the lar crystals — those from which the w evaporated first, or those from whic evaporated more slowly?

4. Could you see how the crystals gan? What is the shape of the per salt crystal? Do the smallest crystals h the same shape as the largest ones?

5. What happens to people who not get salt to eat?

6. How are dairy salt and table salt tained? What is rock salt? What are licks? Where are the salt mines fou Why is the ocean called "the b deep"?

7. Name and locate some salt la Why are some lakes salt? Why is ocean salt?

QUARTZ

Quartz is the least destructible and is one of the most abundant materials in the crust of the earth as we know it. It is made up of two elements chemically united — the solid silicon and the gas oxygen. It is the chief material of most sand and sandstones, and it occurs, mixed with grains of other minerals, in granite, gneiss, and many lavas; it also occurs in the form of veins, and sometimes in crystals ornamenting the walls of cavities in rocks. Subterranean waters often contain a small amount of silica, the substance of quartz, in solution; from such solutions it may be deposited in fissures or cracks in the rock, thus forming bodies called "veins." Other materials are often deposited at the same time, and in this way the ores of the precious metals came to be associated with quartz. Sometimes silica is deposited from hot springs or geysers, forming a spongy substance called geyserite. In this case, some of the water is combined with the silica, making what is called opal. Quartz will cut glass.

Quartz occurs in many varieties: (a) In crystals or masses like glass. If colorless and transparent it is called rock crysta smoky brown, it is called smoky qua if purple, amethyst. (b) In crystals masses, glassy but not transparent. white, it is milky quartz; if pink, quartz. (c) As a compact crystal structure without luster, waxy or opaque or translucent, when polished bright red, it is carnelian; if brownish sard; if in various colors in bands, ag if dull red or brown, jasper; if green w red spots, bloodstone; if smoky or g breaking with small, shell-like, or choidal fractures, flint.

Rock crystals are used in jewelry especially are made to imitate diamo The amethyst is much prized as a se precious stone. Carnelian, bloodstone, agate are also used in jewelry; agate is u also in making many ornamental obje and to make little mortars and pestles grinding hard substances.

One of the marvels of the world is petrified forest of Arizona, now set a by the government as a national rese Great trees have been changed to a and flint, the silica having permeated nt tissue so that the texture of the od is preserved.

When our country was first settled, flint s used to start fires by striking it with el and letting the sparks fly into dry, e material, called tinder. It was also d in guns before the invention of cartdges, and the guns were called flintks. The Indians used flint to make chets and for tips to their arrows. The king of flint implements dates far back o prehistoric times; it was probably one the first steps upward which man ieved in his long, hard climb from a el with the brute creation to the ghts attained by our present civilizan.

Quartz sand is used in making glass. It nelted with soda or potash or lead, and glass varies in hardness according to minerals added. Quartz is also used for dpaper; and, ground to a fine powder, s combined with japans and oils and d as a finish for wood surfaces. Much neral wool is now made from glass, and widely used for insulation in the walls houses. Quartz combined with sodium potassium and water forms a liquid led water-glass, which is used for wateror surfaces; it is also fireproof to a cern degree. Water-glass is the best subnce in which to preserve eggs; one part commercial water-glass to ten parts of ter makes a proper solution for this pose.

LESSON 213

Quartz

Leading Thought — Quartz is one of the most common of minerals. It occurs in many forms. As a crystal it is six-sided, and the ends terminate in a six-sided pyramid. It is very hard and will scratch and cut glass. When broken, it has a glassy luster and it does not break smoothly, but shows an uneven surface.

Method — The pupils should have before them as many varieties of quartz as possible; at least they should have rock crystal, amethyst, rose and smoky quartz, and flint.

Observations — 1. What is the shape of quartz crystals? Are the sides all of the same size? Has the pyramid-shaped end the same number of plane surfaces as the sides?

2. What is the luster of quartz? Is this luster the same in all the different colored kinds of quartz?

3. Can you scratch quartz with the point of a knife? Can you scratch glass with a corner or piece of the quartz? Can you cut glass with quartz?

4. Describe the following kinds of quartz and their uses: amethyst, agate, flint.

5. How many varieties of quartz do you know? What has quartz to do with the formation of the petrified forests of Arizona?

FELDSPAR

We most commonly see feldspar as the kish portion of granite. This does not an that feldspar is always pink, for it y be the lime-soda form known as labraite, which is dark gray, brown or greenbrown, or white; or it may be the sodae feldspar called oligoclase, which is yish green, grayish white, or white; the most common feldspar of all is potash feldspar — orthoclase — which y be white, nearly transparent, or pink. Orthoclase is different from other feldrs in that, when it splits, its plane surfaces form right angles. Feldspar is next in the scale of hardness to quartz, and will with effort and perseverance scratch glass but will not cut it; it can be scratched with a steel point. Its luster is glassy and often somewhat pearly.

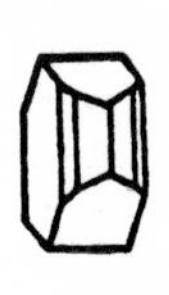
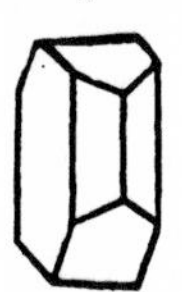
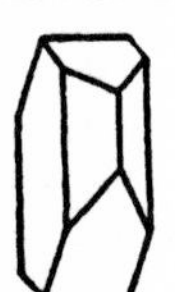

Forms of feldspar crystals

FOSSILS

In very early times fossils were considered more or less as freak relics of an ancient past; but now a fossil may be defined as an organism or anything indicating the former presence of an organism which has been preserved in any natural deposit in or on the earth's crust. In fact, any vestige of life of a former age may be considered a fossil.

The types of fossils vary greatly in their nature and in their completeness. There are instances of animals having been preserved as unaltered remains. *Such is the case of the rare mammoths found in Siberia and the insects caught in the tree resin which we now find as amber.* Petrifactions *are fossils in which some of the original portions of the organisms have been replaced at least in part by a mineral. A* mold *is an interesting type of fossil; it is the impression of a plant or animal left formerly in soft mud. After the body decayed or was removed in some manner the impression still remained and became permanently preserved. A shell formerly buried in rock and later dissolved by water leaves a cavity bearing the shape of the shell; if later that cavity becomes filled with some mineral substance the result is a fossil called a* cast. Trails *of marine animals,* tracks *of dinosaurs, or* burrows *of worms are all considered as fossils.*

The sea is the most favorable place for the burial of organisms; many forms of life are present, much sediment is available to cover the dead bodies, and decay is checked by the salt water. There are few places where good preservation of land animals has been possible; but fossils of land animals have been found in caves, under lava flows, and in swamps; in some instances the bodies have been washed out to sea and preserved there.

By means of fossils man has been able to unravel much interesting earth history. In fact some fossils and combinations of fossils have come to be known as "guide fossils," and these can be used to determine very definitely the geologic age of the rock strata in which they are found. From fossils much can be learned about the factors of the environment in which the plant or animal lived — whether the atmosphere was moist or dry, cool or warm, whether the water was fresh or salty.

1. HYPOHIPPUS *skeleton found in rocks in Colorado.*

2. BRACHIOPODS, Lingula. *These soft-bod animals had two shells not quite equal in s See No. 7 and No. 9.*

3. CRANE FLY, Tipula. *We can see that t insect was very similar to the present- crane fly.*

4. TRILOBITES, Phacops. *These crustace were among the first fossils to attract the tention of naturalists and are used as "gu fossils" in the rock formations of the Ca brian period. The trilobites varied in len from a fraction of an inch to two feet.*

5. CYCADS, Otozamites. *These plants w similar to pines and spruces in structure their palmlike leaves were somewhat on order of ferns. A few tropical forms of cyc are to be found living today.*

6. CRINOID OR SEA LILY, Taxocrinus. *T animal was named from its resemblance t lily. It had a long stem, at the upper en which was a cluster of plumelike arms shown in the picture.*

7 AND 9. BRACHIOPOD. *Pictured here are ferent aspects of this very abundant fo The animal has an "upper" and "low valve rather than a "right" and a "left' have the oyster, clam, and mussel. More t* 200 *kinds of living brachiopods are kno but in the Paleozoic time there were m than* 2500 *known forms in what is now No America. They do not move about but fixed to one place in the sea by a stalk f the lower valve.*

8. DINOSAUR TRACKS. *These tracks w found on the brownstone of Connecticut V ley; the fossil shown in this picture is the collection at Amherst College. Dinosa were reptiles which ran about on their h legs as do birds. They were so abundant were such extraordinary animals that the t in which they lived has been called the of Reptiles. These creatures were at least much diversified in size, form, and ada tions as are the mammals of today. Some birdlike feet with great claws; others w lived in swampy areas had huge ducklike b Still other dinosaurs reached a length o feet or more and weighed as much as* 40 *t*

Photographs by courtesy of the American Museum of Natural History and the Buffalo Museum of Science

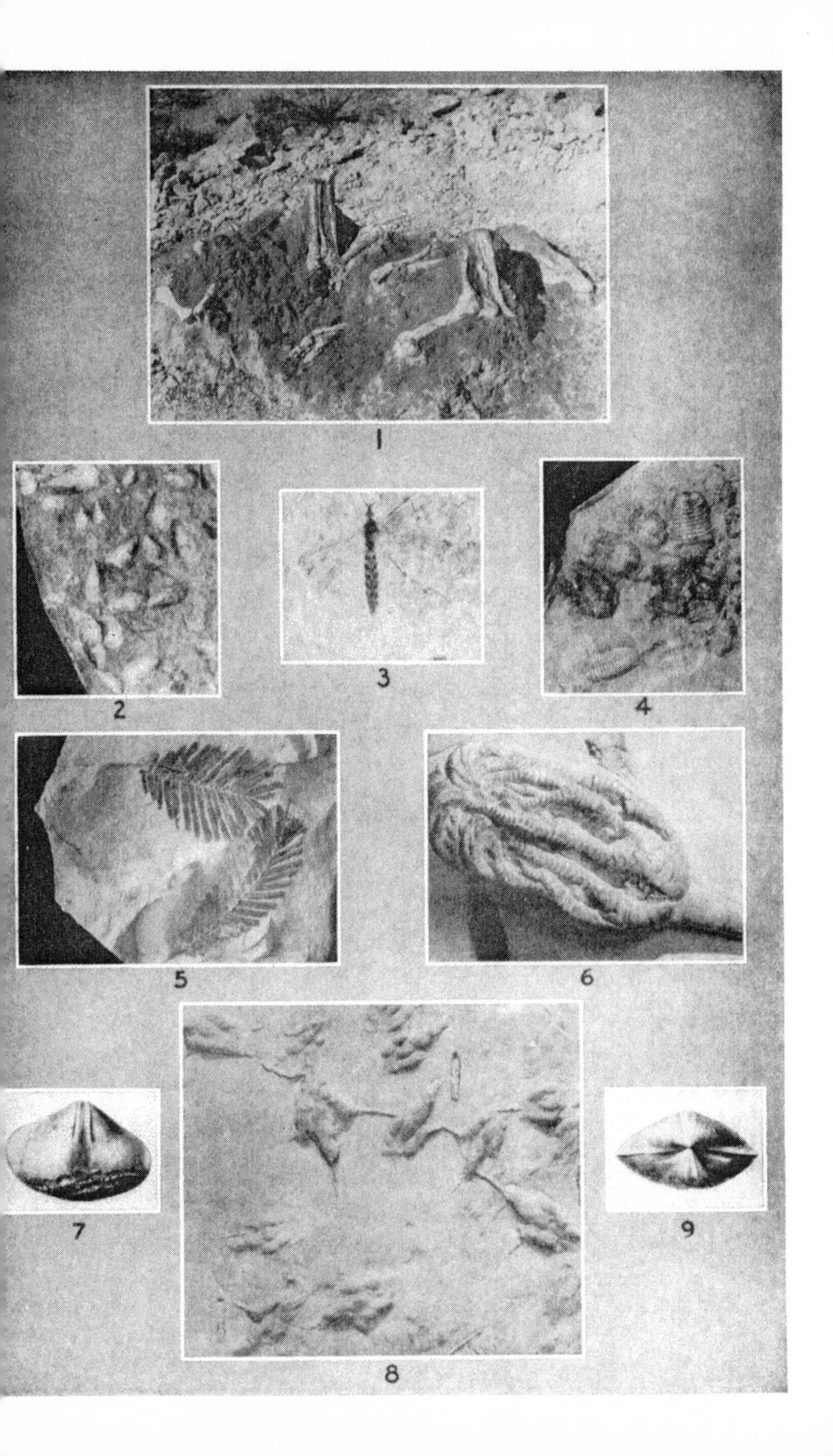
1
2
3
4
5
6
7
8
9

North Carolina leads all other states in the production of feldspar; but considerable quantities are produced in South Dakota, New Hampshire, and Colorado. It is quarried, crushed, and ground to powder as fine as flour to use with the clay from which china and most kinds of white pottery are made. Kaolin, which has been used so extensively in making the finest porcelain, is the purest of all clays, and is formed of weathered feldspar. Moonstone is clean soda-lime feldspar, whitish in color, and with a reflection something like an opal.

LESSON 214

Feldspar

Leading Thought — Feldspar is an exceedingly common mineral in some rocks. When broken, orthoclase feldspar splits in two directions nearly at right angles to each other, making pieces that are obliquely brick-shaped. It is next in hardness to quartz; it will scratch glass but will not cut it.

Method — If possible have the co
mon feldspar (orthoclase), and some
the less common kinds like oligoclase a
labradorite.

Observations — 1. What is the sha
of the feldspar crystal?

2. What colors are your specimens
feldspar? How many kinds have you?

3. What is the luster of feldspar?

4. Can you scratch feldspar with t
point of a knife? Can you scratch it w
quartz? Can you scratch glass with it?

5. When you scratch feldspar w
steel what is the color of the streak l
upon it?

6. If feldspar is broken, does it bre
along certain lines, leaving smooth fac
At what angles do these smooth fa
stand to each other?

7. How can you tell feldspar fr
quartz? Write a comparison of felds
and quartz, giving clearly the charact
istics of both.

8. Hunt over the pebbles found in
sand-bank. Which ones are quartz? I
you find any of feldspar?

MICA

The mica crystal when perfect has six sides and flat ends, because it splits very easily at angles to the sides. In color, mica varies through shades of brown, from a pale smoked pearl to black. Its luster is pearly, and it can be scratched with the thumbnail. Its distinguishing characteristic is that the thin layers into which it splits bend without breaking. When mica flakes decay, they take on a golden luster and are frequently mistaken for gold.

Mica was used in antiquity for windows. Because it is transparent and not affected by heat, it has been used in the doors of stoves and furnaces and for lamp chimneys. Powdered mica is the artificial snow that is scattered over cotton batting for the decoration of Christmas trees.

Mica mines are scarce in this country; but the mining of mica is important in North Carolina and New Hampshire. India and Canada are also sources of supply. The entire production of this mine
in the United States for the year 1936 w
valued at almost a million dollars. M
of this output was used in the electri
industries, since mica is one of the b
insulating materials known.

LESSON 215

Mica

Leading Thought — Mica is a crys
which flakes off in thin scales paral
with the base of the crystal. We rarely
a complete mica crystal but simply t
thin plates which have split off. The or
nary mica is light colored, but there i
black form.

Method — If it is not possible to
tain a mica crystal, get a thick piece
mica which the pupils may split off in
layers.

Observations—1. Describe your pi

mica. Pull off a layer with the point our knife. See if you can separate this r into two layers or more.

Can you see through mica? Can you d it? Does it break easily? What is the r of your specimen? What is its luster? you cut it with a knife? Can you scratch it with the thumbnail? What color is the streak left by scratching it with steel?

3. What are some of the uses of mica? How is it especially fitted for some uses?

4. Write a theme on how and where mica is obtained.

THE SOIL

REVISED BY H. O. BUCKMAN
Professor of Soil Technology, Cornell University

Arthur Keith, U. S. Geological S

The brook mill even at low water grinds ceaselessly, sorting out the finer products a carrying them away to serve as soil material

The soil is the sepulcher and the resurrection of all life in the past. The greater sepulcher the greater the resurrection. The greater the resurrection the greater growth. The life of yesterday seeks the earth to-day that new life may come from i morrow. The soil is composed of stone flour and organic matter (humus) mixed; greater the store of organic matter the greater the fertility. — JOHN WALTON SPEN

While the coarser burden of streams is of great consequence, as the preceding sections have shown, the finer materials so carried are of even greater human importance. Few people realize the sig cance of the soil and the part that it p in the life of man. Because a child, a making mud pies, is told that his fac ty, he naturally concludes that soil is re dirt. But it is only when out of place t soil is dirt; for in place and perform- its normal and natural functions, it the home of miracles — the seat of intricate chemical and biochemical nges that make possible the nourish- nt of higher plants on which all animal depends. The study of soil is a funda- ntal introduction to agriculture.

W. C. Alden, U. S. Geological Survey

Boulders, sand, and rock flour were carried southward by the glacial ice. Climatic agencies have since changed the finer of such materials into soils, some of them the most fertile in the United States. Even this great boulder is slowly yielding to the attacks of the weather

Soil Material

f we should go back to the very be- ning, we should find that soil forma- n is initiated by rock fragments of vari- kinds — some coarse, some fine, some y fine. In our study of the brook, it noted that certain stones with sharp ners were just entering the water mill ile others had been reduced to gravel, d, or even rock flour. We saw how this ding action is done, why it is so effec- , and how the mineral grist is sorted l sifted as it is carried along. And lly we saw it deposited ready for the t step in soil formation.

p of the United States, showing the southern extent of glaciation

t must not be inferred that running er is the only grinding and carrying ncy engaged in the preparation of soil terial. The United States, north of a roughly traced by the Ohio and Mis- ri rivers, was at one time covered by immense ice sheet that pushed over lands from the north. This great ice ntle, many hundreds of feet in thick- s, scoured the bedrocks, tearing, rend- , and grinding, often to the fineness of powder. Millions of tons of rock material, varying in size from boulders to gravel, sand, and clay, were carried southward for miles, often hundreds of miles, finally to be dumped indiscriminately as the ice melted away. Perhaps one-fifth of the United States was covered by such rock debris ready to be changed by weathering agencies into soil.

Soil material, once it is sufficiently fine, is also subject to transportation by wind. In fact, thousands of square miles in our Middle West are covered by such finely divided materials, the result of ancient and violent "dust storms." This silty deposit, often many feet in thickness, is called

Agronomy Dept. Cornell U.

Boulders left by a glacier

A. F. Gustafson

A soil formed by the weathering of the wind-carried " loess " of Illinois. The fertile land on either side is being eroded by the stream

" loess " and has produced some of our most fertile soils.

But there are other agencies besides running water, glacial ice, and restless winds that help grind the bedrock into soil material. If we visit some rocky cliff, we are sure to find at its base heaps of stones, which the geologist calls " talus." These we know were pried loose by temperature changes aided by freezing water — Jack Frost and his ice wedges. This stone-cracking goes on everywhere in regions where the temperature drops below the freezing point, and not only furnishes soil material in place but also aids the scouring of the winds and the grinding, mill-like action of ice and water.

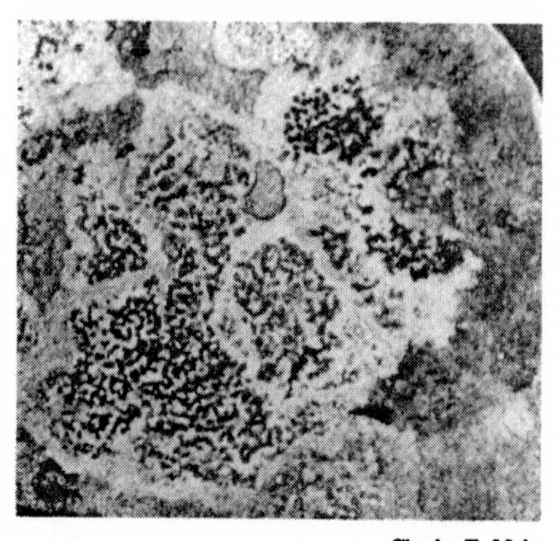

Charles E. Mohr

A common rock lichen, Lecidia albocærulescens

Soil Formation

The mere fining of rock materi whether in place or transported, does r produce a soil; far from it. Other and mo complex changes must occur. The acti of atmospheric gases, especially oxyg and carbon dioxide, is particularly eff tive. We know how iron rusts and fa away, and how limestone slowly etch and dissolves. In a similar way rock m

G. K. Gilbert, U. S. Geological Sur

A rock split by the roots of a tree. The log the foreground is also being reduced to s

terials decompose and form fine earth, t mother substance of our soil.

As this decay progresses another and very different material gradually appe — organic matter. First, perhaps, liche gain a foothold on the soil material. Th higher plants appear. And as they die a regenerate, their tissue is left mixed wi the decomposing mineral fragments. T organic matter, acted on by bacter molds, and other micro-organisms, cays, and " humus," the dark coloring m ter of soil, is produced. One of the esse tial differences between a fertile soil a mere mass of rock fragments lies in the ganic content of the former. This point ould be kept constantly in mind, es- cially when soil productivity is the ue.

Humus intensifies the chemical proc- es already described, stimulates the life thin the soil, and initiates certain ochemical changes essential to higher ants. Gradually the raw soil material pplies a more suitable foothold for gher plants and provides nutrients more undantly for their growth. Thus a soil slowly evolved from the lifeless rock d the residues of living matter — a soil at should present, if properly handled, e loose, mellow seed-bed that brings joy a farmer's heart.

Kinds of Soil

Soils may be divided for convenience to four groups, according to the pre- minant sizes of mineral particles. Thus readily recognize "gravel," "sands," oams," and "clay." Gravel soil is very arse and not of great value in growing ants. Sand soils are loose and open and sy to till. Water drains through such il very rapidly and its moisture-holding pacity is usually low. It is likely to be oughty. But when it contains plenty of mus it is a very satisfactory soil, espe- lly for certain vegetables.

Clay soils are sticky and cohesive when t and are likely to be cloddy when dry. is often difficult to create a suitable d-bed on such soils; besides, they drain ry slowly. A loam, which combines the sirable properties of both a sand and lay without their disadvantages, is per- ps the ideal soil for general purposes. ost field soils are loams of some kind.

LESSON 216

The Soil

Leading Thought — The soil usually composed of a mixture of different sizes mineral particles (sand, silt. and clay) with variable amounts of humus depending on circumstances. Soil, to supply most plants satisfactorily, should be well drained and porous so that roots may penetrate it easily, and readily obtain from it sufficient water and nutrients.

Method — The children should bring in as many different samples of soil as possible. Then classify them as gravel, sand, loam, or clay as the case may be. Now try to find some loams that are especially sandy — they should be called "sandy loams." In like manner identify some "clay loams" — that is, soils that contain more fine material than a typical loam but less than a real clay.

Now select a soil that is quite sandy and one that is decidedly clayey. Wet both moderately and knead them with the fingers. Add more water if necessary. Note the differences in the feel and other physical properties. The clay, if rightly selected and properly moistened, should be sticky and plastic. When dry it becomes hard. Try making marbles with this soil. The sand, on the other hand, is hardly sticky at all and a marble made of it usually falls apart when dried. Now the class is ready for further observations.

Observations — 1. Examine a sandy soil under a hand lens and tell why you think that it contains different sizes of mineral particles. The more numerous mineral fragments that you see are probably quartz.

2. Examine this soil, or some other soil more suitable, for humus. Humus is quite dark and acts as a coloring matter by coating the sand particles and by mixing with clay. It furnishes plant food and improves the physical condition of soils in which it is present.

3. Compare the sand and the clay that were used to make marbles under the hand lens. Describe the differences most apparent.

4. Take a piece of fresh rock such as shale or soft limestone and pound it into fine pieces. Does the fine material look like soil? Would it grow plants very satisfactorily? In what respects does a soil differ from fresh rock powder?

5. How does water grind up rocks and

G. K. Gilbert, U. S. Geological Survey

Jack Frost is busy here with his ice wedges, flaking, scaling, and cracking the rocks, adding little by little to the talus slope at their base. His effect in the soil itself is even more marked, especially in the spring when freezing and thawing occur in rapid succession

help make soil material? Is the brook a good rock mill?

6. What part does wind perform in supplying soil material? Where is " loess " found?

7. How does Jack Frost take a hand in soil formation? Why are his ice wedges so effective?

8. If there is a cliff or a gorge in your neighborhood, look for the work of Jack Frost. Find a " talus."

9. Have you ever noticed old headstones in a cemetery that are crumbling to pieces or are so worn that the carving can hardly be seen? This is the work of the gases of the air — oxygen and carbon dioxide. Examine such tombstones again carefully. The gases of the air affect soil minerals in just the same way.

10. Find a road cut where a suitable soil is exposed for a considerable depth. The dark surface layer on top is called " surface " soil. Below, extending perha to a depth of three feet, is the " subsoil Underneath is " soil material " and pe haps farther down may be seen " be rock." Examine the various layers. Th show the changes that occur as " soil m terial " is changed to soil.

11. Now explain in your own wor how soil material is prepared and how is changed to a fertile soil. This would a good subject for a short essay.

EXPERIMENT 1. *To show what kind soil drains most readily and which hol the most water.*

Take two straight glass lamp chimne or pieces of tubing six or eight inches lo (see sketch), tie cheesecloth over the bo toms, and trim it neatly. Then fill o with a dry, sandy soil and the other wi a fine, dry, clayey one. Compact the so by jarring. Then set the chimneys (s sketch) so that any water coming throu the soils may be caught in glasses or pa

Now carefully pour water from a me ured quantity on the sandy soil just f enough to keep the soil surface nice covered. Consult a watch and note h long it takes for the water to run throu and drip from the bottom of the colum At this time cease adding water and ma note of the total amount of water adde

Do the same with the clay. Compa

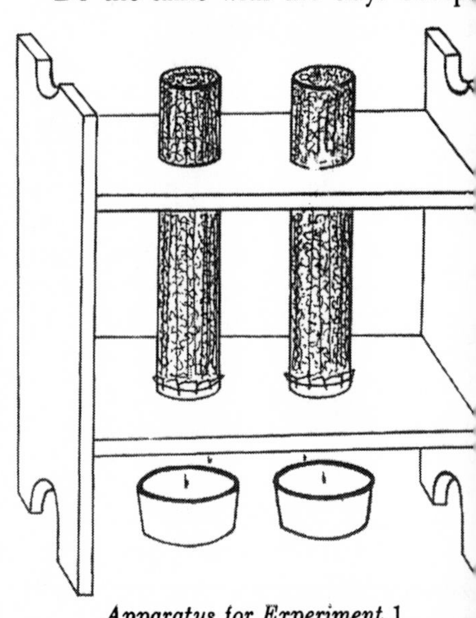

Apparatus for Experiment 1

time necessary for drainage to occur he two cases.

fter water has ceased dripping through soils, measure the amount caught in a case. Now see if you can determine ch soil held more water. Ease of drain- and moisture-holding capacity are h important in a practical way.

Iint to teacher — Water usually passes ough sand much more rapidly than ough clayey soils. But clayey soils have uch greater water capacity, especially umus is present in sufficient amounts. show this difference in moisture capac- the water added and that coming ough should be carefully measured. ands are well-drained soils, while clays n give difficulty in this respect. This nportant in the spring when an early l-bed is necessary. But clays resist ught better because of their high water acity. These practical points should brought out as the object lessons of experiment.

XPERIMENT 2. *To show that soil can water from below.*

ill two chimneys as already directed Experiment 1, but instead of pouring er on the soils, set the chimneys in a llow pan of water (see sketch). Watch t happens. In which soil does the er rise more rapidly? In which does water rise higher after several days?

Iint to teacher — Water rises through ndy soil more rapidly than through a ey one, but if time enough is given, upward distance will be greater in the . It would seem, therefore, that clay move water farther for the use of nts and convey more because of its ter moisture capacity. The object of experiment obviously is to learn the acity of soils to supply crops with mois- e. Apply this thought as practically as sible.

XPERIMENT 3. *To show the effect of anic matter on the physical condition clayey soil.*

o to the woods, scrape away the sur- e accumulation of leaves and other un- ayed matter, and get some of the dark nus (leaf mold) below. Mix this with the heaviest clay soil that has been collected. Use one part of humus to three parts of clay.

First, take some of the original clay, add water slowly, and work the soil into the very best condition possible for plant growth. Apply just the right amount of water. Treat the soil just as though you were going to pot it and use it for growing plants.

Now work up the humus-treated clay in the same way. Which soil works up better

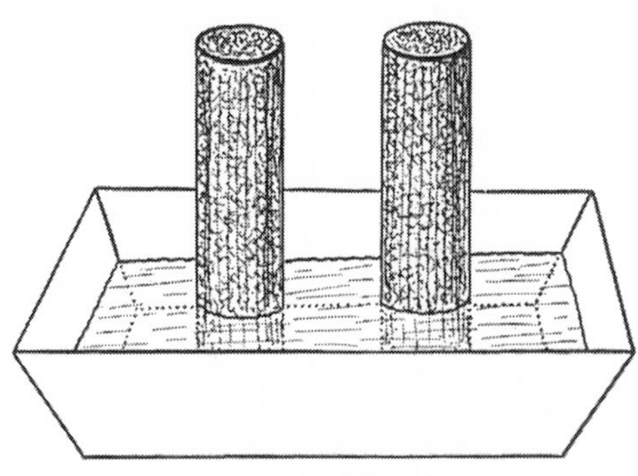

Apparatus for Experiment 2

and more easily? Compare the two samples. Why should a good supply of organic matter be kept in a field soil? Can you guess how humus affects the water-holding power of soil? Will it lower or raise it?

Hint to teacher — If leaf mold is not available for this exercise, use well decomposed organic matter of any kind. Humus makes soils easier to work and lessens the labor of seed-bed preparation. It also increases the water-holding capacity of soils and renders land less susceptible to drought. These are the practical points that should be stressed in this experiment.

Beside the moist clods the slender flags arise filled with the sweetness of the earth. Out of the darkness — under that darkness which knows no day save when the ploughshare opens its chinks — they have come to the light. To the light they have brought a colour which will attract the sunbeams from now till harvest.

— RICHARD JEFFERIES

Here is a problem, a wonder for all to see.
Look at this marvelous thing I hold in my hand!
This is a magic surprising, a mystery
Strange as a miracle, harder to understand.

What is it? Only a handful of dust: to your touch
A dry, rough powder you trample beneath your feet,
Dark and lifeless; but think for a moment, how much
It hides and holds that is beautiful, bitter, or sweet.

Think of the glory of color! The red of the rose,
Green of the myriad leaves and the fields of grass,
Yellow as bright as the sun where the daffodil blows,
Purple where violets nod as the breezes pass.

Strange, that this lifeless thing gives v[illegible] flower, tree,
Color and shape and character, [illegible]grance too;
That the timber that builds the ho[illegible] the ship for the sea,
Out of this powder its strength and toughness drew!

— From "Dust," Celia Thax[illegible]

Some years ago there was received [illegible] Cornell University a letter from a [illegible] working upon a farm in Canada. In [illegible] letter he said:

"*I have read your leaflet entitled, '*[illegible] *Soil, What It Is,' and as I trudged up* [illegible] *down the furrows every stone, every lu*[illegible] *of earth, every shady knoll, every sod* [illegible]*low had for me a new interest. The* [illegible] *passed, the work was done, and I at le*[illegible] *had had a rich experience.*"

HOW VALUABLE SOIL IS LOST

By A. F. Gustafson

Professor of Soil Technology, Cornell University

Were the soil indestructible and everlasting, as so many people imagine, its study would be of general interest only. Unfortunately, however, our lands are subject to ravages and losses so extensive and far-reaching that not only is their crop producing capacity much reduced but also they oftentimes are threatened with total destruction. Because man formerly did not realize that this was true, he took no steps to prevent such losses; and when he finally became conscious of the danger, much damage had already been done.

Many years ago, when the white man came to this country, he found the eastern part of what is now the United States covered with forest trees. In the central Mississippi Valley area there were forests along many of the larger streams, and tall-growing prairie grasses on the wide open spaces between them. To the westward on the Great Plains, where the rainfall was less, the land was covered with short grasses. In the mountains farther west [illegible] along the western coast, trees grew [illegible] lower elevations wherever the rainfall [illegible] sufficient for them. Thus, in nature, [illegible] land was covered, protected, and held [illegible] place by vegetation; and that form of ve[illegible]tation for the growth of which conditi[illegible] were most favorable predominated.

The trees covered the soil somew[illegible] like leaky umbrellas. Rain fell on [illegible] leaves, twigs, and branches; thus the [illegible] of the raindrops was broken and some [illegible] the water ran down the branches [illegible] trunks of the trees directly into the s[illegible] which held part of it for the use of [illegible] trees. Likewise, the rain fell on the pra[illegible] grasses and ran down into the soil [illegible] much as it did in the forest.

The leaf and twig litter in the fo[illegible] caught the water, so that much of it co[illegible] be absorbed by the soil. The old d[illegible] grasses on the prairies and plains h[illegible] water in much the same way. Under b[illegible]

W. C. Mendenhall, U. S. Geological Survey

ıe work of the wind. Wind erosion is irresistible. The wind has covered forests, farmsteads, and even cities with sand

s and trees the soil was loose and open. aying roots left openings in the soil. remains of leaves and grasses were ɛen down by earthworms and other ınisms living in the soil; as these anis moved about, they left many open- in the soil. Moreover, the decaying ɪr kept the soil in a loose condition, so enabled it to absorb the rain rather dly. The litter itself also absorbed conrable water, so that less was lost as off to the streams. The old dead grass the growing grass kept the water from ıing off until much of the rainfall ɛed into the soil. The absorbed rain ɛr came back to the surface of the soil ɔwer elevations, in the form of springs. ing long periods between rains, the ngs supplied water for man and for his stock; the excess, then as now, flowed ɔ form streams which in turn fed the ɛr bodies of water.

'he white man cut down the forest trees and then plowed the land; a little later he broke the sod on the prairies. Once Nature's protecting cover for the soil was plowed under it soon rotted and was lost. Immediately after the forest was cleared, good yields of wheat, corn, and other farm crops were produced even on rather steep slopes. But when the roots of trees and grasses and the other organic matter in the soil had decayed and disappeared, the supply could not be quickly renewed; and as a result, the soil was no longer loose and open but became hard and closely packed. In this condition it would not readily absorb water, which consequently ran off the fields into the brooks.

When the topsoil was thus left without protection, the raindrops fell directly on the bare surface and churned it into a thin mud. This mud ran down the slopes and filled up the small openings in the soil called pores. In the forest or un-

U. S. Soil Conservation Service (N. C.)

The sides of these old gullies have been seeded and mulched with pine needles; this helps keep the soil moist and encourages the growth of the young seedlings. Vegetation will soon cover the soil and protect it from further washing. Pine trees are usually planted in the mulch

der the prairie grasses these pores and wormholes were loosely covered with litter, and being open they permitted the rain to enter the soil freely. But after cultivation and rains had clogged the openings in the soil, much rain water ran off the sloping fields. It is the running off of this surface water that causes erosion or the loss of soil. In the farmer's sloping, clean-cultivated fields, the water collects

J. S. Cutler, U. S. Soil Conservation Service (Ohio)

Rows of cultivated crops running up and down the slope often permit loss of soil. The stubble in the foreground stopped the wasting of the soil. Strip cropping usually prevents such loss

between the rows, and if these run do hill, much water and soil are lost. As m water runs in one place, it runs faster a faster; this gives it added cutting a carrying power, so that the top soil r be readily carried away. Often slight pressions such as wheel tracks or furr become rills and even small gullies dur a single rain.

Heavy rains cause more loss of soil th do light showers. More soil is lost fr steep slopes than from gentle ones heavy rain falls on wet soils, they can t up but little of it and most of the

W. C. Lowdermilk, U. S. Forest Se

Badly gullied slope, Oak Creek drai area, California. This washing resulted f the Tehockapi cloudburst, October, 1 *Note that one of the men in the foregro is up to his waist in one of the smaller gu*

must run off over the surface. Much m erosion occurs, therefore, if rain falls wet soil than if it falls on dry soil. soils, clean-cultivated orchard soils, or s growing cultivated crops such as veg bles, corn, cotton, and tobacco have li protection and suffer greater losses washing than do soils protected by for or pasture or hay grasses.

Soils that are well supplied with of the plant foods and that are in g condition in every way produce ds. Large yields which are accompanied a thick thrifty growth help protect and d the soil. Moreover, soil in good tilth pen and porous and thus takes up rain ter, so that less runs off carrying soil and nt food away with it. Water absorbed the soil is saved for future use; that ich runs off the surface causes the ero- n. Anything, therefore, that slows down flow of water over cultivated land ecks the loss of soil by erosion.

During dry periods, the finest material bare soils, especially sandy ones, may carried away by the wind. It drifts into ds and ditches and onto farmsteads and ps so as to cause untold damage.

U. S. Soil Conservation Service (N. C.)

Sheet erosion and gullies in a North Carolina pasture. Severe sheet erosion usually precedes this type of gullying. Active erosion in places of this sort may usually be checked by means of grass and legumes and woody vines or brushy shrubs. Immediate attention is needed here

Soil Erosion, an Old Problem

As far back as colonial days, Washing- and Jefferson as farmers recognized sion on their lands in Virginia. Geolo- ts have long believed that soil losses ough erosion are so serious as to eaten mankind with starvation at some e in the future. Farmers and workers experiment stations have recognized menace of soil erosion for more than f a century; as long ago as 1885 Priestly Mangum built on his own North Caro- a farm his first terrace, modeled on thods of terracing already in use in orgia. (For further discussion of the ngum terrace, as it is called after its in- tor, see p. 774.) Bulletins discussing erosion were published soon after 1890 Tennessee and Arkansas.

Not until 1934, however, did the fed- l government make a systematic at- npt to control erosion. In that year the il Erosion Service was established in United States Department of the In- ior; it has since been broadened into Soil Conservation Service and trans- red to the Department of Agriculture. . H. H. Bennett was called on to head s service; he was well qualified for this sition by his long interest in and experi- ce with soil erosion in the South, and his recognition of the seriousness of e injury done by erosion to the fertile tton lands of the South, the corn soils of the Middle West, and the wheat and orchard lands of the Far West.

One of the first tasks confronting the Soil Erosion Service was the making of a survey to learn the extent and seriousness of soil erosion by wind and water throughout the United States. The results of the survey were alarming. It was found that all but 30 per cent of the land area of the United States had been injured by erosion. Of the total area 45 per cent or 855,000,000 acres had lost from one-fourth to three-fourths of the top six inches of soil. Of this depleted portion 10 per cent had lost more than three-fourths of its topsoil. Wind erosion had damaged 233,000,000 acres or 17 per cent of the country. This survey was made in 1934; and, of course, much additional damage has occurred during 1935, 1936, 1937, and 1938; nearly 90,000,000 acres or one-twentieth of the country have been severely damaged or completely destroyed for agriculture by wind erosion. By far the most of the total damage to the soil has been done by the somewhat uniform removal of surface soil, known as sheet erosion; but almost one-half of the country has been damaged by gullying, and on the land actually occupied by gullies, they do greater injury than does sheet erosion; gullies too large to cross with ordinary farm implements make farming difficult and have greatly increased the cost of growing some crops.

J. S. Cutler, U. S. Soil Conservation Service (Mich.)

The topsoil has all been washed away from the hill shown in the background. Now the under soil is washing down upon good soil and covering it up — a common occurrence in many parts of the United States

LESSON 217

How Valuable Soil Is Lost

Leading Thought — Soil on a sloping surface that lacks a protecting cover of some type of vegetation is easily washed away. This washing away of soil is called erosion.

Method — Erosion may be studied any time, but perhaps the best time is i mediately after a hard rain, in a pla where there is some soil that is not co ered with a protecting crop.

Observations — 1. When the wh man came to our country, what conditio existed in the eastern and western se tions? Where did great grassland are exist?

2. What, in a general way, becomes much of the rain that falls on land co ered by trees or heavy grass? Why is t soil under trees and grass usually loose a open? How do earthworms serve in hel ing the soil to remain open? See T Earthworm, page 422.

3. Once rain water has been absorb by the soil, under what conditions m we expect to see it again?

4. Why did the early settlers remove many forest trees from the land? Wh changes occurred in the soil after seve crops had been produced? Why do worn-out soil become hard and packed gether?

5. Why does much of the water th falls on the bare soil of sloping fields r off? What is erosion?

6. Why does more erosion occur if ra falls on soil that is already wet, than it falls on dry soil? Why do bare fields s fer more loss than do those that are co ered by forests or grass? How do trees a grass conserve water and soil?

7. How long ago was erosion noticed our country?

8. What has the federal governme done in an effort to control erosion? H much of the land area of the United Stat has been damaged by erosion? How mu has been damaged by wind erosion? what ways do sheet erosion and gullyi differ?

HOW TO CONSERVE OUR SOIL

By A. F. Gustafson

As already shown, sloping fields on which are grown clean-cultivated farm or garden crops and clean-cultivated orchards or groves are subject to extensi soil erosion, which may result either fro rains or from water running off after t

H. A. Daniel, Oklahoma Agricultural Experiment Station

n Oklahoma country road filled with sand after a wind storm in April, 1936. *The grass and weeds hold the sand in drifts*

J. S. Cutler, U. S. Soil Conservation Service (Ohio)

Strip-cropping. On sloping land clean-cultivated crops are alternated with close-growing ·ops; all are farmed on the contour. This helps to prevent erosion, and to keep water on ·e land

thawing of snow. How may this loss of valuable soil be prevented?

We may learn much about conservation from Nature. Seldom does Nature permit much of the soil to remain bare or exposed long in areas that receive rain enough for fairly good growth of crops. Land that is not under a crop, if not cultivated, is soon covered with weeds and grasses, and these plants help to hold the soil against washing. Grasses, however, give much better protection than do most weeds. In forests, or even in pastures or meadows that are making reasonably good growth, the vegetation protects and holds the soil. Keeping the soil covered and protected by close-growing vegetation, then, is one important way of controlling soil erosion by both wind and water.

W. R. Mattoon, U. S. Forest Service (Tenn.)

Before. Large gully with banks sloped, being planted to black or common locust, oaks, and Japanese honeysuckle vines. The brush dam in the foreground helps to hold soil until the plantings get well started

Of course, we cannot grow cotton, corn, or vegetable crops and at the same time keep the soil covered with grass. We must therefore develop ways and means of managing the land so that it will continue to produce food for man and his livestock and materials for man's clothing. Vari methods have been developed and u during the past half-century.

Fertilization. To fertilize the soil whatever extent is economical for production of relatively large crop yie is a first step. Thrifty crops protect the better, and they leave on it more resi materials such as stalks of corn or cott potato tops, and wheat, oat, or bar

W. R. Mattoon, U. S. Forest Service (Te

After. The gully shown in the oppo picture at the end of the first season. In other year or two the plantings will h checked erosion entirely

stubble to protect it somewhat, u such materials are plowed under. Th residue materials help to hold the soil gether and upon decaying supply pl nutrients for the crops that follow. some areas, economical fertilization c sists of the addition of phosphorus alo in such forms as superphosphate, ba slag, or rock phosphate; in other areas, pecially on sandy soils, potash salts needed in addition to phosphorus; a vegetables in general require the appli tion of nitrogen as well as phospho and potash, or they may require a co plete fertilizer.

Growing legumes. In all extensive cr ping systems, legumes are needed to h keep up the yield of grain and grass cro Such legumes as red and alsike clov alfa, lespedeza, and sweet clover can be d. Wherever the soil is too low in e a dressing of an economical form of e is needed to enable these legumes protect the soil and to produce good lds.

Crop rotations. The growing of crops rotation is usually good farm practice. tation means the growing of crops regular order such as cotton, corn, po-oes, or other vegetable crop the first r, barley, oats, wheat, or other grain p the second year, and a legume or grass third year. Grass for hay the fourth r often follows. On lands that wash her easily, grass for hay for several ad-ional years makes a longer and better ation. Thus the land is kept in a culti-ed crop a smaller proportion of the e than if a three- or four-year rotation followed.

On land which is steep and easily ded, long-term meadows or pasture e good protection to the soil and should the same time provide fair income for h lands. Both meadows and pastures uire fertilization, at least with phos-orus and often with lime, in order that y may produce good yields and inci-ntally that they may provide suitable l protection.

Planting trees. Steeper, more easily oded, shallower, and less productive d may well be reforested. Care should taken to make certain that erosion is ought under control before planting, cause several years usually pass before e trees that are planted become large ough to hold the soil in place. Once ung trees are well established, however, ey afford excellent protection for the il.

Contour farming. Many advantages ac-mpany the carrying out of all tillage, ding, and most harvesting operations on e contour, or crosswise on all the main pes. To begin with, it is easier to plow ross than up and down slopes. The cross-se depressions left by the plow catch d hold water until it soaks into the soil. the soil is well plowed, seed-bed prepa-tion may best be done on the contour, for the same reason that the plowing may best be done in this way.

W. R. Mattoon, U. S. Forest Service (Tenn.)

Locusts not only hold the soil, but enrich it; they also produce useful wood

Seeding crops across the slopes has many advantages. Any depressions and ridges left by seeding implements check the flow of water over the surface and give the soil more opportunity to absorb the water. Up- and downhill cultivation, on the other hand, often leads to heavy loss of soil both by sheet erosion and by gullying. When grains and grasses are seeded on the contour the plants are in rows across the slope. Thus much more water is held by the rows than would have been held if the rows had been seeded up- and downhill.

Contour harvesting is advantageous in that it requires less horse or tractor power than up- and downhill harvesting. Mak-ing wheel tracks down slopes should be avoided because these lead to the collec-

W. A. Rockie and P. C. McGrew, Washington Agricultural Experiment Station

One heavy rain caused this washing of soil. Note that the land under grass and shrubs to the left has not lost its productive topsoil

tion of water and often result in gully formation. Contour tillage saves much water for use in dry periods and thus often increases crop yields materially.

Contour strip cropping. Entire slopes are often plowed and planted all at the same time to a clean-tilled crop such as corn or cotton. Rain water falling on the slope collects in low places, and as it passes down over the surface, the streams grow larger and flow faster. Soon they attain great cutting and carrying power. Under these conditions the result of a heavy rain may be gullying as well as severe sheet washing.

Nowadays many long slopes are broken up into a number of strips laid out across the slope so that crop rows are on the level. Grass is alternated with clean-tilled crops on part of the sloping land. On the other land grain is alternated with clover or another legume, thus completing a four-year rotation. Some sheet washing is bound to take place on bare slopes under heavy beating rains, but the grass strips check the current and cause sedimentation. (For a fuller explanation of this process, see the lesson on the brook, p. 736.) The checking of the flow of water by the grass strips thus tends to prevent gully formation. The best width of strip to use varies with such factors as soil, type of rainfall, crops, and the steepness of slope.

Terracing. About 1885, Mr. Priestly Mangum, who lived near Wake For North Carolina, after observing the eff of various hillside ditches, developed a race with a broad-bottomed channel laid out as to give the channel a sli slope toward the outlet. The principles veloped by Mangum have been extensi adopted: terraces are now usually laid in accord with a definite plan for field to be protected. Each terrace has own channel, which is in fact a hills ditch. These channels break up long slo into a number of small watersheds. water instead of "running off" rap in the usual way and causing severe sh washing and gullying is made to "wa slowly along the contour of the land.

In parts of the South, strip croppin practiced on terraced land. One n study all the conditions and then emp such erosion-control measures as will duce the loss of soil as much as poss. and at the same time produce the cr needed by man for himself and his l stock.

Controlling wind erosion. Keeping soil covered with vegetation, rota crops, keeping the soil rough rather t

A. F. Gust

American beach grass planted on blo beach sand on Long Island, New York. E such weeds as the lowly cocklebur in the f ground help hold the soil against blowing

ooth, strip cropping and seeding crops oss the prevailing wind direction, mak- ; furrows across the slopes on the level holding water, and planting tall crops trees as windbreaks — all these help to eck the blowing away of valuable soil, d the drifting of sands on to crops, ds, and farmsteads. The level furrows ld water that is badly needed for crops. e conservation of water, therefore, pro- ces more plant growth and better cover the protection of the soil; and, more- er, the water so saved keeps the soil ist longer and thus helps greatly in the ntrol of wind erosion.

LESSON 218

How to Conserve Our Soils

Leading Thought — Soils may be con- ved by fertilization and liming, the wing of legumes, rotation of crops, orestation, contour cultivation, strip pping, and the building of terraces.

Method — In most sections of the ited States, it will be possible for pu- s to locate some land that has sufficient pe to show the effects of erosion on e soil after a hard rain.

Observations — 1. Can you find a clean-cultivated field on a slope that shows any loss of soil particles? What does Nature do to land from which a cultivated crop has been taken? In what way do meadows, pastures, and forests help prevent erosion?

2. In what way can fertilizers applied to the soil influence the amount of erosion that takes place?

3. How may legumes serve in the prevention of erosion?

4. What is meant by rotation of crops? Does this practice have any influence in preventing erosion?

5. If a farmer desires to prevent erosion is it better for him to plow, cultivate, and harvest his fields up- and downhill or on the contour?

6. What is strip cropping? How does this method tend to check erosion?

7. Where did Mr. Priestly Mangum build his first terrace? Describe a Mangum terrace. How do terraces check the rate of flow of run-off water? How does this lessen erosion? What kind of farming can be practiced on terraced land?

8. How can the blowing of soil, or wind erosion, be controlled?

9. Are some methods of control effective for both wind and water erosion?

THE MAGNET

Until comparatively recent times, the power of the magnet was so inexplicable that it was regarded as the working of magic. The tale of the Great Black Mountain Island magnet described in the *Arabian Nights Entertainments* — the story of the island that pulled the nails from passing ships and thus wrecked them — was believed by the mariners of the Middle Ages. Professor George L. Burr assures me that this mountain of lodestone and the fear which it inspired were potent factors in the development of medieval navigation. Even yet, with all our scientific knowledge, the magnet is a mystery. We know what it does, but we do not know what it is. That a force unseen by us is flowing off the ends of a bar magnet, the force flowing from one end attracted to the force flowing from the other and repellent to a force similar to itself, we perceive clearly. We also know that there is less of this force at a point in the magnet halfway between the poles; and we know that the force of the magnet acts more strongly if we offer it more surface to act upon, as is shown in the experiment of drawing a needle to a magnet by trying to attract it first at its point and then along its length. The child likes to demonstrate that this force extends out beyond the ends of the magnet by seeing across how wide a space the magnet, without touching the objects, can draw to it iron filings or tacks. That the magnet can impart this force to iron objects is demonstrated with curious interest, as the child takes up a chain of tacks at the end of the magnet; and yet the tacks when removed from the magnet have no such power of cohesion. That some magnets are stronger than others is shown in the favorite game of "stealing tacks," the stronger magnet taking them away from the weaker; it can also be demonstrated by a competition between magnets, noting how many ta each will hold.

One of the most interesting thi about a magnet is that like poles repel a opposite poles attract each other. H hard must we pull to separate two m nets that have the south pole of c against the north pole of the other! Ev more interesting is the repellent power two similar poles, which is shown by a proaching a suspended magnetized nee with a magnet. These attractive and rep lent forces are most interestingly dem strated by the experiment in question of the lesson. These needles floating cork join the magnet or flee from it, cording to which pole is presented them.

Not only does this power reside in t magnet, but it can be imparted to oth objects of iron and steel. By rubbing c pole of the magnet over a needle seve times, always in the same direction, convert the needle into a magnet. If suspend such a needle by a bit of thre from its center, and the needle is n affected by the nearness of a magnet other metal, it will soon arrange its nearly north and south. It is well to thr the needle through a cork, so it will ha horizontally, and then suspend the cork a thread. The magnetized needle will n point exactly north, for the magnet po of the earth do not quite coincide with t poles of the earth's axis.

The direction assumed by the m netized needle may be explained by t fact that the earth is a great magnet, b the south pole of the great earth magn lies near the north pole of the earth. Th a magnet on the earth's surface, if allow to move freely, will turn its north p toward the south pole of the great ea magnet. Then, we might ask, why call the earth's magnetic pole that rest our North Pole its north magnetic e? That is merely a matter of conven- ce for us. We see that the compass dle points north and south, and the of the needle which points north we veniently call its north pole.

he above experiment with a suspended dle shows how the mariner's com- s is made. This most useful instrument said to have been invented by the inese at least as early as 1400 B.C., and haps even longer ago. It was used by m to guide armies over the great plains, the needle was made of lodestone. e compass was introduced into Europe ut 1300 A.D., and has been used by riners ever since. To "box the com- s" is to tell all the points on the com- s dial, and is an exercise which the dren will enjoy.

Ve are able to tell the direction of lines of force flowing from a magnet placing fine iron filings on a pane of s or a sheet of paper and holding one both poles of a magnet close beneath; tantly the filings assume certain lines. he two ends of a horseshoe magnet are d, we can see the direction of the lines force that flow from one pole to the er.

The action of the magnetic force of the th on the electrons streaming from the produces the auroral streamers called rora Borealis in North latitudes and rora Australis in high Southern lati- es.

Magnets made from lodestone are ed natural magnets. A bar magnet or orseshoe magnet has received its mag- ism from some other magnet or from trical sources. An electromagnet is of t iron, and is only a magnet when ler the influence of a coil of wire rged with electricity. As soon as the rent is shut off, the iron immediately ses to be a magnet.

LESSON 219

The Magnet

Leading Thought — Any substance t will attract iron is called a magnet, and the force which enables it to attract iron is called magnetism. This force resides chiefly at the ends of magnets, called the poles. The forces residing at the opposite ends of a magnet act in opposite directions; in two magnets the like poles repel and the unlike poles attract each other. The needle of the mariner's compass points north and south, because the earth is a great magnet which has its south pole as a magnet near the north pole of the world.

Method — Cheap toy horseshoe magnets are sufficiently good for this lesson, but the teacher should have a bar magnet, also a cheap toy compass, and a specimen of lodestone, which can be procured from any dealer in minerals. In addition, there should be nails, iron filings, and tacks of both iron and brass, pins, darning needles or knitting needles, pens, etc. Each child, during play time, should have a chance to test the action of the magnets on these objects, and thus be able to answer for himself the questions, which should be given a few at a time.

Observations — 1. How do we know that an object is a magnet? How many kinds of magnets do you know? Of what substance are the objects which the magnets can pick up made? Does a magnet pick up as many iron filings at its middle as at its ends? What does this show?

2. How far away from a needle must one end of the magnet be before the needle leaps toward it? Does it make any difference in this respect, if the magnet approaches the needle toward the point or along its length? Does this show that the magnetic force extends out beyond the magnet? Does it show that the magnetic force works more strongly where it has more surface to act upon?

3. Take a tack and see if it will pick up iron filings or another tack. Place a tack on one end of the magnet; does the tack pick up iron filings now? What do you think is the reason for this difference in the powers of the tack?

4. Are some magnets stronger than others? Will some magnets pull the iron filings off from others? In the game of

"stealing tacks," which can be played with two magnets, does each end of the magnet work equally well in pulling the tacks away from the other magnet?

5. Pick up a tack with a magnet. Hang another tack to this one end to end. How many tacks will it thus hold? Can you hang more tacks to some magnets than to others? Will the last tack picked up attract iron filings as strongly as the first next to the magnet? Why? Pull off the tack which is next to the magnet. Do the other tacks continue to hold together? Why? Instead of placing the tacks end to end, pick up one tack with the magnet and place others around it. Will it hold more tacks in this way? Why? If a magnet is covered with iron filings will it hold as many tacks without dropping the filings?

6. Take two horseshoe magnets and bring their ends together. Then turn one over and again bring the ends together. Will they cling to each other more or less strongly than before? Bring two ends of two bar magnets together; do they hold fast to each other? Change ends with one; now do the two magnets cling more or less closely than before? Does this show that the forces in the two ends of a magnet are different in character?

7. Magnetize a metal knitting needle or a long sewing needle by rubbing one end of a magnet along its length twelve times, always in the same direction, *and not back and forth*. Does a needle thus treated pick up iron filings? Why?

8. Suspend this magnetized needle by a thread from some object where it can swing clear, or, better, float a magnetized sewing needle on the surface of a glass of water. When it finally rests, does it point north and south or east and west?

9. Bring one end of a bar magnet or of a horseshoe magnet near to the north end of the suspended needle; what happens? Bring the other end of the magnet near the north end of the needle; what happens?

10. Magnetize two needles so that their eyes point in the same direction when they are suspended. Then bring the point of one of these needles toward the eye of the other; what happens? Bring the of one toward the eye of the other; w happens? When a needle is thus m netized the end which turns toward north is called the north pole, and the e pointing south is called the south pole.

11. Try this same experiment by thr ing the needles through the top of a c and floating them on a pan of water. the north poles of these needles attract repel each other? Do the south poles these needles attract or repel each oth If you place the north pole of one nee at the south pole of the other do they j and make one long magnet pointing no and south?

12. Take a pocket compass; place north end of one of the magnetized n dles near the north arm of the comp needle; what happens? Place the so pole of the needle near the north arm the compass needle; what happens? C you tell by the action of your mag upon the compass needle which end your magnet is the north pole and wh the south pole?

13. Magnetize several long sewing n dles by rubbing some of them with magnet from the point toward the and some from the eye toward the po Take some small corks, cut them in cr sections about one-fourth inch thick, a thrust a needle down through the cen of each leaving only the eye above cork. Then set them afloat on a pan water. How do they act toward each oth Try them with a bar magnet first with end and then with the other; how do th act?

14. Describe how the needle in mariner's compass is used in navigatio

15. Place fine iron filings on a pane glass or on a stiff paper. Pass a magnet derneath; what forms do the filings sume? Do they make a picture of the rection of the lines of force which co from the magnet? Describe or sketch direction of these lines of force, when poles of a horseshoe magnet are placed low the filings. Place two similar poles a bar magnet beneath the filings; w form do they take now?

6. What is lodestone? Why is it so
ed?

7. What is the difference between
estone and a bar magnet? What is an
tromagnet?

8. Write an English theme on "The
covery and Early Use of the Mariner's
mpass."

w, chief of all, the magnet's power I sing,
d from what laws the attractive functions spring;
e magnet's name the observing Grecians drew
m the magnetic regions where it grew;
viewless potent virtues men surprise,
strange effects they view with wondering eyes,
ien, without aid of hinges, links, or springs,
endant chain we hold of steely rings
Dropt from the stone — the stone the binding source, —
Ring cleaves to ring, and owes magnetic force:
Those held superior, those below maintain,
Circle 'neath circle downward draws in vain,
Whilst free in air disports the oscillating chain.

— "DE RERUM NATURA," LUCRETIUS, 93–55 B.C.

CLIMATE AND WEATHER

By Wilford M. Wilson

Late Section Director, U. S. Weather Bureau, and Professor of Meteorology in Cornell University.

Lightning flash behind a cloud

The atmosphere, at the bottom of which we live, may be compared to a great ocean of air, about two hundred miles deep, resting upon the earth. The changes and movements that take place in this ocean of air, the storms that invade it, the clouds that float in it, the sunshine, the rain, the dew, the sleet, the frost, the snow, and the hail are termed " weather."

Let us suppose we have just returned from a trip, of two or three months, to some distant part of the country. We can tell of the people we saw, the cities we visited, and the weather we found in the various places; but we cannot tell, from personal experience, about the climate of the places we visited. The weather is the condition of the atmosphere at the moment, while climate is the sum total of weather conditions over a period of several years. One season may be very while another may be very wet, one be exceedingly cold, and the next ma unusually hot; but climate is a term wh includes all of these variations.

A study of weather quite naturally sults in a study of climate, since clim includes, in addition to all the reg daily, monthly, seasonal, or annual a ages, all the extreme departures from th general conditions. We live in weat we partake of its moods; we reflect its shine and shadows; it invades the every affairs of life, influences every busi and social activity, and molds the c acter of nations; and yet nearly everyth we know about the weather has b learned within the lifetime of the pre generation. Not that the weather did interest men of early times, but the p ı appeared to be so complicated and so ɴplex that it baffled their utmost en- ıvors.

ɪᴇ Tower of the Winds at Athens

Ꭲhe Tower of the Winds, erected prob- .y before 35 B.C., indicates the knowl- ge of the weather possessed by the cient Greeks. This tower is a little octa- n, the eight sides of which face the ;ht principal winds. On each of its eight es is a human figure cut in the marble, nbolizing the kind of weather the wind m that particular direction brought to hens.

Boreas, the cold north wind, is rep- ented by the figure of an old man aring a thick mantle, high buskins oots), and blowing on a "wreathed rn." Cæcias, the northeast wind, which ɔught, and still brings to Athens, cold, ɔw, sleet, and hail, is symbolized by a ın with a severe countenance who is lding a dish of olives, because this nd shakes down the olives in Attica.

Apeliotes, the east wind, which brought ather favorable to the growth of vegeta- n, is shown by the figure of a beautiful uth bearing fruit and flowers in his :ked-up mantle.

J. H. Comstock

The Tower of the Winds at Athens

Notus, the warm south wind, brought rain, and he is about to pour the water over the earth from the jar which he carries.

Lips, the southwest wind, beloved of the Greek sailors, drives a ship before him,

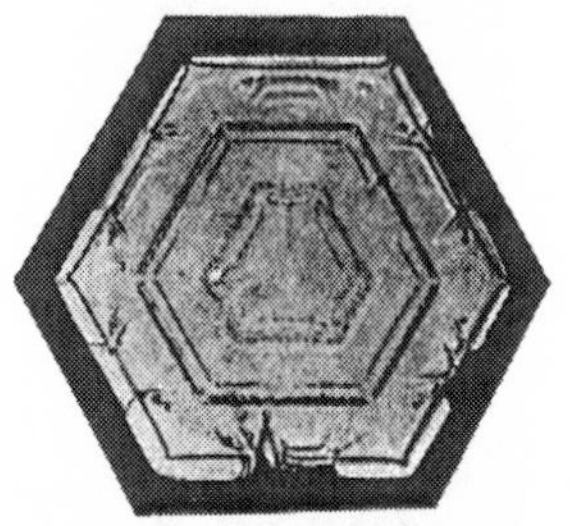

Photomicrograph by W. A. Bentley

Snow crystal

while Zephyrus, the gentle west wind, is represented by a youth lightly clad, scattering flowers as he goes.

Sciron, the northwest wind, which brought dry and usually cold weather to Athens, is symbolized in the figure of a man holding a vessel of charcoal in his hands, because this wind parched the vegetation. Thus, the character of the weather brought by each separate wind is fixed in stone, and from this record we learn that, even with the lapse of twenty centuries, there has come no material change.

Historical

There is no record of any rational progress having been made in the study of the weather until about the middle of the seventeenth century, when Torricelli discovered the principles of the barometer. This was a most important discovery and marks the beginning of the modern science of meteorology. Soon after Torricelli's discovery of the barometer his great teacher, Galileo, discovered the thermometer, and thus made possible the collection of data upon which all meteorological investigations are based. About one hundred years after the discovery of the barometer, Benjamin Franklin made a discovery of equal importance. He demonstrated that storms were eddies in the atmosphere, and that they progressed or moved as a whole, along the surface of the earth.

It might be interesting to learn how Franklin made this discovery. Franklin, being interested at that time in astronomy, had arranged with a friend in Boston to take observations of a lunar eclipse at the same time that he, himself, was to take observations at Philadelphia. On the night of the eclipse a terrific northeast wind and rain storm set in at Philadelphia, and Franklin was unable to make any observations. He reasoned, that as the wind blew from the northeast, the storm must have been experienced in Boston before it reached Philadelphia. But imagine his surprise, when he heard from his friend in Boston that the night had been clear and favorable for observation, but that a fierce wind and rain storm set in on the following morning. Franklin determined to investigate. He sent out letters of inquiry to all surrounding mail stations, asking for the time of the beginning and ending of the storm, the direction and strength of the wind, etc. When the information contained in the replies was charted on a map it showed that, at all places to the southwest of Philadelphia, the beginning of the storm was earlier than at Philadelphia, while at all places to the northeast of Philadelphia the beginning of the storm was later than at Philadelphia. Likewise, the ending was earlier to the southwest and later to the northeast of Philadelphia than at Philadelphia. He also found that the winds in every instance passed through a regular sequence, setting in from some easterly point and veering to the south as the storm progressed, then to the southeast and finally to the west or northwest as the storm passed away and the weather cleared.

A further study of these facts convinced Franklin that the storm was an eddy in the atmosphere, that the eddy moved as a whole from the southwest toward the northeast, and that the winds blew from all directions toward the center of the eddy, impelled by what he termed suction.

Franklin was so far in advance of his time that his ideas about storms made little impression on his contemporaries, a so it remained for Redfield, Espy, Loon Henry and Maury, and other Americ meteorologists, a hundred years later, show that Franklin had gained the fi essentially correct and adequate conc tion of the structure and movement storms.

During the first half of the ninetee century, considerable progress was ma in the study of storms, principally American meteorologists, among wh was William Redfield of New York, w first demonstrated that storms had bot rotary and a progressive movement. Jan Espy followed Redfield in the constr tion of weather maps, although he had ready published much on meteorolog subjects before the latter entered the fie

Professor Joseph Henry, secretary of Smithsonian Institution at Washingt was the first to prepare a daily weat map from observations collected by te graph. He made no attempt to make fc casts, but used his weather map to de onstrate to members of Congress feasibility of a national weather service

An incident occurred during the C mean War that gave meteorology a gr impetus, especially in Europe. On N vember 10, 1854, while the French fl was at anchor in the Black Sea, a sto of great intensity occurred which pra cally destroyed its effectiveness against enemy. The investigation that follow showed that the storm came from west Europe, and that if there had been a quate means of communication and character and direction of progress be known, it would have been possible warn the fleet of its approach and th afford an opportunity for its protection

This report created a profound imp sion among scientific men, and act measures were taken at once, which sulted in the organization of weat services in the principal countries of rope between 1855 and 1860.

The work of Professor Henry Ab and others in this country would, dou less, have resulted in such an organizat in the United States in the early 60's,

ses as the distance from the center of earth increases, so there is a point at a ain distance above the earth where the forces just balance each other, and gas will expand upward to that point will not rise beyond it. Therefore, if know the expansive force of a gas and rate at which gravity decreases, it is sible to calculate the height to which different gases that compose the air rise.

n this way it has been determined that on dioxide, which is one of the vier gases, extends upward about 10 es, water vapor about 12 miles, oxygen ut 30 miles, and nitrogen about 35 es, while hydrogen and helium, the test gases known, do not appear at the ace at all, but probably exist at a height rom 30 miles to possibly 200 miles.

here are other ways in which we are to gain some idea of the approximate ght at which there is an appreciable atsphere. When the rays of light from sun enter our atmosphere they are ken up or scattered — diffracted — so t the atmosphere is partially lighted for ne time before sunrise and after sunset. is is called twilight. If there were no osphere, there would be no twilight, darkness would fall the instant the passed below the horizon. Twilight, ich is caused by the sun shining on upper atmosphere, is perceptible until sun is about 16° below the horizon.

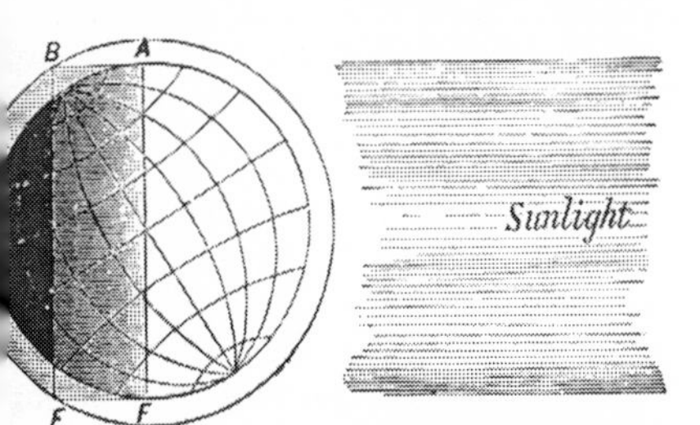

Todd's New Astronomy

The zone of twilight in midwinter

m this it is calculated that the atmosre has sufficient density at a height of miles to scatter, or diffract, sunlight.

Observations of meteors, commonly led shooting stars, indicate that there is an appreciable atmosphere at a height of nearly 200 miles. Meteors are solid bodies flying with great velocity through space. Occasionally they enter our atmosphere.

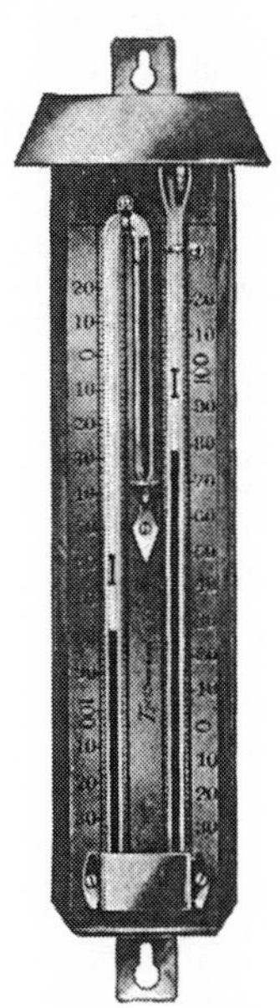

Taylor Instrument Companies

Maximum and minimum thermometer. The index, a miniature glass bottle with a piece of steel wire inside, is left at the highest and lowest points recorded; it can be pulled down with a magnet

Their velocity is so great that the slight resistance offered by the air generates enough heat by friction, or by the compression of the air in the path of the meteor, to make it red hot or to burn it up before it reaches the bottom of the atmosphere. Only the largest meteors reach the earth.

When a meteor is observed by two or more persons at a known distance from each other, and the angle which the line of vision makes with the horizon is noted by each, it is a simple matter to calculate the distance from the earth where the lines of vision intersect, and thus determine the height of the meteor. In this way, reliable observations have given the height at which there is sufficient density in the atmosphere to render meteors luminous as 188 miles.

Temperature of the Atmosphere

The condition of the atmosphere with respect to its temperature is determined by means of the thermometer. This instrument is in such common use that a detailed description is not necessary. It might be interesting to note that the in-

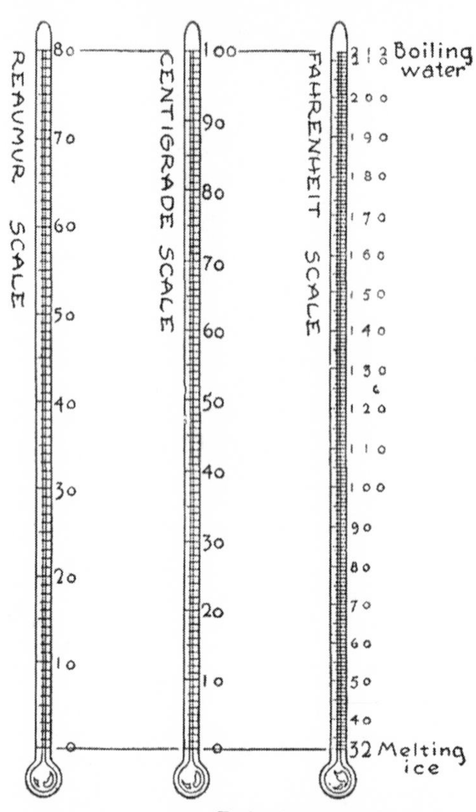

Taylor Instrument Companies

The three standard thermometer scales

strument invented by Galileo was very different from those now in use. Galileo's original thermometer was what is know as an air thermometer, and its operation when subjected to different degrees of heat or cold depended upon the expansion and contraction of air instead of mercury or alcohol. It had one serious defect, viz., the length of a column of air is affected by pressure as well as by temperature and it was therefore necessary, when using this thermometer, to obtain the pressure of the atmosphere by means of the barometer before the temperature could be determined. This is obviated in the mod thermometer by the use of mercury or cohol in a vacuum tube. Mercury is used when very low temperatures must registered, because it congeals at about degrees below zero Fahrenheit.

Thermometer Scales in Use

There are three systems in common for marking the degrees on the scale, Fahrenheit, Centigrade, and Reaumur

The Fahrenheit scale was the invent of a German by that name, but it is wor of note that this scale is used princip by English-speaking nations and is no common use in Germany. Fahren found that by mixing snow and salt was able to obtain a very low temperat and believing that the temperature t obtained was the lowest possible started his scale at that point, which called zero. He then fixed the freez temperature of water 32 degrees ab this zero, and the boiling point of w at 212 degrees. There are, therefore, divisions or degrees between the freez and boiling point of water on the Fah heit scale.

The Centigrade scale starts with zer the freezing point of water and makes boiling point 100. Thus 180 degrees on Fahrenheit scale equal 100 degrees the Centigrade. The Fahrenheit degre therefore, only a little more than hal large, to be exact five-ninths of a degree a degree on the Centigrade scale. Centigrade scale is in common use France and is used almost exclusively all scientific work throughout the wo

The Reaumur scale is not so comn but is used in some parts of Europe. this scale the zero is placed at the free point of water and the boiling point 80 degrees. The divisions are, theref larger than those of the Centigrade s and more than twice as large as the F renheit. The general use of these differ scales has led to endless confusion made the comparison of records diffic so that even at the present time w making a temperature record it is ne sary to indicate the scale in use.

)istribution of the Temperature and Pressure

[he heat received on the earth from sun is the controlling factor in all ither conditions. If the earth were com-ed of all land or all water, and the ount of heat received were everywhere same throughout the year, there would no winds, no storms, and probably no uds and no rain, because the force of vity, which acts on everything on the th's surface and on the air as well, ıld soon settle all differences and the ıosphere would become perfectly still. t the earth is composed of land and er, and the land heats up more rapidly ler sunshine than the water and also :s off — " radiates " — its heat more idly than water. As a result, the air over land is warmer in summer than the over the water. During the winter this eversed, and the air over the oceans is mer than the air over the land. The great ocean currents, by carrying the heat from the equatorial regions toward the poles, and by bringing the cold from the polar regions toward the equator, assist in maintaining a constant difference in temperature between the continents and the adjacent oceans.

Furthermore, the facts that the path of the earth about the sun is not a circle but an ellipse, and that the axis of the earth is not perpendicular to the plane of its orbit, result in an unequal distribution of heat over the surface. It is always warmer near the equator than at the poles, and warmer in summer than in winter. All these differences in temperature cause corresponding differences in density, which, in turn, cause differences in weight or pressure over various parts of the earth's surface. These changes are in no way the result of chance but are determined by the operation of fixed natural laws, and with this in mind we may now take up the study of the winds of the world.

THE WINDS OF THE WORLD

[he general circulation of the atmos-ɜre may be best studied by disregard-those smaller differences of tempera-e and pressure that result from local ɪses and by viewing the earth and its ıosphere as a whole, considering only ›se larger differences which are in con-nt operation. In the great oceans of : world we find the water constantly ›ving in a very systematic manner, and call this system of movements ocean rents. The Gulf Stream, the Equato-. Current, the Japan Current, some-ıes called Kuro Siwo Current, and ıers may be likened to great rivers of ter moving systematically on their ırses in the ocean.

There are greater rivers of air in the nosphere than any in the oceans, and :y move on their courses with equally tematic precision and in obedience to ɜd laws, which we may in a measure derstand.

The air river at the bottom of which we live is broad and deep, extending in width from Florida northward nearly to the North Pole. It flows from west to east circling the globe and its name is the Prevailing Westerlies. The other air river in this hemisphere extends southward from latitude about 35° nearly to the equator. Its name is the Northeast Trade Winds.

In the southern hemisphere are two similar air rivers, one extending southward from latitude about 30° nearly to the South Pole with its current, like its counterpart in the northern hemisphere, flowing from west to east, circling the globe. It is also called the Prevailing Westerlies. The other air river in the southern hemisphere extends from about latitude 30° northward nearly to the equator and flows from the southeast toward the northwest, hence the name Southeast Trade Winds. The dividing line, or bank, between the air rivers in each hemisphere belts the earth at about 35° north and 30° south of the equator. Why does the air move.

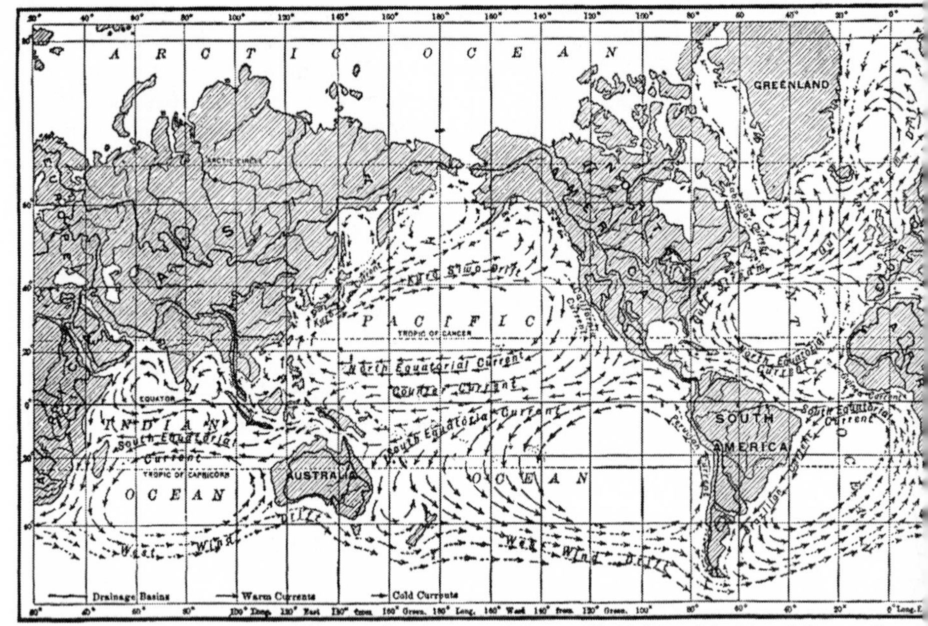

Ellsworth Huntington and Sumner W. Cushing, *Principles of Human Geogra*
John Wiley and Sons,

Ocean currents

and why does it move in such a regular, systematic manner? To answer these questions we will rely upon gravity, the heat from the sun, and the effect of the rotation of the earth on moving wind currents.

Everyone knows that water flows down hill because of the force of gravity. Grav-

J. Russell Smith, *Human Geography*,
The John C. Winston Co.

The earth's prevailing winds and air circulation

ity is nature's great peacemaker. It is ways trying to settle disturbances, e things up, smooth them over. If th were no winds to bring rain to the l or to stir up the ocean, gravity wo soon run all the water into the lakes the seas, and then smooth them out sheets of glass; and if there were noth to stir up the winds, gravity would s settle all differences in the atmosph and the air would become perfectly qu So gravity is kept busy trying to smo out the water which the wind stirs up the same time trying to quiet the wi which are stirred up by the heat of sun.

Tyndall says that heat is a mode motion; that when heat is imparted a substance, the molecules of which composed are set into very rapid vi tion. They are continually trying to away from each other and usually succ in getting more space, and thus incr the size or volume of the substance, in other words, expand it. Iron, brass, c per, water, and many other substances pand under heat. Air is a gas and expa y rapidly when heated. One cubic foot :old air becomes two cubic feet when ıted. Now gravity pulls things down 'ard the center of the earth in accord- :e with their weight-density, and a cubic t of cold air, being more dense and ıs heavier than an equal volume of :m air, is pulled down with greater ce. We therefore say that warm air ighter than cold air, and if lighter it l rise. What it actually does is to press ıally in all directions, and when a place ound where there is less resistance than :where it moves in that direction. So en heat causes air to expand and be- ne lighter than the surrounding cool it moves, and air in motion is *wind*. 'igure No. 1 represents a section of : atmosphere over a broad, level plain h the air at rest and pressing down ıally on every part of the surface. The :ted line H represents the top of the iet atmosphere. Such a condition oc- 's frequently at night after the heat

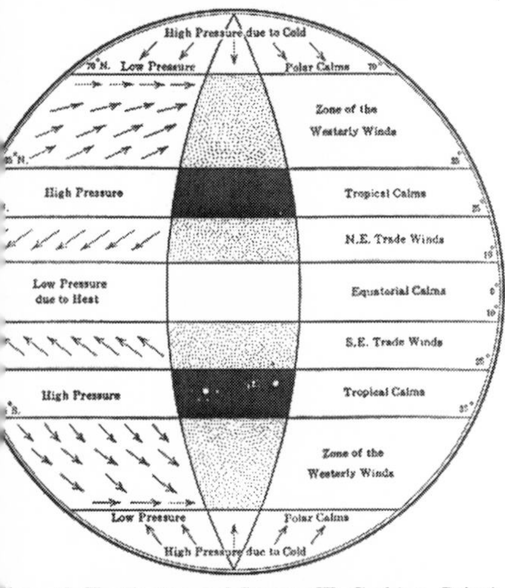

llsworth Huntington and Sumner W. Cushing, *Principles of Human Geography*, John Wiley and Sons, Inc.

Pressure belts on a simplified globe

m the sun is withdrawn and gravity ; settled the atmosphere. When the s of the sun fall on the earth upon ich this quiet air rests they warm the th first, and then the layer of air im- :diately in contact with the surface, so : atmosphere is heated from the bottom ward. We will assume that the layer of air between the earth and the dotted line G is thus heated to a higher temperature than the air above it. It will, therefore, expand. It cannot expand downward because of the earth. It cannot expand

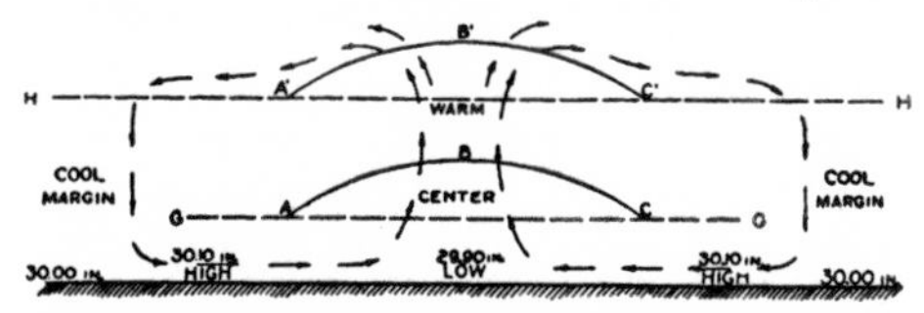

Fig. 1. *Diagram showing air currents set up by sun's heat*

much laterally because it is pressed upon by air that is also seeking more space. It therefore expands upward as represented by the line A B C. Now in expanding upward it lifts all the air above it, and the line H, representing the top of the atmosphere, will become bowed upward also as indicated by the line A′ B′ C′. As a result, the air at the top of the atmosphere over the warm center slides down the slopes on either side toward the cool margins. As soon as the flow of air away from the warm center begins, just that instant the pressure upon the heated layer at the surface is relieved and the warm air is pushed upward and the whole circulation, as indicated by the arrows, begins. It must be remembered that gravity is the really active force in maintaining this movement, because it pulls down the denser, heavier air at the cool margins with greater force than the warm, expanded, light air at the warm center. The descent of the cool air actually lifts the warm air.

The normal pressure, or weight, of the atmosphere at sea level is about 14.7 pounds on each square inch of surface. It is customary, however, to express the weight of the atmosphere in terms of inches of mercury instead of in pounds and ounces. A column of air one inch square from sea level to the top of the atmosphere will just counterbalance a column of mercury 30.00 inches high in a barometer tube of the same size. (See this type of barometer in the sketch shown on p. 787.) We, therefore, say that the normal pressure of the atmosphere at sea level is about 30.00 inches. If, for any reason, the atmosphere becomes heavier than normal, it will raise the column of mercury above the 30-inch mark, and we say that the pressure is "high." If the atmosphere becomes lighter than normal, we say that the pressure is "low." So high pressure means a heavy atmosphere and low pressure a light atmosphere.

At the beginning we assumed that the atmosphere over the broad, level plain

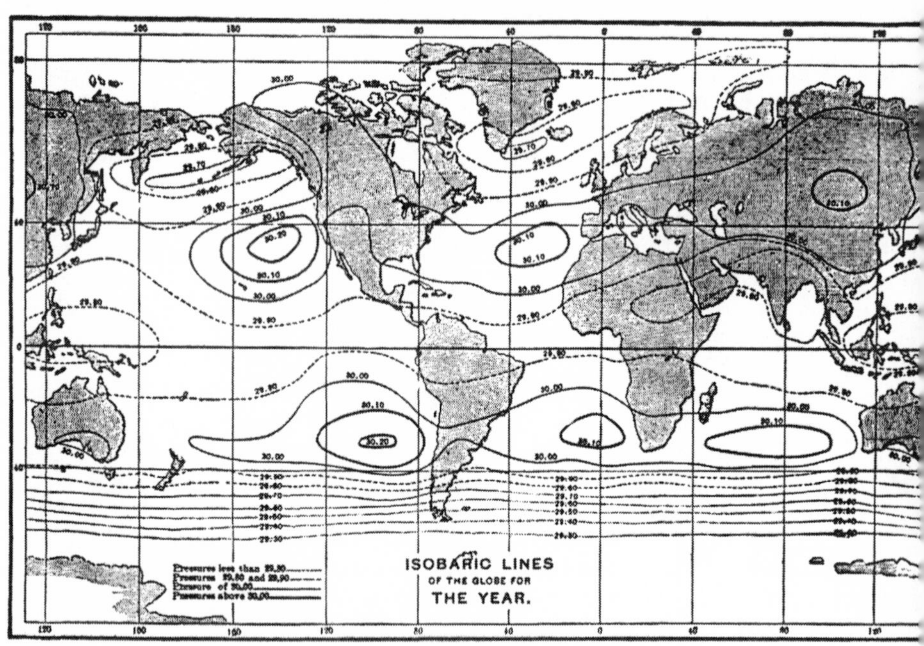

Fig. 2. Isobars of the world

was quiet and that it pressed down equally on every part of the surface. We will now assume that the pressure was normal, or 30.00 inches, and note the changes in pressure that result from the interchange of air between the warm center and the cool margins. So long as none of the air raised by the expanding layer at the surface moved away toward the cool margins, no change in pressure occurred; but the instant the air began to glide down the slopes away from the warm center, then the pressure at the surface decreased, because, some air having moved away, there was less to press down than before. The pressure at the warm center, therefore, became less than 30.00 inches, or in other words, low. Likewise, the air as it mo
away from the warm center, having
much of its heat during its ascent,
gradually pulled down by gravity beca
of its greater density, thus increasing
pressure over the cool margins. We th
fore have low pressure at the warm cen
29.90 inches; and we have high press
30.10 inches, at the cool margins. Fr
this illustration we obtain the six pri
ples of convectional circulation, viz.:

1. Low pressure at warm center.
2. High pressure at cool margins.
3. Ascending currents at warm cen
4. Descending currents at cool margi
5. Surface winds from high pressure low pressure.
6. Upper currents from low pressure high pressure.

Now we all know that the temperat
of air is much higher at the equator th
at the poles, and we may, therefore,
Fig. 1 represent a section of the atm
phere along any meridian from the No
to the South Pole. The equator wo
then become the warm center and
poles the cool margins. We would then

t to find a belt of low pressure around world near the equator because of the h temperature, and high pressure at poles because of the low temperature. e would, also, expect to find ascending rents at the equator; upper currents ving from the equator toward the poles; cending currents at the poles; and sure winds blowing from the poles toward equator. Let us now test our theory actual facts and see how far they are accord.

The chart, Fig. 2, represents the normal, average, pressure at sea level for the rld, and if our theory is in accord with facts, we should find a belt of low ssure all around the world near the

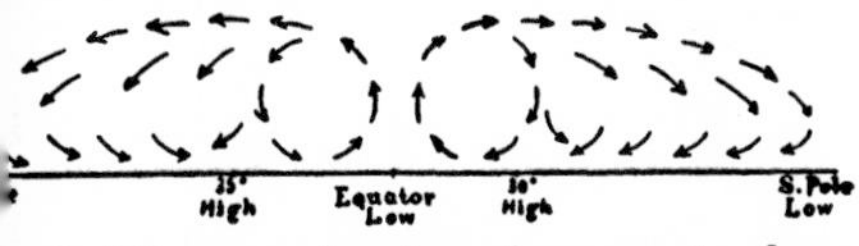

. 3. *Diagram showing air currents along any meridian*

uator, with areas of high pressure at poles. Let us examine the chart. Beinning at the equator, and bearing in nd that the normal pressure is about .00 inches, we find irregular lines, repenting pressures of 29.90 inches — ghtly below normal — around the world both sides of the equator. Between ese lines we find pressure as low as 29.80. is, therefore, evident that there is a lt of low pressure around the world near e equator, as anticipated. Let us look the high pressure at the poles. We ve comparatively few observations near e poles, but the line nearest the South le is marked 29.30 inches, a surprisingly v pressure, much lower even than the v belt at the equator, and just the rerse of what we expected to find. When look at the North Pole we find that e pressure is not so low as at the uth Pole, but still below normal and out as low as at the equator. Going rth and south from the equator we d that the pressure increases gradually up to about latitude 35° in the northn hemisphere and to about latitude 30° the southern, after which it decreases toward the poles. So there are two well-marked belts of high pressure circling the globe; the one about 35° north, and the other about 30° south of the equator. May it not be significant that these belts of high pressure coincide so nearly with the margins, or banks, of the air rivers mentioned on page 791?

Thus far our theory does not accord very well with the facts. True, we found the low pressure at the equator as anticipated; but we also found low pressure at the poles, where the reverse was expected; and the high pressure that we anticipated at the poles, we found not far north and south of the equator. We will, therefore, have to discard our theory, or reconstruct it to accord with the facts. Let us reconstruct Fig. 1, and mark the pressure on the line representing the earth's surface along any meridian to accord with the facts as they appear on Fig. 2.

The diagram shown above now represents the true pressure along any meridian, as determined by actual observations, and we cannot escape the conviction that the requirements as to temperature and pressure at the warm center are fulfilled by the high temperature and low pressure found at the equator. Furthermore, the temperature decreases north and south from the equator, and thus the belts of high pressure near the tropics may be taken to represent the conditions at the cool margins. The first and second principles of a convectional circulation, viz., low pressure at the warm center and a high pressure at the cool margins, are thus fulfilled. To satisfy the remaining conditions, we should find ascending currents near the equator, upper currents flowing from the equator toward the tropical belts of high pressure, descending currents at the tropics, and surface winds blowing from the tropics toward the equator. Let us now examine the surface winds of the world as illustrated by the diagram on page 792.

On either side of the equator and blowing toward it, we find the famous trade winds — the most constant and steady winds of the world. Their northern and

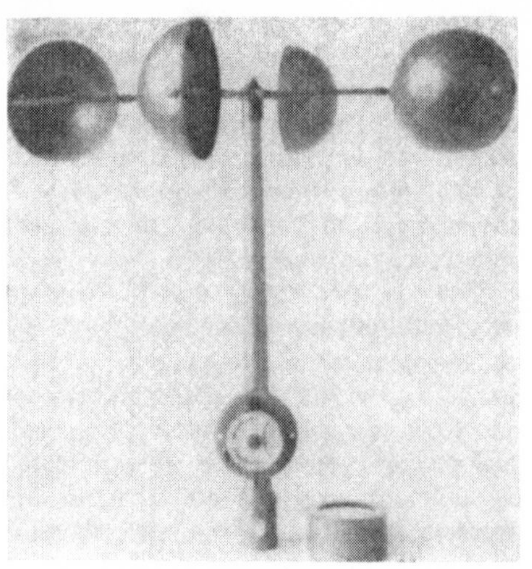

Weather Bureau, U. S. D. A.

Cup anemometer. The dial cover is removed to show the mechanism

southern margins coincide with the tropical belts of high pressure. They blow from high pressure to low pressure and we cannot doubt that they act in obedience to the fifth principle of convectional circulation. From observation of the lofty cirrus clouds in the trade wind belts, we have abundant evidence of upper currents, flowing away from the equator toward the tropical belts of high pressure; thus the sixth principle is satisfied. The torrential rains and violent thunderstorms, characteristic of the equatorial regions, bear evidence to the rapid cooling of the ascending currents near the equator; while the clear, cool weather and light winds of the Horse Latitudes clearly indicate the presence of descending currents at the tropics. Thus, the six principles of a convectional circulation are satisfied, and the evidence is conclusive that the trade winds form a part of a convectional circulation between the tropical belts of high pressure and the equatorial belt of low pressure.

You have doubtless observed that the trade winds do not blow directly toward the equator but are turned to the west so that they blow from the northeast in the Northern Hemisphere, and from the southeast in the Southern. This peculiarity is not in strict accord with our ideas of a simple convectional circulation and suggests at least the presence of some outs influence. If we turn to Ferre's treatise the winds, we find a demonstration of following principle: a free moving bo such as air, in moving over the surface a rotating globe, such as the earth, scribes a path on the surface that tu to the right of the direction of motion the Northern Hemisphere and to the in the Southern. The curvature of the p increases with the latitude, being zero the equator and greatest at the poles, a is independent of direction. With this mind, if we take position at the north limit of the trade winds in the North Hemisphere and face the equator (see 792) we find that the winds moving ward the equator turn to our right; l wise, if we face the equator from the sou ern limit of the southeast trades, we f them turning to our left. Observation upper clouds in the trade wind belts sh that the upper currents also turn to right in the Northern Hemisphere, and the left in the Southern. It is, theref clear that the systematic turning of trade winds from the meridian is due the rotation of the earth. The value c force at various latitudes and for vari velocities that would cause a body to t away from a straight line is purely a pr lem in mathematics, and for the benefi those versed in the science the formul given. The amount of such a force is pressed by 2 MVW sin D, where M is mass, V the velocity, W the angular r tion of the earth, and D the latitude.

Not all of us may be able to solve problem, but we may understand so thing of the effect of the rotation of earth on moving wind currents. It is a w known principle of physics that if a b be given a motion in any direction, it continue to move in a straight line reason of its inertia, without reference north, south, east, or west. A personal perience of this principle may be gai in a street car while it is rounding a cu

In the diagram shown on the next p we have a view of the Northern He sphere. The direction of the rotation is dicated by the curved arrows outside le representing the equator. Suppose a wind starts from the equator, mov- along the meridian A directly toward North Pole. It is clear that it cannot tinue to move along the meridian, be- se the direction of the meridian with rence to space is continually chang- and the inertia of the wind compels it nove in a straight line without refer- e to the points of the compass. So n the meridian A has been moved 3 by the rotation of the earth, the wind, ough it maintains its original direc- , no longer points toward the pole to the right of the pole. Likewise, a d starting from the pole toward the ator also turns to the right of the me- ans and becomes a northeast wind as pproaches the equator. A wind moving or west also turns to the right of the allels for the same reason. So a wind ting out from the equator with the t possible intention of hitting the pole, all the while continuing in the same ight line, will miss the pole by many es, and always on the right side in Northern and on the left side in the thern Hemisphere. Thus, the oblique vement of both the trade winds and prevailing westerlies is accounted for. t now remains to consider the cause the unexpected low pressure found at poles, and the reason for the belts of

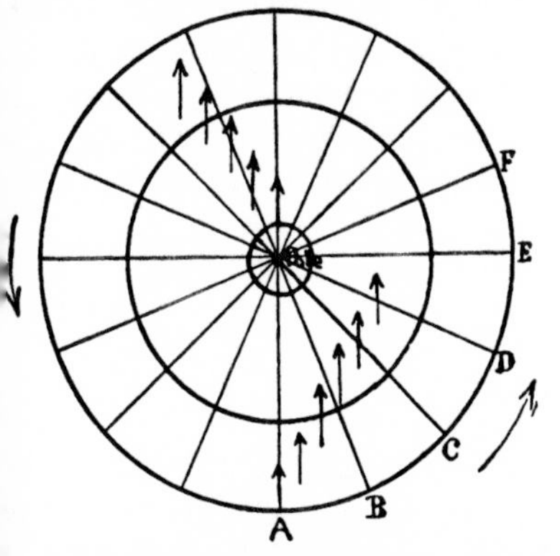

gram showing the effect of the earth's rotation on the atmosphere

h pressure at the tropics. If we refer Fig. 2, page 794, we see that not all air t ascends at the equator descends at the tropics, else there would be an absence of air at the higher latitudes, which is manifestly not the case. On the other

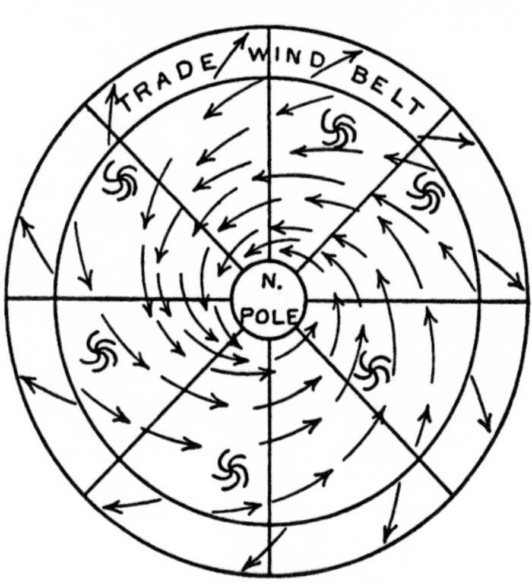

The circumpolar whirl

hand, it is equally impossible that all the air ascending at the equator should move to the poles, because the space it could occupy decreases rapidly from a maximum at the equator to zero at the poles. Only a part of the air that ascends at the equator is, therefore, involved in the trade wind circulation and a part passes over the tropics, and moves on toward the low pressure at the poles. Furthermore, some of the air that descends at the tropics moves along the surface toward the poles, obeying the law that impels air to move from high pressure to low pressure. Now every particle of air that passes over the tropics, every particle that moves northward along the surface, turns to the right in the Northern and to the left in the Southern Hemisphere. All, therefore, miss the poles — on the right side in the Northern and on the left side in the Southern Hemisphere. The result is that two great whirlpools develop in the atmosphere; one whirling about the North and the other whirling about the South Pole. The outer margins of these whirlpools coincide with the tropical belts of high pressure.

As an example of a whirlpool we may take a basin having a vent at the center of the bottom. If the basin is filled with water, the plug withdrawn, and the water given a slight rotary motion, its velocity will increase as it approaches the center and the rapid whirling will develop sufficient centrifugal force to open an empty core. Those who have visited the great whirlpool at Niagara have undoubtedly noticed that the whirling waters are held away from the center and piled up around the margins by the centrifugal force developed. Let us suppose that air starting from the equator moves without friction or other resistances toward the pole. Its velocity must increase as its radius shortens, because the law of the conservation of areas requires that the radius must always sweep over equal areas in a given unit of time. (See law of conservation of areas.) At the equator, the air has an easterly motion equal to the eastward motion of the earth, which is 1,000 miles per hour. At latitude 60° the radius will have decreased one-half and the velocity, therefore, doubled; but at latitude 60° the eastward motion of the earth is only 500 miles per hour, so the air would be moving 1,500 miles per hour faster than the earth. At a distance of 40 miles from the pole the wind would attain an easterly velocity of 100,000 miles per hour, and moving on so short a radius would develop sufficient centrifugal force to hold all the air away from the pole and thus form a vacuum. That the supposed case of no friction is far from the truth is evidenced by the fact that the pressure at the North Pole is but little less than at the equator; but the centrifugal force developed by the gyration winds, in thus withdrawing the air from the poles and piling it up at tropics, may be fairly taken as suffic cause for the low pressure found at poles and the belts of high pressure at tropics.

The questions that remain to be c sidered are: (1) the low pressure at South Pole as compared with the press at the North Pole, and (2) the unec distance of the tropical belts of high p sure from the equator. These questi may be considered together.

It is to be remembered that the Sou ern Hemisphere is the water hemisph and that the prevailing westerlies, in g ing over the smooth water surface, are little retarded by friction and, theref attain a higher velocity than the co sponding winds of the Northern He sphere, where the rougher surface terially retards their movement. As consequence, the circumpolar whirl of Southern Hemisphere is stronger, and velops a greater centrifugal force, t holding a larger quantity of air away fr the South Pole and reducing the press to a greater degree than is brought ab by the weaker winds of the North Hemisphere.

Since the circumpolar whirl of Southern Hemisphere is the stronger the two, it withdraws the air to a grea distance from the pole than does weaker counterpart of the North Hemisphere, and piles it up in the tro cal belt of high pressure about five degr nearer the equator than do the wea forces that operate in the Northern He sphere.

STORMS

Having gained a comprehensive view of the general planetary wind system, we may now undertake the study of local disturbances that arise within the general circulation and are known as "storms."

Storms are simply eddies in the atmosphere. They may be compared to the eddies that are often seen floating along with the current of a river or creek. In th eddies the water is seen to move rapi around a central vertex, developing su cient centrifugal force to hold some the water away from the center, t forming a well-marked depression, quently of considerable depth. The wh circulation of the eddy is quite indepe of the current of the stream which ies it along its course, and while its eral direction and velocity of move- t coincide with that of the current, e are times when it will be seen to e quickly from side to side and again n it will remain nearly stationary for me or take on a rapid movement.

he eddies or storms in the atmos- re act in much the same way. They carried along by the general currents he river of air in which they exist. ir general direction coincides with the ction of the current in which they are ting, and their rate of movement con- ns in a general way to its velocity; but, the eddies in the river, they do not ays move in straight lines or at a form rate of speed.

here is one important respect in which eddies in the air differ from eddies in er. The water may revolve in either direction, depending upon the direction in which the initial force was applied, but the storm eddies in the atmosphere al-

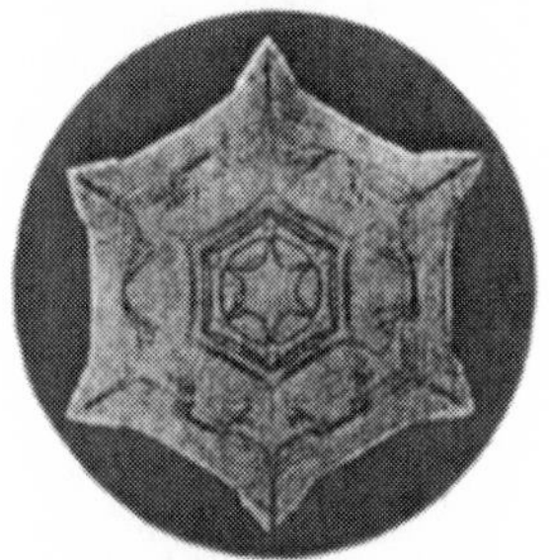

Photomicrograph by W. A. Bentley

Snow crystal

ways revolve counterclockwise in the Northern Hemisphere, and clockwise in the Southern.

This is due to the deflecting force of the earth's rotation, which is fully explained on pages 796–98.

WEATHER MAPS

 weather map is a sort of flashlight tograph of a section of the bottom of or more of these great rivers of air. rings into view the whole meteorologi-

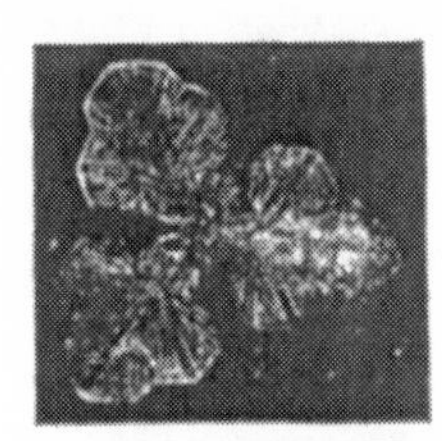

Photomicrograph by W. A. Bentley

Snow crystal

situation over a large territory at a en instant of time; and, while a single p conveys no indication of the move- nts continually taking place in the at- sphere, a series of maps, like a moving ture, shows not only the whirling ed- s, the hurrying clouds, and the fast- ving winds, but the ceaseless on-flow the great river of air in which they at. Our present knowledge of the move- ments of the atmosphere has been gained chiefly from a study of weather maps; they form the basis of the modern system of weather forecasting, and their careful study is essential to any adequate under- standing of the problems presented by the atmosphere. (See pp. 801–6.)

The Principles of Weather Forecasting

The forecasting of the weather has been made possible by the electric telegraph. It is based upon a perfectly simple, rational

Photomicrograph by W. A. Bentley

Snow crystal

process constantly employed in everyday affairs. We go to a railway station and ask the operator about a certain train. He tells us that it will arrive in an hour. We accept his statement without question, because we are confident that he knows the speed at which the train is approaching; a few clicks of his telegraph instrument have told him just where it is, and the time it will arrive, barring accidents, is

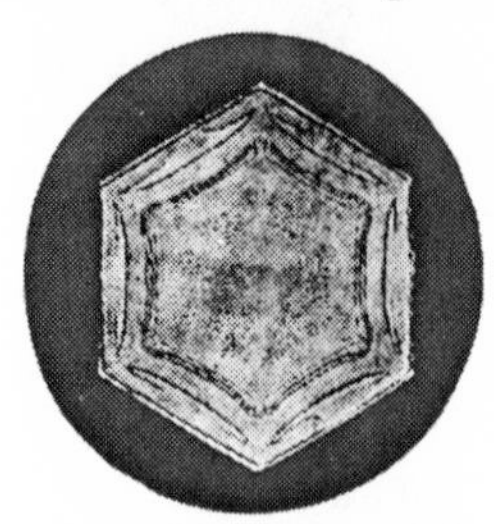

Photomicrograph by W. A. Bentley

Snow crystal

a simple calculation. Information of coming weather changes is obtained in a similar manner. Although storms do not run on steel rails like a train, nevertheless their movements may be foreseen with a reasonable degree of accuracy, depending chiefly upon the size of the territory from which telegraphic reports are received and the experience and skill of the forecaster. As a rule, the larger the territory brought under observation, especially in its longitudinal extent (the general currents carry storms of the middle latitudes eastward around the world and those of the tropics westward), the earlier advancing changes may be recognized and the more accurately their movements foreseen.

Forecasts Based on Weather Maps

The forecasts issued by the United States Weather Bureau are based on weather maps, prepared from observations taken at 7:30 A.M. and 7:30 P.M. Eastern Standard Time, throughout the country, at about 200 observatories. In addition to the reports received by telegraph by the Central Office at Washington, the several forecast centers, and other designated stations from observatories or stations in the United States, a system of interchange with Canada, Mexico, the West Inc and other island outposts in the Atla and Pacific give to the forecaster daily photographs of the weather co tions over a territory embracing nearly whole of the inhabited part of the W ern Hemisphere north of the equator. sort of disturbance within this vast re is photographed at once upon the wea map. If it be a West Indies hurrican other destructive storm, its characte recognized instantly, its rate and direc determined, and information of the p able time of its arrival sent to those pl that lie in its path. The method is fectly simple. Anyone with a weather and a little experience can forecast weather with some degree of accuracy at least, gain an intelligent understan of the conditions upon which the f casts that accompany the map are base

Maps, Where Published and How tained

Weather maps are published in s daily papers, and in somewhat larger fo and more in detail, at Weather Bu stations in some of the largest cities. T may usually be obtained for school use applying to the Chief of the Weather reau at Washington, D. C.

The forecasts that accompany the m are simply an expression on the par the official forecaster as to the wea changes he expects to occur in vari parts of the country within the time sp fied, usually within 36 to 48 hours. opinion is based upon the conditi shown by the map. He has no se source of information. You may accept conclusions, or, if in your opinion t are not justified, you have all the infor tion necessary to make a forecast for y self. Weather maps are published so tensively with a view to thus stimula an intelligent interest in the problen weather forecasting, and also that one see at a glance what the temperature, r fall, wind, and weather are in any par the country in which he may be interes The friends of the weather service those who best understand its work.

The Value of the Weather Service

No one knows so well as the forecaster that the changes that appear most certain to come sometimes fail, or come too late; but taking all in all, about 85 out of 100 forecasts are correct. Of those that fail, probably not more than three or four per cent fail because severe changes come unannounced. Most forecasters predict too much, and their forecasts fail because the expected changes come after the time specified or not at all. It is fortunate that this is so; for it is better to be prepared for the change though it be late in coming than to have it come without warning.

The value of the weather service to the agriculture and commerce of the United States cannot be questioned seriously. That the appropriations for its support have been increased year by year from $1,500 in 1871 to nearly $4,400,000 in 1929 is evidence of its value and efficiency. A conservative estimate places the value of property saved by the warnings issued by the Weather Bureau at many millions of dollars annually.

HOW TO READ WEATHER MAPS

Weather maps may be obtained by writ-
ıg to the Chief of the Weather Bureau, Vashington, D. C., stating that you wish › post the maps in a public place. A sup-
ly of maps for three successive days for se in these lessons may be obtained at 20 ents per hundred. Sometimes they are ›nt free, if it is stated that they are to be sed for school purposes.

The words *isobar* and *isotherm* have een bogies which have frightened many teacher from undertaking to teach about eather maps, and yet how simple are the eanings of these two words. Isobar is ade up of two Greek words, *isos* mean-
g equal and *baros* meaning weight. herefore, as isobar means equal weight, ıd on a map one of these continuous ıes means that, wherever it passes, the mosphere there has equal weight, and e barometer stands at equal height. ıe isobar of 30 means that the mercury the barometer stands 30 inches in ight in all the regions where that line sses.

Isotherm comes from two Greek words, s meaning equal and *therme* meaning at. Therefore, on the map the dotted es show the region where the tempera-
e is the same. If at the end of the ›tted line you find 60 it means that, erever that line passes, the thermome-
stands at 60 degrees.

Highs and Lows

Many of the "highs" and "lows" enter the United States from the Pacific Ocean about the latitude of Washington State or southwest British Columbia; however, by far the greater number enter from the

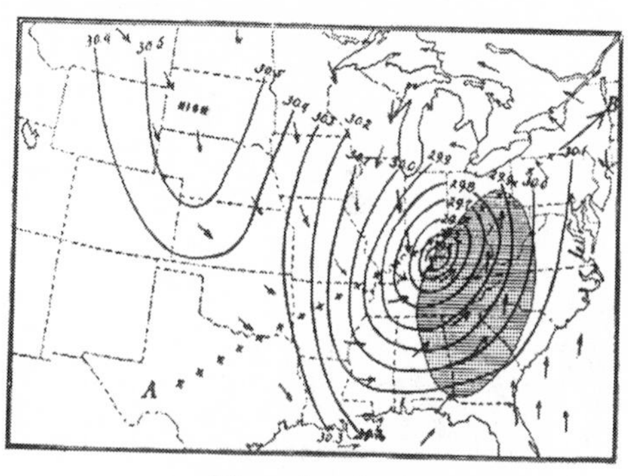

Map of a storm

Canadian Northwest. They follow one another alternately, crossing the continent in the general direction of west to east in a path which curves somewhat to the north, and they leave the United States in the latitude of Maine or New Brunswick. If they enter by way of Lower California, they pass over to the Atlantic Ocean farther south. The time for the passage of a high or low across the continent averages about three and one-half days, sometimes a little more. These areas are usually more

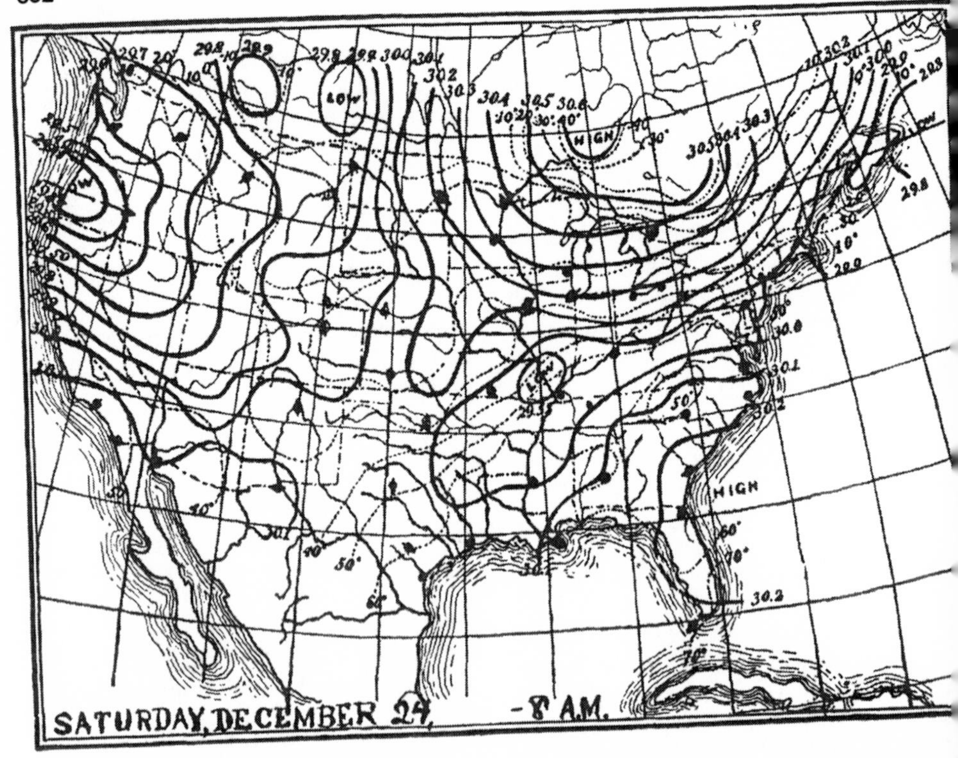

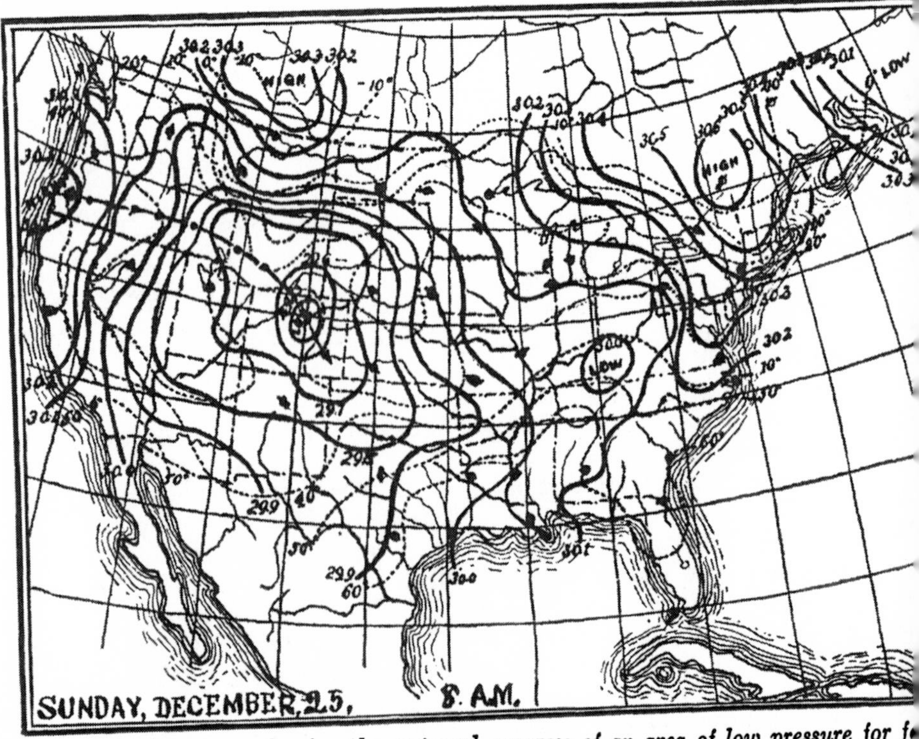

U. S. weather maps, showing the eastward progress of an area of low pressure for f
indicated by the line of d

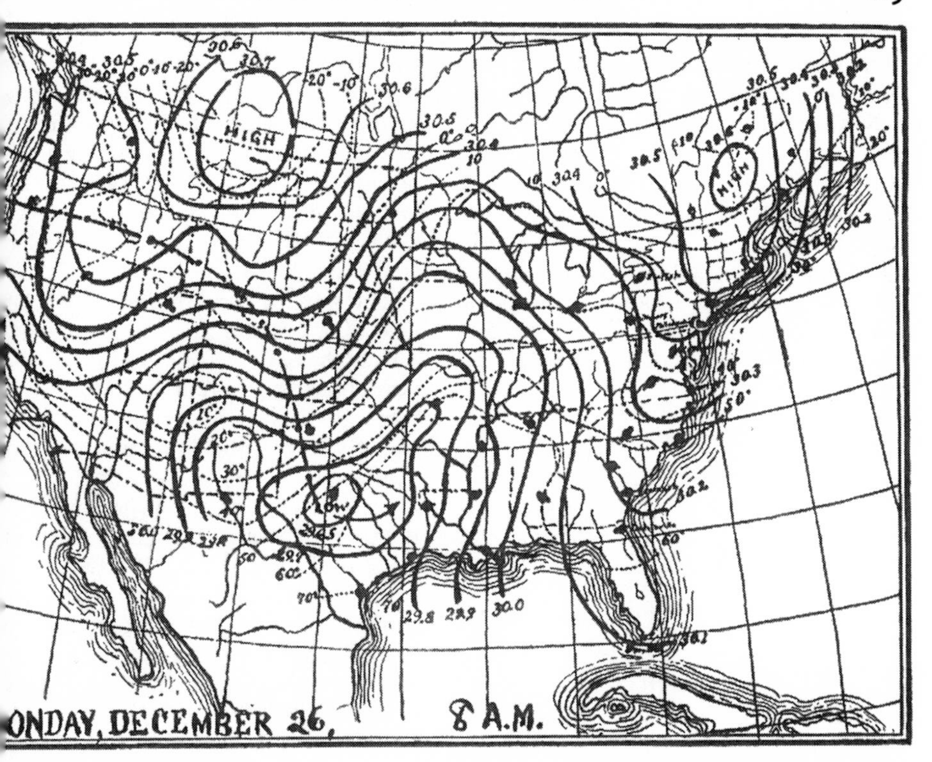

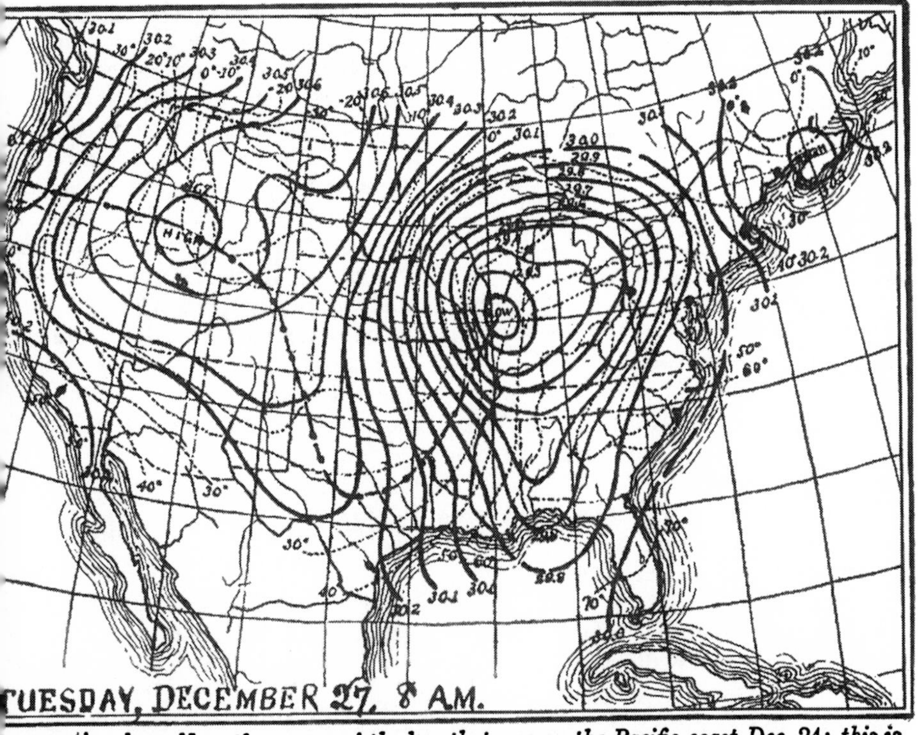

consecutive days. Note the course of the low that was on the Pacific coast Dec. 24; this is
nd dashes on the later maps

marked in winter, and wind storms are more marked and more regular.

A low area is called a cyclone and a high area an anticyclone. The destructive winds, popularly called cyclones, which occur in certain regions, should be called tornadoes instead, although in fact they are simply small and violent cyclones. But a cyclone, when used in a meteorological sense, extends over thousands of square miles and is not violent; while a tornado may be only a few rods in diameter and may be very destructive. The little whirl-

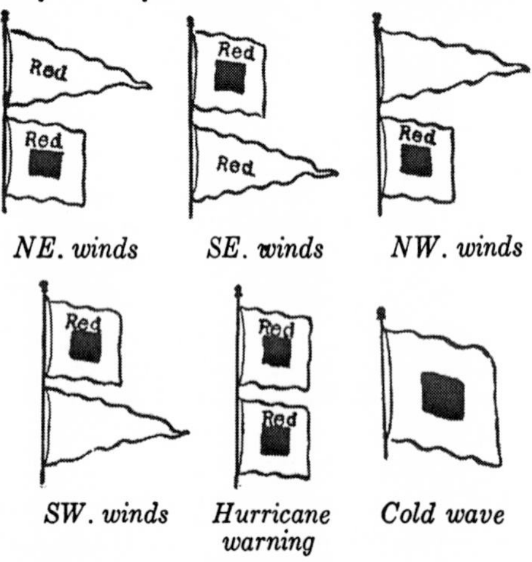

Explanation of storm and hurricane signals

Storm warning — A red flag with a black center indicates that a storm of marked violence is expected.

The pennants displayed with the flags indicate the direction of the wind; red, easterly (from northeast to south); white, westerly (from southwest to north). The pennant above the flag indicates that the wind is expected to blow from the northerly quadrants; below from the southerly quadrants.

By night a red light indicates easterly winds and a white light below a red light westerly winds.

Hurricane warning — Two red flags with black centers displayed one above the other indicates the expected approach of a tropical hurricane or one of those extremely severe and dangerous storms which occasionally move across the Lakes and northern Atlantic coast.

No night hurricane warnings are displayed.

winds which lift the dust in the roads are rotary winds also, but merely the eddies of a gentle wind.

In a cyclone or "low," and also in a tornado, the air blows from *all* sides spirally inward *toward* the center where there is a column of *ascending* air.

In an anticyclone or "high" the air blows outward in every direction in curved lines *from* a column of *descending* air.

In the map (page 801), the curved lines are isobars; the line of crosses, A to B, indicates the course of the storm; the rows indicate the direction of the wind note that it is moving counterclockw around the area of low pressure; shaded area indicates the region where i raining or snowing — note that this is area where the warm, moist air from Gulf and the ocean meets the colder of the north.

The weather conditions during the p sage of a cyclone are briefly as follo Small, changing wisps of cirrus clouds pear about twenty-four hours before r these gradually become larger and co the whole sky, making a nimbus clo The wind changes from northeast to e or southeast to south. The barometer fa the thermometer rises; that is, air press is less to the square inch, and the temp ture of the atmosphere is warmer. An curate record of the temperature ra can be had from the maximum and m mum thermometers (page 789). R begins and falls for a time, varying from hour to a day or more. After the rain th appear breaks in the great nimbus clo and finally the blue sky conquers u there are only a few or no clouds. T wind changes to southwest and west; barometer rises, the temperature fa The rain ceases, the sun shines brightly. The low has passed and the h is approaching, to last about three days

Formerly, the Weather Bureau use series of flags, displayed in public pla to indicate approaching weather co tions; but that practice in general has b discontinued. Some local authorities s maintain the system at their own expe The storm and hurricane warnings the cold wave signal are still in use.

LESSON 222

How to Read Weather Maps

Leading Thought — Weather maps made with great care by the Weather reau experts. Each map is the result many telegraphic communications fr all parts of the country. Every intellig person should be able to understand weather maps.

METHOD — Get several weather maps ·m a nearby Weather Bureau station. ey should be maps for successive days, d there should be enough so that each pil can have three maps, showing the ather conditions for three successive ys.

OBSERVATIONS — 1. Take the map of e earliest date of the three. Where was ur map used? What is its date? How ny kinds of lines are there on your map? e there explanatory notes on the lower t-hand corner of your map? Explain at the continuous lines mean. Find an bar of 30; to what does this figure refer? nd all the towns on your map where the rometer stands at 30 inches. Is there re than one isobar on your map where e barometer stands at 30?

2. Where is the greatest air pressure on ur map? How high does the barometer nd there? How are the isobars arranged th reference to this region? What word printed in the center of this series of bars?

3. What do the arrows indicate? What the circles attached to the arrows indi- te?

4. In general, what is the direction of e winds with reference to this high cen- ?

5. Is the air rising or sinking at the cen- of this area? If the wind is blowing in directions from a center marked high, at sort of weather must the places just st of the high be having? Do the arrows th their circles indicate this?

6. Find a center marked low. How high es the barometer stand there? Does the pressure increase or diminish away m the center marked low, as indicated the isobars? Do the winds blow toward is center or away from it?

7. What must the weather in the region st east of the low be? Why? Do the ar- ws and circles indicate this?

8. Is there a shaded area on your map? so, what does this show?

9. Compare the map of the next date th the one you have just studied. Are e highs and lows in just the same posi- n that they were the day before? Where

Weather Bureau, U. S. D. A.

A rain gauge dismantled to show parts. By means of this instrument the amount of rain- fall is measured by inches

are the centers high and low now? In what directions have they moved?

10. Look at the third map and compare the three maps. Where do the high and low centers seem to have originated? How long does it take a high or low to cross the United States? How far north and south does a high or low, with all its isobars, ex- tend?

11. What do the dotted lines on your map mean? Do they follow exactly the isobars?

12. What is the greatest isotherm on your map? Through or near what towns does it pass?

13. Do the regions of high air pressure have the highest temperature or the low- est? Do high temperatures accompany low pressures? Why?

14. What is the condition of the sky just east of a low center? What is its con- dition just west of low?

15. If the isobars are near together in a low, it means that the wind is moving rather fast and that there will be a well- marked storm. Look at the column giving wind velocity. Was the wind blowing to- ward the center of the low on the map? If so, does that mean it is coming fast or slow? How does this fact correspond with the indications shown by the distance be- tween the isobars?

16. Describe the weather accompany- ing the approach and passage of a low in

the region where your town is situated. What sort of clouds would you have, what winds, what change of the barometer and thermometer?

Note: The amount of rainfall that has been recorded in representative areas will be indicated in a table printed below the map.

How to Find the General Direction and Average Rate of Motion of Highs and Lows

Observations — 1. On the first map of the series of three given, put an X in red pencil or crayon at the center of the high and a blue one at the center of the low; or if you do not have the colored pencils, use some other distinguishing marks for the two. If there are two highs and two lows, use a different mark for each one.

2. Mark the position of each center on this map for the following day with the same mark that you first used for that area. Do this for each of the highs and lows until it leaves the map or until your maps have been used. All the marks of one kind can be joined by a line, using a red line for the red marks and a blue line for the blue marks.

3. What do you find to be the general direction of the movement of the highs and lows?

4. Examine the scale marked statute miles at the bottom of the map. How many miles are represented by one inch on the scale?

5. With your ruler find out how many miles one area of high or low has moved in twenty-four hours; in three days. Divide the distance which the area has moved in three days by three and this will give the average velocity for one day.

6. In the same way find the average velocity of each of the areas on your map for three days and write down all your answers. From all your results find the average weekly velocity; that is, how many miles per hour and the general direction which has characterized the movement of the high and low areas.

How to Keep a Daily Weather M

The pupils should keep a daily weat map record for at least six months. T observations should be made twice ea day and always at the same hours. Wh it would be better if these records co be made at 7:30 o'clock in the morn and again at 7:30 o'clock in the eveni this is hardly practicable and they shou therefore, be made at 9 o'clock and at The accompanying chart may be dra enlarged. Sheets of manila paper are of used, so that one chart may cover the servations for a month.

Few schools are able to have a work barometer, but observations of tempe ture and sky should be made in ev school. Almost any boy can make weather vane, which should be placed a high building or tree where the wind not be deflected from its true direct when striking it. A thermometer sho be placed on the north side of a post a on a level with the eyes; it should not hung from a building, as the temperat of the building might affect it.

The direction of the wind and cloudiness of the day may be indicated the chart, as it is on the weather maps, a circle attached to an arrow which poi in the direction in which the wind blowing. See weather maps for expla tion of symbols.

Observations Concerning the Weather

It is an interesting hobby to really serve the weather. Of course, we all t about the weather if rain or snow is f ing; some people even remark about wind. Let us make it a daily habit to g a thought to weather conditions: the w directions; the presence or absence of d during the hours of evening, night, or ea morning; and the readings of the rometer, thermometer, and the weat maps if any are available.

There are many "weather signs" common circulation; some have a lutely no foundation and others have entific basis. The latter can usually be pended upon, and, in many instances, te interesting to study in an effort to l a reason why they are good signs. To ke a collection of all the weather signs t one can learn from friends or find various books is another interesting bby; the next thing, for an inquiring nd, is to attempt to find out how many merely sayings and how many are lly good signs. Some of the books listed the bibliography will be found quite ful in this field of inquiry.

ANY WEATHER PROVERBS ARE BASED ON SCIENTIFIC FACTS

There follows a short list of weather overbs or sayings that are based on scitific facts:

Evening red and morning gray,
Set the traveler on his way;
Evening gray and morning red,
Bring down rain upon his head.

Rainbow in the morning, sailors take warning,
Rainbow at night, sailor's delight.

Mackerel scales and mare's tails
Make lofty ships to carry low sails.

A mackerel sky,
Not twenty-four hours dry.

When walls are unusually damp, rain is to be expected.

Clouds flying against the wind indicate rain.

CHART FOR SCHOOL WEATHER-RECORDS

ate	Hour	Temp.	Barometer	Direction of Wind	Cloudiness. Fogs	Dew or Frost	Rain or Snow	Remarks
eekly Summary								

WATER FORMS

Gayle Pic

Water in its various changing forms, liquid, gas, and solid, is an example of another overworked miracle — so common that we fail to see the miraculous in it. We cultivate the imagination of our children by tales of the prince who became invisible when he put on his cap of darkness, and who made far journeys through the air on his magic carpet. And yet no cap of darkness ever wrought more astonishing disappearances than occur when this most common of our earth's elements disappears from under our very eyes, dissolving into thin air. We cloak the miracle by saying " water evaporates," but think once of the travels of one of these drops of water in its invisible cap! It may be a drop caught and clogged in a towel hung on the line after washing, but as soon as it dons its magic cap, it flies off in the atmosphere invisible to our eyes; and the next time any of its parts are evident our senses, they may occur as a port of the white masses of cloud sailing ac the blue sky, the cloud which Shelley p sonifies:

I am the daughter of Earth and Wa
And the nursling of the Sky;
I pass through the pores of the ocean a
shores;
I change, but I cannot die.

We have, however, learned the myst ous key word which brings back the va spirit to our sight and touch. This w is " cold." For if our drop of water, in cap of darkness, meets in its travels object which is cold, straightway the falls off and it becomes visible. If it b stratum of cold air that meets the invisi wanderer, it becomes visible as a clo s mist, or as rain. If the cold object be ice pitcher, then it appears as drops its surface, captured from the air and ined as "flowing tears" upon its cold ace. And again, if it be the cooling ace of the earth at night that captures wanderer, it appears as dew.

ut the story of the water magic is only told. The cold brings back the in-ble water vapor, forming it into visible ps; but if it is cold enough to freeze, n we behold another miracle, for the ps are changed to crystals. The cool dowpane at evening may be dimmed h mist caught from the air of the room; e examine the mist with a lens we find omposed of tiny drops of water. But

Weather Bureau, U. S. D. A.

Hailstones as large as hens' eggs" is no re of speech as applied to these stones that at Girard, Ill., in 1929. Note the three eggs ront of the right-hand pile of hailstones

he night be very cold, we find next rning upon the windowpane exquisite is, or stars, or trees, all formed of the stals grown from the mist which was re the night before. Moreover, the ps of mist have been drawn together crystal magic, leaving portions of the ss dry and clear.

f we examine the grass during a cool ning of October we find it pearled with v, wrung from the atmosphere by the meating coolness of the surface of the und. If the following night be freez-cold, the next morning we find the ss blades covered with the beautiful stals of hoar frost.

f a rain cloud encounters a stratum of cold enough to freeze, then what would ve been rain or mist comes down to us sleet, hail, or snowflakes; and of all the ms of water crystals, that of snow in perfection is the most beautiful; it is,

Weather Bureau, U. S. D. A.

After an ice storm. After this storm of Nov. 26, 1929, *at Worcester, Mass., the ice on the wires was three inches in diameter and weighed* 800 *pounds per wire from pole to pole*

indeed, the most beautiful of all crystals that we know. Why should water freezing freely in the air so demonstrate geometry by forming, as it does, a star with six rays, each set to another, at an angle of 60 degrees? And as if to prove geometry divine beyond cavil, sometimes the rays are only three in number — a factor of six — and include angles of twice 60 degrees. More-

Photomicrograph by W. A. Bentley

Snow crystal formed in high clouds

over, the rays are decorated, making thou sands of intricate and beautiful forms; but if one ray of the six is ornamented with additional crystals the other five are decorated likewise. Those snow crystals formed in the higher clouds and, therefore, in cooler regions may be more solid in form, the spaces in the angles being built out to the tips of the rays, and including air spaces set in symmetrical patterns; and some of the crystals may be columnar in

Photomicrograph by W. A. Bentley

Composite snow crystal; the center formed in a high cloud and the margins in a lower cloud

form, the column being six-sided. Those snow crystals formed in the lower currents of air, and therefore in warmer regions, on the other hand, show their six rays marvelously ornamented. The reason why the snow crystals are so much more beautiful and perfect than the crystals of hoar frost or ice, is that they are formed from water vapor, and grow freely in the regions of the upper air. Mr. W. A. Bentley, who spent many years photographing the snow crystals, found nearly 5,000 distinct designs.

The high clouds are composed of ice crystals formed from the cloud mists; such ice clouds form a halo when veiling the sun or the moon.

When the water changes to vapor and is absorbed into the atmosphere, we call the process evaporation. The water left in an open saucer will evaporate more rapidly than that in a covered saucer, because it comes in contact with more air. The clothes which are hung on the line wet, dry more rapidly if the air is dry and not damp; for if the air is damp, it already has almost as much water in it as it can hold. The clothes will dry more rapidly when the air is hot, because hot air takes up moisture more readily and holds more of it than does cold air. The clothes will dry more rapidly on a windy day, because more air moves over them and comes in contact with them than a still day.

If we observe a boiling teakettle, can see a clear space of perhaps an i or less in front of the spout. This spac filled with steam, which is hot air sa rated with hot water vapor. But what call "steam" from a kettle is this sa water vapor condensed back into t drops of water or mist by coming i contact with the cooler air of the roo When the atmosphere is dry, water v boil away much more rapidly than wh the air is damp.

The breath of a horse, or our o breath, is invisible during a warm day; b during a cold day, it is condensed to m as soon as it is expelled from the nost and comes in contact with the cold a A person who wears spectacles finds th unclouded during warm days; but in w ter the glasses become cold out-of-doo and as soon as they are brought into co tact with the warmer, damp atmosph of a room, they are covered with a mi In a like manner, the windowpane in w ter, cooled by the outside temperatu condenses on its inner surface the m from the damp air of the room.

The water vapor in the atmosphere invisible, and it moves with the air c rents until it is wrung out by coming in contact with the cold. The air thus fil

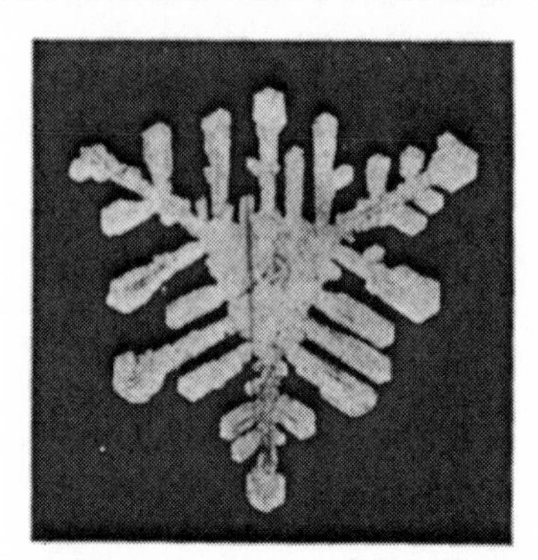

Photomicrograph by W. A. Bentley

Blizzard type of snow crystal formed in l cloud

with water vapor may be entirely cle near the surface of the earth; but, as rises, it comes in contact with cooler a d discharges its vapor in the form of st, which we call clouds; and if there enough vapor in the air when it meets old current, it is discharged as rain and ls back to the earth. Thus, when it is y cloudy, we think it will rain, because uds consist of mist or fog; and if they subjected to a colder temperature, the st is condensed to rain. Thus, often mountainous regions, the fog may be n streaming and boiling over a moun-n peak, and yet always disappears at a tain distance below it. This is because temperature around the peak is cold d condenses the water vapor as fast as wind brings it along, but the mist sses over and soon meets a warm cur-t below and, presto, it disappears! It then taken back into the atmosphere. e level base of a cumulus cloud has a atum of warmer air below it, and marks level of condensation.

At the end of the day, the surface of the ound cools more quickly than the air ove it. If it becomes sufficiently cold d the air is damp, then the water from s condensed, and dew is formed during night. However, all dew is not always ndensed from the atmosphere, since ne of it is moisture given off by the nts, which could not evaporate in the

Weather Bureau, U. S. D. A.

Cumulus clouds

ld night air. On windy nights, the stra-m of air cooled by the surface of the rth is moved along and more air takes

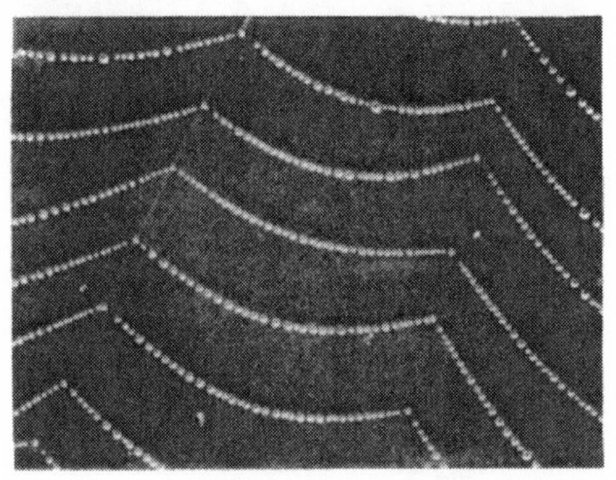

W. A. Bentley

Dew on a spider's web

its place, and it therefore does not become cold enough to be obliged to yield up its water vapor as dew. If the weather during a dewy night becomes very cold, the dew becomes crystallized into hoar frost. The crystals of hoar frost are often very beautiful and are well worth our study.

The ice on the surface of a still pond usually begins to form around the edges first, and fine, lancelike needles of ice are sent out across the surface. It is a very interesting experience to watch the ice crystals form on a shallow pond of water. This may easily be seen during cold winter weather. It is equally interesting to watch the formation of the ice crystals in a glass bottle or jar. Water, in crystallizing, expands, and requires more room than it does as a fluid; therefore, as the water changes to ice it must have more room, and often presses so hard against the sides of the bottle as to break it. The ice in the surface soil of the wheat fields expands and buckles, holding fast in its grip the leaves of the young wheat and tearing them loose from their roots; this "heaving" is one cause for the winter-killing of wheat. Sleet consists of rain crystallized in the form of sharp needles. Hail consists of ice and snow compacted together, making the hard, more or less globular hailstones.

W. A. Bentley

In transpiring, plants give off moisture. " The magic of the cold " has held this moisture in the form of drops on a strawberry leaf

LESSON 223

Water Forms

Leading Thought — Water occurs as an invisible vapor in the air and also as mist and rain; and when subjected to freezing, it crystallizes into ice and frost and snow.

Method — The answers to the questions of this lesson should, as far as possible, be given in the form of a demonstration. All of the experiments suggested should be tried, and the pupils should think the matter out for themselves. In the study of the snow crystals a compound microscope is a great help, but a hand lens will do. This part of the work must be done out of doors. The most advantageous time for studying the perfect snow crystals is when the snow is falling in small, hard flakes; since, when the snow is soft, there are many crystals massed together into great fleecy flakes, and they have lost their original form. The lessons on frost or dew may be given best in the autumn or spring.

Observations — 1. Place a saucer filled with water near a stove or radiator; do not cover it or disturb it. Place anoth saucer filled with water near this but cov it with a tight box. From which sauc does the water evaporate more rapidl Why?

2. We hang the clothes, after they a washed, out-of-doors to dry; what becom of the water that was in them? Will th dry more rapidly during a clear or duri a damp day? Why? Will they dry mo rapidly during a still or during a win day? Why? Will they dry more rapic during hot or cold weather? Why?

3. Watch a teakettle of water as it boiling. Notice that near its spout th is no mist, but what we call steam formed beyond this. Why is this so? Wh is steam? Why does water boil away? kettles boil dry sooner on some days th on others? Why?

4. If the water disappears in the atm phere where does it go? Why do we s " the weather is damp "? What force it that wrings the water out of the atm phere?

5. Why does the breath of a horse sh as a mist on a cold day? Why do perso who wear spectacles find their glasses co ered with mist as soon as they enter warm room after having been out in t cold? Why do the windowpanes becom covered with mist during cold weath Is it the mist on the outside or on t

W. A. Ben

Frost crystals on a windowpane

ide? Why does steam show as a white st? Why does the ice pitcher, on a rm day, become covered on the outside h drops of water? Would this happen a cold day? Why not?

. Why, when the water is invisible in atmosphere, does it become visible as uds? What causes the lower edges of nulus clouds to be so level? What is ? Why do clouds occur on mountain ks? What causes rain?

. What causes dew to form? When the ss is covered with dew, are the leaves the higher trees likewise covered? Why t? What kind of weather must we have order to have dewy nights? What must

Photomicrograph by W. A. Bentley

Forms of hoar frost

the atmosphere of the air in relation that of the ground in order to condense dew? Does dew form on windy nights? hy not? Does all dew come from the , or does some of it come from the und through the plants? Why is not s water, pumped up by the plants, evap- ited?

8. What happens to the dew if the ather becomes freezing during the ght? What is hoar frost? Why should ter change form when it is frozen? How ny forms of frost crystals can you find the grass on a frosty morning?

9. When a pond begins freezing over, at part of it freezes first? Describe how e first layer of ice is formed over the rface.

10. Place a bottle of water out of doors

Marjorie Ruth Ross

Hoar frost on a tree

in freezing weather. How does the ice appear in it at first? What happens later? Why does the bottle break? How is it that water which has filled the crevices of rocks scales off pieces of the rock in cold weather? Why does winter wheat "win- ter-kill" on wet soil?

11. Why does frost form on a window- pane? How many different figures can you trace on a frosted pane? Are there any long, needle-like forms? Are there star forms? Can you find forms that resemble ferns and trees? Do you sometimes see, on boards or on the pavement, frost in forms like those on the windowpane?

12. When there is a fine, dry snow falling, take a piece of dark flannel and catch some flakes upon it. Examine them with a lens, being careful not to breathe upon them. How many forms of snow crystals can you find? How many rays are there in the star-shaped snow crystals? Do

Photomicrograph by W. A. Bentley

High cloud snow crystal

you find any solid crystals? Can you find any crystals that are triangular? When the snow is falling in large, feathery flakes, can you find the crystals? Why not?

13. What is the difference between a hailstone and a snow crystal? What is sleet?

When in the night we wake and hear the
rain
Which on the white bloom of the orchard
falls,
And on the young, green wheat-blades,
where thought recalls
How in the furrow stands the rusting plow,
Then fancy pictures what the day will
see —
The ducklings paddling in the puddled
lane,
Sheep grazing slowly up the emerald slope,
Clear bird-notes ringing, and the droning
bee
Among the lilac's bloom — enchanting
hope —
How fair the fading dreams we entertain,
When in the night we wake and hear the
rain! — Robert Burns Wilson

The thin snow now driving from
north and lodging on my coat consist
those beautiful star crystals, not cott
and chubby spokes, but thin and pa
transparent crystals. They are abou
tenth of an inch in diameter, perfect li
wheels with six spokes without a tire,
rather with six perfect little leaflets, f
like, with a distinct straight and slen
midrib, raying from the center. On e
side of each midrib there is a transpar
thin blade with a crenate edge. How
of creative genius is the air in which th
are generated! I should hardly adm
more if real stars fell and lodged on
coat. Nature is full of genius, full of di
ity. Nothing is cheap and coarse, neit
dewdrops nor snowflakes.

A divinity must have stirred wit
them before the crystals did thus sh
and set. Wheels of storm-chariots. T
same law that shapes the earth-star sha
the snow-stars. As surely as the petals
a flower are fixed, each of these count
snow-stars comes whirling to earth, p
nouncing thus, with emphasis, the n
ber six. — Thoreau's Journal

THE SKIES

REVISED BY S. L. BOOTHROYD
Professor of Astronomy in Cornell University

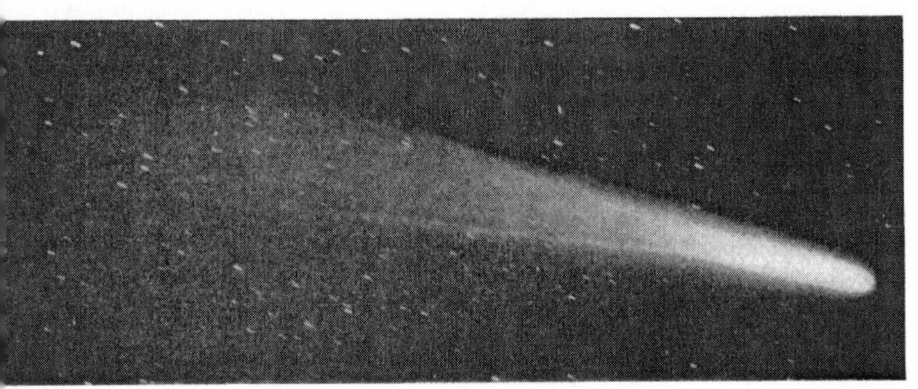

Lick Observatory

Halley's Comet, May 7, 1910

THE STORY OF THE STARS

Why did not somebody teach me the constellations and make me at home in the rry heavens, which are always overhead, and which I don't half know to this day.
— THOMAS CARLYLE

'or many reasons aside from the mere wledge acquired, children should be ght to know something of the stars. s an investment for future years; the s are a constant reminder to us of the usands of worlds outside our own, and king at them intelligently lifts us out urselves in wonder and admiration for infinity of the universe, and serves to ke our own cares and trials seem trivial. e author has not a wide knowledge of stars; a dozen constellations were ght to her as a little child by her ther, who loved the sky as well as the th; but perhaps nothing she has ever rned has been to her such a constant rce of satisfaction and pleasure as this lity to call a few stars by the names y have borne since the men of ancient es first mapped the heavens. It has en her a sense of friendliness with the ht sky that can only be understood by se who have had a similar experience.

There are three ways by which the mysteries of the skies are made plain to us: first, by our own eyes; second, by the telescope; and third, by the spectroscope and other physical instruments. These instruments help us to interpret the messages brought by the light coming from the heavenly bodies. The spectroscope is an instrument which tells us, by analyzing the light of stars, not only the chemical elements which compose them, but something of the state in which the gases exist in the stars, planets, and nebulæ. It also makes possible the measurement of the rate at which a heavenly body is approaching or receding from us. Further still it gives information which assists in determining the temperature of stars as well as in measuring their sizes.

Thus, we have learned many things about the stars; we know that every shining star is a great blazing sun, and there is no reason to doubt that many of these suns have worlds like the earth spinning around them, although, of course, so far away as to be invisible to us; for our world could not be seen at all from even the nearest star.

The telescope early revealed to Galileo that the Milky Way or Galaxy is not a nebulous band of hazy light around the sky, as it appears to the unaided eye, but is composed of myriads of faint stars, too faint to be seen individually without telescopic aid.

We also know that many of the stars which seem single to us are really double — made up of two vast suns swinging around a common center; and although they may be millions of miles apart, they are so far away that they seem to us as one star. The telescope reveals many of these double stars and shows that they circle around their orbits in various periods of time, the most rapid making the circuit in five years, another in sixteen years, another in forty-six years; while there is at least one lazy pair which seems to require fully sixteen hundred years to complete one journey around their elongated, oval orbit. And the spectroscope has revealed to us that many of the stars which seem single through the largest telescope are really double, and some of these great suns race around each other in the period of a few days, at a rate of speed we could hardly imagine.

Astronomers have been able to measure the distance from us to many of the stars, but when this distance is expressed in miles it is too much for us to grasp. Consequently, they have come to express distance to heavenly bodies in terms of the time it takes light to reach us from them. Light travels 186,300 miles a second or about six trillion miles a year; this distance is called a light-year. Thus a star whose distance is such that it takes eight years for its light to reach us, is said to be removed from us eight light-years. Light reaches us from the sun in eight and one-third minutes; but it takes more than four years for a ray to reach us from the nearest star. It adds new interest to the Polestar to know that the light which reaches our eyes left that star almost half a cent ago, and that the light we get from Pleiades started on its journey bef America was discovered. Most of the s are so far away that we cannot meas the distance.

Although stars seem stationary, they all moving through space just as our c sun is doing; but the stars are so far a that even if one moved a million mil day, it would require years of observat to detect that it moved at all, except that component of its motion whic directly toward or away from us. We kn the stars are in motion — just as pla are in motion in our solar system. T problem of determining these motions longs in the realm of advanced astrono cal study, and is too difficult to consi here.

The spectroscope reveals the life cy of stars; when young they are compose thin gases shining red and are giant st when older and more condensed t shine yellow; when still more conden they shine white and may shine blue. T condition marks the end of their infa then they decline through these colors reverse order. When on the decline, t become yellowish, white stars; after they change very little in size and sh with a constant light through milli upon millions of years; this stage co tutes the solar stage of their life and cupies the major part of their life hist

Scattered through the skies are obj which look like clouds; these are the called nebulæ. Some of these have alre been found to consist of stars arrange a globular system of suns. Others h been found to be whole Milky-Way tems — the so-called spiral nebulæ; ot are found to be large volumes of glow gas and yet others are immense areas interstellar dust illumined by giant st

Only two nebulæ can be seen with unaided eye. The telescope, by the aid photography, reveals planetary neb and both bright and dark irregular neb within our own Milky-Way system of s and in the Magellanic clouds (those n est external galaxies beyond the Mi ay system known to man). Photographs by the largest telescopes reveal mil-ns of spiral nebulæ which are distant ilky-Way systems of stars, each contain-; all the types of stars, planetary nebulæ, d irregular nebulæ which are found in r own Milky-Way system.

The planetary nebulæ are so named be-use they have a fairly definite boundary d may appear like a luminous disc or minous ring; they always have at their nter a hot star, whose radiation is rich-: in the ultraviolet rays.

The number of stars that can be seen the whole sky with the unaided eye is tween six and seven thousand; since we : only half of the sky at once, and since e region near the horizon is obscured a more or less dense haze, we can sel-m see more than about two thousand ırs at a given time. With the help of the escope, about eight hundred thousand ırs have been seen individually, classi-d, and catalogued, while photography of e skies reveals thousands of millions.

The Milky Way or Galaxy, that great ite band across the heavens, is made up stars which are so far away that we can-t see them individually, but see only eir diffused light. It is well called a river of stars" flowing in a circle around r whole sky; and during the early hours night, except in the months of spring, e-half of it is seen directly above us ile the other half is hidden below us. one observes the skies at a late hour of e night in early spring, a portion of e Galaxy can be seen then; since it is t visible in early evening hours, most servers do not see it during the spring onths. The place of the Milky Way in e heavens seems fixed and eternal; any r within its borders is always seen at e same point. When the Northern Cross ts itself toward the zenith we are able see that near that constellation the star river divides into three streams with long, blue islands between.

HOW TO BEGIN STAR STUDY: THE CIRCUMPOLAR CONSTELLATIONS

THE POLESTAR AND THE DIPPERS

The way to begin star study is to learn to know the Big Dipper, and through its pointers to distinguish the Polestar; for whenever we try to find any star we have to find the Big Dipper and Polestar first, so as to have some fixed point to start from. There are four stars in the bowl of the Big Dipper and three in the curved handle. A line drawn through the outer two stars of the bowl, if extended, will touch the North Star, or Polestar. It is very important for us to know the Polestar, because the northern end of the earth's axis is directed toward it, and it is therefore situated in the heavens almost directly above our North Pole. For those of us who live in the Northern Her sphere, the North Star never sets, but always in our sky. Of course, the Nor Star has nothing to do with the axis our earth any more than the figure on t blackboard has to do with the pointer; simply happens to lie in the direction ward which the northern end of the eart axis points. In the southern skies, the is no bright star which lies directly abo the South Pole, so there is no South Po star.

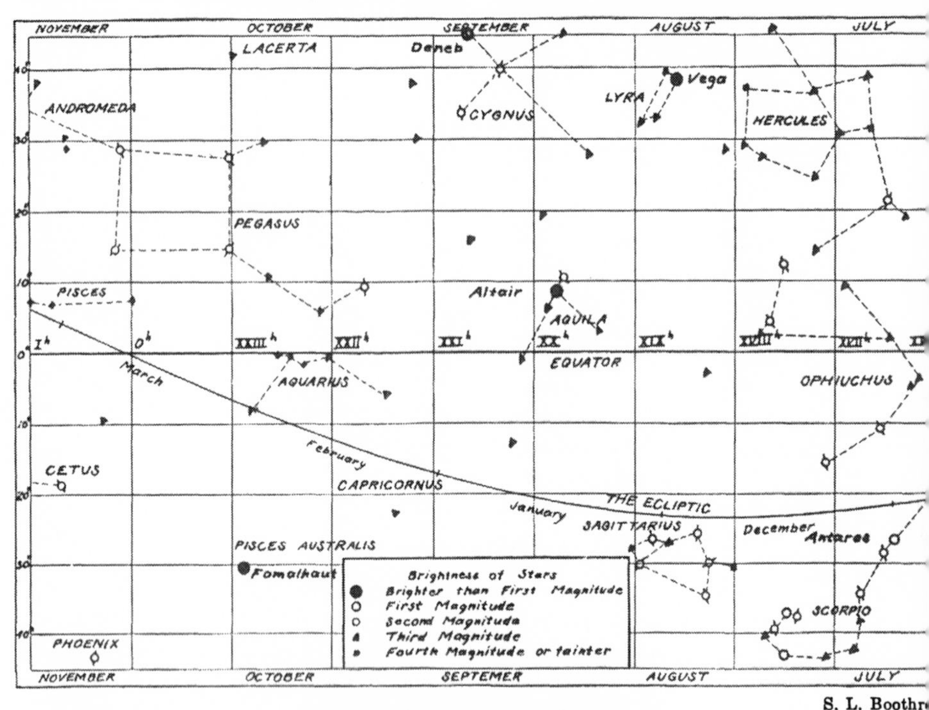

S. L. Boothr

Stars of late summer and autumn

Key map to the sky as the observer in the Northern Hemisphere faces south. An observer in the South Hemisphere would need to face north and hold the map upside down.

On Sept. 1, 10 P.M., Sept. 15, 9 P.M., Oct. 1, 8 P.M., Oct. 15, 7 P.M., Nov. 1, 6 P.M., the regions shown in center of the map are due south of an observer in the Northern Hemisphere and due north of an observer the Southern Hemisphere. Use the map that represents the **date nearest the time the observations are being made**

The Polestar cannot be seen from t Southern Hemisphere; but if we shou start from Florida, on a journey towa Baffin's Bay, we should discover that ea night this star would seem higher in t sky. And if we should succeed in reachi

North Pole, we would find the Pole-
r directly over our heads, and what
onderful sight the stars would be from
s point! For none of the stars we could
would rise or set, but would move
und us in circles parallel to the horizon.
The Big Dipper shows us the Polestar,
l seems to revolve around it counter-
ckwise every twenty-three hours and
y-six minutes; but of course this ap-
rance is caused by the fact that we
selves are revolving from west to east.

Thus, the Big Dipper and the other polar constellations are the night clock of the sailors of the Northern Hemisphere; for though this great polar clock has its hands moving around the wrong way, it gains time with such regularity that anyone who understands it is able to compute exact time by it.

The Little Dipper lies much nearer the Polestar than does the Big Dipper; in fact, the Polestar itself is the end of the handle of the Little Dipper. Besides the Polestar,

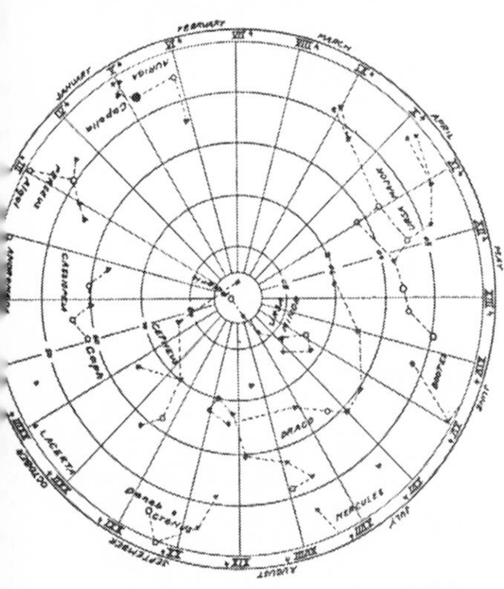

S. L. Boothroyd

North circumpolar chart. For observations
out 9 P.M. Hold the map in such a way that
name of the month in which the observa-
n is made is at the top of the map

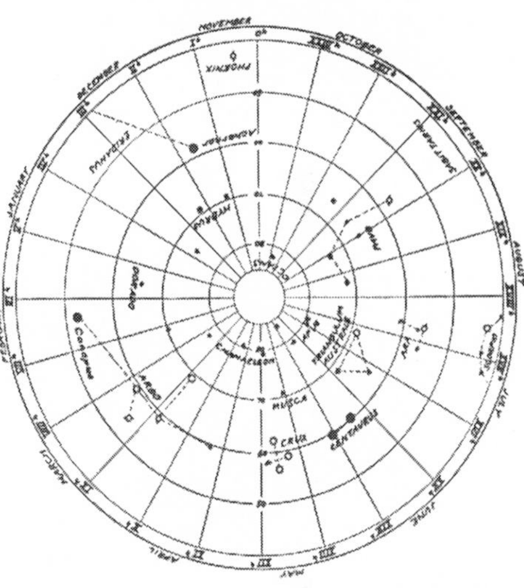

S. L. Boothroyd

South circumpolar chart. For observations about 9 P.M. *Hold the map in such a way that the month in which the observation is made occurs at the top of the map*

erefore, the stars seem to revolve
unterclockwise about the Polestar as
face it. The fact that the sky makes
e revolution in nearly four minutes less
an twenty-four hours is due to the anal motion of the earth around the sun.
ne Big Dipper is seen east of the Pole-
r with handle down early in the eveng in January, and it is seen west of the
lestar with handle up at the same time
evening in July; but the time of year
at a certain star reaches a given point
th reference to our horizon is so invaria-
e that if we know star time, or sidereal
me as it is called, we can tell just what
our of night it is when a star is at this
int.

there are two more stars in the handle of the Little Dipper, and of the four stars which make the bowl, the two that form the outer edge are much the brighter. The bowl of the Little Dipper is above or below the Polestar according to the hour of the evening and the night of the year, for it apparently revolves about the Polestar as does the Big Dipper. The two Dippers open toward each other, and as someone has said, "They pour into each other."

The Big Dipper is a part of a constellation called *Ursa Major*, the Great Bear; and the Little Dipper is the Little Bear, the handle of the dipper being the bear's tail.

There is an ancient myth telling the

story of the Big and Little Bears: A beautiful mother called Callisto had a little son whom she named Arcas. Callisto was so beautiful that she awakened the anger of Juno, who changed her to a bear; and when her son grew up he became a hunter, and one day would have killed his transformed mother; but Jupiter seeing the danger of this crime caught the two up into the heavens, and set them there as shining stars. But Juno was still vindictive,

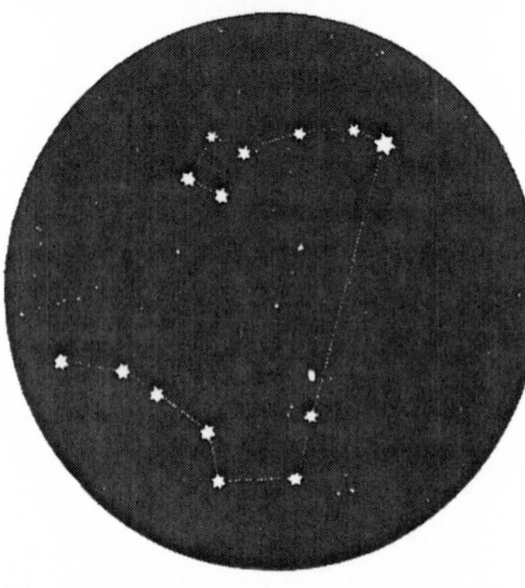

The Polestar and the Big and Little Dippers

so she wrought a spell which never allowed these stars to rise and set like other stars, but kept them always moving around and around.

LESSON 224

The Two Dippers

Leading Thought — The North Star or Polestar can always be found by the stars known as the pointers in the Big Dipper; the stars of the Big Dipper seem to revolve around the Polestar once in twenty-three hours and fifty-six minutes.

Method — The time to begin these servations is when the moon is in its quarter, so that the moonlight will pale the stars in early evening. Draw u the blackboard, from the chart sho opposite, the Big Dipper and the P star, with a line extending through pointers. Say to the pupils that this Dipper is above or below or at one s of the Polestar, and that you wish th to observe for themselves where it is tell you about it the next day. After t surely know the Big Dipper, ask the lowing questions.

Observations — 1. Can you find Big Dipper among the stars?

2. Is it in the north, south, east, or w

3. Which stars are the "pointers" the Dipper, and why are they so called

4. Make a drawing showing how can always find the Polestar, if you see the Big Dipper.

5. How many stars make the bowl the Dipper?

6. How many stars in the handle?

7. Is the handle straight or is it curv

8. Does the Big Dipper open tow the Polestar, or away from it?

9. On the night of your observat was it above or below the Polestar at ei o'clock in the evening, or at the right the left of it?

10. Does the Big Dipper remain in same direction from the Polestar night? Look at it at seven o'clock a again at nine o'clock and see whethe has changed position.

11. Do you think it moves around Polestar approximately every twenty-f hours? In which direction? How co you tell the time of night by the Big D per and the Polestar?

12. Does the Big Dipper ever rise a set?

13. The Big Dipper is part of the Gr Bear. Can you find the stars which m the bear's head and front legs?

After the pupils surely know the Dipper and Polestar, draw the compl diagram upon the board to show the Li Dipper and where it may be found, a call attention to the fact that the end

: Little Dipper's handle is the Polestar
elf and that its bowl is not flaring, like
at of the Big Dipper, and that the two
ur into each other. Let the pupils find
: Little Dipper in the sky for themselves
d ask the following questions.

14. Is the Little Dipper nearer or far-
er from the Polestar than the Big Dip-
r?

15. How many stars in the handle of
: Little Dipper?

16. How many stars make the bowl of
: Little Dipper? Which of these stars
: the brightest? Is the bowl of the Little
pper above or below the Polestar?

17. Does the Little Dipper extend in
: same direction in relation to the Pole-
r all night?

18. Make observations on the relation
each other of the two dippers at eight
lock in the evenings of January, Febru-
, March, and April.

After the above lessons are well learned,
e the following questions about Polaris
ne North Star) and try to have the
pils think out the answers.

19. How many names has the Polestar?
n the Polestar be seen from the South-
1 Hemisphere? If not, why not?

20. If you should start from southern
orida and travel straight north, how
uld the Polestar seem to change posi-
n each succeeding night?

21. If you could stand at the North
le, where would the Polestar seem
be?

22. If you were at the North Pole,
uld any of the stars rise and set? In
at direction would the stars seem to
ve and why?

23. How does the North Star help the
lors to navigate the seas and why?

24. How do astronomers reckon dis-
ces between us and the stars? What
a light-year?

TOPICS FOR ENGLISH LESSON — (a)
hat a star is. (b) What a constellation
(c) How the stars and constellations
eived their names in ancient times. In
cient times the constellations which
ntain the Big and Little Dippers were
med the Big and Little Bears, and those are their Latin names to this day. Write a story about what the ancient Greeks told about these Bears and how they came to be in the sky.

CASSIOPEIA'S CHAIR, CEPHEUS, AND THE DRAGON

There are other constellations besides the two Dippers which never rise and set in the latitude of central New York, because they are so near to the Polestar that, when revolving around it, they do not fall below the horizon. There is one very brilliant star, called Capella, which almost belongs to the polar constellations in this latitude but not quite, for it is far enough

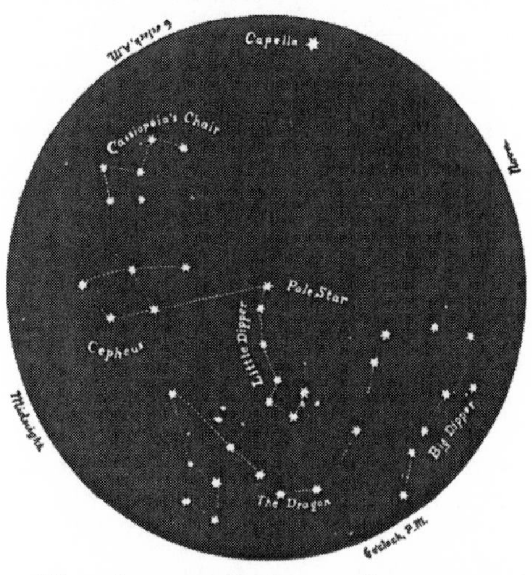

The polar constellations as they appear at about 8 P.M. *on January* 20, *the Dragon being below the Polestar. By revolving this chart as indicated, the positions of the stars are shown for* 6 P.M., midnight, 6 A.M., *and* noon *of January* 20

away from Polaris to dip below the horizon for a few hours during each circuit around the Pole.

Queen Cassiopeia's Chair is on the opposite side of the Polestar from the Big Dipper and at about equal distance from it. It consists of five brilliant stars that form a W with the top toward Polaris, one-half of the W being wider than the other. There is a less brilliant sixth star which finishes out half of the W into a chair seat, making of the figure a very uneasy looking throne for a poor queen to sit upon.

King Cepheus is Queen Cassiopeia's husband, and he sits with one foot on the Polestar quite near to his royal spouse. His constellation is marked by five stars, four of which form a lozenge, and a line connecting the two stars on the side of the lozenge farthest from Cassiopeia, if extended, will reach the Polestar as surely as a line from the Big Dipper pointers. Cepheus is not such a shining light in the heavens as is his wife, for his stars are not so brilliant. Perhaps this is because he was only incidentally put in the skies. He was merely the consort of Queen Cassiopeia, who being a vain and jealous lady boasted that she and her daughter Andromeda were far more beautiful than any goddesses that ever were, and thus incurred the wrath of Juno and Jupiter who set the whole family " sky high " and quite out of the way, a punishment which has its compensations since they are where the world of men may look at and admire them for all ages.

Lying between the Big and Little Dippers and extending beyond the latter is a straggling line of stars, which, if connected by a line, make a very satisfactory dragon. Nine stars form his body and three his head, the two brighter ones being the eyes.

LESSON 225

Cassiopeia's Chair, Cepheus, and the Dragon

Leading Thought — To learn to know and to map the constellations which are so near the Polestar that they never rise or set in our latitude, but seem to swing around the North Star once in twenty-three hours and fifty-six minutes.

Method — Place on the blackboard the diagram given showing the Polestar, Big and Little Dippers, and Cassiope Chair, and ask for observations a sketches showing their position in skies the following evening. After the pils have observed the Chair and know add to your diagram first Cepheus a then the Dragon. After you are sure pupils know these constellations, give following lesson. The observations sho be made early and late in the same even and at different times of the month, that pupils will in every case note the parent movement of these stars arou the Polestar.

Observations — 1. How many s form Cassiopeia's Chair? Make a dr ing showing them and their relation to Polestar.

2. Is the Queen's Chair on the sa side of the Polestar as the Big Dipp Is the top or the bottom of the " V which forms Cassiopeia's Chair turned ward the Polestar?

3. Does Cassiopeia's Chair m around the Polestar, like the Big Dipp

4. How many stars mark the cons lation of Cepheus?

5. Make a sketch of these stars a show the two which are pointers tow the North Star.

6. Does Cepheus also move around Polestar, and in which direction?

7. Describe where the Dragon lies, a where his tail and his head are in relat to the two Dippers. Make a sketch of Dragon.

8. Why do all the polar constellati seem to move around the Polestar ev twenty-three hours and fifty-six minu and why do they seem to go in a direct opposite the movement of the hands c clock? What do we mean by " polar c stellations "?

Topics for English Themes — T Story of Queen Cassiopeia, King Cephe and their Daughter, Andromeda; T Story of the Dragon.

THE WINTER STARS

The natural time for beginning star ıdy is in the autumn when the days are ortening and the early evenings give us portunity for observation. After the lar constellations are learned, we are en ready for further study in the still rlier evenings of winter, when the clear mosphere makes the stars seem more ive, more sparkling, and more beautiful an at any other period of the year. One the first lessons should be to instruct the pils how to draw an imaginary straight ıe from one star to another, and to perive the angles which such lines make ıen they meet at a given star. A rule, what is just as effective, a postal card or me other piece of stiff paper which ows right-angled corners, is very useful this work. It should be held between e eyes and the stars which we wish to nnect, and thus make us certain of a aight line and a right angle.

ORION (*o-ry'on*)

During the evenings of January, Februy, and March the splendid constellation Orion takes possession of the southern lf of the heavens; and so striking is it at we find other stars by referring to it stead of to the Polestar. Orion is a conellation which almost everyone knows; ree stars in a row outline his belt, and a rving line of stars, set obliquely below e belt, outlines the sword. Above the lt as the constellation is seen in the eveng sky of middle northern latitudes we n see the splendid red star Betelgeuse *et'el-jooz*), and below the belt, at about equal distance, is the white star Rigel *y'jel*). West of the red star above, and st of the white star below, are two fainter ırs, and if these four stars are connected lines, an irregular four-sided figure relts, which includes the belt and the ord. In this constellation the ancients w Orion, the great hunter, with his belt d his sword; Betelgeuse was set like a glowing ruby on his shoulder, and the white star Rigel was set like a spur on his heel. Thus stood the great hunter in the sky, with his club raised to keep off the plunging bull whose eye is the red Aldebaran (*al-deb'a-ran*). And beyond him follows the Great Dog with the bright blue white star Sirius (*sir'i-us*) in his mouth, and the Little Dog branded by the white star Procyon (*pro'si-on*). However, our New England ancestors did not see this grand figure in the sky; they called the constellation the Yard-ell or the Ell-yard.

The three beautiful stars which make Orion's belt are all double stars; the belt is just three degrees long and is a good unit for sky measurement. The sword is not merely the three stars which we ordinarily see, but is really a curved line of five stars; and what seems to be a hazy star, third from the tip of the sword is in fact a great nebula. Through the telescope this nebula seems a splash of light with six beautiful stars within it. Near latitude 40 degrees north, the first star in Orion to appear above the horizon is red

Orion, the three large stars in a line forming the belt, the curved line of smaller stars below forming the sword, Betelgeuse above to the left, Rigel below the belt, forming with Betelgeuse and the three stars of the belt a long narrow diamond in the sky

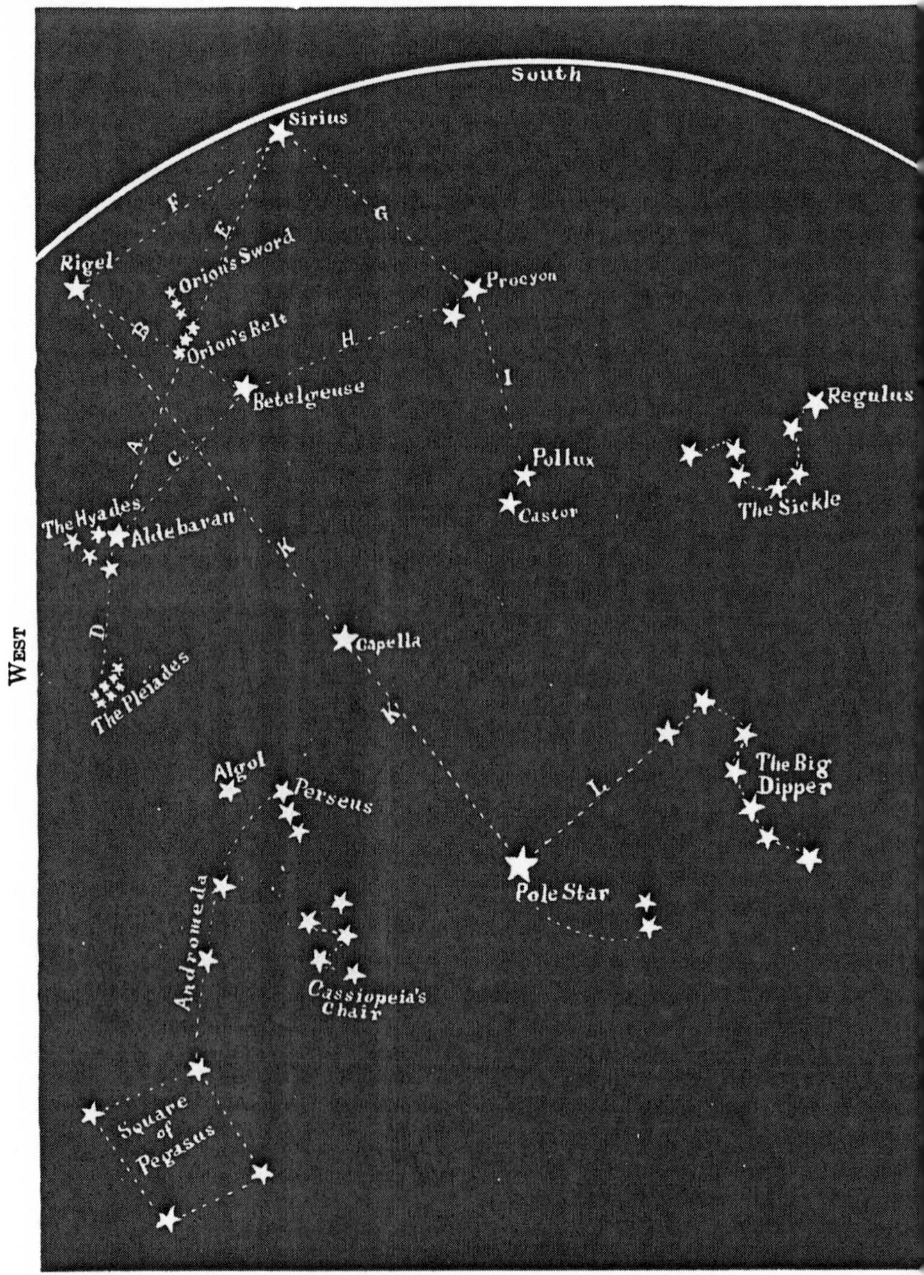

A diagram of the principal stars of winter as seen in early evening late in February

Betelgeuse, a blushing young giant just starting on its career as a star; it is composed of a gas much thinner than our air. Its diameter is 300,000,000 miles, which is more than one and one-half times that of the Earth's orbit. It is 200 light-years away from us. About fifteen minutes af Betelgeuse rises, and after the belt a sword are in sight, a white sparkling s appears 10 degrees to the south of the b This is Rigel, at a distance of 550 lig years. Seventeen thousand of our s

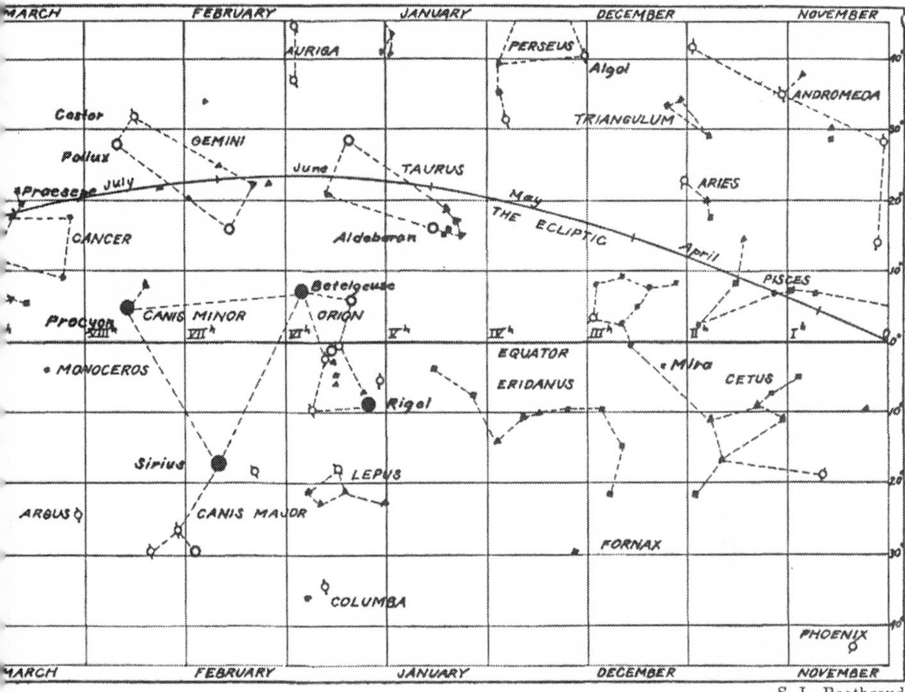

Stars of late autumn and winter

Key map to the sky as the observer in the Northern Hemisphere faces south. An observer in the Southern nisphere would need to face north and hold the map upside down

On Nov. 30, 12 P.M., Dec. 15, 11 P.M., Jan. 1, 10 P.M., Jan. 15, 9 P.M., Feb. 1, 8 P.M., Feb. 15, 7 P.M., March 1, M., the regions shown in the center of the map are due south of an observer in the Northern Hemisphere and due th of an observer in the Southern Hemisphere. Use the map that represents the date nearest the time the ervations are being made

uld be required to send us the same ount of light if they were as far away as s lovely star.

LESSON 226
ORION

LEADING THOUGHT — Orion is one of most beautiful constellations in the avens. It is especially marked by the ee stars which form Orion's belt, and line of stars below the belt which form sword.

METHOD — Place on the blackboard the tline of Orion as given in the diagram. k the pupils to make the following ob- vations in the evening and give their ort the next day.

OBSERVATIONS — 1. Where is Orion in relation to the Polestar?

2. How many stars in the belt of Orion? How many stars in the sword? Can you see plainly the third star from the bottom of the sword?

3. Notice above the belt, about three times its length, a bright star; this is Betelgeuse. What is the color of this star? What do we know about the age of a star if it is red?

4. Look below the belt and observe another bright star at about the same distance below that Betelgeuse is above. What is the color of this star? What does its color signify? The name of this is Rigel.

5. Note that west of the red star above and east of the white star below are two fainter stars. If we connect these four stars by lines we shall make an irregular four-sided figure, fencing in the belt and sword.

Sketch this figure with the belt and sword, and write on your diagram the name of the red star above and the white star below and also the name of the constellation.

6. Which star of the constellation rises first in the evening? Which last?

7. Write a story about Orion, the great hunter.

ALDEBARAN AND THE PLEIADES

Almost in a line with the belt of Orion, up in the skies northwest from it, is the rosy star Aldebaran. This star, which is also a ruddy young giant, marks the end of the lower arm of a V-shaped cluster

Aldebaran in the V-shaped cluster called the Hyades. This is a part of the constellation Taurus

composed of this and four other stars. This cluster is the Hyades (*hy'a-deez*). The Hyades is a part of the constellation called by the ancients Taurus, the bull, and is the head of the infuriated animal. Aldebaran is a comparatively near neighbor of ours, since it takes light only fifty-seven years to pass from it to us. It gives off about one hundred times as much light as does our sun; it lies in the path traversed by the moon as it crosses the sky, and is often thus hidden from our view when the moon occults the star.

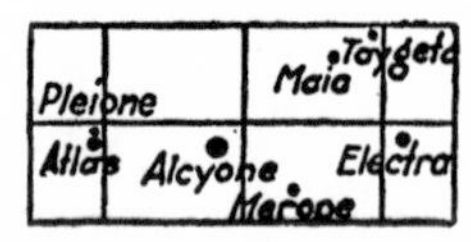

The Pleiades, a well-known group of stars, with the visible stars named

Although we are attracted by many bright stars in the winter sky, yet there is a little misty group of stars, which has ever held the human attention enthralled, and of which the poets of all the ages have sung. These stars are called the Pleia (*plee'ya-deez*); most eyes can count (six stars in the cluster. There are nine s large enough to be seen through the f glass, and which have been given nan but sky photography has revealed to that there are more than two thous stars in this little group. Perhaps no s in the heavens give us such a feeling the infinity of the universe as do Pleiades; for they form a great star syst known as an open cluster. These s which look so close together to us really so far apart that our own sun and its planets could roll in between them never be noticed. It would require sev years for light to travel from one of th stars in the Pleiades to another. The P ades are so far from us that it takes li three hundred years to reach us fr them. There is a mythical story that o the unaided eye could see seven instead six stars in the Pleiades, and much po imagining has been developed to acco for the "lost Pleiad." This myth is pro bly founded on fact.

LESSON 227

Aldebaran and the Pleiades

Leading Thought — The Pleia seem to be a little misty group of six st but instead there are in it two thousa stars. Half way between the Pleiades a Orion's belt is Aldebaran, an adolesc ruddy star.

Method — Draw the diagram (p. 8 on the blackboard showing Orion, debaran, and the Pleiades, and the li B, C, D. Give an outline of the obser tions to be made by the pupils, and them work out the answers when th have opportunity. Each pupil should p pare a chart of these constellations.

Observations — 1. Imagine a l drawn from Rigel to Betelgeuse and th another line just as long extending to west of the latter at a little less than a rig angle, and it will end in a bright, rosy st not so red as Betelgeuse.

2. What is the name of this star? Wr it on your chart.

3. Can you see the figure V formed by [Al]debaran and four fainter stars? Sketch [the] V and show where in it Aldebaran be[lo]ngs. This V-shaped cluster is called the [Hy]ades.

4. Imagine a line drawn from Orion's [bel]t to Aldebaran and extend it to not [qu]ite an equal length beyond it, and it [wil]l end near a " fuzzy little bunch " of [sta]rs which are called the Pleiades. Place [the] Pleiades on your chart.

5. How many stars can you see in the [Ple]iades?

6. Why are they called the seven [sis]ters?

7. How many stars in the Pleiades [ar]e named, and how many does photog[ra]phy show that there really are in the [gro]up?

8. How far apart from each other are [th]e nearest neighbors of the Pleiades?

[T]HE TWO DOG STARS, SIRIUS AND PROCYON

If a line from Aldebaran is passed [th]rough the belt of Orion and is extended [ab]out as far on the other side, it will reach [th]e Great Dog Star, following at Orion's [he]els. This is Sirius (*sir'i-us*) the most [br]illiant of all the stars in our skies, glint[in]g with ever changing colors, sometimes [bl]ue, at others rosy or white. It must have [be]en of this star that Browning wrote:

All that I know
Of a certain star
Is, it can throw
(Like the angled spar)
Now a dart of red,
Now a dart of blue.

Sirius has reached the blue white stage [of] star development. Although it is larger [th]an our sun, and gives twenty-six times [as] much light as our sun, its superior bril[lia]nce is due to its nearness to us; it is only [ei]ght and three-fourths light-years away [fr]om us. It is the most celebrated star in [lit]erature. The ancients knew it, the Egyp[tia]ns worshipped it, Homer sang of it, and [it] has had its place in the poetry of all ages.

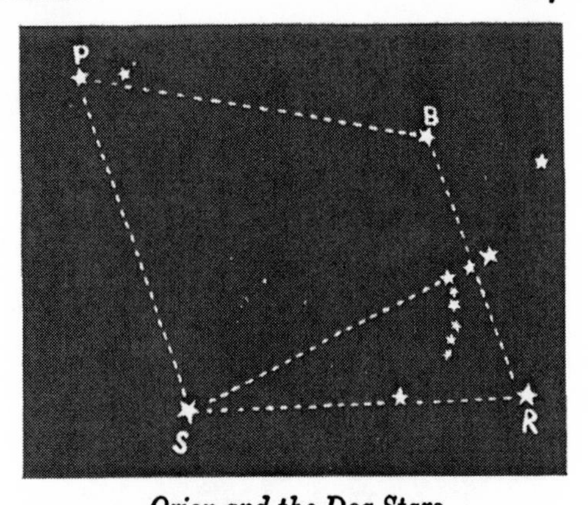

Orion and the Dog Stars

B, Betelgeuse; R, Rigel; S, Sirius, the Great Dog Star; P, Procyon, the Little Dog Star

Procyon (*pro'si-on*) the Little Dog Star was so called perhaps because it trots up the eastern skies a little ahead of the magnificent Great Dog Star; it gives out five times as much light as our sun, and is only ten light-years away from us.

LESSON 228

The Two Dog Stars

Leading Thought — The Great Dog Star, Sirius, is the most famous of all stars in the literature of the ages. The two Dog Stars were supposed by the ancients to be following the great hunter, Orion.

Method — Draw upon the board from the chart shown on this page, the constellation of Orion with Sirius and Procyon. Ask the pupils to note that after Orion is well up in the sky a straight line drawn through Orion's belt and dropping down toward the eastern horizon ends in a beautiful blue white star, which is Sirius; and that if we draw a line from Betelgeuse to Rigel, from Rigel to Sirius, and then draw lines to complete a quadrangle, we shall find our lines meet at a bright star just a little too far away to make the figure a square, but making it somewhat kite-shaped instead. This is the Little Dog Star, Procyon, and it has a twin which can be seen near it. After giving these directions let the children make the following observations.

Observations — 1. How do you find Sirius? Which rises first, Orion or Sirius?

2. What color is Sirius? Judging from its color what stage of development do you think it is in?

3. Try to find out how large Sirius is compared with our sun and how near it is to us.

4. Why is Sirius called the Great Dog Star? Is the Little Dog Star nearer to the North Star than Sirius? Which is the brighter, the Great Dog Star or the Little Dog Star? Can you see any fainter star near Procyon? What direction is it from Procyon?

5. Why is Procyon called the Little Dog Star?

6. Make a chart showing Orion and the two Dog Stars.

CAPELLA AND THE HEAVENLY TWINS

Capella is nearer to the North Star than any other of the bright stars, and in the latitude of northern New York it is a circumpolar star. Its light very much resembles that of our sun, as does that of all

Capella in the constellation Auriga

the bright yellow white stars; but it is a much larger star. Capella is always a beautiful feature of the northern skies, being almost in the zenith during the evenings of January and February. It is in a brilliant shield-shaped constellation known as Auriga.

Capella is a double star; its two components give off 150 times as much light as our sun and it is forty-eight light-years away from us. If our sun were where Capella is, it would be barely visible to the unaided eye on a very clear night. These two components, which make up Capella as we know it, are removed from ea other about the same distance as the ea and the sun. One revolves about the ot in a period of 104 days. The attraction tween these two massive suns is m powerful than the attraction between earth and the sun; hence they race th and one-half times as fast in their or as does the earth.

During the winter evenings we see t stars set like glowing eyes almost in zenith, and in a region of the sky wh

Gemini, the heavenly twins: the larger on Pollux, and the smaller, Castor

there are no other bright stars. These t stars are set just a little closer toget than are the pointers of the Big Dipp To this brilliant pair the ancients gave names of Castor and Pollux. Pollux is brighter of the two and is the more sou ward in situation. Pollux and Castor w two beautiful twin boys who loved ea other so much that, after they were de they were placed in the skies where t could always be near each other. Althou Castor and Pollux seem so near toget in the sky, they are separated by a dista of eleven light-years, Castor being t much farther away from us than Poll

The twin stars are supposed to exe benign influence on oceans and seas a are, therefore, beloved by sailors. Wh a boy says "By Jimminy," he does not alize that he may be using an ancient pletive "By Gemini," which is the La name of these twin stars and was a favo ancient oath, especially with sailors.

Castor is easily seen as a double star a three-inch telescope. Each star of pair is really two stars, as revealed by spectroscope. There are also three ot faint stars in the system, so Castor is re seven stars — a most remarkable syste

LESSON 229

Capella and the Heavenly Twins

Leading Thought — There are, during evenings of January and February, ee brilliant stars almost directly overad. One of these is Capella; the other o are the Heavenly Twins.

Method — Place on the board the part the chart (p. 824) showing the Big pper, Polestar, Capella, and the Twins. aw a line, L, from the pointers of the g Dipper, and extend it to the Polestar. aw another line, K, from the Polestar right angles to the line L, and on the e away from the Big Dipper's handle, d it will pass through a large, brilliant, low star which is Capella. Ask the pus to imagine similar lines drawn across sky when they are making their obvations, and thus find these stars, and place them on their charts, making the lowing observations.

Observations — 1. Is Capella as near to the Polestar as the Big Dipper? Is it near enough so that it never sets where you live?

2. Can you see the shieldlike constellation of which Capella is a part? Do you know the name of this constellation?

3. How do you find the Heavenly Twins after you have found Capella?

4. Why are these stars called the Heavenly Twins? What is their Latin name? What are the names of the two stars?

5. How can you tell the Heavenly Twins from the Little Dog Star and its companion?

6. Read in the books all that you can find about the Heavenly Twins. Try to find whether they are the same age, whether they are as near together as they seem, and whether they are going in the same direction. What sort of influence did the ancient sailors attribute to these twin stars?

THE STARS OF SUMMER

To us, who dwell in a world of change, stars give the comfort of abidingness; y remain ever the same to our eyes and teacher should make much of this. hen we once come to know a star, we ow exactly where to find it in the heavs, wherever we may be. A star which a rson knows during childhood will, in er life and in other lands, seem a unch friend and a bond, drawing n back to his early home and associans.

The summer is an inviting season for king the acquaintance of eight of the een brightest stars visible in northern itudes. Few midsummer entertainnts rival that of lying on one's back on grass of some open space which comnds a wide view of the heavens. There h a planisphere and an intermittently hted flashlight with which to consult it, rn by sight, by name, and by heart se brilliant stars which will ever after et our uplifted eyes with friendly greet. To teach the children in a true informing way about the stars, the teacher should know them, and nowhere in Nature's realm is there a more thought-awakening lesson.

LESSON 230

The Bright Stars of Summer

Leading Thought — The stars we see shining during summer evenings are not the same ones that we see during the winter evenings, except those in the circumpolar constellations. There are eight of the brilliant summer stars which we should be able to distinguish and call by name.

Method — Begin by the middle of May when the Big Dipper is well above the Polestar in the early evening, and when, therefore, Regulus, Spica, Arcturus, and the Crown are high in the sky. The others may be learned in June, although July is the best month for observing them. In teaching the pupils how to find the stars, again instruct them how to draw an imaginary straight line from one star to another and to observe the angles made by such lines connecting three or four stars.

Place upon the blackboard the figures from the chart below, as indicated, leav-

this line lies a group of stars called t Sickle, and the stars that form it outli this implement. The Sickle has a je at the end of the handle, which is a wh

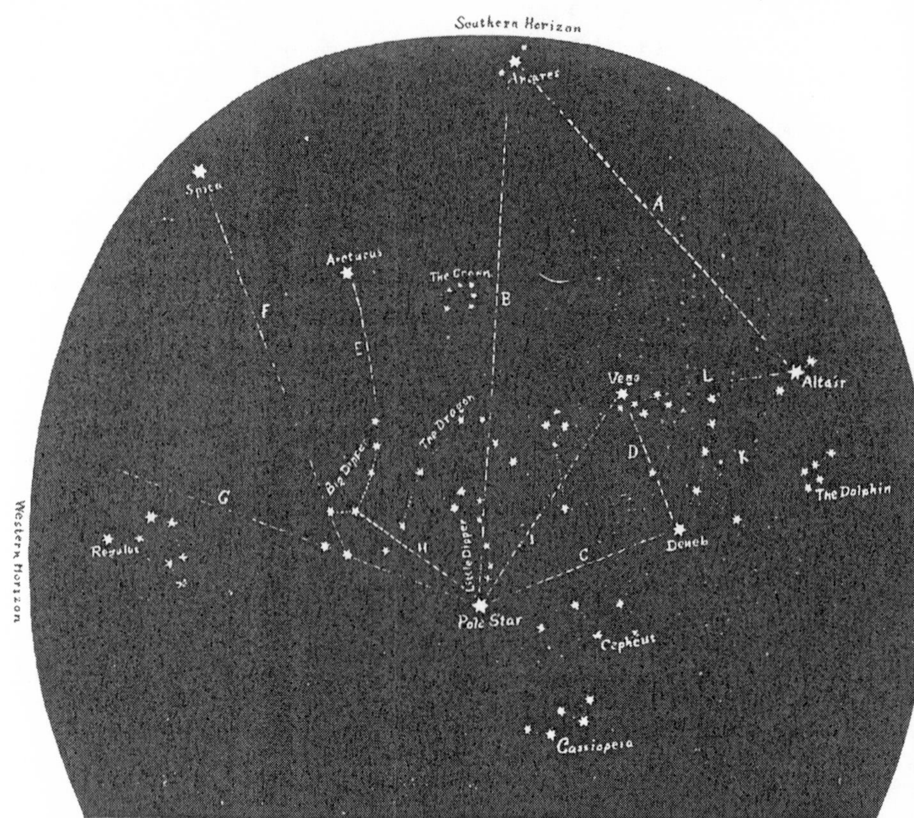

A chart of the brightest stars of summer, showing their positions in early evenings of Jun

ing each one there until the pupils have observed and learned it. Then erase it and put on another figure. In each case try to get the pupils interested in what we know about each star, a brief summary of which is given. Note that the observations given in the lessons are for early in the evenings of the last of May, of June, and of early July.

Regulus (reg′yu-lus)

Draw upon the blackboard from the above chart the Polestar, the Big Dipper, the line G, and the Sickle, shown just below the outer end of the line. Extend the line that passes through the pointers of the Big Dipper to the North Star backward into the western skies; just west of

and diamond-glittering star called Re lus. It is a great sun giving out seve times as much light as our own sun, a this light reaches us in about fifty-sev years. The Sickle is part of a constellati

Regulus, the large star in the handle of sickle

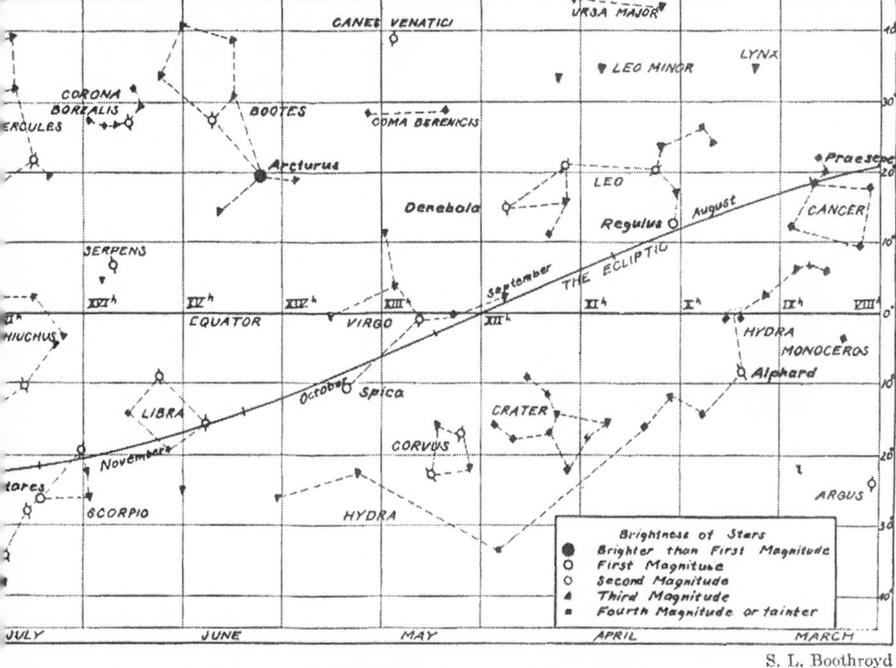

S. L. Boothroyd

Stars of spring and summer

ey map to the sky as the observer in the Northern Hemisphere faces south. An observer in the Southern isphere would need to face north and hold the map upside down
ı April 1, 12 P.M., April 15, 11 P.M., May 1, 10 P.M., May 15, 9 P.M., June 1, 8 P.M., the regions shown in the r of the map are due south of an observer in the Northern Hemisphere and due north of an observer in Southern Hemisphere. Use the map that represents the date nearest the time the observations are being made

ed the Lion, from which comes the wer of meteors which we see on the ings from November 11 — 15. Regu- is seen best in the evening skies of ng.

Arcturus (ark-tu'rus)

lace on the blackboard the Big Dipper, Polestar and the line E, Arcturus, and Crown. Extend the handle of the Big

Arcturus and the Big Dipper

per following its own curve for about wn length and it will end in a beauti- orange star, the only very bright one at region. It is one hundred times as bright as our own sun, but its light does not reach us for thirty-eight years after it is given off. Arcturus is a giant sun, having

The Northern Crown

a diameter of nineteen million miles. During the latter part of June and July it is almost overhead in the early evening.

The Crown

Between Arcturus and Vega, but much nearer the former, is a circle of smaller stars that is called the Northern Crown, which because of its form is quite noticeable.

Spica (*spy'ka*)

Place on the blackboard the Big Dipper, Polestar, line F (Fig. p. 830), and Spica. To find Spica draw a line through the star on the outer edge of the top of the bowl of

Vega and her train of five stars

the Big Dipper, through the star at the bottom of the bowl next the handle, and extend this line far over to the southwest, during the evenings of June and July. (See p. 830.) Spica is a white star, and is the only bright one in that part of the sky. It is over two hundred light-years distant; 1,500 of our suns would be required to equal its brilliance at that distance. Spica is in the constellation called the Virgin.

Vega (*vee'ga*)

Place on the blackboard the Polestar, the Big Dipper, the lines H and I (Fig. p. 830), and Vega with her five attendant stars, as shown in the chart. Teach that these stars are the chief ones in the constellation called the Lyre. To find Vega, draw a line from the Polestar to the star in the Big Dipper which joins the bowl to the handle. Then draw a line at right angles to this (see chart lines H, I) and extend line I a little farther from the North than is the end star of the Dipper han this line will reach a bright star, bluis color, which can always be identified four smaller attendant stars that lie it and outline a parallelogram with sl ing ends. Vega is the most brilliant s mer star that we see in the North Hemisphere. It is a very large sun, gi out fifty times as much light as our su is so far away that it requires twenty years for a ray of light to reach us fron Vega's chief interest for us, aside from beauty, is that toward it our sun and al planets, including our earth, are movin the rate of thirteen miles per second.

Antares (*an-ta'rees*)

Add to the last diagram on the bl board the line E (Fig. p. 830), to Arctu the line B, and Antares. To find this s draw a line half way between Arcturus Vega from the Polestar straight across sky to the south, and just above the so ern horizon it will point to the glow star, Antares, in the constellation of Scorpion. Also a line drawn at right an to the line connecting Altair with its c panions and extending toward the so will reach Antares. Late June and

Antares, a brilliant star in the southern s

about ten o'clock in the evening is best for viewing this beautiful star. An teresting thing about Antares is that the greatest of the young giant stars measured; it has a diameter of 400,000, miles.

Deneb or Arided (*den'eb; a'ri-ded*)

Erase from the last diagram Antares the line B. Add to it the lines C and making a right angle at Deneb, and oss — the head of which is Deneb, the t ending near the letter on line L. is star is at the head of the Northern oss, which is a very shaky looking cross l appears in the eastern skies during the nings of June and July, with its upright n nearly horizontal as seen in a middle rthern latitude. Deneb is white in color

e Northern Cross, in the constellation of the Swan

d is a very large sun, because it seems us a bright star although it is so far away m us that the distance has never been ely measured; but it has been estimated at a ray of light would need at least six ndred years to reach us from Deneb. It d the cross are a part of the constellation Cygnus, or the Swan.

Altair (al'ta-ir)

Add to the last diagram on the board e lines L, K, Altair and its two attendant rs, and the Dolphin. Emphasize the t that Altair marks the constellation of uila, or the Eagle. This beautiful star easily distinguished because of the small r on either side, all three being in a line. The three belong to a constellation called the Eagle, and may be seen in early evening from June to December. Altair,

Altair in the constellation of the Eagle

Deneb, and Vega form a triangle with the most acute angle at Altair. (See diagr. L, K.) Just northeast of Altair is a little diamond-shaped cluster of stars called the Dolphin, which is a good name for it, since it looks like a Dolphin, the fifth star forming the tail. It is also called Job's Coffin, but the reason for this is uncertain, unless Job's trials extended to a coffin which could not possibly fit him. If the line C on the chart drawn from the Polestar to

The Dolphin or Job's Coffin

Deneb be extended, it will touch the Dolphin. Altair is always low in the sky; it is a great sun giving off nearly ten times as much light as our own sun; light reaches us from it in fifteen years.

THE SUN

If, only once in a century, there came us from our great sun light and heat inging the power to awaken dormant e, to lift the plant from the seed and othe the earth with verdure, then it ould indeed be a miracle. But the sun by ining every day cheapens its miracles in e eyes of the thoughtless. While it rdly comes within the province of the nature-study teacher to make a careful study of the sun, yet she may surely stimulate in her pupils a desire to know something of this great luminous center of our system.

Our sun is a great shining globe about one hundred and ten times as thick through as the earth, and more than a million times as large. If we look at the sun in a clear sky, it is so brilliant that it hurts our eyes. Thus, it is better to look at it through a smoked glass, or when the atmosphere is very hazy. If we should see the sun through a telescope, we should find that its surface is not one great glare of light but is mottled, looking like a plate of rice soup, and at times there are dark spots to be seen upon its surface. Some of these spots are so large that during very "smoky weather" we can see them with the naked eye. In September, 1908, a sunspot was plainly visible; it was fifty thousand miles across, and our whole world could have been dropped into it with twelve thousand miles to spare all around it. We do not know the cause of these sunspots, but we know they appear in greater numbers in certain regions of the sun, above and below the equator. And since each sunspot retains its place on the surface of the sun, just as a hole dug in the surface of our earth would retain its place, we have been able to tell by the apparent movement of these spots how rapidly and in which direction the sun is turning on its axis; it revolves once in about twenty-six days and, since the sun is so much larger than our earth, a spot on the equator travels at a rate of more than a mile a second. There is a queer thing about the outside surface of the sun — the equator rotates more rapidly that the parts lying nearer the poles; this shows that the sun is a gaseous or liquid body, for if it were solid, like our earth, all its parts would have to rotate at the same rate. At periods of eleven years the greatest number of spots appear upon the sun.

Another interesting feature of the sun is the tremendous explosions of hydrogen gas mixed with the vapors of calcium and magnesium, which shoot out flames from twenty-five thousand to three hundred thousand miles high, at a rate of speed two hundred times as swift as a rifle bullet travels. Think what fireworks one might see from the sun's surface all the time! These great, explosive flames can be seen by the telescope when the moon eclipses the sun, and may be seen with the aid of a spectroscope at any time. Besi
these magnificent explosions, there is s
rounding the sun a glow which is brigh
near the sun's surface and paler at t
edges; it is a magnificent solar halo, so
of its streamers being millions of mi
long. This halo is called the Corona, a
is visible only during total eclipses.
means of the spectroscope we know th
there are about seventy chemical e
ments in the sun, which are the same
those we find upon our earth. The e
ment helium which gets its name fr
Helios, the sun, was discovered in t
sun by means of the spectroscope befor
was isolated on the earth.

The sun weighs 330,000 times as mu
as the earth; the force of gravity upon
surface is twenty-seven and two-thi
times as much as it is here. A letter wh
weighs an ounce here would weigh alm
a pound and three-quarters on the s
and a man of ordinary size in this wo
would weigh more than two tons the
and would be crushed to death by
own weight. Find how much your wat
your book, your pencil, your baseball, a
your football would weigh on the sun.

Our Sun and Its Family — The Solar System

First of all we shall have to acknowled
that our great, blazing sun is simply
medium-sized star, not nearly so large
Vega nor even so large as the Polest
but it happens to be our own particul
star and so is of the greatest importan
to us. The sun has several other worl
more or less like our own, called plane
these planets revolve around the sun
almost the same level or plane in whi
our world revolves, but some of the
worlds are much nearer the sun and oth

much farther away than ours. See the gram showing the orbits of planets, ow, and page 836.

Ɔne peculiar thing about all the planets the sun's family and all their moons is t they shine by reflecting the light of : sun, and none of them is hot enough give off light independently; but these er worlds of ours are so near us that y often seem larger and brighter than : stars which are true suns and give off ch more light than our own sun. After ittle experience the young astronomer rns to distinguish the planets from the e stars: the planets always closely fol- v the path of the sun and the moon ough the sky; they often seem larger l brighter than the true stars and do not inkle so much. The so-called morning l evening stars are planets of our sun's nily and are not stars at all.

To determine which planets are morn- ; or evening stars at a given time con- t an almanac for the current year, or h a publication as *The Monthly Eve- ıg Sky Map.*

The planets in order of their relative tance from the sun are *Mercury, Venus, rth, Mars, Jupiter, Saturn, Uranus, ptune,* and *Pluto.*

The planet nearest the sun is *Mercury;* was so named by the ancients because travels very rapidly about the sun, and : god Mercury was very fleet-footed be- ıse of his winged slippers. It makes a cle about the sun in eighty-eight days; ıt means its year is eighty-eight days ıg. The amount of heat it receives is ır and three-fifths to ten and one-fourth ıes greater than the amount received r unit area outside of the earth's atmos- ere. Because Mercury follows the sun closely, it is very difficult to observe; t it is possible to see it sometimes, quite ar the horizon, as a morning or evening r.

Venus is the next planet in order of tance from the sun, and is called the in sister of the earth; it is a little aller than the earth, but has about the ne amount of atmosphere, of a different mposition. Venus is, to earth dwellers, the most brilliant object in the sky, with the exception of the sun and moon. It receives about twice as much heat and light from the sun as does the earth. The silvery light from Venus is so strong that on moonless nights it often casts shadows. Sirius, the brightest star in the sky, is only about one-twelfth as bright as Venus is at its brightest. This planet makes a circuit about the sun in a period of 225 days.

Our *earth,* with its year of 365¼ days, comes next. If we could view Earth from some other nearby planet, it too would shine by reflected light from the sun. If the earth and moon could be seen from Venus, when Venus and the earth are nearest each other, the moon would appear as bright as Venus at its best, and earth would be about eighty times as brilliant.

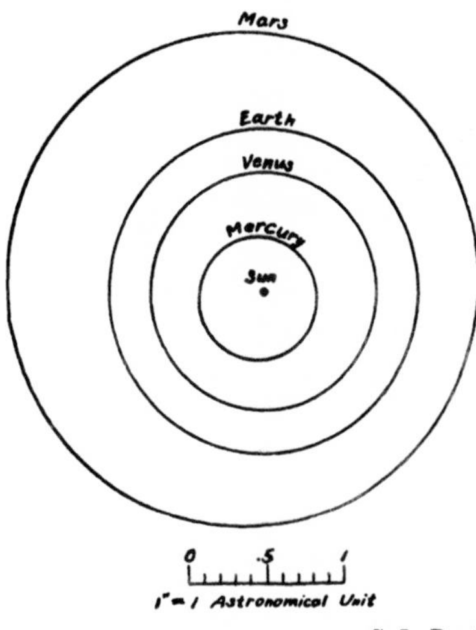

S. L. Boothroyd

The orbits of the Inner Planets. Note that each planet has an orbit which is not circular but is very nearly so

Mars is next in relative distance from the sun, but is so much farther from the sun that it receives only about one-half as much heat and light as does the earth, per unit of area outside of the earth's atmosphere. Polar caps may be seen on Mars, similar to those around our North and South Poles. A year on Mars is equal to 687 of our days or 669⅗ of its own. While Mercury and Venus have no satellites and the earth has only one, Mars has two; these satellites are named Phobos and Deimos. They are quite small; Phobos has a diameter of seven miles, and goes around Mars every seven hours, while Deimos has a diameter of sixteen miles and circles Mars every thirty hours. The diameter of our moon is 2163 miles, and it circles the earth every twenty-seven and one-third days. If you were on Mars, you would see Phobos rise in the west and set in the east, while Deimos would rise in the east and set in the west.

Mercury, Venus, Earth, and Mars as a group are relatively near the sun. Since

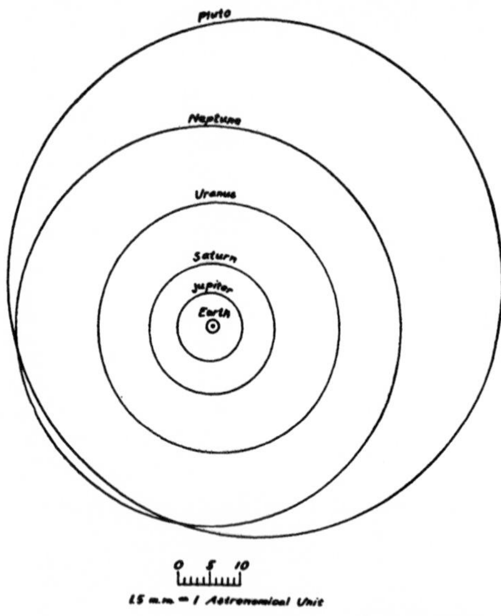

S. L. Boothroyd

The Outer Planets and their orbits. The orbits of these planets also are not circular but are nearly so

they are nearer the sun than other planets, they are called the Inner Planets. Because they are not greatly different in size from the earth and have earthlike surfaces they are also called Terrestrial Planets.

The remaining planets are larger than the ones just discussed, and as a group they are called the Major Planets. They are much farther from the sun and for that reason they are often called the Outer Planets. In contrast to the Terrestrial Planets, all except Pluto have immensely deep, cloudy atmospheres, and so we see only the outer layer of clouds, hundreds miles deep, which completely hide th surfaces.

Beyond the Terrestrial Planets, th is a tremendous space in our solar syste in this space are to be found several hu dred smaller bodies, called planetoids asteroids. The next planet is *Jupit* found at a great distance from Mars. J piter is the largest of all the planets; its ameter is eleven times and its volu thirteen hundred times those of the ear In reality, it is larger in mass and volu than all the other eight planets put gether. About twelve years are requir for the journey of Jupiter around the su The atmosphere of Jupiter seems ve dense with vapors of ammonia and m thane. Jupiter has eleven satellites, a three of them go around the planet in t opposite direction from the other eig satellites, which go around him fro west to east as does our moon about t earth.

Saturn is located at a point nine a one-half times the earth's distance fro the sun, and requires about twenty-ni and one-half years to go around the su This is the last planet that can be eas seen with the naked eye, and it appe about as bright as Arcturus; it has a de nite orange tinge. This planet is s rounded by nine satellites and by mul tudes of tiny particles which revolve circular orbits, and are so numerous th seen from the earth they appear like f rings encircling the planet. With even small telescope the rings are quite evide except when presented edgewise or nea so.

Soon after *Uranus* had been discovere the astronomers calculated its orbit a thereby found where it should be fro day to day; but the planet did not beha as they thought it should. The astron mers decided there was a far-distant bo pulling the planet Uranus from its co puted orbit; they undertook to calcula the position of the body that must causing the disturbance, and in 1846 tw different men, working independentl discovered a new planet, *Neptune*,

position calculated. Neptune requires
rly 165 years to complete a journey
ind the sun. One satellite, similar in
to our moon, accompanies Neptune
ts long journey.
ven after Neptune had been discov-
l, Uranus still failed to follow the new
t computed for it. A long series of
putations followed, and in 1930 a new
iet was located near the position pre-
ed from the remaining discrepancies
Jranus' orbit. The name *Pluto*, sug-
ed by a child in England, has been
n to this planet. Its trip around the
has been calculated to require about
years; it is about forty times as far
n the sun as is the earth, and is quite
ly one of several planets that may ex-
in the remote portions of the solar
em. Careful photographic search,
ong the millions of faint stars, may
al more distant planets in the future.
verful spectrographs, aided by the light
p of some of the great reflectors now
oon to be available, may reveal new
interesting wonders about the sun
his attendant planets.
Dr. Simon Newcomb in his delightful
k, *Astronomy for Everybody*, gives the
illustration to make us understand
place of our sun and its planets and
relation to the stars in space. He ex-
ns that if here in the Atlantic States
should make a model of our solar sys-
by putting an apple of four and one-
inches diameter down in a field to
resent the sun, our earth could be
esented by a mustard seed forty feet
y revolving around the apple; and
otune could be represented as a small
circling around the apple at the disce of a quarter of a mile. Now that a
iet more distant than Neptune has
n discovered, the size of our solar
em has been definitely increased and
shall need to consider Pluto in con-
cting our model. Pluto is forty times
her away from the sun than is the
h; so we shall need to represent Pluto
an object placed at a point 1600 feet
n the apple which we have used to
resent the sun. But to find the star nearest to our earth, the star that is only four and one-half light-years away from us, we should have to travel from this field across the whole of North America to California, and then take steamer and go out into the Pacific Ocean before we should reach our nearest star neighbor, which would be another sun like our own represented by a pair of apples.

A HAPPY FAMILY

The Sun, a great father, is hanging in space
With his children all playing around,
And each child is careful to play as it should,
Without commotion or sound.

Little Mercury stays near his father's side
And hangs on his every smile,
While he kicks up his tiny impertinent heels
And speeds over mile after mile.

Next to Mercury the beauteous one,
Venus, her father's delight,
Unrivaled reigns, without sceptre or crown,
The glorious queen of the night.

And Venus's twin, our own Mother Earth,
Though not considered so fair,
Must be great to observe on a clear night in June,
With the moonbeams astream through her hair.

Then next to the Earth comes little red Mars
With Deimos and Phobos at hand,
They swing into place and scamper around
Within the Zodiac band.

After Mars comes Jupiter, largest of all;
His father looks at him with pride;
And with the big giant his ten satellites
Come tripping in side by side.

Then Saturn rolls in his three pretty rings
While around him nine moons swarm;
He's next to his brother Jupiter in size
And his sister Venus in charm.

And out beyond Saturn are still other twins,
Uranus and Neptune, so far
That to either of them, astronomers say
Their father looks like a big star.

And Pluto finally comes out into view,
After aeons of hiding away;
Although he is quiet, secretive, and shy,
He merrily joins in the play.

But the Sun and his family will all sta
place
Right on to the end of time;
Nor does it disturb them one tittle or
This spinning them into a rhyme.

"MONTHLY EVENING SKY MA
NANCY L. MOOREF
(April 19

COMETS AND METEORS

COMETS

Besides planets and stars there are in space other bodies moving around our great star, the sun, and following paths shaped quite differently from those followed by our earth and its sister planets. We move around the sun nearly in a circle with the sun at the center, but these other heavenly bodies move in narrow oval orbits, the sun being near one end of the ellipse and the other end being much farther out in space, in some cases beyond our farthest planet. These bodies do not revolve around the sun in the same plane as our world and the other planets; indeed they often move in quite the opposite direction. The most noticeable of these bodies whose race-track around the sun is long instead of circular are the comets, and we know that some of these almost brush the sun when turning at the sunward end of their course. The astronomers have been able to measure the length of the race-tracks of some of the comets and thus tell when they will come back. Encke's comet, named after the German astronomer, makes its course in three and one-half years and this is the shortest period any we know. When nearest the sun, i just within the orbit of Mercury, a when farthest, it is about one hundred a ten million miles nearer the sun th Jupiter.

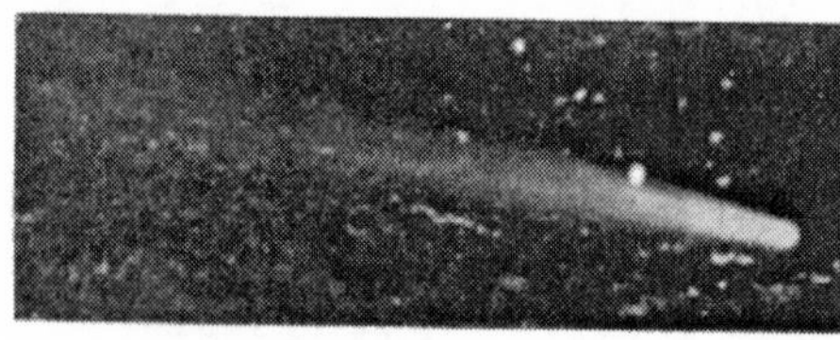

Lick Observatory

Halley's Comet, May 6, 1910. This photograph was taken in Chile

There are about five hundred com whose courses have been thus determin the longest accurately known period longs to Halley's comet, which makes s a long trip that it comes back only o in seventy-six years; but there are ot comets which travel such long routes t they come back only once in hundreds even thousands of years. About tw hundred comets have been discover many of them so small that they can o be seen with the aid of the telescope; it has been found that in one instance least, three comets are racing around sun on the same track.

A comet is a beautiful object, usu having a head which is a point of brilli light and a long, flaring tail of fainter lig which always extends out from it on side opposite the sun. The head o comet must be nearly twice as th through as the earth in order to be la enough for our telescopes to discover Some of the comet heads have been me ured; one was thirty-one times, and other one hundred and fifty times as w as our earth. If the heads are this la imagine how long the tails must Some of them are far longer than distance from our earth to the sun. T comet head decreases as it approac the sun. The head of a comet is suppo be a swarm of meteors with some gas, wing by the reflected light of the sun. hen in the end of the orbit near the , the gas which the comet contains sorbs the energy of the ultraviolet radi- on of the sun and re-emits it as visible ht; thus at such times the comet ap- rs to be partly self-luminous. In fact, s gas has so little weight that light can sh it; one would never believe that light ıld push anything, because we cannot l it strike against us; but the physicists ve found that it does push, and by push- ; against the particles of the gas of com- it sends them streaming away from the ı, just as the heat appears to push out aring cloud of steam from the spout of eakettle.

Comets have played an important part history; they were formerly considered ns of the approval or wrath of God. The urn of Halley's Comet appearing in 66 struck terror to the Saxons and pre- ;ed the Norman conquest of England. e comet of 1811 was thought to warn us the war of 1812 and Napoleon of his ming defeat. This was a wonderful met illuminating our skies for a year d a half; its rosy head was veiled in a seous sphere, which with the head was ger than the sun. Some comets, which ve failed to appear when expected, have ir orbits marked by swarms of meteors.

Shooting Stars

When we look up during an evening lk and see a star falling through space, netimes leaving a track of light behind we wonder which of the beautiful stars heaven has fallen. But astronomers tell that these so-called shooting stars are all pieces of solid material which are veling around the sun in an orbit that ersects the orbit of the earth. Arriving the point of intersection of the orbits, en the earth is there, they hit the upper nosphere and become luminous. The sh of light, which we call the *shooting star*, is due to the heat resulting from the impact, just as a bullet melts when it hits a big rock. The difference between the small dust particle, which in reality is a meteor, and the bullet striking a big rock, is that the meteor strikes a lot of air mole- cules; one after another, the molecules become luminous by the impacts. The molecules of air become luminous along the path traveled by the meteor through a distance of some thirty to sixty miles through the rare upper atmosphere of the earth. For some time after the bright flash of light has vanished, one sees the numerous particles of the meteor left be- hind, and also the glowing air molecules which cause the luminous train as they persist in the path of a bright shooting star.

Meteorites sometimes weigh hundreds of pounds; one in the Yale Museum weighs 1,635 pounds. If it were not for the air, which wraps our globe like a great kindly blanket, and by its friction heats the meteors and reduces them to micro- scopic dust particles, no one could live on this earth; the meteors would pelt us to death. It is reliably estimated that during every twenty-four hours our world meets hundreds of millions of these meteors; some of them are no larger than fine shot and others weigh a few ounces. Occa- sionally meteorites which weigh from a few pounds to many tons do reach the earth. The origin of these is not certainly known; but some may be the larger parti- cles which make up the nucleus of small comets. Others almost certainly come to earth from interstellar space.

The Relation between Comets and Meteors

Before we see the meteor as a shooting star, it is traveling around the sun in an orbit which intersected the orbit of the earth. It is very interesting to know that many meteors travel in swarms made up of a scattered assemblage of particles of matter; this matter once formed part of the head of a comet. Whenever the orbit of a comet intersects or comes very near the orbit of the earth, we get a shower of meteors if we are in that part of our orbit which is near the orbit of the comet. For instance, on May sixth the earth passes near the orbit of Halleys comet; we always get many meteors near that date whose paths seem to radiate from a point near the star Eta Aquarii. These meteors are known as Eta Aquarids.

In the same way the orbit of Tuttle's comet of 1862 intersects the orbit of the earth near the point where our planet is on about August 10. Since much of the original material of this comet is widely dispersed around its orbit, we get meteors of this swarm from about July 12 to September 1 of each year. As the paths of these shooting stars seem to radiate from a point in the constellation Perseus, they are called Perseid Meteors. The orbits of these meteors lie in a plane nearly at right angles to the earth's orbit plane. It is a very elongated orbit; at its farthest point from the sun it reaches nearly to the orbit of Pluto. Tuttle's comet and its associated meteors take 123 years to complete one circuit of the orbit. This is an unusually good meteor shower to observe, because the meteors are fairly abundant every year; many of them are bright enough to be seen even in the presence of the full moon, and besides this the August nights are comfortable nights for outdoor observations.

The most notable meteor shower is undoubtedly the Leonid shower; the paths of these meteors seem to come from a point near the curve of the Sickle, in the constellation Leo. These meteors are associated with Temple's comet of 1866; its orbit crosses the orbit of the earth at the place where we are about November 14. As the main meteor swarm is very compact, some spectacular meteor showers often occur when the earth encounters the main swarm, every thirty-three years. Some Leonid meteors are seen every ye near the middle of November. Some sp tacular showers of Leonids occurred 1799, 1833, and 1866, when meteors w counted by several observers, each wat ing small areas of the sky, at rates as h as 300,000 per hour. Historical records spectacular showers of these meteors c be traced in Chinese annals, almost New Testament times.

We are now nearly certain that all m teors whose paths as shooting stars see to radiate from a small area on the s were traveling in elliptical orbits about t sun, and hence were members of the so system even before becoming a part the earth. Such meteor showers take th names from the star or constellation in t area from which they seem to radiate.

There are, however, many meteors a pearing all the time whose paths have radiant, or central point of origin; rece observations indicate that all of the come to us from interstellar space. Th are continually streaming through t solar system along hyperbolic orbits; the whose orbits intersect the orbit of t earth and arrive at the point of inters tion when the earth is there, produce t flash of light we see as a shooting st They thereby lose their identity as in pendent bodies and become a part of t earth. Judging by the number of mete that strike the earth yearly, the numb streaming through the solar system mu run into millions of millions daily.

We thus learn from observation a the study of meteors that interstellar spa is far from being completely devoid matter. Occasionally one of these piec of matter is so large that it is not co pletely disintegrated in its passage throu our atmosphere, and solid pieces of it f on the surface of the earth. Such an eve is called a fall of meteorites and the piec which fall are called meteorites.

THE RELATION BETWEEN THE TROPIC OF CANCER AND THE PLANTING OF THE GARDEN

By John W. Spencer

In years gone by, many farmers had a vorite phase of the moon when they anted certain crops, usually spoken of the "dark" or the "light" of the oon. I once knew a woman who picked r geese by the "sign of the moon." Hogs re butchered in the "light" of the oon, and then the pork would not "fry vay" so much in the skillet. It is true me pork from some hogs wastes faster an that of others, but the difference is e to the kind of food given the hogs. lany farmers hold to those old supersti- ons yet, but the number is much less w than twenty-five years ago. I wish I ight impress on you young agriculturists at the moon has no influence on plant e, or pork, or geese, but the position of e sun most decidedly has. We have me plants that had best be planted when e sun's rays strike the State of New York antingly, which means in early spring or e fall. We have other plants that should t be put in the open ground until the ys of the sun strike the state more direct ows, which means the hotter weather of mmer. If I were in close touch with you pils, I should be glad to tell some things at happen to three young friends of ine, hoping that thereby my statement ight give the boys and girls an interest three geographical lines concerning the opics, and lead them to find their loca- on on the map, particularly when later ey learn what happens to my three ung friends. There is one in Quito, cuador, of whom we will speak as Equa- r Shem; the one on the Island of Cuba named Tropic of Cancer Ham; and the ther in São Paulo, Brazil, answers to the ame of Tropic of Capricorn Japhet.

What happens to these three boys, hem, Ham, and Japhet, is this. At cer- in times of the year they have no shadow hen they go home for dinner at noon. This state of affairs is no fault of theirs. It is not because they are too thin to make shadows. It is due to the position of the sun. If the boys should look for that luminary at noon, they would find it as directly over their heads as a plumb line. It is a case of direct or straight blows from rays of the sun, and, oh, how hot — hotter than any Fourth of July the oldest inhabitant can remember! These three boys are not hit squarely on the head on one and the same day. Each is hit three months after the other. The first boy to be hit this year in the above manner will be Equator Shem. The time will be during the last half of March. Can any of my young friends in this grade tell me the exact day of March that Equator Shem has no shadow? If no one of you can answer that question at this time, you had best talk it over with your friends, and bring your answers tomorrow. It happens at a time when our days and nights are of about equal length.

Another thing about this particular day is that our almanacs call it the first day of spring. All because no boy or anything else has a shadow on the equator at noon time. People and bluebirds and robins in the State of New York will see squalls of snow about that time, and there will be some freezing nights. But after the first day of spring the cold storms do not last so long as during December, January, and early February, when the sun's rays hit us with very glancing blows. Watch to see how much faster the sun melts the snow on the last days of March than it did at Christmas time. The light is also stronger and brighter, and plants in greenhouses and our homes have more life, and are not so shiftless, so to speak. Even the hens feel the influence, for they begin to lay more eggs and cackle, and down goes the price of eggs. Do not forget to learn what day in March spring begins, when the Equator boy finds it so hot that he would like to take off his flesh, and sit in his bones. After a few days, Equator Shem will find he again has a shadow at noon — a short one it is true, but it will get longer and longer each day. Now his shadow will be on the south side of him. Is this a queer thing to happen? On which side of you is your noontime shadow? I will give every one of you a red apple that finds it anywhere but on the north side of him at twelve o'clock. Every time the sun shines at noon, watch to find your old uncle in the wrong, and thereby get the apple. Each day that the shadow of Equator Shem becomes longer and longer, the noonday shadow of Tropic of Cancer Ham, living on the Island of Cuba, will be getting shorter and shorter, until at last there comes a day during the last of June when he, too, will have no shadow, and the almanac says that that day is the beginning of summer.

Now it will be the turn of the Tropic of Cancer Ham, on the Island of Cuba, to say the weather is hotter than two Fourths of July beat into one, and he too will wish that he could take off his flesh, and sit in his bones. Everybody in the State of New York will say that the first summer day is the longest day of the year. It is on this day that Equator Shem will have as long a shadow as *he* ever had in his life. No United States boy will ever be without a shadow at noon so long as he remains in his own country. When the eight o'clock curfew bell says it is time for boys and girls to go to bed, it will yet be light enough to read the papers. The sun not only sets late on that first summer day, but it appears early next morning. What a beautiful spectacle a sunrise in June is! Men of wealth will pay thousands of dollars for pictures showing its glory, yet I suppose that not one boy in five hundred ever saw the beauty of the birth of a new day in the sixth month of the year, and with no price of admission that.

For only one day do the sun's rays f directly on top of the head of Tropic Cancer Ham, who lives on the Island Cuba — just for one day, after which t up and down rays travel back towards t Equator Shem. On the twenty-third September Shem again has no shadow noon, and the almanac makers say th is the last day of summer, and tomorro will be the first day of autumn. Again it very hot where Shem lives, but the al gators and monkeys and the parrots do n seem to mind it. Where do the up an down rays of the sun go next? They ke going south, hunting for the boy nam Tropic of Capricorn Japhet, to warm *hi* up, as was the case with the boys in Cu and at the Equator. The up and down ra do not find the top of the head of the l in the City of São Paulo, Brazil, until t last part of December, just three days b fore Christmas, and then the almanac sa this is the beginning of winter, and t shorter days of the year, when we in t State of New York light the lamp at fi o'clock in the afternoon. Now, my bo and girls, do you understand why we ha a change of seasons? Do you understa that the sun changes his manner of pitc ing his rays at us? That in winter, wh he is over the head of the Tropic of Cap corn Japhet in São Paulo, and maki summer on that part of the earth, to people in the north, in the State of Ne York, he pitches only slanting rays th do not hit us hard, and have but litt power? Thus you will see that the rays the sun that strike the earth direct blow swing back and forth like a pendulu year after year, and century after centu coming north as far as Tropic of Canc Shem, but no farther, and then swingi south as far as the boy named Tropic Capricorn Japhet, and no farther, ju stopping and swinging back again towar the north.

THE ECLIPTIC AND THE ZODIAC

By S. L. Boothroyd

.ong before man began to write his-
', he noticed that the sun appeared to
ve all the way around the sky in a year.
noticed also that the yearly path
ong the stars was always the same. If
moon happened to be on the sun's
ı at the time of new moon, there was
ays an eclipse of the sun; and if the
on was on the same line at full moon
re was always an eclipse of the moon.
ıat was more natural than to call this
the Ecliptic? Since it was found by
ervation that the moon and the wan-
ng stars which we now call planets,
e always quite near this same line, it
perfectly natural that this band of the
traversed by the sun, moon, and plan-
should seem especially important to
y man. This region of the sky, a band
wide, 8° on either side of the Ecliptic,
alled the Zodiac and the stars in it
e, long ago, divided into twelve groups
ed constellations.

'hese constellations, in order, around
Ecliptic are: *Aries*, the Ram; *Taurus*,
Bull; *Gemini*, the Twins; *Cancer*, the
b; *Leo*, the Lion; *Virgo*, the Virgin;
ra, the Scales; *Scorpio*, the Scorpion;
ittarius, the Archer; *Capricornus*, the
Goat; *Aquarius*, the Water Bearer;
Pisces, the Fishes. What a collection
zoological specimens this is!

f one were to go from California to
w York across the United States, he
ıld pass through many states; as he left
: state he would pass into the state to
east of the one he had just left. In the
ıe way the sun in its annual eastward
rney around the sky as he leaves one
iacal constellation enters another.

. little more than two thousand years
, when these zodiacal constellations
e adopted, substantially as they are
wn to us, the sun just as it crosses the
ıator, going north on about March 21,
entering Aries. We still call this point the first of Aries, and Aries the first sign of the Zodiac; but due to an effect we call "precession of the equinoxes" this point is now in the constellation Pisces.

At present, the sun is in the same constellation about a month later than it was when the zodiacal constellations were adopted; and furthermore, owing to slight changes in the boundaries of the constellations as we know them, the dates are not the same as given in the almanac.

Referring to star maps, pages 818, 825, and 831, the line marked Ecliptic is the sun's apparent annual path around the sky. On the maps the approximate time when the sun is in each of the constellations along the Ecliptic is given. If desired, more accurate dates may be found in the following table.

At present, the sun is in the given constellations during approximately the times indicated:

1. Aries	April 18	to	May 14
2. Taurus	May 14	"	June 21
3. Gemini	June 21	"	July 20
4. Cancer	July 20	"	Aug. 10
5. Leo	Aug. 10	"	Sept. 16
6. Virgo	Sept. 16	"	Oct. 31
7. Libra	Oct. 31	"	Nov. 23
8. Scorpio and Ophiuchus	Nov. 23	"	Dec. 18
9. Sagittarius	Dec. 18	"	Jan. 19
10. Capricornus	Jan. 19	"	Feb. 16
11. Aquarius	Feb. 16	"	Mar. 12
12. Pisces	Mar. 12	"	Apr. 18

It may be difficult for the pupils to learn to know all these constellations, as some of them are not very well marked; however, if they wish to learn them they can do so by the use of the planisphere. Some of the constellations of the Zodiac are marked by brilliant stars which have already been learned. Regulus is the heart of Leo, the Lion; Spica which means "ear" is the ear of wheat which the Virgin is holding in the constellation Virgo. Red Antares lies in the Scorpion; and the Milk Dipper, which is shaped like the Big Dipper, but smaller, marks Sagittarius. Red Aldebaran is the fiery eye of Taurus, the Bull, while Gemini, or the Twins, are the most conspicuous of the stars high overhead in the evening skies of February and March.

In almanacs one may see a table indicating the signs of the Zodiac; for a certain stated period the sun is said to be in a definite sign which corresponded to the region traversed by the sun at the time these Zodiacal constellations w adopted. Each of the twelve Zodiacal c stellations constitutes a sign. The ti given in these signs are those that w used by the ancients.

The following lines will aid one in coming familiar with the relative positi of the zodiac:

The Ram, the Bull, the Heavenly Tw
And next the Crab, the Lion shines,
The Virgin and the Scales.
The Scorpion, the Archer and He-goa
The Man that holds the Watering-po
The Fish with glittering tails.

THE SKY CLOCK

By S. L. Boothroyd

Since the sky, to a northern observer, appears as a vast sphere which turns at a uniform rate about the line joining the observer's eye to Polaris it is evident that the turning of the sky may be used to measure the passage of time. Of course, the stars appear as fixed points upon this sphere.

Everywhere on the earth, north of 33° north latitude, the five bright stars of Cassiopeia and all the stars of the Big Dipper, except the two farthest from Polaris, always above the observer's horizon.

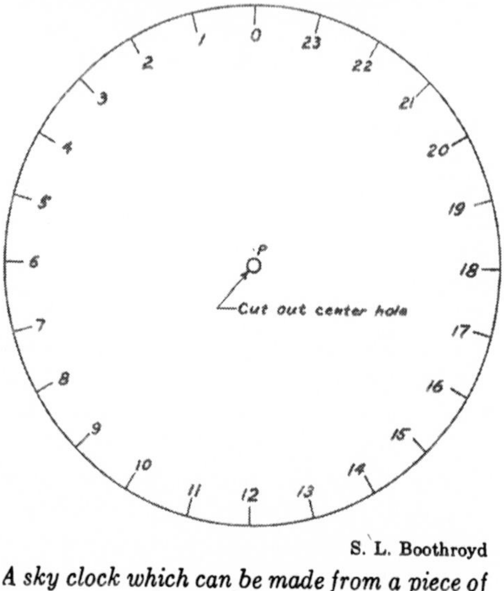

S. L. Boothroyd

A sky clock which can be made from a piece of heavy firm paper

All the stars on the sky appear to m as though they were bright points on surface of an immense sphere. To a p son facing Polaris the sky seems to be tating counterclockwise about the l joining his eye to Polaris. It rotates such a rate that the line joining Pol to any star appears to revolve through degrees in 23 hours 56 minutes and seconds. This is the rate of rotation of earth on its axis. Since this apparent tation of the sky about the line joining observer's eye to Polaris takes place a perfectly uniform rate, we may regard sky about the North Star as the face c great sky clock.

It is most convenient to consider hour hand of this sky clock as the l joining Polaris to the star Caph (caf) Cassiopeia. See the north circumpolar s map. The dial of this sky clock must considered as a circle drawn around Pol as a center, and the figures on the cl face go around the dial in the oppo direction from those on an ordinary clo Also the clock has 24 divisions or hours its face instead of 12. The 0 or 24 hour vision is straight above Polaris and the

ır division is straight below the same
r. The left half of the dial is then num-
ed from 0 (or 24) at the top to 12 at
lowest point straight below Polaris,
l the right half of the dial is numbered
m 12 at the lowest point to 24 (or 0)
the top.

One then imagines this dial printed on
sky, and to get the star time simply
tes where the line from Polaris to Caph
sses the face of the dial. As an aid to
serving the star time with fair accuracy,
ke a dial on a piece of stiff white bristol
ırd, cut in a circle one foot across. Make
ole one-half inch in diameter at the
ter of the dial. Divide the circumfer-
e into 24 equal spaces and number the
isions, with black waterproof ink, from
o 23. Fasten a piece of black string 8
hes long on a line from the center of
dial to a point about one inch above
center and fasten to the other end a
llet or a washer so the string will hang
mb when the dial is held up to the
server as he faces the Polestar. Hold
dial up so that Polaris can be seen
ough the hole in the center; then turn
dial around the line leading from the
to Polaris until the plumb line falls di-
tly over the line from the center to the
hour mark on the dial. At this mont, note where the line from Polaris to
ph crosses the dial; the reading of the
l at this point is *star time*, called *side-
l time* by the astronomer. This time is
d also by sailors and others who often
d to calculate time without the use
the customary timepieces.

Suppose that on an August evening one
found (with the aid of the dial just
cribed) the time to be 19 hours. The
server will now locate the line marked
X on the equator of the Equatorial Star
ıp and the line marked XIX extending
m the center to the boundary of the
rcumpolar Star Chart. These lines in-
ate the line on the sky which at that
ment passes from Polaris directly overad to the south part of the horizon.
ain, if the star time read on the dial is
hours, then the line from Polaris to
X on the boundary of the Circumpolar
Star Chart, and the line marked XX on the equator of the Equatorial Star Chart, are at that moment the line on the sky which passes from Polaris through the zenith to the south part of the horizon. It makes no difference what the day of the year happens to be; a person can, by using his dial, observe the star time and tell just what part of the sky as shown on the star map is on his celestial meridian, as the line from Polaris through his zenith to the south part of the horizon is called. The ability to get star time thus enables one to use his star maps more effectively as aids in learning to know the constellations and stars.

We must bear in mind that the time we use is standard time. This time is the local mean time for a given meridian, usually a meridian that is a multiple of 15 degrees from Greenwich, England. The Eastern standard time is the mean time for the 75° meridian west of Greenwich. It is therefore 5 hours slower than Greenwich time. If a standard meridian were 75 degrees east of Greenwich its time would be 5 hours faster than that of Greenwich.

Unless one lives exactly on the meridian that is the basis for the time that is used in his particular time belt, he is not using the *exact* time indicated by the position of the sun. He is using *standard time* and the time indicated by the mean sun is known as *local mean solar time*. It is easy to find the *local mean solar time* for any location after one has determined, by means of the sky clock, the star time. The following rule is used:

Subtract from the star time observed, a number of hours which is twice the number of months since March 23 and the remainder will be the number of hours since the observer's preceding local mean noon. If the observed star time is less than twice the number of months since March 23, add 24 hours to the observed time before subtracting.

Following are examples to illustrate the use of the sky clock:

1. Suppose that on the night of August 18, at Ithaca, New York, the sky clock was read. The part of the sky between the

lines marked XIX and XX on the star maps was on the celestial meridian and the star time was observed to be 19 hours and 40 minutes.

August 18 is 4 months and 25 days after March 23, or 4 25/30 months. This number of months multiplied by 2 equals 9 2/3.

This number expressed as hours and minutes equals 9 hours and 40 minutes and when this number is subtracted from the observed star time, 19 hours and 40 minutes, the remainder is 10. This figure represents the number of hours since the observer's local mean noon, or 10 P.M. by local mean solar time.

2. Suppose on the night of January 27 at Boston, Massachusetts, the sky clock was read, and the part of the sky between the lines marked VII and VIII on the sky maps was on the celestial meridian. The star time was observed to be 7 hours and 20 minutes.

January 27 is 10 months and 4 days after March 23 or 10 4/30 months after March 23. This number of months multiplied by 2 equals 20 4/15. This number expressed as hours and minutes equals 20 hours and 16 minutes. Since 20 hours and 16 minutes is larger than the observed star time, 7 hours and 20 minutes, 24 hours must be added to 7 hours and 20 minutes making 31 hours and 20 minutes; 31 hours and 20 minutes minus 20 hours and 16 minutes equals 11 hours and 4 minutes. Hence the observation was made 4 minutes after 11 on the night of January 27, local mean solar time.

If the observer wishes to express this "local mean solar time" in the "standard time" for the belt in which he lives, one more step must be taken. It is necessary to determine the longitude of the place where the observations are being made. The longitude is expressed in degrees; it can be easily determined by use of a map. By consulting a map which shows standard time belts, determine what ridian is used as the standard meridian the time belt in which the observati are being made; next determine how east or west of the standard meridian observer is located. For each degree th will be a difference of 4 minutes of tir that is, if the observer is 1 degree east the standard meridian, his local me solar time will be 4 minutes faster th standard time. If he is 1 degree west the standard meridian, his local me solar time will be 4 minutes slower th standard time. Therefore, if local me solar time has been found, the stand time can be easily found provided o knows the longitude of the place wh the observations were made. Four minu will be added to the standard time each degree the observer is located east the standard meridian; 4 minutes will subtracted from standard time for ea degree the observer is located west the standard meridian. Thus, it will seen that the local mean solar ti will be faster than standard time if e of a standard meridian and slower th standard time if west of a standard ridian.

Let us refer to the examples given this and the preceding page; note that t reading of the sky clock in Example 1 w taken at Ithaca, New York. This town h a longitude of 1½ degrees west of the 75 meridian; it is in the Eastern standa time belt, whose standard meridian is degrees. Therefore, since Ithaca is 1½ grees west of the standard meridian, local mean solar time would be 6 minut slower than standard time. Boston, which point the sky clock was read Example 2, is at a point 4 degrees ea of the 75th meridian; therefore its loc mean solar time would be 16 minut faster than standard time.

THE EQUATORIAL STAR FINDER

By S. L. Boothroyd

The line marked Equator on the star ıps is the line in which the plane of : earth's equator extended outward :ets the sky. By means of a very sim- : homemade device, which we call an ıatorial star finder we can locate visible rs and the equator of the sky, or the :lestial Equator as it is called.

RECTIONS FOR MAKING AND OPERATING THE EQUATORIAL STAR FINDER

Cut pieces A, B, and C from a piece plank 2 by 8 inches which has been ıned and smoothed off on all sides; it ɔuld then be about 7½ inches wide and ¾ inches thick. Pieces A and B are each out 2 feet long with the ends accurately ıared. Piece C is cut square on one end d on a bevel on the other end, the face contact with B being about 11 inches ıg and the upper face being about 9½ :hes long. Pieces A and B are hinged gether at one end so that their inner :es will be everywhere in contact when : hinges are closed. Block C is nailed screwed to piece B, as shown in the ıstration. A ⅞ inch hole is bored into at right angles to its upper surface and :ending to within ¼ inch of the lower :face of piece B. The center of this hole ıst be 6 inches from the upper edge the upper face of C and at the center the face. The use of a press drill will ure the hole being at right angles to : face of C and will also prevent the le being bored entirely through B.

Into this hole is inserted the polar axis, of the star finder. This can be made ım a piece of broom handle which is ry straight, and can be filed or sand- pered to accurately fit the hole in C, that it will not wobble when pushed to e bottom of the hole and yet will be free turn on its axis without undue effort.

Disc E is a circular piece of masonite, 7⅛ inches in diameter, in the center of which is bored a hole, ⅞ inch in diameter. The polar axis will just go through this hole and allow the disc to be turned on this axis, and yet be tight enough not to turn of its own accord. On the top face of disc E is glued a specially gradu-

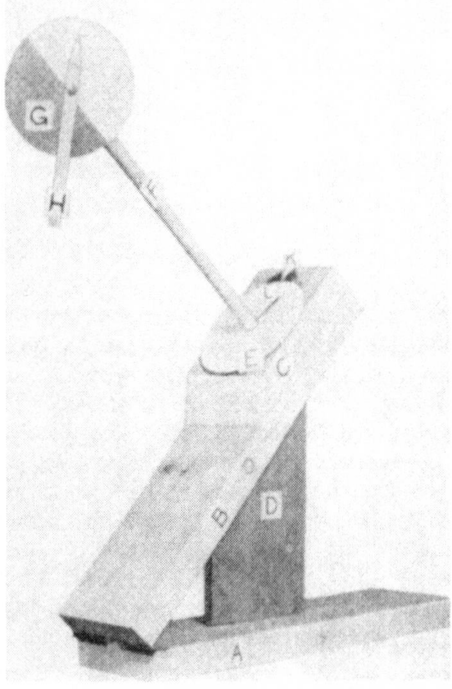

S. L. Boothroyd

The equatorial star finder. See pp. 847–50 for instructions for making and using it

ated protractor on which are printed the positions of 40 bright stars as projected on the equator plane of the celestial sphere. This protractor sheet is shown in the first illustration on page 848. When this protractor sheet is glued to the disc above mentioned, we shall call disc E the "hour disc" of the star finder.

Flatten the upper 6 inches of one side of the polar axis, F, cutting a maximum depth of ⅛ inch from the round stick. G is another disc of masonite of the same size as disc E, 7⅛ inches in diameter. G is nailed to the polar axis so that a diameter of the disc lies accurately along

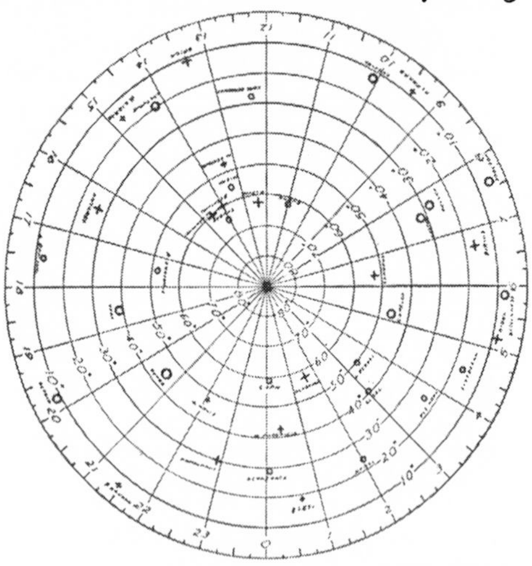

S. L. Boothroyd

Hour disc, E

the flattened face of the upper end of the polar axis, F. Now bore a ¼ inch hole through the center of disc G and through the axis of the polar axis, F, and at right angles thereto. Glue to disc G the declination protractor shown in the illustration opposite. Be sure to glue this to disc G so that the arrow marked *to visible pole* points upward when the polar axis, F, is vertical. This disc, G, with its declination protractor glued to it will be called the "declination disc."

Next, prepare the star pointer, H. Use a piece of wood ⅝ inch wide, ⅛ inch thick and 10 inches long; symmetrically point one end and bore a ⅛ inch hole three inches from the point and in the center of the piece. A ⅛ inch bolt through this and through the ⅛ inch hole bored through the declination disc, G, and the polar axis, F, will pivot this pointer, H, to the upper end of the polar axis, F, as shown in the illustration. Put washers under the head of the bolt, under the nut, and between the star pointer, H, and the declination disc, G. Screw the nut tight enough so that the star pointer, H, will remain in any position, and so t it is not too hard to turn. It is well to pu lock washer on to prevent the nut comi unscrewed. Insert screw eyes on the c ter line of this pointer, near its ends, a the pointer, F, is complete.

Next, rotate the star pointer, H, on axis until it points at 0 degrees on declination disc, G. Then rotate the po axis, F, until the star pointer is parallel the long side of piece B, and is pointi away from the hinges. Now hold the po axis, F, so that it will not turn on its a and rotate the hour disc, E, until the hour-to-12 hour line is parallel to the s pointer, H, with the 0 hour point un the point of the star pointer, H. C an arrow point, K, from a piece of du ble paper and glue it to a small piece masonite to bring it on a level with t graduated face of the hour disc, E. Pl the arrow point at the outer edge of t hour disc, E, with the point of the arr at the end of the 0 hour line and pointi to the center of the hour disc, E. N bore a hole in the polar axis, F, radia in from where the 0 hour line of the h disc, E, meets the polar axis when all been adjusted as has been explained. T hole should be a little smaller than a

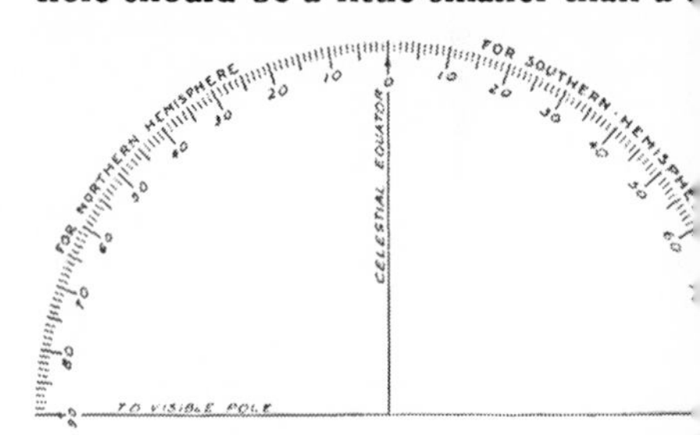

S. L. Booth

Declination disc, G

penny finishing nail and should be ab ⅜ inch deep. A nail when inserted this hole, with the instrument adjus as directed, will be over the 0 hour l of the hour disc, E. It will be well to off the head of the nail to a blunt poi after the nail has been snugly driven i the hole which was bored to receive We shall call the nail the *right ascens pointer, L.*

astly, prepare a piece of ⅞ inch thick :d for piece *D*, in illustration page 847. : preparation of this is best explained howing a side view of it with all the essary dimensions and angles given on diagram below.

he degrees in angle W should be equal he latitude of the locality where the rument is to be used. Angles Y and Z

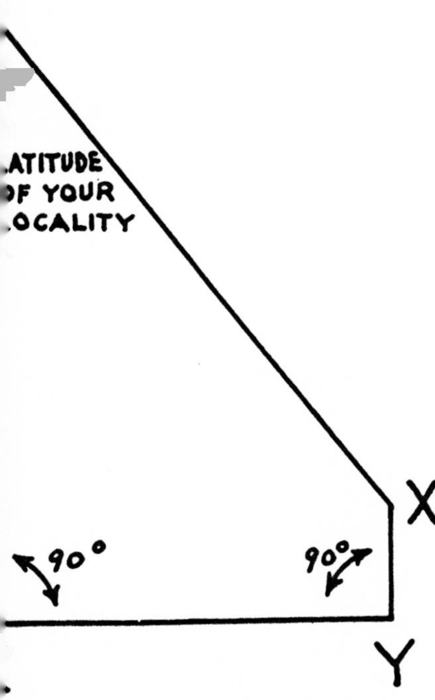

The equatorial star finder, piece D

always 90°. Angle X is equal to 180° ıus latitude of the locality where the rument is to be used. Line ZY is always ıches. Line XY may be made about 4 hes; then the other lines are fully demined if the angles are laid out as cified.

Vith piece *A* on a horizontal surface, ate *B* about the hinges until *D* can be ced with side ZY in contact with the ɔer face of piece A and side WX in tact with the lower face of piece *B*. ce *D*, when accurately constructed for tudes between 30° and 90°, will stay place without anything but friction been the surfaces to hold it in place. For latitudes below 30°, piece *D* will need to be nailed or screwed into place. Piece *D* will be called *the latitude board.*

The equatorial star finder is now ready to be put in the place where it is to be used for observation. It must be set on a level surface, at least 2½ feet long and about 1 foot wide, with its long dimension north and south. The surface should be about as high as a table or perhaps a little higher.

With a compass, if you know the compass variation, or by means of the shadow of a vertical stick at apparent noon, mark a north and south line on the top of your table or slab. The line must be exactly north and south and the top of the table must be level if your star finder is to point accurately to the stars you wish to locate.

Having everything ready, go out with the star finder and place the long side of piece A of the star finder on the north and south line of the level surface, and with the polar axis, *F*, pointing towards the visible pole of the heavens. For an observer in the Northern Hemisphere, it will point at Polaris very nearly. You are now ready to use your star finder. Some star which you already know can be used as a helper to aid you in finding an unknown star. It is best, however, to select a helper that is located some distance from Polaris. Suppose you know the star Vega and wish to find the star Capella. Turn the polar axis, *F*, about its axis, and the star pointer, *H*, about its pivot until you can see Vega through the sight line determined by the center line of the two screw eyes in the star pointer, *H*. Now hold the polar axis, *F*, from turning and rotate the hour disc, *E*, until the star Vega shown on it is under the right ascension pointer, *L*. Now hold the hour disc, *E*, from turning and rotate the polar axis, *F*, until the right ascension pointer is over the star, Capella. Note, on the hour disc, *E*, the declination of Capella, which is its angular distance from the celestial equator. Now move the point of the star pointer, *H*, over the protractor on disc G until it is on the degree mark corresponding to the declination of Capella, as read off the hour disc,

E, and the star pointer, *H*, will now be pointing at Capella, providing you have performed these operations promptly.

By following the above procedure, one can point the star finder at any of the forty stars shown on the hour disc, *E*, of the instrument, whether the star is above or below the horizon. Of course, you can see only those stars which are above your horizon. *Do not expect to find all the stars by one pointing on* Vega, as in the example illustrated; but set on Vega, or some other known star each time, and adjust the hour disc, *E*, as explained above, before setting on the star to be found.

Another way to find an unknown star is to find the star time using the sky clock explained on page 844. Immediately after you have obtained the star time, rotate the hour disc, *E*, of the star finder until the reading on the hour disc which is under the time arrow, *K*, is the same as the star time. Now hold the hour disc, *E*, until the right ascension pointer, *L*, is over the star to be found. Read its declination from the hour disc, *E*, adjust the star pointer, *H*, to this reading on its protractor, *G*, and the star pointer will point at the star in question.

It should, from the above, be seen that one can use the star finder to obtain the star time. For example, suppose one knows where to find the star Vega. Rotate the polar axis and adjust the star pointer until Vega is seen along the star pointer. Now hold the polar axis and rotate the hour disc, *E*, until Vega is under the right ascension pointer, *L*. The star time is the reading on the hour disc opposite the time arrow, *K*. To find other stars than those shown on the hour disc, consult the star maps on pages 818, 825, and 831 and find the right ascension and declination of each star. Then after adjusting the hour disc, *E*, by one of the methods already explained, hold the hour disc, *E*, and rotate the polar axis, *F*, until the right ascension pointer, *L*, points to the right ascension of the star to be found. Then adjust the star poir *H*, until its pointer is over the declina of said star, and the star pointer wil pointed at the star to be identified.

An ingenious boy will use the work an old alarm clock to cause the hour to turn around at the proper rate, so when once set with the time arrow at time, it will continue to be so as lon the clock runs. One can then point one star after another without first adjusting the hour disc. The clock ke the hour disc, *E*, adjusted, when once People who are more deeply intere in the subject will make the whole paratus a little heavier than indicate the instructions given here. A small t scope can then be mounted on the pointer; this makes possible the obse tion of many objects that are not ea visible to the naked eye.

To find planets, obtain, from some liable source, the right ascension declination of the planet and follow instructions for pointing at stars of knc right ascension and declination.

The hour disc, *E*, shown here is use in the Northern Hemisphere.

The outer edge of the hour disc, represents the Celestial Equator. The ures appearing near the outer edge of disc indicate right ascension. For ex ple, to find the right ascension of V imagine a line drawn from the cente the disc through Vega until it inters the graduated outer edge of the disc. T reading at this point in hours and n utes is the *right ascension* of Vega. I about 18 hours and 30 minutes.

The concentric circles on the disc resent the parallels of declination wh are similar to parallels of latitude on earth. To find the *declination* of V note that it lies between the parallel declination of 30° and 40°. The decl tion being about 38½°. Stars which represented by circles (O) have no declination and those represented crosses (+) have south declination.

THE RELATIONS OF THE SUN TO THE EARTH

Whether we look or whether we listen,
We hear life murmur or see it glisten.
— Lowell

ll this murmuring and glistening life our earth planet has its source in the at sun which swings through our skies ly, sending to us his messages of light warmth — messages that kindle life the seed and perfect the existence of ry living organism, whether it be the ed in the field or the king on his throne.

t sunrise this heat which the sun ds out equally at all times of day and ht is tempered when it reaches us, be- se it passes obliquely through our at- sphere-blanket, and thus traverses a ater distance in the cooling air. The e is true at sunset; but at noon, when sun is most directly over our heads, its s pass through the least possible dis- ce of the atmosphere-blanket and refore lose less heat on the way. It is e that often about three o'clock in the ernoon is the hottest period of the day, this is because the air-blanket has be- ne thoroughly heated; it is still true t we receive the most heat directly m the sun at noon.

The variations in the time of the rising l the setting of the sun may be made a st interesting investigation on the part the pupils. They should keep a record a month in the winter, and with this a basis use the almanac to complete the son. Thus each one may learn for him- f which is the shortest and which the gest day of the year. There is a slight iation in different years; for a person in itude 45° north the shortest day of year when this lesson was written, as mputed from a current almanac, was cember 22; the day was eight hours d forty-six minutes long. The longest y of the year was June 22, and it s fifteen hours and thirty-seven min- es in duration. On the longest day of the year the sun reaches its farthest point north and is, therefore, most nearly above us at midday. On the shortest day of the year, the sun reaches its farthest point south and is, therefore, farther from the point directly above us at midday than during any other day of the year.

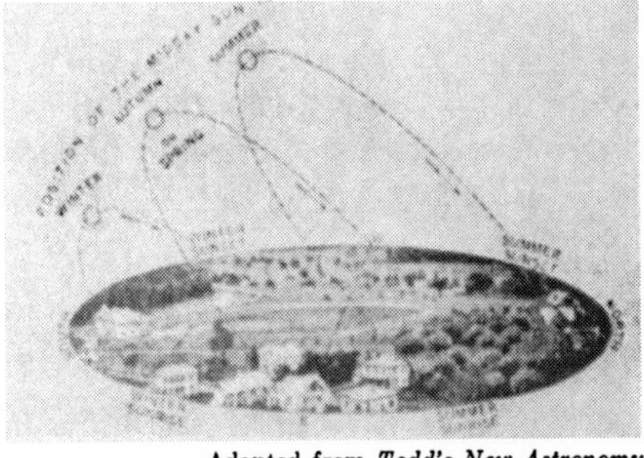

Adapted from *Todd's New Astronomy*

Path of the sun across the sky of an observer in about latitude 40 *degrees north, on June* 22, *on March* 21 *or September* 23, *and on December* 22

The movement of the sun north and south is an interesting subject for personal investigation, as suggested in the lesson. Through quite involuntary observation, I have become so accustomed to the arc traversed by the points of sunrise as seen from my home, that I can tell what month of the year it is by simply noting the place where the sun rises. When it first peeps at us over a certain pine tree far to the south, it is December; when it rises over the reservoir it is February or October; and when it rises over Beebe Lake it is July. Only at the equinox of spring and fall does it rise exactly in the east and set directly in the west. Equinox means equal nights, that is, the length of the night is equal to that of the day.

Because of the vast weight of the sun,

the force of gravity upon its surface is so great that even if it were not for the white-hot fireworks so constantly active there, we could not live upon it, for our own weight would crush us to death. But this multiplying the weight of common objects by twenty-seven and two-thirds to find how much they would weigh on the sun is an interesting diversion for the pupils, and incidentally teaches them how to weigh objects, and something about that mysterious force called gravity; and it is also an excellent lesson in fractions.

LESSON 231

The Relation of the Sun to the Earth

Leading Thought — The sun, which is the source of all our light and heat and therefore of all life on our globe, travels a path that is higher across the sky in June

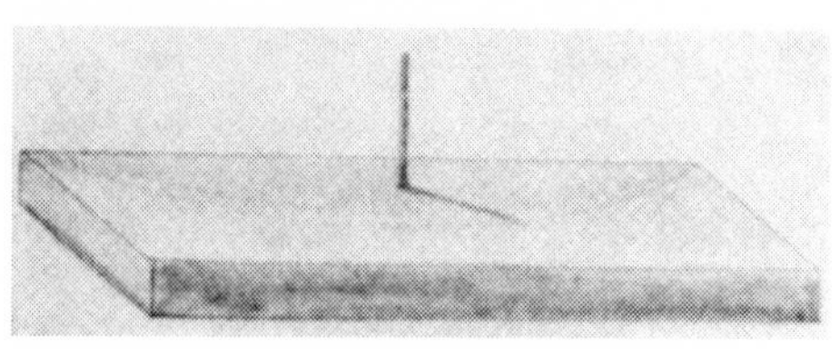

A shadow stick

than the path which it follows in December, and hence we experience changes of seasons. The lesson should be given to the pupils of the upper grades and should be correlated with reading and arithmetic.

Observations — 1. What does the sun do for us?

2. At what time of the day after the sun rises do we get the least heat from it? At what hour of the day do we get the most heat from it?

3. Is the sun equally hot all day? Why does it seem hotter to us at one time of the day than at another?

4. At what hour does the sun rise and set on the first of the following months: February, March, April, May, and June?

5. Which is the shortest day of the year, and how long is it?

6. Which is the longest day of the year, and how many hours and minu are there in it?

7. On what day of the year is the nearest a point directly over our head midday?

8. On which day of the year is the at midday farthest from the point dire above our heads? Explain why this is

9. Standing in a certain place, mark some building, tree, or other object where the sun rises in the east and in the west on the first of February. serve the rising and setting of the from the same place on the first day March and again on the first of Ap Does it rise and set in the same place ways or does its place of rising and sett move northward or southward?

10. Is the sun farthest south on shortest day of the year? If so, is it fa est north on the longest day of the ye

11. At what time of the year does sun rise due east and set due west?

12. The sun is so much more mass than the earth that, in spite of its grea size, its force of gravity is twenty-se and two-thirds times that of the ear How much would your watch weigh if were living on the sun? How much wo you yourself weigh if you were there?

13. Experiment. A Shadow Stick Place a peg two or three inches high right in a level board and place the bo lengthwise on a sill of a south window where it will be in the sunlight, at le from 9 A.M. to 3 P.M. Mark the shad cast by the peg at half-hour intervals d ing a sunny day and draw a line with pe cil or chalk outlining the tip of shadow of the stick. Make a similar o line a month later, and again a mon later and note whether the shadow tra the same line during each of these d of observation. Note especially the leng of the shadow at noon, on March 21, Ju 22, September 23, and December 22, as near these dates as possible.

A measurement that is even more ex than the one just described can be tained by means of a *gnomon pin* plac in a board.

\nother excellent observation lesson for ching the fact that the sun travels ther south in the winter, is to measure : shadow of a tree on the school grounds noonday once a month during the ıool year. The length of the tree shadow ı be measured from the base of the tree .nk, a memorandum being made of it.

14. When does the stick or tree cast its ıgest shadow at noon — in December February? February or April? April or ıe? Why?

Topics for English Themes — The :e and Distance of the Sun. The Heat the Sun and Its Effect upon the Earth. hat We Know about the Sun Spots. ır Path around the Sun.

How to Make a Sundial

Method — The diagram for the dial is esson in mechanical drawing. Each pu- should construct a gnomon (*no′mon*) cardboard, and should make a drawing the face of the dial upon paper. Then e sundial may be constructed by the lp of the more skillful in the class. It ould be made and set up by the pupils. sundial in the school grounds may be ade a center of interest and an object of :auty as well.

Materials — For the gnomon a piece board a half inch thick and six inches uare is required. It should be given sev- al coats of white paint so that it will not

A sundial made by pupils

warp. For the dial, take a board about 14 inches square and an inch or more thick. The lower edge may be bevelled if de-

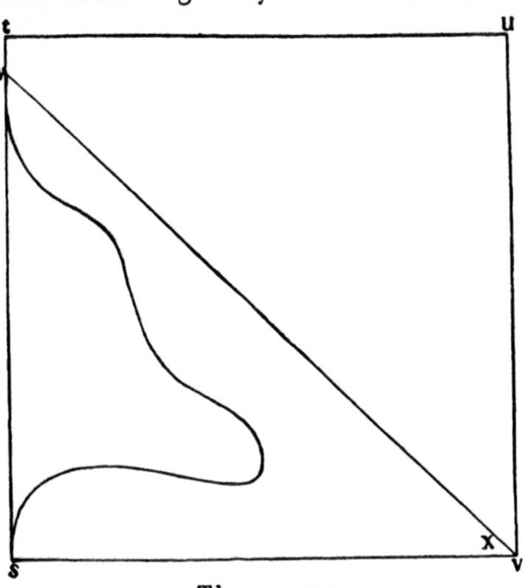

The gnomon

sired. This should be given three coats of white paint, so that it will not warp and check.

To Make the Gnomon — The word gnomon is a Greek word meaning " one that knows." It is the hand of the sundial, which throws its shadow on the face of the dial, indicating the hour. Take a piece of board six inches square, and be sure its angles are right angles. Let s, t, u, v represent the four angles; draw on it a quarter of a circle from s to u with a radius equal to the line vs. Then with a cardboard protractor, costing fifteen cents, or by working it out without any help except knowing that a right angle is 90°, draw the line vw making the angle at x the same as the degree of latitude where the sundial is to be placed. At Ithaca the latitude is 42°, 27′ and the angle at x measures 42°, 27′. Then the board should be cut off at the line vw, and later the edge sw may be cut in some ornamental pattern.

To Make the Dial — Take the painted board 14 inches square and find its exact center, y. Draw on it with a pencil the line AA″ a foot long and one-fourth inch at the left of the center. Then draw the line BB″ exactly parallel to the line AA″ and a half inch to the right of it. These lines should be one-half inch apart — which is just the thickness of the gnomon. If the gnomon were only one-fourth inch thick, then these lines should be one-fourth inch apart, etc.

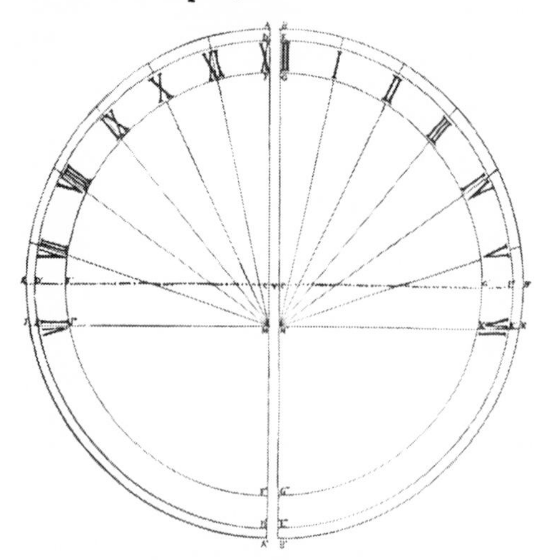

The face of the sundial

With a compass, or a pencil fastened to a string, draw the half-circle AA′A″ with a radius of six inches, with the point c for its center. Draw a similar half-circle BB′B″ opposite, with c′ for its center. Then draw the half-circle from DD′D″, from c with a radius of five and three-quarter inches. Then draw similarly from c′ the half-circle EE′E″. Then draw from c the half-circle FF′F″ with a radius of five inches, and a similar half-circle GG′G″ from c′ as a center.

Find the points M, M′ just six inches from the points F, G; draw the line JK through M, M′ exactly at right angles to the line AA″. This will mark the six o'clock points; so the figures VI may be placed on it in the space between the two inner circles. The noon mark XII should be placed as indicated (the " X " at D, F, the " II " at E, G). With black paint outline all the semicircles and figures.

To Set Up the Sundial — Fasten the base of the gnomon by screws or brads to the dial with the point s of the gnomon at F, G, and the point v of the gnomon M, M′, so that the point W is up in t air. Set the dial on some perfectly le standard with the line AA″ extendi exactly north and south. If no comp is available, wait until noon by the s and set the dial so that the shad from W will fall exactly between t points A, B, and this will mean that t dial is set exactly right. Then with a go watch note the points on the arc E on which the shadow falls at one, tw three, four, and five o'clock by sun tin and in the morning the points on the a J′D on which the shadow falls at seve eight, nine, ten, and eleven o'clock by s time. Draw lines from M to these poin and lines from M′ to the point on t arc EK′. Then place the figures on t dial as indicated in the spaces betwe the two inner circles. The space betwe the two outer circles may be marked wi lines indicating the half and quarter hou The figures should be outlined in pen and then painted with black paint, carved in the wood and then painted.

Sundial on the author's lawn

THE MOON

'he moon is in more senses than one lluminating object for both the earth the skies. As a beginning for earth ly it is an object lesson, illustrating t air and water do for our world and dentally for us; while as the beginning he study of astronomy, it appears as the est and brightest object seen in the at night; and since it lies nearest us, the first natural step from our world outer space.

'he moon is a little dead world that les around our earth with one face ays toward us, just as a hat-pin thrust an apple would keep the same side its head always toward the apple no tter how rapidly the apple was twirled. we study the face of the moon, thus ays turned toward us, we see that it dark in some places and shining in ers, and some uninformed people have ught that the dark places are oceans the light places, land. But the dark tions are simply areas of darker rocks, ile the lighter portions are yellowish or itish rocks. The dark portions are of h a form that people have imagined m to represent the eyes, nose, and uth of a man's face; but a far prettier ture is that of a woman's uplifted face profile. The author has a personal feel- on this point, for as a child she saw the n's face always and thought it very ugly, l, moreover, concluded that he chewed acco; but after she had been taught to d the face of the lady, the moon was ays a beautiful object to her.

The moon is a member of our sun's nily, his granddaughter we might call if the earth be his daughter; and since moon has no fires or light of its own, it nes by light reflected from the sun and erefore one-half of it is always in shadow. hen we see the whole surface of the hted half we say the moon is full; but en we see only half of the lighted side rned toward us, we say the moon is in quarter, because all we can see is one-half of one-half, which is one-quarter; and when the lighted side is almost entirely turned away from us we say it is a crescent moon; and when the lighted side is entirely turned away from us we say there is no moon, although it is always there just the same. Thus, although we can never see the other side of the moon, we can understand that the sun shines on all sides of it.

Lick Observatory

The Moon, age 14.9 *days. This and the following photograph were made with the* 36-*inch refractor of the Lick Observatory*

Our earth, like the moon, shines always by reflected light and is almost four times as wide as the moon. When we see the old moon in the new moon's arms, the dark outline of the moon within the bright crescent is visible because of the earthshine which illumines it, part of which is reflected from the moon back again to us. Sometimes pupils confuse this appearance of the moon with a partial eclipse; but the former is the old moon, which is one edge of the moon shining in the sunlight, the remainder faintly illumined by earth light, while an eclipse must always occur at the full of the moon when the earth passes between the sun and the moon, almost completely hiding the latter in its shadow.

It is approximately a month from one new moon to the next, since it takes twenty-nine and one-half days for the moon to complete its cycle around the earth with respect to the sun, and thus turn once around in the sunshine. Therefore, each moon day is fourteen and three-

Lick Observatory

The moon, age 22.06 *days*

quarter days long and the night is the same length. The moon always rises in the eastern sky and sets in the western sky. The full moon rises at sunset and sets at sunrise, but owing to the movement of the earth around the sun the moon rises about fifty minutes later each evening; however, this time varies with the different phases of the moon and at different times of the year. This difference in the time of rising is so shortened at full moon in August and September, that we have sev nights when the full moon lengthens day; and it is called the "harvest mo because in northern Europe in ea times it was customary to work and in the harvest fields until late at nigh

A Visit to the Moon

If we could be shot out from a J Verne cannon and make a visit to moon, it would be a strange experie First, we should find on this little wo which is only as thick through as the tance from Boston to Salt Lake C mountains rising from its surface m than thirty thousand feet high, which twice as high as Mt. Blanc and a thous feet higher than the tallest peak of Himalayas; and these moon mounta are so steep that no one could climb th Besides ranges of these tremend mountains, there are great craters or cular spaces enclosed with steep rock w many thousand feet high. Sometimes the center of the crater there is a p lifting itself up thousands of feet, a sometimes the space within the cra circle is relatively level. Thirty-three th sand of these craters have been disc ered. And, too, on the moon, there great plains and chasms; and all these tures of the moon have been pho graphed, measured, and mapped by p ple on our earth. For a boy study geometry, the measuring of the height the mountains of the moon is an int esting story.

But we could never in our pres bodies visit the moon, because of one rible fact — the moon has no air surrou ing it. No air! What does that mean a world? First of all, as we know life, living thing — animal or plant — co exist there, for living beings must h air. Neither is there water on the mo for if there were water there would h to be air. And without water no gr thing can be grown, and the surface the moon is simply naked, barren ro If we were on the moon, we could turn our eyes toward the sun, for w air to veil it, its fierce light would blind and the sky is as black at midday as nidnight, since there is no atmospheric t to scatter the blue rays of light, leav- the beautiful blue in the sky; nor is re a glow at sunset because there is no prism to separate the rays of light and clouds to reflect or refract them. The s could be seen in the black skies of lday as well as in the black skies of ht and they would be simply points of t and could not twinkle, since there is air to diffuse the sun's light and thus tain the stars by day and cause them to nkle at night. The shadows on the on are, for the same reason, as black as lnight and as sharply defined; and if we uld step into the shadow of a rock at lday we should be hidden, although ne light reflected from the rocks around might reach us. Hiding in such a dow would be like putting on the in- ible cloak of fairy lore. And because re would be no layers of air to make an ial perspective, a mountain a hundred les away would seem as close to us as e a mile away.

ince there is no atmosphere on the on to act as a blanket to prevent radia- n of heat to outer space and to shield m the direct radiation of the sun, temperature of the moon reaches ve boiling point at noon and near olute zero at midnight. This great nge of temperature between sunlight d darkness is the only force on the on to change the shape of its rocks, the expansion under heat and con- ction under cold must break and crum- even the firmest rock. Our rocks broken by the freezing of water that eps into every crevice, but there is no ter to act on the moon's mountains in s fashion or to wear them away by dash- over their surface. However, the rocks d mountains of the moon may be anged in shape by the battering of me- orites, which pelt into the moon by the llion, since the moon has no air to them afire and make them into harm- s shooting stars, burning up before they ike. But though a meteorite weighing thousands of tons should crash into a moon mountain and shatter it to atoms there would be no sound, since sound is carried only by the atmosphere.

Imagine this barren, dead world, chained to our earth by links forged from unbreakable gravity, with never a breath of air, a drop of rain or flake of snow, with no streams, or seas; graced by not a green thing — not even a blade of grass or a tree, or by the presence of any living creature! Out there in space it whirls its dreary round, with its stupendous moun- tains cutting the black skies with their jagged peaks above, and casting their inky shadows below; heated to a terrific tem- perature, by the sun's rays, then suddenly immersed into cold that would freeze our air solid, its only companion the terrific rain of meteoric stones driven against it with a force far beyond that of cannon balls, and yet with never a sound as loud as a whisper to break the terrible stillness which envelops it.

LESSON 232

THE MOON

LEADING THOUGHT — The moon always has the same side turned toward us, so we do not know what is on the other side. The moon shines by reflected light from the sun, and is always half in light and half in shadow. The moon has neither air nor water on its surface, and what we call the moon phases depend on how much of the lighted surface we see.

METHOD — Have the pupils observe the moon as often as possible for a month, beginning with the full moon. After the suggested experiment, the questions which follow may be given a few at a time.

EXPERIMENT — Darken the room as much as possible; use a lighted lamp or electric light for the sun, which is, of course, stationary. Take a large apple to represent the earth and a small one to represent the moon. Thrust a piece of stiff wire, at least one foot long, through the big apple to represent the axis of the earth and also the axis about which the moon revolves. Tie a string about a foot long to the stem of the moon apple and make fast the other end to the piece of wire just above the earth apple. Hold the

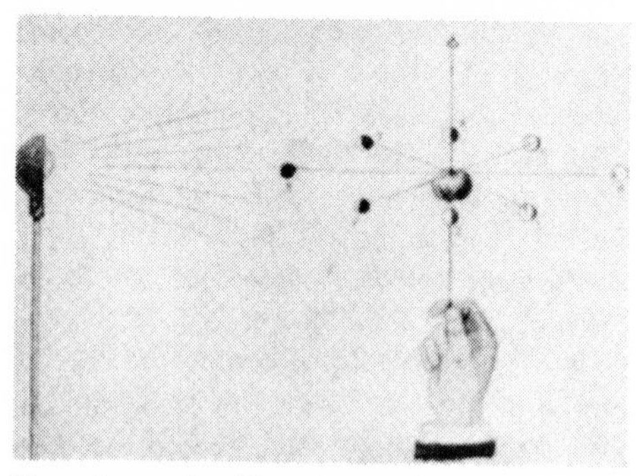

Experiment for illustrating the phases of the moon

wire in one hand and revolve the apple representing the moon slowly with the other hand, letting the children see that if they were living on the earth apple the following things would be true:

1. Moving from right to left when it is between the earth and the sun the moon reflects no light.

2. Moving a little to the left a crescent appears.

3. Moving a quarter around shows the first quarter.

4. When just opposite the lamp, it shows its whole face lighted and turned toward the earth.

5. Another quarter around shows a half disc, which is the third quarter.

6. When almost between the sun and the earth the crescent of the old moon appears.

7. The moon always keeps one face toward the earth.

8. Note that the new moon crescent is the lighted edge of one side of the moon, while the old moon crescent is the lighted edge of the opposite side.

9. Make an eclipse of the moon by letting the shadow of the earth fall upon it, and an eclipse of the sun by revolving moon apple between the sun and earth. The earth's orbit and the moo orbit are such that this relative positi of the two bodies occurs but seldom.

OBSERVATIONS — 1. Describe how moon looks when it is full.

2. What do you think you see in moon?

3. Describe the difference in appe ance between the new moon and the f moon, and explain this difference.

4. Where does the new moon rise a where does it set?

5. When does it rise and when d it set?

6. Where and when does the full mo rise and where and when does it set?

7. How does the old moon look?

8. Could the crescent moon which seen in early evening be the old moon stead of the new; and, if not, why not?

9. When and where do we ordinar see the old moon when it is cresce shaped?

10. Does the moon rise earlier or la on succeeding nights? What is appro mately the difference in time of moonr on two successive nights?

11. Do you think we always look at t same side of the moon? If so, why?

12. Is more than one side of the mo lighted by the sun? Why?

13. How many days from one n moon until the next?

14. How long is the day on the mo and how long the night?

15. How many times does the mo go around the earth in a year?

16. What is the difference between t disappearance of the old moon and eclipse of the moon?

THE PHYSICAL GEOGRAPHY OF THE MO

QUESTIONS FOR THE PUPILS TO TH ABOUT AND ANSWER IF THEY CAN

17. Since it has been found that there no air or water on the moon, could th be any life there?

18. Supposing you could do without or water and should be able to visit t

›on what would you find to be the color the sky there?

19. Would there be a red glow before ırise or beautiful colors at sunset?

20. Would the sun appear to have rays? uld you look at the sun without being nded?

21. Would the stars appear to twinkle? uld you see the stars in the daytime?

22. How would the shadows look? If u could step into the shadow of a rock midday, could you be seen?

23. Could you tell by looking at it ether a mountain was far or near?

24. Why is it so much hotter and lder on the moon than upon the earth?

25. If you could shout on the moon, how would it sound? If one hundred cannons should be fired at once on the moon, how would it sound?

26. Is there any rain or snow on the moon? Are there any clouds there? If there is no air or water on the moon, would the intense heat and the powerful cold affect the soils or rocks, as freezing and thawing affect our rocks?

27. The moon is so small that the force of gravity on its surface is one-sixth that on the earth's surface. If a man can carry seventy-five pounds on his back here, how many pounds could he carry on the moon?

INDEX

INDEX

CPSIA information can be obtained
at www.ICGtesting.com
Printed in the USA
LVOW11s0004301117
557892LV00009B/550/P